SPECIAL INDIAN EDITION

SHIGLEY'S MECHANICAL ENGINEERING DESIGN

SI Units

Twelfth Edition

SPECIAL INDIAN EDITION

SHIGLEY'S MECHANICAL ENGINEERING DESIGN

SI Units

Twelfth Edition

J. Keith Nisbett

Associate Professor of Mechanical Engineering,
Missouri University of Science and Technology

Adapted to SI Units by:

Gianluca Chiappini

Associate Professor of Mechanical Engineering Design and Machine Design
Università eCAMPUS

Edoardo Mancini

Associate Professor of Mechanical Engineering Design and Machine Design
Università degli Studi dell'Aquila

McGraw Hill Education (India) Private Limited

Published by McGraw Hill Education (India) Private Limited
Registered Office: Anjana Complex No: 5/90 A, Butt Road, St. Thomas Mount, Chennai 600 016

Special Indian Edition (SIE) 2026
This edition can be exported from India only by the publisher
Sales Territories: India, Pakistan, Nepal, Bangladesh, Sri Lanka and Bhutan

Print Edition:
ISBN-13: 978-93-6444-073-8
ISBN-10: 93-6444-073-0

E-Book Edition:
ISBN-13: 978-93-6444-421-7
ISBN-10: 93-6444-421-3

1 2 3 4 5 6 7 8 9 7091297 30 29 28 27 26

Printed and bound in India

Typeset at Pee-Gee Graphics, A-6/95, Sector 15, Rohini, Delhi 110089, and printed at
Rajkamal Electric Press, Plot No. 2, Phase – IV, Kundli, Sonepat, Haryana.

Cover Designer: Neeraj Dayal Cover
Image Source: Shutterstock

Visit us at: www.mheducation.co.in
Write to us at: support.india@mheducation.com
CIN: U80302TN2010PTC111532
Toll Free Number: 1800 103 5875

Dedication

To Richard Budynas, in appreciation of his mentorship, his attention to detail, and his contributions to the engineering profession.

J. Keith Nisbett

Joseph Edward Shigley (1909–1994) is undoubtedly one of the most well-known and respected contributors in machine design education. He authored or coauthored eight books, including *Theory of Machines and Mechanisms* (with John J. Uicker, Jr.), and *Applied Mechanics of Materials.* He was coeditor-in-chief of the well-known *Standard Handbook of Machine Design.* He began *Machine Design* as sole author in 1956, and it evolved into *Mechanical Engineering Design,* setting the model for such textbooks. He contributed to the first five editions of this text, along with coauthors Larry Mitchell and Charles Mischke. Uncounted numbers of students across the world got their first taste of machine design with Shigley's textbook, which has literally become a classic. Nearly every mechanical engineer for the past half century has referenced terminology, equations, or procedures as being from "Shigley." McGraw Hill is honored to have worked with Professor Shigley for more than 40 years, and as a tribute to his lasting contribution to this textbook, its title officially reflects what many have already come to call it—*Shigley's Mechanical Engineering Design.*

Having received a bachelor's degree in Electrical and Mechanical Engineering from Purdue University and a master of science in Engineering Mechanics from the University of Michigan, Professor Shigley pursued an academic career at Clemson College from 1936 through 1954. This led to his position as professor and head of Mechanical Design and Drawing at Clemson College. He joined the faculty of the Department of Mechanical Engineering of the University of Michigan in 1956, where he remained for 22 years until his retirement in 1978.

Professor Shigley was granted the rank of Fellow of the American Society of Mechanical Engineers in 1968. He received the ASME Mechanisms Committee Award in 1974, the Worcester Reed Warner Medal for outstanding contribution to the permanent literature of engineering in 1977, and the ASME Machine Design Award in 1985.

Joseph Edward Shigley indeed made a difference. His legacy shall continue.

J. Keith Nisbett is an Associate Professor of Mechanical Engineering at the Missouri University of Science and Technology. He has more than 40 years of experience with using and teaching from this classic textbook. He was awarded the BS, MS, and PhD of the University of Texas at Arlington.

Part 1

Part 2

Part 3

Part 4

Preface xvi

Part **1**

Basics 2

Chapter **1**

Introduction to Mechanical Engineering Design 3

Chapter **2**

Materials 39

Preface

Objectives

This text is intended for students beginning the study of mechanical engineering design. The focus is on blending fundamental development of concepts with practical specification of components. Students of this text should find that it inherently directs them into familiarity with both the basis for decisions and the standards of industrial components. For this reason, as students transition to practicing engineers, they will find that this text is indispensable as a reference text. The objectives of the text are to:

- Cover the basics of machine design, including the design process, engineering mechanics and materials, failure prevention under static and variable loading, and characteristics of the principal types of mechanical elements.
- Offer a practical approach to the subject through a wide range of real-world applications and examples.
- Encourage readers to link design and analysis.
- Encourage readers to link fundamental concepts with practical component specification.

New to This Release

Enhancements and modifications to this release are described in the following summaries:

- Chapter 3, *Load and Stress Analysis*. This foundational topic is often assumed to be understood from a previous course in Mechanics of Materials. However, students benefit greatly from a review that focuses their attention on the conceptual basis behind the math, and on being able to home in on the most needed aspects for machine design. This core topic is given a fresh presentation. The goal is to properly identify critical stress elements. This requires clarity in using free-body diagrams, identifying internal stresses, visualizing a stress element, and determining extreme stresses from the stress element.

- Chapter 4, *Deflection and Stiffness*. The coverage of singularity functions in Section 4–6 has been expanded to cover the method sufficiently without assuming prior exposure. The coverage from Section 3–3 has been merged into an expanded consolidated coverage in Section 4–6. The Euler–Bernoulli beam theory is more thoroughly explained in Section 4–3.

- Chapter 14, *Spur and Helical Gears*. The AGMA safety factors are now treated as design factors (as defined by AGMA) to account for unquantifiable aspects of the application. New factors of safety are defined to represent the typical comparison of stress to allowable stress. Also, new factors of safety are introduced to allow linear comparison of load-based failure.

- Chapter 10, *Mechanical Springs*. For fatigue loading of helical compression springs, the Zimmerli experiments and the recommendations based on them are better explained.
- Chapter 1, *Introduction to Mechanical Engineering Design*. This chapter has been generally edited for cleaner presentation. Notable changes are found in Sections 1–9, 1–10, and 1–11 with regard to stress, strength, uncertainty, and factor of safety.

The following sections received notable improvements in presentation:

Section 1–5 The Design Engineer's Professional Responsibilities
Section 1–9 Stress and Strength
Section 1–10 Uncertainty
Section 1–11 Design Factor and Factor of Safety
Section 3–1 Equilibrium and Free-Body Diagrams
Section 3–2 Shear Force and Bending Moments in Beams
Section 3–4 Stress
Section 3–5 Cartesian Stress Components
Section 3–6 Mohr's Circle for Plane Stress
Section 3–7 General Three-Dimensional Stress
Section 3–9 Uniformly Distributed Stresses (direct shear)
Section 3–11 Shear Stresses for Beams in Bending (new Example 3–9)
Section 3–12 Torsion (Noncircular Cross Sections; Power Relations)
Section 3–14 Stresses in Pressurized Cylinders
Section 3–15 Stresses in Rotating Rings
Section 4–3 Deflection Due to Bending
Section 4–6 Beam Deflections by Singularity Functions
Section 5–3 Failure Theories
Section 6–3 Crack Nucleation and Propagation (additional failure examples)
Section 8–2 The Mechanics of Power Screws
Section 10–4 Compression Springs (presetting)
Section 10–9 Fatigue Loading of Helical Compression Springs
Section 13–7 Interference (new interference tables)
Section 14–8 Overload Factor K_o
Section 14–17 Safety Factors S_F and S_H
Section 14–18 Analysis (Factors of safety; updated example problems; updated summary pages)
Section 14–19 Design of a Gear Mesh (updated example problem)

In keeping with the well-recognized accuracy and consistency within this text, minor improvements and corrections are made throughout with each release. Many of these are in response to the diligent feedback from the community of users.

Instructor Supplements

Additional offerings available in the Instructor Resources include:

- *Solutions manual.* The instructor's manual contains solutions to most end-of-chapter nondesign problems.
- *PowerPoint® slides.* Slides outlining the content of the text are provided in PowerPoint format for instructors to use as a starting point for developing lecture presentation materials. The slides include all figures, tables, and equations from the text.

Acknowledgments

The authors would like to acknowledge those who have contributed to this text for over 50 years. Input from instructors, students, and engineers is instrumental in keeping this text current and clean from errors.

Reviewers of This Release

Milo Rice, *California State University, Fullerton*
Mark Hapstack, *Greenville Technical College*
Matthew Brady, *Greenville Technical College*
Naga Aditya Musunuri, *Indiana Institute of Technology*
James Gerard Freygang, *Ivy Tech Community College*
Valmiki Sooklal, *Kennesaw State University*
Shane A. Brauer, *Mississippi State University*
Rungun Nathan, *Penn State Berks*
Zhongming Liang, *Purdue University Fort Wayne*
Nuri Zeytinoglu, *Purdue University Northwest*
William Palm, *Roger Williams University*
Keivan Davami, *University of Alabama*
Todd Nelson, *University of Southern Indiana*
Brendan O'Toole, *University of Nevada, Las Vegas*
Louie Yaw, *Walla Walla University*
Gene Liao, *Wayne State University*
Richard Meyer, *Western Michigan University*

Publisher's Note

McGraw Hill Education (India) invites suggestions and comments from you, all of which can be sent to *support.india@mheducation.com* (kindly mention the title and author name in the subject line). Piracy related issues may also be reported

This is a list of common symbols used in machine design and in this book. Specialized use in a subject-matter area often attracts fore and post subscripts and superscripts. To make the table brief enough to be useful, the symbol kernels are listed. See Table 14–1 for spur and helical gearing symbols, and Table 15–1 for bevel-gear symbols.

A	Area, coefficient
a	Distance
B	Coefficient, bearing length
Bhn	Brinell hardness
b	Distance, fatigue strength exponent, Weibull shape parameter, width
C	Basic load rating, bolted-joint constant, center distance, coefficient of variation, column end condition, correction factor, specific heat capacity, spring index, radial clearance
c	Distance, fatigue ductility exponent, radial clearance
COV	Coefficient of variation
D	Diameter, helix diameter
d	Diameter, distance
E	Modulus of elasticity, energy, error
e	Distance, eccentricity, efficiency, Naperian logarithmic base
F	Force, fundamental dimension force
f	Coefficient of friction, frequency, function
fom	Figure of merit
G	Torsional modulus of elasticity
g	Acceleration due to gravity, function
H	Heat, power
H_B	Brinell hardness
HRC	Rockwell C-scale hardness
h	Distance, film thickness
$\hbar_{CR}$	Combined overall coefficient of convection and radiation heat transfer
I	Integral, linear impulse, mass moment of inertia, second moment of area
i	Index
$\mathbf{i}$	Unit vector in x-direction
J	Mechanical equivalent of heat, polar second moment of area, geometry factor
$\mathbf{j}$	Unit vector in the y-direction
K	Service factor, stress-concentration factor, stress-augmentation factor, torque coefficient
k	Marin endurance limit modifying factor, spring rate
$\mathbf{k}$	Unit vector in the z-direction
L	Length, life, fundamental dimension length

$\mathscr{L}$	Life in hours
l	Length
M	Fundamental dimension mass, moment
$\mathbf{M}$	Moment vector, mobility vector
m	Mass, slope, strain-strengthening exponent
N	Normal force, number, rotational speed, number of cycles
n	Load factor, rotational speed, factor of safety
n_d	Design factor
P	Force, pressure, diametral pitch
PDF	Probability density function
p	Pitch, pressure, probability
Q	First moment of area, imaginary force, volume
q	Distributed load, notch sensitivity
R	Radius, reaction force, reliability, Rockwell hardness, stress ratio, reduction in area
$\mathbf{R}$	Vector reaction force
r	Radius
$\mathbf{r}$	Distance vector
S	Sommerfeld number, strength
s	Distance, sample standard deviation, stress
T	Temperature, tolerance, torque, fundamental dimension time
$\mathbf{T}$	Torque vector
t	Distance, time, tolerance
U	Strain energy
u	Strain energy per unit volume
V	Linear velocity, shear force
v	Linear velocity
W	Cold-work factor, load, weight
w	Distance, gap, load intensity
X	Coordinate, truncated number
x	Coordinate, true value of a number, Weibull parameter
Y	Coordinate
y	Coordinate, deflection
Z	Coordinate, section modulus, viscosity
z	Coordinate, dimensionless transform variable for normal distributions
α	Coefficient, coefficient of linear thermal expansion, end-condition for springs, thread angle
β	Bearing angle, coefficient
Δ	Change, deflection
δ	Deviation, elongation
ϵ	Eccentricity ratio
ε	Engineering strain
$\tilde{\varepsilon}$	True or logarithmic strain
$\tilde{\varepsilon}_f$	True fracture strain
ε'_f	Fatigue ductility coefficient
Γ	Gamma function, pitch angle
γ	Pitch angle, shear strain, specific weight
λ	Slenderness ratio for springs
μ	Absolute viscosity, population mean
ν	Poisson ratio

ω	Angular velocity, circular frequency
ϕ	Angle, wave length
ψ	Slope integral
ρ	Radius of curvature, mass density
σ	Normal stress
σ_a	Alternating stress, stress amplitude
σ_{ar}	Completely reversed alternating stress
σ_m	Mean stress
σ_0	Nominal stress, strength coefficient or strain-strengthening coefficient
σ'_f	Fatigue strength coefficient
$\tilde{\sigma}$	True stress
$\tilde{\sigma}_f$	True fracture strength
σ'	Von Mises stress
$\hat{\sigma}$	Standard deviation
τ	Shear stress
θ	Angle, Weibull characteristic parameter
¢	Cost per unit weight
$	Cost

SHIGLEY'S

MECHANICAL ENGINEERING DESIGN

SI Units

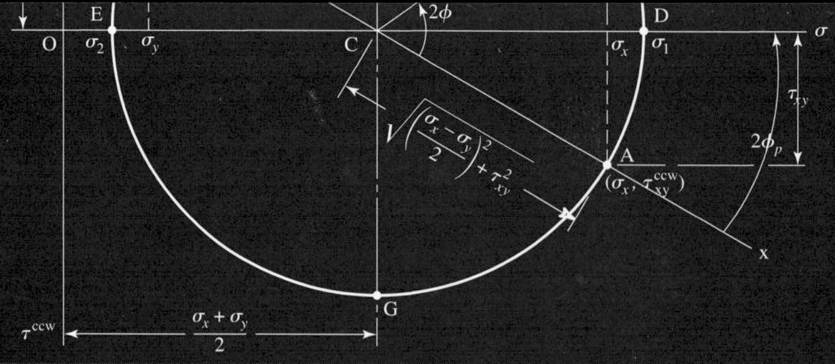

part 1

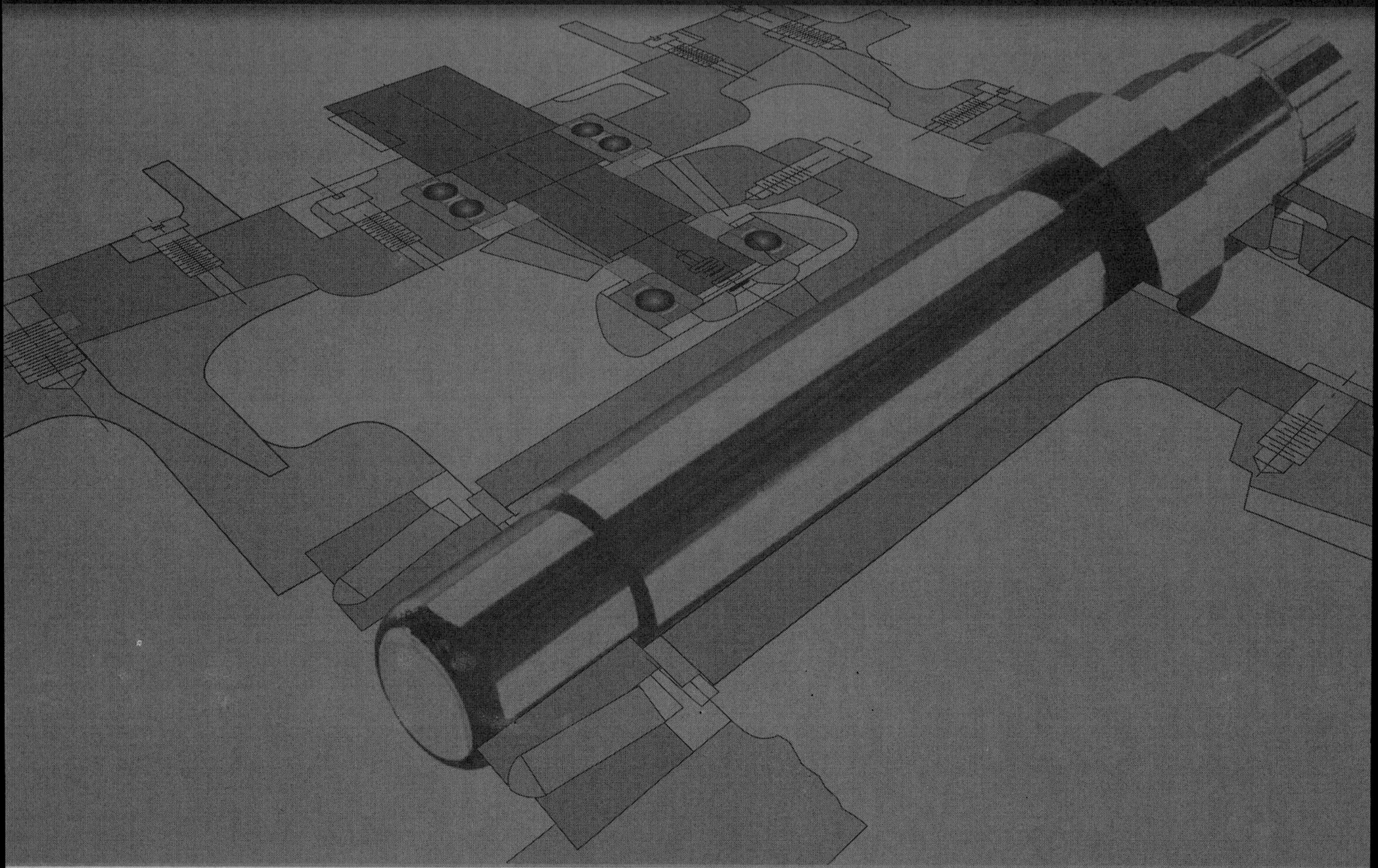

Courtesy of Dee Dehokenanan

Basics

1 Introduction to Mechanical Engineering Design

Monty Rakusen/Cultura/Getty Images

Chapter Outline

Mechanical engineering design is a complex process. No single textbook can adequately address all of the aspects that are interwoven into the topic. Some design topics that warrant special attention include developing creativity, design methodology, and mathematical tools for analysis and optimization. This textbook focuses on physical systems, and the practical specification of components of these systems. This requires the incorporation of many prerequisite topics, such as physics, materials, metallurgy, statics, dynamics, and mechanics of materials. The goal is to use fundamental principles to guide the design and specification of mechanical systems that will perform the desired function for the specified life of the part.

This book consists of four parts. Part 1, *Basics,* begins by introducing some fundamental approaches to design. It continues with three chapters reviewing material properties, stress analysis, and stiffness and deflection analysis.

Part 2, *Failure Prevention,* consists of two chapters on the mechanisms of failure of parts subjected to static and dynamic loading conditions.

In Part 3, *Design of Mechanical Elements,* the concepts of Parts 1 and 2 are applied to the analysis, selection, and design of specific mechanical elements such as shafts, fasteners, weldments, springs, bearings, gears, belts, chains, and wire ropes.

Part 4, *Special Topics,* provides introductions to two important methods used in mechanical design—finite element analysis and geometric dimensioning and tolerancing.

1–1 Design

To design is to formulate a plan for the satisfaction of a specified need. It involves a mixture of creativity, organization, estimation, analysis, and optimization. It often leans heavily on the designer's experience and exposure to other designs.

Design is an innovative and highly iterative process. It is also a decision-making process. Decisions sometimes have to be made with too little information, occasionally with just the right amount of information, or with an excess of partially contradictory information. Decisions are sometimes made tentatively, with the right reserved to adjust as more becomes known. The point is that the engineering designer has to be personally comfortable with a decision-making, problem-solving role.

A designer's personal resources of creativity, communicative ability, and problem-solving skill are intertwined with the knowledge of technology and first principles. Engineering tools (such as mathematics, statistics, analysis software, simulation software, and graphics) are combined to produce a plan that, when carried out, produces a product that is *functional, safe, reliable, competitive, usable, manufacturable, and marketable*.

1–2 Mechanical Engineering Design

Mechanical engineering design involves all the disciplines of mechanical engineering, as well as foundational topics such as mechanics of solids and fluids, mass and momentum transport, manufacturing processes, and electrical and information theory.

Real problems resist compartmentalization. For example, a simple journal bearing involves fluid flow, heat transfer, friction, energy transport, material selection, thermomechanical treatments, statistical descriptions, and so on. There are phrases such as machine design, machine-element design, machine-component design, systems design, and fluid-power design. All of these phrases are somewhat more focused *examples* of mechanical engineering design. They all draw on the same bodies of knowledge, are similarly organized, and require similar skills.

1–3 Phases and Interactions of the Design Process

The complete design process, from start to finish, is often outlined in a simplistic form as in Figure 1–1. The process begins with an identification of a need and a decision to do something about it. After many iterations, the process ends with the documentation and presentation of the plans for satisfying the need. The development of specific strategies and tools to efficiently and effectively carry out this process is called *design methodology*, for which many resources are available for in-depth study. For the purposes of this text, an overview of the fundamental tasks is sufficient.

Identification of Need

Identification of need generally starts the design process. Recognition of the need and phrasing the need often constitute a highly creative act, because the need may be only a vague discontent, a feeling of uneasiness, or a sensing that something is not right. The need is often not evident at all; recognition can be triggered by a particular adverse circumstance or a set of random circumstances that arises almost simultaneously. For example, the need to do something about a food-packaging machine may be indicated by the noise level, by a variation in package weight, and by slight but perceptible variations in the quality of the packaging or wrap.

Definition of Problem

There is a distinct difference between the identification of the need and the definition of the problem. The *definition of problem* is more specific and must include all the specifications for the object that is to be designed. The specifications are the input and output quantities, the characteristics and dimensions of the space the object must occupy, and all the limitations on these quantities. We can regard the object to be designed as something in a black box. In this case we must specify the inputs and outputs of the box, together with their characteristics and limitations. The specifications define the cost, the number to be manufactured, the expected life, the range, the

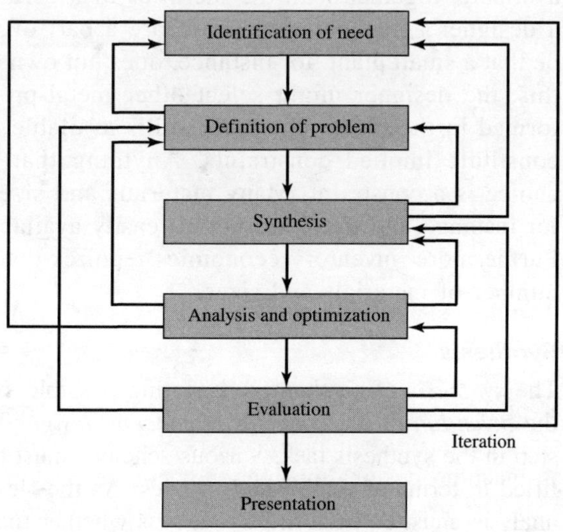

Figure 1–1

The phases in design, acknowledging the many feedbacks and iterations.

operating temperature, and the reliability. Specified characteristics can include the speeds, feeds, temperature limitations, maximum range, expected variations in the variables, dimensional and weight limitations, and more. An example of a specification is given in Section 1–18. This example is a case study of a power transmission that is presented in detail in Chapter 18, and referenced throughout this text.

The problem definition must be sufficiently complete to allow comparison and measurement of the performance of all competing designs. Anything that is not specified in the specifications does not need to be considered when evaluating a proposed design. Any important aspect of the problem that must be considered is referred to as a *design consideration.* Some design considerations that are often of importance for an engineering problem are as follows:

1	Functionality	**14**	Noise
2	Strength/stress	**15**	Styling
3	Distortion/deflection/stiffness	**16**	Shape
4	Wear	**17**	Size
5	Corrosion	**18**	Control
6	Safety	**19**	Thermal properties
7	Reliability	**20**	Surface
8	Manufacturability	**21**	Lubrication
9	Utility	**22**	Marketability
10	Cost	**23**	Maintenance
11	Friction	**24**	Volume
12	Weight	**25**	Liability
13	Life	**26**	Remanufacturing/resource recovery

Some design considerations are obvious, such as the need for a component to not break. However, some design considerations are more elusive and require careful thought to identify aspects that are important to a final solution. There are many implied specifications that result either from the designer's particular environment or from the nature of the problem itself. The manufacturing processes that are available, together with the facilities of a certain plant, constitute restrictions on a designer's freedom, and hence are a part of the implied specifications. It may be that a small plant, for instance, does not own cold-working machinery. Knowing this, the designer might select other metal-processing methods that can be performed in the plant. The labor skills available and the competitive situation also constitute implied constraints. Anything that limits the designer's freedom of choice is a constraint. Many materials and sizes are listed in supplier's catalogs, for instance, but these are not all easily available and shortages frequently occur. Furthermore, inventory economics requires that a manufacturer stock a minimum number of materials and sizes.

Synthesis

The *synthesis* of a scheme connecting possible system elements is sometimes called the *invention of the concept or concept design.* This is the first and most important step in the synthesis task. Various schemes must be proposed, investigated, and quantified in terms of established metrics. As the fleshing out of the scheme progresses, analyses must be performed to assess whether the system performance is satisfactory or better, and, if satisfactory, just how well it will perform. In the early phases of synthesis, the analysis may be based largely on estimations and rough calculations, with the goal of determining the merits of the potential design concepts. System schemes that do not survive analysis are revised, improved, or discarded. Those with potential are revised to improve the potential for satisfying the design specifications. Competing schemes are compared so that the path leading to the most competitive product can be chosen.

Analysis and Optimization

When a design concept has been identified as having good potential, the *analysis and optimization* stage is engaged with greater rigor. In this stage, estimations are replaced with mathematical models. These models may be straightforward equations, such as from a free-body diagram to represent the forces, or from the mechanics of materials to represent the stress situation. Or the models may be complex computer models, such as a motion simulation of the physical system. Using the mathematical models, we flesh out specific details of the design (e.g., dimensions, power requirements, component selection, etc.). We move beyond rough estimation to a specific design that can be manufactured, with a basis to justify that it will satisfy the design specifications.

Design is an iterative process in which we proceed through several steps, evaluate the results, and then return to an earlier phase of the procedure as needed. Thus, we may synthesize several components of a system, analyze and optimize them, and return to synthesis to see what effect this has on the remaining parts of the system. For example, the design of a system to transmit power requires attention to the design and selection of individual components (e.g., gears, bearings, shaft). However, as is often the case in design, these components are not independent. In order to design the shaft for stress and deflection, it is necessary to know the applied forces. If the forces are transmitted through gears, it is necessary to know the gear specifications in order to determine the forces that will be transmitted to the shaft. But stock gears come with certain bore sizes, requiring knowledge of the necessary shaft diameter. Clearly, rough estimates will need to be made in order to proceed through the process, refining and iterating until a final design is obtained that is satisfactory for each individual component as well as for the overall design specifications. Throughout the text we will elaborate on this process for the case study of a power transmission design.

Evaluation

As indicated in Figure 1–1, *evaluation* is a significant phase of the total design process. Evaluation is the final proof of a successful design and often involves the testing of a prototype in the laboratory. Here we wish to discover if the design really satisfies the needs. Is it reliable? Will it compete successfully with similar products? Is it economical to manufacture and to use? Is it easily maintained and adjusted? Can a profit be made from its sale or use? How likely is it to result in product-liability lawsuits? Is insurance easily and cheaply obtained? Is it likely that recalls will be needed to replace defective parts or systems? The project designer or design team will need to address a myriad of engineering and non-engineering questions.

Presentation

Communicating the design to others is the final, vital *presentation* step in the design process. Presentation is a selling job. The engineer, when presenting a new solution to administrative, management, or supervisory persons, is attempting to prove to them that their solution is a better one. Unless this can be done successfully, the time and effort spent on obtaining the solution have been largely wasted.

1–4 Design Tools and Resources

Today, the engineer has a great variety of tools and resources available to assist in the solution of design problems. Robust computer software packages provide tools of immense capability for the design, analysis, and simulation of mechanical components. In addition to these tools, the engineer always needs technical information, either in the form of basic science/engineering behavior or the characteristics of specific off-the-shelf components. The resources can range from science/engineering textbooks to manufacturers' websites and product catalogs.

Computational Tools

Computer-aided design (CAD) software allows the development of three-dimensional (3-D) designs from which conventional two-dimensional orthographic views with automatic dimensioning can be produced. Manufacturing tool paths can be generated from the computer 3-D models, and in many cases, parts can be created directly from the 3-D database using rapid prototyping additive methods referred to as *3-D printing* or STL (*stereolithography*). Another advantage of a 3-D database is that it allows rapid and accurate calculation of mass properties such as mass, location of the center of gravity, and mass moments of inertia. Other geometric properties such as areas and distances between points are likewise easily obtained. There are a great many CAD software packages available such as CATIA, AutoCAD, Autodesk Inventor, IRONCAD, NX, MicroStation, Solid Edge, SolidWorks, and Creo.[1]

The term *computer-aided engineering* (CAE) generally applies to all computer-related engineering applications. With this definition, CAD can be considered as a subset of CAE. Some computer software packages perform specific engineering analysis and/or simulation tasks that assist the designer. Such software fits into two categories: engineering-based and non-engineering-specific. Some examples of engineering-based software for mechanical engineering applications—software that might also be integrated within a CAD system—include finite-element analysis (FEA) programs (see Chapter 19) for analysis of stress and deflection, vibration, and heat transfer (e.g., ABAQUS, ALGOR, ANSYS, MSC Nastran, etc.); computational fluid dynamics (CFD) programs for fluid-flow analysis and simulation (e.g., CFD++, Star-CCM+, Fluent, etc.); and programs for simulation of dynamic force and motion (e.g., Adams, LMS Virtual.Lab Motion, Simulink, Working Model, etc.).

Examples of non-engineering-specific computer-aided applications include software for word processing, spreadsheet, and mathematical solvers (e.g., Maple, MathCad, MATLAB, Mathematica, TKsolver, etc.).

Computer software is no substitute for the human thought process. The engineer is the driver here; the computer is the vehicle to assist on the journey to a solution. Numbers generated by a computer can be far from the truth if incorrect data are entered, if the application or the output of the program is misinterpreted, if the

[1]The commercial softwares mentioned in this section are representative of the available options and are not meant to be endorsements. The resources are continually evolving and no list can ever be exhaustive or remain entirely current.

program contains bugs, etc. It is the engineer's responsibility to assure the validity of the results, by checking the application and results carefully, and by performing benchmark testing on problems with known solutions.

Acquiring Technical Information

A design engineer will need to keep abreast of past and current developments in the field of work. Some sources of information are provided as a starting point.

- **Libraries (community, university, and private).** Engineering dictionaries and encyclopedias, textbooks, monographs, handbooks, indexing and abstract services, journals, translations, technical reports, patents, and business sources/brochures/catalogs.
- **Government sources.** Departments of Defense, Commerce, Energy, and Transportation; NASA; Government Printing Office; U.S. Patent and Trademark Office; National Technical Information Service; and National Institute for Standards and Technology.
- **Professional societies.** American Society of Mechanical Engineers, Society of Manufacturing Engineers, Society of Automotive Engineers, American Society for Testing and Materials, and American Welding Society.
- **Commercial vendors.** Catalogs, technical literature, test data, samples, and cost information.
- **Internet.** Online search tools for engineering products, vendors, tools, and patents.[2]

1–5 The Design Engineer's Professional Responsibilities

In general, the design engineer is required to satisfy the needs of customers (management, clients, consumers, etc.) and is expected to do so in a competent, responsible, ethical, and professional manner. Much of engineering course work and practical experience focuses on competence, but a competent engineer should be very attentive to the broader aspects of professional responsibility. Certainly, the basic principles of professional responsibility (e.g., good work ethic, integrity, consideration of public welfare, etc.) apply to everyone regardless of profession. But it is particularly important to embrace the importance of professional responsibility in professions which have a direct impact on the safety, health, and welfare of society. The National Society of Professional Engineers (NSPE) is an excellent resource, providing guidance for engineers to solidify their awareness and commitment to professional behavior in the engineering profession. The six fundamental canons from the *NSPE Code of Ethics for Engineers* are reproduced here.[3]

Engineers, in the fulfillment of their professional duties, shall:

1. *Hold paramount the safety, health, and welfare of the public.*
2. *Perform services only in areas of their competence.*
3. *Issue public statements only in an objective and truthful manner.*

---

[2]Some helpful resources include www.globalspec.com, www.engnetglobal.com, www.efunda.com, www.thomasnet.com, and www.uspto.gov.

[3]The fundamental canons are reprinted with permission of the National Society of Professional Engineers from the NSPE Code of Ethics for Engineers, available at www.nspe.org/resources/ethics/code-ethics. Many other resources are available at www.nspe.org.

4. *Act for each employer or client as faithful agents or trustees.*
5. *Avoid deceptive acts.*
6. *Conduct themselves honorably, responsibly, ethically, and lawfully so as to enhance the honor, reputation, and usefulness of the profession.*

1–6 Standards and Codes

Engineers must be cognizant of applicable standards and codes. A *standard* is a set of specifications for parts, materials, or processes intended to achieve uniformity, efficiency, and a specified quality. Standards may be generated within a company, across an industry, within a country, or internationally. Standards are intended to define a recognized good practice, or an agreed upon uniformity, or a minimum level of acceptability. One of the important purposes of a standard is to limit the multitude of variations that can arise from the arbitrary creation of a part, material, or process. For example, components such as bolts, bearings, and gears are universally manufactured according to standards that have been set forth by appropriate organizations. It is almost always desirable to incorporate existing standards into designs to optimize cost, interchangeability, and efficiency.

A *code*, or *design code*, is a set of specifications for the analysis, design, manufacture, and construction of something. The purpose of a code is to achieve a specified degree of safety, efficiency, and performance or quality. Codes have been established in many application areas in which the safety of the public is at stake. By designing according to the applicable codes, the design engineer is able to take advantage of the extensive knowledge, experience, and testing that have been used to specify the codes. In many industries, adherence to applicable codes is mandated by law. A search for applicable codes is a necessary first step in engaging in a new design. Usually, the engineering department of the company will have experiential awareness of applicable standards and will provide guidance to an engineer who is new to the field. It is important to observe that safety codes do not imply absolute safety. In fact, absolute safety is impossible to obtain. Sometimes the unexpected event really does happen. Designing a building to withstand a 120 mi/h wind does not mean that the designers think a 140 mi/h wind is impossible; it simply means that they think it is highly improbable.

All of the organizations and societies listed here have established specifications for standards and safety or design codes. The name of the organization provides a clue to the nature of the standard or code. Some of the standards and codes, as well as addresses, can be obtained in most technical libraries or on the Internet. Some of the organizations of particular interest to mechanical engineers are:

Aluminum Association (AA)
American Bearing Manufacturers Association (ABMA)
American Gear Manufacturers Association (AGMA)
American Institute of Steel Construction (AISC)
American Iron and Steel Institute (AISI)
American National Standards Institute (ANSI)
American Society of Heating, Refrigerating and Air-Conditioning Engineers (ASHRAE)
American Society of Mechanical Engineers (ASME)
American Society of Testing and Materials (ASTM)

American Welding Society (AWS)
ASM International
British Standards Institution (BSI)
Industrial Fasteners Institute (IFI)
Institute of Transportation Engineers (ITE)
Institution of Mechanical Engineers (IMechE)
International Bureau of Weights and Measures (BIPM)
International Federation of Robotics (IFR)
International Standards Organization (ISO)
National Association of Power Engineers (NAPE)
National Institute for Standards and Technology (NIST)
Society of Automotive Engineers (SAE)

1–7 Economics

The consideration of cost plays such an important role in the design decision process that we could easily spend as much time in studying the cost factor as in the study of the entire subject of design. Here we introduce only a few general concepts and simple rules.

First, observe that nothing can be said in an absolute sense concerning costs. Materials and labor usually show an increasing cost from year to year. But the costs of processing the materials can be expected to exhibit a decreasing trend because of the use of automated machine tools and robots. The cost of manufacturing a single product will vary from city to city and from one plant to another because of overhead, labor, taxes, and freight differentials and the inevitable slight manufacturing variations.

Standard Sizes

The use of standard or stock sizes is a first principle of cost reduction. An engineer who specifies an AISI 1020 bar of hot-rolled steel 53 mm square has added cost to the product, provided that a bar 50 or 60 mm square, both of which are preferred sizes, would do equally well. The 53-mm size can be obtained by special order or by rolling or machining a 60-mm square, but these approaches add cost to the product. To ensure that standard or preferred sizes are specified, designers must have access to stock lists of the materials they employ.

A further word of caution regarding the selection of preferred sizes is necessary. Although a great many sizes are usually listed in catalogs, they are not all readily available. Some sizes are used so infrequently that they are not stocked. A rush order for such sizes may add to the expense and delay. Thus you should also have access to a list such as those in Table A–15 for preferred inch and millimeter sizes.

There are many purchased parts, such as motors, pumps, bearings, and fasteners, that are specified by designers. In the case of these, too, you should make a special effort to specify parts that are readily available. Parts that are made and sold in large quantities usually cost somewhat less than the odd sizes. The cost of rolling bearings, for example, depends more on the quantity of production by the bearing manufacturer than on the size of the bearing.

Large Tolerances

Among the effects of design specifications on costs, tolerances are perhaps most significant. Tolerances, manufacturing processes, and surface finish are interrelated

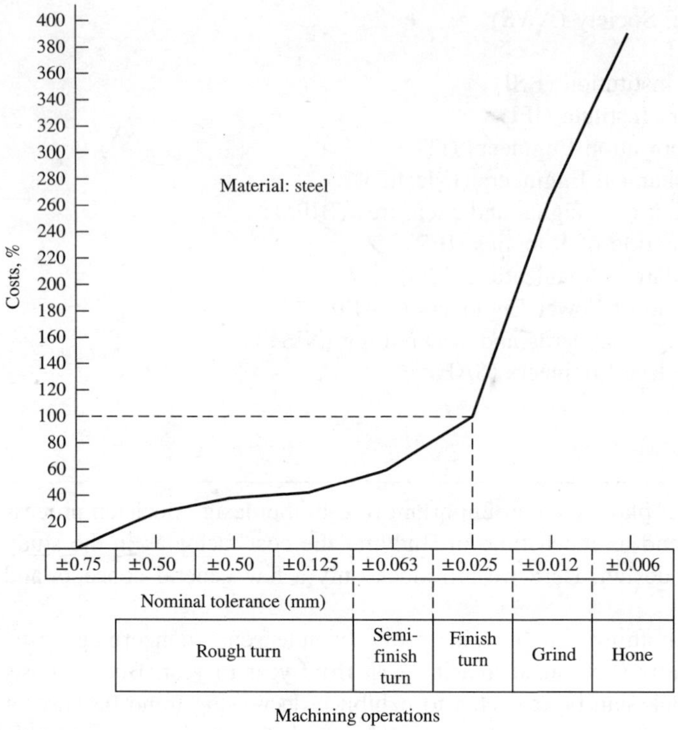

Figure 1–2

Cost versus tolerance/machining process. (*Source: From Ullman, David G.* The Mechanical Design Process. *3rd ed. New York: McGraw Hill, 2003.*)

and influence the producibility of the end product in many ways. Close tolerances may necessitate additional steps in processing and inspection or even render a part completely impractical to produce economically. Tolerances cover dimensional variation and surface-roughness range and also the variation in mechanical properties resulting from heat treatment and other processing operations.

Because parts having large tolerances can often be produced by machines with higher production rates, costs will be significantly smaller. Also, fewer such parts will be rejected in the inspection process, and they are usually easier to assemble. A plot of cost versus tolerance/machining process is shown in Figure 1–2, and illustrates the drastic increase in manufacturing cost as tolerance diminishes with finer machining processing.

Breakeven Points

Sometimes it happens that, when two or more design approaches are compared for cost, the choice between the two depends on a set of conditions such as the quantity of production, the speed of the assembly lines, or some other condition. There then occurs a point corresponding to equal cost, which is called the *breakeven point.*

As an example, consider a situation in which a certain part can be manufactured at the rate of 25 parts per hour on an automatic screw machine or 10 parts per hour on a hand screw machine. Let us suppose, too, that the setup time for the automatic is 3 h and that the labor cost for either machine is $20 per hour, including overhead. Figure 1–3 is a graph of cost versus production by the two methods. The breakeven point for this example corresponds to 50 parts. If the desired production is greater than 50 parts, the automatic machine should be used.

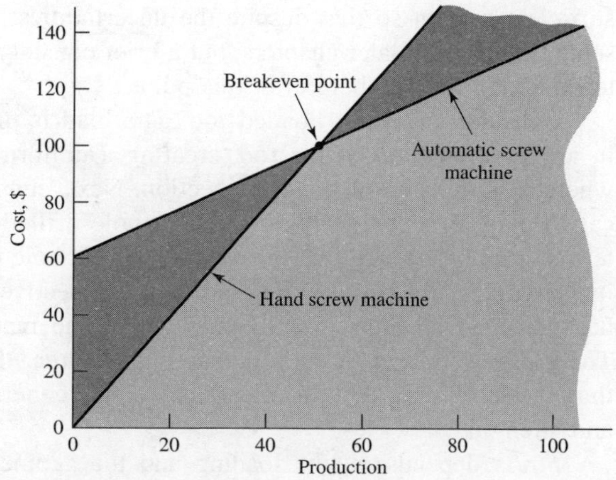

Figure 1–3

A breakeven point.

Cost Estimates

There are many ways of obtaining relative cost figures so that two or more designs can be roughly compared. A certain amount of judgment may be required in some instances. For example, we can compare the relative value of two automobiles by comparing the dollar cost per pound of weight. Another way to compare the cost of one design with another is simply to count the number of parts. The design having the smaller number of parts is likely to cost less. Many other cost estimators can be used, depending upon the application, such as area, volume, horsepower, torque, capacity, speed, and various performance ratios.

1–8 Safety and Product Liability

The *strict liability* concept of product liability generally prevails in the United States. This concept states that the manufacturer of an article is liable for any damage or harm that results because of a defect. And it doesn't matter whether the manufacturer knew about the defect, or even could have known about it. For example, suppose an article was manufactured, say, 10 years ago. And suppose at that time the article could not have been considered defective on the basis of all technological knowledge then available. Ten years later, according to the concept of strict liability, the manufacturer is still liable. Thus, under this concept, the plaintiff needs only to prove that the article was defective and that the defect caused some damage or harm. Negligence of the manufacturer need not be proved.

The best approaches to the prevention of product liability are good engineering in analysis and design, quality control, and comprehensive testing procedures. Advertising managers often make glowing promises in the warranties and sales literature for a product. These statements should be reviewed carefully by the engineering staff to eliminate excessive promises and to insert adequate warnings and instructions for use.

1–9 Stress and Strength

The survival of many products depends on how the designer adjusts the maximum stresses in a component to be less than the component's strength at critical locations. The designer must allow the maximum stress to be less than the strength by a sufficient margin so that despite the uncertainties, failure is rare. This topic is presented in depth in later chapters, but a brief consideration of some basic concepts and terminology may be helpful at this point.

Consider an axially loaded rod. The load is uniformly distributed to all points in any cross section of the rod, creating a uniform stress distribution of $\sigma = F/A$, where A is the area of the cross section. Next, consider an identical rod with identical loading, except that this rod has a hole drilled laterally through it. In layman's terms, it might be said that the addition of the hole has "weakened" the bar, reducing its "strength." But from an engineering perspective, we will recognize that the hole has not changed any strength property of the material; it has changed the stress. The cross section at the hole has a smaller area, thus resulting in a larger stress at that cross section. Let us then note some general observations regarding stress and strength.

Stress depends on the loading and the geometry. Stress is independent of the material. Note that familiar stress equations are functions of load and geometry, and do not contain any material property terms (e.g., F/A, My/I, Tr/J). Stress is, in general, not uniform throughout a part. In fact, we will often evaluate the stress state at discrete locations (called stress elements) within a part, seeking locations where the stress is more critical than other locations. The loads are often externally applied and readily determined from a free-body diagram. But loading may also be a result of a body force, such as weight, or an internal loading from restrained deformation due to temperature change or a manufacturing operation.

Strengths are the magnitudes of stresses at which something of interest occurs, such as the proportional limit, 0.2 percent-offset yielding, or fracture (see Section 2–1). In many cases, such events represent the stress level at which a particular loss of function occurs.

Strength is a property of a material or of a mechanical element. The strength of an element depends on the composition, the treatment, and the processing of the material. Strength is not dependent on the loading or the geometry of the part. Consider, for example, a shipment of springs. We can associate a strength with a specific spring. When this spring is incorporated into a machine, external forces are applied that result in load-induced stresses in the spring, the magnitudes of which depend on its geometry and are independent of the material and its processing. If the spring is removed from the machine undamaged, the stress due to the external forces will return to zero. But the strength remains as one of the properties of the spring. Remember, then, that *strength is an inherent property of a part,* a property built into the part because of the use of a particular material and process.

Various metalworking and heat-treating processes, such as forging, rolling, and cold forming, cause variations in the strength from point to point throughout a part. The spring cited previously is quite likely to have a strength on the outside of the coils different from its strength on the inside because the spring has been formed by a cold winding process, and the two sides may not have been deformed by the same amount. Remember, too, therefore, that a strength value given for a part may apply to only a particular point or set of points on the part.

In this book we shall use the capital letter S to denote *strength,* with appropriate subscripts to denote the type of strength. Thus, S_y is a yield strength, S_u an ultimate strength, S_{sy} a shear yield strength, and S_e an endurance strength. A review of material properties and many of the common strengths is given in Chapter 2.

We shall employ the Greek letters σ (sigma) and τ (tau) to designate normal and shear *stresses,* respectively. Again, various subscripts will indicate some special characteristic. For example, σ_1 is a principal normal stress, σ_y a normal stress component in the y direction, and σ_r a normal stress component in the radial direction. A review of the common stress situations is given in Chapter 3.

1–10 Uncertainty

Uncertainties in machinery design abound. Examples of uncertainties concerning stress and strength include

- Composition of material and the effect of variation on properties.
- Variations in properties from place to place within a bar of stock.
- Effect of processing locally, or nearby, on properties.
- Effect of nearby assemblies such as weldments and shrink fits on stress conditions.
- Effect of thermomechanical treatment on properties.
- Intensity and distribution of loading.
- Validity of mathematical models used to represent reality.
- Intensity of stress concentrations.
- Influence of time on strength and geometry.
- Effect of corrosion.
- Effect of wear.
- Uncertainty as to the length of any list of uncertainties.

Engineers must accommodate uncertainty. Uncertainty always accompanies change. Material properties, load variability, fabrication fidelity, and validity of mathematical models are among concerns to designers.

There are mathematical models to address uncertainties. The primary techniques are the deterministic and the stochastic methods. The *deterministic method*, also known as the *factor of safety method*, attempts to determine a value for a "strength" parameter and a value for a "stress" parameter and provide sufficient margin between the two to predict successful functionality. The *stochastic method*, also known as the *reliability method*, is based on the measured or predicted statistical nature of the "strength" and "stress" parameters, allowing a prediction of the percent probability of success. These methods are described in the following sections.

1–11 Design Factor and Factor of Safety

The deterministic factor of the safety method is intuitive to implement, yet it requires significant judgment to use effectively. Let's take a simple example. Suppose we want to use a rope to suspend a child's swing. We know that we want a rope that can support the weight of a child without breaking. If we know that the child weighs 25 kg and we have a rope that breaks when loaded with 50 kg, then we may believe it to be safe to use the rope for this application since there is a factor of two between the child's weight and the rope's capacity. We have intuitively applied the deterministic method to determine the applied load, the load capacity, and the factor of safety as the ratio, such that

$$\text{factor of safety} = \frac{\text{load capacity}}{\text{applied load}} = \frac{50 \text{ kg}}{25 \text{ kg}} = 2$$

A factor of safety of 2 may be perfectly suitable if you are in the backyard one day and decide to tie a rope to the tree branch to let a child swing on it. But, let us consider several factors.

Both the load capacity and the applied load have uncertainties. On what basis did we conclude that the rope capacity is 50 kg? Did we test it? Was this labeled on the rope packaging as a load rating? Do we know what a reported load rating is based on? What do we know about the distribution of the rope's capacity if different sections of the rope are tested? Is the rope new or worn? Then consider the applied load. How accurately do we know the weight? Is it estimated or measured? Will the child be bouncing, climbing, and swinging to an extent that a static applied load is not sufficient, and dynamic loading should be determined? What about unexpected uses of the swing? Do we need to account for a heavier person using the swing? Another consideration is the ramifications of failure. A small tumble onto soft grass is much different than a fall onto concrete from a high swinging arc. How well do we know the situation?

The point here is that the deterministic approach is clearly simple, but its usefulness depends heavily on how well we can identify and make judgments about the uncertainties. The better we are able to model the situation, the better we can have confidence in our ability to determine a meaningful factor of safety that is at an appropriate level for the needs of the design problem.

Let us generalize the method and more formally define it. The factor of safety, sometimes called the design factor, is the ratio of a loss-of-function parameter to an applied parameter.

$$\text{factor of safety} = n = \frac{\text{loss-of-function parameter}}{\text{applied parameter}} \tag{1–1}$$

The parameter to be compared in this ratio can be load, stress, deflection, etc. The loss-of-function parameter is a limiting value of the parameter at which it is no longer able to sustain the desired function of that parameter. Some examples include a load at which the rope breaks, a level of stress causing the part to permanently yield, or a deflection that is unacceptable. Perhaps the most classical representation of the factor of safety is by comparing strength to stress.

$$n = \frac{Strength}{Stress} = \frac{S}{\sigma} \tag{1–2}$$

where the strength term may be any limiting value of stress, such as yield strength or ultimate strength.

The factor of safety parameters must be applicable at the same location in the part, and of the same type (e.g., both normal stresses or both shear stresses or both deflections), and of the same units. There will potentially be multiple factors of safety for the same design problem, taking into account multiple locations at which failure might be expected, as well as multiple failure modes that are possible. The smallest factor of safety indicates the critical point that is predicted to be most likely to fail.

A factor of safety less than unity predicts failure. A factor of safety greater than unity predicts a successful design. In neither case is there any information about a statistical percentage of failures. A higher factor of safety may increase our confidence in achieving a greater reliability, but we are unable to quantify that confidence with this deterministic method. We rely on the designer's assessment of the model and its accuracy, as well as good judgment, to have a sense of the anticipated reliability of a design. It is crucial that this be well documented, for a numerical value of factor of safety has useful meaning only when accompanied with the documented rationale behind it.

It is common to start a design process with a factor of safety that it is desired to achieve. In this case, the desired goal is usually called a *design factor*. The design factor n_d may be used in place of the factor of safety n in Equation (1–1) to solve for the maximum allowable value of the applied parameter.

$$\text{maximum allowable parameter} = \frac{\text{loss-of-function parameter}}{n_d} \tag{1–3}$$

The design factor and the factor of safety are essentially the same thing. But in practice, the design factor is specified up front as a design goal, while the factor of safety is calculated from the realized design. The actual design will usually incorporate standard sizes for dimensions and components, such that the manufactured dimensions and selected components slightly over-achieve the calculated requirements. Accordingly, the factor of safety will typically be slightly larger than the original design factor.

EXAMPLE 1–1

A rod with a cross-sectional area of A and loaded in tension with an axial force of $P = 1000$ kg undergoes a stress of $\sigma = P/A$. Using a material strength of 170 MPa and a *design factor* of 3.0, determine the minimum diameter of a solid circular rod. Using Table A–15, select a preferred fractional diameter and determine the rod's *factor of safety*.

Solution

Since $A = \pi d^2/4$, $\sigma = P/A$, and from Equation (1–3), $\sigma = S/n_d$, then

$$\sigma = \frac{P}{A} = \frac{P}{\pi d^2/4} = \frac{S}{n_d}$$

Solving for d yields

Answer
$$d = \left(\frac{4Pn_d}{\pi S}\right)^{1/2} = \left(\frac{4(10\ 000)3}{\pi(24\ 000)}\right)^{1/2} = 15 \text{ mm}$$

From Table A–15, the next higher preferred size is 170 MPa = 16 mm. Thus, when n_d is replaced with n in the equation developed above, the factor of safety n is

Answer
$$n = \frac{\pi S d^2}{4P} = \frac{\pi(170)16^2}{4(10\ 000)} = 3.42$$

Thus, rounding the diameter has provided a factor of safety that is slightly larger than the original design factor.

Guidance on selecting an appropriate design factor is nebulous because it depends on a judicial consideration of a multitude of issues. The considerations can be summarized in the following three categories:

- Accuracy of the prediction of the situation, including such things as loads, strengths, wear, environment, manufacturing quality, and maintenance
- Cost of overachieving the requirements
- Consequences of failure

The first category has already been discussed. Basically, design factors can be lowered by better understanding of the situation, accurately modeling the behavior, and acquiring better data from testing. The second category reflects the ever-present economic factor. We may use a design factor of 20 on a bolt, while using a design factor of 1.5 on an airplane part for which extra weight is costly. Of course, to be able to use a relatively low design factor for the airplane part, we will put significant effort into the first category of accurate prediction. The third category—consequences of failure—accounts for the need for higher reliabilities when failure has high consequences. When safety is at stake, there will usually be relevant industry codes that specify the design factor, along with the accompanying details regarding how that design factor is to be determined. These codes may be mandated by law. An engineer should seek guidance on applicable codes and standards from the collected knowledge and experience within the company and the industry.

1–12 Reliability and Probability of Failure

The *reliability method* of design is one in which we obtain the distribution of stresses and the distribution of strengths and then relate these two in order to achieve an acceptable success rate. The statistical measure of the probability that a mechanical element will not fail in use is called the *reliability* of that element and is related to the *probability of failure*, p_f.

Probability of Failure

The probability of failure, p_f, is obtained from the *probability density function* (PDF), which represents the distribution of events within a given range of values. A number of standard discrete and continuous probability distributions are commonly applicable to engineering problems. The two most important continuous probability distributions for our use in this text are the *Gaussian* (*normal*) *distribution* and the *Weibull distribution*. The normal distribution is described in this section and in Section 2–2. The Weibull distribution is widely used in rolling-contact bearing design and is described in Chapter 11.

The continuous Gaussian (normal) distribution is an important one whose *probability density function* (PDF) is expressed in terms of its mean, μ_x, and its standard deviation[4] $\hat{\sigma}_x$ as

$$f(x) = \frac{1}{\hat{\sigma}_x\sqrt{2\pi}}\exp\left[-\frac{1}{2}\left(\frac{x-\mu_x}{\hat{\sigma}_x}\right)^2\right] \tag{1–4}$$

Plots of Equation (1–4) are shown in Figure 1–4 for small and large standard deviations. The bell-shaped curve is taller and narrower for small values of $\hat{\sigma}$ and shorter and broader for large values of $\hat{\sigma}$. Note that the area under each curve is unity. That is, the probability of all events occurring is one (100 percent).

To obtain values of p_f, integration of Equation (1–4) is necessary. This can come easily from a table if the variable x is placed in dimensionless form. This is done using the transform

$$z = \frac{x-\mu_x}{\hat{\sigma}_x} \tag{1–5}$$

[4]The symbol σ is normally used for the standard deviation. However, in this text σ is used for stress. Consequently, we will use $\hat{\sigma}$ for the standard deviation.

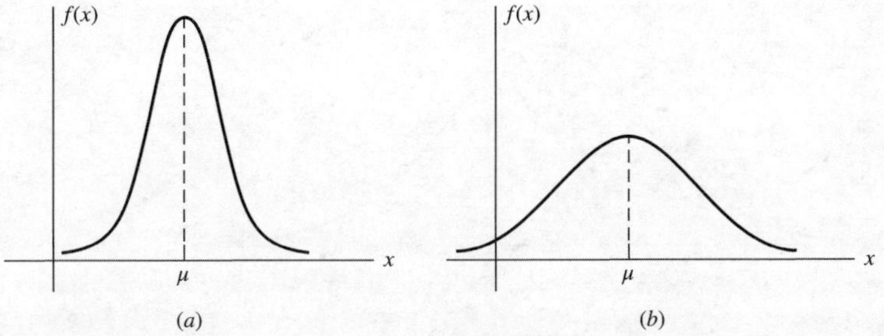

Figure 1–4

The shape of the normal distribution curve: (*a*) small $\hat{\sigma}$; (*b*) large $\hat{\sigma}$.

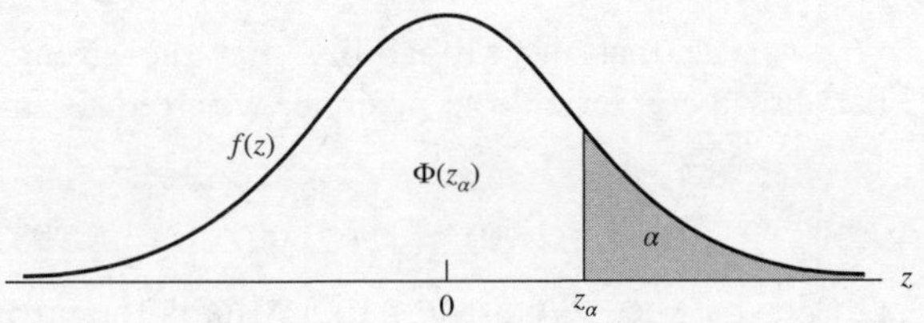

Figure 1–5

Transformed normal distribution function of Table A–10.

The integral of the transformed normal distribution is tabulated in Table A–10, where α is defined, and is shown in Figure 1–5. The value of the normal density function is used so often, and manipulated in so many equations, that it has its own particular symbol, $\Phi(z)$. The transform variant z has a mean value of zero and a standard deviation of unity. In Table A–10, the probability of an observation less than z is $\Phi(z)$ for negative values of z and $1 - \Phi(z)$ for positive values of z.

EXAMPLE 1–2

In a shipment of 250 connecting rods, the mean tensile strength is found to be $\overline{S} = 315$ MPa and has a standard deviation of $\hat{\sigma}_S = 35$ MPa.
(*a*) Assuming a normal distribution, how many rods can be expected to have a strength less than $S = 276.5$ MPa?
(*b*) How many are expected to have a strength between 276.5 MPa and 416.5 MPa?

Solution
(*a*) Substituting in Equation (1–5) gives the transform z variable as

$$z_{276.5} = \frac{x - \mu_x}{\hat{\sigma}_x} = \frac{S - \overline{S}}{\hat{\sigma}_S} = \frac{276.5 - 315}{35} = -1.10$$

The probability that the strength is less than 276.5 MPa can be designated as $F(z) = \Phi(z_{276.5}) = \Phi(-1.10)$. Using Table A–10, and referring to Figure 1–6, we find $\Phi(z_{276.5}) = 0.1357$. So the number of rods having a strength less than 276.5 MPa is,

Answer $$N\Phi(z_{276.5}) = 250(0.1357) = 33.9 \approx 34 \text{ rods}$$

because $\Phi(z_{276.5})$ represents the *proportion* of the population N having a strength less than 276.5 MPa.

Figure 1–6

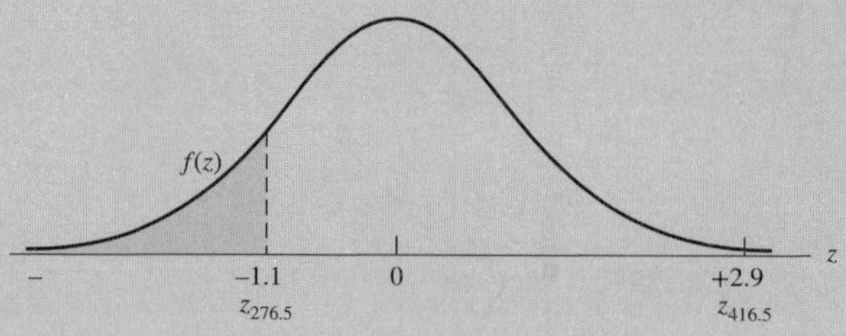

(*b*) Corresponding to $S = 416.5$ MPa, we have

$$z_{416.5} = \frac{416.5 - 315}{35} = 2.90$$

Referring again to Figure 1–6, we see that the probability that the strength is less than 416.5 MPa is $F(z) = \Phi(z_{416.5}) = \Phi(2.90)$. Because the z variable is positive, we need to find the value complementary to unity. Thus, from Table A–10

$$\Phi(2.90) = 1 - \Phi(-2.90) = 1 - 0.001\ 87 = 0.998\ 13$$

The probability that the strength lies between 276.5 and 416.5 MPa is the area between the ordinates at $z_{276.5}$ and $z_{416.5}$ in Figure 1–6. This probability is found to be

$$\begin{aligned} p &= \Phi(z_{416.5}) - \Phi(z_{276.5}) = \Phi(2.90) - \Phi(-1.10) \\ &= 0.998\ 13 - 0.1357 = 0.862\ 43 \end{aligned}$$

Therefore the number of rods expected to have strengths between 276.5 and 416.5 MPa is

Answer $$Np = 250(0.862) = 215.5 \approx 216 \text{ rods}$$

Events typically arise as *discrete distributions,* which can be approximated by continuous distributions. Consider N samples of events. Let x_i be the value of an event ($i = 1, 2, \ldots k$) and f_i is the class frequency or number of times the event x_i occurs within the class frequency range. The *discrete* mean, $\bar{x}$, and standard deviation, defined as s_x, are given by

$$\bar{x} = \frac{1}{N}\sum_{i=1}^{k} f_i x_i \tag{1–6}$$

$$s_x = \sqrt{\frac{\sum_{i=1}^{k} f_i x_i^2 - N\bar{x}^2}{N - 1}} \tag{1–7}$$

EXAMPLE 1–3

Five tons of 50 mm round rods of 1030 hot-rolled steel have been received for workpiece stock. Nine standard-geometry tensile test specimens have been machined from random locations in various rods. In the test report, the ultimate tensile strength was given in MPa. The data in the ranges 427–462, 462–469, 469–504, and 504–511 MPa are given in histographic form as follows:

S_{ut} (MPa)	444.5	465.5	486.5	507.5
f	2	2	3	2

where the values of S_{ut} are the midpoints of each range. Find the mean and standard deviation of the data.

Solution

Table 1–1 provides a tabulation of the calculations for the solution.

Table 1–1

Class Midpoint x, MPa	Class Frequency f	Extension fx	fx^2
444.5	2	889	395 160.50
465.5	2	931	433 380.50
486.5	3	1459.5	710 046.75
507.5	2	1015	515 112.50
	Σ 9	4294.50	2 053 700.25

From Equation (1–6),

Answer $$\bar{x} = \frac{1}{N}\sum_{i=1}^{k} f_i x_i = \frac{1}{9}(4294.50) = 477.2 \text{ MPa}$$

From Equation (1–7),

Answer $$s_x = \sqrt{\frac{\sum_{i=1}^{k} f_i x_i^2 - N\bar{x}^2}{N-1}} = \sqrt{\frac{2\ 053\ 700.25 - 9(477.2^2)}{9-1}}$$
$$= 23.74 \text{ MPa}$$

Reliability

The reliability R can be expressed by

$$R = 1 - p_f \tag{1–8}$$

where p_f is the *probability of failure,* given by the number of instances of failures per total number of possible instances. The value of R falls in the range $0 \le R \le 1$. A reliability of $R = 0.90$ means there is a 90 percent chance that the part will perform its proper function without failure. The failure of 6 parts out of every 1000 manufactured, $p_f = 6/1000$, might be considered an acceptable failure rate for a certain class of products. This represents a reliability of $R = 1 - 6/1000 = 0.994$ or 99.4 percent.

In the *reliability method of design,* the designer's task is to make a judicious selection of materials, processes, and geometry (size) so as to achieve a specific reliability goal. Thus, if the objective reliability is to be 99.4 percent, as shown, what combination of materials, processing, and dimensions is needed to meet this goal?

If a mechanical system fails when any one component fails, the system is said to be a *series system.* If the reliability of component i is R_i in a series system of n components, then the reliability of the system is given by

$$R = \prod_{i=1}^{n} R_i \tag{1–9}$$

For example, consider a shaft with two bearings having reliabilities of 95 percent and 98 percent. From Equation (1–9), the overall reliability of the shaft system is then

$$R = R_1 R_2 = 0.95(0.98) = 0.93$$

or 93 percent.

Analyses that lead to an assessment of reliability address uncertainties, or their estimates, in parameters that describe the situation. Stochastic variables such as stress, strength, load, or size are described in terms of their means, standard deviations, and distributions. If bearing balls are produced by a manufacturing process in which a diameter distribution is created, we can say upon choosing a ball that there is uncertainty as to size. If we wish to consider weight or moment of inertia in rolling, this size uncertainty can be considered to be *propagated* to our knowledge of weight or inertia. There are ways of estimating the statistical parameters describing weight and inertia from those describing size and density. These methods are variously called *propagation of error, propagation of uncertainty,* or *propagation of dispersion.* These methods are integral parts of analysis or synthesis tasks when probability of failure is involved.

It is important to note that good statistical data and estimates are essential to perform an acceptable reliability analysis. This requires a good deal of testing and validation of the data. In many cases, this is not practical and a deterministic approach to the design must be undertaken.

1–13 Relating Design Factor to Reliability

Reliability is the statistical probability that machine systems and components will perform their intended function satisfactorily without failure. Stress and strength are statistical in nature and very much tied to the reliability of the stressed component. Consider the probability density functions for stress and strength, σ and S, shown in Figure 1–7*a*. The mean values of stress and strength are $\bar{\sigma} = \mu_\sigma$ and $\bar{S} = \mu_S$, respectively. Here, the "average" design factor is

$$\bar{n}_d = \frac{\mu_S}{\mu_\sigma} \qquad (a)$$

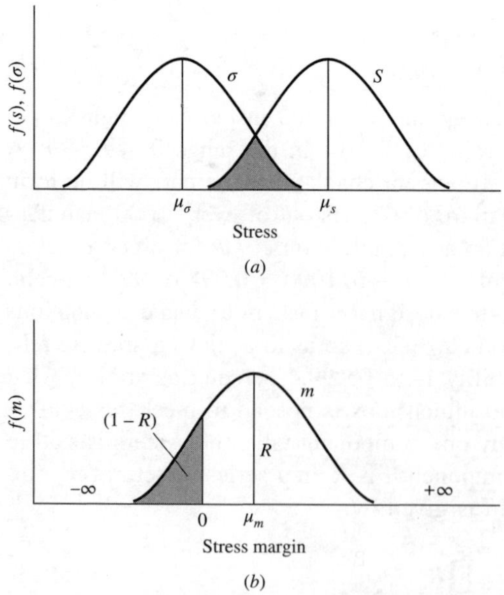

Figure 1–7

Plots of density functions showing how the interference of S and σ is used to explain the stress margin m. (*a*) Stress and strength distributions. (*b*) Distribution of interference; the reliability R is the area of the density function for $m > 0$; the interference is the area $(1 - R)$.

The *margin of safety* for any value of stress σ and strength S is defined as

$$m = S - \sigma \tag{b}$$

The average of the margin of safety is $\bar{m} = \mu_S - \mu_\sigma$. However, for the overlap of the distributions shown by the shaded area in Figure 1–7*a*, the stress exceeds the strength. Here, the margin of safety is negative, and these parts are expected to fail. This shaded area is called the *interference* of σ and S.

Figure 1–7*b* shows the distribution of m, which obviously depends on the distributions of stress and strength. The reliability that a part will perform without failure, R, is the area of the margin of safety distribution for $m > 0$. The interference is the area, $1 - R$, where parts are expected to fail. Assuming that σ and S each have a normal distribution, the stress margin m will also have a normal distribution. Reliability is the probability p that $m > 0$. That is,

$$R = p(S > \sigma) = p(S - \sigma > 0) = p(m > 0) \tag{c}$$

To find the probability that $m > 0$, we form the z variable of m and substitute $m = 0$. Noting that $\mu_m = \mu_S - \mu_\sigma$, and[5] $\hat{\sigma}_m = (\hat{\sigma}_S^2 + \hat{\sigma}_\sigma^2)^{1/2}$, use Equation (1–5) to write

$$z = \frac{m - \mu_m}{\hat{\sigma}_m} = \frac{0 - \mu_m}{\hat{\sigma}_m} = -\frac{\mu_m}{\hat{\sigma}_m} = -\frac{\mu_S - \mu_\sigma}{\left(\hat{\sigma}_S^2 + \hat{\sigma}_\sigma^2\right)^{1/2}} \tag{1–10}$$

Comparing Figure 1–7*b* with Table A–10, we see that

$$\begin{aligned} R &= 1 - \Phi(z) \qquad z \le 0 \\ &= \Phi(z) \qquad z > 0 \end{aligned} \tag{d}$$

To relate to the design factor, $\bar{n}_d = \mu_S/\mu_\sigma$, divide each term on the right side of Equation (1–10) by μ_σ and rearrange as shown:

$$\begin{aligned} z &= -\frac{\dfrac{\mu_S}{\mu_\sigma} - 1}{\left[\dfrac{\hat{\sigma}_S^2}{\mu_\sigma^2} + \dfrac{\hat{\sigma}_\sigma^2}{\mu_\sigma^2}\right]^{1/2}} = -\frac{\bar{n}_d - 1}{\left[\dfrac{\hat{\sigma}_S^2}{\mu_\sigma^2}\dfrac{\mu_S^2}{\mu_S^2} + \dfrac{\hat{\sigma}_\sigma^2}{\mu_\sigma^2}\right]^{1/2}} \\ &= -\frac{\bar{n}_d - 1}{\left[\dfrac{\mu_S^2}{\mu_\sigma^2}\dfrac{\hat{\sigma}_S^2}{\mu_S^2} + \dfrac{\hat{\sigma}_\sigma^2}{\mu_\sigma^2}\right]^{1/2}} = -\frac{\bar{n}_d - 1}{\left[\bar{n}_d^2\dfrac{\hat{\sigma}_S^2}{\mu_S^2} + \dfrac{\hat{\sigma}_\sigma^2}{\mu_\sigma^2}\right]^{1/2}} \end{aligned} \tag{e}$$

Introduce the terms $C_S = \hat{\sigma}_S/\mu_S$ and $C_\sigma = \hat{\sigma}_\sigma/\mu_\sigma$, called the *coefficients of variance* for strength and stress, respectively. Equation (*e*) is then rewritten as

$$z = -\frac{\bar{n}_d - 1}{\sqrt{\bar{n}_d^2 C_S^2 + C_\sigma^2}} \tag{1–11}$$

[5] *Note:* If a and b are normal distributions, and $c = a \pm b$, then c is a normal distribution with a mean of $\mu_c = \mu_a \pm \mu_b$, and a standard deviation of $\hat{\sigma}_c = \left(\hat{\sigma}_a^2 + \hat{\sigma}_b^2\right)^{1/2}$. Tabular results for means and standard deviations for simple algebraic operations can be found in R. G. Budynas and J. K. Nisbett, *Shigley's Mechanical Engineering Design*, 9th ed., McGraw Hill, New York, 2011, Table 20-6, p. 993.

Squaring both sides of Equation (1–11) and solving for $\bar{n}_d$ results in

$$\bar{n}_d = \frac{1 \pm \sqrt{1 - (1 - z^2C_S^2)(1 - z^2C_\sigma^2)}}{1 - z^2C_S^2} \tag{1–12}$$

The plus sign is associated with $R > 0.5$, and the minus sign with $R \leq 0.5$.

Equation (1–12) is remarkable in that it relates the design factor $\bar{n}_d$ to the reliability goal R (through z) and the coefficients of variation of the strength and stress.

EXAMPLE 1–4

A round cold-drawn 1018 steel rod has 0.2 percent mean yield strength $\bar{S}_y = 540.5$ MPa with a standard deviation of 41 MPa. The rod is to be subjected to a mean static axial load of $\bar{P} = 225$ kN with a standard deviation of 18 kN. Assuming the strength and load have normal distributions, what value of the design factor $\bar{n}_d$ corresponds to a reliability of 0.999 against yielding? Determine the corresponding diameter of the rod.

Solution

For strength, $C_S = \hat{\sigma}_S/\mu_S = 41/540.5 = 0.0759$. For stress,

$$\sigma = \frac{P}{A} = \frac{4P}{\pi d^2}$$

Since the tolerance on the diameter will be an order of magnitude less than that of the load or strength, the diameter will be treated deterministically. Thus, statistically, the stress is linearly proportional to the load, and $C_\sigma = C_P = \hat{\sigma}_P/\mu_P = 18/225 = 0.08$. From Table A–10, for $R = 0.999$, $z = -3.09$. Then, Equation (1–12) gives

Answer

$$\bar{n}_d = \frac{1 + \sqrt{1 - [1 - (-3.09)^2(0.0759)^2][1 - (-3.09)^2(0.08)^2]}}{1 - (-3.09)^2(0.0759)^2} = 1.413$$

The diameter is found deterministically from

$$\bar{\sigma} = \frac{4\bar{P}}{\pi d^2} = \frac{\bar{S}_y}{\bar{n}_d}$$

Solving for d gives

Answer

$$d = \sqrt{\frac{4\bar{P}\bar{n}_d}{\pi\bar{S}_y}} = \sqrt{\frac{4(225\ 000)(1.413)}{\pi(540.5)}} = 27.37 \text{ mm}$$

1–14 Dimensions and Tolerances

Part of a machine designer's task is to specify the parts and components necessary for a machine to perform its desired function. Early in the design process, it is usually sufficient to work with nominal dimensions to determine function, stresses, deflections, and the like. However, eventually it is necessary to get to the point of specificity that every component can be purchased and every part can be manufactured. For a part to be manufactured, its essential shape, dimensions, and tolerances must be communicated to the manufacturers. This is usually done by means of a machine drawing, which may either be a multiview drawing on paper, or digital data from a CAD file. Either way, the drawing represents a legal document between the parties involved in the design and manufacture of the part. It is essential that the part be defined precisely and completely so that it can only be interpreted in one way. The designer's intent must be conveyed in such a way that any manufacturer can make the part and/or component to the satisfaction of any inspector.

Common Dimensioning Terminology

We will define a few terms commonly used in dimensioning.

- **Nominal size.** The size we use in speaking of an element. For example, we may specify a 40 mm pipe or a 12 mm bolt. Either the theoretical size or the actual measured size may be quite different. The theoretical size of a 40 mm pipe is 47.5 mm for the outside diameter. And the diameter of the 12 mm bolt may actually measure 11.8 mm.
- **Limits.** The stated maximum and minimum dimensions.
- **Tolerance.** The difference between the two limits.
- **Bilateral tolerance.** The variation in both directions from the basic dimension. That is, the basic size is between the two limits, for example, 25 ± 0.05 mm. The two parts of the tolerance need not be equal.
- **Unilateral tolerance.** The basic dimension is taken as one of the limits, and variation is permitted in only one direction, for example,

$$25\,^{+0.05}_{-0.00}\ \text{mm}$$

- **Clearance.** A general term that refers to the mating of cylindrical parts such as a bolt and a hole. The word *clearance* is used only when the internal member is smaller than the external member. The *diametral clearance* is the measured difference in the two diameters. The *radial clearance* is the difference in the two radii.
- **Interference.** The opposite of clearance, for mating cylindrical parts in which the internal member is larger than the external member (e.g., press-fits).
- **Allowance.** The minimum stated clearance or the maximum stated interference for mating parts.
- **Fit.** The amount of clearance or interference between mating parts. See Section 7–8 for a standardized method of specifying fits for cylindrical parts, such as gears and bearings onto a shaft.
- **GD&T.** Geometric Dimensioning and Tolerancing (GD&T) is a comprehensive system of symbols, rules, and definitions for defining the nominal (theoretically perfect) geometry of parts and assemblies, along with the allowable variation in size, location, orientation, and form of the features of a part. See Chapter 20 for an overview of GD&T.

Choice of Tolerances

The choice of tolerances is the designer's responsibility and should not be made arbitrarily. Tolerances should be selected based on a combination of considerations including functionality, fit, assembly, manufacturing process ability, quality control, and cost. While there is need for balancing these considerations, functionality must not be compromised. If the functionality of the part or assembly cannot be achieved with a reasonable balance of the other considerations, the entire design may need to be reconsidered. The relationship of tolerances to functionality is usually associated with the need to assemble multiple parts. For example, the diameter of a shaft does not generally need a tight tolerance, except for the portions that must fit with components like bearings or gears. The bearings need a particular press fit in order to function properly. Section 7–8 addresses this issue in detail.

Manufacturing methods evolve over time. The manufacturer is free to use any manufacturing process, as long as the final part meets the specifications. This allows the manufacturer to take advantage of available materials and tools, and to specify the least expensive manufacturing methods. Excessive precision by the designer may seem like an easy way to achieve functionality, but it is actually a poor design choice in that it limits the manufacturing options and drives up the cost. In a competitive manufacturing environment, the designer must embrace the idea that less expensive manufacturing methods *should* be selected, even though the parts may be less than perfect. Since tight tolerances usually correlate to higher production costs, as shown in Figure 1–2, the designer should generally be thinking in terms of loosening the tolerances as much as possible, while still achieving the desired functionality.

Choice of Dimensions

Dimensioning a part is a designer's responsibility, since the choice of which dimensions to specify can make a difference in the functionality of the part. A properly dimensioned part will include just enough information, with no extraneous information that can lead to confusion or multiple interpretations. For example, the part shown in Figure 1–8*a* is over-specified in its length dimensions. Note, in machine drawings, the units for the dimensions are typically specified in an overall note on the drawing, and are not shown with the dimensions. If all the dimensions were theoretically perfect, there would be no inconsistency in the over-specified dimensions. But in reality every dimension can only be manufactured to some less-than-perfect level of accuracy. Suppose every dimension in Figure 1–8*a* is specified with a tolerance of +/− 1. It would be possible to manufacture the part such that some dimensions were within the specified tolerance, while forcing related redundant dimensions to be out of tolerance. For example, in Figure 1–8*b*, three of the dimensions are within the +/− 1 tolerance, but they force the other two dimensions to be out of tolerance. In this example, only three length dimensions should be specified. The designer should determine which three are most important to the functioning and assembly of the part.

Figure 1–9 shows four different choices of how the length dimensions might be specified for the same part. None of them are incorrect, but they are not all equivalent in terms of satisfying a particular function. For example, if the two holes are to mate with a pair of corresponding features from another part, the distance between the holes is critical. The choice of dimensions in Figure 1–9*c* would not be a good choice in this case. Even if the part is manufactured within the specified tolerances of +/− 1, the distance between the holes could range anywhere from 47 to 53, an effective

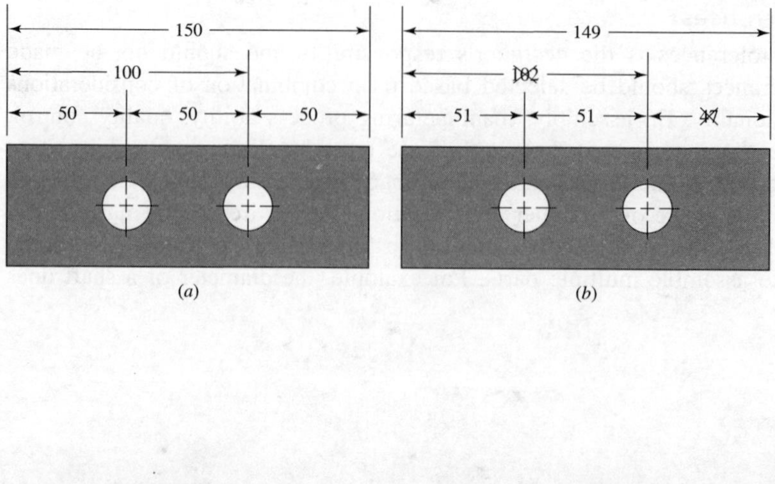

Figure 1–8

Example of over-specified dimensions. (*a*) Five nominal dimensions specified. (*b*) With +/− 1 tolerances, two dimensions are incompatible.

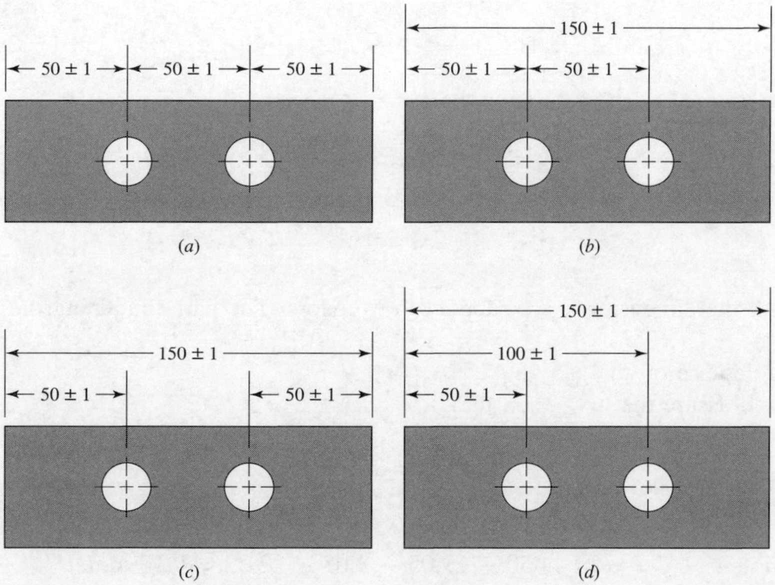

Figure 1–9

Examples of choice of which dimensions to specify.

tolerance of +/− 3. Choosing dimensions as shown in Figure 1–9*a* or 1–9*b* would serve the purpose better to limit the dimension between the holes to a tolerance of +/− 1. For a different application, the distance of the holes to one or both edges might be important, while the overall length might be critical for another application. The point is, the designer should make this determination, not the manufacturer.

Tolerance Stack-up

Note that while there are always choices of which dimensions to specify, the cumulative effect of the individual specified tolerances must be allowed to accumulate *somewhere*. This is known as *tolerance stack-up*. Figure 1–9*a* shows an example of *chain dimensioning,* in which several dimensions are specified in series such that the tolerance stack-up can become large. In this example, even though the individual tolerances are all +/− 1, the total length of the part has an implied tolerance of +/− 3 due to the tolerance stack-up. A common method of minimizing a large tolerance stack-up is to dimension from a common *baseline,* as shown in Figure 1–9*d*.

The tolerance stack-up issue is also pertinent when several parts are assembled. A gap or interference will occur, and will depend on the dimensions and tolerances of the individual parts. An example will demonstrate the point.

EXAMPLE 1–5

A shouldered screw contains three hollow right circular cylindrical parts on the screw before a nut is tightened against the shoulder. To sustain the function, the gap w must equal or exceed 0.08 mm. The parts in the assembly depicted in Figure 1–10 have dimensions and tolerances as follows:

$$a = 44.50 \pm 0.08 \text{ mm} \qquad b = 19.05 \pm 0.02 \text{ mm}$$

$$c = 3.05 \pm 0.13 \text{ mm} \qquad d = 22.23 \pm 0.02 \text{ mm}$$

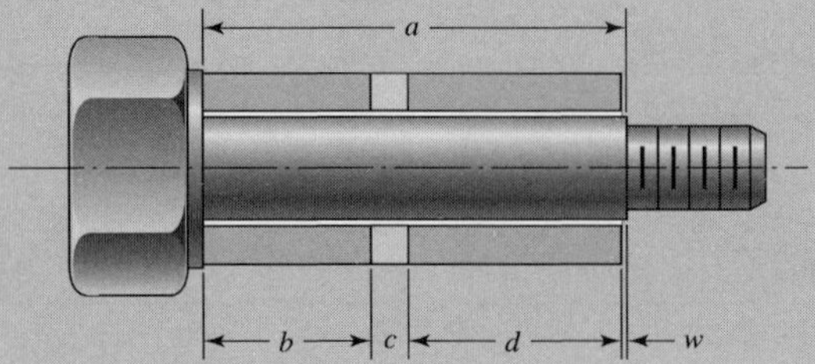

Figure 1–10
An assembly of three cylindrical sleeves of lengths *b*, *c*, and *d* on a shoulder bolt shank of length *a*. The gap *w* is of interest.

All parts except the part with the dimension d are supplied by vendors. The part containing the dimension d is made in-house.
(*a*) Estimate the mean and tolerance on the gap w.
(*b*) What basic value of d will assure that $w \geq 0.08$ mm?

Solution
(*a*) The mean value of w is given by

Answer
$$\bar{w} = \bar{a} - \bar{b} - \bar{c} - \bar{d} = 44.50 - 19.05 - 3.05 - 22.23 = 0.17 \text{ mm}$$

For equal bilateral tolerances, the tolerance of the gap is

Answer
$$t_w = \sum_{\text{all}} t = 0.08 + 0.02 + 0.13 + 0.02 = 0.25 \text{ mm}$$

Then, $w = 0.17 \pm 0.25$ mm, and

$$w_{max} = \bar{w} + t_w = 0.17 + 0.25 = 0.42 \text{ mm}$$
$$w_{min} = \bar{w} - t_w = 0.17 - 0.25 = -0.08 \text{ mm}$$

Thus, both clearance and interference are possible.
(*b*) If w_{min} is to be 0.08 in, then, $\bar{w} = w_{min} + t_w = 0.08 + 0.25 = 0.33$ mm. Thus,

Answer
$$\bar{d} = \bar{a} - \bar{b} - \bar{c} - \bar{w} = 44.50 - 19.05 - 3.05 - 0.33 = 27.07 \text{ mm}$$

1–15 Units

In the symbolic units equation for Newton's second law, $F = ma$,

$$F = MLT^{-2} \tag{1–13}$$

F stands for force, M for mass, L for length, and T for time. Units chosen for *any* three of these quantities are called *base* units. The first three having been chosen, the fourth unit is called a *derived* unit. When force, length, and time are chosen as base units, the mass is the derived unit and the system that results is called a *gravitational system of units*. When mass, length, and time are chosen as base units, force is the derived unit and the system that results is called an *absolute system of units*.

In some English-speaking countries, the *U.S. customary foot-pound-second system* (fps) and the *inch-pound-second system* (ips) are the two standard gravitational systems most used by engineers. In the fps system the unit of mass is

$$M = \frac{FT^2}{L} = \frac{(\text{pound-force})(\text{second})^2}{\text{foot}} = \text{lbf} \cdot \text{s}^2/\text{ft} = \text{slug} \tag{1–14}$$

Thus, length, time, and force are the three base units in the fps gravitational system.

The unit of force in the fps system is the pound, more properly the *pound-force*. We shall often abbreviate this unit as lbf; the abbreviation lb is permissible however, since we shall be dealing only with the U.S. customary gravitational system. In some branches of engineering it is useful to represent 1000 lbf as a kilopound and to abbreviate it as kip. *Note:* In Equation (1–14) the derived unit of mass in the fps gravitational system is the lbf · s²/ft and is called a *slug;* there is no abbreviation for slug.

The unit of mass in the ips gravitational system is

$$M = \frac{FT^2}{L} = \frac{(\text{pound-force})(\text{second})^2}{\text{inch}} = \text{lbf} \cdot \text{s}^2/\text{in} \tag{1–15}$$

The mass unit lbf · s^2/in has no official name.

The *International System of Units* (SI) is an absolute system. The base units are the meter, the kilogram (for mass), and the second. The unit of force is derived by using Newton's second law and is called the *newton*. The units constituting the newton (N) are

$$F = \frac{ML}{T^2} = \frac{(\text{kilogram})(\text{meter})}{(\text{second})^2} = \text{kg} \cdot \text{m/s}^2 = \text{N} \tag{1–16}$$

The weight of an object is the force exerted upon it by gravity. Designating the weight as W and the acceleration due to gravity as g, we have

$$W = mg \tag{1–17}$$

With SI units, standard gravity is 9.806 or about 9.81 m/s. Thus, the weight of a 1-kg mass is

$$W = (1\ \text{kg})(9.81\ \text{m/s}^2) = 9.81\ \text{N}$$

In the fps system, standard gravity is $g = 32.1740$ ft/s^2. For most cases this is rounded off to 32.2. Thus the weight of a mass of 1 slug in the fps system is

$$W = mg = (1\ \text{slug})(32.2\ \text{ft/s}^2) = 32.2\ \text{lbf}$$

In the ips system, standard gravity is 386.088 or about 386 in/s^2. Thus, in this system, a unit mass weighs

$$W = (1\ \text{lbf} \cdot \text{s}^2/\text{in})(386\ \text{in/s}^2) = 386\ \text{lbf}$$

A series of names and symbols to form multiples and submultiples of SI units has been established to provide an alternative to the writing of powers of 10. Table A–1 includes these prefixes and symbols.

Numbers having four or more digits are placed in groups of three and separated by a space instead of a comma. However, the space may be omitted for the special case of numbers having four digits. A period is used as a decimal point. These recommendations avoid the confusion caused by certain European countries in which a comma is used as a decimal point, and by the English use of a centered period. Examples of correct and incorrect usage are as follows:

1924 or 1 924 but not 1,924
0.1924 or 0.192 4 but not 0.192,4
192 423.618 50 but not 192,423.618 50

The decimal point should always be preceded by a zero for numbers less than unity.

1–16 Calculations and Significant Figures

The discussion in this section applies to real numbers, not integers. The accuracy of a real number depends on the number of significant figures describing the number. Usually, but not always, three or four significant figures are necessary for engineering accuracy. Unless otherwise stated, *no less* than three significant figures should be used in your calculations. The number of significant figures is usually inferred by the number of figures given (except for leading zeros). For example, 706, 3.14, and 0.002 19 are assumed to be numbers with three significant figures. For trailing zeros, a little more clarification is necessary. To display 706 to four significant figures insert a trailing zero and display either 706.0, 7.060×10^2, or 0.7060×10^3. Also, consider a number such as 91 600. Scientific notation should be used to clarify the accuracy. For three significant figures express the number as 91.6×10^3. For four significant figures express it as 91.60×10^3.

Computers and calculators display calculations to many significant figures. However, you should never report a number of significant figures of a calculation any greater than the smallest number of significant figures of the numbers used for the calculation. Of course, you should use the greatest accuracy possible when performing a calculation. For example, determine the circumference of a solid shaft with a diameter of $d = 12$ mm. The circumference is given by $C = \pi d$. Since d is given with two significant figures, C should be reported with only two significant figures. Now if we used only two significant figures for π our calculator would give $C = 3.1\ (12) = 37.2$ mm. This rounds off to two significant figures as $C = 37$ mm. However, using $\pi = 3.141\ 592\ 654$ as programmed in the calculator, $C = 3.141\ 592\ 654\ (12) = 37.699\ 111\ 84$ mm. This rounds off to $C = 38$ mm, which is 2.7 percent higher than the first calculation. Note, however, since d is given with two significant figures, it is implied that the range of d is 12 ± 0.5. This means that the calculation of C is only accurate to within $\pm 0.5/12 = \pm 0.0417 = \pm 4.17\%$. The calculation could also be one in a series of calculations, and rounding each calculation separately may lead to an accumulation of greater inaccuracy. Thus, it is considered good engineering practice to make all calculations to the greatest accuracy possible and report the results within the accuracy of the given input.

1–17 Design Topic Interdependencies

One of the characteristics of machine design problems is the interdependencies of the various elements of a given mechanical system. For example, a change from a spur gear to a helical gear on a drive shaft would add axial components of force, which would have implications on the layout and size of the shaft, and the type and size of the bearings. Further, even within a single component, it is necessary to consider many different facets of mechanics and failure modes, such as excessive deflection, static yielding, fatigue failure, contact stress, and material characteristics. However, in order to provide significant attention to the details of each topic, most machine design textbooks focus on these topics separately and give end-of-chapter problems that relate only to that specific topic.

To help the reader see the interdependence between the various design topics, this textbook presents many ongoing and interdependent problems in the end-of-chapter problem sections. Each row of Table 1–2 shows the problem numbers that apply to the same mechanical system that is being analyzed according to the topics

Table 1–2 Problem Numbers for Linked End-of-Chapter Problems*

3–1	4–50	4–81											
3–39	5–78	5–79											
3–76	4–23	4–29	4–35	5–50	6–37	7–12	11–16						
3–77	4–24	4–30	4–36	5–51	6–38	7–13	11–17						
3–78	4–25	4–31	4–37	5–52	6–39	7–14	11–18						
3–79	4–26	4–32	4–38	5–53	6–40	7–15	11–19						
3–80	4–27	4–33	4–39	5–54	6–41	7–16	7–24	7–25	7–39	11–29	11–30	13–44	14–37
3–81	4–28	4–34	4–40	5–55	6–42	7–17	7–26	7–27	7–40	11–31	11–32	13–45	14–38
3–82	5–56	6–43	7–18	11–47	13–48								
3–84	5–57	6–44	7–19	11–48	13–48								
3–85	5–58	6–45	7–20	11–20	13–46	14–39							
3–87	5–59	6–46	7–21	11–21	13–47	14–40							
3–88	4–41	4–78	5–60	6–47									
3–89	5–61	6–48											
3–90	5–62	6–49											
3–91	5–63	6–50											
3–92	4–43	4–80	5–64	5–67	6–51								
3–93	5–65	6–52											
3–94	5–66	6–53											
3–95	5–67												

*Each row corresponds to the same mechanical component repeated for a different design concept.

being presented in that particular chapter. For example, in the second row, Problems 3–39, 5–78, and 5–79 correspond to a pin in a knuckle joint that is to be analyzed for stresses in Chapter 3 and then for static failure in Chapter 5. This is a simple example of interdependencies, but as can be seen in the table, other systems are analyzed with as many as 10 separate problems. It may be beneficial to work through some of these continuing sequences as the topics are covered to increase your awareness of the various interdependencies.

In addition to the problems given in Table 1–2, Section 1–18 describes a power transmission case study where various interdependent analyses are performed throughout the book, when appropriate in the presentation of the topics. The final results of the case study are then presented in Chapter 18.

1–18 Power Transmission Case Study Specifications

We will consider a case study incorporating the many facets of the design process for a power transmission speed reducer throughout this textbook. The problem will be introduced here with the definition and specification for the product to be designed. Further details and component analysis will be presented in subsequent chapters. Chapter 18 provides an overview of the entire process, focusing on the design sequence, the interaction between the component designs, and other details pertinent to transmission of power. It also contains a complete case study of the power transmission speed reducer introduced here.

Many industrial applications require machinery to be powered by engines or electric motors. The power source usually runs most efficiently at a narrow range of rotational speed. When the application requires power to be delivered at a slower speed than supplied by the motor, a speed reducer is introduced. The speed reducer should transmit the power from the motor to the application with as little energy loss as practical, while reducing the speed and consequently increasing the torque. For example, assume that a company wishes to provide off-the-shelf speed reducers in various capacities and speed ratios to sell to a wide variety of target applications. The marketing team has determined a need for one of these speed reducers to satisfy the following customer requirements.

Design Requirements

Power to be delivered: 15 kW
Input speed: 1750 rev/min
Output speed: 84 rev/min
Targeted for uniformly loaded applications, such as conveyor belts, blowers, and generators
Output shaft and input shaft in-line
Base mounted with 4 bolts
Continuous operation
6-year life, with 8 hours/day, 5 days/wk
Low maintenance
Competitive cost
Nominal operating conditions of industrialized locations
Input and output shafts standard size for typical couplings

In reality, the company would likely design for a whole range of speed ratios for each power capacity, obtainable by interchanging gear sizes within the same overall design. For simplicity, in this case study we will consider only one speed ratio.

Notice that the list of customer requirements includes some numerical specifics, but also includes some generalized requirements, e.g., low maintenance and competitive cost. These general requirements give some guidance on what needs to be considered in the design process, but are difficult to achieve with any certainty. In order to pin down these nebulous requirements, it is best to further develop the customer requirements into a set of product specifications that are measurable. This task is usually achieved through the work of a team including engineering, marketing, management, and customers. Various tools may be used to prioritize the requirements, determine suitable metrics to be achieved, and to establish target values for each metric. The goal of this process is to obtain a product specification that identifies precisely what the product must satisfy. The following product specifications provide an appropriate framework for this design task.

Design Specifications

Power to be delivered: 15 kW
Power efficiency: >95%
Steady state input speed: 1750 rev/min
Maximum input speed: 2400 rev/min
Steady-state output speed: 82–88 rev/min

Usually low shock levels, occasional moderate shock
Input and output shafts extend 4 in outside gearbox
Input and output shaft diameter tolerance: ±0.025 mm
Input and output shafts in-line: concentricity ±0.125 mm, alignment ±0.001 rad
Maximum allowable loads on input shaft: axial, 220 N; transverse, 440 N
Maximum allowable loads on output shaft: axial, 220 N; transverse, 2220 N
Maximum gearbox size: 350 × 350 mm base, 550 mm height
Base mounted with 4 bolts
Mounting orientation only with base on bottom 100% duty cycle
100% duty cycle
Maintenance schedule: lubrication check every 2000 hours; change of lubrication every 8000 hours of operation; gears and bearing life >12 000 hours; infinite shaft life; gears, bearings, and shafts replaceable
Access to check, drain, and refill lubrication without disassembly or opening of gasketed joints
Manufacturing cost per unit: <$300
Production: 10 000 units per year
Operating temperature range: −23° to 65°C
Sealed against water and dust from typical weather
Noise: <85 dB from 1 meter

PROBLEMS

1–1 Select a mechanical component from Part 3 of this book (roller bearings, springs, etc.), go to your university's library or the appropriate Internet website, and, using the *Thomas Register of American Manufacturers* (www.thomasnet.com), report on the information obtained on five manufacturers or suppliers.

1–2 Select a mechanical component from Part 3 of this book (roller bearings, springs, etc.), go to the Internet, and, using a search engine, report on the information obtained on five manufacturers or suppliers.

1–3 Select an organization listed in Section 1–6, go to the Internet, and list what information is available on the organization.

1–4 Go to the Internet and connect to the NSPE website (www.nspe.org/ethics). Read the history of the *Code of Ethics* and briefly discuss your reading.

1–5 Go to the Internet and connect to the NSPE website (www.nspe.org/ethics). Read the complete *NSPE Code of Ethics for Engineers* and briefly discuss your reading.

1–6 Go to the Internet and connect to the NSPE website (www.nspe.org/ethics). Go to *Ethics Resources* and review one or more of the topics given. Briefly discuss your reading.

1–7 Estimate how many times more expensive it is to grind a steel part to a tolerance of ±0.0125 mm versus turning it to a tolerance of ±0.075 mm.

1–8 The costs to manufacture a part using methods A and B are estimated by $C_A = 10 + 0.8\ P$ and $C_B = 60 + 0.8\ P - 0.005\ P^2$ respectively, where the cost C is in dollars and P is the number of parts. Estimate the breakeven point.

1–9 A cylindrical part of diameter d is loaded by an axial force P. This causes a stress of P/A, where $A = \pi d^2/4$. If the load is known with an uncertainty of ± 10 percent, the diameter is known within ± 5 percent (tolerances), and the stress that causes failure (strength) is known within ± 15 percent, determine the minimum design factor that will guarantee that the part will not fail.

1–10 When one knows the true values x_1 and x_2 and has approximations X_1 and X_2 at hand, one can see where errors may arise. By viewing error as something to be added to an approximation to attain a true value, it follows that the error e_i is related to X_i and x_i as $x_i = X_i + e_i$.

(*a*) Show that the error in a sum $X_1 + X_2$ is

$$(x_1 + x_2) - (X_1 + X_2) = e_1 + e_2$$

(*b*) Show that the error in a difference $X_1 - X_2$ is

$$(x_1 - x_2) - (X_1 - X_2) = e_1 - e_2$$

(*c*) Show that the error in a product $X_1 X_2$ is

$$x_1 x_2 - X_1 X_2 \approx X_1 X_2 \left(\frac{e_1}{X_1} + \frac{e_2}{X_2}\right)$$

(*d*) Show that in a quotient X_1/X_2 the error is

$$\frac{x_1}{x_2} - \frac{X_1}{X_2} \approx \frac{X_1}{X_2}\left(\frac{e_1}{X_1} - \frac{e_2}{X_2}\right)$$

1–11 Use the true values $x_1 = \sqrt{7}$ and $x_2 = \sqrt{8}$

(*a*) Demonstrate the correctness of the error equation from Problem 1–10 for addition if X_1 and X_2 are obtained by truncating x_1 and x_2 to three digits.

(*b*) Demonstrate the correctness of the error equation for addition if X_1 and X_2 are obtained by rounding x_1 and x_2 to three significant figures.

1–12 A solid circular rod of diameter d undergoes a bending moment $M = 110$ N · m inducing a stress $\sigma = 32M/(\pi d^3)$. Using a material strength of 170 MPa and a *design factor* of 2.5, determine the minimum diameter of the rod. Using Table A–15, select a preferred fractional diameter and determine the resulting *factor of safety*.

1–13 A fatigue test is performed on rotating beam specimens where, for each rotation cycle, the specimens experience tensile and compressive stresses of equal magnitude. The cycles-to-failure experience with 69 specimens of 5160H steel from 30-mm hexagonal bar stock was as follows:

L	60	70	80	90	100	110	120	130	140	150	160	170	180	190	200	210
f	2	1	3	5	8	12	6	10	8	5	2	3	2	1	0	1

where L is the life in thousands of cycles, and f is the class frequency of failures.

(*a*) Estimate the mean and standard deviation of the life for the population from which the sample was drawn.

(*b*) Presuming the distribution is normal, how many specimens are predicted to fail at less than 115 kcycles?

1–14 Determinations of the ultimate tensile strength S_{ut} of stainless-steel sheet (17-7PH, condition TH 1050), in sizes from 0.5 to 1.5 mm, in 197 tests combined into seven classes were

S_{ut} (MPa)	1200	1255	1310	1365	1420	1475	1530
f	6	9	44	67	53	12	6

where S_{ut} is the class midpoint and f is the class frequency. Estimate the mean and standard deviation.

1–15 The lives of parts are often expressed as the number of cycles of operation that a specified percentage of a population will exceed before experiencing failure. The symbol L is used to designate this definition of life. Thus, we can speak of L_{10} life as the number of cycles to failure exceeded by 90 percent of a population of parts. Given a normal distribution model, with a mean of $\bar{L} = 122.9$ kilocycles and standard deviation of $s_L = 30.3$ kilocycles, estimate the corresponding L_{10} life.

1–16 The tensile 0.2 percent offset yield strength of AISI 1137 cold-drawn steel bars up to 25 mm in diameter from 2 mills and 25 heats is reported as follows:

S_y	651	665	679	693	707	721	735	749	763	777
f	19	25	38	17	12	10	5	4	4	2

where S_y is the class midpoint in MPa and f is the number in each class. Presuming the distribution is normal, what is the yield strength exceeded by 99 percent of the population?

1–17 A mechanical system comprises three subsystems in series with reliabilities of 98, 96, and 94 percent. What is the overall reliability of the system?

1–18 The maximum shear stress in a solid round bar of diameter, d, due to an applied torque, T, is given by $\tau_{max} = 16\,T/(\pi d^3)$. A round, cold-drawn 1018 steel rod is subjected to a mean torsional load of $\bar{T} = 1.5$ kN · m with a standard deviation of 145 N · m. The rod material has a mean shear yield strength of $\bar{S}_{sy} = 312$ MPa with a standard deviation of 23.5 MPa. Assuming the strength and load have normal distributions, what value of the design factor $\bar{n}_d$ corresponds to a reliability of 0.99 against yielding? Determine the corresponding diameter of the rod.

1–19 A round cold-drawn 1045 steel rod has a mean strength $\bar{S}_y = 650$ MPa with a standard deviation of $\hat{\sigma}_{S_y} = 45$ MPa. The rod is to be subjected to a mean static axial load of $\bar{P} = 300$ kN with a standard deviation of $\hat{\sigma}_P = 20$ kN. Assuming the strength and load have normal distributions, determine the reliabilities corresponding to the design factors of (*a*) 1.2, (*b*) 1.5. Also, determine the diameter corresponding to each case.

1–20 A beam subjected to axial loading will experience an axial stress, σ_a. If, in addition, the beam is subjected to a bending load, a bending stress, σ_b, will also occur at the outer fibers of the beam. The maximum stress at the outer fibers of the beam will be $\sigma_{max} = \sigma_a + \sigma_b$. Assume that σ_a and σ_b are independent and $\bar{\sigma}_a = 90$ MPa, $\hat{\sigma}_{\sigma a} = 8.4$ MPa, $\bar{\sigma}_b = 383$ MPa, $\hat{\sigma}_{\sigma b} = 22.3$ MPa. The rod is made of a steel with $\bar{S}_y = 553$ MPa and $\hat{\sigma}_{S_y} = 42.7$ MPa. Assuming the strength and load have normal distributions, determine the design factor and the reliability guarding against yielding.

1–21 Three blocks A, B, and C and a grooved block D have dimensions a, b, c, and d as follows:

$a = 37.5 \pm 0.025$ mm $\quad b = 50 \pm 0.075$ mm

$c = 75 \pm 0.1$ mm $\quad d = 163 \pm 0.25$ mm

Problem 1–21

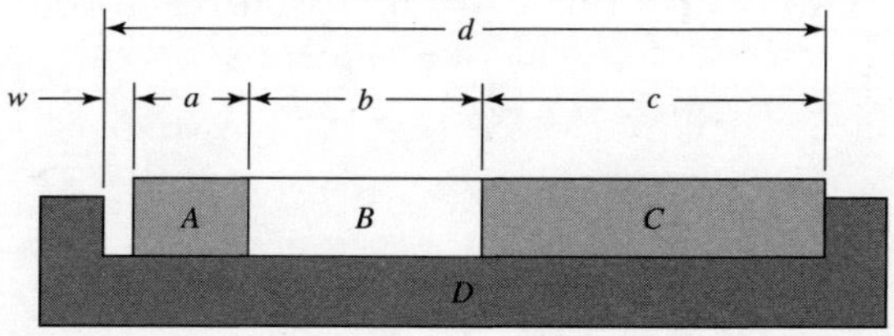

(*a*) Determine the mean gap $\bar{w}$ and its tolerance.
(*b*) Determine the mean size of d that will assure that $w \geq 0.25$ mm.

1–22 The volume of a rectangular parallelepiped is given by $V = xyz$. If $x = a \pm \Delta a$, $y = b \pm \Delta b$, $z = c \pm \Delta c$, show that

$$\frac{\Delta V}{\bar{V}} \approx \frac{\Delta a}{\bar{a}} + \frac{\Delta b}{\bar{b}} + \frac{\Delta c}{\bar{c}}$$

Use this result to determine the bilateral tolerance on the volume of a rectangular parallelepiped with dimensions

$$a = 37.5 \pm 0.05 \text{ mm} \qquad b = 47 \pm 0.075 \text{ mm} \qquad c = 75 \pm 0.1 \text{ mm}$$

1–23 A pivot in a linkage has a pin in the figure whose dimension $a \pm t_a$ is to be established. The thickness of the link clevis is 37.5 ± 0.125 mm. The designer has concluded that a gap of between 0.1 and 1.25 mm will satisfactorily sustain the function of the linkage pivot. Determine the dimension a and its tolerance.

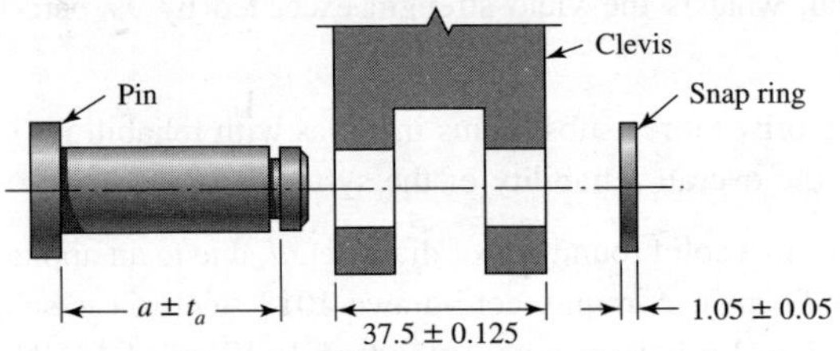

Problem 1–23
Dimensions in inches.

1–24 A circular cross-section O ring has the dimensions shown in the figure. In particular, an AS 568A standard No. 240 O ring has an inside diameter D_i and a cross-section diameter d of

$$D_i = 84.4 \pm 0.7 \text{ mm} \qquad d_i = 3.5 \pm 0.1 \text{ mm}$$

Estimate the mean outside diameter $\bar{D}_o$ and its bilateral tolerance.

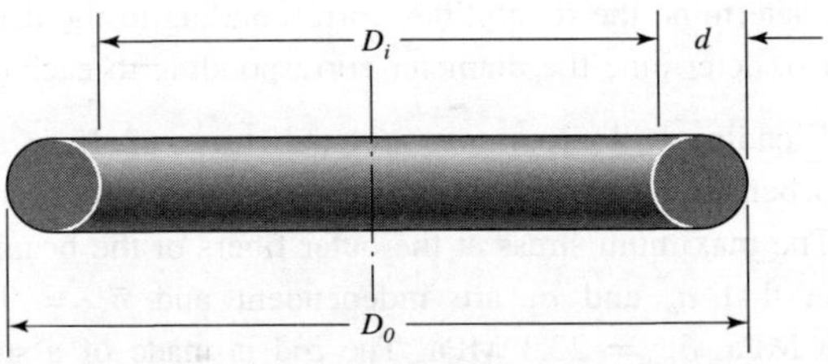

Problem 1–24

1–25 to 1–28 For the table given, repeat Problem 1–24 for the following O rings, given the AS 568A standard number. Solve Problems 1–25 and 1–26 using SI units. Solve Problems 1–27 and 1–28 using ips units. *Note:* The solutions require research.

Problem number	1–25	1–26	1–27	1–28
AS 568A No.	110	220	160	320

1–29 Generally, final design results are rounded to or fixed to three digits because the given data cannot justify a greater display. In addition, prefixes should be selected so as to limit number strings to no more than four digits to the left of the decimal point. Using these rules, as well as those for the choice of prefixes, solve the following relations:
(*a*) $\sigma = M/Z$, where $M = 200$ N · m and $Z = 15.3 \times 10^3$ mm^3.
(*b*) $\sigma = F/A$, where $F = 42$ kN and $A = 600$ mm^2.
(*c*) $y = Fl^3/3EI$, where $F = 1200$ N, $l = 800$ mm, $E = 207$ GPa, and $I = 64 \times 10^3$ mm^4.
(*d*) $\theta = Tl/GJ$, where $T = 1100$ N · m, $l = 250$ mm, $G = 79.3$ GPa, and $J = \pi d^4/32$ with $d = 25$ mm.

1–30 Repeat Problem 1–31 for the following:
(*a*) $\sigma = F/wt$, where $F = 1$ kN, $w = 25$ mm, and $t = 5$ mm.
(*b*) $I = bh^3/12$, where $b = 10$ mm and $h = 25$ mm.
(*c*) $I = \pi d^4/64$, where $d = 25.4$ mm.
(*d*) $\tau = 16\,T/\pi d^3$, where $T = 25$ N · m, and $d = 12.7$ mm.

1–31 Repeat Problem 1–31 for:
(*a*) $\tau = F/A$, where $A = \pi d^2/4$, $F = 120$ kN, and $d = 20$ mm.
(*b*) $\sigma = 32\,Fa/\pi d^3$, where $F = 800$ N, $a = 800$ mm, and $d = 32$ mm.
(*c*) $Z = \pi(d_o^4 - d_i^4)/(32\,d_o)$ for $d_o = 36$ mm and $d_i = 26$ mm.
(*d*) $k = (d^4\,G)/(8\,D^3\,N)$, where $d = 1.6$ mm, $G = 79.3$ GPa, $D = 19.2$ mm, and $N = 32$ (a dimensionless number).

2 Materials

Glow Images

Chapter Outline

The selection of a material for a machine part or a structural member is one of the most important decisions the designer is called on to make. The decision is usually made before the dimensions of the part are established. After choosing the process of creating the desired geometry and the material (the two cannot be divorced), the designer can proportion the member so that loss of function can be avoided or the chance of loss of function can be held to an acceptable risk.

In Chapters 3 and 4, methods for estimating stresses and deflections of machine members are presented. These estimates are based on the properties of the material from which the member will be made. For deflections and stability evaluations, for example, the elastic (stiffness) properties of the material are required, and evaluations of stress at a critical location in a machine member require a comparison with the strength of the material at that location in the geometry and condition of use. This strength is a material property found by testing and is adjusted to the geometry and condition of use as necessary.

As important as stress and deflection are in the design of mechanical parts, the selection of a material is not always based on these factors. Many parts carry no loads on them whatever. Parts may be designed merely to fill up space or for aesthetic qualities. Members must frequently be designed to also resist corrosion. Sometimes temperature effects are more important in design than stress and strain. So many other factors besides stress and strain may govern the design of parts that the designer must have the versatility that comes only with a broad background in materials and processes.

2–1 Material Strength and Stiffness

Stress-Strain Relationships from the Tensile Test

The standard tensile test is used to obtain a variety of material characteristics and strengths that are used in design. Figure 2–1 illustrates a typical tension-test specimen and its characteristic dimensions.[1] The original diameter d_0 and the gauge length l_0, used to measure the deflections, are recorded before the test is begun. The specimen is then mounted in the test machine and slowly loaded in tension while the load P and deflection are observed. The load is converted to stress by the calculation

$$\sigma = \frac{P}{A_0} \tag{2–1}$$

where $A_0 = \frac{1}{4}\pi d_0^2$ is the original area of the specimen.

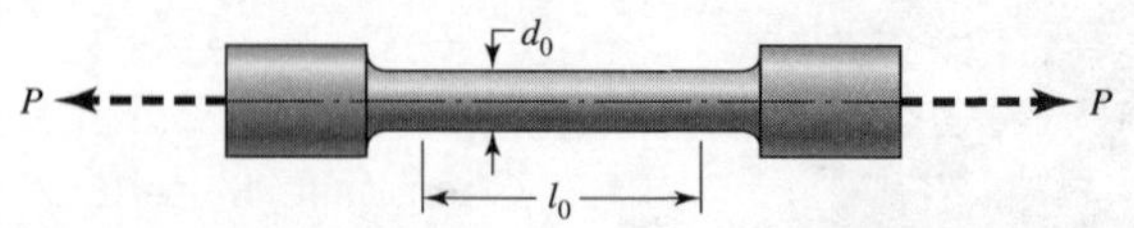

Figure 2–1

A typical tension-test specimen. Some of the standard dimensions used for d_0 are 2.5, 6.25, and 12.5 mm and 0.505 in, but other sections and sizes are in use. Common gauge lengths l_0 used are 10, 25, and 50 mm and 1 and 2 in.

[1]See ASTM standards E8 and E-8 m for standard dimensions.

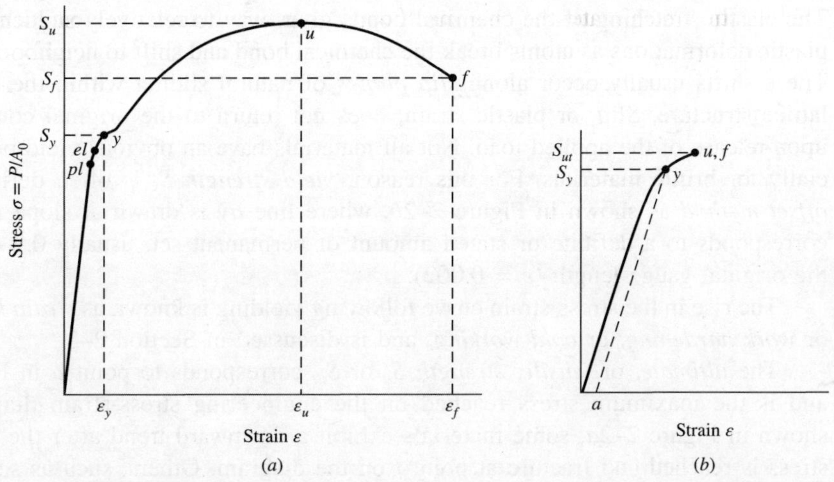

Figure 2–2

Typical engineering stress-strain diagram obtained from the standard tensile test (*a*) Ductile material; (*b*) brittle material.

The deflection, or extension of the gauge length, is given by $l - l_0$ where l is the gauge length corresponding to the load P. The normal strain is calculated from

$$\varepsilon = \frac{l - l_0}{l_0} \tag{2–2}$$

The results are plotted as an *engineering stress-strain diagram.* Figure 2–2 depicts typical stress-strain diagrams for ductile and brittle materials. Ductile materials deform much more than brittle materials.

Point pl in Figure 2–2a is called the *proportional limit.* This is the point at which the curve first begins to deviate from a straight line. In the linear range, the uniaxial stress-strain relation is given by *Hooke's law* as

$$\sigma = E\varepsilon \tag{2–3}$$

where the constant of proportionality E, the slope of the linear part of the stress-strain curve, is called *Young's modulus* or the *modulus of elasticity.* E is a measure of the stiffness of a material, and since strain is dimensionless, the units of E are the same as stress. An interesting characteristic of Young's modulus is that it is very nearly constant for each type of material (e.g., steel, aluminum, copper). Steel, for example, has a modulus of elasticity of about 207 GPa *regardless of heat treatment, carbon content, or alloying.* This characteristic is a consequence of elastic deformations being primarily governed by the stretching of chemical bonds that are determined by the particular crystal structure of each material. Values of E for a variety of materials can be found in Appendix A–5.

Point el in Figure 2–2a is called the *elastic limit*. If the specimen is loaded beyond this point, the material will take on a permanent set when the load is removed. Loading to the left of the elastic limit produces what is referred to as *elastic deformation,* while loading to the right of the elastic limit produces *plastic deformation*. Between pl and el the diagram is not a perfectly straight line, even though the specimen is elastic. For many materials the difference between points el and pl is insignificant.

During the tension test, many materials reach a point at which the strain begins to increase very rapidly without a corresponding increase in stress. This point is called the *yield point,* labeled y in Figure 2–2a. Many ductile materials have a clear "knee" in the curve where the stress flattens or even decreases as the strain increases rapidly.

The elastic stretching of the chemical bonds gives way to relatively sudden and large plastic deformations as atoms break the chemical bond and shift to neighboring atoms. These shifts usually occur along *slip planes* of natural sliding within the crystalline lattice structure. Slip, or plastic strain, does not return to the original configuration upon release of the applied load. Not all materials have an obvious yield point, especially for brittle materials. For this reason, *yield strength* S_y is often defined by an *offset method* as shown in Figure 2–2*b*, where line *ay* is drawn at slope *E*. Point *a* corresponds to a definite or stated amount of permanent set, usually 0.2 percent of the original gauge length ($\varepsilon = 0.002$).

The rise in the stress-strain curve following yielding is known as *strain hardening,* or *work hardening,* or *cold working,* and is discussed in Section 2–3.

The *ultimate,* or *tensile, strength* S_u or S_{ut} corresponds to point *u* in Figure 2–2 and is the maximum stress reached on the engineering stress-strain diagram.[2] As shown in Figure 2–2*a*, some materials exhibit a downward trend after the maximum stress is reached and fracture at point *f* on the diagram. Others, such as some of the cast irons and high-strength steels, fracture while the stress-strain trace is still rising, as shown in Figure 2–2*b*, where points *u* and *f* are identical.

True Stress and True Strain

The stresses and strains in the engineering stress-strain diagram of Figure 2–2 are based on the original undeformed geometry, such as in Equations (2–1) and (2–2). In reality, as the load is applied the area reduces and the specimen lengthens so that the actual stresses and strains are larger than the calculated engineering stresses and strains. Consequently, the engineering stresses and strains are only accurate when the overall specimen deformation is small. This is very appropriate for the small strains realized in the elastic region, and even acceptable for practical purposes up to the ultimate strength when necking begins. However, it is sometimes useful, particularly when evaluating the plastic region, to work with stresses and strains that are calculated from the instantaneous deformed geometry rather than the original gauge dimensions. This requires the continual measurement of the cross-sectional area and length of the specimen while the load increases.

The *true stress* is given by

$$\tilde{\sigma} = \frac{P}{A} \tag{2–4}$$

where *A* is the cross-sectional area measured simultaneously with the applied load *P*.

From a comparison with Equation (2–1), it is seen that the true stress is related to the engineering stress by

$$\tilde{\sigma} = \sigma\left(\frac{A_0}{A}\right) \tag{2–5}$$

The engineering strain given by Equation (2–2) is based on the net change in length from the *original* length. The *true strain* is the sum of incremental elongations Δl_i divided by the *current* gauge length l_i corresponding to the current load P_i, thus

$$\tilde{\varepsilon} = \sum \frac{\Delta l_i}{l_i} \tag{2–6}$$

[2]Usage varies. For a long time engineers used the term *ultimate strength,* hence the subscript *u* in S_u or S_{ut}. However, in material science and metallurgy the term *tensile strength* is used.

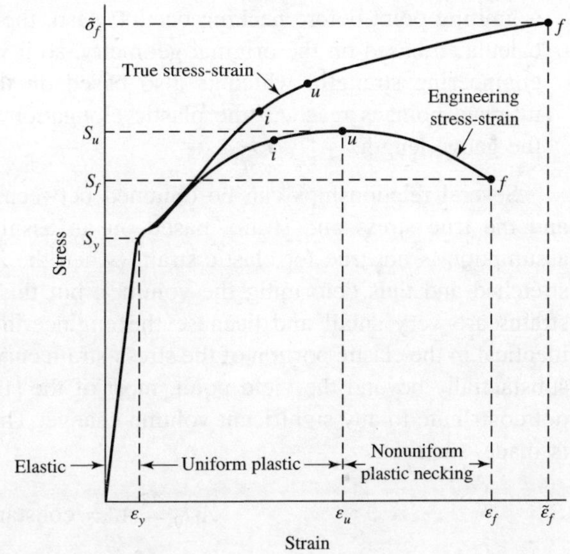

Figure 2–3

Engineering stress-strain and true stress-strain diagrams for a ductile material.

For very small measurement increments, this can be stated in integral form as

$$\tilde{\varepsilon} = \int_{l_0}^{l} \frac{dl}{l} = \ln \frac{l}{l_0} \tag{2–7}$$

where l is the current gauge length and l_0 is the original gauge length.

A representative comparison of engineering stress-strain and true stress-strain for a ductile material is shown in Figure 2–3. Several observations are worth noting.

- For small strains, less than about 2 percent, there is practically no difference between the two curves. For any situation that is working below the yield point, which is very often a design constraint, engineering stress and strain are entirely suitable. Consequently, it is customary for designs working in the elastic region to simply refer to stress and strain without including the distinguishing terms "engineering" or "true."
- For large strains, the difference in the two curves is substantial. This is especially important in the study of localized plastic deformation at a location of stress concentration, particularly at the tip of a crack. Large strains are also prevalent with forming processes such as drawing, rolling, and forging. In these situations, use of true stress and true strain is essential.
- The engineering stress decreases after reaching the ultimate strength at point u. This is due to a characteristic "necking" of ductile materials, where the area reduces dramatically, as shown in Figure 2–4. The true stress takes into account the reduced necked area, resulting in an accurate representation of the continuously increasing true stress all the way to fracture.
- The ultimate tensile strength, S_{ut}, is best determined from the engineering stress-strain curve at the peak point u. Even though this is not the actual (true) stress magnitude, it is of great practical value as a strength property because it represents

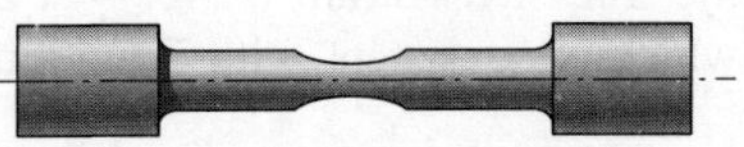

Figure 2–4

Tension specimen after necking.

a limiting point before necking occurs. Also, the stress in an actual part is usually calculated based on the original geometry, so it is appropriate to compare it to the engineering strength, which is also based on the original geometry. Before the ultimate point is reached, the plastic elongation is essentially uniform throughout the gauge length.

Several relationships can be obtained between the engineering stress and strain and the true stress and strain, based on an assumption of constant volume. This assumption is not true for elastic strain (when the atomic bonds are being elastically stretched and thus deforming the volume), but this is of little consequence as these strains are very small and because the engineering and true values are essentially identical in the elastic portion of the stress-strain curve. Once the strains have increased substantially beyond the yield point, most of the strain is inelastic (plastic) and does not contribute to any significant volume change. Thus, a constant volume assumption is made, such that

$$A_0 l_0 = Al = \text{constant} \tag{a}$$

From this, and applying Equation (2–2),

$$\frac{A_0}{A} = \frac{l}{l_0} = \frac{l_0 + \Delta l}{l_0} = \frac{l_0 + (l - l_0)}{l_0} = 1 + \varepsilon \tag{b}$$

From this, engineering strain can be expressed in terms of areas,

$$\varepsilon = \frac{A_0 - A}{A} \tag{2–8}$$

Experimentally, it is usually best to obtain engineering strain from the lengths, using Equation (2–2), before necking occurs, since the change in area is nearly negligible. Then, switch to Equation (2–8) after necking occurs, where the area change is more pronounced and the change in length is no longer uniform throughout the gauge length.

Applying the definition of true strain from Equation (2–7) with Equation (*b*) gives two useful relationships.

$$\tilde{\varepsilon} = \ln \frac{l}{l_0} = \ln \frac{A_0}{A} \tag{2–9}$$

$$\tilde{\varepsilon} = \ln (1 + \varepsilon) \qquad \varepsilon \leq \varepsilon_u \tag{2–10}$$

Incorporating Equation (2–5) with Equation (2–8), we obtain an expression for true stress in terms of engineering stress and engineering strain.

$$\tilde{\sigma} = \sigma(1 + \varepsilon) \qquad \varepsilon \leq \varepsilon_u \tag{2–11}$$

Prior to reaching the ultimate point, the plastic strain is reasonably uniform throughout the gauge length. Once the ultimate point is passed, the plastic elongation is nonuniform and much more localized in the necking region. Since engineering strain is based on an average deformation across the gauge length, it is not a valid measurement for determining true stress or true strain once necking has begun. Consequently, Equations (2–10) and (2–11) are only valid up to the beginning of necking. Before necking, notice in Figure 2–3 that for specific points on the engineering stress-strain curve, such as points labeled i and u, that the corresponding point on the true stress-strain curve is at a larger value of stress and smaller value of strain, which is consistent with Equations (2–10) and (2–11). After necking, the true strain may increase rapidly and exceed the engineering strain. The true stress will always continue to increase monotonically.

The *true fracture strength* and *true fracture strain,* that is, the true stress and true strain at the point of fracture, are important material properties and are designated as $\tilde{\sigma}_f$ and $\tilde{\varepsilon}_f$, respectively. They are readily obtained by applying the area at fracture in Equations (2–4) and (2–9).

Stress-Strain Relationships from Compression and Torsion Tests

Compression tests are more difficult to conduct, and the geometry of the test specimens differs from the geometry of those used in tension tests. The reason for this is that the specimen may buckle during testing or it may be difficult to distribute the stresses evenly. Other difficulties occur because ductile materials will bulge after yielding. However, the results can be plotted on a stress-strain diagram also, and the same strength definitions can be applied as used in tensile testing. For most ductile materials the compressive strengths are about the same as the tensile strengths. When substantial differences occur between tensile and compressive strengths, however, as is the case with the cast irons, the tensile and compressive strengths should be stated separately, S_{ut}, S_{uc}, where S_{uc} is reported as a *positive* quantity.

Torsional strengths are found by twisting solid circular bars and recording the torque and the twist angle. The results are then plotted as a *torque-twist diagram.* The shear stresses in the specimen are linear with respect to radial location, being zero at the center of the specimen and maximum at the outer radius r (see Chapter 3). The maximum shear stress $\tau_{\max}$ is related to the angle of twist θ by

$$\tau_{\max} = \frac{Gr}{l_0}\theta \tag{2–12}$$

where θ is in radians, r is the radius of the specimen, l_0 is the gauge length, and G is the material stiffness property called the *shear modulus* or the *modulus of rigidity.* The maximum shear stress is also related to the applied torque T as

$$\tau_{\max} = \frac{Tr}{J} \tag{2–13}$$

where $J = \frac{1}{2}\pi r^4$ is the polar second moment of area of the cross section.

The torque-twist diagram will be similar to Figure 2–2, and, using Equations (2–12) and (2–13), the modulus of rigidity can be found as well as the elastic limit and the *torsional yield strength* S_{sy}. The maximum point on a torque-twist diagram, corresponding to point u on Figure 2–2, is T_u. The equation

$$S_{su} = \frac{T_u r}{J} \tag{2–14}$$

defines the *modulus of rupture* for the torsion test. Note that it is incorrect to call S_{su} the ultimate torsional strength, as the outermost region of the bar is in a plastic state at the torque T_u and the stress distribution is no longer linear.

Energy Absorption Properties

In addition to providing strength values for a material, the stress-strain diagram provides insight into the energy-absorbing characteristics of a material. This is because the stress-strain diagram involves both loads and deflections, which are directly related to energy. The capacity of a material to absorb energy within its elastic range is called *resilience.* The *modulus of resilience* u_R of a material is defined as the energy absorbed per unit volume without permanent deformation, and is equal to the area under the stress-strain curve up to the elastic limit. The elastic limit is often approximated by the yield point, since it is more readily determined, giving

$$u_R \approx \int_0^{\varepsilon_y} \sigma d\varepsilon \tag{2–15}$$

where ε_y is the strain at the yield point. If the stress-strain is linear to the yield point, then the area under the curve is simply a triangular area; thus

$$u_R \approx \frac{1}{2} S_y \varepsilon_y = \frac{1}{2}(S_y)(S_y/E) = \frac{S_y^2}{2E} \tag{2–16}$$

This relationship indicates that for two materials with the same yield strength, the less stiff material (lower E), will have a greater resilience, that is, an ability to absorb more energy without yielding.

The capacity of a material to absorb energy without fracture is called *toughness.* The *modulus of toughness* u_T of a material is defined as the energy absorbed per unit volume without fracture, which is equal to the total area under the stress-strain curve up to the fracture point, or

$$u_T = \int_0^{\varepsilon_f} \sigma d\varepsilon \tag{2–17}$$

where ε_f is the strain at the fracture point. This integration is often performed graphically from the stress-strain data, or a rough approximation can be obtained by using the average of the yield and ultimate strengths and the strain at fracture to calculate an area; that is,

$$u_T \approx \left(\frac{S_y + S_{ut}}{2}\right)\varepsilon_f \tag{2–18}$$

The units of toughness and resilience are energy per unit volume (J/m^3), which are numerically equivalent to Pa. These definitions of toughness and resilience assume the low strain rates that are suitable for obtaining the stress-strain diagram. For higher strain rates, see Section 2–6 for impact properties.

2–2 The Statistical Significance of Material Properties

There is some subtlety in the ideas presented in the previous section that should be pondered carefully before continuing. Figure 2–2 depicts the result of a *single* tension test (*one* specimen, now fractured). It is common for engineers to consider these important *stress* values (at points *pl*, *el*, *y*, *u*, and *f*) as properties and to denote them as strengths with a special notation, uppercase S, in lieu of lowercase sigma σ, with subscripts added: S_{pl} for proportional limit, S_y for yield strength, S_u for ultimate tensile strength (S_{ut} or S_{uc}, if tensile or compressive sense is important).

If there were 1000 nominally identical specimens, the values of strength obtained would be distributed between some minimum and maximum values. It follows that the description of strength, a material property, is distributional and thus is statistical in nature.

Consider the following example.

EXAMPLE 2–1

One thousand specimens of 1020 steel were tested to rupture, and the ultimate tensile strengths were reported as follows:

Range midpoint S_{ut} (MPa)	389	396	403	410	417	424	431	438	445	452	459	466	473	480	487	494
Frequency, f_i	2	18	23	31	83	109	138	151	139	130	82	49	28	11	4	2

where the *range* of stress for each midpoint entry is $w = 7$ MPa.

Plot a histogram of the data (a bar chart of f versus S_{ut}). Presuming the distribution is normal, plot Equation (1–4). Compare the plots.

Solution

The worksheet for the data is

Range Midpoint x_i (MPa)	Frequency f_i	$x_i f_i$	$x_i^2 f_i$	Observed PDF* $f_i/(Nw)$†	Normal PDF* $f(x)$
389	2	778	302 642	0.000 286	0.000 506
396	18	7 128	2 822 688	0.002 571	0.001 353
403	23	9 269	3 735 407	0.003 286	0.003 121
410	31	12 710	5 211 100	0.004 429	0.006 206
417	83	34 611	14 432 787	0.011 857	0.010 634
424	109	46 216	19 595 584	0.015 571	0.015 707
431	138	59 478	25 635 018	0.019 714	0.019 997
438	151	66 138	28 968 444	0.021 571	0.021 943
445	139	61 855	27 525 475	0.019 857	0.020 754
452	130	58 760	26 559 520	0.018 571	0.016 919
459	82	37 638	17 275 842	0.011 714	0.011 888
466	49	22 834	10 640 644	0.007 000	0.007 200
473	28	13 244	6 264 412	0.004 000	0.003 758
480	11	5 280	2 534 400	0.001 571	0.001 691
487	4	1 948	948 676	0.000 571	0.000 656
494	2	988	488 072	0.000 286	0.000 219
	Σ 1000	438 875	192 940 711		

*PDF refers to Probability Density Function (see Section 1–12).

†To compare discrete frequency data with continuous density functions, f_i must be divided by Nw. Here, N = sample size = 1000, and w = width of range interval = 7 MPa.

From Equation (1–6),

$$\bar{x} = \frac{1}{N}\sum_{i=1}^{k} f_i x_i = \frac{1}{1000}(438\ 875) = 438.875\ \text{MPa}$$

From Equation (1–7),

$$s_x = \sqrt{\frac{\sum_{i=1}^{k} f_i x_i^2 - N\bar{x}^2}{N-1}} = \sqrt{\frac{192\ 940\ 711 - 1000(438.875^2)}{1000 - 1}}$$
$$= 18.2 \text{ MPa}$$

From Equation (1–4), with $\mu_x = \bar{x}$ and $\hat{\sigma}_x = s_x$, the PDF for a normal density function is

$$f(x) = \frac{1}{\hat{\sigma}_x\sqrt{2\pi}} \exp\left[-\frac{1}{2}\left(\frac{x-\mu_x}{\hat{\sigma}_x}\right)^2\right]$$
$$= \frac{1}{18.2\sqrt{2\pi}} \exp\left[-\frac{1}{2}\left(\frac{x-438.875}{18.2}\right)^2\right]$$

For example, $f(438) = 0.021\ 943$.

The bar chart shown in Figure 2–5 depicts the histogram of the PDF of the discrete data. A plot of the continuous normal PDF, $f(x)$, is also included.

Figure 2–5

Histogram for 1000 tensile tests on a 1020 steel from a single heat.

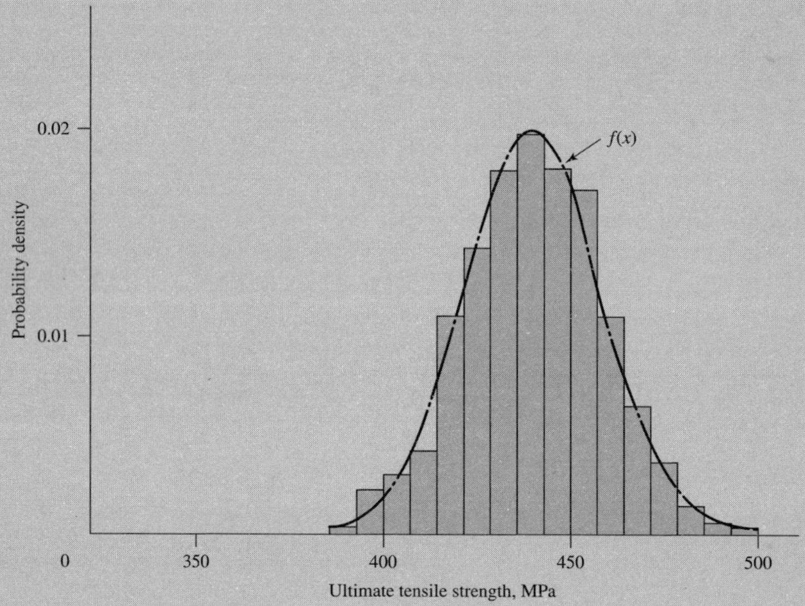

Note, in Example 2–1, the test program has described 1020 property S_{ut}, for only one heat of one supplier. Testing is an involved and expensive process. Tables of properties are often prepared to be helpful to other persons. A statistical quantity is described by its mean, standard deviation, and distribution type. Many tables display a single number, which is often the mean, minimum, or some percentile, such as the 99th percentile. Always read the footnotes to the table.

2–3 Plastic Deformation and Cold Work

Many applications are designed to operate entirely in the elastic region of the stress-strain diagram, with the stresses controlled so as not to exceed the yield point. However, there are two common circumstances that warrant the study of a material's behavior

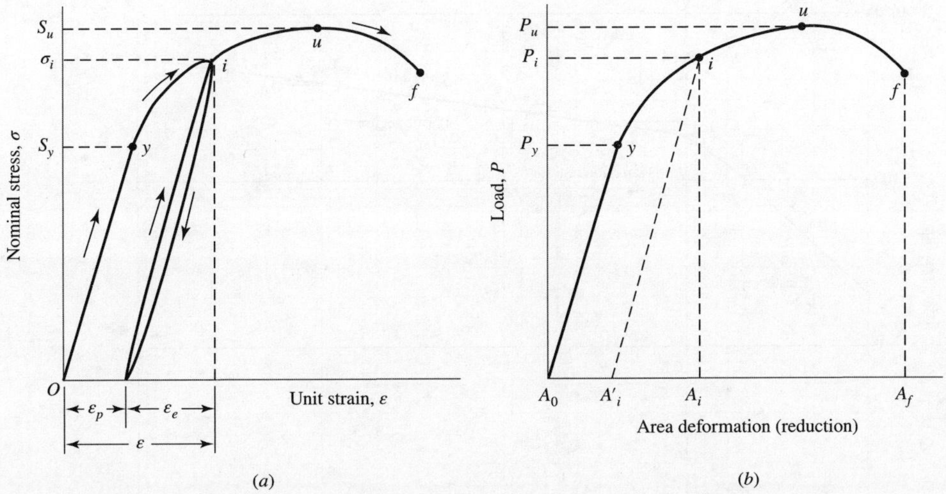

Figure 2–6

(*a*) Stress-strain diagram showing unloading and reloading at point *i* in the plastic region; (*b*) analogous load-deformation diagram.

in the plastic region. The first is that even when the nominal stress is in the elastic region, it is possible for localized stresses to exceed the yield strength. This can be from stress concentrations due to such things as discontinuities in the geometry, cracks, internal material flaws, or localized thermal stresses. For static loading, this type of localized plastic strain may not be detrimental to the functioning of the part. But for repetitive loading, understanding the plastic strain behavior is crucial to understanding fatigue failures (see Chapter 6). The second situation that operates in the plastic deformation zone is material processing and manufacturing operations that deliberately deform the material or part into the plastic zone in order to modify the geometry or the material properties.

Consider a stress-strain diagram of a ductile material, such as in Figure 2–6*a*. As a material is loaded, it first experiences the elastic region. Recall that this region is dominated by stretching of the chemical bonds, resulting in a high stiffness, a linear stress-strain relationship, and a recoverable dimension upon release of the load. Upon passing the elastic limit, roughly near the yield point *y*, the atoms begin to slip along slip planes within the crystalline lattice structure and permanently shift their locations within the lattice. At first, many materials experience this as a significant increase in strain with little increase, and sometimes even a decrease, in stress. But as the load continues to increase, the slope of the stress-strain curve increases. This is due to the proliferation of new slip planes that create defects in the crystalline lattice known as *dislocations*. The material is increasingly saturated with new dislocations, which resist additional slipping, hence the observed strengthening. This process is known as *strain hardening,* and when done deliberately to affect the strength properties of a material it is commonly referred to as *work hardening* or *cold working*.

Cold working gets its name from physically working the material by plastic straining while staying below the recrystallization temperature. Materials can also be deformed plastically by the application of heat, as in forging or hot rolling, but the resulting mechanical properties are significantly dictated by the heat treatment and are quite different from those obtained by strain hardening.

Returning to Figure 2–6*a*, assume a material is stressed beyond the yield strength at *y* to some point *i*, in the plastic region, and then the load removed. At this point the material has a permanent plastic deformation ε_p. If the load corresponding to

Figure 2–7

True stress–true strain diagram with log-log scale, typical for a mild steel.

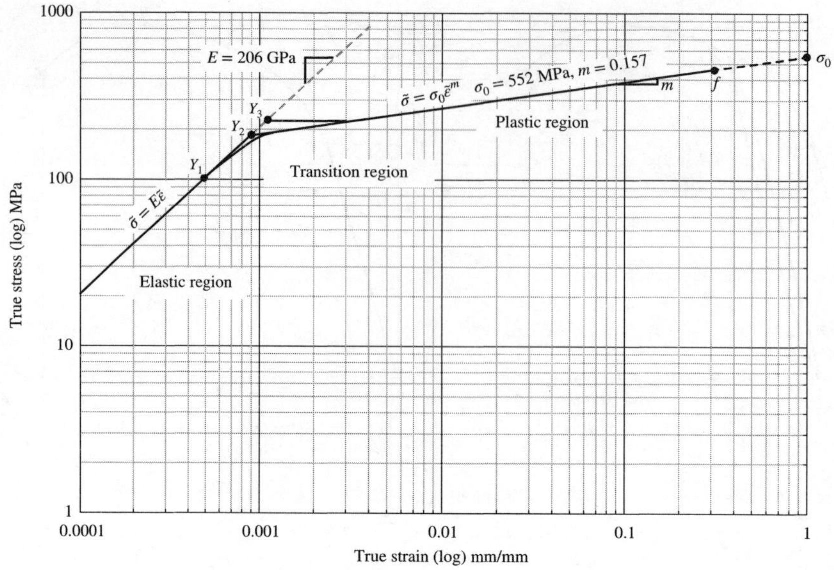

point i is now reapplied, the material will be elastically deformed by the amount ε_e. Thus at point i the total unit strain consists of the two components ε_p and ε_e and is given by the equation

$$\varepsilon = \varepsilon_p + \varepsilon_e \tag{2–19}$$

This material can be unloaded and reloaded any number of times from and to point i, and it is found that the action always occurs along the straight line that is approximately parallel to the initial elastic line Oy. Thus

$$\varepsilon_e = \frac{\sigma_i}{E} \tag{2–20}$$

The material now has a higher yield point, is less ductile as a result of a reduction in strain capacity, and is said to be *strain-hardened*. If the process is continued, increasing ε_p, the material can become brittle and exhibit sudden fracture.

For most metals, several interesting observations can be made by plotting the true stress–true strain diagram on log-log coordinates, such as in Figure 2–7. The elastic region still plots as a straight line. Remember that in this region the strains are very small, such that true stresses and strains are very nearly equal to engineering stresses and strains. Therefore, the elastic region can equally well be expressed by Hooke's law in Equation (2–3), or in terms of true stress and true strain as

$$\tilde{\sigma} = E\tilde{\varepsilon} \tag{2–21}$$

The linear relationship, slope, and intercept on the log-log scale can be readily recognized by taking the logarithm of both sides of Equation (2–21),

$$\log \tilde{\sigma} = \log E + (1) \log \tilde{\varepsilon} \tag{a}$$

On the log-log scale, this line has a unit slope for all materials, and is positioned vertically such that, if extended, its intercept with the ordinate through $\tilde{\varepsilon} = 1$ ($\log \tilde{\varepsilon} = 0$) is at E.

In the plastic strain-strengthening region of Figure 2–7, most metals exhibit a linear relationship. This is generally valid for strains greater than about 0.01, such that the plastic strain is significantly larger than the elastic strain, making the elastic strain negligible. Datsko[3] describes the linear relationship in the plastic strain region by the *strain-strengthening equation,*

$$\tilde{\sigma} = \sigma_0 \tilde{\varepsilon}^m \tag{2–22}$$

Technically, the strain in Equation (2–22) is the plastic component of the true strain, and could be expressed as $\tilde{\varepsilon}_p$. However, for larger strains, the true plastic strain $\tilde{\varepsilon}_p$ can be replaced by the total true strain $\tilde{\varepsilon}$ with negligible difference. Taking the logarithm of both sides of Equation (2–22), the linear relationship is again apparent.

$$\log \tilde{\sigma} = \log \sigma_0 + m \log \tilde{\varepsilon} \tag{b}$$

The term σ_0 is the *strength coefficient* or *strain-strengthening coefficient,* and is the value of the true stress correlating to a true plastic strain of unity (that is, the ordinate intercept of the line when $\tilde{\varepsilon}_p = 1(\log \tilde{\varepsilon} = 0)$). The exponent m is the slope of the line, and is called the *strain-strengthening exponent,* or sometimes the *strain-hardening exponent*. It is a measure of the rate of strain hardening and is thus an indicator of a material's work-hardening behavior. For materials that have an engineering stress-strain curve that exhibits a peak at the ultimate strength, it can be shown[4] that the strain-strengthening exponent is equal to the true strain at the ultimate point, that is,

$$m = \tilde{\varepsilon}_u \tag{2–23}$$

Values for σ_0 and m for several materials are given in Table A–21.

In the neighborhood of the yield point, there is often a transition region between the purely elastic line and the purely plastic line. The shape of this elastic-plastic curve varies for different types of materials. The three points Y_1, Y_2, and Y_3 describe three possible yield points. The intersection of the elastic line and plastic line at Y_2 would be characteristic of an ideal material that behaves as purely elastic until it yields and becomes purely plastic. Most engineering materials are said to *overyield* to Y_3, because they have a yield strength greater than the ideal value. The alloys of the steels, coppers, brasses, and nickels have this characteristic. For most metals, this transition region corresponds roughly to a line of zero slope at the yield strength. This region represents a combination of elastic and plastic strain, with little or no strain strengthening, before the fully plastic strain-strengthening region begins. A few materials, such as a fully annealed aluminum alloy, have the characteristic of *underyielding* at point Y_1.

A single equation to represent both the elastic and plastic behavior can be obtained by adding the elastic true strain and the plastic true strain, as in Equation (2–19). The elastic true strain is obtained from the elastic line of Equation (2–21), and the plastic true strain is obtained from the plastic line of Equation (2–22), by noting that the strain is properly plastic strain. Thus, the total true strain is

$$\tilde{\varepsilon} = \frac{\tilde{\sigma}}{E} + \left(\frac{\tilde{\sigma}}{\sigma_0}\right)^{1/m} \tag{2–24}$$

This is known as the Ramberg-Osgood relationship.

---

[3]Joseph Datsko, "Solid Materials," chapter 32 in Joseph E. Shigley, Charles R. Mischke, and Thomas H. Brown, Jr. (eds.), *Standard Handbook of Machine Design*, 3rd ed., McGraw Hill, New York, 2004. See also Joseph Datsko, "New Look at Material Strength," *Machine Design*, vol. 58, no. 3, Feb. 6, 1986, pp. 81–85.

[4]See section 5–2, J. E. Shigley and C. R. Mischke, *Mechanical Engineering Design*, 6th ed., McGraw Hill, New York, 2001.

EXAMPLE 2–2

The first three columns of Table 2–1 list the results obtained from a tensile test of A–40 annealed titanium
(*a*) Plot the engineering and true stress-strain diagrams.
(*b*) Find the modulus of elasticity, the yield strength, and the ultimate strength.
(*c*) Find the plastic strain-strengthening coefficient and exponent.

Table 2–1 Results of a Tensile Test of Annealed A–40 Titanium as Reported by Datsko (Specimen size is d_0 = 12.8 mm, l_0 = 50.8 mm)

Observed Test Results				Analytical Results			
	Gauge			Engineering		True	
Load P, N (1)	Length l, mm (2)	Diameter d, mm (3)	Area A, mm^2 (4)	Stress P/A_0, MPa (5)	Strain ε, mm/mm (6)	Stress P/A, MPa (7)	Strain $\tilde{\varepsilon}$, mm/mm (8)
0	50.8	12.8	129.22	0	0	0	0
4 448.22	50.815			34.42	0.000 30	34.4	0.000 30
8 896.44	50.830			68.84	0.000 60	68.9	0.000 60
13 344.66	50.846			103.27	0.000 90	103.4	0.000 90
17 792.88	50.861			137.69	0.001 20	137.9	0.001 20
22 241.10	50.889			172.11	0.001 75	172.4	0.001 75
26 689.32	50.912			206.53	0.002 20	207.0	0.002 20
31 137.54	50.945			240.95	0.002 85	241.6	0.002 85
35 585.76	50.978			275.38	0.003 50	276.3	0.003 49
40 033.98	51.039	12.8	128.71	309.80	0.004 70	311.3	0.004 69
44 482.20	51.156			344.22	0.007 00	346.6	0.006 98
53 378.64		12.7	127.18	413.07	0.016 03	419.8	0.015 90
62 275.08		12.5	123.15	481.91	0.049 27	505.6	0.048 10
64 499.19		12.3	119.68	499.12	0.079 72	539.0	0.076 70
66 500.89	58.674	11.9	111.93	514.61	0.154 48	594.1	0.143 65
64 499.19		11.2	98.99	499.12	0.305 38	651.7	0.266 50
62 275.08		10.8	91.52	481.91	0.411 90	680.3	0.344 94
51 154.53	62.992	8.9	62.78	395.85	1.058 25	814.9	0.721 85

Source: Data from Datsko, Joseph. Materials in Design and Manufacturing. *Ann Arbor, Michigan: CRC Press Inc, 1977, Chap 5.*

Solution

(*a*) The engineering stress-strain diagram is plotted in Figure 2–8 from the data in columns 5 and 6 of Table 2–1. Note that two strain scales are used in Figure 2–8 to fit all data points in a compact plot. The first 11 values in column 6 are obtained from Equation (2–2). The remaining values in column 6 are found from Equation (2–8). Note that the strain for the last entry in column 6 does not really mean that the test specimen elongated 1.058 58 mm/mm, because necking has occurred and the result is based on a change of areas. As shown in the table, the final length is actually 62.992 mm.

(*b*) Columns 7 and 8 of Table 2–1 are the computed values of the true stress and strain which are plotted in Figure 2–9. The true strain is obtained using Equation (2–10).

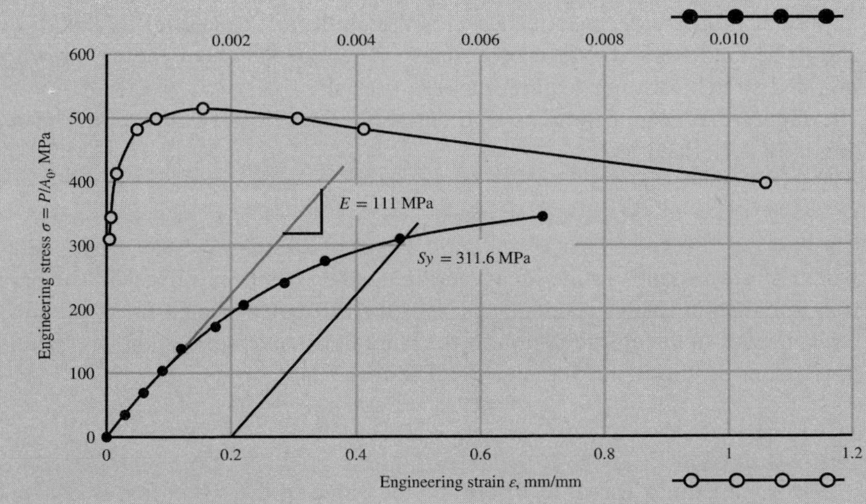

Figure 2–8

Engineering stress-strain diagram for Example 2–2.

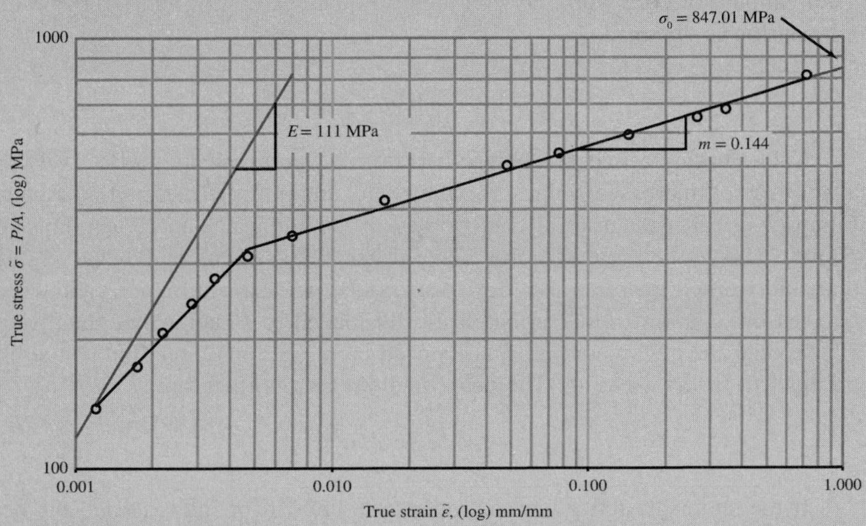

Figure 2–9

True stress-strain in log-log scale for Example 2–2.

The ultimate strength is $S_{ut} = 514.61$ MPa; it is the value of the engineering stress in column 5 corresponding to the ultimate load $P = 66\ 500.89$ N.

(*c*) Figure 2–9 shows that the plastic strain-strengthening coefficient σ_0 is found at the intersection of the stress-strain line and the ordinate corresponding to $\tilde{\varepsilon} = 1$, that is, $\log \tilde{\varepsilon} = 0$.

The strain-strengthening exponent is found to be $m = 0.144$, from Equation (2–23), and corresponds to the true strain at the ultimate load $P = 66\ 500.89$ N .

The ability of a material to be cold-worked can be quantified. The stress-strain loading of Figure 2–6*a* is shown in terms of area deformation in Figure 2–6*b*. The *reduction in area* corresponding to the load P_f, at fracture, is defined as

$$R = \frac{A_0 - A_f}{A_0} = 1 - \frac{A_f}{A_0} \tag{2–25}$$

where A_0 is the original area and A_f is the area at fracture. The quantity R in Equation (2–25) is usually expressed in percent and tabulated in lists of mechanical properties as a measure of *ductility*. See Appendix Table A–19, for example. Ductility is an important property because it measures the ability of a material to absorb overloads and to be cold-worked. Metal processing operations such as bending, drawing, heading, and stretch forming require ductile materials.

Figure 2–6*b* can also be used to define the quantity of cold work. The *cold-work factor W* is defined as

$$W = \frac{A_0 - A_i'}{A_0} \approx \frac{A_0 - A_i}{A_0} \tag{2–26}$$

where A_i' corresponds to the area after the load P_i has been released. The approximation in Equation (2–26) results because of the difficulty of measuring the small diametral changes in the elastic region. If the amount of cold work is known, then Equation (2–26) can be solved for the area A_i'. The result is

$$A_i' = A_0(1 - W) \tag{2–27}$$

It is sometimes useful to determine the value of the strain that is equivalent to a given amount of cold work. By combining Equations (2–26) and (2–9) the following relationship is obtained.

$$\tilde{\varepsilon} = \ln\left(\frac{1}{1 - W}\right) \tag{2–28}$$

Cold working a material before it is put into an application effectively produces a new set of increased values for the yield and ultimate strengths. Returning to Figure 2–6*b*, if a material is cold-worked by applying a load P_i to point i between points y and u, then released, the material has been strain-strengthened. Considering this cold-worked material as a fresh material, it will have a higher yield strength for two reasons. First, it now can be elastically loaded to P_i before yielding, rather than P_y. Second, the new material has a reduced cross-sectional area, but can still achieve the load P_i before yielding. The new yield strength is therefore

$$S_y' = \frac{P_i}{A_i'} \qquad P_i \le P_u \tag{2–29}$$

If the strain-strengthening parameters are known for the material, the new yield strength can be estimated from the amount of strain by applying Equation (2–22),

$$S_y' = \sigma_0 \tilde{\varepsilon}_i^m \qquad \tilde{\varepsilon}_i \le \tilde{\varepsilon}_u \tag{2–30}$$

In determining a new ultimate strength for the cold-worked material, note that the ultimate load has not changed, but the area is reduced. Thus,

$$S_u' = \frac{P_u}{A_i'} \tag{2–31}$$

Since $P_u = S_u A_0$, we find, with Equation (2–27), that

$$S_u' = \frac{S_u A_0}{A_0(1 - W)} = \frac{S_u}{1 - W} \qquad \tilde{\varepsilon}_i \le \tilde{\varepsilon}_u \tag{2–32}$$

Pairing Equations (2–28) and (2–30), the new yield strength is seen to be a function of the amount of cold work and the strain-strengthening properties of the material. From Equation (2–32), however, the new ultimate strength depends only on the amount of cold work and not on the strain-strengthening properties. This is because the new ultimate strength is due only to the reduced area. The consequence is that cold working will increase the yield strength more than the ultimate strength, causing the yield strength to approach the ultimate strength. With continued cold working, the plastic region between the yield point and ultimate point is reduced; thus, the material's ductility is reduced. Severely cold working a ductile material can make it into a brittle material.

EXAMPLE 2–3

An annealed AISI 1018 steel (see Table A–21) is given 15 percent cold work. Find the new values for the yield strength and the ultimate strength.

Solution

From Table A–21, $S_y = 220$ MPa, $S_u = 341$ MPa, $\sigma_0 = 620$ MPa, $m = 0.25$. From Equation (2–28), the true strain that is induced by 15 percent cold work is

$$\tilde{\varepsilon}_{15} = \ln\left(\frac{1}{1 - W}\right) = \ln\left(\frac{1}{1 - 0.15}\right) = 0.1625$$

From Equation (2–23), the true strain at the ultimate point is equal to m. That is, $\tilde{\varepsilon}_u = m = 0.25$. Therefore, since $\tilde{\varepsilon}_{15} < \tilde{\varepsilon}_u$, the strain from 15 percent cold work has not reached the strain at the ultimate point, and Equations (2–30) and (2–32) are valid.

Answer Equation (2–30): $S_y' = \sigma_0 \tilde{\varepsilon}_{15}^m = 620(0.1625)^{0.25} = 393.6$ MPa

Answer Equation (2–32): $S_u' = \dfrac{S_u}{1 - W} = \dfrac{341}{1 - 0.15} = 401.2$ MPa

Notice that the original yield strength was not needed to find the new yield strength.

2–4 Cyclic Stress-Strain Properties

A typical stress-strain diagram is said to be *monotonic,* because it is generated with steadily increasing loads, either entirely in tension or compression. The monotonic stress-strain diagram discussed in the previous section demonstrates how material properties, such as yield strength, can change due to loading in the plastic region. Additional effects are observed when the loading is cycled between tension and compression.

Figure 2–10 illustrates the behavior of a polycrystalline metal that is loaded in tension into the plastic region to point B, followed by loading in compression into the plastic region to point E. The yield point in compression, point D, often has a lower magnitude than the original yield point in tension, point A. This is known as the *Bauschinger effect* and demonstrates a reduction in yield strength due to even a single reversal of direction of plastic strain. As shown, the compressive yield point D often occurs when the stress change from the unloading point B is about twice the monotonic yield strength, that is, $2S_y$. Apparently, in the first yielding, some grains will slip and some will not, depending on the orientation of their natural slip planes. Upon unloading, the grains that did not slip try to elastically return to their initial geometry, while the slipped (yielded) grains naturally retain their plastic deformation. This

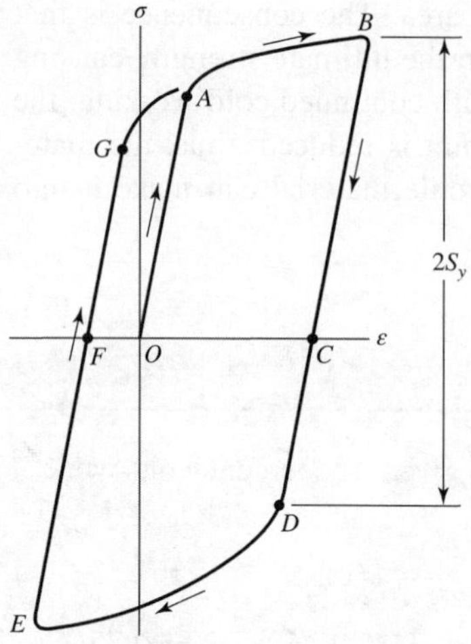

Figure 2–10

Stress-strain diagram showing the Bauschinger effect.

results in residual stresses that are favorable to new loads in the original direction, but which tend to yield sooner when loaded in the opposite direction.

Returning to Figure 2–10, reversing the direction of loading from compression at point E to the original tensile direction can result in a lower tensile yield strength, at point G, than the original yield strength, at point A. This is referred to as *cyclic softening,* and is characteristic of cold-worked metals. On the other hand, annealed metals can exhibit a *cyclic hardening* characteristic, in which the yield strength increases with reversed plastic cycling.

The Bauschinger effect makes it apparent that a single reversal of plastic loading can change the stress-strain behavior from the monotonic behavior. Noteworthy points are that gains in tensile yield strength from strength hardening can be at the expense of compressive yield strength, and can be partially lost by repetitive cycling. Thus, care should be taken with using published strength data for cold-worked metals in applications subject to reversed plastic loading directions. In many situations, the noted effects of plastic behavior will be localized due to the mechanics of the cold-working operation, and to the stress gradient, such as from bending or torsion. This localized behavior may have little consequence to the gross geometry of the part. Consequently, the cyclic stress-strain properties are often of most interest at locations where fatigue cracks are likely to begin and grow, as will be explored more fully in Chapter 6.

With continued cycling between controlled levels of reversed strain, cyclic hardening or softening is usually more pronounced with the first several cycles, but then gradually stabilizes to an equilibrium, as shown in Figure 2–11. The stable cyclic

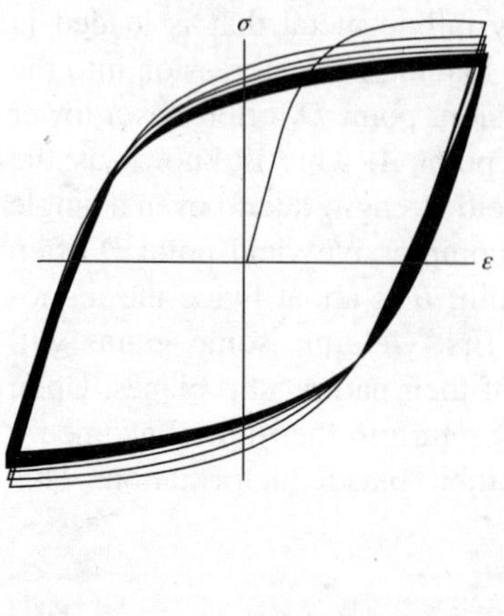

Figure 2–11

A cyclic stress-strain diagram showing strain softening, converging to a stable condition.

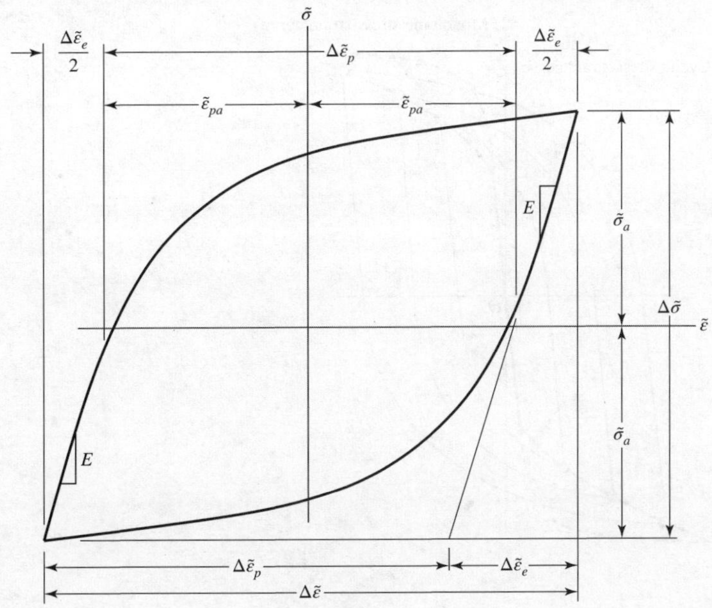

Figure 2–12

A stable cyclic hysteresis loop.

stress-strain diagram is known as a *stable cyclic hysteresis loop,* and is shown in Figure 2–12. From the figure, it can be seen that the true strain range, $\Delta\tilde{\varepsilon}$, is the sum of the true elastic strain and the true plastic strain, given by

$$\Delta\tilde{\varepsilon} = \Delta\tilde{\varepsilon}_e + \Delta\tilde{\varepsilon}_p = \frac{\Delta\tilde{\sigma}}{E} + \Delta\tilde{\varepsilon}_p \tag{2–33}$$

where $\Delta\tilde{\sigma}$ is the true stress range, and E is Young's modulus.

In engineering metals, the stable hysteresis loop is typically close to symmetrical with respect to tension and compression. This is an indication that continued cycling tends to overcome the Bauschinger effect as the tensile and compressive behaviors become symmetric in the stable condition. The area within a hysteresis loop is the energy of plastic work per unit volume done during a cycle, which is dissipated in heat.

Similar to a monotonic stress-strain curve, a cyclic stress-strain curve can be generated for a material to observe the relationship between stress and strain ranges when subject to reversed cyclic loading. A series of stable cyclic hysteresis loops can be generated, each at a different level of reversed cyclic strain. The stable hysteresis loops are superimposed, as shown in Figure 2–13. A curve from the origin passing through the tips of the hysteresis loops, such as *O-A-B-C*, is called a *cyclic stress-strain curve*. For most engineering metals, this curve is symmetric for compression and tension. The cyclic stress-strain curve represents the relationship between the stress amplitude and strain amplitude for cyclic loading. Figure 2–14 shows how a monotonic and cyclic stress-strain curve might typically compare for cyclic hardening and cyclic softening materials. Clearly, use of monotonic material properties in a cyclic loading situation can lead to poor prediction of such things as the yield point, or the amount of strain to be expected for a given stress amplitude.

Similar to the monotonic stress-strain curve, the plastic cyclic stress-strain curve exhibits a linear relationship on a log-log scale. Adapting Equation (2–22) for true

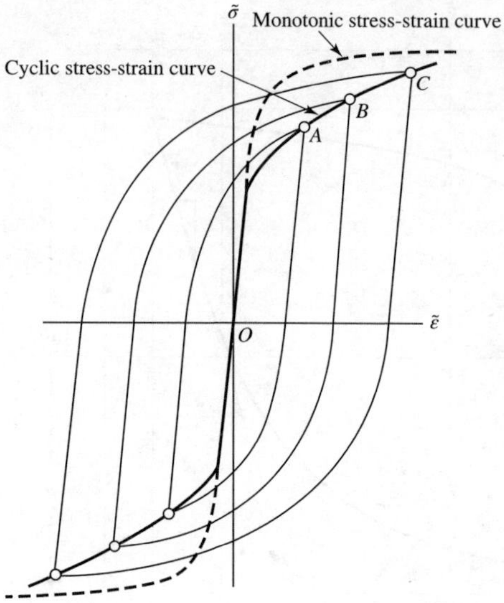

Figure 2–13

A cyclic stress-strain curve generated from superimposing a series of stable hysteresis loops.

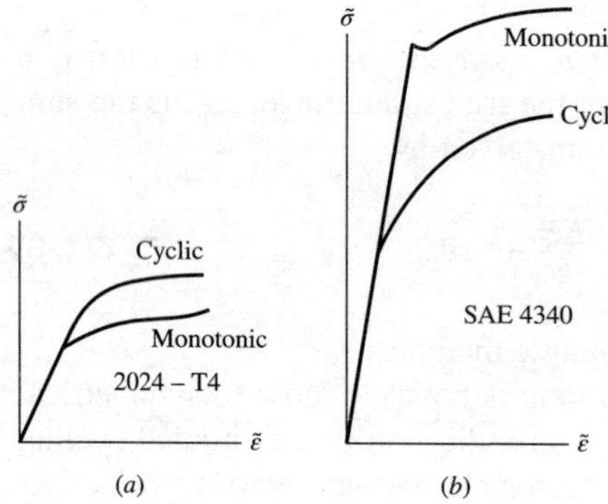

Figure 2–14

Comparison of monotonic and cyclic stress-strain curves for (*a*) a cyclic hardening material and (*b*) a cyclic softening material. *(Source: Adapted from Landgraf, R. W. et al. "Determination of the Cyclic Stress-Strain Curve."* Journal of Materials, American Society for Testing and Materials *4, no. 1 (March 1969).)*

stress amplitude, $\tilde{\sigma}_a = \Delta\tilde{\sigma}/2$, and true plastic strain amplitude, $\tilde{\varepsilon}_{pa} = \Delta\tilde{\varepsilon}_p/2$, the linear relationship is expressed by the power function

$$\tilde{\sigma}_a = \sigma'_0(\tilde{\varepsilon}_{pa})^{m'} \tag{2–34}$$

where σ'_0 is the *cyclic strength coefficient,* and m' is the *cyclic strain strengthening exponent,* with values for select materials in Table A–21. The *cyclic yield strength,* S'_y is defined by a 0.2 percent offset on the cyclic stress-strain curve. This can be estimated by using a plastic strain amplitude of 0.002 in Equation (2–34).

The combined elastic and plastic cyclic stress-strain curve can be represented by a single equation, using a parallel development to Equation (2–24), obtaining the *cyclic Ramberg-Osgood* relationship

$$\tilde{\varepsilon}_a = \frac{\tilde{\sigma}_a}{E} + \left(\frac{\tilde{\sigma}_a}{\sigma'_0}\right)^{1/m'} \tag{2–35}$$

2–5 Hardness

The resistance of a material to penetration by a pointed tool is called *hardness*. Though there are many hardness-measuring systems, we shall consider here only the two in greatest use.

Rockwell hardness tests are described by ASTM standard hardness method E–18 and measurements are quickly and easily made, they have good reproducibility, and the test machine for them is easy to use. In fact, the hardness number is read directly from a dial. Rockwell hardness scales are designated as *A*, *B*, *C*, . . . , etc. The indenters are described as a diamond, a 1.5875 ± 0.003 mm diameter ball, and a diamond for scales *A*, *B*, and *C*, respectively, where the load applied is either 60, 100, or 150 kg. Thus the Rockwell *B* scale, designated R_B, uses a 100-kg load and a No. 2 indenter, which is a 1.5875 ± 0.003 mm diameter ball. The Rockwell *C* scale R_C uses a diamond cone, which is the No. 1 indenter, and a load of 150 kg. Hardness numbers so obtained are relative. Therefore a hardness $R_C = 50$ has meaning only in relation to another hardness number using the same scale.

The *Brinell hardness* is another test in very general use. In testing, the indenting tool through which force is applied is a ball and the hardness number H_B is found as a number equal to the applied load divided by the spherical surface area of the indentation. Thus the units of H_B are the same as those of stress, though they are seldom used. Brinell hardness testing takes more time, since H_B must be computed from the test data. The primary advantage of both methods is that they are nondestructive in most cases. Both are empirically and directly related to the ultimate strength of the material tested. This means that the strength of parts could, if desired, be tested part by part during manufacture.

Hardness testing provides a convenient and nondestructive means of estimating the strength properties of materials. The Brinell hardness test is particularly well known for this estimation, since for many materials the relationship between the minimum ultimate strength and the Brinell hardness number is roughly linear. The constant of proportionality varies between classes of materials, and is also dependent on the load used to determine the hardness. There is a wide scatter in the data, but for rough approximations for *steels*, the relationship is generally accepted as

$$S_u = 3.4\ H_B \quad \text{MPa} \tag{2–36}$$

Similar relationships for *cast iron* can be derived from data supplied by Krause.[5] The minimum strength, as defined by the ASTM, is found from these data to be

$$S_u = 1.58\ H_B - 86\ \text{MPa} \tag{2–37}$$

Walton[6] shows a chart from which the SAE minimum strength can be obtained, which is more conservative than the values obtained from Equation (2–37).

[5]D. E. Krause, "Gray Iron—A Unique Engineering Material," ASTM Special Publication 455, 1969, pp. 3–29, as reported in Charles F. Walton (ed.), *Iron Castings Handbook*, Iron Founders Society, Inc., Cleveland, 1971, pp. 204, 205.

[6]Ibid.

EXAMPLE 2–4

It is necessary to ensure that a certain part supplied by a foundry always meets or exceeds ASTM No. 20 specifications for cast iron (see Table A–23). What hardness should be specified?

Solution

From Equation (2–37), with $(S_u)_{min} = 138$ MPa, we have

Answer

$$H_B = \frac{S_u + 86}{1.58} = \frac{138 + 86}{1.58} = 141.8$$

If the foundry can control the hardness within 20 points, routinely, then specify $145 < H_B < 165$. This imposes no hardship on the foundry and assures the designer that ASTM grade 20 will always be supplied at a predictable cost.

2–6 Impact Properties

An external force applied to a structure or part is called an *impact load* if the time of application is less than one-third the lowest natural period of vibration of the part or structure. Otherwise it is called simply a *static load.*

The *Charpy* (commonly used) and *Izod* (rarely used) *notched-bar tests* utilize bars of specified geometries to determine brittleness and impact strength. These tests are helpful in comparing several materials and in the determination of low-temperature brittleness. In both tests the specimen is struck by a pendulum released from a fixed height, and the energy absorbed by the specimen, called the *impact value,* can be computed from the height of swing after fracture, but is read from a dial that essentially "computes" the result.

The effect of temperature on impact values is shown in Figure 2–15 for a material showing a ductile-brittle transition. Not all materials show this transition. Notice the narrow region of critical temperatures where the impact value increases very rapidly. In the low-temperature region the fracture appears as a brittle, shattering type, whereas the appearance is a tough, tearing type above the critical-temperature region. The critical temperature seems to be dependent on both the material and the geometry of the notch. For this reason designers should not rely too heavily on the results of notched-bar tests.

The average strain rate used in obtaining the stress-strain diagram is about 0.001 mm/(mm · s) or less. When the strain rate is increased, as it is under impact conditions, the strengths increase, as shown in Figure 2–16. In fact, at very high strain rates the yield strength seems to approach the ultimate strength as a limit. But note

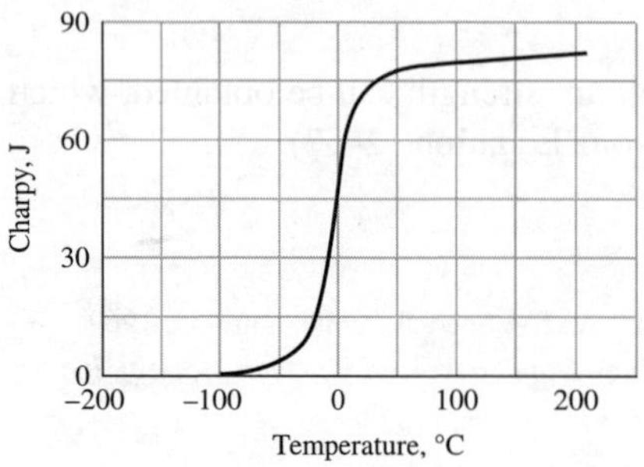

Figure 2–15

A mean trace shows the effect of temperature on impact values. The result of interest is the brittle-ductile transition temperature, often defined as the temperature at which the mean trace passes through the 20 J level. The critical temperature is dependent on the geometry of the notch, which is why the Charpy V notch is closely defined.

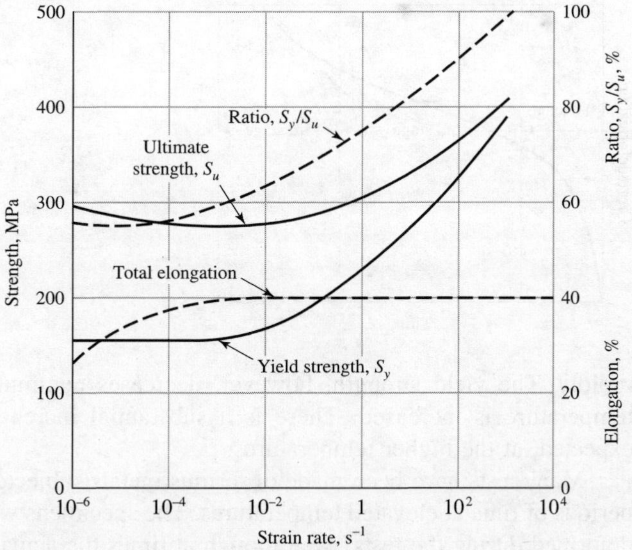

Figure 2–16

Influence of strain rate on tensile properties.

that the curves show little change in the elongation. This means that the ductility remains about the same. Also, in view of the sharp increase in yield strength, a mild steel could be expected to behave elastically throughout practically its entire strength range under impact conditions.

The Charpy and Izod tests really provide toughness data under dynamic, rather than static, conditions. It may well be that impact data obtained from these tests are as dependent on the notch geometry as they are on the strain rate. For these reasons it may be better to use the concepts of notch sensitivity, fracture toughness, and fracture mechanics, discussed in Chapters 5 and 6, to assess the possibility of cracking or fracture.

2–7 Temperature Effects

Strength and ductility, or brittleness, are properties affected by the temperature of the operating environment.

The effect of temperature on the static properties of steels is typified by the strength versus temperature chart of Figure 2–17. Note that the tensile strength changes only a small amount until a certain temperature is reached. At that point it falls off

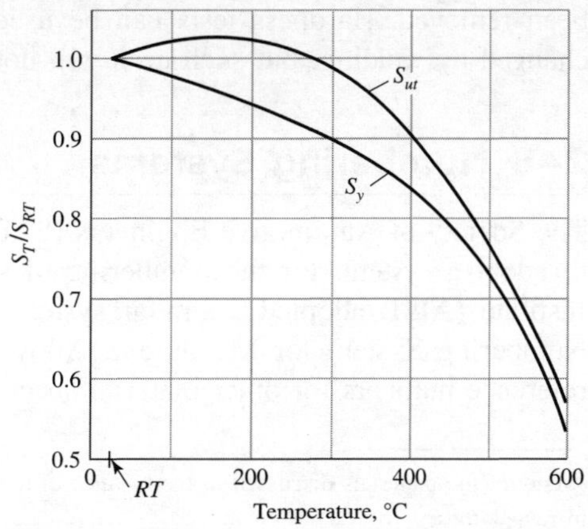

Figure 2–17

A plot of the results of 145 tests of 21 carbon and alloy steels showing the effect of operating temperature on the yield strength S_y and the ultimate strength S_{ut}. The ordinate is the ratio of the strength at the operating temperature to the strength at room temperature. The standard deviations were $0.0442 \leq \hat{\sigma}_{Sy} \leq 0.152$ for S_y and $0.099 \leq \hat{\sigma}_{Sut} \leq 0.11$ for S_{ut}. *(Source: Data from Brandes, E. A., (ed.),* Smithells Metals Reference Book, *6th ed., Butterworth, London, 1983, 22–128 to 22–131.)*

Figure 2–18

Creep-time curve.

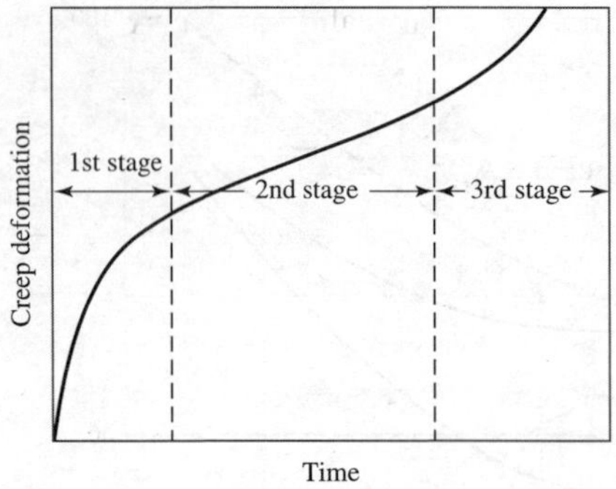

rapidly. The yield strength, however, decreases continuously as the environmental temperature is increased. There is a substantial increase in ductility, as might be expected, at the higher temperatures.

Many tests have been made of ferrous metals subjected to constant loads for long periods of time at elevated temperatures. The specimens were found to be permanently deformed during the tests, even though at times the actual stresses were less than the yield strength of the material obtained from short-time tests made at the same temperature. This continuous deformation under load is called *creep*.

One of the most useful tests to have been devised is the long-time creep test under constant load. Figure 2–18 illustrates a curve that is typical of this kind of test. The curve is obtained at a constant stated temperature. A number of tests are usually run simultaneously at different stress intensities. The curve exhibits three distinct regions. In the first stage are included both the elastic and the plastic deformation. This stage shows a decreasing creep rate, which is due to the strain hardening. The second stage shows a constant minimum creep rate caused by the annealing effect. In the third stage the specimen shows a considerable reduction in area, the true stress is increased, and a higher creep eventually leads to fracture.

When the operating temperatures are lower than the transition temperature (Figure 2–15), the possibility arises that a part could fail by a brittle fracture. This subject is discussed in Chapter 5.

Of course, heat treatment, as will be shown, is used to make substantial changes in the mechanical properties of a material.

Heating due to electric and gas welding also changes the mechanical properties. Such changes may be due to clamping during the welding process, as well as heating; the resulting stresses then remain when the parts have cooled and the clamps have been removed. Hardness tests can be used to learn whether the strength has been changed by welding, but such tests will not reveal the presence of residual stresses.

2–8 Numbering Systems

The Society of Automotive Engineers (SAE) was the first to recognize the need, and to adopt a system, for the numbering of steels. Later the American Iron and Steel Institute (AISI) adopted a similar system. In 1975 the SAE published the Unified Numbering System for Metals and Alloys (UNS); this system also contains cross-reference numbers for other material specifications.[7] The UNS uses a letter prefix to

[7]Many of the materials discussed in the balance of this chapter are listed in the Appendix tables. Be sure to review these.

designate the material, as, for example, G for the carbon and alloy steels, A for the aluminum alloys, C for the copper-base alloys, and S for the stainless or corrosion-resistant steels. For some materials, not enough agreement has as yet developed in the industry to warrant the establishment of a designation.

For the steels, the first two numbers following the letter prefix indicate the composition, excluding the carbon content. The various compositions used are as follows:

G10	Plain carbon	G46	Nickel-molybdenum
G11	Free-cutting carbon steel with more sulfur or phosphorus	G48	Nickel-molybdenum
		G50	Chromium
G13	Manganese	G51	Chromium
G23	Nickel	G52	Chromium
G25	Nickel	G61	Chromium-vanadium
G31	Nickel-chromium	G86	Chromium-nickel-molybdenum
G33	Nickel-chromium	G87	Chromium-nickel-molybdenum
G40	Molybdenum	G92	Manganese-silicon
G41	Chromium-molybdenum	G94	Nickel-chromium-molybdenum
G43	Nickel-chromium-molybdenum		

The second number pair refers to the approximate carbon content. Thus, G10 400 is a plain carbon steel with a nominal carbon content of 0.40 percent (0.37 to 0.44 percent). The fifth number following the prefix is used for special situations. For example, the old designation AISI 52 100 represents a chromium alloy with about 100 points of carbon. The UNS designation is G52 986.

The UNS designations for the stainless steels, prefix S, utilize the older AISI designations for the first three numbers following the prefix. The next two numbers are reserved for special purposes. The first number of the group indicates the approximate composition. Thus 2 is a chromium-nickel-manganese steel, 3 is a chromium-nickel steel, and 4 is a chromium alloy steel. Sometimes stainless steels are referred to by their alloy content. Thus S30 200 is often called an 18-8 stainless steel, meaning 18 percent chromium and 8 percent nickel.

The prefix for the aluminum group is the letter A. The first number following the prefix indicates the processing. For example, A9 is a wrought aluminum, while A0 is a casting alloy. The second number designates the main alloy group as shown in Table 2–2. The third number in the group is used to modify the original alloy or to designate the impurity limits. The last two numbers refer to other alloys used with the basic group.

Table 2–2 Aluminum Alloy Designations

Aluminum 99.00% pure and greater	Ax1xxx
Copper alloys	Ax2xxx
Manganese alloys	Ax3xxx
Silicon alloys	Ax4xxx
Magnesium alloys	Ax5xxx
Magnesium-silicon alloys	Ax6xxx
Zinc alloys	Ax7xxx

The American Society for Testing and Materials (ASTM) numbering system for cast iron is in widespread use. This system is based on the tensile strength. Thus ASTM A18 speaks of classes; e.g., 30 cast iron has a minimum tensile strength of 30 kpsi (around 200 MPa). Note from Appendix A-23, however, that the *typical* tensile strength is 31 kpsi (213 MPa). You should be careful to designate which of the two values is used in design and problem work because of the significance of factor of safety.

2–9 Sand Casting

Sand casting is a basic low-cost process, and it lends itself to economical production in large quantities with practically no limit to the size, shape, or complexity of the part produced.

In sand casting, the casting is made by pouring molten metal into sand molds. A pattern, constructed of metal or wood, is used to form the cavity into which the molten metal is poured. Recesses or holes in the casting are produced by sand cores introduced into the mold. The designer should make an effort to visualize the pattern and casting in the mold. In this way the problems of core setting, pattern removal, draft, and solidification can be studied. Castings to be used as test bars of cast iron are cast separately and properties may vary.

Steel castings are the most difficult of all to produce, because steel has the highest melting temperature of all materials normally used for casting. This high temperature aggravates all casting problems.

The following rules will be found quite useful in the design of any sand casting:

1. All sections should be designed with a uniform thickness.
2. The casting should be designed so as to produce a gradual change from section to section where this is necessary.
3. Adjoining sections should be designed with generous fillets or radii.
4. A complicated part should be designed as two or more simple castings to be assembled by fasteners or by welding.

Steel, gray iron, brass, bronze, and aluminum are most often used in castings. The minimum wall thickness for any of these materials is about 5 mm, though with particular care, thinner sections can be obtained with some materials.

2–10 Shell Molding

The shell-molding process employs a heated metal pattern, usually made of cast iron, aluminum, or brass, which is placed in a shell-molding machine containing a mixture of dry sand and thermosetting resin. The hot pattern melts the plastic, which, together with the sand, forms a shell about 5 to 10 mm thick around the pattern. The shell is then baked at from 400 to 700°F for a short time while still on the pattern. It is then stripped from the pattern and placed in storage for use in casting.

In the next step the shells are assembled by clamping, bolting, or pasting; they are placed in a backup material, such as steel shot; and the molten metal is poured into the cavity. The thin shell permits the heat to be conducted away so that solidification takes place rapidly. As solidification takes place, the plastic bond is burned and the mold collapses. The permeability of the backup material allows the gases to escape and the casting to air-cool. All this aids in obtaining a fine-grain, stress-free casting.

Shell-mold castings feature a smooth surface, a draft that is quite small, and close tolerances. In general, the rules governing sand casting also apply to shell-mold casting.

2–11 Investment Casting

Investment casting uses a pattern that may be made from wax, plastic, or other material. After the mold is made, the pattern is melted out. Thus a mechanized method of casting a great many patterns is necessary. The mold material is dependent upon the melting point of the cast metal. Thus a plaster mold can be used for some materials while others would require a ceramic mold. After the pattern is melted out, the mold is baked or fired; when firing is completed, the molten metal may be poured into the hot mold and allowed to cool.

If a number of castings are to be made, then metal or permanent molds may be suitable. Such molds have the advantage that the surfaces are smooth, bright, and accurate, so that little, if any, machining is required. *Metal-mold castings* are also known as *die castings* and *centrifugal castings*.

2–12 Powder-Metallurgy Process

The powder-metallurgy process is a quantity-production process that uses powders from a single metal, several metals, or a mixture of metals and nonmetals. It consists essentially of mechanically mixing the powders, compacting them in dies at high pressures, and heating the compacted part at a temperature less than the melting point of the major ingredient. The particles are united into a single strong part similar to what would be obtained by melting the same ingredients together. The advantages are (1) the elimination of scrap or waste material, (2) the elimination of machining operations, (3) the low unit cost when mass-produced, and (4) the exact control of composition. Some of the disadvantages are (1) the high cost of dies, (2) the lower physical properties, (3) the higher cost of materials, (4) the limitations on the design, and (5) the limited range of materials that can be used. Parts commonly made by this process are oil-impregnated bearings, incandescent lamp filaments, cemented-carbide tips for tools, and permanent magnets. Some products can be made only by powder metallurgy: surgical implants, for example. The structure is different from what can be obtained by melting the same ingredients.

2–13 Hot-Working Processes

By *hot working* are meant such processes as rolling, forging, hot extrusion, and hot pressing, in which the metal is heated above its recrystallization temperature.

Hot rolling is usually used to create a bar of material of a particular shape and dimension. Figure 2–19 shows some of the various shapes that are commonly produced by the hot-rolling process. All of them are available in many different sizes as well as in different materials. The materials most available in the hot-rolled bar sizes are steel, aluminum, magnesium, and copper alloys.

Tubing can be manufactured by hot-rolling strip or plate. The edges of the strip are rolled together, creating seams that are either butt-welded or lap-welded. Seamless tubing is manufactured by roll-piercing a solid heated rod with a piercing mandrel.

Extrusion is the process by which great pressure is applied to a heated metal billet or blank, causing it to flow through a restricted orifice. This process is more

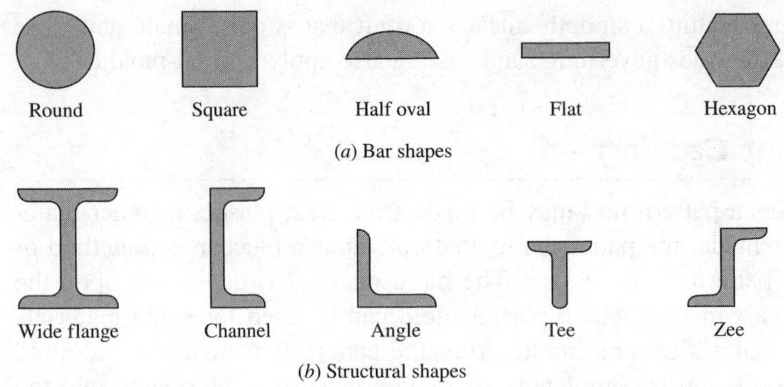

Figure 2–19

Common shapes available through hot rolling.

common with materials of low melting point, such as aluminum, copper, magnesium, lead, tin, and zinc. Stainless steel extrusions are available on a more limited basis.

Forging is the hot working of metal by hammers, presses, or forging machines. In common with other hot-working processes, forging produces a refined grain structure that results in increased strength and ductility. Compared with castings, forgings have greater strength for the same weight. In addition, drop forgings can be made smoother and more accurate than sand castings, so that less machining is necessary. However, the initial cost of the forging dies is usually greater than the cost of patterns for castings, although the greater unit strength rather than the cost is usually the deciding factor between these two processes.

2–14 Cold-Working Processes

By *cold working* is meant the forming of the metal while at a low temperature (usually room temperature). In contrast to parts produced by hot working, cold-worked parts have a bright new finish, are more accurate, and require less machining.

Cold-finished bars and shafts are produced by rolling, drawing, turning, grinding, and polishing. Of these methods, by far the largest percentage of products are made by the cold-rolling and cold-drawing processes. Cold rolling is now used mostly for the production of wide flats and sheets. Practically all cold-finished bars are made by cold drawing but even so are sometimes mistakenly called "cold-rolled bars." In the drawing process, the hot-rolled bars are first cleaned of scale and then drawn by pulling them through a die that reduces the size about 0.8 to 1.6 mm. This process does not remove material from the bar but reduces, or "draws" down, the size. Many different shapes of hot-rolled bars may be used for cold drawing.

Cold rolling and cold drawing have the same effect upon the mechanical properties. The cold-working process does not change the grain size but merely distorts it. Cold working results in a large increase in yield strength, an increase in ultimate strength and hardness, and a decrease in ductility. In Figure 2–20 the properties of a cold-drawn bar are compared with those of a hot-rolled bar of the same material.

Heading is a cold-working process in which the metal is gathered, or upset. This operation is commonly used to make screw and rivet heads and is capable of producing a wide variety of shapes. *Roll threading* is the process of rolling threads by squeezing and rolling a blank between two serrated dies. *Spinning* is the operation of working sheet material around a rotating form into a circular shape. *Stamping* is the term used to describe punch-press operations such as *blanking, coining, forming,* and *shallow drawing*.

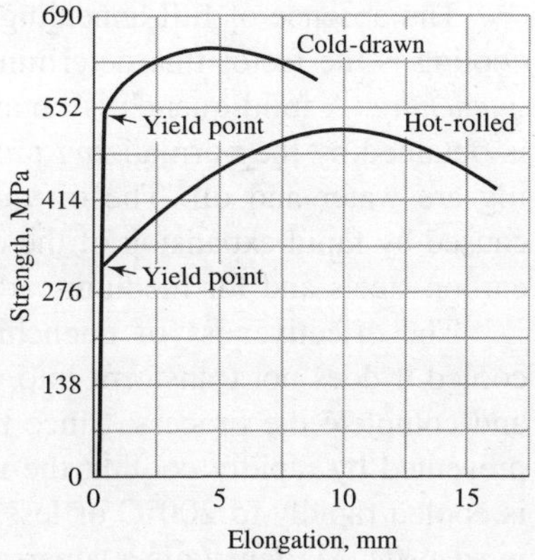

Figure 2–20
Stress-strain diagram for hot-rolled and cold-drawn UNS G10 350 steel.

2–15 The Heat Treatment of Steel

Heat treatment of steel refers to time- and temperature-controlled processes that relieve residual stresses and/or modifies material properties such as hardness (strength), ductility, and toughness. Other mechanical or chemical operations are sometimes grouped under the heading of heat treatment. The common heat-treating operations are annealing, quenching, tempering, and case hardening.

Annealing

When a material is cold- or hot-worked, residual stresses are built in, and, in addition, the material usually has a higher hardness as a result of these working operations. These operations change the structure of the material so that it is no longer represented by the equilibrium diagram. Full annealing and normalizing is a heating operation that permits the material to transform according to the equilibrium diagram. The material to be annealed is heated to a temperature that is approximately 100°F above the critical temperature. It is held at this temperature for a time that is sufficient for the carbon to become dissolved and diffused through the material. The object being treated is then allowed to cool slowly, usually in the furnace in which it was treated. If the transformation is complete, then it is said to have a full anneal. Annealing is used to soften a material and make it more ductile, to relieve residual stresses, and to refine the grain structure.

The term *annealing* includes the process called *normalizing*. Parts to be normalized may be heated to a slightly higher temperature than in full annealing. This produces a coarser grain structure, which is more easily machined if the material is a low-carbon steel. In the normalizing process the part is cooled in still air at room temperature. Since this cooling is more rapid than the slow cooling used in full annealing, less time is available for equilibrium, and the material is harder than fully annealed steel. Normalizing is often used as the final treating operation for steel. The cooling in still air amounts to a slow quench.

Quenching

Eutectoid steel that is fully annealed consists entirely of pearlite, which is obtained from austenite under conditions of equilibrium. A fully annealed hypoeutectoid steel would consist of pearlite plus ferrite, while hypereutectoid steel in the fully annealed condition would consist of pearlite plus cementite. The hardness of steel of a given carbon content depends upon the structure that replaces the pearlite when full annealing is not carried out.

The absence of full annealing indicates a more rapid rate of cooling. The rate of cooling is the factor that determines the hardness. A controlled cooling rate is called *quenching*. A mild quench is obtained by cooling in still air, which, as we have seen, is obtained by the normalizing process. The two most widely used media for quenching are water and oil. The oil quench is quite slow but prevents quenching cracks caused by rapid expansion of the object being treated. Quenching in water is used for carbon steels and for medium-carbon, low-alloy steels.

The effectiveness of quenching depends upon the fact that when austenite is cooled it does not transform into pearlite instantaneously but requires time to initiate and complete the process. Since the transformation ceases at about 430°C, it can be prevented by rapidly cooling the material to a lower temperature. When the material is cooled rapidly to 200°C or less, the austenite is transformed into a structure called *martensite*. Martensite is a supersaturated solid solution of carbon in ferrite and is the hardest and strongest form of steel.

If steel is rapidly cooled to a temperature between 200 and 430°C and held there for a sufficient length of time, the austenite is transformed into a material that is generally called *bainite*. Bainite is a structure intermediate between pearlite and martensite. Although there are several structures that can be identified between the temperatures given, depending upon the temperature used, they are collectively known as bainite. By the choice of this transformation temperature, almost any variation of structure may be obtained. These range all the way from coarse pearlite to fine martensite.

Tempering

When a steel specimen has been fully hardened, it is very hard and brittle and has high residual stresses. The steel is unstable and tends to contract on aging. This tendency is increased when the specimen is subjected to externally applied loads, because the resultant stresses contribute still more to the instability. These internal stresses can be relieved by a modest heating process called *stress relieving,* or a combination of stress relieving and softening called *tempering* or *drawing*. After the specimen has been fully hardened by being quenched from above the critical temperature, it is reheated to some temperature below the critical temperature for a certain period of time and then allowed to cool in still air. The temperature to which it is reheated depends upon the composition and the degree of hardness or toughness desired.[8] This reheating operation releases the carbon held in the martensite, forming carbide crystals. The structure obtained is called *tempered martensite*. It is now essentially a superfine dispersion of iron carbide(s) in fine-grained ferrite.

The effect of heat-treating operations upon the various mechanical properties of a low alloy steel is shown graphically in Figure 2–21.

Case Hardening

The purpose of case hardening is to produce a hard outer surface on a specimen of low-carbon steel while at the same time retaining the ductility and toughness in the core. This is done by increasing the carbon content at the surface. Either solid, liquid, or gaseous carburizing materials may be used. The process consists of introducing the part to be carburized into the carburizing material for a stated time and at a stated

[8]For the quantitative aspects of tempering in plain carbon and low-alloy steels, see Charles R. Mischke, "The Strength of Cold-Worked and Heat-Treated Steels," chapter 33 in Joseph E. Shigley, Charles R. Mischke, and Thomas H. Brown, Jr. (eds.), *Standard Handbook of Machine Design*, 3rd ed., McGraw Hill, New York, 2004.

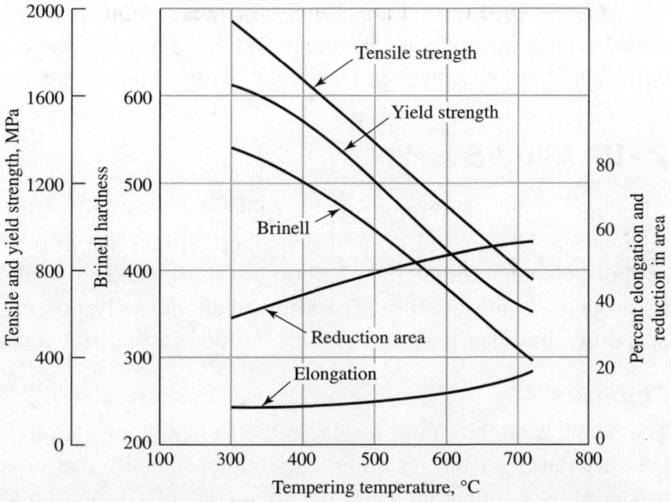

Figure 2–21

The effect of thermal-mechanical history on the mechanical properties of AISI 4340 steel. *(Source: Adapted from the International Nickel Company.)*

Condition	Tensile Strength, MPa	Yield Strength, MPa	Reduction in Area, %	Elongation in 50 mm, %	Brinell Hardness, Bhn
Normalized	1380	1000	20	10	410
As rolled	1300	990	18	9	380
Annealed	830	680	43	18	228

temperature, depending upon the depth of case desired and the composition of the part. The part may then be quenched directly from the carburization temperature and tempered, or in some cases it must undergo a double heat treatment in order to ensure that both the core and the case are in proper condition. Some of the more useful case-hardening processes are pack carburizing, gas carburizing, nitriding, cyaniding, induction hardening, and flame hardening. In the last two cases carbon is not added to the steel in question, generally a medium carbon steel, for example SAE/AISI 1144.

Quantitative Estimation of Properties of Heat-Treated Steels

Courses in metallurgy (or material science) for mechanical engineers usually present the addition method of Crafts and Lamont for the prediction of heat-treated properties from the Jominy test for plain carbon steels.[9] If this has not been in your prerequisite experience, then refer to the *Standard Handbook of Machine Design,* where the addition method is covered with examples.[10] If this book is a textbook for a machine elements course, it is a good class project (many hands make light work) to study the method and report to the class.

For low-alloy steels, the multiplication method of Grossman[11] and Field[12] is explained in the *Standard Handbook of Machine Design* (Sections 29.6 and 33.6).

[9]W. Crafts and J. L. Lamont, *Hardenability and Steel Selection*, Pitman and Sons, London, 1949.

[10]Charles R. Mischke, chapter 33 in Joseph E. Shigley, Charles R. Mischke, and Thomas H. Brown, Jr. (eds.), *Standard Handbook of Machine Design*, 3rd ed., McGraw Hill, New York, 2004, p. 33.9.

[11]M. A. Grossman, *AIME*, February 1942.

[12]J. Field, *Metals Progress*, March 1943.

Modern Steels and Their Properties Handbook explains how to predict the Jominy curve by the method of Grossman and Field from a ladle analysis and grain size.[13] Bethlehem Steel has developed a circular plastic slide rule that is convenient to the purpose.

2–16 Alloy Steels

Although a plain carbon steel is an alloy of iron and carbon with small amounts of manganese, silicon, sulfur, and phosphorus, the term *alloy steel* is applied when one or more elements other than carbon are introduced in sufficient quantities to modify its properties substantially. The alloy steels not only possess more desirable physical properties but also permit a greater latitude in the heat-treating process.

Chromium

The addition of chromium results in the formation of various carbides of chromium that are very hard, yet the resulting steel is more ductile than a steel of the same hardness produced by a simple increase in carbon content. Chromium also refines the grain structure so that these two combined effects result in both increased toughness and increased hardness. The addition of chromium increases the critical range of temperatures and moves the eutectoid point to the left. Chromium is thus a very useful alloying element.

Nickel

The addition of nickel to steel also causes the eutectoid point to move to the left and increases the critical range of temperatures. Nickel is soluble in ferrite and does not form carbides or oxides. This increases the strength without decreasing the ductility. Case hardening of nickel steels results in a better core than can be obtained with plain carbon steels. Chromium is frequently used in combination with nickel to obtain the toughness and ductility provided by the nickel and the wear resistance and hardness contributed by the chromium.

Manganese

Manganese is added to all steels as a deoxidizing and desulfurizing agent, but if the sulfur content is low and the manganese content is over 1 percent, the steel is classified as a manganese alloy. Manganese dissolves in the ferrite and also forms carbides. It causes the eutectoid point to move to the left and lowers the critical range of temperatures. It increases the time required for transformation so that oil quenching becomes practicable.

Silicon

Silicon is added to all steels as a deoxidizing agent. When added to very-low-carbon steels, it produces a brittle material with a low hysteresis loss and a high magnetic permeability. The principal use of silicon is with other alloying elements, such as manganese, chromium, and vanadium, to stabilize the carbides.

Molybdenum

While molybdenum is used alone in a few steels, it finds its greatest use when combined with other alloying elements, such as nickel, chromium, or both. Molybdenum forms carbides and also dissolves in ferrite to some extent, so that it adds both hardness and toughness. Molybdenum increases the critical range of temperatures and substantially lowers the transformation point. Because of this lowering of the transformation point,

---

[13]*Modern Steels and Their Properties*, 7th ed., Handbook 2757, Bethlehem Steel, 1972, pp. 46–50.

molybdenum is most effective in producing desirable oil-hardening and air-hardening properties. Except for carbon, it has the greatest hardening effect, and because it also contributes to a fine grain size, this results in the retention of a great deal of toughness.

Vanadium

Vanadium has a very strong tendency to form carbides; hence, it is used only in small amounts. It is a strong deoxidizing agent and promotes a fine grain size. Since some vanadium is dissolved in the ferrite, it also toughens the steel. Vanadium gives a wide hardening range to steel, and the alloy can be hardened from a higher temperature. It is very difficult to soften vanadium steel by tempering; hence, it is widely used in tool steels.

Tungsten

Tungsten is widely used in tool steels because the tool will maintain its hardness even at red heat. Tungsten produces a fine, dense structure and adds both toughness and hardness. Its effect is similar to that of molybdenum, except that it must be added in greater quantities.

2–17 Corrosion-Resistant Steels

Iron-base alloys containing at least 12 percent chromium are called *stainless steels*. The most important characteristic of these steels is their resistance to many, but not all, corrosive conditions. The four types available are the ferritic chromium steels, the austenitic chromium-nickel steels, and the martensitic and precipitation-hardenable stainless steels.

The ferritic chromium steels have a chromium content ranging from 12 to 27 percent. Their corrosion resistance is a function of the chromium content, so that alloys containing less than 12 percent still exhibit some corrosion resistance, although they may rust. The quench-hardenability of these steels is a function of both the chromium and the carbon content. The very high carbon steels have good quench hardenability up to about 18 percent chromium, while in the lower carbon ranges it ceases at about 13 percent. If a little nickel is added, these steels retain some degree of hardenability up to 20 percent chromium. If the chromium content exceeds 18 percent, they become difficult to weld, and at the very high chromium levels the hardness becomes so great that very careful attention must be paid to the service conditions. Since chromium is expensive, the designer will choose the lowest chromium content consistent with the corrosive conditions.

The chromium-nickel stainless steels retain the austenitic structure at room temperature; hence, they are not amenable to heat treatment. The strength of these steels can be greatly improved by cold working. They are not magnetic unless cold-worked. Their work hardenability properties also cause them to be difficult to machine. All the chromium-nickel steels may be welded. They have greater corrosion-resistant properties than the plain chromium steels. When more chromium is added for greater corrosion resistance, more nickel must also be added if the austenitic properties are to be retained.

2–18 Casting Materials

Gray Cast Iron

Of all the cast materials, gray cast iron is the most widely used. This is because it has a very low cost, is easily cast in large quantities, and is easy to machine. The principal objections to the use of gray cast iron are that it is brittle and that it is weak in tension. In addition to a high carbon content (over 1.7 percent and usually greater than 2 percent), cast iron also has a high silicon content, with low percentages of sulfur, manganese, and phosphorus. The resultant alloy is composed of pearlite, ferrite, and graphite, and under certain conditions the pearlite may decompose into graphite and ferrite. The resulting product then contains all ferrite and graphite. The graphite, in the form of thin flakes distributed evenly throughout the structure, darkens it; hence, the name *gray cast iron*.

Gray cast iron is not readily welded, because it may crack, but this tendency may be reduced if the part is carefully preheated. Although the castings are generally used in the as-cast condition, a mild anneal reduces cooling stresses and improves the machinability. The tensile strength of gray cast iron varies from 100 to 400 MPa, and the compressive strengths are 3 to 4 times the tensile strengths. The modulus of elasticity varies widely, with values extending all the way from 75 to 150 GPa.

Ductile and Nodular Cast Iron

Because of the lengthy heat treatment required to produce malleable cast iron, engineers have long desired a cast iron that would combine the ductile properties of malleable iron with the ease of casting and machining of gray iron and at the same time would possess these properties in the as-cast conditions. A process for producing such a material using magnesium-containing material seems to fulfill these requirements.

Ductile cast iron, or *nodular cast iron,* as it is sometimes called, is essentially the same as malleable cast iron, because both contain graphite in the form of spheroids. However, ductile cast iron in the as-cast condition exhibits properties very close to those of malleable iron, and if a simple 1-h anneal is given and is followed by a slow cool, it exhibits even more ductility than the malleable product. Ductile iron is made by adding MgFeSi to the melt; since magnesium boils at this temperature, it is necessary to alloy it with other elements before it is introduced.

Ductile iron has a high modulus of elasticity (172 GPa) as compared with gray cast iron, and it is elastic in the sense that a portion of the stress-strain curve is a straight line. Gray cast iron, on the other hand, does not obey Hooke's law, because the modulus of elasticity steadily decreases with increase in stress. Like gray cast iron, however, nodular iron has a compressive strength that is higher than the tensile strength, although the difference is not as great. In 40 years it has become extensively used.

White Cast Iron

If all the carbon in cast iron is in the form of cementite and pearlite, with no graphite present, the resulting structure is white and is known as *white cast iron*. This may be produced in two ways. The composition may be adjusted by keeping the carbon and silicon content low, or the gray-cast-iron composition may be cast against chills in order to promote rapid cooling. By either method, a casting with large amounts of cementite is produced, and as a result the product is very brittle and hard to machine but also very resistant to wear. A chill is usually used in the production of gray-iron castings in order to provide a very hard surface within a particular area of the casting, while at the same time retaining the more desirable gray structure within the remaining portion. This produces a relatively tough casting with a wear-resistant area.

Malleable Cast Iron

If white cast iron within a certain composition range is annealed, a product called *malleable cast iron* is formed. The annealing process frees the carbon so that it is present as graphite, just as in gray cast iron but in a different form. In gray cast iron the graphite is present in a thin flake form, while in malleable cast iron it has a nodular form and is known as *temper carbon*. A good grade of malleable cast iron may have a tensile strength of over 350 MPa, with an elongation of as much as 18 percent. The percentage elongation of a gray cast iron, on the other hand, is seldom over 1 percent. Because of the time required for annealing (up to 6 days for large and heavy castings), malleable iron is necessarily somewhat more expensive than gray cast iron.

Alloy Cast Irons

Nickel, chromium, and molybdenum are the most common alloying elements used in cast iron. Nickel is a general-purpose alloying element, usually added in amounts up to 5 percent. Nickel increases the strength and density, improves the wearing qualities, and raises the machinability. If the nickel content is raised to 10 to 18 percent, an austenitic structure with valuable heat- and corrosion-resistant properties results. Chromium increases the hardness and wear resistance and, when used with a chill, increases the tendency to form white iron. When chromium and nickel are both added, the hardness and strength are improved without a reduction in the machinability rating. Molybdenum added in quantities up to 1.25 percent increases the stiffness, hardness, tensile strength, and impact resistance. It is a widely used alloying element.

Cast Steels

The advantage of the casting process is that parts having complex shapes can be manufactured at costs less than fabrication by other means, such as welding. Thus the choice of steel castings is logical when the part is complex and when it must also have a high strength. The higher melting temperatures for steels do aggravate the casting problems and require closer attention to such details as core design, section thicknesses, fillets, and the progress of cooling. The same alloying elements used for the wrought steels can be used for cast steels to improve the strength and other mechanical properties. Cast-steel parts can also be heat-treated to alter the mechanical properties, and, unlike the cast irons, they can be welded.

2–19 Nonferrous Metals

Aluminum

The outstanding characteristics of aluminum and its alloys are their strength-weight ratio, their resistance to corrosion, and their high thermal and electrical conductivity. The density of aluminum is about 2770 kg/m^3, compared with 7750 kg/m^3 for steel. Pure aluminum has a tensile strength of about 90 MPa, but this can be improved considerably by cold working and also by alloying with other materials. The modulus of elasticity of aluminum, as well as of its alloys, is 71.7 GPa, which means that it has about one-third the stiffness of steel.

Considering the cost and strength of aluminum and its alloys, they are among the most versatile materials from the standpoint of fabrication. Aluminum can be processed by sand casting, die casting, hot or cold working, or extruding. Its alloys can be machined, press-worked, soldered, brazed, or welded. Pure aluminum melts at 660°C, which makes it very desirable for the production of either permanent or sand-mold castings. It is commercially available in the form of plate, bar, sheet, foil, rod, and tube and in structural and extruded shapes. Certain precautions must be taken in joining aluminum by soldering, brazing, or welding; these joining methods are not recommended for all alloys.

The corrosion resistance of the aluminum alloys depends upon the formation of a thin oxide coating. This film forms spontaneously because aluminum is inherently very reactive. Constant erosion or abrasion removes this film and allows corrosion to take place. An extra-heavy oxide film may be produced by the process called *anodizing*. In this process the specimen is made to become the anode in an electrolyte, which may be chromic acid, oxalic acid, or sulfuric acid. It is possible in this process to control the color of the resulting film very accurately.

The most useful alloying elements for aluminum are copper, silicon, manganese, magnesium, and zinc. Aluminum alloys are classified as *casting alloys* or *wrought alloys*. The casting alloys have greater percentages of alloying elements to facilitate casting, but this makes cold working difficult. Many of the casting alloys, and some of the wrought alloys, cannot be hardened by heat treatment. The alloys that are heat-treatable use an alloying element that dissolves in the aluminum. The heat treatment consists of heating the specimen to a temperature that permits the alloying element to pass into solution, then quenching so rapidly that the alloying element is not precipitated. The aging process may be accelerated by heating slightly, which results in even greater hardness and strength. One of the better-known heat-treatable alloys is duralumінum, or 2017 (4 percent Cu, 0.5 percent Mg, 0.5 percent Mn). This alloy hardens in 4 days at room temperature. Because of this rapid aging, the alloy must be stored under refrigeration after quenching and before forming, or it must be formed immediately after quenching. Other alloys (such as 5053) have been developed that age-harden much more slowly, so that only mild refrigeration is required before forming. After forming, they are artificially aged in a furnace and possess approximately the same strength and hardness as the 2024 alloys. Those alloys of aluminum that cannot be heat-treated can be hardened only by cold working. Both work hardening and the hardening produced by heat treatment may be removed by an annealing process.

Magnesium

The density of magnesium is about 1800 kg/m^3, which is two-thirds that of aluminum and one-fourth that of steel. Since it is the lightest of all commercial metals, its greatest use is in the aircraft and automotive industries, but other uses are now being found for it. Although the magnesium alloys do not have great strength, because of their light weight the strength-weight ratio compares favorably with the stronger aluminum and steel alloys. Even so, magnesium alloys find their greatest use in applications where strength is not an important consideration. Magnesium will not withstand elevated temperatures; the yield point is definitely reduced when the temperature is raised to that of boiling water.

Magnesium and its alloys have a modulus of elasticity of 45 GPa in tension and in compression, although some alloys are not as strong in compression as in tension. Curiously enough, cold working reduces the modulus of elasticity. A range of cast magnesium alloys are also available.

Titanium

Titanium and its alloys are similar in strength to moderate-strength steel but weigh half as much as steel. The material exhibits very good resistance to corrosion, has low thermal conductivity, is nonmagnetic, and has high-temperature strength. Its modulus of elasticity is between those of steel and aluminum at 114 GPa. Because of its many advantages over steel and aluminum, applications include: aerospace and military aircraft structures and components, marine hardware, chemical tanks and processing equipment, fluid handling systems, and human internal replacement devices. The disadvantages of titanium are its high cost compared to steel and aluminum and the difficulty of machining it.

Copper-Base Alloys

When copper is alloyed with zinc, it is usually called *brass*. If it is alloyed with another element, it is often called *bronze*. Sometimes the other element is specified too, as, for example, *tin bronze* or *phosphor bronze*. There are hundreds of variations in each category.

Brass with 5 to 15 Percent Zinc

The low-zinc brasses are easy to cold work, especially those with the higher zinc content. They are ductile but often hard to machine. The corrosion resistance is good. Alloys included in this group are *gilding brass* (5 percent Zn), *commercial bronze* (10 percent Zn), and *red brass* (15 percent Zn). Gilding brass is used mostly for jewelry and articles to be gold-plated; it has the same ductility as copper but greater strength, accompanied by poor machining characteristics. Commercial bronze is used for jewelry and for forgings and stampings, because of its ductility. Its machining properties are poor, but it has excellent cold-working properties. Red brass has good corrosion resistance as well as high-temperature strength. Because of this it is used a great deal in the form of tubing or piping to carry hot water in such applications as radiators or condensers.

Brass with 20 to 36 Percent Zinc

Included in the intermediate-zinc group are *low brass* (20 percent Zn), *cartridge brass* (30 percent Zn), and *yellow brass* (35 percent Zn). Since zinc is cheaper than copper, these alloys cost less than those with more copper and less zinc. They also have better machinability and slightly greater strength; this is offset, however, by poor corrosion resistance and the possibility of cracking at points of residual stresses. Low brass is very similar to red brass and is used for articles requiring deep-drawing operations. Of the copper-zinc alloys, cartridge brass has the best combination of ductility and strength. Cartridge cases were originally manufactured entirely by cold working; the process consisted of a series of deep draws, each draw being followed by an anneal to place the material in condition for the next draw, hence the name cartridge brass. Although the hot-working ability of yellow brass is poor, it can be used in practically any other fabricating process and is therefore employed in a large variety of products.

When small amounts of lead are added to the brasses, their machinability is greatly improved and there is some improvement in their abilities to be hot-worked. The addition of lead impairs both the cold-working and welding properties. In this group are *low-leaded brass* (32.5 percent Zn, 0.5 percent Pb), *high-leaded brass* (34 percent Zn, 2 percent Pb), and *free-cutting brass* (35.5 percent Zn, 3 percent Pb). The low-leaded brass is not only easy to machine but has good cold-working properties. It is used for various screw-machine parts. High-leaded brass, sometimes called *engraver's brass,* is used for instrument, lock, and watch parts. Free-cutting brass is also used for screw-machine parts and has good corrosion resistance with excellent mechanical properties.

Admiralty metal (28 percent Zn) contains 1 percent tin, which imparts excellent corrosion resistance, especially to saltwater. It has good strength and ductility but only fair machining and working characteristics. Because of its corrosion resistance it is used in power-plant and chemical equipment. *Aluminum brass* (22 percent Zn) contains 2 percent aluminum and is used for the same purposes as admiralty metal, because it has nearly the same properties and characteristics. In the form of tubing or piping, it is favored over admiralty metal, because it has better resistance to erosion caused by high-velocity water.

Brass with 36 to 40 Percent Zinc

Brasses with more than 38 percent zinc are less ductile than cartridge brass and cannot be cold-worked as severely. They are frequently hot-worked and extruded. *Muntz metal* (40 percent Zn) is low in cost and mildly corrosion-resistant. *Naval brass* has the same composition as Muntz metal except for the addition of 0.75 percent tin, which contributes to the corrosion resistance.

Bronze

Silicon bronze, containing 3 percent silicon and 1 percent manganese in addition to the copper, has mechanical properties equal to those of mild steel, as well as good corrosion resistance. It can be hot- or cold-worked, machined, or welded. It is useful wherever corrosion resistance combined with strength is required.

Phosphor bronze, made with up to 11 percent tin and containing small amounts of phosphorus, is especially resistant to fatigue and corrosion. It has a high tensile strength and a high capacity to absorb energy, and it is also resistant to wear. These properties make it very useful as a spring material.

Aluminum bronze is a heat-treatable alloy containing up to 12 percent aluminum. This alloy has strength and corrosion-resistance properties that are better than those of brass, and in addition, its properties may be varied over a wide range by cold working, heat treating, or changing the composition. When iron is added in amounts up to 4 percent, the alloy has a high endurance limit, a high shock resistance, and excellent wear resistance.

Beryllium bronze is another heat-treatable alloy, containing about 2 percent beryllium. This alloy is very corrosion resistant and has high strength, hardness, and resistance to wear. Although it is expensive, it is used for springs and other parts subjected to fatigue loading where corrosion resistance is required.

With slight modification most copper-based alloys are available in cast form.

2–20 Plastics

The term *thermoplastics* is used to mean any plastic that flows or is moldable when heat is applied to it; the term is sometimes applied to plastics moldable under pressure. Such plastics can be remolded when heated.

A *thermoset* is a plastic for which the polymerization process is finished in a hot molding press where the plastic is liquefied under pressure. Thermoset plastics cannot be remolded.

Table 2–3 lists some of the most widely used thermoplastics, together with some of their characteristics and the range of their properties. Table 2–4, listing some of the thermosets, is similar. These tables are presented for information only and should not be used to make a final design decision. The range of properties and characteristics that can be obtained with plastics is very great. The influence of many factors, such as cost, moldability, coefficient of friction, weathering, impact strength, and the effect of fillers and reinforcements, must be considered. Manufacturers' catalogs will be found quite helpful in making possible selections.

Table 2–3 The Thermoplastics

Name	S_u MPa	E GPa	Hardness Rockwell	Elongation %	Dimensional Stability	Heat Resistance	Chemical Resistance	Processing
ABS group	14–55	0.7–2.5	60–110R	3–50	Good	*	Fair	EMST
Acetal group	55–70	2.8–3.6	80–94M	40–60	Excellent	Good	High	M
Acrylic	35–70	1.4–3.3	92–110M	3–75	High	*	Fair	EMS
Fluoroplastic group	4–50	. . .	50–80D	100–300	High	Excellent	Excellent	MPR†
Nylon	55–95	1.2–3.1	112–120R	10–200	Poor	Poor	Good	CEM
Phenylene oxide	50–125	2.4–6.3	115R, 106L	5–60	Excellent	Good	Fair	EFM
Polycarbonate	55–110	2.3–5.9	62–91M	10–125	Excellent	Excellent	Fair	EMS
Polyester	55–125	1.9–11	65–90M	1–300	Excellent	Poor	Excellent	CLMR
Polyimide	40–345	. . .	88–120M	Very low	Excellent	Excellent	Excellent†	CLMP
Polyphenylene sulfide	95–130	0.8	122R	1.0	Good	Excellent	Excellent	M
Polystyrene group	18–85	0.1–4.1	10–90M	0.5–60	. . .	Poor	Poor	EM
Polysulfone	70	2.5	120R	50–100	Excellent	Excellent	Excellent†	EFM
Polyvinyl chloride	10–50	2.4–4.1	65–85D	40–450	. . .	Poor	Poor	EFM

*Heat-resistant grades available.
†With exceptions.
C Coatings L Laminates R Resins E Extrusions M Moldings S Sheet F Foams P Press and sinter methods T Tubing
Source: This data has been obtained from the *Machine Design Materials Reference Issue,* published by Penton/IPC, Cleveland. These reference issues are published about every 2 years and constitute an excellent source of data on a great variety of materials.

Table 2–4 The Thermosets

Name	S_u MPa	E GPa	Hardness Rockwell	Elongation %	Dimensional Stability	Heat Resistance	Chemical Resistance	Processing
Alkyd	20–60	0.3–2	99M*	. . .	Excellent	Good	Fair	M
Allylic	30–70	. . .	105–120M	. . .	Excellent	Excellent	Excellent	CM
Amino group	35–55	0.9–1.6	110–120M	0.30–0.90	Good	Excellent*	Excellent*	LR
Epoxy	35–140	0.2–2*	80–120M	1–10	Excellent	Excellent	Excellent	CMR
Phenolics	35–60	0.7–1.7	70–95E	. . .	Excellent	Excellent	Good	EMR
Silicones	35–40	. . .	80–90M	. . .	. . .	Excellent	Excellent	CLMR

*With exceptions.
C Coatings L Laminates R Resins E Extrusions M Moldings S Sheet F Foams P Press and sinter methods T Tubing
Source: This data has been obtained from the *Machine Design Materials Reference Issue,* published by Penton/IPC, Cleveland. These reference issues are published about every 2 years and constitute an excellent source of data on a great variety of materials.

2–21 Composite Materials[14]

Composite materials are formed from two or more dissimilar materials, each of which contributes to the final properties. Unlike metallic alloys, the materials in a composite remain distinct from each other at the macroscopic level.

Most engineering composites consist of two materials: a reinforcement called a *filler* and a *matrix*. The filler provides stiffness and strength; the matrix holds the material together and serves to transfer load among the discontinuous reinforcements. The most common reinforcements, illustrated in Figure 2–22, are continuous fibers, either straight or woven, short chopped fibers, and particulates. The most common matrices are various plastic resins although other materials including metals are used.

Metals and other traditional engineering materials are uniform, or isotropic, in nature. This means that material properties, such as strength, stiffness, and thermal conductivity, are independent of both position within the material and the choice of coordinate system. The discontinuous nature of composite reinforcements, though, means that material properties can vary with both position and direction. For example, an epoxy resin reinforced with continuous graphite fibers will have very high strength and stiffness in the direction of the fibers, but very low properties normal or transverse to the fibers. For this reason, structures of composite materials are normally constructed of multiple plies (laminates) where each ply is oriented to achieve optimal structural stiffness and strength performance.

High strength-to-weight ratios, up to five times greater than those of high-strength steels, can be achieved. High stiffness-to-weight ratios can also be obtained, as much as eight times greater than those of structural metals. For this reason, composite materials are becoming very popular in automotive, marine, aircraft, and spacecraft applications where weight is a premium.

The directionality of properties of composite materials increases the complexity of structural analyses. Isotropic materials are fully defined by two engineering constants: Young's modulus E and Poisson's ratio ν. A single ply of a composite material, however, requires four constants, defined with respect to the ply coordinate system. The constants are two Young's moduli (the longitudinal modulus in the direction of the fibers, E_1, and the transverse modulus normal to the fibers, E_2), one Poisson's ratio (ν_{12}, called the major Poisson's ratio), and one shear modulus (G_{12}). A fifth constant, the minor Poisson's ratio, ν_{21}, is determined through the reciprocity relation, $\nu_{21}/E_2 = \nu_{12}/E_1$. Combining this with multiple plies oriented at different angles makes structural analysis of complex structures unapproachable by manual techniques. For this reason, computer software is available to calculate the properties of a laminated composite construction.

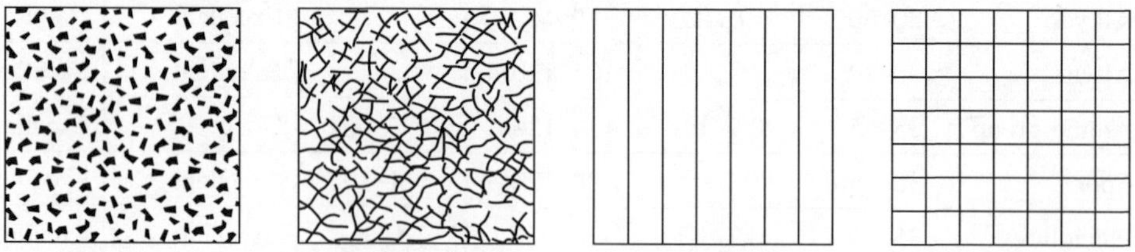

Figure 2–22

Composites categorized by type of reinforcement.

[14]For references see I. M. Daniel and O. Ishai, *Engineering Mechanics of Composite Materials*, Oxford University Press, 1994, and *ASM Engineered Materials Handbook: Composites*, ASM International, Materials Park, OH, 1988.

2–22 Materials Selection

As stated earlier, the selection of a material for a machine part or structural member is one of the most important decisions the designer is called on to make. Up to this point in this chapter we have discussed many important material physical properties, various characteristics of typical engineering materials, and various material production processes. The actual selection of a material for a particular design application can be an easy one, say, based on previous applications (1020 steel is always a good candidate because of its many positive attributes), or the selection process can be as involved and daunting as any design problem with the evaluation of the many material physical, economical, and processing parameters. There are systematic and optimizing approaches to material selection. Here, for illustration, we will only look at how to approach some material properties. One basic technique is to list all the important material properties associated with the design, e.g., strength, stiffness, and cost. This can be prioritized by using a weighting measure depending on what properties are more important than others. Next, for each property, list all available materials and rank them in order beginning with the best material; e.g., for strength, high-strength steel such as 4340 steel should be near the top of the list. For completeness of available materials, this might require a large source of material data. Once the lists are formed, select a manageable amount of materials from the top of each list. From each reduced list select the materials that are contained within every list for further review. The materials in the reduced lists can be graded within the list and then weighted according to the importance of each property.

M. F. Ashby has developed a powerful systematic method using *materials selection charts*.[15] This method has also been implemented in a software package called CES Edupack.[16] The charts display data of various properties for the families and classes of materials listed in Table 2–5. For example, considering material stiffness properties, a simple bar chart plotting Young's modulus E on the y axis is shown in

Table 2–5 Material Families and Classes

Family	Classes	Short Name
Metals	Aluminum alloys	Al alloys
(the metals and alloys of engineering)	Copper alloys	Cu alloys
	Lead alloys	Lead alloys
	Magnesium alloys	Mg alloys
	Nickel alloys	Ni alloys
	Carbon steels	Steels
	Stainless steels	Stainless steels
	Tin alloys	Tin alloys
	Titanium alloys	Ti alloys
	Tungsten alloys	W alloys
	Lead alloys	Pb alloys
	Zinc alloys	Zn alloys

(Continued)

[15]M. F. Ashby, *Materials Selection in Mechanical Design*, 3rd ed., Elsevier Butterworth-Heinemann, Oxford, 2005.

[16]Produced by Granta Design Limited. See www.grantadesign.com.

Table 2–5 (*Continued*)

Family	Classes	Short Name
Ceramics	Alumina	Al_2O_3
Technical ceramics (fine ceramics capable of load-bearing application)	Aluminum nitride	AlN
	Boron carbide	B_4C
	Silicon carbide	SiC
	Silicon nitride	Si_3N_4
	Tungsten carbide	WC
Nontechnical ceramics (porous ceramics of construction)	Brick	Brick
	Concrete	Concrete
	Stone	Stone
Glasses	Soda-lime glass	Soda-lime glass
	Borosilicate glass	Borosilicate glass
	Silica glass	Silica glass
	Glass ceramic	Glass ceramic
Polymers	Acrylonitrile butadiene styrene	ABS
(the thermoplastics and thermosets of engineering)	Cellulose polymers	CA
	Ionomers	Ionomers
	Epoxies	Epoxy
	Phenolics	Phenolics
	Polyamides (nylons)	PA
	Polycarbonate	PC
	Polyesters	Polyester
	Polyetheretherkeytone	PEEK
	Polyethylene	PE
	Polyethylene terephalate	PET or PETE
	Polymethylmethacrylate	PMMA
	Polyoxymethylene (Acetal)	POM
	Polypropylene	PP
	Polystyrene	PS
	Polytetrafluorethylene	PTFE
	Polyvinylchloride	PVC
Elastomers	Butyl rubber	Butyl rubber
(engineering rubbers, natural and synthetic)	EVA	EVA
	Isoprene	Isoprene
	Natural rubber	Natural rubber
	Polychloroprene (Neoprene)	Neoprene
	Polyurethane	PU
	Silicon elastomers	Silicones
Hybrids	Carbon-fiber reinforced polymers	CFRP
Composites	Glass-fiber reinforced polymers	GFRP
	SiC reinforced aluminum	Al-SiC
Foams	Flexible polymer foams	Flexible foams
	Rigid polymer foams	Rigid foams
Natural materials	Cork	Cork
	Bamboo	Bamboo
	Wood	Wood

Source: From Ashby, M. F., *Materials Selection in Mechanical Design.* 3rd ed., Oxford: Elsevier Butterworth-Heinemann, 2005. Table 4–1, 49–50.

Figure 2–23

Young's modulus E for various materials. *(Source: Adapted from figure by Prof. Mike Ashby, Granta Design, Cambridge, U.K.)*

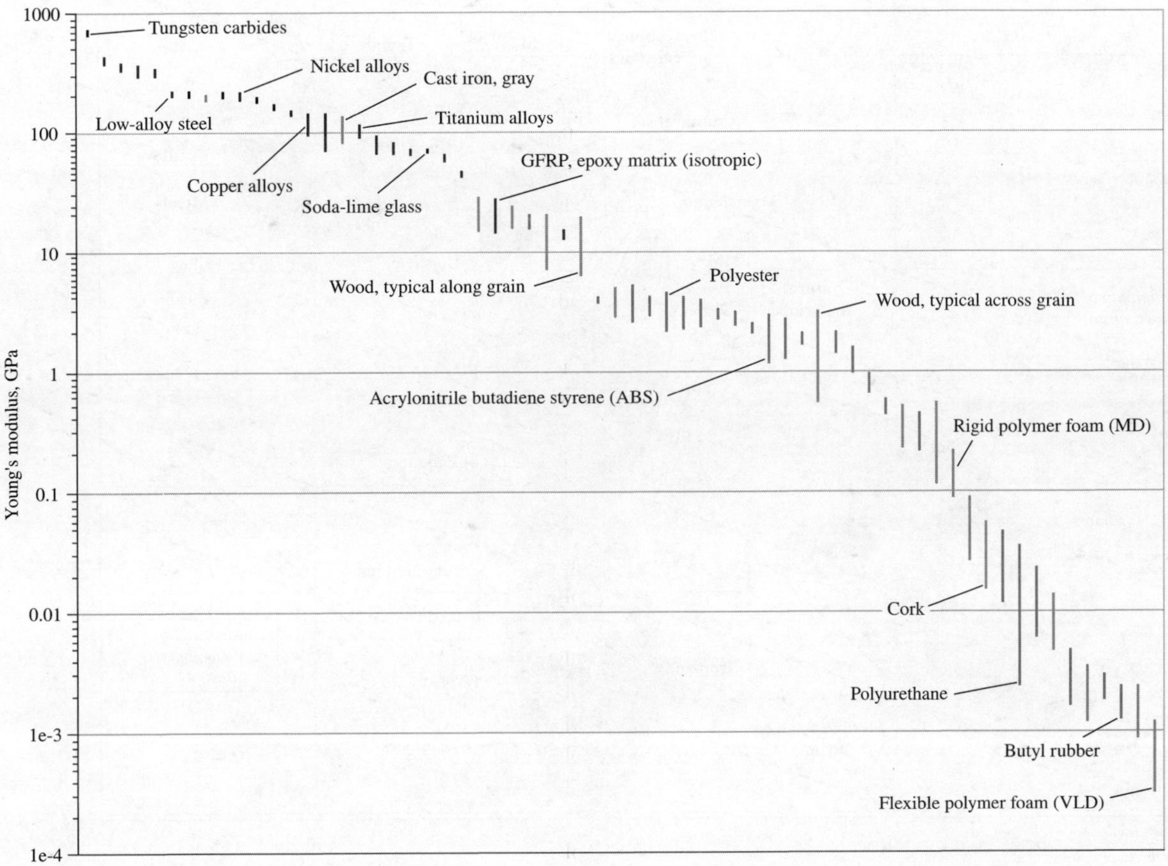

Figure 2–23. Each vertical line represents the range of values of E for a particular material. Only some of the materials are labeled. Now, more material information can be displayed if the x axis represents another material property, say density. Figure 2–24, called a "bubble" chart, represents Young's modulus E plotted against density ρ. The line ranges for each material property plotted two-dimensionally now form ellipses, or bubbles. Groups of bubbles outlined according to the material families of Table 2–5 are also shown. This plot is more useful than the two separate bar charts of each property. Now, we also see how stiffness/weight for various materials relate. The ratio of Young's modulus to density, E/ρ, is known as the *specific modulus,* or *specific stiffness.* This ratio is of particular interest when it is desired to minimize weight where the primary design limitation is deflection, stiffness, or natural frequency, rather than strength. Machine parts made from materials with higher specific modulus will exhibit lower deflection, higher stiffness, and higher natural frequency.

In the lower right corner of the chart in Figure 2–24, dotted lines indicate ratios of E^{β}/ρ. Several parallel dotted lines are shown for $\beta = 1$ that represent different values of the specific modulus E/ρ. This allows simple comparison of the specific modulus between materials. It can be seen, for example, that some woods and

Figure 2–24

Young's modulus E versus density ρ for various materials. *(Source: Adapted from figure by Prof. Mike Ashby, Granta Design, Cambridge, U.K.)*

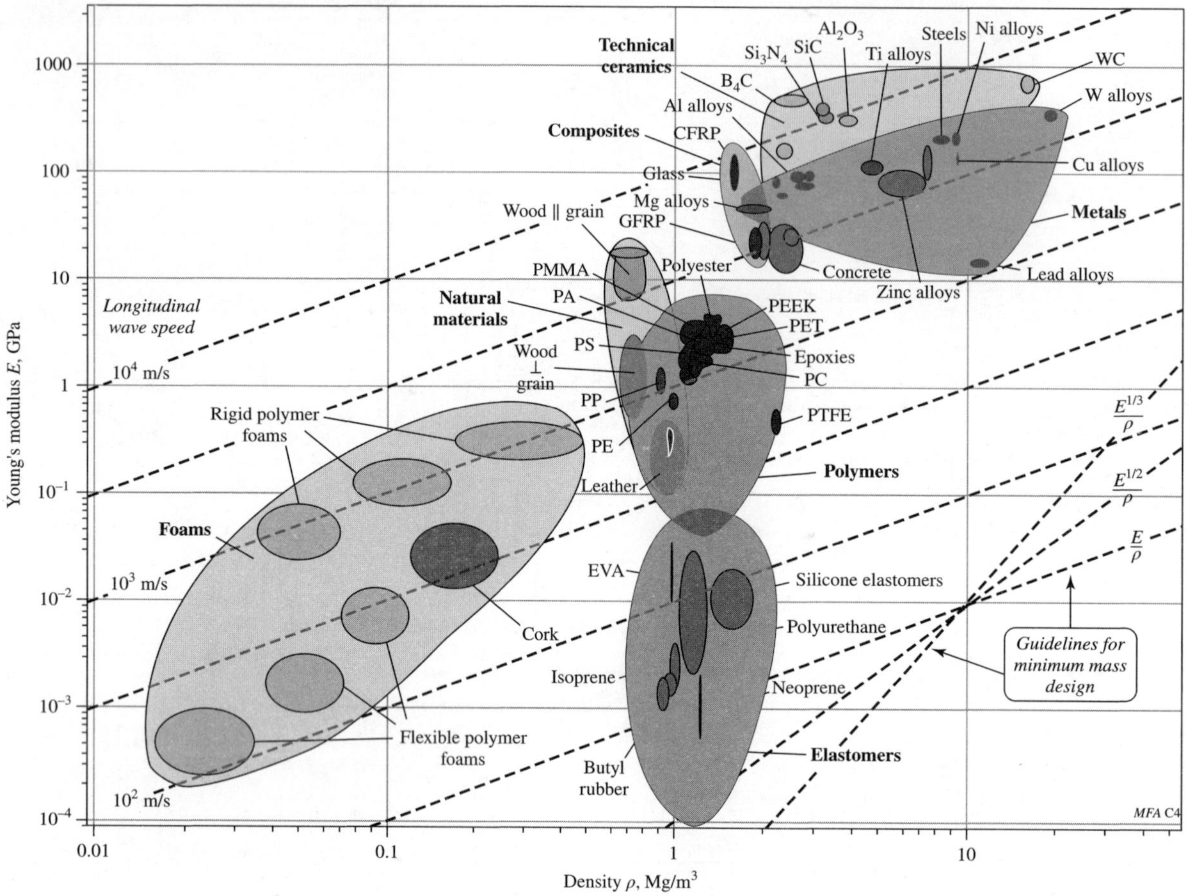

aluminum alloys have about the same specific modulus as steels. Different values of β allow comparisons for various relationships between stiffness and weight, such as in different loading conditions. The relationship is linear ($\beta = 1$) for axial loading, but nonlinear ($\beta = 1/2$) for bending loading [see Equation (2–46) and its development]. Since the plot is on a log-log scale, the exponential functions still plot as straight lines. The $\beta = 1$ lines can also be used to represent constant values of the speed of sound in a material, since the relationship between E and ρ is linear in the equation for the speed of sound in a material, $c = (E/\rho)^{1/2}$. The same can be shown for natural frequency, which is a function of the ratio of stiffness to mass.

To see how β fits into the mix, consider the following. The performance metric P of a structural element depends on (1) the functional requirements, (2) the geometry, and (3) the material properties of the structure. That is,

$$P = \left[\begin{pmatrix}\text{functional}\\ \text{requirements } F\end{pmatrix}, \begin{pmatrix}\text{geometric}\\ \text{parameters } G\end{pmatrix}, \begin{pmatrix}\text{material}\\ \text{properties } M\end{pmatrix}\right]$$

or, symbolically,

$$P = f(F, G, M) \tag{2–38}$$

If the function is *separable,* which it often is, we can write Equation (2–38) as

$$P = f_1(F) \cdot f_2(G) \cdot f_3(M) \tag{2–39}$$

For optimum design, we desire to maximize or minimize P. With regards to material properties alone, this is done by maximizing or minimizing $f_3(M)$, called the *material efficiency coefficient*.

For illustration, say we want to design a light, stiff, end-loaded cantilever beam with a circular cross section. For this we will use the mass m of the beam for the performance metric to minimize. The stiffness of the beam is related to its material and geometry. The stiffness of a beam is given by $k = F/\delta$, where F and δ are the end load and deflection, respectively (see Chapter 4). The end deflection of an end-loaded cantilever beam is given in Table A–9, beam 1, as $\delta = y_{\text{max}} = (Fl^3)/(3EI)$, where E is Young's modulus, I the second moment of the area, and l the length of the beam. Thus, the stiffness is given by

$$k = \frac{F}{\delta} = \frac{3EI}{l^3} \tag{2–40}$$

From Table A–16, the second moment of the area of a circular cross section is

$$I = \frac{\pi D^4}{64} = \frac{A^2}{4\pi} \tag{2–41}$$

where D and A are the diameter and area of the cross section, respectively. Substituting Equation (2–41) in (2–40) and solving for A, we obtain

$$A = \left(\frac{4\pi k l^3}{3E}\right)^{1/2} \tag{2–42}$$

The mass of the beam is given by

$$m = Al\rho \tag{2–43}$$

Substituting Equation (2–42) into (2–43) and rearranging yields

$$m = 2\sqrt{\frac{\pi}{3}}\,(k^{1/2})(l^{5/2})\left(\frac{\rho}{E^{1/2}}\right) \tag{2–44}$$

Equation (2–44) is of the form of Equation (2–39). The term $2\sqrt{\pi/3}$ is simply a constant and can be associated with any function, say $f_1(F)$. Thus, $f_1(F) = 2\sqrt{\pi/3}(k^{1/2})$ is the functional requirement, stiffness; $f_2(G) = (l^{5/2})$, the geometric parameter, length; and the material efficiency coefficient

$$f_3(M) = \frac{\rho}{E^{1/2}} \tag{2–45}$$

is the material property in terms of density and Young's modulus. To minimize m we want to minimize $f_3(M)$, or maximize

$$M = \frac{E^{1/2}}{\rho} \tag{2–46}$$

where M is called the *material index,* and $\beta = \frac{1}{2}$. Returning to Figure 2–24, draw lines of various values of $E^{1/2}/\rho$ as shown in Figure 2–25. Lines of increasing M move up and to the left as shown. Thus, we see that good candidates for a light, stiff, end-loaded cantilever beam with a circular cross section are certain woods, composites, and ceramics.

Figure 2–25

A schematic E versus ρ chart showing a grid of lines for various values the material index $M = E^{1/2}/\rho$. *(Source: Adapted from Ashby, M. F.,* Materials Selection in Mechanical Design, *3rd ed., Oxford: Elsevier Butterworth-Heinemann, 2005.)*

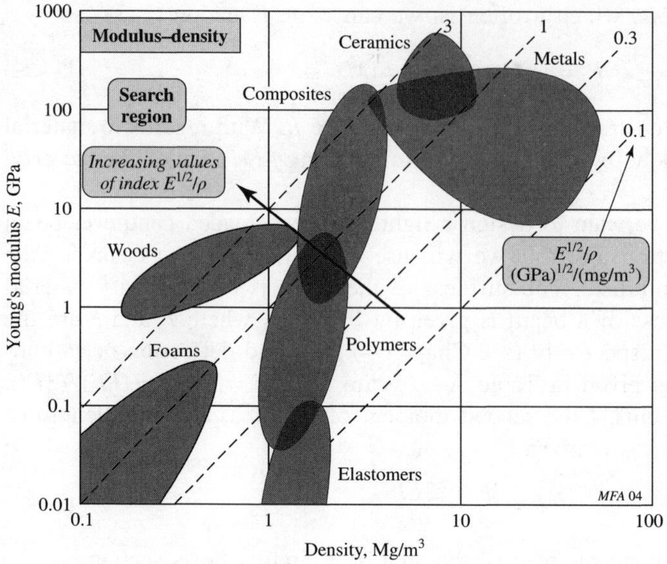

Other limits/constraints may warrant further investigation. Say, for further illustration, the design requirements indicate that we need a Young's modulus greater than 50 GPa. Figure 2–26 shows how this further restricts the search region. This eliminates woods as a possible material.

Another commonly useful chart, shown in Figure 2–27, represents strength versus density for the material families. The ratio of strength to density is known as *specific strength,* and is particularly useful when it is desired to minimize weight where the primary design limitation is strength, rather than deflection. The guidelines in the lower right corner represent different relationships between strength and density, in the form of S^{β}/ρ. Following an approach similar to that used before, it can be shown that for axial loading, $\beta = 1$, and for bending loading, $\beta = 2/3$.

Figure 2–26

The search region of Figure 2–24 further reduced by restricting $E \geq 50$ GPa. *(Source: Adapted from Ashby, M. F.,* Materials Selection in Mechanical Design, *3rd ed., Oxford: Elsevier Butterworth-Heinemann, 2005.)*

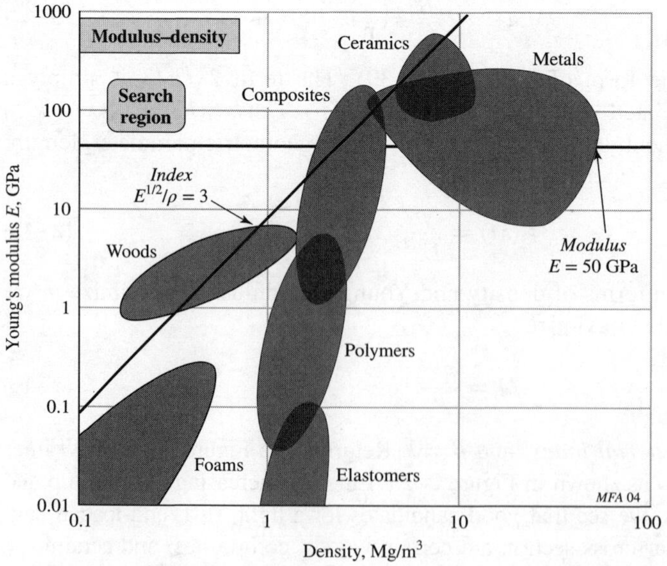

Figure 2–27

Strength S versus density ρ for various materials. For *metals,* S is the 0.2 percent offset yield strength. For *polymers,* S is the 1 percent yield strength. For *ceramics and glasses,* S is the compressive crushing strength. For *composites,* S is the tensile strength. For *elastomers,* S is the tear strength. *(Source: Adapted from figure by Prof. Mike Ashby, Granta Design, Cambridge, U.K.)*

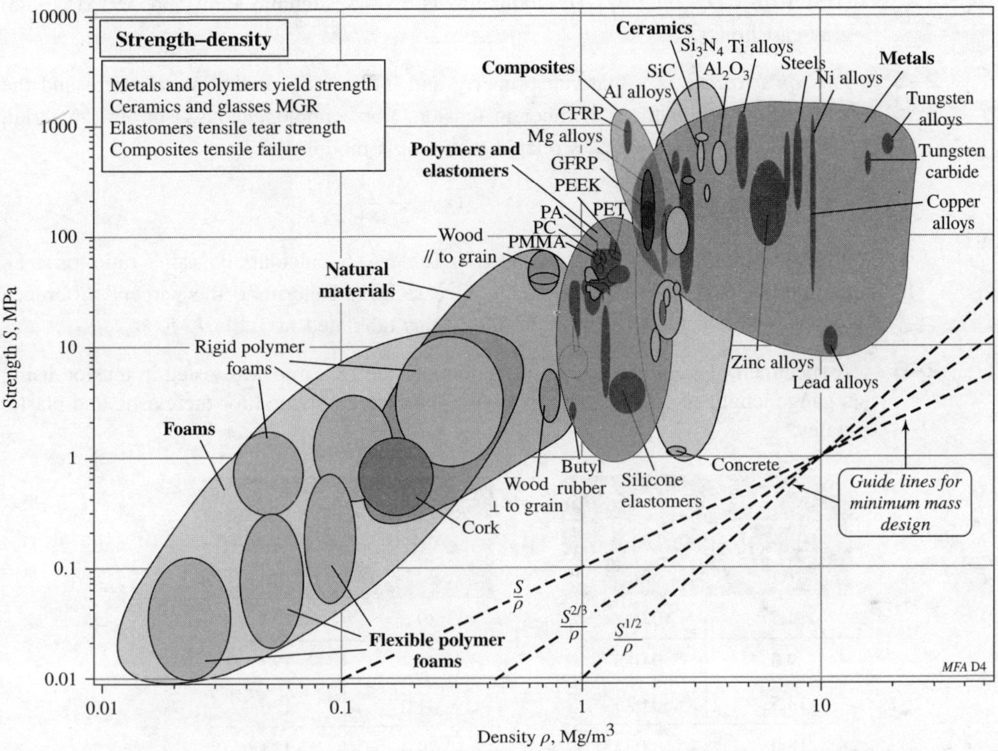

Certainly, in a given design exercise, there will be other considerations such as environment, cost, availability, and machinability, and other charts may be necessary to investigate. Also, we have not brought in the material process selection part of the picture. If done properly, material selection can result in a good deal of bookkeeping. This is where software packages such as CES Edupack become very effective.

PROBLEMS

2–1 Determine the tensile and yield strengths for the following materials:

(*a*) UNS G10 200 hot-rolled steel.

(*b*) SAE 1050 cold-drawn steel.

(*c*) AISI 1141 steel quenched and tempered at 540°C.

(*d*) 2024-T4 aluminum alloy.

(*e*) Ti-6Al-4V annealed titanium alloy.

2–2 Assume you were specifying an AISI 1060 steel for an application. Using Table A–20,

(*a*) how would you specify it if you desired to maximize the yield strength?

(*b*) how would you specify it if you desired to maximize the ductility?

2–3 Determine the yield strength-to-density ratios (specific strength) in units of kN • m/kg for AISI 1018 CD steel, 2011-T6 aluminum, Ti-6Al-4V titanium alloy, and ASTM No. 40 gray cast iron.

2–4 Determine the stiffness-to-weight density ratios (specific modulus) in units of meters for AISI 1018 CD steel, 2011-T6 aluminum, Ti-6Al-4V titanium alloy, and ASTM No. 40 gray cast iron.

2–5 *Poisson's ratio* ν is a material property and is the ratio of the lateral strain and the longitudinal strain for a member in tension. For a homogeneous, isotropic material, the modulus of rigidity G is related to Young's modulus as

$$G = \frac{E}{2(1 + \nu)}$$

Using the tabulated values of G and E in Table A–5, calculate Poisson's ratio for steel, aluminum, beryllium copper, and gray cast iron. Determine the percent difference between the calculated values and the values tabulated in Table A–5.

2–6 A specimen of steel having an initial diameter of 12.8 mm was tested in tension using a gauge length of 50 mm. The following data were obtained for the elastic and plastic states:

Elastic State	
Load P kN	Elongation mm
4.5	0.01
9.0	0.015
13.5	0.025
18.0	0.0325
31.5	0.0575
37.8	0.07
39.6	0.09
41.4	0.2225

Plastic State	
Load P kN	Area A_i mm^2
39.6	128
41.4	127
41.0	126
59.4	124
68.4	121
76.5	101
73.8	84
66.6	70

Note that there is some overlap in the data.

(*a*) Plot the engineering or nominal stress-strain diagram using two scales for the unit strain ε, one scale from zero to about 0.02 mm/mm and the other scale from zero to maximum strain.

(*b*) From this diagram find the modulus of elasticity, the 0.2 percent offset yield strength, the ultimate strength, and the percent reduction in area.

(*c*) Characterize the material as ductile or brittle. Explain your reasoning.

(*d*) Identify a material specification from Table A–19 that has a reasonable match to the data.

2–7 Compute the true stress and the true strain using the data of Problem 2–6 and plot the results on log-log paper. Then find the strain-strengthening coefficient σ_0 and the strain-strengthening exponent m. Find also the yield strength and the ultimate strength after the specimen has had 20 percent cold work.

2–8 The stress-strain data from a tensile test on a cast-iron specimen are

Engineering stress, MPa	35	70	112	133	182	224	280	332	343	378
Engineering strain, $\varepsilon \cdot 10^{-3}$ mm/mm	0.20	0.44	0.80	1.0	1.5	2.0	2.8	3.4	4.0	5.0

Plot the stress-strain locus and find the 0.1 percent offset yield strength, and the tangent modulus of elasticity at zero stress and at 140 Mpa.

2–9 A part made from annealed AISI 1018 steel undergoes a 20 percent cold-work operation.
(*a*) Obtain the yield strength and ultimate strength before and after the cold-work operation. Determine the percent increase in each strength.
(*b*) Determine the ratios of ultimate strength to yield strength before and after the cold-work operation. What does the result indicate about the change of ductility of the part?

2–10 Repeat Problem 2–9 for a part made from hot-rolled AISI 1212 steel.

2–11 Repeat Problem 2–9 for a part made from 2024-T4 aluminum alloy.

2–12 A steel member has a Brinell of $H_B = 275$. Estimate the ultimate strength of the steel in MPa.

2–13 A gray cast iron part has a Brinell hardness number of $H_B = 200$. Estimate the ultimate strength of the part in MPa. Make a reasonable assessment of the likely grade of cast iron by comparing both hardness and strength to material options in Table A–23.

2–14 A part made from 1040 hot-rolled steel is to be heat treated to increase its strength to approximately 700 MPa. What Brinell hardness number should be expected from the heat-treated part?

2–15 Brinell hardness tests were made on a random sample of 10 steel parts during processing. The results were H_B values of 230, 232(2), 234, 235(3), 236(2), and 239. Estimate the mean and standard deviation of the ultimate strength in MPa.

2–16 Repeat Problem 2–15 assuming the material to be cast iron.

2–17 For the material in Problem 2–6: (*a*) Determine the modulus of resilience, and (*b*) Estimate the modulus of toughness, assuming that the last data point corresponds to fracture.

2–18 Some commonly used plain carbon steels are AISI 1010, 1018, and 1040. Research these steels and provide a comparative summary of their characteristics, focusing on aspects that make each one unique for certain types of application. Product application guides provided on the Internet by steel manufacturers and distributors are one source of information.

2–19 Repeat Problem 2–18 for the commonly used alloy steels, AISI 4130 and 4340.

2–20 An application requires the support of an axial load of 100 kips with a round rod without exceeding the yield strength of the material. Assume the current cost per pound for round stock is given in the table below for several materials that are being considered. Material properties are available in Tables A–5, A–19, A–20, and A–23. Select one of the materials for each of the following additional design goals.
(*a*) Minimize diameter.
(*b*) Minimize weight.
(*c*) Minimize cost.
(*d*) Minimize axial deflection.

Material	Cost/N
1020 HR	\$0.06
1020 CD	\$0.07
1040 Q&T @800°F	\$0.08
4140 Q&T @800°F	\$0.18
Wrought Al 2024 T3	\$0.24
Titanium alloy (Ti-6Al-4V)	\$1.56

2–21 to 2–23 A 25 mm diameter rod, 1 m long, of unknown material is found in a machine shop. A variety of inexpensive nondestructive tests are readily available to help determine the material, as described below:

(*a*) Visual inspection.
(*b*) Scratch test: Scratch the surface with a file; observe color of underlying material and depth of scratch.
(*c*) Check if it is attracted to a magnet.
(*d*) Measure weight (± 0.25 N).
(*e*) Inexpensive bending deflection test: Clamp one end in a vise, leaving 24 in cantilevered. Apply a force of 450 N (± 4.5 N). Measure deflection of the free end (within ± 0.8 mm).
(*f*) Brinell hardness test.

Choose which tests you would actually perform, and in what sequence, to minimize time and cost, but to determine the material with a reasonable level of confidence. The table below provides results that would be available to you if you choose to perform a given test. Explain your process, and include any calculations. You may assume the material is one listed in Table A–5. If it is carbon steel, try to determine an approximate specification from Table A–19.

Test	Results if test were made	
Problem 2–21	**Problem 2–22**	**Problem 2–23**
(*a*) Dark gray, rough surface finish, moderate scale	Silvery gray, smooth surface finish, slightly tarnished	Reddish-brown, tarnished, smooth surface finish
(*b*) Metallic gray, moderate scratch	Silvery gray, deep scratch	Shiny brassy color, deep scratch
(*c*) Magnetic	Not magnetic	Not magnetic
(*d*) $W = 36$ N	$W = 14$ N	$W = 40$ N
(*e*) $\delta = 8$ mm	$\delta = 22$ mm	$\delta = 14$ mm
(*f*) $H_B = 200$	$H_B = 95$	$H_B = 70$

2–24 Research the material Inconel, briefly described in Table A–5. Compare it to various carbon and alloy steels in stiffness, strength, ductility, and toughness. What makes this material so special?

2–25 Consider a rod transmitting a tensile force. The following materials are being considered: tungsten carbide, zinc alloy, polycarbonate polymer, and aluminum alloy. Using the Ashby charts, recommend the best material for a design situation in which failure is by exceeding the strength of the material, and it is desired to minimize the weight.

2–26 Repeat Problem 2–25, except that the design situation is failure by excessive deflection, and it is desired to minimize the weight.

2–27 Consider a cantilever beam that is loaded with a transverse force at its tip. The following materials are being considered: tungsten carbide, high-carbon heat-treated steel, polycarbonate polymer, and aluminum alloy. Using the Ashby charts, recommend the best material for a design situation in which failure is by exceeding the strength of the material and it is desired to minimize the weight.

2–28 Repeat Problem 2–27, except that the design situation is failure by excessive deflection, and it is desired to minimize the weight.

2–29 For an axially loaded rod, prove that $\beta = 1$ for the E^{β}/ρ guidelines in Figure 2–24.

2–30 For an axially loaded rod, prove that $\beta = 1$ for the S^{β}/ρ guidelines in Figure 2–27.

2–31 For a cantilever beam loaded in bending, prove that $\beta = 1/2$ for the E^{β}/ρ guidelines in Figure 2–24.

2–32 For a cantilever beam loaded in bending, prove that $\beta = 2/3$ for the S^{β}/ρ guidelines in Figure 2–27.

2–33 Consider a tie rod transmitting a tensile force F. The corresponding tensile stress is given by $\sigma = F/A$, where A is the area of the cross section. The deflection of the rod is given by $\delta = (Fl)/(AE)$, where l is the length of the rod. Using the Ashby charts of Figures 2–24 and 2–27, explore what ductile materials are best suited for a light, stiff, and strong tie rod. *Hint:* Consider stiffness and strength separately.

2–34 Repeat Problem 1–13. Does the data reflect the number found in part (*b*)? If not, why? Plot a histogram of the data. Presuming the distribution is normal, plot Equation (1–4) and compare it with the histogram.

3 Load and Stress Analysis

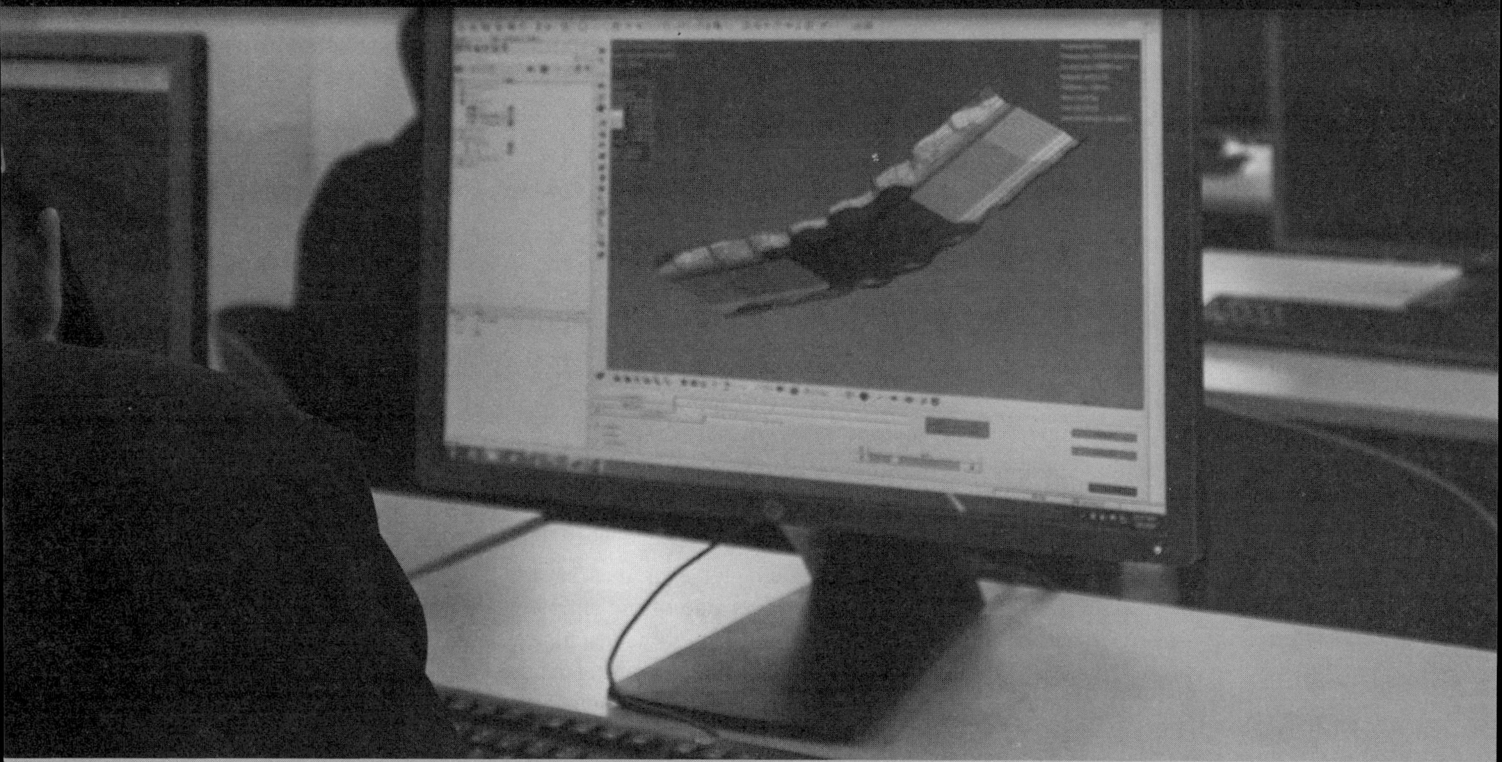

mofaez/Shutterstock

Chapter Outline

One of the main objectives of this book is to design specific machine components so that they function safely without failing structurally. One mode of failure is by the stress at a critical location exceeding the material strength at that location. This chapter addresses the determination of the stress situation in a part.

The coverage here assumes prior knowledge of statics, dynamics, and mechanics of material. Though some review is provided, the emphasis is on extending a basic knowledge of these topics into the practical application in machine design. For thorough derivations and background, the reader is referred to basic textbooks on these subjects.

The material in this chapter is organized as follows:

- This chapter begins with a consideration of force analysis. This includes a review of free-body diagrams, along with techniques to evaluate the shear forces and bending moments in beams.
- A review is provided of the concept of stress, the stress element, and the normalization of the stress element by means of the Mohr's circle to obtain principal stresses and maximum shear stresses. This review will focus on helping the reader who has already had a first exposure to the topic to grasp the concepts at a deeper level.
- The primary stress situations (e.g., axial, bending, torsion, transverse shear, and direct shear) are summarized.
- Pulling together the review material, we will address how to evaluate a mechanical part to determine locations where the stress situations are potentially critical failure points.
- The latter portion of the chapter examines several situations for which the stress distribution does not fall under one of the simple categories (e.g., bending, torsion, etc.), but which are common enough in machine design to warrant special consideration (e.g., pressurized cylinders, rotating rings, press and shrink fits, thermal stresses, curved beams in bending, and contact stresses).

3–1 Equilibrium and Free-Body Diagrams

A first step in analyzing or designing a mechanical part is to determine the forces acting on and within it. The techniques for doing so are covered in courses and textbooks on statics and dynamics. Here, we will focus only on a few highlights regarding how this step fits within a machine design process.

The starting point for a force analysis is the *free-body diagram* (FBD). The FBD provides a means of visualizing the basic geometry of the part(s), the interconnections within a system of parts, the applied forces, and the reaction forces. Note that the term "force" here is used in a general sense to include all types of loading, including forces, bending moments, and torsional moments. Determining internal forces within the geometry is necessary to evaluate how the forces are distributed as stress throughout the geometry. Determining reaction forces at the boundaries of the system, such as where a machine part attaches to the frame of the machine, is necessary for properly sizing the components (e.g., bearings, bolts, gears, and welds).

For each FBD, equations are generated for Newton's second law.

$$\sum F = ma \tag{3–1}$$

$$\sum M = I\alpha \tag{3–2}$$

The force and acceleration terms are vectors. Commonly, separate equations are generated for the forces along three orthogonal axes (e.g., x, y, and z), and for the moments around three orthogonal axes. Of course, some of these directions may not have any non-zero contributions, and can then be ignored. The inertia terms on the right-hand side of these equations may be zero if either the mass or the acceleration is negligibly small or zero.

A few tips regarding FBDs may be helpful to solidify the understanding gained from a first exposure in a statics course.

- An FBD can be drawn for any body, or part of a body, or grouping of connected bodies. Draw as many as are needed to provide the information required.
- Forces and moments for three orthogonal axes are applied anywhere an FBD is isolated from where it touches its surroundings. Some of these may have magnitude of zero.
- It is often useful to cut through a body to expose a cross section, allowing the forces transmitted through the cross section to be determined. At the cut section, the FBD should include consideration of forces along the three axes directions and moments around the three axes. Any forces or moments are to be understood to be applied to the entire cross section of the cut.

EXAMPLE 3–1

Figure 3–1*a* shows a simplified rendition of a gear reducer where the input and output shafts *AB* and *CD* are rotating at constant speeds ω_i and ω_o, respectively. The input and output torques (torsional moments) are $T_i = 27$ N · m and T_o, respectively. The shafts are supported in the housing by bearings at *A*, *B*, *C*, and *D*. The pitch radii of gears G_1 and G_2 are $r_1 = 20$ mm and $r_2 = 40$ mm, respectively. Draw the free-body diagrams of each member and determine the net reaction forces and moments at all points.

Solution

First, we will list all simplifying assumptions.

1. Gears G_1 and G_2 are simple spur gears with a standard pressure angle $\phi = 20°$ (see Section 13–5).
2. The bearings are self-aligning and the shafts can be considered to be simply supported.
3. The weight of each member is negligible.
4. Friction is negligible.
5. The mounting bolts at *E*, *F*, *H*, and *I* are the same size.

The separate free-body diagrams of the members are shown in Figures 3–1*b*–*d*. Note that Newton's third law, called *the law of action and reaction,* is used extensively where each member mates. The force transmitted between the spur gears is not tangential but at the pressure angle ϕ. Thus, $N = F \tan \phi$.

Summing moments about the x axis of shaft *AB* in Figure 3–1*c* gives

$$\sum M_x = F(20) - 27 = 0$$

$$F = 1350 \text{ N}$$

The normal force is $N = 1350 \tan 20° = 491.36$ N.

Using the equilibrium equations for Figures 3–1*c* and *d*, the reader should verify that: $R_{Ay} = 830.8$ N, $R_{Az} = 302.4$ N, $R_{By} = 519.2$ N, $R_{Bz} = 188.9$ N, $R_{Cy} = 830.8$ N, $R_{Cz} = 302.4$ N, $R_{Dy} = 519.2$ N, $R_{Dz} = 188.9$ N, and $T_o = 54$ N · m. The direction of the output torque T_o is opposite ω_o because it is the resistive load on the system opposing the motion ω_o.

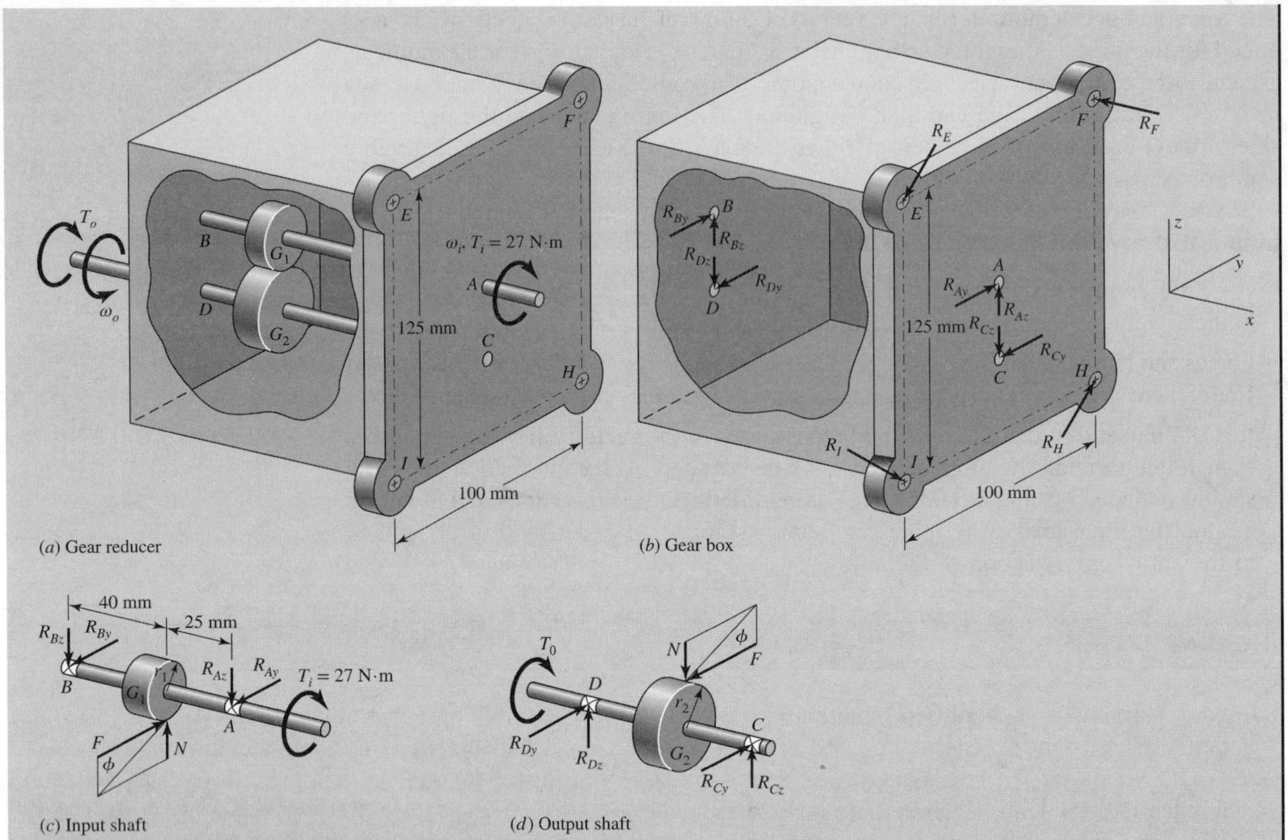

Figure 3–1

(*a*) Gear reducer; (*b–d*) free-body diagrams. Diagrams are not drawn to scale.

Note in Figure 3–1*b* the net force from the bearing reactions is zero whereas the net moment about the *x* axis is (40 + 20)(830.8) + (40 + 20)(519.2) = 81 N · m. This value is the same as $T_i + T_o = 27 + 54 = 81$ N · m, as shown in Figure 3–1*a*. The reaction forces R_E, R_F, R_H, and R_I, from the mounting bolts cannot be determined from the equilibrium equations as there are too many unknowns. Only three equations are available, $\Sigma F_y = \Sigma F_z = \Sigma M_x = 0$. In case you were wondering about assumption 5, here is where we will use it (see Section 8–12). The gear box tends to rotate about the *x* axis because of a pure torsional moment of 81 N · m. The bolt forces must provide an equal but opposite torsional moment. The center of rotation relative to the bolts lies at the centroid of the bolt cross-sectional areas. Thus if the bolt areas are equal: the center of rotation is at the center of the four bolts, a distance of $\sqrt{(100/2)^2 + (125/2)^2} = 80.03$ mm from each bolt; the bolt forces are equal ($R_E = R_F = R_H = R_I = R$), and each bolt force is perpendicular to the line from the bolt to the center of rotation. This gives a net torque from the four bolts of $4R(0.080\ 03) = 81$. Thus, $R_E = R_F = R_H = R_I = 253.02$ N.

3–2 Shear Force and Bending Moments in Beams

A *beam* is a long slender structural member that supports loads at various points along its length. Examples include shafts and structural elements (e.g., trusses, joists, girders, etc.). The beam is so common in machine design that its force analysis receives special consideration. Beams are often loaded with concentrated and distributed loads that are transverse to the axis of the beam. Beams may also carry axial and torsional loads. Note that a torsional load is a moment around the long axis of the beam which creates a twist in the beam, as contrasted with a moment around a transverse axis which creates a bending along the long axis of the beam.

For force analysis of a beam, the axial and torsional loads tend to be constant along the length of a beam between points of interest, and are therefore easily handled with standard free-body diagram evaluation. However, the transverse loads create bending moments that change along the axis of the beam, and therefore require more attention to determine where the bending moments are greatest.

Consider the simply supported beam in Figure 3–2*a*, along with its free-body diagram in Figure 3–2*b* where the ground reaction forces have been determined from applying the summation of forces and moments equations. If the beam is cut open at some location along its length, say at 0.2 meters from the left end, the internal shear force V and bending moment M carried by the beam at that cross section can be determined by applying the equilibrium equations for the cut beam, as in Figure 3–2*c*. This could be repeated at any location of interest along the beam. The problem is that it is not obvious which location along the beam will carry the largest load. Consequently, it is advantageous to generate plots of the shear force and bending moment with respect to the axis of the beam. This will allow the entire beam to be evaluated for potentially critical locations.

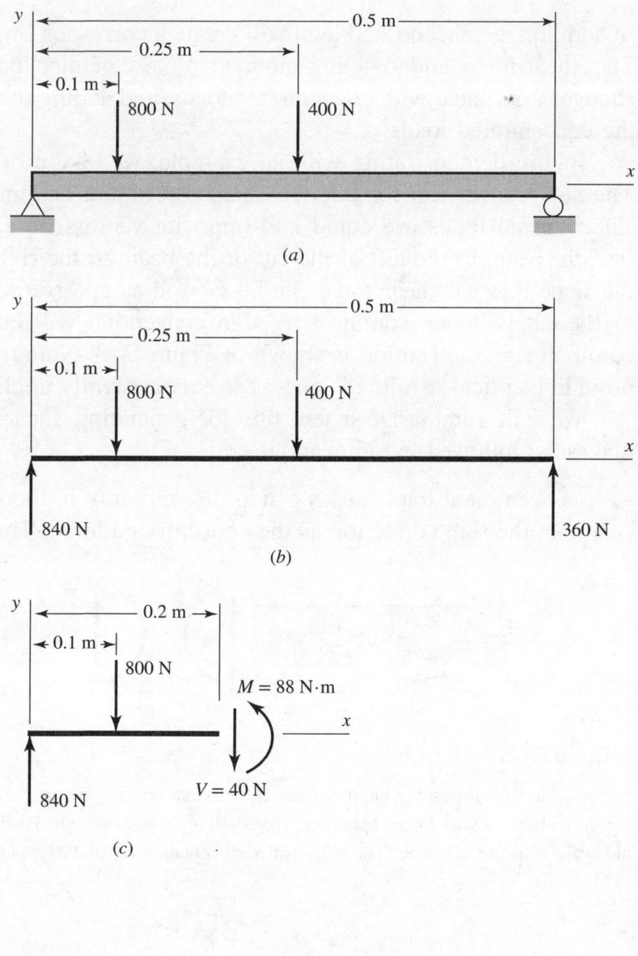

Figure 3–2

(*a*) Simply-supported beam; (*b*) Free-body diagram with ground reaction forces; (*c*) Internal shear force and bending moment at cut section.

A simple technique for generating these plots as a whole rather than point by point is obtained by recalling from mechanics of materials the relationships between the transverse load q, internal shear force V, and internal bending moment M,

$$q = \frac{dV}{dx} \tag{3–3}$$

$$V = \frac{dM}{dx} \tag{3–4}$$

Integrating these equations between any two points A and B, with the condition that q and V remain continuous between these points, results in

$$\int_{x_A}^{x_B} q\,dx = \int_{V_A}^{V_B} dV = V_B - V_A \tag{3–5}$$

$$\int_{x_A}^{x_B} V\,dx = \int_{M_A}^{M_B} dM = M_B - M_A \tag{3–6}$$

These relationships between load, shear force, and bending moment extend further to slope and deflection. This is developed more thoroughly in Section 4–3.

From the nature of integrals, Equations (3–5) and (3–6) indicate the following:

- The change in shear force V from A to B is equal to the area under the load curve between A and B.
- The change in moment M from A to B is equal to the area under the shear-force curve between A and B.

In addition, a concentrated load will create a corresponding step change in shear force. The shear-force and bending-moment plots can then be obtained by plotting step changes associated with concentrated loads, and adding the areas under curves between the concentrated loads.

Before demonstrating with an example, we take a brief look at sign convention. The sign convention for internal shear forces and bending moments is not intuitive. The internal loads are equal and opposite vectors on each side of the cut section (i.e., the beam to the left of the cut or the beam to the right of the cut). Consequently, the same internal shear force can be viewed as up or down, depending on which side of the cut is being examined. A sign convention will be needed for consistency. A common sign convention is shown in Figure 3–3. Note that *any* sign convention will provide identical results as long as it is consistently applied.

We will summarize a few tips for generating shear-force and bending-moment diagrams, followed by an example.

- A concentrated load causes a step discontinuity in the shear-force plot. The step in V is in the same direction as the concentrated force. This is consistent with the sign

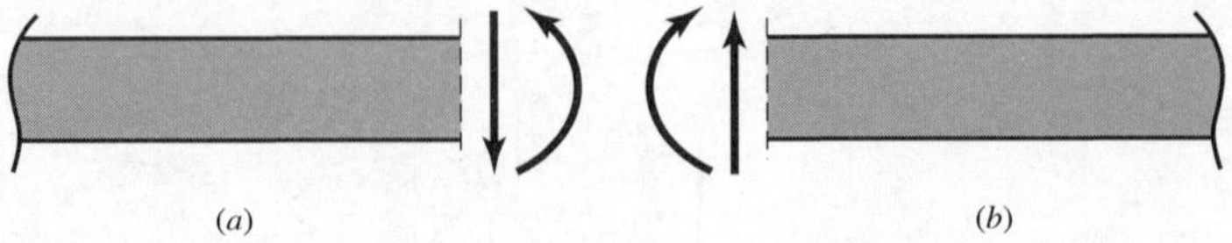

Figure 3–3

Sign convention for positive internal shear force V and bending moment M, as viewed from both sides of a cut in the beam. Other forces applied to the beam are not shown. (*a*) Positive directions for V and M when the beam is to the left of the cut. (*b*) Positive directions for V and M when the beam is to the right of the cut.

convention shown in Figure 3–3*a*, in which an upward applied force will be balanced by a downward shear force in the cut beam, which is defined by sign convention as positive.

- A uniformly distributed load results in a linear change in V, with the sign of the change in V being the same as the sign of the uniform load. The magnitude of the change of V is equal to the area under the distributed load plot.
- A concentrated bending moment causes a step discontinuity in the bending-moment plot. Consistent with the sign convention shown in Figure 3–3*a*, a CW external moment causes a positive step in the bending-moment plot.
- The change in bending moment between any two points is equal to the area under the shear diagram between those two points.
- The slope of the moment diagram at any point is equal to the shear force at that point.
- The shear-force and bending-moment diagrams should start and end at zero. If they don't, an error has been made in either the plots or in the equilibrium calculations to find the ground reactions.

EXAMPLE 3–2

For the beam in Figure 3–4*a*, generate plots of the shear force and the bending moment within the beam.

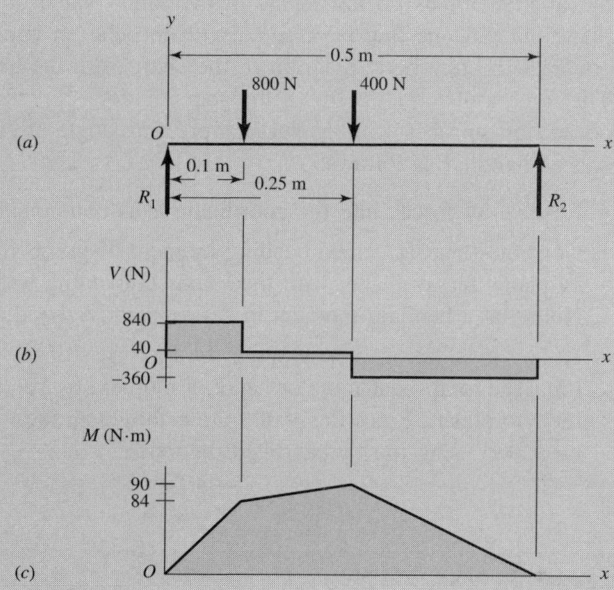

Figure 3–4

(*a*) Loading diagram; (*b*) Shear-force diagram; (*c*) Bending-moment diagram.

Solution

The loading diagram given in Figure 3–4*a* also serves as the free-body diagram for determining the ground reaction forces. From summation of forces in the y direction and summation of moments around the z axis at any convenient point, we can readily find the ground reaction forces to be $R_1 = 840$ N and $R_2 = 360$ N.

Beginning at the left end of the beam, the concentrated force of R_1 = 840 N is plotted on the shear-force plot as a positive value, since from the sign convention in Figure 3–3 the shear force counteracting R_1 is acting downward (positive by convention). Moving to the right, the shear force needed within the beam to maintain equilibrium remains the same until x = 0.1 m, where the concentrated 800 N creates a step change in the shear force. Continuing this process, the shear force plot is obtained as shown in Figure 3–4*b*. Note that the plot closes to zero at the right end of the beam, when the concentrated force of R_2 = 360 N is applied.

Applying Equation (3–6) between points x = 0 m and x = 0.1 m, the bending moment must change by the amount of the area under the shear curve between those points, thus at x = 0.1 m, M = (840 N)(0.1 m) = 84 N · m. Continuing this process between each of the concentrated forces, the bending moment plot is obtained as shown in Figure 3–4*c*. Again, note that the plot closes to zero at the right end of the beam, which is a good indicator that the plot is correct.

Many common beam configurations are given in Appendix A–9, with plots and equations for shear force, bending moment, and deflection. Superposition of the beam configurations is valid, and is given thorough treatment in Section 4–5.

For the beam in Example 3–2, all of the forces in the loading diagram lie in the same plane. This conveniently led to a plot of shear forces in the *xy* plane, and a plot of bending moments around the *z* axis. In many machine design situations, however, the applied forces do not all lie in the same plane, or perhaps they are in a single plane but not one that is oriented with the chosen coordinate axes. As an example, consider the free-body diagram of the shaft from the gearbox shown in Figure 3–1*c*. The force transmitted between the gears is along a pressure line that is normal to the contacting surfaces of the gears, at a pressure angle of 20° to the *y* axis. This situation can be handled as follows:

- Resolve all forces into the coordinate axes components.
- Generate shear-force and bending-moment diagrams for two orthogonal planes (e.g., *xy* plane and *xz* plane, with the *x* axis coinciding with the long axis of the beam). Note that a bending moment in the *xy* plane is bending around the *z* axis.
- At any location of interest along the long axis of the beam (e.g., the *x* axis), determine the total shear force or bending moment by vectorially adding the values from the two planes. Since the planes are orthogonal, the vectors are orthogonal, and can be added using the Pythagorean theorem.

EXAMPLE 3–3

The shaft shown in Figure 3–5*a* is simply supported by ball bearings at *A* and *D*, and supports gears at *B* and *C*. The gears have a standard pressure angle of 20°, such that the transmitted forces are 20° from the *y* axis. The gear forces are F_B = 5499 N and F_C = 3437 N. The effective diameters of the gears for force transmission purposes are their pitch diameters, d_B = 280 mm and d_C = 448 mm.

(*a*) Generate shear-force and bending-moment diagrams for the shaft. Make one set for the *xy* plane and another set for the *xz* plane.

(*b*) Determine the location and the magnitude of the maximum bending moment along the shaft.

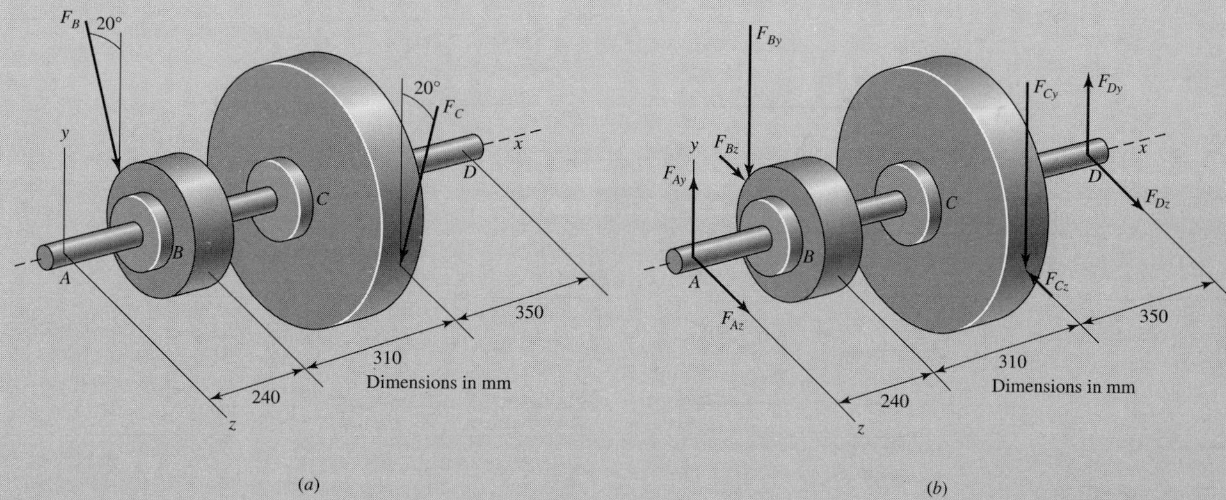

Figure 3–5

(*a*) Shaft with two gears. (*b*) Free-body diagram with forces shown as components in the coordinate system.

Solution

Determine the *y* and *z* components of the gear forces, where the direction of each force is determined by inspection of the directions given on the free-body diagram.

$$F_{By} = F_B \cos 20° = 5499 \cos 20° = 5167 \text{ N (negative } y \text{ direction)}$$
$$F_{Bz} = F_B \sin 20° = 5499 \sin 20° = 1881 \text{ N (positive } z \text{ direction)}$$
$$F_{Cy} = F_C \cos 20° = 3437 \cos 20° = 3230 \text{ N (negative } y \text{ direction)}$$
$$F_{Cz} = F_C \sin 20° = 3437 \sin 20° = 1176 \text{ N (negative } z \text{ direction)}$$

Using the free-body diagram in Figure 3–5*b*, determine the ground reaction forces at the bearings, using the *y* and *z* components of the bearing forces.

$$\sum F_y = 0 \qquad F_{Ay} - F_{By} - F_{Cy} + F_{Dy} = 0$$
$$\sum F_z = 0 \qquad F_{Az} + F_{Bz} - F_{Cz} + F_{Dz} = 0$$
$$\sum M_{Az} = 0 \qquad -(240)F_{By} - (240 + 310)F_{Cy} + (240 + 310 + 350)F_{Dy} = 0$$
$$\sum M_{Ay} = 0 \qquad -(240)F_{Bz} + (240 + 310)F_{Cz} - (240 + 310 + 350)F_{Dz} = 0$$

Substitute the values for the gear forces and solve the four equations for the four ground reaction force components, where the signs are with respect to the positive *y* and *z* axes.

$$F_{Ay} = 5045 \text{ N}$$
$$F_{Az} = -922 \text{ N}$$
$$F_{Dy} = 3352 \text{ N}$$
$$F_{Dz} = 217 \text{ N}$$

Draw the load diagram for the *xy* and *xz* planes, shown in Figure 3–6. Note that the load diagram for the *xz* plane is plotting the positive *z* axis in the downward direction in order to have the viewpoint from the "top" looking down the *y* axis. From the load diagrams, generate the shear-force and bending-moment plots for each plane, also shown in Figure 3–6.

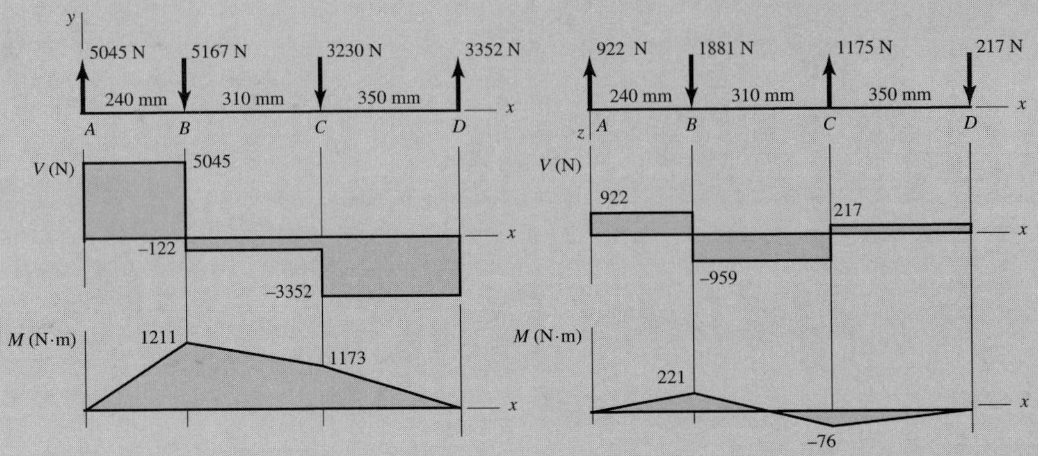

Figure 3–6

Shear-force and bending-moment diagrams for the *xy* and *xz* planes.

The bending moments in each plane have peaks at *B* and *C*, so the maximum total bending moment will be at one of those points. Find the total bending moment at each point by adding the bending moment vectors. In particular, considering point *B*, the bending moment in the *xy* plane (around the *z* axis) can be treated as a vector with magnitude 1211 N · m along the *z* axis. Similarly, the bending moment in the *xz* plane (around the *y* axis) can be treated as a vector with magnitude 221 N · m along the *y* axis. The total bending moment at *B* is the vector sum of these two moment vectors, thus

$$M_B = \sqrt{1211^2 + 221^2} = 1231 \text{ N}$$

$$M_C = \sqrt{1173^2 + (-76)^2} = 1175 \text{ N}$$

Answer

The maximum bending moment is $M_B = 1231$ N at point *B*.

3–3 Singularity Functions

An extremely useful tool in handling discontinuous and complex loading of beams employs a mathematical tool called *Macaulay functions* or *singularity functions*. In addition to handling shear force and bending moment, this tool is even more commonly used when extended further to slope and deflection. Rather than splitting the topic between two chapters, it is comprehensively developed in Section 4–6.

3–4 Stress

When loads are applied to a physical object, the laws of static and dynamic equilibrium allow us to determine the forces and moments experienced by the part. With a theoretical slice through the object, the forces and moments that exist at the exposed cross section are revealed. This is a macro view that essentially treats the body as if it is rigid. If we hypothetically look closer, the material is not rigid, but is elastically

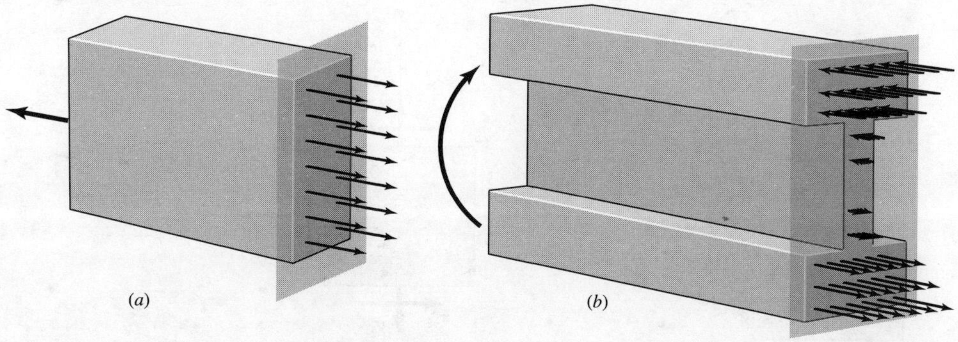

Figure 3–7

(*a*) Axial load is uniformly distributed to all particles in the cross section. (*b*) The bending moment is non-uniformly distributed throughout the cross section.

deforming due to the loads. It may be helpful to think of the continuum of the material as tiny particles that are each being pushed by, and are pushing on, neighboring particles as necessary to transmit the loads. This imagery is one way to understand the concept of stress. *Stress* expresses the internal forces that neighboring particles of a continuous material exert on each other. Stress is therefore a measure of the distribution of macro-level loads carried by a body into the internal forces within the body. To continue the imagery, imagine that each particle is bonded to neighboring particles. To accommodate deformation within the material, each particle will stretch or compress or distort when the bonds of neighboring particles pull or push on it.

Stress is a measure of the distribution of a force over an area, so it has units of pressure. In SI units, stress is in newtons per square meter (N/m^2) which is equal to pascal (Pa), and will often practically be at the level of MPa. In U.S. Customary units this will be pounds per square inch (psi), and in practical engineering problems will often be at the level of kpsi.

The distribution of a load over an area is sometimes uniform, such as when an axial load is distributed over a cross section that is perpendicular to the load, as depicted in Figure 3–7*a*.

But the distribution is not always uniform, such as when a bending load is distributed in a cross section of a beam, as shown in Figure 3–7*b*. The elements on the top of the beam are compressed, and the elements on the bottom are stretched, with a linear distribution based on the distance from the neutral plane. Specific details for these cases are provided in later sections.

3–5 Cartesian Stress Components

As an aid to articulating the potentially complex stress situation at points within a body, the concept of a stress element is introduced. The *stress element* is a cube that is infinitesimally small, but drawn to a scale large enough for us to visualize its six faces and the stresses acting on those faces. Essentially, the stress element replaces the formerly referenced "particles" in the continuum with tiny cubes. The stress elements have the advantage of conveniently representing an orthogonal set of axes. Figure 3–8*a* shows the stresses acting on the surfaces of a stress element oriented with faces aligning with an *xyz* coordinate axes. A stress acting normal to a face is a *normal stress*. A stress acting parallel to a surface is a *shear stress*. The normal stresses are easily understood from the prior discussion of particles pushing and pulling on neighboring particles. Any neighboring stress element with faces aligned with this stress element can push or pull on an adjacent surface face with a normal stress.

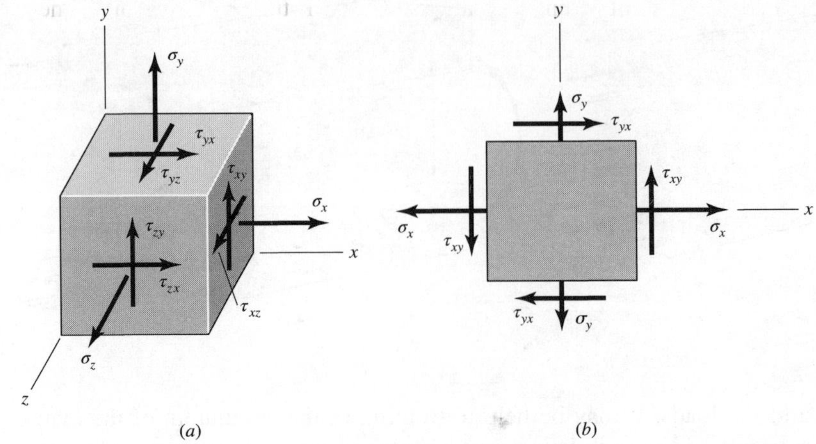

Figure 3–8

(*a*) General three-dimensional stress. (*b*) Plane stress.

To understand the concept of shear stress, imagine a neighboring stress element with a face that is trying to slide parallel to a face of this stress element. The bond between the faces of the two elements will create a stress on the faces, causing the elements to skew rather than to stretch or compress.

Figure 3–8 gives the nomenclature for the stresses. The normal stresses are labeled with the symbol σ with a subscript indicating the direction of the surface normal. Normal stresses have positive values when stretching the element in tension, and negative values when compressing the element in compression. The net shear stress on a face is resolved into two component directions. Shear stresses are labeled with the symbol τ with two subscripts. The first subscript indicates the normal of the face on which the stress is acting; the second subscript indicates the direction of the shear stress. As the dimension of the cube approaches zero, which is required by the definition of a stress element, the stresses on opposite faces are equal in magnitude and opposite in direction to each other. Further, to maintain rotational equilibrium, it is necessary for "cross shears" to have equal magnitudes, hence,

$$\tau_{yx} = \tau_{xy} \qquad \tau_{zy} = \tau_{yz} \qquad \tau_{xz} = \tau_{zx} \tag{3–7}$$

A common state of stress occurs when the stresses on one surface are zero. When this occurs the state of stress is called *plane stress*. Figure 3–8*b* shows a state of plane stress, arbitrarily assuming that the normal for the stress-free surface is the z direction such that $\sigma_z = \tau_{zx} = \tau_{zy} = 0$. It is important to note that the element in Figure 3–8*b* is still a three-dimensional cube.

A state of plane stress typically occurs under two conditions. The first condition is on *free surfaces* where no stresses exist perpendicular to the surface, or the second, on thin, flat parts only loaded perpendicular to the thickness plane.

3–6 Mohr's Circle for Plane Stress

Transformation of Coordinates

It is important to remember that the stress element is infinitesimally small. The cubic shape, the faces, and the orientation of the axes are only specified for convenience of identifying the stresses in a manner consistent with the orientations of the loads and the geometry of the part. But the stress situation at a point can just as well be represented by a stress element oriented in any direction. It turns out that often there is reason to use a different orientation of the axes. The stress situation at a point is unchanged by a change of orientation of the stress element. But, as with any change of coordinate system, the stresses resolved into different coordinate axes will have different magnitudes.

The transformation of coordinates for plane stress is readily derived, serves to demonstrate basic concepts, and is suitable for the majority of practical problems in machine design.

Suppose the *dx dy dz* element of the plane stress element in Figure 3–8*b* is cut by an oblique plane with a normal *n* at an arbitrary angle ϕ counterclockwise from the *x* axis as shown in Figure 3–9. Here, we are concerned with the stresses σ and τ that act upon this oblique plane. The equations derived from cut at an oblique plane are also applicable for rotating the entire stress element and its coordinates around the *z* axis by the angle ϕ. By summing the forces caused by all the stress components to zero, the stresses σ and τ are found to be

$$\sigma = \frac{\sigma_x + \sigma_y}{2} + \frac{\sigma_x - \sigma_y}{2} \cos 2\phi + \tau_{xy} \sin 2\phi \tag{3–8}$$

$$\tau = -\frac{\sigma_x - \sigma_y}{2} \sin 2\phi + \tau_{xy} \cos 2\phi \tag{3–9}$$

Equations (3–8) and (3–9) are called the *plane-stress transformation equations.*

We will apply the transformation equations to a coordinate system rotation of ϕ around the *z* axis, from *xy* coordinates as shown in Figure 3–8*b* to *x'y'* coordinates as shown in Figure 3–10. Directly apply Equations (3–8) and (3–9) to obtain σ'_x and τ'_{xy}, and use a rotation angle of $(\phi + \pi/2)$ to obtain σ'_y. This gives

$$\sigma'_x = \frac{\sigma_x + \sigma_y}{2} + \frac{\sigma_x - \sigma_y}{2} \cos 2\phi + \tau_{xy} \sin 2\phi \tag{3–10}$$

$$\sigma'_y = \frac{\sigma_x + \sigma_y}{2} - \frac{\sigma_x - \sigma_y}{2} \cos 2\phi - \tau_{xy} \sin 2\phi \tag{3–11}$$

$$\tau'_{xy} = \tau'_{yx} = -\frac{\sigma_x - \sigma_y}{2} \sin 2\phi + \tau_{xy} \cos 2\phi \tag{3–12}$$

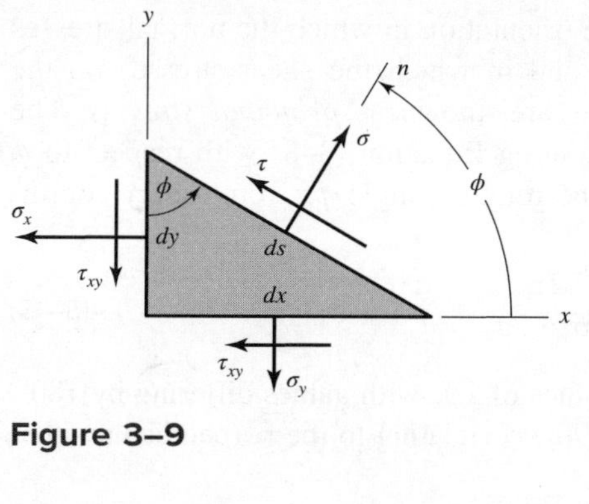

Figure 3–9

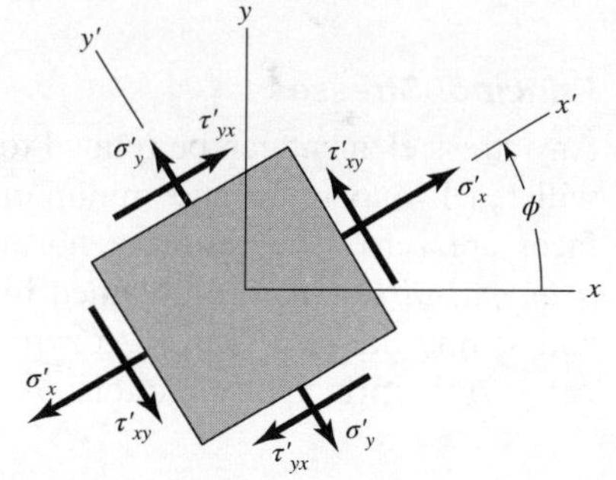

Figure 3–10

Rotation of a stress element by angle ϕ from *xy* coordinates to *x'y'* coordinates.

EXAMPLE 3–4

A stress element is shown in Figure 3–11. Rotate the stress element 35° CCW to new coordinates $x'y'$. Determine the stresses acting on the stress element in the new coordinates.

Solution

The stresses in the xy coordinate system as shown in Figure 3–11 are $\sigma_x = 120$ MPa, $\sigma_y = -200$ MPa, and $\tau_{xy} = -60$ MPa. Note that σ_y is compressive and therefore negative. Applying Equations (3–10) through (3–12),

Answer

$$\sigma'_x = \frac{120 + (-200)}{2} + \frac{120 - (-200)}{2} \cos(2 \cdot 35°) + (-60) \sin(2 \cdot 35°) = -41.7 \text{ MPa}$$

$$\sigma'_y = \frac{120 + (-200)}{2} - \frac{120 - (-200)}{2} \cos(2 \cdot 35°) - (-60) \sin(2 \cdot 35°) = -38.3 \text{ MPa}$$

$$\tau'_{xy} = -\frac{120 - (-200)}{2} \sin(2 \cdot 35°) + (-60) \cos(2 \cdot 35°) = -170.9 \text{ MPa}$$

The stresses in the $x'y'$ coordinates are shown on the stress element in Figure 3–11*b*, where the signs have been incorporated in the directions of the arrows.

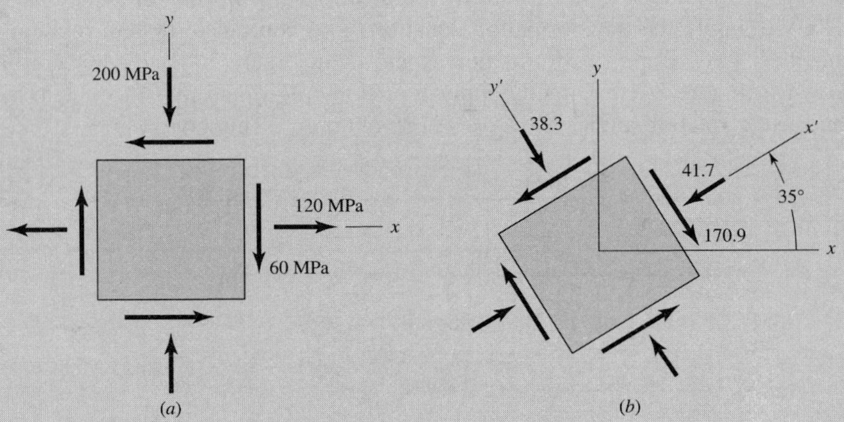

Figure 3–11

(*a*) Stress state in xy coordinates.
(*b*) The same stress state rotated 35° to $x'y'$ coordinates.

Principal Stresses

Any stress element can be rotated to a unique orientation in which the normal stresses will reach maximum and minimum values, and in which the shear stresses on the faces are zero. The resulting normal stresses are known as *principal stresses*. The *principal directions* are obtained by differentiating Equation (3–8) with respect to ϕ and setting the result equal to zero, giving the rotation angle ϕ_p from the xy coordinates to the principal directions as

$$\tan 2\phi_p = \frac{2\tau_{xy}}{\sigma_x - \sigma_y} \tag{3–13}$$

Obtaining the inverse tangent will give two values of $2\phi_p$ with values differing by 180°. Thus there are two values of ϕ_p, differing by 90°, correlating to the perpendicular faces containing the principal stresses. Substituting ϕ_p into Equation (3–8), the principal stresses are

$$\sigma_1, \sigma_2 = \frac{\sigma_x + \sigma_y}{2} \pm \sqrt{\left(\frac{\sigma_x - \sigma_y}{2}\right)^2 + \tau_{xy}^2} \tag{3–14}$$

For plane stress, recall that there is also a third principal stress equal to zero on the face oriented along the z axis. Substituting ϕ_p into Equation (3–9) results in shear stresses of zero on the principal planes, consistent with the definition of principal stress.

A similar approach is used to obtain the orientation of a stress element with the extreme magnitudes of shear stress. Differentiating Equation (3–9) and setting the resulting equation equal to zero gives the rotation ϕ_s,

$$\tan 2\phi_s = -\frac{\sigma_x - \sigma_y}{2\tau_{xy}} \tag{3–15}$$

Comparing Equations (3–15) and (3–13), we see that $\tan 2\phi_s$ is the negative reciprocal of $\tan 2\phi_p$ which leads to

$$\phi_s = \phi_p \pm 45° \tag{3–16}$$

So, the planes of maximum shear stress occur at 45° to the principal planes.

Substituting ϕ_s into Equation (3–9) gives the equation for the extreme shear stresses in the plane of principal axes 1 and 2 as,

$$\tau_{12} = \pm\sqrt{\left(\frac{\sigma_x - \sigma_y}{2}\right)^2 + \tau_{xy}^2} = \frac{\sigma_1 - \sigma_2}{2} \tag{3–17}$$

Due to the $\pm$ sign, there are two extreme values, of equal magnitude and opposite directions. We usually only need the absolute value, which is referred to as the *principal shear stress*. Note that there are two more principal shear stresses in planes defined by principal axes 1 and 3, and by principal axes 2 and 3. See Section 3–7 for further discussion of this, and for determination of the absolute maximum shear stress.

We will summarize a few of the key points for principal stresses in plane stress.

- Principal stresses occur at the unique orientation of the stress element where the faces have only normal stresses and no shear stresses.
- There are always three principal stresses. For a plane stress situation, one principal stress is equal to zero.
- The maximum and minimum values of normal stresses that can be obtained for any orientation of a stress element will be principal stresses.
- The principal shear stresses occur on faces rotated 45° from the principal faces.
- On the principal shear stress faces, the normal stresses are equal to the average of the principal stresses, which is also the average of the normal stresses on any orientation of the stress element.

Mohr's Circle

A graphical method for expressing the relations developed in this section, called *Mohr's circle diagram,* is a very effective means of visualizing the stress state at a point and keeping track of the directions of the various components associated with plane stress. Equations (3–8) and (3–9) can be shown to be a set of parametric equations for σ and τ, where the parameter is 2ϕ. The parametric relationship between σ

and τ is that of a circle plotted in the σ, τ plane, where the center of the circle is located at $C = (\sigma, \tau) = [(\sigma_x + \sigma_y)/2, 0]$ and has a radius of $R = \sqrt{[(\sigma_x - \sigma_y)/2]^2 + \tau_{xy}^2}$.

A problem arises in the sign of the shear stress. The transformation equations are based on a positive ϕ being counterclockwise, as shown in Figure 3–9. If a positive τ were plotted above the σ axis, points would rotate clockwise on the circle 2ϕ in the opposite direction of rotation on the element. It would be convenient if the rotations were in the same direction.

This is accomplished by use of the following convention in drawing Mohr's circle:

- Shear stresses tending to rotate the element clockwise (cw) are plotted *above* the σ axis.
- Shear stresses tending to rotate the element counterclockwise (ccw) are plotted *below* the σ axis.

This traditional convention does not treat the shear axis as positive and negative, but as clockwise and counterclockwise. It turns out that for most machine design purposes we are interested in magnitudes of the stresses and not orientations of the stress element, so the sign convention could be reversed with no loss of functionality.

In Figure 3–12 we create a coordinate system with normal stresses plotted along the abscissa and shear stresses plotted as the ordinates. On the abscissa, tensile (positive) normal stresses are plotted to the right of the origin O and compressive (negative) normal stresses to the left. On the ordinate, clockwise (cw) shear stresses are plotted up; counterclockwise (ccw) shear stresses are plotted down.

Using the stress state of Figure 3–8*b*, we plot Mohr's circle, Figure 3–12, by first looking at the right surface of the element containing σ_x to establish the sign of σ_x and the cw or ccw direction of the shear stress. The right face is called the *x face* where $\phi = 0°$. If σ_x is positive and the shear stress τ_{xy} is ccw as shown in Figure 3–8*b*, we can establish point A with coordinates $(\sigma_x, \tau_{xy}^{ccw})$ in Figure 3–12. Next, we look at the top *y face*, where $\phi = 90°$, which contains σ_y, and repeat the process to obtain point

Figure 3–12

Mohr's circle diagram.

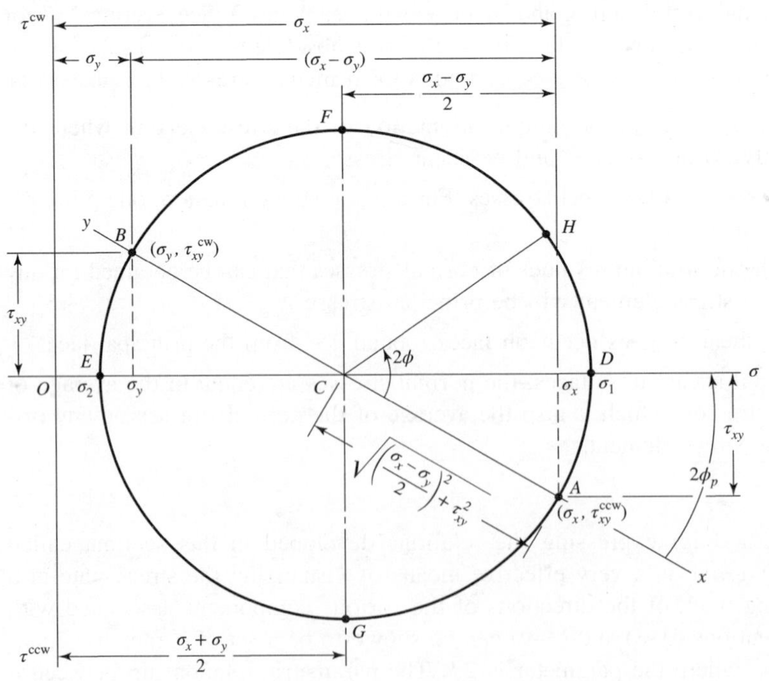

B with coordinates $(\sigma_y, \tau_{xy}^{\text{cw}})$ as shown in Figure 3–12. The two states of stress for the element are $\Delta\phi = 90°$ from each other on the element so they will be $2\Delta\phi = 180°$ from each other on Mohr's circle. Points A and B are the same vertical distance from the σ axis. Thus, AB must be on the diameter of the circle, and the center of the circle C is where AB intersects the σ axis. With points A and B on the circle, and center C, the complete circle can then be drawn. Note that the extended ends of line AB are labeled x and y as references to the normals to the surfaces for which points A and B represent the stresses.

The entire Mohr's circle represents the state of stress at a *single* point in a structure. Each point on the circle represents the stress state for a *specific* surface intersecting the point in the structure. Each pair of points on the circle 180° apart represent the state of stress on an element whose surfaces are 90° apart. Once the circle is drawn, the states of stress can be visualized for various surfaces intersecting the point being analyzed. For example, the principal stresses σ_1 and σ_2 are points D and E, respectively, and their values obviously agree with Equation (3–14). We also see that the shear stresses are zero on the surfaces containing σ_1 and σ_2. The two extreme-value shear stresses, one clockwise and one counterclockwise, occur at F and G with magnitudes equal to the radius of the circle. The surfaces at F and G each also contain normal stresses of $(\sigma_x + \sigma_y)/2$ which is the average of the normal stresses for any stress element orientation. Finally, the state of stress on an arbitrary surface located at an angle ϕ counterclockwise from the x face is point H.

At one time, Mohr's circle was used graphically where it was drawn to scale very accurately and values were measured by using a scale and protractor. Here, we are strictly using Mohr's circle as a visualization aid and will use a semigraphical approach, calculating values from the properties of the circle. This is illustrated by the following example.

EXAMPLE 3–5

A plane stress element has $\sigma_x = 80$ MPa, $\sigma_y = 0$ MPa, and $\tau_{xy} = 50$ MPa cw, as shown in Figure 3–13*a*.
(*a*) Using Mohr's circle, find the principal stresses and directions, and show these on a stress element correctly aligned with respect to the *xy* coordinates. Draw another stress element to show τ_1 and τ_2, find the corresponding normal stresses, and label the drawing completely.
(*b*) Calculate the principal stresses and the extreme shear stresses using Equations (3–14) and (3–16). Compare to the results of part (*a*).

Solution

(*a*) In the semigraphical approach used here, we first make an approximate freehand sketch of Mohr's circle and then use the geometry of the figure to obtain the desired information.

Draw the σ and τ axes first (Figure 3–13*b*) and from the x face locate $\sigma_x = 80$ MPa along the σ axis. On the x face of the element, we see that the shear stress is 50 MPa in the cw direction. Thus, for the x face, this establishes point A $(80, 50^{\text{cw}})$ MPa. Corresponding to the y face, the stress is $\sigma = 0$ and $\tau = 50$ MPa in the ccw direction. This locates point B $(0, 50^{\text{ccw}})$ MPa. The line AB forms the diameter of the required circle, which can now be drawn. The intersection of the circle with the σ axis defines σ_1 and σ_2 as shown. Now, noting the triangle ACD, indicate on the sketch the length of the legs AD and CD as 50 and 40 MPa, respectively. The length of the hypotenuse AC is

Answer
$$\tau_1 = \sqrt{(50)^2 + (40)^2} = 64.0 \text{ MPa}$$

and this should be labeled on the sketch too. Since intersection C is 40 MPa from the origin, the principal stresses are now found to be

Answer $\sigma_1 = 40 + 64 = 104$ MPa and $\sigma_2 = 40 - 64 = -24$ MPa

The angle 2ϕ from the x axis cw to σ_1 is

Answer $$2\phi_p = \tan^{-1}\frac{50}{40} = 51.3°$$

To draw the principal stress element (Figure 3–13c), sketch the x and y axes parallel to the original axes. The angle ϕ_p on the stress element must be measured in the *same* direction as is the angle $2\phi_p$ on the Mohr circle. Thus, from x measure 25.7° (half of 51.3°) clockwise to locate the σ_1 axis. The σ_2 axis is 90° from the σ_1 axis and the stress element can now be completed and labeled as shown. Note that there are *no* shear stresses on this element.

The two maximum shear stresses occur at points E and F in Figure 3–13b. The two normal stresses corresponding to these shear stresses are each 40 MPa, as indicated. Point E is 38.7° ccw from point A on Mohr's circle. Therefore, in Figure 3–13d, draw a stress element oriented 19.3° (half of 38.7°) ccw from x. The element should then be labeled with magnitudes and directions as shown.

Figure 3–13

All stresses in MPa.

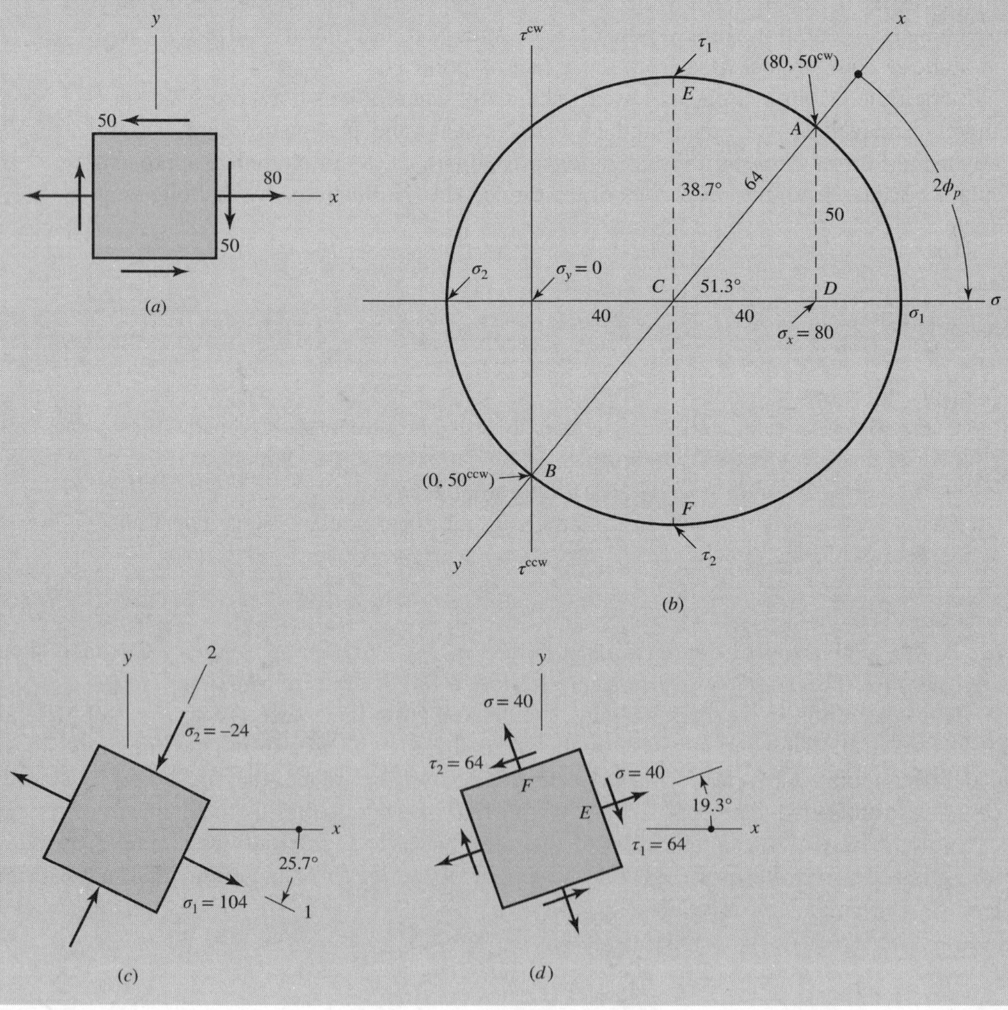

Answer

(*b*) Using Equations (3–14) and (3–16) we can calculate the principal stresses and principal shear stresses directly.

$$\sigma_1, \sigma_2 = \frac{80 + 0}{2} \pm \sqrt{\left(\frac{80 - 0}{2}\right)^2 + (-50)^2}$$

$$\sigma_1 = 104.0 \text{ MPa}$$

$$\sigma_2 = -24.0 \text{ MPa}$$

$$\tau_{12} = \pm\sqrt{\left(\frac{\sigma_x - \sigma_y}{2}\right)^2 + \tau_{xy}^2} = \pm\sqrt{\left(\frac{80 - 0}{2}\right)^2 + (-50)^2} = \pm 64 \text{ MPa}$$

For purposes of machine design, we often only need a rough sketch of the Mohr's circle and the magnitudes of the principal stresses and principal shear stresses. Don't let the details explained in part (*a*) deter you from quickly sketching the circle. It only requires plotting two points *A* and *B* correlating to the normal and shear stresses on the *x* face and the *y* face. Then draw the circle through these two points. Identify the principal stresses where the circle crosses the σ axis, and the principal shear stresses at the top and bottom of the circle. Calculating the magnitudes of these is simple, and comparing to the quick sketch of the circle is confirming. The tedious part is keeping up with sign conventions and orientations. Fortunately, we often don't actually need to do this tedious part.

3–7 General Three-Dimensional Stress

As in the case of plane stress, a particular orientation of a stress element occurs in space for which all shear-stress components are zero. When an element has this particular orientation, the normals to the faces are mutually orthogonal and correspond to the principal directions, and the normal stresses associated with these faces are the principal stresses. Since there are three faces, there are three principal directions and three principal stresses σ_1, σ_2, and σ_3. For plane stress, the stress-free surface contains the third principal stress, which is zero.

In our studies of plane stress we were able to specify any stress state σ_x, σ_y, and τ_{xy} and find the principal stresses and principal directions. But six components of stress are required to specify a general state of stress in three dimensions, and the problem of determining the principal stresses and directions is more difficult. In design, three-dimensional transformations are rarely performed since most maximum stress states occur under plane stress conditions. One notable exception is contact stress, which is not a case of plane stress, where the three principal stresses are given in Section 3–19. In fact, *all* states of stress are truly three-dimensional, where they might be described one- or two-dimensionally with respect to *specific* coordinate axes. Here it is most important to understand the relationship among the *three* principal stresses. The process in finding the three principal stresses from the six stress components σ_x, σ_y, σ_z, τ_{xy}, τ_{yz}, and τ_{zx}, involves finding the roots of the cubic equation[1]

$$\begin{aligned}\sigma^3 - (\sigma_x + \sigma_y + \sigma_z)\sigma^2 + \left(\sigma_x\sigma_y + \sigma_x\sigma_z + \sigma_y\sigma_z - \tau_{xy}^2 - \tau_{yz}^2 - \tau_{zx}^2\right)\sigma \\ - \left(\sigma_x\sigma_y\sigma_z + 2\tau_{xy}\tau_{yz}\tau_{zx} - \sigma_x\tau_{yz}^2 - \sigma_y\tau_{zx}^2 - \sigma_z\tau_{xy}^2\right) = 0\end{aligned} \tag{3–18}$$

[1]For development of this equation and further elaboration of three-dimensional stress transformations see: Richard G. Budynas, *Advanced Strength and Applied Stress Analysis*, 2nd ed., McGraw Hill, New York, 1999, pp. 46–78.

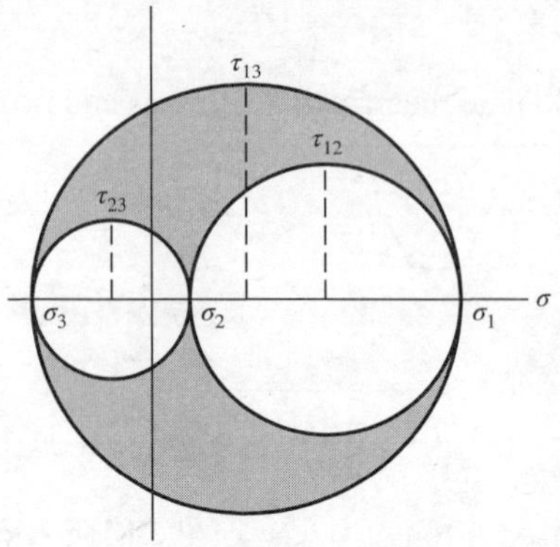

Figure 3–14

Mohr's circles for three-dimensional stress.

In plotting Mohr's circles for three-dimensional stress, the principal normal stresses are typically ordered so that $\sigma_1 \geq \sigma_2 \geq \sigma_3$. Then the result appears as in Figure 3–14.

Note, however, that for plane stress it is customary to label the two in-plane stresses as σ_1 and σ_2 and the principal stress at zero as σ_3.

Principal Shear Stresses and the Maximum Shear Stress

There are always three circles on the Mohr's circle plot, though we often are only concerned with the largest one. The radius of each circle corresponds to a principal shear stress. Each principal shear stress is the maximum shear stress for the in-plane stress case defined by a pair of the principal stresses. The relationships are shown in Figure 3–15. For example, in Figure 3–15*a*, principal stresses σ_1 and σ_2 define a plane. Principal shear stress τ_{12} acts on a surface that is 45° from the σ_1 and σ_2 surfaces, and is the maximum shear stress acting in plane stress defined by σ_1 and σ_2. The three in-plane principal shear stresses are given by

$$\tau_{12} = \frac{\sigma_1 - \sigma_2}{2} \qquad \tau_{23} = \frac{\sigma_2 - \sigma_3}{2} \qquad \tau_{13} = \frac{\sigma_1 - \sigma_3}{2} \tag{3–19}$$

To more clearly show the in-plane relationships for the principal shear stresses, Figure 3–15 does not show the third principal stress or the normal stress acting on the surface of the principal shear stress.

The *absolute maximum shear stress*, often simply referred to as the *maximum shear stress*, is the greatest of the three in-plane principal shear stresses. This is the largest value of shear stress that can be realized within the stress element for any orientation. It turns out to be significant as a means of comparing the overall stress situation between stress elements. See Section 5–4 for use of the maximum shear stress in a failure theory. If the normal principal stresses are ordered ($\sigma_1 > \sigma_2 > \sigma_3$), then $\tau_{max} = \tau_{13}$. If they are not ordered, then it is necessary to determine which principal shear stress is maximum.

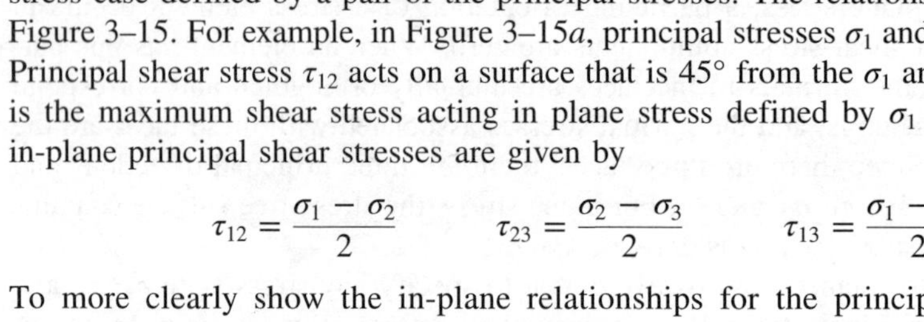

Figure 3–15

Principal shear stresses for the in-plane principal stress cases.
(*a*) Principal shear stress τ_{12} in the $\sigma_1 - \sigma_2$ plane
(*b*) Principal shear stress τ_{23} in the $\sigma_2 - \sigma_3$ plane
(*c*) Principal shear stress τ_{13} in the $\sigma_1 - \sigma_3$ plane

A word of caution is in order regarding the maximum shear stress for a plane stress case. It is easy to complacently assume Equation (3–17) provides the maximum shear stress, as it often does in practical design situations. However, in the plane stress situation where both non-zero principal stresses are positive, or where both are negative, then the principal shear stress defined in the plane of those two principal stresses is not the largest. The following example demonstrates such a case:

EXAMPLE 3–6

A plane stress element has $\sigma_x = 50$ MPa, $\sigma_y = 120$ MPa, and $\tau_{xy} = 40$ MPa cw, as shown in Figure 3–16*a*. Determine the absolute maximum shear stress.

Solution

A quick sketch of the Mohr's circles is helpful to visualize the situation. In Figure 3–16*b*, points *A* and *B* are plotted from the (σ, τ) coordinates of the *x*-face and *y*-face of the stress element. With only a rough sketch approximately to scale, it is apparent that principal stresses σ_1 and σ_2 are both positive. With σ_3 at zero for this plane stress element, we sketch in the other two Mohr's circles between σ_1 and σ_3, and between σ_2 and σ_3. The absolute maximum shear stress will be the radius of the largest circle, corresponding to τ_{13}. Obtaining the largest principal stress from Equation (3–14),

$$\sigma_1 = \frac{\sigma_x + \sigma_y}{2} + \sqrt{\left(\frac{\sigma_x - \sigma_y}{2}\right)^2 + \tau_{xy}^2} = \frac{50 + 120}{2} + \sqrt{\left(\frac{50 - 120}{2}\right)^2 + 40^2} = 138 \text{ MPa}$$

Thus, from Equation (3–19),

Answer
$$\tau_{max} = \tau_{13} = \frac{\sigma_1 - \sigma_3}{2} = \frac{\sigma_1 - 0}{2} = \frac{\sigma_1}{2} = \frac{164.0}{2} = 69 \text{ MPa}$$

For comparison, obtain the shear stress in the *xy* plane from Equation (3–17).

$$\tau_{12} = \sqrt{\left(\frac{\sigma_x - \sigma_y}{2}\right)^2 + \tau_{xy}^2} = \sqrt{\left(\frac{50 - 120}{2}\right)^2 + 40^2} = 53 \text{ MPa}$$

Clearly, the maximum shear stress in the *xy* plane, τ_{12}, is not the absolute maximum shear stress for this case.

It is easy to determine the absolute maximum shear stress using Equation (3–19), but it is necessary to remember to do so. This is where a quick sketch of the Mohr's circles clarifies the overall stress situation, which can avoid mistakes.

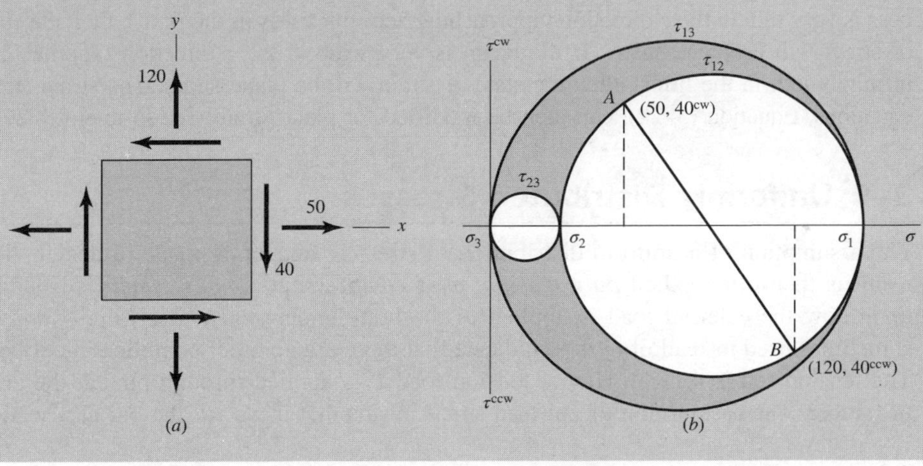

Figure 3–16
(*a*) Plane stress element; (*b*) Mohr's circles for plane stress element in which both non-zero principal stresses are positive.

3–8 Elastic Strain

Normal strain ε is defined and discussed in Section 2–1 for the tensile specimen and is given by Equation (2–2) as $\varepsilon = \delta/l$, where δ is the total elongation of the bar within the length l. Hooke's law for the tensile specimen is given by Equation (2–3) as

$$\sigma = E\varepsilon \tag{3–20}$$

where the constant E is called *Young's modulus* or the *modulus of elasticity.*

When a material is placed in tension, there exists not only an axial strain, but also negative strain (contraction) perpendicular to the axial strain. Assuming a linear, homogeneous, isotropic material, this lateral strain is proportional to the axial strain. If the axial direction is x, then the lateral strains are $\varepsilon_y = \varepsilon_z = -\nu\varepsilon_x$. The constant of proportionality ν is called *Poisson's ratio,* which is about 0.3 for most structural metals. See Table A–5 for values of ν for common materials.

If the axial stress is in the x direction, then from Equation (3–20)

$$\varepsilon_x = \frac{\sigma_x}{E} \qquad \varepsilon_y = \varepsilon_z = -\nu\frac{\sigma_x}{E} \tag{3–21}$$

For a stress element undergoing σ_x, σ_y, and σ_z simultaneously, the normal strains are given by

$$\begin{aligned}\varepsilon_x &= \frac{1}{E}[\sigma_x - \nu(\sigma_y + \sigma_z)]\\ \varepsilon_y &= \frac{1}{E}[\sigma_y - \nu(\sigma_x + \sigma_z)]\\ \varepsilon_z &= \frac{1}{E}[\sigma_z - \nu(\sigma_x + \sigma_y)]\end{aligned} \tag{3–22}$$

Shear strain γ is the change in a right angle of a stress element when subjected to pure shear stress, and Hooke's law for shear is given by

$$\tau = G\gamma \tag{3–23}$$

where the constant G is the *shear modulus of elasticity* or *modulus of rigidity.*

It can be shown for a linear, isotropic, homogeneous material, the three elastic constants are related to each other by

$$E = 2G(1 + \nu) \tag{3–24}$$

A state of strain, called *plane strain,* occurs when all strains in a given direction are zero. For example, if all the strains in the z direction were zero, then $\varepsilon_z = \gamma_{yz} = \gamma_{zx} = 0$. As an example of plane strain, consider the beam shown in Figure 3–20. If the beam thickness is very thin in the z direction (approaching zero thickness in the limit), then the state of stress will be plane stress. If the beam is very wide in the z direction (approaching infinite width in the limit), then the state of strain will be plane strain. The strain-stress equations, Equation (3–22), can then be modified for plane strain (see Problem 3–29).

3–9 Uniformly Distributed Stresses

The assumption of a uniform distribution of stress is frequently made in design. The result is then often called *pure tension, pure compression,* or *pure shear,* depending upon how the external load is applied to the body under study. The word *simple* is sometimes used instead of *pure* to indicate that there are no other complicating effects. The tension rod is typical. Here a tension load F is applied through pins at the ends of the bar. The assumption of uniform stress means that if we cut the bar at a section remote from the ends and remove one piece, we can replace its effect by applying a uniformly distributed force of magnitude σA to the cut end, as depicted in Figure 3–7*a*. So the stress σ is said to be uniformly distributed. It is calculated from the equation

$$\sigma = \frac{F}{A} \tag{3–25}$$

This assumption of uniform stress distribution requires that:

- The bar be straight and of a homogeneous material
- The line of action of the force contains the centroid of the section
- The section be taken remote from the ends and from any discontinuity or abrupt change in cross section

For simple compression, Equation (3–25) is applicable with F normally being considered a negative quantity. Also, a slender bar in compression may fail by buckling. See Section 4–11 for consideration of this failure mode.

Another type of loading that assumes a uniformly distributed stress is known as *direct shear*. This occurs when there is a shearing action with no bending. An example is the action on a piece of sheet metal caused by the two blades of tin snips. Think of a cantilever beam with a force pushing down on it. Now move the force all the way up to the wall so there is no bending moment, just a force trying to shear the beam off the wall. This is direct shear. Direct shear is usually assumed to be uniform across the cross section, and is given by

$$\tau = \frac{V}{A} \tag{3–26}$$

where V is the shear force and A is the area of the cross section that is being sheared.

Bolts and pins that are loaded in shear often are modeled as direct shear with uniform shear stress over the cross section. In reality, the direct shear stress is an average stress for the cross section, and the stress distribution is complex. In the immediate vicinity of the applied force the stress is higher than the average. Also, due to elasticity of the members and clearance fits between mating members, some bending will be present, which affects the distribution of the shear stress. For these reasons, direct shear should be understood to be a simplified model. This model is mostly useful when failure is from ultimate failure of the entire cross section, such as shearing a pin, rather than by yielding at a critical stress element within the cross section. Localized yielding causes localized deflection which tends to distribute the localized higher stress into neighboring stress elements. Consequently, for ultimate failure of the entire cross section, the direct shear model turns out to be acceptable. See Section 8–12 for further analysis of shear-loaded bolts and pins. See Section 3–11 for a more detailed model of the shear stress distribution over a cross section, particularly in the presence of bending.

3–10 Normal Stresses for Beams in Bending

The equations for the normal bending stresses in straight beams are based on the following assumptions.

- The beam is subjected to pure bending. This means that the shear force is zero, and that no torsion or axial loads are present. (However, for most engineering applications it is assumed that these loads affect the bending stresses minimally.)
- The material is isotropic and homogeneous.

- The material obeys Hooke's law.
- The beam is initially straight with a cross section that is constant throughout the beam length.
- The cross section of the beam has an axis of symmetry in the plane of bending.
- The proportions of the beam are such that it would fail by bending rather than by crushing, wrinkling, or sidewise buckling.
- Plane cross sections of the beam remain plane during bending.

In Figure 3–17 we visualize a portion of a straight beam acted upon by a positive bending moment M shown by the curved arrow showing the physical action of the moment together with a straight, double-headed, arrow indicating the moment vector. The x axis is coincident with the *neutral axis* of the section, and the xz plane, which contains the neutral axes of all cross sections, is called the *neutral plane*. Elements of the beam coincident with this plane have zero bending stress. The location of the neutral axis with respect to the cross section is coincident with the *centroidal axis* of the cross section.

The bending stress varies linearly with the perpendicular distance, y, from the neutral plane and is given by

$$\sigma_x = -\frac{My}{I} \tag{3–27}$$

where I is the *second-area moment* about the z axis. That is,

$$I = \int y^2 dA \tag{3–28}$$

Equations for I for standard cross sections are given in Table A–16. The stress distribution given by Equation (3–27) is shown in Figure 3–18, as well as depicted in Figure 3–7*b*. The maximum magnitude of the bending stress will occur where y has the greatest magnitude. Designating $\sigma_{\max}$ as the maximum *magnitude* of the bending stress, and c as the maximum *magnitude* of y

$$\sigma_{\max} = \frac{Mc}{I} \tag{3–29a}$$

Equation (3–27) can still be used to ascertain whether $\sigma_{\max}$ is tensile or compressive.

Equation (3–29*a*) is often written as

$$\sigma_{\max} = \frac{M}{Z} \tag{3–29b}$$

where $Z = I/c$ is called the *section modulus*.

Figure 3–17

Straight beam in positive bending.

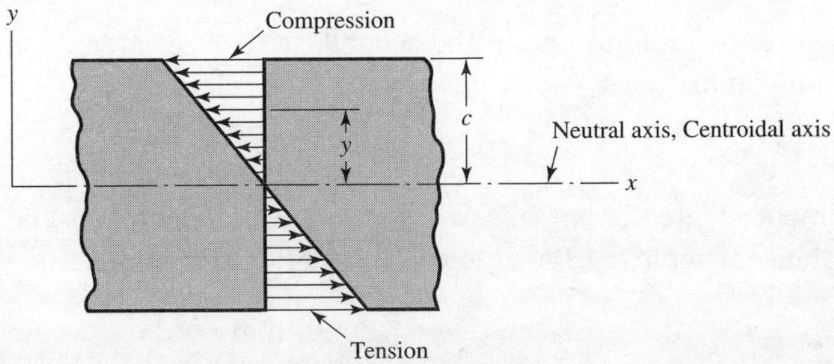

Figure 3–18
Bending stresses according to Equation (3–27).

EXAMPLE 3–7

A beam having a T section with the dimensions shown in Figure 3–19 is subjected to a bending moment of 1600 N · m, about the negative z axis, that causes tension at the top surface. Locate the neutral axis and find the maximum tensile and compressive bending stresses.

Solution

Dividing the T section into two rectangles, numbered 1 and 2, the total area is $A = 12(75) + 12(88) = 1956\ \text{mm}^2$. Summing the area moments of these rectangles about the top edge, where the moment arms of areas 1 and 2 are 6 mm and $(12 + 88/2) = 56$ mm respectively, we have

$$1956c_1 = 12(75)(6) + 12(88)(56)$$

and hence $c_1 = 32.99$ mm. Therefore $c_2 = 100 - 32.99 = 67.01$ mm.

Next we calculate the second moment of area of each rectangle about its own centroidal axis. Using Table A–16, we find for the top rectangle

$$I_1 = \frac{1}{12}bh^3 = \frac{1}{12}(75)12^3 = 1.080 \times 10^4\ \text{mm}^4$$

For the bottom rectangle, we have

$$I_2 = \frac{1}{12}(12)88^3 = 6.815 \times 10^5\ \text{mm}^4$$

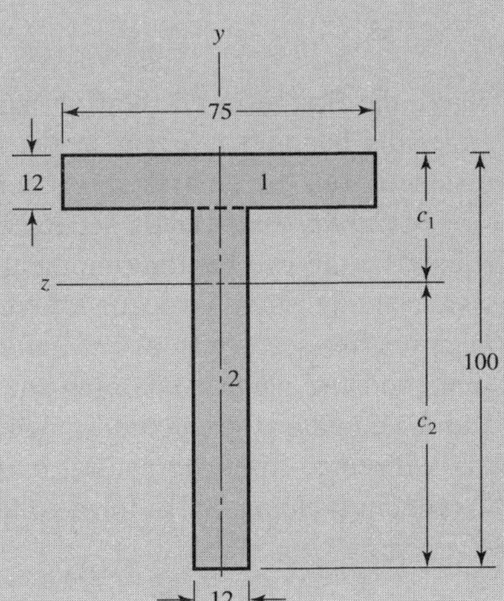

Figure 3–19
Dimensions in millimeters.

We now employ the *parallel-axis theorem* to obtain the second moment of area of the composite figure about its own centroidal axis. This theorem states

$$I_z = I_{ca} + Ad^2$$

where I_{ca} is the second moment of area about its own centroidal axis and I_z is the second moment of area about any parallel axis a distance d removed. For the top rectangle, the distance is

$$d_1 = 32.99 - 6 = 26.99 \text{ mm}$$

and for the bottom rectangle,

$$d_2 = 67.01 - \frac{88}{2} = 23.01 \text{ mm}$$

Using the parallel-axis theorem for both rectangles, we now find that

$$\begin{aligned} I &= [1.080 \times 10^4 + 12(75)26.99^2] + [6.815 \times 10^5 + 12(88)23.01^2] \\ &= 1.907 \times 10^6 \text{ mm}^4 \end{aligned}$$

Finally, the maximum tensile stress, which occurs at the top surface, is found to be

Answer
$$\sigma = \frac{Mc_1}{I} = \frac{1600(32.99)10^{-3}}{1.907(10^{-6})} = 27.68(10^6) \text{ Pa} = 27.68 \text{ MPa}$$

Similarly, the maximum compressive stress at the lower surface is found to be

Answer
$$\sigma = -\frac{Mc_2}{I} = -\frac{1600(67.01)10^{-3}}{1.907(10^{-6})} = -56.22(10^6) \text{ Pa} = -56.22 \text{ MPa}$$

Two-Plane Bending

Quite often, in mechanical design, bending occurs in both xy and xz planes. Considering cross sections with one or two planes of symmetry only, the bending stresses are given by

$$\sigma_x = -\frac{M_z y}{I_z} + \frac{M_y z}{I_y} \tag{3–30}$$

where the first term on the right side of the equation is identical to Equation (3–27), M_y is the bending moment in the xz plane (moment vector in y direction), z is the distance from the neutral y axis, and I_y is the second area moment about the y axis.

For *noncircular* cross sections, Equation (3–30) is the superposition of stresses caused by the two bending moment components. The maximum tensile and compressive bending stresses occur where the summation gives the greatest positive and negative stresses, respectively. For solid *circular* cross sections, all lateral axes are the same and the plane containing the moment corresponding to the vector sum of M_z and M_y contains the maximum bending stresses. For a beam of diameter d the maximum distance from the neutral axis is $d/2$, and from Table A–16, $I = \pi d^4/64$. The maximum bending stress for a solid circular cross section is then

$$\sigma_m = \frac{Mc}{I} = \frac{(M_y^2 + M_z^2)^{1/2}(d/2)}{\pi d^4/64} = \frac{32}{\pi d^3}(M_y^2 + M_z^2)^{1/2} \tag{3–31}$$

EXAMPLE 3–8

As shown in Figure 3–20*a*, beam *OC* is loaded in the *xy* plane by a uniform load of 9 kN/m, and in the *xz* plane by a concentrated force of 0.4 kN at end *C*. The beam is 0.2 m long.
(*a*) For the cross section shown determine the maximum tensile and compressive bending stresses and where they act.
(*b*) If the cross section was a solid circular rod of diameter, $d = 30$ mm, determine the magnitude of the maximum bending stress.

Solution

(*a*) The reactions at *O* and the bending-moment diagrams in the *xy* and *xz* planes are shown in Figures 3–20*b* and *c*, respectively. The maximum moments in both planes occur at *O* where

$$(M_z)_O = -\frac{1}{2}(9)0.2^2 = -180 \text{ N}\cdot\text{m} \qquad (M_y)_O = 0.4(0.2) = 80 \text{ N}\cdot\text{m}$$

The second moments of area in both planes are

$$I_z = \frac{1}{12}(0.02)0.04^3 = 106.7(10)^{-9} \text{ m}^4 \qquad I_y = \frac{1}{12}(0.04)0.02^3 = 26.7(10)^{-9} \text{ m}^4$$

The maximum tensile stress occurs at point *A*, shown in Figure 3–20*a*, where the maximum tensile stress is due to both moments. At *A*, $y_A = 0.02$ m and $z_A = 0.01$ m. Thus, from Equation (3–30)

Answer $$(\sigma_x)_A = -\frac{(-180 \times 0.02)}{(106.7)(10)^{-9}} + \frac{(80 \times 0.01)}{(26.7)(10)^{-9}} = 63.7(10)^6 \text{ Pa} = 63.7 \text{ MPa}$$

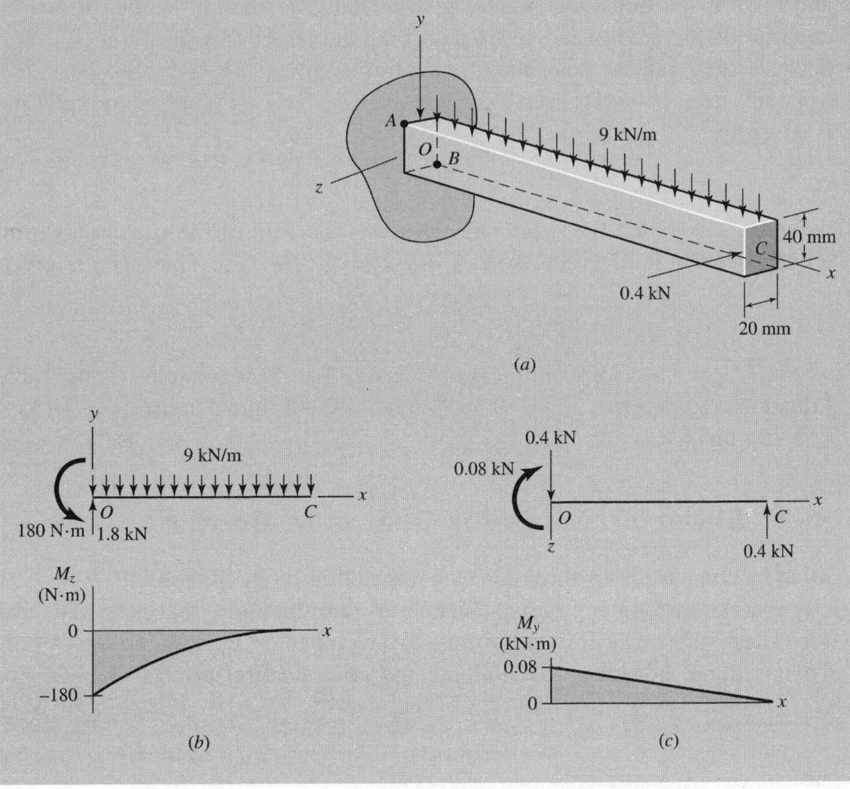

Figure 3–20

(*a*) Beam loaded in two planes; (*b*) loading and bending-moment diagrams in *xy* plane; (*c*) loading and bending-moment diagrams in *xz* plane.

The maximum compressive bending stress occurs at point B, where $y_B = -0.02$ m and $z_B = -0.01$ m. Thus

Answer
$$(\sigma_x)_B = -\frac{(-180)(0.02)}{(106.7)(10)^{-9}} + \frac{(80)(0.01)}{(26.7)(10)^{-9}} = -63.7(10)^6 \text{ Pa} = -63.7 \text{ MPa}$$

(*b*) For a solid circular cross section of diameter, $d = 30$ mm, the maximum bending stress at end O is given by Equation (3–31) as

Answer
$$\sigma_m = \frac{32}{\pi(0.3)^3}[80^2 + (-180)^2]^{1/2} = 74.3(10)^6 \text{ Pa} = 74.3 \text{ MPa}$$

Beams with Asymmetrical Sections[2]

The bending stress equations, given by Equations (3–27) and (3–30), can also be applied to beams having asymmetrical cross sections, provided the planes of bending coincide with the *area principal axes* of the section. The method for determining the orientation of the area principal axes and the values of the corresponding *principal second-area moments* can be found in any statics book. If a section has an axis of symmetry, that axis and its perpendicular axis are the area principal axes.

For example, consider a beam in bending, using an equal leg angle as shown in Table A–6. Equation (3–30) cannot be used if the bending moments are resolved about axis 1–1 and/or axis 2–2. However, Equation (3–30) can be used if the moments are resolved about axis 3–3 and its perpendicular axis (let us call it, say, axis 4–4). Note, for this cross section, axis 4–4 is an axis of symmetry. Table A–6 is a standard table, and for brevity, does not directly give all the information needed to use it. The orientation of the area principal axes and the values of I_{2-2}, I_{3-3}, and I_{4-4} are not given because they can be determined as follows. Since the legs are equal, the principal axes are oriented ±45° from axis 1–1, and $I_{2-2} = I_{1-1}$. The second-area moment I_{3-3} is given by

$$I_{3-3} = A(k_{3-3})^2 \tag{a}$$

where k_{3-3} is called the *radius of gyration*. The sum of the second-area moments for a cross section is invariant, so $I_{1-1} + I_{2-2} = I_{3-3} + I_{4-4}$. Thus, I_{4-4} is given by

$$I_{4-4} = 2I_{1-1} - I_{3-3} \tag{b}$$

where $I_{2-2} = I_{1-1}$. For example, consider a 80 × 80 × 6 angle. Using Table A–6 and Equations (*a*) and (*b*), $I_{3-3} = 9.35\ (1.57)^2 = 23.05$ mm^4, and $I_{4-4} = 2\ (55.8) - 23.05 = 88.5$ mm^4.

3–11 Shear Stresses for Beams in Bending

Most beams have both shear forces and bending moments present. It is only occasionally that we encounter beams subjected to pure bending, that is to say, beams having zero shear force. The flexure formula is developed on the assumption of pure bending. This is done, however, to eliminate the complicating effects of shear force in the

[2]For further discussion, see Section 5.3, Richard G. Budynas, *Advanced Strength and Applied Stress Analysis*, 2nd ed., McGraw Hill, New York, 1999.

development of the formula. For engineering purposes, the flexure formula is a reasonable model no matter whether a shear force is present or not. For this reason, we shall utilize the same normal bending-stress distribution [Equations (3–27) and (3–29)] when shear forces are also present.

In Figure 3–21*a* we show a beam segment of constant cross section subjected to a shear force V and a bending moment M at x. Because of external loading and V, the shear force and bending moment change with respect to x. At $x + dx$ the shear force and bending moment are $V + dV$ and $M + dM$, respectively. Considering forces in the x direction only, Figure 3–21*b* shows the stress distribution σ_x due to the bending moments. If dM is positive, with the bending moment increasing, the stresses on the right face, for a given value of y, are larger in magnitude than the stresses on the left face. If we further isolate the element by making a slice at $y = y_1$ (see Figure 3–21*b*), the net *force* in the x direction will be directed to the left with a value of

$$\int_{y_1}^{c} \frac{(dM)y}{I} dA$$

as shown in the rotated view of Figure 3–21*c*. For equilibrium, a shear force on the bottom face, directed to the right, is required. This shear force gives rise to a shear stress τ, where, if assumed uniform, the force is $\tau b\, dx$. Thus

$$\tau b\, dx = \int_{y_1}^{c} \frac{(dM)y}{I} dA \tag{a}$$

The term dM/I can be removed from within the integral and $b\, dx$ placed on the right side of the equation; then, from Equation (3–4) with $V = dM/dx$ Equation (*a*) becomes

$$\tau = \frac{V}{Ib} \int_{y_1}^{c} y\, dA \tag{3–32}$$

In this equation, the integral is the first moment of the area A' with respect to the neutral axis (see Figure 3–21*c*). This integral is usually designated as Q. Thus

$$Q = \int_{y_1}^{c} y\, dA = \bar{y}'A' \tag{3–33}$$

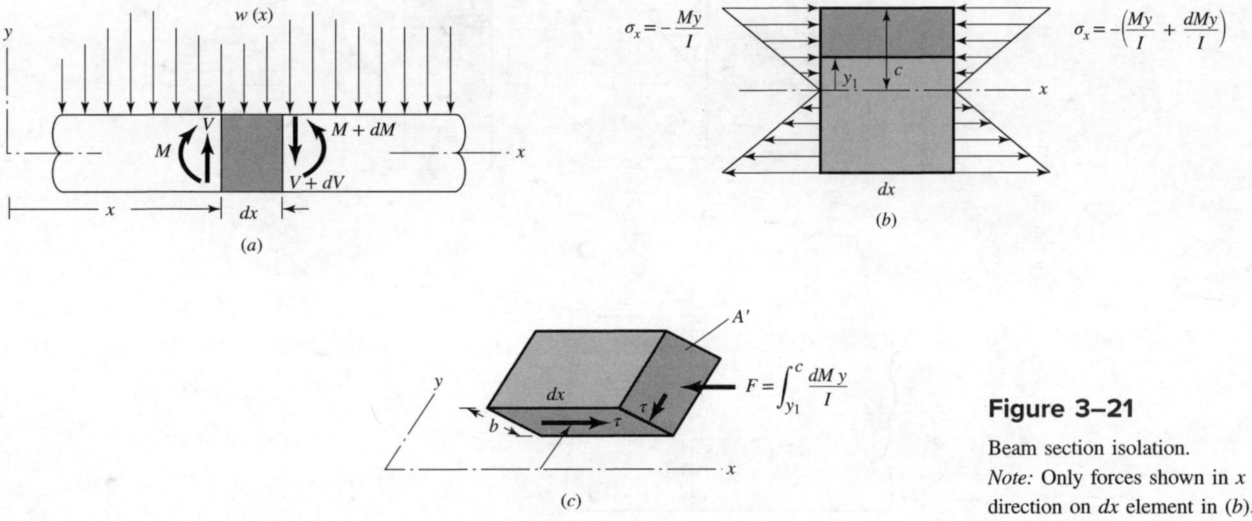

Figure 3–21

Beam section isolation.

Note: Only forces shown in x direction on dx element in (*b*).

where, for the isolated area y_1 to c, $\bar{y}'$ is the distance in the y direction from the neutral plane to the centroid of the area A'. With this, Equation (3–32) can be written as

$$\tau = \frac{VQ}{Ib} \tag{3–34}$$

This stress is known as the *transverse shear stress*. It is always accompanied with bending stress.

In using this equation, note that b is the width of the section at $y = y_1$. Also, I is the second moment of area of the entire section about the neutral axis.

Because cross shears are equal, and area A' is *finite,* the shear stress τ given by Equation (3–34) and shown on area A' in Figure 3–21c occurs only at $y = y_1$. The shear stress on the lateral area varies with y, normally maximum at $y = 0$ (where $\bar{y}'A'$ is maximum) and zero at the outer fibers of the beam where $A' = 0$.

The shear stress distribution in a beam depends on how Q/b varies as a function of y_1. Here we will show how to determine the shear stress distribution for a beam with a rectangular cross section and provide results of maximum values of shear stress for other standard cross sections. Figure 3–22 shows a portion of a beam with a rectangular cross section, subjected to a shear force V and a bending moment M. As a result of the bending moment, a normal stress σ is developed on a cross section such as A–A, which is in compression above the neutral axis and in tension below. To investigate the shear stress at a distance y_1 above the neutral axis, we select an element of area dA at a distance y above the neutral axis. Then, $dA = b\ dy$, and so Equation (3–33) becomes

$$Q = \int_{y_1}^{c} y\, dA = b \int_{y_1}^{c} y\, dy = \left.\frac{by^2}{2}\right|_{y_1}^{c} = \frac{b}{2}(c^2 - y_1^2) \tag{b}$$

Substituting this value for Q into Equation (3–34) gives

$$\tau = \frac{V}{2I}(c^2 - y_1^2) \tag{c}$$

Figure 3–22

Transverse shear stresses in a rectangular beam.

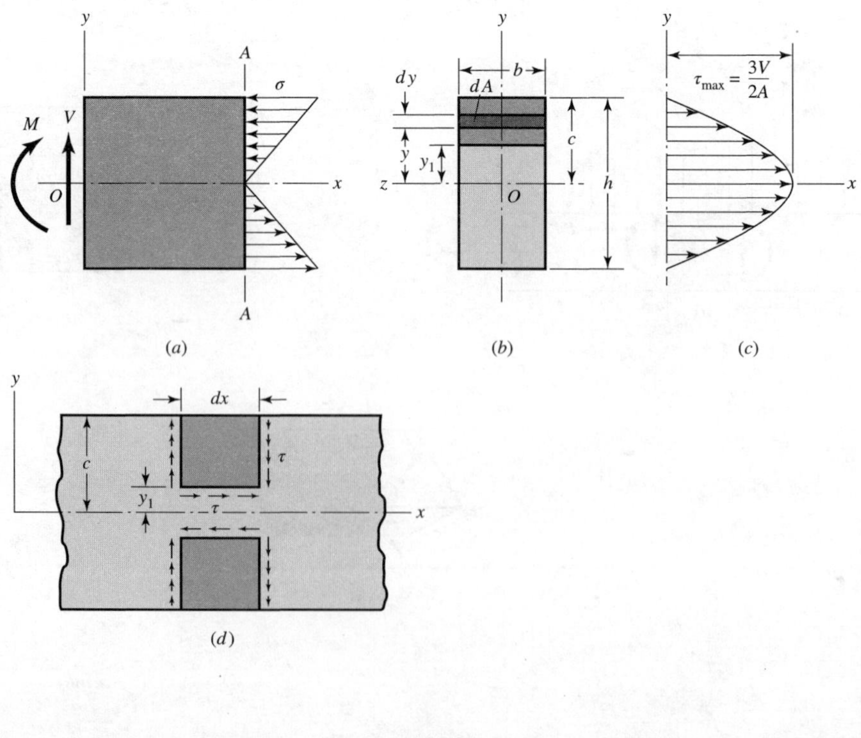

This is the general equation for shear stress in a rectangular beam. To learn something about it, let us make some substitutions. From Table A–16, the second moment of area for a rectangular section is $I = bh^3/12$; substituting $h = 2c$ and $A = bh = 2bc$ gives

$$I = \frac{Ac^2}{3} \tag{d}$$

If we now use this value of I for Equation (c) and rearrange, we get

$$\tau = \frac{3V}{2A}\left(1 - \frac{y_1^2}{c^2}\right) \tag{e}$$

We note that the maximum shear stress exists when $y_1 = 0$, which is at the bending neutral axis. Thus

$$\tau_{\max} = \frac{3V}{2A} \tag{f}$$

for a rectangular section. As we move away from the neutral axis, the shear stress decreases parabolically until it is zero at the outer surfaces where $y_1 = \pm c$, as shown in Figure 3–22c. Horizontal shear stress is always accompanied by vertical shear stress of the same magnitude, and so the distribution can be diagrammed as shown in Figure 3–22d. Figure 3–22c shows that the shear τ on the vertical surfaces varies with y. We are almost always interested in the horizontal shear, τ in Figure 3–22d, which is nearly uniform over dx with constant $y = y_1$. The maximum horizontal shear occurs where the vertical shear is largest. This is usually at the neutral axis but may not be if the width b is smaller somewhere else. Furthermore, if the section is such that b can be minimized on a plane not horizontal, then the horizontal shear stress occurs on an inclined plane. For example, with tubing, the horizontal shear stress occurs on a radial plane and the corresponding "vertical shear" is not vertical, but tangential.

The distributions of transverse shear stresses for several commonly used cross sections are shown in Table 3–1. The profiles represent the VQ/Ib relationship, which is a function of the distance y from the neutral axis. For each profile, the formula for the maximum value at the neutral axis is given. Note that the expression given for the I beam is a commonly used approximation that is reasonable for a standard I beam with a thin web. Also, the profile for the I beam is idealized. In reality the transition from the web to the flange is quite complex locally, and not simply a step change.

It is significant to observe that the transverse shear stress in each of these common cross sections is maximum on the neutral axis, and zero on the outer surfaces. Since this is exactly the opposite of where the bending and torsional stresses have their maximum and minimum values, the transverse shear stress is often not critical from a design perspective.

Table 3–1 **Formulas for Maximum Transverse Shear Stress from VQ/Ib**

Beam Shape	Formula	Beam Shape	Formula
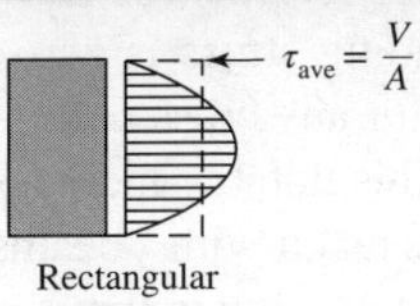 Rectangular	$\tau_{\max} = \frac{3V}{2A}$	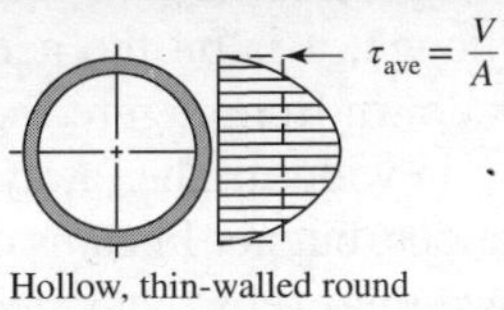 Hollow, thin-walled round	$\tau_{\max} = \frac{2V}{A}$
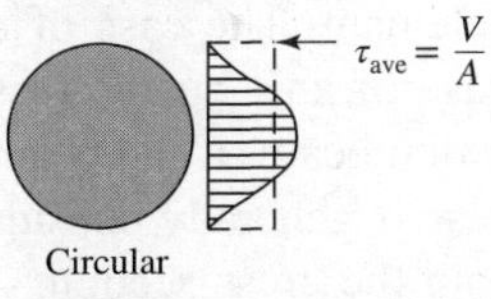 Circular	$\tau_{\max} = \frac{4V}{3A}$	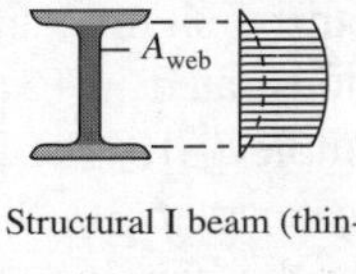 Structural I beam (thin-walled)	$\tau_{\max} \approx \frac{V}{A_{web}}$

Let us examine the significance of the transverse shear stress, using as an example a cantilever beam of length L, with rectangular cross section $b \times h$, loaded at the free end with a transverse force F. At the wall, where the bending moment is the largest, at a distance y from the neutral axis, a stress element will include both bending stress and transverse shear stress. In Section 5–4 it will be shown that a good measure of the combined effects of multiple stresses on a stress element is the maximum shear stress. The bending stress is given by $\sigma = My/I$, where, for this case, $M = FL$, $I = bh^3/12$, and $h = 2c$. This gives

$$\sigma = \frac{My}{I} = \frac{12FLy}{bh^3} = \frac{3F}{2bh} \cdot 8\left(\frac{L}{h}\right)\left(\frac{y}{2c}\right) = \frac{3F}{2bh} \cdot 4\left(\frac{L}{h}\right)\left(\frac{y}{c}\right) \tag{g}$$

The shear stress given by Equation (e), with $V = F$ and $A = bh$, is

$$\tau = \frac{3F}{2bh}\left[1 - \left(\frac{y}{c}\right)^2\right] \tag{h}$$

Substituting Equations (g) and (h) into Equation (3–17) we obtain a general equation for the maximum shear stress in a cantilever beam with a rectangular cross section. The result is

$$\tau_{max} = \sqrt{\left(\frac{\sigma}{2}\right)^2 + \tau^2} = \frac{3F}{2bh}\sqrt{4(L/h)^2(y/c)^2 + [1 - (y/c)^2]^2} \tag{i}$$

Defining a normalized maximum shear stress, $\boldsymbol{\tau}_{max}$ as τ_{max} divided by the maximum shear stress due to bending, $(\sigma/2)|_{y=c} = 6FLc/(bh^3) = 3FL/(bh^2)$, Equation ($i$) can be rewritten as

$$\boldsymbol{\tau}_{max} = \frac{1}{2(L/h)}\sqrt{4(L/h)^2(y/c)^2 + [1 - (y/c)^2]^2} \tag{j}$$

To investigate the significance of transverse shear stress, we plot $\boldsymbol{\tau}_{max}$ as a function of L/h for several values of y/c, as shown in Figure 3–23. Since F and b appear only as linear multipliers outside the radical, they will only serve to scale the plot in the vertical direction without changing any of the relationships. Notice that at the neutral axis where $y/c = 0$, $\boldsymbol{\tau}_{max}$ is constant for any length beam, since the bending stress is zero at the neutral axis and the transverse shear stress is independent of L. On the other hand, on the outer surface where $y/c = 1$, $\boldsymbol{\tau}_{max}$ increases linearly with L/h because of the bending moment. For y/c between zero and one, $\boldsymbol{\tau}_{max}$ is nonlinear for low values of L/h, but behaves linearly as L/h increases, displaying the dominance of the bending stress as the moment arm increases. We can see from the graph that the critical stress element (the largest value of $\boldsymbol{\tau}_{max}$) will always be either on the outer surface ($y/c = 1$) or at the neutral axis ($y/c = 0$), and never between. Thus, for the rectangular cross section, the transition between these two locations occurs at $L/h = 0.5$ where the line for $y/c = 1$ crosses the horizontal line for $y/c = 0$. The critical stress element is either on the outer surface where the transverse shear is zero, or if L/h is small enough, it is on the neutral axis where the bending stress is zero.

The conclusions drawn from Figure 3–23 are generally similar for any cross section that does not increase in width farther away from the neutral axis. This notably includes solid round cross sections, but not I beams or channels. Care must be taken with I beams and channels that have thin webs that extend far enough from the neutral axis that the bending and shear may both be significant on the same stress element. The case of a round cross section is demonstrated in Example 3–9, and the case of an I beam cross section is examined in Example 3–10. For any common cross section beam, if the beam length to height ratio is greater than 10, the transverse shear stress is generally considered negligible compared to the bending stress at any point within the cross section.

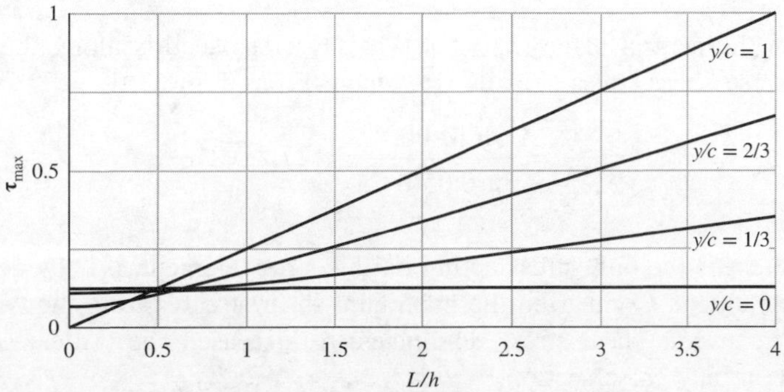

Figure 3–23

Plot of dimensionless maximum shear stress, τ_{max}, for a rectangular cantilever beam, combining the effects of bending and transverse shear stresses.

EXAMPLE 3–9

A short cantilevered round peg is 50 mm in diameter and extends 10 mm from the wall. It is loaded with a force $F = 40$ kN applied at the tip in the negative y direction as shown in Figure 3–24. Determine the critical stress element which has the largest absolute principal shear stress.

Solution

The ratio of length to height of the peg is 1/5, which is definitely in the range for which transverse shear stress could be significant. The round cross section qualifies for the same conclusions as for the rectangular cross section in Figure 3–23. Specifically, the critical stress element will either be at the top (or bottom) where the bending stress is maximum, or at the neutral plane where the transverse shear is maximum.

At the top, at the wall where the bending moment is maximum, the stress element is in tension due to the bending stress,

$$\sigma_x = \frac{Mc}{I} = \frac{40\ 000(0.010)(0.050/2)}{\pi(0.050)^4/64} = 32.6 \text{ MPa}$$

The transverse shear stress is zero at this stress element at the top, because it is the maximum distance from the neutral plane, where Q is zero. The absolute principal shear stress, or maximum shear stress, for the stress element at the top can be obtained from Equation (3–19),

$$\tau_{\max} = \sqrt{\left(\frac{\sigma_x - \sigma_y}{2}\right)^2 + \tau_{xy}^2} = \sqrt{\left(\frac{32.6 - 0}{2}\right)^2 + 0^2} = 16.3 \text{ MPa}$$

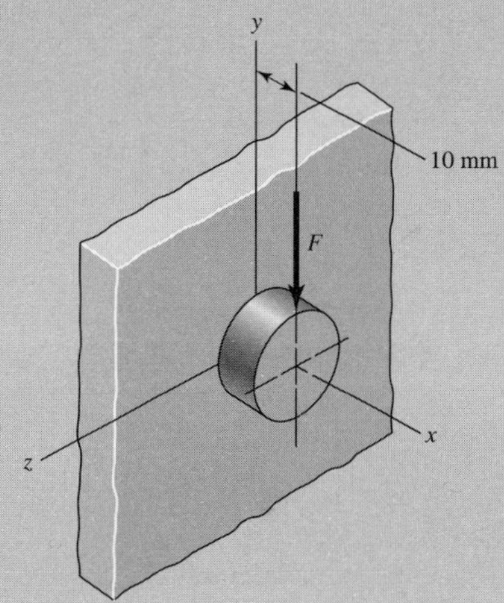

Figure 3–24

For any stress element on the neutral plane (xz plane), at any cross section along the x axis, the bending stress is zero, and the transverse shear stress is at its maximum. From Table 3–1,

$$\tau_{\max} = \frac{4V}{3A} = \frac{4(40\ 000)}{3\pi(0.050)^2/4} = 27.2 \text{ MPa}$$

Since this transverse shear stress is the only stress acting on this stress element, it is by definition the absolute principal shear stress for the element. Comparing the maximum shear stresses from the two potentially critical locations, we find that the transverse shear stress dominates in this case. The critical stress element is any element on the neutral plane, at any cross section.

EXAMPLE 3–10

A simply supported beam, 0.3 m long, is to support a load of 2 kN acting 80 mm from the left support, as shown in Figure 3–25a. The beam is an I beam with the cross-sectional dimensions shown. To simplify the calculations, assume a cross section with square corners, as shown in Figure 3–25c. Points of interest are labeled (a, b, c, and d) at distances y from the neutral axis of 0 mm, 32^- mm, 32^+ mm, and 40 mm (Figure 3–25c). At the critical axial location along the beam, find the following information.

Figure 3–25

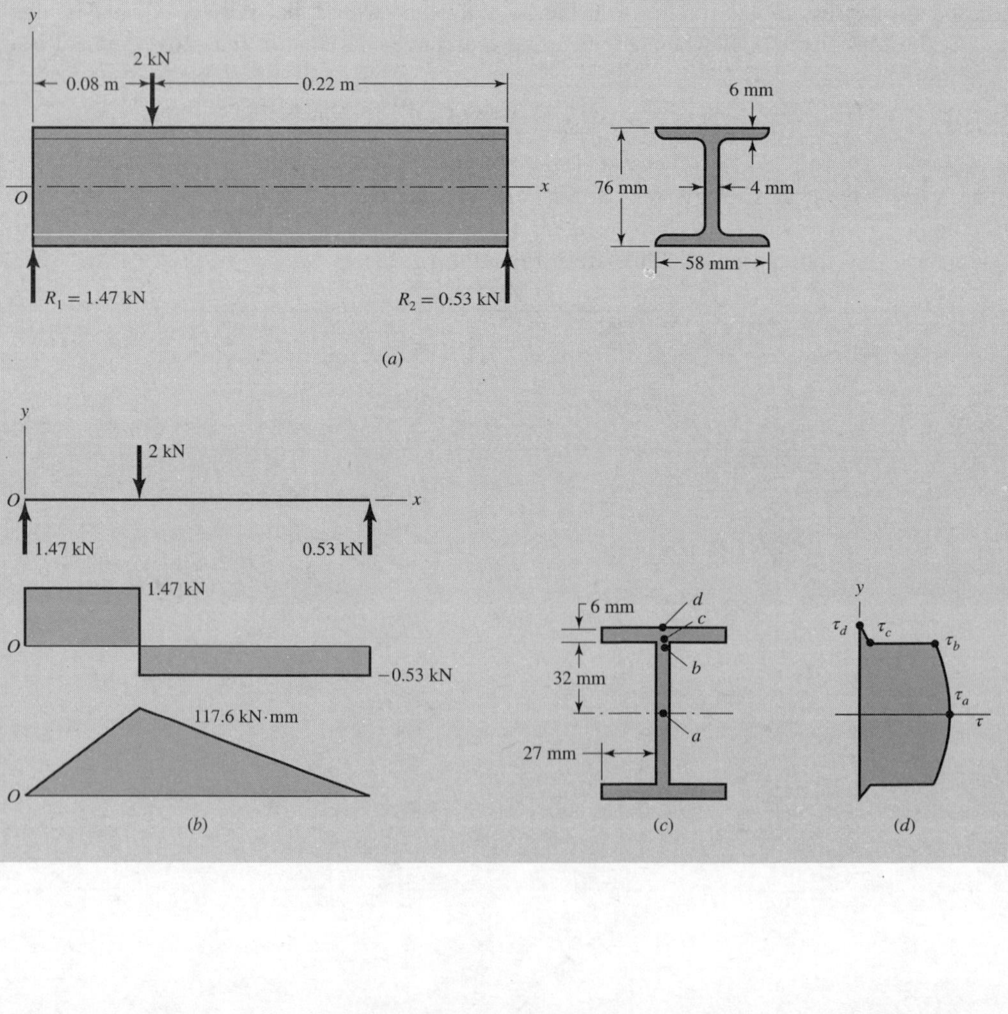

(*a*) Determine the profile of the distribution of the transverse shear stress, obtaining values at each of the points of interest.
(*b*) Determine the bending stresses at the points of interest.
(*c*) Determine the maximum shear stresses at the points of interest, and compare them.

Solution

First, we note that the transverse shear stress is not likely to be negligible in this case since the beam length to height ratio is much less than 10, and since the thin web and wide flange will allow the transverse shear to be large. The loading, shear-force, and bending-moment diagrams are shown in Figure 3–25*b*. The critical axial location is at $x = 80^-$ where the shear force and the bending moment are both maximum.
(*a*) We obtain the area moment of inertia I by evaluating I for a solid 80 mm × 58 mm rectangular area, and then subtracting the two rectangular areas that are not part of the cross section.

$$I = \frac{(58)(80)^3}{12} - 2\left[\frac{(27)(64)^3}{12}\right] = 1\ 295\ 019\ \text{mm}^4$$

Finding Q at each point of interest using Equation (3–33) gives

$$Q_a = \left(\sum \bar{y}'A'\right)\Big|_{y=0}^{y=40} = \left(32 + \frac{6}{2}\right)[(58)(6)] + \left(\frac{32}{2}\right)[(32)(4)] = 14\ 228\ \text{mm}^3$$

$$Q_b = Q_c = \left(\sum \bar{y}'A'\right)\Big|_{y=32}^{y=40} = \left(32 + \frac{6}{2}\right)[(58)(6)] = 12\ 180\ \text{mm}^3$$

$$Q_d = \left(\sum \bar{y}'A'\right)\Big|_{y=40}^{y=40} = (40)(0) = 0\ \text{mm}^3$$

Applying Equation (3–34) at each point of interest, with V and I constant for each point, and b equal to the width of the cross section at each point, shows that the magnitudes of the transverse shear stresses are

Answer

$$\tau_a = \frac{VQ_a}{Ib_a} = \frac{(1.47)(14\ 228)}{(1\ 295\ 019)(4)} = 4\ \text{MPa}$$

$$\tau_b = \frac{VQ_b}{Ib_b} = \frac{(1.47)(12\ 180)}{(1\ 295\ 019)(4)} = 3.5\ \text{MPa}$$

$$\tau_c = \frac{VQ_c}{Ib_c} = \frac{(1.47)(12\ 180)}{(1\ 295\ 019)(58)} = 0.24\ \text{MPa}$$

$$\tau_d = \frac{VQ_d}{Ib_d} = \frac{(1.47)(0)}{(1\ 295\ 019)(58)} = 0\ \text{MPa}$$

The magnitude of the idealized transverse shear stress profile through the beam depth will be as shown in Figure 3–25*d*.
(*b*) The bending stresses at each point of interest are

Answer

$$\sigma_a = \frac{My_a}{I} = \frac{(117\ 600)(0)}{(1\ 295\ 019)} = 0\ \text{MPa}$$

$$\sigma_b = \sigma_c = -\frac{My_b}{I} = -\frac{(117\ 600)(32)}{(1\ 295\ 019)} = -2.9\ \text{MPa}$$

$$\sigma_d = -\frac{My_d}{I} = -\frac{(117\ 600)(40)}{(1\ 295\ 019)} = -3.6\ \text{MPa}$$

(*c*) Now at each point of interest, consider a stress element that includes the bending stress and the transverse shear stress. The maximum shear stress for each stress element can be determined by Mohr's circle, or analytically by Equation (3–17) with $\sigma_y = 0$,

$$\tau_{\max} = \sqrt{\left(\frac{\sigma}{2}\right)^2 + \tau^2}$$

Thus, at each point

$$\tau_{\max,a} = \sqrt{0 + (4)^2} = 4 \text{ MPa}$$

$$\tau_{\max,b} = \sqrt{\left(\frac{-2.9}{2}\right)^2 + (3.5)^2} = 3.8 \text{ MPa}$$

$$\tau_{\max,c} = \sqrt{\left(\frac{-2.9}{2}\right)^2 + (0.24)^2} = 1.47 \text{ MPa}$$

$$\tau_{\max,d} = \sqrt{\left(\frac{-3.6}{2}\right)^2 + 0} = 3.24 \text{ MPa}$$

Answer

Interestingly, the critical location is at point *a* where the maximum shear stress is the largest, even though the bending stress is zero. The next critical location is at point *b* in the web, where the thin web thickness dramatically increases the transverse shear stress compared to points *c* or *d*. These results are counterintuitive, since both points *a* and *b* turn out to be more critical than point *d*, even though the bending stress is maximum at point *d*. The thin web and wide flange increase the impact of the transverse shear stress. If the beam length to height ratio were increased, the critical point would move from point *a* to point *b*, since the transverse shear stress at point *a* would remain constant, but the bending stress at point *b* would increase. The designer should be particularly alert to the possibility of the critical stress element not being on the outer surface with cross sections that get wider farther from the neutral axis, particularly in cases with thin web sections and wide flanges. For rectangular and circular cross sections, however, the maximum bending stresses at the outer surfaces will dominate, as was shown in Figure 3–23.

3–12 Torsion

Any moment vector that is collinear with an axis of a mechanical part is called a *torque vector,* because the moment causes the part to be twisted about that axis. A bar subjected to such a moment is said to be in *torsion*.

As shown in Figure 3–26, the torque T applied to a bar can be designated by drawing arrows on the surface of the bar to indicate direction or by drawing torque-vector arrows along the axes of twist of the bar. Torque vectors are the hollow arrows shown on the x axis in Figure 3–26. Note that they conform to the right-hand rule for vectors.

The *angle of twist,* in radians, for a solid round bar is

$$\theta = \frac{Tl}{GJ} \tag{3–35}$$

where T = torque
l = length
G = modulus of rigidity
J = polar second moment of area

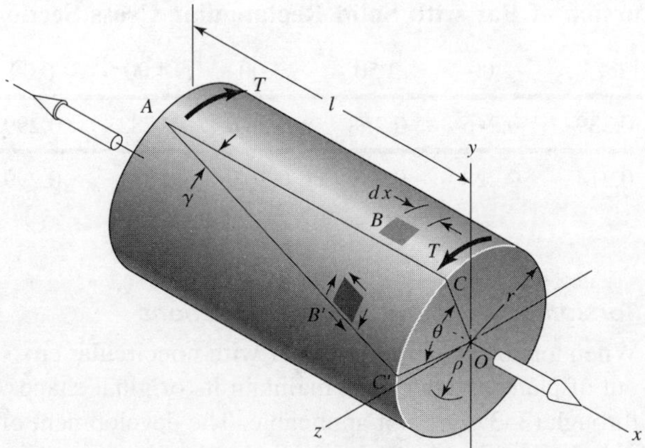

Figure 3–26

Shear stresses develop throughout the cross section. For a round bar in torsion, these stresses are proportional to the radius ρ and are given by

$$\tau = \frac{T\rho}{J} \tag{3–36}$$

Designating r as the radius to the outer surface, we have

$$\tau_{\max} = \frac{Tr}{J} \tag{3–37}$$

The assumptions used in the analysis are:

- The bar is acted upon by a pure torque, and the sections under consideration are remote from the point of application of the load and from a change in diameter.
- The material obeys Hooke's law.
- Adjacent cross sections originally plane and parallel remain plane and parallel after twisting, and any radial line remains straight.

The last assumption depends upon the axisymmetry of the member, so it does not hold true for noncircular cross sections. Consequently, Equations (3–35) through (3–37) apply *only* to circular sections. For a solid round section,

$$J = \frac{\pi d^4}{32} \tag{3–38}$$

where d is the diameter of the bar. For a hollow round section,

$$J = \frac{\pi}{32}\left(d_o^4 - d_i^4\right) \tag{3–39}$$

where the subscripts o and i refer to the outside and inside diameters, respectively.

Table 3–2 Parameters for Torsion of Bar with Solid Rectangular Cross Section

b/c	1.00	1.50	1.75	2.00	2.50	3.00	4.00	6.00	8.00	10	∞
α	0.208	0.231	0.239	0.246	0.258	0.267	0.282	0.299	0.307	0.313	0.333
β	0.141	0.196	0.214	0.228	0.249	0.263	0.281	0.299	0.307	0.313	0.333

Torsion in Noncircular Cross Sections

When torsion is applied to a bar with noncircular cross section, the cross section warps out of plane and does not maintain its original shape. Consequently, Equations (3–35) through (3–37) are not applicable. The development of equations for stress and deflection for torsional loading of noncircular cross sections can be obtained from the mathematical theory of elasticity. In general, the shear stress does not vary linearly with the distance from the axis, and depends on the specific cross section. In fact, for a rectangular section bar the shear stress is zero at the corners where the distance from the axis is the largest. The maximum shearing stress in a rectangular $b \times c$ section bar occurs in the middle of the *longest* side b and is of the magnitude

$$\tau_{\max} = \frac{T}{\alpha b c^2} \approx \frac{T}{bc^2}\left(3 + \frac{1.8}{b/c}\right) \tag{3–40}$$

where b is the width (longer side) and c is the thickness (shorter side). They can *not* be interchanged. The parameter α is a factor that is a function of the ratio b/c as shown in Table 3–2.[3] The angle of twist is given by

$$\theta = \frac{Tl}{\beta b c^3 G} \tag{3–41}$$

where β is a function of b/c, as shown in Table 3–2.

Equation (3–40) is also approximately valid for equal-sided angles; these can be considered as two rectangles, each of which is capable of carrying half the torque.[4]

Torque and Power

Application of torque to a shaft is frequently associated with the transmission of power. For a rotating shaft, the relationship between power H, torque T, and angular velocity ω is given by

$$H = T\omega \tag{3–42}$$

This relationship is valid for any units, and the user is responsible for any unit conversions to put the results into desired units. If standard metric units are used, then torque is in N · m, angular velocity is in rad/s, and power is in watts.

[3] S. Timoshenko, *Strength of Materials*, Part I, 3rd ed., D. Van Nostrand Company, New York, 1955, p. 290.

[4] For other sections see W. C. Young, R. G. Budynas, and A. M. Sadegh, *Roark's Formulas for Stress and Strain*, 8th ed., McGraw Hill, New York, 2012.

EXAMPLE 3–11

Figure 3–27 shows a crank loaded by a force $F = 1.35$ kN that causes twisting and bending of a 19 mm-diameter shaft fixed to a support at the origin of the reference system. In actuality, the support may be an inertia that we wish to rotate, but for the purposes of a stress analysis we can consider this a statics problem.
(*a*) Draw separate free-body diagrams of the shaft *AB* and the arm *BC*, and compute the values of all forces, moments, and torques that act. Label the directions of the coordinate axes on these diagrams.
(*b*) For member *BC*, compute the maximum bending stress and the maximum shear stress associated with the applied torsion and transverse loading. Indicate where these stresses act.
(*c*) Locate a stress element on the top surface of the shaft at *A*, and calculate all the stress components that act upon this element.
(*d*) Determine the maximum principal stress and the maximum shear stress for the stress element at *A*.

Solution

(*a*) The two free-body diagrams are shown in Figure 3–28. The results are

At end *C* of arm *BC*: $\mathbf{F} = -1350\mathbf{j}$ N, $\mathbf{T}_C = -51\ 300\mathbf{k}$ N · mm
At end *B* of arm *BC*: $\mathbf{F} = 1350\mathbf{j}$ N, $\mathbf{M}_1 = 135\ 000\mathbf{i}$ N · mm, $\mathbf{T}_1 = 51\ 300\mathbf{k}$ N · mm
At end *B* of shaft *AB*: $\mathbf{F} = -1350\mathbf{j}$ N, $\mathbf{T}_2 = -135\ 000\mathbf{i}$ N · mm, $\mathbf{M}_2 = -51\ 300\mathbf{k}$ N · mm
At end *A* of shaft *AB*: $\mathbf{F} = 1350\mathbf{j}$ N, $\mathbf{M}_A = 220\ 050\mathbf{k}$ N · mm, $\mathbf{T}_A = 135\ 000\mathbf{i}$ N · mm

(*b*) For arm *BC*, the bending moment will reach a maximum near the shaft at *B*. If we assume this is 135 000 N · m, then the bending stress for a rectangular section will be

Answer
$$\sigma = \frac{M}{I/c} = \frac{6M}{bh^2} = \frac{6(135\ 000)}{6(31)^2} = 140.5 \text{ MPa}$$

Of course, this is not exactly correct, because at *B* the moment is actually being transferred into the shaft, probably through a weldment.

For the torsional stress, use Equation (3–40). Thus

$$\tau_T = \frac{T}{bc^2}\left(3 + \frac{1.8}{b/c}\right) = \frac{51\ 300}{31(6^2)}\left(3 + \frac{1.8}{31/6}\right) = 153.9 \text{ MPa}$$

This stress occurs on the outer surfaces at the middle of the 31 mm side. In addition to the torsional shear stress, a transverse shear stress exists at the same location and in *the same direction* on the visible side of *BC*, hence they are additive. On the opposite side, they subtract. The maximum transverse shear stress, from Table 3–1, with $V = F$, is

$$\tau_V = \frac{3F}{2A} = \frac{(3)(1350)}{(2)\ (31)\ (6)} = 11 \text{ MPa}$$

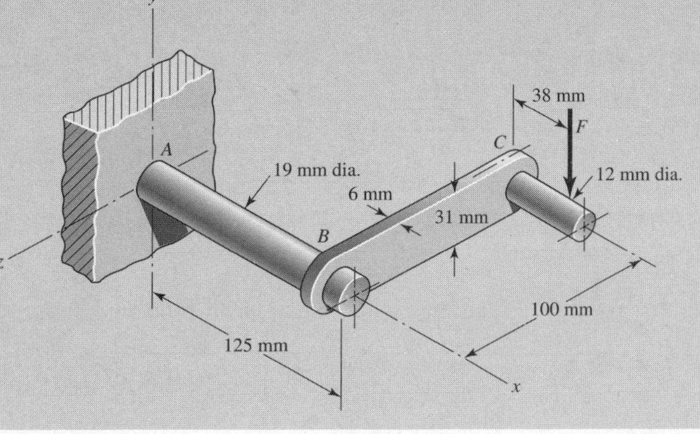

Figure 3–27

Figure 3–28

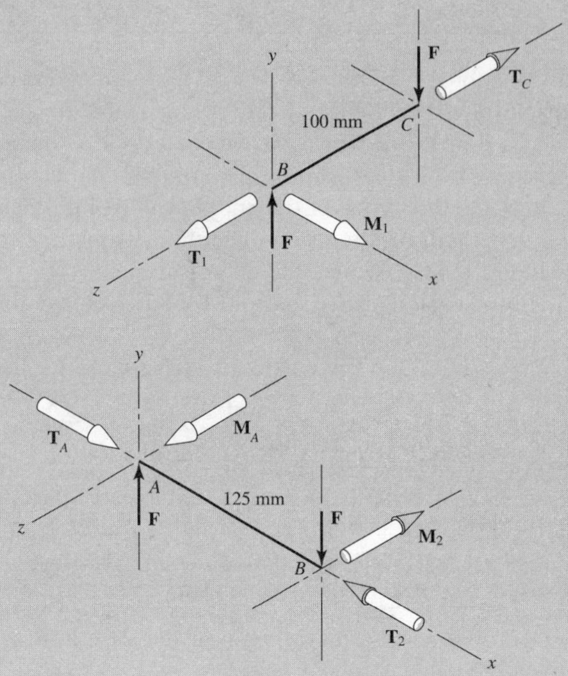

Adding this to τ_T gives the maximum shear stress of

Answer
$$\tau_{\max} = \tau_T + \tau_V = 153.9 + 11 = 164.9 \text{ MPa}$$

This stress occurs on the outer-facing surface at the middle of the $1\frac{1}{4}$-in side.
(*c*) For a stress element at *A*, the bending stress is tensile and is

Answer
$$\sigma_x = \frac{M}{I/c} = \frac{32M}{\pi d^3} = \frac{32(220\ 050)}{\pi(19)^3} = 326.9 \text{ MPa}$$

The torsional stress is

Answer
$$\tau_{xz} = \frac{-T}{J/c} = \frac{-16T}{\pi d^3} = \frac{-16(135\ 000)}{\pi(9)^3} = -100.3 \text{ MPa}$$

where the negative sign accounts for the direction of τ_{xz}.
(*d*) Point *A* is in a state of plane stress where the stresses are in the *xz* plane. Thus, the principal stresses are given by Equation (3–14) with subscripts corresponding to the *x*, *z* axes.

Answer
The maximum normal stress is then given by

$$\sigma_1 = \frac{\sigma_x + \sigma_z}{2} + \sqrt{\left(\frac{\sigma_x - \sigma_z}{2}\right)^2 + \tau_{xz}^2}$$
$$= \frac{326.9 + 0}{2} + \sqrt{\left(\frac{326.9 - 0}{2}\right)^2 + (-100.3)^2} = 355.2 \text{ MPa}$$

Answer
The maximum shear stress is given by Equation (3–17), again with modified subscripts, and is given by

$$\tau_{12} = \sqrt{\left(\frac{\sigma_x - \sigma_z}{2}\right)^2 + \tau_{xz}^2} = \sqrt{\left(\frac{326.9 - 0}{2}\right)^2 + (-100.3)^2} = 191.7 \text{ MPa}$$

EXAMPLE 3–12

The 38 mm diameter solid steel shaft shown in Figure 3–29*a* is simply supported at the ends. Two pulleys are keyed to the shaft where pulley B is of diameter 100 mm and pulley C is of diameter 200 mm. Considering bending and torsional stresses only, determine the locations and magnitudes of the greatest tensile, compressive, and shear stresses in the shaft.

Solution

Figure 3–29*b* shows the net forces, reactions, and torsional moments on the shaft. Although this is a three-dimensional problem and vectors might seem appropriate, we will look at the components of the moment vector by performing a two-plane analysis. Figure 3–29*c* shows the loading in the *xy* plane, as viewed down

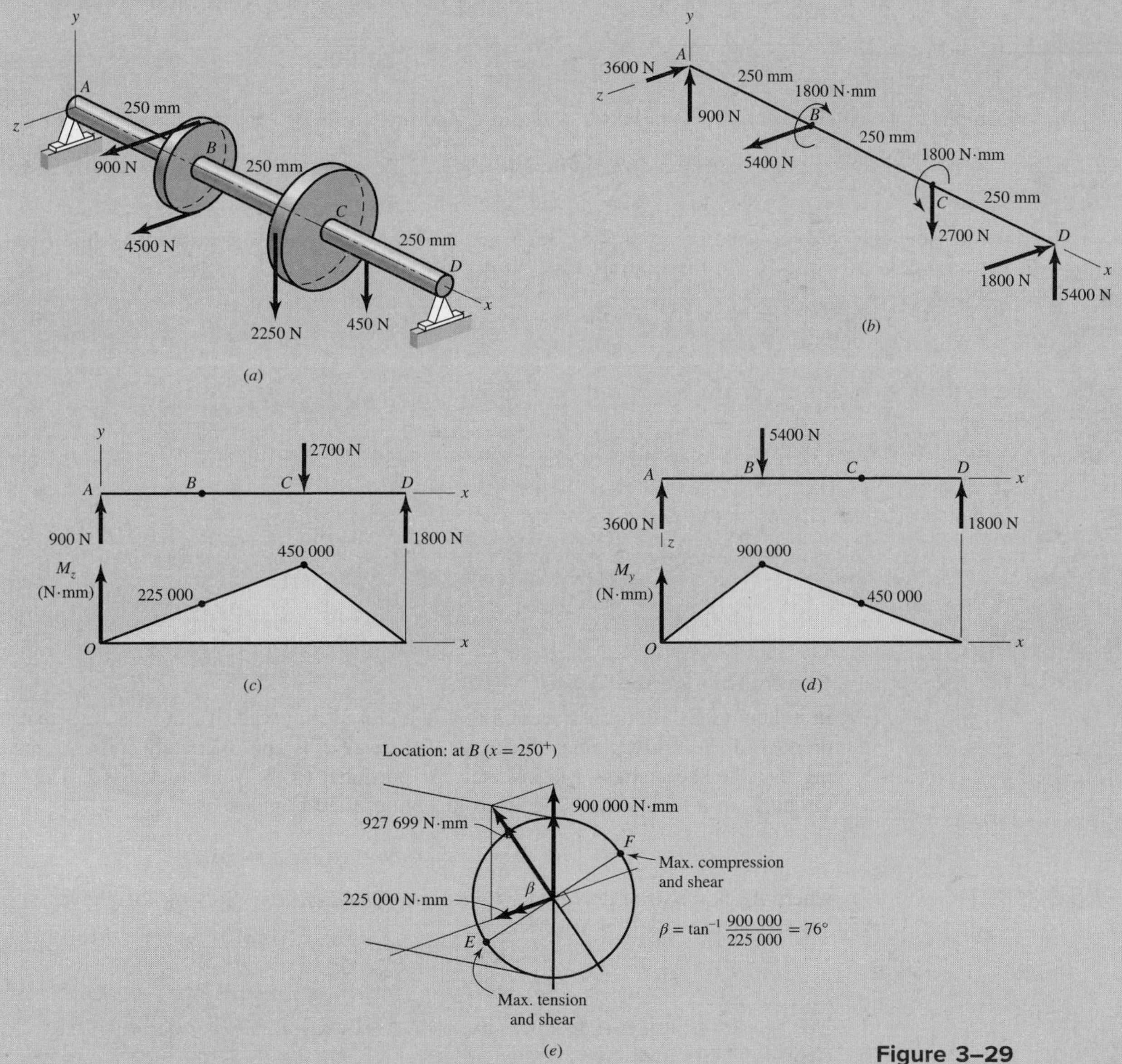

Figure 3–29

the z axis, where bending moments are actually vectors in the z direction. Thus we label the moment diagram as M_z versus x. For the xz plane, we look down the y axis, and the moment diagram is M_y versus x as shown in Figure 3–29*d*.

The net moment on a section is the vector sum of the components. That is,

$$M = \sqrt{M_y^2 + M_z^2} \tag{1}$$

At point B,

$$M_B = \sqrt{225\ 000^2 + 900\ 000^2} = 927\ 669 \text{ N}\cdot\text{mm}$$

At point C,

$$M_C = \sqrt{450\ 000^2 + 450\ 000^2} = 636\ 396 \text{ N}\cdot\text{mm}$$

Thus the maximum bending moment is 927 669 N · mm and the maximum bending stress at pulley B is

$$\sigma = \frac{Md/2}{\pi d^4/64} = \frac{32M}{\pi d^3} = \frac{32(927\ 669)}{\pi(38^3)} = 172.3 \text{ MPa}$$

The maximum torsional shear stress occurs between B and C and is

$$\tau = \frac{Td/2}{\pi d^4/32} = \frac{16T}{\pi d^3} = \frac{16(180\ 000)}{\pi(38^3)} = 16.7 \text{ MPa}$$

The maximum bending and torsional shear stresses occur just to the right of pulley B at points E and F as shown in Figure 3–29*e*. At point E, the maximum tensile stress will be σ_1 given by

Answer
$$\sigma_1 = \frac{\sigma}{2} + \sqrt{\left(\frac{\sigma}{2}\right)^2 + \tau^2} = \frac{172.3}{2} + \sqrt{\left(\frac{172.3}{2}\right)^2 + 16.7^2} = 173.8 \text{ MPa}$$

At point F, the maximum compressive stress will be σ_2 given by

Answer
$$\sigma_2 = \frac{-\sigma}{2} - \sqrt{\left(\frac{-\sigma}{2}\right)^2 + \tau^2} = \frac{-172.3}{2} - \sqrt{\left(\frac{-172.3}{2}\right)^2 + 16.7^2} = -173.8 \text{ MPa}$$

The extreme shear stress also occurs at E and F and is

Answer
$$\tau_{12} = \sqrt{\left(\frac{\pm\sigma}{2}\right)^2 + \tau^2} = \sqrt{\left(\frac{\pm 172.3}{2}\right)^2 + 16.7^2} = 87.7 \text{ MPa}$$

Closed Thin-Walled Tubes[5]

In hollow tubes with cross section that is a closed thin wall, it can be shown that the product of shear stress times thickness of the wall τt is approximately constant, meaning that the shear stress τ is inversely proportional to the wall thickness t. The total torque T on a tube such as depicted in Figure 3–30 is given by

$$T = \int \tau t r\, ds = (\tau t)\int r\, ds = \tau t(2A_m) = 2A_m t\tau$$

where A_m is the *area enclosed by the section median line*. Solving for τ gives

$$\tau = \frac{T}{2A_m t} \tag{3–43}$$

[5]See Section 3–13, F. P. Beer, E. R. Johnston, and J. T. De Wolf, *Mechanics of Materials*, 5th ed., McGraw Hill, New York, 2009.

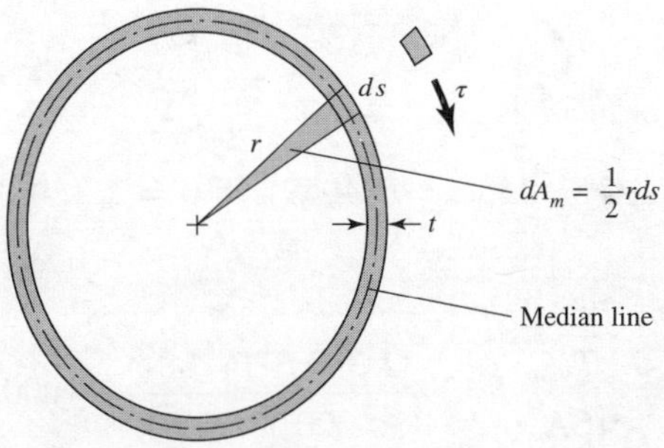

Figure 3–30

The depicted cross section is elliptical, but the section need not be symmetrical nor of constant thickness.

For constant wall thickness t, the angular twist (radians) per unit of length of the tube θ_1 is given by

$$\theta_1 = \frac{TL_m}{4GA_m^2 t} \tag{3–44}$$

where L_m is the *length of the section median line*. These equations presume the buckling of the tube is prevented by ribs, stiffeners, bulkheads, and so on, and that the stresses are below the proportional limit.

EXAMPLE 3–13

A welded steel tube is 1 m long, has a 3 mm wall thickness, and a 60 mm by 90 mm rectangular cross section as shown in Figure 3–31. Assume an allowable shear stress of 80 MPa and a shear modulus of 80 GPa.
(*a*) Estimate the allowable torque T.
(*b*) Estimate the angle of twist due to the torque.

Solution
(*a*) Within the section median line, the area enclosed is

$$A_m = (60 - 3)(90 - 0.3) = 4959 \text{ mm}^2$$

and the length of the median perimeter is

$$L_m = 2[(60 - 3) + (90 - 3)] = 288 \text{ mm}$$

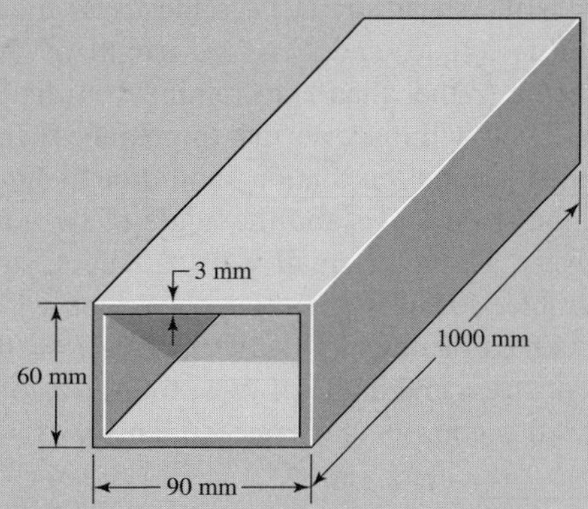

Figure 3–31

A rectangular steel tube produced by welding.

Answer

From Equation (3–43) the torque T is

$$T = 2A_m t\tau = 2(4959)3(80) = 2\ 380\ 320\ \text{N} \cdot \text{mm}$$

Answer

(*b*) The angle of twist θ from Equation (3–44) is

$$\theta = \theta_1 l = \frac{TL_m}{4GA_m^2 t} l = \frac{2\ 380\ 320(288)}{4(80\ 000^6)(4959^2)(3)}(1000) = 0.029\ \text{rad} = 1.66°$$

EXAMPLE 3–14

Compare the shear stress on a circular cylindrical tube with an outside diameter of 25 mm and an inside diameter of 22 mm, predicted by Equation (3–37), to that estimated by Equation (3–43).

Solution

From Equation (3–37),

$$\tau_{\max} = \frac{Tr}{J} = \frac{Tr}{(\pi/32)\left(d_o^4 - d_i^4\right)} = \frac{T(12.5)}{(\pi/32)(25^4 - 22^4)} = 8.14(10)^{-4}T$$

From Equation (3–43),

$$\tau = \frac{T}{2A_m t} = \frac{T}{2(\pi 23.5^2/4)1.5} = 7.69(10)^{-4}T$$

Taking Equation (3–37) as correct, the error in the thin-wall estimate is 5.6%.

Open Thin-Walled Sections

When the median wall line of the cross section of a long member is not closed, the section is said to be an *open section*. Figure 3–32 presents some examples. Open sections in torsion, where the wall is thin, have relations derived from the membrane analogy theory[6] resulting in:

$$\tau = G\theta_1 c = \frac{3T}{Lc^2} \tag{3–45}$$

where τ is the shear stress, G is the shear modulus, θ_1 is the angle of twist per unit length, T is torque, and L is the length of the median line. The wall thickness is designated c (rather than t) to remind you that you are in open sections. By studying Table 3–2 you will discover that membrane theory presumes $b/c \to \infty$. Note that open thin-walled sections in torsion should be avoided in design. As indicated in Equation (3–45), the shear stress and the angle of twist are inversely proportional to c^2 and c^3, respectively. Thus, for small wall thickness, stress and twist can become quite large. For example, consider the thin round tube with a slit in Figure 3–32. For a ratio of wall thickness of outside diameter of $c/d_o = 0.1$, the open section has greater magnitudes of stress and angle of twist by factors of 12.3 and 61.5, respectively, compared to a closed section of the same dimensions.

---

[6]See S. P. Timoshenko and J. N. Goodier, *Theory of Elasticity*, 3rd ed., McGraw Hill, New York, 1970, Section 109.

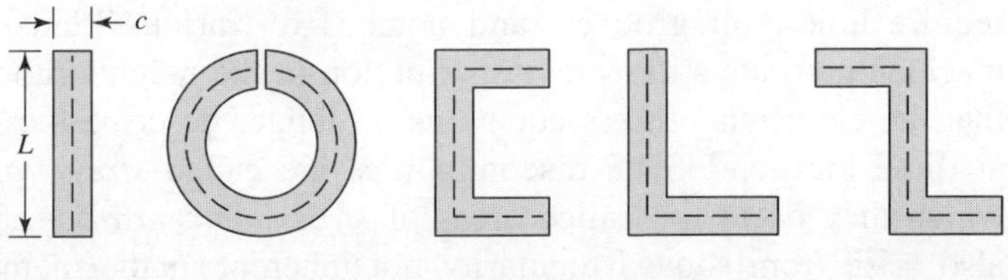

Figure 3–32

Some open thin-walled sections.

EXAMPLE 3–15

A 0.3 m-long strip of steel is 3 mm thick and 25 mm wide, as shown in Figure 3–33. If the allowable shear stress is 80 MPa and the shear modulus is 80 GPa, find the torque corresponding to the allowable shear stress and the angle of twist, in degrees, (*a*) using Equation (3–45) and (*b*) using Equations (3–40) and (3–41).

Solution

(*a*) The length of the median line is 25 mm. From Equation (3–45),

$$T = \frac{Lc^2\tau}{3} = \frac{(25)(3)^2 80}{3} = 6000 \text{ N} \cdot \text{mm}$$

$$\theta = \theta_1 l = \frac{\tau l}{Gc} = \frac{80(300)}{80\ 000(3)} = 0.1 \text{ rad} = 5.73°$$

A torsional spring rate k_t can be expressed as T/θ:

$$k_t = 6000/0.1 = 60\ 000 \text{ N} \cdot \text{mm/rad}$$

(*b*) From Equation (3–40),

$$T = \frac{\tau_{\max} bc^2}{3 + 1.8/(b/c)} = \frac{80(25)(3)^2}{3 + 1.8/(25/3)} = 5597 \text{ N} \cdot \text{mm}$$

From Equation (3–41), with $b/c = 25/3 = 8.3$

$$\theta = \frac{Tl}{\beta bc^3 G} = \frac{5597(300)}{0.307(25)3^3(80\ 000)} = 0.1013 \text{ rad} = 5.8°$$

$$k_t = 5597/0.1013 = 55\ 252 \text{ N} \cdot \text{mm/rad}$$

The cross section is not thin, where b should be greater than c by at least a factor of 10. In estimating the torque, Equation (3–45) provided a value of 7.5 percent higher than Equation (3–40), and 8.5 percent higher than Table 3–2.

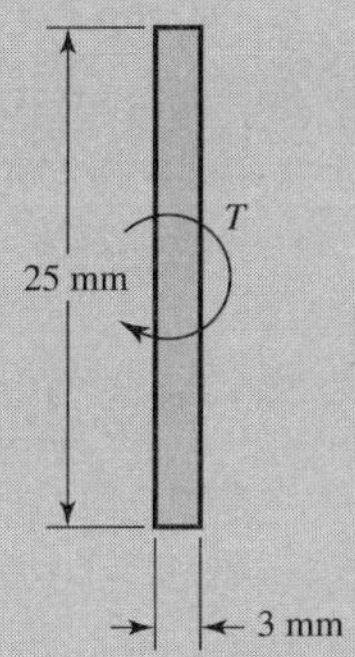

Figure 3–33

The cross section of a thin strip of steel subjected to a torsional moment T.

3–13 Stress Concentration

In the development of the basic stress equations for tension, compression, bending, and torsion, it was assumed that no geometric irregularities occurred in the member under consideration. But it is quite difficult to design a machine without permitting some changes in the cross sections of the members. Rotating shafts must have shoulders designed on them so that the bearings can be properly seated and so that they will take thrust loads; and the shafts must have key slots machined into them for securing pulleys and gears. A bolt has a head on one end and screw threads on the other end, both of which account for abrupt changes in the cross section. Other parts require holes, oil grooves, and notches of various kinds. Any discontinuity in a machine part alters the stress distribution in the neighborhood of the discontinuity so that the elementary stress equations no longer describe the state of stress in the part at these locations. Such discontinuities are called *stress raisers,* and the regions in which they occur are called areas of *stress concentration*. Stress concentrations can also arise from some irregularity not inherent in the member, such as tool marks, holes, notches, grooves, or threads.

A *theoretical,* or *geometric, stress-concentration factor* K_t or K_{ts} is used to relate the actual maximum stress at the discontinuity to the *nominal stress.* The factors are defined by the equations

$$K_t = \frac{\sigma_{\max}}{\sigma_0} \qquad K_{ts} = \frac{\tau_{\max}}{\tau_0} \tag{3–46}$$

where K_t is used for normal stresses and K_{ts} for shear stresses. The nominal stress σ_0 or τ_0 is the stress calculated by using the elementary stress equations and the net area, or net cross section. Sometimes the gross cross section is used instead, and so it is always wise to double check the source of K_t or K_{ts} before calculating the maximum stress.

The stress-concentration factor depends for its value only on the *geometry* of the part. That is, the particular material used has no effect on the value of K_t. This is why it is called a *theoretical* stress-concentration factor.

The analysis of geometric shapes to determine stress-concentration factors is a difficult problem, and not many solutions can be found. Most stress-concentration factors are found by using experimental techniques.[7] Though the finite-element method has been used, the fact that the elements are indeed finite prevents finding the exact maximum stress. Experimental approaches generally used include photoelasticity, grid methods, brittle-coating methods, and electrical strain-gauge methods. Of course, the grid and strain-gauge methods both suffer from the same drawback as the finite-element method.

Stress-concentration factors for a variety of geometries may be found in Tables A–13 and A–14.

An example is shown in Figure 3–34, that of a thin plate loaded in tension where the plate contains a centrally located hole.

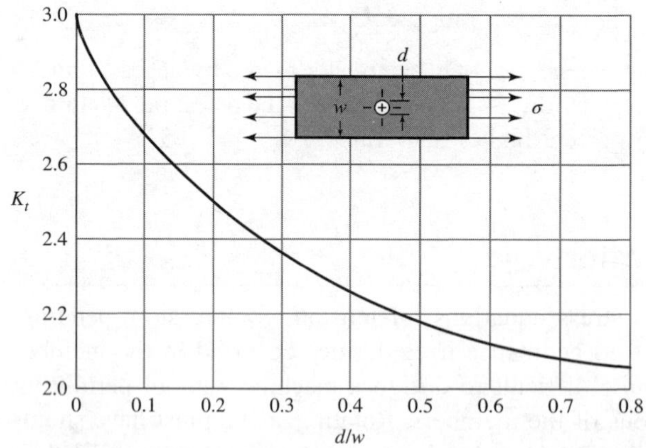

Figure 3–34

Thin plate in tension or simple compression with a transverse central hole. The net tensile force is $F = \sigma wt$, where t is the thickness of the plate. The nominal stress is given by

$$\sigma_0 = \frac{F}{(w-d)t} = \frac{w}{(w-d)}\sigma.$$

[7]W. D. Pilkey and D. F. Pilkey, *Peterson's Stress Concentration Factors*, 3rd ed., John Wiley & Sons, New York, 2008.

In *static loading,* stress-concentration factors are applied as follows. In *ductile materials* ($\varepsilon_f \geq 0.05$), the stress-concentration factor *is not* usually applied to predict the critical stress, because plastic strain in the region of the stress concentration is localized and has a strengthening effect. In *brittle materials* ($\varepsilon_f < 0.05$), the geometric stress-concentration factor K_t is applied to the nominal stress before comparing it with strength. Gray cast iron has so many inherent stress raisers that the stress raisers introduced by the designer have only a modest (but additive) effect.

Consider a part made of a ductile material and loaded by a gradually applied static load such that the stress in an area of a stress concentration goes beyond the yield strength. The yielding will be restricted to a very small region, and the permanent deformation as well as the residual stresses after the load is released will be insignificant and normally can be tolerated. If yielding does occur, the stress distribution changes and tends toward a more uniform distribution. In the region where yielding occurs, there is little danger of fracture of a ductile material, but if the possibility of a brittle fracture exists, the stress concentration must be taken seriously. Brittle fracture is not just limited to brittle materials. Materials often thought of as being ductile can fail in a brittle manner under certain conditions, e.g., any single application or combination of cyclic loading, rapid application of static loads, loading at low temperatures, and parts containing defects in their material structures (see Section 5–12). The effects on a ductile material of processing, such as hardening, hydrogen embrittlement, and welding, may also accelerate failure. Thus, care should always be exercised when dealing with stress concentrations.

For *dynamic loading,* the stress concentration effect is significant for *both* ductile and brittle materials and must always be taken into account (see Section 6–10).

EXAMPLE 3–16

The 2-mm-thick bar shown in Figure 3–35 is loaded axially with a constant force of 10 kN. The bar material has been heat treated and quenched to raise its strength, but as a consequence it has lost most of its ductility. It is desired to drill a hole through the center of the 40-mm face of the plate to allow a cable to pass through it. A 4-mm hole is sufficient for the cable to fit, but an 8-mm drill is readily available. Will a crack be more likely to initiate at the larger hole, the smaller hole, or at the fillet?

Solution

Since the material is brittle, the effect of stress concentrations near the discontinuities must be considered. Dealing with the hole first, for a 4-mm hole, the nominal stress is

$$\sigma_0 = \frac{F}{A} = \frac{F}{(w-d)t} = \frac{10\ 000}{(40-4)2} = 139\text{ MPa}$$

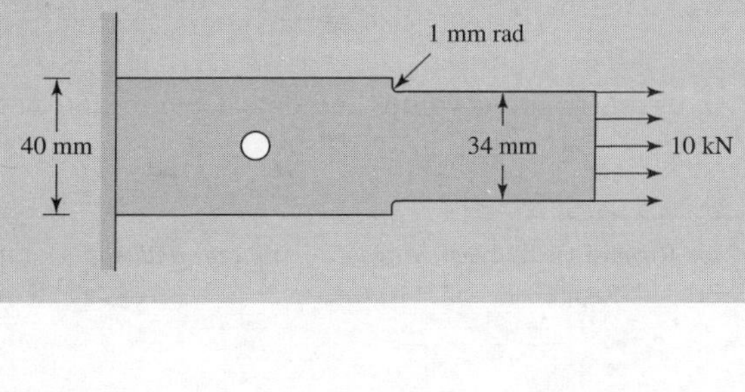

Figure 3–35

The theoretical stress concentration factor, from Figure A–13–1, with $d/w = 4/40 = 0.1$, is $K_t = 2.7$. The maximum stress is

Answer $$\sigma_{\max} = K_t\sigma_0 = 2.7(139) = 380 \text{ MPa}$$

Similarly, for an 8-mm hole,

$$\sigma_0 = \frac{F}{A} = \frac{F}{(w-d)t} = \frac{10\ 000}{(40-8)2} = 156 \text{ MPa}$$

With $d/w = 8/40 = 0.2$, then $K_t = 2.5$, and the maximum stress is

Answer $$\sigma_{\max} = K_t\sigma_0 = 2.5(156) = 390 \text{ MPa}$$

Though the stress concentration is higher with the 4-mm hole, in this case the increased nominal stress with the 8-mm hole has more effect on the maximum stress.

For the fillet,

$$\sigma_0 = \frac{F}{A} = \frac{10\ 000}{(34)2} = 147 \text{ MPa}$$

From Table A–13–5, $D/d = 40/34 = 1.18$, and $r/d = 1/34 = 0.026$. Then $K_t = 2.5$.

Answer $$\sigma_{\max} = K_t\sigma_0 = 2.5(147) = 368 \text{ MPa}$$

Answer
The crack will most likely occur with the 8-mm hole, next likely would be the 4-mm hole, and least likely at the fillet.

3–14 Stresses in Pressurized Cylinders

Cylindrical pressure vessels, hydraulic cylinders, gun barrels, and pipes carrying fluids at high pressures develop both radial and tangential stresses with values that depend upon the radius of the element under consideration. In determining the radial stress σ_r and the tangential stress σ_t, we make use of the assumption that the longitudinal elongation is constant around the circumference of the cylinder. In other words, a right section of the cylinder remains plane after stressing.

Referring to Figure 3–36, we designate the inside radius of the cylinder by r_i, the outside radius by r_o, the internal pressure by p_i, and the external pressure by p_o. Then it can be shown that the tangential and radial stresses are[8]

$$\sigma_t = \frac{p_i r_i^2 - p_o r_o^2 - r_i^2 r_o^2(p_o - p_i)/r^2}{r_o^2 - r_i^2}$$
$$\sigma_r = \frac{p_i r_i^2 - p_o r_o^2 + r_i^2 r_o^2(p_o - p_i)/r^2}{r_o^2 - r_i^2} \qquad (3\text{–}47)$$

As usual, positive values indicate tension and negative values, compression.

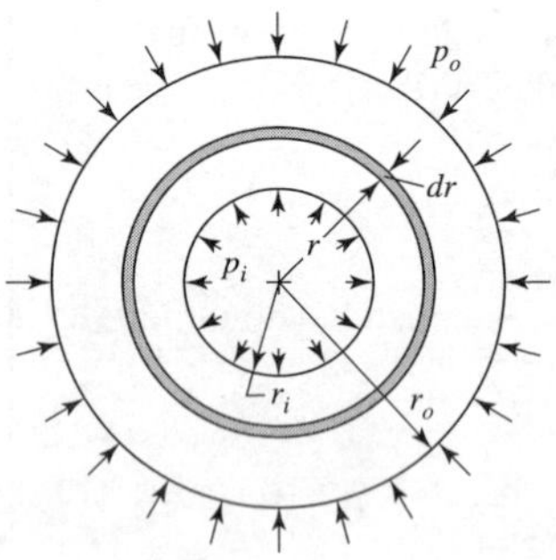

Figure 3–36

A cylinder subjected to both internal and external pressure.

[8]See Richard G. Budynas, *Advanced Strength and Applied Stress Analysis*, 2nd ed., McGraw Hill, New York, 1999, pp. 348–352.

The tangential and radial stresses are orthogonal, and at any location of interest can be modeled as principal stresses on a stress element. Chapter 5 deals with methods to evaluate a multiaxial stress element.

Thick-Walled Cylinders Subjected to Internal Pressure Only

Equation (3–47) applies for cylinders with both internal and external pressure. We shall further develop the case of cylinders subject to internal pressure only, in order to better understand the stress situation. A similar analysis can be done for the case of external pressure. For the special case of $p_o = 0$, Equation (3–47) gives

$$\sigma_t = \frac{r_i^2 p_i}{r_o^2 - r_i^2}\left(1 + \frac{r_o^2}{r^2}\right)$$
$$\sigma_r = \frac{r_i^2 p_i}{r_o^2 - r_i^2}\left(1 - \frac{r_o^2}{r^2}\right) \qquad (3\text{–}48)$$

The equations of set (3–48) are plotted in Figure 3–37 to show the distribution of stresses over the wall thickness. The tangential and radial stresses are greatest in magnitude at the inside radius. Substituting $r = r_i$ into Equation (3–48) yields

$$(\sigma_t)_{\max} = p_i\left(\frac{r_o^2 + r_i^2}{r_o^2 - r_i^2}\right)$$
$$(\sigma_r)_{\max} = -p_i \qquad (3\text{–}49)$$

If the cylinder is closed on the ends, the internal pressure creates a force on the ends, equal to the product of the pressure and the projected head area on which it acts. Dividing this force by the area of the cylinder's cross section, assuming the force is carried uniformly over the cylinder's cross section, the longitudinal stress is found to be

$$\sigma_l = \frac{p_i r_i^2}{r_o^2 - r_i^2} \qquad (3\text{–}50)$$

The critical stress element will be at the inner radius, where $(\sigma_t)_{\max}$, $(\sigma_r)_{\max}$, and σ_l are all at peak values. These three stresses are orthogonal, and are therefore principal stresses. $(\sigma_t)_{\max}$ and σ_l are positive and $(\sigma_r)_{\max}$ is negative, with $(\sigma_t)_{\max} > \sigma_l > 0 > (\sigma_r)_{\max}$.

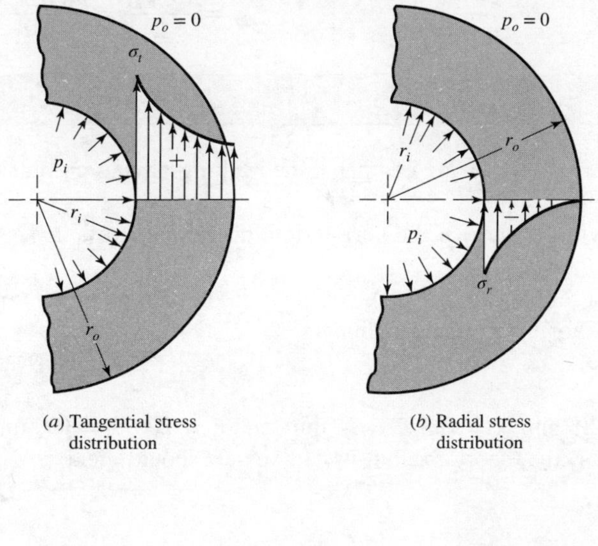

Figure 3–37

Distribution of stresses in a thick-walled cylinder subjected to internal pressure.

Consequently, sketching the Mohr's circles for this situation shows that the largest Mohr's circle is determined by $(\sigma_t)_{\max}$ and $(\sigma_r)_{\max}$. Therefore, the maximum principal shear stress is independent of σ_l, and given by

$$\tau_{\max} = \frac{(\sigma_t)_{\max} - (\sigma_r)_{\max}}{2} = \frac{p_i r_o^2}{r_o^2 - r_i^2} \tag{3–51}$$

The longitudinal stress is positive, in the axial direction, and constant for any radius. Equations (3–47), (3–48), (3–49), (3–50), and (3–51) apply only to sections taken a significant distance from the ends and away from any areas of stress concentration.

Thin-Walled Vessels Subjected to Internal Pressure Only

When the wall thickness of a cylindrical pressure vessel is about one-tenth, or less, of its radius, the radial stress that results from pressurizing the vessel is quite small compared with the tangential stress. Under these conditions the tangential stress, called the *hoop stress,* can be considered uniform throughout the cylinder wall. From Equation (3–48), the average tangential stress is given by

$$(\sigma_t)_{\text{av}} = \frac{\int_{r_i}^{r_o} \sigma_t dr}{r_o - r_i} = \frac{\int_{r_i}^{r_o} \frac{r_i^2 p_i}{r_o^2 - r_i^2}\left(1 + \frac{r_o^2}{r_i^2}\right) dr}{r_o - r_i} = \frac{p_i r_i}{r_o - r_i} = \frac{p_i d_i}{2t} \tag{a}$$

where d_i is the inside diameter. So far, this is applicable as an average tangential stress regardless of the wall thickness.

For a thin-walled pressure vessel, an approximation to the maximum tangential stress is obtained by replacing the inside radius $(d_i/2)$ with the average radius $(d_i + t)/2$, giving

$$(\sigma_t)_{\max} = \frac{p_i(d_i + t)}{2t} \tag{3–52}$$

If the thin-walled cylinder has closed ends, the longitudinal stress from Equation (3–50) can be simplified to

$$\sigma_l = \frac{p_i d_i}{4t} \tag{3–53}$$

Since both the tangential and longitudinal stresses are positive, the absolute maximum principal shear stress will be in the plane of the tangential stress and the third principal stress at zero, thus

$$\tau_{\max} = \frac{(\sigma_t)_{\max}}{2} = \frac{p_i(d_i + t)}{4t} \tag{3–54}$$

EXAMPLE 3–17

An aluminum-alloy pressure vessel is made of open-ended tubing having an outside diameter of 200 mm and a wall thickness of 6 mm.

(*a*) What pressure can the cylinder carry if the permissible tangential stress is 82 MPa and the theory for thin-walled vessels is assumed to apply?

(*b*) On the basis of the pressure found in part (*a*), compute the maximum stress components and the maximum shear stress using the theory for thick-walled cylinders.

Solution

(*a*) Here $d_i = 200 - 2(6) = 188$ mm, $r_i = 188/2 = 94$ mm, and $r_o = 200/2 = 100$ mm. Then $t/r_i = 6/94 = 0.064$. Since this ratio is less than 0.1, the theory for thin-walled vessels should yield safe results.

We first solve Equation (3–52) to obtain the allowable pressure. This gives

Answer
$$p = \frac{2t(\sigma_t)_{\max}}{d_i + t} = \frac{2(6)82}{188 + 6} = 5.07 \text{ MPa}$$

(*b*) The maximum tangential and radial stresses will occur at the inside radius, where Equation (3–49) is applicable, giving

Answer
$$(\sigma_t)_{\max} = p_i \frac{r_o^2 + r_i^2}{r_o^2 - r_i^2} = 5.07 \frac{(100^2 + 94^2)}{(100^2 - 94^2)} = 82.04 \text{ MPa}$$

Answer
$$(\sigma_r)_{\max} = -p_i = -5.07 \text{ MPa}$$

From Equation (3–51), the maximum shear stress for the stress element at the inside radius is

Answer
$$\tau_{\max} = \frac{p_i r_o^2}{r_o^2 - r_i^2} = \frac{5.07(100)^2}{100^2 - 94^2} = 43.56 \text{ MPa}$$

For comparison, with the thin-walled approximation of Equation (3–54), which essentially ignores the radial stress,

$$\tau_{\max} = \frac{p_i(d_i + t)}{4t} = \frac{5.07(188 + 6)}{(4)(6)} = 40.98 \text{ MPa}$$

Thin-walled pressure vessels are often rated for permissible tangential stress. Note that there is essentially no difference in the tangential stress from parts (*a*) and (*b*), so the thin-wall theory is satisfactory. If the pressure vessel is being designed based on the maximum shear stress in the critical stress element, the thin-wall estimation is about 6 percent low in this case.

3–15 Stresses in Rotating Rings

Many rotating elements, such as flywheels and blowers, can be simplified to a rotating ring to determine the stresses. When this is done it is found that the same tangential and radial stresses exist as in the theory for thick-walled cylinders except that they are caused by inertial forces acting on all the particles of the ring. The tangential and radial stresses so found are subject to the following restrictions:

- The outside radius of the ring, or disk, is large compared with the axial thickness, t, that is, $r_o \geq 10t$.
- The axial thickness of the ring or disk is constant.
- The stresses are constant over the axial thickness.

The stresses are[9]

$$\sigma_t = \rho\omega^2\left(\frac{3 + \nu}{8}\right)\left(r_i^2 + r_o^2 + \frac{r_i^2 r_o^2}{r^2} - \frac{1 + 3\nu}{3 + \nu}r^2\right)$$
$$\sigma_r = \rho\omega^2\left(\frac{3 + \nu}{8}\right)\left(r_i^2 + r_o^2 - \frac{r_i^2 r_o^2}{r^2} - r^2\right) \tag{3–55}$$

[9]Ibid, pp. 348–357.

where r is the radius to the stress element under consideration, ρ is the mass density, and ω is the angular velocity of the ring in radians per second. For a rotating disk with no central hole, use $r_i = 0$ in these equations.

The tangential stress is tensile and maximum at the inner radius. The radial stress is zero at the inner radius and at the outer radius, and tensile everywhere else. See Figure 3–38 from Example 3–18 for an example of the distribution of these stresses. The two stresses are orthogonal, and they are principal stresses. Since both of these principal stresses are positive, the third principal stress with magnitude of zero is utilized to determine the absolute maximum principal shear stress from Equation (3–19),

$$\tau_{13} = \frac{\sigma_1 - \sigma_3}{2} = \frac{\sigma_t - 0}{2} = \frac{\sigma_t}{2} \tag{a}$$

The critical stress element is at the inner radius where σ_t is maximum. Setting $r = r_i$ in Equation (3–55), and incorporating Equation (*a*), the maximum shear stress at the critical stress element is

$$\tau_{\max} = \frac{(\sigma_t)_{\max}}{2} = \frac{(\sigma_t)_{r=r_i}}{2} \tag{3–56}$$

A disk spinning at high speed can have an appreciable change in outer diameter. The strain in the tangential direction is given by Equation (3–22), where the subscripts on the σ's are $x = r$, $y = t$, and $z = z$. The longitudinal stress is $\sigma_z = 0$. Thus, the tangential strain is $\varepsilon_t = \sigma_t/E$. The length of the tangential line at the outer radius, the circumference C, increases by $\Delta C = C\varepsilon_t = 2\pi r_o \varepsilon_t$. The change in circumference is also equal to $2\pi \Delta r_o$. As a result, $2\pi r_o \varepsilon_t = 2\pi \Delta r_o$, or

$$\Delta r_o = r_o \varepsilon_t \tag{3–57}$$

This is an important, but not obvious, equation for cylindrical problems.

EXAMPLE 3–18

A steel flywheel 30 mm thick with an outer diameter of 1 m and an inner diameter of 0.2 m is rotating at 5000 rpm.
(*a*) Determine the radial and tangential stress distributions as functions of the radial position, r. Also, determine the maxima, and plot the stress distributions.
(*b*) Determine the maximum shear stress for the critical stress element at the inner radius.
(*c*) From the tangential strain, determine the radial deflection of the outer radius of the flywheel.

Solution

(*a*) From Table A–5, $\nu = 0.292$, and $\gamma = 76.5$ kN/m^3. The mass density is $\rho = (76.5 \times 1000)/9.81 = 7798$ kg/m^3, and the speed is

$$\omega = 2\pi N/60 = 2\pi(5000)/60 = 523.6 \text{ rad/s}$$

Answer

From Equation (3–55)

$$\begin{aligned}\sigma_t &= 7798 \times 10^{(-6)}(523.6)^2\left(\frac{3.292}{8}\right)\left[0.1^2 + 0.5^2 + \frac{0.1^2(0.5^2)}{r^2} - \left(\frac{0.1 + 3(0.292)}{3 + 0.292}\right)r^2\right] \\ &= 879.73\left(0.26 + \frac{0.0025}{r^2} - 0.56\,987r^2\right)\end{aligned} \tag{1}$$

$$\sigma_r = 7798 \times 10^{(-6)}(523.6)^2\left(\frac{3.292}{8}\right)\left[0.1^2 + 0.5^2 - \frac{0.1^2(0.5^2)}{r^2} - r^2\right]$$
$$= 879.73\left(0.26 - \frac{0.0025}{r^2} - r^2\right) \qquad (2)$$

The maximum tangential stress occurs at the inner radius

Answer $$(\sigma_t)_{\max} = 879.73\left(0.26 + \frac{0.0025}{4^2} - 0.56\,987\,(0.1^2)\right) = 443.65 \text{ MPa}$$

(*b*) The critical stress element is at the inner radius. From Equation (3–56), the maximum shear stress for the critical stress element is

Answer $$\tau_{\max} = \frac{(\sigma_t)_{\max}}{2} = \frac{443.65}{2} = 221.8 \text{ MPa}$$

The location of the maximum radial stress is found from evaluating $d\sigma_r/dr = 0$ in Equation (3–53). This occurs at $r = \sqrt{r_i r_o} = \sqrt{0.1(0.5)} = 22.36$ cm. Thus,

Answer $$(\sigma_r)_{\max} = 879.73\left(0.26 - \frac{0.0025}{22.36^2} - 22.36^2\right) = 140.76 \text{ MPa}$$

A plot of the distributions is given in Figure 3–38.

(*c*) At $r = 50$ cm, $\sigma_r = 0$, and $\sigma_t = 112.20$ MPa. The tangential strain is $\varepsilon_t = \sigma_t/E = 112.2/[207\,000] = 5.42(10^{-4})$ cm/cm. From Equation (3–57),

Answer $$\Delta r_o = (50)\,[5.42(10^{-4})] = 0.271 \text{ mm}$$

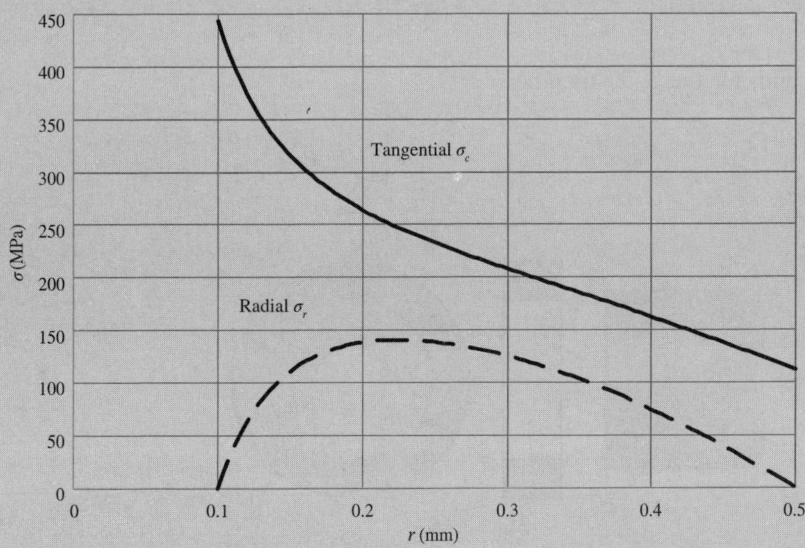

Figure 3–38

Tangential and radial stress distribution for Example 3–18.
$(\sigma_t)_{\max} = 443.65$ MPa
$(\sigma_r)_{\max} = 140.76$ MPa

3–16 Press and Shrink Fits

When two cylindrical parts are assembled by shrinking or press fitting one part upon another, a contact pressure is created between the two parts. The stresses resulting from this pressure may easily be determined with the equations of the preceding sections.

Figure 3–39 shows two cylindrical members that have been assembled with a shrink fit. Prior to assembly, the outer radius of the inner member was larger than the inner radius of the outer member by the *radial interference* δ. After assembly, an interference contact pressure p develops between the members at the nominal radius R, causing radial stresses $\sigma_r = -p$ in each member at the contacting surfaces. This pressure is given by[10]

$$p = \frac{\delta}{R\left[\dfrac{1}{E_o}\left(\dfrac{r_o^2 + R^2}{r_o^2 - R^2} + \nu_o\right) + \dfrac{1}{E_i}\left(\dfrac{R^2 + r_i^2}{R^2 - r_i^2} - \nu_i\right)\right]} \tag{3–58}$$

where the subscripts o and i on the material properties correspond to the outer and inner members, respectively. If the two members are of the same material with $E_o = E_i = E$, and $\nu_o = \nu_i$, the relation simplifies to

$$p = \frac{E\delta}{2R^3}\left[\frac{(r_o^2 - R^2)(R^2 - r_i^2)}{r_o^2 - r_i^2}\right] \tag{3–59}$$

For Equations (3–58) and (3–59), diameters can be used in place of R, r_i, and r_o, provided δ is the diametral interference (twice the radial interference).

With p, Equation (3–47) can be used to determine the radial and tangential stresses in each member. For the inner member, $p_o = p$ and $p_i = 0$. For the outer member, $p_o = 0$ and $p_i = p$. For example, the magnitudes of the tangential stresses at the transition radius R are maximum for both members. For the inner member

$$(\sigma_t)_i\Big|_{r=R} = -p\frac{R^2 + r_i^2}{R^2 - r_i^2} \tag{3–60}$$

and, for the outer member

$$(\sigma_t)_o\Big|_{r=R} = p\frac{r_o^2 + R^2}{r_o^2 - R^2} \tag{3–61}$$

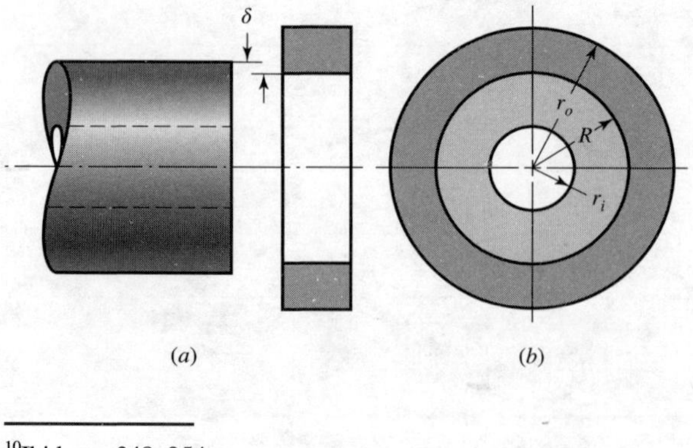

Figure 3–39

Notation for press and shrink fits. (*a*) Unassembled parts; (*b*) after assembly.

[10]Ibid, pp. 348–354.

It is assumed that both members have the same length. In the case of a hub that has been press-fitted onto a shaft, this assumption would not be true, and there would be an increased pressure at each end of the hub. It is customary to allow for this condition by employing a stress-concentration factor. The value of this factor depends upon the contact pressure and the design of the female member, but its theoretical value is seldom greater than 2.

See Section 7–8 for further discussion of press fits.

3–17 Temperature Effects

When the temperature of an unrestrained body is uniformly increased, the body expands, and the normal strain is

$$\epsilon_x = \epsilon_y = \epsilon_z = \alpha(\Delta T) \tag{3-62}$$

where α is the coefficient of thermal expansion and ΔT is the temperature change, in degrees. In this action the body experiences a simple volume increase with the components of shear strain all zero.

If a straight bar is restrained at the ends so as to prevent lengthwise expansion and then is subjected to a uniform increase in temperature, a compressive stress will develop because of the axial constraint. The stress is

$$\sigma = -\epsilon E = -\alpha(\Delta T)E \tag{3-63}$$

In a similar manner, if a uniform flat plate is restrained at the edges and also subjected to a uniform temperature rise, the compressive stress developed is given by the equation

$$\sigma = -\frac{\alpha(\Delta T)E}{1 - \nu} \tag{3-64}$$

The stresses expressed by Equations (3–63) and (3–64) are called *thermal stresses.* They arise because of a temperature change in a clamped or restrained member. Such stresses, for example, occur during welding, since parts to be welded must be clamped before welding. Table 3–3 lists approximate values of the coefficients of thermal expansion.

Table 3–3 Coefficients of Thermal Expansion (Linear Mean Coefficients for the Temperature Range 0–100°C)

Material	Celsius Scale (°C^{-1})
Aluminum	$23.9(10)^{-6}$
Brass, cast	$18.7(10)^{-6}$
Carbon steel	$10.8(10)^{-6}$
Cast iron	$10.6(10)^{-6}$
Magnesium	$25.2(10)^{-6}$
Nickel steel	$13.1(10)^{-6}$
Stainless steel	$17.3(10)^{-6}$
Tungsten	$4.3(10)^{-6}$

3–18 Curved Beams in Bending[11]

The distribution of stress in a curved flexural member is determined by using the following assumptions:

- The cross section has an axis of symmetry in the plane of bending.
- Plane cross sections remain plane after bending.
- The modulus of elasticity is the same in tension as in compression.

We shall find that the neutral axis and the centroidal axis of a curved beam, unlike the axes of a straight beam, are not coincident and also that the stress does not vary linearly from the neutral axis. The notation shown in Figure 3–40 is defined as follows:

r_o = radius of outer fiber
r_i = radius of inner fiber
h = depth of section
c_o = distance from neutral axis to outer fiber
c_i = distance from neutral axis to inner fiber
r_n = radius of neutral axis
r_c = radius of centroidal axis
$e = r_c - r_n$, distance from centroidal axis to neutral axis
M = bending moment; positive M decreases curvature

Figure 3–40 shows that the neutral and centroidal axes are not coincident. The location of the neutral axis with respect to the center of curvature O is given by the equation

$$r_n = \frac{A}{\int \frac{dA}{r}} \tag{3–65}$$

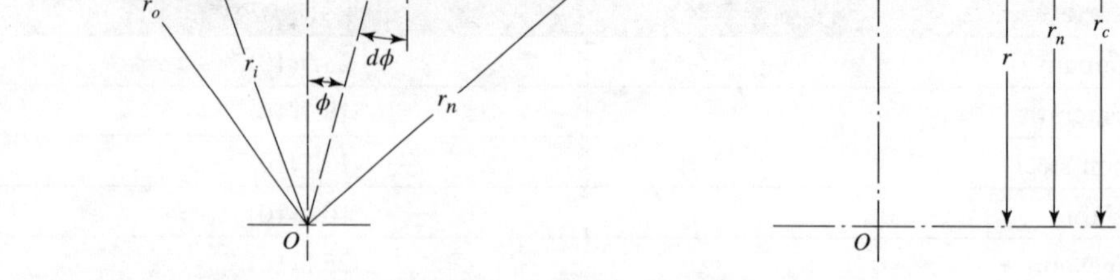

Figure 3–40

Note that y is positive in the direction toward the center of curvature, point O.

[11]For a complete development of the relations in this section, see Richard G. Budynas, *Advanced Strength and Applied Stress Analysis*, 2nd ed., McGraw Hill, New York, 1999, pp. 309–317.

Furthermore, it can be shown that the stress distribution is given by

$$\sigma = \frac{My}{Ae(r_n - y)} \tag{3–66}$$

where M is positive in the direction shown in Figure 3–40. The stress distribution given by Equation (3–66) is *hyperbolic* and not linear as is the case for straight beams. The critical stresses occur at the inner and outer surfaces where $y = c_i$ and $y = -c_o$, respectively, and are

$$\sigma_i = \frac{Mc_i}{Aer_i} \qquad \sigma_o = -\frac{Mc_o}{Aer_o} \tag{3–67}$$

These equations are valid for pure bending. In the usual and more general case, such as a crane hook, the U frame of a press, or the frame of a C clamp, the bending moment is due to a force acting at a distance from the cross section under consideration. Thus, the cross section transmits a bending moment *and* an axial force. The axial force is located at *the centroidal axis* of the section and the bending moment is then computed at this location. The tensile or compressive stress due to the axial force is then added to the bending stresses given by Equations (3–66) and (3–67) to obtain the resultant stresses acting on the section.

EXAMPLE 3–19

Plot the distribution of stresses across section A–A of the crane hook shown in Figure 3–41*a*. The cross section is rectangular, with $b = 18$ mm and $h = 100$ mm, and the load is $F = 22$ kN.

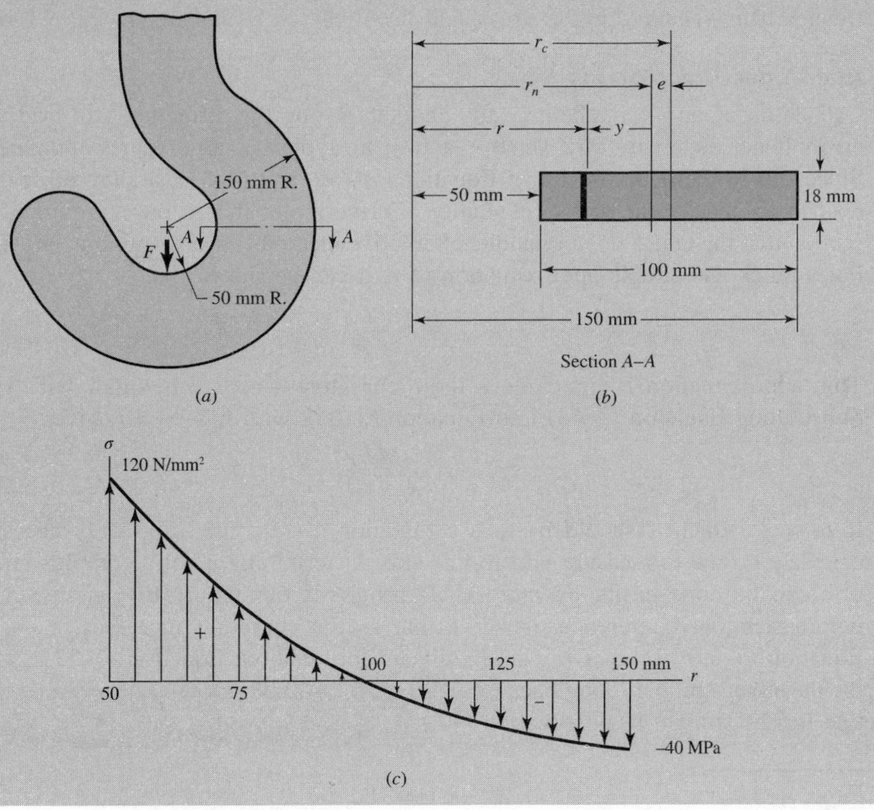

Figure 3–41

(*a*) Plan view of crane hook; (*b*) cross section and notation; (*c*) resulting stress distribution. There is no stress concentration.

Solution

Since $A = bh$, we have $dA = b\,dr$ and, from Equation (3–65),

$$r_n = \frac{A}{\int \frac{dA}{r}} = \frac{bh}{\int_{r_i}^{r_o} \frac{b}{r} dr} = \frac{h}{\ln \frac{r_o}{r_i}} \tag{1}$$

From Figure 3–41b, we see that $r_i = 50$ mm, $r_o = 150$ mm, $r_c = 100$ mm, and $A = 1900$ mm^2. Thus, from Equation (1),

$$r_n = \frac{h}{\ln(r_o/r_i)} = \frac{100}{\ln(150/50)} = 91.02 \text{ mm}$$

and the eccentricity is $e = r_c - r_n = 100 - 91.02 = 8.98$ mm. The moment M is positive and is $M = Fr_c =$ 22 000(100) = 2 200 000 N · mm^2. Adding the axial component of stress to Equation (3–66) gives

$$\sigma = \frac{F}{A} + \frac{My}{Ae(r_n - y)} = \frac{22\,000}{1900} + \frac{(2\,200\,000)(91.02 - r)}{1900(8.98)r} \tag{2}$$

Substituting values of r from 50 to 150 mm results in the stress distribution shown in Figure 3–41c. The stresses at the inner and outer radii are found to be 120 MPa and −40 MPa, respectively, as shown.

Note in the hook example, the symmetrical rectangular cross section causes the maximum tensile stress to be 3 times greater than the maximum compressive stress. If we wanted to design the hook to use material more effectively we would use more material at the inner radius and less material at the outer radius. For this reason, trapezoidal, T, or unsymmetric I, cross sections are commonly used. Sections most frequently encountered in the stress analysis of curved beams are shown in Table 3–4.

Alternative Calculations for e

Calculating r_n and r_c mathematically and subtracting the difference can lead to large errors if not done carefully, since r_n and r_c are typically large values compared to e. Since e is in the denominator of Equations (3–66) and (3–67), a large error in e can lead to an inaccurate stress calculation. Furthermore, if you have a complex cross section that the tables do not handle, alternative methods for determining e are needed. For a quick and simple approximation of e, it can be shown that[12]

$$e \approx \frac{I}{r_c A} \tag{3–68}$$

This approximation is good for a large curvature where e is small with $r_n \approx r_c$. Substituting Equation (3–68) into Equation (3–66), with $r_n - y = r$, gives

$$\sigma \approx \frac{My}{I}\frac{r_c}{r} \tag{3–69}$$

If $r_n \approx r_c$, which it should be to use Equation (3–69), then it is only necessary to calculate r_c, and to measure y from this axis. Determining r_c for a complex cross section can be done easily by most CAD programs or numerically as shown in the before-mentioned reference. Observe that as the curvature increases, $r \to r_c$, and Equation (3–69) becomes the straight-beam formulation, Equation (3–27). Note that the negative sign is missing because y in Figure 3–40 is vertically downward, opposite that for the straight-beam equation.

---

[12]Ibid., pp. 317–321. Also presents a numerical method.

Table 3–4 Formulas for Sections of Curved Beams

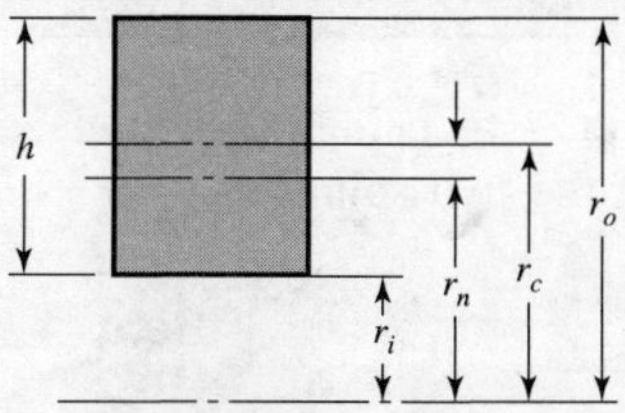

$$r_c = r_i + \frac{h}{2}$$

$$r_n = \frac{h}{\ln(r_o/r_i)}$$

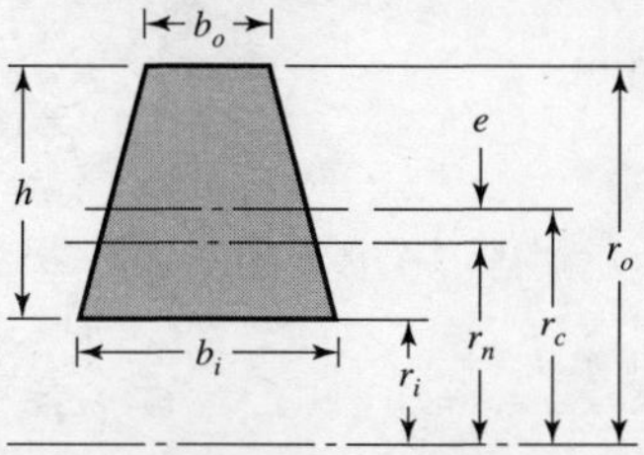

$$r_c = r_i + \left(\frac{h}{3}\right)\frac{b_i + 2b_o}{b_i + b_o}$$

$$r_n = \frac{A}{b_o - b_i + [(b_i r_o - b_o r_i)/h]\ln(r_o/r_i)}$$

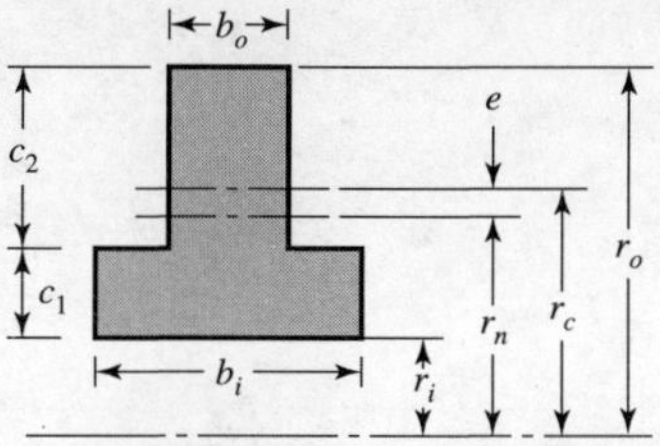

$$r_c = r_i + \frac{b_i c_1^2 + 2b_o c_1 c_2 + b_o c_2^2}{2(b_o c_2 + b_i c_1)}$$

$$r_n = \frac{b_i c_1 + b_o c_2}{b_i \ln[(r_i + c_i)/r_i] + b_o \ln[r_o/(r_i + c_1)]}$$

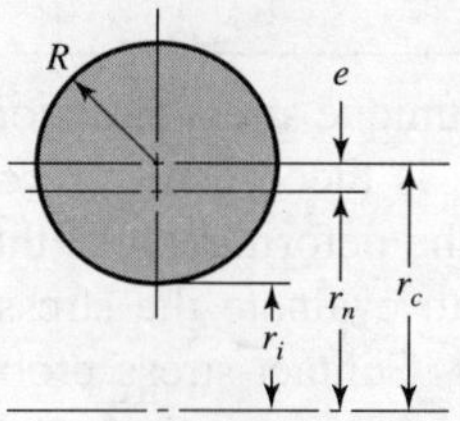

$$r_c = r_i + R$$

$$r_n = \frac{R^2}{2\left(r_c - \sqrt{r_c^2 - R^2}\right)}$$

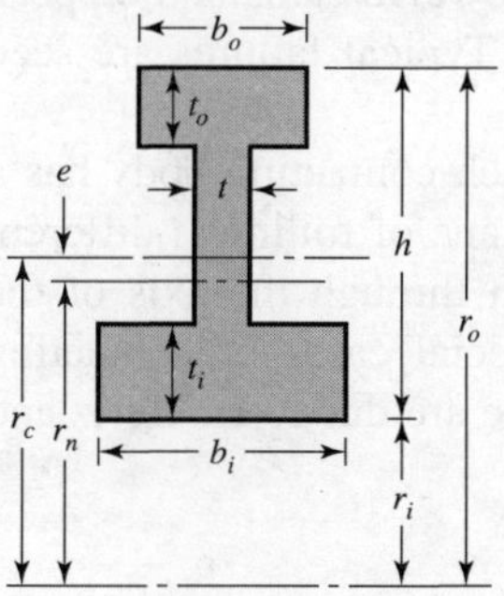

$$r_c = r_i + \frac{\frac{1}{2}h^2 t + \frac{1}{2}t_i^2(b_i - t) + t_o(b_o - t)(h - t_o/2)}{t_i(b_i - t) + t_o(b_o - t) + ht}$$

$$r_n = \frac{t_i(b_i - t) + t_o(b_o - t) + ht}{b_i \ln\dfrac{r_i + t_i}{r_i} + t\ln\dfrac{r_o - t_o}{r_i + t_i} + b_o\ln\dfrac{r_o}{r_o - t_o}}$$

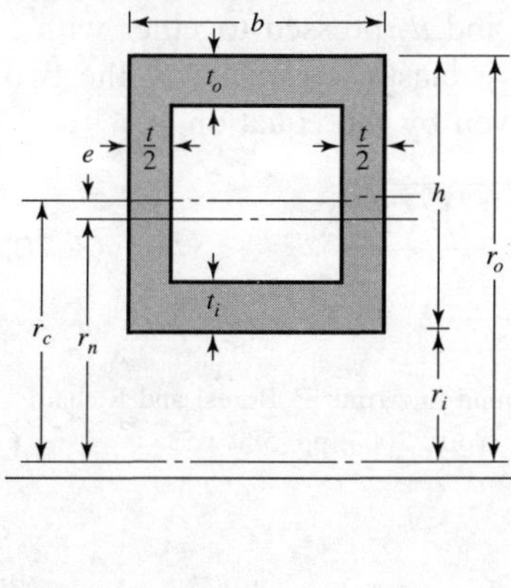

$$r_c = r_i + \frac{\frac{1}{2}h^2 t + \frac{1}{2}t_i^2(b - t) + t_o(b - t)(h - t_o/2)}{ht + (b - t)(t_i + t_o)}$$

$$r_n = \frac{(b - t)(t_i + t_o) + ht}{b\left(\ln\dfrac{r_i + t_i}{r_i} + \ln\dfrac{r_o}{r_o - t_o}\right) + t\ln\dfrac{r_o - t_o}{r_i + t_i}}$$

EXAMPLE 3–20

Consider the circular section in Table 3–4 with $r_c = 70$ mm and $R = 25$ mm. Determine e by using the formula from the table and approximately by using Equation (3–68). Compare the results of the two solutions.

Solution

Using the formula from Table 3–4 gives

$$r_n = \frac{R^2}{2\left(r_c - \sqrt{r_c^2 - R^2}\right)} = \frac{25^2}{2\left(70 - \sqrt{70^2 - 25}\right)} = 67.69 \text{ mm}$$

This gives an eccentricity of

Answer $$e = r_c - r_n = 70 - 67.69 = 2.31 \text{ mm}$$

The approximate method, using Equation (3–68), yields

Answer $$e \approx \frac{I}{r_c A} = \frac{\pi R^4/4}{r_c(\pi R^2)} = \frac{R^2}{4r_c} = \frac{25^2}{4(70)} = 2.23 \text{ mm}$$

This differs from the exact solution by −2.9 percent.

3–19 Contact Stresses

When two bodies having curved surfaces are pressed together, a unique stress situation arises due to the point or line contact. The contact surface area is theoretically zero, which would imply an infinite stress. It is necessary to include the deformation of the contacting surfaces, creating a finite area of contact, in order to evaluate the stress. The stresses developed in the two bodies are three-dimensional. Contact-stress problems arise in the contact of a wheel and a rail, in automotive valve cams and tappets, in mating gear teeth, and in the action of rolling bearings. Typical failures are seen as cracks, pits, or flaking in the surface material.

The most general case of contact stress occurs when each contacting body has a double radius of curvature; that is, when the radius in the plane of rolling is different from the radius in a perpendicular plane, both planes taken through the axis of the contacting force. Here we shall consider only the two special cases of contacting spheres and contacting cylinders.[13] The results presented here are due to H. Hertz and so are frequently known as *Hertzian stresses*.

Spherical Contact

Figure 3–42 shows two solid spheres of diameters d_1 and d_2 pressed together with a force F. Specifying E_1, ν_1 and E_2, ν_2 as the respective elastic constants of the two spheres, the radius a of the circular contact area is given by the equation

$$a = \sqrt[3]{\frac{3F}{8} \frac{\left(1 - \nu_1^2\right)/E_1 + \left(1 - \nu_2^2\right)/E_2}{1/d_1 + 1/d_2}} \tag{3–70}$$

[13]A more comprehensive presentation of contact stresses may be found in Arthur P. Boresi and Richard J. Schmidt, *Advanced Mechanics of Materials*, 6th ed., Wiley, New York, 2003, pp. 589–623.

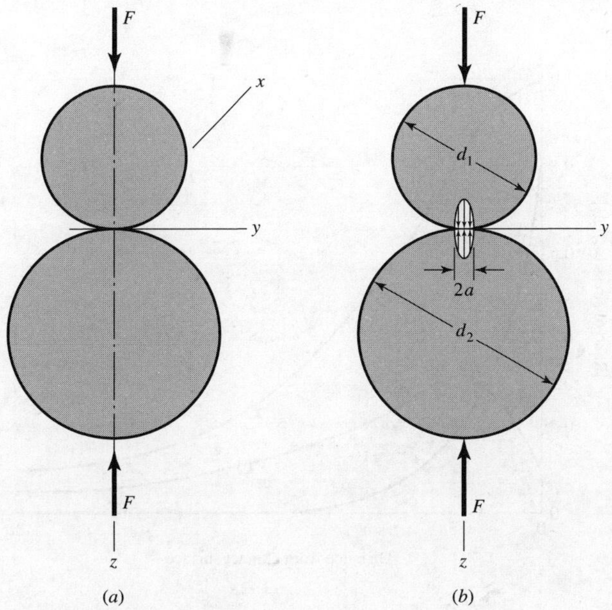

Figure 3–42

(*a*) Two spheres held in contact by force *F*; (*b*) contact stress has a hemispherical distribution across contact zone of diameter $2a$.

The pressure distribution within the contact area of each sphere is hemispherical, as shown in Figure 3–42*b*. The maximum pressure occurs at the center of the contact area and is

$$p_{\max} = \frac{3F}{2\pi a^2} \tag{3–71}$$

Equations (3–70) and (3–71) are perfectly general and also apply to the contact of a sphere and a plane surface or of a sphere and an internal spherical surface. For a plane surface, use $d = \infty$. For an internal surface, the diameter is expressed as a negative quantity.

The maximum stresses occur on the z axis, and these are principal stresses. Their values are

$$\sigma_1 = \sigma_2 = \sigma_x = \sigma_y = -p_{\max}\left[\left(1 - \left|\frac{z}{a}\right| \tan^{-1}\frac{1}{|z/a|}\right)(1+\nu) - \frac{1}{2\left(1 + \dfrac{z^2}{a^2}\right)}\right] \tag{3–72}$$

$$\sigma_3 = \sigma_z = \frac{-p_{\max}}{1 + \dfrac{z^2}{a^2}} \tag{3–73}$$

These equations are valid for either sphere, but the value used for Poisson's ratio must correspond with the sphere under consideration. The equations are even more complicated when stress states off the z axis are to be determined, because here the x and y coordinates must also be included. But these are not required for design purposes, because the maxima occur on the z axis.

Mohr's circles for the stress state described by Equations (3–72) and (3–73) are a point and two coincident circles. Since $\sigma_1 = \sigma_2$, we have $\tau_{12} = 0$ and

$$\tau_{\max} = \tau_{13} = \tau_{23} = \frac{\sigma_1 - \sigma_3}{2} = \frac{\sigma_2 - \sigma_3}{2} \tag{3–74}$$

Figure 3–43

Magnitude of the stress components below the surface as a function of the maximum pressure of contacting spheres. Note that the maximum shear stress is slightly below the surface at $z = 0.48a$ and is approximately $0.3p_{max}$. The chart is based on a Poisson ratio of 0.30. Note that the normal stresses are all compressive stresses.

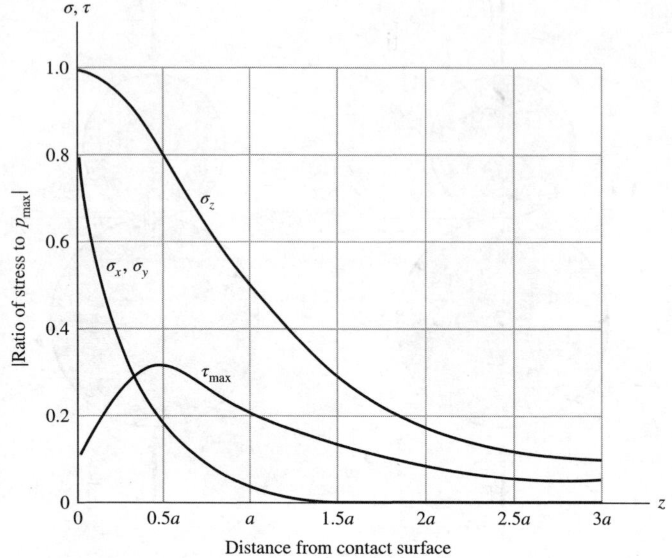

Figure 3–43 is a plot of Equations (3–72), (3–73), and (3–74) for a distance to $3a$ below the surface. Note that the shear stress reaches a maximum value slightly below the surface. It is the opinion of many authorities that this maximum shear stress is responsible for the surface fatigue failure of contacting elements. The explanation is that a crack originates at the point of maximum shear stress below the surface and progresses to the surface and that the pressure of the lubricant wedges the chip loose.

Cylindrical Contact

Figure 3–44 illustrates a similar situation in which the contacting elements are two cylinders of length l and diameters d_1 and d_2. As shown in Figure 3–44*b*, the area of

Figure 3–44

(*a*) Two right circular cylinders held in contact by forces F uniformly distributed along cylinder length l. (*b*) Contact stress has an elliptical distribution across the contact zone of width $2b$.

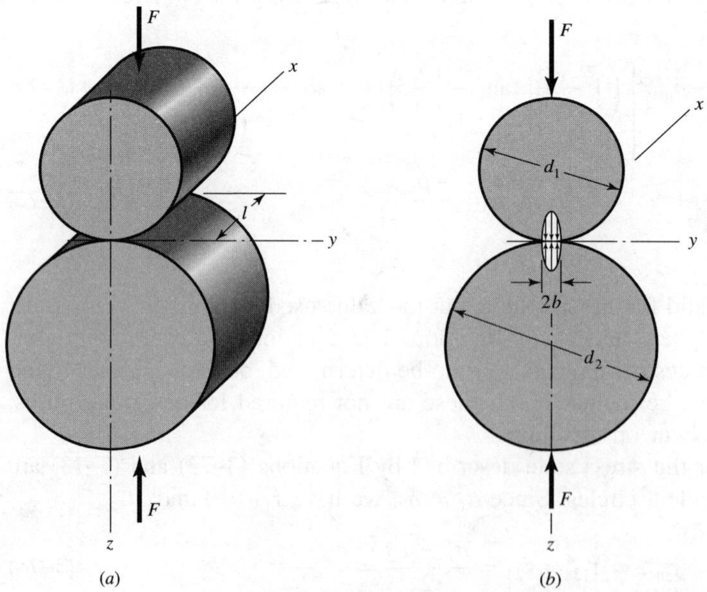

contact is a narrow rectangle of width $2b$ and length l, and the pressure distribution is elliptical. The half-width b is given by the equation

$$b = \sqrt{\frac{2F}{\pi l}\frac{(1-\nu_1^2)/E_1 + (1-\nu_2^2)/E_2}{1/d_1 + 1/d_2}} \tag{3–75}$$

The maximum pressure is

$$p_{\max} = \frac{2F}{\pi b l} \tag{3–76}$$

Equations (3–75) and (3–76) apply to a cylinder and a plane surface, such as a rail, by making $d = \infty$ for the plane surface. The equations also apply to the contact of a cylinder and an internal cylindrical surface; in this case d is made negative for the internal surface.

The stress state along the z axis is given by the equations

$$\sigma_x = -2\nu p_{\max}\left(\sqrt{1+\frac{z^2}{b^2}} - \left|\frac{z}{b}\right|\right) \tag{3–77}$$

$$\sigma_y = -p_{\max}\left(\frac{1+2\dfrac{z^2}{b^2}}{\sqrt{1+\dfrac{z^2}{b^2}}} - 2\left|\frac{z}{b}\right|\right) \tag{3–78}$$

$$\sigma_3 = \sigma_z = \frac{-p_{\max}}{\sqrt{1+z^2/b^2}} \tag{3–79}$$

These three equations are plotted in Figure 3–45 up to a distance of $3b$ below the surface. For $0 \le z \le 0.436b$, $\sigma_1 = \sigma_x$, and $\tau_{\max} = (\sigma_1 - \sigma_3)/2 = (\sigma_x - \sigma_z)/2$. For $z \ge 0.436b$, $\sigma_1 = \sigma_y$, and $\tau_{\max} = (\sigma_y - \sigma_z)/2$. A plot of $\tau_{\max}$ is also included in Figure 3–45, where the greatest value occurs at $z/b = 0.786$ with a value of $0.30\ p_{\max}$.

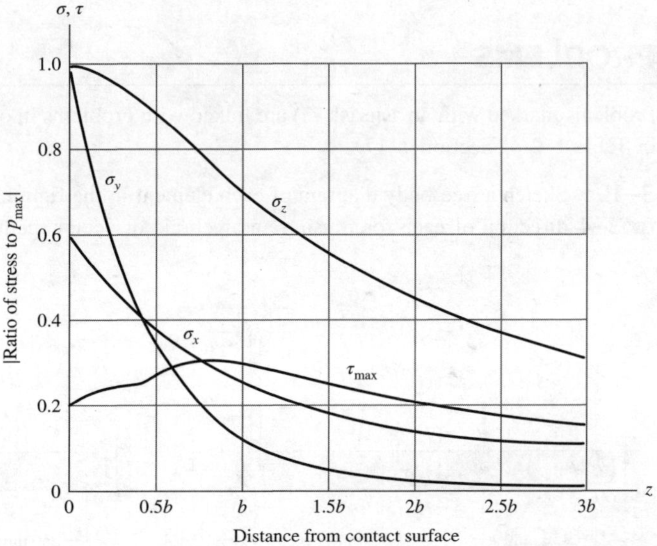

Figure 3–45

Magnitude of the stress components below the surface as a function of the maximum pressure for contacting cylinders. The largest value of $\tau_{\max}$ occurs at $z/b = 0.786$. Its maximum value is $0.30p_{\max}$. The chart is based on a Poisson ratio of 0.30. Note that all normal stresses are compressive stresses.

Hertz (1881) provided the preceding mathematical models of the stress field when the contact zone is free of shear stress. Another important contact stress case is *line of contact* with friction providing the shearing stress on the contact zone. Such shearing stresses are small with cams and rollers, but in cams with flatfaced followers, wheel-rail contact, and gear teeth, the stresses are elevated above the Hertzian field. Investigations of the effect on the stress field due to normal and shear stresses in the contact zone were begun theoretically by Lundberg (1939), and continued by Mindlin (1949), Smith-Liu (1949), and Poritsky (1949) independently. For further detail, see the reference cited in footnote 13.

3–20 Summary

The ability to quantify the stress condition at a critical location in a machine element is an important skill of the engineer. Why? Whether the member fails or not is assessed by comparing the stress at a critical location with the corresponding material strength at this location. This chapter has addressed the description of stress.

Stresses can be estimated with reasonable precision where the geometry is sufficiently simple that theory provides the necessary quantitative relationships. In other cases, approximations are used. There are numerical approximations such as finite element analysis (FEA, see Chapter 19), whose results tend to converge on the true values. There are experimental measurements, strain gauging, for example, allowing *inference* of stresses from the measured strain conditions. Whatever the method(s), the goal is a robust description of the stress condition at a critical location.

Methods such as FEA are attractive, but a solid foundational understanding of stress is essential. An engineer should avoid the trap of thinking of stresses as simply "high" or "low" at a given location. To understand the stress situation, and to properly evaluate a part for failure, it is necessary to move through the steps from free-body diagram to determine forces, evaluation of the internal forces within members, determination of the types of stress carried by a member, and visualization of the 3D stress element. From here, it is a straight forward process to determine principal stresses and maximum shear stresses in order to hone in on the critical stress element(s). Chapters 5 and 6 will provide the means to evaluate critical stress elements for static and dynamic failure. But it starts with the ability to find and analyze the critical stress elements.

PROBLEMS

Problems marked with an asterisk (*) are linked with problems in other chapters, as summarized in Table 1–2 of Section 1–17.

3–1* to 3–4 Sketch a free-body diagram of each element in the figure. Compute the magnitude and direction of each force using an algebraic or vector method, as specified.

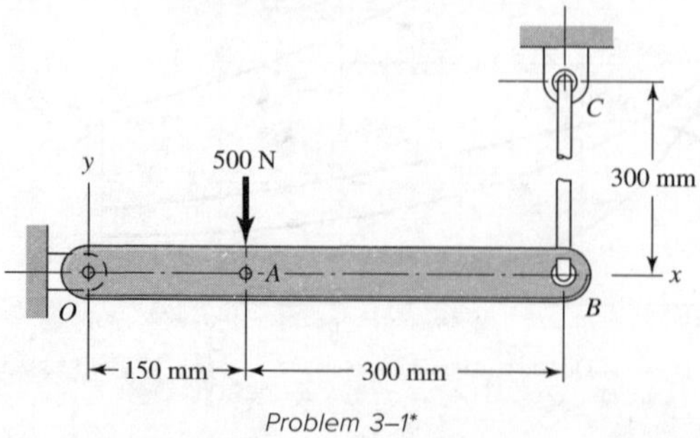

Problem 3–1*

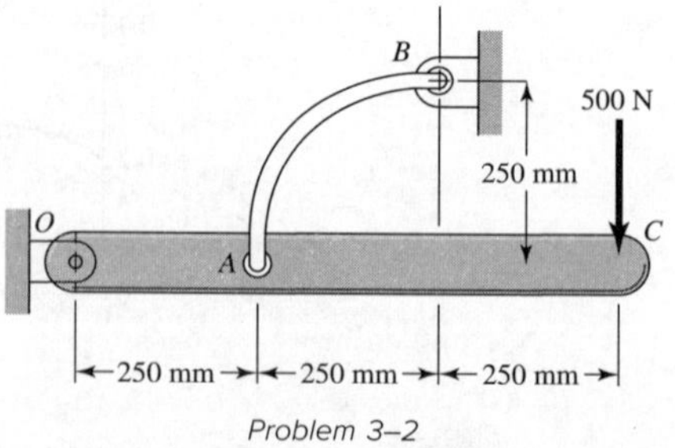

Problem 3–2

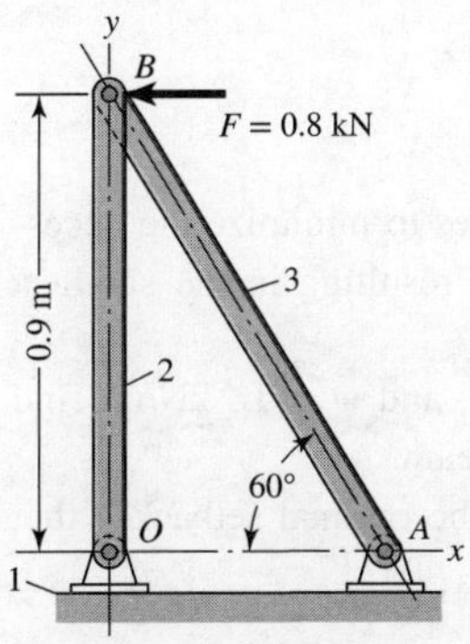

Problem 3–3

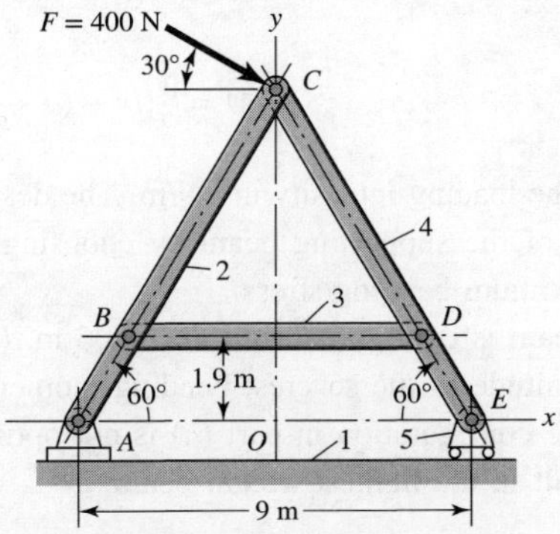

Problem 3–4

3–5 to 3–8 For the beam shown, find the reactions at the supports and plot the shear-force and bending-moment diagrams. Label the diagrams properly and provide values at all key points.

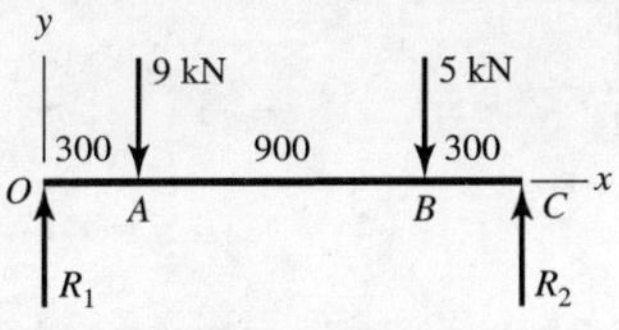

Problem 3–5
Dimensions in millimeters

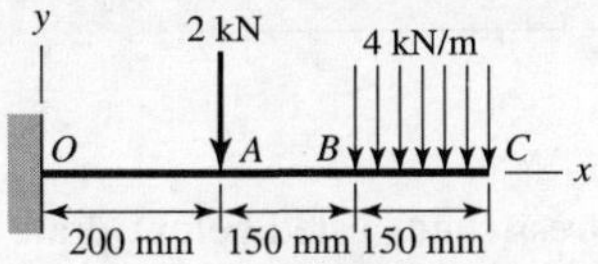

Problem 3–6

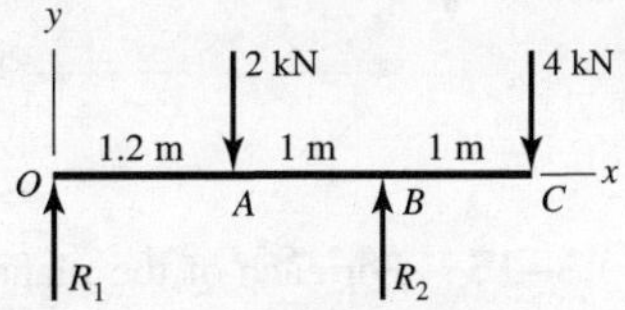

Problem 3–7

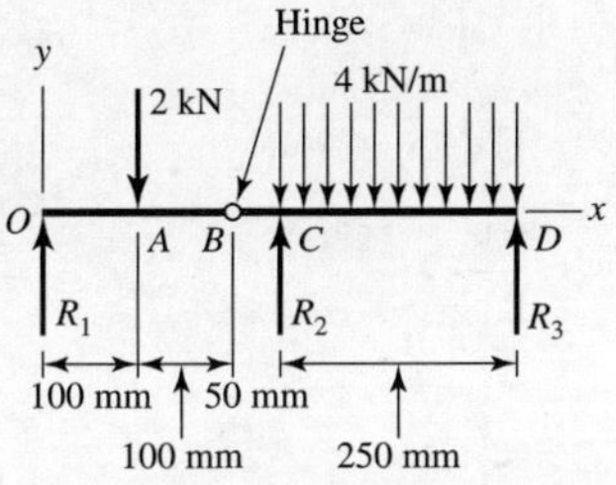

Problem 3–8

3–9 Repeat Problem 3–5 using singularity functions exclusively (including reactions). Singularity functions are covered comprehensively in Section 4–6 in the context of deflection analysis, but are very suitable for solving for shear force and bending moments in beams.

3–10 Repeat Problem 3–6 using singularity functions exclusively (including reactions). Singularity functions are covered comprehensively in Section 4–6 in the context of deflection analysis, but are very suitable for solving for shear force and bending moments in beams.

3–11 Repeat Problem 3–7 using singularity functions exclusively (including reactions). Singularity functions are covered comprehensively in Section 4–6 in the context of deflection analysis, but are very suitable for solving for shear force and bending moments in beams.

3–12 Repeat Problem 3–8 using singularity functions exclusively (including reactions). Singularity functions are covered comprehensively in Section 4–6 in the context of deflection analysis, but are very suitable for solving for shear force and bending moments in beams.

3–13 For a beam from Table A–9, as specified by your instructor, find general expressions for the loading, shear-force, bending-moment, and support reactions. Use the method specified by your instructor.

3–14 A beam carrying a uniform load is simply supported with the supports set back a distance a from the ends as shown in the figure. The bending moment at x can be found from summing moments to zero at section x:

$$\sum M = M + \frac{1}{2}w(a + x)^2 - \frac{1}{2}wlx = 0$$

or

$$M = \frac{w}{2}[lx - (a + x)^2]$$

where w is the loading intensity in kN/m. The designer wishes to minimize the necessary weight of the supporting beam by choosing a setback resulting in the smallest possible maximum bending stress.

(*a*) If the beam is configured with $a = 0.06$ m, $l = 0.25$ m, and $w = 18$ kN/m, find the magnitude of the severest bending moment in the beam.

(*b*) Since the configuration in part (*a*) is not optimal, find the optimal setback a that will result in the lightest-weight beam.

Problem 3–14

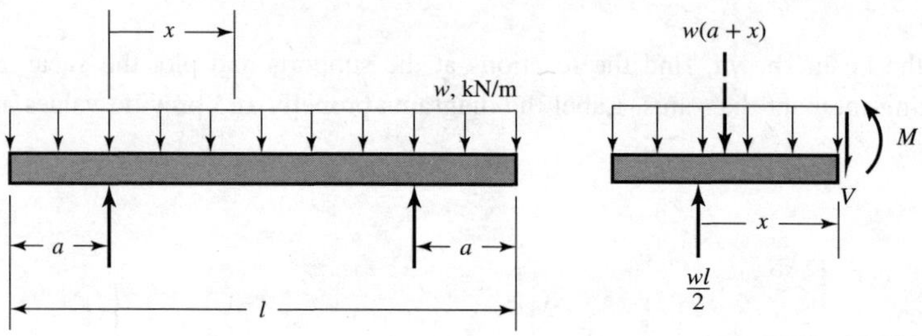

3–15 For each of the plane stress states listed below, draw a Mohr's circle diagram properly labeled, find the principal normal and shear stresses, and determine the angle from the x axis to σ_1. Draw stress elements as in Figure 3–13*c* and *d* and label all details.

(*a*) $\sigma_x = 20$ MPa, $\sigma_y = -10$ MPa, $\tau_{xy} = 8$ MPa cw
(*b*) $\sigma_x = 16$ MPa, $\sigma_y = 9$ MPa, $\tau_{xy} = 5$ MPa ccw
(*c*) $\sigma_x = 10$ MPa, $\sigma_y = 24$ MPa, $\tau_{xy} = 6$ MPa ccw
(*d*) $\sigma_x = -12$ MPa, $\sigma_y = 22$ MPa, $\tau_{xy} = 12$ MPa cw

3–16 Repeat Problem 3–15 for:

(*a*) $\sigma_x = -8$ MPa, $\sigma_y = 7$ MPa, $\tau_{xy} = 6$ MPa cw
(*b*) $\sigma_x = 9$ MPa, $\sigma_y = -6$ MPa, $\tau_{xy} = 3$ MPa cw
(*c*) $\sigma_x = -4$ MPa, $\sigma_y = 12$ MPa, $\tau_{xy} = 7$ MPa ccw
(*d*) $\sigma_x = 6$ MPa, $\sigma_y = -5$ MPa, $\tau_{xy} = 8$ MPa ccw

3–17 Repeat Problem 3–15 for:

(*a*) $\sigma_x = 12$ MPa, $\sigma_y = 6$ MPa, $\tau_{xy} = 4$ MPa, cw
(*b*) $\sigma_x = 30$ MPa, $\sigma_y = -10$ MPa, $\tau_{xy} = 10$ MPa ccw
(*c*) $\sigma_x = -10$ MPa, $\sigma_y = 18$ MPa, $\tau_{xy} = 9$ MPa cw
(*d*) $\sigma_x = 9$ MPa, $\sigma_y = 19$ MPa, $\tau_{xy} = 8$ MPa cw

3–18 For each of the stress states listed below, find all three principal normal and shear stresses. Draw a complete Mohr's three-circle diagram and label all points of interest.

(*a*) $\sigma_x = -80$ MPa, $\sigma_y = -30$ MPa, $\tau_{xy} = 20$ MPa cw
(*b*) $\sigma_x = 30$ MPa, $\sigma_y = -60$ MPa, $\tau_{xy} = 30$ MPa cw
(*c*) $\sigma_x = 40$ MPa, $\sigma_z = -30$ MPa, $\tau_{xy} = 20$ MPa ccw
(*d*) $\sigma_x = 50$ MPa, $\sigma_z = -20$ MPa, $\tau_{xy} = 30$ MPa cw

3–19 Repeat Problem 3–18 for:

(*a*) $\sigma_x = 10$ MPa, $\sigma_y = -4$ MPa
(*b*) $\sigma_x = 10$ MPa, $\tau_{xy} = 4$ MPa ccw

(c) $\sigma_x = -2$ MPa, $\sigma_y = -8$ MPa, $\tau_{xy} = 4$ MPa cw
(d) $\sigma_x = 10$ MPa, $\sigma_y = -30$ MPa, $\tau_{xy} = 10$ MPa ccw

3–20 The state of stress at a point is $\sigma_x = -6$, $\sigma_y = 18$, $\sigma_z = -12$, $\tau_{xy} = 9$, $\tau_{yz} = 6$, and $\tau_{zx} = -15$ MPa. Determine the principal stresses, draw a complete Mohr's three-circle diagram, labeling all points of interest, and report the maximum shear stress for this case.

3–21 Repeat Problem 3–20 with $\sigma_x = 20$, $\sigma_y = 0$, $\sigma_z = 20$, $\tau_{xy} = 40$, $\tau_{yz} = -20\sqrt{2}$, and $\tau_{zx} = 0$ MPa.

3–22 Repeat Problem 3–20 with $\sigma_x = 10$, $\sigma_y = 40$, $\sigma_z = 40$, $\tau_{xy} = 20$, $\tau_{yz} = -40$, and $\tau_{zx} = -20$ MPa.

3–23 A 12 mm diameter steel tension rod is 1.8 m long and carries a load of 9 kN. Find the tensile stress, the total deformation, the unit strains, and the change in the rod diameter.

3–24 Repeat Problem 3–23 except change the rod to aluminum and the load to 1.8 kN.

3–25 A 30-mm-diameter copper rod is 1 m long with a yield strength of 70 MPa. Determine the axial force necessary to cause the diameter of the rod to reduce by 0.01 percent, assuming elastic deformation. Check that the elastic deformation assumption is valid by comparing the axial stress to the yield strength.

3–26 A diagonal aluminum alloy tension rod of diameter d and initial length l is used in a rectangular frame to prevent collapse. The rod can safely support a tensile stress of σ_{allow}. If $d = 16$ mm, $l = 3$ m, and $\sigma_{\text{allow}} = 140$ MPa, determine how much the rod must be stretched to develop this allowable stress.

3–27 Electrical strain gauges were applied to a notched specimen to determine the stresses in the notch. The results were $\varepsilon_x = 0.0019$ and $\varepsilon_y = -0.000\ 72$. Find σ_x and σ_y if the material is carbon steel.

3–28 Repeat Problem 3–27 for a material of aluminum.

3–29 For plane strain, prove that:

$$\sigma_z = \nu(\sigma_x + \sigma_y) \tag{3–80}$$

$$\varepsilon_x = \frac{1+\nu}{E}[(1-\nu)\sigma_x - \nu\sigma_y] \tag{3–81a}$$

$$\varepsilon_y = \frac{1+\nu}{E}[(1-\nu)\sigma_y - \nu\sigma_x] \tag{3–81b}$$

3–30 The Roman method for addressing uncertainty in design was to build a copy of a design that was satisfactory and had proven durable. Although the early Romans did not have the intellectual tools to deal with scaling size up or down, you do. Consider a simply supported, rectangular-cross-section beam with a concentrated load F, as depicted in the figure.
(a) Show that the stress-to-load equation is

$$F = \frac{\sigma b h^2 l}{6ac}$$

(b) Subscript every parameter with m (for model) and divide into the above equation. Introduce a scale factor, $s = a_m/a = b_m/b = c_m/c$ etc. Since the Roman method

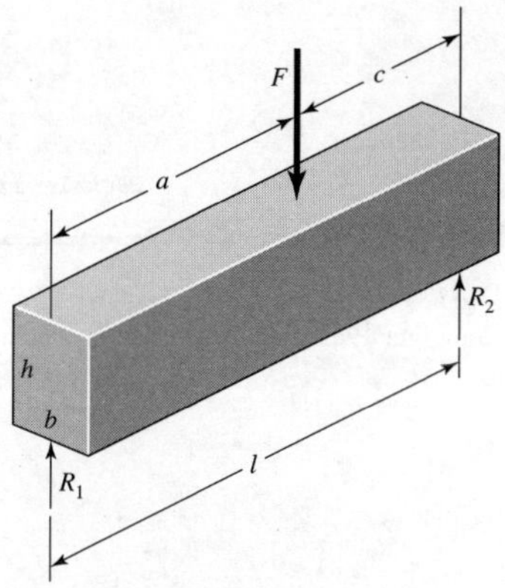

Problem 3–30

was to not "lean on" the material any more than the proven design, set $\sigma_m/\sigma = 1$. Express F_m in terms of the scale factors and F, and comment on what you have learned.

3–31 Using our experience with concentrated loading on a simple beam, Problem 3–30, consider a uniformly loaded simple beam (Table A–9–7).

(*a*) Show that the stress-to-load equation for a rectangular-cross-section beam is given by

$$W = \frac{4}{3}\frac{\sigma b h^2}{l}$$

where $W = wl$.

(*b*) Subscript every parameter with m (for model) and divide the model equation into the prototype equation. Introduce the scale factor s as in Problem 3–30, setting $\sigma_m/\sigma = 1$. Express W_m and w_m in terms of the scale factor, and comment on what you have learned.

3–32 Many years ago, the Chicago North Shore & Milwaukee Railroad was an electric railway running between the cities in its corporate title. It had passenger cars as shown in the figure, which weighed 460 kN, had 10 m, 0.2-m truck centers, 2-m-wheelbase trucks, and a coupled length of 17 m, 20 mm. Consider the case of a single car on a 30-m-long, simply supported deck plate girder bridge.

(*a*) What was the largest bending moment in the bridge?

(*b*) Where on the bridge was the moment located?

(*c*) What was the position of the car on the bridge?

(*d*) Under which axle is the bending moment?

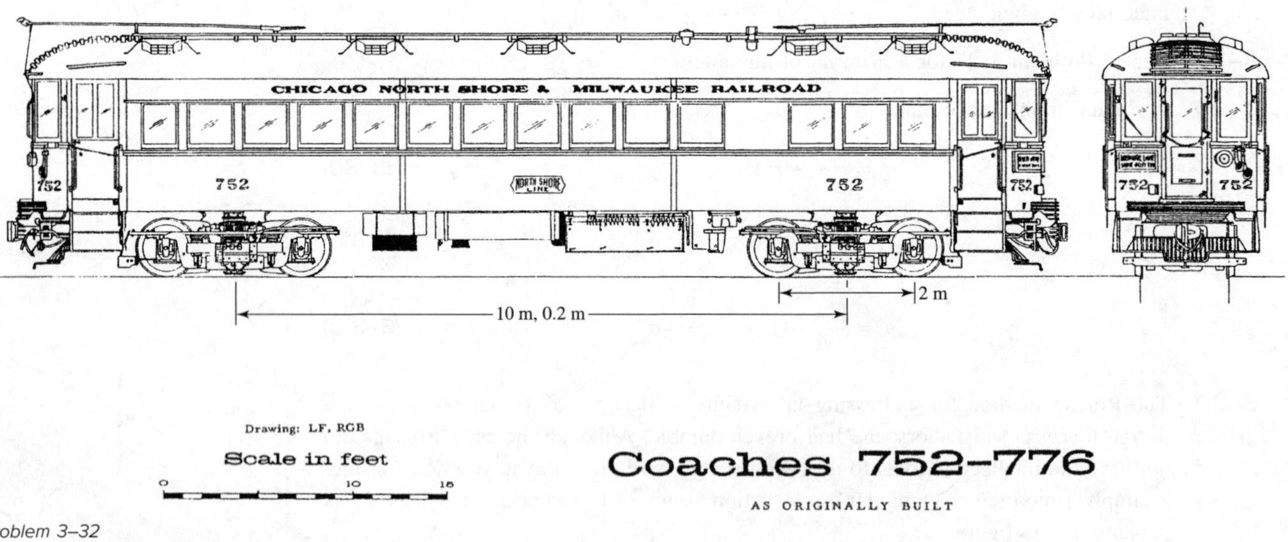

Problem 3–32

3–33 For each section illustrated, find the second moment of area, the location of the neutral axis, and the distances from the neutral axis to the top and bottom surfaces. Consider that the section is transmitting a positive bending moment about the z

axis, $M_z = 1.13$ kN · m. Determine the resulting stresses at the top and bottom surfaces and at every abrupt change in the cross section.

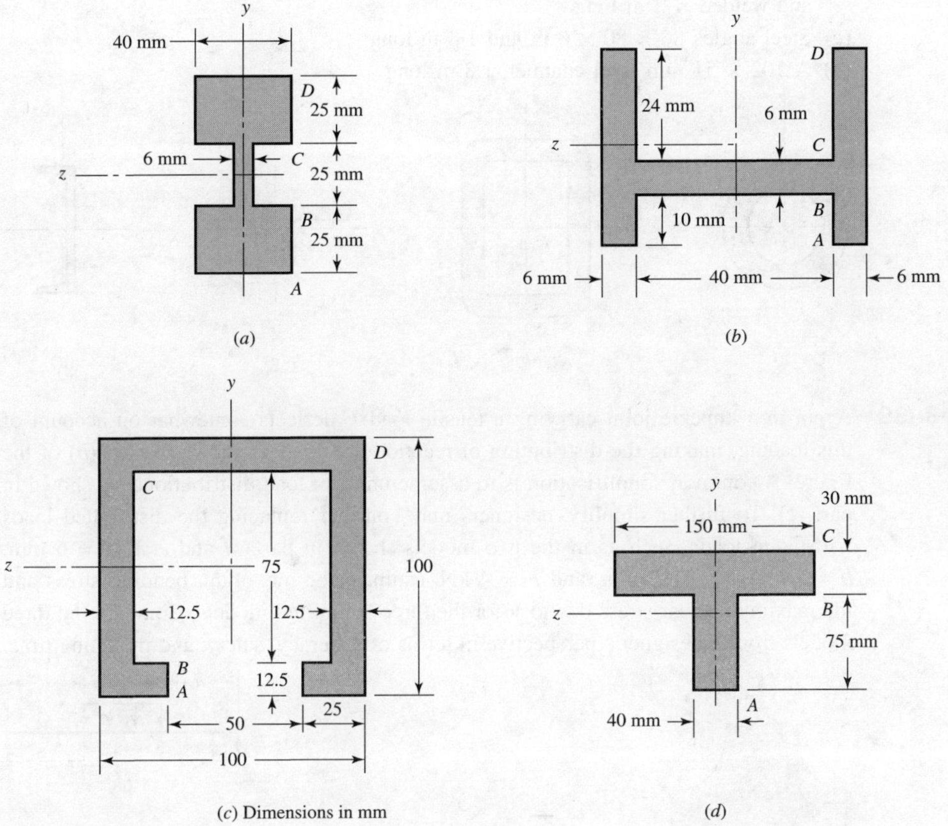

Problem 3–33

3–34 to 3–37 For the beam illustrated in the figure, find the locations and magnitudes of the maximum tensile bending stress due to M and the maximum shear stress due to V.

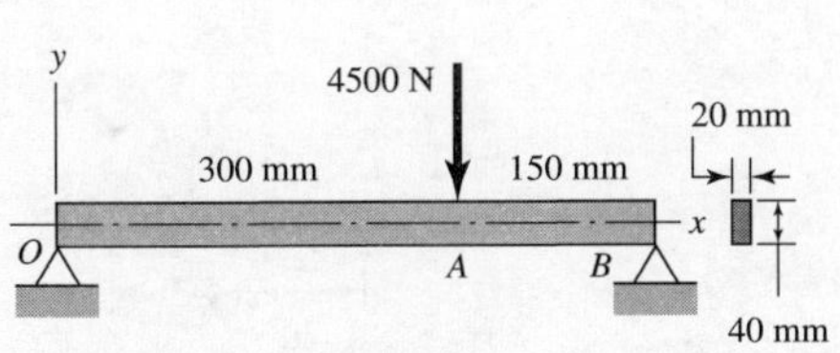

Problem 3–34

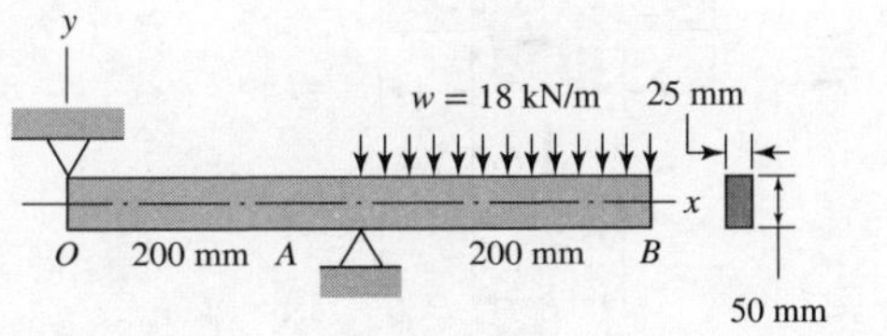

Problem 3–35

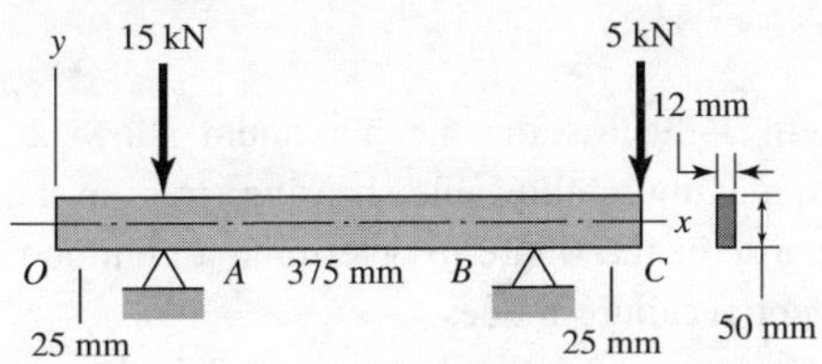

Problem 3–36

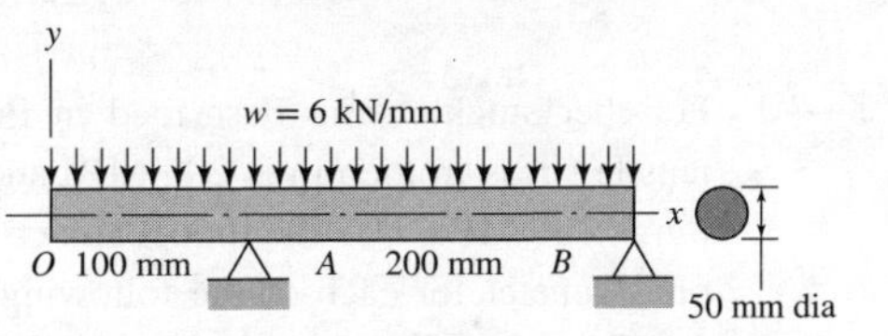

Problem 3–37

3–38 The figure illustrates a number of beam sections. Use an allowable bending stress of 80 MPa for steel and find the maximum safe uniformly distributed load that each beam can carry if the given lengths are between simple supports.

(*a*) Standard 50 mm × 5 mm tube, 3.6 m long
(*b*) Hollow steel tube 75 by 50 mm, outside dimensions, formed from 5 mm material and welded, 1.2 m long
(*c*) Steel angles 80 × 80 × 6 m and 1.8 m long
(*d*) A 102 × 51 mm steel channel, 1.8 m long

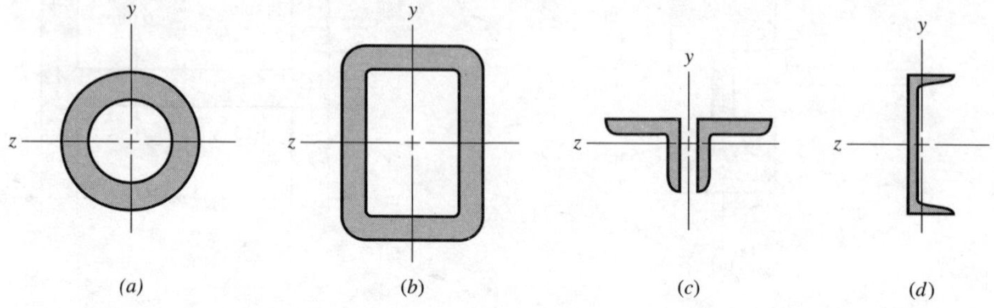

Problem 3–38

3–39* A pin in a knuckle joint carrying a tensile load F deflects somewhat on account of this loading, making the distribution of reaction and load as shown in part (*b*) of the figure. A common simplification is to assume uniform load distributions, as shown in part (*c*). To further simplify, designers may consider replacing the distributed loads with point loads, such as in the two models shown in parts *d* and *e*. If $a = 6$ mm, $b = 18$ mm, $d = 12$ mm, and $F = 4$ kN, estimate the maximum bending stress and the maximum shear stress due to V for the three simplified models. Compare the three models from a designer's perspective in terms of accuracy, safety, and modeling time.

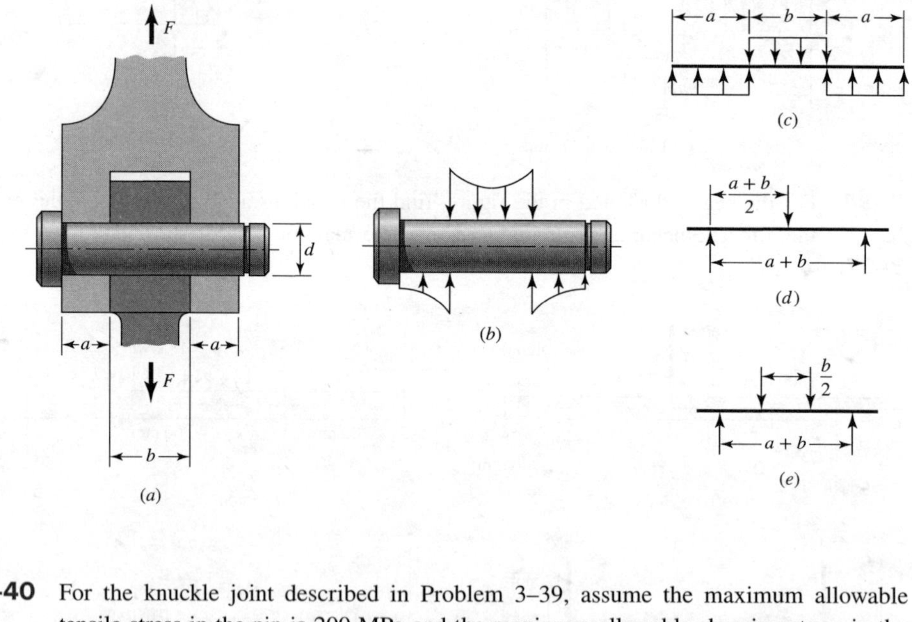

Problem 3–39*

3–40 For the knuckle joint described in Problem 3–39, assume the maximum allowable tensile stress in the pin is 200 MPa and the maximum allowable shearing stress in the pin is 100 MPa. Use the model shown in part *c* of the figure to determine a minimum pin diameter for each of the following potential failure modes.

(*a*) Consider failure based on bending at the point of maximum bending stress in the pin.
(*b*) Consider failure based on the average shear stress on the pin cross section at the interface plane of the knuckle and clevis.
(*c*) Consider failure based on shear at the point of the maximum transverse shear stress in the pin.

3–41 The figure illustrates a pin tightly fitted into a hole of a substantial member. A usual analysis is one that assumes concentrated reactions R and M at distance l from F. Suppose the reaction is distributed linearly along distance a. Is the resulting moment reaction larger or smaller than the concentrated reaction? What is the loading intensity q? What do you think of using the usual assumption?

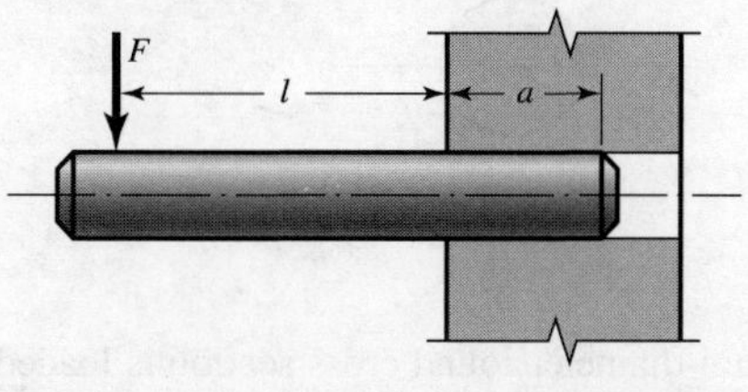

Problem 3–41

3–42 For the beam shown, determine (*a*) the maximum tensile and compressive bending stresses, (*b*) the maximum shear stress due to V, and (*c*) the maximum shear stress in the beam.

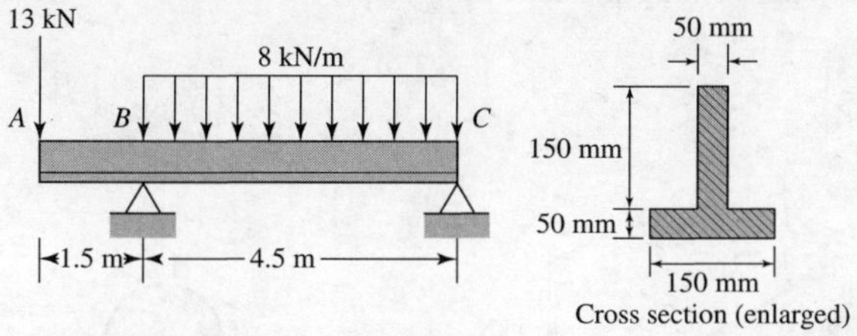

Problem 3–42

3–43 A cantilevered beam is loaded as shown. The cross section at the wall is shown, with points of interest A (at the top), B (at the center), and C (midway between A and B).

(*a*) Sketch the stress elements at points A, B, and C. Calculate the magnitudes of the stresses acting on the stress elements. Do not neglect transverse shear stress.

(*b*) Find the *maximum shear stresses* (i.e., from Mohr's circle) for the stress elements at A, B, and C.

(*c*) Assuming the *critical* stress element is the one with the largest *maximum shear stress* from part (*b*), which point is critical?

(*d*) If the length of the beam is varied from longer to shorter, the critical stress element will switch from point A to point B. At what length does this happen?

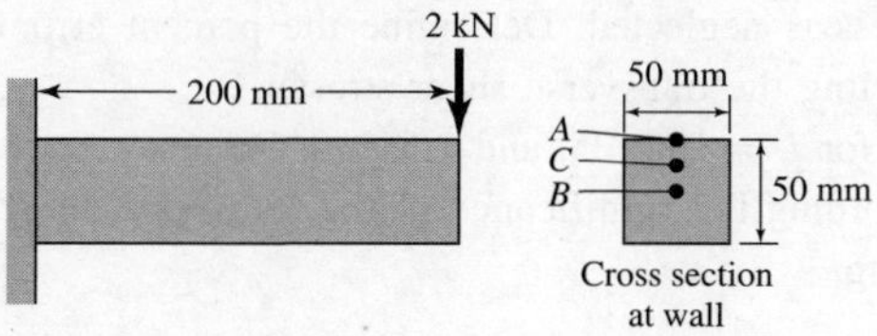

Problem 3–43

3–44 A short, cantilevered round peg is 40 mm in diameter and extends 10 mm from the wall. It is loaded, as shown, with a force $F = 50$ kN applied at the tip in the negative z direction, and a torque $T = 800$ N · m. The cross section at the wall is also shown, with points of interest A, B, and C.

(*a*) Sketch the stress elements at points A, B, and C. Calculate the magnitudes of the stresses acting on the stress elements. Do not neglect transverse shear stress.

(*b*) Find the maximum shear stresses (i.e., from Mohr's circle) for the stress elements at A, B, and C.

(*c*) Assuming the critical stress element is the one with the largest maximum shear stress from part (*b*), which point is critical?

Problem 3–44

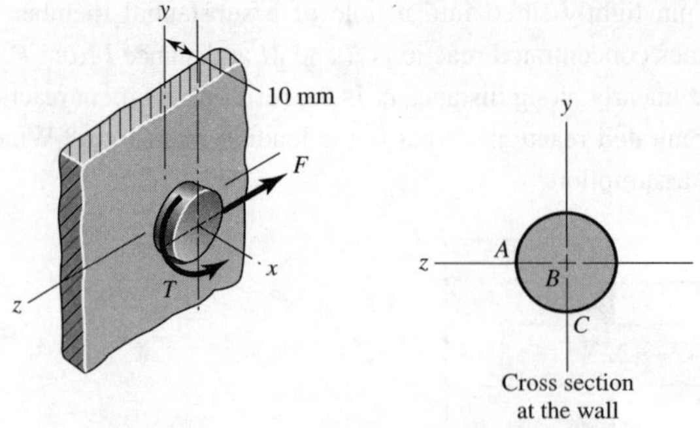

3–45 A cantilever beam with a 25-mm-diameter round cross section is loaded at the tip with a transverse force of 5 kN, as shown in the figure. The cross section at the wall is also shown, with labeled points A at the top, B at the center, and C at the midpoint between A and B. Study the significance of the transverse shear stress in combination with bending by performing the following steps.

Problem 3–45

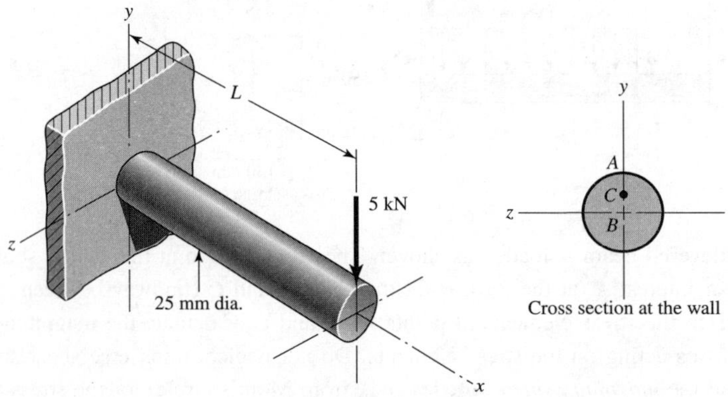

(*a*) Assume $L = 250$ mm. For points A, B, and C, sketch three-dimensional stress elements, labeling the coordinate directions and showing all stresses. Calculate magnitudes of the stresses on the stress elements. Do not neglect transverse shear stress. Calculate the maximum shear stress for each stress element.

(*b*) For each stress element in part (*a*), calculate the maximum shear stress if the transverse shear stress is neglected. Determine the percent error for each stress element from neglecting the transverse shear stress.

(*c*) Repeat the problem for $L = 100$, 25, and 2.5 mm. Compare the results and state any conclusions regarding the significance of the transverse shear stress in combination with bending.

3–46 Consider a simply supported beam of rectangular cross section of constant width b and variable depth h, so proportioned that the maximum stress σ_x at the outer surface due to bending is constant, when subjected to a load F at a distance a from the left support and a distance c from the right support. Show that the depth h at location x is given by

$$h = \sqrt{\frac{6Fcx}{lb\sigma_{\max}}} \qquad 0 \le x \le a$$

3–47 In Problem 3–46, $h \to 0$ as $x \to 0$, which cannot occur. If the maximum shear stress $\tau_{\max}$ due to direct shear is to be constant in this region, show that the depth h at location x is given by

$$h = \frac{3}{2}\frac{Fc}{lb\tau_{\max}} \qquad 0 \le x \le \frac{3}{8}\frac{Fc\sigma_{\max}}{lb\tau_{\max}^2}$$

3–48 and 3–49 The beam shown is loaded in the xy and xz planes.

(*a*) Find the y- and z-components of the reactions at the supports.

(*b*) Plot the shear-force and bending-moment diagrams for the xy and xz planes. Label the diagrams properly and provide the values at key points.

(*c*) Determine the net shear-force and bending-moment at the key points of part (*b*).

(*d*) Determine the maximum tensile bending stress. For Problem 3–48, use the cross section given in Problem 3–33, part (*a*). For Problem 3–49, use the cross section given in Problem 3–38, part (*b*).

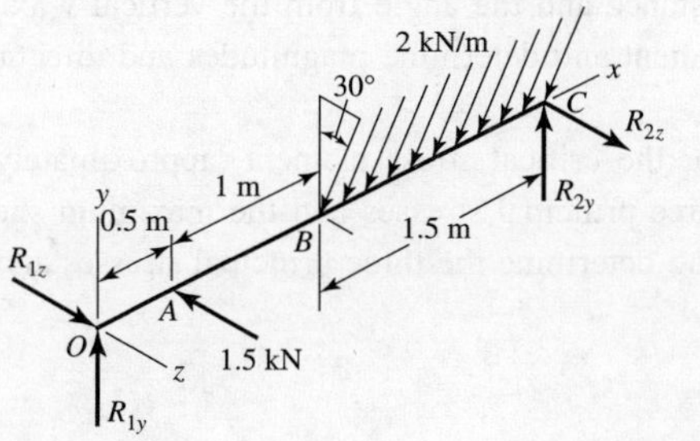

Problem 3–48

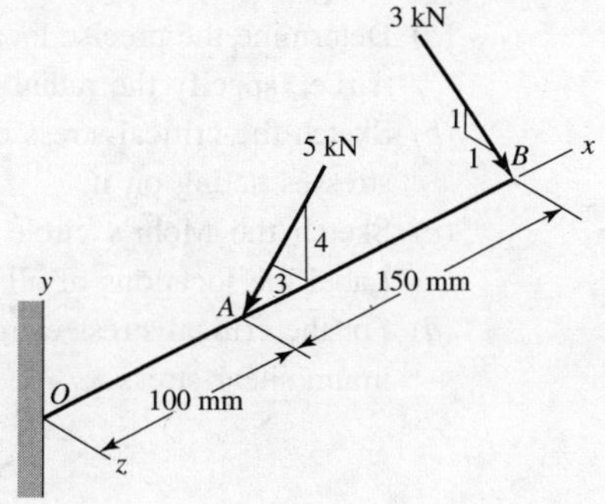

Problem 3–49

3–50 The part shown is loaded at point C with 300 N in the positive x direction and at point E with 200 N in the positive y direction. The diameter of the bar ABD is 12 mm. Evaluate the likelihood of failure in section AB by providing the following information:

(*a*) Determine the precise location of the critical stress element at the cross section at A (i.e., specify the radial distance and the angle from the vertical y axis).

(*b*) Sketch the critical stress element and determine magnitudes and directions for all stresses acting on it.

(*c*) Sketch the Mohr's circle for the critical stress element, approximately to scale. Label the locations of all three principal stresses and the maximum shear stress.

(*d*) For the critical stress element, determine the three principal stresses and the maximum shear stress.

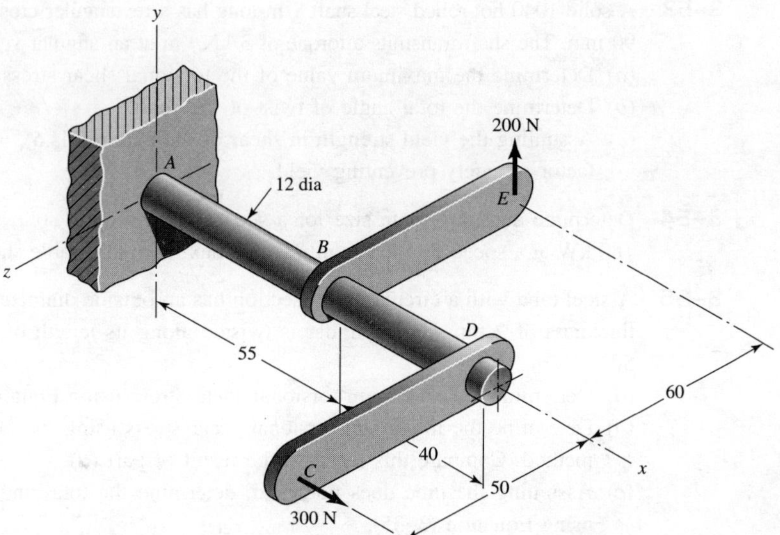

Problem 1–50

Dimensions in mm.

3–51 Repeat Problem 3–47, except replace the 200 N force at E with a 200 N force in the positive x direction, and replace the 300 N force at C with a 300 N force in the positive y direction.

3–52 The part shown is loaded at point B with 1.2 kN in the negative y direction, at point C with 0.8 kN in the negative z direction, and at point D with a force with components of 0.3 kN in the positive x direction and 0.4 kN in the positive y direction. The diameter of the bar AB is 20 mm. Evaluate the likelihood of failure in bar AB by providing the following information:

(*a*) Determine the precise location of the critical stress element at the cross section at A (i.e., specify the radial distance and the angle from the vertical y axis).

(*b*) Sketch the critical stress element and determine magnitudes and directions for all stresses acting on it.

(*c*) Sketch the Mohr's circle for the critical stress element, approximately to scale. Label the locations of all three principal stresses and the maximum shear stress.

(*d*) For the critical stress element, determine the three principal stresses and the maximum shear stress.

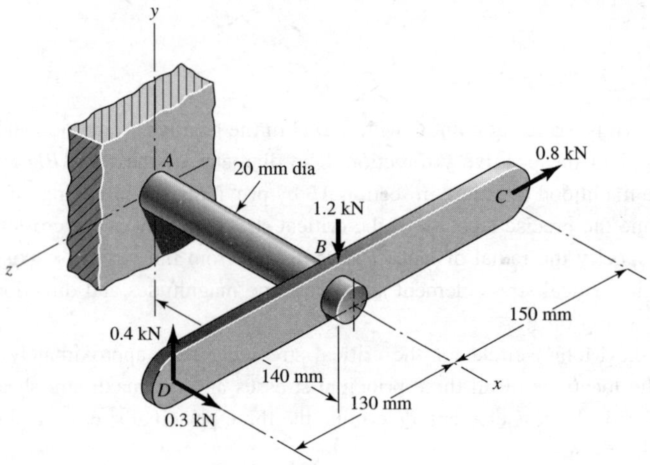

Problem 3–52

3–53 A solid 1040 hot-rolled steel shaft 1 m long has a rectangular cross section of 60 mm × 90 mm. The shaft transmits a torque of 3 kN · m at an angular velocity of 500 rpm.

(*a*) Determine the maximum value of the torsional shear stress.

(*b*) Determine the total angle of twist of the shaft.

(*c*) Assuming the yield strength in shear of the material is $S_{sy} = 0.5\,S_y$, determine the factor of safety preventing yield.

3–54 Determine an appropriate size for a square cross-section solid steel shaft to transmit 180 kW at a speed of 540 rev/min if the maximum allowable shear stress is 100 MPa.

3–55 A steel tube with a circular cross section has an outside diameter of 300 mm and wall thickness of 2 mm. The cylinder is twisted along its length of 2 m with a torque of 50 kN · m.

(*a*) Determine the maximum torsional shear stress using Equation (3–37).

(*b*) Determine the maximum torsional shear stress using the closed thin-walled tube method. Compare this result to the result of part (*a*).

(*c*) Assuming the tube does not yield, determine the total angle of twist in degrees using Equation (3–35).

(*d*) Assuming the tube does not yield, determine the total angle of twist in degrees using the closed thin-walled tube method. Compare this result to the result of part (*c*).

3–56 A power-take-off (PTO) shaft to provide power from a tractor to an implement (e.g., a mower) is made of two parts that can slide with respect to each other to provide a variable length. The inner part is a solid bar with a rectangular cross section with dimensions 35 × 50 mm. The outer part is a tube with a rectangular cross section with nominal dimensions to loosely fit over the solid bar, and has a wall thickness of 3 mm. The material for both parts is 1035 CD steel. The maximum operating speed of the shaft is 540 rev/min. Assume the yield strength in shear is $S_{sy} = 0.5 S_y$.

(*a*) Determine the maximum power (in kW) that the solid bar can deliver. Use a factor of safety of 2.

(*b*) Determine the maximum power that the hollow tube can deliver, using the closed thin-walled method. Use a factor of safety of 2.

3–57 A steel tube, 1 m long, has a rectangular cross section with outer dimensions of 20 × 30 mm, and a uniform wall thickness of 1 mm. The tube is twisted along its length with a torque *T*. The tube material is 1018 CD steel, and assume the yield strength in shear is $S_{sy} = 0.5\ S_y$.

(*a*) Estimate the maximum torque *T* that can be applied without yielding.

(*b*) Estimate the necessary torque *T* to obtain a total angle of twist over the length of the tube of 3°.

3–58 Two steel thin-wall tubes in torsion of equal length are to be compared. The first is of square cross section, side length *b*, and wall thickness *t*. The second is a round of diameter *b* and wall thickness *t*. The largest allowable shear stress is τ_{all} and is to be the same in both cases.

(*a*) Determine the ratio of maximum torque for the square tube versus the round tube.

(*b*) Determine the ratio of the angle of twist per unit length for the square tube versus the round tube.

3–59 Consider a 24 mm-square steel thin-walled tube loaded in torsion. The tube has a wall thickness t = 1.5 mm, is 1 m long, and has a maximum allowable shear stress of 80 MPa. Determine the maximum torque that can be applied and the corresponding angle of twist of the tube.

(*a*) Assume that the internal radius at the corners $r_i = 0$.

(*b*) Assume that the internal radius at the corners is more realistically r_i =3 mm.

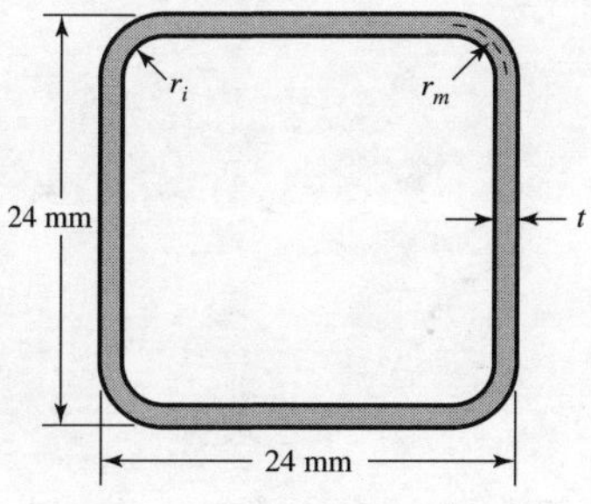

Problem 3–59

3–60 The thin-walled open cross section shown is transmitting torque *T*. The angle of twist per unit length of each leg can be determined separately using Equation (3–45) and is given by

$$\theta_1 = \frac{3T_i}{GL_i c_i^3}$$

where for this case, $i = 1, 2, 3$, and T_i represents the torque in leg *i*. Assuming that the angle of twist per unit length for each leg is the same, show that

$$T = \frac{G\theta_1}{3}\sum_{i=1}^{3} L_i c_i^3 \quad \text{and} \quad \tau_{max} = G\theta_1 c_{max}$$

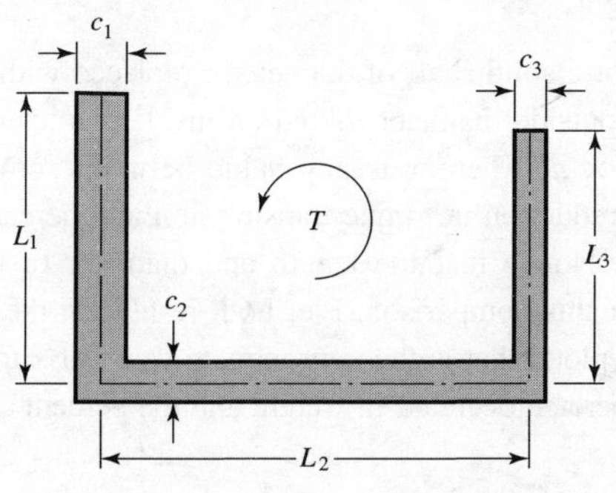

Problem 3–60

3–61 to 3–63 Using the results from Problem 3–60, consider a steel section with $\tau_{allow} = 80$ MPa.
(*a*) Determine the torque transmitted by each leg and the torque transmitted by the entire section.
(*b*) Determine the angle of twist per unit length.

Problem Number	c_1	L_1	c_2	L_2	c_3	L_3
3–61	2 mm	20 mm	3 mm	30 mm	0	0
3–62	1.5 mm	18 mm	18 mm	25 mm	1.5 mm	15 mm
3–63	2 mm	20 mm	3 mm	30 mm	2 mm	25 mm

3–64 Two 300-mm-long rectangular steel strips are placed together as shown. Using a maximum allowable shear stress of 80 MPa, determine the maximum torque and angular twist, and the torsional spring rate. Compare these with a single strip of cross section 30 mm by 4 mm. Solve the problem two ways: (*a*) using Equations (3–40) and (3–41), and (*b*) using Equation (3–45). Compare and discuss your results

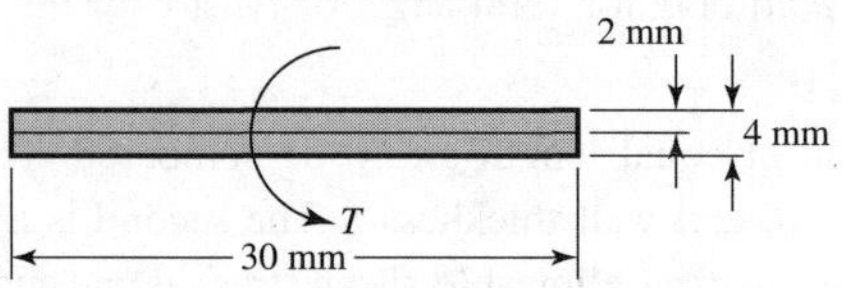

Problem 3–64

3–65 Using a maximum allowable shear stress of 70 MPa, find the shaft diameter needed to transmit 40 kW when
(*a*) The shaft speed is 2500 rev/min.
(*b*) The shaft speed is 250 rev/min.

3–66 Repeat Problem 3–65 with an allowable shear stress of 140 MPa and a power of 37 kW.

3–67 Using an allowable shear stress of 50 MPa, determine the power that can be transmitted at 2000 rpm through a shaft with a 30-mm diameter.

3–68 A 20-mm-diameter steel bar is to be used as a torsion spring. If the torsional stress in the bar is not to exceed 110 MPa when one end is twisted through an angle of 15°, what must be the length of the bar?

3–69 A 0.6-m-long steel bar with a 20 mm diameter is to be used as a torsion spring. If the torsional stress in the bar is not to exceed 210 MPa, what is the maximum angle of twist of the bar?

3–70 A 40-mm-diameter solid steel shaft, used as a torque transmitter, is replaced with a hollow shaft having a 40-mm OD and a 36-mm ID. If both materials have the same strength, what is the percentage reduction in torque transmission? What is the percentage reduction in shaft weight?

3–71 Generalize Problem 3–70 for a solid shaft of diameter d replaced with a hollow shaft of the same material with an outside diameter d, and an inside diameter that is a fraction of the outside diameter, $x \times d$, where x is any value between zero and one. Obtain expressions for percentage reduction in torque transmission and percentage reduction in weight in terms of only x. Notice that the length and diameter of the shaft, and the material, are not needed for this comparison. Plot both results on the same axis for the range $0 < x < 1$. From the plot, what is the approximate value of x to obtain the greatest difference between the percent decrease in weight and the percent decrease in torque?

3–72 A hollow steel shaft is to transmit 4200 N · m of torque and is to be sized so that the torsional stress does not exceed 120 MPa.

(*a*) If the inside diameter is 70 percent of the outside diameter, what size shaft should be used? Use preferred sizes.

(*b*) What is the stress on the inside of the shaft when full torque is applied?

3–73 The figure shows an endless-belt conveyor drive roll. The roll has a diameter of 120 mm and is driven at 10 rev/min by a geared-motor source rated at 1.5 kW. Determine a suitable shaft diameter d_c for an allowable torsional stress of 80 MPa.

(*a*) What would be the stress in the shaft you have sized if the motor starting torque is twice the running torque?

(*b*) Is bending stress likely to be a problem? What is the effect of different roll lengths *B* on bending?

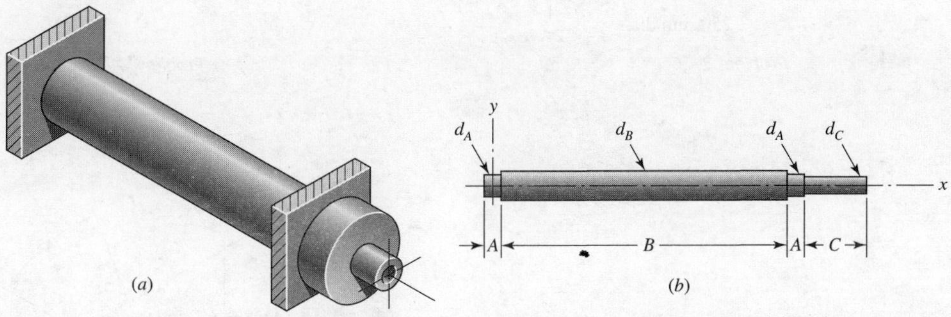

Problem 3–73

3–74 The conveyer drive roll in the figure for Problem 3–73 is 150 mm in diameter and is driven at 8 rev/min by a geared-motor source rated at 1 kW. Find a suitable shaft diameter d_c from the preferred decimal sizes in Table A–15, based on an allowable torsional stress of 75 Mpa.

3–75 Consider two shafts in torsion, each of the same material, length, and cross-sectional area. One shaft has a solid square cross section and the other shaft has a solid circular section.

(*a*) Which shaft has the greater maximum shear stress and by what percentage?

(*b*) Which shaft has the greater angular twist θ and by what percentage?

3–76* to 3–79* A countershaft carrying two V-belt pulleys is shown in the figure. Pulley *A* receives power from a motor through a belt with the belt tensions shown. The power is transmitted through the shaft and delivered to the belt on pulley *B*. Assume the belt tension on the loose side at *B* is 15 percent of the tension on the tight side.

(*a*) Determine the tensions in the belt on pulley *B*, assuming the shaft is running at a constant speed.

(*b*) Find the magnitudes of the bearing reaction forces, assuming the bearings act as simple supports.

(*c*) Draw shear-force and bending-moment diagrams for the shaft. If needed, make one set for the horizontal plane and another set for the vertical plane.

(*d*) At the point of maximum bending moment, determine the bending stress and the torsional shear stress.

(*e*) At the point of maximum bending moment, determine the principal stresses and the maximum shear stress.

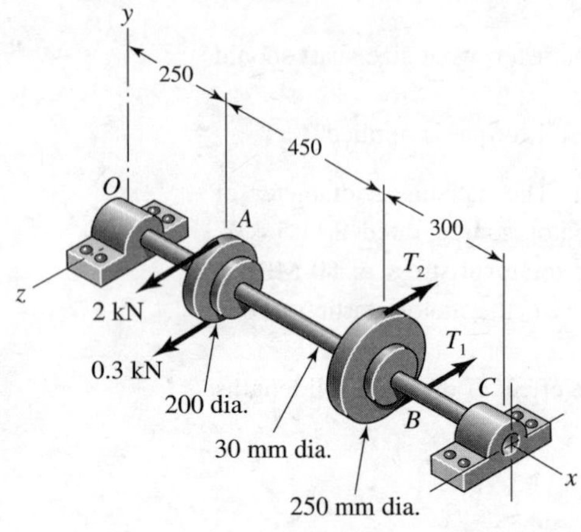

*Problem 3–76**

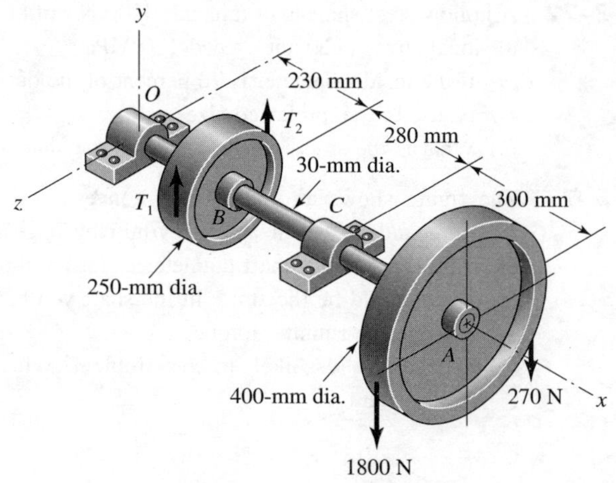

*Problem 3–77**

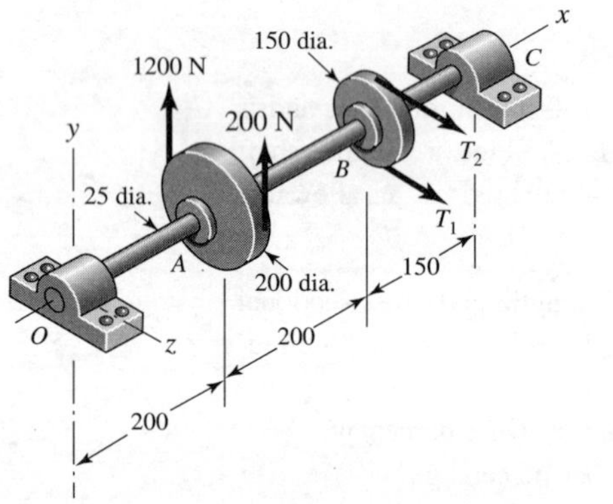

*Problem 3–78**

Dimensions in millimeters.

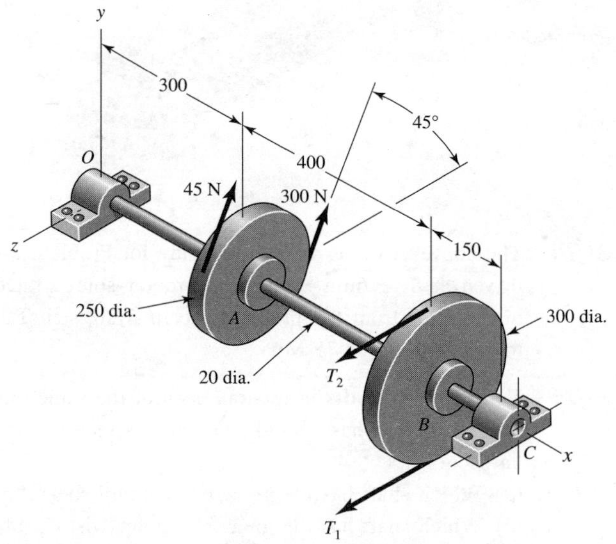

*Problem 3–79**

Dimensions in millimeters.

3–80* to 3–81* A gear reduction unit uses the countershaft shown in the figure. Gear A receives power from another gear with the transmitted force F_A applied at the 20° pressure angle as shown. The power is transmitted through the shaft and delivered through gear B through a transmitted force F_B at the pressure angle shown.

(*a*) Determine the force F_B, assuming the shaft is running at a constant speed.

(*b*) Find the bearing reaction forces, assuming the bearings act as simple supports.

(*c*) Draw shear-force and bending-moment diagrams for the shaft. If needed, make one set for the horizontal plane and another set for the vertical plane.

(*d*) At the point of maximum bending moment, determine the bending stress and the torsional shear stress.

(*e*) At the point of maximum bending moment, determine the principal stresses and the maximum shear stress.

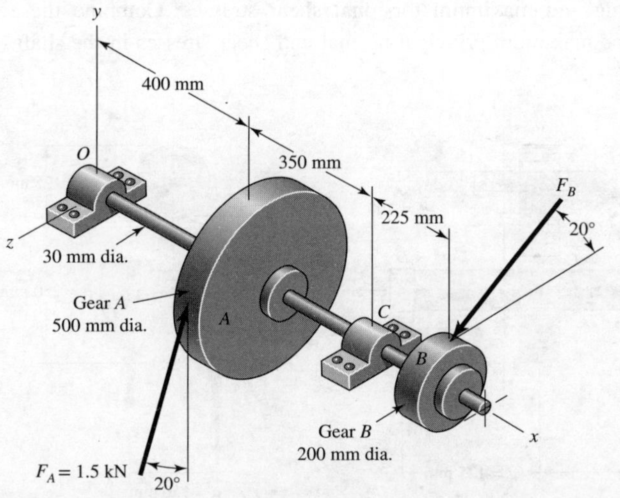

Problem 3–80*

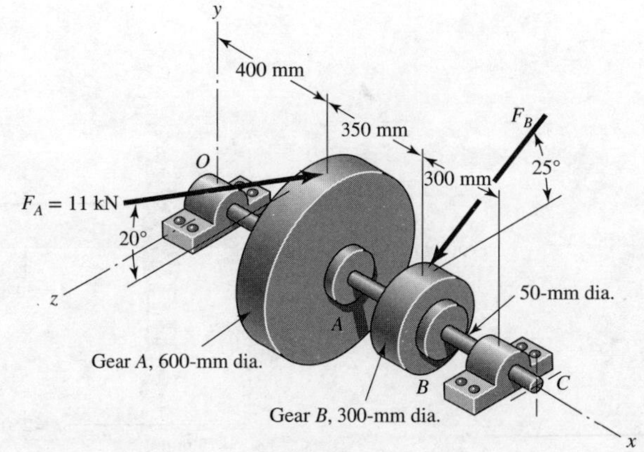

Problem 3–81*

3–82* In the figure, shaft *AB* transmits power to shaft *CD* through a set of bevel gears contacting at point *E*. The contact force at *E* on the gear of shaft *CD* is determined to be $(\mathbf{F}_E)_{CD} = -400\mathbf{i} - 1600\mathbf{j} + 3600\mathbf{k}$ N. For shaft *CD*: (*a*) draw a free-body diagram and determine the reactions at *C* and *D* assuming simple supports (assume also that bearing *C* carries the thrust load), (*b*) draw the shear-force and bending-moment diagrams, (*c*) for the critical stress element, determine the torsional shear stress, the bending stress, and the axial stress, and (*d*) for the critical stress element, determine the principal stresses and the maximum shear stress.

Problem 3–82*

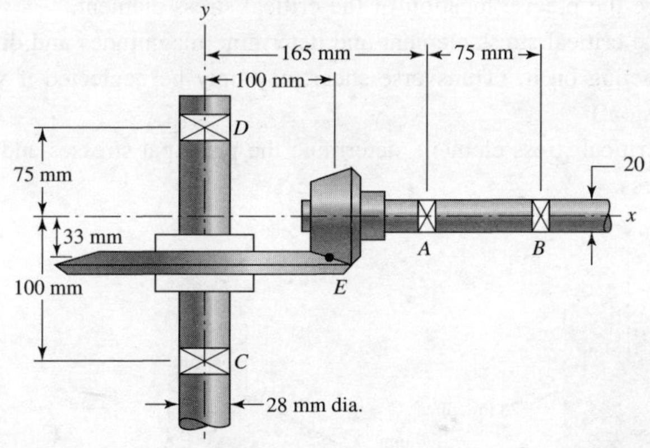

3–83 Repeat Problem 3–82 except for a contact force at *E* of $(\mathbf{F}_E)_{CD} = -200\mathbf{i} - 600\mathbf{j} + 1800\mathbf{k}$ N and a shaft diameter of 25 mm.

3–84* Repeat the analysis of Problem 3–82 for shaft *AB*. Assume that bearing *A* carries the thrust load.

3–85* A torque $T = 100$ N · m is applied to the shaft *EFG*, which is running at constant speed and contains gear *F*. Gear *F* transmits torque to shaft *ABCD* through gear *C*, which drives the chain sprocket at *B*, transmitting a force *P* as shown. Sprocket *B*, gear *C*, and gear *F* have pitch diameters of $a = 150$, $b = 250$, and $c = 125$ mm, respectively. The contact force between the gears is transmitted through the pressure angle $\phi = 20°$. Assuming no frictional losses and considering the bearings at *A*, *D*, *E*, and *G* to be simple supports, locate the point on shaft *ABCD* that contains the maximum tensile bending and maximum torsional shear stresses. Combine these stresses and determine the maximum principal normal and shear stresses in the shaft.

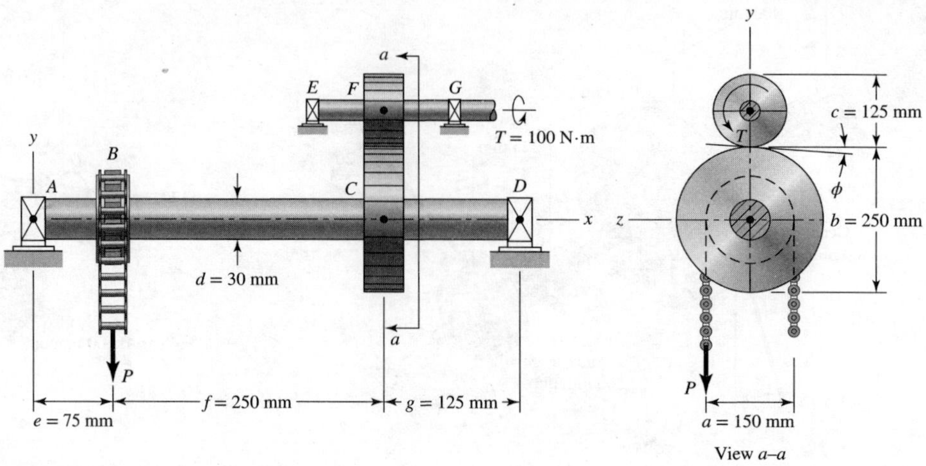

Problem 3–85*

3–86 Repeat Problem 3–85 with the chain parallel to the z axis with P in the positive z direction.

3–87* Repeat Problem 3–85 with $T = 100$ N · m, $a = 150$ mm, $b = 125$ mm, $c = 250$ mm, $d = 35$ mm, $e = 100$ mm, $f = 250$ mm, and $g = 150$ mm.

3–88* The cantilevered bar in the figure is made from a ductile material and is statically loaded with $F_y = 800$ N and $F_x = F_z = 0$. Analyze the stress situation in rod AB by obtaining the following information.

(*a*) Determine the precise location of the critical stress element.

(*b*) Sketch the critical stress element and determine magnitudes and directions for all stresses acting on it. (Transverse shear may only be neglected if you can justify this decision.)

(*c*) For the critical stress element, determine the principal stresses and the maximum shear stress.

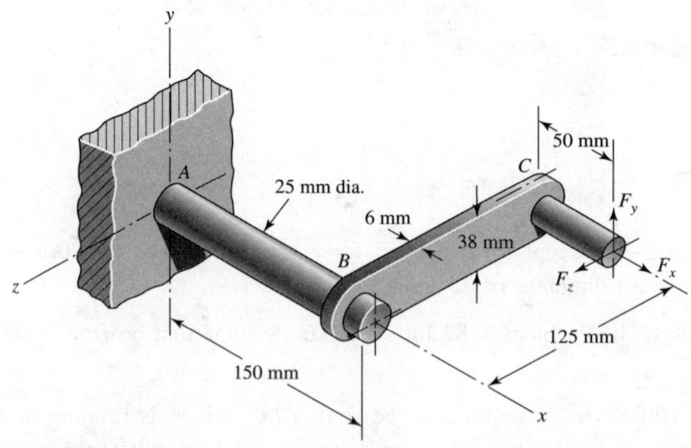

Problem 3–88*

3–89* Repeat Problem 3–88 with $F_x = 0$, $F_y = 700$ N, and $F_z = 400$ N.

3–90* Repeat Problem 3–88 with $F_x = 300$ N, $F_y = -800$ N, and $F_z = 400$ N.

3–91* For the handle in Problem 3–88, one potential failure mode is twisting of the flat plate BC. Determine the maximum value of the shear stress due to torsion in the main section of the plate, ignoring the complexities of the interfaces at B and C.

3–92* The cantilevered bar in the figure is made from a ductile material and is statically loaded with $F_y = 1$ kN and $F_x = F_z = 0$. Analyze the stress situation in the small diameter at the shoulder at *A* by obtaining the following information.

(*a*) Determine the precise location of the critical stress element at the cross section at *A*.

(*b*) Sketch the critical stress element and determine magnitudes and directions for all stresses acting on it. (Transverse shear may be neglected if you can justify this decision.)

(*c*) For the critical stress element, determine the principal stresses and the maximum shear stress.

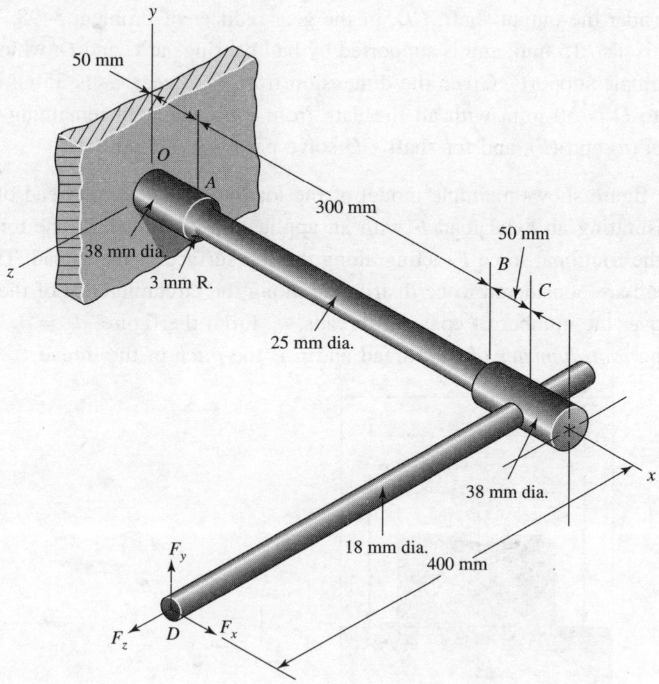

Problem 3–92*

3–93* Repeat Problem 3–92 with $F_x = 1.2$ kN, $F_y = 1$ kN, and $F_z = 0$.

3–94* Repeat Problem 3–92 with $F_x = 1.2$ kN, $F_y = 1$ kN, and $F_z = -0.4$ kN.

3–95* Repeat Problem 3–92 for a brittle material, requiring the inclusion of stress concentration in the fillet radius.

3–96 Repeat Problem 3–92 with $F_x = 1.2$ kN, $F_y = 1$ kN, and $F_z = 0$, and for a brittle material, requiring the inclusion of stress concentration in the fillet radius.

3–97 Repeat Problem 3–92 with $F_x = 1.2$ kN, $F_y = 1$ kN, and $F_z = -0.4$ kN, and for a brittle material, requiring the inclusion of stress concentration in the fillet radius.

3–98 A gear reducer similar to Figure 3–1(*a*), transmits power from input shaft *AB* to output shaft *CD*. The input torque and constant speed are $T_i = 20$ N · m and $\omega_i = 60$ rev/min, respectively. The output load torque and speed are T_o and ω_o, respectively. Shaft *AB*, with diameter 15 mm, is supported by ball bearings at *A* and *B*, which can be treated as simple supports. For this shaft, the dimension from *A* to gear G_1 is 35 mm, and gear G_1 to *B* is 50 mm. The pitch radii of the gears are $r_1 = 25$ mm and $r_2 = 60$ mm. For the spur gears, the pressure angle, ϕ, is 20°.

(*a*) Determine the power transmitted.

(*b*) Determine the output torque.

(*c*) Find the y and z components of the force at each bearing. Also determine the magnitude of the total force at each bearing.

(*d*) Draw the shear-force and bending-moment diagrams for the shaft. Make one set for the horizontal plane and another set for the vertical plane.

(*e*) For a stress element on the outer surface of the shaft, at the point along the shaft where the maximum bending moment occurs, determine the bending stress and the torsional shear stress.

(*f*) For the stress element of part (*e*), determine the principal stresses and the maximum shear stress.

3–99 Consider the output shaft, *CD*, of the gear reducer of Problem 3–98. The shaft diameter is also 15 mm, and is supported by ball bearings at *C* and *D*, which can be treated as simple supports. Given the dimension from *C* to gear G_2 is 35 mm, and from gear G_2 to *D* is 50 mm, with all the data from Problem 3–98 remaining the same, solve parts (*a*) and (*b*), and for shaft *CD* solve parts (*c*) through (*f*).

3–100 The figure shows a simple model of the loading of a square thread of a power screw transmitting an axial load F with an application of torque T. The torque is balanced by the frictional force F_f acting along the top surface of the thread. The forces on the thread are considered to be distributed along the circumference of the *mean diameter* d_m over the number of engaged threads, n_t. From the figure, $d_m = d_r + p/2$, where d_r is the *root diameter* of the thread and p is the *pitch* of the thread.

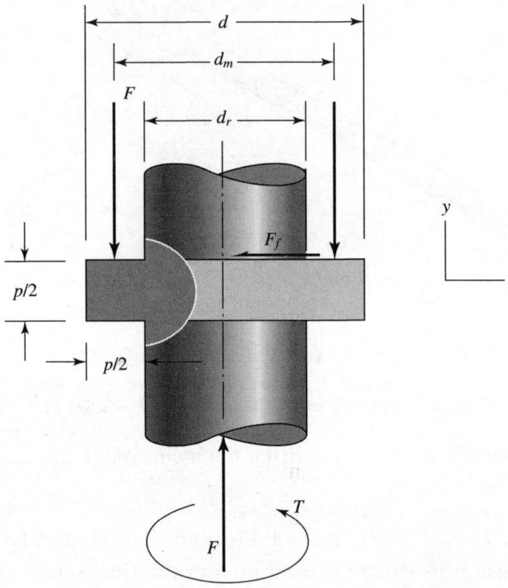

Problem 3–100

(*a*) Considering the thread to be a cantilever beam as shown in the cutaway view, show that the nominal bending stress at the root of the thread can be approximated by

$$\sigma_b = +\frac{6F}{\pi d_r n_t p}$$

(*b*) Show that the axial and maximum torsional shear stresses in the body of the shaft can be approximated by

$$\sigma_a = -\frac{4F}{\pi d_r^2} \quad \text{and} \quad \tau_t = \frac{16T}{\pi d_r^3}$$

(*c*) For the stresses of parts (*a*) and (*b*) show a three-dimensional representation of the state of stress on an element located at the intersection of the lower thread root base and the thread body. Using the given coordinate system label the stresses using the notation given in Figure 3–8a.

(*d*) A square-thread power screw has an outside diameter $d = 38$ mm, pitch $p = 6$ mm, and transmits a load $F = 6$ KN through the application of a torque $T = 26$ N · m. If $n_t = 2$, determine the key stresses and the corresponding *principal stresses* (normal and shear).

3–101 Develop the formulas for the maximum radial and tangential stresses in a thick-walled cylinder due to internal pressure only.

3–102 Repeat Problem 3–101 where the cylinder is subject to external pressure only. At what radii do the maximum stresses occur?

3–103 Develop the equations for the principal stresses in a thin-walled spherical pressure vessel of inside diameter d_i, thickness t, and with an internal pressure p_i. You may wish to follow a process similar to that used for a thin-walled cylindrical pressure vessel as discussed in Section 3–14.

3–104 to 3–106 A pressure cylinder has an outer diameter d_o, wall thickness t, internal pressure p_i, and maximum allowable shear stress τ_{max}. In the table given, determine the appropriate value of x.

Problem Number	d_o	t	p_i	τ_{max}
3–104	150 mm	6 mm	x_{max}	70 MPa
3–105	200 mm	x_{min}	4 MPa	25 MPa
3–106	200 mm	6 mm	3 MPa	x

3–107 to 3–109 A pressure cylinder has an outer diameter d_o, wall thickness t, external pressure p_o, and maximum allowable shear stress τ_{max}. In the table given, determine the appropriate value of x.

Problem Number	d_o	t	p_o	τ_{max}
3–107	150 mm	6 mm	x_{max}	70 MPa
3–108	200 mm	x_{min}	4 MPa	25 MPa
3–109	200 mm	6 mm	3 MPa	x

3–110 An AISI 1040 cold-drawn steel tube has an OD = 50 mm and wall thickness 6 mm. What maximum external pressure can this tube withstand if the largest principal normal stress is not to exceed 80 percent of the minimum yield strength of the material?

3–111 Repeat Problem 3–110 with an OD of 48 mm and wall thickness of 6 mm.

3–112 Repeat Problem 3–110 with an internal pressure.

3–113 Repeat Problem 3–111 with an internal pressure.

3–114 A thin-walled cylindrical steel water storage tank 9 m in diameter and 18 m long is oriented with its longitudinal axis vertical. The tank is topped with a hemispherical steel dome. The wall thickness of the tank and dome is 18 mm. If the tank is unpressurized and contains water 16 m above its base, and considering the weight of the tank, determine the maximum state of stress in the tank and the corresponding principal stresses (normal and shear). The weight density of water is 9810 N/m^3.

3–115 Repeat Problem 3–114 with the tank being pressurized to 350 kPa.

3–116 For mass-induced stresses in rotating rings, the tangential and radial stresses are given in Equation (3–55).

(*a*) Sketch stress profiles (similar to Figure 3–37). The sketch needs to be detailed enough only to indicate generally what to expect on the inside and outside radii (e.g., zero/nonzero, at a maximum, positive/negative, etc.). Don't worry about the inflections of the profile shapes.

(*b*) Evaluate where the critical stress element(s) will be, and develop useful equations for the values of the tangential and radial stresses at critical location(s) of interest.

3–117 Find the maximum shear stress in a 140-mm-diameter circular saw blade if it runs idle at 5000 rev/min. The saw is 2 mm steel and is used on a 16-mm-diameter arbor. The thickness is uniform. What is the maximum radial component of stress?

3–118 The maximum recommended speed for a 250-mm-diameter abrasive grinding wheel is 2000 rev/min. Assume that the material is isotropic; use a bore of 20 mm, $\nu = 0.24$, and a mass density of 3320 kg/m^3, and find the maximum tensile stress at this speed.

3–119 An abrasive cutoff wheel has an outer diameter of 120 mm, is 1.5 mm thick, and has a 18 mm bore. The wheel weighs 140 g and runs at 12 000 rev/min. The wheel material is isotropic, with a Poisson's ratio of 0.2, and has an ultimate strength of 80 MPa.

(*a*) Determine and justify the location of the critical stress element.

(*b*) Determine the stresses on the stress element at the critical location and the factor of safety corresponding to fracture.

3–120 A rotary lawnmower blade rotates at 3500 rev/min. The steel blade has a uniform cross section 3 mm thick by 30 mm wide, and has a 12-mm-diameter hole in the center as shown in the figure. Estimate the nominal tensile stress at the central section due to rotation.

Problem 3–120

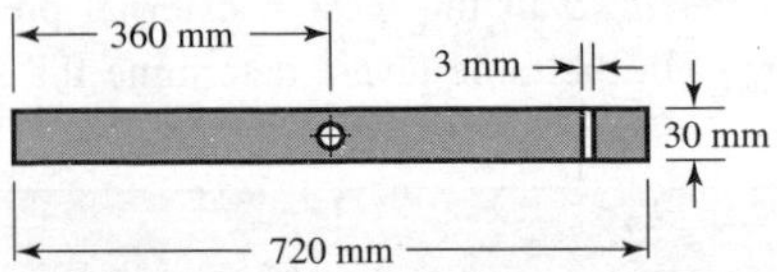

3–121 to 3–126 The table lists the maximum and minimum hole and shaft dimensions for a variety of standard press and shrink fits. The materials are both hot-rolled steel. Find the maximum and minimum values of the radial interference and the corresponding interface pressure. Use a collar diameter of 100 mm and 80 mm for even and odd numbered problems, respectively.

Problem Number	Fit Designation†	Basic Size	Hole		Shaft	
			D_{max}	D_{min}	d_{max}	d_{min}
3–121	50H7/p6	50 mm	50.025	50.000	50.042	50.026
3–122	40H7/p6	40 mm	40.025	40	40.042	40.026
3–123	50H7/s6	50 mm	50.025	50.000	50.059	50.043
3–124	40H7/s6	40 mm	40.025	40	40.059	40.043
3–125	50H7/u6	50 mm	50.025	50.000	50.086	50.070
3–126	40H7/u6	40 mm	40.025	40	40.076	40.060

†Note: See Table 7–9 for description of fits.

3–127 to 3–130 The table gives data concerning the shrink fit of two cylinders of differing materials and dimensional specification in millimeters. Elastic constants for different materials may be found in Table A–5. Identify the radial interference δ, then find the interference pressure p, and the tangential normal stress on both sides of the fit surface. If dimensional tolerances are given at fit surfaces, repeat the problem for the highest and lowest stress levels.

Problem Number	Inner Cylinder			Outer Cylinder		
	Material	d_i	d_O	Material	D_i	D_O
3–127	Steel	0	50.05	Steel	50	75
3–128	Steel	0	50.05	Cast iron	50	75
3–129	Steel	0	25.43/25.48	Steel	25.4/25.43	50
3–130	Aluminum	0	50.88/50.93	Steel	50.8/50.85	50

3–131 A utility hook was formed from a round rod of diameter $d = 20$ mm into the geometry shown in the figure. What are the stresses at the inner and outer surfaces at section A–A if $F = 4$ kN, $L = 250$ mm, and $D_i = 75$ mm?

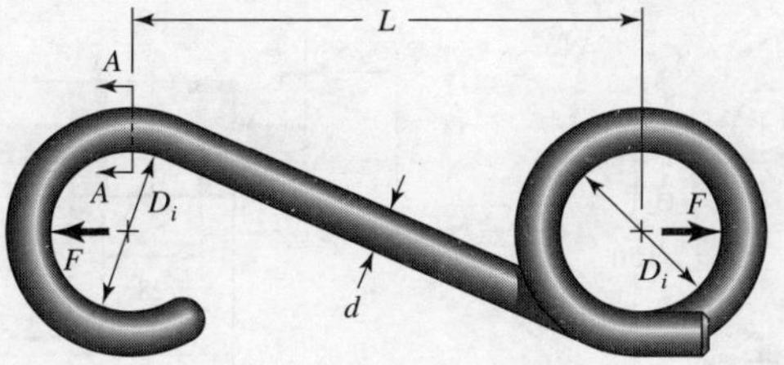

Problem 3–131

3–132 The steel eyebolt shown in the figure is loaded with a force $F = 300$ N. The bolt is formed from wire of diameter $d = 6$ mm to a radius $R_i = 10$ mm in the eye and at the shank. Estimate the stresses at the inner and outer surfaces at section A–A.

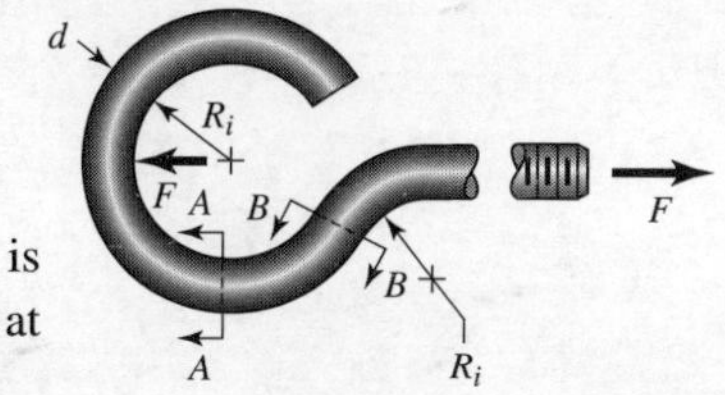

Problem 3–132

3–133 For Problem 3–132 estimate the stresses at the inner and outer surfaces at section B–B, located along the line between the radius centers.

3–134 Repeat Problem 3–132 with $d = 6$ mm, $R_i = 12$ mm, and $F = 300$ N.

3–135 Repeat Problem 3–133 with $d = 6$ mm, $R_i = 12$ mm, and $F = 300$ N.

3–136 Shown in the figure is a 2.8 by 28 latching spring that supports a load of $F = 100$ N. The inside radius of the bend is 3 mm.

(*a*) Using straight-beam theory, determine the stresses at the top and bottom surfaces immediately to the right of the bend.

(*b*) Using curved-beam theory, determine the stresses at the inner and outer surfaces at the bend.

(*c*) By comparing the stresses at the bend with the nominal stresses before the bend, estimate effective stress concentration factors for the inner and outer surfaces.

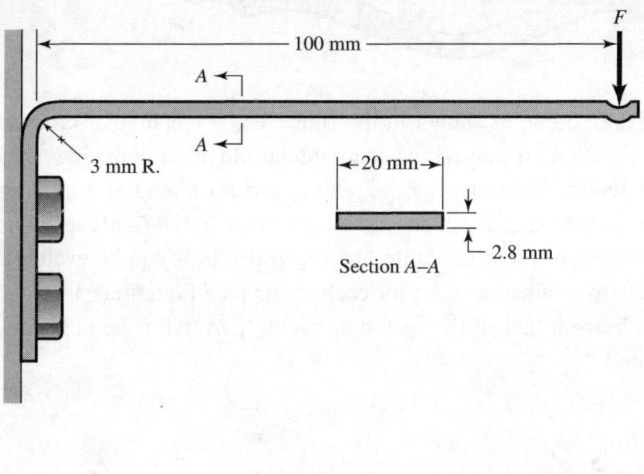

Problem 3–136

3–137 Repeat Problem 3–136 with a 3.6 mm material thickness.

3–138 Repeat Problem 3–136 with a bend radius of 6 mm.

3–139 The cast-iron bell-crank lever depicted in the figure is acted upon by forces F_1 of 2.4 kN and F_2 of 3.2 kN. The section *A–A* at the central pivot has a curved inner surface with a radius of $r_i = 25$ mm. Estimate the stresses at the inner and outer surfaces of the curved portion of the lever.

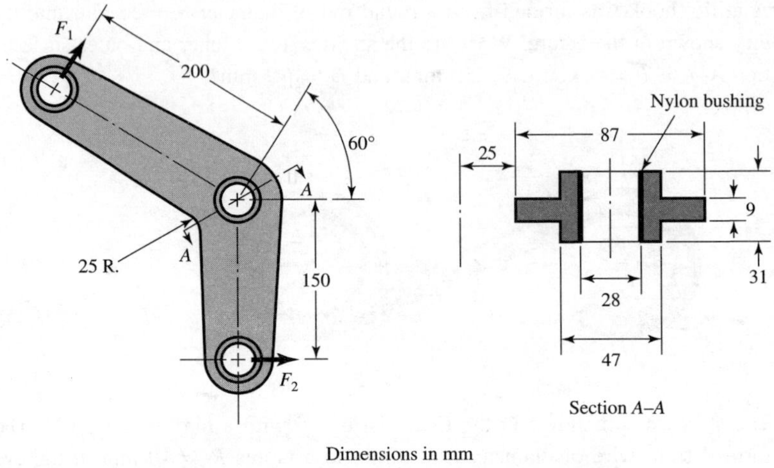

Problem 3–139

3–140 The crane hook depicted in Figure 3–41 has a 18-mm-diameter hole in the center of the critical section. For a load of 30 kN, estimate the bending stresses at the inner and outer surfaces at the critical section.

3–141 An offset tensile link is shaped to clear an obstruction with a geometry as shown in the figure. The cross section at the critical location is elliptical, with a major axis of 100 mm and a minor axis of 50 mm. For a load of 90 kN, estimate the stresses at the inner and outer surfaces of the critical section.

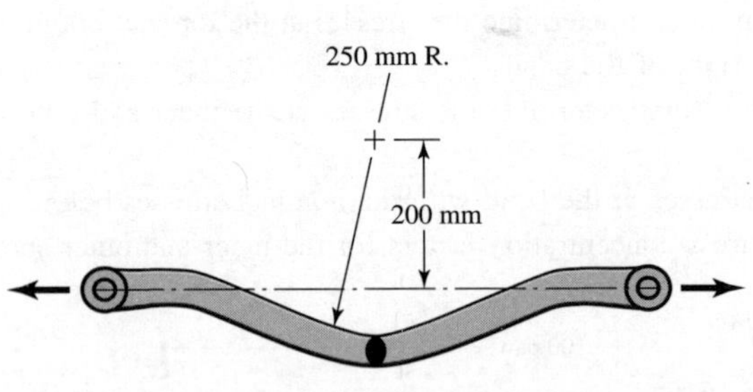

Problem 3–141

3–142 A cast-steel C frame as shown in the figure has a rectangular cross section of 25 mm by 40 mm, with a 10 mm-radius semicircular notch on both sides that forms midflank fluting as shown. Estimate A, r_c, r_n, and e, and for a load of 13 kN, estimate the inner and outer surface stresses at the throat *C*. *Note:* Table 3–4 can be used to determine r_n for this section. From the table, the integral $\int dA/r$ can be evaluated for a rectangle and a circle by evaluating A/r_n for each shape [see Equation (3–66)]. Subtracting A/r_n of the circle from that of the rectangle yields $\int dA/r$ for the C frame, and r_n can then be evaluated.

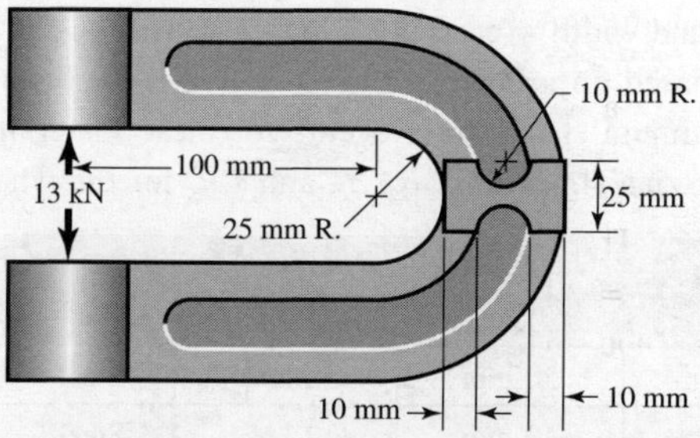

Problem 3–142

3–143 A glass sphere with a 30-mm diameter is pressed against a flat carbon steel plate with a force of 5 N. Evaluate the following for the glass sphere.
(*a*) Determine the maximum pressure at the contact surface.
(*b*) Calculate the principal stresses and maximum shear stress, as a function of the depth z. Use an increment of 0.01 mm for z, over the range $0 < z < a$. Plot your results.
(*c*) Determine the depth at which the maximum shear stress occurs, rounded to the nearest 0.01 mm. Do not assume Figure 3–43 is applicable, with its underlying assumption of a Poisson ratio of 0.3.
(*d*) Determine the maximum shear stress.
(*e*) For comparison, determine the maximum shear stress that is predicted by Figure 3–43.

3–144 Two carbon steel balls, each 30 mm in diameter, are pressed together by a force F. In terms of the force F in Newtons, find the maximum values of the principal stress, and the maximum shear stress, in MPa.

3–145 A carbon steel ball with 25-mm diameter is pressed together with an aluminum ball with a 40-mm diameter by a force of 10 N. Determine the maximum shear stress, and the depth at which it will occur for the aluminum ball. Assume Figure 3–43, which is based on a typical Poisson's ratio of 0.3, is applicable to estimate the depth at which the maximum shear stress occurs for these materials.

3–146 Repeat Problem 3–145 but determine the maximum shear stress and depth for the steel ball.

3–147 A carbon steel ball with a 30-mm diameter is pressed against a flat carbon steel plate with a force of 20 N. Determine the maximum shear stress, and the depth in the plate at which it will occur.

3–148 An AISI 1018 steel ball with 25-mm diameter is used as a roller between a flat plate made from 2024 T3 aluminum and a flat table surface made from ASTM No. 30 gray cast iron. Determine the maximum amount of weight that can be stacked on the aluminum plate without exceeding a maximum shear stress of 140 MPa in any of the three pieces. Assume Figure 3–43, which is based on a typical Poisson's ratio of 0.3, is applicable to estimate the depth at which the maximum shear stress occurs for these materials.

3–149 An aluminum alloy cylindrical roller with diameter 25 mm and length 50 mm rolls on the inside of a cast-iron ring having an inside radius of 100 mm, which is 50 mm thick. Find the maximum contact force F that can be used if the shear stress is not to exceed 28 MPa.

3–150 A pair of mating steel spur gears with a 18 mm face width transmits a load of 200 N. For estimating the contact stresses, make the simplifying assumption that the teeth profiles can be treated as cylindrical with instantaneous radii at the contact point of interest of 11 mm and 15 mm, respectively. Estimate the maximum contact pressure and the maximum shear stress experienced by either gear.

3–151 to 3–153 A wheel of diameter d and width w carrying a load F rolls on a flat rail. Assume that Figure 3–45, which is based on a Poisson's ratio of 0.3, is applicable to estimate the depth at which the maximum shear stress occurs for these materials. At this critical depth, calculate the Hertzian stresses σ_x, σ_y, σ_z, and τ_{max} for the wheel.

Problem Number	d	w	F	Wheel Material	Rail Material
3–151	125 mm	50 mm	3 kN	Steel	Steel
3–152	150 mm	40 mm	2 kN	Steel	Cast iron
3–153	42 mm	30 mm	1 kN	Cast iron	Cast iron

4 Deflection and Stiffness

Pixtal/age fotostock

Chapter Outline

All real bodies deform under load, either elastically or plastically. Controlling this deflection is a necessary part of machine design. Excess deflection is a failure mode that should be considered. The size of a load-bearing component is often determined based on deflections, rather than limits on stress.

Sometimes deflection is a desired part of the design. A snap ring, or retaining ring, must be flexible enough to be bent without permanent deformation and assembled with other parts, and then it must be rigid enough to hold the assembled parts together. In a transmission, the gears must be supported by a rigid shaft. If the shaft bends too much, that is, if it is too flexible, the teeth will not mesh properly, and the result will be excessive impact, noise, wear, and early failure. In rolling sheet or strip steel to prescribed thicknesses, the rolls must be crowned, that is, curved, so that the finished product will be of uniform thickness. Thus, to design the rolls it is necessary to know exactly how much they will bend when a sheet of steel is rolled between them. Sometimes mechanical elements must be designed to have a particular force-deflection characteristic. The suspension system of an automobile, for example, must be designed within a very narrow range to achieve an optimum vibration frequency for all conditions of vehicle loading, because the human body is comfortable only within a limited range of frequencies.

This chapter considers distortion of single bodies due to geometry (shape) and loading, then, briefly, the behavior of groups of bodies.

4–1 Spring Rates

Elasticity is that property of a material that enables it to regain its original configuration after having been deformed. A *spring* is a mechanical element that exerts a force when deformed. Figure 4–1*a* shows a straight beam of length l simply supported at the ends and loaded by the transverse force F. The deflection y is linearly related to the force, as long as the elastic limit of the material is not exceeded, as indicated by the graph. This beam can be described as a *linear spring.*

In Figure 4–1*b* a straight beam is supported on two cylinders such that the length between supports decreases as the beam is deflected by the force F. A larger force is required to deflect a short beam than a long one, and hence the more this beam is deflected, the stiffer it becomes. Also, the force is not linearly related to the deflection, and hence this beam can be described as a *nonlinear stiffening spring.*

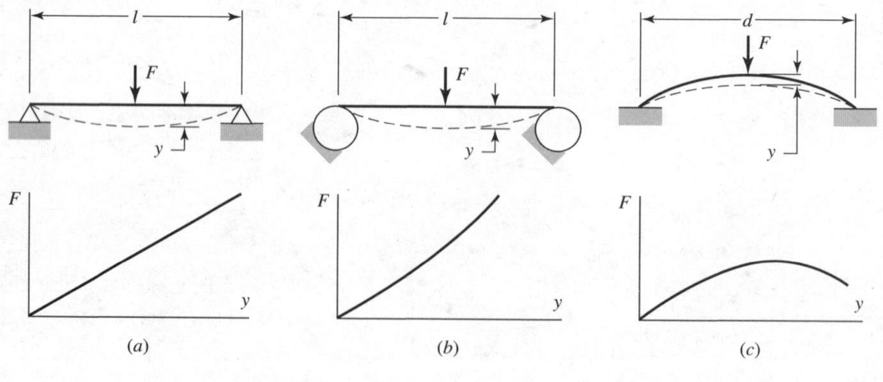

Figure 4–1

(*a*) A linear spring;
(*b*) a stiffening spring;
(*c*) a softening spring.

Figure 4–1*c* is an edge-view of a dish-shaped round disk. The force necessary to flatten the disk increases at first and then decreases as the disk approaches a flat configuration, as shown by the graph. Any mechanical element having such a characteristic is called a *nonlinear softening spring*.

If we designate the general relationship between force and deflection by the equation

$$F = F(y) \tag{a}$$

then *spring rate* is defined as

$$k(y) = \lim_{\Delta y \to 0} \frac{\Delta F}{\Delta y} = \frac{dF}{dy} \tag{4–1}$$

where y must be measured in the direction of F and at the point of application of F. Most of the force-deflection problems encountered in this book are linear, as in Figure 4–1*a*. For these, k is a constant, also called the *spring constant;* consequently Equation (4–1) is written

$$k = \frac{F}{y} \tag{4–2}$$

We might note that Equations (4–1) and (4–2) are quite general and apply equally well for torques and moments, provided angular measurements are used for y. For linear displacements, the units of k are often newtons per meter, and for angular displacements, newton-meters per radian.

4–2 Tension, Compression, and Torsion

The total extension or contraction of a uniform bar in pure tension or compression, respectively, is given by

$$\delta = \frac{Fl}{AE} \tag{4–3}$$

This equation does not apply to a *long* bar loaded in compression if there is a possibility of buckling (see Sections 4–11 to 4–15). Using Equations (4–2) and (4–3) with $\delta = y$, we see that the spring constant of an axially loaded bar is

$$k = \frac{AE}{l} \tag{4–4}$$

The angular deflection of a uniform solid or hollow round bar subjected to a twisting moment T was given in Equation (3–35), and is

$$\theta = \frac{Tl}{GJ} \tag{4–5}$$

where θ is in radians.

Equation (4–5) can be rearranged to give the torsional spring rate as

$$k = \frac{T}{\theta} = \frac{GJ}{l} \tag{4–6}$$

Equations (4–5) and (4–6) apply *only* to circular cross sections. Torsional loading for bars with noncircular cross sections is discussed in Section 3–12. For the angular twist of rectangular cross sections, closed thin-walled tubes, and open thin-walled sections, refer to Equations (3–41), (3–44), and (3–45), respectively.

4–3 Deflection Due to Bending

Beam Theory

A beam is a long slender structural member that supports loads at points along its length. In addition to structural elements, components such as shafts, axles, cranks, levers, springs, and brackets are modeled as beams. The subject is covered extensively in mechanics of materials. In this text, the goal is to provide a summary and to focus attention on some of the more practical aspects for machine design.

The Euler–Bernoulli beam theory is a simple and useful model for determining beam deflection. The primary assumptions of the theory are as follows:

- The beam has one dimension, known as its axis, that is significantly larger than its other two dimensions.
- Plane sections remain plane and perpendicular to the neutral axis.
- The deformations, including lateral deflections and slopes, are small.
- The beam material is isotropic and is linearly elastic.

The curvature of a beam subjected to a bending moment M is given by

$$\frac{1}{\rho} = \frac{M}{EI} \tag{4–7}$$

where ρ is the radius of curvature. From studies in mathematics, the curvature of a plane curve is given by the equation

$$\frac{1}{\rho} = \frac{d^2y/dx^2}{[1 + (dy/dx)^2]^{3/2}} \tag{4–8}$$

where y is the lateral deflection of the centroidal axis of the beam at any point x along its length. The slope of the beam at any point x is

$$\theta = \frac{dy}{dx} \tag{a}$$

For many problems in bending, the slope is very small, and for these the denominator of Equation (4–8) can be taken as unity. Equation (4–7) can then be written

$$\frac{M}{EI} = \frac{d^2y}{dx^2} \tag{b}$$

Noting Equations (3–3) and (3–4) and successively differentiating Equation (*b*) yields

$$\frac{V}{EI} = \frac{d^3y}{dx^3} \tag{c}$$

$$\frac{q}{EI} = \frac{d^4y}{dx^4} \tag{d}$$

It is convenient to display these relations in a group as follows:

$$\begin{aligned}
\frac{q}{EI} &= \frac{d^4y}{dx^4}\\
\frac{V}{EI} &= \frac{d^3y}{dx^3}\\
\frac{M}{EI} &= \frac{d^2y}{dx^2}\\
\theta &= \frac{dy}{dx}\\
y &= f(x)
\end{aligned} \tag{4–9}$$

This remarkable set of equations characterizes the Euler–Bernoulli relations, showing a pattern of derivative relationships from deflection, to slope, to bending moment, to shear force, to load, with each being proportional to the rate of change of the prior one. The middle equation of the set is particularly referred to as the Euler–Bernoulli equation of bending theory, describing the relationship between the bending moment in a beam and the deflection of the beam. Typically, in application, we start with a given load function q and solve sequentially to get to deflection. Consequently, it is convenient to reverse the order of these equations and show them in integral form.

$$q = \text{distributed load function} \tag{4–10}$$

$$V = \int q\, dx \tag{4–11}$$

$$M = \int V\, dx \tag{4–12}$$

$$\theta = \int \frac{M}{EI}\, dx \tag{4–13}$$

$$y = \int \theta\, dx \tag{4–14}$$

It is also valid and common to start a problem by obtaining the bending moment from standard free-body diagram equilibrium methods, then integrate twice to get to the deflection equation.

The relationships are illustrated by the beam of Figure 4–2. Here, a beam of length l is loaded by the uniform load w. The x axis is positive to the right, and the y axis positive upward. All quantities—loading, shear, moment, slope, and deflection—have the same sense as y; they are positive if upward, negative if downward.

Note that each curve represents the slope of the next curve, as predicted by the differential relationship equations.

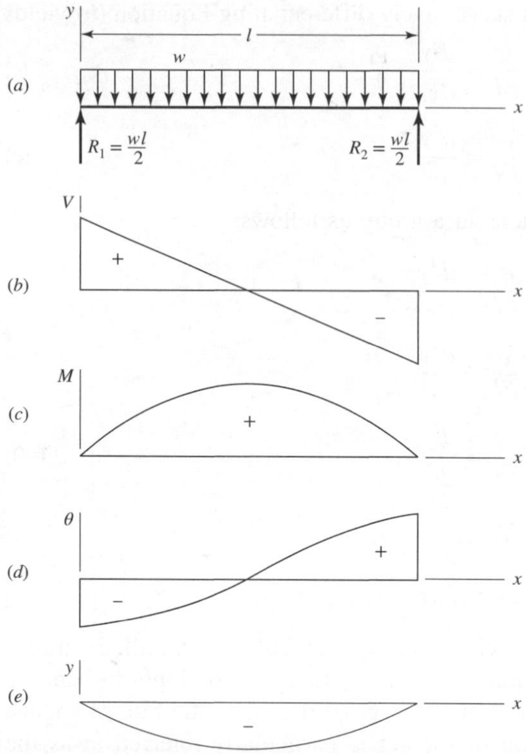

Figure 4–2

Simply-supported beam with uniform load, illustrating the integral relationships of Equations (4–15) through (4–19).

EXAMPLE 4–1

For the beam in Figure 4–2*a*, generate equations and plots of shear force, bending moment, slope, and deflection. Also determine the slopes at the ends of the beam, and the maximum deflection.

Solution

Start with the distributed load function as given in Figure 4–2*a*,

$$q(x) = -w \tag{1}$$

The negative sign is because q is defined as positive in the positive y direction, and the applied load w is downward.

Apply Equation (4–11).

$$V = \int q\,dx = \int -w\,dx = -wx + C_1$$

To determine the constant of integration C_1, apply a boundary condition for shear load, that is, a known shear force at some value of x. At the left end of the beam, just to the right of the support, so at $x = 0^+$, $V = R_1 = wl/2$, which leads to $C_1 = wl/2$. Thus,

Answer

$$V = -wx + \frac{wl}{2} \tag{2}$$

This equation for shear force is plotted in Figure 4–2*b*.

Apply Equation (4–12) to integrate Equation (2) to obtain the moment equation.

$$M = \int V\,dx = \int \left(-wx + \frac{wl}{2}\right) dx = \frac{-wx^2}{2} + \frac{wlx}{2} + C_2$$

To determine the constant of integration C_2, apply the known boundary condition for bending moment from the simple support, that is, at $x = 0$, $M = 0$, which results in $C_2 = 0$.

Answer
$$M = \frac{-wx^2}{2} + \frac{wlx}{2} \tag{3}$$

Equation (3) for bending moment is plotted in Figure 4–2*c*. Note that the bending moment in Equation (3) could have been obtained as a starting point simply by breaking into the beam and performing a moment balance around that point. It is often simpler to start with the bending moment rather than with the load, thereby skipping the work of the first two integrations.

Continuing, apply Equation (4–13).

$$\theta = \int \frac{M}{EI}\,dx = \frac{1}{EI}\int\left(\frac{-wx^2}{2} + \frac{wlx}{2}\right)dx = \frac{1}{EI}\left(\frac{-wx^3}{6} + \frac{wlx^2}{4} + C_3\right) \tag{4}$$

To solve for C_3, we could impose that the slope is zero at the midspan of the beam, since the beam and loading are symmetric relative to the midspan. However, this is not technically a boundary condition of the problem, but an expected result. We have two boundary conditions for displacement, one at each end of the beam, so we will proceed with the next integration and then use the two boundary conditions to solve for C_3 and C_4.

Apply Equation (4–14).

$$y = \int \theta\,dx = \int\left[\frac{1}{EI}\left(\frac{-wx^3}{6} + \frac{wlx^2}{4} + C_3\right)\right]dx = \frac{1}{EI}\left(\frac{-wx^4}{24} + \frac{wlx^3}{12} + C_3x + C_4\right) \tag{5}$$

The boundary conditions for the simply supported beam are zero deflection at each end. Applying the boundary condition at the left end, $y = 0$ at $x = 0$, to Equation (5) results in $C_4 = 0$. Applying the boundary condition at the right end, $y = 0$ at $x = l$, to Equation (5) results in $C_3 = -wl^3/24$. Substituting C_3 and C_4 back into Equations (4) and (5) and collecting common terms yields

Answer
$$\theta = \frac{w}{24EI}(-4x^3 + 6lx^2 - l^3) \tag{6}$$

Answer
$$y = \frac{w}{24EI}(-x^4 + 2lx^3 - l^3x) \tag{7}$$

Equations (4) and (5) are plotted in Figures 4–2*d* and 4–2*e*, respectively.

Comparing Equation (7) with that given in Table A–9, beam 7, we see complete agreement. For the slope at the left end, substituting $x = 0$ into Equation (6) yields

Answer
$$\theta|_{x=0} = -\frac{wl^3}{24EI}$$

and at $x = l$,

Answer
$$\theta|_{x=l} = \frac{wl^3}{24EI}$$

At the midspan, substituting $x = l/2$ into Equation (6) gives $\theta = 0$, as earlier suspected.

The maximum deflection occurs where $\theta = dy/dx = 0$. Substituting $x = l/2$ into Equation (7) yields

Answer
$$y_{\max} = -\frac{5wl^4}{384EI}$$

which again agrees with Table A–9–7.

The approach used in the example is fine for simple beams with continuous loading. However, for beams with discontinuous loading and/or geometry changes, such as a stepped shaft to support gears, flywheels, pulleys, etc., the problem can be quite complex. The following section discusses bending deflections in general and the techniques that are provided in this chapter.

4–4 Beam Deflection Methods

Equations (4–10) through (4–14) are the basis for relating the intensity of loading q, vertical shear V, bending moment M, slope of the neutral surface θ, and the transverse deflection y. Beams have intensities of loading that range from q = constant (uniform loading), variable intensity $q(x)$, to Dirac delta functions (concentrated loads).

The intensity of loading usually consists of piecewise contiguous zones, the expressions for which are integrated through Equations (4–10) to (4–14) with varying degrees of difficulty. Another approach is to represent the deflection $y(x)$ as a Fourier series, which is capable of representing single-valued functions with a finite number of finite discontinuities, then differentiating through Equation (4–9), and stopping at some level where the Fourier coefficients can be evaluated. A complication is the piecewise continuous nature of some beams (shafts) that are stepped-diameter bodies.

All of the above constitute, in one form or another, formal integration methods, which, with properly selected problems, result in solutions for q, V, M, θ, and y. These solutions may be

- Closed-form, or
- Represented by infinite series, which amount to closed form if the series are rapidly convergent, or
- Approximations obtained by evaluating the first or the first and second terms.

The series solutions can be made equivalent to the closed-form solution by the use of a computer. Roark's[1] formulas are committed to commercial software and can be used on a personal computer.

There are many techniques employed to solve the integration problem for beam deflection. Some of the popular methods include:

- Superposition (see Section 4–5)
- The moment-area method[2]
- Singularity functions (see Section 4–6)
- Numerical integration[3]

The two methods described in this chapter are easy to implement and can handle a large array of problems.

[1]Warren C. Young, Richard G. Budynas, and Ali M. Sadegh, *Roark's Formulas for Stress and Strain,* 8th ed., McGraw Hill, New York, 2012.

[2]See Chapter 9, F. P. Beer, E. R. Johnston Jr., and J. T. DeWolf, *Mechanics of Materials,* 5th ed., McGraw Hill, New York, 2009.

[3]See Section 4–4, J. E. Shigley and C. R. Mischke, *Mechanical Engineering Design,* 6th ed., McGraw Hill, New York, 2001.

There are methods that do not deal with Equations (4–10) to (4–14) directly. An energy method, based on Castigliano's theorem, is quite powerful for problems not suitable for the methods mentioned earlier and is discussed in Sections 4–7 to 4–10. Finite element programs are also quite useful for determining beam deflections.

4–5 Beam Deflections by Superposition

The results of many simple load cases and boundary conditions have been solved and are available. Table A–9 provides a limited number of cases. Roark's[4] provides a much more comprehensive listing. *Superposition* resolves the effect of combined loading on a structure by determining the effects of each load separately and adding the results algebraically. Superposition may be applied provided

1 each effect is linearly related to the load that produces it;

2 a load does not create a condition that affects the result of another load; and

3 the deformations resulting from any specific load are not large enough to appreciably alter the geometric relations of the parts of the structural system.

The following examples are illustrations of the use of superposition.

EXAMPLE 4–2

Consider the uniformly loaded beam with a concentrated force as shown in Figure 4–3. Using superposition, determine the reactions and the deflection as a function of x.

Solution

Considering each load state separately, we can superpose beams 6 and 7 of Table A–9. For the reactions we find

Answer $$R_1 = \frac{Fb}{l} + \frac{wl}{2}$$

Answer $$R_2 = \frac{Fa}{l} + \frac{wl}{2}$$

The loading of beam 6 is discontinuous and separate deflection equations are given for regions AB and BC. Beam 7 loading is not discontinuous so there is only one equation. Superposition yields

Answer $$y_{AB} = \frac{Fbx}{6EIl}(x^2 + b^2 + l^2) + \frac{wx}{24EI}(2lx^2 - x^3 - l^3)$$

Answer $$y_{BC} = \frac{Fa(l - x)}{6EIl}(x^2 + a^2 - 2lx) + \frac{wx}{24EI}(2lx^2 - x^3 - l^3)$$

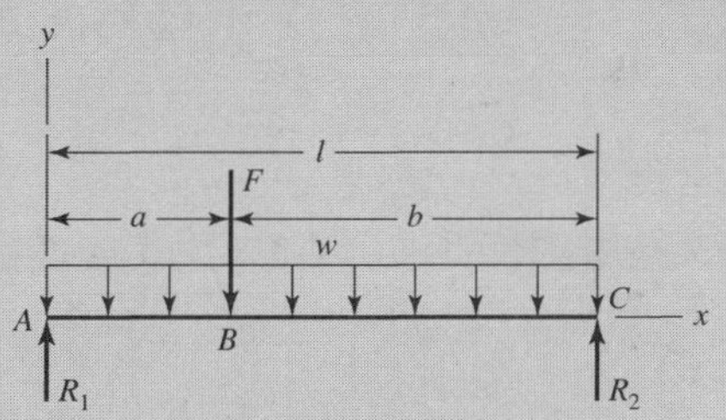

Figure 4–3

[4]Young, op. cit.

If the maximum deflection of a beam is desired, it will occur either where the slope is zero or at the end of the overhang if the beam has a free end. In the previous example, there is no overhang, so setting $dy/dx = 0$ will yield the equation for x that locates where the maximum deflection occurs. In the example there are two equations for y where only one will yield a solution. If $a = l/2$, the maximum deflection would obviously occur at $x = l/2$ because of symmetry. However, if $a < l/2$, where would the maximum deflection occur? It can be shown that as F moves toward the left support, the maximum deflection moves toward the left support also, but not as much as F (see Problem 4–55). Thus, we would set $dy_{BC}/dx = 0$ and solve for x. If $a > l/2$, then we would set $dy_{AB}/dx = 0$. For more complicated problems, plotting the equations using numerical data is the simplest approach to finding the maximum deflection.

Sometimes it may not be obvious that we can use superposition with the tables at hand, as demonstrated in the next example.

EXAMPLE 4–3

Consider the beam in Figure 4–4*a* and determine the deflection equations using superposition.

Solution

For region *AB* we can superpose beams 7 and 10 of Table A–9 to obtain

Answer

$$y_{AB} = \frac{wx}{24EI}(2lx^2 - x^3 - l^3) + \frac{Fax}{6EIl}(l^2 - x^2)$$

For region *BC*, how do we represent the uniform load? Considering the uniform load *only*, the beam deflects as shown in Figure 4–4*b*. Region *BC* is straight since there is no bending moment due to w. The slope of the beam at B is θ_B and is obtained by taking the derivative of y given in the table with respect to x and setting $x = l$. Thus,

$$\frac{dy}{dx} = \frac{d}{dx}\left[\frac{wx}{24EI}(2lx^2 - x^3 - l^3)\right] = \frac{w}{24EI}(6lx^2 - 4x^3 - l^3)$$

Substituting $x = l$ gives

$$\theta_B = \frac{w}{24EI}(6ll^2 - 4l^3 - l^3) = \frac{wl^3}{24EI}$$

The deflection in region *BC* due to w is $\theta_B(x - l)$, and adding this to the deflection due to F, in *BC*, yields

Answer

$$y_{BC} = \frac{wl^3}{24EI}(x - l) + \frac{F(x - l)}{6EI}[(x - l)^2 - a(3x - l)]$$

Figure 4–4

(*a*) Beam with uniformly distributed load and overhang force; (*b*) deflections due to uniform load only.

EXAMPLE 4–4

Figure 4–5*a* shows a cantilever beam with an end load. Normally we model this problem by considering the left support as rigid. After testing the rigidity of the wall it was found that the translational stiffness of the wall was k_t force per unit vertical deflection, and the rotational stiffness was k_r moment per unit angular (radian) deflection (see Figure 4–5*b*). Determine the deflection equation for the beam under the load F.

Solution

Here we will superpose the *modes* of deflection. They are: (1) translation due to the compression of spring k_t, (2) rotation of the spring k_r, and (3) the elastic deformation of beam 1 given in Table A–9. The force in spring k_t is $R_1 = F$, giving a deflection from Equation (4–2) of

$$y_1 = -\frac{F}{k_t} \tag{1}$$

The moment in spring k_r is $M_1 = Fl$. This gives a clockwise rotation of $\theta = Fl/k_r$. Considering this mode of deflection only, the beam rotates rigidly clockwise, leading to a deflection equation of

$$y_2 = -\frac{Fl}{k_r}x \tag{2}$$

Finally, the elastic deformation of beam 1 from Table A–9 is

$$y_3 = \frac{Fx^2}{6EI}(x - 3l) \tag{3}$$

Adding the deflections from each mode yields

Answer

$$y = \frac{Fx^2}{6EI}(x - 3l) - \frac{F}{k_t} - \frac{Fl}{k_r}x$$

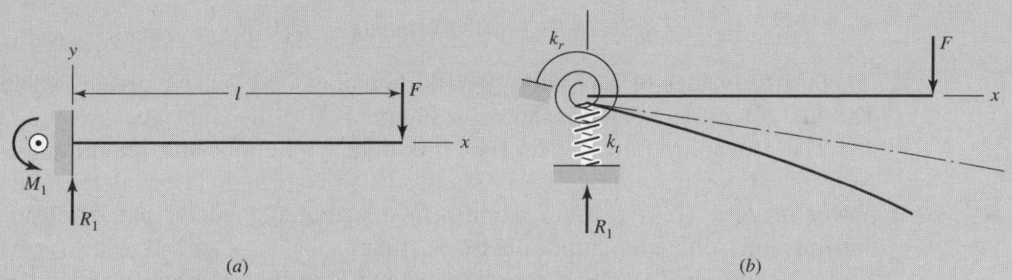

Figure 4–5

4–6 Beam Deflections by Singularity Functions

As shown in Section 4–3, determination of beam deflection is fundamentally a matter of successive integration of the load function. Direct integration is straightforward when the load function is continuous for the entire length of the beam, such as in Example 4–1. Using superposition allows for more complexity of loading, but is still limiting for more complex loading and for changes in geometry. Beams will often have changes in loading and geometry along the length, requiring each unique section of the beam to be handled with a separate load function. Further, when a load function contains discontinuities such as discrete loads and moments, integration at the discontinuity is not possible due to the integral being undefined at a singularity. This requires separate integrations, with separate constants of integration, for each unique section. The constants of integration must be determined with the application of boundary conditions for each section, accounting for the discontinuous load function at the boundaries between the sections.

A well-known method to simplify integration of complex and discontinuous loading problems is *Macaulay's Method*, also known as the method of *singularity function*. This method employs a mathematical representation for the discontinuities that allows the entire beam loading to be modeled with a single equation. Incorporating rules for integration of the singularity functions does away with the need for separate sections, and most importantly for the need for separate boundary conditions for each section. Each integration of the Euler–Bernoulli relations requires only one boundary condition for the entire beam to solve for a constant of integration, rather than one boundary condition for each unique section of the beam and for each integration. This is a significant savings in time for setup and computation. The method is ideally suited for computer-programmed solutions.

Macaulay functions are defined to represent quantities that begin at a specified distance a from the origin of coordinates x. In general form this is represented as

$$\langle x - a\rangle^n = \begin{cases} 0 & \text{when } x < a \\ (x - a)^n & \text{when } x \geq a \end{cases} \qquad \text{for } n \geq 0 \tag{4–15}$$

where x represents the location of a point on the beam, a is the location where the loading begins, and n is a positive integer exponent indicating the degree of the function. The special bracket is used to indicate the Macaulay function. The essence of the function is that it is zero until it is activated when x reaches a, then it continues active for the rest of the beam. Another way to think of the Macaulay bracket is that the equation in the bracket is active if positive, and zero otherwise. For programming a Macaulay bracket, it is convenient to use a *max* function,

$$\langle x - a\rangle^n = [\max(x - a, 0)]^n \tag{4–16}$$

Positive values of n provide specific functions of x^n once activated when $x > \alpha$. The first three functions are shown in Figure 4–6.

The Macaulay function for $n = 0$ is a step function with magnitude of unity. Note that once it is active, it is simply $(x - a)^0$ which is equal to 1 due to the power of zero. To represent a uniform distributed load w that starts at $x = a$, multiply this unit step function by w, that is, $w\ \langle x - a\rangle^0$. If a uniform load is applied between $x = a$ and $x = b$, a second step function can be included in the overall load function to "turn off," or cancel, the uniform load when both functions are active for $x \geq b$. See Figure 4–7. Note that once a function is active, it continues indefinitely as x increases, though it is irrelevant once the end of the beam is reached.

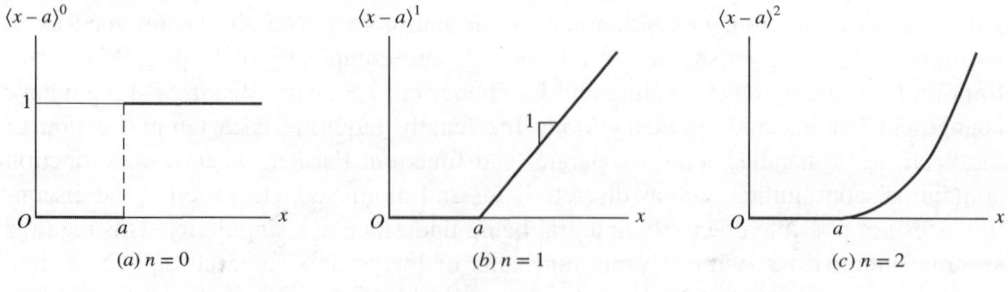

Figure 4–6

Graphs of Macaulay functions for $n = 0$, 1, and 2
(*a*) Unit step function;
(*b*) Unit ramp function;
(*c*) Parabolic ramp function

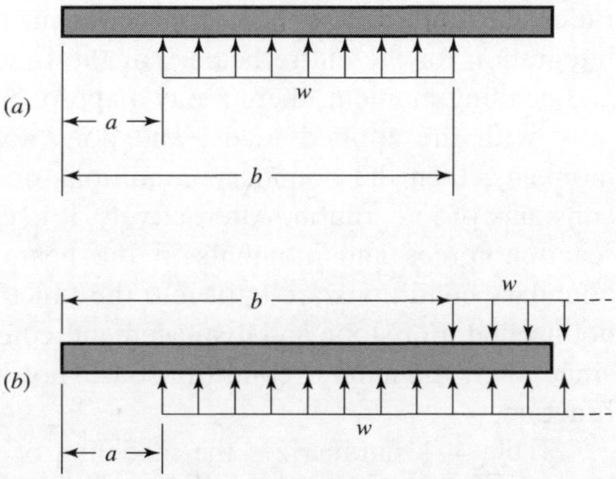

$$q = w\,\langle x - a\rangle^0 - w\,\langle x - b\rangle^0$$

Figure 4–7

(*a*) Distributed load between $x = a$ and $x = b$; (*b*) Use of a second step function to cancel the first step function at $x = b$

Sometimes the Macaulay functions are referred to as singularity functions. However, we will distinguish the two terms to clarify a different behavior between them. Macaulay functions have exponent $n \geq 0$, and *singularity functions* have $n < 0$. A singularity function is singular at a, and zero everywhere else. Mathematically, singularity functions are defined as

$$\langle x - a\rangle^n = \begin{cases} 0 & \text{when } x \neq a \\ \text{undefined} & \text{when } x = a \end{cases} \qquad \text{for } n < 0 \tag{4–17}$$

Two singularity functions are useful for beam analysis. When $n = -1$, the function at $x = a$ is defined as a unit impulse that is infinitely tall and infinitesimally narrow with an enclosed area of unity. A concentrated force P can be modeled by multiplying the unit impulse function by P. Even though the impulse function is singular at $x = a$, upon integration the resulting shear force will be the area of the impulse (unity) times the force P.

Similarly, when $n = -2$, the function at $x = a$ is defined as a unit moment. It can be thought of as a couple of forces that are infinitely large and infinitesimally separated, creating a moment of unit magnitude. The unit moment function can model a concentrated moment M_0 by multiplying the function by M_0.

Integration of the Macaulay functions and singular functions are defined as follows:

$$\text{Macaulay function, } n \geq 0: \qquad \int \langle x - a\rangle^n\, dx = \frac{\langle x - a\rangle^{n+1}}{(n+1)} \tag{4–18}$$

$$\text{Singularity function, } n < 0: \qquad \int \langle x - a\rangle^n\, dx = \langle x - a\rangle^{n+1} \tag{4–19}$$

Note that when $n \geq 0$, the exponent behaves mathematically as an exponent and follows normal rules of integration. But when $n < 0$, the exponent is an indication of degree of singularity and does not behave mathematically as an exponent. Thus, its integration rule is uniquely defined as shown in Equation (4–19).

A clarification regarding constants of integration is appropriate here. The beam loading function will consist of multiple functions. At each integration of the

Euler–Bernoulli relationship for the beam, there will be one collective constant of integration for the entire beam. For the first two integrations to obtain shear force and bending moment, there are two approaches. If the loading function is generated only with the applied loads, and does not include ground reaction forces and moments, then the boundary conditions on V and M are applied to solve for the constants of integration. Alternatively, it is convenient to simply include the ground reaction forces and moments in the beam loading function, in which case the boundary conditions are built into the function and the constants of integration are not needed. For slope and displacement, constants of integration are always needed, since their boundary conditions are not directly built into the beam loading function.

Table 4–1 summarizes the modeling of beam loading functions with Macaulay and singularity functions. It is sometimes convenient to start a problem by skipping directly to the bending moment equation for the beam rather than starting with the loading function. To aid in this, the first two integrations of the loading functions are shown in Table 4–1.

Macaulay's method is actually straightforward. A few examples will help at this point. We will start with a simple example to establish the method, then progressively move to more complex examples where this method has advantage.

Table 4–1 Summary of Modeling Beam Loads with Macaulay and Singularity Functions

Loading	Loading function $q = q(x)$	Shear $V = \int q\,dx$	Moment $M = \int V\,dx$
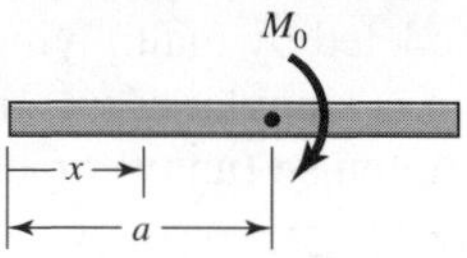	$q = M_0\langle x - a\rangle^{-2}$	$V = M_0\langle x - a\rangle^{-1}$	$M = M_0\langle x - a\rangle^{0}$
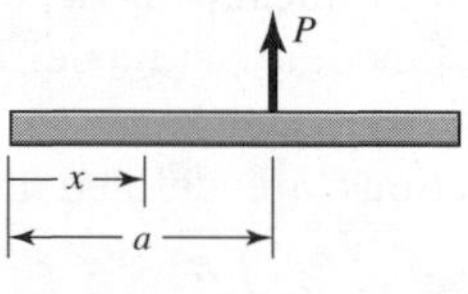	$q = P\langle x - a\rangle^{-1}$	$V = P\langle x - a\rangle^{0}$	$M = P\langle x - a\rangle^{1}$
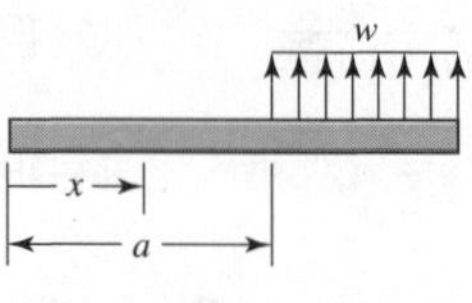	$q = w\langle x - a\rangle^{0}$	$V = w\langle x - a\rangle^{1}$	$M = \frac{w}{2}\langle x - a\rangle^{2}$
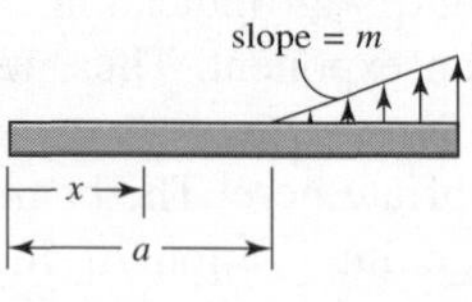	$q = m\langle x - a\rangle^{1}$	$V = \frac{m}{2}\langle x - a\rangle^{2}$	$M = \frac{m}{6}\langle x - a\rangle^{3}$

EXAMPLE 4–5

Use Macaulay's method to develop the deflection equation for the beam in Figure 4–8.

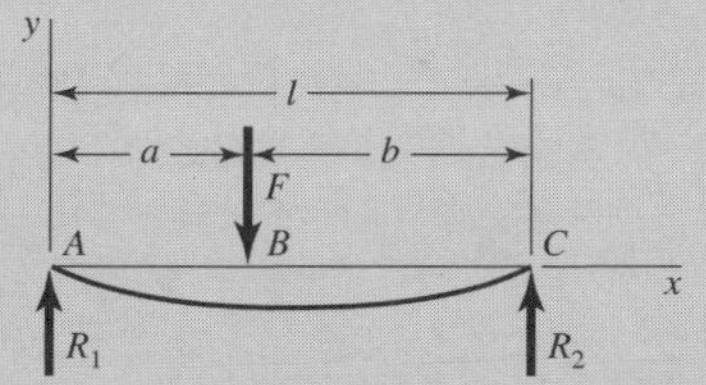

Figure 4–8

Simply-supported beam with concentrated load.

Solution

As a preliminary step, determine the ground reaction forces from standard free-body diagram equilibrium equations. This gives $R_1 = Fb/l$ and $R_2 = Fa/l$. To keep the integration equations cleaner, we will incorporate these after the integrations are completed.

Write the loading function equation from the free-body diagram, including the ground reaction forces.

$$q = R_1 \langle x \rangle^{-1} - F\langle x - a \rangle^{-1} + R_2 \langle x - l \rangle^{-1}$$

All three forces are implemented as impulse singularity functions, thus with $n = -1$. R_1 is active starting at $x = 0$, F is active starting at $x = a$, and R_2 is active starting at $x = l$. Here, we can take advantage of an opportunity to simplify the loading function. Note that R_2 is only active for $x > l$, but the beam doesn't exist beyond $x = l$. Consequently, the R_2 term will never be active and can be eliminated. This is a general principle of Macaulay's method, that any load at the extreme end of the beam will not contribute to the load function or its integrals. Note, however, that R_2 was included in the determination of the ground reaction forces from free-body equilibrium, so its effect on the beam is included in the value of R_1. Deleting the R_2 term, the load function is

$$q = R_1 \langle x \rangle^{-1} - F\langle x - a \rangle^{-1} \tag{1}$$

Applying Equations (4–11) and (4–12), using the singularity integration rules of Equations (4–18) and (4–19), the shear force and bending moment equations are obtained.

$$V = \int q\, dx = R_1 \langle x \rangle^{0} - F\langle x - a \rangle^{0} \tag{2}$$

$$M = \int V\, dx = R_1 \langle x \rangle^{1} - F\langle x - a \rangle^{1} \tag{3}$$

Since the ground reaction forces and moments are included in the beam load function of Equation (1), constants of integration are not needed. The next two integrations will include constants of integration to account for the necessary boundary conditions on slope and/or deflection that are not built into the original loading function.

At this point, notice that we could have taken a shortcut directly to Equation (3) by building the moment equation directly from the last column of Table 4–1.

Continuing from Equation (3), apply Equations (4–13) and (4–14), with the integration rule of Equation (4–18),

$$\theta = \int \frac{M}{EI}\, dx = \frac{1}{EI}\left(\frac{R_1}{2}\langle x \rangle^2 - \frac{F}{2}\langle x - a \rangle^2 + C_1\right) \tag{4}$$

$$y = \int \theta\, dx = \frac{1}{EI}\left(\frac{R_1}{6}\langle x \rangle^3 - \frac{F}{6}\langle x - a \rangle^3 + C_1 x + C_2\right) \tag{5}$$

To solve for the constants of integration, apply the boundary conditions for zero deflection at each of the simple supports. With $y = 0$ at $x = 0$, Equation (5) yields $C_2 = 0$. With $y = 0$ at $x = l$, Equation (5) is

$$0 = \frac{R_1}{6}(l)^3 - \frac{F}{6}(l - a)^3 + C_1 l$$

Noting that $(l - a) = b$, solving for C_1,

$$C_1 = \frac{R_1 l^2}{6} - \frac{Fb^3}{6l}$$

Now substituting the ground reaction force previously obtained as $R_1 = Fb/l$,

$$C_1 = -\frac{Fb}{6l}(l^2 - b^2)$$

Finally, substitute C_1 and R_1 into Equation (5). Also, note that $\langle x \rangle^3 = (x)^3$ since it is active for the entire length of the beam.

$$y = \frac{1}{EI}\left(\frac{Fb}{6l}\langle x \rangle^3 - \frac{F}{6}\langle x - a \rangle^3 - \frac{Fb}{6l}(l^2 - b^2)x\right)$$

$$y = \frac{F}{6EIl}[bx(x^2 + b^2 - l^2) - l\langle x - a \rangle^3] \tag{6}$$

Solution for this beam loading case is provided in the beam table of Table A-9, case 6, where two equations are given for the two sections of the beam. The results are easily shown to be identical with only slight rearrangement and incorporating $b = (l - a)$.

EXAMPLE 4–6

Figure 4–9 shows the free-body diagram for a beam cantilevered at A with a uniform load of 4 kN/m acting on the portion 50 mm $\leq x \leq$ 150 mm, and a concentrated counterclockwise moment of 25 N · m at $x = 200$ mm. The beam is steel, with a rectangular 3 mm × 15 mm cross section. Use Macaulay's method to obtain equations for shear force, bending moment, slope, and deflection. Program the equations to generate plots.

Solution

From standard summation of forces and moments on the free-body diagram, determine the ground reaction force and moment to be $R_1 = +400$ N and $M_1 = +15$ N · m CCW.

Generate the beam loading function, using Table 4–1.

$$q = -15\ 000\langle x \rangle^{-2} + 400\langle x \rangle^{-1} - 4\langle x - 50 \rangle^0 + 4\langle x - 150 \rangle^0 - 25\ 000\langle x - 200 \rangle^{-2} \tag{1}$$

The first term is from the ground reaction moment. The moment M_1 on the free-body diagram is CCW and therefore positive according to standard sign convention. However, when applied as a concentrated moment

Figure 4–9

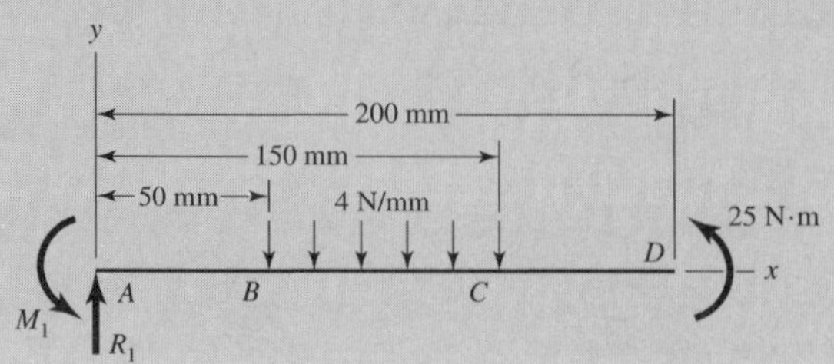

function, the sign convention used in Table 4–1 is that CW is positive, thus it is entered in the loading function as a negative step moment. This may seem odd, but it is to be consistent with the positive bending moment within the beam as specified in Figure 3–3. This is a tricky point, so pay particular attention to the sign convention when applying a moment function. The second term in Equation (1) is the impulse point load for the ground reaction force, positive upward. The distributed load is given by the third and fourth terms in Equation (1), applying a negative (downward) step function of 4 N · m for $x > 50$ mm, and then canceling the distributed load with a positive (upward) step function of 4 N/mm for $x > 150$ mm. The last term of Equation (1) is the concentrated moment at the tip of the beam. Since this function will never be active (zero for the length of the beam, and non-existent past the end of the beam), we will delete it, giving the beam loading function as

$$q = -15\ 000\langle x\rangle^{-2} + 400\langle x\rangle^{-1} - 4\langle x - 50\rangle^0 + 4\langle x - 150\rangle^0 \tag{2}$$

Applying the Euler–Bernoulli relationships of Equations (4–10) through (4–14), we take successive integrals using the integration rules of Equations (4–18) and (4–19).

$$V = \int q\, dx = -15\ 000\langle x\rangle^{-1} + 400\langle x\rangle^0 - 4\langle x - 50\rangle^1 + 4\langle x - 150\rangle^1 \tag{3}$$

$$M = \int V\, dx = -15\ 000\langle x\rangle^0 + 400\langle x\rangle^1 - \frac{4}{2}\langle x - 50\rangle^2 + \frac{4}{2}\langle x - 150\rangle^2 \tag{4}$$

$$\theta = \int \frac{M}{EI}\, dx = \frac{1}{EI}\left(-15\ 000\langle x\rangle^1 + \frac{400}{2}\langle x\rangle^2 - \frac{4}{6}\langle x - 50\rangle^3 + \frac{4}{6}\langle x - 150\rangle^3 + C_1\right) \tag{5}$$

$$y = \int \theta\, dx = \frac{1}{EI}\left(-\frac{15\ 000}{2}\langle x\rangle^2 + \frac{400}{6}\langle x\rangle^3 - \frac{4}{24}\langle x - 50\rangle^4 + \frac{4}{24}\langle x - 150\rangle^4 + C_1x + C_2\right) \tag{6}$$

Since the ground reaction forces and moments are included in the original loading function of Equation (2), the boundary conditions for shear force and bending moment are already incorporated into Equations (3) and (4) without the need for constants of integration.

Apply boundary conditions for slope and deflection to solve for constants of integration C_1 and C_2. The slope is zero at $x = 0$, so Equation (5) gives $C_1 = 0$. The deflection is zero at $x = 0$, so Equation (6) gives $C_2 = 0$.

We now restate Equations (3) through (6), simplifying terms. In particular, we can set $\langle x\rangle^0 = 1$, and, for Macaulay functions we can replace $\langle x\rangle^n$ with $(x)^n$ since these functions are applicable for the entire beam. Note that in Equation (3) there is a singularity function with $n = -1$, which is applicable only at the singularity point of $x = 0$, so it must keep its bracket.

Answer

$$V = -15\ 000\langle x\rangle^{-1} + 400 - 4\langle x - 50\rangle^1 + 4\langle x - 150\rangle^1 \tag{7}$$

$$M = -15\ 000 + 400x - 2\langle x - 50\rangle^2 + 2\langle x - 150\rangle^2 \tag{8}$$

$$\theta = \frac{1}{EI}\left(-15\ 000x + 200x^2 - \frac{2}{3}\langle x - 50\rangle^3 + \frac{2}{3}\langle x - 150\rangle^3\right) \tag{9}$$

$$y = \frac{1}{EI}\left(-7500x^2 + \frac{200}{3}x^3 - \frac{1}{6}\langle x - 50\rangle^4 + \frac{1}{6}\langle x - 150\rangle^4\right) \tag{10}$$

These equations are readily implemented for plotting. For example, in a spreadsheet, the cell implementation for each equation could be something like the following:

Cells for V: = 400 – 4 * MAX(A5-50,0) + 4 * MAX(A5-150,0)

Cells for M: = –15 000 + 400 * A5 – 2 * MAX(A5-50,0)^2 + 2 * MAX(A5-50,0)^2

Cells for q: = 1/(E * I) * (–15 000 * A5 + 200 * A5^2 – 2/3 * MAX(A5-50,0)^3 + 2/3 * MAX(A5-150,0)^3)

Cells for y: = 1/(E * I) * (–7500 * A5^2 + 200/3 * A5^3 – 1/6 * MAX(A5-50,0)^4 + 1/6 * MAX(A5-150,0)^4)

Each equation is valid for the entire range of x, making it very easy to implement. References here to cell A5 are to column A which contains the values of x. E and I are references to defined cells containing their values of $E = 2.1(10)^6$ MPa and $I = bh^3/12 = 843.75$ mm^4. The MAX function is used to evaluate the Macaulay functions to set the value to the maximum of the function value or zero. In implementing Equation (7), the first term is a singularity function which is by definition undefined at $x = 0$ and zero everywhere else. This function is not included in the implementation of V since it is an undefined impulse that cannot be calculated, and it also represents a moment which is not shown on a shear-force plot. The plots obtained directly from a spreadsheet implementation of these equations are shown in Figure 4–10.

Figure 4–10

Plots generated from a spreadsheet for Example 4–6.

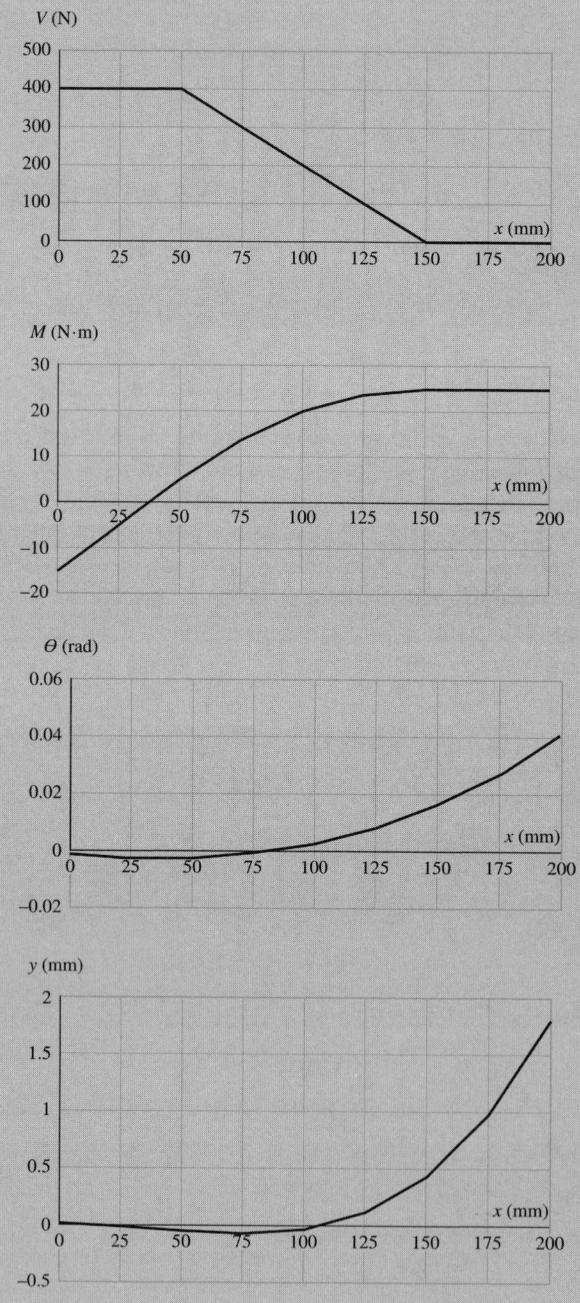

EXAMPLE 4–7

The steel step shaft shown in Figure 4–11*a* is mounted in bearings at *A* and *F*. A pulley is centered at *C* where a total radial force of 2.6 kN is applied. Using singularity functions evaluate the shaft displacements at 12.7 mm increments. Assume the shaft is simply supported.

Solution

The reactions are found to be $R_1 = 1.56$ kN and $R_2 = 1.04$ kN. Ignoring R_2 since it occurs at the far right, using singularity functions, the moment equation is

$$M = 1560x - 2600\langle x - 0.2\rangle^1 \tag{1}$$

This is plotted in Figure 4–11*b*.

For simplification, we will consider only the step at *D*. That is, we will assume section *AB* has the same diameter as *BC* and section *EF* has the same diameter as *DE*. Since these sections are short and at the supports, the size reduction will not add much to the deformation. We will examine this simplification later. The second area moments for *BC* and *DE* are

$$I_{BC} = \frac{\pi}{64}0.038^4 = 102.35\ 10^{-9}\ \text{m}^4 \qquad I_{DE} = \frac{\pi}{64}0.045^4 = 201.29\ 10^{-9}\ \text{m}^4$$

A plot of M/I is shown in Figure 4–11*c*. The values at points *b* and *c*, and the step change are

$$\left(\frac{M}{I}\right)_b = \frac{302}{(102.35\ 10^{-9})} = 2.95\ 10^9\ \text{N/m}^3 \qquad \left(\frac{M}{I}\right)_c = \frac{302}{(201.29\ 10^{-9})} = 1.50\ 10^9\ \text{N/m}^3$$

$$\Delta\left(\frac{M}{I}\right) = 1.5\ 10^9 - 2.95\ 10^9 = -1.45\ 10^9\ \text{N/m}^3$$

The slopes for *ab* and *cd*, and the change are

$$m_{ab} = \frac{302\ 000 - 312\ 000}{102.35\ 10^9(10)} = -9.77\ 10^9\ \text{N/m}^4 \qquad m_{cd} = \frac{1.50\ 10^9}{0.29} = -5.17\ 10^9\ \text{N/m}^4$$

$$\Delta m = -5.17\ 10^9 - (-9.77\ 10^9) = 4.6\ 10^9\ \text{N/m}^4$$

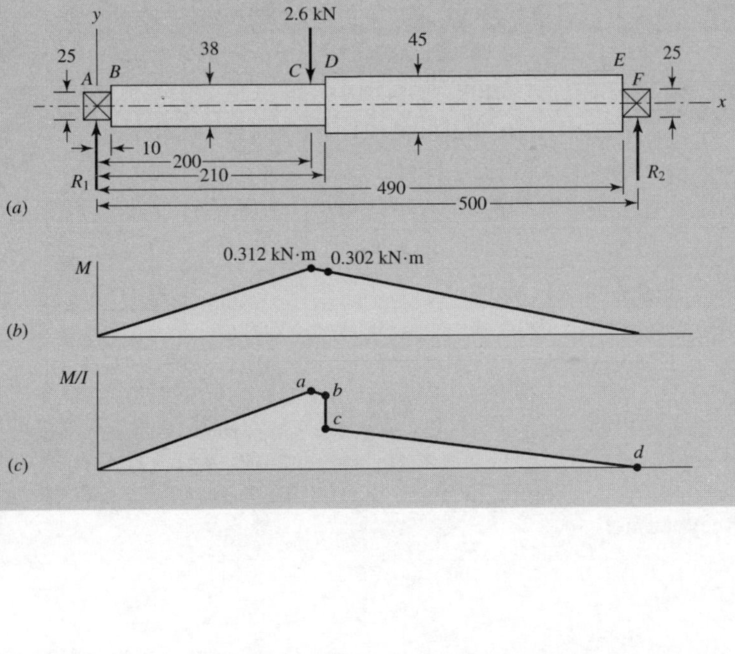

Figure 4–11
Dimensions in millimeters

Dividing Equation (1) by I_{BC} and, at $x = 0.21$ mm, adding a step of $-1.45\ 10^9$ N/m^3 and a ramp of slope $4.66\ 10^9$ N/m^4, gives

$$\frac{M}{I} = 1.52\text{E+}10x - 2.54\text{E+}10\langle x - 0.2\rangle^1 - 1.45\ 10^9\langle x - 0.21\rangle^0 + 4.66\ 10^9\langle x - 0.21\rangle^1 \tag{2}$$

Integration gives

$$E\frac{dy}{dx} = 7.62\text{E+}09x^2 - 1.27\text{E+}10\langle x - 0.2\rangle^2 - 1.45\ 10^9\langle x - 0.21\rangle^1 + 2.33\text{E+}09\langle x - .21\rangle^2 + C_1 \tag{3}$$

Integrating again yields

$$Ey = 2.54\text{E+}09x^3 - 4.23\text{E+}09\langle x - 0.2\rangle^3 - 7.25\text{E+}08\langle x - 0.21\rangle^2 + 7.77\text{E+}08\langle x - 0.21\rangle^3 + C_1x + C_2 \tag{4}$$

At $x = 0$, $y = 0$. This gives $C_2 = 0$ (remember, singularity functions do not exist until the argument is positive). At $x = 0.5$ m, $y = 0$, and

$$0 = 2.54\text{E+}09(0.5)^3 - 4.23\text{E+}09(0.5 - 0.2)^3 - 7.25\text{E+}08(0.5 - 0.21)^2 + 7.77\text{E+}08(0.5 - 0.21)^3 + C_1(0.5)$$

Solving, gives $C_1 = -0.323(10^9)$ N/m^2. Thus, Equation (4) becomes, with $E = 200(10^9)$ Pa

$$y = \frac{1}{200(10^9)}(2.54(10^9) - 4.23(10^9)\langle x - 0.2\rangle^3 - 0.725(10^9)\langle x - 0.21\rangle^2 + 0.777(10^9)\langle x - 0.21\rangle^3 - 0.323(10^9)x) \tag{5}$$

When using a spreadsheet, program the following equations:

$$y = \frac{1}{200(10^9)}(2.54(10^9)x^3 - 0.323(10^9)x) \qquad 0 \le x \le 0.2 \text{ m}$$

$$y = \frac{1}{200(10^9)}[2.54(10^9)x^3 - 4.23(10^9)(x - 0.2)^3 - 0.323(10^9)x] \qquad 0.2 \le x \le 0.21 \text{ m}$$

$$y = \frac{1}{200(10^9)}[2.54(10^9)x^3 - 4.23(10^9)(x - 0.2)^3 - 0.725(10^9)(x - 0.21)^2 + 0.777(10^9)(x - 0.21)^3 - 0.323(10^9)x] \qquad 0.21 \le x \le 0.5 \text{ m}$$

The following table results.

x (m)	y (mm)	x (m)	y (mm)	x (m)	y (mm)
0	0	0.18	–0.2162	0.34	–0.1599
0.02	–0.0322	0.20	–0.2209	0.36	–0.1431
0.04	–0.0637	0.21	–0.2211	0.38	–0.1250
0.06	–0.0940	0.22	–0.2201	0.4	–0.1057
0.08	–0.1225	0.24	–0.2160	0.42	–0.0855
0.1	–0.1486	0.26	–0.2092	0.44	–0.0646
0.12	–0.1716	0.28	–0.2000	0.46	–0.0433
0.14	–0.1909	0.30	–0.1886	0.48	–0.0216
0.16	–0.2060	0.32	–0.1751	0.5	0

where x is in meters and y in millimeters. We see that the greatest deflection is at $x = 210$ mm, where $y = -0.2211$ mm.

Substituting C_1 into Equation (3) the slopes at the supports are found to be $\theta_A = 1.626(10^{-3})$ rad $= 0.09\,316$ deg, and $\theta_F = 1.134(10^{-3})$ rad $= 0.06\,497$ deg. You might think these to be insignificant deflections, but as you will see in Chapter 7, on shafts, they are not.

A finite-element analysis was performed for the same model and resulted in

$$y|_{x=210\text{ mm}} = -0.2299\text{ mm} \qquad \theta_A = -0.09\,293° \qquad \theta_F = 0.0652°$$

Virtually the same answer save some round-off error in the equations.

If the steps of the bearings were incorporated into the model, more equations result, but the process is the same. The solution to this model is

$$y|_{x=210\text{ mm}} = -0.2241\text{ mm} \qquad \theta_A = -0.0941° \qquad \theta_F = 0.0657°$$

The largest difference between the models is of the order of 1.2 percent. Thus the simplification was justified.

In Section 4–9, we will demonstrate the usefulness of singularity functions in solving statically indeterminate problems.

4–7 Strain Energy

The external work done on an elastic member in deforming it is transformed into *strain,* or *potential, energy*. If the member is deformed a distance y, and if the force-deflection relationship is linear, this energy is equal to the product of the average force and the deflection, or

$$U = \frac{F}{2}y = \frac{F^2}{2k} \tag{4–20}$$

This equation is general in the sense that the force F can also mean torque, or moment, provided, of course, that consistent units are used for k. By substituting appropriate expressions for k, strain-energy formulas for various simple loadings may be obtained. For tension and compression, for example, we employ Equation (4–4) and obtain

$$\left.\begin{aligned} U &= \frac{F^2 l}{2AE} \\ \text{or}\quad U &= \int \frac{F^2}{2AE}\,dx \end{aligned}\right\} \text{tension and compression} \tag{4–21}$$

where the first equation applies when all the terms are constant throughout the length, and the more general integral equation allows for any of the terms to vary through the length.

Similarly, from Equation (4–6), the strain energy for torsion is given by

$$\left.\begin{aligned} U &= \frac{T^2 l}{2GJ} \\ \text{or}\quad U &= \int \frac{T^2}{2GJ}\,dx \end{aligned}\right\} \text{torsion} \tag{4–22}$$

To obtain an expression for the strain energy due to direct shear, consider the element with one side fixed in Figure 4–12*a*. The force F places the element in pure shear, and the work done is $U = F\delta/2$. Since the shear strain is $\gamma = \delta/l = \tau/G = F/AG$, we have

$$\left.\begin{aligned} U &= \frac{F^2 l}{2AG} \\ \text{or}\quad U &= \int \frac{F^2}{2AG}\,dx \end{aligned}\right\} \text{direct shear} \tag{4–23}$$

Figure 4–12

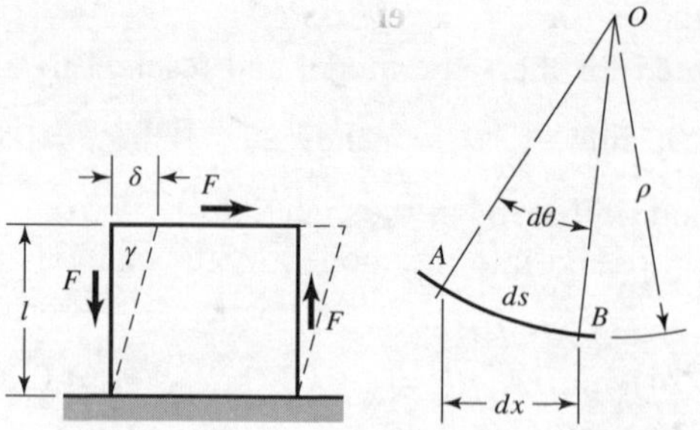

(*a*) Pure shear element (*b*) Beam bending element

The strain energy stored in a beam or lever by bending may be obtained by referring to Figure 4–12*b*. Here *AB* is a section of the elastic curve of length *ds* having a radius of curvature ρ. The strain energy stored in this element of the beam is $dU = (M/2)d\theta$. Since $\rho\ d\theta = ds$, we have

$$dU = \frac{M\ ds}{2\rho} \tag{a}$$

We can eliminate ρ by using Equation (4–7), $\rho = EI/M$. Thus

$$dU = \frac{M^2\ ds}{2EI} \tag{b}$$

For small deflections, $ds \approx dx$. Then, for the entire beam

$$U = \int dU = \int \frac{M^2}{2EI} dx \tag{c}$$

The integral equation is commonly needed for bending, where the moment is typically a function of *x*. Summarized to include both the integral and nonintegral form, the strain energy for bending is

$$\left.\begin{aligned} U &= \frac{M^2 l}{2EI} \\ \text{or} \quad U &= \int \frac{M^2}{2EI} dx \end{aligned}\right\} \text{bending} \tag{4–24}$$

Equation (4–24) is exact only when a beam is subject to pure bending. Even when transverse shear is present, these equations continue to give quite good results, except for very short beams. The strain energy due to shear loading of a beam is a complicated problem. An approximate solution can be obtained by using Equation (4–23) with a correction factor whose value depends upon the shape of the cross section. If we use *C* for the correction factor and *V* for the shear force, then the strain energy due to shear in bending is

$$\left.\begin{aligned} U &= \frac{CV^2 l}{2AG} \\ \text{or} \quad U &= \int \frac{CV^2}{2AG} dx \end{aligned}\right\} \text{transverse shear} \tag{4–25}$$

Values of the factor *C* are listed in Table 4–2.

Table 4–2 Strain-Energy Correction Factors for Transverse Shear

Beam Cross-Sectional Shape	Factor C
Rectangular	1.2
Circular	1.11
Thin-walled tubular, round	2.00
Box sections†	1.00
Structural sections†	1.00

†Use area of web only.

Source: Budynas, Richard G. *Advanced Strength and Applied Stress Analysis.* 2nd ed. New York: McGraw Hill Education, 1999. Copyright © 1999 The McGraw Hill Companies.

EXAMPLE 4–8

A cantilever beam with a round cross section has a concentrated load F at the end, as shown in Figure 4–13*a*. Find the strain energy in the beam.

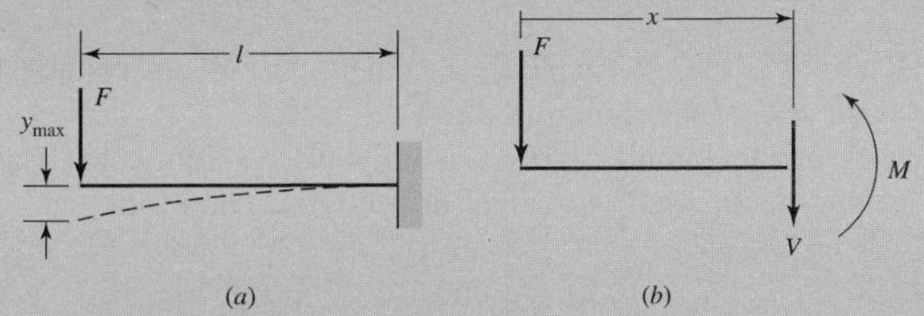

Figure 4–13

Solution

To determine what forms of strain energy are involved with the deflection of the beam, we break into the beam and draw a free-body diagram to see the forces and moments being carried within the beam. Figure 4–13*b* shows such a diagram in which the transverse shear is $V = -F$, and the bending moment is $M = -Fx$. The variable x is simply a variable of integration and can be defined to be measured from any convenient point. The same results will be obtained from a free-body diagram of the right-hand portion of the beam with x measured from the wall. Using the free end of the beam usually results in reduced effort since the ground reaction forces do not need to be determined.

For the transverse shear, using Equation (4–25) with the correction factor $C = 1.11$ from Table 4–2, and noting that V is constant through the length of the beam,

$$U_{\text{shear}} = \frac{CV^2 l}{2AG} = \frac{1.11F^2 l}{2AG}$$

For the bending, since M is a function of x, Equation (4–24) gives

$$U_{\text{bend}} = \int \frac{M^2 dx}{2EI} = \frac{1}{2EI}\int_0^l (-Fx)^2\, dx = \frac{F^2 l^3}{6EI}$$

The total strain energy is

Answer
$$U = U_{\text{bend}} + U_{\text{shear}} = \frac{F^2 l^3}{6EI} + \frac{1.11F^2 l}{2AG}$$

Note, except for very short beams, the shear term (of order l) is typically small compared to the bending term (of order l^3). This will be demonstrated in the next example.

4–8 Castigliano's Theorem

A most unusual, powerful, and often surprisingly simple approach to deflection analysis is afforded by an energy method called *Castigliano's theorem*. It is a unique way of analyzing deflections and is even useful for finding the reactions of indeterminate structures. Castigliano's theorem states that *when forces act on elastic systems subject to small displacements, the displacement corresponding to any force, in the direction of the force, is equal to the partial derivative of the total strain energy with respect to that force.* The terms *force* and *displacement* in this statement are broadly interpreted to apply equally to moments and angular displacements. Mathematically, the theorem of Castigliano is

$$\delta_i = \frac{\partial U}{\partial F_i} \tag{4–26}$$

where δ_i is the displacement of the point of application of the force F_i in the direction of F_i. For rotational displacement Equation (4–26) can be written as

$$\theta_i = \frac{\partial U}{\partial M_i} \tag{4–27}$$

where θ_i is the rotational displacement, in radians, of the beam where the moment M_i exists and in the direction of M_i.

As an example, apply Castigliano's theorem using Equations (4–21) and (4–22) to get the axial and torsional deflections. The results are

$$\delta = \frac{\partial}{\partial F}\left(\frac{F^2 l}{2AE}\right) = \frac{Fl}{AE} \tag{a}$$

$$\theta = \frac{\partial}{\partial T}\left(\frac{T^2 l}{2GJ}\right) = \frac{Tl}{GJ} \tag{b}$$

Compare Equations (*a*) and (*b*) with Equations (4–3) and (4–5).

EXAMPLE 4–9

The cantilever of Example 4–8 is a carbon steel bar 250-mm long with a 25-mm diameter and is loaded by a force $F = 400$ N.
(*a*) Find the maximum deflection using Castigliano's theorem, including that due to shear.
(*b*) What error is introduced if shear is neglected?

Solution
(*a*) From Example 4–8, the total energy of the beam is

$$U = \frac{F^2 l^3}{6EI} + \frac{1.11F^2 l}{2AG} \tag{1}$$

Then, according to Castigliano's theorem, the deflection of the end is

$$y_{\text{max}} = \frac{\partial U}{\partial F} = \frac{Fl^3}{3EI} + \frac{1.11Fl}{AG} \tag{2}$$

We also find that

$$I = \frac{\pi d^4}{64} = \frac{\pi(25)^4}{64} = 19\ 174.76\ \text{mm}^4$$

$$A = \frac{\pi d^2}{4} = \frac{\pi(25)^2}{4} = 490.87\ \text{mm}^2$$

Substituting these values, together with $F = 400$ N, $l = 250$ mm, $E = 210\ 000$ N/mm^2, and $G = 80\ 000$ N/mm^2, in Equation (2) gives

Answer

$$y_{\max} = 0.517 + 0.0028 = 0.52\ \text{mm}$$

Note that the result is positive because it is in the *same* direction as the force F.

Answer

(*b*) The error in neglecting shear for this problem is $(0.52 - 0.517)/0.52 = 0.0054 = 0.54$ percent.

The relative contribution of transverse shear to beam deflection decreases as the length-to-height ratio of the beam increases, and is generally considered negligible for $l/d > 10$. Note that the deflection equations for the beams in Table A–9 do not include the effects of transverse shear.

Castigliano's theorem can be used to find the deflection at a point even though no force or moment acts there. The procedure is as follows:

1 Set up the equation for the total strain energy U by including the energy due to a fictitious force or moment Q acting at the point whose deflection is to be found.

2 Find an expression for the desired deflection δ, in the direction of Q, by taking the derivative of the total strain energy with respect to Q.

3 Since Q is a fictitious force, solve the expression obtained in step 2 by setting Q equal to zero. Thus, the displacement at the point of application of the fictitious force Q is

$$\delta = \left.\frac{\partial U}{\partial Q}\right|_{Q=0} \tag{4–28}$$

In cases where integration is necessary to obtain the strain energy, it is more efficient to obtain the deflection directly without explicitly finding the strain energy, by moving the partial derivative inside the integral. For the example of the bending case,

$$\delta_i = \frac{\partial U}{\partial F_i} = \frac{\partial}{\partial F_i}\left(\int \frac{M^2}{2EI}\,dx\right) = \int \frac{\partial}{\partial F_i}\left(\frac{M^2}{2EI}\right)dx = \int \frac{2M\dfrac{\partial M}{\partial F_i}}{2EI}\,dx = \int \frac{1}{EI}\left(M\frac{\partial M}{\partial F_i}\right)dx$$

This allows the derivative to be taken before integration, simplifying the mathematics. This method is especially helpful if the force is a fictitious force Q, since it can be set to zero as soon as the derivative is taken. The expressions for the common cases in Equations (4–21), (4–22), and (4–24) are rewritten as

$$\delta_i = \frac{\partial U}{\partial F_i} = \int \frac{1}{AE}\left(F\frac{\partial F}{\partial F_i}\right)dx \qquad \text{tension and compression} \tag{4–29}$$

$$\theta_i = \frac{\partial U}{\partial M_i} = \int \frac{1}{GJ}\left(T\frac{\partial T}{\partial M_i}\right)dx \qquad \text{torsion} \tag{4–30}$$

$$\delta_i = \frac{\partial U}{\partial F_i} = \int \frac{1}{EI}\left(M\frac{\partial M}{\partial F_i}\right)dx \qquad \text{bending} \tag{4–31}$$

EXAMPLE 4–10

Using Castigliano's method, determine the deflections of points A and B due to the force F applied at the end of the step shaft shown in Figure 4–14. The second area moments for sections AB and BC are I_1 and $2I_1$, respectively.

Figure 4–14

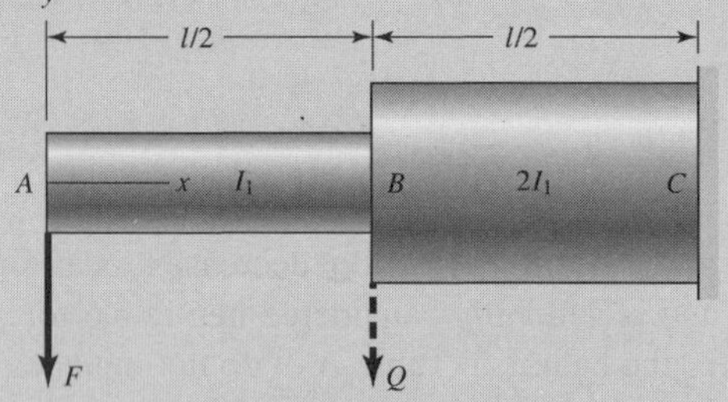

Solution

To avoid the need to determine the support reaction forces, define the origin of x at the left end of the beam as shown. For $0 \le x \le l$, the bending moment is

$$M = -Fx \tag{1}$$

Since F is at A and in the direction of the desired deflection, the deflection at A from Equation (4–31) is

$$\delta_A = \frac{\partial U}{\partial F} = \int_0^l \frac{1}{EI}\left(M\frac{\partial M}{\partial F}\right)dx \tag{2}$$

Substituting Equation (1) into Equation (2), noting that $I = I_1$ for $0 \le x \le l/2$, and $I = 2I_1$ for $l/2 \le x \le l$, we get

$$\delta_A = \frac{1}{E}\left[\int_0^{l/2} \frac{1}{I_1}(-Fx)(-x)\,dx + \int_{l/2}^{l} \frac{1}{2I_1}(-Fx)(-x)\,dx\right]$$

Answer

$$= \frac{1}{E}\left[\frac{Fl^3}{24I_1} + \frac{7Fl^3}{48I_1}\right] = \frac{3}{16}\frac{Fl^3}{EI_1}$$

which is positive, as it is in the direction of F.

For B, a fictitious force Q is necessary at the point. Assuming Q acts down at B, and x is as before, the moment equation is

$$M = -Fx \qquad 0 \le x \le l/2$$

$$M = -Fx - Q\left(x - \frac{l}{2}\right) \qquad l/2 \le x \le l \tag{3}$$

For Equation (4–31), we need $\partial M/\partial Q$. From Equation (3),

$$\frac{\partial M}{\partial Q} = 0 \qquad 0 \le x \le l/2$$

$$\frac{\partial M}{\partial Q} = -\left(x - \frac{l}{2}\right) \qquad l/2 \le x \le l \tag{4}$$

Once the derivative is taken, Q can be set to zero, so Equation (4–31) becomes

$$\delta_B = \left[\int_0^l \frac{1}{EI}\left(M\frac{\partial M}{\partial Q}\right) dx\right]_{Q=0}$$

$$= \frac{1}{EI_1}\int_0^{l/2}(-Fx)(0)\,dx + \frac{1}{E(2I_1)}\int_{l/2}^{l}(-Fx)\left[-\left(x - \frac{l}{2}\right)\right] dx$$

Evaluating the last integral gives

Answer

$$\delta_B = \frac{F}{2EI_1}\left(\frac{x^3}{3} - \frac{lx^2}{4}\right)\Bigg|_{l/2}^{l} = \frac{5}{96}\frac{Fl^3}{EI_1}$$

which again is positive, in the direction of Q.

EXAMPLE 4–11

For the wire form of diameter d shown in Figure 4–15a, determine the deflection of point B in the direction of the applied force F (neglect the effect of transverse shear).

Solution

Figure 4–15b shows free body diagrams where the body has been broken in each section, and internal balancing forces and moments are shown. The sign convention for the force and moment variables is positive in the directions shown. With energy methods, sign conventions are arbitrary, so use a convenient one. In each section, the

Figure 4–15

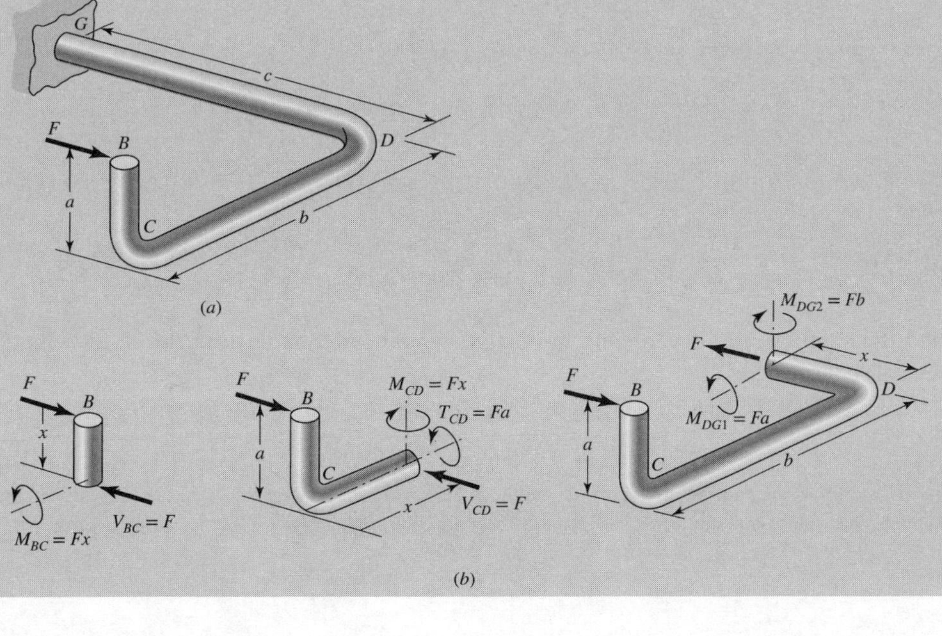

variable x is defined with its origin as shown. The variable x is used as a variable of integration for each section independently, so it is acceptable to reuse the same variable for each section. For completeness, the transverse shear forces are included, but the effects of transverse shear on the strain energy (and deflection) will be neglected.

Element BC is in bending only so from Equation (4–31),[5]

$$\frac{\partial U_{BC}}{\partial F} = \frac{1}{EI}\int_0^a (Fx)(x)\,dx = \frac{Fa^3}{3EI} \tag{1}$$

Element CD is in bending and in torsion. The torsion is constant so Equation (4–30) can be written as

$$\frac{\partial U}{\partial F_i} = \left(T\frac{\partial T}{\partial F_i}\right)\frac{l}{GJ}$$

where l is the length of the member. So for the torsion in member CD, $F_i = F$, $T = Fa$, and $l = b$. Thus,

$$\left(\frac{\partial U_{CD}}{\partial F}\right)_{\text{torsion}} = (Fa)(a)\frac{b}{GJ} = \frac{Fa^2 b}{GJ} \tag{2}$$

For the bending in CD,

$$\left(\frac{\partial U_{CD}}{\partial F}\right)_{\text{bending}} = \frac{1}{EI}\int_0^b (Fx)(x)\,dx = \frac{Fb^3}{3EI} \tag{3}$$

Member DG is axially loaded and is bending in two planes. The axial loading is constant, so Equation (4–29) can be written as

$$\frac{\partial U}{\partial F_i} = \left(F\frac{\partial F}{\partial F_i}\right)\frac{l}{AE}$$

where l is the length of the member. Thus, for the axial loading of DG, $F_i = F$, $l = c$, and

$$\left(\frac{\partial U_{DG}}{\partial F}\right)_{\text{axial}} = \frac{Fc}{AE} \tag{4}$$

The bending moments in each plane of DG are constant along the length, with $M_{DG2} = Fb$ and $M_{DG1} = Fa$. Considering each one separately in the form of Equation (4–31) gives

$$\left(\frac{\partial U_{DG}}{\partial F}\right)_{\text{bending}} = \frac{1}{EI}\int_0^c (Fb)(b)\,dx + \frac{1}{EI}\int_0^c (Fa)(a)\,dx$$

$$= \frac{Fc(a^2 + b^2)}{EI} \tag{5}$$

Adding Equations (1) to (5), noting that $I = \pi d^4/64$, $J = 2I$, $A = \pi d^2/4$, and $G = E/[2(1 + \nu)]$, we find that the deflection of B in the direction of F is

Answer $$(\delta_B)_F = \frac{4F}{3\pi E d^4}\left[16(a^3 + b^3) + 48c(a^2 + b^2) + 48(1 + \nu)a^2 b + 3cd^2\right]$$

Now that we have completed the solution, see if you can physically account for each term in the result using an independent method such as superposition.

[5] It is very tempting to mix techniques and try to use superposition also, for example. However, some subtle things can occur that you may visually miss. It is highly recommended that if you are using Castigliano's theorem on a problem, you use it for all parts of the problem.

4–9 Deflection of Curved Members

Machine frames, springs, clips, fasteners, and the like frequently occur as curved shapes. The determination of stresses in curved members has already been described in Section 3–18. Castigliano's theorem is particularly useful for the analysis of deflections in curved parts too.[6] Consider, for example, the curved frame of Figure 4–16*a*. We are interested in finding the deflection of the frame due to F and in the direction of F. Unlike straight beams, the bending moment and axial force are coupled for curved beams, creating an additional energy term.[7] The energy due to the moment alone is

$$U_1 = \int \frac{M^2 d\theta}{2AeE} \tag{4–32}$$

In this equation, the eccentricity e is

$$e = R - r_n \tag{4–33}$$

where r_n is the radius of the neutral axis as defined in Section 3–18 and shown in Figure 3–40.

Analogous to Equation (4–21), the strain energy component due to the axial force F_θ alone is

$$U_2 = \int \frac{F_\theta^2 R\, d\theta}{2AE} \tag{4–34}$$

The additional coupling term between M and F_θ is

$$U_3 = -\int \frac{MF_\theta\, d\theta}{AE} \tag{4–35}$$

The negative sign of Equation (4–35) can be appreciated by referring to both parts of Figure 4–16. Note that the moment M tends to decrease the angle $d\theta$. On the other hand, F_θ tends to increase $d\theta$. Thus U_3 is negative. If F_θ had been acting in the opposite direction, then both M and F_θ would tend to decrease the angle $d\theta$.

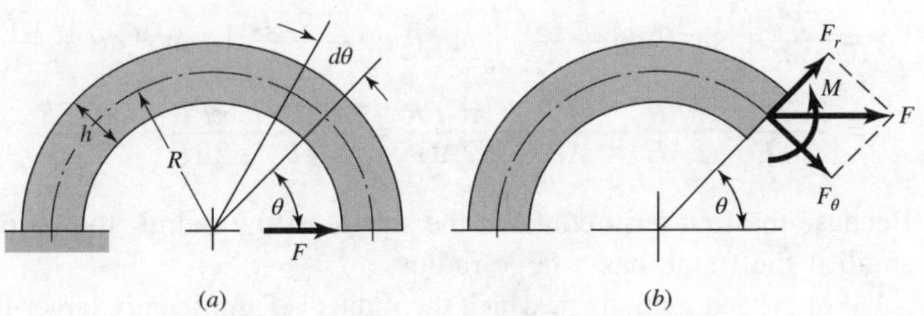

Figure 4–16

(*a*) Curved bar loaded by force F. R = radius to centroidal axis of section; h = section thickness. (*b*) Diagram showing forces acting on section taken at angle θ. $F_r = V$ = shear component of F; F_θ is component of F normal to section; M is moment caused by force F.

[6]For more solutions than are included here, see Joseph E. Shigley, "Curved Beams and Rings," Chapter 38 in Joseph E. Shigley, Charles R. Mischke, and Thomas H. Brown, Jr. (eds.), *Standard Handbook of Machine Design,* 3rd ed., McGraw Hill, New York, 2004.

[7]See Richard G. Budynas, *Advanced Strength and Applied Stress Analysis,* 2nd ed., Section 6.7, McGraw Hill, New York, 1999.

The fourth and last term is the transverse shear energy due to F_r. Adapting Equation (4–25) gives

$$U_4 = \int \frac{CF_r^2\, R\, d\theta}{2AG} \tag{4–36}$$

where C is the correction factor of Table 4–2.

Combining the four terms gives the total strain energy

$$U = \int \frac{M^2 d\theta}{2AeE} + \int \frac{F_\theta^2 R\, d\theta}{2AE} - \int \frac{MF_\theta d\theta}{AE} + \int \frac{CF_r^2 R\, d\theta}{2AG} \tag{4–37}$$

The deflection produced by the force F can now be found. It is

$$\delta = \frac{\partial U}{\partial F} = \int \frac{M}{AeE}\left(\frac{\partial M}{\partial F}\right) d\theta + \int \frac{F_\theta R}{AE}\left(\frac{\partial F_\theta}{\partial F}\right) d\theta$$
$$- \int \frac{1}{AE}\,\frac{\partial (MF_\theta)}{\partial F}\, d\theta + \int \frac{CF_r R}{AG}\left(\frac{\partial F_r}{\partial F}\right) d\theta \tag{4–38}$$

This equation is general and may be applied to any section of a thick-walled circular curved beam with application of appropriate limits of integration.

For the specific curved beam in Figure 4–16*b*, the integrals are evaluated from 0 to π. Also, for this case we find

$$M = FR \sin\theta \qquad \frac{\partial M}{\partial F} = R \sin\theta$$
$$F_\theta = F \sin\theta \qquad \frac{\partial F_\theta}{\partial F} = \sin\theta$$
$$MF_\theta = F^2 R \sin^2\theta \qquad \frac{\partial (MF_\theta)}{\partial F} = 2FR \sin^2\theta$$
$$F_r = F \cos\theta \qquad \frac{\partial F_r}{\partial F} = \cos\theta$$

Substituting these into Equation (4–38) and factoring yields

$$\delta = \frac{FR^2}{AeE}\int_0^\pi \sin^2\theta\, d\theta + \frac{FR}{AE}\int_0^\pi \sin^2\theta\, d\theta - \frac{2FR}{AE}\int_0^\pi \sin^2\theta\, d\theta + \frac{CFR}{AG}\int_0^\pi \cos^2\theta\, d\theta$$
$$= \frac{\pi FR^2}{2AeE} + \frac{\pi FR}{2AE} - \frac{\pi FR}{AE} + \frac{\pi CFR}{2AG} = \frac{\pi FR^2}{2AeE} - \frac{\pi FR}{2AE} + \frac{\pi CFR}{2AG} \tag{4–39}$$

Because the first term contains the square of the radius, the second two terms will be small if the frame has a large radius.

For curved sections in which the radius is significantly larger than the thickness, say $R/h > 10$, the effect of the eccentricity is negligible, so that the strain energies can be approximated directly from Equations (4–21), (4–24), and (4–25) with a substitution of $R\, d\theta$ for dx. Further, as R increases, the contributions to deflection from the normal force and tangential force becomes negligibly small compared to the bending component. Therefore, an approximate result can be obtained for a thin circular curved member as

$$U \approx \int \frac{M^2}{2EI} R\, d\theta \qquad R/h > 10 \tag{4–40}$$

$$\delta = \frac{\partial U}{\partial F} \approx \int \frac{1}{EI}\left(M \frac{\partial M}{\partial F}\right) R\, d\theta \qquad R/h > 10 \tag{4–41}$$

EXAMPLE 4–12

The cantilevered hook shown in Figure 4–17*a* is formed from a round steel wire with a diameter of 2 mm. The hook dimensions are $l = 40$ and $R = 50$ mm. A force P of 1 N is applied at point C. Use Castigliano's theorem to estimate the deflection at point D at the tip.

Solution

Since l/d and R/d are significantly greater than 10, only the contributions due to bending will be considered. To obtain the vertical deflection at D, a fictitious force Q will be applied there. Free-body diagrams are shown in Figures 4–17*b*, *c*, and *d*, with breaks in sections AB, BC, and CD, respectively. The normal and shear forces, N and V respectively, are shown but are considered negligible in the deflection analysis.

For section AB, with the variable of integration x defined as shown in Figure 4–17*b*, summing moments about the break gives an equation for the moment in section AB,

$$M_{AB} = P(R + x) + Q(2R + x) \tag{1}$$

$$\partial M_{AB}/\partial Q = 2R + x \tag{2}$$

Since the derivative with respect to Q has been taken, we can set Q equal to zero. From Equation (4–31), inserting Equations (1) and (2),

$$(\delta_D)_{AB} = \left[\int_0^l \frac{1}{EI}\left(M_{AB}\frac{\partial M_{AB}}{\partial Q}\right)dx\right]_{Q=0} = \frac{1}{EI}\int_0^l P(R + x)(2R + x)\,dx$$

$$= \frac{P}{EI}\int_0^l (2R^2 + 3Rx + x^2)\,dx = \frac{P}{EI}\left(2R^2 l + \frac{3}{2}l^2 R + \frac{1}{3}l^3\right) \tag{3}$$

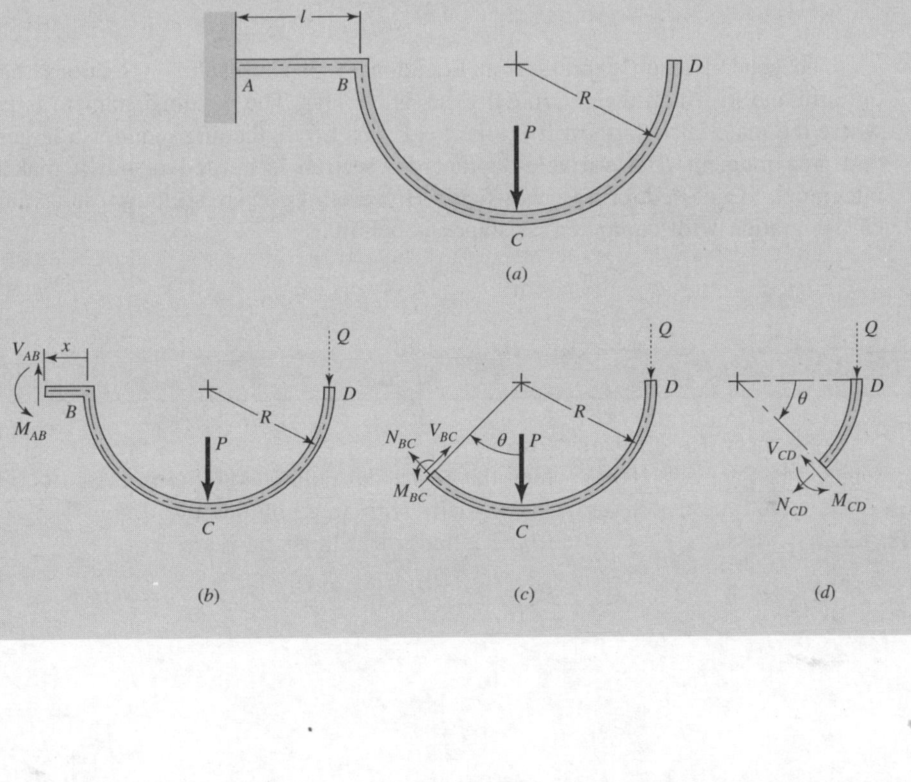

Figure 4–17

For section BC, with the variable of integration θ defined as shown in Figure 4–17c, summing moments about the break gives the moment equation for section BC.

$$M_{BC} = Q(R + R\sin\theta) + PR\sin\theta \tag{4}$$

$$\partial M_{BC}/\partial Q = R(1 + \sin\theta) \tag{5}$$

From Equation (4–41), inserting Equations (4) and (5) and setting $Q = 0$, we get

$$(\delta_D)_{BC} = \left[\int_0^{\pi/2} \frac{1}{EI}\left(M_{BC}\frac{\partial M_{BC}}{\partial Q}\right)R\,d\theta\right]_{Q=0} = \frac{R}{EI}\int_0^{\pi/2}(PR\sin\theta)[R(1+\sin\theta)]\,d\theta$$

$$= \frac{PR^3}{EI}\left(1 + \frac{\pi}{4}\right) \tag{6}$$

Noting that the break in section CD contains nothing but Q, and after setting $Q = 0$, we can conclude that there is no actual strain energy contribution in this section. Combining terms from Equations (3) and (6) to get the total vertical deflection at D,

$$\delta_D = (\delta_D)_{AB} + (\delta_D)_{BC} = \frac{P}{EI}\left(2R^2 l + \frac{3}{2}l^2 R + \frac{1}{3}l^3\right) + \frac{PR^3}{EI}\left(1 + \frac{\pi}{4}\right)$$

$$= \frac{P}{EI}(1.785R^3 + 2R^2 l + 1.5\,Rl^2 + 0.333l^3) \tag{7}$$

Substituting values, and noting $I = \pi d^4/64$, and $E = 207$ GPa for steel, we get

Answer $$\delta_D = \frac{1}{207(10^9)[\pi(0.002^4)/64]}[1.785(0.05^3) + 2(0.05^2)0.04 + 1.5(0.05)0.04^2 + 0.333(0.04^3)]$$

$$= 3.47(10^{-3})\text{ m} = 3.47\text{ mm}$$

The general result expressed in Equation (4–39) is useful in sections that are uniform and in which the centroidal locus is circular. The bending moment is largest where the material is farthest from the load axis. Strengthening requires a larger second area moment I. A variable-depth cross section is attractive, but it makes the integration to a closed form very difficult. However, if you are seeking results, numerical integration with computer assistance is helpful.

EXAMPLE 4–13

Deflection in a Variable-Cross-Section Punch-Press Frame

Consider the steel C frame depicted in Figure 4–18a in which the centroidal radius is 800 mm, the cross section at the ends is 50 mm × 50 mm, and the depth varies sinusoidally with an amplitude of 50 mm. The load is 4 kN. It follows that $C = 1.2$, $G = 80$ GPa, $E = 210$ GPa. The outer and inner radii are

$$R_{\text{out}} = 33 + 2\sin\theta \qquad R_{\text{in}} = 31 - 2\sin\theta$$

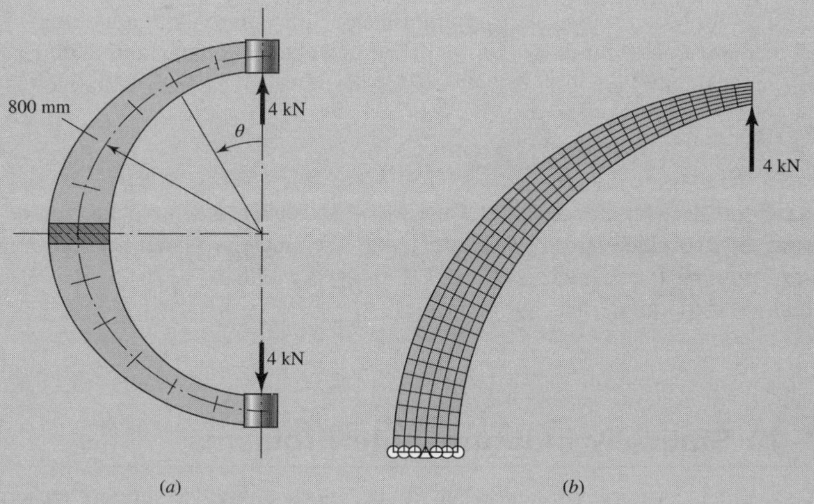

Figure 4–18

(*a*) A steel punch press has a *C* frame with a varying-depth rectangular cross section depicted. The cross section varies sinusoidally from 50 mm × 50 mm at $\theta = 0°$ to 50 mm × 150 mm at $\theta =$ 90°, and back to 50 mm × 50 mm at $\theta = 180°$. Of immediate interest to the designer is the deflection in the load axis direction under the load. (*b*) Finite-element model.

The remaining geometrical terms are

$$h = R_{\text{out}} - R_{\text{in}} = 2(1 + 2\sin\theta)$$

$$A = bh = 4(1 + 2\sin\theta)$$

$$r_n = \frac{h}{\ln(R_{\text{out}}/R_{\text{in}})} = \frac{2(1 + 2\sin\theta)}{\ln[(33 + 2\sin\theta)/(31 - 2\sin\theta)]}$$

$$e = R - r_n = 32 - r_n$$

Note that

$$M = FR\sin\theta \qquad \partial M/\partial F = R\sin\theta$$

$$F_\theta = F\sin\theta \qquad \partial F_\theta/\partial F = \sin\theta$$

$$MF_\theta = F^2R\sin^2\theta \qquad \partial M\,F_\theta/\partial F = 2FR\sin^2\theta$$

$$F_r = F\cos\theta \qquad \partial F_r/\partial F = \cos\theta$$

Substitution of the terms into Equation (4–38) yields three integrals

$$\delta = I_1 + I_2 + I_3 \tag{1}$$

where the integrals are

$$I_1 = 8.5333(10^{-3})\int_0^{\pi} \frac{\sin^2\theta\, d\theta}{(1 + 2\sin\theta)\left[32 - \dfrac{2(1 + 2\sin\theta)}{\ln\left(\dfrac{33 + 2\sin\theta}{31 - 2\sin\theta}\right)}\right]} \tag{2}$$

$$I_2 = -2.6667(10^{-4})\int_0^{\pi} \frac{\sin^2\theta\, d\theta}{1 + 2\sin\theta} \tag{3}$$

$$I_3 = 8.3478(10^{-4})\int_0^{\pi} \frac{\cos^2\theta\, d\theta}{1 + 2\sin\theta} \tag{4}$$

The integrals may be evaluated in a number of ways: by a program using Simpson's rule integration,[8] by a program using a spreadsheet, or by mathematics software. Using MathCad and checking the results with Excel gives the integrals as $I_1 = 1.91\ 5375$, $I_2 = -0.003\ 975$, and $I_3 = 0.019\ 325$. Substituting these into Equation (1) gives

Answer $$\delta = 1.93 \text{ mm}$$

Finite-element (FE) programs are also very accessible. Figure 4–18*b* shows a simple half-model, using symmetry, of the press consisting of 216 plane-stress (2-D) elements. Creating the model and analyzing it to obtain a solution took minutes. Doubling the results from the FE analysis yielded $\delta = 1.95$ mm, a less than 1 percent variation from the results of the numerical integration.

4–10 Statically Indeterminate Problems

A system is *overconstrained* when it has more unknown support (reaction) forces and/or moments than static equilibrium equations. Such a system is said to be *statically indeterminate* and the extra constraint supports are called *redundant supports.* In addition to the static equilibrium equations, a deflection equation is required for *each* redundant support reaction in order to obtain a solution. For example, consider a beam in bending with a wall support on one end and a simple support on the other, such as beam 12 of Table A–9. There are three support reactions and only two static equilibrium equations are available. This beam has *one* redundant support. To solve for the three unknown support reactions we use the two equilibrium equations and *one* additional deflection equation. For another example, consider beam 15 of Table A–9. This beam has a wall on both ends, giving rise to *two* redundant supports requiring *two* deflection equations in addition to the equations from statics. The purpose of redundant supports is to provide additional safety and reduce deflection.

A simple example of a statically indeterminate problem is furnished by the nested helical springs in Figure 4–19*a*. When this assembly is loaded by the compressive force F, it deforms through the distance δ. What is the compressive force in each spring?

Only one equation of static equilibrium can be written. It is

$$\sum F = F - F_1 - F_2 = 0 \tag{a}$$

which simply says that the total force F is resisted by a force F_1 in spring 1 plus the force F_2 in spring 2. Since there are two unknowns and only one static equilibrium equation, the system is statically indeterminate.

To write another equation, note the deformation relation in Figure 4–19*b*. The two springs have the same deformation. Thus, we obtain the second equation as

$$\delta_1 = \delta_2 = \delta \tag{b}$$

If we now substitute Equation (4–2) in Equation (*b*), we have

$$\frac{F_1}{k_1} = \frac{F_2}{k_2} \tag{c}$$

[8]See Case Study 4, p. 203, J. E. Shigley and C. R. Mischke, *Mechanical Engineering Design,* 6th ed., McGraw Hill, New York, 2001.

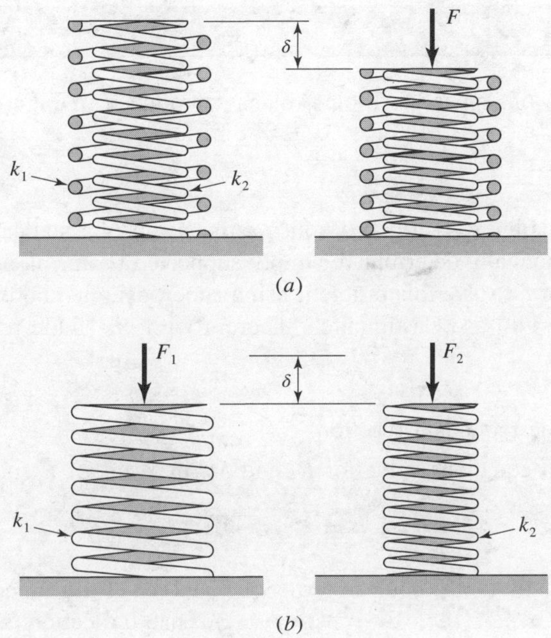

Figure 4–19

Now we solve Equation (*c*) for F_1 and substitute the result in Equation (*a*). This gives

$$F - \frac{k_1}{k_2}F_2 - F_2 = 0 \qquad \text{or} \qquad F_2 = \frac{k_2 F}{k_1 + k_2} \tag{d}$$

Substituting F_2 into Equation (*c*) gives $F_1 = k_1 F/(k_1 + k_2)$ and so $\delta = \delta_1 = \delta_2 = F/(k_1 + k_2)$. Thus, for two springs in parallel, the overall spring constant is $k = F/\delta = k_1 + k_2$.

In the spring example, obtaining the necessary deformation equation was very straightforward. However, for other situations, the deformation relations may not be as easy. A more structured approach may be necessary. Here we will show two basic procedures for general statically indeterminate problems.

Procedure 1

1. Choose the redundant reaction(s). There may be alternative choices (see Example 4–14).
2. Write the equations of static equilibrium for the remaining reactions in terms of the applied loads and the redundant reaction(s) of step 1.
3. Write the deflection equation(s) for the point(s) at the locations of the redundant reaction(s) of step 1 in terms of the applied loads and the redundant reaction(s) of step 1. Normally the deflection(s) is (are) zero. If a redundant reaction is a moment, the corresponding deflection equation is a rotational deflection equation.
4. The equations from steps 2 and 3 can now be solved to determine the reactions.

In step 3 the deflection equations can be solved in any of the standard ways. Here we will demonstrate the use of superposition and Castigliano's theorem on a beam problem.

EXAMPLE 4–14

The indeterminate beam 11 of Appendix Table A–9 is reproduced in Figure 4–20. Determine the reactions using procedure 1.

Solution

The reactions are shown in Figure 4–20*b*. Without R_2 the beam is a statically determinate cantilever beam. Without M_1 the beam is a statically determinate simply supported beam. In either case, the beam has only *one* redundant support. We will first solve this problem using superposition, choosing R_2 as the redundant reaction. For the second solution, we will use Castigliano's theorem with M_1 as the redundant reaction.

Solution 1

1 Choose R_2 at B to be the redundant reaction.

2 Using static equilibrium equations solve for R_1 and M_1 in terms of F and R_2. This results in

$$R_1 = F - R_2 \qquad M_1 = \frac{Fl}{2} - R_2 l \tag{1}$$

3 Write the deflection equation for point B in terms of F and R_2. Using superposition of beam 1 of Table A–9 with $F = -R_2$, and beam 2 of Table A–9 with $a = l/2$, the deflection of B, at $x = l$, is

$$\delta_B = -\frac{R_2 l^2}{6EI}(l - 3l) + \frac{F(l/2)^2}{6EI}\left(\frac{l}{2} - 3l\right) = \frac{R_2 l^3}{3EI} - \frac{5Fl^3}{48EI} = 0 \tag{2}$$

4 Equation (2) can be solved for R_2 directly. This yields

Answer

$$R_2 = \frac{5F}{16} \tag{3}$$

Next, substituting R_2 into Equation (1) completes the solution, giving

Answer

$$R_1 = \frac{11F}{16} \qquad M_1 = \frac{3Fl}{16} \tag{4}$$

Note that the solution agrees with what is given for beam 11 in Table A–9.

Solution 2

1 Choose M_1 at O to be the redundant reaction.

2 Using static equilibrium equations solve for R_1 and R_2 in terms of F and M_1. This results in

$$R_1 = \frac{F}{2} + \frac{M_1}{l} \qquad R_2 = \frac{F}{2} - \frac{M_1}{l} \tag{5}$$

Figure 4–20

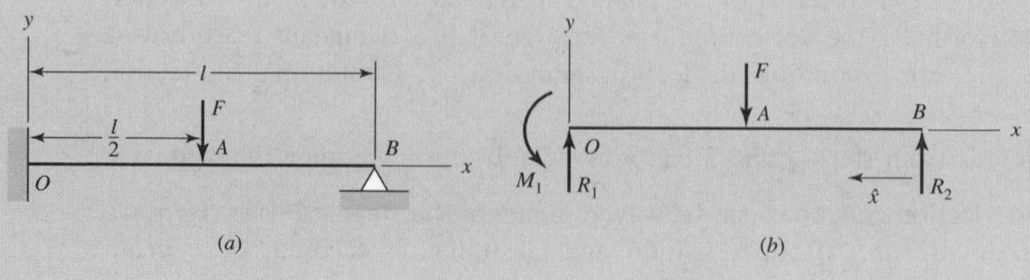

3 Since M_1 is the redundant reaction at O, write the equation for the angular deflection at point O. From Castigliano's theorem this is

$$\theta_O = \frac{\partial U}{\partial M_1} \tag{6}$$

We can apply Equation (4–31), using the variable x as shown in Figure 4–20*b*. However, simpler terms can be found by using a variable $\hat{x}$ that starts at B and is positive to the left. With this and the expression for R_2 from Equation (5) the moment equations are

$$M = \left(\frac{F}{2} - \frac{M_1}{l}\right)\hat{x} \qquad 0 \le \hat{x} \le \frac{l}{2} \tag{7}$$

$$M = \left(\frac{F}{2} - \frac{M_1}{l}\right)\hat{x} - F\left(\hat{x} - \frac{l}{2}\right) \qquad \frac{l}{2} \le \hat{x} \le l \tag{8}$$

For both equations

$$\frac{\partial M}{\partial M_1} = -\frac{\hat{x}}{l} \tag{9}$$

Substituting Equations (7) to (9) in Equation (6), using the form of Equation (4–31) where $F_i = M_1$, gives

$$\theta_O = \frac{\partial U}{\partial M_1} = \frac{1}{EI}\left\{\int_0^{l/2}\left(\frac{F}{2} - \frac{M_1}{l}\right)\hat{x}\left(-\frac{\hat{x}}{l}\right)d\hat{x} + \int_{l/2}^{l}\left[\left(\frac{F}{2} - \frac{M_1}{l}\right)\hat{x} - F\left(\hat{x} - \frac{l}{2}\right)\right]\left(-\frac{\hat{x}}{l}\right)d\hat{x}\right\} = 0$$

Canceling $1/EIl$, and combining the first two integrals, simplifies this quite readily to

$$\left(\frac{F}{2} - \frac{M_1}{l}\right)\int_0^l \hat{x}^2\, d\hat{x} - F\int_{l/2}^{l}\left(\hat{x} - \frac{l}{2}\right)\hat{x}\, d\hat{x} = 0$$

Integrating gives

$$\left(\frac{F}{2} - \frac{M_1}{l}\right)\frac{l^3}{3} - \frac{F}{3}\left[l^3 - \left(\frac{l}{2}\right)^3\right] + \frac{Fl}{4}\left[l^2 - \left(\frac{l}{2}\right)^2\right] = 0$$

which reduces to

$$M_1 = \frac{3Fl}{16} \tag{10}$$

4 Substituting Equation (10) into (5) results in

$$R_1 = \frac{11F}{16} \qquad R_2 = \frac{5F}{16} \tag{11}$$

which again agrees with beam 11 of Table A–9.

For some problems even procedure 1 can be a task. Procedure 2 eliminates some tricky geometric problems that would complicate procedure 1. We will describe the procedure for a beam problem.

Procedure 2

1 Write the equations of static equilibrium for the beam in terms of the applied loads and unknown restraint reactions.

2 Write the deflection equation for the beam in terms of the applied loads and unknown restraint reactions.

3 Apply boundary conditions to the deflection equation of step 2 consistent with the restraints.

4 Solve the equations from steps 1 and 3.

EXAMPLE 4–15

The rods *AD* and *CE* shown in Figure 4–21*a* each have a diameter of 10 mm. The second-area moment of beam *ABC* is $I = 62.5(10^3)$ mm^4. The modulus of elasticity of the material used for the rods and beam is $E = 200$ GPa. The threads at the ends of the rods are single-threaded with a pitch of 1.5 mm. The nuts are first snugly fit with bar *ABC* horizontal. Next the nut at *A* is tightened one full turn. Determine the resulting tension in each rod and the deflections of points *A* and *C*.

Solution

There is a lot going on in this problem; a rod shortens, the rods stretch in tension, and the beam bends. Let's try the procedure!

1 The free-body diagram of the beam is shown in Figure 4–21*b*. Summing forces, and moments about *B*, gives

$$F_B - F_A - F_C = 0 \tag{1}$$

$$4F_A - 3F_C = 0 \tag{2}$$

2 Using singularity functions, we find the moment equation for the beam is

$$M = -F_A x + F_B \langle x - 0.2 \rangle^1$$

where x is in meters. Integration yields

$$EI\frac{dy}{dx} = -\frac{F_A}{2}x^2 + \frac{F_B}{2}\langle x - 0.2 \rangle^2 + C_1$$

$$EI\,y = -\frac{F_A}{6}x^3 + \frac{F_B}{6}\langle x - 0.2 \rangle^3 + C_1 x + C_2 \tag{3}$$

The term $EI = 200(10^9)\,62.5(10^{-9}) = 1.25(10^4)$ N · m^2.

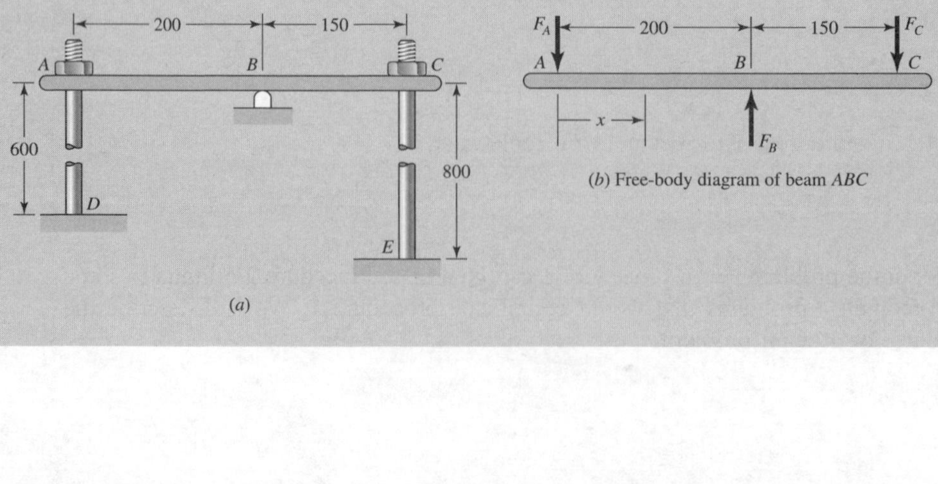

Figure 4–21

Dimensions in mm.

3 The upward deflection of point A is $(Fl/AE)_{AD} - Np$, where the first term is the elastic stretch of AD, N is the number of turns of the nut, and p is the pitch of the thread. Thus, the deflection of A in meters is

$$\begin{aligned} y_A &= \frac{F_A(0.6)}{\frac{\pi}{4}(0.010)^2(200)(10^9)} - (1)(0.0015) \\ &= 3.8197(10^{-8})F_A - 1.5(10^{-3}) \end{aligned} \tag{4}$$

The upward deflection of point C is $(Fl/AE)_{CE}$, or

$$y_C = \frac{F_C(0.8)}{\frac{\pi}{4}(0.010)^2(200)(10^9)} = 5.093(10^{-8})F_C \tag{5}$$

Equations (4) and (5) will now serve as the boundary conditions for Equation (3). At $x = 0$, $y = y_A$. Substituting Equation (4) into (3) with $x = 0$ and $EI = 1.25(10^4)$, noting that the singularity function is zero for $x = 0$, gives

$$-4.7746(10^{-4})F_A + C_2 = -18.75 \tag{6}$$

At $x = 0.2$ m, $y = 0$, and Equation (3) yields

$$-1.3333(10^{-3})F_A + 0.2C_1 + C_2 = 0 \tag{7}$$

At $x = 0.35$ m, $y = y_C$. Substituting Equation (5) into (3) with $x = 0.35$ m and $EI = 1.25(10^4)$ gives

$$-7.1458(10^{-3})F_A + 5.625(10^{-4})F_B - 6.3662(10^{-4})F_C + 0.35C_1 + C_2 = 0 \tag{8}$$

Equations (1), (2), (6), (7), and (8) are five equations in F_A, F_B, F_C, C_1 and C_2. Written in matrix form, they are

$$\begin{bmatrix} -1 & 1 & -1 & 0 & 0 \\ 4 & 0 & -3 & 0 & 0 \\ -4.7746(10^{-4}) & 0 & 0 & 0 & 1 \\ -1.3333(10^{-3}) & 0 & 0 & 0.2 & 1 \\ -7.1458(10^{-3}) & 5.625(10^{-4}) & -6.3662(10^{-4}) & 0.35 & 1 \end{bmatrix} \begin{Bmatrix} F_A \\ F_B \\ F_C \\ C_1 \\ C_2 \end{Bmatrix} = \begin{Bmatrix} 0 \\ 0 \\ -18.75 \\ 0 \\ 0 \end{Bmatrix}$$

Solving these equations yields

Answer $\quad F_A = 2988$ N $\quad F_B = 6971$ N $\quad F_C = 3983$ N

$C_1 = 106.54\ \text{N} \cdot \text{m}^2 \quad C_2 = -17.324\ \text{N} \cdot \text{m}^3$

Equation (3) can be reduced to

$$y = -(39.84x^3 - 92.95\langle x - 0.2\rangle^3 - 8.523x + 1.386)(10^{-3})$$

Answer $\quad$ At $x = 0$, $y = y_A = -1.386(10^{-3})$ m $= -1.386$ mm

Answer $\quad$ At $x = 0.35$m, $y = y_C = -[39.84(0.35)^3 - 92.95(0.35 - 0.2)^3 - 8.523(0.35) + 1.386](10^{-3}) = 0.203(10^{-3})$ m $= 0.203$ mm

Note that we could have easily incorporated the stiffness of the support at B if we were given a spring constant.

4–11 Compression Members—General

The analysis and design of compression members can differ significantly from that of members loaded in tension or in torsion. If you were to take a long rod or pole, such as a meterstick, and apply gradually increasing compressive forces at each end, very small axial deflections would happen at first, but then the stick would bend (buckle), and very quickly bend so much as to possibly fracture. Try it. The other extreme would occur if you were to saw off, say, a 5-mm length of the meterstick and perform the same experiment on the short piece. You would then observe that the failure exhibits itself as a mashing of the specimen, that is, a simple compressive failure. For these reasons it is convenient to classify compression members according to their length and according to whether the loading is central or eccentric. The term *column* is applied to all such members except those in which failure would be by simple or pure compression. Columns can be categorized then as:

1 Long columns with central loading
2 Intermediate-length columns with central loading
3 Columns with eccentric loading
4 Struts or short columns with eccentric loading

Classifying columns as above makes it possible to develop methods of analysis and design specific to each category. Furthermore, these methods will also reveal whether or not you have selected the category appropriate to your particular problem. The four sections that follow correspond, respectively, to the four categories of columns listed above.

4–12 Long Columns with Central Loading

Figure 4–22 shows long columns with differing end (boundary) conditions. If the axial force P shown acts along the centroidal axis of the column, simple compression of the member occurs for low values of the force. However, under certain conditions, when P reaches a specific value, the column becomes *unstable* and bending as shown in Figure 4–22 develops rapidly. This force is determined by writing the bending deflection equation for the column, resulting in a differential equation where when the boundary conditions are applied, results in the *critical load* for unstable bending.[9] The critical force for the pin-ended column of Figure 4–22*a* is given by

$$P_{\mathrm{cr}} = \frac{\pi^2 EI}{l^2} \tag{4–42}$$

which is called the *Euler column formula*. Equation (4–42) can be extended to apply to other end-conditions by writing

$$P_{\mathrm{cr}} = \frac{C\pi^2 EI}{l^2} \tag{4–43}$$

where the constant C depends on the end conditions as shown in Figure 4–22.

[9]See F. P. Beer, E. R. Johnston, Jr., and J. T. DeWolf, *Mechanics of Materials,* 5th ed., McGraw Hill, New York, 2009, pp. 610–613.

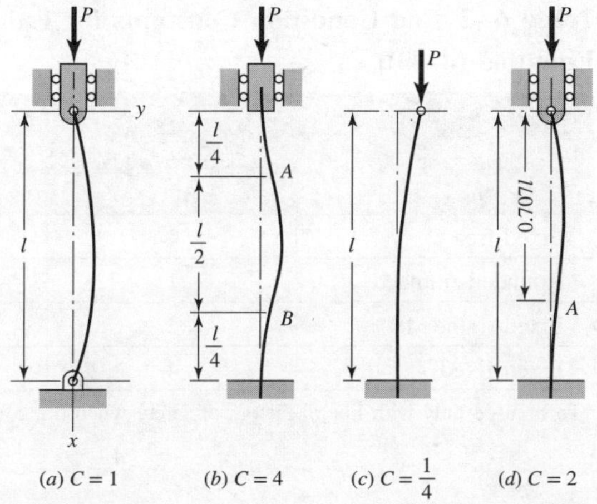

Figure 4–22

(*a*) Both ends rounded or pivoted; (*b*) both ends fixed; (*c*) one end free and one end fixed; (*d*) one end rounded and pivoted, and one end fixed.

Using the relation $I = Ak^2$, where A is the area and k the *radius of gyration,* enables us to rearrange Equation (4–43) into the more convenient form

$$\frac{P_{cr}}{A} = \frac{C\pi^2 E}{(l/k)^2} \tag{4–44}$$

where l/k is called the *slenderness ratio.* This ratio, rather than the actual column length, will be used in classifying columns according to length categories.

The quantity P_{cr}/A in Equation (4–44) is the *critical unit load*. It is the load per unit area necessary to place the column in a condition of *unstable equilibrium.* In this state any small crookedness of the member, or slight movement of the support or load, will cause the column to begin to collapse. The unit load has the same units as strength, but this is the strength of a specific column, not of the column material. Doubling the length of a member, for example, will have a drastic effect on the value of P_{cr}/A but no effect at all on, say, the yield strength S_y of the column material itself.

Equation (4–44) shows that the critical unit load depends only upon the end conditions, the modulus of elasticity, and the slenderness ratio. Thus a column obeying the Euler formula made of high-strength alloy steel is *no stronger* than one made of low-carbon steel, since E is the same for both.

The factor C is called the *end-condition constant,* and it may have any one of the theoretical values $\frac{1}{4}$, 1, 2, and 4, depending upon the manner in which the load is applied. In practice it is difficult, if not impossible, to fix the column ends so that the factor $C = 2$ or $C = 4$ would apply. Even if the ends are welded, some deflection will occur. Because of this, some designers never use a value of C greater than unity. However, if liberal factors of safety are employed, and if the column load is accurately known, then a value of C not exceeding 1.2 for both ends fixed, or for one end rounded and one end fixed, is not unreasonable, since it supposes only partial fixation. Of course, the value $C = \frac{1}{4}$ must always be used for a column having one end fixed and one end free. These recommendations are summarized in Table 4–3.

When Equation (4–44) is solved for various values of the unit load P_{cr}/A in terms of the slenderness ratio l/k, we obtain the curve PQR shown in Figure 4–23. Since the yield strength of the material has the same units as the unit load, the horizontal line through S_y and Q has been added to the figure. This would appear to make the figure $S_y\,QR$ cover the entire range of compression problems from the shortest to the longest compression member. Thus it would appear that any compression member

Table 4–3 End-Condition Constants for Euler Columns [to Be Used with Equation (4–43)]

	End-Condition Constant C		
Column End Conditions	Theoretical Value	Conservative Value	Recommended Value*
Fixed-free	$\frac{1}{4}$	$\frac{1}{4}$	$\frac{1}{4}$
Rounded-rounded	1	1	1
Fixed-rounded	2	1	1.2
Fixed-fixed	4	1	1.2

*To be used only with liberal factors of safety when the column load is accurately known.

Figure 4–23

Euler curve plotted using Equation (4–43) with $C = 1$.

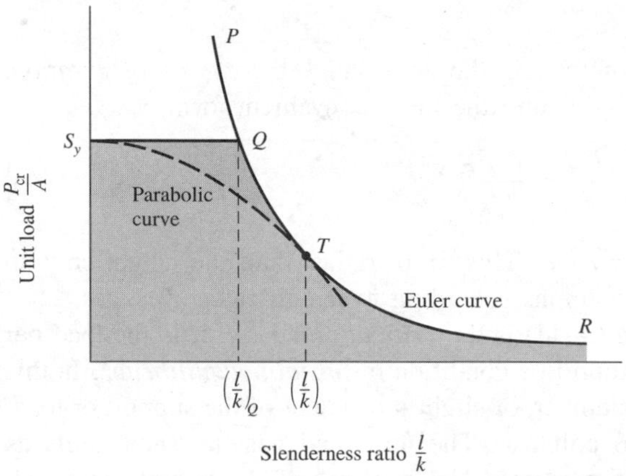

having an l/k value less than $(l/k)_Q$ should be treated as a pure compression member while all others are to be treated as Euler columns. Unfortunately, this is not true.

In the actual design of a member that functions as a column, the designer will be aware of the end conditions shown in Figure 4–22, and will endeavor to configure the ends, using bolts, welds, or pins, for example, so as to achieve the required ideal end conditions. In spite of these precautions, the result, following manufacture, is likely to contain defects such as initial crookedness or load eccentricities. The existence of such defects and the methods of accounting for them will usually involve a factor-of-safety approach or a stochastic analysis. These methods work well for long columns and for simple compression members. However, tests show numerous failures for columns with slenderness ratios below and in the vicinity of point Q, as shown in the shaded area in Figure 4–23. These have been reported as occurring even when near-perfect geometric specimens were used in the testing procedure.

A column failure is always sudden, total, unexpected, and hence dangerous. There is no advance warning. A beam will bend and give visual warning that it is overloaded, but not so for a column. For this reason neither simple compression methods nor the Euler column equation should be used when the slenderness ratio is near $(l/k)_Q$. Then what should we do? The usual approach is to choose some point T on the Euler curve of Figure 4–23. If the slenderness ratio is specified as $(l/k)_1$ corresponding to point T, then use the Euler equation only when the actual slenderness ratio is greater than $(l/k)_1$. Otherwise, use one of the methods in the sections that follow. See Examples 4–17 and 4–18.

Most designers select point T such that $P_{cr}/A = S_y/2$. Using Equation (4–43), we find the corresponding value of $(l/k)_1$ to be

$$\left(\frac{l}{k}\right)_1 = \left(\frac{2\pi^2 CE}{S_y}\right)^{1/2} \tag{4–45}$$

Euler Equations for Specific Cross Sections

Equations can be developed for simple round or rectangular cross sections.

Round Cross Sections

Given a column of diameter d, the area is $A = \pi d^2/4$ and the radius of gyration is $k = (I/A)^{1/2} = [(\pi d^4/64)/(\pi d^2/4)]^{1/2} = d/4$. Substituting these into Equation (4–44) gives

$$\frac{P_{cr}}{\pi d^2/4} = \frac{C\pi^2 E}{[l/(d/4)]^2}$$

Solving for d yields

$$d = \left(\frac{64 P_{cr} l^2}{\pi^3 CE}\right)^{1/4} \tag{4–46}$$

Rectangular Cross Sections

Consider a column of cross section $h \times b$ with the restriction that $h \leq b$. If the end conditions are the same for buckling in both directions, then buckling will occur about the axis of the least thickness. Therefore, $I = bh^3/12$, $A = bh$, and $k^2 = I/A = h^2/12$. Substituting these into Equation (4–44) gives

$$\frac{P_{cr}}{bh} = \frac{C\pi^2 E}{[l/(h/\sqrt{12})]^2}$$

Solving for b or h yields

$$b = \frac{12 P_{cr} l^2}{\pi^2 CEh^3} \tag{4–47a}$$

$$h = \left(\frac{12 P_{cr} l^2}{\pi^2 CEb}\right)^{1/3} \tag{4–47b}$$

Note, however, that rectangular columns do not generally have the same end conditions in both directions.

4–13 Intermediate-Length Columns with Central Loading

Over the years there have been a number of column formulas proposed and used for the range of l/k values for which the Euler formula is not suitable. Many of these are based on the use of a single material; others, on a so-called safe unit load rather than the critical value. Most of these formulas are based on the use of a linear relationship between the slenderness ratio and the unit load. However, the nonlinear *parabolic* or *J. B. Johnson formula* now seems to be the preferred one among designers in the machine, automotive, aircraft, and structural-steel construction fields.

The general form of the parabolic formula is

$$\frac{P_{cr}}{A} = a - b\left(\frac{l}{k}\right)^2 \tag{a}$$

where a and b are constants that are evaluated by fitting a parabola to the Euler curve of Figure 4–23 as shown by the dashed line ending at T. If the parabola is begun at S_y, then $a = S_y$. If point T is selected as previously noted, then Equation (4–45) gives the value of $(l/k)_1$ and the constant b is found to be

$$b = \left(\frac{S_y}{2\pi}\right)^2 \frac{1}{CE} \tag{b}$$

Upon substituting the known values of a and b into Equation (a), we obtain, for the parabolic equation,

$$\frac{P_{cr}}{A} = S_y - \left(\frac{S_y}{2\pi}\frac{l}{k}\right)^2 \frac{1}{CE} \qquad \frac{l}{k} \le \left(\frac{l}{k}\right)_1 \tag{4–48}$$

Parabolic Equations for Specific Cross Sections

The J. B. Johnson Equations can be developed for simple round or rectangular cross sections.

Round Cross Sections

Given a column of diameter d, the area is $A = \pi d^2/4$ and the radius of gyration is $k = (I/A)^{1/2} = [(\pi d^4/64)/(\pi d^2/4)]^{1/2} = d/4$. Substituting these into Equation (4–48) gives

$$\frac{P_{cr}}{(\pi d^2/4)} = S_y - \left(\frac{S_y l}{2\pi(d/4)}\right)^2 \frac{1}{CE}$$

Solving for d yields

$$d = 2\left(\frac{P_{cr}}{\pi S_y} + \frac{S_y l^2}{\pi^2 CE}\right)^{1/2} \tag{4–49}$$

Rectangular Cross Sections

Consider a column of cross section $h \times b$ with the restriction that $h \le b$. If the end conditions are the same for buckling in both directions, then buckling will occur about the axis of the least thickness. Therefore, $I = bh^3/12$, $A = bh$, and $k^2 = I/A = h^2/12$. Substituting these into Equation (4–48) gives

$$\frac{P_{cr}}{bh} = S_y - \left(\frac{S_y l}{2\pi(h/\sqrt{12})}\right)^2 \frac{1}{CE}$$

Solving for b or h yields

$$b = \frac{P_{cr}}{hS_y\left(1 - \dfrac{3l^2 S_y}{\pi^2 CEh^2}\right)} \tag{4–50a}$$

$$h = \frac{P_{cr}}{2bS_y} + \left[\left(\frac{P_{cr}}{2bS_y}\right)^2 + \frac{3l^2 S_y}{\pi^2 CE}\right]^{1/2} \tag{4–50b}$$

Note, however, that rectangular columns do not generally have the same end conditions in both directions.

4–14 Columns with Eccentric Loading

We have noted before that deviations from an ideal column, such as load eccentricities or crookedness, are likely to occur during manufacture and assembly. Though these deviations are often quite small, it is still convenient to have a method of dealing with them. Frequently, too, problems occur in which load eccentricities are unavoidable.

Figure 4–24*a* shows a column in which the line of action of the column forces is separated from the centroidal axis of the column by the eccentricity e. From Figure 4–24*b*, $M = -P(e + y)$. Substituting this into Equation (4–9), $d^2y/dx^2 = M/EI$, results in the differential equation

$$\frac{d^2y}{dx^2} + \frac{P}{EI}y = -\frac{Pe}{EI} \tag{a}$$

The solution of Equation (*a*), using the boundary conditions, $y = 0$ at $x = 0$ and l is

$$y = e\left[\tan\left(\frac{l}{2}\sqrt{\frac{P}{EI}}\right)\sin\left(\sqrt{\frac{P}{EI}}\,x\right) + \cos\left(\sqrt{\frac{P}{EI}}\,x\right) - 1\right] \tag{b}$$

By substituting $x = l/2$ in Equation (*b*) and using a trigonometric identity, we obtain

$$\delta = e\left[\sec\left(\sqrt{\frac{P}{EI}}\,\frac{l}{2}\right) - 1\right] \tag{4–51}$$

The magnitude of the maximum bending moment also occurs at midspan and is

$$M_{\max} = P(e + \delta) = Pe\sec\left(\frac{l}{2}\sqrt{\frac{P}{EI}}\right) \tag{4–52}$$

The magnitude of the maximum *compressive* stress at midspan is found by superposing the axial component and the bending component. This gives

$$\sigma_c = \frac{P}{A} + \frac{Mc}{I} = \frac{P}{A} + \frac{Mc}{Ak^2} \tag{c}$$

Substituting $M_{\max}$ from Equation (4–52) yields

$$\sigma_c = \frac{P}{A}\left[1 + \frac{ec}{k^2}\sec\left(\frac{l}{2k}\sqrt{\frac{P}{EA}}\right)\right] \tag{4–53}$$

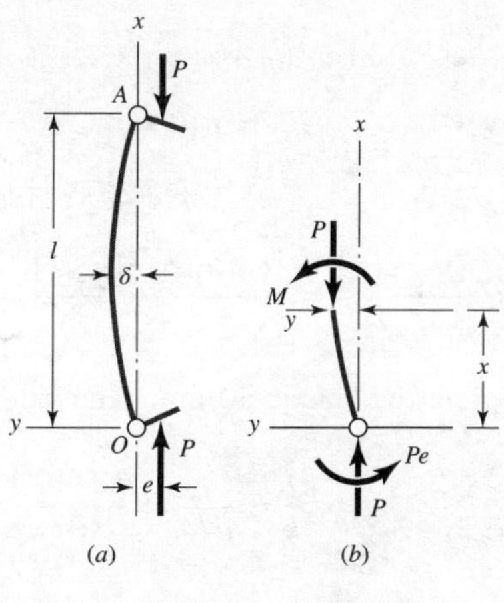

Figure 4–24

Notation for an eccentrically loaded column.

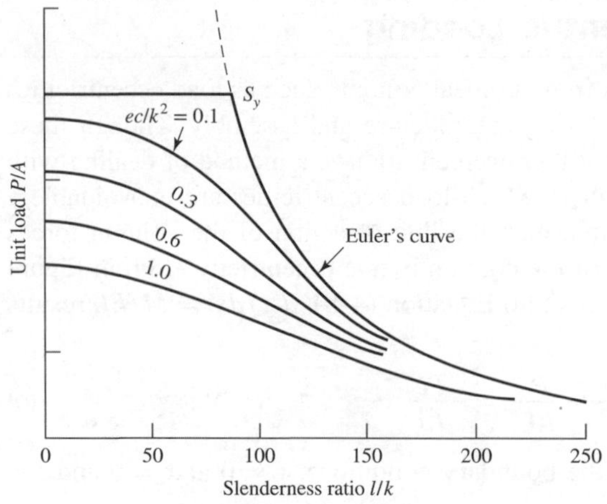

Figure 4–25

Comparison of secant and Euler equations for steel with $S_y = 288$ MPa.

By imposing the compressive yield strength S_{yc} as the maximum value of σ_c, we can write Equation (4–53) in the form

$$\frac{P}{A} = \frac{S_{yc}}{1 + (ec/k^2)\sec\left[\left(\frac{l}{2k}\right)\sqrt{P/AE}\right]} \tag{4–54}$$

This is called the *secant column formula.* The term ec/k^2 is called the *eccentricity ratio.* Figure 4–25 is a plot of Equation (4–54) for a steel having a compressive (and tensile) yield strength of 288 MPa. Note how the P/A contours asymptotically approach the Euler curve as l/k increases.

Equation (4–54) cannot be solved explicitly for the load P. Design charts, in the fashion of Figure 4–25, can be prepared for a single material if much column design is to be done. Otherwise, a root-finding technique using numerical methods must be used.

EXAMPLE 4–16

Specify the diameter of a round column 1.5 m long that is to carry a maximum load estimated to be 22 kN. Use a design factor $n_d = 4$ and consider the ends as pinned (rounded). The column material selected has a minimum yield strength of 500 MPa and a modulus of elasticity of 207 GPa.

Solution

We shall design the column for a critical load of

$$P_{cr} = n_d P = 4(22) = 88 \text{ kN}$$

Then, using Equation (4–46) with $C = 1$ (see Table 4–3) gives

$$d = \left(\frac{64P_{cr}l^2}{\pi^3 CE}\right)^{1/4} = \left[\frac{64(88)10^3(1.5)^2}{\pi^3(1)207(10^9)}\right]^{1/4} = 0.0375 \text{ m} = 37.5 \text{ mm}$$

Table A–15 shows that the preferred size is 40 mm. The slenderness ratio for this size is

$$\frac{l}{k} = \frac{l}{d/4} = \frac{1.5(10^3)}{40/4} = 150$$

To be sure that this is an Euler column, we use Equation (4–45) and obtain

$$\left(\frac{l}{k}\right)_1 = \left(\frac{2\pi^2 CE}{S_y}\right)^{1/2} = \left[\frac{2\pi^2(1)207(10^9)}{500(10^6)}\right]^{1/2} = 90.4$$

where $l/k > (l/k)_1$ indicates that it is indeed an Euler column. So select

Answer $d = 40$ mm

EXAMPLE 4–17

Repeat Example 4–16 for a fixed-fixed column with a rectangular cross section with $b = 4h$. Round up the results to the next higher whole millimeter.

Solution

From Example 4–16, $P_{cr} = 88$ kN. From Table 4–3, we select the recommended value of $C = 1.2$. Setting $b = 4h$ into Equation (4–47*a*) and solving for h gives

$$h = \left(\frac{3P_{cr}l^2}{\pi^2 CE}\right)^{1/4} = \left[\frac{3(88)10^3(1.5)^2}{\pi^2(1.2)(207)10^9}\right]^{1/4} = 0.0222 \text{ m} = 22.2 \text{ mm}$$

Rounding up the next higher whole millimeter gives $h = 23$ mm and $b = 92$ mm. The slenderness ratio for this size is

$$\frac{l}{k} = \frac{\sqrt{12}\, l}{h} = \frac{\sqrt{12}\,(1.5)}{0.023} = 226$$

From Equation (4–45) and using the recommended value of $C = 1.2$ from Table 4–3

$$\left(\frac{l}{k}\right)_1 = \left(\frac{2\pi^2 CE}{S_y}\right)^{1/2} = \left[\frac{2\pi^2(1.2)207(10^9)}{500(10^6)}\right]^{1/2} = 99.03$$

Since $(l/k) > (l/k)_1$, the column is an Euler column. So select

Answer $h = 23$ mm, $b = 92$ mm

EXAMPLE 4–18

Repeat Example 4–16 with $l = 375$ mm.

Solution

Following the same procedure as Example 4–16,

$$d = \left(\frac{64P_{cr}l^2}{\pi^3 CE}\right)^{1/4} = \left(\frac{64(88)10^3(0.375)^2}{\pi^3(1)207(10^9)}\right)^{1/4} = 0.0187 \text{ m} = 18.7 \text{ mm}$$

From Table A–15, $d = 20$ mm.

$$\frac{l}{k} = \frac{4l}{d} = \frac{4(375)}{20} = 75$$

From Example 4–16, $(l/k)_1 = 90.4$. Since $(l/k) < (l/k)_1$, the column is not a Euler column and the parabolic equation, Equation (4–49), must be used. Thus,

$$d = 2\left(\frac{P_{\text{cr}}}{\pi S_y} + \frac{S_y l^2}{\pi^2 CE}\right)^{1/2} = 2\left[\frac{88(10^3)}{\pi(500)10^6} + \frac{500(10^6)(0.375)^2}{\pi^2(1)207(10^9)}\right]^{1/2}$$

$$= 0.0190 \text{ m} = 19.0 \text{ mm}$$

From Table A–15 select

Answer $\qquad d = 20$ mm

The answer is not substantially different from what was obtained from the Euler formula since we were close to point T of Figure 4–23.

EXAMPLE 4–19

Choose a set of dimensions for a rectangular link that is to carry a maximum compressive load of 22.5 kN. The material selected has a minimum yield strength of 540 MPa and a modulus of elasticity $E = 210$ GPa. Use a design factor of 4 and an end condition constant $C = 1$ for buckling in the weakest direction, and design for (*a*) a length of 375 mm, and (*b*) a length of 200 mm with a minimum thickness of 12.5 mm.

Solution

(*a*) Using Equation (4–45), we find the limiting slenderness ratio to be

$$\left(\frac{l}{k}\right)_1 = \left(\frac{2\pi^2 CE}{S_y}\right)^{1/2} = \left[\frac{2\pi^2(1)(210)(10^9)}{540(10)^6}\right]^{1/2} = 87.6$$

By using $P_{\text{cr}} = n_d P = 4(22\ 500) = 90$ kN, Equations (4–47*a*) and (4–50*a*) are solved, using various values of h, to form Table 4–4. The table shows that a cross section of 15.6 by 19 mm, which is marginally suitable, gives the least area.

(*b*) An approach similar to that in part (*a*) is used with $l = 200$ mm. All trial computations are found to be in the J. B. Johnson region of l/k values. A minimum area occurs when the section is a near square. Thus a cross section of 12.5 by 18.75 mm is found to be suitable and safe.

Table 4–4 Table Generated to Solve Example 4–19, part (*a*)

h	b	A	l/k	Type	Eq. No.
9.4	86.5	813.1	139	Euler	(4–47*a*)
12.5	36.5	456.25	104	Euler	(4–47*a*)
15.6	19	296.4	83	Johnson	(4–50*a*)
14.1	25.8	363.8	92	Euler	(4–47*a*)

4–15 Struts or Short Compression Members

A short bar loaded in pure compression by a force P acting along the centroidal axis will shorten in accordance with Hooke's law, until the stress reaches the elastic limit of the material. At this point, permanent set is introduced and usefulness as a machine member may be at an end. If the force P is increased still more, the material either becomes "barrel-like" or fractures. When there is eccentricity in the loading, the elastic limit is encountered at smaller loads.

A *strut* is a *short compression member* such as the one shown in Figure 4–26. The magnitude of the maximum compressive stress in the x direction at point B in an intermediate section is the sum of a simple component P/A and a flexural component Mc/I; that is,

$$\sigma_c = \frac{P}{A} + \frac{Mc}{I} = \frac{P}{A} + \frac{PecA}{IA} = \frac{P}{A}\left(1 + \frac{ec}{k^2}\right) \tag{4–55}$$

where $k = (I/A)^{1/2}$ and is the radius of gyration, c is the coordinate of point B, and e is the eccentricity of loading.

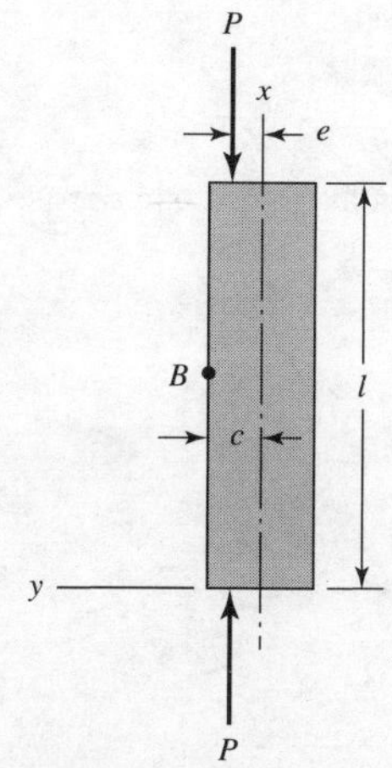

Figure 4–26

Eccentrically loaded strut.

Note that the length of the strut does not appear in Equation (4–55). In order to use the equation for design or analysis, we ought, therefore, to know the range of lengths for which the equation is valid. In other words, how long is a short member?

The difference between the secant formula Equations (4–54) and Equation (4–55) is that the secant equation, unlike Equation (4–55), accounts for an increased bending moment due to bending deflection. Thus the secant equation shows the eccentricity to be magnified by the bending deflection. This difference between the two formulas suggests that one way of differentiating between a "secant column" and a strut, or short compression member, is to say that in a strut, the effect of bending deflection must be limited to a certain small percentage of the eccentricity. If we decide that the limiting percentage is to be 1 percent of e, then, from Equation (4–44), the limiting slenderness ratio turns out to be

$$\left(\frac{l}{k}\right)_2 = 0.282\left(\frac{AE}{P}\right)^{1/2} \tag{4–56}$$

This equation then gives the limiting slenderness ratio for using Equation (4–55). If the actual slenderness ratio is greater than $(l/k)_2$, then use the secant formula; otherwise, use Equation (4–55).

EXAMPLE 4–20

Figure 4–27*a* shows a workpiece clamped to a milling machine table by a bolt tightened to a tension of 4500 N. The clamp contact is offset from the centroidal axis of the strut by a distance $e = 2.5$ mm, as shown in part *b* of the figure. The strut, or block, is steel, 25 mm square and 100 mm long, as shown. Determine the maximum compressive stress in the block.

Solution

First we find $A = bh = 25(25) = 625\text{ mm}^2$, $I = bh^3/12 = 25(25)^3/12 = 32\,552.1\text{ mm}^4$, $k^2 = I/A = 52.1\text{ mm}^2$, and $l/k = 100/(52.1)^{1/2} = 13.9$. Equation (4–56) gives the limiting slenderness ratio as

$$\left(\frac{l}{k}\right)_2 = 0.282\left(\frac{AE}{P}\right)^{1/2} = 0.282\left[\frac{625(210\,000)}{4500}\right]^{1/2} = 48.2$$

Thus the block could be as long as

$$l = 48.2k = 48.2(52.1)^{1/2} = 348\text{ mm}$$

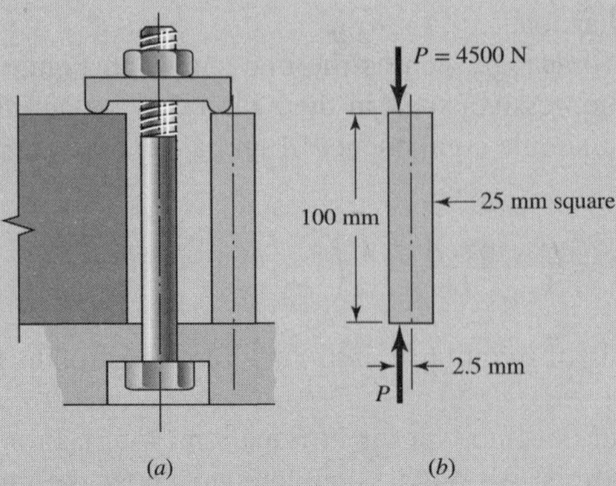

Figure 4–27

A strut that is part of a workpiece clamping assembly.

before it need be treated by using the secant formula. So Equation (4–55) applies and the maximum compressive stress is

Answer

$$\sigma_c = \frac{P}{A}\left(1 + \frac{ec}{k^2}\right) = \frac{4500}{625}\left[1 + \frac{2.5(12.5)}{52.1}\right] = 11.52 \text{ MPa}$$

4–16 Elastic Stability

Section 4–12 presented the conditions for the unstable behavior of long, slender columns. *Elastic instability* can also occur in structural members other than columns. *Compressive loads/stresses within any long, thin structure can cause structural instabilities* (buckling). The compressive stress may be elastic or inelastic and the instability may be global or local. Global instabilities can cause *catastrophic* failure, whereas local instabilities may cause permanent deformation and function failure but not a catastrophic failure. The buckling discussed in Section 4–12 was global instability. However, consider a wide flange beam in bending. One flange will be in compression, and if thin enough, can develop localized buckling in a region where the bending moment is a maximum. Localized buckling can also occur in the web of the beam, where transverse shear stresses are present at the beam centroid. Recall, for the case of pure shear stress τ, a stress transformation will show that at 45°, a *compressive stress* of $\sigma = -\tau$ exists. If the web is sufficiently thin where the shear force V is a maximum, localized buckling of the web can occur. For this reason, additional support in the form of bracing is typically applied at locations of high shear forces.[10]

Thin-walled beams in bending can buckle in a torsional mode as illustrated in Figure 4–28. Here a cantilever beam is loaded with a lateral force, F. As F increases from zero, the end of the beam will deflect in the negative y direction normally according to the bending equation, $y = -FL^3/(3EI)$. However, if the beam is long enough and the ratio of b/h is sufficiently small, there is a critical value of F for which the beam will collapse in a twisting mode as shown. This is due to the *compression* in the bottom fibers of the beam that cause the fibers to buckle sideways (z direction).

[10]See C. G. Salmon, J. E. Johnson, and F. A. Malhas, *Steel Structures: Design and Behavior,* 5th ed., Prentice Hall, Upper Saddle River, NJ, 2009.

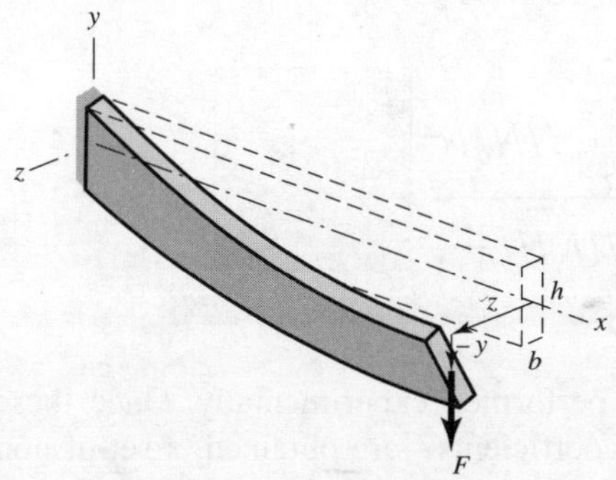

Figure 4–28

Torsional buckling of a thin-walled beam in bending.

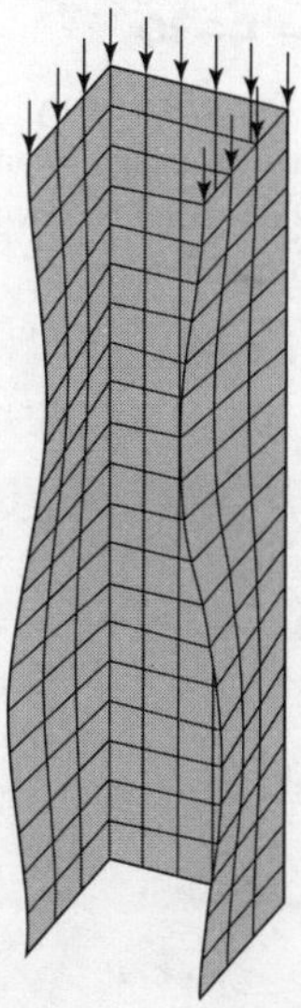

Figure 4–29

Finite-element representation of flange buckling of a channel in compression.

There are a great many other examples of unstable structural behavior, such as thin-walled pressure vessels in compression or with outer pressure or inner vacuum, thin-walled open or closed members in torsion, thin arches in compression, frames in compression, and shear panels. Because of the vast array of applications and the complexity of their analyses, further elaboration is beyond the scope of this book. The intent of this section is to make the reader aware of the possibilities and potential safety issues. The key issue is that the designer should be aware that if any *unbraced* part of a structural member is *thin,* and/or *long,* and in *compression* (directly or *indirectly*), the possibility of buckling should be investigated.[11]

For unique applications, the designer may need to revert to a numerical solution such as using finite elements. Depending on the application and the finite-element code available, an analysis can be performed to determine the critical loading (see Figure 4–29).

4–17 Shock and Impact

Impact refers to the collision of two masses with initial relative velocity. In some cases it is desirable to achieve a known impact in design; for example, this is the case in the design of coining, stamping, and forming presses. In other cases, impact occurs because of excessive deflections, or because of clearances between parts, and in these cases it is desirable to minimize the effects. The rattling of mating gear teeth in their tooth spaces is an impact problem caused by shaft deflection and the clearance between the teeth. This impact causes gear noise and fatigue failure of the tooth surfaces. The clearance space between a cam and follower or between a journal and its bearing may result in crossover impact and also cause excessive noise and rapid fatigue failure.

Shock is a more general term that is used to describe any suddenly applied force or disturbance. Thus the study of shock includes impact as a special case.

Figure 4–30 represents a highly simplified mathematical model of an automobile in collision with a rigid obstruction. Here m_1 is the lumped mass of the engine. The displacement, velocity, and acceleration are described by the coordinate x_1 and its time derivatives. The lumped mass of the vehicle less the engine is denoted by m_2, and its motion by the coordinate x_2 and its derivatives. Springs k_1, k_2, and k_3 represent the linear and nonlinear stiffnesses of the various structural elements that compose the vehicle. Friction and damping can and should be included, but is not shown in this model. The determination of the spring rates for such a complex structure will

[11]See S. P. Timoshenko and J. M. Gere, *Theory of Elastic Stability,* 2nd ed., McGraw Hill, New York, 1961. See also, Z. P. Bazant and L. Cedolin, *Stability of Structures,* Oxford University Press, New York, 1991.

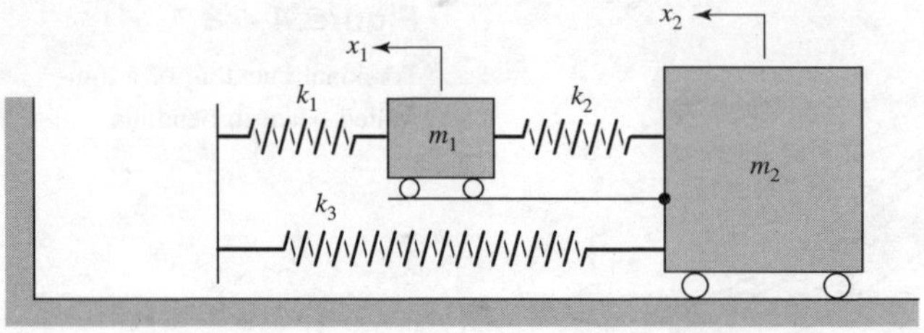

Figure 4–30

Two-degree-of-freedom mathematical model of an automobile in collision with a rigid obstruction.

almost certainly have to be performed experimentally. Once these values—the k's, m's, damping and frictional coefficients—are obtained, a set of nonlinear differential equations can be written and a computer solution obtained for any impact velocity. For sake of illustration, assuming the springs to be linear, isolate each mass and write their equations of motion. This results in

$$\begin{aligned} m\ddot{x}_1 + k_1x_1 + k_2(x_1 - x_2) &= 0 \\ m\ddot{x}_2 + k_3x_2 - k_2(x_1 - x_2) &= 0 \end{aligned} \tag{4–57}$$

The analytical solution of the Equation (4–57) pair is harmonic and is studied in a course on mechanical vibrations.[12] If the values of the m's and k's are known, the solution can be obtained easily using a program such as MATLAB.

Suddenly Applied Loading

A simple case of impact is illustrated in Figure 4–31*a*. Here a weight W falls a distance h and impacts a cantilever of stiffness EI and length l. We want to find the maximum deflection and the maximum force exerted on the beam due to the impact.

Figure 4–31*b* shows an abstract model of the system considering the beam as a simple spring. For beam 1 of Table A–9, we find the spring rate to be $k = F/y = 3EI/l^3$. The beam mass and damping can be accounted for, but for this example will be considered negligible. If the beam is considered massless, there is no momentum transfer, only energy. If the maximum deflection of the spring (beam) is considered to be δ, the drop of the weight is $h + \delta$, and the loss of potential energy is $W(h + \delta)$. The resulting increase in potential (strain) energy of the spring is $\frac{1}{2}k\delta^2$. Thus, for energy conservation, $\frac{1}{2}k\delta^2\ W(h + \delta)$. Rearranging this gives

$$\delta^2 - 2\frac{W}{k}\delta - 2\frac{W}{k}h = 0 \tag{a}$$

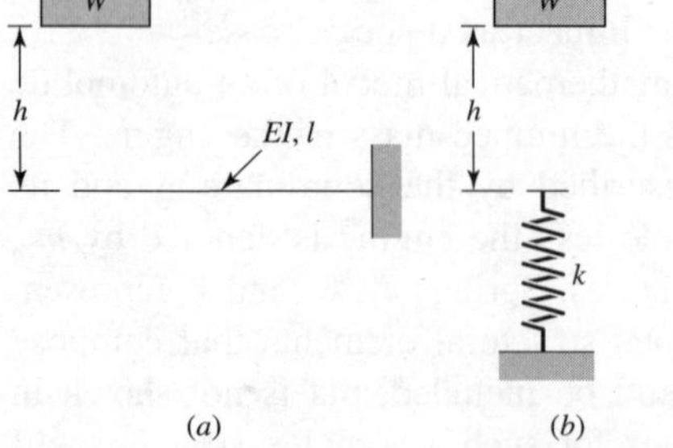

Figure 4–31

(*a*) A weight free to fall a distance h to free end of a beam. (*b*) Equivalent spring model.

[12]See William T. Thomson and Marie Dillon Dahleh, *Theory of Vibrations with Applications,* 5th ed., Prentice Hall, Upper Saddle River, NJ, 1998.

Solving for δ yields

$$\delta = \frac{W}{k} \pm \frac{W}{k}\left(1 + \frac{2hk}{W}\right)^{1/2} \tag{b}$$

The negative solution is possible only if the weight "sticks" to the beam and vibrates between the limits of Equation (*b*). Thus, the maximum deflection is

$$\delta = \frac{W}{k} + \frac{W}{k}\left(1 + \frac{2hk}{W}\right)^{1/2} \tag{4–58}$$

The maximum force acting on the beam is now found to be

$$F = k\delta = W + W\left(1 + \frac{2hk}{W}\right)^{1/2} \tag{4–59}$$

Note, in this equation, that if $h = 0$, then $F = 2W$. This says that when the weight is released while in contact with the spring but is not exerting any force on the spring, the largest force is double the weight.

Most systems are not as ideal as those explored here, so be wary about using these relations for nonideal systems.

PROBLEMS

Problems marked with an asterisk (*) are linked to problems in other chapters, as summarized in Table 1–2 of Section 1–17.

4–1 The figure shows a torsion bar *OA* fixed at *O*, simply supported at *A*, and connected to a cantilever *AB*. The spring rate of the torsion bar is k_T, in newton-meters per radian, and that of the cantilever is k_l in newtons per meter. What is the overall spring rate based on the deflection *y* at point *B*?

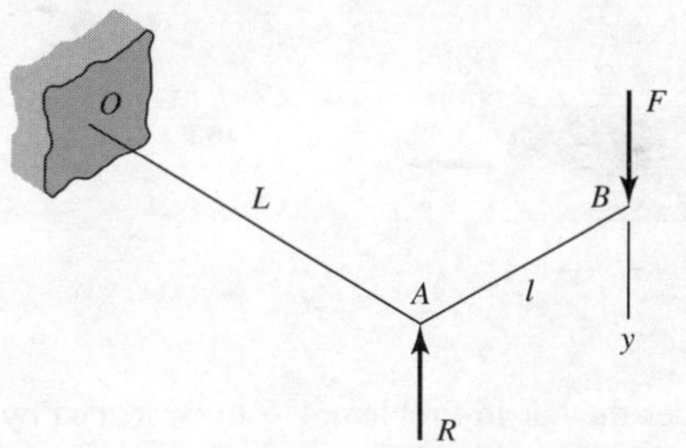

Problem 4–1

4–2 For Problem 4–1, if the simple support at point *A* were eliminated and the cantilever spring rate of *OA* is given by k_L, determine the overall spring rate of the bar based on the deflection of point *B*.

4–3 A torsion-bar spring consists of a prismatic bar, usually of round cross section, that is twisted at one end and held fast at the other to form a stiff spring. An engineer needs a stiffer one than usual and so considers building in both ends and applying the torque somewhere in the central portion of the span, as shown in the figure. This effectively creates two springs in parallel. If the bar is uniform in diameter, that is, if $d = d_1 = d_2$, (*a*) determine how the spring rate and the end reactions depend on the location *x* at which the torque is applied, (*b*) determine the spring rate, the end reactions, and the maximum shear stress, if $d = 12$ mm, $x = 120$ mm, $l = 240$ mm, $T = 150$ N · m, and $G = 79$ GPa.

Problem 4–3

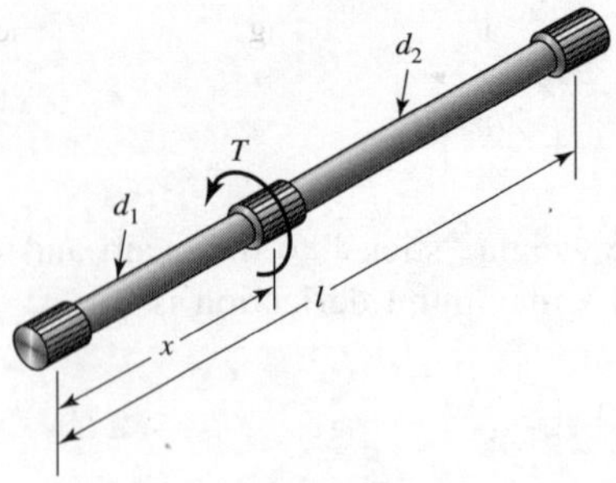

4–4 An engineer is forced by geometric considerations to apply the torque on the spring of Problem 4–3 at the location $x = 0.4l$. For a uniform-diameter spring, this would cause one leg of the span to be underutilized when both legs have the same diameter. For optimal design the diameter of each leg should be designed such that the maximum shear stress in each leg is the same. This problem is to redesign the spring of part (*b*) of Problem 4–3. Using $x = 0.4l$, $l = 240$ mm, $T = 150$ N · m, and $G = 79$ GPa, design the spring such that the maximum shear stresses in each leg are equal and the spring has the same spring rate (angle of twist) as part (*b*) of Problem 4–3. Specify d_1, d_2, the spring rate k, and the torque and the maximum shear stress in each leg.

4–5 A bar in tension has a circular cross section and includes a tapered portion of length l, as shown.

(*a*) For the tapered portion, use Equation (4–3) in the form of $\delta = \int_0^l [F/(AE)]\, dx$ to show that

$$\delta = \frac{4}{\pi}\frac{Fl}{d_1 d_2 E}$$

(*b*) Determine the elongation of each portion if $d_1 = 12$ mm, $d_2 = 18$ mm, $l = l_1 = l_2 = 48$ mm, $E = 207$ GPa, and $F = 4$ kN.

Problem 4–5

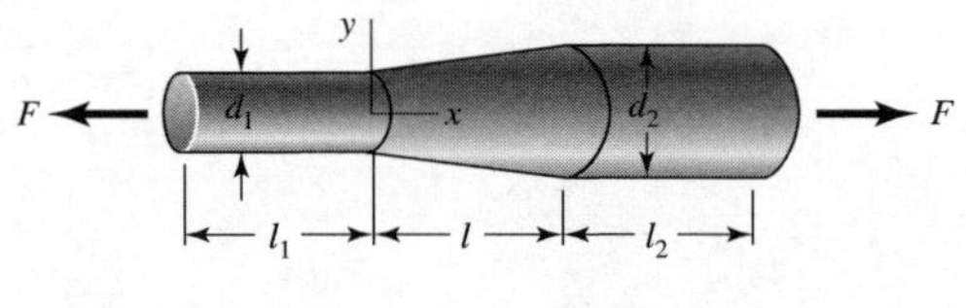

4–6 Instead of a tensile force, consider the bar in Problem 4–5 to be loaded by a torque T.

(*a*) Use Equation (4–5) in the form of $\theta = \int_0^l [T/(GJ)]\, dx$ to show that the angle of twist of the tapered portion is

$$\theta = \frac{32}{3\pi}\frac{Tl(d_1^2 + d_1 d_2 + d_2^2)}{G d_1^3 d_2^3}$$

(*b*) Using the same geometry as in Problem 4–5*b* with $T = 150$ N · m and $G = 79$ GPa, determine the angle of twist in degrees for each portion.

4–7 When a vertically suspended hoisting cable is long, the weight of the cable itself contributes to the elongation. If a 150 m steel cable has an effective diameter of 12 mm and lifts a load of 20 kN, determine the total elongation and the percent of the total elongation due to the cable's own weight.

4–8 Derive the equations given for beam 2 in Table A–9 using statics and the double-integration method.

4–9 Derive the equations given for beam 5 in Table A–9 using statics and the double-integration method.

4–10 The figure shows a cantilever consisting of steel angles size 100 × 100 × 12 mm mounted back to back. Using superposition, find the deflection at B and the maximum stress in the beam.

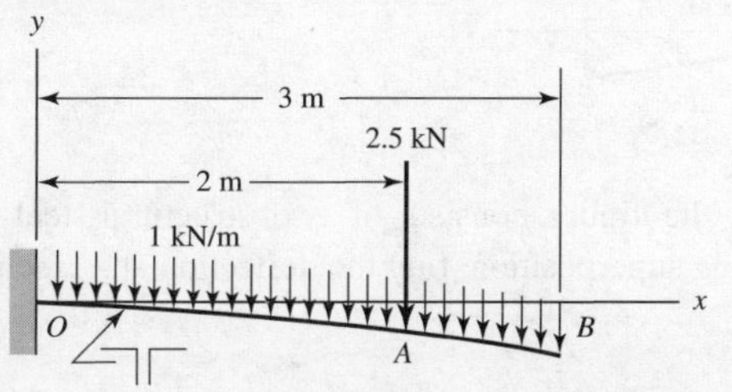

Problem 4–10

4–11 A simply supported beam loaded by two forces is shown in the figure. Select a pair of structural steel channels mounted back to back to support the loads in such a way that the deflection at midspan will not exceed 1.6 mm and the maximum stress will not exceed 40 MPa. Use superposition.

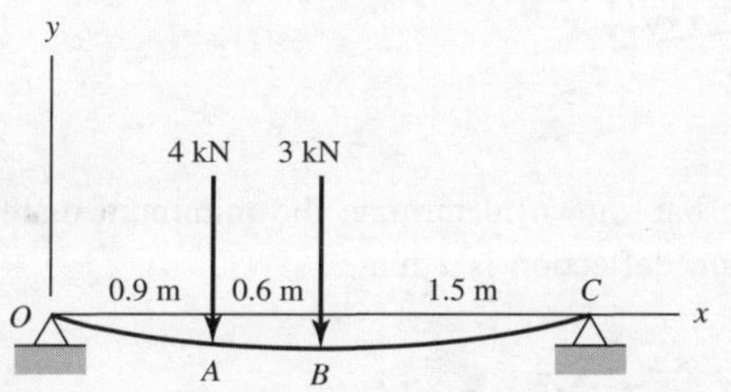

Problem 4–11

4–12 Using superposition, find the deflection of the steel shaft at A in the figure. Find the deflection at midspan. By what percentage do these two values differ?

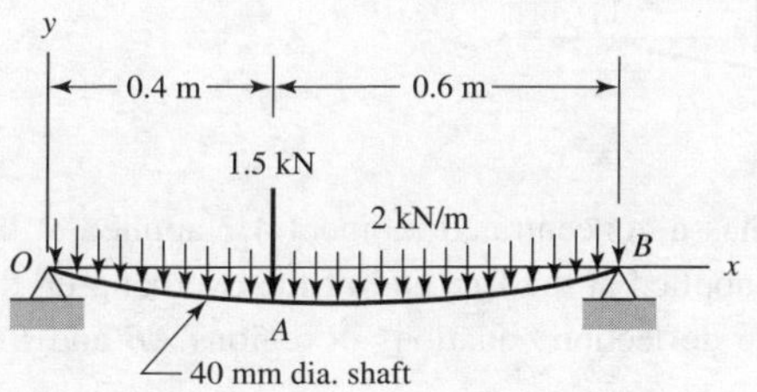

Problem 4–12

4–13 A rectangular steel bar supports the two overhanging loads shown in the figure. Using superposition, find the deflection at the ends and at the center.

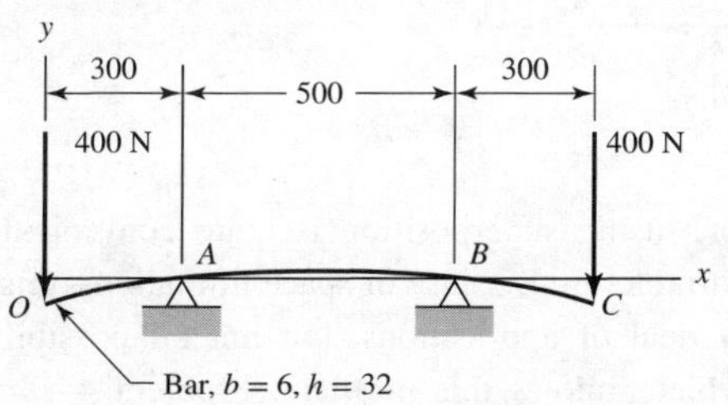

Problem 4–13

Dimensions in millimeters.

4–14 An aluminum tube with outside diameter of 50 mm and inside diameter of 38 mm is cantilevered and loaded as shown. Using the formulas in Appendix Table A–9 and superposition, find the deflection at *B*.

Problem 4–14

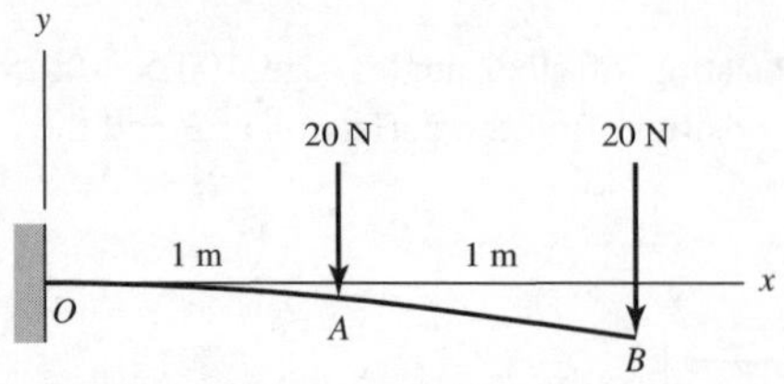

4–15 The cantilever shown in the figure consists of two structural-steel channels size 76 × 38 mm, 2 kN/m. Using superposition, find the deflection at *A*. Include the weight of the channels.

Problem 4–15

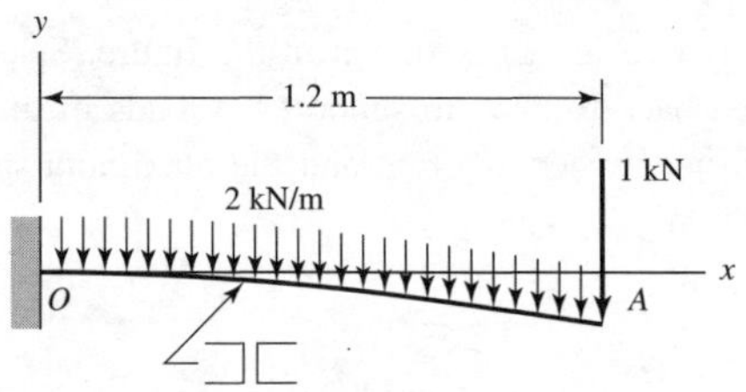

4–16 Using superposition for the bar shown, determine the minimum diameter of a steel shaft for which the maximum deflection is 2 mm.

Problem 4–16
Dimensions in millimeters.

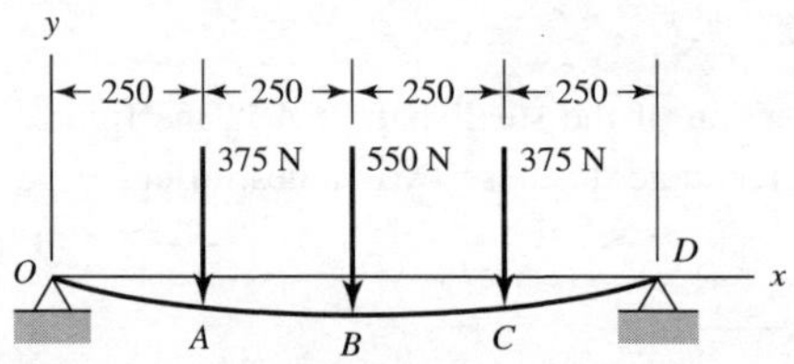

4–17 A simply supported beam has a concentrated moment M_A applied at the left support and a concentrated force *F* applied at the free end of the overhang on the right. Using superposition, determine the deflection equations in regions *AB* and *BC*.

Problem 4–17

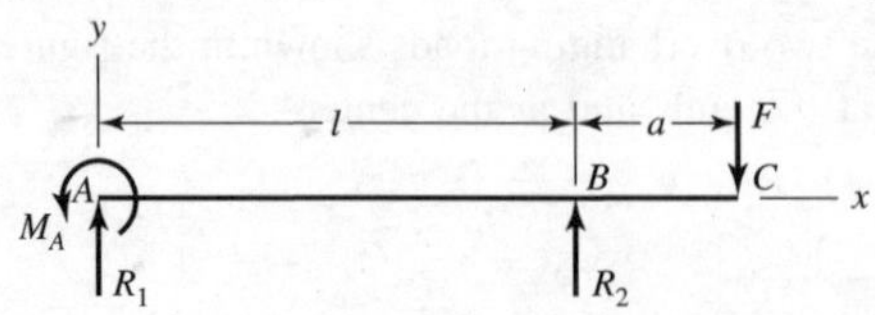

4–18 Calculating beam deflections using superposition is quite convenient provided you have a comprehensive table to refer to. Because of space limitations, this book provides a table that covers a great deal of applications, but not all possibilities. Take for example, Problem 4–19, which follows this problem. Problem 4–19 is not directly solvable from Table A–9, but with the addition of the results of this problem, it is. For the beam shown, using statics and double integration, show that

$$R_1 = \frac{wa}{2l}(2l - a) \qquad R_2 = \frac{wa^2}{2l} \qquad V_{AB} = \frac{w}{2l}[2l(a - x) - a^2] \qquad V_{BC} = -\frac{wa^2}{2l}$$

$$M_{AB} = \frac{wx}{2l}(2al - a^2 - lx) \qquad M_{BC} = \frac{wa^2}{2l}(l - x)$$

$$y_{AB} = \frac{wx}{24EIl}[2ax^2(2l - a) - lx^3 - a^2(2l - a)^2] \qquad y_{BC} = y_{AB} + \frac{w}{24EI}(x - a)^4$$

Problem 4–18

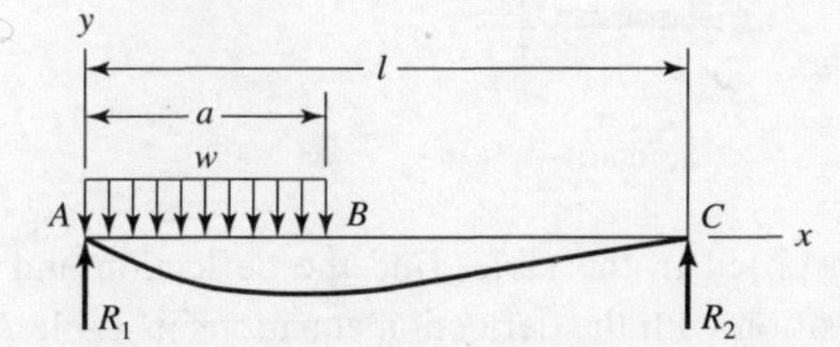

4–19 Using the results of Problem 4–18, use superposition to determine the deflection equations for the three regions of the beam shown.

Problem 4–19

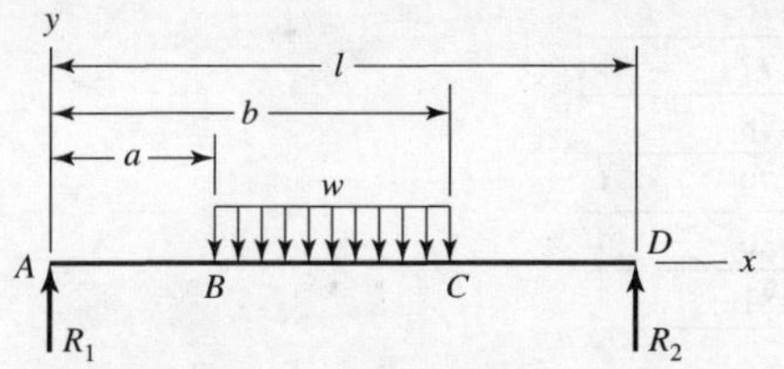

4–20 Like Problem 4–18, this problem provides another beam to add to Table A–9. For the simply supported beam shown with an overhanging uniform load, use statics and double integration to show that

$$R_1 = \frac{wa^2}{2l} \qquad R_2 = \frac{wa}{2l}(2l + a) \qquad V_{AB} = -\frac{wa^2}{2l} \qquad V_{BC} = w(l + a - x)$$

$$M_{AB} = -\frac{wa^2}{2l}x \qquad M_{BC} = -\frac{w}{2}(l + a - x)^2$$

$$y_{AB} = \frac{wa^2x}{12EIl}(l^2 - x^2) \qquad y_{BC} = -\frac{w}{24EI}[(l + a - x)^4 - 4a^2(l - x)(l + a) - a^4]$$

Problem 4–20

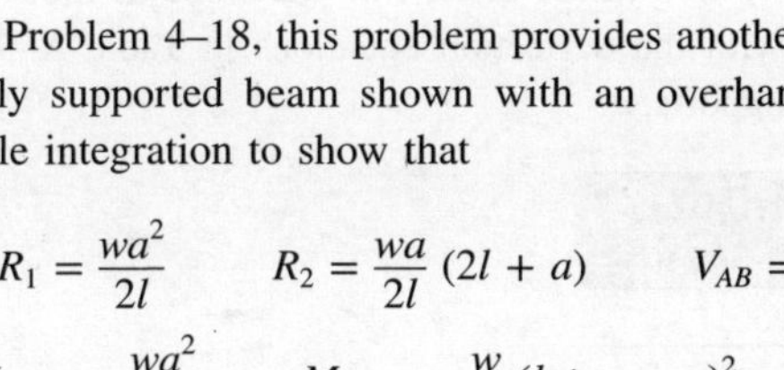

4–21 Consider the uniformly loaded simply supported steel beam with an overhang as shown. The second-area moment of the beam is $I = 19\ 500\ \text{mm}^4$. Use superposition (with Table A–9 and the results of Problem 4–20) to determine the reactions and the deflection equations of the beam. Plot the deflections.

Problem 4–21

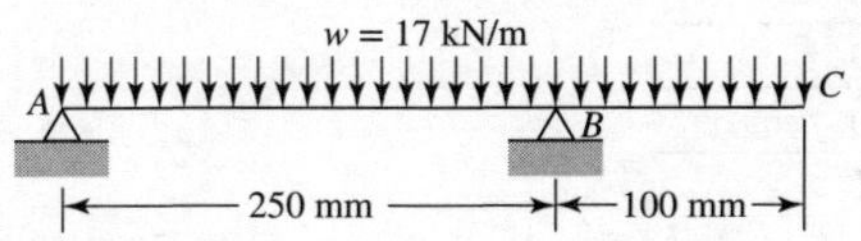

4–22 Illustrated is a rectangular steel bar with simple supports at the ends and loaded by a force F at the middle; the bar is to act as a spring. The ratio of the width to the thickness is to be about $b = 10h$, and the desired spring scale is 300 kN/m.

(*a*) Find a set of cross-section dimensions, using preferred fractional sizes from Table A–15.

(*b*) What deflection would cause a permanent set in the spring if this is estimated to occur at a normal stress of 400 Mpa?

Problem 4–22

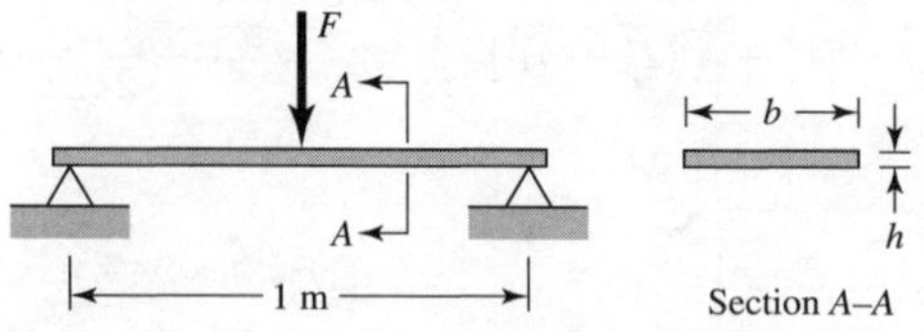

4–23* to 4–28* For the steel countershaft specified in the table, find the deflection and slope of the shaft at point A. Use superposition with the deflection equations in Table A–9. Assume the bearings constitute simple supports.

Problem Number	Problem Number Defining Shaft
4–23*	3–76
4–24*	3–77
4–25*	3–78
4–26*	3–79
4–27*	3–80
4–28*	3–81

4–29* to 4–34* For the steel countershaft specified in the table, find the slope of the shaft at each to bearing. Use superposition with the deflection equations in Table A–9. Assume the bearings constitute simple supports.

Problem Number	Problem Number Defining Shaft
4–29*	3–76
4–30*	3–77
4–31*	3–78
4–32*	3–79
4–33*	3–80
4–34*	3–81

4–35* to 4–40* For the steel countershaft specified in the table, assume the bearings have a maximum slope specification of 0.06° for good bearing life. Determine the minimum shaft diameter.

Problem Number	Problem Number Defining Shaft
4–35*	3–76
4–36*	3–77
4–37*	3–78
4–38*	3–79
4–39*	3–80
4–40*	3–81

4–41* The cantilevered handle in the figure is made from mild steel that has been welded at the joints. For $F_y = 800$ N, $F_x = F_z = 0$, determine the vertical deflection (along the y axis) at the tip. Use superposition. See the discussion on Equation (3–41) for the twist in the rectangular cross section in section BC.

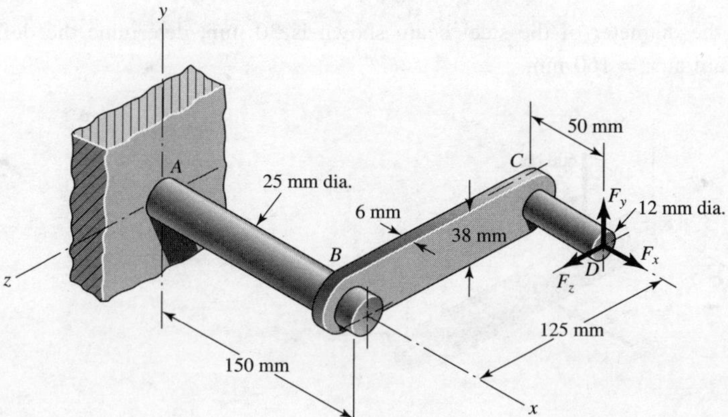

Problem 4–41

4–42 For the cantilevered handle in Problem 4–41, let $F_x = -600$ N, $F_y = 0$ N, $F_z = -400$ N. Find the deflection at the tip along the x axis.

4–43* The cantilevered handle in Problem 3–95 is made from mild steel. Let $F_y = 1$ kN, $F_x = F_z = 0$. Determine the angle of twist in bar OC, ignoring the fillets but including the changes in diameter along the 400 mm effective length. Compare the angle of twist if the bar OC is simplified to be all of uniform 24 mm diameter. Use superposition to determine the vertical deflection (along the y axis) at the tip, using the simplified bar OC.

4–44 A flat-bed trailer is to be designed with a curvature such that when loaded to capacity the trailer bed is flat. The load capacity is to be 45 kN/m between the axles, which are 7 m apart, and the second-area moment of the steel structure of the bed is $I = 190 \times 10^6$ mm^4. Determine the equation for the curvature of the unloaded bed and the maximum height of the bed relative to the axles.

4–45 The designer of a shaft usually has a slope constraint imposed by the bearings used. This limit will be denoted as ξ. If the shaft shown in the figure is to have a uniform diameter d except in the locality of the bearing mounting, it can be approximated as a uniform beam with simple supports. Show that the minimum diameters to meet the slope constraints at the left and right bearings are, respectively,

$$d_L = \left|\frac{32Fb(l^2 - b^2)}{3\pi El\xi}\right|^{1/4} \qquad d_R = \left|\frac{32Fa(l^2 - a^2)}{3\pi El\xi}\right|^{1/4}$$

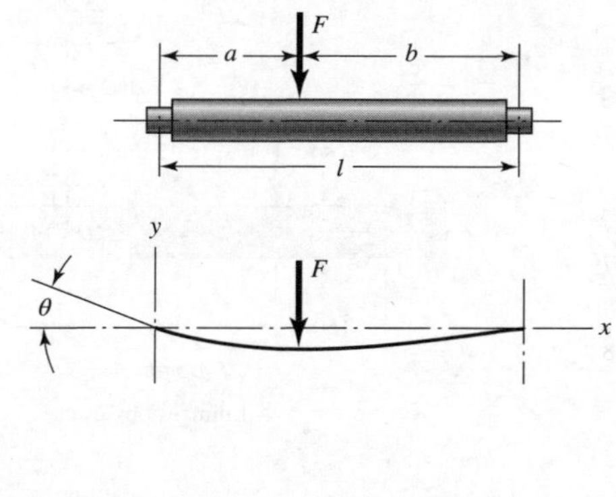

Problem 4–45

4–46 A steel shaft is to be designed so that it is supported by roller bearings. The basic geometry is shown in the figure from Problem 4–45, with $l = 300$ mm, $a = 100$ mm, and $F = 3$ kN. The allowable slope at the bearings is 0.001 mm/mm without bearing life penalty. For a design factor of 1.28, what uniform-diameter shaft will support the load without penalty? Determine the maximum deflection of the shaft.

4–47 If the diameter of the steel beam shown is 30 mm, determine the deflection of the beam at $x = 160$ mm.

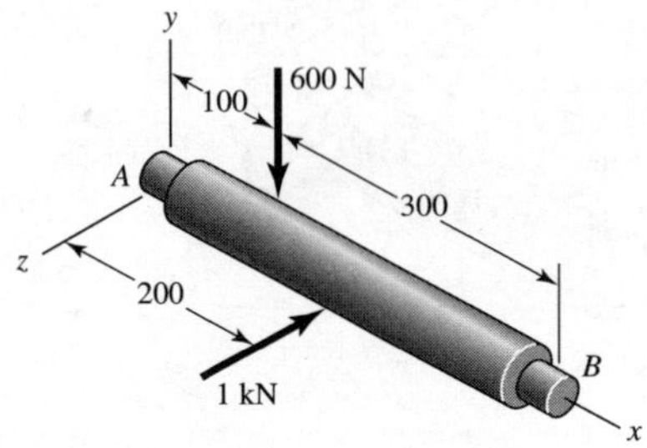

Problem 4–47
Dimensions in millimeters.

4–48 For the beam of Problem 4–47, plot the *magnitude* of the displacement of the beam in 2 mm increments. Approximate the maximum displacement and the value of x where it occurs.

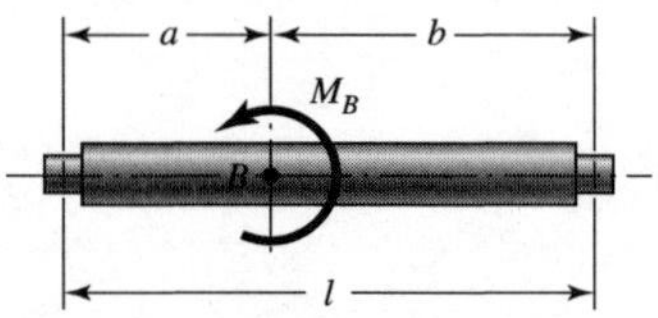

Problem 4–49

4–49 Shown in the figure is a uniform-diameter shaft with bearing shoulders at the ends; the shaft is subjected to a concentrated moment $M = 130$ N · m. The shaft is of carbon steel and has $a = 80$ mm and $l = 200$ mm. The slope at the ends must be limited to 0.002 rad. Find a suitable diameter d.

4–50* to 4–53 The figure shows an aluminum beam *OB* with rectangular cross section, pinned to the ground at one end, and supported by a round steel rod with hooks formed on the ends. A load is applied as shown. Use superposition to determine the vertical deflection at point *B*.

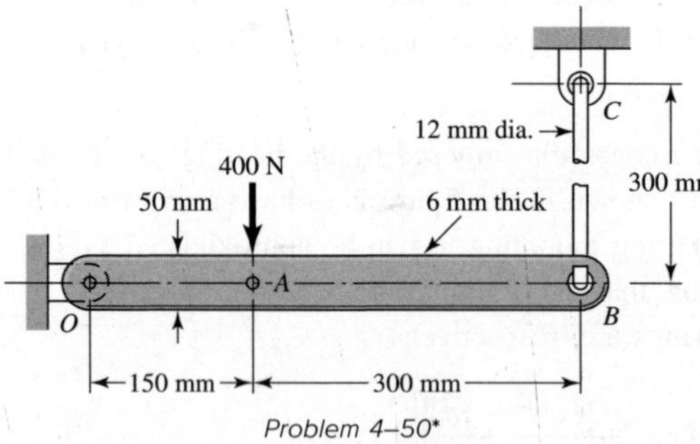

*Problem 4–50**

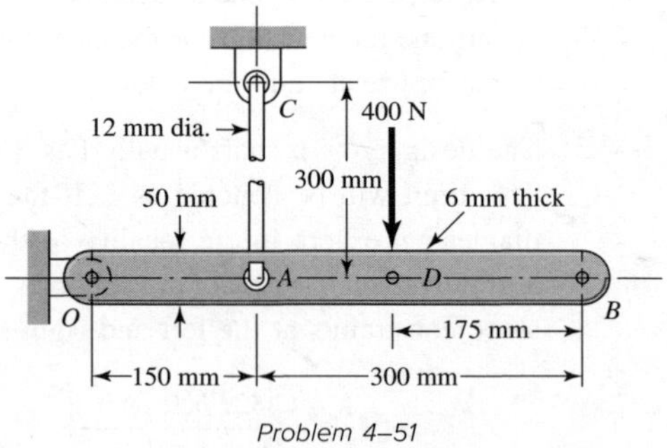

Problem 4–51

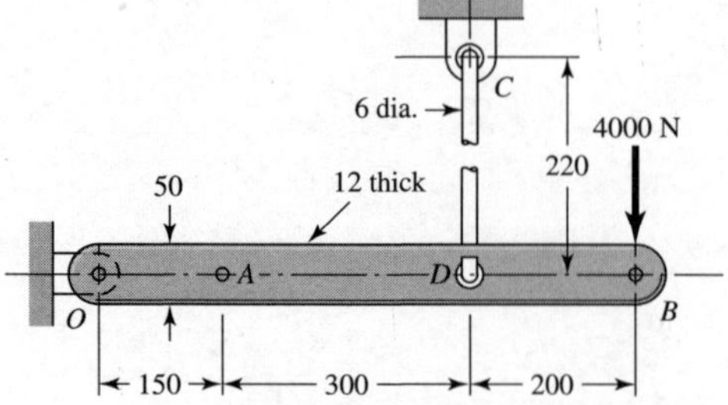

Problem 4–52
All Dimensions in mm.

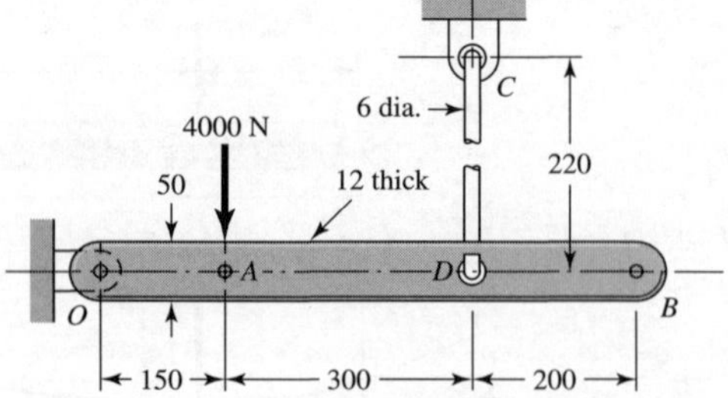

Problem 4–53
All dimensions in mm.

4–54 Solve Problem 4–50 for the vertical deflection at point *A*.

4–55 Solve Problem 4–51 for the vertical deflection at point *A*.

4–56 Solve Problem 4–52 for the vertical deflection at point *A*.

4–57 Solve Problem 4–53 for the vertical deflection at point *A*.

4–58 The figure illustrates a stepped torsion-bar spring *OA* with an actuating cantilever *AB*. Both parts are of carbon steel. Use superposition and find the spring rate *k* corresponding to a force *F* acting at *B*.

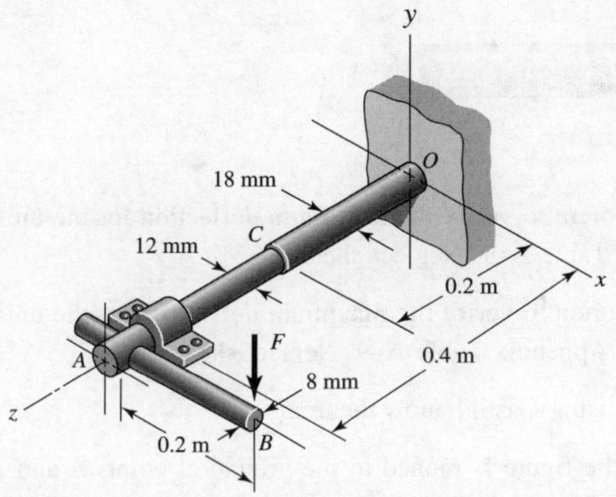

Problem 4–58

4–59 Consider the simply supported beam 5 with a center load in Appendix A–9. Determine the deflection equation if the stiffness of the left and right supports are k_1 and k_2, respectively.

4–60 Consider the simply supported beam 10 with an overhanging load in Appendix A–9. Determine the deflection equation if the stiffness of the left and right supports are k_1 and k_2, respectively.

4–61 Prove that for a uniform-cross-section beam with simple supports at the ends loaded by a single concentrated load, the location of the maximum deflection will never be outside the range of $0.423l \leq x \leq 0.577l$ regardless of the location of the load along the beam. The importance of this is that you can always get a quick estimate of y_{max} by using $x = l/2$.

4–62 Solve Problem 4–10 using singularity functions. Use statics to determine the reactions.

4–63 Solve Problem 4–11 using singularity functions. Use statics to determine the reactions.

4–64 Solve Problem 4–12 using singularity functions. Use statics to determine the reactions.

4–65 Solve Problem 4–21 using singularity functions to determine the deflection equation of the beam. Use statics to determine the reactions.

4–66 Solve Problem 4–13 using singularity functions. Since the beam is symmetric, only write the equation for half the beam and use the slope at the beam center as a boundary condition. Use statics to determine the reactions.

4–67 Solve Problem 4–17 using singularity functions. Use statics to determine the reactions.

4–68 Solve Problem 4–19 using singularity functions to determine the deflection equation of the beam. Use statics to determine the reactions.

4–69 Using singularity functions, write the deflection equation for the steel beam shown. Since the beam is symmetric, write the equation for only half the beam and use the slope at the beam center as a boundary condition. Plot your results and determine the maximum deflection.

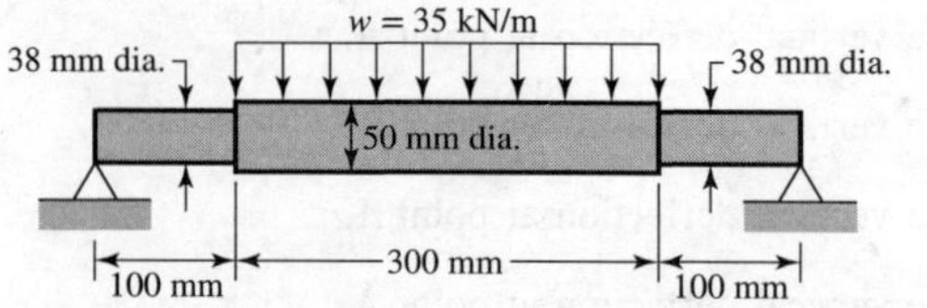

Problem 4–69

4–70 Determine the deflection equation for the cantilever beam shown using singularity functions. Evaluate the deflections at B and C and compare your results with Example 4–10.

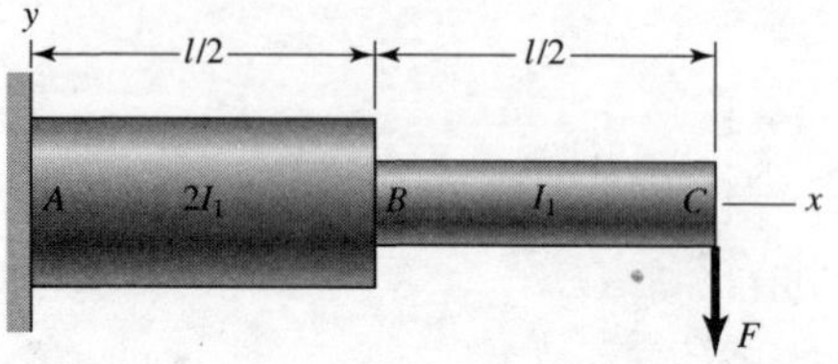

Problem 4–70

4–71 Use Castigliano's theorem to verify the maximum deflection for the uniformly loaded beam 7 of Appendix Table A–9. Neglect shear.

4–72 Use Castigliano's theorem to verify the maximum deflection for the uniformly loaded cantilever beam 3 of Appendix Table A–9. Neglect shear.

4–73 Solve Problem 4–15 using Castigliano's theorem.

4–74 The beam shown in the figure is pinned to the ground at points A and B, and loaded by a force P at point C.

(*a*) Using Castigliano's method, derive an expression for the vertical deflection at D.

(*b*) Using Table A–9, determine the vertical deflection at D, and compare your results with part (*a*).

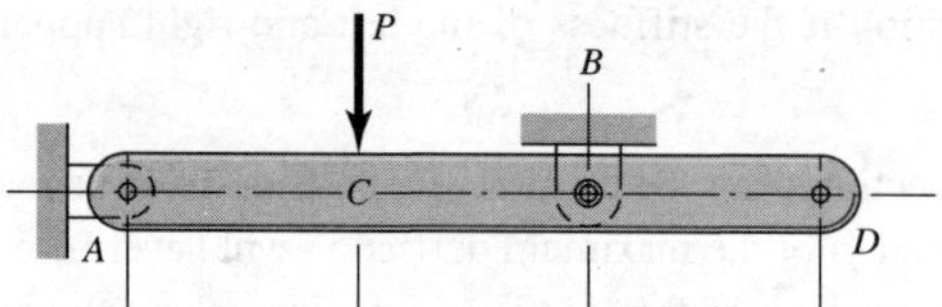

Problem 4–74

4–75 Solve Problem 4–58 using Castigliano's theorem.

4–76 Determine the deflection at midspan for the beam of Problem 4–69 using Castigliano's theorem.

4–77 Using Castigliano's theorem, determine the deflection of point B in the direction of the force F for the steel bar shown.

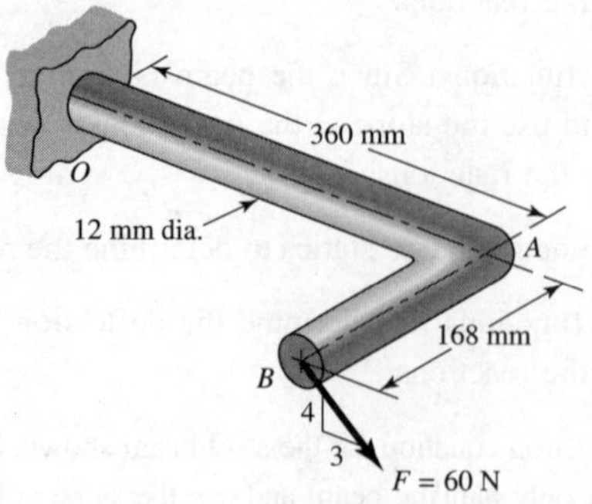

Problem 4–77

4–78* Solve Problem 4–41 using Castigliano's theorem. Since Equation (4–22) for torsional strain energy was derived from the angular displacement for circular cross sections, it is not applicable for section *BC*. You will need to obtain a new strain energy equation for the rectangular cross section from Equations (4–22) and (3–41).

4–79 Solve Problem 4–42 using Castigliano's theorem.

4–80* The cantilevered handle in Problem 3–95 is made from mild steel. Let $F_y = 1$ kN and $F_x = F_z = 0$. Using Castigliano's theorem, determine the vertical deflection (along the *y* axis) at the tip. Repeat the problem with shaft *OC* simplified to a uniform diameter of 1 in for its entire length. What is the percent error from this simplification?

4–81* Solve Problem 4–50 using Castigliano's theorem.

4–82 Solve Problem 4–51 using Castigliano's theorem.

4–83 Solve Problem 4–52 using Castigliano's theorem.

4–84 Solve Problem 4–53 using Castigliano's theorem.

4–85 Solve Problem 4–54 using Castigliano's theorem.

4–86 Solve Problem 4–55 using Castigliano's theorem.

4–87 The figure shows a rectangular member *OB*, made from 0.5-cm thick aluminum plate, pinned to the ground at one end and supported by a coil spring at point *A*. The spring has been measured to deflect 1 mm with an applied force of 10 kN. Find the deflection at point *B*. Use Castigliano's method directly without using superposition or indirect geometric extrapolations. Organize your work in the following manner:

(*a*) Determine the component of the deflection at *B* due to the energy in section *OA*.

(*b*) Determine the component of the deflection at *B* due to the energy in section *AB*.

(*c*) Determine the component of the deflection at *B* due to the energy in section *AC*. You may find it helpful to consider the definition of strain energy in Equation (4–20), and apply it to the spring.

(*d*) Determine the total deflection at *B* due to all of the components combined.

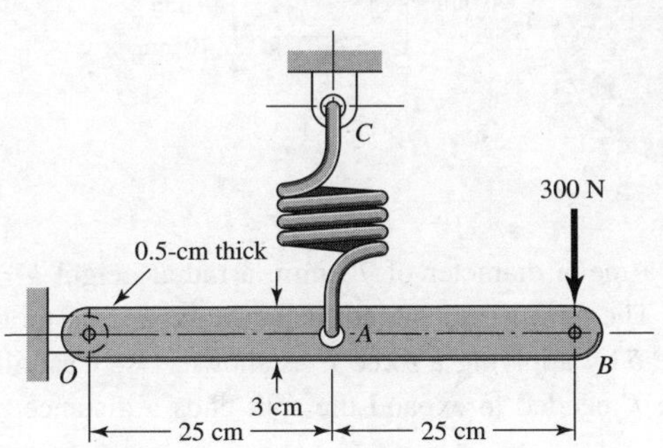

Problem 4–87

4–88 and 4–89 The part shown in the figure is made from cold-drawn AISI 1020 steel. Use Castigliano's method to directly find the deflection of point *D* in the *y* direction. Organize your work in the following manner:

(*a*) Determine the component of the deflection at *D* due to the energy in section *AB*.

(*b*) Determine the component of the deflection at *D* due to the energy in section *BC*.

(*c*) Determine the component of the deflection at *D* due to the energy in section *BD*.

(*d*) Determine the total deflection at *D* due to all of the components combined.

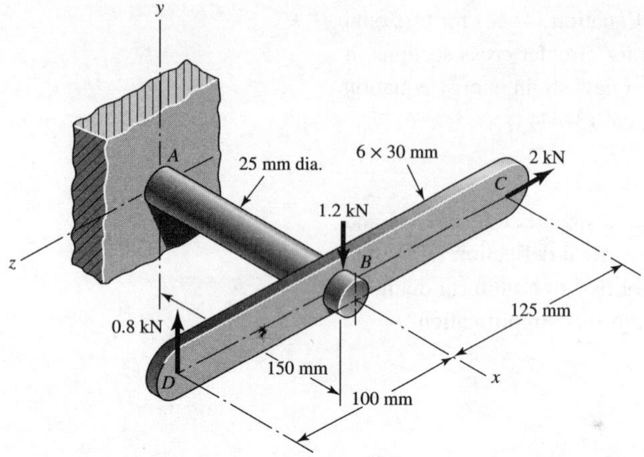

Problem 4–88

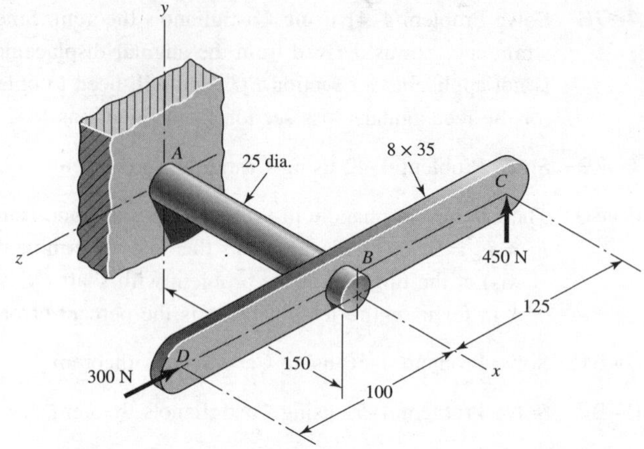

Problem 4–89

Dimensions in mm.

4–90 For Problem 4–88, determine the displacement of point D in the z direction.

4–91 For Problem 4–89, determine the displacement of point D in the z direction.

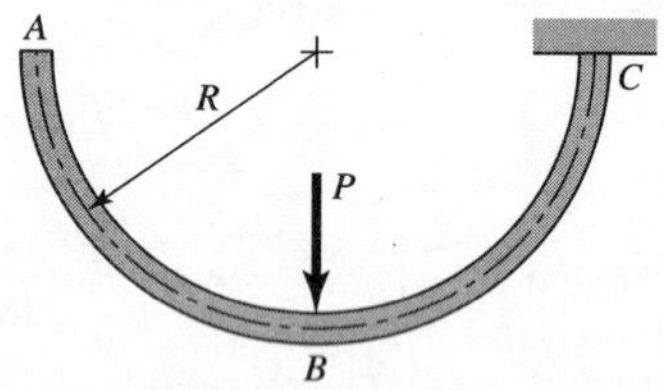

Problem 4–92

4–92 The steel curved bar shown has a rectangular cross section with a radial height $h = 6$ mm, and a thickness $b = 4$ mm. The radius of the centroidal axis is $R = 40$ mm. A force $P = 10$ N is applied as shown. Find the vertical deflection at B. Use Castigliano's method for a curved flexural member, and since $R/h < 10$, do not neglect any of the terms.

4–93 Repeat Problem 4–92 to find the vertical deflection at A.

4–94 For the curved steel beam shown, $F = 30$ kN. Determine the relative deflection of the applied forces.

Problem 4–94

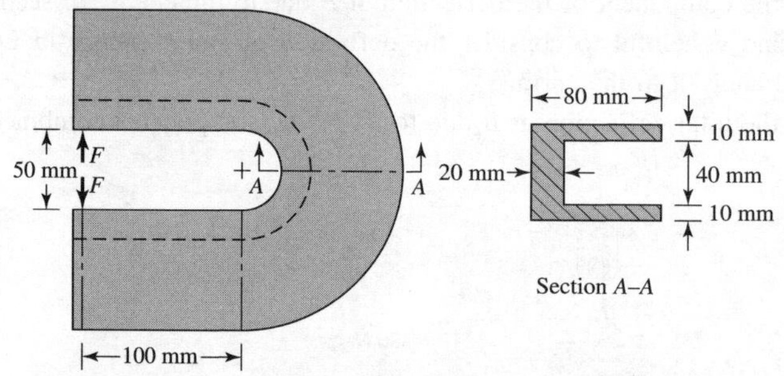

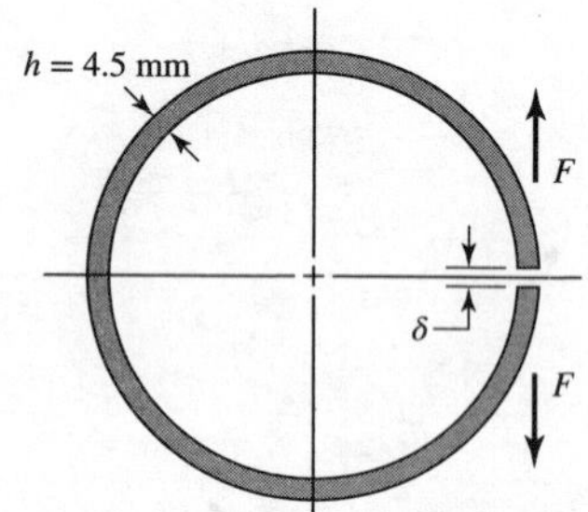

Problem 4–95

4–95 A steel piston ring has a mean diameter of 70 mm, a radial height $h = 4.5$ mm, and a thickness $b = 3$ mm. The ring is assembled using an expansion tool that separates the split ends a distance δ by applying a force F as shown. Use Castigliano's theorem and determine the force F needed to expand the split ends a distance $\delta = 1$ mm.

4–96 For the steel wire form shown, use Castigliano's method to determine the horizontal reaction forces at A and B and the deflection at C.

Problem 4–96

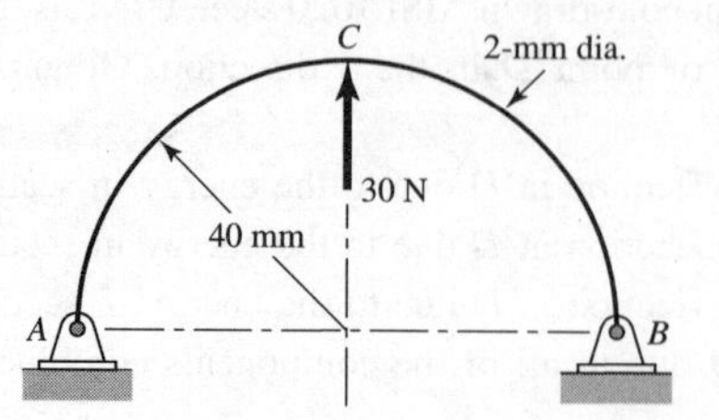

4–97 to 4–100 The part shown is formed from a 3 mm diameter steel wire, with $R = 125$ mm and $l = 100$ mm. A force is applied with $P = 5$ N. Use Castigliano's method to estimate the horizontal deflection at point A. Justify any components of strain energy that you choose to neglect.

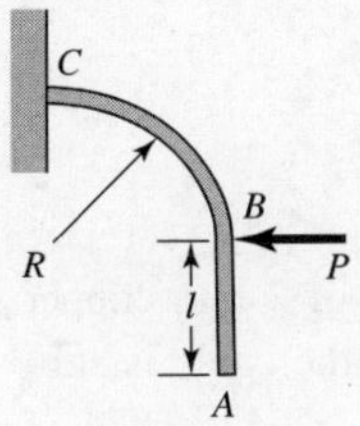

Problem 4–97

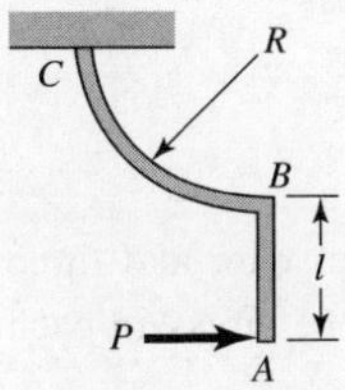

Problem 4–98

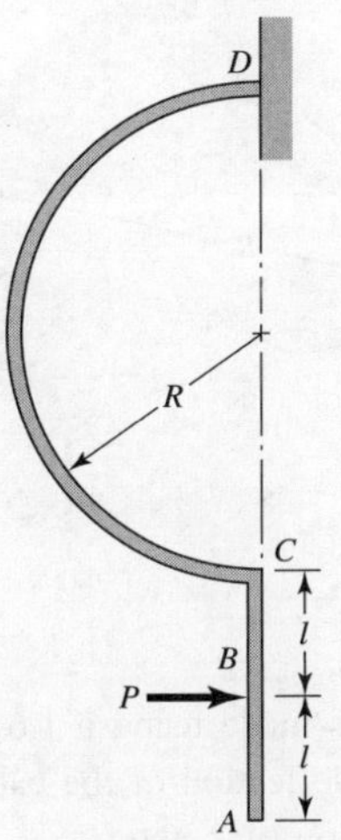

Problem 4–99

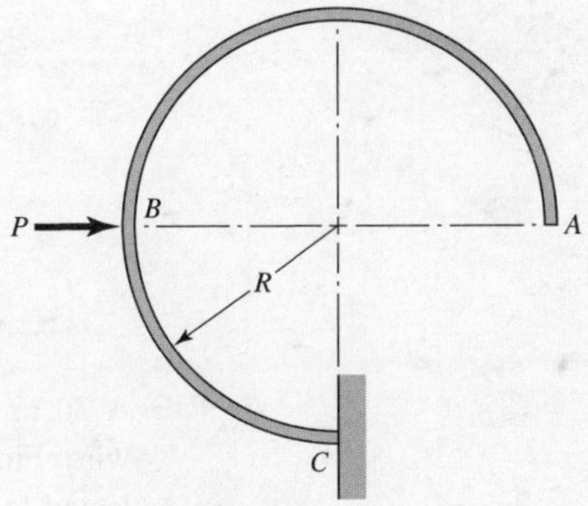

Problem 4–100

4–101 Repeat Problem 4–97 for the vertical deflection at point A.

4–102 Repeat Problem 4–98 for the vertical deflection at point A.

4–103 Repeat Problem 4–99 for the vertical deflection at point A.

4–104 Repeat Problem 4–100 for the vertical deflection at point A.

4–105 A hook is formed from a 2-mm-diameter steel wire and fixed firmly into the ceiling as shown. A 1-kg mass is hung from the hook at point D. Use Castigliano's theorem to determine the vertical deflection of point D.

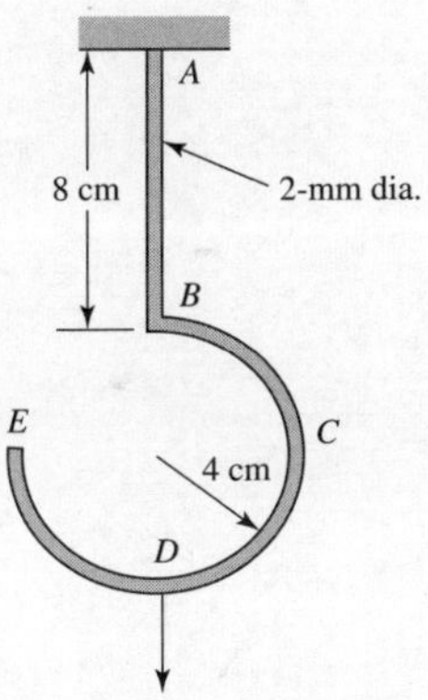

Problem 4–105

4–106 The figure shows a rectangular member OB, made from 6 mm thick aluminum plate, pinned to the ground at one end, and supported by a 12 mm diameter round steel rod that is formed into an arc and pinned to the ground at C. A load of 500 N is applied at B. Use Castigliano's theorem to determine the vertical deflection at point B. Justify any choices to neglect any components of strain energy.

Problem 4–106

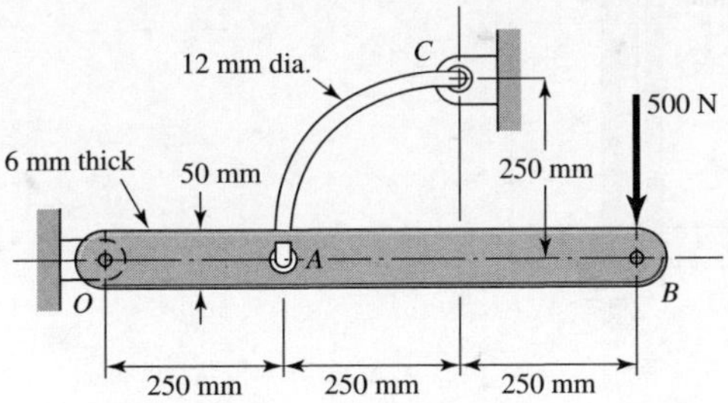

4–107 Repeat Problem 4–106 for the vertical deflection at point A.

4–108 For the wire form shown, determine the deflection of point A in the y direction. Assume $R/h > 10$ and consider the effects of bending and torsion only. The wire is steel with $E = 200$ GPa, $\nu = 0.29$, and has a diameter of 6 mm. Before application of the 250-N force the wire form is in the xz plane where the radius R is 80 mm.

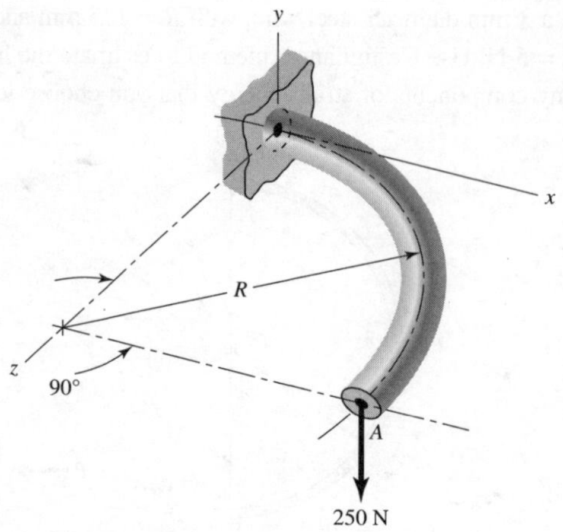

Problem 4–108

4–109 A 30 m cable is made using a 1.6 mm steel wire and three strands of 2 mm copper wire. Find the deflection of the cable and the stress in each wire if the cable is subjected to a tension of 2 kN.

4–110 The figure shows a steel pressure cylinder of diameter 100 mm that uses six SAE grade 5 steel bolts having a grip of 300 mm. These bolts have a proof strength (see Chapter 8) of 580 MPa. Suppose the bolts are tightened to 75 percent of this strength.

(*a*) Find the tensile stress in the bolts and the compressive stress in the cylinder walls.

(*b*) Repeat part (*a*), but assume now that a fluid under a pressure of 4 MPa is introduced into the cylinder.

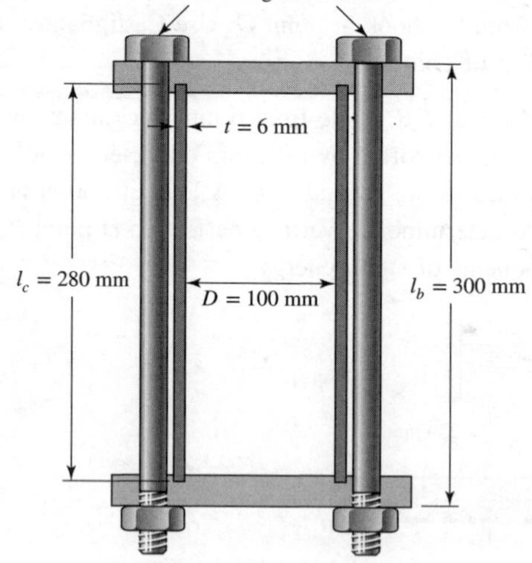

Problem 4–110

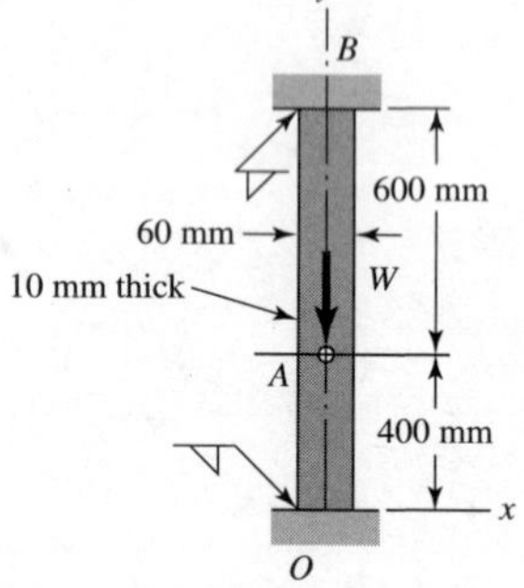

Problem 4–112

4–111 A torsion bar of length L consists of a round core of stiffness $(GJ)_c$ and a shell of stiffness $(GJ)_s$. If a torque T is applied to this composite bar, what percentage of the total torque is carried by the shell?

4–112 A rectangular aluminum bar 10 mm thick and 60 mm wide is welded to fixed supports at the ends, and the bar supports a load $W = 4$ kN, acting through a pin as shown. Find the reactions at the supports and the deflection of point A.

4–113 Solve Problem 4–112 using Castigliano's method and procedure 1 from Section 4–10.

4–114 An aluminum step bar is loaded as shown. (*a*) Verify that end *C* deflects to the rigid wall, and (*b*) determine the wall reaction forces, the stresses in each member, and the deflection of *B*.

Problem 4–114
(Not drawn to scale)

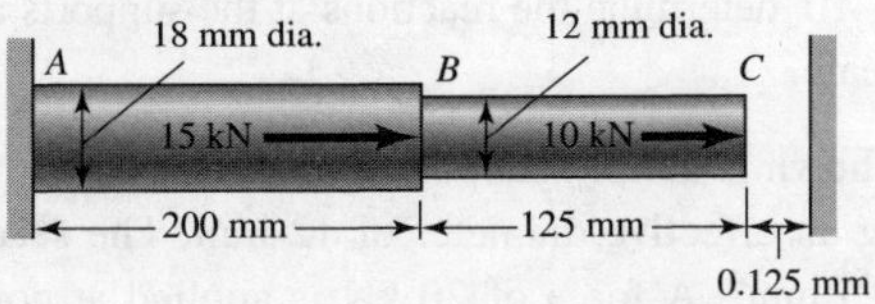

4–115 The steel shaft shown in the figure is subjected to a torque of 5 N · m applied at point *A*. Find the torque reactions at *O* and *B*; the angle of twist at *A*, in degrees; and the shear stress in sections *OA* and *AB*.

Problem 4–115

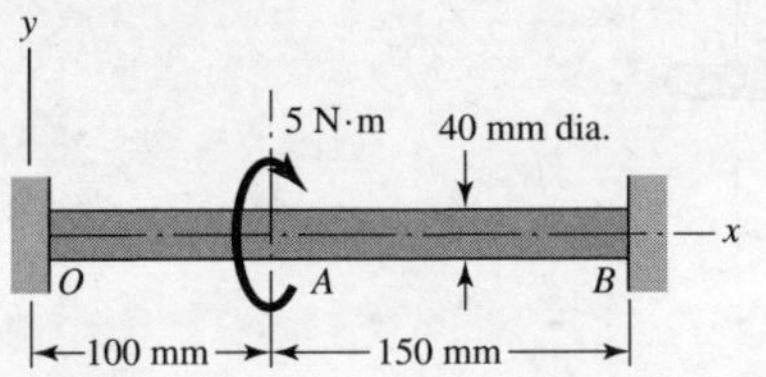

4–116 Repeat Problem 4–115 with the diameters of section *OA* being 12 mm and section *AB* being 18 mm.

4–117 The figure shows a 10 by 40 mm rectangular steel bar welded to fixed supports at each end. The bar is axially loaded by the forces $F_A = 60$ kN and $F_B = 30$ kN acting on pins at *A* and *B*. Assuming that the bar will not buckle laterally, find the reactions at the fixed supports, the stress in section *AB*, and the deflection of point *A*. Use procedure 1 from Section 4–10.

Problem 4–117

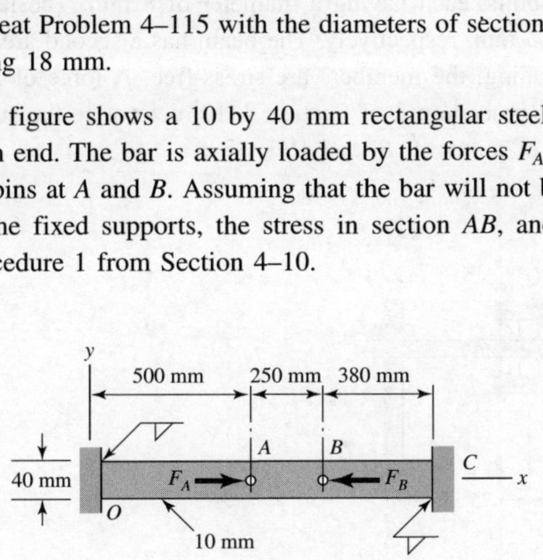

4–118 For the beam shown, determine the support reactions using superposition and procedure 1 from Section 4–10.

Problem 4–118

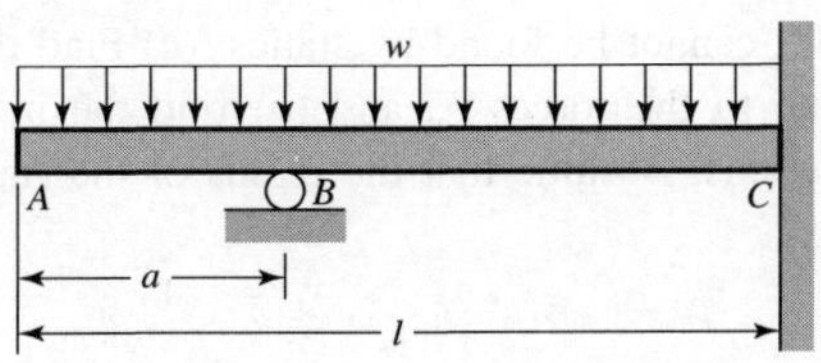

4–119 Solve Problem 4–118 using Castigliano's theorem and procedure 1 from Section 4–10.

4–120 Consider beam 13 in Table A–9, but with flexible supports. Let $w = 7$ kN/m, $l = 0.6$ m, $E = 207$ GPa, and $I = 330(10^3)$ mm^4. The support at the left end has a translational spring constant of $k_1 = 255(10^6)$ N/m and a rotational spring constant of $k_2 = 280$ kN · m. The right support has a translational spring constant of $k_3 = 340(10^6)$ N/m. Using procedure 2 of Section 4–10, determine the reactions at the supports and the deflection at the midpoint of the beam.

4–121 The steel beam *ABCD* shown is simply supported at *A* and supported at *B* and *D* by steel cables, each having an effective diameter of 12 mm. The second area moment of the beam is $I = 8(10^5)$ mm^4. A force of 20 kN is applied at point *C*. Using procedure 2 of Section 4–10 determine the stresses in the cables and the deflections of *B*, *C*, and *D*.

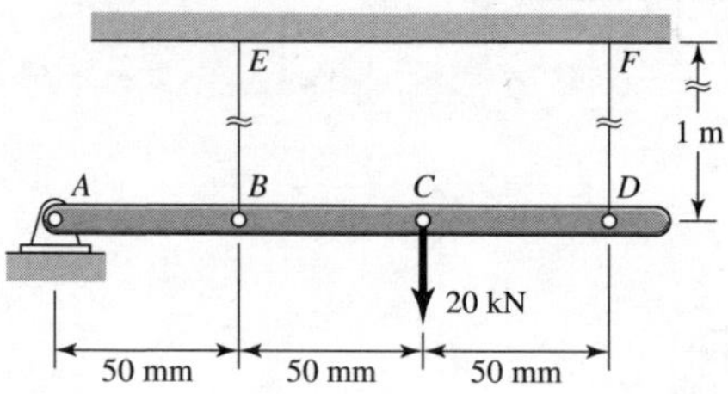

Problem 4–121

4–122 The steel beam *ABCD* shown is simply supported at *C* as shown and supported at *B* and *D* by shoulder steel bolts, each having a diameter of 8 mm. The lengths of *BE* and *DF* are 50 mm and 65 mm, respectively. The beam has a second area moment of $21(10^3)$ mm^4. Prior to loading, the members are stress-free. A force of 2 kN is then applied at point *A*. Using procedure 2 of Section 4–10, determine the stresses in the bolts and the deflections of points *A*, *B*, and *D*.

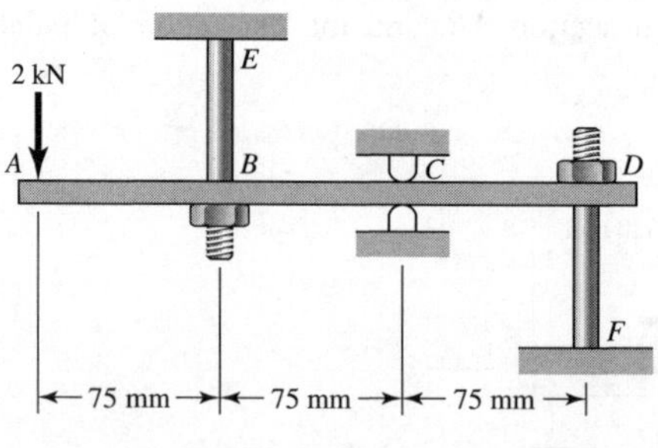

Problem 4–122

4–123 Repeat Example 4–15 except consider the ground support at point *B* to be elastic with a spring constant of $k_B = 40$ MN/m.

4–124 A thin ring is loaded by two equal and opposite forces *F* in part *a* of the figure. A free-body diagram of one quadrant is shown in part *b*. This is a statically indeterminate problem, because the moment M_A cannot be found by statics. (*a*) Find the maximum bending moment in the ring due to the forces *F*, and (*b*) find the increase in the diameter of the ring along the *y* axis. Assume that the radius of the ring is large so that Equation (4–41) can be used.

Problem 4–124

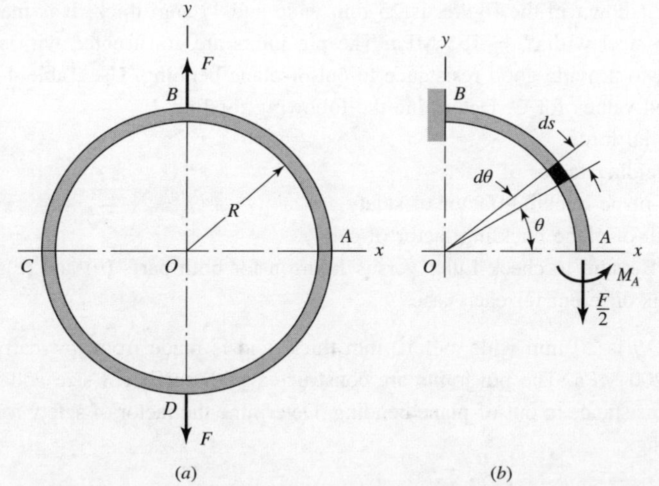

4–125 A round tubular column has outside and inside diameters of D and d, respectively, and a diametral ratio of $K = d/D$. Show that buckling will occur when the outside diameter is

$$D = \left[\frac{64P_{cr}l^2}{\pi^3 CE(1 - K^4)}\right]^{1/4}$$

4–126 For the conditions of Problem 4–125, show that buckling according to the parabolic formula will occur when the outside diameter is

$$D = 2\left[\frac{P_{cr}}{\pi S_y(1 - K^2)} + \frac{S_y l^2}{\pi^2 CE(1 + K^2)}\right]^{1/2}$$

4–127 Link 2, shown in the figure, is 25 mm wide, has 12-mm-diameter bearings at the ends, and is cut from low-carbon steel bar stock having a minimum yield strength of 165 MPa. The end-condition constants are $C = 1$ and $C = 1.2$ for buckling in and out of the plane of the drawing, respectively.

(*a*) Using a design factor $n_d = 4$, find a suitable thickness for the link.

(*b*) Are the bearing stresses at O and B of any significance?

Problems 4–127 and 4–128

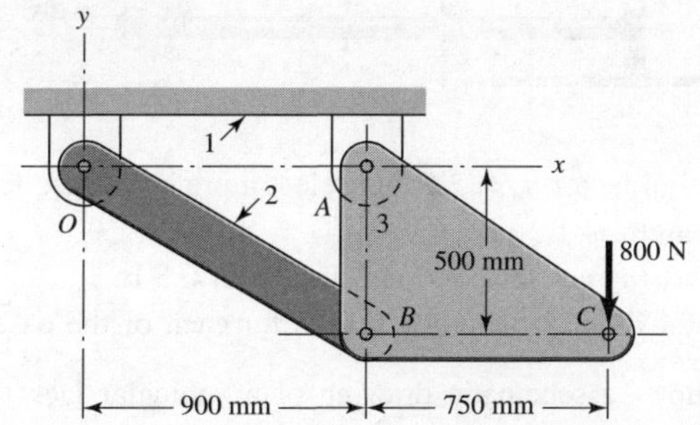

4–128 Link 2, shown in the figure, is 25 mm wide and 11 mm thick. It is made from low-carbon steel with S_y = 165 MPa. The pin joints are constructed with sufficient size and fit to provide good resistance to out-of-plane bending. Use Table 4–3 for recommended values for C. Determine the following for link 2.

(*a*) Axial force

(*b*) Yielding factor of safety

(*c*) In-plane buckling factor of safety

(*d*) Out-of-plane buckling factor of safety

Hint: Be sure to check Euler versus Johnson for both parts (*b*) and (*c*), as the $(l/k)_1$ point is different for each case.

4–129 Link OB is 20 mm wide and 10 mm thick, and is made from low-carbon steel with S_y = 200 MPa. The pin joints are constructed with sufficient size and fit to provide good resistance to out-of-plane bending. Determine the factor of safety for out-of-plane buckling.

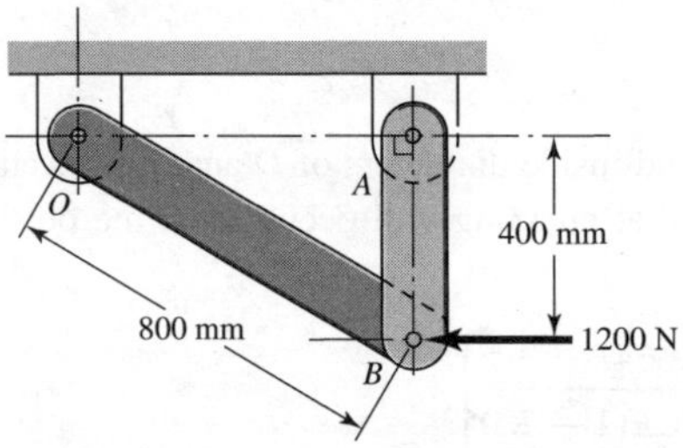

Problem 4–129

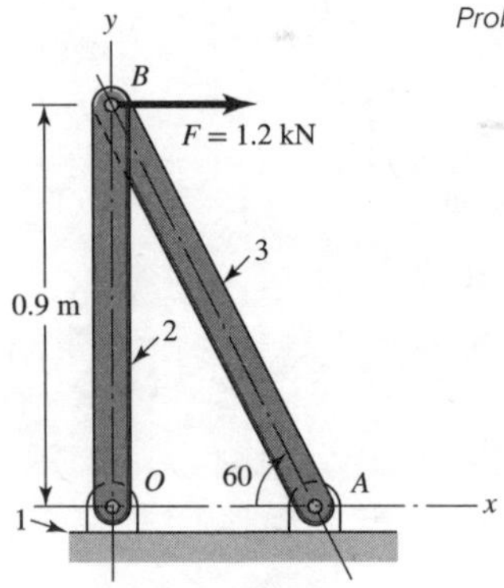

Problem 4–130

4–130 Link 3, shown schematically in the figure, acts as a brace to support the 1.2kN load. For buckling in the plane of the figure, the link may be regarded as pinned at both ends. For out-of-plane buckling, the ends are fixed. Select a suitable material and a method of manufacture, such as forging, casting, stamping, or machining, for casual applications of the brace in oil-field machinery. Specify the dimensions of the cross section as well as the ends so as to obtain a strong, safe, well-made, and economical brace.

4–131 The hydraulic cylinder shown in the figure has a 50 mm bore and is to operate at a pressure of 10 MPa. With the clevis mount shown, the piston rod should be sized as a column with both ends rounded for any plane of buckling. The rod is to be made of forged AISI 1030 steel without further heat treatment.

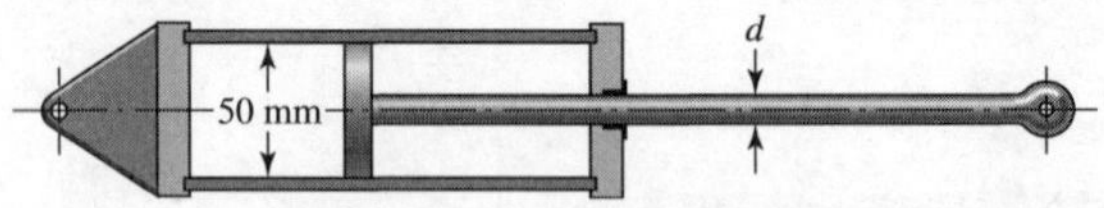

Problem 4–131

(*a*) Use a design factor n_d = 2.5 and select a preferred size for the rod diameter if the column length is 1.2 m.

(*b*) Repeat part (*a*) but for a column length of 0.5 m.

(*c*) What factor of safety actually results for each of the cases above?

4–132 The figure shows a schematic drawing of a vehicular jack that is to be designed to support a maximum mass of 300 kg based on the use of a design factor n_d = 3.50. The opposite-handed threads on the two ends of the screw are cut to allow the link angle θ to vary from 15 to 70°. The links are to be machined from AISI 1010 hot-rolled steel bars. Each of the four links is to consist of two bars, one on each side of the central bearings. The bars are to be 350 mm long and have a bar width of $w = 30$ mm. The pinned ends are to be designed to secure an end-condition constant of at least $C = 1.4$ for out-of-plane buckling. Find a suitable preferred thickness and the resulting factor of safety for this thickness.

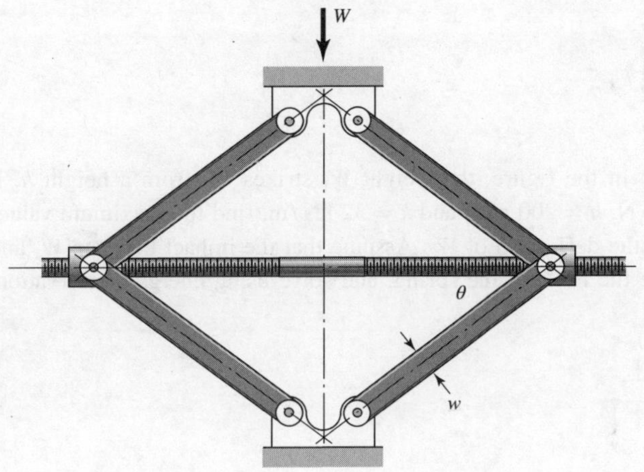

Problem 4–132

4–133 If drawn, a figure for this problem would resemble that for Problem 4–110. A strut that is a standard hollow right circular cylinder has an outside diameter of 100 mm and a wall thickness of 10 mm and is compressed between two circular end plates held by four bolts equally spaced on a bolt circle of 140 mm diameter. All four bolts are hand-tightened, and then bolt A is tightened to a tension of 10 kN and bolt C, diagonally opposite, is tightened to a tension of 50 kN. The strut axis of symmetry is coincident with the center of the bolt circles. Find the maximum compressive load, the eccentricity of loading, and the largest compressive stress in the strut.

4–134 Design link CD of the hand-operated toggle press shown in the figure. Specify the cross-section dimensions, the bearing size and rod-end dimensions, the material, and the method of processing.

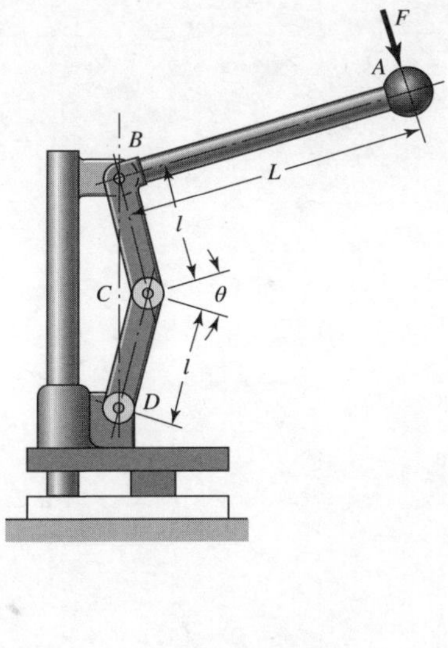

Problem 4–134
$L = 300$ mm, $l = 100$ mm, $\theta_{min} = 0°$.

4–135 Find the maximum values of the spring force and deflection of the impact system shown in the figure if $W = 150$ N, $k = 17$ kN/m, and $h = 50$ mm. Ignore the mass of the spring and solve using energy conservation.

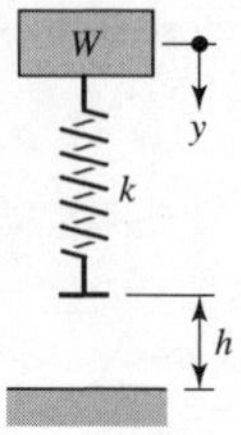

Problem 4–135

4–136 As shown in the figure, the weight W_1 strikes W_2 from a height h. If $W_1 = 40$ N, $W_2 = 400$ N, $h = 200$ mm, and $k = 32$ kN/m, find the maximum values of the spring force and the deflection of W_2. Assume that the impact between W_1 and W_2 is *inelastic,* ignore the mass of the spring, and solve using energy conservation.

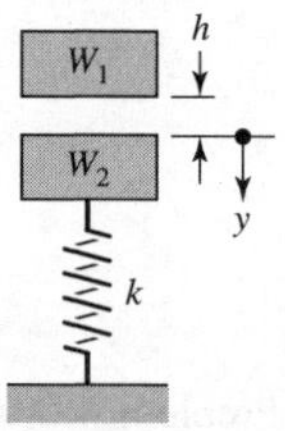

Problem 4–136

4–137 Part a of the figure shows a weight W mounted between two springs. If the free end of spring k_1 is suddenly displaced through the distance $x = a$, as shown in part b, determine the maximum displacement y of the weight. Let $W = 20$ N, $k_1 = 1.7$ kN/m, $k_2 = 3.4$ kN/m, and $a = 6$ mm. Ignore the mass of each spring and solve using energy conservation.

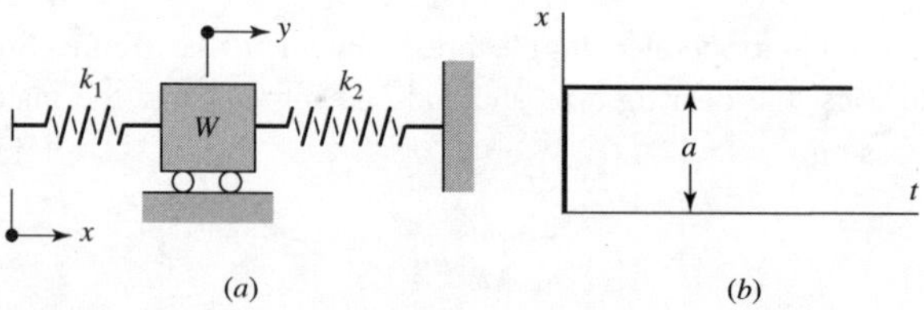

Problem 4–137

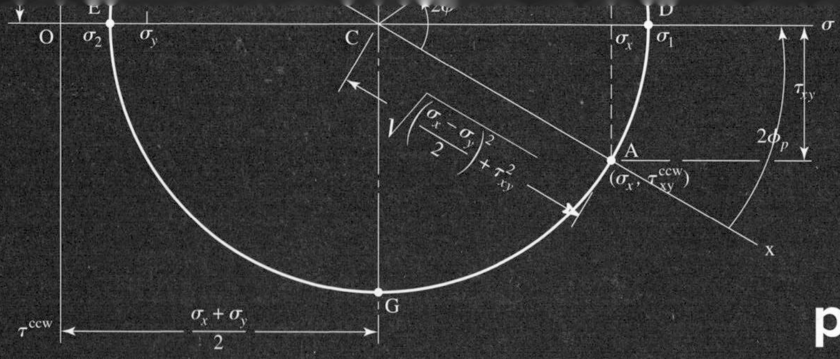

part 2

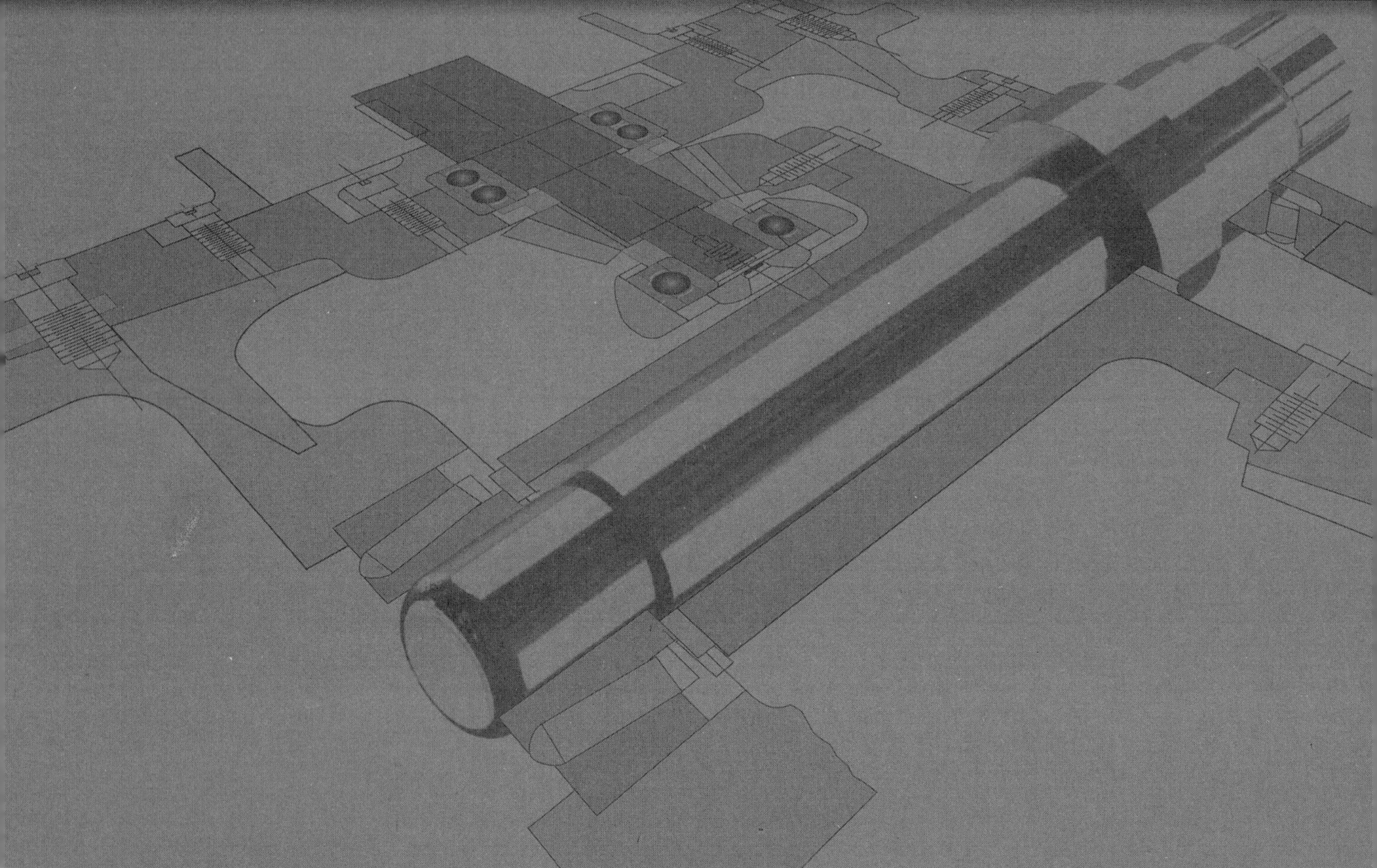

Courtesy of Dee Dehokenanan

Failure Prevention

5 Failures Resulting from Static Loading

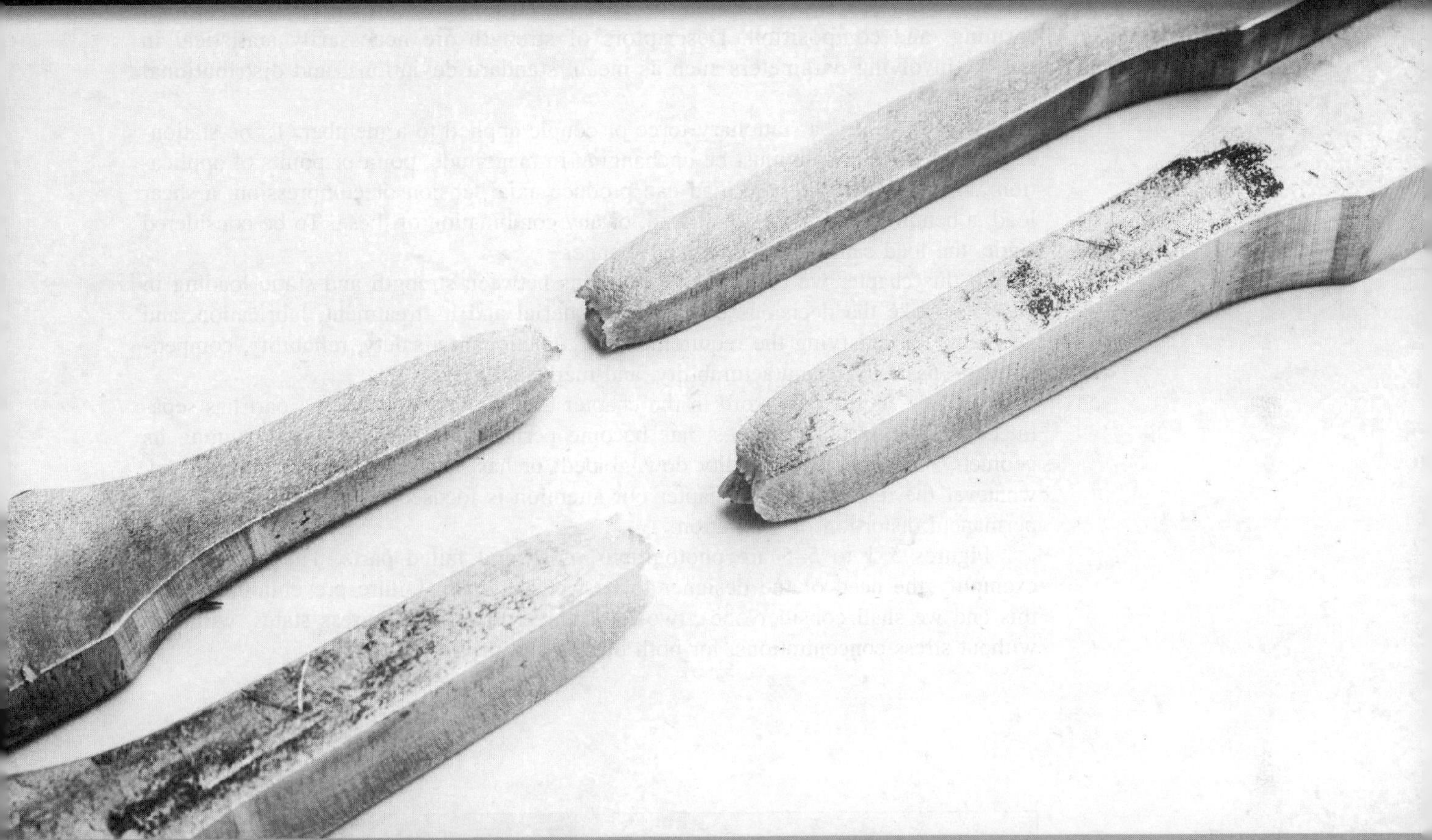

arhendrix/123RF

Chapter Outline

In Chapter 1 we learned that *strength is a property or characteristic of a mechanical element*. This property results from the material identity, the treatment and processing incidental to creating its geometry, and the loading, and it is at the controlling or critical location.

In addition to considering the strength of a single part, we must be cognizant that the strength of a mass-produced part will be somewhat different from the others in the collection or ensemble because of variations in dimensions, machining, forming, and composition. Descriptors of strength are necessarily statistical in nature, involving parameters such as mean, standard deviations, and distributional identification.

A *static load* is a stationary force or couple applied to a member. To be stationary, the force or couple must be unchanging in magnitude, point or points of application, and direction. A static load can produce axial tension or compression, a shear load, a bending load, a torsional load, or any combination of these. To be considered static, the load cannot change in any manner.

In this chapter we consider the relations between strength and static loading in order to make the decisions concerning material and its treatment, fabrication, and geometry for satisfying the requirements of functionality, safety, reliability, competitiveness, usability, manufacturability, and marketability.

"Failure" is the first word in the chapter title. Failure can mean a part has separated into two or more pieces; has become permanently distorted, thus ruining its geometry; has had its reliability downgraded; or has had its function compromised, whatever the reason. In this chapter our attention is focused on the predictability of permanent distortion or separation.

Figures 5–1 to 5–5 are photographs of several failed parts. The photographs exemplify the need of the designer to be well-versed in failure prevention. Toward this end we shall consider one-, two-, and three-dimensional stress states, with and without stress concentrations, for both ductile and brittle materials.

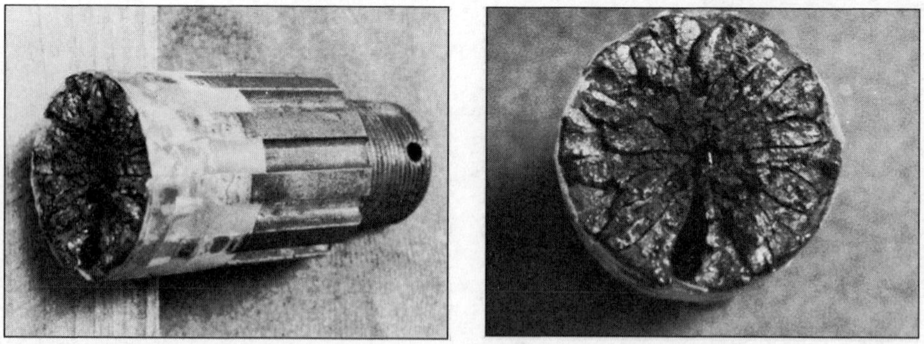

(*a*) (*b*)

Figure 5–1

(*a*) Failure of a truck drive-shaft spline due to corrosion fatigue. Note that it was necessary to use clear tape to hold the pieces in place. (*b*) Direct end view of failure. *(For permission to reprint Figs. 5-1 through 5-5, the authors are grateful for the personal photographs of Larry D. Mitchell, co-author of* Mechanical Engineering Design, *4th ed., McGraw Hill, New York, 1983.)*

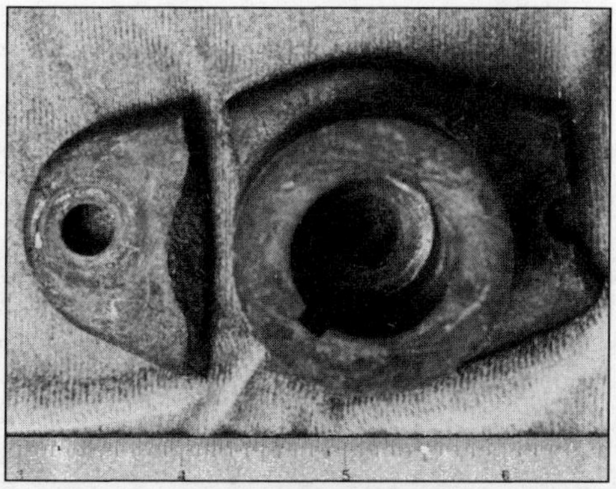

Figure 5–2

Impact failure of a lawn-mower blade driver hub. The blade impacted a surveying pipe marker. *(For permission to reprint Figs. 5-1 through 5-5, the authors are grateful for the personal photographs of Larry D. Mitchell, co-author of* Mechanical Engineering Design, *4th ed., McGraw Hill, New York, 1983.)*

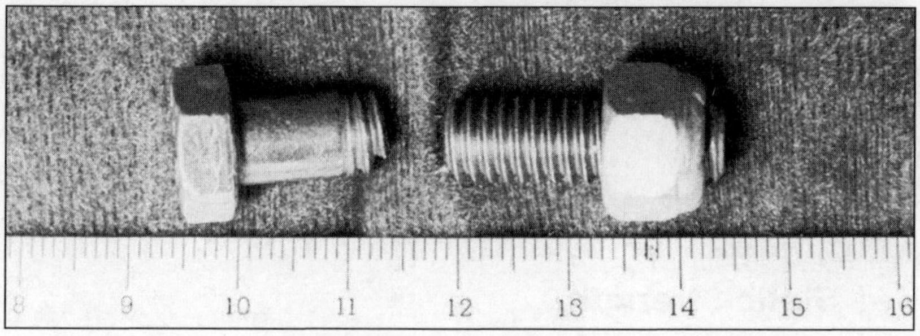

Figure 5–3

Failure of an overhead-pulley retaining bolt on a weightlifting machine. A manufacturing error caused a gap that forced the bolt to take the entire moment load. *(For permission to reprint Figs. 5-1 through 5-5, the authors are grateful for the personal photographs of Larry D. Mitchell, co-author of* Mechanical Engineering Design, *4th ed., McGraw Hill, New York, 1983.)*

(*a*) (*b*)

Figure 5–4

Chain test fixture that failed in one cycle. To alleviate complaints of excessive wear, the manufacturer decided to case-harden the material. (*a*) Two halves showing fracture; this is an excellent example of brittle fracture initiated by stress concentration. (*b*) Enlarged view of one portion to show cracks induced by stress concentration at the support-pin holes. *(For permission to reprint Figs. 5-1 through 5-5, the authors are grateful for the personal photographs of Larry D. Mitchell, co-author of* Mechanical Engineering Design, *4th ed., McGraw Hill, New York, 1983.)*

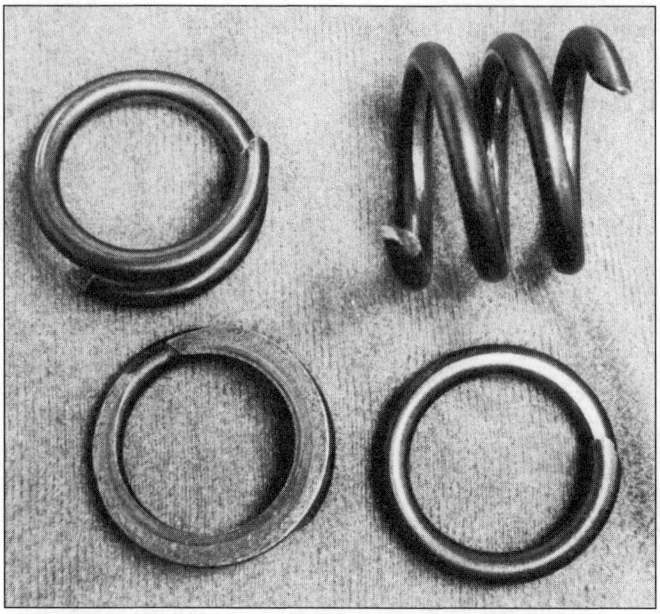

Figure 5–5

Valve-spring failure caused by spring surge in an oversped engine. The fractures exhibit the classic 45° shear failure. *(For permission to reprint Figs. 5-1 through 5-5, the authors are grateful for the personal photographs of Larry D. Mitchell, co-author of* Mechanical Engineering Design, *4th ed., McGraw Hill, New York, 1983.)*

5–1 Static Strength

Ideally, in designing any machine element, the engineer should have available the results of a great many strength tests of the particular material chosen. These tests should be made on specimens having the same heat treatment, surface finish, and size as the element the engineer proposes to design; and the tests should be made under exactly the same loading conditions as the part will experience in service. This means that if the part is to experience a bending load, it should be tested with a bending load. If it is to be subjected to combined bending and torsion, it should be tested under combined bending and torsion. If it is made of heat-treated AISI 1040 steel drawn at 500°C with a ground finish, the specimens tested should be of the same material prepared in the same manner. Such tests will provide very useful and precise information. Whenever such data are available for design purposes, the engineer can be assured of doing the best possible job of engineering.

The cost of gathering such extensive data prior to design is justified if failure of the part may endanger human life or if the part is manufactured in sufficiently large quantities. Refrigerators and other appliances, for example, have very good reliabilities because the parts are made in such large quantities that they can be thoroughly tested in advance of manufacture. The cost of making these tests is very low when it is divided by the total number of parts manufactured.

You can now appreciate the following four design categories:

1 Failure of the part would endanger human life, or the part is made in extremely large quantities; consequently, an elaborate testing program is justified during design.

2 The part is made in large enough quantities that a moderate series of tests is feasible.

3 The part is made in such small quantities that testing is not justified at all; or the design must be completed so rapidly that there is not enough time for testing.

4 The part has already been designed, manufactured, and tested and found to be unsatisfactory. Analysis is required to understand why the part is unsatisfactory and what to do to improve it.

More often than not it is necessary to design using only published values of yield strength, ultimate strength, percentage reduction in area, and percentage elongation, such as those listed in Appendix A. How can one use such meager data to design against both static and dynamic loads, two- and three-dimensional stress states, high and low temperatures, and very large and very small parts? These and similar questions will be addressed in this chapter and those to follow.

5–2 Stress Concentration

Stress concentration (see Section 3–13) is a highly localized effect. In some instances it may be due to a surface scratch. If the material is ductile and the load static, the design load may cause yielding in the critical location in the notch. This yielding can involve strain strengthening of the material and an increase in yield strength at the small critical notch location. Since the loads are static and the material is ductile, that part can carry the loads satisfactorily with no general yielding. In these cases the designer sets the geometric (theoretical) stress-concentration factor K_t to unity.

The rationale can be expressed as follows. The worst-case scenario is that of an idealized non–strain-strengthening material shown in Figure 5–6. The stress-strain curve rises linearly to the yield strength S_y, then proceeds at constant stress, which is equal to S_y. Consider a filleted rectangular bar as depicted in Figure A–13–5, where the cross-section area of the small shank is 645 mm^2. If the material is ductile, with a yield point of 280 MPa, and the theoretical *stress-concentration factor* (SCF) K_t is 2,

- A load of 90 kN induces a nominal tensile stress of 140 MPa in the shank as depicted at point *A* in Figure 5–6. At the critical location in the fillet the stress is 280 MPa, and the SCF is $K = \sigma_{\text{max}}/\sigma_{\text{nom}} = 280/140 = 2$.
- A load of 135 kN induces a nominal tensile stress of 210 MPa in the shank at point *B*. The fillet stress is still 280 MPa (point *D*), and the SCF $K = \sigma_{\text{max}}/\sigma_{\text{nom}} = S_y/\sigma = 280/210 = 1.33$.
- At a load of 180 kN the induced tensile stress (point *C*) is 280 MPa in the shank. At the critical location in the fillet, the stress (at point *E*) is 280 MPa. The SCF $K = \sigma_{\text{max}}/\sigma_{\text{nom}} = S_y/\sigma = 280/280 = 1$.

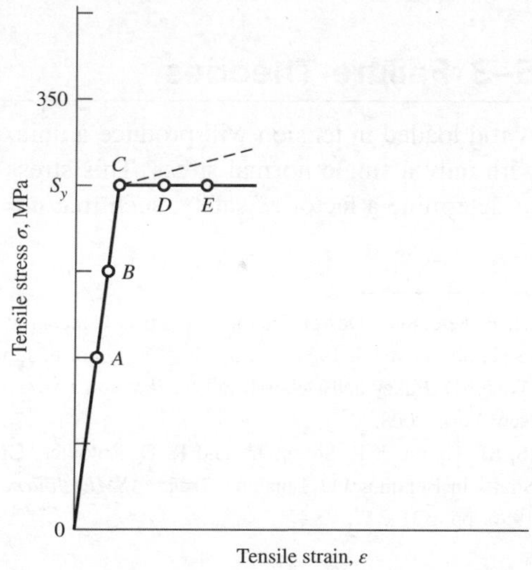

Figure 5–6

An idealized stress-strain curve. The dashed line depicts a strain-strengthening material.

For materials that strain-strengthen, the critical location in the notch has a higher S_y. The shank area is at a stress level a little below 280 MPa, is carrying load, and is very near its failure-by-general-yielding condition. This is the reason designers do not apply K_t in *static loading* of a *ductile material* loaded elastically, instead setting $K_t = 1$.

When using this rule for ductile materials with static loads, be careful to assure yourself that the material is not susceptible to brittle fracture (see Section 5–12) in the environment of use. The usual definition of geometric (theoretical) stress-concentration factor for normal stress K_t and shear stress K_{ts} is given by Equation pair (3–46) as

$$\sigma_{\max} = K_t \sigma_{\text{nom}} \tag{a}$$

$$\tau_{\max} = K_{ts} \tau_{\text{nom}} \tag{b}$$

As shown in Figure 2–2*b*, brittle materials do not exhibit a plastic range. The stress-concentration factor given by Equation (*a*) or (*b*) could raise the stress to a level to cause fracture to initiate at the stress raiser, and initiate a catastrophic failure of the member.

An exception to this rule is a brittle material that inherently contains microdiscontinuity stress concentration, worse than the macrodiscontinuity that the designer has in mind. Sand molding introduces sand particles, air, and water vapor bubbles. The grain structure of cast iron contains graphite flakes (with little strength), which are literally cracks introduced during the solidification process. When a tensile test on a cast iron is performed, the strength reported in the literature *includes* this stress concentration. In such cases K_t or K_{ts} need not be applied.

An important source of stress-concentration factors is R. E. Peterson, who compiled them from his own work and that of others.[1] Peterson developed the style of presentation in which the stress-concentration factor K_t is multiplied by the nominal stress σ_{nom} to estimate the magnitude of the largest stress in the locality. His approximations were based on photoelastic studies of two-dimensional strips (Hartman and Levan, 1951; Wilson and White, 1973), with some limited data from three-dimensional photoelastic tests of Hartman and Levan. A contoured graph was included in the presentation of each case. Filleted shafts in tension were based on two-dimensional strips. Table A–13 provides many charts for the theoretical stress-concentration factors for several fundamental load conditions and geometry. Additional charts are also available from Peterson.[2]

Finite element analysis (FEA) can also be applied to obtain stress-concentration factors. Improvements on K_t and K_{ts} for filleted shafts were reported by Tipton, Sorem, and Rolovic.[3]

5–3 Failure Theories

A rod loaded in tension will produce a uniaxial stress element, that is, a stress element with only a single normal stress. This stress can be compared directly with a strength to determine a factor of safety, such that n = Strength/Stress. But, a more complicated

[1]R. E. Peterson, "Design Factors for Stress Concentration," *Machine Design*, vol. 23, no. 2, February 1951; no. 3, March 1951; no. 5, May 1951; no. 6, June 1951; no. 7, July 1951.

[2]Walter D. Pilkey and Deborah Pilkey, *Peterson's Stress-Concentration Factors*, 3rd ed., John Wiley & Sons, New York, 2008.

[3]S. M. Tipton, J. R. Sorem Jr., and R. D. Rolovic, "Updated Stress-Concentration Factors for Filleted Shafts in Bending and Tension," *Trans. ASME, Journal of Mechanical Design*, vol. 118, September 1996, pp. 321–327.

loading situation will produce a multi-axial stress element, such as shown in Figure 3–8, with multiple normal and shear stresses acting on the faces of the stress element. It is not immediately clear how to compare this stress situation to a single value of strength, such as a yield strength, that was obtained from a simple tension test. Similarly, how do we compare one stress element to another to determine which is critical, that is, closer to failure? Various theories may be proposed as to how to do such comparisons. For example, the stress element could be oriented into its principal stress directions, eliminating the shear stresses. Then, a theory could be proposed that the largest of the three principal stresses is to be compared to the yield strength to obtain a factor of safety for yielding of the stress element. This theory could be compared with experimental testing to determine if it is a good representation of what happens in reality. In fact, it does not match the experimental data for many stress states.

Several failure theories have been proposed and examined and found to be useful to the designer. Usually, each theory is based on some hypothesis of what aspect of the stress state is critical to causing failure. It is interesting to note that we really don't know the answer to this. This is why multiple theories have been proposed. Due to the inherent scatter of experimental data, it is difficult to say that a given theory is "correct." We instead take note of the usefulness of a theory due to how it fits within the scatter of the experimental data.

The following theories have generally been accepted as useful and will be presented in the following sections.

Ductile materials (yield criteria)

- Maximum shear stress (MSS), Section 5–4
- Distortion energy (DE), Section 5–5
- Ductile Coulomb-Mohr (DCM), Section 5–6

Brittle materials (fracture criteria)

- Maximum normal stress (MNS), Section 5–8
- Brittle Coulomb-Mohr (BCM), Section 5–9
- Modified Mohr (MM), Section 5–9

Structural metal behavior is typically classified as being ductile or brittle, although under special situations, a material normally considered ductile can fail in a brittle manner (see Section 5–12). Ductile materials are normally classified such that $\varepsilon_f \geq 0.05$ and have an identifiable yield strength that is often the same in compression as in tension ($S_{yt} = S_{yc} = S_y$). Brittle materials, $\varepsilon_f < 0.05$, do not exhibit an identifiable yield strength, and are typically classified by ultimate tensile and compressive strengths, S_{ut} and S_{uc}, respectively (where S_{uc} is given as a positive quantity).

5–4 Maximum-Shear-Stress Theory for Ductile Materials

The *maximum-shear-stress* (MSS) *theory* predicts that *yielding begins when the maximum shear stress in any stress element equals or exceeds the maximum shear stress in a tension-test specimen of the same material when that specimen begins to yield.* The MSS theory is also referred to as the *Tresca* or *Guest theory*.

Many theories are postulated on the basis of the consequences seen from tensile tests. As a strip of a ductile material is subjected to tension, slip lines (called *Lüders lines*) form at approximately 45° with the axis of the strip. These slip lines are the beginning of yield, and when loaded to fracture, fracture lines are also seen at angles approximately 45° with the axis of tension. Since the shear stress is maximum at 45° from the axis of tension, it makes sense to think that this is the mechanism of failure. It will be shown in the next section that there is a little more going on than this. However, it turns out the MSS theory is an acceptable but conservative predictor of failure; and since engineering design often benefits from being conservative, it is quite often used.

Recall that for simple tensile stress, $\sigma = P/A$, and the maximum shear stress occurs on a surface 45° from the tensile surface with a magnitude of $\tau_{max} = \sigma/2$. So the maximum shear stress at yield is $\tau_{max} = S_y/2$. For a general state of stress, three principal stresses can be determined and ordered such that $\sigma_1 \geq \sigma_2 \geq \sigma_3$. The maximum shear stress is then $\tau_{max} = (\sigma_1 - \sigma_3)/2$ (see Figure 3–14). Thus, for a general state of stress, the maximum-shear-stress theory predicts yielding when

$$\tau_{max} = \frac{\sigma_1 - \sigma_3}{2} \geq \frac{S_y}{2} \quad \text{or} \quad \sigma_1 - \sigma_3 \geq S_y \tag{5–1}$$

Note that this implies that the yield strength in shear is given by

$$S_{sy} = 0.5S_y \tag{5–2}$$

which, as we will see later is about 15 percent low (conservative).

For design purposes, Equation (5–1) can be modified to incorporate a factor of safety, n. Thus,

$$\tau_{max} = \frac{S_y}{2n} \quad \text{or} \quad \sigma_1 - \sigma_3 = \frac{S_y}{n} \tag{5–3}$$

Plane stress is a very common state of stress in design. However, it is extremely important to realize that plane stress is a *three-dimensional* state of stress. Plane stress transformations in Section 3–6 are restricted to the in-plane stresses only, where the in-plane principal stresses are given by Equation (3–14) and labeled as σ_1 and σ_2. It is true that these are the principal stresses in the *plane of analysis,* but out of plane there is a third principal stress and it is *always zero* for plane stress. This means that if we are going to use the convention of ordering $\sigma_1 \geq \sigma_2 \geq \sigma_3$ for three-dimensional analysis, upon which Equation (5–1) is based, we cannot arbitrarily call the in-plane principal stresses σ_1 and σ_2 until we relate them with the third principal stress of zero. To illustrate the MSS theory graphically for plane stress, we will first label the principal stresses given by Equation (3–14) as σ_A and σ_B, and then order them with the zero principal stress according to the convention $\sigma_1 \geq \sigma_2 \geq \sigma_3$. Assuming that $\sigma_A \geq \sigma_B$, there are three cases to consider when using Equation (5–1) for plane stress:

Case 1: $\sigma_A \geq \sigma_B \geq 0$. For this case, $\sigma_1 = \sigma_A$ and $\sigma_3 = 0$. Equation (5–1) reduces to a yield condition of

$$\sigma_A \geq S_y \tag{5–4}$$

Case 2: $\sigma_A \geq 0 \geq \sigma_B$. Here, $\sigma_1 = \sigma_A$ and $\sigma_3 = \sigma_B$, and Equation (5–1) becomes

$$\sigma_A - \sigma_B \geq S_y \tag{5–5}$$

Case 3: $0 \geq \sigma_A \geq \sigma_B$. For this case, $\sigma_1 = 0$ and $\sigma_3 = \sigma_B$, and Equation (5–1) gives

$$\sigma_B \leq -S_y \tag{5–6}$$

Equations (5–4) to (5–6) are represented in Figure 5–7 by the three lines indicated in the σ_A, σ_B plane. The remaining unmarked lines are cases for $\sigma_B \geq \sigma_A$, which completes the *stress yield envelope* but are not normally used. The maximum-shear-stress theory predicts yield if a stress state is outside the shaded region bordered by the stress yield

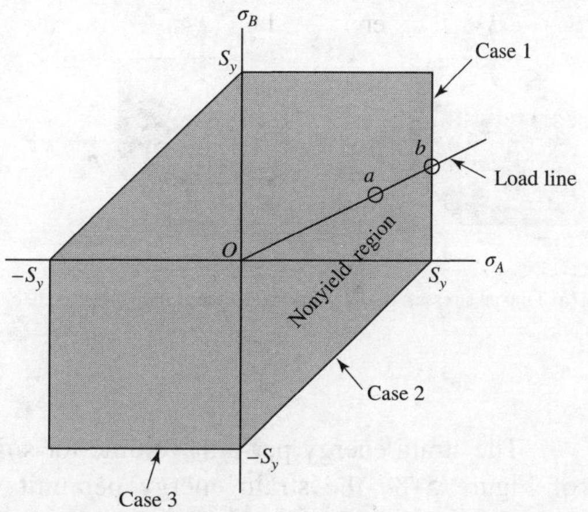

Figure 5–7

The maximum-shear-stress (MSS) theory yield envelope for plane stress, where σ_A and σ_B are the two nonzero principal stresses.

envelope. In Figure 5–7, suppose point *a* represents the stress state of a critical stress element of a member. If the load is increased, it is typical to assume that the principal stresses will increase proportionally along the line from the origin through point *a*. Such a *load line* is shown. If the stress situation increases along the load line until it crosses the stress failure envelope, such as at point *b*, the MSS theory predicts that the stress element will yield. The factor of safety guarding against yield at point *a* is given by the ratio of strength (distance to failure at point *b*) to stress (distance to stress at point *a*), that is $n = Ob/Oa$.

Note that the first part of Equation (5–3), $\tau_{max} = S_y/2n$, is sufficient for design purposes provided the designer is careful in determining τ_{max}. For plane stress, Equation (3–17) *does not always predict* τ_{max}. However, consider the special case when one normal stress is zero in the plane, say σ_x and τ_{xy} have values and $\sigma_y = 0$. It can be easily shown that this is a Case 2 problem, and the shear stress determined by Equation (3–17) *is* τ_{max}. Shaft design problems typically fall into this category where a normal stress exists from bending and/or axial loading, and a shear stress arises from torsion.

5–5 Distortion-Energy Theory for Ductile Materials

The *distortion-energy theory* predicts that *yielding occurs when the distortion strain energy per unit volume reaches or exceeds the distortion strain energy per unit volume for yield in simple tension or compression of the same material.*

The distortion-energy (DE) theory originated from the observation that ductile materials stressed hydrostatically (equal principal stresses) exhibited yield strengths greatly in excess of the values given by the simple tension test. Therefore it was postulated that yielding was not a simple tensile or compressive phenomenon at all, but, rather, that it was related somehow to the angular distortion of the stressed element. To develop the theory, note, in Figure 5–8*a*, the unit volume subjected to any three-dimensional stress state designated by the stresses σ_1, σ_2, and σ_3. The stress state shown in Figure 5–8*b* is one of hydrostatic normal stresses due to the stresses σ_{av} acting in each of the same principal directions as in Figure 5–8*a*. The formula for σ_{av} is simply

$$\sigma_{av} = \frac{\sigma_1 + \sigma_2 + \sigma_3}{3} \qquad (a)$$

Thus, the element in Figure 5–8*b* undergoes pure volume change, that is, no angular distortion. If we regard σ_{av} as a component of σ_1, σ_2, and σ_3, then this component can be subtracted from them, resulting in the stress state shown in Figure 5–8*c*. This element is subjected to pure angular distortion, that is, no volume change.

Figure 5–8

(*a*) Element with triaxial stresses; this element undergoes both volume change and angular distortion. (*b*) Element under hydrostatic normal stresses undergoes only volume change. (*c*) Element has angular distortion without volume change.

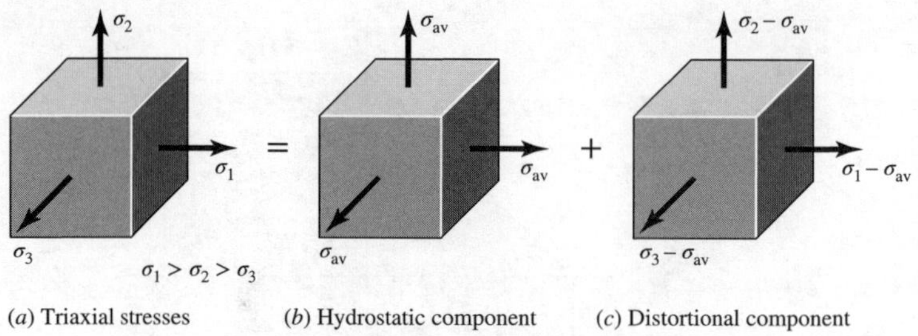

The strain energy per unit volume for simple tension is $u = \frac{1}{2}\epsilon\sigma$. For the element of Figure 5–8*a* the strain energy per unit volume is $u = \frac{1}{2}[\epsilon_1\sigma_1 + \epsilon_2\sigma_2 + \epsilon_3\sigma_3]$. Substituting Equation (3–22) for the principal strains gives

$$u = \frac{1}{2E}\left[\sigma_1^2 + \sigma_2^2 + \sigma_3^2 - 2\nu(\sigma_1\sigma_2 + \sigma_2\sigma_3 + \sigma_3\sigma_1)\right] \tag{b}$$

The strain energy for producing only volume change u_v can be obtained by substituting σ_{av} for σ_1, σ_2, and σ_3 in Equation (*b*). The result is

$$u_v = \frac{3\sigma_{av}^2}{2E}(1 - 2\nu) \tag{c}$$

If we now substitute the square of Equation (*a*) in Equation (*c*) and simplify the expression, we get

$$u_v = \frac{1 - 2\nu}{6E}\left(\sigma_1^2 + \sigma_2^2 + \sigma_3^2 + 2\sigma_1\sigma_2 + 2\sigma_2\sigma_3 + 2\sigma_3\sigma_1\right) \tag{5–7}$$

Then the distortion energy is obtained by subtracting Equation (5–7) from Equation (*b*). This gives

$$u_d = u - u_v = \frac{1 + \nu}{3E}\left[\frac{(\sigma_1 - \sigma_2)^2 + (\sigma_2 - \sigma_3)^2 + (\sigma_3 - \sigma_1)^2}{2}\right] \tag{5–8}$$

Note that the distortion energy is zero if $\sigma_1 = \sigma_2 = \sigma_3$.

For the simple tensile test, at yield, $\sigma_1 = S_y$ and $\sigma_2 = \sigma_3 = 0$, and from Equation (5–8) the distortion energy is

$$u_d = \frac{1 + \nu}{3E} S_y^2 \tag{5–9}$$

So for the general state of stress given by Equation (5–8), yield is predicted if Equation (5–8) equals or exceeds Equation (5–9). This gives

$$\left[\frac{(\sigma_1 - \sigma_2)^2 + (\sigma_2 - \sigma_3)^2 + (\sigma_3 - \sigma_1)^2}{2}\right]^{1/2} \geq S_y \tag{5–10}$$

If we had a simple case of tension σ, then yield would occur when $\sigma \geq S_y$. Thus, the left of Equation (5–10) can be thought of as a *single, equivalent,* or *effective stress* for the entire general state of stress given by σ_1, σ_2, and σ_3. This effective stress is usually called the *von Mises stress,* σ', named after Dr. R. von Mises, who contributed to the theory. Thus Equation (5–10), for yield, can be written as

$$\sigma' \geq S_y \tag{5–11}$$

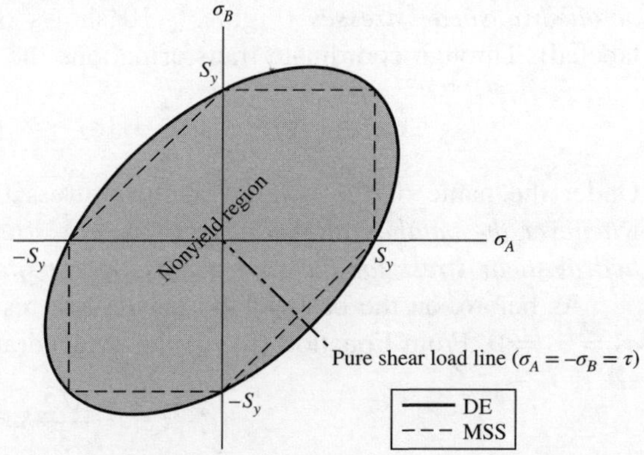

Figure 5–9

The distortion-energy (DE) theory yield envelope for plane stress states. This is a plot of points obtained from Equation (5–13) with $\sigma' = S_y$.

where the von Mises stress is

$$\sigma' = \left[\frac{(\sigma_1 - \sigma_2)^2 + (\sigma_2 - \sigma_3)^2 + (\sigma_3 - \sigma_1)^2}{2}\right]^{1/2} \tag{5–12}$$

For plane stress, the von Mises stress can be represented by the principal stresses σ_A, σ_B, and zero. Then from Equation (5–12), we get

$$\sigma' = \left(\sigma_A^2 - \sigma_A\sigma_B + \sigma_B^2\right)^{1/2} \tag{5–13}$$

Equation (5–13) is a rotated ellipse in the σ_A, σ_B plane, as shown in Figure 5–9 with $\sigma' = S_y$. The dotted lines in the figure represent the MSS theory, which can be seen to be more restrictive, hence, more conservative.[4]

Using *xyz* components of three-dimensional stress, the von Mises stress can be written as

$$\sigma' = \frac{1}{\sqrt{2}}\left[(\sigma_x - \sigma_y)^2 + (\sigma_y - \sigma_z)^2 + (\sigma_z - \sigma_x)^2 + 6(\tau_{xy}^2 + \tau_{yz}^2 + \tau_{zx}^2)\right]^{1/2} \tag{5–14}$$

and for plane stress,

$$\sigma' = \left(\sigma_x^2 - \sigma_x\sigma_y + \sigma_y^2 + 3\tau_{xy}^2\right)^{1/2} \tag{5–15}$$

The distortion-energy theory is also called:

- The von Mises or von Mises–Hencky theory
- The shear-energy theory
- The octahedral-shear-stress theory

Understanding octahedral shear stress will shed some light on why the MSS is conservative. Consider an isolated element in which the normal stresses on each surface are equal to the hydrostatic stress σ_{av}. There are eight surfaces symmetric to the principal directions that contain this stress. This forms an octahedron as shown in Figure 5–10. The shear stresses on these surfaces are equal and are called the

[4]The three-dimensional equations for DE and MSS can be plotted relative to three-dimensional σ_1, σ_2, σ_3, coordinate axes. The failure surface for DE is a circular cylinder with an axis inclined at 45° from each principal stress axis, whereas the surface for MSS is a hexagon inscribed within the cylinder. See Arthur P. Boresi and Richard J. Schmidt, *Advanced Mechanics of Materials*, 6th ed., John Wiley & Sons, New York, 2003, Section 4.4.

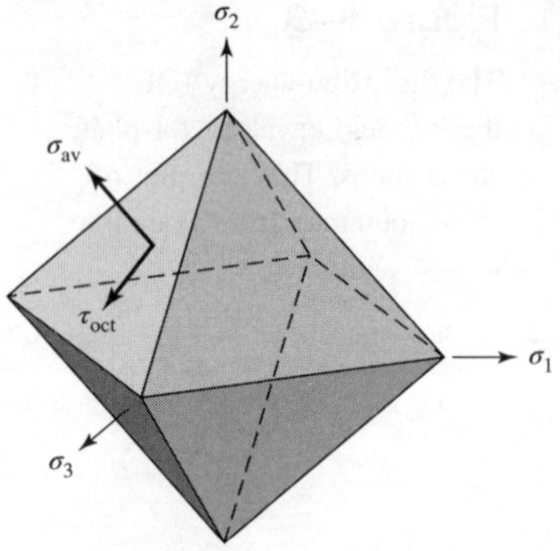

Figure 5–10

Octahedral surfaces.

octahedral shear stresses (Figure 5–10 shows only one of the octahedral surfaces labeled). Through coordinate transformations the octahedral shear stress is given by[5]

$$\tau_{oct} = \frac{1}{3}\left[(\sigma_1 - \sigma_2)^2 + (\sigma_2 - \sigma_3)^2 + (\sigma_3 - \sigma_1)^2\right]^{1/2} \tag{5–16}$$

Under the name of the octahedral-shear-stress theory, *failure is assumed to occur whenever the octahedral shear stress for any stress state equals or exceeds the octahedral shear stress for the simple tension-test specimen at failure.*

As before, on the basis of the tensile test results, yield occurs when $\sigma_1 = S_y$ and $\sigma_2 = \sigma_3 = 0$. From Equation (5–16) the octahedral shear stress under this condition is

$$\tau_{oct} = \frac{\sqrt{2}}{3} S_y \tag{5–17}$$

When, for the general stress case, Equation (5–16) is equal or greater than Equation (5–17), yield is predicted. This reduces to

$$\left[\frac{(\sigma_1 - \sigma_2)^2 + (\sigma_2 - \sigma_3)^2 + (\sigma_3 - \sigma_1)^2}{2}\right]^{1/2} \geq S_y \tag{5–18}$$

which is identical to Equation (5–10), verifying that the maximum-octahedral-shear-stress theory is equivalent to the distortion-energy theory.

The model for the MSS theory ignores the contribution of the normal stresses on the 45° surfaces of the tensile specimen. However, these stresses are $P/2A$, and *not* the hydrostatic stresses which are $P/3A$. Herein lies the difference between the MSS and DE theories.

The mathematical manipulation involved in describing the DE theory might tend to obscure the real value and usefulness of the result. The equations given allow the most complicated stress situation to be represented by a single quantity, the von Mises stress, which then can be compared against the yield strength of the material through Equation (5–11). This equation can be expressed as a design equation by

$$\sigma' = \frac{S_y}{n} \tag{5–19}$$

The distortion-energy theory predicts no failure under hydrostatic stress and agrees well with all data for ductile behavior. Hence, it is the most widely used theory for ductile materials and is recommended for design problems unless otherwise specified.

One final note concerns the shear yield strength. Consider a case of pure shear τ_{xy}, where for plane stress $\sigma_x = \sigma_y = 0$. For yield, Equation (5–11) with Equation (5–15) gives

$$\left(3\tau_{xy}^2\right)^{1/2} = S_y \quad \text{or} \quad \tau_{xy} = \frac{S_y}{\sqrt{3}} = 0.577 S_y \tag{5–20}$$

Thus, the shear yield strength predicted by the distortion-energy theory is

$$S_{sy} = 0.577 S_y \tag{5–21}$$

which as stated earlier, is about 15 percent greater than the 0.5 S_y predicted by the MSS theory. For pure shear, τ_{xy} the principal stresses from Equation (3–14) are $\sigma_A = -\sigma_B = \tau_{xy}$. The load line for this case is in the fourth quadrant at an angle of 45° from the σ_A, σ_B axes shown in Figure 5–9.

[5]For a derivation, see Arthur P. Boresi, op. cit., pp. 36–37.

EXAMPLE 5–1

A hot-rolled steel has a yield strength of $S_{yt} = S_{yc} = 700$ MPa and a true strain at fracture of $\varepsilon_f = 0.55$. Estimate the factor of safety for the following principal stress states:
(*a*) $\sigma_x = 500$ MPa, $\sigma_y = 500$ MPa, $\tau_{xy} = 0$ MPa
(*b*) $\sigma_x = 200$ MPa, $\sigma_y = 500$ MPa, $\tau_{xy} = 0$ MPa
(*c*) $\sigma_x = 0$ MPa, $\sigma_y = 500$ MPa, $\tau_{xy} = -200$ MPa
(*d*) $\sigma_x = 0$ MPa, $\sigma_y = -200$ MPa, $\tau_{xy} = -500$ MPa
(*e*) $\sigma_x = 200$ MPa, $\sigma_y = 200$ MPa, $\tau_{xy} = 200$ MPa

Solution

Since $\varepsilon_f > 0.05$ and S_{yt} and S_{yc} are equal, the material is ductile and both the distortion-energy (DE) theory and maximum-shear-stress (MSS) theory apply. Both will be used for comparison. Note that cases *a* to *d* are plane stress states.

(*a*) Since there is no shear stress on this stress element, the normal stresses are equal to the principal stresses. The ordered principal stresses are $\sigma_A = \sigma_1 = 500$, $\sigma_B = \sigma_2 = 500$, $\sigma_3 = 0$ MPa.

DE From Equation (5–13),

$$\sigma' = [500^2 - 500(500) + 500^2]^{1/2} = 500 \text{ MPa}$$

From Equation (5–19),

Answer
$$n = \frac{S_y}{\sigma'} = \frac{700}{500} = 1.40$$

MSS Noting that the two nonzero principal stresses are equal, τ_{max} will be from the largest Mohr's circle, which will incorporate the third principal stress at zero. From Equation (3–19),

$$\tau_{max} = \frac{\sigma_1 - \sigma_3}{2} = \frac{500 - 0}{2} = 250 \text{ MPa}$$

From Equation (5–3),

Answer
$$n = \frac{S_y/2}{\tau_{max}} = \frac{700/2}{250} = 1.40$$

(*b*) From Equation (3–14), the nonzero principal stresses are

$$\sigma_A, \sigma_B = \frac{200 + 500}{2} \pm \sqrt{\left(\frac{200 - 500}{2}\right)^2 + (0)^2} = 500, 200 \text{ MPa}$$

The ordered principal stresses are $\sigma_A = \sigma_1 = 500$, $\sigma_B = \sigma_2 = 200$, $\sigma_3 = 0$ MPa.

DE
$$\sigma' = [500^2 - 500(200) + 200^2]^{1/2} = 436 \text{ MPa}$$

Answer
$$n = \frac{S_y}{\sigma'} = \frac{700}{436} = 1.60$$

MSS Noting that the two nonzero principal stresses are both positive, τ_{max} will be from the largest Mohr's circle which will incorporate the third principal stress at zero. From Equation (3–19),

$$\tau_{max} = \frac{\sigma_1 - \sigma_3}{2} = \frac{500 - 0}{2} = 250 \text{ MPa}$$

Answer
$$n = \frac{S_y/2}{\tau_{max}} = \frac{700/2}{250} = 1.40$$

(*c*) This time, we shall obtain the factors of safety directly from the *xy* components of stress.

DE From Equation (5–15),

$$\sigma' = \left(\sigma_x^2 - \sigma_x\sigma_y + \sigma_y^2 + 3\tau_{xy}^2\right)^{1/2} = \left[(500)^2 + 3(-200)^2\right]^{1/2} = 608 \text{ MPa}$$

Answer
$$n = \frac{S_y}{\sigma'} = \frac{700}{608} = 1.15$$

MSS Taking care to note from a quick sketch of Mohr's circle that one nonzero principal stress will be positive while the other one will be negative, τ_{max} can be obtained from the extreme-value shear stress given by Equation (3–17) without finding the principal stresses.

$$\tau_{max} = \sqrt{\left(\frac{\sigma_x - \sigma_y}{2}\right)^2 + \tau_{xy}^2} = \sqrt{\left(\frac{0 - 500}{2}\right)^2 + -200^2} = 320 \text{ MPa}$$

Answer
$$n = \frac{S_y/2}{\tau_{max}} = \frac{700/2}{320} = 1.09$$

For graphical comparison purposes later in this problem, the nonzero principal stresses can be obtained from Equation (3–14) to be 570 MPa and −70 MPa.
The ordered principal stresses are $\sigma_1 = 0$, $\sigma_A = \sigma_2 = -200$, $\sigma_B = \sigma_3 = -500$ MPa.

DE
$$\sigma' = \left[(-500)^2 - (-500)(-200) + (-200)^2\right]^{1/2} = 436 \text{ MPa}$$

Answer
$$n = \frac{S_y}{\sigma'} = \frac{700}{436} = 1.61$$

MSS From Equation (3–19),

$$\tau_{max} = \frac{\sigma_1 - \sigma_3}{2} = \frac{0 - (-500)}{2} = 250 \text{ MPa}$$

Answer
$$n = \frac{S_y/2}{\tau_{max}} = \frac{700/2}{250} = 1.40$$

(*e*) The ordered principal stresses are $\sigma_1 = 200$, $\sigma_2 = 200$, $\sigma_3 = 200$ MPa.

DE From Equation (5–12),

$$\sigma' = \left[\frac{(200 - 200)^2 + (200 - 200)^2 + (200 - 200)^2}{2}\right]^{1/2} = 0 \text{ MPa}$$

Answer
$$n = \frac{S_y}{\sigma'} = \frac{700}{0} \rightarrow \infty$$

MSS From Equation (5–3),

Answer

$$n = \frac{S_y}{\sigma_1 - \sigma_3} = \frac{700}{200 - 200} \rightarrow \infty$$

A tabular summary of the factors of safety is included for comparisons.

	(a)	(b)	(c)	(d)	(e)
DE	1.40	1.60	1.15	1.61	∞
MSS	1.40	1.40	1.09	1.40	∞

Since the MSS theory is on or within the boundary of the DE theory, it will always predict a factor of safety equal to or less than the DE theory, as can be seen in the table. For each case, except case (*e*), the coordinates and load lines in the σ_A, σ_B plane are shown in Figure 5–11. Case (*e*) is not plane stress. Note that the load line for case (*a*) is the only plane stress case given in which the two theories agree, thus giving the same factor of safety.

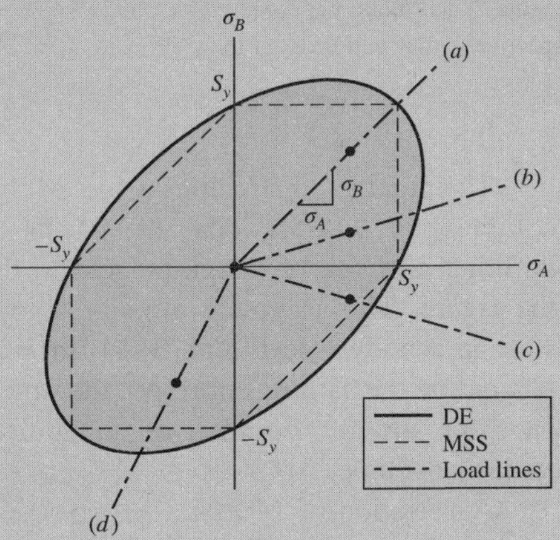

Figure 5–11

Load lines for Example 5–1.

5–6 Coulomb-Mohr Theory for Ductile Materials

Not all materials have compressive strengths equal to their corresponding tensile values. For example, the yield strength of magnesium alloys in compression may be as little as 50 percent of their yield strength in tension. The ultimate strength of gray cast irons in compression varies from 3 to 4 times greater than the ultimate tensile strength. So, in this section, we are primarily interested in those theories that can be used to predict failure for materials whose strengths in tension and compression are not equal.

Historically, the Mohr theory of failure dates to 1900, a date that is relevant to its presentation. There were no computers, just slide rules, compasses, and French curves. Graphical procedures, common then, are still useful today for visualization. The idea of Mohr is based on three "simple" tests: tension, compression, and shear, to yielding if the material can yield, or to rupture. It is easier to define shear yield strength as S_{sy} than it is to test for it.

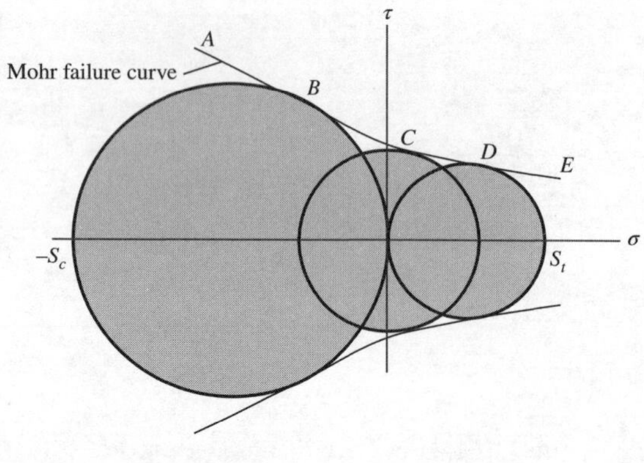

Figure 5–12

Three Mohr circles, one for the uniaxial compression test, one for the test in pure shear, and one for the uniaxial tension test, are used to define failure by the Mohr hypothesis. The strengths S_c and S_t are the compressive and tensile strengths, respectively; they can be used for yield or ultimate strength.

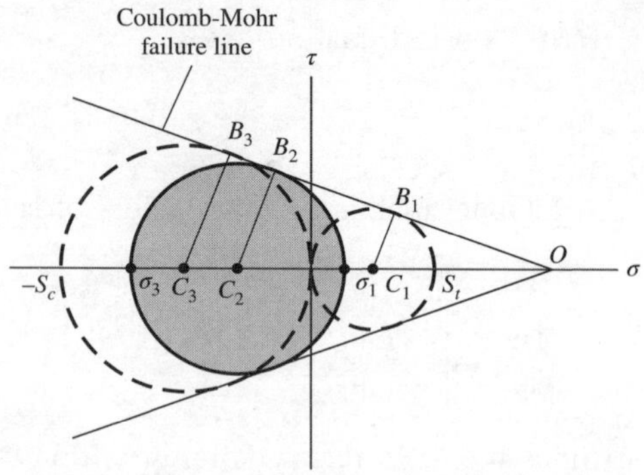

Figure 5–13

Mohr's largest circle for a general state of stress.

The practical difficulties aside, Mohr's hypothesis was to use the results of tensile, compressive, and torsional shear tests to construct the three circles of Figure 5–12 defining a failure envelope tangent to the three circles, depicted as curve *ABCDE* in the figure. The argument amounted to the three Mohr circles describing the stress state in a body (see Figure 3–14) growing during loading until one of them became tangent to the failure envelope, thereby defining failure. Was the form of the failure envelope straight, circular, or quadratic? A compass or a French curve defined the failure envelope.

A variation of Mohr's theory, called the *Coulomb-Mohr theory* or the *internal-friction theory,* assumes that the boundary *BCD* in Figure 5–12 is straight. With this assumption only the tensile and compressive strengths are necessary. Consider the conventional ordering of the principal stresses such that $\sigma_1 \geq \sigma_2 \geq \sigma_3$. The largest circle connects σ_1 and σ_3, as shown in Figure 5–13. The centers of the circles in Figure 5–13 are C_1, C_2, and C_3. Triangles OB_iC_i are similar, therefore

$$\frac{B_2C_2 - B_1C_1}{OC_2 - OC_1} = \frac{B_3C_3 - B_1C_1}{OC_3 - OC_1}$$

or,

$$\frac{B_2C_2 - B_1C_1}{C_1C_2} = \frac{B_3C_3 - B_1C_1}{C_1C_3}$$

where $B_1C_1 = S_t/2$, $B_2C_2 = (\sigma_1 - \sigma_3)/2$, and $B_3C_3 = S_c/2$, are the radii of the right, center, and left circles, respectively. The distance from the origin to C_1 is $S_t/2$, to C_3 is $S_c/2$, and to C_2 (in the *positive* σ direction) is $(\sigma_1 + \sigma_3)/2$. Thus,

$$\frac{\dfrac{\sigma_1 - \sigma_3}{2} - \dfrac{S_t}{2}}{\dfrac{S_t}{2} - \dfrac{\sigma_1 + \sigma_3}{2}} = \frac{\dfrac{S_c}{2} - \dfrac{S_t}{2}}{\dfrac{S_t}{2} + \dfrac{S_c}{2}}$$

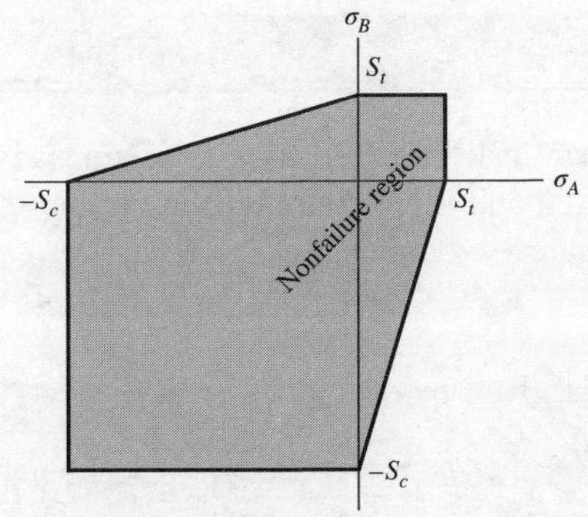

Figure 5–14

Plot of the Coulomb-Mohr theory failure envelope for plane stress states.

Canceling the 2 in each term, cross-multiplying, and simplifying reduces this equation to

$$\frac{\sigma_1}{S_t} - \frac{\sigma_3}{S_c} = 1 \tag{5–22}$$

where either yield strength or ultimate strength can be used.

For plane stress, when the two nonzero principal stresses are $\sigma_A \geq \sigma_B$, we have a situation similar to the three cases given for the MSS theory, Equations (5–4) to (5–6). That is, the failure conditions are

Case 1: $\sigma_A \geq \sigma_B \geq 0$. For this case, $\sigma_1 = \sigma_A$ and $\sigma_3 = 0$. Equation (5–22) reduces to

$$\sigma_A \geq S_t \tag{5–23}$$

Case 2: $\sigma_A \geq 0 \geq \sigma_B$. Here, $\sigma_1 = \sigma_A$ and $\sigma_3 = \sigma_B$, and Equation (5–22) becomes

$$\frac{\sigma_A}{S_t} - \frac{\sigma_B}{S_c} \geq 1 \tag{5–24}$$

Case 3: $0 \geq \sigma_A \geq \sigma_B$. For this case, $\sigma_1 = 0$ and $\sigma_3 = \sigma_B$, and Equation (5–22) gives

$$\sigma_B \leq -S_c \tag{5–25}$$

A plot of these cases, together with the normally unused cases corresponding to $\sigma_B \geq \sigma_A$, is shown in Figure 5–14.

For design equations, incorporating the factor of safety n, divide all strengths by n. For example, Equation (5–22) as a design equation can be written as

$$\frac{\sigma_1}{S_t} - \frac{\sigma_3}{S_c} = \frac{1}{n} \tag{5–26}$$

Since for the Coulomb-Mohr theory we do not need the torsional shear strength circle we can deduce it from Equation (5–22). For pure shear τ, $\sigma_1 = -\sigma_3 = \tau$. The torsional yield strength occurs when $\tau_{\max} = S_{sy}$. Substituting $\sigma_1 = -\sigma_3 = S_{sy}$ into Equation (5–22) and simplifying gives

$$S_{sy} = \frac{S_{yt}S_{yc}}{S_{yt} + S_{yc}} \tag{5–27}$$

EXAMPLE 5–2

A 25-mm-diameter shaft is statically torqued to 230 N · m. It is made of cast 195-T6 aluminum, with a yield strength in tension of 160 MPa and a yield strength in compression of 170 MPa. It is machined to final diameter. Estimate the factor of safety of the shaft.

Solution

The maximum shear stress is given by

$$\tau = \frac{16T}{\pi d^3} = \frac{16(230)}{\pi[25(10^{-3})]^3} = 75(10^6)\ \text{N/m}^2 = 75\ \text{MPa}$$

The two nonzero principal stresses are 75 and −75 MPa, making the ordered principal stresses $\sigma_1 = 75$, $\sigma_2 = 0$, and $\sigma_3 = -75$ MPa. From Equation (5–26), for yield,

Answer
$$n = \frac{1}{\sigma_1/S_{yt} - \sigma_3/S_{yc}} = \frac{1}{75/160 - (-75)/170} = 1.10$$

Alternatively, from Equation (5–27),

$$S_{sy} = \frac{S_{yt}S_{yc}}{S_{yt} + S_{yc}} = \frac{160(170)}{160 + 170} = 82.4\ \text{MPa}$$

and $\tau_{max} = 75$ MPa. Thus,

Answer
$$n = \frac{S_{sy}}{\tau_{max}} = \frac{82.4}{75} = 1.10$$

5–7 Failure of Ductile Materials Summary

Having studied some of the various theories of failure, we shall now evaluate them and show how they are applied in design and analysis. In this section we limit our studies to materials and parts that are known to fail in a ductile manner. Materials that fail in a brittle manner will be considered separately because these require different failure theories.

To help decide on appropriate and workable theories of failure, Marin[6] collected data from many sources. Some of the data points used to select failure theories for ductile materials are shown in Figure 5–15.[7] Mann also collected many data for copper and nickel alloys; if shown, the data points for these would be mingled with those already diagrammed. Figure 5–15 shows that either the maximum-shear-stress theory or the distortion-energy theory is acceptable for design and analysis of materials that would fail in a ductile manner.

The selection of one or the other of these two theories is something that the engineer must decide. For design purposes the maximum-shear-stress theory is easy, quick to use, and conservative. If the problem is to learn *why* a part failed, then the

[6]Joseph Marin was one of the pioneers in the collection, development, and dissemination of material on the failure of engineering elements. He has published many books and papers on the subject. Here the reference used is Joseph Marin, *Engineering Materials*, Prentice Hall, Englewood Cliffs, N.J., 1952.

[7]Note that some data in Figure 5–15 are displayed along the top horizontal boundary where $\sigma_B \geq \sigma_A$. This is often done with failure data to thin out congested data points by plotting on the mirror image of the line $\sigma_B = \sigma_A$.

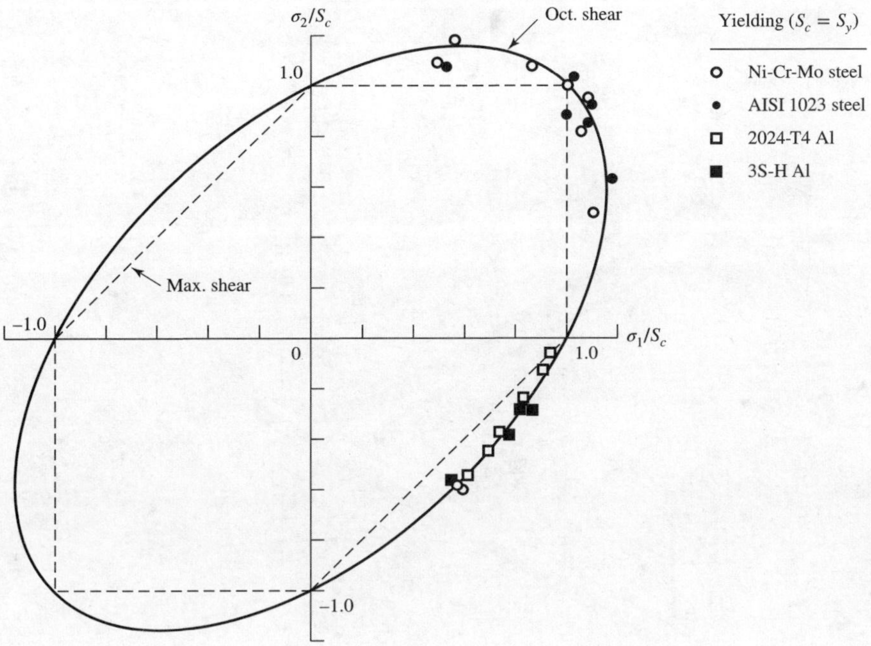

Figure 5–15

Experimental data superposed on failure theories. *(Source: Adapted from Fig. 7.11, p. 257,* Mechanical Behavior of Materials, *2nd ed., N. E. Dowling, Prentice Hall, Englewood Cliffs, N.J., 1999. Modified to show only ductile failures.)*

distortion-energy theory may be the best to use; Figure 5–15 shows that the plot of the distortion-energy theory passes closer to the central area of the data points, and thus is generally a better predictor of failure. However, keep in mind that though a failure curve passing through the center of the experimental data is *typical* of the data, its *reliability* from a statistical standpoint is about 50 percent. For design purposes, a larger factor of safety may be warranted when using such a failure theory.

For ductile materials with unequal yield strengths, S_{yt} in tension and S_{yc} in compression, the Mohr theory is the best available. However, the theory requires the results from three separate modes of tests, graphical construction of the failure locus, and fitting the largest Mohr's circle to the failure locus. The alternative to this is to use the Coulomb-Mohr theory, which requires only the tensile and compressive yield strengths and is easily dealt with in equation form.

EXAMPLE 5–3

This example illustrates the use of a failure theory to determine the strength of a mechanical element or component. The example may also clear up any confusion existing between the phrases *strength of a machine part, strength of a material, and strength of a part at a point.*

A certain force F applied at D near the end of the 380 mm lever shown in Figure 5–16, which is quite similar to a socket wrench, results in certain stresses in the cantilevered bar $OABC$. This bar ($OABC$) is of AISI 1035 steel, forged and heat-treated so that it has a minimum (ASTM) yield strength of 560 MPa. We presume that this component would be of no value after yielding. Thus the force F required to initiate yielding can be regarded as the strength of the component part. Find this force.

Solution

In this problem we focus on bar $OABC$. Bar CD requires its own examination for failure, which is not presented here. A 1035 steel, heat-treated, will have a reduction in area of 50 percent or more and hence is a ductile material at normal temperatures. This also means that stress concentration at shoulder A need not be considered. A stress

Figure 5–16

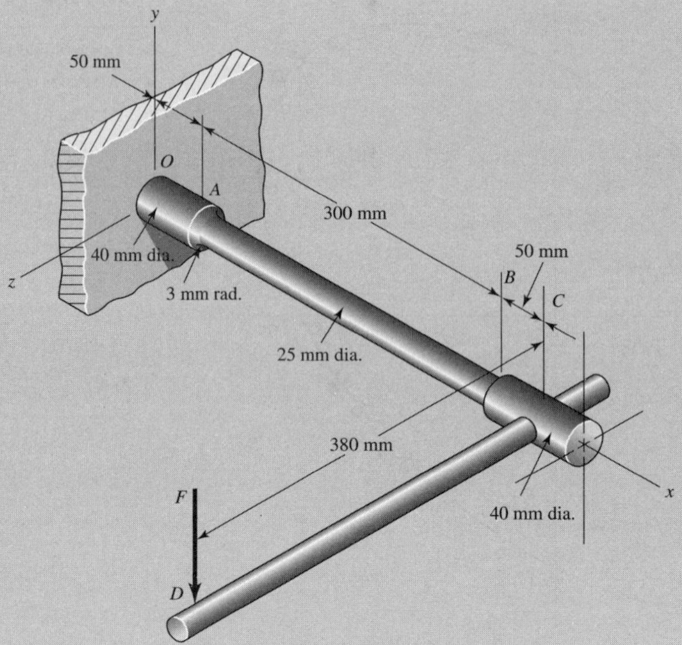

element at A on the top surface will be subjected to a tensile bending stress and a torsional stress. This point, on the 25-mm-diameter section, is the weakest section, and governs the strength of the assembly. The two stresses are

$$\sigma_x = \frac{M}{I/c} = \frac{32M}{\pi d^3} = \frac{32(350F)}{\pi(25^3)} = 0.228F$$

$$\tau_{zx} = \frac{Tr}{J} = \frac{16T}{\pi d^3} = \frac{16(380F)}{\pi(25^3)} = 0.124F$$

Employing the distortion-energy theory, we find, from Equation (5–15), that

$$\sigma' = \left(\sigma_x^2 + 3\tau_{zx}^2\right)^{1/2} = \left[(0.228F)^2 + 3(0.124F)^2\right]^{1/2} = 0.313F$$

Equating the von Mises stress to S_y, we solve for F and get

Answer
$$F = \frac{S_y}{194.5} = \frac{560}{0.313} = 1789 \text{ N}$$

In this example the strength of the material at point A is $S_y = 560$ MPa. The strength of the assembly or component is $F = 1789$ N.

Let us apply the MSS theory for comparison. For a point undergoing plane stress with only one nonzero normal stress and one shear stress, the two nonzero principal stresses will have opposite signs, and hence the maximum shear stress is obtained from the Mohr's circle between them. From Equation (3–17)

$$\tau_{\max} = \sqrt{\left(\frac{\sigma_x}{2}\right)^2 + \tau_{zx}^2} = \sqrt{\left(\frac{0.228F}{2}\right)^2 + (0.124F)^2} = 0.168F$$

Setting this equal to $S_y/2$, from Equation (5–3) with $n = 1$, and solving for F, we get

$$F = \frac{560/2}{0.168} = 1667 \text{ N}$$

which is about 7 percent less than found for the DE theory. As stated earlier, the MSS theory is more conservative than the DE theory.

EXAMPLE 5–4

The cantilevered tube shown in Figure 5–17 is to be made of 2014 aluminum alloy treated to obtain a specified minimum yield strength of 276 MPa. We wish to select a stock-size tube from Table A–8 using a design factor $n_d = 4$. The bending load is $F = 1.75$ kN, the axial tension is $P = 9.0$ kN, and the torsion is $T = 72$ N · m. What is the realized factor of safety?

Solution

The critical stress element is at point A on the top surface at the wall, where the bending moment is the largest, and the bending and torsional stresses are at their maximum values. The critical stress element is shown in Figure 5–17b. Since the axial stress and bending stress are both in tension along the x axis, they are additive for the normal stress, giving

$$\sigma_x = \frac{P}{A} + \frac{Mc}{I} = \frac{9}{A} + \frac{120(1.75)(d_o/2)}{I} = \frac{9}{A} + \frac{105d_o}{I} \tag{1}$$

where, if millimeters are used for the area properties, the stress is in gigapascals.

The torsional stress at the same point is

$$\tau_{zx} = \frac{Tr}{J} = \frac{72(d_o/2)}{J} = \frac{36d_o}{J} \tag{2}$$

For accuracy, we choose the distortion-energy theory as the design basis. The von Mises stress from Equation (5–15) is

$$\sigma' = \left(\sigma_x^2 + 3\tau_{zx}^2\right)^{1/2} \tag{3}$$

On the basis of the given design factor, the goal for σ' is

$$\sigma' \le \frac{S_y}{n_d} = \frac{0.276}{4} = 0.0690 \text{ GPa} \tag{4}$$

where we have used gigapascals in this relation to agree with Equations (1) and (2).

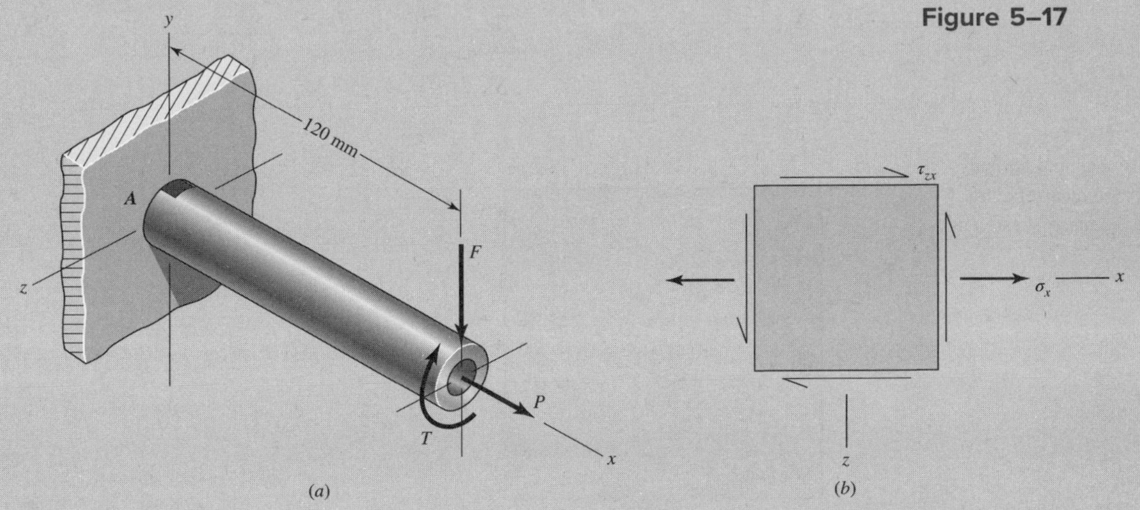

Figure 5–17

Programming Equations (1) to (3) on a spreadsheet and entering metric sizes from Table A–8 reveals that a 42 × 5-mm tube is satisfactory. The von Mises stress is found to be $\sigma' = 0.060\ 43$ GPa for this size. Thus the realized factor of safety is

Answer
$$n = \frac{S_y}{\sigma'} = \frac{0.276}{0.060\ 43} = 4.57$$

For the next size smaller, a 42 × 4-mm tube, $\sigma' = 0.071\ 05$ GPa giving a factor of safety of

$$n = \frac{S_y}{\sigma'} = \frac{0.276}{0.071\ 05} = 3.88$$

5–8 Maximum-Normal-Stress Theory for Brittle Materials

The maximum-normal-stress (MNS) theory states that *failure occurs when one of the three principal stresses equals or exceeds the strength.* Again we arrange the principal stresses for a general stress state in the ordered form $\sigma_1 \geq \sigma_2 \geq \sigma_3$. This theory then predicts that failure occurs whenever

$$\sigma_1 \geq S_{ut} \qquad \text{or} \qquad \sigma_3 \leq -S_{uc} \tag{5–28}$$

where S_{ut} and S_{uc} are the ultimate tensile and compressive strengths, respectively, given as positive quantities.

For plane stress, with the principal stresses given by Equation (3–14), with $\sigma_A \geq \sigma_B$, Equation (5–28) can be written as

$$\sigma_A \geq S_{ut} \qquad \text{or} \qquad \sigma_B \leq -S_{uc} \tag{5–29}$$

which is plotted in Figure 5–18.

As before, the failure criteria equations can be converted to design equations. We can consider two sets of equations where $\sigma_A \geq \sigma_B$ as

$$\sigma_A = \frac{S_{ut}}{n} \qquad \text{or} \qquad \sigma_B = -\frac{S_{uc}}{n} \tag{5–30}$$

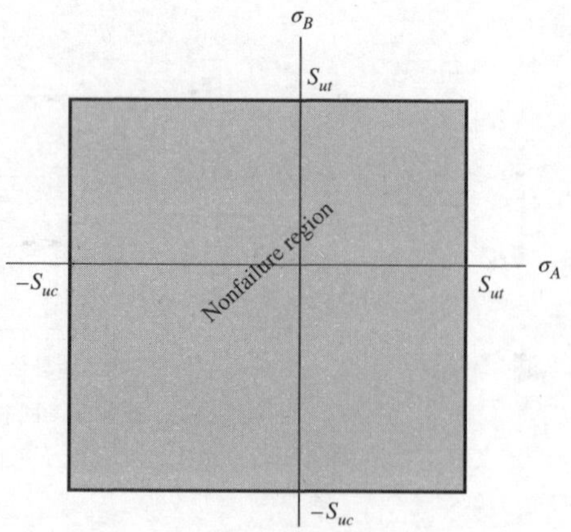

Figure 5–18

Graph of maximum-normal-stress (MNS) theory failure envelope for plane stress states.

As will be seen later, the maximum-normal-stress theory is not very good at predicting failure in the second and fourth quadrants of the σ_A, σ_B plane. Thus, we will not recommend the theory for use. It has been included here mainly for historical reasons.

5–9 Modifications of the Mohr Theory for Brittle Materials

We will discuss two modifications of the Mohr theory for brittle materials: the Brittle-Coulomb-Mohr (BCM) theory and the modified Mohr (MM) theory. The equations provided for the theories will be restricted to plane stress and be of the design type incorporating the factor of safety.

The Coulomb-Mohr theory was discussed earlier in Section 5–6 with Equations (5–23) to (5–25). Written as design equations for a brittle material, they are:

Brittle-Coulomb-Mohr

$$\sigma_A = \frac{S_{ut}}{n} \qquad \sigma_A \geq \sigma_B \geq 0 \tag{5–31a}$$

$$\frac{\sigma_A}{S_{ut}} - \frac{\sigma_B}{S_{uc}} = \frac{1}{n} \qquad \sigma_A \geq 0 \geq \sigma_B \tag{5–31b}$$

$$\sigma_B = -\frac{S_{uc}}{n} \qquad 0 \geq \sigma_A \geq \sigma_B \tag{5–31c}$$

On the basis of observed data for the fourth quadrant, the modified Mohr theory expands the fourth quadrant with the solid lines shown in the second and fourth quadrants of Figure 5–19 (where the factor of safety, n, is set to one).

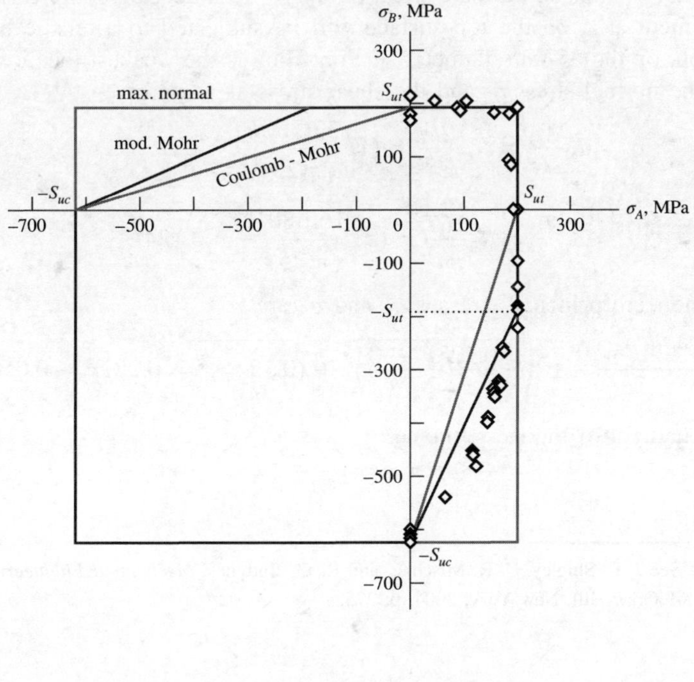

Figure 5–19

Biaxial fracture data of gray cast iron compared with various failure criteria. *(Source: Data points from Grassi, R. C. and I. Cornet. "Fracture of Gray Cast Iron Tubes under Biaxial Stress."* Journal of Applied Mechanics *16 (June 1949): 178–182.)*

Modified Mohr

$$\sigma_A = \frac{S_{ut}}{n} \qquad \sigma_A \geq \sigma_B \geq 0$$

$$\sigma_A \geq 0 \geq \sigma_B \text{ and } \left|\frac{\sigma_B}{\sigma_A}\right| \leq 1 \tag{5–32a}$$

$$\frac{(S_{uc} - S_{ut})\sigma_A}{S_{uc}S_{ut}} - \frac{\sigma_B}{S_{uc}} = \frac{1}{n} \qquad \sigma_A \geq 0 \geq \sigma_B \text{ and } \left|\frac{\sigma_B}{\sigma_A}\right| > 1 \tag{5–32b}$$

$$\sigma_B = -\frac{S_{uc}}{n} \qquad 0 \geq \sigma_A \geq \sigma_B \tag{5–32c}$$

Data are still outside this extended region. The straight line introduced by the modified Mohr theory, for $\sigma_A \geq 0 \geq \sigma_B$ and $|\sigma_B/\sigma_A| > 1$, can be replaced by a parabolic relation which can more closely represent some of the data.[8] However, this introduces a nonlinear equation for the sake of a minor correction, and will not be presented here.

EXAMPLE 5–5

Consider the wrench in Example 5–3, Figure 5–16, as made of cast iron, machined to dimension. The force F required to fracture this part can be regarded as the strength of the component part. If the material is ASTM grade 30 cast iron, find the force F with
(*a*) Coulomb-Mohr failure model.
(*b*) Modified Mohr failure model.

Solution

In this problem we focus on bar *OABC*. Bar *CD* requires its own examination for failure, which is not presented here. Since grade 30 cast iron is a brittle material *and* cast iron, the stress-concentration factors K_t and K_{ts} are set to unity. From Table A–23, the tensile ultimate strength is 210 MPa and the compressive ultimate strength is 750 MPa. The stress element at A on the top surface will be subjected to a tensile bending stress and a torsional stress. This location, on the 25-mm-diameter section fillet, is the weakest location, and it governs the strength of the assembly. The normal stress σ_x and the shear stress at A are given by

$$\sigma_x = K_t\frac{M}{I/c} = K_t\frac{32M}{\pi d^3} = (1)\frac{32(350F)}{\pi(25)^3} = 0.228F$$

$$\tau_{xy} = K_{ts}\frac{Tr}{J} = K_{ts}\frac{16T}{\pi d^3} = (1)\frac{16(380F)}{\pi(25)^3} = 0.124F$$

From Equation (3–13) the nonzero principal stresses σ_A and σ_B are

$$\sigma_A,\ \sigma_B = \frac{0.228F + 0}{2} \pm \sqrt{\left(\frac{0.228F - 0}{2}\right)^2 + (0.124F)^2} = 0.282F,\ -0.054F$$

This puts us in the fourth-quadrant of the σ_A, σ_B plane.

[8]See J. E. Shigley, C. R. Mischke, and R. G. Budynas, *Mechanical Engineering Design*, 7th ed., McGraw Hill, New York, 2004, p. 275.

(*a*) For BCM, Equation (5–31*b*) applies with $n = 1$ for failure.

$$\frac{\sigma_A}{S_{ut}} - \frac{\sigma_B}{S_{uc}} = \frac{0.282F}{210} - \frac{(-0.054F)}{750} = 1$$

Solving for F yields

Answer $$F = 707 \text{ N}$$

(*b*) For MM, the slope of the load line is $|\sigma_B/\sigma_A| = 0.054/0.282 = 0.191 < 1$. Obviously, Equation (5–32*a*) applies.

$$\frac{\sigma_A}{S_{ut}} = \frac{0.282F}{210} = 1$$

Answer $$F = 745 \text{ N}$$

As one would expect from inspection of Figure 5–19, Coulomb-Mohr is more conservative.

5–10 Failure of Brittle Materials Summary

We have identified failure or strength of brittle materials that conform to the usual meaning of the word *brittle,* relating to those materials whose true strain at fracture is 0.05 or less. We also have to be aware of normally ductile materials that for some reason may develop a brittle fracture or crack if used below the transition temperature. Figure 5–20 shows data for a nominal grade 30 cast iron taken under biaxial stress conditions, with several brittle failure hypotheses shown, superposed. We note the following:

- In the first quadrant the data appear on both sides and along the failure curves of maximum-normal-stress, Coulomb-Mohr, and modified Mohr. All failure curves are the same, and data fit well.

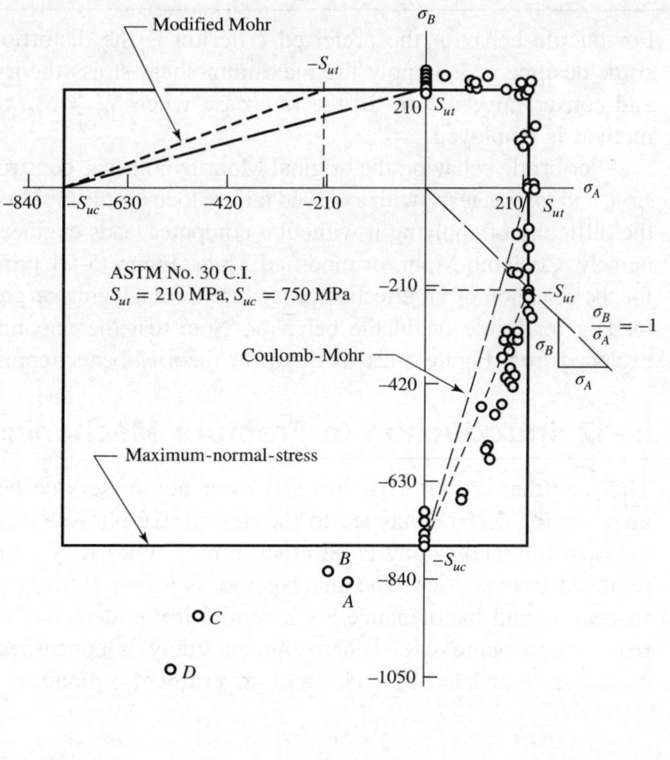

Figure 5–20

A plot of experimental data points obtained from tests on cast iron. Shown also are the graphs of three failure theories of possible usefulness for brittle materials. Note points *A*, *B*, *C*, and *D*. To avoid congestion in the first quadrant, points have been plotted for $\sigma_A > \sigma_B$ as well as for the opposite sense. *(Source: Data from Walton, F. Charles (ed.).* Iron Castings Handbook. *Cleveland, Ohio: Iron Founders' Society, 1971, 215, 216.)*

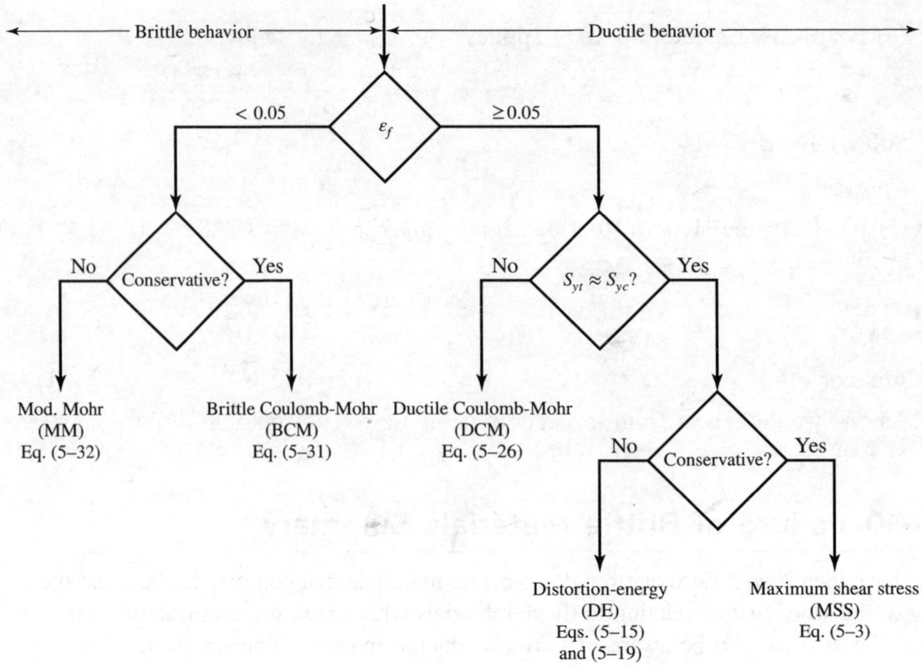

Figure 5–21

Failure theory selection flowchart.

- In the fourth quadrant the modified Mohr theory represents the data best, whereas the maximum-normal-stress theory does not.
- In the third quadrant the points *A*, *B*, *C*, and *D* are too few to make any suggestion concerning a fracture locus.

5–11 Selection of Failure Criteria

For ductile behavior the preferred criterion is the distortion-energy theory, although some designers also apply the maximum-shear-stress theory because of its simplicity and conservative nature. In the rare case when $S_{yt} \neq S_{yc}$, the ductile Coulomb-Mohr method is employed.

For brittle behavior, the original Mohr hypothesis, constructed with tensile, compression, and torsion tests, with a curved failure locus is the best hypothesis we have. However, the difficulty of applying it without a computer leads engineers to choose modifications, namely, Coulomb Mohr, or modified Mohr. Figure 5–21 provides a summary flowchart for the selection of an effective procedure for analyzing or predicting failures from static loading for brittle or ductile behavior. Note that the maximum-normal-stress theory is excluded from Figure 5–21 as the other theories better represent the experimental data.

5–12 Introduction to Fracture Mechanics

The idea that cracks exist in parts even before service begins, and that cracks can grow during service, has led to the descriptive phrase "damage-tolerant design." The focus of this philosophy is on crack growth until it becomes critical, and the part is removed from service. The analysis tool is *linear elastic fracture mechanics* (LEFM). Inspection and maintenance are essential in the decision to retire parts before cracks reach catastrophic size. Where human safety is concerned, periodic inspections for cracks are mandated by codes and government ordinance.

We shall now briefly examine some of the basic ideas and vocabulary needed for the potential of the approach to be appreciated. The intent here is to make the reader aware of the dangers associated with the sudden brittle fracture of so-called ductile materials. The topic is much too extensive to include in detail here and the reader is urged to read further on this complex subject.[9]

The use of elastic stress-concentration factors provides an indication of the average load required on a part for the onset of plastic deformation, or yielding. These factors are also useful for analysis of the loads on a part that will cause fatigue fracture. However, stress-concentration factors are limited to structures for which all dimensions are precisely known, particularly the radius of curvature in regions of high stress concentration. When there exists a crack, flaw, inclusion, or defect of unknown small radius in a part, the elastic stress-concentration factor approaches infinity as the root radius approaches zero, thus rendering the stress-concentration factor approach useless. Furthermore, even if the radius of curvature of the flaw tip is known, the high local stresses there will lead to local plastic deformation surrounded by a region of elastic deformation. Elastic stress-concentration factors are no longer valid for this situation, so analysis from the point of view of stress-concentration factors does not lead to criteria useful for design when very sharp cracks are present.

By combining analysis of the gross elastic changes in a structure or part that occur as a sharp brittle crack grows with measurements of the energy required to produce new fracture surfaces, it is possible to calculate the average stress (if no crack were present) that will cause crack growth in a part. Such calculation is possible only for parts with cracks for which the elastic analysis has been completed, and for materials that crack in a relatively brittle manner and for which the fracture energy has been carefully measured. The term *relatively brittle* is rigorously defined in the test procedures,[10] but it means, roughly, *fracture without yielding occurring throughout the fractured cross section.*

Thus glass, hard steels, strong aluminum alloys, and even low-carbon steel below the ductile-to-brittle transition temperature can be analyzed in this way. Fortunately, ductile materials blunt sharp cracks, as we have previously discovered, so that fracture occurs at average stresses of the order of the yield strength, and the designer is prepared for this condition. The middle ground of materials that lie between "relatively brittle" and "ductile" is now being actively analyzed, but exact design criteria for these materials are not yet available.

Quasi-Static Fracture

Many of us have had the experience of observing brittle fracture, whether it is the breaking of a cast-iron specimen in a tensile test or the twist fracture of a piece of blackboard chalk. It happens so rapidly that we think of it as instantaneous, that is, the cross section simply parting. Fewer of us have skated on a frozen pond in the

---

[9]References on brittle fracture include:

H. Tada, P. C. Paris, and G. R. Irwin, *The Stress Analysis of Cracks Handbook*, 3rd ed., ASME Press, New York, 2000.

D. Broek, *Elementary Engineering Fracture Mechanics*, 4th ed., Martinus Nijhoff, London, 1985.

D. Broek, *The Practical Use of Fracture Mechanics*, Kluwer Academic Publishers, London, 1988.

David K. Felbeck and Anthony G. Atkins, *Strength and Fracture of Engineering Solids*, 2nd ed., Prentice Hall, Englewood Cliffs, N.J., 1995.

Kåre Hellan, *Introduction to Fracture Mechanics*, McGraw Hill, New York, 1984.

[10]BS 5447:1977 and ASTM E399-78.

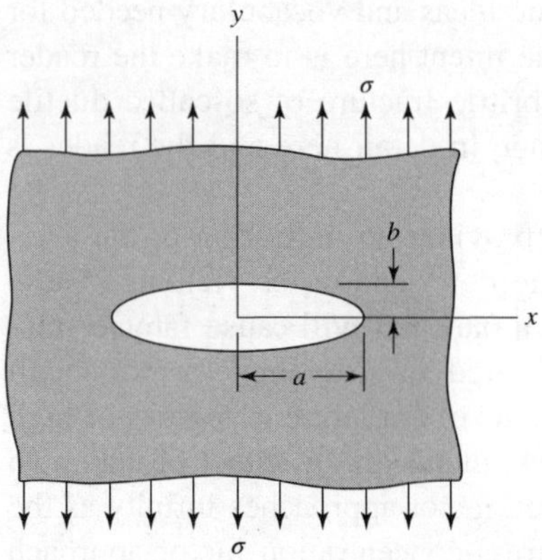

Figure 5–22

spring, with no one near us, heard a cracking noise, and stopped to observe. The noise is due to cracking. The cracks move slowly enough for us to see them run. The phenomenon is not instantaneous, since some time is necessary to feed the crack energy from the stress field to the crack for propagation. Quantifying these things is important to understanding the phenomenon "in the small." In the large, a static crack may be stable and will not propagate. Some level of loading can render the crack unstable, and the crack propagates to fracture.

The foundation of fracture mechanics was first established by Griffith in 1921 using the stress field calculations for an elliptical flaw in a plate developed by Inglis in 1913. For the infinite plate loaded by an applied uniaxial stress σ in Figure 5–22, the maximum stress occurs at $(\pm a, 0)$ and is given by

$$(\sigma_y)_{\max} = \left(1 + 2\frac{a}{b}\right)\sigma \tag{5–33}$$

Note that when $a = b$, the ellipse becomes a circle and Equation (5–33) gives a stress-concentration factor of 3. This agrees with the well-known result for an infinite plate with a circular hole (see Table A–13–1). For a fine crack, $b/a \to 0$, and Equation (5–33) predicts that $(\sigma_y)_{\max} \to \infty$. However, on a microscopic level, an infinitely sharp crack is a hypothetical abstraction that is physically impossible, and when plastic deformation occurs, the stress will be finite at the crack tip.

Griffith showed that the crack growth occurs when the energy release rate from applied loading is greater than the rate of energy for crack growth. Crack growth can be stable or unstable. Unstable crack growth occurs when the *rate* of change of the energy release rate relative to the crack length is equal to or greater than the *rate* of change of the crack growth rate of energy. Griffith's experimental work was restricted to brittle materials, namely glass, which pretty much confirmed his surface energy hypothesis. However, for ductile materials, the energy needed to perform plastic work at the crack tip is found to be much more crucial than surface energy.

Crack Modes and the Stress Intensity Factor

Three distinct modes of crack propagation exist, as shown in Figure 5–23. A tensile stress field gives rise to mode I, the *opening crack propagation mode,* as shown in Figure 5–23*a*. This mode is the most common in practice. Mode II is the *sliding mode,* is due to in-plane shear, and can be seen in Figure 5–23*b*. Mode III is the *tearing mode,* which arises from out-of-plane shear, as shown in Figure 5–23*c*. Combinations of these modes can also occur. Since mode I is the most common and important mode, the remainder of this section will consider only this mode.

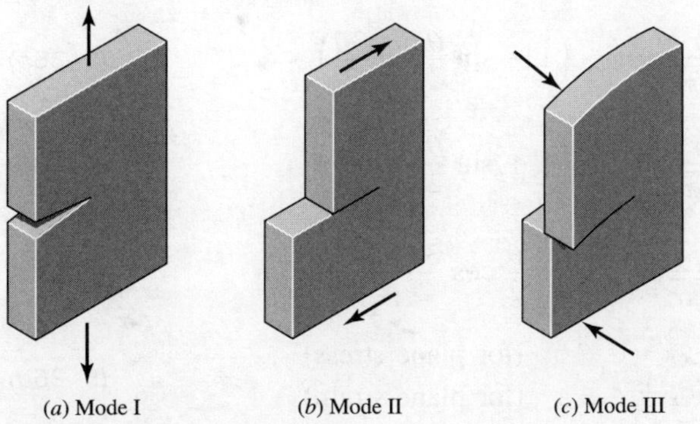

Figure 5–23

Crack propagation modes.

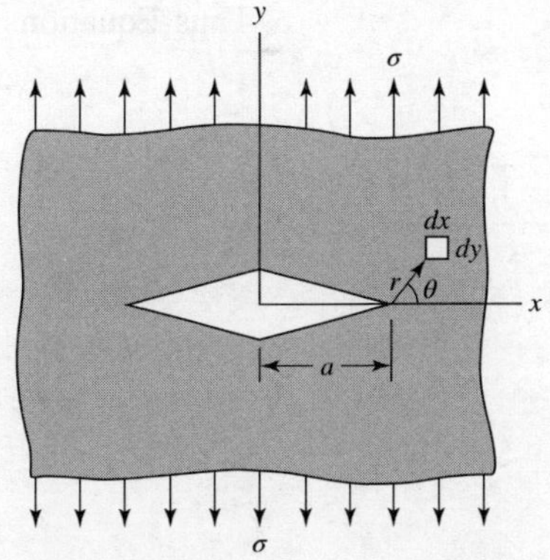

Figure 5–24

Mode I crack model.

Consider a mode I crack of length $2a$ in the infinite plate of Figure 5–24. By using complex stress functions, it has been shown that the stress field on a $dx\, dy$ element in the vicinity of the crack tip is given by

$$\sigma_x = \sigma\sqrt{\frac{a}{2r}}\cos\frac{\theta}{2}\left(1 - \sin\frac{\theta}{2}\sin\frac{3\theta}{2}\right) \tag{5–34a}$$

$$\sigma_y = \sigma\sqrt{\frac{a}{2r}}\cos\frac{\theta}{2}\left(1 + \sin\frac{\theta}{2}\sin\frac{3\theta}{2}\right) \tag{5–34b}$$

$$\tau_{xy} = \sigma\sqrt{\frac{a}{2r}}\sin\frac{\theta}{2}\cos\frac{\theta}{2}\cos\frac{3\theta}{2} \tag{5–34c}$$

$$\sigma_z = \begin{cases} 0 & \text{(for plane stress)} \\ \nu(\sigma_x + \sigma_y) & \text{(for plane strain)} \end{cases} \tag{5–34d}$$

where plane stress and plane strain are defined in Sections 3–5 and 3–8, respectively. The stress σ_y near the tip, with $\theta = 0$, is

$$\sigma_y|_{\theta=0} = \sigma\sqrt{\frac{a}{2r}} \tag{a}$$

As with the elliptical crack, we see that $\sigma_y|_{\theta=0} \to \infty$ as $r \to 0$, and again the concept of an infinite stress concentration at the crack tip is inappropriate. The quantity $\sigma_y|_{\theta=0}\sqrt{2r} = \sigma\sqrt{a}$, however, does remain constant as $r \to 0$. It is common practice to define a factor K called the *stress intensity factor* given by

$$K = \sigma\sqrt{\pi a} \tag{b}$$

where the units are $\text{MPa}\sqrt{\text{m}}$. Since we are dealing with a mode I crack, Equation (*b*) is written as

$$K_I = \sigma\sqrt{\pi a} \tag{5–35}$$

The stress intensity factor is *not* to be confused with the static stress-concentration factors K_t and K_{ts} defined in Sections 3–13 and 5–2.

Thus Equations (5–34) can be rewritten as

$$\sigma_x = \frac{K_I}{\sqrt{2\pi r}} \cos\frac{\theta}{2}\left(1 - \sin\frac{\theta}{2}\sin\frac{3\theta}{2}\right) \tag{5–36a}$$

$$\sigma_y = \frac{K_I}{\sqrt{2\pi r}} \cos\frac{\theta}{2}\left(1 + \sin\frac{\theta}{2}\sin\frac{3\theta}{2}\right) \tag{5–36b}$$

$$\tau_{xy} = \frac{K_I}{\sqrt{2\pi r}} \sin\frac{\theta}{2}\cos\frac{\theta}{2}\cos\frac{3\theta}{2} \tag{5–36c}$$

$$\sigma_z = \begin{cases} 0 & \text{(for plane stress)} \\ \nu(\sigma_x + \sigma_y) & \text{(for plane strain)} \end{cases} \tag{5–36d}$$

The stress intensity factor is a function of geometry, size and shape of the crack, and the type of loading. For various load and geometric configurations, Equation (5–35) can be written as

$$K_I = \beta\sigma\sqrt{\pi a} \tag{5–37}$$

where β is the *stress intensity modification factor*. Tables for β are available in the literature for basic configurations.[11] Figures 5–25 to 5–30 present a few examples of β for mode I crack propagation.

Fracture Toughness

When the magnitude of the mode I stress intensity factor reaches a critical value, K_{Ic}, crack propagation initiates. The *critical stress intensity factor* K_{Ic} is a material property that depends on the material, crack mode, processing of the material, temperature, loading rate, and the state of stress at the crack site (such as plane stress versus plane strain). The critical stress intensity factor K_{Ic} is also called the *fracture toughness* of the material. The fracture toughness for plane strain is normally lower than that for plane stress. For this reason, the term K_{Ic} is typically defined as the *mode I, plane strain fracture toughness*. Fracture toughness K_{Ic} for engineering metals lies in the range $20 \leq K_{Ic} \leq 200$ MPa · $\sqrt{\text{m}}$; for engineering polymers and ceramics, $1 \leq K_{Ic} \leq 5$ MPa · $\sqrt{\text{m}}$. For a 4340 steel, where the yield strength due to heat treatment ranges from 800 to 1600 MPa, K_{Ic} *decreases* from 190 to 40 MPa · $\sqrt{\text{m}}$.

Table 5–1 gives some approximate typical room-temperature values of K_{Ic} for several materials. As previously noted, the fracture toughness depends on many factors and the table is meant only to convey some typical magnitudes of K_{Ic}. For an actual application, it is recommended that the material specified for the application be certified using standard test procedures [see the American Society for Testing and Materials (ASTM) standard E399].

---

[11]See, for example:

H. Tada, P. C. Paris, and G. R. Irwin, *The Stress Analysis of Cracks Handbook*, 3rd ed., ASME Press, New York, 2000.

G. C. Sib, *Handbook of Stress Intensity Factors for Researchers and Engineers*, Institute of Fracture and Solid Mechanics, Lehigh University, Bethlehem, Pa., 1973.

Y. Murakami, ed., *Stress Intensity Factors Handbook*, Pergamon Press, Oxford, U.K., 1987.

W. D. Pilkey, *Formulas for Stress, Strain, and Structural Matrices*, 2nd ed., John Wiley & Sons, New York, 2005.

Table 5–1 Values of K_{Ic} for Some Engineering Materials at Room Temperature

Material	K_{Ic}, MPa $\sqrt{m}$	S_y, MPa
Aluminum		
2024	26	455
7075	24	495
7178	33	490
Titanium		
Ti-6Al-4V	115	910
Ti-6Al-4V*	55	1035
Steel		
4340	99	860
4340*	60	1515
52 100	14	2070

*Heat treated to a higher strength.

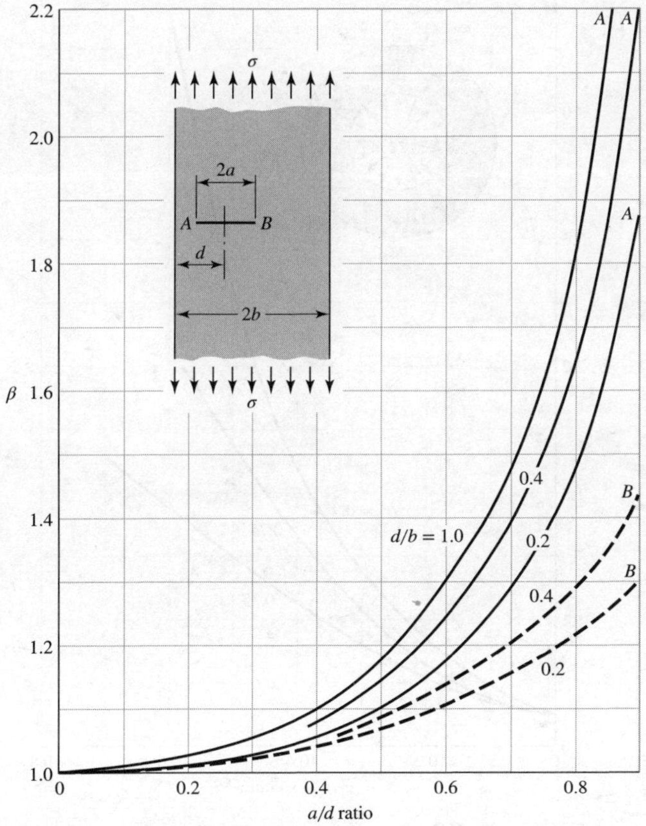

Figure 5–25

Off-center crack in a plate in longitudinal tension; solid curves are for the crack tip at *A*; dashed curves are for the tip at *B*.

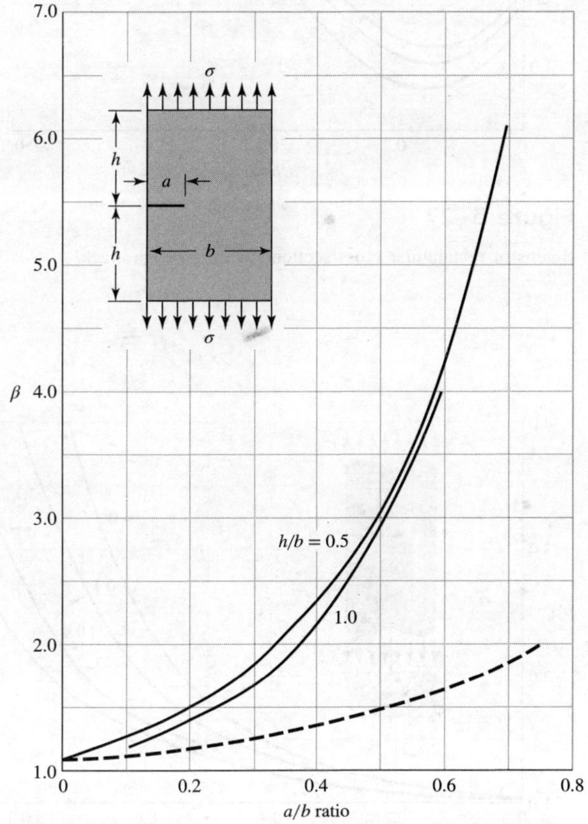

Figure 5–26

Plate loaded in longitudinal tension with a crack at the edge; for the solid curve there are no constraints to bending; the dashed curve was obtained with bending constraints added.

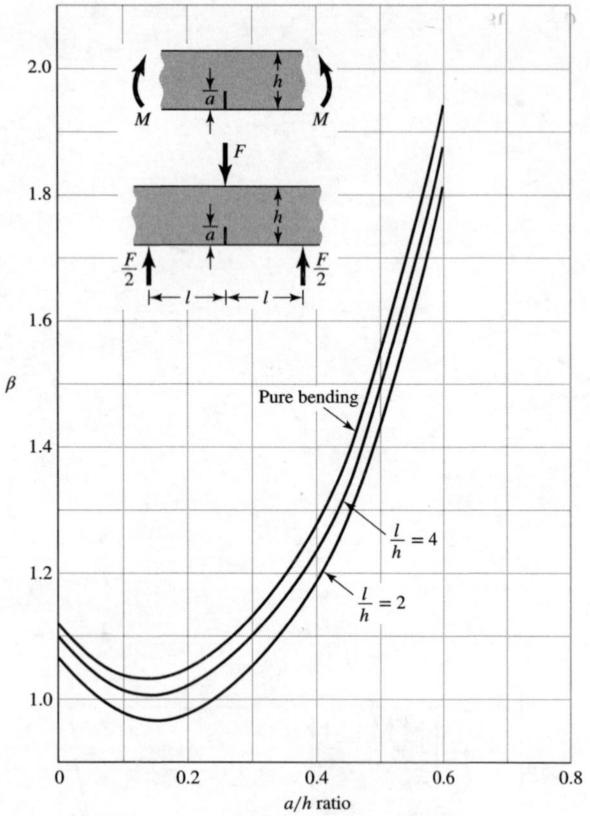

Figure 5–27

Beams of rectangular cross section having an edge crack.

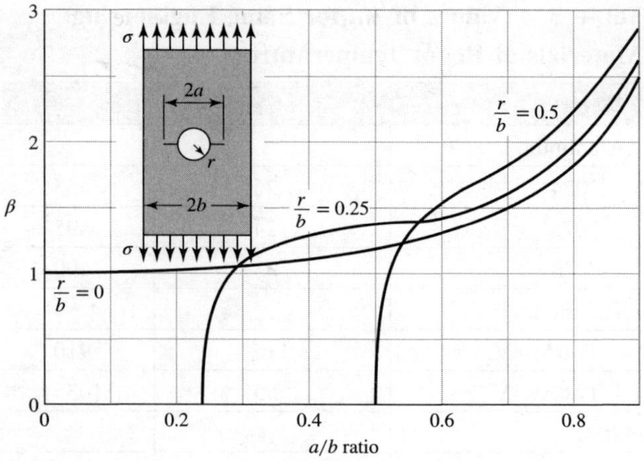

Figure 5–28

Plate in tension containing a circular hole with two cracks.

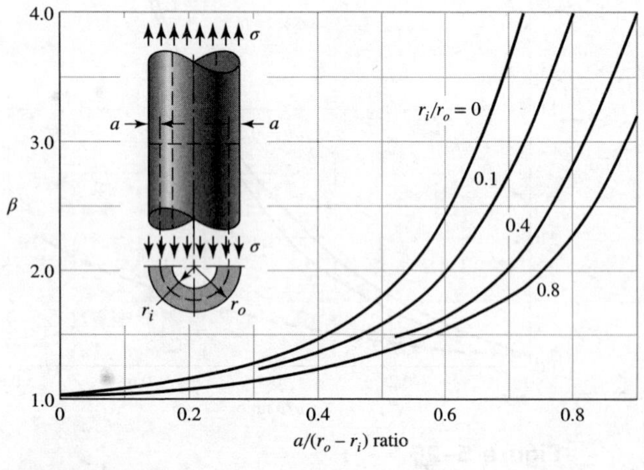

Figure 5–29

A cylinder loading in axial tension having a radial crack of depth a extending completely around the circumference of the cylinder.

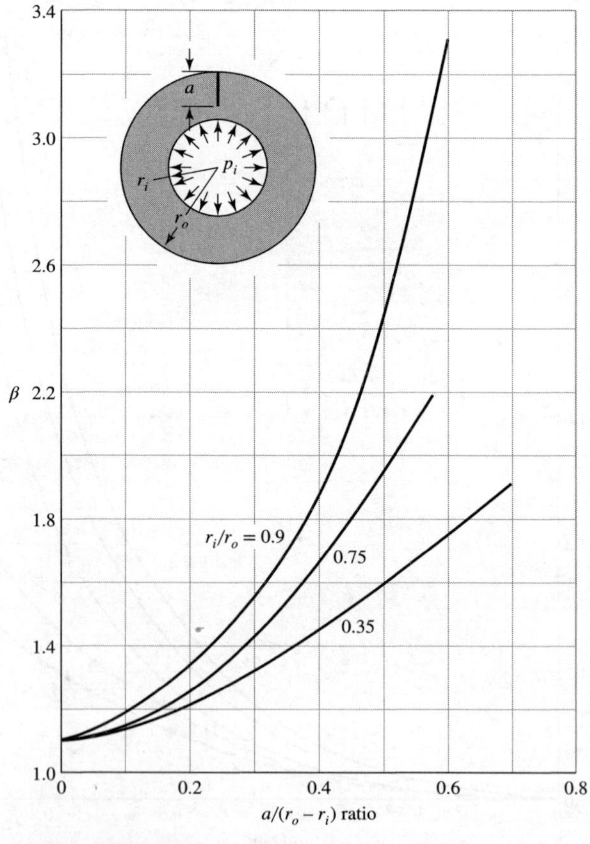

Figure 5–30

Cylinder subjected to internal pressure p_i, having a radial crack in the longitudinal direction of depth a. Use Equation (3–48) for the tangential stress at $r = r_0$.

One of the first problems facing the designer is that of deciding whether the conditions exist, or not, for a brittle fracture. Low-temperature operation, that is, operation below room temperature, is a key indicator that brittle fracture is a possible failure mode. Tables of transition temperatures for various materials have not been published, possibly because of the wide variation in values, even for a single material. Thus, in many situations, laboratory testing may give the only clue to the possibility of a brittle fracture. Another key indicator of the possibility of fracture is the ratio of the yield strength to the ultimate strength. A high ratio of S_y/S_u indicates there is only a small ability to absorb energy in the plastic region and hence there is a likelihood of brittle fracture.

The strength-to-stress ratio K_{Ic}/K_I can be used as a factor of safety as

$$n = \frac{K_{Ic}}{K_I} \tag{5–38}$$

EXAMPLE 5–6

A steel ship deck plate is 30 mm thick and 12 m wide. It is loaded with a nominal uniaxial tensile stress of 50 MPa. It is operated below its ductile-to-brittle transition temperature with K_{Ic} equal to 28.3 MPa. If a 65-mm-long central transverse crack is present, estimate the tensile stress at which catastrophic failure will occur. Compare this stress with the yield strength of 240 MPa for this steel.

Solution

For Figure 5–25, with $d = b$, $2a = 65$ mm and $2b = 12$ m, so that $d/b = 1$ and $a/d = 65/12(10^3) = 0.005\ 42$. Since a/d is so small, $\beta = 1$, so that

$$K_I = \sigma\sqrt{\pi a} = 50\sqrt{\pi(32.5 \times 10^{-3})} = 16.0 \text{ MPa } \sqrt{\text{m}}$$

From Equation (5–38),

$$n = \frac{K_{Ic}}{K_I} = \frac{28.3}{16.0} = 1.77$$

The stress at which catastrophic failure occurs is

Answer
$$\sigma_c = \frac{K_{Ic}}{K_I}\sigma = \frac{28.3}{16.0}(50) = 88.4 \text{ MPa}$$

The yield strength is 240 MPa, and catastrophic failure occurs at $88.4/240 = 0.37$, or at 37 percent of yield. The factor of safety in this circumstance is $K_{Ic}/K_I = 28.3/16 = 1.77$ and *not* $240/50 = 4.8$.

EXAMPLE 5–7

A plate of width 1.4 m and length 2.8 m is required to support a tensile force in the 2.8-m direction of 4.0 MN. Inspection procedures will detect only through-thickness edge cracks larger than 2.7 mm. The two Ti-6AL-4V alloys in Table 5–1 are being considered for this application, for which the safety factor must be 1.3 and minimum weight is important. Which alloy should be used?

Solution

(*a*) We elect first to estimate the thickness required to resist yielding. Because $\sigma = P/wt$, we have $t = P/w\sigma$. For the weaker alloy, we have, from Table 5–1, $S_y = 910$ MPa. Thus,

$$\sigma_{\text{all}} = \frac{S_y}{n} = \frac{910}{1.3} = 700 \text{ MPa}$$

Thus,

$$t = \frac{P}{w\sigma_{\text{all}}} = \frac{4.0(10)^3}{1.4(700)} = 4.08 \text{ mm or greater}$$

For the stronger alloy, we have, from Table 5–1,

$$\sigma_{\text{all}} = \frac{1035}{1.3} = 796 \text{ MPa}$$

and so the thickness is

Answer
$$t = \frac{P}{w\sigma_{\text{all}}} = \frac{4.0(10)^3}{1.4(796)} = 3.59 \text{ mm or greater}$$

(*b*) Now let us find the thickness required to prevent crack growth. Using Figure 5–26, we have

$$\frac{h}{b} = \frac{2.8/2}{1.4} = 1 \qquad \frac{a}{b} = \frac{2.7}{1.4(10^3)} = 0.001\ 93$$

Corresponding to these ratios we find from Figure 5–26 that $\beta \approx 1.1$, and $K_I = 1.1\sigma\sqrt{\pi a}$.

$$n = \frac{K_{Ic}}{K_I} = \frac{115\sqrt{10^3}}{1.1\sigma\sqrt{\pi a}} \qquad \sigma = \frac{K_{Ic}}{1.1n\sqrt{\pi a}}$$

From Table 5–1, $K_{Ic} = 115$ MPa$\sqrt{\text{m}}$ for the weaker of the two alloys. Solving for σ with $n = 1$ gives the fracture stress

$$\sigma = \frac{115}{1.1\sqrt{\pi(2.7)(10)^{-3}}} = 1135 \text{ MPa}$$

which is greater than the yield strength of 910 MPa, and so yield strength is the basis for the geometry decision. For the stronger alloy $S_y = 1035$ MPa, with $n = 1$ the fracture stress is

$$\sigma = \frac{K_{Ic}}{nK_I} = \frac{55}{1(1.1)\sqrt{\pi(2.7)(10)^{-3}}} = 542.9 \text{ MPa}$$

which is less than the yield strength of 1035 MPa. The thickness t is

$$t = \frac{P}{w\sigma_{\text{all}}} = \frac{4.0(10^3)}{1.4(542.9/1.3)} = 6.84 \text{ mm or greater}$$

This example shows that the fracture toughness K_{Ic} limits the geometry when the stronger alloy is used, and so a thickness of 6.84 mm or larger is required. When the weaker alloy is used the geometry is limited by the yield strength, giving a thickness of only 4.08 mm or greater. Thus the weaker alloy leads to a thinner and lighter weight choice since the failure modes differ.

5–13 Important Design Equations

The following equations are provided as a summary. *Note for plane stress:* The principal stresses in the following equations that are labeled σ_A and σ_B represent the principal stresses determined from the *two-dimensional* Equation (3–14).

Maximum Shear Theory

$$\tau_{\max} = \frac{\sigma_1 - \sigma_3}{2} = \frac{S_y}{2n} \tag{5–3}$$

Distortion-Energy Theory

Von Mises stress

$$\sigma' = \left[\frac{(\sigma_1 - \sigma_2)^2 + (\sigma_2 - \sigma_3)^2 + (\sigma_3 - \sigma_1)^2}{2}\right]^{1/2} \tag{5–12}$$

$$\sigma' = \frac{1}{\sqrt{2}}\left[(\sigma_x - \sigma_y)^2 + (\sigma_y - \sigma_z)^2 + (\sigma_z - \sigma_x)^2 + 6\left(\tau_{xy}^2 + \tau_{yz}^2 + \tau_{zx}^2\right)\right]^{1/2} \tag{5–14}$$

Plane stress

$$\sigma' = \left(\sigma_A^2 - \sigma_A\sigma_B + \sigma_B^2\right)^{1/2} \tag{5–13}$$

$$\sigma' = \left(\sigma_x^2 - \sigma_x\sigma_y + \sigma_y^2 + 3\tau_{xy}^2\right)^{1/2} \tag{5–15}$$

Yield design equation

$$\sigma' = \frac{S_y}{n} \tag{5–19}$$

Shear yield strength

$$S_{sy} = 0.577\ S_y \tag{5–21}$$

Coulomb-Mohr Theory

$$\frac{\sigma_1}{S_t} - \frac{\sigma_3}{S_c} = \frac{1}{n} \tag{5–26}$$

where S_t is tensile yield (ductile) or ultimate tensile (brittle), and S_c is compressive yield (ductile) or ultimate compressive (brittle) strengths.

Modified Mohr (Plane Stress)

$$\sigma_A = \frac{S_{ut}}{n} \qquad \sigma_A \ge \sigma_B \ge 0 \tag{5–32a}$$

$$\sigma_A \ge 0 \ge \sigma_B \quad \text{and} \quad \left|\frac{\sigma_B}{\sigma_A}\right| \le 1$$

$$\frac{(S_{uc} - S_{ut})\sigma_A}{S_{uc}S_{ut}} - \frac{\sigma_B}{S_{uc}} = \frac{1}{n} \qquad \sigma_A \ge 0 \ge \sigma_B \quad \text{and} \quad \left|\frac{\sigma_B}{\sigma_A}\right| > 1 \tag{5–32b}$$

$$\sigma_B = -\frac{S_{uc}}{n} \qquad 0 \ge \sigma_A \ge \sigma_B \tag{5–32c}$$

Failure Theory Flowchart

Figure 5–21

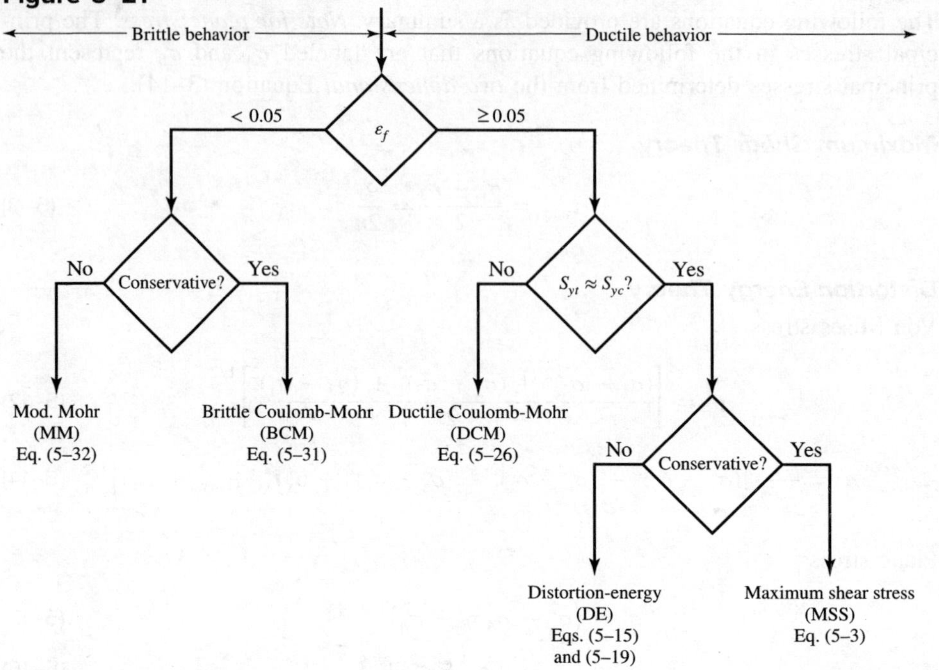

Fracture Mechanics

$$K_I = \beta\sigma\sqrt{\pi a} \tag{5–37}$$

where β is found in Figures 5–25 to 5–30

$$n = \frac{K_{Ic}}{K_I} \tag{5–38}$$

where K_{Ic} is found in Table 5–1

PROBLEMS

Problems marked with an asterisk (*) are linked to problems in other chapters, as summarized in Table 1–2 of Section 1–17.

5–1 A ductile hot-rolled steel bar has a minimum yield strength in tension and compression of 350 MPa. Using the distortion-energy and maximum-shear-stress theories determine the factors of safety for the following plane stress states:

(a) $\sigma_x = 100$ MPa, $\sigma_y = 100$ MPa
(b) $\sigma_x = 100$ MPa, $\sigma_y = 50$ MPa
(c) $\sigma_x = 100$ MPa, $\tau_{xy} = -75$ MPa
(d) $\sigma_x = -50$ MPa, $\sigma_y = -75$ MPa, $\tau_{xy} = -50$ MPa
(e) $\sigma_x = 100$ MPa, $\sigma_y = 20$ MPa, $\tau_{xy} = -20$ MPa

5–2 Repeat Problem 5–1 with the following principal stresses obtained from Equation (3–14):

(a) $\sigma_A = 100$ MPa, $\sigma_B = 100$ MPa
(b) $\sigma_A = 100$ MPa, $\sigma_B = -100$ MPa
(c) $\sigma_A = 100$ MPa, $\sigma_B = 50$ MPa
(d) $\sigma_A = 100$ MPa, $\sigma_B = -50$ MPa
(e) $\sigma_A = -50$ MPa, $\sigma_B = -100$ MPa

5–3 Repeat Problem 5–1 for a bar of AISI 1030 hot-rolled steel and:
(*a*) $\sigma_x = 175$ MPa, $\sigma_y = 105$ MPa
(*b*) $\sigma_x = 105$ MPa, $\sigma_y = -105$ MPa
(*c*) $\sigma_x = 140$ MPa, $\tau_{xy} = -70$ MPa
(*d*) $\sigma_x = -84$ MPa, $\sigma_y = 105$ MPa, $\tau_{xy} = -63$ MPa
(*e*) $\sigma_x = -168$ MPa, $\sigma_y = -168$ MPa, $\tau_{xy} = -105$ MPa

5–4 Repeat Problem 5–1 for a bar of AISI 1015 cold-drawn steel with the following principal stresses obtained from Equation (3–14):
(*a*) $\sigma_A = 210$ MPa, $\sigma_B = 210$ MPa
(*b*) $\sigma_A = 210$ MPa, $\sigma_B = -210$ MPa
(*c*) $\sigma_A = 210$ MPa, $\sigma_B = 105$ MPa
(*d*) $\sigma_A = -210$ MPa, $\sigma_B = -105$ MPa
(*e*) $\sigma_A = -350$ MPa, $\sigma_B = 70$ MPa

5–5 Repeat Problem 5–1 by first plotting the failure loci in the σ_A, σ_B plane to scale; then, for each stress state, plot the load line and by graphical measurement estimate the factors of safety.

5–6 Repeat Problem 5–3 by first plotting the failure loci in the σ_A, σ_B plane to scale; then, for each stress state, plot the load line and by graphical measurement estimate the factors of safety.

5–7 to 5–11 An AISI 1018 steel has a yield strength, $S_y = 295$ MPa. Using the distortion-energy theory for the given state of plane stress, (*a*) determine the factor of safety, (*b*) plot the failure locus, the load line, and estimate the factor of safety by graphical measurement.

Problem Number	σ_x (MPa)	σ_y (MPa)	τ_{xy} (MPa)
5–7	75	−35	0
5–8	−100	30	0
5–9	100	0	−25
5–10	−30	−65	40
5–11	−80	30	−10

5–12 A ductile material has the properties $S_{yt} = 420$ MPa and $S_{yc} = 525$ MPa. Using the ductile Coulomb-Mohr theory, determine the factor of safety for the states of plane stress given in Problem 5–3.

5–13 Repeat Problem 5–12 by first plotting the failure loci in the σ_A, σ_B plane to scale; then for each stress state, plot the load line and by graphical measurement estimate the factor of safety.

5–14 to 5–18 A material at 427°C exhibits $S_{yt} = 235$ MPa, $S_{yc} = 285$ MPa, and $\varepsilon_f = 0.07$. For the given state of plane stress, (*a*) determine the factor of safety, (*b*) plot the failure locus and the load line, and estimate the factor of safety by graphical measurement.

Problem Number	σ_x (MPa)	σ_y (MPa)	τ_{xy} (MPa)
5–14	150	−50	0
5–15	−150	50	0
5–16	125	0	−75
5–17	−80	−125	50
5–18	125	80	−75

5–19 A brittle material has the properties $S_{ut} = 210$ MPa and $S_{uc} = 630$ MPa. Using the brittle Coulomb-Mohr and modified-Mohr theories, determine the factor of safety for the following states of plane stress.

(*a*) $\sigma_x = 175$ MPa, $\sigma_y = 105$ MPa
(*b*) $\sigma_x = 105$ MPa, $\sigma_y = -105$ MPa
(*c*) $\sigma_x = 140$ MPa, $\tau_{xy} = -70$ MPa
(*d*) $\sigma_x = -105$ MPa, $\sigma_y = -70$ MPa, $\tau_{xy} = -105$ MPa
(*e*) $\sigma_x = -140$ MPa, $\sigma_y = -140$ MPa, $\tau_{xy} = -105$ MPa

5–20 Repeat Problem 5–19 by first plotting the failure loci in the σ_A, σ_B plane to scale; then for each stress state, plot the load line and by graphical measurement estimate the factor of safety.

5–21 to 5–25 For an ASTM 30 cast iron, (*a*) find the factors of safety using the BCM and MM theories, (*b*) plot the failure diagrams in the σ_A, σ_B plane to scale and locate the coordinates of the stress state, and (*c*) estimate the factors of safety from the two theories by graphical measurements along the load line.

Problem Number	σ_x (MPa)	σ_y (MPa)	τ_{xy} (MPa)
5–21	105	70	0
5–22	105	−350	0
5–23	105	0	−70
5–24	−70	−175	−70
5–25	−245	90	−70

5–26 to 5–30 A cast aluminum 195-T6 exhibits $S_{ut} = 252$ MPa, $S_{uc} = 245$ MPa, and $\varepsilon_f = 0.045$. For the given state of plane stress, (*a*) using the Coulomb-Mohr theory, determine the factor of safety, (*b*) plot the failure locus and the load line, and estimate the factor of safety by graphical measurement.

Problem Number	σ_x (MPa)	σ_y (MPa)	τ_{xy} (MPa)
5–26	105	−70	0
5–27	−105	70	0
5–28	84	0	−56
5–29	−70	−105	10
5–30	105	56	−56

5–31 to 5–35 Repeat Problems 5–26 to 5–30 using the modified-Mohr theory.

Problem number	5–31	5–32	5–33	5–34	5–35
Repeat problem	5–26	5–27	5–28	5–29	5–30

5–36 This problem illustrates that the factor of safety for a machine element depends on the particular point selected for analysis. Here you are to compute factors of safety, based upon the distortion-energy theory, for stress elements at *A* and *B* of the member shown in the figure. This bar is made of AISI 1006 cold-drawn steel and is loaded by the forces $F = 0.55$ kN, $P = 4.0$ kN, and $T = 25$ N · m.

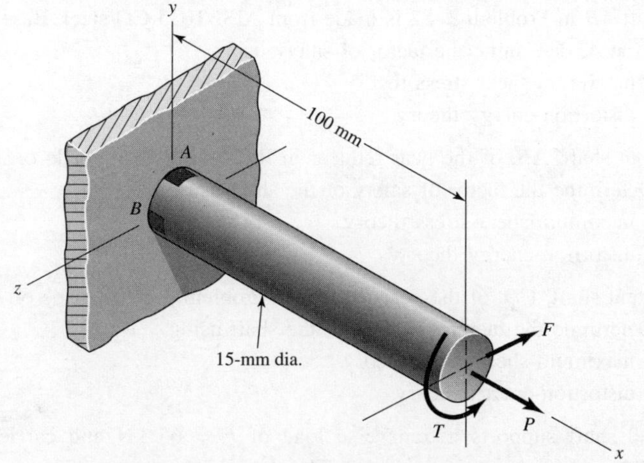

Problem 5–36

5–37 For the beam in Problem 3–42 determine the minimum yield strength that should be considered to obtain a minimum factor of safety of 2 based on the distortion-energy theory.

5–38 A 1020 CD steel shaft is to transmit 15 kW while rotating at 1750 rpm. Determine the minimum diameter for the shaft to provide a minimum factor of safety of 3 based on the maximum-shear-stress theory.

5–39 A 30-mm-diameter shaft, made of AISI 1018 HR steel, transmits 10 kW of power while rotating at 200 rev/min. Assume any bending moments present in the shaft to be negligibly small compared to the torque. Determine the static factor of safety based on
(*a*) the maximum-shear-stress failure theory.
(*b*) the distortion-energy failure theory.

5–40 A 20-mm-diameter steel shaft, made of AISI 1035 HR steel, transmits power while rotating at 400 rev/min. Assume any bending moments in the shaft to be relatively small compared to the torque. Determine how much power, in units of kW, the shaft can transmit with a static factor of safety of 1.5 based on
(*a*) the maximum-shear-stress theory.
(*b*) distortion-energy theory.

5–41 The short cantilevered peg in Problem 3–44 is made from AISI 1040 cold-drawn steel. Evaluate the state of stress at points *A*, *B*, and *C*, determine the critical case, and evaluate the minimum factor of safety based on
(*a*) the maximum-shear-stress theory.
(*b*) the distortion-energy theory.

5–42 The shaft *ABD* in Problem 3–50 is made from AISI 1040 CD steel. Based on failure at the wall at *A*, determine the factor of safety using
(*a*) the maximum-shear-stress theory.
(*b*) the distortion-energy theory.

5–43 The shaft *ABD* in Problem 3–51 is made from AISI 1020 CD steel. Based on failure at the wall at *A*, determine the factor of safety using
(*a*) the maximum-shear-stress theory.
(*b*) the distortion-energy theory.

5–44 The shaft *AB* in Problem 3–52 is made from AISI 1035 CD steel. Based on failure at the wall at *A*, determine the factor of safety using
(*a*) the maximum-shear-stress theory.
(*b*) the distortion-energy theory.

5–45 The input shaft, *AB*, of the gear reducer in Problem 3–98 is made of AISI 1040 CD steel. Determine the factor of safety of the shaft using
(*a*) the maximum-shear-stress theory.
(*b*) the distortion-energy theory.

5–46 The output shaft, *CD*, of the gear reducer in Problem 3–99 is made of AISI 1040 CD steel. Determine the factor of safety of the shaft using
(*a*) the maximum-shear-stress theory.
(*b*) the distortion-energy theory.

5–47 A round shaft supports a transverse load of $F = 65$ kN and carries a torque of $T = 800$ N · m, as shown in the figure. The shaft does not rotate. The shaft is machined from AISI 4140 steel, quenched and tempered at 200°C. Document the location of the critical stress element and how you determined that element to be critical. Do not assume transverse shear stress is negligible without proving it. Then,
(*a*) determine the minimum factor of safety based on yielding according to the maximum-shear-stress theory.
(*b*) determine the minimum factor of safety based on yielding according to the distortion-energy theory.

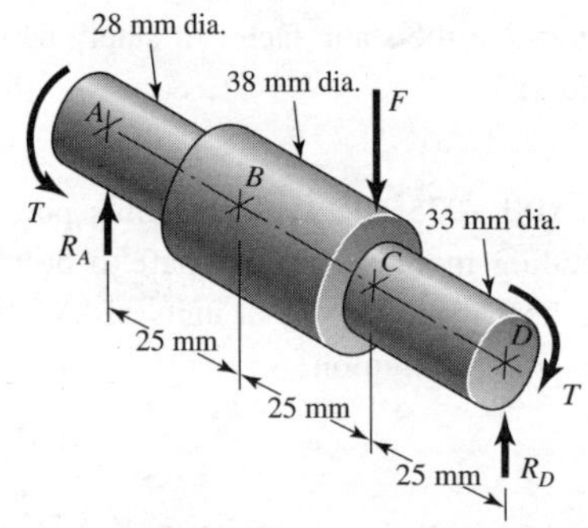

Problem 5–47

5–48 A 25-mm, constant diameter shaft, is loaded with forces at *A* and *B* as shown, with ground reaction forces at *O* and *C*. The shaft also transmits a torque of 150 N · m throughout the length of the shaft. The shaft has a tensile yield strength of 350 MPa. Determine the minimum static factor of safety using
(*a*) the maximum-shear-stress failure theory.
(*b*) the distortion-energy failure theory.

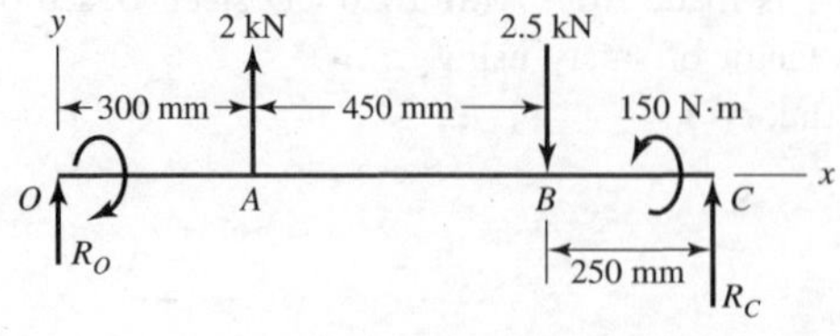

Problem 5–48

5–49 Cantilevered rod *OA* is 0.5 m long, and made from AISI 1010 hot-rolled steel. A constant force and torque are applied as shown. Determine the minimum diameter, *d*, for the rod that will achieve a minimum static factor of safety of 2
(*a*) using the maximum-shear-stress failure theory.
(*b*) using the distortion-energy failure theory.

Problem 5–49

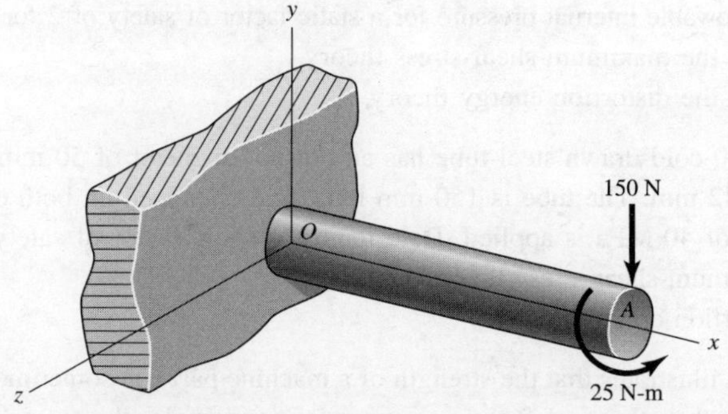

5–50* to 5–66* For the problem specified in the table, build upon the results of the original problem to determine the minimum factor of safety for yielding. Use both the maximum-shear-stress theory and the distortion-energy theory, and compare the results. The material is 1018 CD steel.

Problem Number	Original Problem Number
5–50*	3–76
5–51*	3–77
5–52*	3–78
5–53*	3–79
5–54*	3–80
5–55*	3–81
5–56*	3–82
5–57*	3–84
5–58*	3–85
5–59*	3–87
5–60*	3–88
5–61*	3–89
5–62*	3–90
5–63*	3–91
5–64*	3–92
5–65*	3–93
5–66*	3–94

5–67* Build upon the results of Problems 3–93 and 3–95 to compare the use of a low-strength, ductile material (1018 CD) in which the stress-concentration factor can be ignored to a high-strength but more brittle material (4140 Q&T @ 204°C) in which the stress-concentration factor should be included. For each case, determine the factor of safety for yielding using the distortion-energy theory.

5–68 Using $F = 1770$ N, design the lever arm *CD* of Figure 5–16 by specifying a suitable size and material.

5–69 A spherical pressure vessel is formed of 1.25 mm cold-drawn AISI 1020 sheet steel. If the vessel has a diameter of 200 mm, use the distortion-energy theory to estimate the pressure necessary to initiate yielding. What is the estimated bursting pressure?

5–70 An AISI 1040 cold-drawn steel tube has an outside diameter of 50 mm and an inside diameter of 42 mm. The tube is 150 mm long and is capped at both ends. Determine the maximum allowable internal pressure for a static factor of safety of 2 for the tube walls

(*a*) based on the maximum-shear-stress theory.

(*b*) based on the distortion-energy theory.

5–71 An AISI 1040 cold-drawn steel tube has an outside diameter of 50 mm and an inside diameter of 42 mm. The tube is 150 mm long, and is capped on both ends. An internal pressure of 40 MPa is applied. Determine the static factor of safety using

(*a*) the maximum-shear-stress theory.

(*b*) the distortion-energy theory.

5–72 This problem illustrates that the strength of a machine part can sometimes be measured in units other than those of force or moment. For example, the maximum speed that a flywheel can reach without yielding or fracturing is a measure of its strength. In this problem you have a rotating ring made of hot-forged AISI 1020 steel; the ring has a 150 mm inside diameter and a 250 mm outside diameter and is 40 mm thick. Using the distortion-energy theory, determine the speed in revolutions per minute that would cause the ring to yield. At what radius would yielding begin? [*Note:* The maximum radial stress occurs at $r = (r_o r_i)^{1/2}$; see Equation (3–55).]

5–73 A light pressure vessel is made of 2024-T3 aluminum alloy tubing with suitable end closures. This cylinder has a 100 mm OD, a 1.5 mm wall thickness, and $\nu = 0.334$. The purchase order specifies a minimum yield strength of 320 MPa. Using the distortion-energy theory, determine the factor of safety if the pressure-release valve is set at 3.5 MPa.

5–74 A cold-drawn AISI 1015 steel tube is 300 mm OD by 200 mm ID and is to be subjected to an external pressure caused by a shrink fit. Using the distortion-energy theory, determine the maximum pressure that would cause the material of the tube to yield.

5–75 What speed would cause fracture of the ring of Problem 5–72 if it were made of grade 30 cast iron?

5–76 The figure shows a shaft mounted in bearings at *A* and *D* and having pulleys at *B* and *C*. The forces shown acting on the pulley surfaces represent the belt tensions. The shaft is to be made of AISI 1035 CD steel. Using a conservative failure theory with a design factor of 2, determine the minimum shaft diameter to avoid yielding.

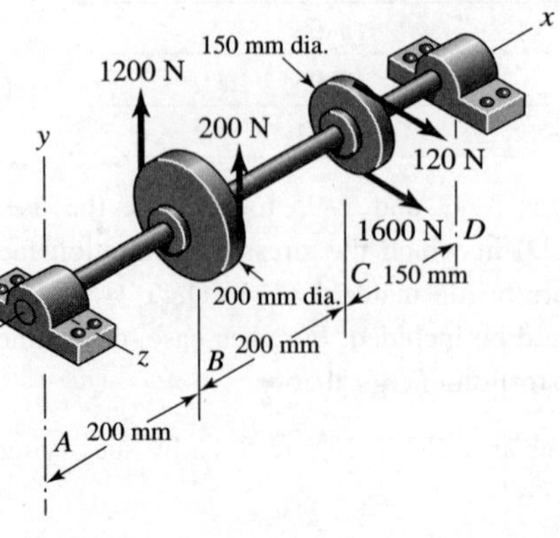

Problem 5–76

5–77 By modern standards, the shaft design of Problem 5–76 is poor because it is so long. Suppose it is redesigned by halving the length dimensions. Using the same material and design factor as in Problem 5–76, find the new shaft diameter.

5–78* Build upon the results of Problem 3–39 to determine the factor of safety for yielding based on the distortion-energy theory for each of the simplified models in parts *c*, *d*, and *e* of the figure for Problem 3–39. The pin is machined from AISI 1018 hot-rolled steel. Compare the three models from a designer's perspective in terms of accuracy, safety, and modeling time.

5–79* For the clevis pin of Problem 3–39, redesign the pin diameter to provide a factor of safety of 2.5 based on a conservative yielding failure theory, and the most conservative loading model from parts *c*, *d*, and *e* of the figure for Problem 3–39. The pin is machined from AISI 1018 hot-rolled steel.

5–80 A split-ring clamp-type shaft collar is shown in the figure. The collar is 50 mm OD by 25 mm ID by 12 mm wide. The screw is designated as M 6 × 1. The relation between the screw tightening torque T, the nominal screw diameter d, and the tension in the screw F_i is approximately $T = 0.2\ F_i d$. The shaft is sized to obtain a close running fit. Find the axial holding force F_x of the collar as a function of the coefficient of friction and the screw torque.

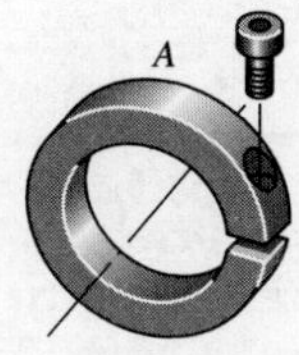

Problem 5–80

5–81 Suppose the collar of Problem 5–80 is tightened by using a screw torque of 20 N · m. The collar material is AISI 1035 steel heat-treated to a minimum tensile yield strength of 450 MPa.

(*a*) Estimate the tension in the screw.

(*b*) By relating the tangential stress to the hoop tension, find the internal pressure of the shaft on the ring.

(*c*) Find the tangential and radial stresses in the ring at the inner surface.

(*d*) Determine the maximum shear stress and the von Mises stress.

(*e*) What are the factors of safety based on the maximum-shear-stress and the distortion-energy theories?

5–82 In Problem 5–80, the role of the screw was to induce the hoop tension that produces the clamping. The screw should be placed so that no moment is induced in the ring. Just where should the screw be located?

5–83 A tube has another tube shrunk over it. The specifications are:

	Inner Member	Outer Member
ID	25 ± 0.05 mm	49.98 ± 0.01 mm
OD	50 ± 0.01 mm	75 ± 0.1 mm

Both tubes are made of a plain carbon steel.

(*a*) Find the nominal shrink-fit pressure and the von Mises stresses at the fit surface.

(*b*) If the inner tube is changed to solid shafting with the same outside dimensions, find the nominal shrink-fit pressure and the von Mises stresses at the fit surface.

5–84 Two steel tubes have the specifications:

	Inner Tube	Outer Tube
ID	20 ± 0.050 mm	39.98 ± 0.008 mm
OD	40 ± 0.008 mm	65 ± 0.10 mm

These are shrink-fitted together. Find the nominal shrink-fit pressure and the von Mises stress in each body at the fit surface.

5–85 Repeat Problem 5–84 for maximum shrink-fit conditions.

5–86 A solid steel shaft has a gear with ASTM grade 20 cast-iron hub (E = 100 GPa) shrink-fitted to it. The shaft diameter is 50 ± 0.01 mm. The specifications for the gear hub are

$$49\begin{array}{l}+0.01\\-0.00\end{array}\text{ mm}$$

ID with an OD of 100 ± 0.8 mm. Using the midrange values and the modified Mohr theory, estimate the factor of safety guarding against fracture in the gear hub due to the shrink fit.

5–87 Two steel tubes are shrink-fitted together where the nominal diameters are 40, 45, and 50 mm. Careful measurement before fitting determined the diametral interference between the tubes to be 0.062 mm. After the fit, the assembly is subjected to a torque of 900 N · m and a bending-moment of 6.75 N · m. Assuming no slipping between the cylinders, analyze the outer cylinder at the inner and outer radius. Determine the factor of safety using distortion energy with S_y = 415 MPa.

5–88 Repeat Problem 5–87 for the inner tube.

5–89 to 5–94 For the problem given in the table, the specifications for the press fit of two cylinders are given in the original problem from Chapter 3. If both cylinders are hot-rolled AISI 1040 steel, determine the minimum factor of safety for the outer cylinder based on the distortion-energy theory.

Problem Number	Original Problem Number
5–89	3–121
5–90	3–122
5–91	3–123
5–92	3–124
5–93	3–125
5–94	3–126

5–95 For Equations (5–36) show that the principal stresses are given by

$$\sigma_1 = \frac{K_I}{\sqrt{2\pi r}}\cos\frac{\theta}{2}\left(1+\sin\frac{\theta}{2}\right)$$

$$\sigma_2 = \frac{K_I}{\sqrt{2\pi r}}\cos\frac{\theta}{2}\left(1-\sin\frac{\theta}{2}\right)$$

$$\sigma_3 = \begin{cases} 0 & \text{(plane stress)} \\ \sqrt{\dfrac{2}{\pi r}}\,\nu K_I\cos\dfrac{\theta}{2} & \text{(plane strain)} \end{cases}$$

5–96 Use the results of Problem 5–95 for plane strain near the tip with $\theta = 0$ and $\nu = \frac{1}{3}$. If the yield strength of the plate is S_y, what is σ_1 when yield occurs?

(*a*) Use the distortion-energy theory.

(*b*) Use the maximum-shear-stress theory. Using Mohr's circles, explain your answer.

5–97 A plate 100 mm wide, 200 mm long, and 12 mm thick is loaded in tension in the direction of the length. The plate contains a crack as shown in Figure 5–26 with the crack length of 16 mm. The material is steel with K_{Ic} = 80 MPa · $\sqrt{\text{m}}$, and S_y = 950 MPa. Determine the maximum possible load that can be applied before the plate (*a*) yields, and (*b*) has uncontrollable crack growth.

5–98 A cylinder subjected to internal pressure p_i has an outer diameter of 350 mm and a 25 mm wall thickness. For the cylinder material, K_{Ic} = 80 MPa · $\sqrt{\text{m}}$, S_y = 1200 MPa, and S_{ut} = 1350 MPa. If the cylinder contains a radial crack in the longitudinal direction of depth 12.5 mm determine the pressure that will cause uncontrollable crack growth.

6 Fatigue Failure Resulting from Variable Loading

From ASM Handbook, Vol. 12: Fractography, 2nd printing, 1992, ASM International, Materials Park, OH 44073-0002, fig. 50, p. 120. Reprinted by permission of ASM International®, www.asminternational.org.

Chapter Outline

In Chapter 5 we considered the analysis and design of parts subjected to static loading. In this chapter we shall examine the failure mechanisms of parts subjected to fluctuating loading conditions.

6–1 Introduction to Fatigue

Prior to the nineteenth century, engineering design was based primarily on static loading. Speeds were relatively slow, loads were light, and factors of safety were large. About the time of the development of the steam engine, this changed. Shafts were expected to run faster, accumulating high repetitions of dynamically cycling stresses. An example is a particular stress element on the surface of a rotating shaft subjected to the action of bending loads that cycles between tension and compression with each revolution of the shaft. In the days of waterwheel-powered machinery, a shaft might experience only a few thousand cycles in a day. The steam engine could rotate at several hundred revolutions per minute, accumulating hundreds of thousands of cycles in a single day.

As material manufacturing processes improved, steels with higher strengths and greater uniformity became available, leading to the confidence to design for higher stresses and lower static factors of safety. Increasingly, machine members failed suddenly and often dramatically, under the action of fluctuating stresses; yet a careful analysis revealed that the maximum stresses were well below the yield strength of the material. The most distinguishing characteristic of these failures was that the stresses had been repeated a large number of times. This led to the notion that the part had simply become "tired" from the repeated cycling, hence the origin of the term *fatigue failure*. However, posttesting of failed parts indicated that the material still had its original properties. In actuality, fatigue failure is not a consequence of changed material properties but from a crack initiating and growing when subjected to many repeated cycles of stress.

Some of the first notable fatigue failures involved railroad axles in the mid-1800s and led to investigations by August Wöhler, who is credited with deliberately studying and articulating some of the basic principles of fatigue failure, as well as design strategies to avoid fatigue failure. The next 150 years saw a steady stream of research to both understand the underlying mechanisms of fatigue and to develop theoretical modeling techniques for practical design use. There have been times in which some postulated theories gained traction, only to later be discarded as false paths. The understanding of fatigue failure has been somewhat elusive, perhaps to be expected due to the fact that the failures are a result of many factors combining, seemingly with a high level of randomness included. With persistence, the field of study has stabilized. With the advancement of technology to examine microstructures at extreme magnification, much of the physical mechanics of the problem has been well established. More recent work in the last few decades has focused on the theoretical modeling to incorporate both the mechanics as well as the empirical evidence.

Historically, even as progress was made, dramatic failures continued to prove the need for continued study. A few notable examples that can be readily researched for details include the following: Versailles railroad axle (1842), Liberty ships (1943), multiple de Havilland Comet crashes (1954), Kielland oil platform collapse (1980), Aloha B737 accident (1988), DC10 Sioux City accident (1989), MD-88 Pensacola engine failure (1996), Eschede railway accident (1998) (See Figure 6–17), GE CF6 engine failure (2016), Denver Colorado Boeing 777 turbine blade break (2021) (See Figure 6–18).

6–2 Chapter Overview

In this chapter, we will present the concepts pertinent to understanding the mechanisms of fatigue failure, as well as a practical approach to estimating fatigue life. The study of fatigue includes the overlaying of several complexities. As an aid to organizing these complexities, a brief outline of the chapter is given here.

I. *Crack Nucleation and Propagation (Section 6–3)*
This section describes the stages of crack nucleation, propagation, and fracture. The description starts with macroscale observations, then moves to examining the reaction of the material at a microscopic level to cyclical strain loading. It provides the conceptual background to understanding why and how fatigue fracture happens.

II. *Fatigue-Life Methods (Section 6–4)*
This section describes and contrasts the three major approaches to predicting the fatigue life of a part, namely, the *linear-elastic fracture mechanics method,* the *strain-life method,* and the *stress-life method*. The purpose of this section is to briefly introduce these topics to provide a big picture of the methods to be described in the following sections.

III. *The Linear-Elastic Fracture Mechanics Method (Section 6–5)*
The *linear-elastic fracture mechanics method* assumes a crack exists and uses principles of fracture mechanics to model the growth of the crack. This section is useful for a general understanding of crack growth. It is intended as an overview and is not essential for understanding the remainder of the chapter.

IV. *The Strain-Life Method (Section 6–6)*
The *strain-life method* uses strain testing to analyze the effects of local yielding at notches. It is particularly useful for situations with high stress and plastic strain, and therefore low life expectancy. Local plastic strain is the driving mechanism for fatigue failures, so a general understanding of the concepts in this section is worthwhile. The method is complex enough that this section does not attempt to cover all of the details. It is intended as an overview and is not essential for understanding the remainder of the chapter.

V. *The Stress-Life Method in Detail (Sections 6–7 through 6–17)*
The stress-life method is fully developed in the remainder of the chapter as a practical approach to estimating the fatigue life of a part. The coverage is organized into three categories of loading: completely reversed, fluctuating, and combined modes.

A. *Completely Reversed Loading (Sections 6–7 through 6–10)*
Completely reversed loading is the simplest type of stress cycling between equal amplitudes of tension and compression. The majority of fatigue experimentation uses this type of loading, so the fundamentals are first addressed with this loading. These sections represent the heart of the stress-life method.

i. *The Stress-Life Method and the S-N Diagram (Section 6–7)*
ii. *The Idealized S-N Diagram for Steels (Section 6–8)*
iii. *Endurance Limit Modifying Factors (Section 6–9)*
iv. *Stress Concentration and Notch Sensitivity (Section 6–10)*

B. *Fluctuating Loading (Sections 6–11 through 6–15)*
Generalized fluctuating stress levels are introduced to model the effect of nonzero mean stress on the fatigue life. Several well-known models are introduced to represent the experimental data.
  i. *Characterizing Fluctuating Stresses (Section 6–11)*
  ii. *The Fluctuating-Stress Diagram (Section 6–12)*
  iii. *Fatigue Failure Criteria (Section 6–13)*
  iv. *Constant-Life Curves (Section 6–14)*
  v. *Fatigue Failure Criterion for Brittle Materials (Section 6–15)*

C. *Combinations of Loading Modes (Section 6–16)*
A procedure based on the distortion-energy theory is presented for analyzing combined fluctuating stress states, such as combined bending and torsion.

D. *Cumulative Fatigue Damage (Section 6–17)*
The fluctuating stress levels on a machine part may be time-varying. Methods are provided to assess the fatigue damage on a cumulative basis.

VI. *Surface Fatigue (Section 6–18)*
Surface fatigue is discussed as an important special case when contact stress is cyclic. It is useful background for the study of fatigue failure in gears and bearings. This section is not essential for a first approach to the fatigue methods.

VII. *Road Maps and Important Design Equations for the Stress-Life Method (Section 6–19)*
The important equations and procedures for the stress-life method are summarized in a concise manner.

6–3 Crack Nucleation and Propagation

Fatigue failure is due to crack nucleation and propagation. A fatigue crack will initiate at a location that experiences repeated applications of locally high stress (and thus high strain). This is often at a discontinuity in the material, such as any of the following:

- Geometric changes in cross section, keyways, holes, and the like where stress concentrations occur, as discussed in Sections 3–13 and 5–2.
- Manufacturing imperfections such as stamp marks, tool marks, scratches, and burrs; poor joint design; improper assembly; and other fabrication faults.
- Composition of the material itself as processed by rolling, forging, casting, extrusion, drawing, heat treatment, and more. Microscopic and submicroscopic surface and subsurface discontinuities arise, such as inclusions of foreign material, alloy segregation, voids, hard precipitated particles, and crystal discontinuities.

A fatigue failure is characterized by three stages. *Stage I* is the initiation of one or more microcracks due to cyclic plastic deformation followed by crystallographic propagation extending from three to ten grains from the origin. Stage I cracks are not normally discernible to the naked eye. *Stage II* progresses from microcracks to macrocracks forming parallel plateau-like fracture surfaces separated by longitudinal ridges. The plateaus are generally smooth and normal to the direction of maximum tensile stress. These surfaces are often characterized by dark and light bands referred to as *beach marks,* as seen in Figure 6–1. During cyclic loading, these cracked surfaces open and close, rubbing together, and the beach mark appearance depends on the

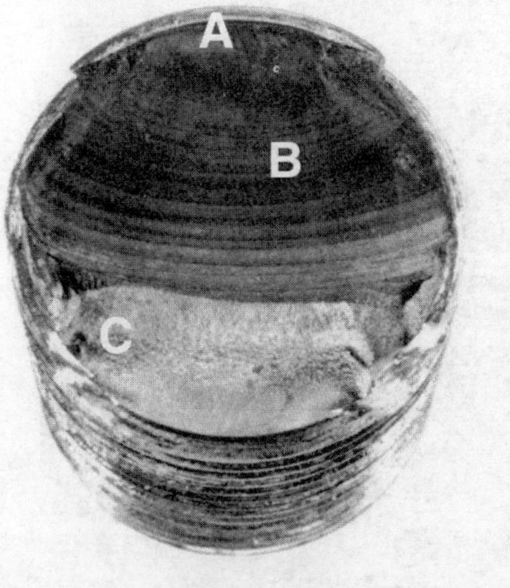

Figure 6–1

Fatigue failure of a bolt due to repeated unidirectional bending. The failure started at the thread root at *A*, propagated across most of the cross section shown by the beach marks at *B*, before final fast fracture at *C*. *(From ASM Handbook, Vol. 12:* Fractography, *2nd printing, 1992, ASM International, Materials Park, OH 44073-0002, fig. 50, p. 120. Reprinted by permission of ASM International®, www.asminternational.org.)*

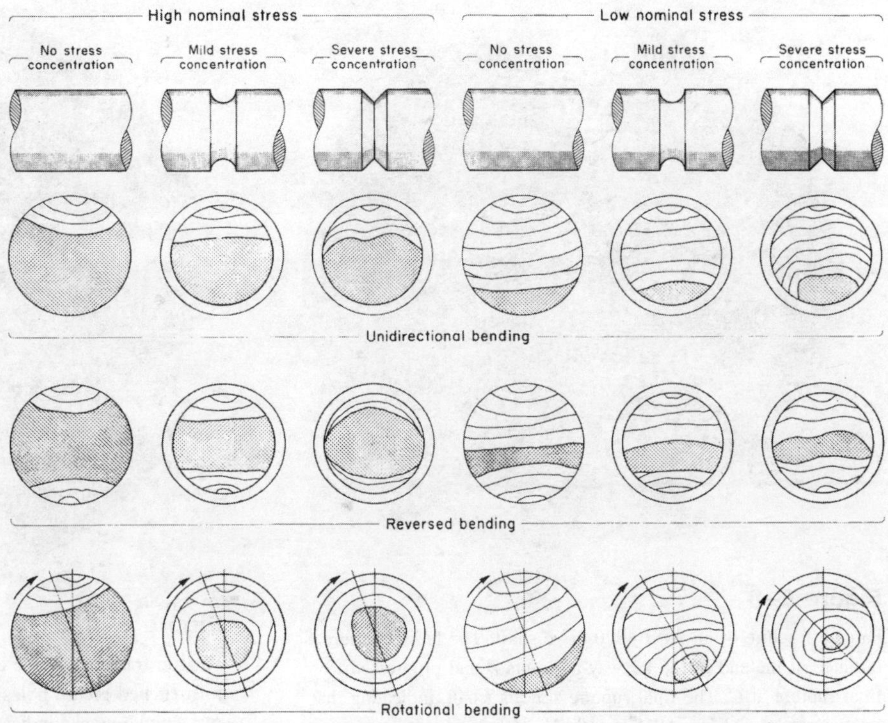

Figure 6–2

Schematics of fatigue fracture surfaces produced in smooth and notched components with round cross sections under various loading conditions and nominal stress levels. *(From ASM Metals Handbook, Vol. 11:* Failure Analysis and Prevention, *1986, ASM International, Materials Park, OH 44073-0002, fig. 18, p. 111. Reprinted by permission of ASM International®, www.asminternational.org.)*

changes in the level or frequency of loading and the corrosive nature of the environment. *Stage III* occurs during the final stress cycle when the material of the remaining cross section cannot support the loads, resulting in a sudden, fast fracture. A stage III fracture can be brittle, ductile, or a combination of both.

There is a good deal that can be observed visually from the fracture patterns of a fatigue failure. Quite often the beach marks, if they exist, and possible patterns in the stage III fracture called *chevron lines,* point toward the origins of the initial cracks. Figure 6–2 shows representations of typical failure surfaces of round cross sections under differing load conditions and levels of stress concentration. Note that the level of stress is evident from the amount of area for the stage III final fracture. The sharper stress concentrations tend to advance the stage II crack growth into a more aggressive and prominent leading edge. Figures 6–3 to 6–8 show a few examples of fatigue failure surfaces. Several of these examples are from the ASM *Metals Handbook,* which is a 21-volume major reference source in the study of fatigue failure.

Crack Nucleation

Crack nucleation occurs in the presence of localized plastic strain. Plastic strain involves the breaking of a limited number of atomic bonds by the movement of

Figure 6–3

Fatigue fracture of an AISI 4320 drive shaft. The fatigue failure initiated at the end of the keyway at points *B* and progressed to final rupture at *C*. The final rupture zone is small, indicating that loads were low. *(From ASM Handbook, Vol. 12:* Fractography, *2nd printing, 1992, ASM International, Materials Park, OH 44073-0002, fig. 51, p. 120. Reprinted by permission of ASM International®, www.asminternational.org.)*

Figure 6–4

Fatigue fracture surface of an AISI 8640 pin. Sharp corners of the mismatched grease holes provided stress concentrations that initiated two fatigue cracks indicated by the arrows. *(From ASM Handbook, Vol. 12:* Fractography, *2nd printing, 1992, ASM International, Materials Park, OH 44073-0002, fig. 520, p. 331. Reprinted by permission of ASM International®, www.asminternational.org.)*

Figure 6–5

Fatigue fracture surface of a forged connecting rod of AISI 8640 steel. The fatigue crack origin is at the left edge, at the flash line of the forging, but no unusual roughness of the flash trim was indicated. The fatigue crack progressed halfway around the oil hole at the left, indicated by the beach marks, before final fast fracture occurred. Note the pronounced shear lip in the final fracture at the right edge. *(From ASM Handbook, Vol. 12:* Fractography, *2nd printing, 1992, ASM International, Materials Park, OH 44073-0002, fig. 523, p. 332. Reprinted by permission of ASM International®, www.asminternational.org.)*

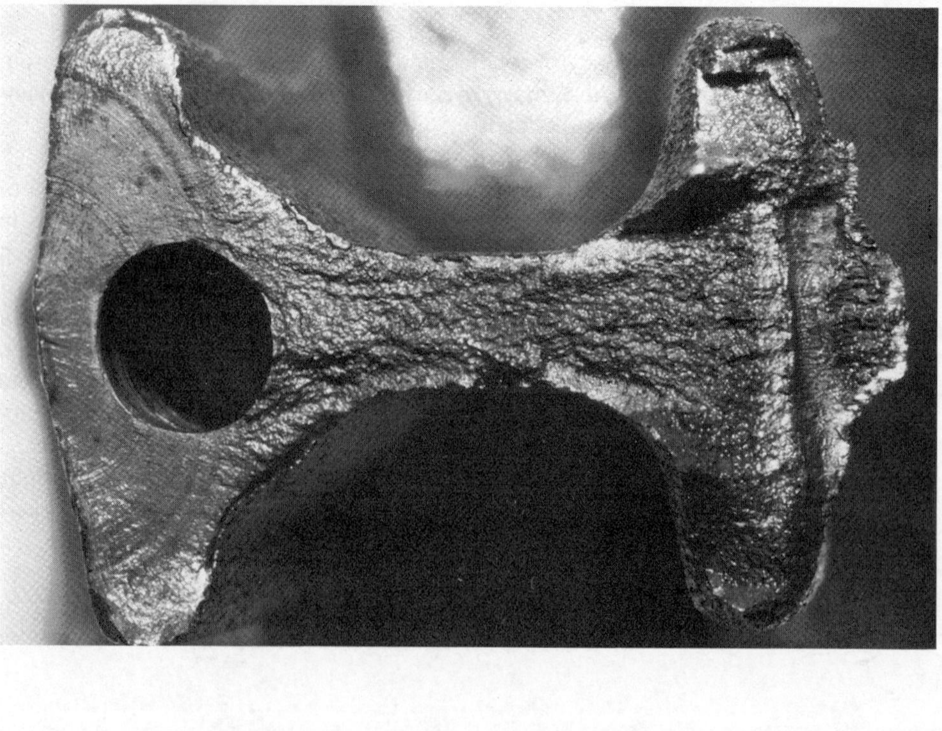

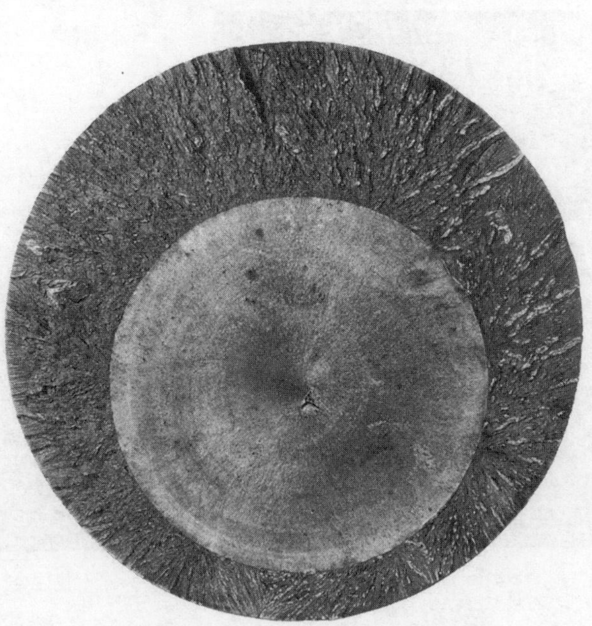

Figure 6–6

Fatigue fracture surface of a 200-mm diameter piston rod of an alloy steel steam hammer used for forging. This is an example of a fatigue fracture caused by pure tension where surface stress concentrations are absent and a crack may initiate anywhere in the cross section. In this instance, the initial crack formed at a forging flake slightly below center, grew outward symmetrically, and ultimately produced a brittle fracture without warning. *(From ASM Handbook, Vol. 12:* Fractography, *2nd printing, 1992, ASM International, Materials Park, OH 44073-0002, fig. 570, p. 342. Reprinted by permission of ASM International®, www.asminternational.org.)*

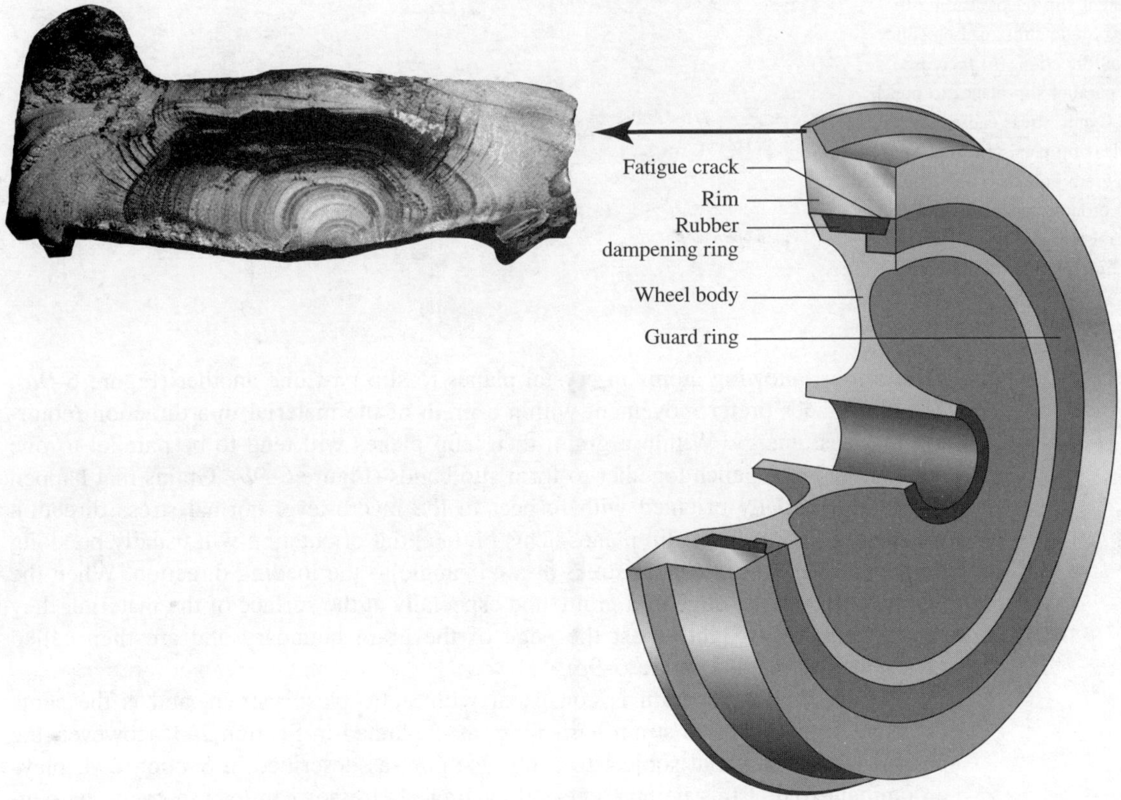

Figure 6–7

Fatigue failure of wheel rim of the Eschede railway accident (1998). The crack origin is near the bottom center in the detail view, with beach marks emanating from it, showing approximately 80% Stage II crack growth before final fracture. *(broken wheel tire): Courtesy of Bernhard Weisse*

Figure 6–8

Fatigue failure of a turbine blade from a Pratt & Whitney engine on a Boeing 777 which failed shortly after takeoff from Denver, Colorado in 2021. (*a*) View of the stub of the hollow turbine blade that broke due to a fatigue crack. The longer broken blade on the left is from collateral damage. (*b*) Close up view of the turbine blade showing the origin of the fatigue crack on the inside of the blade. *(a): National Transportation Safety Board; (b): Photo courtesy of Pratt & Whitney, illustrated by NTSB.*

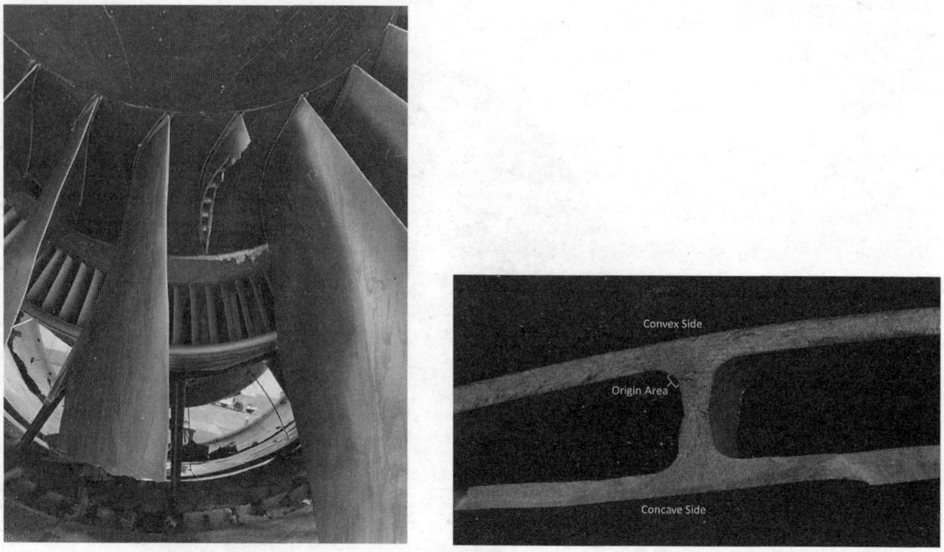

(*a*) (*b*)

Figure 6–9

Formation of persistent slip bands leading to crack nucleation. (*a*) Static stress causes slip planes 45° from load direction. (*b*) Increased static stress causes parallel slip planes to bunch into slip bands. (*c*) Cyclic stress causes slipping in both directions, leaving persistent slip bands protruding past the grain boundary. (*d*) Continued cyclic stress causes pronounced extrusions and intrusions of the persistent slip bands, which are prime locations for microcracks to nucleate.

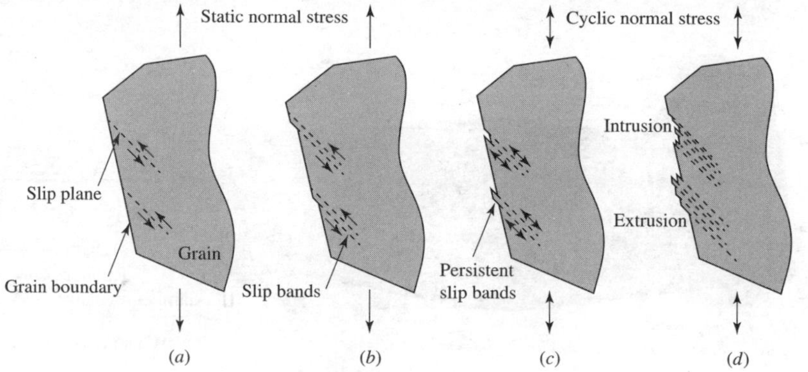

dislocations, allowing atoms in crystal planes to slip past one another (Figure 6–9*a*). The slip planes prefer movement within a grain of the material in a direction requiring the least energy. Within a grain, then, slip planes will tend to be parallel to one another, and will bunch together to form slip bands (Figure 6–9*b*). Grains that happen to be preferentially oriented with respect to the macro-level normal stress direction will be the first to form slip planes. This preferential orientation will usually be along the plane of maximum shear stress at a 45° angle to the loading direction. When the slip bands reach the edge of a grain, and especially at the surface of the material, they can extrude very slightly past the edge of the grain boundary and are then called persistent slip bands (Figure 6–9*c*).

So far, this explanation is consistent with static plastic strain, and is the same mechanism that leads to strain hardening, as explained in Section 2–3. However, the properties of a material subject to cyclic loading, as described in Section 2–4, plays a dominant role at this point. Repeated cycling of stresses can lead to cyclic hardening or cyclic softening, depending on the original condition of the material. The cyclic stress-strain properties of the material are of more significance in dictating the plastic yield strength than the monotonic stress-strain properties. Strain hardening and cyclic hardening will resist the slip planes, slowing or halting the crack nucleation.

Continued cyclic loading of sufficient level eventually causes the persistent slip bands to slide back and forth with respect to one another, forming both extrusions and intrusions at the grain boundary on the order of 1 to 10 microns (Figure 6–9*d*). This leaves tiny steps in the surface that act as stress concentrations that are prone to nucleating a microcrack. Conditions that can accelerate crack nucleation include residual tensile stresses, elevated temperatures, temperature cycling, and a corrosive environment.

Microcrack nucleation is much more likely to occur at the free surface of a part. The stresses are often highest at the outer part of the cross section; stress concentrations are often at the surface; surface roughness creates local stress concentrations; oxidation and corrosion at the surface accelerate the process; the plastic deformation necessary for persistent slip bands to form has less resistance at the surface.

Crack Propagation

After a microcrack nucleates, it begins stage I crack growth by progressively breaking the bonds between slip planes across a single grain. The growth rate is very slow, on the order of 1 n · m per stress cycle. When a microcrack reaches the grain boundary, it may halt, or eventually it may transfer into the adjacent grain, especially if that grain is also preferentially oriented such that its maximum shear plane is near 45° to the direction of loading. After repeating this process across approximately three to ten grains, the crack is sufficiently large to form a stress concentration at its tip that forms a tensile plastic zone. If several microcracks are in near vicinity to one another, they may join together, increasing the size of the tensile plastic zone. At this point, the crack is vulnerable to being "opened" by a tensile normal stress. This causes the crack to change from growing along slip planes to growing in a direction perpendicular to the direction of the applied load, beginning stage II crack growth. Figure 6–10 shows a schematic of the stage I and II transcrystalline growth of a crack. A compressive normal stress does not open the crack, so the change from stage I to stage II may not be as distinct. Later we will see that compressive stresses have much less effect on fatigue life than tensile stress.

Figure 6–11 shows a representative graph of the crack nucleation, growth, and fracture on a plot of cycling stress amplitude versus the fatigue life in number of cycles.

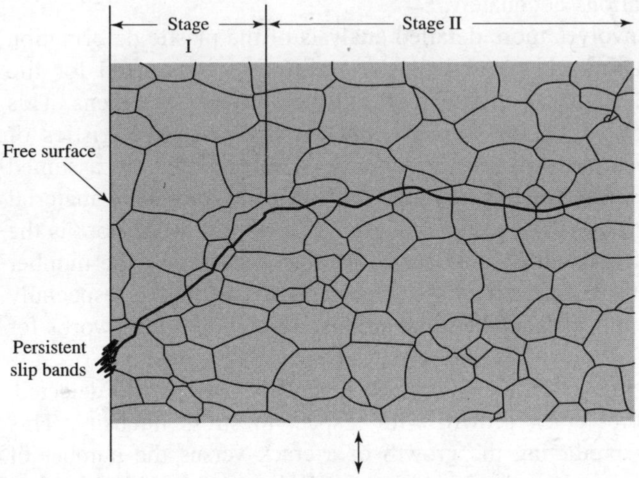

Figure 6–10

Schematic of stages I (shear mode) and II (tensile mode) transcrystalline microscopic fatigue crack growth.

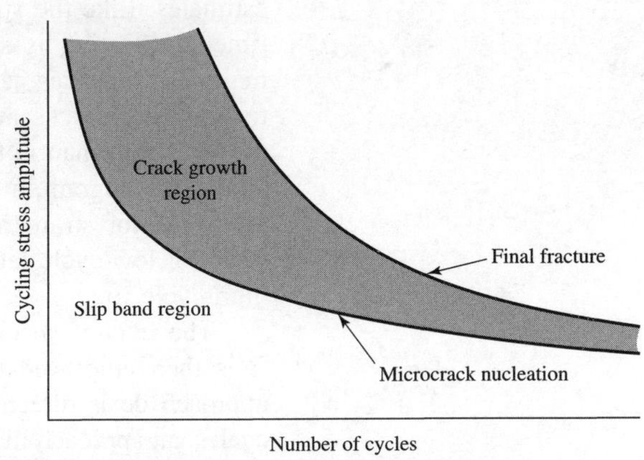

Figure 6–11

Crack nucleation and growth as a portion of total fatigue life.

The graph shows that at higher stress levels a crack initiates quickly, and most of the fatigue life is used in growing the crack. This lengthy crack growth at high stress levels is well modeled by methods of fracture mechanics, described in Section 6–5. At lower stress levels, a large fraction of the fatigue life is spent to nucleate a crack, followed by a quick crack growth. If the stress level is low enough, it is possible that a crack never nucleates, or that a nucleated crack never grows to fracture. This last phenomenon is characteristic of some materials, including steels, and is one of the early discoveries of Wöhler that allowed for designing for long or infinite life.

6–4 Fatigue-Life Methods

The three major fatigue-life methods used in design and analysis are the *stress-life method,* the *strain-life method,* and the *linear-elastic fracture mechanics (LEFM) method.* These methods attempt to predict the life in number of cycles to failure, N, for a specific level of loading. Broadly speaking, the strain-life method focuses on crack nucleation (stage I), the LEFM method focuses on crack propagation (stage II), and the stress-life method merges all three stages together with an empirical view to estimating fatigue life based on comparison to experimental test specimens. When cyclic loads are relatively low, stresses and strains are mostly elastic, and long lives are achieved (say, greater than about 10^4), the domain is referred to as *high-cycle fatigue*. On the other hand, for high cyclic loads with mostly plastic stresses and strains, and short lives, the domain is *low-cycle fatigue.*

The stress-life method estimates the life to complete fracture, ignoring the details of crack initiation and propagation. It is based on nominal stress levels only, applying stress concentration factors at notches, with no accounting for local plastic strain. Consequently, it is not useful for conditions with high stresses, plastic strains, and low cycles (i.e., the low-cycle fatigue domain). It is almost entirely based on empirical data, with little theoretical basis, with the goal of developing a stress-life diagram, a plot of the cycling stress levels versus the number of cycles before failure. Though it is the least accurate approach, it is the most traditional method, since it is the easiest to implement for a wide range of design applications, has ample supporting data, and represents high-cycle applications adequately.

The strain-life method involves more detailed analysis of the plastic deformation at localized regions where both elastic and plastic strains are considered for life estimates. Like the stress-life method, this method is based on test specimens. This time, the testing is strain-based, taking into account the cyclic characteristics of material properties at the localized level in the vicinity of notches that are assumed to be where cracks nucleate. The method requires a significant amount of material property information from cyclic stress-strain and strain-life curves. A key tool is the strain-life diagram, a plot of both elastic and plastic strain levels versus the number of cycles (or strain reversals) before a crack nucleates. This method is especially good for low-cycle fatigue applications where the strains are high, but also works for high-cycle life.

The fracture mechanics method assumes a crack is already present and detected. It is then employed to predict crack growth with respect to stress intensity. This approach deals directly with predicting the growth of a crack versus the number of cycles, an approach that is unique from the other two approaches, which do not address the actual crack. The method is useful for predicting fatigue life when the stress levels are high and a large fraction of the fatigue life is spent in the slow growth of a crack, as indicated in the discussion of Figure 6–11.

All three methods have their place in fatigue design. It is not that one is better than the others. Each one is the prime tool for the application and domain in which it is strong, but cumbersome or completely incapable for applications and domain in which it is weak. For applications that need to monitor the actual growth rate of a crack, LEFM is the prime tool. For the low-cycle domain in the presence of a notch, the strain-life method is optimal. For high-cycle domain, both the strain-life and stress-life methods apply. The strain-life method is more accurate but requires significantly more overhead. From a practical perspective, the stress-life method is by far the easiest for beginning engineers. It requires less conceptual mastery, less need to acquire material properties, and less mathematical analysis. It allows rough life estimates to be made in the high-cycle domain with minimal effort. It provides a path to observe the factors that have been proven to affect fatigue life. In this text, the focus is on first exposure to fatigue, so we focus on the stress-life method. The brief coverage of LEFM and strain-life is to encourage familiarity with the concepts, and to provide a good starting point for digging deeper when greater investment is warranted for the strengths of these methods.

Fatigue Design Criteria

As the understanding of fatigue failure mechanisms has developed, design philosophies have also evolved to provide strategies for safe designs. There are four design approaches, each with an appropriate application for use.

The *infinite-life design* is the oldest strategy, originating in the mid-1800s when Wöhler discovered that steels have a stress level, referred to as a fatigue limit or endurance limit, below which fatigue cracks do not ever grow. This design approach uses materials for which this characteristic is true and designs for stresses that never exceed the endurance limit. This approach is most suitable for applications that need to sustain over a million stress cycles before the end of service life is reached. Because the endurance limit is significantly below the yield strength, this approach requires very low stress levels, such that plastic strain is almost entirely avoided. Consequently, this approach is often used with the stress-life method.

The *safe-life design* criterion designs for a finite life, typically less than a million cycles. This approach is appropriate for applications that experience a limited number of stress cycles. It is not possible to be precise with estimating fatigue life, as the actual failures exhibit a large scatter for the same operating conditions. Significant safety margins are used. Both the strain-life and stress-life approach are used for safe-life designs, though the safe-life approach is not very accurate for lower numbers of cycles where the stress levels are high enough to produce locally plastic strains. When warranted, test data from actual simulated operating conditions on the actual parts can reduce the necessary factor of safety. Standardized parts like bearings are tested sufficiently to specify a reliability for a given load and life.

The *fail-safe design* criterion requires that the overall design of the system is such that if one part fails, the system does not fail. It uses load paths, load transfer between members, crack stoppers, and scheduled inspections. The philosophy originated in the aircraft industry that could not tolerate the added weight from large factors of safety, nor the potentially catastrophic consequences from low factors of safety. This approach does not attempt to entirely prevent crack initiation and growth.

The *damage-tolerant design* criterion is a refinement of the fail-safe criterion. It assumes the existence of a crack, due to material processing, manufacturing, or nucleation. It uses the linear-elastic fracture mechanics method to predict the growth of a hypothetical crack, in order to dictate an inspection and replacement schedule. Materials that exhibit slow crack growth and high fracture toughness are best for this criterion.

6–5 The Linear-Elastic Fracture Mechanics Method

Fracture mechanics is the field of mechanics that studies the propagation of cracks. Linear-elastic fracture mechanics is an analytical approach to evaluating the stress field at the tip of a crack, with the assumptions that the material is isotropic and linear elastic, and that the plastic deformation at the tip of a crack is small compared to the size of the crack. The stress field is evaluated at the crack tip using the theory of elasticity. When the stresses near the crack tip exceed the material fracture toughness, the crack is predicted to grow. The fundamentals of fracture mechanics are developed in Section 5–12, where it is applied to quasi-static loading situations and brittle materials. Here, we will extend the concepts for dynamically loaded applications.

The fracture mechanics approach is useful first of all for studying and understanding the fracture mechanism. Secondly, for a certain class of problems it is effective in predicting fatigue life. In particular, it works well for situations in which most of the fatigue life consists of a slow propagation of a crack. From Figure 6–11, we can see that this is often the case for high stresses in which the crack nucleates quickly. In fact, the method is often used when it is appropriate to assume the crack exists from the beginning, either in the form of a prior flaw or a quickly nucleated crack at a sharp stress concentration. This is the approach used for the damage-tolerant design criterion (Section 6–4), and is prominent in the aircraft industry.

Fatigue cracks nucleate and grow when stresses vary and there is some tension in each stress cycle. Consider the stress to be fluctuating between the limits of $\sigma_{\min}$ and $\sigma_{\max}$, where the stress range is defined as $\Delta\sigma = \sigma_{\max} - \sigma_{\min}$. From Equation (5–37) the stress intensity is given by $K_I = \beta\sigma\sqrt{\pi a}$. Thus, for $\Delta\sigma$, the stress intensity range per cycle is

$$\Delta K_I = \beta(\sigma_{\max} - \sigma_{\min})\sqrt{\pi a} = \beta\Delta\sigma\sqrt{\pi a} \tag{6–1}$$

To develop fatigue strength data, a number of specimens of the same material are tested at various levels of $\Delta\sigma$. Cracks nucleate at or very near a free surface or large discontinuity. Assuming an initial crack length of a_i, crack growth as a function of the number of stress cycles N will depend on $\Delta\sigma$, that is, ΔK_I. For ΔK_I below some threshold value $(\Delta K_I)_{th}$ a crack will not grow. Figure 6–12 represents the crack length a as a function of N for three stress levels $(\Delta\sigma)_3 > (\Delta\sigma)_2 > (\Delta\sigma)_1$, where $(\Delta K_I)_3 > (\Delta K_I)_2 > (\Delta K_I)_1$ for a given crack size. Notice the effect of the higher stress range in Figure 6–12 in the production of longer cracks at a particular cycle count.

When the rate of crack growth per cycle, da/dN in Figure 6–12, is plotted as shown in Figure 6–13, the data from all three stress range levels superpose to give a

Figure 6–12

The increase in crack length a from an initial length of a_i as a function of cycle count for three stress ranges, $(\Delta\sigma)_3 > (\Delta\sigma)_2 > (\Delta\sigma)_1$.

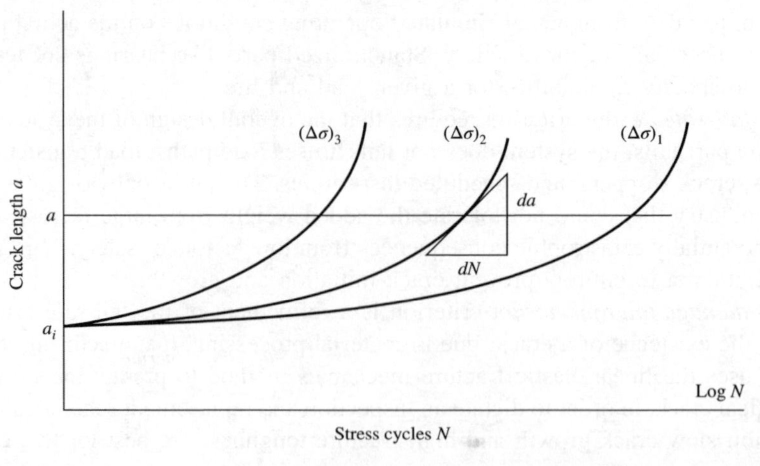

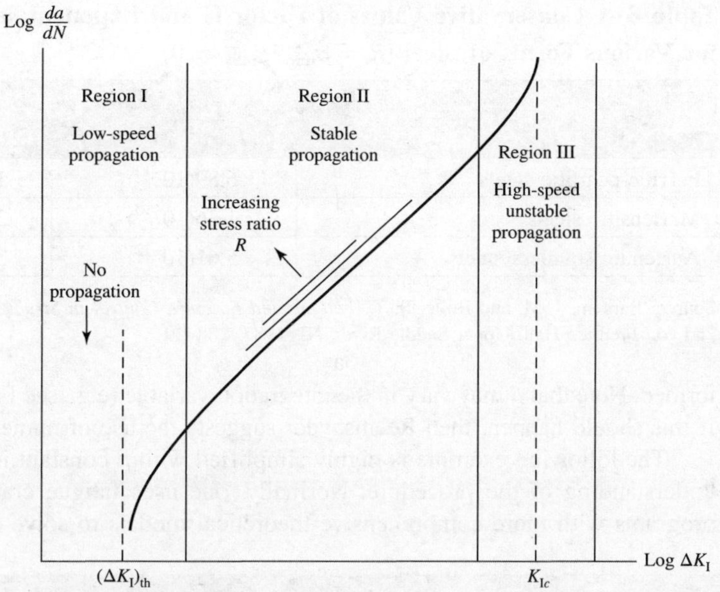

Figure 6–13

Crack growth rate versus stress intensity factor range, showing three regions of crack development.

sigmoidal curve. A group of similar curves can be generated by changing the stress ratio $R = \sigma_{min}/\sigma_{max}$ of the experiment. Three unique regions of crack development are observable from the curve. Region I is known as the near threshold region. If the stress intensity factor range is less than a threshold value $(\Delta K_I)_{th}$, the crack does not grow. However, this threshold level is very low, so it is not usually a practical option to design to stay below this level. Above this threshold, the crack growth rate, though still very small, begins to increase rapidly. Region II is characterized by stable crack propagation with a linear relationship between crack growth and stress intensity factor range. In region III, the crack growth rate is very high and rapidly accelerates to instability. When ΔK_I exceeds the critical stress intensity factor ΔK_{Ic} (also known as the fracture toughness, and defined in Section 5–12), then there is a sudden, complete fracture of the remaining cross section.

Here we present a simplified procedure for estimating the remaining life of a cyclically stressed part after discovery of a crack. This requires the assumption that plane strain conditions prevail.[1] Assuming a crack is discovered early in region II, the crack growth in region II of Figure 6–13 can be approximated by the *Paris equation,* which is of the form

$$\frac{da}{dN} = C(\Delta K_I)^m \tag{6–2}$$

where C and m are empirical material constants and ΔK_I is given by Equation (6–1). Representative, but conservative, values of C and m for various classes of steels are listed in Table 6–1. Substituting Equation (6–1) and integrating gives

$$\int_0^{N_f} dN = N_f = \frac{1}{C} \int_{a_i}^{a_f} \frac{da}{(\beta \Delta\sigma \sqrt{\pi a})^m} \tag{6–3}$$

Here a_i is the initial crack length, a_f is the final crack length corresponding to failure, and N_f is the estimated number of cycles to produce a failure *after* the initial crack is

[1]Recommended references are: Dowling, op. cit.; J. A. Collins, *Failure of Materials in Mechanical Design,* John Wiley & Sons, New York, 1981; R. Stephens, A. Fatemi, R. Stephens, and H. Fuchs, *Metal Fatigue in Engineering,* 2nd ed., John Wiley & Sons, New York, 2001; and Harold S. Reemsnyder, "Constant Amplitude Fatigue Life Assessment Models," *SAE Trans. 820 688,* vol. 91, Nov. 1983.

Table 6–1 Conservative Values of Factor C and Exponent m in Equation (6–2) for Various Forms of Steel ($R = \sigma_{min}/\sigma_{max} \approx 0$)

Material	C, $\frac{\text{m/cycle}}{(\text{MPa}\sqrt{\text{m}})^m}$	m
Ferritic-pearlitic steels	$6.89(10^{-12})$	3.00
Martensitic steels	$1.36(10^{-10})$	2.25
Austenitic stainless steels	$5.61(10^{-12})$	3.25

Source: Barsom, J. M. and Rolfe, S. T., *Fatigue and Fracture Control in Structures*, 2nd ed., Prentice Hall, Upper Saddle River, NJ, 1987, 288–291.

formed. Note that β may vary in the integration variable (e.g., see Figures 5–25 to 5–30). If this should happen, then Reemsnyder suggests the use of numerical integration.[2]

The following example is highly simplified with β constant in order to give some understanding of the procedure. Normally, one uses fatigue crack growth computer programs with more comprehensive theoretical models to solve these problems.

EXAMPLE 6–1

The bar shown in Figure 6–14 is subjected to a repeated moment $0 \le M \le 135$ N · m. The bar is AISI 4430 steel with $S_{ut} = 1300$ MPa, $S_y = 1200$ MPa, and $K_{Ic} = 81$ MPa $\sqrt{\text{m}}$. Material tests on various specimens of this material with identical heat treatment indicate worst-case constants of $C = 114(10^{-15})(\text{m/cycle})/(\text{MPa}\sqrt{\text{m}})^m$ and $m = 3.0$. As shown, a nick of size 0.1 mm has been discovered on the bottom of the bar. Estimate the number of cycles of life remaining.

Solution

The stress range $\Delta\sigma$ is always computed by using the nominal (uncracked) area. Thus

$$\frac{I}{c} = \frac{bh^2}{6} = \frac{6(13)^2}{6} = 169 \text{ mm}^3$$

Therefore, before the crack initiates, the stress range is

$$\Delta\sigma = \frac{\Delta M}{I/c} = \frac{135\,000}{169} = 800 \text{ MPa}$$

which is below the yield strength. As the crack grows, it will eventually become long enough such that the bar will completely yield or undergo a brittle fracture. For the ratio of S_y/S_{ut} it is highly unlikely that the bar will reach complete yield. For brittle fracture, designate the crack length as a_f. If $\beta = 1$, then from Equation (5–37) with $K_I = K_{Ic}$, we approximate a_f as

$$a_f = \frac{1}{\pi}\left(\frac{K_{Ic}}{\beta\sigma_{max}}\right)^2 \approx \frac{1}{\pi}\left(\frac{81}{800}\right)^2 = 3.26 \text{ mm}$$

Figure 6–14

[2]Op. cit.

From Figure 5–27, we compute the ratio a_f/h as

$$\frac{a_f}{h} = \frac{3.26}{13} = 0.251$$

Thus, a_f/h varies from near zero to approximately 0.251. From Figure 5–27, for this range β is nearly constant at approximately 1.07. We will assume it to be so, and re-evaluate a_f as

$$a_f = \frac{1}{\pi}\left(\frac{81}{1.07(800)}\right)^2 = 2.85 \text{ mm}$$

Thus, from Equation (6–3), the estimated remaining life is

Answer

$$N_f = \frac{1}{C}\int_{a_i}^{a_f} \frac{da}{(\beta\Delta\sigma\sqrt{\pi a})^m} = \frac{1}{114(10^{-15})}\int_{0.1}^{2.85} \frac{da}{[1.07(800)\sqrt{\pi a}]^3}$$

$$= -\frac{151(10^3)}{\sqrt{a}}\bigg|_{0.1}^{2.85} = 388(10^3)\,\text{cycles}$$

6–6 The Strain-Life Method

A fatigue failure almost always begins at a local discontinuity such as a notch, crack, or other area of stress concentration. When the stress at the discontinuity exceeds the elastic limit, plastic strain occurs. If a fatigue fracture is to occur, cyclic plastic strains must exist. The strain-life method is based on evaluation of the plastic and elastic strains in the localized regions where a crack begins, usually at a notch. The method is particularly useful for situations involving local yielding, which is often the case when stresses and strains are high and there is a relatively short life expectancy. Because the strain-life method also includes elastic strains, it is also applicable for long life expectancy, when the stresses and strains are lower and mostly elastic. The method is therefore comprehensive, and is widely viewed as the best method for serious work when the fatigue life needs to be predicted with reasonable reliability. The method requires a relatively high learning curve and a significant investment to obtain the necessary strain-based material behaviors. In this section, the goal is to present the fundamentals to allow the reader to be conversant and to have a framework for further study.[3]

Unlike the fracture mechanics approach, the strain-life method does not specifically analyze the crack growth but predicts life based on comparison to the experimentally measured behavior of test specimens. The assumption is that the life of the notched part will be the same as the life of a small unnotched specimen cycled to the same strains as the material at the notch root. The test specimens are loaded in tension to a given strain level, followed by loading in compression to the same strain level. The number of load reversals is used in the evaluations, which is twice the number of complete cycles that is used for the stress-life method.

[3]N. Dowling, *Mechanical Behavior of Materials,* 4th ed., Pearson Education, Upper Saddle River, N.J., 2013, chapter 14.

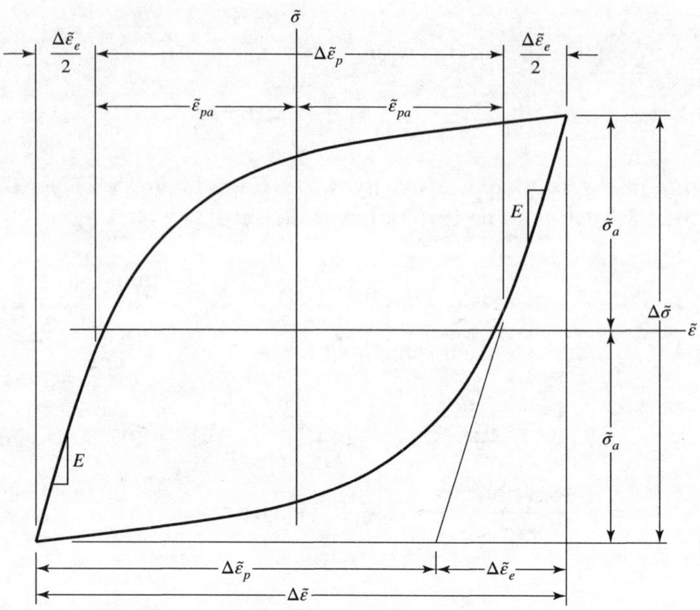

Figure 6–15

A stable cyclic hysteresis loop.

Initially, the strain cycling may lead to changes in the monotonic material properties, such as cyclic hardening or softening, as described in Section 2–4. With relatively few cycles, though, the material response settles into a stable cyclic hysteresis loop, such as in Figure 6–15, and fully described in Section 2–4.

The total true strain amplitude $\Delta\tilde{\varepsilon}/2$ is resolved into elastic and plastic components, $\Delta\tilde{\varepsilon}_e/2$ and $\Delta\tilde{\varepsilon}_p/2$, respectively, as defined in Figure 6–15. This means that a strain-life fatigue test produces two data points, elastic and plastic, for each strain reversal. These strain amplitudes are plotted on a log-log scale versus the number of strain reversals $2N$, such as for the example in Figure 6–16. Both elastic and plastic components have a linear relationship with strain reversals on the log-log scale. The equations for these two lines are well known and named, and are given below, followed by definitions of the terms.

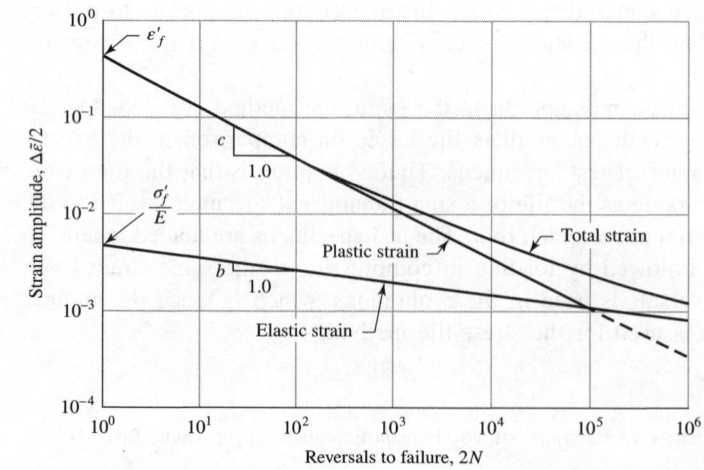

Figure 6–16

A strain-life plot for hot-rolled SAE 1020 steel. The total strain-life curve is the sum of the plastic-strain and elastic-strain lines (on a log-log scale).

Plastic-strain Manson-Coffin equation: $\frac{\Delta\tilde{\varepsilon}_p}{2} = \varepsilon'_f\,(2N)^c$ (6–4)

Elastic-strain Basquin equation: $\frac{\Delta\tilde{\varepsilon}_e}{2} = \frac{\sigma'_f}{E}(2N)^b$ (6–5)

- ***Fatigue ductility coefficient*** ε'_f is the ordinate intercept (at 1 reversal, $2N = 1$) of the plastic-strain line. It is approximately equal to the true fracture strain.
- ***Fatigue strength coefficient*** σ'_f is approximately equal to the true fracture strength. σ'_f/E is the ordinate intercept of the elastic-strain line.
- ***Fatigue ductility exponent c*** is the slope of the plastic-strain line.
- ***Fatigue strength exponent b*** is the slope of the elastic-strain line.

All four of these parameters are considered empirical material properties, with some representative values tabulated in Table A–22. The point of specimen strain testing is to obtain these parameters. Cyclic strain testing is carried out to obtain stable cyclic hysteresis loops, from which elastic and plastic strain data points are obtained. This is repeated over a broad range of strain amplitudes to generate enough data points to plot both the lines for the strain-life plot. This can be rather expensive, and sometimes approximations are used, such as those identified in the aforementioned parameter definitions.

The total strain amplitude is the sum of elastic and plastic components, giving

$$\frac{\Delta\tilde{\varepsilon}}{2} = \frac{\Delta\tilde{\varepsilon}_e}{2} + \frac{\Delta\tilde{\varepsilon}_p}{2} \tag{6–6}$$

Therefore, the total-strain amplitude is

$$\frac{\Delta\tilde{\varepsilon}}{2} = \frac{\sigma'_f}{E}(2N)^b + \varepsilon'_f(2N)^c \tag{6–7}$$

This equation is termed the *strain-life relation* and is the basis of the strain-life method. For high strain amplitudes, the strain-life curve approaches the plastic-strain Manson-Coffin line. At low strain amplitudes, the curve approaches the elastic-strain Basquin equation. In the stress-life method, a similar relationship is developed between stress and life, plotted on the *S-N* diagram, though it only includes an elastic term. The Basquin equation is essentially the same as the stress-life line. This makes sense, because at low strain levels the strain is almost entirely elastic, in which case stress and strain are linearly related. The uniqueness and beauty of the strain-life curve is that it includes both elastic and plastic influences on the life.

The strain-life relation is based on completely reversed tensile and compressive strain of equal amplitudes. There is much research and discussion regarding the effect of nonzero mean strain. Strain-controlled cycling with a mean strain results in a relaxation of the mean stress for large strains, due to plastic deformation. The stress, then, tends to move toward centering around zero mean stress, even as the strain continues to cycle with a mean strain. In this case, the mean strain has little effect. But if the strain level is not sufficiently high, the mean stress does not completely relax, which has a detrimental effect on fatigue life for positive mean stress, and positive effect for negative mean stress. Dowling provides a good discussion of modeling the mean stress effects.[4]

---

[4]N. Dowling, "Mean Stress Effects in Strain-Life Fatigue," *Fatigue Fract. Eng. Mater. Struct.*, vol. 32, pp. 1004–1019, 2009.

6–7 The Stress-Life Method and the *S-N* Diagram

The stress-life method relies on studies of test specimens subjected to controlled cycling between two stress levels, known as *constant amplitude loading*. Figure 6–17*a* shows a general case of constant amplitude loading between a minimum and maximum stress. Load histories of actual parts are often much more diverse, with variable amplitude loading, but many cases can be reasonably modeled with the constant amplitude approach. This is especially true for rotating equipment that experiences repetitive loading with each revolution. Testing with constant amplitude loading also provides a controlled environment to study the nature of fatigue behavior and material fatigue properties.

The constant amplitude stress situation is characterized by the following terminology:

$$\sigma_{\min} = \text{minimum stress}$$
$$\sigma_{\max} = \text{maximum stress}$$
$$\sigma_m = \text{mean stress, or midrange stress}$$
$$\sigma_a = \text{alternating stress, or stress amplitude}$$
$$\sigma_r = \text{stress range}$$

The following relations are evident from Figure 6–17*a*,

$$\sigma_a = \left|\frac{\sigma_{\max} - \sigma_{\min}}{2}\right| \tag{6–8}$$

$$\sigma_m = \frac{\sigma_{\max} + \sigma_{\min}}{2} \tag{6–9}$$

The mean stress can have positive or negative values, while the alternating stress is always a positive magnitude representing the amplitude of the stress that alternates above and below the mean stress.

Two special cases are common enough to warrant special attention. Figure 6–17*b* shows what is called a *repeated stress,* in which the stress cycles between a minimum stress of zero to a maximum stress. Figure 6–17*c* shows a *completely reversed stress,* in which the stress alternates between equal magnitudes of tension and compression, with a mean stress of zero. We shall add the subscript r to the alternating stress, that is σ_{ar}, when it is advantageous to clarify that an alternating stress is completely reversed. Most fatigue testing is done with completely reversed stresses; then the modifying effect of nonzero mean stress is considered separately.

One of the first and most widely used fatigue-testing devices is the R. R. Moore high-speed rotating-beam machine. This machine subjects the test specimen to pure bending by loading as shown in Figure 6–18. A quick sketch of transverse shear and

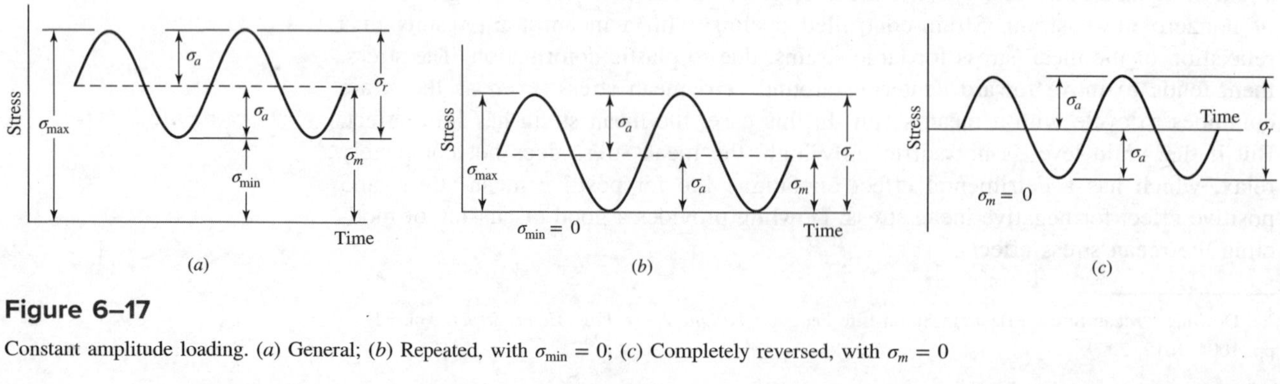

Figure 6–17

Constant amplitude loading. (*a*) General; (*b*) Repeated, with $\sigma_{\min} = 0$; (*c*) Completely reversed, with $\sigma_m = 0$

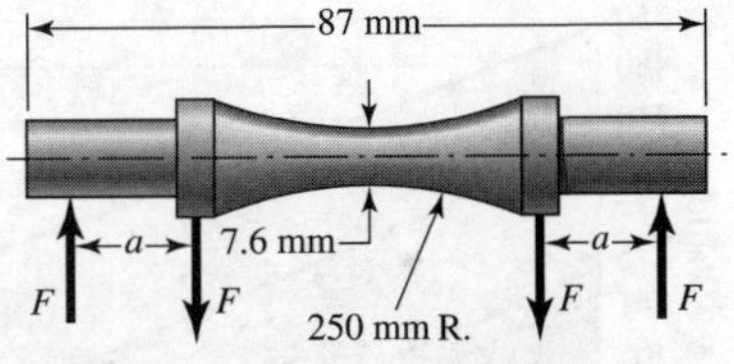

Figure 6–18

Test-specimen geometry for the R. R. Moore rotating-beam machine.

bending moment diagrams will show that throughout the center section of the specimen the transverse shear is zero and the bending moment is constant. The test specimen is very carefully machined and polished, with a final polishing in an axial direction to avoid circumferential scratches. The goal is to eliminate as many extraneous effects as possible to study just the fatigue behavior of the material. Other fatigue-testing machines are available for applying fluctuating or reversed axial stresses, torsional stresses, or combined stresses to the test specimens.

The S-N Diagram

For the rotating-beam test, each revolution of the specimen causes a stress element on the surface to cycle between equal magnitudes of tension and compression, thus achieving completely reversed stress cycling. A large number of tests are necessary because of the statistical nature of fatigue. Many specimens are tested to failure at each level of completely reversed stress. A plot of the completely reversed alternating stress versus the life in cycles to failure is made on a semi-log or log-log scale, such as in Figure 6–19. This plot is called a *Wöhler curve,* a *stress-life diagram,* or an *S-N diagram.* The curve typically passes through the median of the test data for each stress level. A value of completely reversed stress on the ordinate is referred to as a *fatigue strength* S_f when accompanied by a statement of the number of cycles N to which it corresponds.

Fatigue failure with less than 1000 cycles is generally classified as low-cycle fatigue, and fatigue failure with greater than 1000 cycles as high-cycle fatigue, as indicated in Figure 6–19. Low-cycle fatigue is associated with high stresses and strains, often in the plastic range, at least locally. Consequently, low-cycle fatigue is

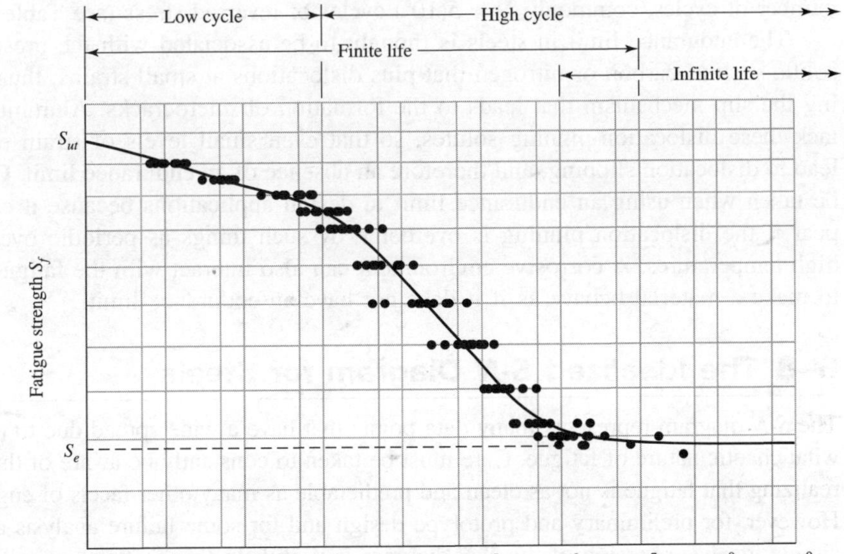

Figure 6–19

A characteristic *S-N* diagram.

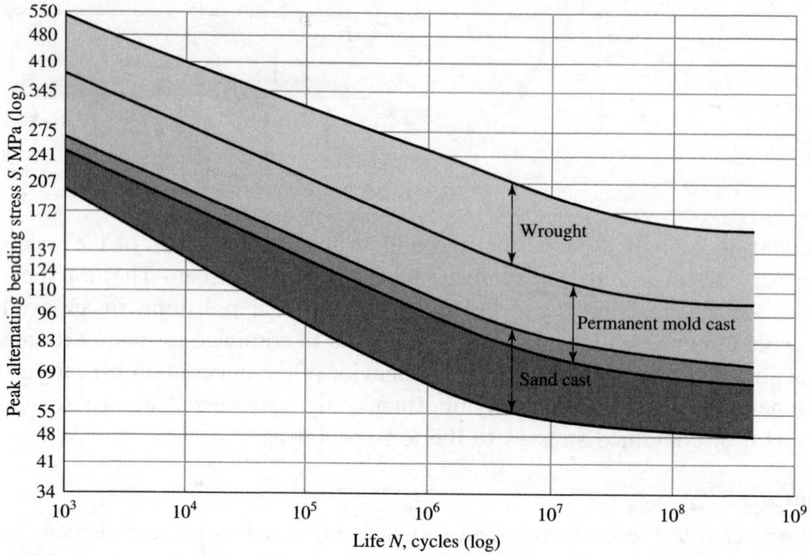

Figure 6–20

S-N bands for representative aluminum alloys.

best modeled with the strain-life approach. Because the stress-life approach is best suited for lower stresses, which are associated with the high-cycle region, a healthy factor of safety with regard to static yielding, including any stress concentration, will usually suffice to avoid failure in the low-cycle region.

In the case of ferrous metals and alloys, a knee occurs in the *S-N* diagram. Reversed stresses below this level do not cause fatigue failure, no matter how great the number of cycles. Accordingly, Figure 6–19 distinguishes a finite-life region and an infinite-life region. The boundary between these regions depends on the specific material, but it lies somewhere between 10^6 and 10^7 cycles for steels. The fatigue strength corresponding to the knee is called the *endurance limit* S_e, or the fatigue limit. Many materials, particularly nonferrous metals and plastics, do not exhibit an endurance limit. For example, Figure 6–20 shows regions for the *S-N* curves for most common aluminum alloys. For materials without an endurance limit, the fatigue strength S_f is reported at a specific number of cycles, commonly $N = 5(10^8)$ cycles of reversed stress (see Table A–23).

The endurance limit in steels is thought to be associated with the presence of a solute such as carbon or nitrogen that pins dislocations at small strains, thus preventing the slip mechanism that leads to the formation of microcracks. Aluminum alloys lack these dislocation-pinning solutes, so that even small levels of strain eventually lead to dislocation slipping, and therefore an absence of an endurance limit. Care must be taken when using an endurance limit in design applications because it can disappear if the dislocation pinning is overcome by such things as periodic overloads or high temperatures. A corrosive environment can also interact with the fatigue process to make a material behave as if it does not have an endurance limit.

6–8 The Idealized *S-N* Diagram for Steels

The *S-N* diagram represents many data points that have a wide spread due to the somewhat chaotic nature of fatigue. Care must be taken to constantly be aware of this scatter, realizing that fatigue is not as clean and predictable as many other facets of engineering. However, for preliminary and prototype design and for some failure analysis as well, a simple idealized version of the *S-N* diagram is useful. In this section, we will develop such a diagram for steels. A similar approach can be used for other materials.

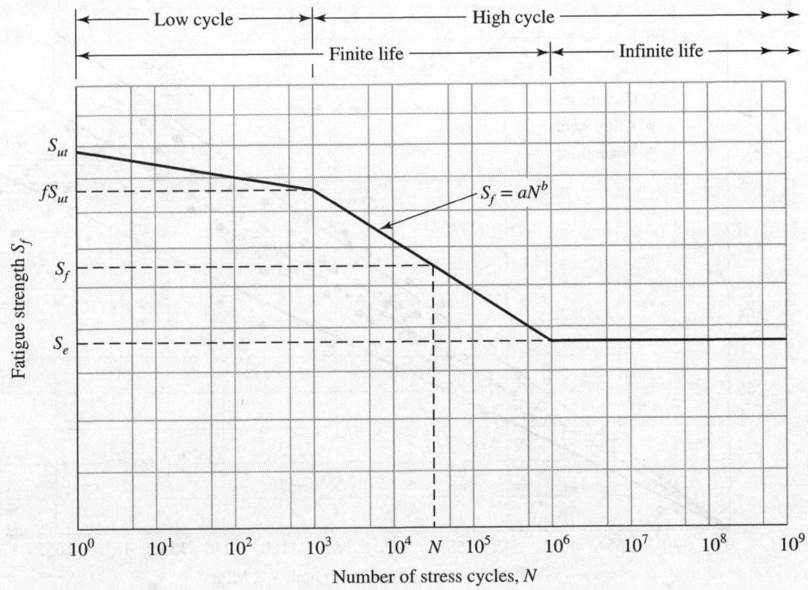

Figure 6–21

An idealized *S-N* diagram for steels.

Figure 6–21 shows an idealized *S-N* diagram for steels, in which the median failure curve in the finite-life region is represented by two lines. In the low-cycle region ($N < 1000$ cycles), the line has a relatively low slope and runs between the ultimate strength to some fraction f of the ultimate strength, where f is generally between 0.8 and 0.9. Between 10^3 and 10^6 cycles, a line of steeper slope represents the median failure curve. At 10^6 cycles, the endurance limit is reached, and the failure curve becomes horizontal. To work with this idealized *S-N* diagram, we will need to determine two points: fS_{ut} at 10^3 cycles and S_e at 10^6 cycles. With these points, the relationship between fatigue strength and life can be approximated for the finite life region, and infinite life can be predicted for stress levels less than the endurance limit.

Estimating the Endurance Limit

Many fatigue tests are run at ever lower stress levels to determine the endurance limit for a particular material. This is very time consuming since close to a million revolutions is necessary for each data point. Consequently, it is desirable for purposes of obtaining an idealized *S-N* diagram to have a quick means of estimating the endurance limit, ideally based on readily available published data. Plotting the experimentally determined endurance limit versus ultimate strength for a large quantity of ferrous metals shows there is a reasonably strong correlation, as shown in Figure 6–22. The plot indicates that the endurance limit generally ranges between 40 to 60 percent of the ultimate strength up to about 1400 MPa. At that point, the scatter increases and the slope flattens.

For steels, simplifying our observation of Figure 6–22, we will estimate the endurance limit as

$$S'_e = \begin{cases} 0.5\,S_{ut} & S_{ut} \le 1400 \text{ MPa} \\ 700 \text{ MPa} & S_{ut} > 1400 \text{ MPa} \end{cases} \tag{6–10}$$

where S_{ut} is the *minimum* tensile strength. The prime mark on S'_e in this equation refers to the *rotating-beam specimen* itself. We wish to reserve the unprimed symbol S_e for the endurance limit of an actual machine element subjected to any kind of loading. Soon we shall learn that the two strengths may be quite different.

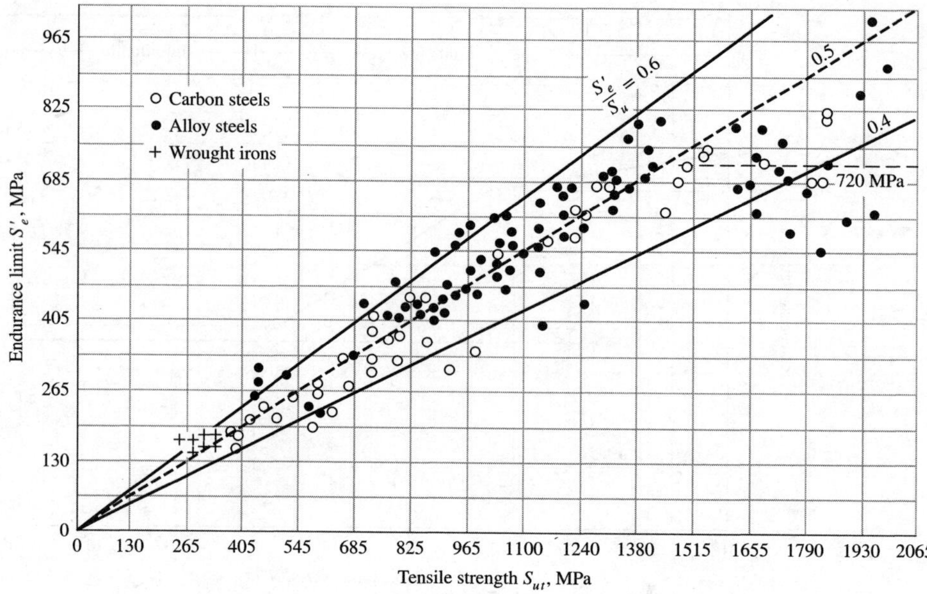

Figure 6–22

Graph of endurance limits versus tensile strengths from actual test results for a large number of wrought irons and steels. Ratios of S'_e/S_{ut} of 0.60, 0.50, and 0.40 are shown by the solid and dashed lines. Note also the horizontal dashed line for $S'_e = 720$ MPa. Points shown having a tensile strength greater than 1450 MPa have a mean endurance limit of $S'_e = 720$ MPa and a standard deviation of 93 MPa. *(Source: Collated from data compiled by Grover, H. J., Gordon, S. A. and Jackson, L. R. In* Fatigue of Metals and Structures, *Bureau of Naval Weapons Document NAVWEPS 00-25-534, 1960; and from* Fatigue Design Handbook, *SAE, 1968, p. 42.)*

Steels treated to give different microstructures have different S'_e/S_{ut} ratios. It appears that the more ductile microstructures have a higher ratio. Martensite has a very brittle nature and is highly susceptible to fatigue-induced cracking; thus the ratio is low. When designs include detailed heat-treating specifications to obtain specific micro structures, it is possible to use an estimate of the endurance limit based on test data for the particular microstructure; such estimates are much more reliable and indeed should be used.

The endurance limits for various classes of cast irons, polished or machined, are given in Table A–23. Aluminum alloys do not have an endurance limit. The fatigue strengths of some aluminum alloys at $5(10^8)$ cycles of reversed stress are given in Table A–23.

Estimating the Fatigue Strength at 10^3 Cycles

The next step in determining the fatigue line in the high-cycle region between 10^3 to 10^6 cycles is to estimate the fatigue strength fS_{ut} at 10^3 cycles. Many references have settled on a value of f of 0.9. Others, noting that some experimental data indicate a lower value is warranted, opt for a more conservative value of f equal to 0.8. The difference is not terribly significant, as it represents a small change on the low-cycle end of a line that is primarily of interest at higher cycles. However, it has been observed that for steels, f is generally a little lower for higher strength materials. A relationship between f and S_{ut} has been developed based on the elastic strain line in the strain-life approach to fatigue analysis.[5] The resulting relationship for steels is plotted in Figure 6–23 and expressed by the curve-fit equations

$$f = 1.06 - 4.1(10^{-4})S_{ut} + 1.5(10^{-7})S_{ut}^2 \qquad 500 < S_{ut} < 1400 \text{ MPa} \quad \text{(6–11)}$$

---

[5]R. G. Budynas and J. K. Nisbett, *Shigley's Mechanical Engineering Design,* 10th ed., McGraw Hill Education, New York, 2015, p. 292.

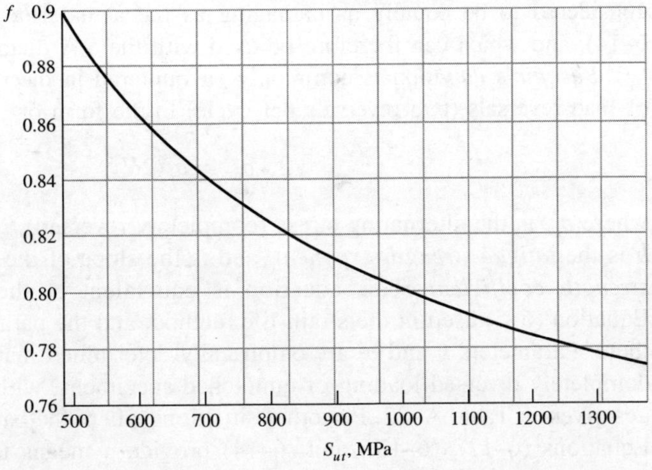

Figure 6–23

Fatigue strength fraction, f, of S_{ut} at 10^3 cycles for steels, with $S_e = S'_e = 0.5S_{ut}$ at 10^6 cycles.

Values for f can be estimated from the plot or equations, noting that they are not experimentally based and are only intended to provide a better estimate than using a fixed value. For values of S_{ut} lower than the limit given, $f = 0.9$ is recommended.

The High-Cycle S-N *Line*

The relationship between fatigue strength and life in the high-cycle finite-life region, that is between 10^3 and 10^6 cycles, is approximately linear on the log-log scale. It thus can be represented on a normal scale by a power function known as *Basquin's equation,*

$$S_f = aN^b \tag{6–12}$$

where S_f is the fatigue strength correlating to a life N in cycles to failure. The constants a and b are the ordinate intercept and the slope of the line in log-log coordinates, which can be readily recognized by taking the logarithm of both sides of Equation (6–12), giving $\log S_f = b \log N + \log a$. To obtain expressions for a and b, substitute into Equation (6–12) for (N, S_f) the two known points $(10^3, fS_{ut})$ and $(10^6, S_e)$. Solving for a and b,

$$a = \frac{(fS_{ut})^2}{S_e} \tag{6–13}$$

$$b = -\frac{1}{3}\log\left(\frac{fS_{ut}}{S_e}\right) \tag{6–14}$$

Equation (6–12) can be solved for the life in cycles correlating to a completely reversed stress, replacing S_f with σ_{ar},

$$N = \left(\frac{\sigma_{ar}}{a}\right)^{1/b} \tag{6–15}$$

The typical *S-N* diagram is only applicable for completely reversed loading. For general fluctuating loading situations, the effect of mean stress must be accounted for (see Section 6–11). This will lead to an equivalent completely reversed stress that is considered to be equally as damaging as the actual fluctuating stress (see Section 6–14), and which can therefore be used with the *S-N* diagram and Equation (6–15).

Basquin's equation is commonly encountered in the research literature in terms of load reversals (two reversals per cycle) in the form of

$$\sigma_{ar} = \sigma'_f (2N)^b \tag{6–16}$$

where σ_{ar} is the alternating stress (completely reversed), N is the number of cycles, b is the *fatigue strength exponent,* and is the slope of the line, and σ'_f is the *fatigue strength coefficient.* This equation is equivalent to the strain-based version of Equation (6–5) used in the strain-life method, and the parameters are fully discussed there. Parameters b and σ'_f are empirically determined material properties, based on completely reversed loading of unnotched specimens, with some representative values given in Table A–22. For other situations where the parameters are not available, Equations (6–12), (6–13), and (6–14) provide a means to estimate the high-cycle *S-N* line.

EXAMPLE 6–2

Given a 1050 HR steel, *estimate*
(*a*) the rotating-beam endurance limit at 10^6 cycles.
(*b*) the endurance strength of a polished rotating-beam specimen corresponding to 10^4 cycles to failure.
(*c*) the expected life of a polished rotating-beam specimen under a completely reversed stress of 385 MPa.

Solution

(*a*) From Table A–19, S_{ut} = 630 MPa. From Equation (6–10),

Answer

$$S'_e = 0.5(630) = 315 \text{ MPa}$$

(*b*) From Figure 6–23, or Equation (6–11), for S_{ut} = 630 MPa, $f \approx 0.86$. From Equation (6–13),

$$a = \frac{[0.86(630)]^2}{315} = 932 \text{ MPa}$$

From Equation (6–14),

$$b = -\frac{1}{3}\log\left[\frac{0.86(630)}{315}\right] = -0.0785$$

Thus, Equation (6–12) is

$$S_f = 932\ N^{-0.0785}$$

Answer

For 10^4 cycles to failure, $S_f = 932(10^4)^{-0.0785} = 452$ MPa

(*c*) From Equation (6–15), with σ_{ar} = 385 MPa,

Answer

$$N = \left(\frac{385}{932}\right)^{1/-0.0785} = 77\,830 = 77.83(10^3)\,\text{cycles}$$

Keep in mind that these are only *estimates,* thus the rounding of the results to fewer significant figures.

6–9 Endurance Limit Modifying Factors

The rotating-beam specimen used in the laboratory to determine endurance limits is prepared very carefully and tested under closely controlled conditions. It is unrealistic to expect the endurance limit of a mechanical or structural member to match the values obtained in the laboratory. Some differences include

- ***Material:*** composition, basis of failure, variability
- ***Manufacturing:*** method, heat treatment, fretting corrosion, surface condition, stress concentration
- ***Environment:*** corrosion, temperature, stress state, relaxation times
- ***Design:*** size, shape, life, stress state, speed, fretting, galling

Marin[6] identified factors that quantified the effects of surface condition, size, loading, temperature, and miscellaneous items. Marin proposed that correction factors for each effect be applied as multipliers to adjust the endurance limit. The multiplicative combination of the various effects has not been thoroughly tested and proven, particularly in capturing the actual impact of interactions between different effects. However, limited testing has shown it to be reasonable for the rough approximations expected from the stress-life approach. Marin's equation is written as

$$S_e = k_a k_b k_c k_d k_e S'_e \tag{6–17}$$

where k_a = surface factor
k_b = size factor
k_c = load factor
k_d = temperature factor
k_e = reliability factor
S'_e = rotary-beam test specimen endurance limit
S_e = endurance limit at the critical location of a machine part in the geometry and condition of use

When endurance tests of parts are not available, estimations are made by applying Marin factors to the endurance limit. This will effectively lower the high-cycle end of the *S-N* line, while not moving the low-cycle end of the line. Accordingly, the modifying effects are applied proportionately through the finite life region. This seems reasonable and consistent with limited experimental data.

Surface Factor k_a

The surface factor is to account for the effect on the endurance limit for actual parts which rarely have, or can maintain, a surface finish as smooth as the polished test specimen. Stresses are often highest at the surface. Surface roughness puts many localized stress concentrations in the area of high stress, often leading to localized plastic strain at the roots of surface imperfections. This creates an environment that is prone to crack initiation. If materials were perfectly homogeneous, elastic, and isotropic, the effect of surface roughness could be predicted from an analysis of geometric stress concentration associated with the surface profile. However, with any real material subject to real manufacturing operations, the surface roughness only adds

[6]Joseph Marin, *Mechanical Behavior of Engineering Materials,* Prentice Hall, Englewood Cliffs, N.J., 1962, p. 224.

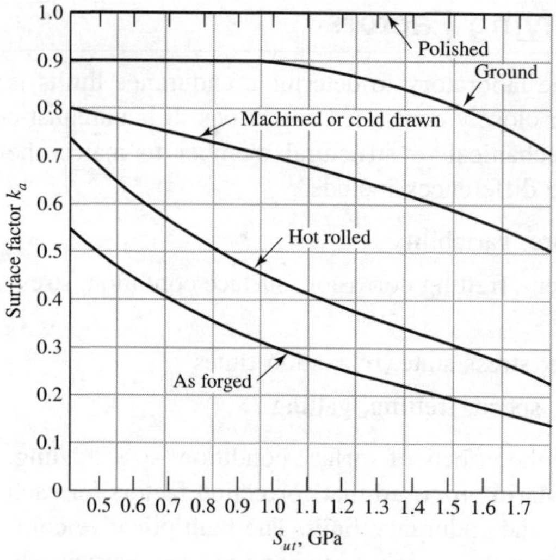

Figure 6–24

Trends for surface factor k_a for steels. *(Source: Generated from data from C. J. Noll and C. Lipson. "Allowable Working Stresses"* Society for Experimental Stress Analysis *vol. 3, no. 2 (1946): 29.)*

stress raisers to those already present in the existing microstructure due to its complicated history of such things as metallurgical treatment, cold working, residual stresses from manufacturing operations, etc. Consequently, from a practical perspective, a surface factor based on purely surface roughness can be misleading when compared to a test specimen that also includes a history of manufacturing processes.

Lipson and Noll collected data from many studies, organizing them into several common commercial surface finishes (ground, machined, hot-rolled, and as-forged).[7] A commonly used adaptation of their data is shown in Figure 6–24. It is clear from the curves that the surface effect is significant, and that it is more detrimental for higher-strength materials. Bringing critical stress locations to at least a machined finish is certainly worth consideration as a cost-effective substantial improvement in fatigue life. Several observations regarding the data for each category is warranted.

The ground surface category includes any surface finish that affects the endurance limit by no more than approximately 10 percent (except for higher strengths), and that has a maximum profilometer reading of 2 μm. This includes ground, honed, and lapped finishes. The test data for higher strengths is significantly scattered. In recognition of this, the declining curve shown for higher strengths is generally conservative and not typical of all the data points.

The machined category includes rough and finish machining operations, as well as unmachined cold-drawn surfaces. The data collected for this curve was very limited for higher strengths, so the curve is extrapolated for strengths above about 1100 MPa.

The hot-rolled class represents surface conditions typical of a hot-rolled manufacturing process and includes surface irregularities, scale defects, oxide, and partial surface decarburization. It is significant to note that this is not entirely a factor of the surface roughness, but also includes metallurgical conditions.

Similar to the hot-rolled category, the as-forged category is heavily influenced by metallurgical conditions, notably that the surface layer is significantly decarburized,

---

[7]C. J. Noll and C. Lipson, "Allowable Working Stresses," *Society for Experimental Stress Analysis,* vol. 3, no. 2, 1946, p. 29. Reproduced by O. J. Horger (ed.), *Metals Engineering Design ASME Handbook,* McGraw Hill, New York, 1953, p. 102.

Table 6–2 Curve Fit Parameters for Surface Factor, Equation (6–18)

Surface Finish	Factor a S_{ut}, MPa	Exponent b
Ground	1.38	−0.067
Machined or cold-drawn	3.04	−0.217
Hot-rolled	38.6	−0.650
As-forged	54.9	−0.758

which is probably as much or more significant than the actual surface roughness. When machining a forged part to improve its surface factor, it is important to machine to a depth that will remove the decarburized layer. More recent investigations by McKelvey and others have noted that the forging process has seen significant improvements since the Lipson and Noll data was collected in the 1940s.[8] Consequently, based on more recent experimental data, McKelvey recommends a surface factor for the as-forged surface that is at least as high as the hot-rolled curve shown in Figure 6–24.

The curves in Figure 6–24 are only intended to capture the broad tendencies. The data came from many studies, gathered under a variety of conditions. Lipson and Noll indicated that they attempted to compensate for other factors, such as the differences in specimen sizes included in the studies. In general, the curves are thought to represent the lower bounds of the spread of the data, and are therefore likely to be conservative compared to what might be proven by specific part testing.

For convenience, the curves of Figure 6–24 are fitted with a power curve equation

$$k_a = aS_{ut}^b \tag{6–18}$$

where S_{ut} is the minimum tensile strength, in units of MPa, and a and b are curve fit parameters tabulated in Table 6–2. The power curve form of Equation (6–18) is not the optimal choice for fitting the shape of the curves, but it is suitable for its simplicity and is sufficient for the level of scatter in the original data. The fitting parameters in Table 6–2 have been updated from previous editions of this text to better fit the curves, and giving less weighting to strengths above 1400 MPa, where there was little test data available.

EXAMPLE 6–3

A steel has a minimum ultimate strength of 520 MPa and a machined surface. Estimate k_a.

Solution

From Table 6–2, $a = 3.04$ and $b = -0.217$. Then, from Equation (6–18)

Answer $$k_a = 3.04(520)^{-0.217} = 0.78$$

[8]S. A. McKelvey and A. Fatemi, "Surface Finish Effect on Fatigue Behavior of Forged Steel," *International Journal of Fatigue*, vol. 36, 2012, pp. 130–145.

Size Factor k_b

The endurance limit of specimens loaded in bending and torsion has been observed to decrease slightly as the specimen size increases. One explanation of this is that a larger specimen size leads to a greater volume of material experiencing the highest stress levels, thus a higher probability of a crack initiating. However, the size effect is not present in axially loaded specimens. This leads to a further hypothesis that the presence of a stress gradient is somehow involved in the size effect.

A size factor has been evaluated by Mischke for bending and torsion of round rotating bars, using a compilation of data from several sources.[9] The data has significant scatter, but curve-fit equations representing the lower edge of the data (and therefore conservative for design purposes) can be given by

$$k_b = \begin{cases} (d/7.62)^{-0.107} = 1.24d^{-0.107} & 7.62 \le d \le 51\,\text{mm} \\ 1.51d^{-0.157} & 51 < d \le 254\,\text{mm} \end{cases} \tag{6–19}$$

For d less than 7.62 mm, the data is quite scattered. Unless more specific data is available to warrant a higher value, a value of $k_b = 1$ is recommended.

For axial loading there is no size effect, so

$$k_b = 1 \tag{6–20}$$

but see k_c.

Equation (6–19) applies to round rotating bars in bending and torsion, in which the highly stressed volume is around the outer circumference. If a round bar is not rotating, the highly stressed volume is the same for torsion (all the way around the circumference), but is much less for bending (e.g., just on opposite sides of the cross section). Kuguel introduced a critical volume theory in which the volume of material experiencing a stress above 95 percent of the maximum stress is considered to be critical.[10] The method employs an *equivalent diameter* d_e obtained by equating the volume of material stressed at and above 95 percent of the maximum stress to the same volume in the rotating-beam specimen. When these two volumes are equated, the lengths cancel, and so we need only consider the areas. For a rotating round section, the 95 percent stress area is the area in a ring having an outside diameter d and an inside diameter of $0.95d$. So, designating the 95 percent stress area $A_{0.95\sigma}$, we have

$$A_{0.95\sigma} = \frac{\pi}{4}[d^2 - (0.95d)^2] = 0.0766d^2 \tag{6–21}$$

This equation is also valid for a rotating hollow round. For nonrotating solid or hollow rounds, the 95 percent stress area is twice the area outside of two parallel chords having a spacing of $0.95d$, where d is the diameter, that is

$$A_{0.95\sigma} = 0.010\,46d^2 \tag{6–22}$$

---

[9]Charles R. Mischke, "Prediction of Stochastic Endurance Strength," *Trans. of ASME, Journal of Vibration, Acoustics, Stress, and Reliability in Design,* vol. 109, no. 1, January 1987, Table 3.

[10]R. Kuguel, "A Relation between Theoretical Stress-Concentration Factor and Fatigue Notch Factor Deduced from the Concept of Highly Stressed Volume," *Proc. ASTM,* vol. 61, 1961, pp. 732–748.

With d_e in Equation (6–21), setting Equations (6–21) and (6–22) equal to each other enables us to solve for the effective diameter. This gives

$$d_e = 0.370d \tag{6–23}$$

as the effective size of a round corresponding to a nonrotating solid or hollow round.

A rectangular section of dimensions $h \times b$ has $A_{0.95\sigma} = 0.05hb$. Using the same approach as before,

$$d_e = 0.808(hb)^{1/2} \tag{6–24}$$

Table 6–3 provides $A_{0.95\sigma}$ areas of common structural shapes undergoing nonrotating bending.

Table 6–3 $A_{0.95\sigma}$ Areas of Common Nonrotating Structural Shapes Loaded in Bending

Shape	Area
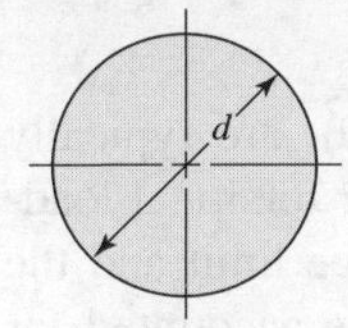	$A_{0.95\sigma} = 0.010\,46d^2$ $d_e = 0.370d$
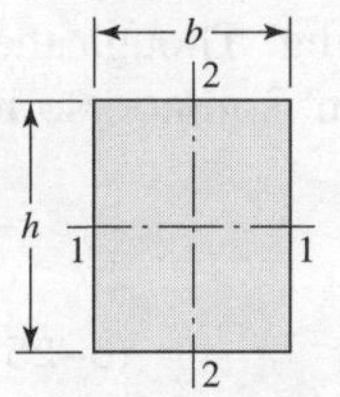	$A_{0.95\sigma} = 0.05hb$ $d_e = 0.808\sqrt{hb}$
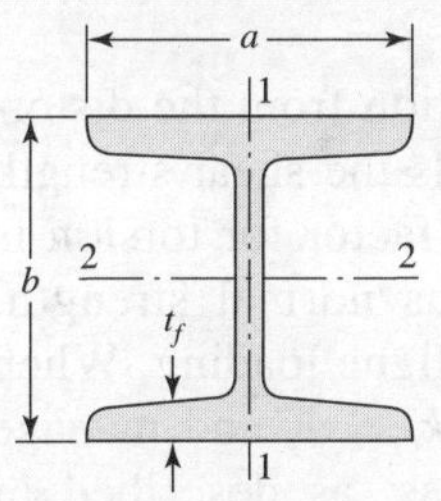	$A_{0.95\sigma} = \begin{cases} 0.10at_f & \text{axis 1-1} \\ 0.05ba \quad t_f > 0.025a & \text{axis 2-2} \end{cases}$
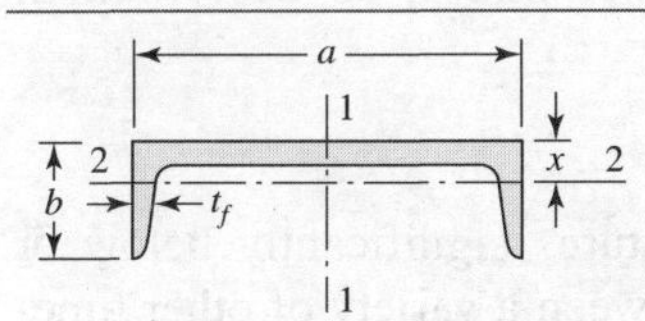	$A_{0.95\sigma} = \begin{cases} 0.05ab & \text{axis 1-1} \\ 0.052xa + 0.1t_f(b - x) & \text{axis 2-2} \end{cases}$

EXAMPLE 6–4

A steel shaft loaded in bending is 32 mm in diameter, abutting a filleted shoulder 38 mm in diameter. Estimate the Marin size factor k_b if the shaft is used in

(*a*) A rotating mode.

(*b*) A nonrotating mode.

Solution

(*a*) From Equation (6–19)

Answer
$$k_b = \left(\frac{d}{7.62}\right)^{-0.107} = \left(\frac{32}{7.62}\right)^{-0.107} = 0.86$$

(*b*) From Table 6–3,

$$d_e = 0.37d = 0.37(32) = 11.84\,\text{mm}$$

From Equation (6–19),

Answer
$$k_b = \left(\frac{11.84}{7.62}\right)^{-0.107} = 0.95$$

Loading Factor k_c

Estimates of endurance limit, such as that given in Equation (6–10), are typically obtained from testing with completely reversed bending. With axial or torsional loading, fatigue tests indicate different relationships between the endurance limit and the ultimate strength for each type of loading.[11] These differences can be accounted for with a load factor to adjust the endurance limit obtained from bending. Though the load factor is actually a function of the ultimate strength, the variation is minor, so it is appropriate to specify average values of the load factor as

$$k_c = \begin{cases} 1 & \text{bending} \\ 0.85 & \text{axial} \\ 0.59 & \text{torsion} \end{cases} \tag{6–25}$$

Note that the load factor for torsion is very close to the prediction from the distortion energy theory, Equation (5–21), where for ductile materials the shear strength is 0.577 times the normal strength. This implies that the load factor for torsion is mainly accounting for the difference in shear strength versus normal strength. Therefore, use the torsion load factor *only* for pure torsional fatigue loading. When torsion is combined with other loading, such as bending, set $k_c = 1$, and manage the loading situation by using the effective von Mises stress, as described in Section 6–16.

Temperature Factor k_d

Fatigue life predictions can be complicated at temperatures significantly below or above room temperature, due to complex interactions between a variety of other time-dependent and material-dependent processes.[12] We shall note some of the issues and trends here, but will provide only very limited quantitative recommendations.

---

[11]H. J. Grover, S. A. Gordon, and L. R. Jackson, *Fatigue of Metals and Structures,* Bureau of Naval Weapons, Document NAVWEPS 00-2500435, 1960; R. G. Budynas and J. K. Nisbett, *Shigley's Mechanical Engineering Design,* 9th ed., McGraw Hill, New York, 2011, pp. 332–333.

[12]A good overview is provided by R. I. Stephens, A. Fatemi, R. R. Stephens, and H. O. Fuchs, *Metal Fatigue in Engineering,* 2nd ed., John Wiley & Sons, 2001, pp. 364–391.

As operating temperatures decrease below room temperature, many materials, particularly steels, show an increase in ultimate and yield strength. However, the ductility decreases, as evidenced by a decrease in the plastic zone between the yield and ultimate strengths. The effect on fatigue is sometimes positive in that the higher ultimate strength also evidences a higher endurance limit. But if strain levels at stress concentrations are high, there may be a greater notch sensitivity due to the reduced ductility. The lower fracture toughness and lower ductility can lead to a smaller crack length necessary before rapid crack growth to fracture. For steels, the overall fatigue performance effect is often improved at lower temperatures, especially in the high-cycle region (and thus low stress and low strain). In the low-cycle region, steels often experience a slight improvement for smooth specimens, but little positive effect, and possibly detrimental effect, for notched specimens. Any impact loading during the dynamic fatigue cycling can have a significant detrimental effect, especially in the presence of sharp notches. In general, then, by avoiding sharp notches and impact loading, the regular fatigue life prediction methods are not unreasonable for low temperatures, especially for the high-cycle region.

High temperature fatigue is primarily a concern for temperatures above about 40 percent of the absolute (Kelvin) melting temperature. At these temperatures, thermally activated processes are involved, including oxidation, creep, relaxation, and metallurgical aspects. Also, time-dependent factors that are of little importance at lower temperatures begin to play an important role, such as frequency of cycling, wave shape, creep, and relaxation. Thermally induced cycling stresses can be significant in applications with frequent start-up and shut-down, or any other form of nonuniform or transient heating. When temperatures exceed about 50 percent of the absolute melting temperature, creep becomes a predominant factor, and the stress-life approach is no longer reasonable.

Crack nucleation and growth rates are accelerated due to high temperature environmental oxidation. Freshly exposed surfaces due to local plasticity are rapidly oxidized, with grain boundaries being particularly attacked. This leads to a greater incidence of intercrystalline crack growth, rather than the usual transcrystalline growth, which can have a faster crack growth rate.

At high temperatures, metals lose the distinct endurance limit, such that the fatigue strength continues to decline with cycles to failure. Further, most metals exhibit a lower long-life fatigue strength (e.g., at 10^8 cycles) with increased temperatures, with the notable exception of an increase for mild carbon steel in the 200° to 400°C range.

For steels operating in steady temperatures less than about 40 percent of the absolute melting point, which is about 380°C, the primary fatigue life mechanism is probably just the temperature effect on the strength properties. Accordingly, for moderate temperature variations from room temperature, it is recommended to simply utilize the appropriate ultimate strength, and if available, the endurance limit or fatigue strength, for the operating temperature, with k_d equal to unity.[13] If the temperature-specific ultimate strength is not available, the graph from Figure 2–17 can be utilized (for steels), represented by the curve-fit polynomials

$$S_T/S_{RT} = 0.99 + 5.9(10^{-4})T_C - 2.1(10^{-6})T_C^2 \tag{6–26}$$

[13]See, for example, W. F. Gale and T. C. Totemeier (eds.), *Smithells Metals Reference Book,* 8th ed., Elsevier, 2004, pp. 22–138 to 22–140, where endurance limits are tabulated for several steels from 100° to 650°C.

where S_T and S_{RT} are the ultimate strengths at the operating temperature and room temperature, respectively, and T_C are the operating temperature in degrees Celsius, Equation (6–26) is for steels, and should be limited to the range 20°C to 550°C. The other factors discussed earlier in this section should be considered as potentially significant for temperatures above about 380°C.

In the case where the endurance limit is known (e.g., by testing, or tabulated material data) at room temperature, it can be adjusted for the operating temperature by applying

$$k_d = S_T/S_{RT} \tag{6–27}$$

where values from Equation (6–26) can be used for steels. Again, though, if the endurance limit is available or being estimated based on the ultimate strength at the operating temperature, it needs no further adjustment and k_d should be set to unity.

EXAMPLE 6–5

A 1035 steel has a tensile strength of 550 MPa and is to be used for a part that operates in a steady temperature of 400°C. Estimate the endurance limit at the operating temperature if
(*a*) only the tensile strength at room temperature is known.
(*b*) the room-temperature endurance limit for the material is found by test to be $(S'_e)_{20°} = 270$ MPa.

Solution
(*a*) Estimate the tensile strength at the operating temperature from Equation (6–26),

$$(S_T/S_{RT})_{400°} = 0.99 + 5.9(10^{-4})(400) - 2.1(10^{-6})(400)^2 = 0.89$$

Thus,

$$(S_{ut})_{400°} = (S_T/S_{RT})_{400°}(S_{ut})_{20°} = 0.89(550) = 489.5 \text{ MPa}$$

From Equation (6–10),

Answer

$$(S_e)_{400°} = 0.5(S_{ut})_{400°} = 0.5(489.5) = 244.7 \text{ MPa}$$

and use $k_d = 1$ since this is already adjusted for the operating temperature.
(*b*) Since the endurance limit is known at room temperature, apply the temperature factor to adjust it to the operating temperature. From Equation (6–27),

$$k_d = (S_T/S_{RT})_{400°} = 0.89$$

Answer

$$(S_e)_{400°} = k_d(S'_e)_{20°} = 0.89(270) = 240 \text{ MPa}$$

Reliability Factor k_e

The reliability factor accounts only for the scatter in the endurance limit fatigue data and is not part of a complete stochastic analysis. The presentation in this text is strictly of a deterministic nature. For a more in-depth stochastic discussion see the previous edition of this text,[14] based on work by Mischke.[15] A true reliability for fatigue life can only be reasonably attempted with actual part testing.

---

[14]R. G. Budynas and J. K. Nisbett, *Shigley's Mechanical Engineering Design*, 9th ed., McGraw Hill Education, New York, 2011, p. 330.

[15]C. R. Mischke, "Prediction of Stochastic Endurance Strength," *Trans. ASME, Journal of Vibration, Acoustics, Stress, and Reliability in Design*, vol. 109, no. 1, January 1987, pp. 113–122.

Table 6–4 Reliability Factors k_e Corresponding to 8 Percent Standard Deviation of the Endurance Limit

Reliability, %	Transformation Variate z_a	Reliability Factor k_e
50	0	1.000
90	1.288	0.897
95	1.645	0.868
99	2.326	0.814
99.9	3.091	0.753
99.99	3.719	0.702

Most endurance strength data are reported as mean values. In Figure 6–22 and Equation (6–10), the endurance limit is experimentally determined to be related to the ultimate strength by $S'_e/S_{ut} = 0.5$. This corresponds roughly to a line through the middle of the scattered data. The reliability factor allows the slope of the line to be adjusted to increase the reliability of data points being included above the line. In turn, this adjusted (lower) endurance limit will account for a higher reliability on the *S-N* diagram of actual failure points being above the predictions of the high-cycle line.

Data presented by Haugen and Wirching[16] show standard deviations of endurance strengths of less than 8 percent. Thus the reliability modification factor to account for this can be written as

$$k_e = 1 - 0.08\, z_a \tag{6–28}$$

where z_a is defined by Equation (1–5) and values for any desired reliability can be determined from Table A–10. Table 6–4 gives reliability factors for some standard specified reliabilities.

Miscellaneous Effects

Any number of additional factors could be included as a multiplier in the Marin equation. Several other effects are known to exist, but they are difficult to quantify in the absence of specific testing. Some of these effects are briefly discussed here. Further investigation of pertinent literature is warranted when dealing with any of these issues.

Residual stresses may either improve the endurance limit or affect it adversely. Generally, if the residual stress in the surface of the part is compression, the endurance limit is improved. Fatigue failures appear to be tensile failures, or at least to be caused by tensile stress, and so anything that reduces tensile stress will also reduce the possibility of a fatigue failure. Operations such as shot peening, hammering, and cold rolling build compressive stresses into the surface of the part and improve the endurance limit significantly. Of course, the material must not be worked to exhaustion.

The endurance limits of parts that are made from rolled or drawn sheets or bars, as well as parts that are forged, may be affected by the so-called *directional characteristics* of the operation. Rolled or drawn parts, for example, have an endurance limit in the transverse direction that may be 10 to 20 percent less than the endurance limit in the longitudinal direction.

---

[16]E. B. Haugen and P. H. Wirsching, "Probabilistic Design," *Machine Design,* vol. 47, no. 12, 1975, pp. 10–14.

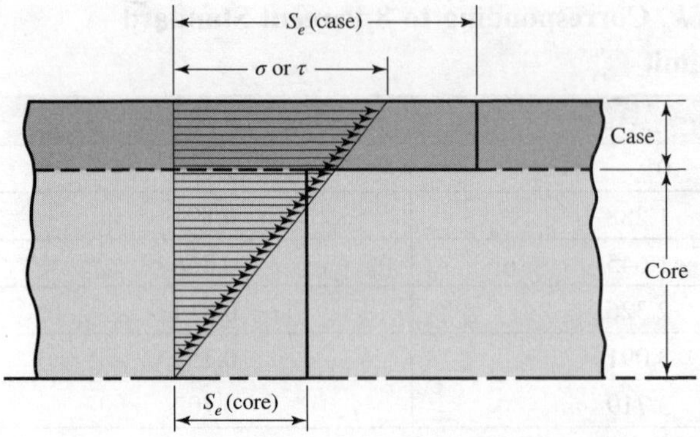

Figure 6–25

The failure of a case-hardened part in bending or torsion. In this example, failure occurs in the core.

Parts that are case-hardened may fail at the surface or at the maximum core radius, depending upon the stress gradient. Figure 6–25 shows the typical triangular stress distribution of a bar under bending or torsion. Also plotted as a heavy line in this figure are the endurance limits S_e for the case and core. For this example the endurance limit of the core rules the design because the figure shows that the stress σ or τ, whichever applies, at the outer core radius, is appreciably larger than the core endurance limit.

Corrosion

It is to be expected that parts that operate in a corrosive atmosphere will have a lowered fatigue resistance. This is true, and it is partly due to the roughening or pitting of the surface by the corrosive material. But the problem is not so simple as the one of finding the endurance limit of a specimen that has been corroded. The reason for this is that the corrosion and the stressing occur at the same time, such that the corrosion has opportunity to interact with and accelerate the crack growth mechanism. Basically, this means that in time any part will fail when subjected to repeated stressing in a corrosive atmosphere. There is no fatigue limit. Thus the designer's problem is to attempt to minimize the factors that affect the fatigue life; these are:

- Mean or static stress
- Alternating stress
- Electrolyte concentration
- Dissolved oxygen in electrolyte
- Material properties and composition
- Temperature
- Cyclic frequency
- Fluid flow rate around specimen
- Local crevices

Electrolytic Plating

Metallic coatings, such as chromium plating, nickel plating, or cadmium plating, reduce the endurance limit by as much as 50 percent. In some cases the reduction by coatings has been so severe that it has been necessary to eliminate the plating process. Zinc plating does not affect the fatigue strength. Anodic oxidation of light alloys reduces bending endurance limits by as much as 39 percent but has no effect on the torsional endurance limit.

Metal Spraying

Metal spraying results in surface imperfections that can initiate cracks. Limited tests show reductions of 14 percent in the fatigue strength.

Cyclic Frequency

If, for any reason, the fatigue process becomes time-dependent, then it also becomes frequency-dependent. Under normal conditions, fatigue failure is independent of frequency. But when corrosion or high temperatures, or both, are encountered, the cyclic rate becomes important. The slower the frequency and the higher the temperature, the higher the crack propagation rate and the shorter the life at a given stress level.

Frettage Corrosion

The phenomenon of frettage corrosion is the result of microscopic motions of tightly fitting parts or structures. Bolted joints, bearing-race fits, wheel hubs, and any set of tightly fitted parts are examples. The process involves surface discoloration, pitting, and eventual fatigue. A frettage factor depends upon the material of the mating pairs and ranges from 0.24 to 0.90.

EXAMPLE 6–6

A 1080 hot-rolled steel bar has been machined to a diameter of 25 mm. It is to be placed in reversed axial loading for 70 000 cycles to failure in an operating environment of 340°C. Using ASTM minimum properties, and a reliability for the endurance limit estimate of 99 percent, estimate the endurance limit and fatigue strength at 70 000 cycles.

Solution

From Table A–19, $S_{ut} = 770$ MPa at 20°C. Since the rotating-beam specimen endurance limit is not known at room temperature, we determine the ultimate strength at the elevated temperature first, using Equation (6–26),

$$(S_T/S_{RT})_{340^\circ} = 0.99 + 5.9(10^{-4})(340) - 2.1(10^{-6})(340)^2 = 0.95$$

The ultimate strength at 340°C is then

$$(S_{ut})_{340^\circ} = (S_T/S_{RT})_{340^\circ}(S_{ut})_{20^\circ} = 0.95(770) = 732 \text{ MPa}$$

The rotating-beam specimen endurance limit at 340°F is then estimated from Equation (6–10) as

$$S'_e = 0.5(732) = 366 \text{ MPa}$$

Next, we determine the Marin factors. For the machined surface, Equation (6–18) with Table 6–2 gives

$$k_a = aS_{ut}^b = 3.04(732)^{-0.217} = 0.73$$

For axial loading, from Equation (6–20), the size factor $k_b = 1$, and from Equation (6–25) the loading factor is $k_c = 0.85$. The temperature factor $k_d = 1$, since we accounted for the temperature in modifying the ultimate strength and consequently the endurance limit. For 99 percent reliability, from Table 6–4, $k_e = 0.814$. The endurance limit for the part is estimated by Equation (6–17) as

Answer

$$S_e = k_a k_b k_c k_d k_e S'_e$$
$$= 0.73(1)(0.85)(1)(0.814)366 = 185 \text{ MPa}$$

For the fatigue strength at 70 000 cycles we need to construct the *S-N* equation. From Equation (6–11), or we could use Figure 6–23,

$$f = 1.06 - 4.1(10^{-4})(732) + 1.5(10^{-7})(732)^2 = 0.84$$

From Equation (6–13),

$$a = \frac{(fS_{ut})^2}{S_e} = \frac{[0.84(732)]^2}{185} = 2044 \text{ MPa}$$

and Equation (6–14)

$$b = -\frac{1}{3}\log\left(\frac{fS_{ut}}{S_e}\right) = -\frac{1}{3}\log\left[\frac{0.84 \times 732}{185}\right] = -0.1739$$

Finally, for the fatigue strength at 70 000 cycles, Equation (6–12) gives

Answer $$S_f = aN^b = 2044(70\,000)^{-0.1741} = 294 \text{ MPa}$$

6–10 Stress Concentration and Notch Sensitivity

In Section 3–13 it was pointed out that the existence of irregularities or discontinuities, such as holes, grooves, or notches, in a part increases the theoretical stresses significantly in the immediate vicinity of the discontinuity. Equation (3–46) defined a stress-concentration factor K_t (or K_{ts}), which is used with the nominal stress to obtain the maximum resulting stress very local to the irregularity or defect. The theoretical stress-concentration factor K_t is defined for static loading conditions. Under dynamic loading conditions leading to fatigue, it turns out that the fatigue strength of a notched specimen is often not affected as much as would be expected from applying the stress-concentrated maximum stress. Consequently, for fatigue purposes, a *fatigue stress-concentration factor,* K_f (also known as the fatigue notch factor), is defined based on the fatigue strengths of notch-free versus notched specimens at long-life conditions (e.g., the endurance limit at 10^6 cycles) under completely reversed loading, such that

$$K_f = \frac{\text{Fatigue strength of notch-free specimen}}{\text{Fatigue strength of notched specimen}} \tag{6–29}$$

Thus, K_f is a reduced version of K_t, taking into account the sensitivity of the actual part to the stress concentrating effects of a notch in a fatigue situation, and is used in place of K_t as a stress increaser of the nominal stress,

$$\sigma_{\max} = K_f\sigma_0 \qquad \text{or} \qquad \tau_{\max} = K_{fs}\tau_0 \tag{6–30}$$

The reason for the reduced sensitivity is complex and not fully understood, and seems to include the interaction of several different phenomena. Perhaps the most fundamental explanation is that the notch stress affecting the fatigue life is not the maximum stress at the notch, but rather an average stress acting over a finite volume of material adjacent to the notch. Additionally, a small crack initiating at a notch will be growing into a region with rapidly decreasing stress levels. Consequently, the stress gradient plays a part. A large notch radius will have less stress gradient, such that the concentrated stress reduces gradually for distances further from the notch. In this case, the average stress near the notch is nearly as high as the maximum stress at the notch, so K_f will be nearly as large as K_t. On the other hand, a small (sharp) notch radius will have a large stress gradient, such that the average stress near the notch is significantly lower than the maximum stress. This leads to an apparent lower sensitivity to the notch. This behavior might be expected due to the fact that at the very small scale, the material is not homogeneous, but is made up of a discrete microstructure of grains which have the effect of equalizing the stress over a small finite distance. In general, materials that are soft, ductile, and low strength with a large grain structure are less sensitive to notches than materials that are hard, brittle, and high strength with a small grain structure. Another factor for notch sensitivity applies particularly when stresses are high enough to cause localized plastic strain at the root of a crack, effectively blunting the tip and reducing the maximum stress to a level that is less than predicted by $K_t\sigma_0$.

To quantify the sensitivity of materials to notches, a *notch sensitivity* q is defined as

$$q = \frac{K_f - 1}{K_t - 1} \qquad \text{or} \qquad q_s = \frac{K_{fs} - 1}{K_{ts} - 1} \tag{6–31}$$

where q is usually between zero and unity. Equation (6–31) shows that if $q = 0$, then $K_f = 1$, and the material has no sensitivity to notches at all. On the other hand, if $q = 1$, then $K_f = K_t$, and the material has full notch sensitivity. In analysis or design work, find K_t first, from the geometry of the part. Then specify the material, find q, and solve for K_f from the equation

$$K_f = 1 + q(K_t - 1) \qquad \text{or} \qquad K_{fs} = 1 + q_s(K_{ts} - 1) \tag{6–32}$$

Notch sensitivities for specific materials are obtained experimentally. Published experimental values are limited, but some values are available for steels and aluminum. Trends for notch sensitivity as a function of notch radius and ultimate strength are shown in Figure 6–26 for reversed bending or axial loading, and Figure 6–27 for reversed torsion. In using these charts it is well to know that the actual test results from which the curves were derived exhibit a large amount of scatter. Because of this scatter it is always safe to use $K_f = K_t$ if there is any doubt about the true value of q. Also, note that q is not far from unity for large notch radii.

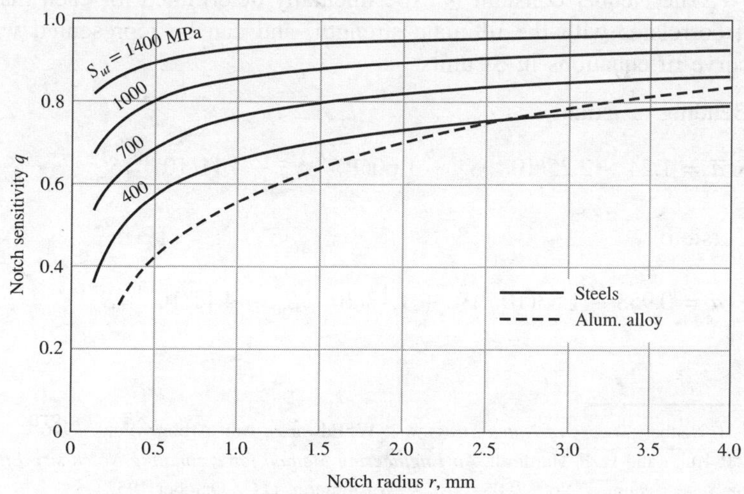

Figure 6–26

Notch-sensitivity charts for steels and UNS A92024-T wrought aluminum alloys subjected to reversed bending or reversed axial loads. For larger notch radii, use the values of q corresponding to the $r = 4$ mm ordinate. *(Source:* Sines, George and Waisman, J. L. (eds.). *Metal Fatigue*. McGraw Hill, New York, 1969.*)*

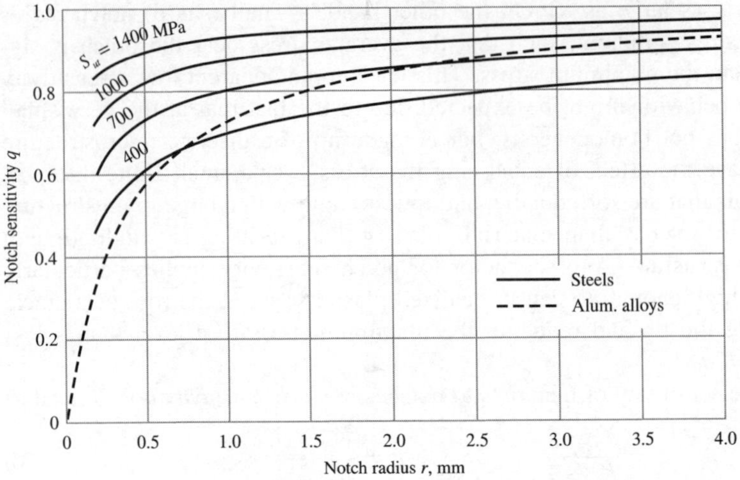

Figure 6–27

Notch-sensitivity curves for materials in reversed torsion. For larger notch radii, use the values of q_s corresponding to $r = 4$ mm.

Figures 6–26 and 6–27 have as their basis the work of Neuber[17] and Kuhn,[18] in which the notch sensitivity is described as a function of the notch radius and a material characteristic length dimension *a*. This characteristic length is roughly several times the size of a single microstructure grain, and can be thought of as near the size of the material's natural internal imperfections. It is often shown in the form of the *Neuber constant* $\sqrt{a}$. The relationship is

$$q = \frac{1}{1 + \frac{\sqrt{a}}{\sqrt{r}}} \tag{6–33}$$

where *r* is the notch radius. For convenience, we can combine Equations (6–32) and (6–33), giving

$$K_f = 1 + \frac{K_t - 1}{1 + \sqrt{a}/\sqrt{r}} \tag{6–34}$$

The Neuber constant is experimentally determined for each material. For steels, it correlates with the ultimate strength, and can be represented with the following curve-fit equations in SI units.

Bending or axial:

$$\sqrt{a} = 1.24 - 2.25(10^{-3})S_{ut} + 1.60(10^{-6})S_{ut}^2 - 4.11(10^{-10})S_{ut}^3 \tag{6–35}$$

Torsion:

$$\sqrt{a} = 0.958 - 1.83(10^{-3})S_{ut} + 1.43(10^{-6})S_{ut}^2 - 4.11(10^{-10})S_{ut}^3 \tag{6–36}$$

[17]H. Neuber, *Theory of Notch Stresses*, J. W. Edwards, Ann Arbor, Mich., 1946.

[18]P. Kuhn and H. F. Hardrath, *An Engineering Method for Estimating Notch-size Effect in Fatigue Tests on Steel.* Technical Note 2805, NACA, Washington, D.C., October 1952.

where the equations apply to steel and S_{ut} is in MPa, the Neuber constant $\sqrt{a}$ has units of $(\text{mm})^{1/2}$, and when used in Equation (6–33), r is in mm. Equation (6–33) used in conjunction with Equation (6–35) or (6–36) is equivalent to Figures (6–26) and (6–27). As with the graphs, the results from the curve fit equations provide only approximations to the experimental data.

The notch sensitivity of cast irons is very low, varying from 0 to about 0.20, depending upon the tensile strength. To be on the conservative side, it is recommended that the value $q = 0.20$ be used for all grades of cast iron.

EXAMPLE 6–7

A steel shaft in bending has an ultimate strength of 690 MPa and a shoulder with a fillet radius of 3 mm connecting a 32-mm diameter with a 38-mm diameter. Estimate K_f using:
(*a*) Figure 6–26.
(*b*) Equations (6–34) and (6–35).

Solution

From Figure A–13–9, using $D/d = 38/32 = 1.1875$, $r/d = 3/32 = 0.093\ 75$, we read the graph to find $K_t = 1.65$.
(*a*) From Figure 6–26, for $S_{ut} = 690$ MPa and $r = 3$ mm, $q = 0.84$. Thus, from Equation (6–32)

Answer

$$K_f = 1 + q(K_t - 1) = 1 + 0.84(1.65 - 1) = 1.55$$

(*b*) From Equation (6–35) with $S_{ut} = 690$ MPa, $\sqrt{a} = 0.314\sqrt{\text{mm}}$. Substituting this into Equation (6–34) with $r = 3$ mm gives

Answer

$$K_f = 1 + \frac{K_t - 1}{1 + \sqrt{a/r}} = 1 + \frac{1.65 - 1}{1 + \dfrac{0.314}{\sqrt{3}}} = 1.55$$

The fatigue stress-concentration factor was defined in Equation (6–29) based on its effect at long-life conditions, such as at 10^6 cycles. At shorter lives, experimental evidence indicates that reduced amounts of K_f are appropriate, particularly for ductile materials. In fact, for ductile materials with static loading (less than 1 cycle), it is standard practice to ignore the stress concentration effect, which we can now understand as using a reduced value of K_f equal to unity. At 10^3 cycles, high strength or brittle materials need nearly the full amount of K_f, while low strength and ductile materials need a concentrating effect that is between 1 and the full amount of K_f. To simplify the matter, a conservative approach will be used in this text, in which the full amount of K_f will be applied even for the low-cycle region. In fact, since the stress-life approach does not attempt to account for localized plastic strain, we will recommend that even the maximum localized stresses at a notch, as represented by $K_f\sigma_0$, be kept below yielding.

Some designers use $1/K_f$ as a Marin factor to reduce S_e, primarily as a convenient method of applying K_f to only the high-cycle end of the *S-N* curve. As has just been explained, this is only appropriate for very ductile materials for which stress concentration is not needed in the low-cycle region. Of course, for simple loading, infinite life problems, it makes no difference whether S_e is reduced by dividing it by K_f or the nominal stress is multiplied by K_f. Furthermore, in Section 6–16, when we consider combining loads, there generally are multiple fatigue stress-concentration factors occurring at a point (e.g., K_f for bending and K_{fs} for torsion). Here, it is only practical to modify the nominal stresses. To be consistent in this text, we will exclusively use the fatigue stress-concentration factor as a multiplier of the nominal stress.

EXAMPLE 6–8

Figure 6–28*a* shows a rotating shaft simply supported in ball bearings at *A* and *D* and loaded by a nonrotating force *F* of 6.8 kN. The shaft is machined from AISI 1050 cold-drawn steel. Estimate the life of the part.

Solution

From Figure 6–28*b* we learn that failure will probably occur at *B* rather than at *C* or at the point of maximum moment. Point *B* has a smaller cross section, a higher bending moment, and a higher stress-concentration factor than *C*, and the location of maximum moment has a larger size and no stress-concentration factor.

We shall solve the problem by first estimating the strength at point *B* and comparing this strength with the stress at the same point.

From Table A–19 we find $S_{ut} = 690$ MPa and $S_y = 580$ MPa. The endurance limit S'_e is estimated as

$$S'_e = 0.5(690) = 345 \text{ MPa}$$

From Equation (6–18) and Table 6–2,

$$k_a = 3.04(690)^{-0.217} = 0.74$$

From Equation (6–19),

$$k_b = (32/7.62)^{-0.107} = 0.86$$

Since $k_c = k_d = k_e = 1$,

$$S_e = 0.74(0.86)345 = 220 \text{ MPa}$$

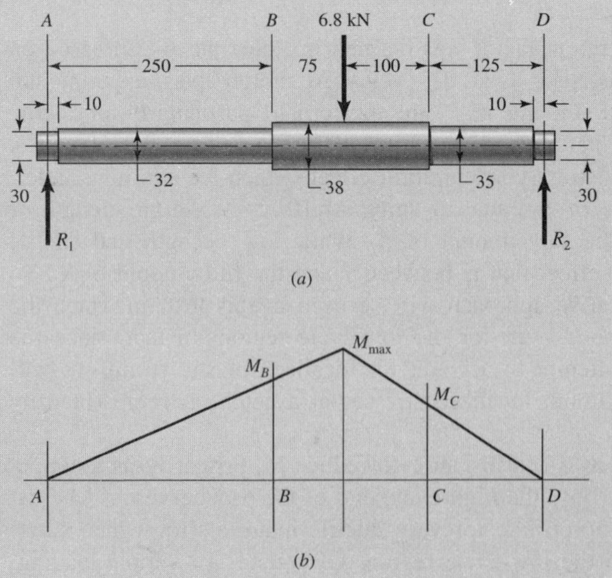

Figure 6–28

(*a*) Shaft drawing showing all dimensions in millimeters; all fillets 3-mm radius.
(*b*) Bending-moment diagram.

To find the geometric stress-concentration factor K_t we enter Figure A–13–9 with $D/d = 38/32 = 1.1875$ and $r/d = 3/32 = 0.09\,375$ and read $K_t = 1.65$. From Equation (6–35a) with $S_{ut} = 690$ MPa, $\sqrt{a} = 0.314\sqrt{\text{mm}}$. Substituting this into Equation (6–34) gives

$$K_f = 1 + \frac{K_t - 1}{1 + \sqrt{a/r}} = 1 + \frac{1.65 - 1}{1 + 0.314/\sqrt{3}} = 1.55$$

The next step is to estimate the bending stress at point B. The bending moment is

$$M_B = R_1 x = \frac{225F}{550}250 = \frac{225(6.8)}{550}250 = 695.5\,\text{N} \cdot \text{m}$$

Just to the left of B the section modulus is $I/c = \pi d^3/32 = \pi 32^3/32 = 3.217\,(10^3)\,\text{mm}^3$. The reversing bending stress is, assuming infinite life,

$$\sigma_{ar} = K_f \frac{M_B}{I/c} = 1.55\frac{695.5}{3.217}(10)^{-6} = 335.1(10^6)\ \text{Pa} = 335.1\ \text{MPa}$$

This stress is greater than S_e and less than S_y. This means we have both finite life and no yielding on the first cycle.

For finite life, we will need to use Equation (6–15). The ultimate strength, $S_{ut} = 690$ MPa. From Figure 6–23, $f = 0.85$. From Equation (6–13)

$$a = \frac{(fS_{ut})^2}{S_e} = \frac{[0.85(690)]^2}{220} = 1564\ \text{MPa}$$

and from Equation (6–14)

$$b = -\frac{1}{3}\log\left(\frac{fS_{ut}}{S_e}\right) = -\frac{1}{3}\log\left[\frac{0.85(690)}{220}\right] = -0.1419$$

From Equation (6–15),

Answer

$$N = \left(\frac{\sigma_{ar}}{a}\right)^{1/b} = \left(\frac{335.1}{1564}\right)^{-1/0.1419} = 52(10^3)\,\text{cycles}$$

6–11 Characterizing Fluctuating Stresses

In Section 6–7 the concept of fluctuating stresses was introduced to define terminology. But so far the focus has been on completely reversed stress situations, with zero mean stress. This has allowed the focus to be on the data predominately generated by completely reversed testing. The next step is to examine the effect of nonzero mean stresses. To do so, we will first consider some general issues characterizing fluctuating stresses and reiterate some of the definitions that will be used in the following sections.

Figure 6–29 illustrates some of the various stress-time traces that occur. Fluctuating stresses in machinery often take the form of a sinusoidal pattern because of the nature of rotating machinery. However, other patterns are common as well. Very irregular or random stress is extremely difficult to model. Simplifications to the stress pattern are sometimes necessary to apply the stress-life method. Tiny ripples in an otherwise large stress variation, such as in Figure 6–29a, can be ignored. Irregularities in the shape or frequency of a wave pattern, such as in Figure 6–29b, do not have any impact. In general, the important thing is the magnitudes of the peaks. For variations in the

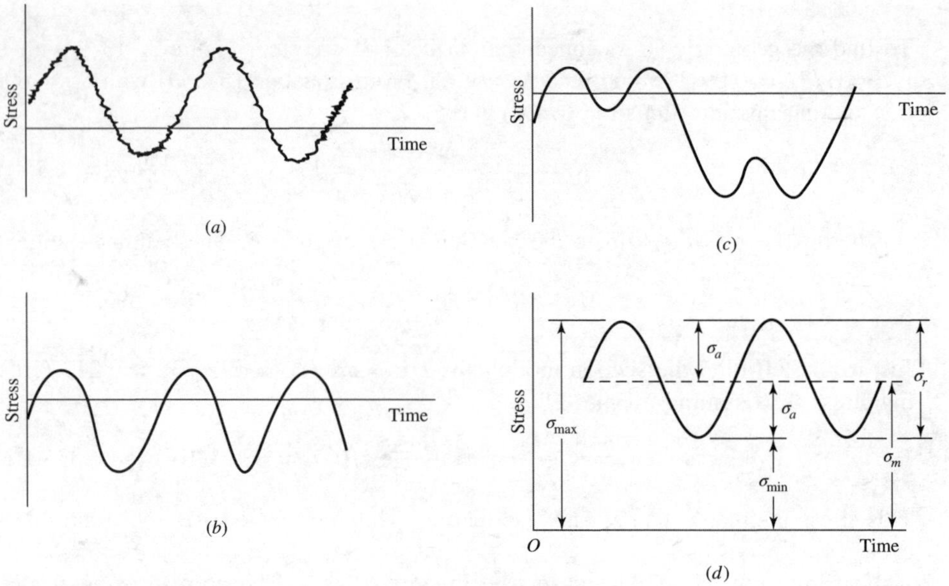

Figure 6–29

Some stress-time relations: (*a*) fluctuating stress with high-frequency ripple; (*b* and *c*) nonsinusoidal fluctuating stress; (*d*) sinusoidal fluctuating stress.

peaks, such as in Figure 6–29*c*, reasonable life estimates may be possible by ignoring the intermediate peaks. To study the effect of mean stress on the fatigue life predictions of the *S-N* diagram, we will focus on the general fluctuating case shown in Figure 6–29*d*, along with the following definitions.

σ_{min} = minimum stress
σ_{max} = maximum stress
σ_a = alternating stress, or stress amplitude
σ_m = mean stress, or midrange stress
σ_r = stress range

The following relations are evident from Figure 6–29*d*. These were defined in Section 6–7 and are repeated here for convenience:

$$\sigma_a = \left|\frac{\sigma_{max} - \sigma_{min}}{2}\right| \tag{6–8}$$

$$\sigma_m = \frac{\sigma_{max} + \sigma_{min}}{2} \tag{6–9}$$

In addition, the *stress ratio* is defined as

$$R = \frac{\sigma_{min}}{\sigma_{max}} \tag{6–37}$$

The stress ratio can have values between -1 and $+1$, and is commonly used to represent with a single value the nature of the stress pattern. For example, $R = -1$ is completely reversed, $R = 0$ is repeated load, $R = 1$ is steady.

Application of Fatigue Stress-Concentration Factor

While we are defining terms, we will take a moment to address the terminology and the method of applying the fatigue stress-concentration factor K_f to the stresses. The stress terms already defined, such as σ_a and σ_m, are to be understood to refer to the stress at the location under scrutiny, including any stress concentration. When it is helpful to be clear that a stress is referring to a nominal stress, without the inclusion of stress concentration, it will be shown with an additional subscript, as in σ_{a0} and σ_{m0}. Therefore, in the presence of a notch, $\sigma_a = K_f\,\sigma_{a0}$, and when no notch is present, $K_f = 1$ and $\sigma_a = \sigma_{a0}$.

When the mean stress is high enough to induce localized notch yielding, the first-cycle local yielding produces plastic strain and strain strengthening. This is occurring at the location where fatigue crack nucleation and growth are most likely. The material properties (S_y and S_{ut}) are new and difficult to quantify. We recommend as a design criterion for the stress-life method that the maximum stress at a notch, including the fatigue stress concentration, not exceed the yield strength. In other words, design to avoid plastic yielding at a notch. This recommendation is because the stress-life method does not address plastic strain and is not particularly suitable for predicting life when plastic strain is included. Consequently, in this text, we will apply the fatigue stress-concentration factor to both the alternating and mean stresses, as well as to the maximum stress when checking for yielding at a notch. Thus,

$$\sigma_a = K_f\,\sigma_{a0} \tag{6–38}$$

$$\sigma_m = K_f\,\sigma_{m0} \tag{6–39}$$

There are other strategies for handling the stress concentration when localized yielding occurs. They mostly take into account that the peak stress is actually reduced when the localized yielding occurs, so the full amount of the concentration factor is not accurate, and can be reduced. This is usually done only for the mean stress. Some of these approaches include the *nominal mean stress* method, the *residual stress* method,[19] and Dowling's method.[20]

6–12 The Fluctuating-Stress Diagram

The *S-N* diagram is normally generated with completely reversed stresses. It is possible to generate *S-N* diagrams with loading situations that include a mean stress. Figure 6–30 shows a characteristic family of *S-N* curves, generated as usual by plotting alternating stresses versus fatigue life, but with each curve corresponding to a

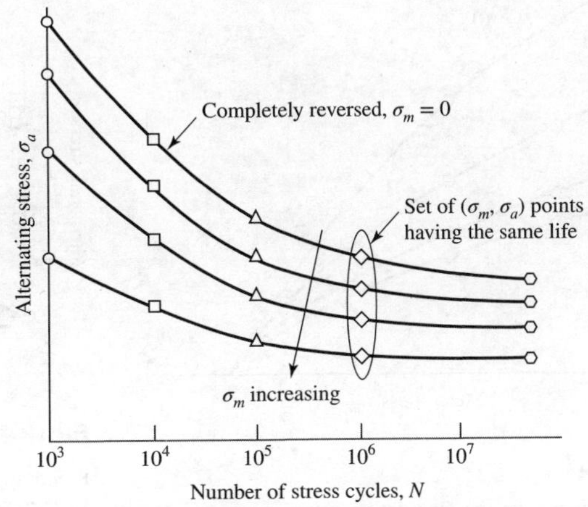

Figure 6–30

Characteristic family of *S-N* curves for increasing levels of mean stress.

---

[19]R. C. Juvinall and K. M. Marshek, *Fundamentals of Machine Component Design,* 4th ed., Wiley, New York, 2006, Section 8.11; M. E. Dowling, *Mechanical Behavior of Materials,* 3rd ed., Prentice Hall, Upper Saddle River, N.J., 2007, Sections 10.2–10.6.

[20]Dowling, op. cit., pp. 486–487.

different mean stress. The top curve is for zero mean stress and corresponds to the usual line on the completely reversed *S-N* diagram. Though details would be different for each material tested, the curves show the usual characteristic that higher mean stresses reduce the endurance limit, and in general reduce the life expectancy for a given alternating stress.

Historically, there have been many ways of plotting the data obtained for general fluctuating stresses, with such names as Goodman diagram, modified-Goodman diagram, master fatigue diagram, and Haigh diagram.[21] Today, probably the simplest and most common means of displaying the fluctuating-stress data is on a plot of alternating stress versus mean stress. We will call this $\sigma_a - \sigma_m$ set of axes a *fluctuating-stress diagram*. From the family of *S-N* curves in Figure 6–30, taking sets of points correlating to the same value of life, combinations of σ_a and σ_m can be determined for each life, and plotted as *constant-life curves* on a fluctuating-stress diagram, as shown in Figure 6–31. Of course, the curves are attempts to fit the scattered data points. Specific equations to "fit" the data are discussed in Section 6–14. This diagram depicts the effect of mean stress on fatigue life, for a specific material from which the curves were generated. Among other things, it is clear that for any desired life, an increase in mean stress must be accompanied by a decrease in alternating stress.

Now, from Figure 6–31, focus in on the data points that correlate to the constant-life curve of 10^6 cycles, shown in Figure 6–32, with many more data points shown to indicate the scatter of data. For steels, with the idealized assumption that the endurance limit corresponds to a life of 10^6 cycles, these data points reflect the predicted boundary between finite life and infinite life. If it is desired to achieve infinite life, the combination of σ_a and σ_m must be inside an envelope bounded by these data points. A straight line between the endurance limit and the ultimate strength is a commonly used design criterion to conservatively represent the infinite life boundary.

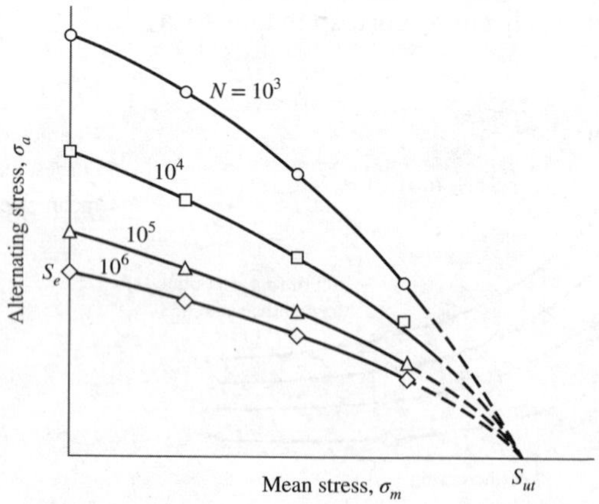

Figure 6–31

Constant-life curves on the fluctuating-stress diagram.

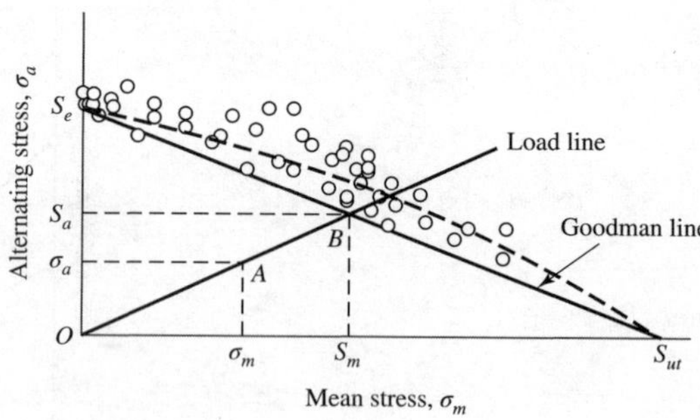

Figure 6–32

Fluctuating-stress diagram showing the Goodman line as an infinite-life fatigue criterion.

[21] W. Schutz, "A History of Fatigue," *Engineering Fracture Mechanics*, vol. 54, no. 2, 1996, pp. 263–300.

The line is well-known and is often called the modified-Goodman line. As time has passed since John Goodman's life (1862–1935), very little reference is ever made to the original Goodman line, so for simplicity, we will refer to the modified-Goodman line as simply the Goodman line. The Goodman line connects two points: the endurance limit when the mean stress is zero (completely reversed) and the ultimate strength when the alternating stress is zero (steady stress). The equation for the Goodman line is

$$\frac{\sigma_a}{S_e} + \frac{\sigma_m}{S_{ut}} = 1 \tag{6–40}$$

To utilize the Goodman line as a design criterion, define a load-line from the origin, through a given stress point (σ_m, σ_a). The default assumption is that if the stress situation is to increase, it will do so along the load line, maintaining the same slope, that is, the ratio of σ_a/σ_m. A factor of safety with respect to achieving infinite life can then be defined as the proportion of the distance along the load line from the origin to the failure criterion that has been achieved by the load stress. In other words, referring to Figure 6–32, define a strength as distance OB, a stress as distance OA, and the factor of safety as strength/stress $= OB/OA$. This can be done by solving for coordinates (S_m, S_a) at point B at the intersection of the failure line and the load line. Then the factor of safety is $n_f = OB/OA = S_a/\sigma_a = S_m/\sigma_m$. Or, for this load line from the origin, it turns out that the factor of safety can be found mathematically by simply applying a fatigue life design factor n_f to each of the stresses in Equation (6–40) and solving for the factor. That is,

$$\frac{(n_f\sigma_a)}{S_e} + \frac{(n_f\sigma_m)}{S_{ut}} = 1 \tag{a}$$

$$n_f = \left(\frac{\sigma_a}{S_e} + \frac{\sigma_m}{S_{ut}}\right)^{-1} \qquad \sigma_m \geq 0 \tag{6–41}$$

This factor of safety based on the Goodman line is commonly used for design situations in which some conservatism is desired (though the level of conservatism is not quantified). In using it, remember that it is being applied to a specific stress element of a specific part made from a specific material. The stresses should both include any applicable fatigue concentration factor, and the endurance limit should be either obtained from actual testing of the part, or estimated and adjusted with appropriate Marin factors. A factor of safety greater than 1 predicts infinite life, and a factor of safety less than 1 predicts a finite life.

So far, the presentation of the Goodman line has been based on hypothetical data from a single material (as in Figures 6–30 and 6–31). To evaluate how well a curve fits the data, it is useful to normalize the fluctuating-stress diagram to allow data from different materials and different testing programs to be plotted on the same plot. Figure 6–33 shows such a plot for a variety of steels, where the ordinate axis is normalized by dividing the alternating stress by the endurance limit, and the abscissa is normalized by dividing the mean stress by the material's ultimate strength.

The plot also includes the negative mean stress range, where it is evident that the failure points do not indicate a reduction in alternating stresses for increasing compressive mean stresses. In fact, tests actually indicate an increase in alternating stress for

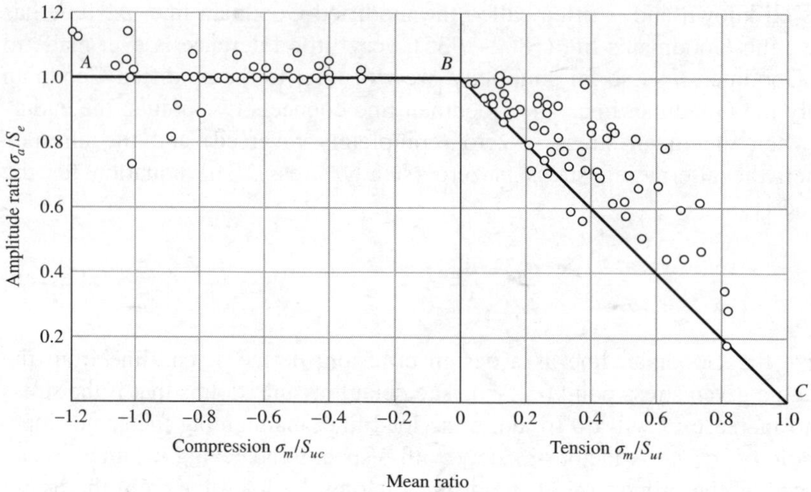

Figure 6–33

Plot of fatigue failures for mean stresses in both tensile and compressive regions. Normalizing the data by using the ratio of steady strength component to tensile strength σ_m/S_{ut}, steady strength component to compressive strength σ_m/S_{uc} and strength amplitude component to endurance limit σ_a/S_e enables a plot of experimental results for a variety of steels. *(Data source: Thomas J. Dolan. "Stress Range." Section 6.2 in O. J. Horger (ed.).* ASME Handbook—Metals Engineering Design. *McGraw Hill, New York, 1953.)*

many materials. This is an interesting phenomenon—that compressive mean stress is not detrimental to fatigue life, and is in fact often beneficial. There are several thoughts as to why this is so, some of which are discussed in Section 6–3 in the crack propagation section. From the conglomeration of test data such as in Figure 6–33, it is confirmed that the Goodman line is a reasonable conservative design tool for positive mean stress. For negative mean stress, a similarly conservative design criterion is a horizontal line from the endurance limit. A fatigue factor of safety based on such a line is

$$n_f = \frac{S_e}{\sigma_a} \qquad \sigma_m < 0 \tag{6–42}$$

It is necessary to check for yielding. So far, since we are only considering a uniaxial stress situation, we can simply compare the maximum stress to the yield strength. The factor of safety guarding against first-cycle yield is given by

$$n_y = \frac{S_y}{\sigma_{\max}} = \frac{S_y}{\sigma_a + |\sigma_m|} \tag{6–43}$$

where the absolute value of the mean stress allows the equation to apply for both positive and negative mean stress. It is helpful to plot the yield condition on the fluctuating-stress diagram to see how it compares with the fatigue criterion. Setting n_y equal to unity, and plotting the equation $\sigma_a + |\sigma_m| = S_y$ produces two diagonal lines as shown in Figure 6–34. The yield lines are known as *Langer lines.* Figure 6–34 provides a summary of the design space for both infinite-life fatigue and yielding. A fluctuating-stress state outside the bounds of the yield lines predicts yielding. Notice there is a zone on the positive mean stress side where the yield line crosses inside the fatigue line. This zone indicates it is possible to yield in the first cycle, even when infinite fatigue life is predicted. For this reason, it is always wise to check for yielding, even on a fatigue-loading situation, especially when the mean stress is large and the alternating stress is small.

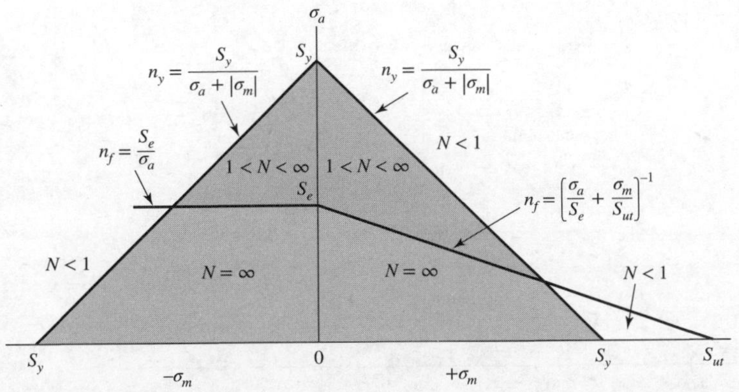

Figure 6–34

Fluctuating-stress diagram with fatigue boundary lines and yield boundary lines defining a design space with zones for infinite life, finite life, and first-cycle yielding.

EXAMPLE 6–9

A steel bar undergoes cyclic loading such that at the critical notch location the nominal stress cycles between $\sigma_{max} = 270$ MPa and $\sigma_{min} = 140$ MPa, and a fatigue stress-concentration factor is applicable with $K_f = 1.2$. For the material, $S_{ut} = 700$ MPa, $S_y = 595$ MPa, and a fully corrected endurance limit of $S_e = 280$ MPa. Estimate
(*a*) the fatigue factor of safety based on achieving infinite life according to the Goodman line.
(*b*) the yielding factor of safety.

Solution

(*a*) From Equations (6–8) and (6–9),

$$\sigma_{a0} = \frac{270 - 140}{2} = 65 \text{ MPa} \qquad \sigma_{m0} = \frac{270 + 140}{2} = 205 \text{ MPa}$$

Applying Equations (6–38) and (6–39),

$$\sigma_a = K_f\, \sigma_{a0} = 1.2(65) = 78 \text{ MPa}$$

$$\sigma_m = K_f\, \sigma_{m0} = 1.2(205) = 246 \text{ MPa}$$

For a positive mean stress, apply Equation (6–41),

Answer
$$n_f = \left(\frac{\sigma_a}{S_e} + \frac{\sigma_m}{S_{ut}}\right)^{-1} = \left(\frac{78}{280} + \frac{246}{700}\right)^{-1} = 1.59$$

Infinite life is predicted.
(*b*) To avoid even localized yielding at the notch, keep K_f applied to the stresses for the yield check. Using Equation (6–43),

Answer
$$n_y = \frac{S_y}{\sigma_a + |\sigma_m|} = \frac{595}{78 + 246} = 1.84$$

No yielding is predicted at the notch at the first stress cycle. Of course, realize that with continued cycling, at the grain level the cyclic stress will eventually lead to very localized plastic strain (see Section 6–3). If there were truly no plastic strain, there would be no fatigue.

Figure 6–35

Fluctuating-stress diagram showing load lines for Examples 6–9, 6–10, and 6–11.

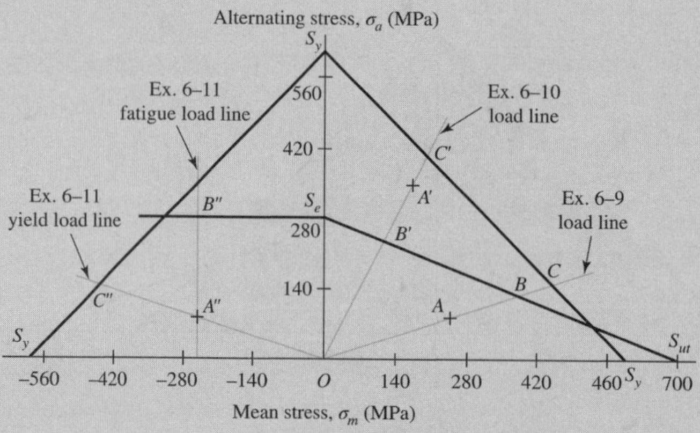

A visual representation of this example is helpful. Figure 6–35 shows this stress situation on the design space of the fluctuating-stress diagram. The load line passes through the fluctuating-stress point at (36, 12), the intersection with the Goodman line at point B, and the intersection with the Langer yield line at point C. Because point A is inside the envelope of both Goodman and Langer, infinite life with no yielding is predicted. Since the load line crosses the Goodman line before crossing the Langer line, fatigue life drives the design. The factor of safety is the proportion of the load line traversed toward failure. For fatigue this is $n_f = OB/OA$, and for yielding this is $n_y = OC/OA$.

EXAMPLE 6–10

Repeat Example 6–9, except for a nominal stress that cycles between $\sigma_{max} = 420$ MPa and $\sigma_{min} = -140$ MPa.

Solution

(*a*) Equations (6–8), (6–9): $\sigma_{a0} = \dfrac{420 - (-140)}{2} = 280$ MPa $\quad \sigma_{m0} = \dfrac{420 + (-140)}{2} = 140$ MPa

Equations (6–38), (6–39): $\sigma_a = K_f \sigma_{a0} = 1.2(280) = 336$ MPa $\quad \sigma_m = K_f \sigma_{m0} = 1.2(140) = 168$ MPa

Answer Equation (6–41):

$$n_f = \left(\frac{\sigma_a}{S_e} + \frac{\sigma_m}{S_{ut}}\right)^{-1} = \left(\frac{336}{280} + \frac{168}{700}\right)^{-1} = 0.69$$

Infinite life is not predicted. In Example 6–15 this problem will be revisited to estimate the predicted finite life.

(*b*) Equation (6–43):

$$n_y = \frac{S_y}{\sigma_a + |\sigma_m|} = \frac{595}{(336 + 168)} = 1.18$$

No yielding is predicted at the notch at the first stress cycle. The stress point, fatigue line intercept, and yield line intercept are plotted as A', B', and C', respectively, on the fluctuating-stress diagram of Figure 6–35.

EXAMPLE 6–11

Repeat Example 6–9, except for a nominal stress that cycles between $\sigma_{max} = -140$ MPa and $\sigma_{min} = -280$ MPa,

Solution

(*a*) Equations (6–8), (6–9): $\sigma_{a0} = \dfrac{-140 - (-280)}{2} = 70$ MPa $\quad \sigma_{m0} = \dfrac{-140 + (-280)}{2} = -210$ MPa

Equations (6–38), (6–39): $\sigma_a = K_f\sigma_{a0} = 1.2(70) = 84$ MPa $\quad \sigma_m = K_f\sigma_{m0} = 1.2(-210) = -252$ MPa

For a negative mean stress, apply Equation (6–42),

Answer $$n_f = \frac{S_e}{\sigma_a} = \frac{280}{84} = 3.3$$

Infinite life is predicted, but with a factor of safety more than double the similar problem in Example 6–9, with the only difference being the negative mean stress.

(*b*) Equation (6–43): $$n_y = \frac{S_y}{\sigma_a + |\sigma_m|} = \frac{595}{84 + |-252|} = 1.8$$

This is the same as in Example 6–9, though it is with regard to compressive yielding in this case. The stress point, fatigue line intercept, and yield line intercept are plotted as A'', B'', and C'', respectively, on the fluctuating-stress diagram of Figure 6–35. Note that the load lines for fatigue and yielding are different this time.

6–13 Fatigue Failure Criteria

The Goodman line is not intended as the best "fit" for the infinite-life data. Indeed, it is not a good fit, as it is almost entirely to one side of the data. Yet it serves useful purposes. Several fatigue failure criteria have been proposed to provide options for various purposes, and are shown in Figure 6–36. Each is briefly described below, along with the equations for the criterion curve and infinite-life fatigue factor of safety.

Goodman

The Goodman line is simple, conservative, and good for design purposes. It is almost entirely to the conservative side of the data, and therefore not good when seeking a criterion that is typical of the data. It is only applicable for

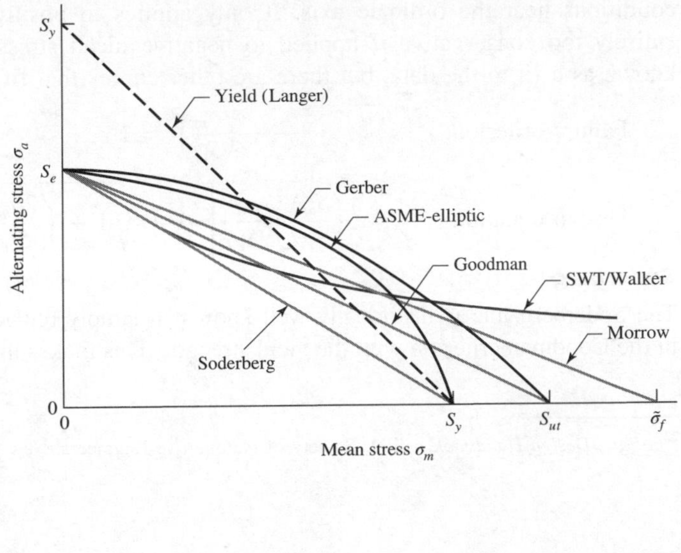

Figure 6–36

Comparison of several infinitelife fatigue failure criteria.

positive mean stress, as it is nonconservative if applied to negative mean stress. The Goodman equations were previously defined and are repeated here for convenience.

$$\text{Failure criterion:} \qquad \frac{\sigma_a}{S_e} + \frac{\sigma_m}{S_{ut}} = 1 \tag{6–40}$$

$$\text{Design equation:} \qquad n_f = \left(\frac{\sigma_a}{S_e} + \frac{\sigma_m}{S_{ut}}\right)^{-1} \qquad \sigma_m \geq 0 \tag{6–41}$$

Morrow

The Morrow line is identical in nature to the Goodman line, except it replaces the ultimate strength as the abscissa intersect with the true fracture strength $\tilde{\sigma}_f$ (defined in Section 2–1) or the fatigue strength coefficient σ_f' (defined in Section 6–6). These two terms are approximately equal for steels, and both are somewhat higher than the ultimate strength. They can be significantly different for aluminum alloys, in which case $\tilde{\sigma}_f$ is the better fit. The Morrow line provides a reasonable fit to the fatigue data on both the positive and negative mean stress side. It tends to be slightly toward the conservative side for positive mean stress, and slightly to the nonconservative side for negative mean stress. It is linear and simple, like the Goodman line, but with a better fit to the data. A primary disadvantage of this criterion is the limited availability of values for $\tilde{\sigma}_f$ and σ_f'. A very crude estimate is given by SAE[22] for the fatigue strength coefficient for steels (HB $\leq$ 500) of

$$\sigma_f' = S_{ut} + 345 \text{ MPa} \tag{6–44}$$

Of course, if the estimate is used, the curve could be anywhere between conservative and nonconservative. The equations for the Morrow criterion are

$$\text{Failure criterion:} \qquad \frac{\sigma_a}{S_e} + \frac{\sigma_m}{\tilde{\sigma}_f} = 1 \qquad \text{or} \qquad \frac{\sigma_a}{S_e} + \frac{\sigma_m}{\sigma_f'} = 1 \tag{6–45}$$

$$\text{Design equation:} \qquad n_f = \left(\frac{\sigma_a}{S_e} + \frac{\sigma_m}{\tilde{\sigma}_f}\right)^{-1} \qquad \text{or} \qquad n_f = \left(\frac{\sigma_a}{S_e} + \frac{\sigma_m}{\sigma_f'}\right)^{-1} \tag{6–46}$$

Gerber

The Gerber curve is a parabolic equation that was one of the early options presented to better pass through the middle of the fatigue points. It certainly fits the data better than the Goodman line, but it can be slightly nonconservative, especially for stress conditions near the ordinate axis. It only applies to positive mean stress, as it is entirely too conservative if applied to negative mean stress. It is historically well known as a fit to the data, but there are other curves that fit better.

$$\text{Failure criterion:} \qquad \frac{\sigma_a}{S_e} + \left(\frac{\sigma_m}{S_{ut}}\right)^2 = 1 \tag{6–47}$$

$$\text{Design equation:} \quad n_f = \frac{1}{2}\left(\frac{S_{ut}}{\sigma_m}\right)^2\left(\frac{\sigma_a}{S_e}\right)\left[-1 + \sqrt{1 + \left(\frac{2\sigma_m S_e}{S_{ut}\sigma_a}\right)^2}\,\right] \qquad \sigma_m \geq 0 \tag{6–48}$$

Soderberg

The Soderberg line is historically well known. It simply replaces the ultimate strength in the Goodman criterion with the yield strength. This makes the line ultra-conservative.

[22] *Fatigue Design Handbook,* vol. 4, Society of Automotive Engineers, New York, 1958, p. 27.

Its sole purpose is to provide a simple, conservative line that checks for infinite-life fatigue and yielding at the same time. It removes the need for a separate yield check. It is perhaps useful for quick first estimates or for situations that can be grossly over-designed.

Failure criterion: $$\frac{\sigma_a}{S_e} + \frac{\sigma_m}{S_y} = 1 \tag{6–49}$$

Design equation: $$n = \left(\frac{\sigma_a}{S_e} + \frac{\sigma_m}{S_y}\right)^{-1} \qquad \sigma_m \geq 0 \tag{6–50}$$

ASME-Elliptic

This criterion uses an elliptic equation to attempt to mix the qualities of the Gerber and the Soderberg criteria, that is, to fit the fatigue data, but check for yielding at the same time. As might be expected, it does both tasks to some extent, but neither very well. It is sometimes conservative and sometimes nonconservative, for both fatigue and yielding. Its primary recognition is that it is the criterion specified in the ANSI/ASME Standard B106.1M-1985 for design of transmission shafting.

Failure criterion: $$\left(\frac{\sigma_a}{S_e}\right)^2 + \left(\frac{\sigma_m}{S_y}\right)^2 = 1 \tag{6–51}$$

Design equation: $$n_f = \left[\left(\frac{\sigma_a}{S_e}\right)^2 + \left(\frac{\sigma_m}{S_y}\right)^2\right]^{-1/2} \qquad \sigma_m \geq 0 \tag{6–52}$$

Smith-Watson-Topper

The Smith-Watson-Topper (SWT) criterion (as well as its closely related Walker criterion) is relatively more recent (1970s) and has gained traction as a good fatigue criterion. It is unique from the previously described criteria in that it has a theoretical basis rather than simply attempting to fit the data.[23] It is primarily associated with the strain-life method, but for high-cycle life where the plastic strain is small, it can be put into terms of stress. It basically asserts that the critical parameters in fatigue-life prediction are the maximum stress and the alternating stress. In equation form, the SWT criterion is

Failure criterion: $$S_e = \sqrt{\sigma_{\max}\sigma_a} = \sqrt{(\sigma_m + \sigma_a)\sigma_a} \tag{6–53}$$

Design equation: $$n_f = \frac{S_e}{\sqrt{(\sigma_m + \sigma_a)\sigma_a}} \tag{6–54}$$

This criterion is not a function of ultimate strength, or any other static strength, so its curve does not intersect the mean stress abscissa. Its range for positive mean stress should be limited by the yield line. It is usually also acceptable for negative mean stress. It is most commonly used for predicting an equivalent completely reversed stress for a fluctuating-stress state that does not predict infinite life. Consequently, it is not usually used to predict a fatigue factor of safety based on achieving infinite life. The SWT criterion is particularly good at fitting data for aluminum and is usually reasonable for steel. For any given combination of (σ_m, σ_a) it could be slightly conservative or nonconservative with respect to the failure data.

[23] K. Smith, P. Watson, and T. Topper, "A Stress-Strain Function for the Fatigue of Metals," *Journal of Materials,* ASTM, vol. 5, no. 4, December 1970, pp. 767–778.

Walker

The Walker criterion is a more generalized version of SWT, in which the square root is replaced by a fitting parameter γ in the form of

$$\text{Failure criterion:} \qquad S_e = \sigma_{\max}^{1-\gamma}\sigma_a^{\gamma} = (\sigma_m + \sigma_a)^{1-\gamma}\,\sigma_a^{\gamma} \tag{6–55}$$

$$\text{Design equation:} \qquad n_f = \frac{S_e}{(\sigma_m + \sigma_a)^{1-\gamma}\sigma_a^{\gamma}} \tag{6–56}$$

The parameter γ is determined by experiment for each material by testing at multiple values of mean stress. The parameter essentially shifts more or less weight between the maximum stress and the alternating stress in order to better fit the experimentally proven behavior of a material. The SWT criterion is a special case of Walker, with $\gamma = 0.5$. The Walker criterion is known to provide good fit for both positive and negative mean stress. The disadvantage is that it is not trivial to obtain the value of the fitting parameter. Dowling[24] has determined an approximate relationship between γ and ultimate strength for steels to be

$$\gamma = -0.0002S_{ut} + 0.8818 \qquad (S_{ut} \text{ in MPa}) \tag{6–57}$$

With this, the simplicity of SWT is achieved with the better fit of the Walker criterion, without testing. Equation (6–57) shows a trend of decreasing γ with increasing S_{ut}. This indicates a shifting of the weighting in Equation (6–55) away from the alternating stress and to the maximum stress, indicating a more brittle behavior for higher strength materials.

Concluding Recommendations

There is not one best infinite-life fatigue criteria, as each has value for different applications. The Goodman criterion is one of the oldest and most well known, and is still a good option for situations that call for simplicity and conservatism. That goal is consistent with the expectations of a single chapter on the topic of fatigue, and especially for when the stress-life approach is being used. So, in this text we will tend to use the Goodman criterion and the conservative fatigue and yield design spaces as modeled in Figure 6–34.

When it is desired to use a fatigue criterion that is more typical of the experimental data, the Morrow criterion is appealing, as it is just as simple as Goodman and is a better fit to the data and can be used for both positive and negative mean stress. If the material properties needed for the Morrow criterion are not available, and estimates have to be used, then it is an acceptable approach, but is perhaps no better than Gerber. The SWT and Walker criteria are more complex in concept for a first exposure to fatigue. They are best known for use in the strain-life method. They are potentially good options but do not seem to be used for an infinite-life factor of safety. They are best for finite-life situations, as discussed in Section 6–14. Their inclusion here is warranted due to their recognition in the current research and application environment.

The Soderberg and ASME-Elliptic criteria may have appropriate applications, but in trying to do two things at once, they suffer some consequences as a result.

---

[24]N. Dowling, C. Calhoun, and A. Arcari, "Mean Stress Effects in Stress-Life Fatigue and the Walker Equation," *Fatigue Fract. Eng. Mater. Struct.*, vol. 32, no. 3, pp. 163–179.

Application to a Pure Shear Case

For the case when the fluctuating stresses are entirely shear stresses, the fluctuating-stress diagram and most of the fatigue failure criteria can be adapted by using shear stresses and shear strengths. Specifically, make the following adjustments:

- Replace normal stresses σ_m and σ_a with shear stresses τ_m and τ_a.
- Apply the load factor $k_c = 0.59$ to the endurance limit.
- Replace S_y with $S_{sy} = 0.577S_y$, based on the relationship predicted by the distortion energy theory.
- Replace S_{ut} with S_{su}.

For most materials, S_{su} ranges from 65 to 80 percent of the ultimate strength. Lacking specific information justifying a higher value, use the conservative estimate of

$$S_{su} = 0.67S_{ut} \tag{6–58}$$

which is consistent with testing on torsional strengths of common spring materials.[25]

Alternatively, the fluctuating-stress diagram and the fatigue failure criterion can be used directly by converting the shear stresses to von Mises stresses using the method in Section 6–16.

EXAMPLE 6–12

For the part shown in Figure 6–37, the 75-mm diameter end is firmly clamped. A force F is repeatedly applied to deflect the tip until it touches the rigid stop, then released. The part is machined from AISI 4130 quenched and tempered to a hardness of approximately 250 HB. Use Table A–22 for material properties. Estimate the fatigue factor of safety based on achieving infinite life, using each of the following criteria. Compare the results.

(*a*) Goodman (*b*) Morrow (*c*) Gerber

Solution

The critical stress location is readily identified as at the fillet radius, on the bottom, where it experiences repeated bending stress in tension. We shall first find the fully modified endurance limit, then the stresses. From Table A–22, the closest material option has S_{ut} = 895 MPa.

Equation (6–18): Machined $k_a = a(S_{ut})^b = 3.04(895)^{-0.217} = 0.70$

Equation (6–23): Nonrotating round $d_e = 0.37d = 0.37(25) = 9.25$ mm

Equation (6–19): $k_b = 1.24d^{-0.107} = 1.24(9.25)^{-0.107} = 0.98$

Equations (6–10) and (6–17): $S_e = (0.70)(0.98)(0.5)(895) = 307$ MPa

$$I = \pi d^4/64 = \pi(25)^4/64 = 19\ 175\ \text{mm}^4$$

Table A–9–1: $$F_{\max} = y_{\max}\frac{3EI}{l^3} = 3\frac{3(2)(10^5)(19\ 175)}{400^3} = 539\ \text{N}$$

$$F_{\min} = 0$$

$$\sigma_{\max} = My/I = 539(400)(12.5)/19\ 175 = 140\ \text{MPa}$$

Equations (6–8), (6–9): $$\sigma_{m0} = \frac{\sigma_{\max} + \sigma_{\min}}{2} = 140/2 = 70\ \text{MPa} = \sigma_{a0}$$

[25]Associated Spring, *Design Handbook: Engineering Guide to Spring Design,* Associated Spring, Barnes Group Inc., Bristol, Conn., 1987.

Figure A–13–9: $r/d = 0.1,\ D/d = 75/25 = 3,\ K_t = 1.8$

Figure 6–26 or Equation (6–34): $q = 0.9$

Equation (6–32): $K_f = 1 + q(K_t - 1) = 1 + 0.9(1.8 - 1) = 1.7$

Equations (6–38), (6–39): $\sigma_m = \sigma_a = K_f\sigma_0 = 1.7(70) = 119$ MPa

(*a*) Goodman

Answer Equation (6–41):

$$n_f = \left(\frac{\sigma_a}{S_e} + \frac{\sigma_m}{S_{ut}}\right)^{-1} = \left(\frac{119}{307} + \frac{119}{895}\right)^{-1} = 1.9$$

(*b*) Morrow
From Table A–22, $\sigma_f' = 1275$ MPa. Note that if σ_f' had not been available for this material, the estimate for steel in Equation 6–44 would have predicted a value of 1240, which would have been quite acceptable to use.

Answer Equation (6–46):

$$n_f = \left(\frac{\sigma_a}{S_e} + \frac{\sigma_m}{\sigma_f'}\right)^{-1} = \left(\frac{119}{307} + \frac{119}{1275}\right)^{-1} = 2.1$$

(*c*) Gerber
Equation (6–48):

$$n_f = \frac{1}{2}\left(\frac{S_{ut}}{\sigma_m}\right)^2\left(\frac{\sigma_a}{S_e}\right)\left[-1 + \sqrt{1 + \left(\frac{2\sigma_m S_e}{S_{ut}\sigma_a}\right)^2}\right]$$

Answer

$$= \frac{1}{2}\left(\frac{895}{119}\right)^2\left(\frac{119}{307}\right)\left[-1 + \sqrt{1 + \left(\frac{2(119)(307)}{895(119)}\right)^2}\right] = 2.3$$

A criterion that predicts a lower factor of safety is predicting that the stress is closer to failure, which sends a message to be careful. From a design perspective, then, a predicted lower factor of safety is more conservative. Ranking the results in order of most conservative to least conservative, gives

Goodman	1.9
Morrow	2.1
Gerber	2.3

Because we had a tabulated value for σ_f' for the material, the Morrow result is probably the most typical of reality. As expected, Goodman is on the conservative side. Gerber is sometimes typical, but for lower mean stress it has a tendency to be a bit nonconservative.

Figure 6–37

The load *F* is repeatedly applied to push the rod to the stop.

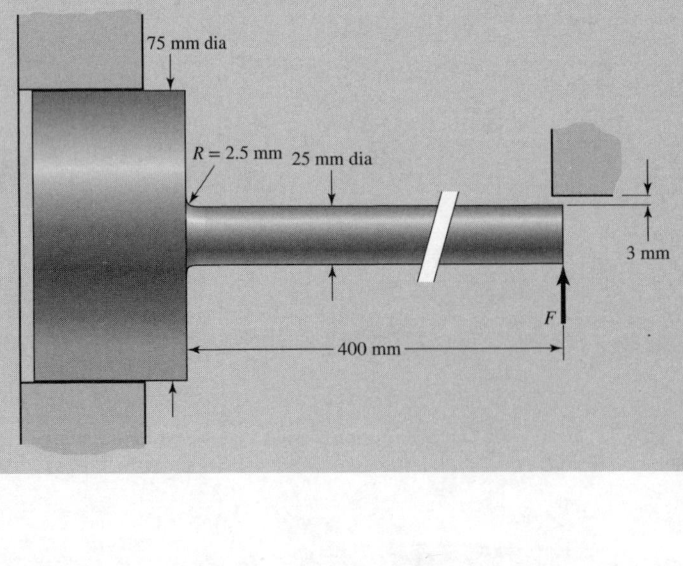

EXAMPLE 6–13

For the problem in Example 6–10, estimate the infinite-life fatigue factor of safety for each of the failure criteria defined in this section. Compare the results.

Solution

From Example 6–10, $\sigma_a = 336$ MPa, $\sigma_m = 168$ MPa, $S_{ut} = 700$ MPa, and $S_e = 280$ MPa.

Goodman:

Equation (6–41) $$n_f = \left(\frac{\sigma_a}{S_e} + \frac{\sigma_m}{S_{ut}}\right)^{-1} = \left(\frac{336}{280} + \frac{168}{700}\right)^{-1} = 0.69$$

Morrow:

Lacking specific material properties, use the estimate for steel in Equation (6–44),

$$\sigma'_f = S_{ut} + 345 = 1045 \text{ MPa}$$

Equation (6–46) $$n_f = \left(\frac{\sigma_a}{S_e} + \frac{\sigma_m}{\sigma'_f}\right)^{-1} = \left(\frac{336}{280} + \frac{168}{1045}\right)^{-1} = 0.73$$

Gerber:

Equation (6–48) $$n_f = \frac{1}{2}\left(\frac{S_{ut}}{\sigma_m}\right)^2\left(\frac{\sigma_a}{S_e}\right)\left[-1 + \sqrt{1 + \left(\frac{2\sigma_m S_e}{S_{ut}\sigma_a}\right)^2}\right]$$

$$= \frac{1}{2}\left(\frac{700}{168}\right)^2\left(\frac{336}{280}\right)\left[-1 + \sqrt{1 + \left(\frac{2(168)(280)}{700(336)}\right)^2}\right] = 0.80$$

Soderberg:

Equation (6–50) $$n_f = \left(\frac{\sigma_a}{S_e} + \frac{\sigma_m}{S_y}\right)^{-1} = \left(\frac{336}{280} + \frac{168}{595}\right)^{-1} = 0.67$$

ASME-Elliptic:

Equation (6–52) $$n_f = \left[\left(\frac{\sigma_a}{S_e}\right)^2 + \left(\frac{\sigma_m}{S_y}\right)^2\right]^{-1/2} = \left[\left(\frac{336}{280}\right)^2 + \left(\frac{168}{595}\right)^2\right]^{-1/2} = 0.81$$

SWT:

Equation (6–54) $$n_f = \frac{S_e}{\sqrt{(\sigma_m + \sigma_a)\sigma_a}} = \frac{280}{\sqrt{(168 + 336)336}} = 0.68$$

Walker:

Lacking specific material test data, use the estimate in Equation (6–57),

$$\gamma = -0.0002S_{ut} + 0.8818 = -0.0002(700) + 0.8818 = 0.74$$

Equation (6–56) $$n_f = \frac{S_e}{(\sigma_m + \sigma_a)^{1-\gamma}\sigma_a^{\gamma}} = \frac{280}{(168 + 336)^{1-0.74}336^{0.74}} = 0.75$$

For comparison, sort all of the results in order of most conservative to least conservative.

Soderberg	0.67
SWT	0.68
Goodman	0.69
Morrow	0.73
Walker	0.75
Gerber	0.80
ASME-Elliptic	0.81

Probably, Walker and Morrow are the most accurate. Gerber and ASME-Elliptic are nonconservative, which they tend to be with low mean stress. Goodman is conservative as expected. SWT is about the same as Goodman, not being a particularly good match for the fitting parameter γ needed for this material (as estimated by Walker). Soderberg is, as always, conservative, though not as much for low mean stress where its line is not much different from Goodman. Which is best? It depends on the goal. Probably, in this case, Walker or Morrow were equally good for a result typical of the data, and Goodman is a good choice for a reasonably predictable amount of conservativeness.

EXAMPLE 6–14

A flat-leaf spring is used to retain an oscillating flat-faced follower in contact with a plate cam. The follower range of motion is 50 mm and fixed, so the alternating component of force, bending moment, and stress are fixed, too. The spring is preloaded to adjust to various cam speeds. The preload must be increased to prevent follower float or jump. For lower speeds the preload should be decreased to obtain longer life of cam and follower surfaces. The spring is a steel cantilever 0.8 m long, 50 mm wide, and 6 mm thick, as seen in Figure 6–38*a*. The spring strengths are $S_{ut} = 1000$ MPa, $S_y = 880$ MPa, and $S_e = 195$ MPa fully corrected. The total cam motion is 50 mm. The designer wishes to preload the spring by deflecting it 50 mm for low speed and 125 mm for high speed.
(*a*) Plot the Gerber failure criterion curve with the load line.
(*b*) What are the strength factors of safety corresponding to 50 mm and 125 mm preload?

Solution

A unique aspect of this problem is that due to the nature of the cam's motion, the alternating force is very defined. The preload can change the mean stress. An appropriate load line is a horizontal line to reflect a steady value of alternating stress. We begin with preliminaries. The second area moment of the cantilever cross section is

$$I = \frac{bh^3}{12} = \frac{50(6)^3}{12} = 900 \text{ mm}^4$$

Since, from Table A–9, beam 1, force F and deflection y in a cantilever are related by $F = 3EIy/l^3$, then stress σ and deflection y are related by

$$\sigma = \frac{Mc}{I} = \frac{800Fc}{I} = \frac{800(3EIy)}{l^3}\frac{c}{I} = \frac{2400Ecy}{l^3} = Ky$$

where $K = \dfrac{2400Ec}{l^3} = \dfrac{2400(207\ 000)3}{800^3} = 2.91$ MPa/mm

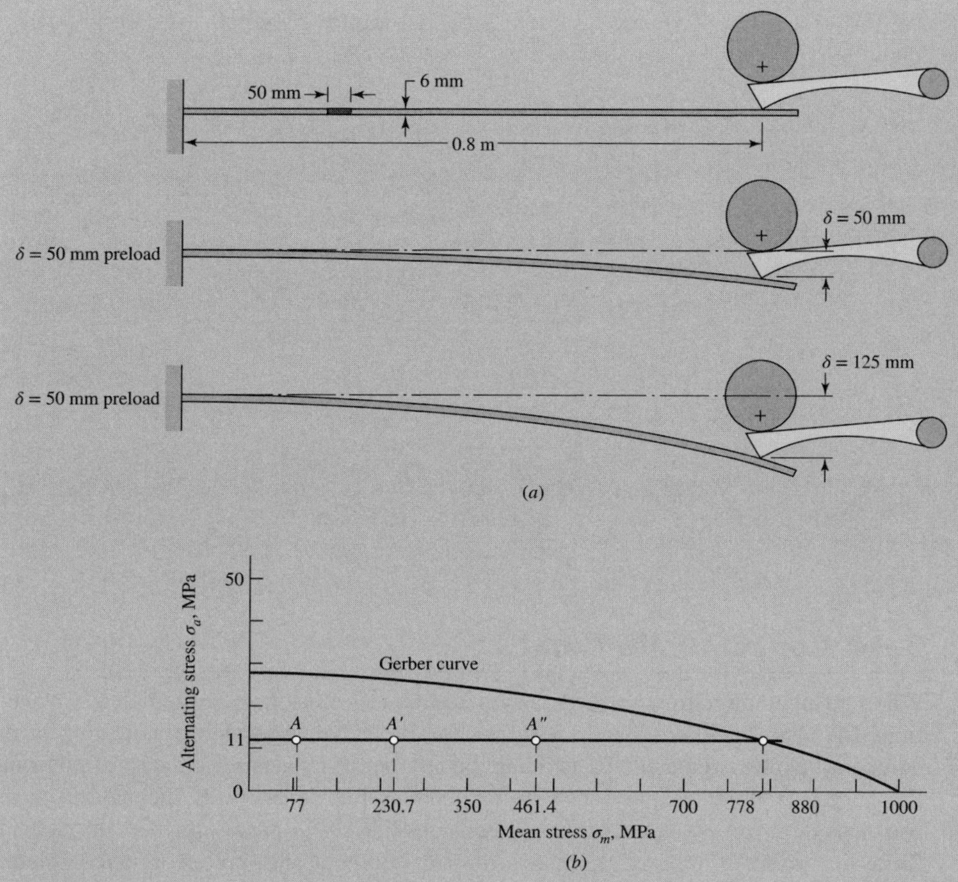

Figure 6–38

Cam follower retaining spring for Example 6–14. (*a*) Geometry; (*b*) Fluctuating-stress diagram.

Now the minimums and maximums of y and σ can be defined by

$$y_{\text{min}} = \delta \qquad y_{\text{max}} = 50 + \delta$$

$$\sigma_{\text{min}} = K\delta \qquad \sigma_{\text{max}} = K(50 + \delta)$$

The stress components are thus

$$\sigma_a = \frac{K(50 + \delta) - K\delta}{2}$$

$$\sigma_m = \frac{K(50 + \delta) + K\delta}{2}$$

For $\delta = 0$, $\quad \sigma_a = \sigma_m = 73$ MPa

For $\delta = 50$ mm, $\quad \sigma_a = 73$ MPa, $\quad \sigma_m = 218$ MPa

For $\delta = 125$ mm, $\quad \sigma_a = 73$ MPa, $\quad \sigma_m = 436.5$ MPa

(*a*) A plot of the Gerber criterion is shown in Figure 6–38*b*. The three preload deflections of 0, 50, and 125 mm are shown as points A, A', and A''. Note that since σ_a is constant at 73 MPa, the load line is horizontal and does not contain the origin. The design equation of Equation (6–48) was derived for the load line from the origin, so it is not applicable in this case. The intersection point (S_m, S_a) between the Gerber line and the load line is found from solving Equation (6–47) for S_m and substituting 73 MPa for S_a:

$$S_m = S_{ut}\sqrt{1 - \frac{S_a}{S_e}} = 1000\sqrt{1 - \frac{73}{195}} = 791 \text{ MPa}$$

(*b*) The factor of safety is found as the proportion of the distance along the load line toward the failure point that the stress point has come. For $\delta = 50$,

Answer $$n_f = \frac{S_m}{\sigma_m} = \frac{791}{218} = 3.63$$

and for $\delta = 125$,

Answer $$n_f = \frac{791}{436.5} = 1.81$$

The problem statement didn't ask for it, but yielding should also be checked in a similar fashion, using the load line and the Langer yield line.

6–14 Constant-Life Curves

When a fluctuating stress exceeds a fatigue failure criterion, it is predicted to have a finite life. The fluctuating-stress diagram does not give any specific indication of the number of cycles predicted. To estimate the number of cycles, the concept of constant-life curves is used, as previously introduced in Figure 6–31. If the nature of the constant-life curves is known, then a constant-life curve can be passed through the finite-life point of interest, and its ordinate intercept interpreted as a completely reversed stress with an equivalent life. This equivalent completely reversed stress can be evaluated on an *S-N* diagram to estimate the life.

A fatigue failure criterion, such as those defined in Section 6–13, can be used to model constant-life curves. Basically, the failure criteria were defined to represent a curve of constant-life of 10^6, and to pass through the ordinate intercept at the endurance limit. Now we will simply adapt the equation to represent a different life, that is, the one associated with the finite-life point (σ_m, σ_a), and take the ordinate intercept as the equivalent completely reversed stress σ_{ar}.

Some of the failure criteria are better predictors than others. In particular, for this purpose, a criterion that is typical of the experimental data is better than a conservative one, or of one that also tries to accommodate the yielding check. Dowling[26] has compared several of the best candidate criteria with studies for steels, aluminum alloys, and one titanium alloy. Some of his conclusions and some accompanying thoughts are given for the Goodman, Morrow, SWT, and Walker criteria. The other criteria are not recommended for this purpose.

Goodman

The Goodman criterion is very inaccurate for the purpose of estimating an equivalent completely reversed stress. Its inaccuracy is almost always in the very conservative

[26]N. Dowling, op. cit.

sense, in that it predicts a higher equivalent completely reversed stress than is validated by experiment. The criterion has been commonly used and will likely continue to be used for cases where excessive design conservatism is acceptable. In the equation for the Goodman line, Equation (6–40), substituting the equivalent completely reversed stress for the endurance limit,

$$\sigma_{ar} = \frac{\sigma_a}{1 - \sigma_m/S_{ut}} \tag{6–59}$$

Morrow

The Morrow criterion is reasonably accurate for steels with either the true fracture strength $\tilde{\sigma}_f$ or the fatigue strength coefficient σ'_f. For aluminum, only the true fracture strength provides good results. Unfortunately, these material properties are not always readily available. The SAE estimate for the fatigue strength coefficient for steels (HB $\leq$ 500) is repeated here for convenience,

$$\sigma'_f = S_{ut} + 345 \text{ MPa} \tag{6–44}$$

Even with such an estimate, the Morrow criterion is probably as good as or better than the Goodman criterion for estimating an equivalent completely reversed stress. From Equation (6–45), the equivalent completely reversed stress is

$$\sigma_{ar} = \frac{\sigma_a}{1 - \sigma_m/\tilde{\sigma}_f} \quad \text{or} \quad \sigma_{ar} = \frac{\sigma_a}{1 - \sigma_m/\sigma'_f} \tag{6–60}$$

Smith-Watson-Topper

The Smith-Watson-Topper (SWT) criterion is very good for aluminum and usually is reasonably good for steels. Obtaining the equivalent completely reversed stress from Equation (6–53),

$$\sigma_{ar} = \sqrt{\sigma_{\max}\sigma_a} = \sqrt{(\sigma_m + \sigma_a)\sigma_a} \tag{6–61}$$

The SWT method has the advantage of simplicity in only needing the two stress components.

Walker

The Walker criterion is considered the best match to experimental predictions for the equivalent completely reversed stress when the material fitting parameter γ is known. From Equation (6–55),

$$\sigma_{ar} = \sigma_{\max}^{1-\gamma}\sigma_a^{\gamma} = (\sigma_m + \sigma_a)^{1-\gamma}\sigma_a^{\gamma} \tag{6–62}$$

For aluminum, γ is close to 0.5, which is equivalent to the SWT criterion.

For steels, the estimate of Equation (6–57) can be used, repeated here for convenience,

$$\gamma = -0.0002S_{ut} + 0.8818 \qquad (S_{ut} \text{ in MPa}) \tag{6–57}$$

With this, the Walker method is very usable for steels and is in most cases as good as or better than any of the other methods.

Concluding Recommendations

The best criteria to use depends on the situation. For aluminum, the SWT method is simple and a good fit and does not require any material properties. For materials other than aluminum, if best empirical fit is desired, then the order of preference is probably Walker, Morrow, SWT, Goodman. If the material properties for Walker and Morrow are not available, then for steels the estimates can be used and the order of preference is still the same. For other materials, if the material properties are not available, either from experiment, table, or estimate, then the order of preference is SWT then Goodman. The primary reason to use Goodman, for any material, is when best fit is not the goal, and a simple and conservative approach is preferred. Of course, these are judgment decisions, and there are certainly other considerations that could change the choice.

EXAMPLE 6–15

For the problem defined in Example 6–10 and extended in Example 6–13, all of the fatigue criteria predicted finite life. For the fatigue criteria of Goodman, Morrow, SWT, and Walker, estimate the equivalent completely reversed stress and the predicted life. Compare the results.

Solution

From Example 6–10, $\sigma_a = 336$ MPa, $\sigma_m = 168$ MPa, $S_{ut} = 700$ MPa, and $S_e = 280$ MPa.

Equation (6–59): Goodman $$\sigma_{ar} = \frac{\sigma_a}{1 - \sigma_m/S_{ut}} = \frac{336}{1 - 168/700} = 442 \text{ MPa}$$

Equation (6–44): $$\sigma_f' = S_{ut} + 345 \text{ MPa} = 1045 \text{ MPa}$$

Equation (6–60): Morrow $$\sigma_{ar} = \frac{\sigma_a}{1 - \sigma_m/\sigma_f'} = \frac{336}{1 - 168/1045} = 400 \text{ MPa}$$

Equation (6–61): SWT $$\sigma_{ar} = \sqrt{(\sigma_m + \sigma_a)\sigma_a} = \sqrt{(168 + 336)336} = 411.5 \text{ MPa}$$

From Example 6–13, $\gamma = 0.74$

Equation (6–62): Walker $$\sigma_{ar} = \sigma_{\max}^{1-\gamma}\sigma_a^{\gamma} = (\sigma_m + \sigma_a)^{1-\gamma}\sigma_a^{\gamma} = (168 + 336)^{1-0.74}336^{0.74} = 373 \text{ MPa}$$

Use the *S-N* diagram equations with these equivalent completely reversed stresses to estimate the life based on each criterion.

Figure 6–23: $$f = 0.84$$

Equation (6–13): $$a = \frac{(fS_{ut})^2}{S_e} = \frac{((0.84)700)^2}{280} = 1235$$

Equation (6–14): $$b = -\frac{1}{3}\log\left(\frac{fS_{ut}}{S_e}\right) = -\frac{1}{3}\log\left(\frac{(0.84)(700)}{280}\right) = -0.1074$$

Equation (6–15): $$N = \left(\frac{\sigma_{ar}}{a}\right)^{1/b}$$

Calculate the estimated life for each criterion. Results are reported in order of lowest life to highest life.

Answer	Goodman	$\sigma_{ar} = 442$ MPa	$N = 14\ 000$ cycles
	SWT	$\sigma_{ar} = 411.5$ MPa	$N = 28\ 000$ cycles
	Morrow	$\sigma_{ar} = 400$ MPa	$N = 36\ 000$ cycles
	Walker	$\sigma_{ar} = 373$ MPa	$N = 69\ 000$ cycles

A higher equivalent completely reversed stress on an *S-N* diagram will predict a shorter life. From a design perspective, a prediction of a shorter life is more conservative. The Walker estimate is expected to be the most typical of reality. The Morrow result is an improvement on Goodman, even with just the estimate for σ'_f. As expected, Goodman is conservative. Though Goodman is in the correct order of magnitude, it is substantially less than the presumably better value from Walker.

6–15 Fatigue Failure Criterion for Brittle Materials

For many *brittle* materials, the first quadrant fatigue failure criterion follows a concave upward Smith-Dolan locus represented by

$$\frac{S_a}{S_e} = \frac{1 - S_m/S_{ut}}{1 + S_m/S_{ut}} \tag{6–63a}$$

or as a design equation,

$$\frac{n\sigma_a}{S_e} = \frac{1 - n\sigma_m/S_{ut}}{1 + n\sigma_m/S_{ut}} \tag{6–63b}$$

For a radial load line of slope r, we substitute S_a/r for S_m in Equation (6–63*a*) and solve for S_a, obtaining the intersect

$$S_a = \frac{rS_{ut} + S_e}{2}\left[-1 + \sqrt{1 + \frac{4rS_{ut}S_e}{(rS_{ut} + S_e)^2}}\right] \tag{6–64}$$

The fatigue diagram for a brittle material differs markedly from that of a ductile material because:

- Yielding is not involved since the material may not have a yield strength.
- Characteristically, the compressive ultimate strength exceeds the ultimate tensile strength severalfold.
- First-quadrant fatigue failure locus is concave-upward (Smith-Dolan), for example, and as flat as Goodman. Brittle materials are more sensitive to midrange stress, being lowered, but compressive midrange stresses are beneficial.
- Not enough work has been done on brittle fatigue to discover insightful generalities, so we stay in the first and a bit of the second quadrant.

The most likely domain of designer use is in the range from $-S_{ut} \le \sigma_m \le S_{ut}$. The locus in the first quadrant is Goodman, Smith-Dolan, or something in between. The portion of the second quadrant that is used is represented by a straight line between the points $-S_{ut}$, S_{ut} and 0, S_e, which has the equation

$$S_a = S_e + \left(\frac{S_e}{S_{ut}} - 1\right)S_m \qquad -S_{ut} \le S_m \le 0 \quad \text{(for cast iron)} \tag{6–65}$$

Table A–23 gives properties of gray cast iron. The endurance limit stated is really $k_a k_b S'_e$ and only corrections k_c, k_d, and k_e need be made. The average k_c for axial and torsional loading is 0.9.

EXAMPLE 6–16

A grade 30 gray cast iron is subjected to a load F applied to a 25 by 10 mm cross-section link with a 6 mm-diameter hole drilled in the center as depicted in Figure 6–39*a*. The surfaces are machined. In the neighborhood of the hole, what is the factor of safety guarding against failure under the following conditions:
(*a*) The load $F = 4450$ N tensile, steady.
(*b*) The load is 4450 N repeatedly applied.
(*c*) The load fluctuates between 4450 N and 1334 N without column action.
Use the Smith-Dolan fatigue locus.

Solution

Some preparatory work is needed. From Table A–23, $S_{ut} = 214$ MPa, $S_{uc} = 752$ MPa, $k_a k_b S'_e = 97$ MPa. Since k_c for axial loading is 0.9, then $S_e = (k_a k_b S'_e)k_c = 97(0.9) = 86$ MPa. From Table A–13–1, $A = t(w - d) = 10(25 - 6) = 190\ \text{mm}^2$, $d/w = 6/25 = 0.24$, and $K_t = 2.45$. The notch sensitivity for cast iron is 0.20 (see Section 6–10), so

$$K_f = 1 + q(K_t - 1) = 1 + 0.20(2.45 - 1) = 1.29$$

(*a*) Since the load is steady, $\sigma_a = 0$, the load is static. Based on the discussion of cast iron in Section 5–2, K_t, and consequently K_f, need not be applied. Thus, $\sigma_m = F_m/A = 4450/190 = 30$ MPa, and

Answer
$$n = \frac{S_{ut}}{\sigma_m} = \frac{214}{30} = 7.13$$

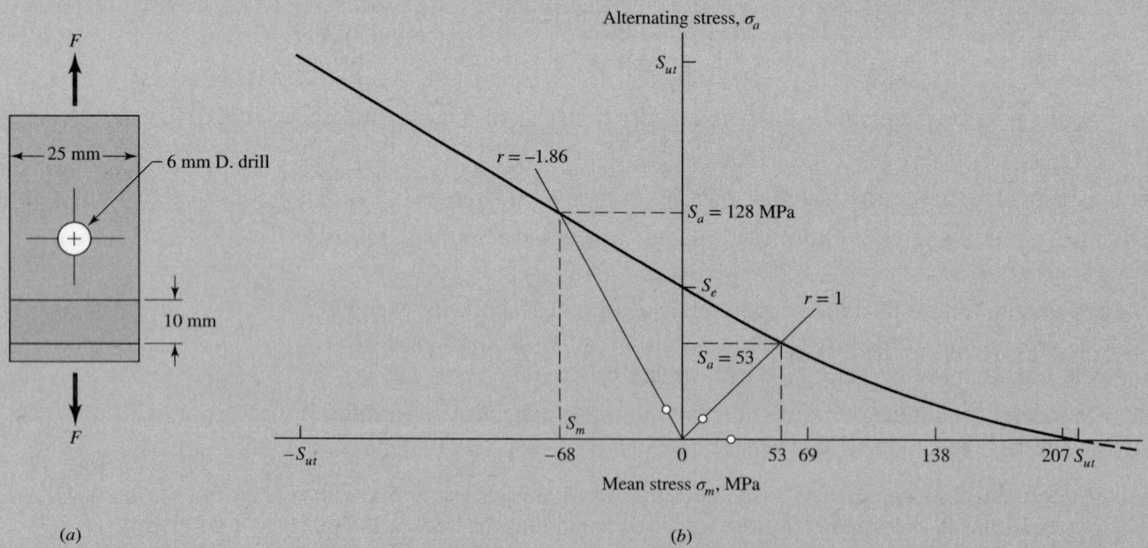

Figure 6–39
The grade 30 cast-iron part in axial fatigue with (*a*) its geometry displayed and (*b*) its fluctuating-stress diagram for the circumstances of Example 6–16.

(*b*)

$$F_a = F_m = \frac{F}{2} = \frac{4450}{2} = 2225\ \text{N}$$

$$\sigma_a = \sigma_m = \frac{K_f F_a}{A} = \frac{1.29(2225)}{190} = 15\ \text{MPa}$$

$$r = \frac{\sigma_a}{\sigma_m} = 1$$

From Equation (6–64),

$$S_a = \frac{(1)214 + 86}{2}\left[-1 + \sqrt{1 + \frac{4(1)214(86)}{[(1)214 + 86]^2}}\right] = 52.25\ \text{MPa}$$

Answer

$$n = \frac{S_a}{\sigma_a} = \frac{52.25}{15} = 3.48$$

(*c*)

$$F_a = \frac{1}{2}|1334 - (-4450)| = 2892\ \text{N} \qquad \sigma_a = \frac{1.29(2892)}{190} = 20\ \text{MPa}$$

$$F_m = \frac{1}{2}[1334 + (-4450)] = -1558\ \text{N} \qquad \sigma_m = \frac{1.29(-1558)}{190} = -11\ \text{MPa}$$

$$r = \frac{\sigma_a}{\sigma_m} = \frac{20}{(-11)} = -1.82$$

From Equation (6–65), $S_a = S_e + (S_e/S_{ut} - 1)S_m$ and $S_m = S_a/r$. It follows that

$$S_a = \frac{S_e}{1 - \dfrac{1}{r}\left(\dfrac{S_e}{S_{ut}} - 1\right)} = \frac{86}{1 - \dfrac{1}{(-1.82)}\left(\dfrac{86}{214} - 1\right)} = 128.1\ \text{MPa}$$

Answer

$$n = \frac{S_a}{\sigma_a} = \frac{128.1}{20} = 6.40$$

Figure 6–39*b* shows the portion of the designer's fatigue diagram that was constructed.

6–16 Combinations of Loading Modes

It may be helpful to think of fatigue problems as being in three categories:

- Completely reversing simple loads
- Fluctuating simple loads
- *Combinations of loading modes*

The simplest category is that of a completely reversed single stress which is handled with the *S-N* diagram, relating the alternating stress to a life. Only one type of loading is allowed here, and the mean stress must be zero. The next category incorporates general fluctuating loads, using a criterion to relate mean and alternating stresses (such as

Goodman, Morrow, or Gerber). Again, only *one* type of loading is allowed at a time. The third category, which we will develop in this section, involves cases where there are combinations of different types of loading, such as combined bending, torsion, and axial.

In Section 6–9 we learned that a load factor k_c is used to modify the endurance limit, based on whether the loading is axial, bending, or torsion. In this section we address how to proceed when the loading is a *mixture* of, say, axial, bending, and torsional loads. This type of loading introduces a few complications in that there may now exist combined normal and shear stresses, each with alternating and mean values, and several of the factors used in determining the endurance limit depend on the type of loading. There may also be multiple stress-concentration factors, one for each mode of loading. The problem of how to deal with combined stresses was encountered when developing static failure theories. The distortion energy failure theory proved to be a satisfactory method of combining the multiple stresses on a stress element of a ductile material into a single equivalent von Mises stress. The same approach will be used here.

The first step is to generate *two* stress elements—one for the alternating stresses and one for the mean stresses. Apply the appropriate fatigue stress concentration factors to each of the stresses; that is, apply $(K_f)_{\text{bending}}$ for the bending stresses, $(K_{fs})_{\text{torsion}}$ for the torsional stresses, and $(K_f)_{\text{axial}}$ for the axial stresses. Next, calculate an equivalent von Mises stress for each of these two stress elements, σ'_a and σ'_m. Finally, select a fatigue failure criterion to complete the fatigue analysis. For the endurance limit, S_e, the only Marin modifiers that are affected by multiple load types are the size factor k_b and the load factor k_c. For k_b, there could be a different value for each load type. A safe and simple approach is to use the lowest value of k_b predicted. The torsional load factor, $k_c = 0.59$, should not be applied as it is already accounted for in the von Mises stress calculation. The load factor for the axial load is sometimes accounted for by dividing the alternating axial stress by the axial load factor of 0.85. However, a simpler approach is to simply use the load factor of 1 unless the axial stress is the dominant stress, in which case use a load factor of 0.85.

Consider the common case of a shaft with bending stresses, torsional shear stresses, and axial stresses. For this case, the von Mises stress is of the form $\sigma' = \left(\sigma_x^2 + 3\tau_{xy}^2\right)^{1/2}$. Considering that the bending, torsional, and axial stresses have alternating and mean components, the von Mises stresses for the two stress elements can be written as

$$\sigma'_a = \{[(K_f)_{\text{bending}}(\sigma_{a0})_{\text{bending}} + (K_f)_{\text{axial}}(\sigma_{a0})_{\text{axial}}]^2 + 3[(K_{fs})_{\text{torsion}}(\tau_{a0})_{\text{torsion}}]^2\}^{1/2} \tag{6–66}$$

$$\sigma'_m = \{[(K_f)_{\text{bending}}(\sigma_{m0})_{\text{bending}} + (K_f)_{\text{axial}}(\sigma_{m0})_{\text{axial}}]^2 + 3[(K_{fs})_{\text{torsion}}(\tau_{m0})_{\text{torsion}}]^2\}^{1/2} \tag{6–67}$$

For first-cycle localized yielding, the maximum von Mises stress is calculated. This would be done by first adding the axial and bending alternating and mean stresses to obtain σ_{max} and adding the alternating and mean shear stresses to obtain τ_{max}. Then substitute σ_{max} and τ_{max} into the equation for the von Mises stress. Actually, this is identical to simply applying the distortion energy theory to the maximum stresses. A convenient and more conservative method is to add Equation (6–66) and Equation (6–67). That is, let $\sigma'_{\text{max}} \approx \sigma'_a + \sigma'_m$.

If the stress components are not in phase but have the same frequency, the maxima can be found by expressing each component in trigonometric terms, using phase angles, and then finding the sum. If two or more stress components have differing frequencies, the problem is difficult; one solution is to assume that the two (or more) components often reach an in-phase condition, so that their magnitudes are additive.

EXAMPLE 6–17

A shaft is made of 42- × 4-mm AISI 1018 cold-drawn steel tubing and has a 6-mm-diameter hole drilled transversely through it. Estimate the factor of safety guarding against fatigue and static failures using the Goodman and Langer failure criteria for the following loading conditions:

(*a*) The shaft is rotating and is subjected to a completely reversed torque of 120 N · m in phase with a completely reversed bending moment of 150 N · m.

(*b*) The shaft is nonrotating and is subjected to a pulsating torque fluctuating from 20 to 160 N · m and a steady bending moment of 150 N · m.

Solution

Here we follow the procedure of estimating the strengths and then the stresses, followed by relating the two.

From Table A–19 we find the minimum strengths to be S_{ut} = 440 MPa and S_y = 370 MPa. The endurance limit of the rotating-beam specimen is 0.5(440) = 220 MPa. The surface factor, obtained from Equation (6–18) and Table 6–2, is

$$k_a = 3.04 S_{ut}^{-0.217} = 3.04(440)^{-0.217} = 0.81$$

From Equation (6–19) the size factor is

$$k_b = \left(\frac{d}{7.62}\right)^{-0.107} = \left(\frac{42}{7.62}\right)^{-0.107} = 0.83$$

The remaining Marin factors are all unity, so the modified endurance strength S_e is

$$S_e = 0.81(0.83)220 = 148 \text{ MPa}$$

(*a*) Theoretical stress-concentration factors are found from Table A–14. Using $a/D = 6/42 = 0.143$ and $d/D = 34/42 = 0.810$, and using linear interpolation, we obtain $A = 0.798$ and $K_t = 2.37$ for bending; and $A = 0.89$ and $K_{ts} = 1.75$ for torsion. Thus, for bending,

$$Z_{\text{net}} = \frac{\pi A}{32D}(D^4 - d^4) = \frac{\pi(0.798)}{32(42)}[(42)^4 - (34)^4] = 3.31(10^3)\,\text{mm}^3$$

and for torsion

$$J_{\text{net}} = \frac{\pi A}{32}(D^4 - d^4) = \frac{\pi(0.89)}{32}[(42)^4 - (34)^4] = 155(10^3)\,\text{mm}^4$$

Next, using Figures 6–26 and 6–27, with a notch radius of 3 mm we find the notch sensitivities to be 0.78 for bending and 0.81 for torsion. The two corresponding fatigue stress concentration factors are obtained from Equation (6–32) as

$$K_f = 1 + q(K_t - 1) = 1 + 0.78(2.37 - 1) = 2.07$$

$$K_{fs} = 1 + 0.81(1.75 - 1) = 1.61$$

The alternating bending stress is now found to be

$$\sigma_{xa} = K_f \frac{M}{Z_{\text{net}}} = 2.07 \frac{150}{3.31(10^{-6})} = 93.8(10^6)\text{Pa} = 93.8 \text{ MPa}$$

and the alternating torsional stress is

$$\tau_{xya} = K_{fs}\frac{TD}{2J_{\text{net}}} = 1.61\frac{120(42)(10^{-3})}{2(155)(10^{-9})} = 26.2(10^6)\text{Pa} = 26.2\ \text{MPa}$$

The mean von Mises component σ'_m is zero. The alternating component σ'_a is given by

$$\sigma'_a = (\sigma_{xa}^2 + 3\tau_{xya}^2)^{1/2} = [93.8^2 + 3(26.2^2)]^{1/2} = 104\ \text{MPa}$$

For this completely reversed loading, the fatigue factor of safety n_f is

Answer $$n_f = \frac{S_e}{\sigma'_a} = \frac{148}{104} = 1.42$$

The first-cycle yield factor of safety is

Answer $$n_y = \frac{S_y}{\sigma'_a} = \frac{370}{104} = 3.56$$

There is no localized yielding; the threat is from fatigue.

(*b*) This part asks us to find the factors of safety when the alternating component is due to pulsating torsion, and a steady component is due to both torsion and bending. We have $T_a = (160 - 20)/2 = 70$ N · m and $T_m = (160 + 20)/2 = 90$ N · m. The corresponding amplitude and steady-stress components are

$$\tau_{xya} = K_{fs}\frac{T_aD}{2J_{\text{net}}} = 1.61\frac{70(42)(10^{-3})}{2(155)(10^{-9})} = 15.3(10^6)\text{Pa} = 15.3\ \text{MPa}$$

$$\tau_{xym} = K_{fs}\frac{T_mD}{2J_{\text{net}}} = 1.61\frac{90(42)(10^{-3})}{2(155)(10^{-9})} = 19.7(10^6)\text{Pa} = 19.7\ \text{MPa}$$

The steady bending stress component σ_{xm} is

$$\sigma_{xm} = K_f\frac{M_m}{Z_{\text{net}}} = 2.07\frac{150}{3.31(10^{-6})} = 93.8(10^6)\text{Pa} = 93.8\ \text{MPa}$$

The von Mises components σ'_a and σ'_m, from Equations (6–66) and (6–67), are

$$\sigma'_a = [3(15.3)^2]^{1/2} = 26.5\ \text{MPa}$$

$$\sigma'_m = [93.8^2 + 3(19.7)^2]^{1/2} = 99.8\ \text{MPa}$$

From Equation (6–41),

Answer $$n_f = \left(\frac{\sigma'_a}{S_e} + \frac{\sigma'_m}{S_{ut}}\right)^{-1} = \left(\frac{26.5}{148} + \frac{99.8}{440}\right)^{-1} = 2.46$$

We will choose to include stress concentration in the yield check, in order to check for even localized yielding. Add the alternating and mean stresses to obtain maximum values for both normal and shear stresses which are used to find the von Mises maximum stress. This is simply applying the Distortion Energy failure theory.

$$\sigma'_{\max} = (\sigma_{\max}^2 + 3\tau_{\max}^2)^{1/2} = [(\sigma_m + \sigma_a)^2 + 3(\tau_m + \tau_a)^2]^{1/2}$$

$$\sigma'_{\max} = [(93.8 + 0)^2 + 3(19.7 + 15.3)^2]^{1/2} = 111.7\ \text{MPa}$$

Answer $$n_y = \frac{S_y}{\sigma'_{\max}} = \frac{370}{111.7} = 3.31$$

Alternatively, a quick conservative check for yielding can be obtained by estimating $\sigma'_{\max} \approx \sigma'_a + \sigma'_m$.

Answer $$n_y \approx \frac{S_y}{\sigma'_a + \sigma'_m} = \frac{370}{26.5 + 99.8} = 2.93$$

There is no notch yielding.

6–17 Cumulative Fatigue Damage

Instead of a single fully reversed stress history block composed of n cycles, suppose a machine part, at a critical location, is subjected to

- A fully reversed stress σ_1 for n_1 cycles, σ_2 for n_2 cycles, . . ., or
- A "wiggly" time line of stress exhibiting many and different peaks and valleys.

What stresses are significant, what counts as a cycle, and what is the measure of damage incurred? Consider a fully reversed cycle with stresses varying 420, 560, 280, and 420 MPa and a second fully reversed cycle −280, −420, −140, and −280 MPa as depicted in Figure 6–40*a*. First, it is clear that to impose the pattern of stress in Figure 6–40*a* on a part it is necessary that the time trace look like the solid lines plus the dashed lines in Figure 6–40*a*. Figure 6–40*b* moves the snapshot to exist beginning with 560 MPa and ending with 560 MPa. Acknowledging the existence of a single stress-time trace is to discover a "hidden" cycle shown as the dashed line in Figure 6–40*b*. If there are 100 applications of the all-positive stress cycle, then 100 applications of the all-negative stress cycle, the hidden cycle is applied but once. If the all-positive stress cycle is applied alternately with the all-negative stress cycle, the hidden cycle is applied 100 times.

To ensure that the hidden cycle is not lost, begin on the snapshot with the largest (or smallest) stress and add previous history to the right side, as was done in Figure 6–40*b*. Characterization of a cycle takes on a max–min–same max (or min–max–same min) form. We identify the hidden cycle first by moving along the dashed-line trace in Figure 6–40*b* identifying a cycle with an 560 MPa max, a −420 MPa min, and returning to 560 MPa. Mentally deleting the used part of the trace (the dashed line) leaves a 280, 420, 280 cycle and a −280, −140, −280 cycle. Since failure loci are expressed in terms of alternating stress σ_a and mean stress σ_m, we use Equations (6–8) and (6–9) to construct the table below:

Cycle Number	$\sigma_{\max}$	$\sigma_{\min}$	σ_a	σ_m
1	560	−420	490	70
2	420	280	70	350
3	−140	−280	70	−210

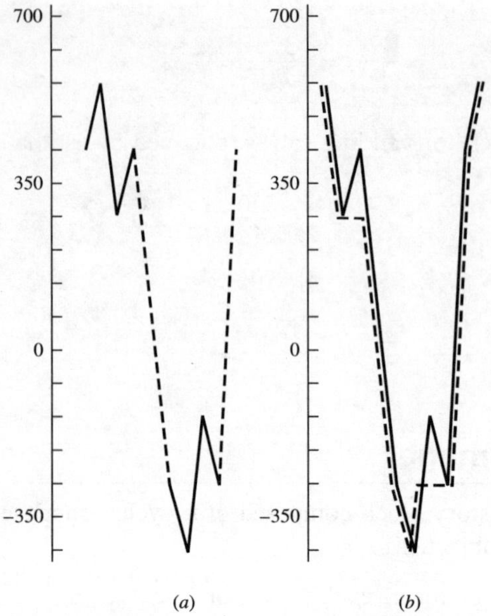

Figure 6–40

Variable stress diagram prepared for assessing cumulative damage.

The most damaging cycle is number 1. It could have been lost.

Methods for counting cycles include:

- Number of tensile peaks to failure.
- All maxima above the waveform mean, all minima below.
- The global maxima between crossings above the mean and the global minima between crossings below the mean.
- All positive slope crossings of levels above the mean, and all negative slope crossings of levels below the mean.
- A modification of the preceding method with only one count made between successive crossings of a level associated with each counting level.
- Each local max–min excursion is counted as a half-cycle, and the associated amplitude is half-range.
- The preceding method plus consideration of the local mean.
- Rain-flow counting technique.

The method used here amounts to a variation of the *rain-flow counting technique*.

The *Palmgren-Miner*[27] *cycle-ratio summation rule,* also called *Miner's rule,* is written

$$\sum \frac{n_i}{N_i} = c \tag{6–68}$$

where n_i is the number of cycles at stress level σ_i and N_i is the number of cycles to failure at stress level σ_i. The parameter c has been determined by experiment; it is usually found in the range $0.7 < c < 2.2$ with an average value near unity.

Using the deterministic formulation as a linear damage rule we write

$$D = \sum \frac{n_i}{N_i} \tag{6–69}$$

where D is the accumulated damage. When $D = c = 1$, failure ensues.

[27] A. Palmgren, "Die Lebensdauer von Kugellagern," *ZVDI,* vol. 68, pp. 339–341, 1924; M. A. Miner, "Cumulative Damage in Fatigue," *J. Appl. Mech.,* vol. 12, *Trans. ASME,* vol. 67, pp. A159–A164, 1945.

EXAMPLE 6–18

Given a steel part with $S_{ut} = 1057$ MPa and at the critical location of the part, $S_e = 472.5$ MPa. For the loading of Figure 6–40, estimate the number of repetitions of the stress-time block in Figure 6–40 that can be made before failure. Use the Morrow criteria.

Solution

From Figure 6–23, for $S_{ut} = 1057$ MPa, $f = 0.795$. From Equation (6–13),

$$a = \frac{(fS_{ut})^2}{S_e} = \frac{[0.795(1057)]^2}{472.5} = 1494.5 \text{ MPa}$$

From Equation (6–14),

$$b = -\frac{1}{3}\log\left(\frac{fS_{ut}}{S_e}\right) = -\frac{1}{3}\log\left[\frac{0.795(1057)}{472.5}\right] = -0.0833$$

From Equation (6–15),

$$N = \left(\frac{\sigma_{ar}}{1494.5}\right)^{-1/0.0833} \tag{1}$$

We prepare to add two columns to the previous table. Lacking specific material information, use the estimate for steel from Equation (6–44),

$$\sigma'_f = S_{ut} + 345 \text{ MPa} = 1057 + 345 = 1402 \text{ MPa}$$

Cycle 1: Check the fatigue factor of safety to see if damage is expected.

Equation (6–46): $$n_f = \left(\frac{\sigma_a}{S_e} + \frac{\sigma_m}{\sigma'_f}\right)^{-1} = \left(\frac{490}{472.5} + \frac{70}{1402}\right)^{-1} = 0.92$$

Since $n_f < 1$, fatigue damage is predicted from cycle 1. Find the equivalent completely reversed stress.

Equation (6–60): $$\sigma_{ar} = \frac{\sigma_a}{1 - \sigma_m/\sigma'_f} = \frac{490}{1 - 70/1402} = 516 \text{ MPa}$$

From Equation (1),

$$N = \left(\frac{\sigma_{ar}}{a}\right)^{1/b} = \left(\frac{516}{1494.5}\right)^{1/-0.0833} = 350(10^3) \text{ cycles}$$

Cycle 2: Repeat the process with the second cycle of stresses.

Equation (6–46): $$n_f = \left(\frac{\sigma_a}{S_e} + \frac{\sigma_m}{\sigma'_f}\right)^{-1} = \left(\frac{70}{472.5} + \frac{350}{1402}\right)^{-1} = 2.51$$

Since $n_f > 1$, no fatigue damage is predicted from cycle 2, so infinite life is predicted.

Cycle 3: This cycle has a negative mean stress. Though the Morrow line can be continued into the negative mean stress region, it is not necessary, as a quick check shows that the alternating stress is well below the endurance limit. No damage is predicted from cycle 3, so infinite life is predicted.

From Equation (6–69) the damage per block is

$$D = \sum\frac{n_i}{N_i} = N\left[\frac{1}{350(10^3)} + \frac{1}{\infty} + \frac{1}{\infty}\right] = \frac{N}{350(10^3)}$$

Answer

Setting $D = 1$ yields $N = 350(10^3)$ cycles.

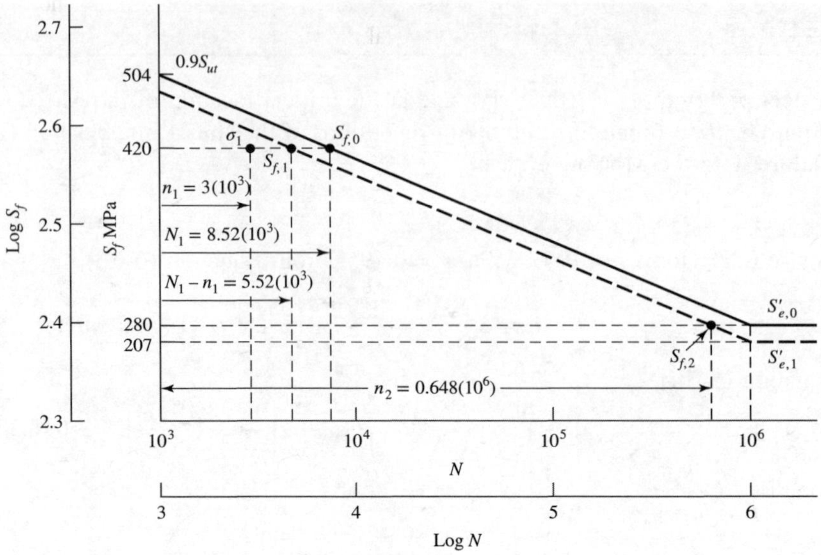

Figure 6–41

Use of the Miner rule to predict the endurance limit of a material that has been overstressed for a finite number of cycles.

To further illustrate the use of the Miner rule, let us consider a steel having the properties $S_{ut} = 560$ MPa, $S'_{e,0} = 280$ MPa, and $f = 0.9$, where we have used the designation $S'_{e,0}$ instead of the more usual S'_e to indicate the endurance limit of the *virgin*, or *undamaged, material*. The log S–log N diagram for this material is shown in Figure 6–41 by the heavy solid line. From Equations (6–13) and (6–14), we find that $a = 907$ MPa and $b = -0.085\,091$. Now apply, say, a reversed stress $\sigma_1 = 420$ MPa for $n_1 = 3000$ cycles. Since $\sigma_1 > S'_{e,0}$, the endurance limit will be damaged, and we wish to find the new endurance limit $S'_{e,1}$ of the damaged material using the Miner rule. The equation of the virgin material failure line in Figure 6–41 in the 10^3 to 10^6 cycle range is

$$S_f = aN^b = 907N^{-0.085\,091}$$

The cycles to failure at stress level $\sigma_1 = 420$ MPa are

$$N_1 = \left(\frac{\sigma_1}{907}\right)^{-1/0.085\,091} = \left(\frac{420}{907}\right)^{-1/0.085\,091} = 8520 \text{ cycles}$$

Figure 6–41 shows that the material has a life $N_1 = 8520$ cycles at 420 MPa, and consequently, after the application of σ_1 for 3000 cycles, there are $N_1 - n_1 = 5520$ cycles of life remaining at σ_1. This locates the finite-life strength $S_{f,1}$ of the damaged material, as shown in Figure 6–41. To get a second point, we ask the question: With n_1 and N_1 given, how many cycles of stress $\sigma_2 = S'_{e,0}$ can be applied before the damaged material fails? This corresponds to n_2 cycles of stress reversal, and hence, from Equation (6–69), we have

$$\frac{n_1}{N_1} + \frac{n_2}{N_2} = 1 \tag{a}$$

Solving for n_2 gives

$$n_2 = (N_1 - n_1)\frac{N_2}{N_1} \tag{b}$$

Then

$$n_2 = [8.52(10^3) - 3(10^3)]\frac{10^6}{8.52(10^3)} = 0.648(10^6) \text{ cycles}$$

This corresponds to the finite-life strength $S_{f,2}$ in Figure 6–41. A line through $S_{f,1}$ and $S_{f,2}$ is the log S–log N diagram of the damaged material according to the Miner rule. Two points, $(N_1 - n_1, \sigma_1)$ and (n_2, σ_2), determine the new equation for the line, $S_f = a'N^{b'}$. Thus, $\sigma_1 = a'(N_1 - n_1)^{b'}$, and $\sigma_2 = a'n_2^{b'}$. Dividing the two equations, taking the logarithm of the results, and solving for b' gives

$$b' = \frac{\log(\sigma_1/\sigma_2)}{\log\left(\dfrac{N_1 - n_1}{n_2}\right)}$$

Substituting n_2 from Equation (b) and simplifying gives

$$b' = \frac{\log(\sigma_1/\sigma_2)}{\log(N_1/N_2)}$$

For the undamaged material, $N_1 = (\sigma_1/a)^{1/b}$ and $N_2 = (\sigma_2/a)^{1/b}$, then

$$b' = \frac{\log(\sigma_1/\sigma_2)}{\log[(\sigma_1/a)^{1/b}/(\sigma_2/a)^{1/b}]} = \frac{\log\ (\sigma_1/\sigma_2)}{(1/b)\log\ (\sigma_1/\sigma_2)} = b$$

This means that the damaged material line has the same slope as the virgin material line, and the two lines are parallel. The value of a' is then found from $a' = S_f/N^b$.

For the case we are illustrating, $a' = 420/[5.52(10)^3]^{-0.085\,091} = 874.286$ MPa, and thus the new endurance limit is $S'_{e,1} = a'N_e^b = 874.286[(10)^6]^{-0.085\,091} = 207$ MPa.

Though the Miner rule is quite generally used, it fails in two ways to agree with experiment. First, note that this theory states that the static strength S_{ut} is damaged, that is, decreased, because of the application of σ_1; see Figure 6–41 at $N = 10^3$ cycles. Experiments fail to verify this prediction.

The Miner rule, as given by Equation (6–69), does not account for the order in which the stresses are applied, and hence ignores any stresses less than $S'_{e,0}$. But we can see in Figure 6–41 that a stress σ_3 in the range $S'_{e,1} < \sigma_3 < S'_{e,0}$ would cause damage if applied after the endurance limit had been damaged by the application of σ_1.

Manson's[28] approach overcomes both of the deficiencies noted for the Palmgren-Miner method; historically it is a much more recent approach, and it is just as easy to use. Except for a slight change, we shall use and recommend the Manson method in this book. Manson plotted the S–log N diagram instead of a log S–log N plot as is recommended here. Manson also resorted to experiment to find the point of convergence of the S–log N lines corresponding to the static strength, instead of arbitrarily selecting the intersection of $N = 10^3$ cycles with $S = 0.9S_{ut}$ as is done here. Of course, it is always better to use experiment, but our purpose in this book has been to use the simple test data to learn as much as possible about fatigue failure.

The method of Manson, as presented here, consists in having all log S–log N lines, that is, lines for both the damaged and the virgin material, converge to the same point, $0.9S_{ut}$ at 10^3 cycles. In addition, the log S–log N lines must be constructed in the same historical order in which the stresses occur.

The data from the preceding example are used for illustrative purposes. The results are shown in Figure 6–42. Note that the strength $S_{f,1}$ corresponding to $N_1 - n_1 = 5.52(10^3)$ cycles is found in the same manner as before. Through this point and through $0.9S_{ut}$ at 10^3 cycles, draw the heavy dashed line to meet $N = 10^6$ cycles

[28]S. S. Manson, A. J. Nachtigall, C. R. Ensign, and J. C. Fresche, "Further Investigation of a Relation for Cumulative Fatigue Damage in Bending," *Trans. ASME, J. Eng. Ind.*, ser. B, vol. 87, no. 1, pp. 25–35, February 1965.

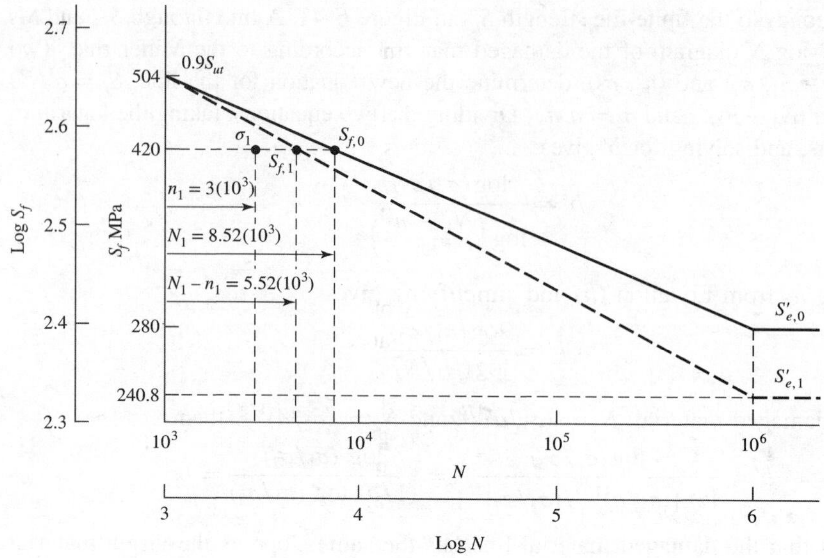

Figure 6–42

Use of the Manson method to predict the endurance limit of a material that has been overstressed for a finite number of cycles.

and define the endurance limit $S'_{e,1}$ of the damaged material. Again, with two points on the line, $b' = [\log(504/420)]/\log[(10^3)/5.52(10^3)] = -0.106\ 722$, and $a' = 420/[5.52(10^3)]^{-0.106\ 722} = 1053.4$ MPa. In this case, the new endurance limit is $S'_{e,1} = a'N_e^{b'} = 1053.4(10^6)^{-0.106\ 722} = 240.8$ MPa, which is somewhat less than that found by the Miner method.

It is now easy to see from Figure 6–42 that a reversed stress $\sigma = 252$ MPa, say, would not harm the endurance limit of the virgin material, no matter how many cycles it might be applied. However, if $\sigma = 252$ MPa should be applied *after* the material was damaged by $\sigma_1 = 420$ MPa, then additional damage would be done.

Both these rules involve a number of computations, which are repeated every time damage is estimated. For complicated stress-time traces, this might be every cycle. Clearly a computer program is useful to perform the tasks, including scanning the trace and identifying the cycles.

Collins said it well: "In spite of all the problems cited, the Palmgren linear damage rule is frequently used because of its simplicity and the experimental fact that other more complex damage theories do not always yield a significant improvement in failure prediction reliability."[29]

6–18 Surface Fatigue Strength

The surface fatigue mechanism is not definitively understood. The contact-affected zone, in the absence of surface shearing tractions, entertains compressive principal stresses. Rotary fatigue has its cracks grown at or near the surface in the presence of tensile stresses that are associated with crack propagation, to catastrophic failure. There are shear stresses in the zone, which are largest just below the surface. Cracks seem to grow from this stratum until small pieces of material are expelled, leaving pits on the surface. Because engineers had to design durable machinery before the surface fatigue phenomenon was understood in detail, they had taken the posture of conducting tests, observing pits on the surface, and declaring failure at an arbitrary projected area of hole, and they related this to the Hertzian contact pressure. This compressive stress did not produce the

[29]J. A. Collins, *Failure of Materials in Mechanical Design,* John Wiley & Sons, New York, 1981, p. 243.

failure directly, but whatever the failure mechanism, whatever the stress type that was instrumental in the failure, the contact stress was an *index* to its magnitude.

Buckingham[30] conducted a number of tests relating the fatigue at 10^8 cycles to endurance strength (Hertzian contact pressure). While there is evidence of an endurance limit at about $3(10^7)$ cycles for cast materials, hardened steel rollers showed no endurance limit up to $4(10^8)$ cycles. Subsequent testing on hard steel shows no endurance limit. Hardened steel exhibits such high fatigue strengths that its use in resisting surface fatigue is widespread.

Our studies thus far have dealt with the failure of a machine element by yielding, by fracture, and by fatigue. The endurance limit obtained by the rotating-beam test is frequently called the *flexural endurance limit,* because it is a test of a rotating beam. In this section we shall study a property of *mating materials* called the *surface endurance shear*. The design engineer must frequently solve problems in which two machine elements mate with one another by rolling, sliding, or a combination of rolling and sliding contact. Obvious examples of such combinations are the mating teeth of a pair of gears, a cam and follower, a wheel and rail, and a chain and sprocket. A knowledge of the surface strength of materials is necessary if the designer is to create machines having a long and satisfactory life.

When two surfaces roll or roll and slide against one another with sufficient force, a pitting failure will occur after a certain number of cycles of operation. Authorities are not in complete agreement on the exact mechanism of the pitting; although the subject is quite complicated, they do agree that the Hertz stresses, the number of cycles, the surface finish, the hardness, the degree of lubrication, and the temperature all influence the strength. In Section 3–19 it was learned that, when two surfaces are pressed together, a maximum shear stress is developed slightly below the contacting surface. It is postulated by some authorities that a surface fatigue failure is initiated by this maximum shear stress and then is propagated rapidly to the surface. The lubricant then enters the crack that is formed and, under pressure, eventually wedges the chip loose.

To determine the surface fatigue strength of mating materials, Buckingham designed a simple machine for testing a pair of contacting rolling surfaces in connection with his investigation of the wear of gear teeth. Buckingham and, later, Talbourdet gathered large numbers of data from many tests so that considerable design information is now available. To make the results useful for designers, Buckingham defined a *load-stress factor,* also called a *wear factor,* which is derived from the Hertz equations. Equations (3–71) and (3–72) for contacting cylinders are found to be

$$b = \sqrt{\frac{2F}{\pi l}\frac{(1-\nu_1^2)/E_1 + (1-\nu_2^2)/E_2}{(1/d_1)+(1/d_2)}} \tag{6–70}$$

$$P_{\max} = \frac{2F}{\pi b l} \tag{6–71}$$

where b = half width of rectangular contact area
F = contact force
l = length of cylinders
ν = Poisson's ratio
E = modulus of elasticity
d = cylinder diameter

[30]Earle Buckingham, *Analytical Mechanics of Gears*, McGraw Hill, New York, 1949.

It is more convenient to use the cylinder radius, so let $2r = d$. If we then designate the length of the cylinders as w (for width of gear, bearing, cam, etc.) instead of l and remove the square root sign, Equation (6–70) becomes

$$b^2 = \frac{4F}{\pi w} \frac{(1 - \nu_1^2)/E_1 + (1 - \nu_2^2)/E_2}{1/r_1 + 1/r_2} \tag{6–72}$$

We can define a *surface endurance strength* S_C using

$$p_{\max} = \frac{2F}{\pi b w} \tag{6–73}$$

as

$$S_C = \frac{2F}{\pi b w} \tag{6–74}$$

which may also be called *contact strength,* the *contact fatigue strength,* or the *Hertzian endurance strength.* The strength is the contacting pressure which, after a specified number of cycles, will cause failure of the surface. Such failures are often called *wear* because they occur over a very long time. They should not be confused with abrasive wear, however. By squaring Equation (6–74), substituting b^2 from Equation (6–72), and rearranging, we obtain

$$\frac{F}{w}\left(\frac{1}{r_1} + \frac{1}{r_2}\right) = \pi S_C^2 \left[\frac{1 - \nu_1^2}{E_1} + \frac{1 - \nu_2^2}{E_2}\right] = K_1 \tag{6–75}$$

The left expression consists of parameters a designer may seek to control independently. The central expression consists of material properties that come with the material and condition specification. The third expression is the parameter K_1, Buckingham's load-stress factor, determined by a test fixture with values F, w, r_1, r_2 and the number of cycles associated with the first tangible evidence of fatigue. In gear studies a similar K factor is used:

$$K_g = \frac{K_1}{4} \sin \phi \tag{6–76}$$

where ϕ is the tooth pressure angle, and the term $[(1 - \nu_1^2)/E_1 + (1 - \nu_2^2)/E_2]$ is defined as $1/(\pi C_P^2)$, so that

$$S_C = C_p \sqrt{\frac{F}{w}\left(\frac{1}{r_1} + \frac{1}{r_2}\right)} \tag{6–77}$$

Buckingham and others reported K_1 for 10^8 cycles and nothing else. This gives only one point on the $S_C N$ curve. For cast metals this may be sufficient, but for wrought steels, heat-treated, some idea of the slope is useful in meeting design goals of other than 10^8 cycles.

Experiments show that K_1 versus N, K_g versus N, and S_C versus N data are rectified by log-log transformation. This suggests that

$$K_1 = \alpha_1 N^{\beta_1} \qquad K_g = aN^b \qquad S_C = \alpha N^\beta$$

The three exponents are given by

$$\beta_1 = \frac{\log(K_{11}/K_{12})}{\log(N_1/N_2)} \qquad b = \frac{\log(K_{g1}/K_{g2})}{\log(N_1/N_2)} \qquad \beta = \frac{\log(S_{C1}/S_{C2})}{\log(N_1/N_2)} \tag{6–78}$$

Data on induction-hardened steel on steel give $(S_C)_{10^7} = 1897$ MPa and $(S_C)_{10^8} = 1673$ MPa, so β, from Equation (6–78), is

$$\beta = \frac{\log(1897/1673)}{\log(10^7/10^8)} = -0.055$$

It may be of interest that the American Gear Manufacturers Association (AGMA) uses $\beta = -0.056$ between $10^4 < N < 10^{10}$ if the designer has no data to the contrary beyond 10^7 cycles.

A long-standing correlation in steels between S_C and H_B at 10^8 cycles is

$$(S_C)_{10^8} = \{2.76H_B - 70 \text{ MPa} \tag{6–79}$$

AGMA uses

$$_{0.99}(S_C)_{10^7} = 0.327H_B + 179 \text{ MPa} \tag{6–80}$$

Equation (6–77) can be used in design to find an allowable surface stress by using a design factor. Since this equation is nonlinear in its stress-load transformation, the designer must decide if loss of function denotes inability to carry the load. If so, then to find the allowable stress, one divides the load F by the design factor n_d:

$$\sigma_C = C_P\sqrt{\frac{F}{wn_d}\left(\frac{1}{r_1} + \frac{1}{r_2}\right)} = \frac{C_P}{\sqrt{n_d}}\sqrt{\frac{F}{W}\left(\frac{1}{r_1} + \frac{1}{r_2}\right)} = \frac{S_C}{\sqrt{n_d}}$$

and $n_d = (S_C/\sigma_C)^2$. If the loss of function is focused on stress, then $n_d = S_C/\sigma_C$. It is recommended that an engineer

- Decide whether loss of function is failure to carry load or stress.
- Define and document the design factor and factor of safety accordingly.

6–19 Road Maps and Important Design Equations for the Stress-Life Method

As stated in Section 6–16, there are three categories of fatigue problems. The important procedures and equations for deterministic stress-life problems are presented here, organized into those three categories.

Completely Reversing Simple Loading

1 Determine S'_e either from test data or

$$S'_e = \begin{cases} 0.5S_{ut} & S_{ut} \le 1400 \text{ MPa} \\ 700 \text{ MPa} & S_{ut} > 1400 \text{ MPa} \end{cases} \tag{6–10}$$

2 Modify S'_e to determine S_e.

$$S_e = k_a k_b k_c k_d k_e S'_e \tag{6–17}$$

$$k_a = aS_{ut}^b \tag{6–18}$$

Table 6–2 Curve Fit Parameters for Surface Factor, Equation (6–18)

Surface Finish	Factor a S_{ut}, MPa	Exponent b
Ground	1.38	−0.067
Machined or cold-drawn	3.04	−0.217
Hot-rolled	38.6	−0.650
As-forged	54.9	−0.758

Rotating shaft. For bending or torsion,

$$k_b = \begin{cases} (d/7.62)^{-0.107} = 1.24d^{-0.107} & 7.62 \le d \le 51 \text{ mm} \\ 1.51d^{-0.157} & 51 < 254 \text{ mm} \end{cases} \tag{6–19}$$

For axial,

$$k_b = 1 \tag{6–20}$$

Nonrotating member. For bending, use Table 6–3 for d_e and substitute into Equation (6–19) for d.

$$k_c = \begin{cases} 1 & \text{bending} \\ 0.85 & \text{axial} \\ 0.59 & \text{torsion} \end{cases} \tag{6–25}$$

$$S_T/S_{RT} = 0.99 + 5.9(10^{-4})T_C - 2.1(10^{-6})T_C^2 \tag{6–26}$$

Either use the ultimate strength from Equation (6–26) to estimate S_e at the operating temperature, with $k_d = 1$, or use the known S_e at room temperature with $k_d = S_T/S_{RT}$ from Equation (6–26).

Table 6–4 Reliability Factor k_e Corresponding to 8 Percent Standard Deviation of the Endurance Limit

Reliability, %	Transformation Variate z_a	Reliability Factor k_e
50	0	1.000
90	1.288	0.897
95	1.645	0.868
99	2.326	0.814
99.9	3.091	0.753
99.99	3.719	0.702

3 Determine fatigue stress-concentration factor, K_f or K_{fs}. First, find K_t or K_{ts} from Table A–13.

$$K_f = 1 + q(K_t - 1) \qquad \text{or} \qquad K_{fs} = 1 + q_s(K_{ts} - 1) \tag{6–32}$$

Obtain q from either Figure 6–26 or 6–27.

Alternatively,

$$K_f = 1 + \frac{K_t - 1}{1 + \sqrt{a/r}} \tag{6–34}$$

Bending or axial:

$$\sqrt{a} = 1.24 - 2.25(10^{-3})S_{ut} + 1.60(10^{-6})S_{ut}^2 - 4.11(10^{-10})S_{ut}^3 \tag{6–35}$$

Torsion:

$$\sqrt{a} = 0.958 - 1.83(10^{-3})S_{ut} + 1.43(10^{-6})S_{ut}^2 - 4.11(10^{-10})S_{ut}^3 \tag{6–36}$$

4 Apply K_f to the nominal completely reversed stress, $\sigma_a = K_f \sigma_{a0}$.

5 Determine f from Figure 6–23 or Equation (6–11). For S_{ut} lower than the range, use $f = 0.9$.

$$f = 1.06 - 4.1(10^{-4})S_{ut} + 1.5(10^{-7})S_{ut}^2 \qquad 500 < S_{ut} < 1400 \text{ MPa} \tag{6–11}$$

$$a = (fS_{ut})^2/S_e \tag{6–13}$$

$$b = -[\log(fS_{ut}/S_e)]/3 \tag{6–14}$$

6 Determine fatigue strength S_f at N cycles, or, N cycles to failure at a reversing stress σ_{ar}.

(*Note*: This only applies to purely reversing stresses where $\sigma_m = 0$.)

$$S_f = aN^b \tag{6–12}$$

$$N = (\sigma_{ar}/a)^{1/b} \tag{6–15}$$

Fluctuating Simple Loading

For S_e, K_f or K_{fs}, see previous subsection.

1 Calculate σ_m and σ_a. Apply K_f to both stresses.

$$\sigma_a = |\sigma_{max} - \sigma_{min}|/2 \qquad \sigma_m = (\sigma_{max} + \sigma_{min})/2 \qquad \text{(6–8), (6–9)}$$

2 Check for infinite life with a fatigue failure criterion. Use Goodman criterion for conservative result, or another criterion from Section 6–13.

$$\sigma_m \geq 0 \qquad n_f = \left(\frac{\sigma_a}{S_e} + \frac{\sigma_m}{S_{ut}}\right)^{-1} \tag{6–41}$$

$$\sigma_m < 0 \qquad n_f = \frac{S_e}{\sigma_a} \tag{6–42}$$

3 Check for localized yielding.

$$n_y = \frac{S_y}{\sigma_{max}} = \frac{S_y}{\sigma_a + |\sigma_m|} \tag{6–43}$$

4 For finite-life, find an equivalent completely reversed stress to use on the *S-N* diagram with Equation (6–15). Select one of the following criterion. Discussion of merits is in Section 6–14.

Goodman: $$\sigma_{ar} = \frac{\sigma_a}{1 - \sigma_m/S_{ut}} \tag{6–59}$$

Morrow: $$\sigma_{ar} = \frac{\sigma_a}{1 - \sigma_m/\tilde{\sigma}_f} \quad \text{or} \quad \sigma_{ar} = \frac{\sigma_a}{1 - \sigma_m/\sigma'_f} \tag{6–60}$$

Estimate for steel: $$\sigma'_f = S_{ut} + 345 \text{ MPa} \tag{6–44}$$

SWT: $$\sigma_{ar} = \sqrt{\sigma_{\max}\sigma_a} = \sqrt{(\sigma_m + \sigma_a)\sigma_a} \tag{6–61}$$

Walker: $$\sigma_{ar} = \sigma_{\max}^{1-\gamma}\sigma_a^{\gamma} = (\sigma_m + \sigma_a)^{1-\gamma}\sigma_a^{\gamma} \tag{6–62}$$

Estimate for steel: $$\gamma = -0.0002S_{ut} + 0.8818 \qquad (S_{ut} \text{ in MPa}) \tag{6–57}$$

For determining the finite life N use the equivalent completely reversed stress in Equation (6–15),

$$N = \left(\frac{\sigma_{ar}}{a}\right)^{1/b}$$

Combination of Loading Modes

See previous subsections for earlier definitions.

1 Calculate von Mises stresses for alternating and mean stress states, σ'_a and σ'_m. When determining S_e, do not use k_c. Apply K_f and/or K_{fs} directly to each specific alternating and mean stress. For the special case of combined bending, torsional shear, and axial stresses

$$\sigma'_a = \{[(K_f)_{\text{bending}}(\sigma_{a0})_{\text{bending}} + (K_f)_{\text{axial}}(\sigma_{a0})_{\text{axial}}]^2 + 3[(K_{fs})_{\text{torsion}}(\tau_{a0})_{\text{torsion}}]^2\}^{1/2} \tag{6–66}$$

$$\sigma'_m = \{[(K_f)_{\text{bending}}(\sigma_{m0})_{\text{bending}} + (K_f)_{\text{axial}}(\sigma_{m0})_{\text{axial}}]^2 + 3[(K_{fs})_{\text{torsion}}(\tau_{m0})_{\text{torsion}}]^2\}^{1/2} \tag{6–67}$$

2 Apply stresses to fatigue criterion [see previous subsection].

3 To check for yielding, apply the distortion energy theory as usual by putting maximum stresses on a stress element and calculating a von Mises stress.

$$n_y = S_y/\sigma'_{\max}$$

Or for a conservative check, apply the yield line with the von Mises stresses for alternating and mean stresses.

$$n_y = \frac{S_y}{\sigma'_a + \sigma'_m} \tag{6–43}$$

PROBLEMS

Problems marked with an asterisk (*) are linked to problems in other chapters, as summarized in Table 1–2 of Section 1–17.

6–1 A 10-mm steel drill rod was heat-treated and ground. The measured hardness was found to be 300 Brinell. Estimate the endurance strength in MPa if the rod is used in rotating bending.

6–2 Estimate S_e' in MPa for the following materials:
(*a*) AISI 1035 CD steel.
(*b*) AISI 1050 HR steel.
(*c*) 2024 T4 aluminum.
(*d*) AISI 4130 steel heat-treated to a tensile strength of 1750 MPa.

6–3 A steel rotating-beam test specimen has an ultimate strength of 840 MPa. Estimate the life of the specimen if it is tested at a completely reversed stress amplitude of 490 MPa.

6–4 A steel rotating-beam test specimen has an ultimate strength of 1600 MPa. Estimate the life of the specimen if it is tested at a completely reversed stress amplitude of 900 MPa.

6–5 A steel rotating-beam test specimen has an ultimate strength of 1610 MPa. Estimate the fatigue strength corresponding to a life of 150 kcycles of stress reversal.

6–6 Repeat Problem 6–5 with the specimen having an ultimate strength of 1100 MPa.

6–7 A steel rotating-beam test specimen has an ultimate strength of 1050 MPa and a yield strength of 945 MPa. It is desired to test low-cycle fatigue at approximately 500 cycles. Check if this is possible without yielding by determining the necessary reversed stress amplitude.

6–8 Estimate the endurance strength of a 32 mm-diameter rod of AISI 1040 steel having a machined finish and heat-treated to a tensile strength of 710 MPa, loaded in rotating bending.

6–9 Two steels are being considered for manufacture of as-forged connecting rods subjected to bending loads. One is AISI 4340 Cr-Mo-Ni steel capable of being heat-treated to a tensile strength of 1820 MPa. The other is a plain carbon steel AISI 1040 with an attainable S_{ut} of 791 MPa. Each rod is to have a size giving an equivalent diameter d_e of 20 mm. Determine the endurance limit for each material. Is there any advantage to using the alloy steel for this fatigue application?

6–10 A *rotating* shaft of 25-mm diameter is simply supported by bearing reaction forces R_1 and R_2. The shaft is loaded with a transverse load of 13 kN as shown in the figure. The shaft is made from AISI 1045 hot-rolled steel. The surface has been machined. Determine
(*a*) the minimum static factor of safety based on yielding.
(*b*) the endurance limit, adjusted as necessary with Marin factors.
(*c*) the minimum fatigue factor of safety based on achieving infinite life.
(*d*) If the fatigue factor of safety is less than 1 (*hint:* it should be for this problem), then estimate the life of the part in number of rotations.

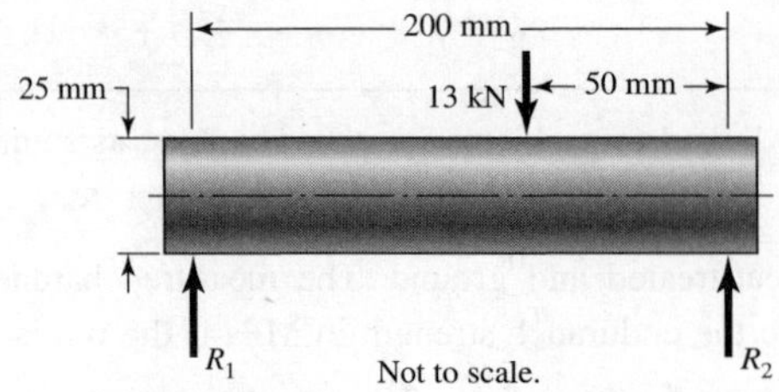

Problem 6–10

6–11 A pin in a knuckle joint is shown in part (*a*) of the figure. The joint is subject to a repeatedly applied and released load of 6000 N in tension. Assume the loading on the pin is modeled as concentrated forces as shown in part (*b*) of the figure. The shaft is made from AISI 1018 hot-rolled steel that has been machined to its final diameter. Based on a stress element on the outer surface at the cross-section *A*, determine a suitable diameter of the pin, rounded up to the next mm increment, to provide at least a factor of safety of 1.5 for both infinite fatigue life and for yielding.

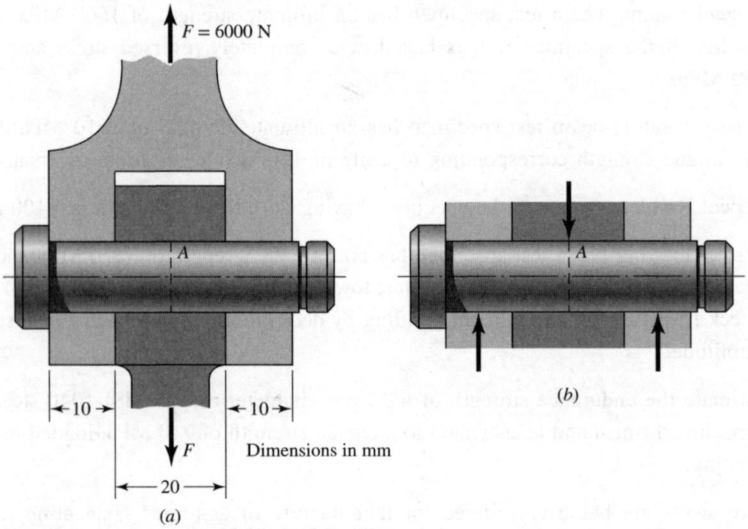

Problem 6–11

6–12 A 25 mm-diameter solid round bar has a groove 2.5 mm deep with a 2.5 mm radius machined into it. The bar is made of AISI 1020 CD steel and is subjected to a purely reversing torque of 200 N · m.

(*a*) Estimate the number of cycles to failure.

(*b*) If the bar is also placed in an environment with a temperature of 750°F, estimate the number of cycles to failure.

6–13 A solid square rod is cantilevered at one end. The rod is 0.6 m long and supports a completely reversing transverse load at the other end of ±2 kN. The material is AISI 1080 hot-rolled steel. If the rod must support this load for 10^4 cycles with a design factor of 1.5, what dimension should the square cross section have? Neglect any stress concentrations at the support end.

6–14 A rectangular bar is cut from an AISI 1020 cold-drawn steel flat. The bar is 60 mm wide by 10 mm thick and has a 12 mm-dia. hole drilled through the center as depicted in Table A–13–1. The bar is concentrically loaded in push-pull fatigue by axial forces F_a, uniformly distributed across the width. Using a design factor of $n_d = 2$, estimate the largest force F_a that can be applied ignoring column action.

6–15 A solid round bar with diameter of 50 mm has a groove cut to a diameter of 45 mm, with a radius of 2.5 mm. The bar is not rotating. The bar is loaded with a repeated bending load that causes the bending moment at the groove to fluctuate between 0 and 2825 N · m. The bar is hot-rolled AISI 1095, but the groove has been machined. Determine the factor of safety for fatigue based on infinite life using the Goodman criterion, and the factor of safety for yielding.

6–16 The rotating shaft shown in the figure is machined from AISI 1020 CD steel. It is subjected to a force of $F = 6$ kN. Find the minimum factor of safety for fatigue based on infinite life. If the life is not infinite, estimate the number of cycles. Be sure to check for yielding.

Problem 6–16
Dimensions in millimeters

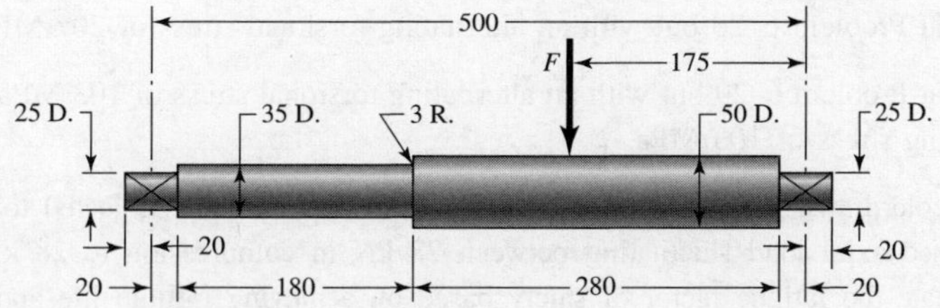

6–17 The shaft shown in the figure is machined from AISI 1040 CD steel. The shaft rotates at 1600 rpm and is supported in rolling bearings at *A* and *B*. The applied forces are $F_1 = 10$ kN and $F_2 = 4$ kN. Determine the minimum fatigue factor of safety based on achieving infinite life. If infinite life is not predicted, estimate the number of cycles to failure. Also check for yielding.

Problem 6–17

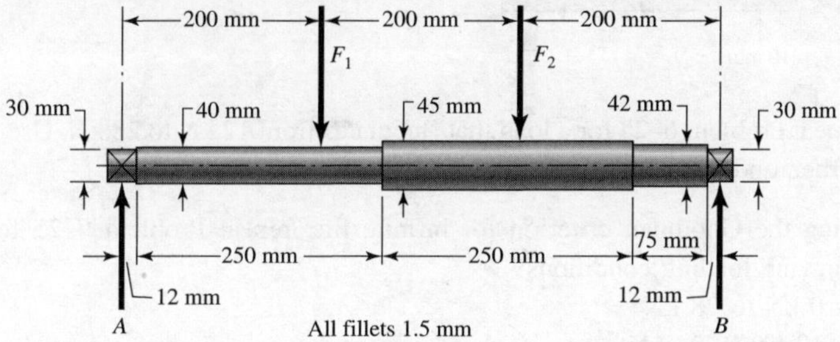

6–18 Solve Problem 6–17 except with forces $F_1 = 4.8$ kN and $F_2 = 9.6$ kN.

6–19 Bearing reactions R_1 and R_2 are exerted on the shaft shown in the figure, which rotates at 950 rev/min and supports an 45 kN bending force. Use a 1095 HR steel. Specify a diameter *d* using a design factor of $n_d = 1.6$ for a life of 10 hr. The surfaces are machined.

Problem 6–19

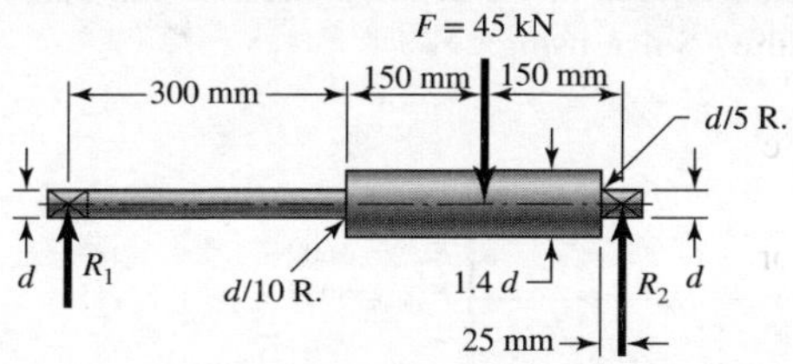

6–20 A bar of steel has the minimum properties $S_e = 276$ MPa, $S_y = 413$ MPa, and $S_{ut} = 551$ MPa. The bar is subjected to a steady torsional stress of 103 MPa and an alternating bending stress of 172 MPa. Find the factor of safety guarding against a static failure, and either the factor of safety guarding against a fatigue failure or the expected life of the part. For the fatigue analysis use:

(*a*) Goodman criterion.
(*b*) Gerber criterion.
(*c*) Morrow criterion.

6–21 Repeat Problem 6–20 but with a steady torsional stress of 138 MPa and an alternating bending stress of 69 MPa.

6–22 Repeat Problem 6–20 but with a steady torsional stress of 103 MPa, an alternating torsional stress of 69 MPa, and an alternating bending stress of 83 MPa.

6–23 Repeat Problem 6–20 but with an alternating torsional stress of 207 MPa.

6–24 Repeat Problem 6–20 but with an alternating torsional stress of 103 MPa and a steady bending stress of 103 MPa.

6–25 The cold-drawn AISI 1040 steel bar shown in the figure is subjected to a completely reversed axial load fluctuating between 28 kN in compression to 28 kN in tension. Estimate the fatigue factor of safety based on achieving infinite life and the yielding factor of safety. If infinite life is not predicted, estimate the number of cycles to failure.

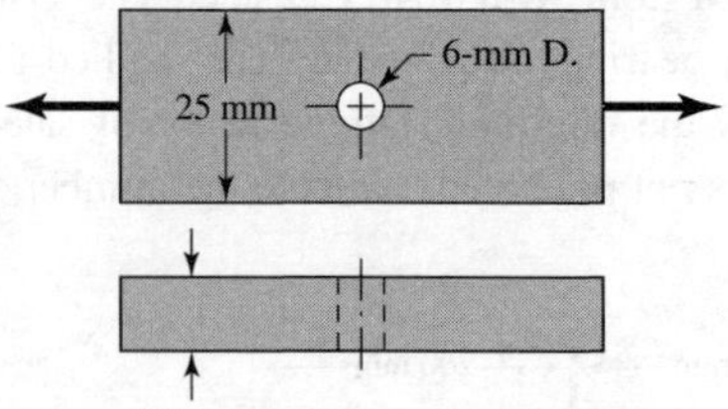

Problem 6–25

6–26 Repeat Problem 6–25 for a load that fluctuates from 12 kN to 28 kN. Use the Goodman, Gerber, and Morrow criteria and compare their predictions.

6–27 Using the Goodman criterion for infinite life, repeat Problem 6–25 for each of the following loading conditions:

(*a*) 0 kN to 28 kN
(*b*) 12 kN to 28 kN
(*c*) −28 kN to 12 kN

6–28 The figure shows a formed round-wire cantilever spring subjected to a varying force. The hardness tests made on 50 springs gave a minimum hardness of 400 Brinell. It is apparent from the mounting details that there is no stress concentration. A visual inspection of the springs indicates that the surface finish corresponds closely to a hot-rolled finish. Ignore curvature effects on the bending stress. What number of applications is likely to cause failure? Solve using:

(*a*) Goodman criterion.
(*b*) Gerber criterion.

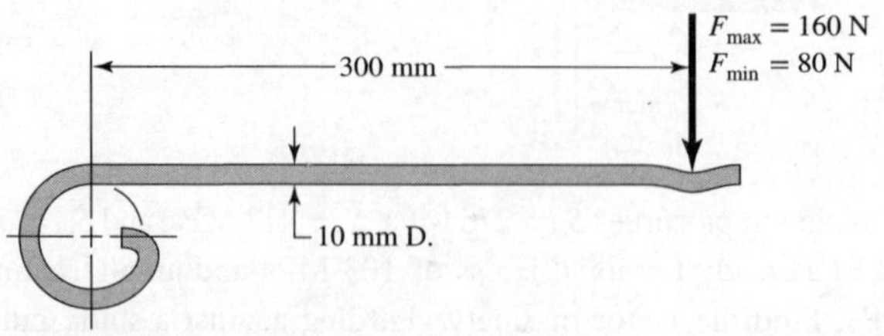

Problem 6–28

6–29 The figure is a drawing of a 4- by 20-mm latching spring. A preload is obtained during assembly by shimming under the bolts to obtain an estimated initial deflection of 2 mm. The latching operation itself requires an additional deflection of exactly 4 mm. The material is ground high-carbon steel, bent then hardened and tempered to a minimum hardness of 490 Bhn. The inner radius of the bend is 4 mm. Estimate the yield strength to be 90 percent of the ultimate strength.

(*a*) Find the maximum and minimum latching forces.

(*b*) Determine the fatigue factor of safety for infinite life, using the Goodman criterion.

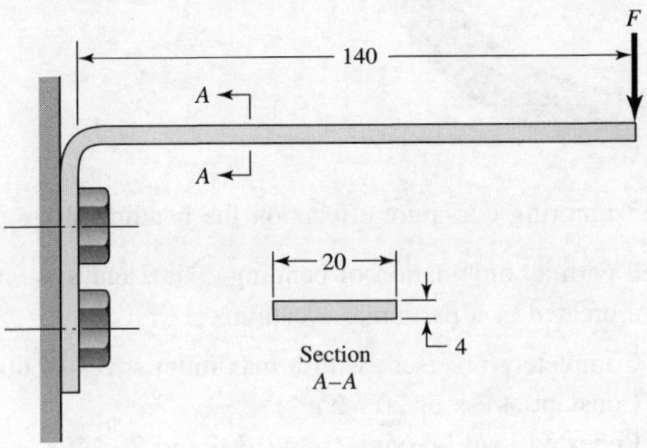

Problem 6–29
Dimensions in millimeters

6–30 The figure shows the free-body diagram of a connecting-link portion having stress concentration at three sections. The dimensions are $r = 6$ mm, $d = 20$ mm, $h = 12$ mm, $w_1 = 90$ mm, and $w_2 = 60$ mm. The forces F fluctuate between a tension of 18 kN and a compression of 72 kN. Neglect column action and find the least factor of safety if the material is cold-drawn AISI 1018 steel.

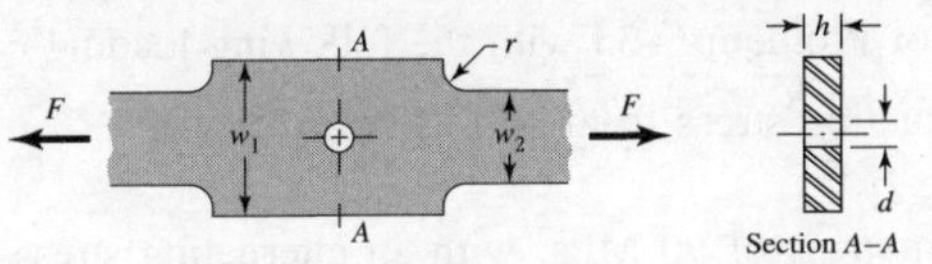

Problem 6–30

6–31 Solve Problem 6–30 except let $w_1 = 60$ mm, $w_2 = 36$ mm, and the force fluctuates between a tension of 72 kN and a compression of 18 kN. Use the Goodman criterion.

6–32 For the part in Problem 6–30, recommend a fillet radius r that will cause the fatigue factor of safety to be the same at the hole and at the fillet.

6–33 The torsional coupling in the figure is composed of a curved beam of square cross section that is welded to an input shaft and output plate. A torque is applied to the shaft and cycles from zero to T. The cross section of the beam has dimensions of 5×5 mm, and the centroidal axis of the beam describes a curve of the form $r = 20 + 10\theta/\pi$, where r and θ are in millimeters and radians, respectively ($0 \le \theta \le 4\pi$). The curved beam has a machined surface with yield and ultimate strength values of 420 and 770 MPa, respectively.

(*a*) Determine the maximum allowable value of T such that the coupling will have an infinite life with a factor of safety, $n = 3$, using the Goodman criterion.

(*b*) Repeat part (*a*) using the Morrow criterion.

(*c*) Using T found in part (*b*), determine the factor of safety guarding against yield.

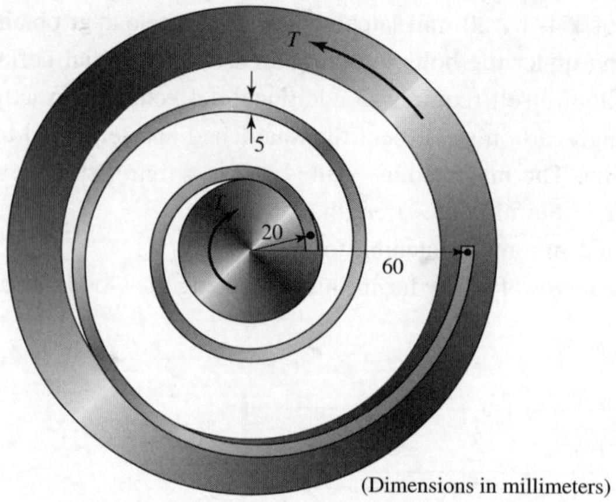

Problem 6–33

6–34 Repeat Problem 6–33 ignoring curvature effects on the bending stress.

6–35 A steel part is loaded with a combination of bending, axial, and torsion such that the following stresses are created at a particular location:

Bending:	Completely reversed, with a maximum stress of 60 MPa
Axial:	Constant stress of 20 MPa
Torsion:	Repeated load, varying from 0 MPa to 70 MPa

Assume the varying stresses are in phase with each other. The part contains a notch such that $K_{f,bending} = 1.4$, $K_{f,axial} = 1.1$, and $K_{f,torsion} = 2.0$. The material properties are $S_y = 300$ MPa and $S_u = 400$ MPa. The completely adjusted endurance limit is found to be $S_e = 160$ MPa. Find the factor of safety for fatigue based on infinite life, using the Goodman criterion. If the life is not infinite, estimate the number of cycles, using the Walker criterion to find the equivalent completely reversed stress. Be sure to check for yielding.

6–36 Repeat the requirements of Problem 6–35 with the following loading conditions:

Bending:	Fluctuating stress from –40 MPa to 150 MPa
Axial:	None
Torsion:	Mean stress of 90 MPa, with an alternating stress of 10 percent of the mean stress

6–37* to 6–46* For the problem specified in the table, build upon the results of the original problem to determine the minimum factor of safety for fatigue based on infinite life, using the Goodman criterion. The shaft rotates at a constant speed, has a constant diameter, and is made from cold-drawn AISI 1018 steel.

Problem Number	Original Problem Number
6–37*	3–76
6–38*	3–77
6–39*	3–78
6–40*	3–79
6–41*	3–80
6–42*	3–81
6–43*	3–82
6–44*	3–84
6–45*	3–85
6–46*	3–87

6–47* to 6–50* For the problem specified in the table, build upon the results of the original problem to determine the minimum factor of safety for fatigue based on infinite life, using the Goodman criterion. If the life is not infinite, conservatively estimate the number of cycles. The force F is applied as a repeated load. The material is AISI 1018 CD steel. The fillet radius at the wall is 2 mm, with theoretical stress concentrations of 1.5 for bending, 1.2 for axial, and 2.1 for torsion.

Problem Number	Original Problem Number
6–47*	3–88
6–48*	3–89
6–49*	3–90
6–50*	3–91

6–51* to 6–53* For the problem specified in the table, build upon the results of the original problem to determine the minimum factor of safety for fatigue at point A, based on infinite life, using the Morrow criterion. If the life is not infinite, estimate the number of cycles. The force F is applied as a repeated load. The material is AISI 1018 CD steel.

Problem Number	Original Problem Number
6–51*	3–92
6–52*	3–93
6–53*	3–94

6–54 Solve Problem 6–17 except include a steady torque of 280 N · m being transmitted through the shaft between the points of application of the forces. Use the Goodman criterion.

6–55 Solve Problem 6–18 except include a steady torque of 250 N · m being transmitted through the shaft between the points of application of the forces. Use the Goodman criterion.

6–56 In the figure shown, shaft A, made of AISI 1020 hot-rolled steel, is welded to a fixed support and is subjected to loading by equal and opposite forces F via shaft B. A theoretical stress-concentration factor K_{ts} of 1.6 is induced in the shaft by the 3 mm weld fillet. The length of shaft A from the fixed support to the connection at shaft B is 1 m. The load F cycles from 0.5 to 2 kN.

(*a*) For shaft A, find the factor of safety for infinite life using the Goodman fatigue failure criterion.

(*b*) Repeat part (*a*) using the Gerber fatigue failure criterion.

Problem 6–56

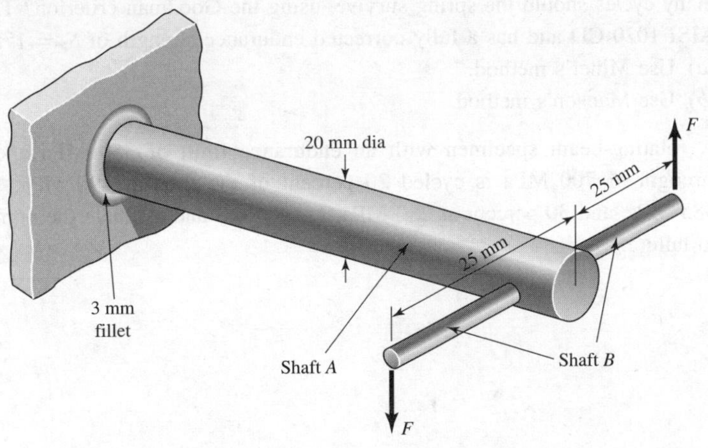

6–57 A schematic of a clutch-testing machine is shown. The steel shaft rotates at a constant speed ω. An axial load is applied to the shaft and is cycled from zero to P. The torque T induced by the clutch face onto the shaft is given by

$$T = \frac{fP(D + d)}{4}$$

where D and d are defined in the figure and f is the coefficient of friction of the clutch face. The shaft is machined with $S_y = 800$ MPa and $S_{ut} = 1000$ MPa. The theoretical stress-concentration factors for the fillet are 3.0 and 1.8 for the axial and torsional loading, respectively.

Assume the load variation P is synchronous with shaft rotation. With $f = 0.3$, find the maximum allowable load P such that the shaft will survive a minimum of 10^6 cycles with a factor of safety of 3. Use the Goodman criterion. Check for yielding.

Problem 6–57

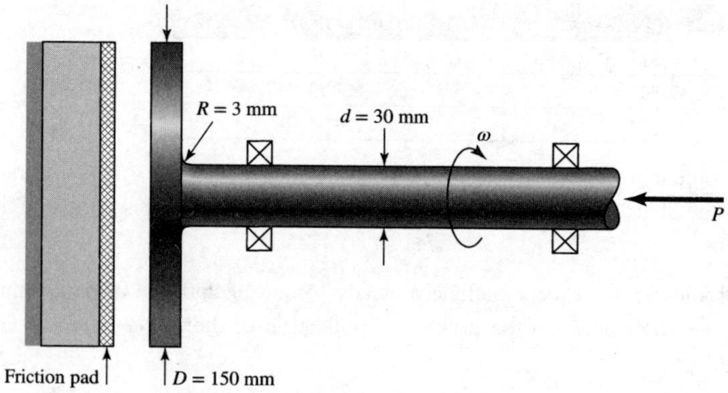

6–58 For the clutch of Problem 6–57, the external load P is cycled between 20 kN and 80 kN. Assuming that the shaft is rotating synchronous with the external load cycle, estimate the number of cycles to failure. Use the Goodman fatigue failure criterion.

6–59 A flat leaf spring has fluctuating stress of $\sigma_{max} = 360$ MPa and $\sigma_{min} = 160$ MPa applied for $8(10^4)$ cycles. If the load changes to $\sigma_{max} = 320$ MPa and $\sigma_{min} = -200$ MPa, how many cycles should the spring survive, using the Goodman criterion? The material is AISI 1020 CD and has a fully corrected endurance strength of $S_e = 175$ MPa.

(*a*) Use Miner's method.

(*b*) Use Manson's method.

6–60 A rotating-beam specimen with an endurance limit of 350 MPa and an ultimate strength of 700 MPa is cycled 20 percent of the time at 490 MPa, 50 percent at 385 MPa, and 30 percent at 280 MPa. Let $f = 0.8$ and estimate the number of cycles to failure.

6–61 A machine part will be cycled at ±350 MPa for $5(10^3)$ cycles. Then the loading will be changed to ±260 MPa for $5(10^4)$ cycles. Finally, the load will be changed to ±225 MPa. How many cycles of operation can be expected at this stress level? For the part, $S_{ut} = 530$ MPa, $f = 0.9$, and has a fully corrected endurance strength of $S_e = 210$ MPa.
(*a*) Use Miner's method.
(*b*) Use Manson's method.

6–62 The material properties of a machine part are $S_{ut} = 595$ MPa, $f = 0.86$, and a fully corrected endurance limit of $S_e = 315$ MPa. The part is to be cycled at $\sigma_a = 245$ MPa and $\sigma_m = 210$ MPa for $12(10^3)$ cycles. Using the Goodman criterion, estimate the new endurance limit after cycling.
(*a*) Use Miner's method.
(*b*) Use Manson's method.

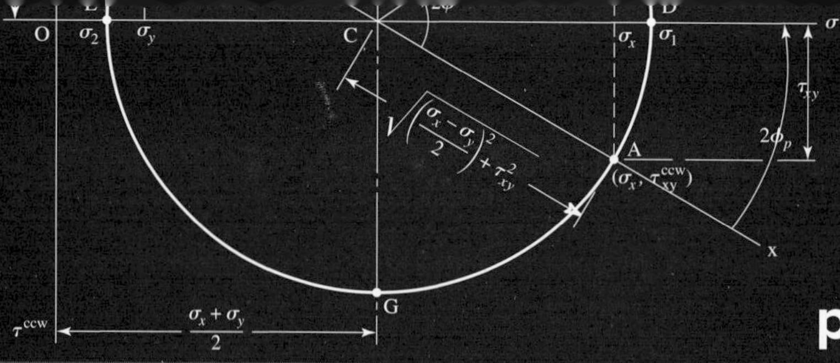

part 3

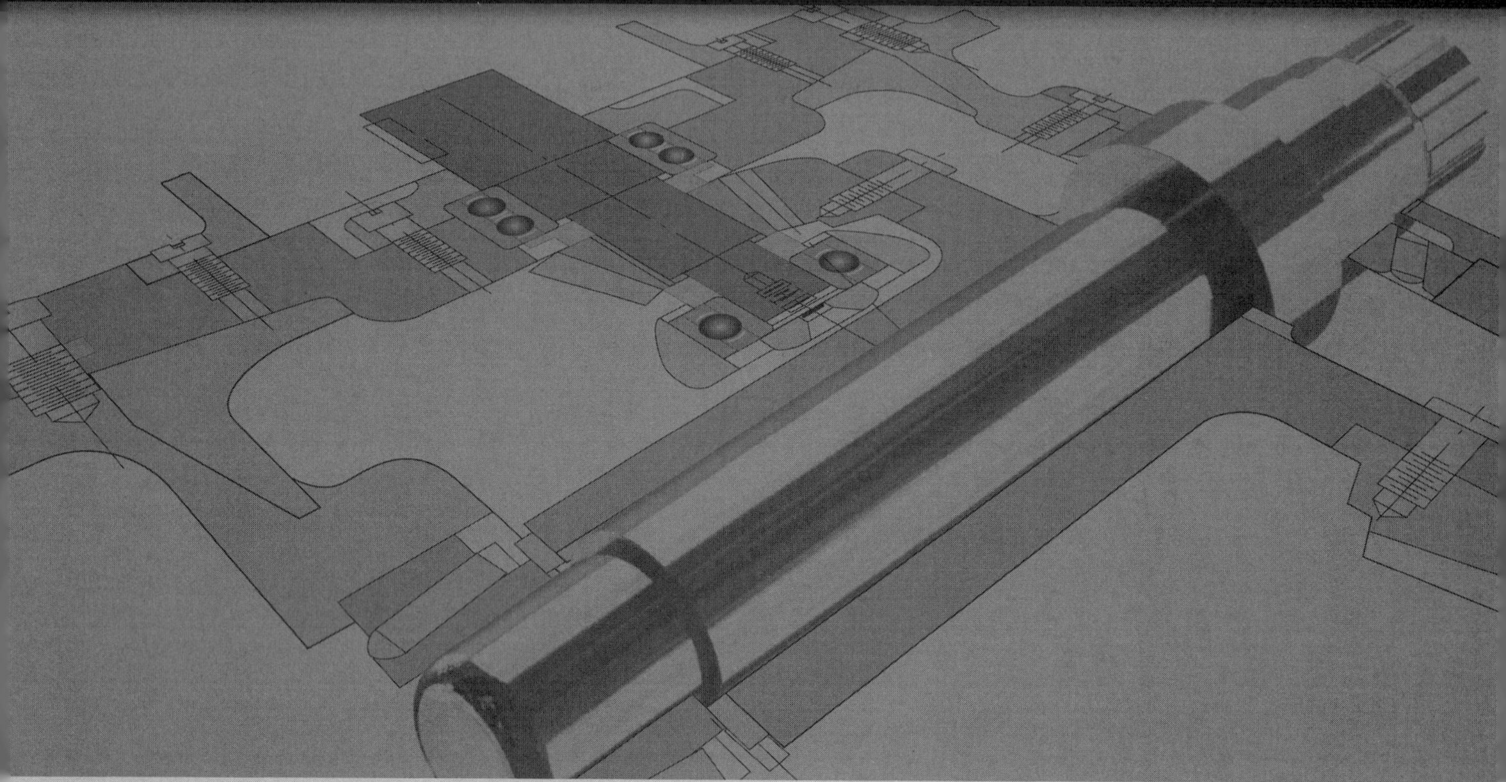

Courtesy of Dee Dehokenanan

Design of Mechanical Elements

7 Shafts and Shaft Components

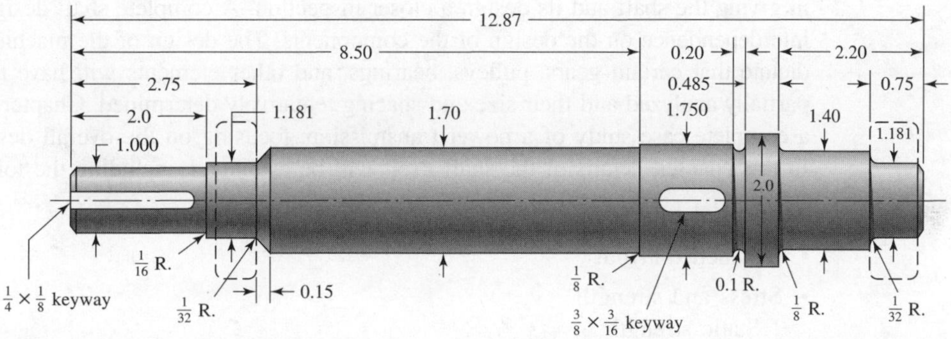

Chapter Outline

7–1 Introduction

A *shaft* is a rotating member, usually of circular cross section, used to transmit power or motion. It provides the axis of rotation, or oscillation, of elements such as gears, pulleys, flywheels, cranks, sprockets, and the like and controls the geometry of their motion. An *axle* is a nonrotating member that carries no torque and is used to support rotating wheels, pulleys, and the like. The automotive axle is not a true axle; the term is a carryover from the horse-and-buggy era, when the wheels rotated on nonrotating members. A nonrotating axle can readily be designed and analyzed as a static beam, and will not warrant the special attention given in this chapter to the rotating shafts, which are subject to fatigue loading.

There is really nothing unique about a shaft that requires any special treatment beyond the basic methods already developed in previous chapters. However, because of the ubiquity of the shaft in so many machine design applications, there is some advantage in giving the shaft and its design a closer inspection. A complete shaft design has much interdependence on the design of the components. The design of the machine itself will dictate that certain gears, pulleys, bearings, and other elements will have at least been partially analyzed and their size and spacing tentatively determined. Chapter 18 provides a complete case study of a power transmission, focusing on the overall design process. In this chapter, details of the shaft itself will be examined, including the following:

- Material selection
- Geometric layout
- Stress and strength
  - Static strength
  - Fatigue strength
- Deflection and rigidity
  - Bending deflection
  - Torsional deflection
  - Slope at bearings and shaft-supported elements
  - Shear deflection due to transverse loading of short shafts
- Vibration due to natural frequency

In deciding on an approach to shaft sizing, it is necessary to realize that a stress analysis at a specific point on a shaft can be made using only the shaft geometry in the vicinity of that point. Thus the geometry of the entire shaft is not needed. In design it is usually possible to locate the critical areas, size these to meet the strength requirements, and then size the rest of the shaft to meet the requirements of the shaft-supported elements.

The deflection and slope analyses cannot be made until the geometry of the entire shaft has been defined. Thus deflection is a function of the geometry *everywhere,* whereas the stress at a section of interest is a function of *local geometry*. For this reason, shaft design allows a consideration of stress first. Then, after tentative values for the shaft dimensions have been established, the determination of the deflections and slopes can be made.

7–2 Shaft Materials

Deflection is not affected by strength, but rather by stiffness as represented by the modulus of elasticity, which is essentially constant within many categories of material, as shown in Figure 2–23 and Appendix A–5. For that reason, for steel shafts, rigidity cannot be controlled by material decisions, but only by geometric decisions.

Necessary strength to resist loading stresses affects the choice of materials and their treatments. Many shafts are made from low-carbon, cold-drawn or hot-rolled steel, such as AISI 1020–1050 steels.

Significant strengthening from heat treatment and high alloy content are often not warranted. Fatigue failure is reduced moderately by increase in strength, and then only to a certain level before adverse effects in endurance limit and notch sensitivity begin to counteract the benefits of higher strength. A good practice is to start with an inexpensive, low or medium carbon steel for the first time through the design calculations. If strength considerations turn out to dominate over deflection, then a higher strength material should be tried, allowing the shaft sizes to be reduced until excess deflection becomes an issue. The cost of the material and its processing must be weighed against the need for smaller shaft diameters. When warranted, typical alloy steels for heat treatment include AISI 1340–50, 3140–50, 4140, 4340, 5140, and 8650.

Shafts usually don't need to be surface hardened unless they serve as the actual journal of a bearing surface. Typical material choices for surface hardening include carburizing grades of AISI 1020, 4320, 4820, and 8620.

Cold-drawn steel is usually used for diameters under about 3 inches. The nominal diameter of the bar can be left unmachined in areas that do not require fitting of components. Hot-rolled steel should be machined all over. For large shafts requiring much material removal, the residual stresses may tend to cause warping. If concentricity is important, it may be necessary to rough machine, then heat treat to remove residual stresses and increase the strength, then finish machine to the final dimensions.

In approaching material selection, the amount to be produced is a salient factor. For low production, turning is the usual primary shaping process. An economic viewpoint may require removing the least material. High production may permit a volume-conservative shaping method (hot or cold forming, casting), and minimum material in the shaft can become a design goal. Cast iron may be specified if the production quantity is high, and the gears are to be integrally cast with the shaft.

Properties of the shaft locally depend on its history—cold work, cold forming, rolling of fillet features, heat treatment, including quenching medium, agitation, and tempering regimen.[1]

Stainless steel may be appropriate for some environments.

7–3 Shaft Layout

The general layout of a shaft to accommodate shaft elements, e.g., gears, bearings, and pulleys, must be specified early in the design process in order to perform a free body force analysis and to obtain shear-moment diagrams. The geometry of a shaft is generally that of a stepped cylinder. The use of shaft shoulders is an excellent means of axially locating the shaft elements and to carry any thrust loads. Figure 7–1 shows an example of a stepped shaft supporting the gear of a worm-gear speed reducer. Each shoulder in the shaft serves a specific purpose, which you should attempt to determine by observation.

The geometric configuration of a shaft to be designed is often simply a revision of existing models in which a limited number of changes must be made. If there is no existing design to use as a starter, then the determination of the shaft layout may

---

[1]See Joseph E. Shigley, Charles R. Mischke, and Thomas H. Brown, Jr. (eds-in-chief), *Standard Handbook of Machine Design*, 3rd ed., McGraw Hill, New York, 2004. For cold-worked property prediction see Chapter 29, and for heat-treated property prediction see Chapters 29 and 33.

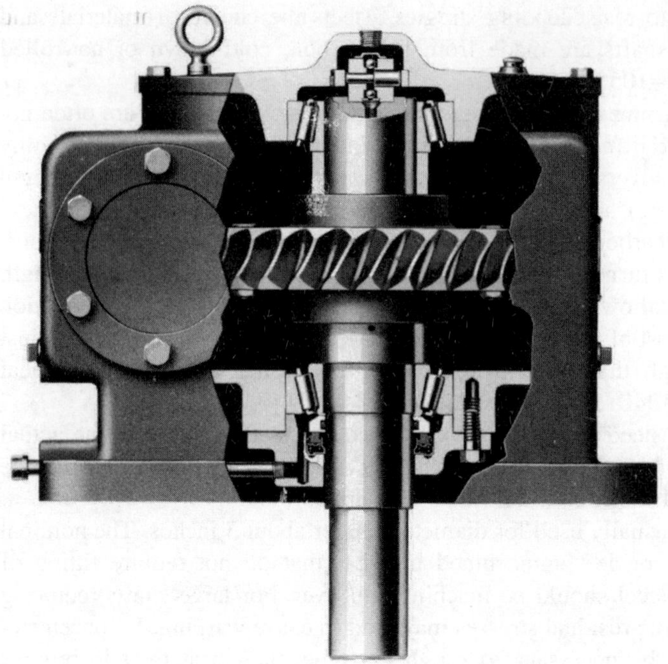

Figure 7–1

A vertical worm-gear speed reducer. *(Courtesy of the Cleveland Gear Company.)*

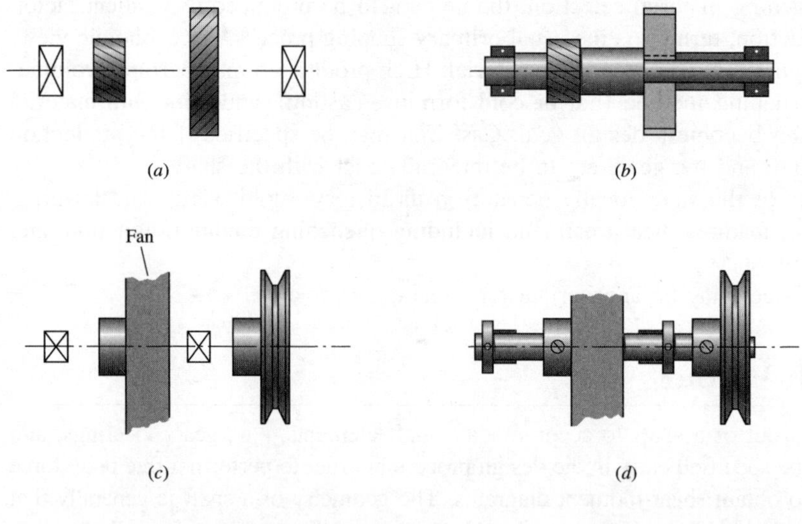

Figure 7–2

(*a*) Choose a shaft configuration to support and locate the two gears and two bearings. (*b*) Solution uses an integral pinion, three shaft shoulders, key and keyway, and sleeve. The housing locates the bearings on their outer rings and receives the thrust loads. (*c*) Choose fan-shaft configuration. (*d*) Solution uses sleeve bearings, a straight-through shaft, locating collars, and setscrews for collars, fan pulley, and fan itself. The fan housing supports the sleeve bearings.

have many solutions. This problem is illustrated by the two examples of Figure 7–2. In Figure 7–2*a* a geared countershaft is to be supported by two bearings. In Figure 7–2*c* a fanshaft is to be configured. The solutions shown in Figure 7–2*b* and 7–2*d* are not necessarily the best ones, but they do illustrate how the shaft-mounted devices are fixed and located in the axial direction, and how provision is made for torque transfer from one element to another. There are no absolute rules for specifying the general layout, but the following guidelines may be helpful.

Axial Layout of Components

The axial positioning of components is often dictated by the layout of the housing and other meshing components. In general, it is best to support load-carrying components between bearings, such as in Figure 7–2*a,* rather than cantilevered outboard of the bearings, such as in Figure 7–2*c*. Pulleys and sprockets often need to be mounted outboard for ease of installation of the belt or chain. The length of the cantilever should be kept short to minimize the deflection.

Only two bearings should be used in most cases. For extremely long shafts carrying several load-bearing components, it may be necessary to provide more than two bearing supports. In this case, particular care must be given to the alignment of the bearings.

Shafts should be kept short to minimize bending moments and deflections. Some axial space between components is desirable to allow for lubricant flow and to provide access space for disassembly of components with a puller. Load-bearing components should be placed near the bearings, again to minimize the bending moment at the locations that will likely have stress concentrations, and to minimize the deflection at the load-carrying components.

The components must be accurately located on the shaft to line up with other mating components, and provision must be made to securely hold the components in position. The primary means of locating the components is to position them against a shoulder of the shaft. A shoulder also provides a solid support to minimize deflection and vibration of the component. Sometimes when the magnitudes of the forces are reasonably low, shoulders can be constructed with retaining rings in grooves, sleeves between components, or clamp-on collars. In cases where axial loads are very small, it may be feasible to do without the shoulders entirely, and rely on press fits, pins, or collars with setscrews to maintain an axial location. See Figures 7–2*b* and 7–2*d* for examples of some of these means of axial location.

Supporting Axial Loads

In cases where axial loads are not trivial, it is necessary to provide a means to transfer the axial loads into the shaft, then through a bearing to the ground. This will be particularly necessary with helical or bevel gears, or tapered roller bearings, as each of these produces axial force components. Often, the same means of providing axial location, e.g., shoulders, retaining rings, and pins, will be used to also transmit the axial load into the shaft.

It is generally best to have only one bearing carry the axial load, to allow greater tolerances on shaft length dimensions, and to prevent binding if the shaft expands due to temperature changes. This is particularly important for long shafts. Figures 7–3 and 7–4 show examples of shafts with only one bearing carrying the axial load against a shoulder, while the other bearing is simply press-fit onto the shaft with no shoulder.

Providing for Torque Transmission

Most shafts serve to transmit torque from an input gear or pulley, through the shaft, to an output gear or pulley. Of course, the shaft itself must be sized to support the torsional stress and torsional deflection. It is also necessary to provide a means of transmitting the torque between the shaft and the gears. Common torque-transfer elements are:

- Keys
- Splines
- Setscrews
- Pins
- Tolerance rings
- Press or shrink fits
- Tapered fits

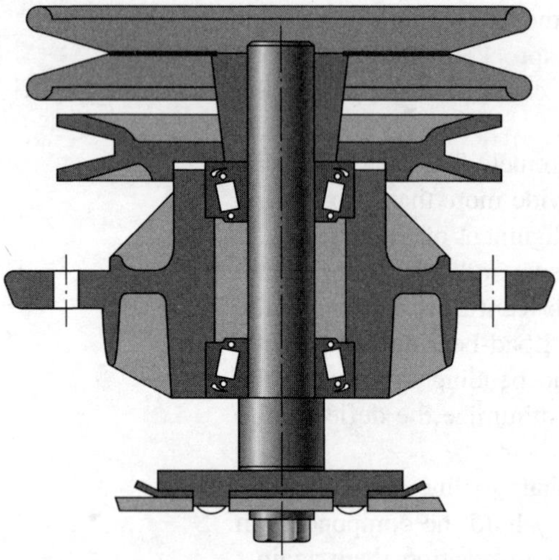

Figure 7–3

Tapered roller bearings used in a mowing machine spindle. This design represents good practice for the situation in which one or more torque-transfer elements must be mounted outboard. *(Source: Redrawn from material furnished by The Timken Company.)*

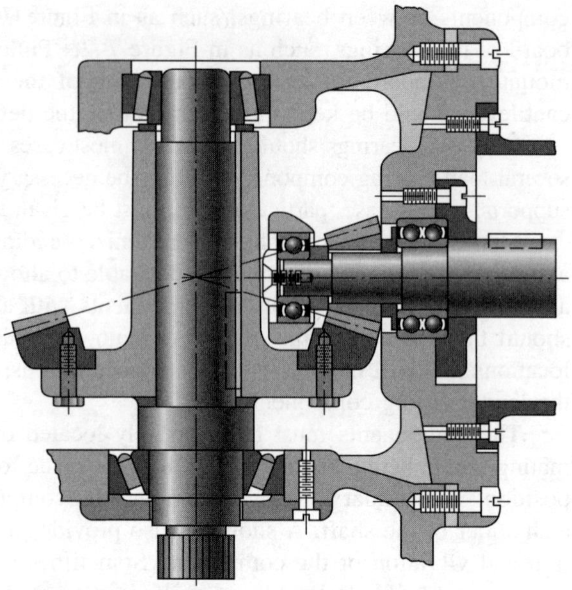

Figure 7–4

A bevel-gear drive in which both pinion and gear are straddle-mounted. *(Source: Redrawn from material furnished by Gleason Machine Division.)*

In addition to transmitting the torque, many of these devices are designed to fail or slip if the torque exceeds acceptable operating limits, protecting more expensive components.

Details regarding hardware components such as *keys, pins,* and *setscrews* are addressed in detail in Section 7–7. One of the most effective and economical means of transmitting moderate to high levels of torque is through a key that fits in a groove in the shaft and gear. Keyed components generally have a slip fit onto the shaft, so assembly and disassembly is easy. The key provides for positive angular orientation of the component, which is useful in cases where phase angle timing is important.

Splines are essentially stubby gear teeth formed on the outside of the shaft and on the inside of the hub of the load-transmitting component. Splines are generally much more expensive to manufacture than keys, and are usually not necessary for simple torque transmission. They are typically used to transfer high torques. One feature of a spline is that it can be made with a reasonably loose slip fit to allow for large axial motion between the shaft and component while still transmitting torque. This is useful for connecting two shafts where relative motion between them is common, such as in connecting a power takeoff (PTO) shaft of a tractor to an implement. SAE and ANSI publish standards for splines. Stress-concentration factors are greatest where the spline ends and blends into the shaft, but are generally quite moderate.

For cases of low-torque transmission, various means of transmitting torque are available. These include pins, setscrews in hubs, tolerance rings, tapered fits, and press fits.

Press and shrink fits for securing hubs to shafts are used both for torque transfer and for preserving axial location. The resulting stress-concentration factor is usually quite small. See Section 7–8 for guidelines regarding appropriate sizing and tolerancing to transmit torque with press and shrink fits. A similar method is to use a split hub with screws to clamp the hub to the shaft. This method allows for disassembly and lateral adjustments. Another similar method uses a two-part hub consisting of a split inner member that fits into a tapered hole. The assembly is then tightened to the shaft with screws, which forces the inner part into the wheel and clamps the whole assembly against the shaft.

Tapered fits between the shaft and the shaft-mounted device, such as a wheel, are often used on the overhanging end of a shaft. Screw threads at the shaft end then permit the use of a nut to lock the wheel tightly to the shaft. This approach is useful because it can be disassembled, but it does not provide good axial location of the wheel on the shaft.

At the early stages of the shaft layout, the important thing is to select an appropriate means of transmitting torque, and to determine how it affects the overall shaft layout. It is necessary to know where the shaft discontinuities, such as keyways, holes, and splines, will be in order to determine critical locations for analysis.

Assembly and Disassembly

Consideration should be given to the method of assembling the components onto the shaft, and the shaft assembly into the frame. This generally requires the largest diameter in the center of the shaft, with progressively smaller diameters toward the ends to allow components to be slid on from the ends. If a shoulder is needed on both sides of a component, one of them must be created by such means as a retaining ring or by a sleeve between two components. The gearbox itself will need means to physically position the shaft into its bearings, and the bearings into the frame. This is typically accomplished by providing access through the housing to the bearing at one end of the shaft. See Figures 7–5 through 7–8 for examples.

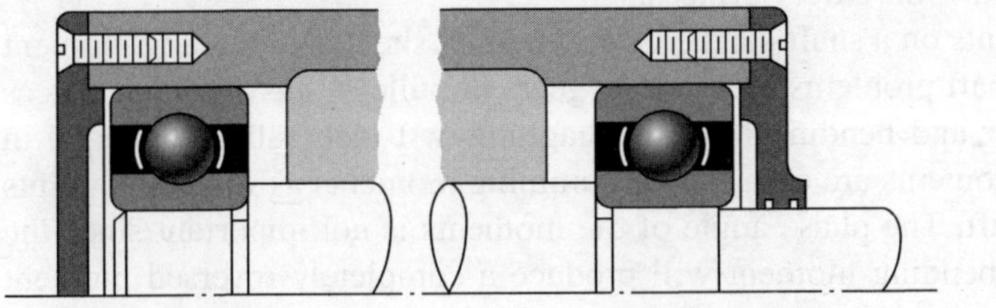

Figure 7–5

Arrangement showing bearing inner rings press-fitted to shaft while outer rings float in the housing. The axial clearance should be sufficient only to allow for machinery vibrations. Note the labyrinth seal on the right.

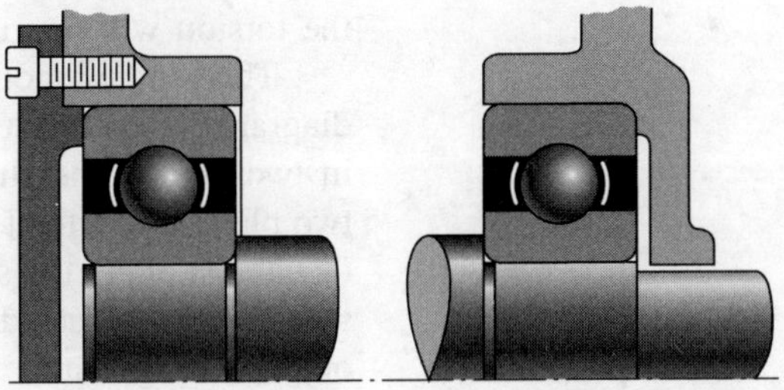

Figure 7–6

Similar to the arrangement of Figure 7–5 except that the outer bearing rings are preloaded.

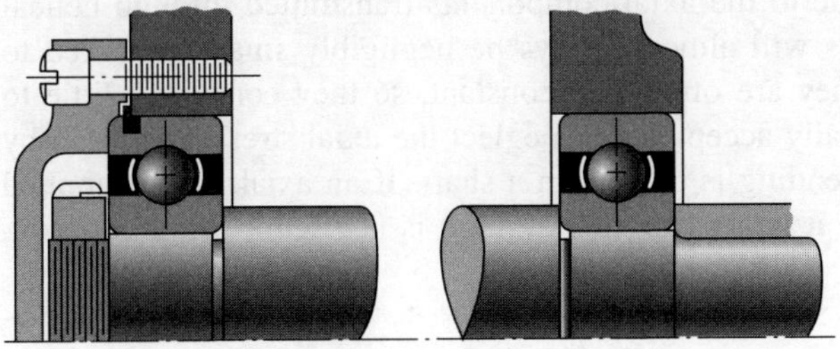

Figure 7–7

In this arrangement the inner ring of the left-hand bearing is locked to the shaft between a nut and a shaft shoulder. The locknut and washer are AFBMA standard. The snap ring in the outer race is used to positively locate the shaft assembly in the axial direction. Note the floating right-hand bearing and the grinding runout grooves in the shaft.

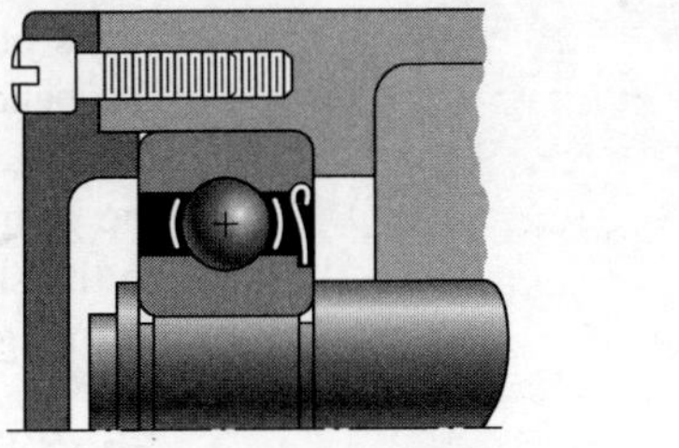

Figure 7–8

This arrangement is similar to Figure 7–7 in that the left-hand bearing positions the entire shaft assembly. In this case the inner ring is secured to the shaft using a snap ring. Note the use of a shield to prevent dirt generated from within the machine from entering the bearing.

When components are to be press-fit to the shaft, the shaft should be designed so that it is not necessary to press the component down a long length of shaft. This may require an extra change in diameter, but it will reduce manufacturing and assembly cost by only requiring the close tolerance for a short length.

Consideration should also be given to the necessity of disassembling the components from the shaft. This requires consideration of issues such as accessibility of retaining rings, space for pullers to access bearings, openings in the housing to allow pressing the shaft or bearings out, etc.

7–4 Shaft Design for Stress

Critical Locations

It is not necessary to evaluate the stresses in a shaft at every point; a few potentially critical locations will suffice. Critical locations will usually be on the outer surface, at axial locations where the bending moment is large, where the torque is present, and where stress concentrations exist. By direct comparison of various points along the shaft, a few critical locations can be identified upon which to base the design. An assessment of typical stress situations will help.

Most shafts will transmit torque through a portion of the shaft. Typically the torque comes into the shaft at one gear and leaves the shaft at another gear. A free body diagram of the shaft will allow the torque at any section to be determined. The torque is often relatively constant at steady state operation. The shear stress due to the torsion will be greatest on outer surfaces.

The bending moments on a shaft can be determined by shear and bending moment diagrams. Since most shaft problems incorporate gears or pulleys that introduce forces in two planes, the shear and bending moment diagrams will generally be needed in two planes. Resultant moments are obtained by summing moments as vectors at points of interest along the shaft. The phase angle of the moments is not important since the shaft rotates. A steady bending moment will produce a completely reversed moment on a rotating shaft, as a specific stress element will alternate from compression to tension in every revolution of the shaft. The normal stress due to bending moments will be greatest on the outer surfaces. In situations where a bearing is located at the end of the shaft, stresses near the bearing are often not critical since the bending moment is small.

Axial stresses on shafts due to the axial components transmitted through helical gears or tapered roller bearings will almost always be negligibly small compared to the bending moment stress. They are often also constant, so they contribute little to fatigue. Consequently, it is usually acceptable to neglect the axial stresses induced by the gears and bearings when bending is present in a shaft. If an axial load is applied to the shaft in some other way, it is not safe to assume it is negligible without checking magnitudes.

Shaft Stresses

Bending, torsion, and axial stresses may be present in both mean and alternating components. For analysis, it is simple enough to combine the different types of stresses into alternating and mean von Mises stresses, as shown in Section 6–16. It is sometimes convenient to customize the equations specifically for shaft applications. Axial loads are usually comparatively very small at critical locations where bending and torsion dominate, so they will be left out of the following equations. The fluctuating stresses due to bending and torsion are given by

$$\sigma_a = K_f \frac{M_a c}{I} \qquad \sigma_m = K_f \frac{M_m c}{I}$$
$$\tau_a = K_{fs} \frac{T_a r}{J} \qquad \tau_m = K_{fs} \frac{T_m r}{J} \tag{7–1}$$

where M_m and M_a are the mean and alternating bending moments, T_m and T_a are the mean and alternating torques, and K_f and K_{fs} are the fatigue stress-concentration factors for bending and torsion, respectively.

Assuming a solid shaft with round cross section, appropriate geometry terms can be introduced for c, I, r, and J resulting in

$$\sigma_a = K_f \frac{32M_a}{\pi d^3} \qquad \sigma_m = K_f \frac{32M_m}{\pi d^3} \tag{7–2}$$

$$\tau_a = K_{fs} \frac{16T_a}{\pi d^3} \qquad \tau_m = K_{fs} \frac{16T_m}{\pi d^3} \tag{7–3}$$

Using the distortion energy failure theory, the von Mises stress is given by Equation (5–15), with $\sigma_x = \sigma$, the bending stress, $\sigma_y = 0$, and $\tau_{xy} = \tau$, the torsional shear stress. Thus, for rotating round solid shafts, neglecting axial loads, the fluctuating von Mises stresses are given by

$$\sigma'_a = (\sigma_a^2 + 3\tau_a^2)^{1/2} = \left[\left(\frac{32K_f M_a}{\pi d^3}\right)^2 + 3\left(\frac{16K_{fs} T_a}{\pi d^3}\right)^2\right]^{1/2} \tag{7–4}$$

$$\sigma'_m = (\sigma_m^2 + 3\tau_m^2)^{1/2} = \left[\left(\frac{32K_f M_m}{\pi d^3}\right)^2 + 3\left(\frac{16K_{fs} T_m}{\pi d^3}\right)^2\right]^{1/2} \tag{7–5}$$

Note that the stress-concentration factors are sometimes considered optional for the mean components with ductile materials, because of the capacity of the ductile material to yield locally at the discontinuity.

Expressions can be obtained for any of the common failure criteria by substituting the von Mises stresses from Equations (7–4) and (7–5) into a failure criterion equation for any of the failure criteria from the fluctuating-stress diagram (Figure 6–36). The resulting equations for several of the commonly used failure curves are summarized below. For design purposes, it is also desirable to solve each equation for the diameter. The names given to each set of equations identifies the significant failure theory, followed by a fatigue failure locus name. For example, DE-Gerber indicates the stresses are combined using the distortion energy (DE) theory, and the Gerber criteria is used for the fatigue failure.

To keep the equations in simpler form, first establish a pair of terms to be used in each of the criteria equations.

$$A = \sqrt{4(K_f M_a)^2 + 3(K_{fs} T_a)^2}$$
$$B = \sqrt{4(K_f M_m)^2 + 3(K_{fs} T_m)^2} \tag{7–6}$$

DE-Goodman

$$n = \frac{\pi d^3}{16}\left(\frac{A}{S_e} + \frac{B}{S_{ut}}\right)^{-1} \tag{7–7}$$

$$d = \left[\frac{16n}{\pi}\left(\frac{A}{S_e} + \frac{B}{S_{ut}}\right)\right]^{1/3} \tag{7–8}$$

DE-Morrow

$$n = \frac{\pi d^3}{16}\left(\frac{A}{S_e} + \frac{B}{\tilde{\sigma}_f}\right)^{-1} \tag{7–9}$$

$$d = \left[\frac{16n}{\pi}\left(\frac{A}{S_e} + \frac{B}{\tilde{\sigma}_f}\right)\right]^{1/3} \tag{7–10}$$

DE-Gerber

$$\frac{1}{n} = \frac{8A}{\pi d^3 S_e}\left\{1 + \left[1 + \left(\frac{2BS_e}{AS_{ut}}\right)^2\right]^{1/2}\right\} \tag{7–11}$$

$$d = \left(\frac{8nA}{\pi S_e}\left\{1 + \left[1 + \left(\frac{2BS_e}{AS_{ut}}\right)^2\right]^{1/2}\right\}\right)^{1/3} \tag{7–12}$$

DE-SWT

$$n = \frac{\pi d^3}{16}\,\frac{S_e}{(A^2 + AB)^{1/2}} \tag{7–13}$$

$$d = \left[\frac{16n}{\pi S_e}(A^2 + AB)^{1/2}\right]^{1/3} \tag{7–14}$$

For a rotating shaft with constant bending and torsion, the bending stress is completely reversed and the torsion is steady. Equations (7–7) through (7–14) can be simplified by setting M_m and T_a equal to 0, which simply drops out some of the terms.

Note that in an analysis situation in which the diameter is known and the factor of safety is desired, as an alternative to using the specialized equations above, it is always still valid to calculate the alternating and mean stresses using Equations (7–4) and (7–5), and substitute them into one of the fatigue failure criterion equations from Section 6–13, and solve directly for n. In a design situation, however, having the equations pre-solved for diameter is quite helpful.

It is always necessary to consider the possibility of static failure in the first load cycle. A von Mises maximum stress is calculated for this purpose.

$$\begin{aligned}\sigma'_{\max} &= \left[(\sigma_m + \sigma_a)^2 + 3(\tau_m + \tau_a)^2\right]^{1/2} \\ &= \left[\left(\frac{32K_f(M_m + M_a)}{\pi d^3}\right)^2 + 3\left(\frac{16K_{fs}(T_m + T_a)}{\pi d^3}\right)^2\right]^{1/2}\end{aligned} \tag{7–15}$$

To check for yielding, this von Mises maximum stress is compared to the yield strength, as usual.

$$n_y = \frac{S_y}{\sigma'_{\max}} \tag{7–16}$$

For a quick, conservative check, an estimate for σ'_{max} can be obtained by simply adding σ'_a and σ'_m. $(\sigma'_a + \sigma'_m)$ will always be greater than or equal to σ'_{max}, and will therefore be conservative.

EXAMPLE 7–1

At a machined shaft shoulder the small diameter d is 28 mm, the large diameter D is 42 mm, and the fillet radius is 2.8 mm. The bending moment is 142.4 N · m and the steady torsion moment is 124.3 N · m. The heat-treated steel shaft has an ultimate strength of S_{ut} = 735 MPa, a yield strength of S_y = 574 MPa, and a true fracture strength of $\tilde{\sigma}_f$ = 1085 MPa. The reliability goal for the endurance limit is 0.99.
(*a*) Determine the fatigue factor of safety of the design using each of the fatigue failure criteria described in this section.
(*b*) Determine the yielding factor of safety.

Solution
(*a*) $D/d = 42/28 = 1.50$, $r/d = 2.8/28 = 0.10$, $K_t = 1.68$ (Figure A–13–9), $K_{ts} = 1.42$ (Figure A–13–8), $q = 0.85$ (Figure 6–26), $q_{shear} = 0.92$ (Figure 6–27).

From Equation (6–32),

$$K_f = 1 + 0.85(1.68 - 1) = 1.58$$

$$K_{fs} = 1 + 0.92(1.42 - 1) = 1.39$$

Equation (6–10): $$S'_e = 0.5(735) = 367.5 \text{ MPa}$$

Equation (6–18): $$k_a = 4.51(735)^{-0.265} = 0.787$$

Equation (6–19): $$k_b = \left(\frac{28}{7.62}\right)^{-0.107} = 0.870$$

$$k_c = k_d = k_f = 1$$

Table 6–4: $$k_e = 0.814$$

$$S_e = 0.787(0.870)0.814(367.5) = 205 \text{ MPa}$$

For a rotating shaft, the constant bending moment will create a completely reversed bending stress.

$$M_a = 142.4 \text{ N} \cdot \text{m} \qquad T_m = 124.3 \text{ N} \cdot \text{m} \qquad M_m = T_a = 0$$

Applying Equations (7–6) and (7–7) for the DE-Goodman criteria gives

Equation (7–6): $$A = \sqrt{4(K_f M_a)^2} = \sqrt{4[(1.58)(142.4)]^2} = 448.7$$

$$B = \sqrt{3(K_{fs} T_m)^2} = \sqrt{3[(1.39)(124.3)]^2} = 299.26$$

Equation (7–7): $$n = \frac{\pi d^3}{16}\left(\frac{A}{S_e} + \frac{B}{S_{ut}}\right)^{-1} = \frac{\pi(0.028)^3}{16}\left(\frac{448.7}{205(10^6)} + \frac{299.26}{735(10^6)}\right)^{-1} = 1.66$$

Answer $n = 1.66$ DE-Goodman

Similarly, applying Equations (7–9), (7–11), and (7–13) for the other failure criteria,

Answer $n = 1.87$ DE-Morrow

Answer $n = 1.87$ DE-Gerber

Answer $n = 1.56$ DE-SWT

For comparison, consider an equivalent approach of calculating the stresses and applying the fatigue failure criteria directly. From Equations (7–4) and (7–5),

$$\sigma_a' = \left[\left(\frac{32 \cdot 1.58 \cdot 142.4}{\pi(0.028)^3}\right)^2\right]^{1/2} = 104.4 \text{ MPa}$$

$$\sigma_m' = \left[3\left(\frac{16 \cdot 1.39 \cdot 124.3}{\pi(0.028)^3}\right)^2\right]^{1/2} = 69.4 \text{ MPa}$$

Taking, for example, the Goodman failure criteria, application of Equation (6–41) gives

$$\frac{1}{n} = \frac{\sigma_a'}{S_e} + \frac{\sigma_m'}{S_{ut}} = \frac{104.4}{205} + \frac{69.4}{735} = 0.604$$

$$n = 1.62$$

which is identical with the previous result. The same process could be used for the other failure criteria.

(*b*) For the yielding factor of safety, determine an equivalent von Mises maximum stress using Equation (7–15).

$$\sigma_{\max}' = \left[\left(\frac{32(1.58)(142.4)}{\pi(0.028)^3}\right)^2 + 3\left(\frac{16(1.37)(124.3)}{\pi(0.028)^3}\right)^2\right]^{1/2} = 125.4 \text{ MPa}$$

Answer

$$n_y = \frac{S_y}{\sigma_{\max}'} = \frac{574}{125.4} = 4.58$$

For comparison, a quick and very conservative check on yielding can be obtained by replacing $\sigma_{\max}'$ with $\sigma_a' + \sigma_m'$. This just saves the extra time of calculating $\sigma_{\max}'$ if σ_a' and σ_m' have already been determined. For this example,

$$n_y = \frac{S_y}{\sigma_a' + \sigma_m'} = \frac{574}{104.4 + 69.4} = 3.30$$

which is quite conservative compared with $n_y = 4.58$.

Estimating Stress Concentrations

The stress analysis process for fatigue is highly dependent on stress concentrations. Stress concentrations for shoulders and keyways are dependent on size specifications that are not known the first time through the process. Fortunately, since these elements are usually of standard proportions, it is possible to estimate the stress-concentration factors for initial design of the shaft. These stress concentrations will be fine-tuned in successive iterations, once the details are known.

Shoulders for bearing and gear support should match the catalog recommendation for the specific bearing or gear. A look through bearing catalogs shows that a typical bearing calls for the ratio of D/d to be between 1.2 and 1.5. For a first approximation, the worst case of 1.5 can be assumed. Similarly, the fillet radius at the shoulder needs to be sized to avoid interference with the fillet radius of the mating component. There is a significant variation in typical bearings in the ratio of fillet radius versus bore diameter, with r/d typically ranging from around 0.02 to 0.06. A quick look at the stress concentration charts (Figures A–13–8 and A–13–9) shows that the stress concentrations for bending and torsion increase significantly in this range. For example, with $D/d = 1.5$ for bending, $K_t = 2.7$ at $r/d = 0.02$, and reduces to $K_t = 2.1$ at

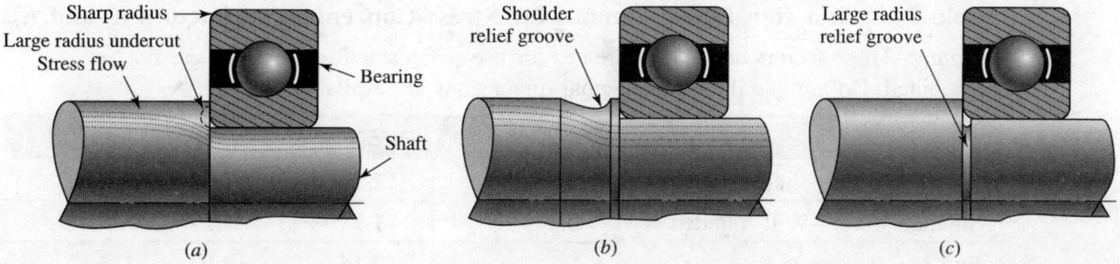

Figure 7–9

Techniques for reducing stress concentration at a shoulder supporting a bearing with a sharp radius. (*a*) Large radius undercut into the shoulder. (*b*) Large radius relief groove into the back of the shoulder. (*c*) Large radius relief groove into the small diameter.

$r/d = 0.05$, and further down to $K_t = 1.7$ at $r/d = 0.1$. This indicates that this is an area where some attention to detail could make a significant difference. Fortunately, in most cases the shear and bending moment diagrams show that bending moments are quite low near the bearings, since the bending moments from the ground reaction forces are small.

In cases where the shoulder at the bearing is found to be critical, the designer should plan to select a bearing with generous fillet radius, or consider providing for a larger fillet radius on the shaft by relieving it into the base of the shoulder as shown in Figure 7–9*a*. This effectively creates a dead zone in the shoulder area that does not carry the bending stresses, as shown by the stress flow lines. A shoulder relief groove as shown in Figure 7–9*b* can accomplish a similar purpose. Another option is to cut a large-radius relief groove into the small diameter of the shaft, as shown in Figure 7–9*c*. This has the disadvantage of reducing the cross-sectional area, but is often used in cases where it is useful to provide a relief groove before the shoulder to prevent the grinding or turning operation from having to go all the way to the shoulder.

For the standard shoulder fillet, for estimating K_t values for the first iteration, an r/d ratio should be selected so K_t values can be obtained. For the worst end of the spectrum, with $r/d = 0.02$ and $D/d = 1.5$, K_t values from the stress concentration charts for shoulders indicate 2.7 for bending, 2.2 for torsion, and 3.0 for axial.

A keyway will produce a stress concentration near a critical point where the load-transmitting component is located. The stress concentration in an end-milled keyseat is a function of the ratio of the radius r at the bottom of the groove and the shaft diameter d. For early stages of the design process, it is possible to estimate the stress concentration for keyways regardless of the actual shaft dimensions by assuming a typical ratio of $r/d = 0.02$. This gives $K_t = 2.14$ for bending and $K_{ts} = 3.0$ for torsion, assuming the key is in place.

Figures A–13–16 and A–13–17 give values for stress concentrations for flat-bottomed grooves such as used for retaining rings. By examining typical retaining ring specifications in vendor catalogs, it can be seen that the groove width is typically slightly greater than the groove depth, and the radius at the bottom of the groove is around 1/10 of the groove width. From Figures A–13–16 and A–13–17, stress-concentration factors for typical retaining ring dimensions are around 5 for bending and axial, and 3 for torsion. Fortunately, the small radius will often lead to a smaller notch sensitivity, reducing K_f.

Table 7–1 summarizes some typical stress-concentration factors for the first iteration in the design of a shaft. Similar estimates can be made for other features. The

Table 7–1 First Iteration Estimates for Stress-Concentration Factors K_t and K_{ts}

Warning: These factors are only estimates for use when actual dimensions are not yet determined. Do *not* use these once actual dimensions are available.

	Bending	Torsional	Axial
Shoulder fillet—sharp ($r/d = 0.02$)	2.7	2.2	3.0
Shoulder fillet—well rounded ($r/d = 0.1$)	1.7	1.5	1.9
End-mill keyseat ($r/d = 0.02$)	2.14	3.0	—
Sled runner keyseat	1.7	—	—
Retaining ring groove	5.0	3.0	5.0

Missing values in the table are not readily available.

point is to notice that stress concentrations are essentially normalized so that they are dependent on ratios of geometry features, not on the specific dimensions. Consequently, by estimating the appropriate ratios, the first iteration values for stress concentrations can be obtained. These values can be used for initial design, then actual values inserted once diameters have been determined.

EXAMPLE 7–2

This example problem is part of a larger case study. See Chapter 18 for the full context.

A double reduction gearbox design has developed to the point that the general layout and axial dimensions of the countershaft carrying two spur gears has been proposed, as shown in Figure 7-10. The gears and bearings are located and supported by shoulders, and held in place by retaining rings. The gears transmit torque through keys. Gears have been specified as shown, allowing the tangential and radial forces transmitted through the gears to the shaft to be determined as follows.

$$W^t_{23} = 3070 \text{ N} \qquad W^t_{54} = -13\ 850 \text{ N}$$

$$W^r_{23} = -1120 \text{ N} \qquad W^r_{54} = -5040 \text{ N}$$

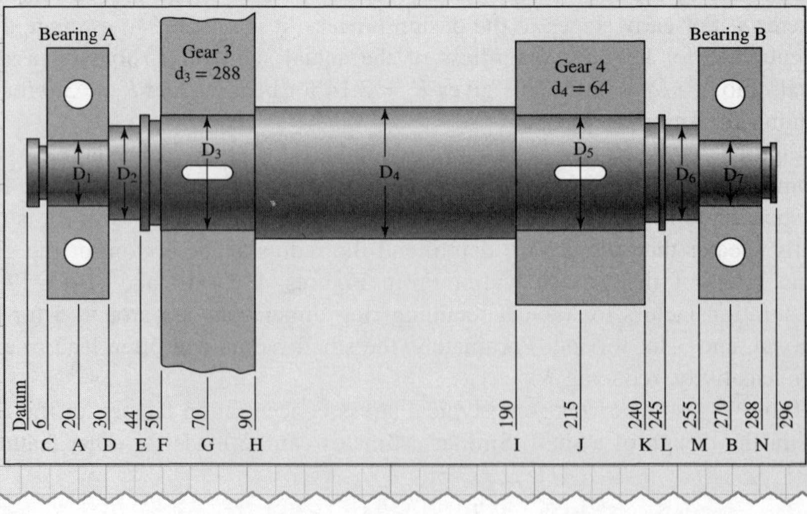

Figure 7–10
Shaft layout for Example 7–2. Dimensions in millimeters.

where the superscripts t and r represent tangential and radial directions, respectively; and the subscripts 23 and 54 represent the forces exerted by gears 2 and 5 (not shown) on gears 3 and 4, respectively.

Proceed with the next phase of the design, in which a suitable material is selected, and appropriate diameters for each section of the shaft are estimated, based on providing sufficient fatigue and static stress capacity for infinite life of the shaft, with minimum safety factors of 1.5.

Solution

Perform free body diagram analysis to get reaction forces at the bearings.

$$R_{Az} = 547 \text{ N}$$

$$R_{Ay} = 1856 \text{ N}$$

$$R_{Bz} = 10\,233 \text{ N}$$

$$R_{By} = 4304 \text{ N}$$

From ΣM_x, find the torque in the shaft between the gears,

$$T = W_{23}^t(d_3/2) = 3070(0.288/2) = 442 \text{ N} \cdot \text{m}.$$

Generate shear-moment diagrams for two planes.

Combine orthogonal planes as vectors to get total moments, e.g., at J, $\sqrt{563^2 + 237^2} = 611 \text{ N} \cdot \text{m}$.

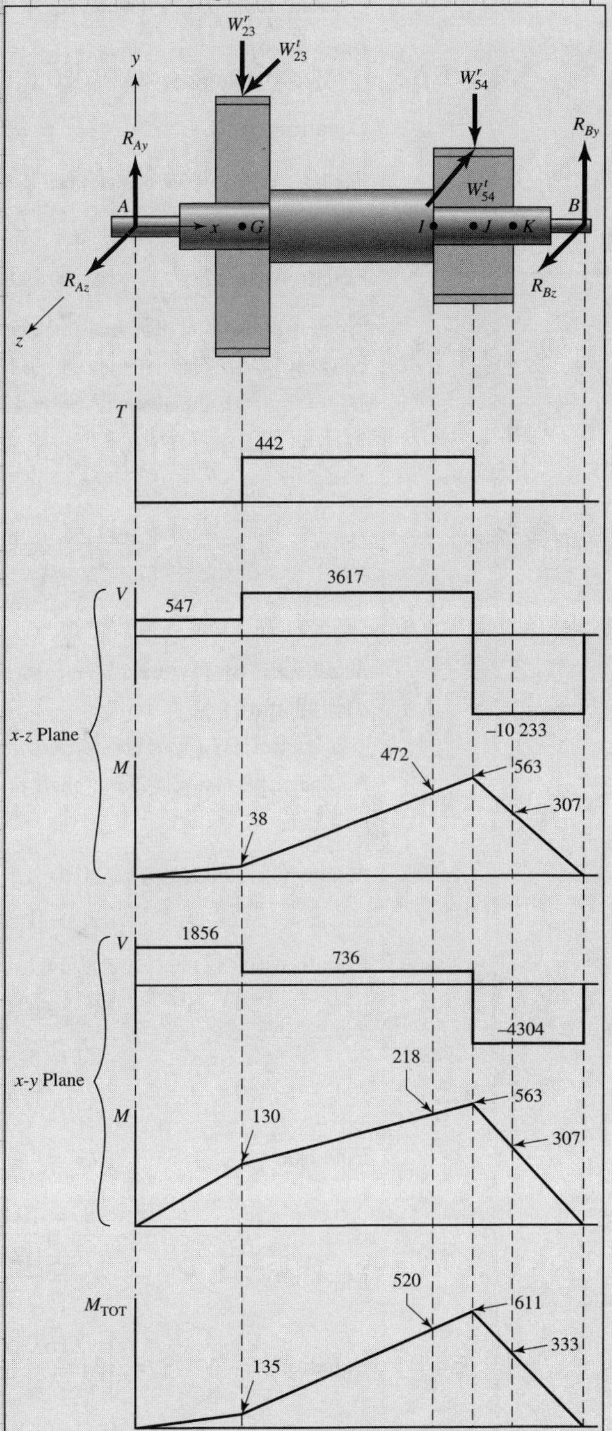

Start with point I, where the bending moment is high, there is a stress concentration at the shoulder, and the torque is present.

$$\text{At } I,\ M_a = 520\ \text{N}\cdot\text{m},\ T_m = 442\ \text{N}\cdot\text{m},\ M_m = T_a = 0$$

Assume generous fillet radius for gear at I.

From Table 7-1, estimate $K_t = 1.7$, $K_{ts} = 1.5$. For quick, conservative first pass, assume $K_f = K_t$, $K_{fs} = K_{ts}$.

Choose inexpensive steel, 1020 CD, with $S_{ut} = 469$ MPa. For S_e,

Equation (6–18) $$k_a = aS_{ut}^b = 4.51(469)^{-0.265} = 0.883$$

Guess $k_b = 0.9$. Check later when d is known.

$$k_c = k_d = k_e = 1$$

Equation (6–17) $$S_e = (0.883)(0.9)(0.5)(469) = 186\ \text{MPa}$$

For first estimate of the small diameter at the shoulder at point I, use the DE-Goodman criterion of Equation (7-8). This criterion is good for the initial design, since it is simple and conservative. With $M_m = T_a = 0$, Equations (7-6) and (7-8) reduce to

$$d = \left\{\frac{16n}{\pi}\left(\frac{2(K_f M_a)}{S_e} + \frac{[3(K_{fs}T_m)^2]^{1/2}}{S_{ut}}\right)\right\}^{1/3}$$

$$d = \left\{\frac{16(1.5)}{\pi}\left(\frac{2(1.7)(520)}{186(10^6)} + \frac{\{3[(1.5)(442)]^2\}^{1/2}}{469(10^6)}\right)\right\}^{1/3}$$

$$d = 45\ \text{mm}$$

All estimates have probably been conservative, so select the next standard size below 45 mm and check, $d = 45$ mm.

A typical D/d ratio for support at a shoulder is $D/d = 1.2$, thus, $D = 1.2(45) = 54$ mm. A nominal 54-mm cold-drawn shaft diameter can be used. Check if estimates were acceptable.

$$D/d = 54/45 = 1.2$$

Assume fillet radius $r = d/10 = 4.5$ mm, $r/d = 0.1$

$$K_t = 1.6\ (\text{Fig. A–13–9}),\ q = 0.82\ (\text{Fig. 6–26})$$

Equation (6–32) $$K_f = 1 + 0.82(1.6 - 1) = 1.49$$

$$K_{ts} = 1.35\ (\text{Fig. A–13–8}),\ q_s = 0.85\ (\text{Fig. 6–27})$$

$$K_{fs} = 1 + 0.85\ (1.35 - 1) = 1.30$$

$$k_a = 0.801\ (\text{no change})$$

Equation (6–19) $$k_b = \left(\frac{45}{7.62}\right)^{-0.107} = 0.827$$

$$S_e = (0.883)(0.827)(0.5)(469) = 171\ \text{MPa}$$

Equation (7–4) $$\sigma'_a = \frac{32K_f M_a}{\pi d^3} = \frac{32(1.49)(520)}{\pi(0.045)^3} = 87\ \text{MPa}$$

Equation (7–5) $$\sigma'_m = \left[3\left(\frac{16K_{fs}T_m}{\pi d^3}\right)^2\right]^{1/2} = \frac{\sqrt{3}(16)(1.30)(442)}{\pi(0.045)^3} = 57\ \text{MPa}$$

Using Goodman criterion

$$\frac{1}{n_f} = \frac{\sigma'_a}{S_e} + \frac{\sigma'_m}{S_{ut}} = \frac{87}{171} + \frac{57}{469} = 0.63$$

$$n_f = 1.59$$

Note that we could have used Equation (7-7) directly.

Check yielding.

$$n_y = \frac{S_y}{\sigma'_{\max}} > \frac{S_y}{\sigma'_a + \sigma'_m} = \frac{393}{87 + 57} = 2.73$$

Also check this diameter at the end of the keyway, just to the right of point I, and at the groove at point K. From moment diagram, estimate M at end of keyway to be $M = 530 \text{ N}\cdot\text{m}$.

Assume the radius at the bottom of the keyway will be the standard $r/d = 0.02$, $r = 0.02d = 0.02(45) = 0.9$ mm.

$$K_t = 2.14 \text{ (Table 7–1)}, \; q = 0.67 \text{ (Figure 6–26)}$$

$$K_f = 1 + 0.67(2.14 - 1) = 1.76$$

$$K_{ts} = 3.0 \text{ (Table 7–1)}, \; q_s = 0.9 \text{ (Figure 6–27)}$$

$$K_{fs} = 1 + 0.9(3 - 1) = 2.8$$

$$\sigma'_a = \frac{32K_f M_a}{\pi d^3} = \frac{32(1.76)(530)}{\pi(0.045)^3} = 104 \text{ MPa}$$

$$\sigma'_m = \sqrt{3}(16)\frac{K_{fs}T_m}{\pi d^3} = \frac{\sqrt{3}(16)(2.8)(442)}{\pi(0.045)^3} = 120 \text{ MPa}$$

$$\frac{1}{n_f} = \frac{\sigma'_a}{S_e} + \frac{\sigma'_m}{S_{ut}} = \frac{104}{171} + \frac{120}{469} = 0.86$$

$$n_f = 1.16$$

The keyway turns out to be more critical than the shoulder. We can either increase the diameter or use a higher strength material. Unless the deflection analysis shows a need for larger diameters, let us choose to increase the strength. We started with a very low strength and can afford to increase it some to avoid larger sizes. Try 1050 CD with $S_{ut} = 690$ MPa.

Recalculate factors affected by S_{ut}, i.e., $k_a \rightarrow S_e$; $q \rightarrow K_f \rightarrow \sigma'_a$

$$k_a = 4.51(690)^{-0.265} = 0.797, \quad S_e = 0.797(0.825)(0.5)(690) = 227 \text{ MPa}$$

$$q = 0.75, \; K_f = 1 + 0.75(2.14 - 1) = 1.86$$

$$\sigma'_a = \frac{32(1.86)(530)}{\pi(0.045)^3} = 110 \text{ MPa}$$

$$\frac{1}{n_f} = \frac{110}{227} + \frac{120}{690} = 0.74$$

$$n_f = 1.35$$

This is slightly under the goal for the design factor to be 1.5. If we round to 2 significant figures, which is actually more realistic for fatigue, then we get 1.5. If the situation calls for a more conservative decision, a higher strength material can be selected.

Check at the groove at K, since K_t for flat-bottomed grooves are often very high. From the torque diagram, note that no torque is present at the groove. From the moment diagram, $M_a = 333 \text{ N}\cdot\text{m}$,

$M_m = T_a = T_m = 0$. To quickly check if this location is potentially critical, just use $K_f = K_t = 5.0$ as an estimate, from Table 7-1.

$$\sigma_a = \frac{32K_fM_a}{\pi d^3} = \frac{32(5)(333)}{\pi(0.045)^3} = 186 \text{ MPa}$$

$$n_f = \frac{S_e}{\sigma_a} = \frac{227}{186} = 1.22$$

This is low. We will look up data for a specific retaining ring to obtain K_f more accurately. With a quick online search of a retaining ring specification using the website www.globalspec.com, appropriate groove specifications for a retaining ring for a shaft diameter of 45 mm are obtained as follows: width, $a = 1.85$ mm; depth, $t = 1.25$ mm; and corner radius at bottom of groove, $r = 0.25$ mm. From Figure A-13-16, with $r/t = 0.25/1.25 = 0.2$, and $a/t = 1.85/1.25 = 1.48$

$$K_t = 4.4,\ q = 0.65 \text{ (Figure 6–26)}$$

$$K_f = 1 + 0.65(4.4 - 1) = 3.21$$

$$\sigma_a = \frac{32K_fM_a}{\pi d^3} = \frac{32(3.21)(333)}{\pi(0.045)^3} = 119 \text{ MPa}$$

$$n_f = \frac{S_e}{\sigma_a} = \frac{227}{119} = 1.91$$

Quickly check if point M might be critical. Only bending is present, and the moment is small, but the diameter is small and the stress concentration is high for a sharp fillet required for a bearing. From the moment diagram, $M_a = 133$ MPa, and $M_m = T_m = T_a = 0$.

Estimate $K_t = 2.7$ from Table 7-1, $d = 30$ mm, and fillet radius r to fit a typical bearing.

$$r/d = 0.02,\ r = 0.02(30) = 0.6 \text{ mm}$$

$$q = 0.7 \text{ (Figure 6–26)}$$

$$K_f = 1 + (0.7)(2.7 - 1) = 2.19$$

$$\sigma_a = \frac{32K_fM_a}{\pi d^3} = \frac{32(2.19)(133)}{\pi(0.03)^3} = 110 \text{ MPa}$$

$$n_f = \frac{S_e}{\sigma_a} = \frac{227}{110} = 2.06$$

This location is more critical than perhaps anticipated. The estimate for stress concentration is likely on the high side, so we will choose to continue and recheck after a specific bearing is selected.

With the diameters specified for the critical locations, fill in trial values for the rest of the diameters, taking into account typical shoulder heights for bearing and gear support.

$$D_1 = D_7 = 30 \text{ mm}$$

$$D_2 = D_6 = 36 \text{ mm}$$

$$D_3 = D_5 = 45 \text{ mm}$$

$$D_4 = 54 \text{ mm}$$

The bending moments are much less on the left end of the shaft, so D_1, D_2, and D_3 could be smaller. However, unless weight is an issue, there is little advantage to requiring more material removal. Also, the extra rigidity may be needed to keep deflections small.

7–5 Deflection Considerations

Deflection analysis at even a single point of interest requires complete geometry information for the entire shaft. For this reason, it is desirable to design the dimensions at critical locations to handle the stresses, and fill in reasonable estimates for all other dimensions, before performing a deflection analysis. Deflection of the shaft, both linear and angular, should be checked at gears and bearings. Allowable deflections will depend on many factors, and bearing and gear catalogs should be used for guidance on allowable misalignment for specific bearings and gears. As a rough guideline, typical ranges for maximum slopes and transverse deflections of the shaft centerline are given in Table 7–2. The allowable transverse deflections for spur gears are dependent on the size of the teeth, as represented by the diametral pitch P, which equals the number of teeth divided by the pitch diameter.

In Section 4–4 several beam deflection methods are described. For shafts, where the deflections may be sought at a number of different points, integration using either singularity functions or numerical integration is practical. In a stepped shaft, the cross-sectional properties change along the shaft at each step, increasing the complexity of integration, since both M and I vary. Fortunately, only the gross geometric dimensions need to be included, as the local factors such as fillets, grooves, and keyways do not have much impact on deflection. Example 4–7 demonstrates the use of singularity functions for a stepped shaft. Many shafts will include forces in multiple planes, requiring either a three-dimensional analysis, or the use of superposition to obtain deflections in two planes which can then be summed as vectors.

A deflection analysis is straightforward, but it is lengthy and tedious to carry out manually, particularly for multiple points of interest. Consequently, practically all shaft deflection analysis will be evaluated with the assistance of software. Any general-purpose finite-element software can readily handle a shaft problem (see Chapter 19). This is practical if the designer is already familiar with using the software and with how to properly model the shaft. Special-purpose software solutions for 3-D shaft analysis are available, but somewhat expensive if only used occasionally. Software requiring very little training is readily available for planar beam analysis, and can be downloaded from the Internet. Example 7–3 demonstrates how to incorporate such a program for a shaft with forces in multiple planes.

Table 7–2 Typical Maximum Ranges for Slopes and Transverse Deflections

Slopes	
Tapered roller	0.0005–0.0012 rad
Cylindrical roller	0.0008–0.0012 rad
Deep-groove ball	0.001–0.003 rad
Spherical ball	0.026–0.052 rad
Self-align ball	0.026–0.052 rad
Uncrowned spur gear	$<$0.0005 rad
Transverse Deflections	
Spur gears with $P < 10$ teeth/cm	0.25 mm
Spur gears with $11 < P < 19$	0.125 mm
Spur gears with $20 < P < 50$	0.075 mm

EXAMPLE 7–3

This example problem is part of a larger case study. See Chapter 18 for the full context.

In Example 7-2, a preliminary shaft geometry was obtained on the basis of design for stress. The resulting shaft is shown in Figure 7-10, with proposed diameters of

$$D_1 = D_7 = 30 \text{ mm}$$
$$D_2 = D_6 = 36 \text{ mm}$$
$$D_3 = D_5 = 45 \text{ mm}$$
$$D_4 = 54 \text{ mm}$$

Check that the deflections and slopes at the gears and bearings are acceptable. If necessary, propose changes in the geometry to resolve any problems.

Solution

A simple planar beam analysis program will be used. By modeling the shaft twice, with loads in two orthogonal planes, and combining the results, the shaft deflections can readily be obtained. For both planes, the material is selected (steel with $E = 210$ GPa), the shaft lengths and diameters are entered, and the bearing locations are specified. Local details like grooves and keyways are ignored, as they will have insignificant effect on the deflections. Then the tangential gear forces are entered in the horizontal xz plane model, and the radial gear forces are entered in the vertical xy plane model. The software can calculate the bearing reaction forces, and numerically integrate to generate plots for shear, moment, slope, and deflection, as shown in Figure 7-11.

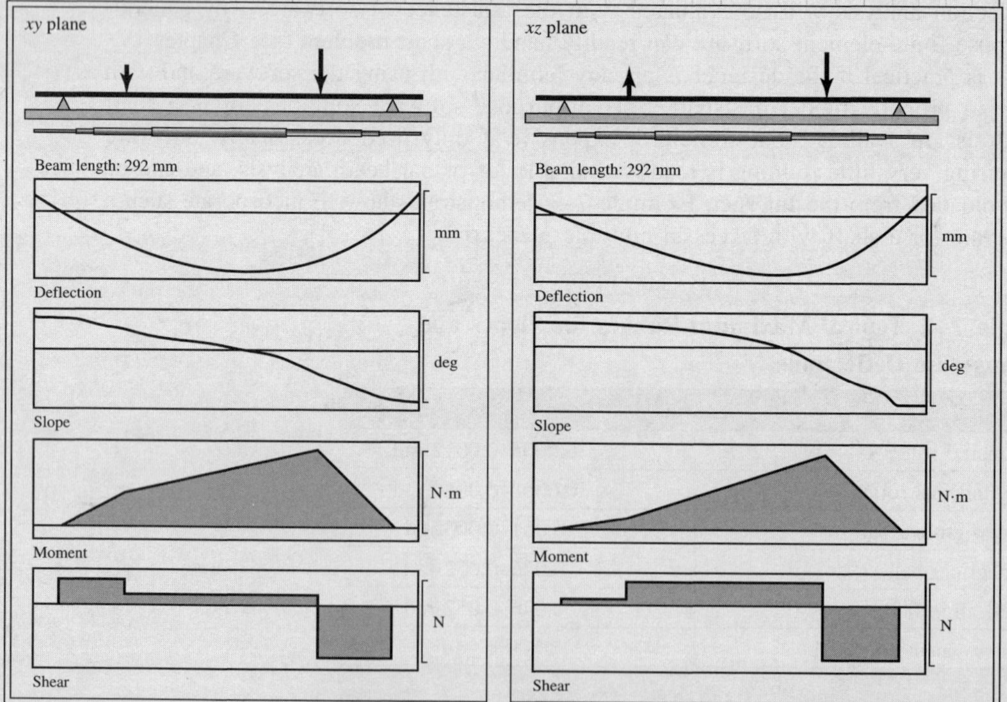

Figure 7–11

Shear, moment, slope, and deflection plots from two planes. *(Source: Beam 2D Stress Analysis, ORAND System, Inc.)*

Table 7–3 Slope and Deflection Values at Key Locations

Point of InteRest	xz Plane	xy Plane	Total
Left bearing slope	0.0193 deg	0.0161 deg	0.0251 deg 0.000 439 rad
Right bearing slope	0.0482 deg	0.0226 deg	0.0533 deg 0.000 93 rad
Left gear slope	0.0177 deg	0.0099 deg	0.0203 deg 0.000 355 rad
Right gear slope	0.0198 deg	0.0104 deg	0.022 deg 0.000 391 rad
Left gear deflection	0.0175 mm	0.0135 mm	0.0221 mm
Right gear deflection	0.0375 mm	0.018 mm	0.0416 mm

The deflections and slopes at points of interest are obtained from the plots, and combined with orthogonal vector addition, that is, $\delta = \sqrt{\delta_{xz}^2 + \delta_{xy}^2}$. Results are shown in Table 7-3.

Whether these values are acceptable will depend on the specific bearings and gears selected, as well as the level of performance expected. According to the guidelines in Table 7-2, all of the bearing slopes are well below typical limits for ball bearings. The right bearing slope is within the typical range for cylindrical bearings. Since the load on the right bearing is relatively high, a cylindrical bearing might be used. This constraint should be checked against the specific bearing specifications once the bearing is selected.

The gear slopes and deflections more than satisfy the limits recommended in Table 7-2. It is recommended to proceed with the design, with an awareness that changes that reduce rigidity should warrant another deflection check.

Once deflections at various points have been determined, if any value is larger than the allowable deflection at that point, a larger shaft diameter is warranted. Since I is proportional to d^4, a new diameter can be found from

$$d_{\text{new}} = d_{\text{old}} \left| \frac{n_d y_{\text{old}}}{y_{\text{all}}} \right|^{1/4} \tag{7–17}$$

where y_{all} is the allowable deflection at that station and n_d is the design factor. Similarly, if any slope is larger than the allowable slope θ_{all}, a new diameter can be found from

$$d_{\text{new}} = d_{\text{old}} \left| \frac{n_d (dy/dx)_{\text{old}}}{(\text{slope})_{\text{all}}} \right|^{1/4} \tag{7–18}$$

where $(\text{slope})_{\text{all}}$ is the allowable slope. As a result of these calculations, determine the largest $d_{\text{new}}/d_{\text{old}}$ ratio, then multiply *all* diameters by this ratio. The tight constraint will be just tight, and all others will be loose. Don't be too concerned about end journal sizes, as their influence is usually negligible. The beauty of the method is that the deflections need to be completed just once and constraints can be rendered loose but for one, with diameters all identified without reworking every deflection.

EXAMPLE 7–4

For the shaft in Example 7–3, it was noted that the slope at the right bearing is near the limit for a cylindrical roller bearing. Determine an appropriate increase in diameters to bring this slope down to 0.0005 rad.

Solution

Applying Equation (7–17) to the deflection at the right bearing gives

$$d_{\text{new}} = d_{\text{old}} \left| \frac{n_d \text{slope}_{\text{old}}}{\text{slope}_{\text{all}}} \right|^{1/4} = 30 \left| \frac{(1)(0.000\ 930)}{(0.0005)} \right|^{1/4} = 35 \text{ mm}$$

Multiplying all diameters by the ratio

$$\frac{d_{\text{new}}}{d_{\text{old}}} = \frac{35}{30} = 1.167$$

gives a new set of diameters,

$$D_1 = D_7 = 35 \text{ mm}$$

$$D_2 = D_6 = 42 \text{ mm}$$

$$D_3 = D_5 = 52.5 \text{ mm}$$

$$D_4 = 63 \text{ mm}$$

Repeating the beam deflection analysis of Example 7–3 with these new diameters produces a slope at the right bearing of 0.0005 rad, with all other deflections less than their previous values.

The transverse shear V at a section of a beam in flexure imposes a shearing deflection, which is superposed on the bending deflection. Usually such shearing deflection is less than 1 percent of the transverse bending deflection, and it is seldom evaluated. However, when the shaft length-to-diameter ratio is less than 10, the shear component of transverse deflection merits attention. There are many short shafts. A tabular method is explained in detail elsewhere,[2] including examples.

For right-circular cylindrical shafts in torsion the angular deflection θ is given in Equation (4–5). For a stepped shaft with individual cylinder length l_i and torque T_i, the angular deflection can be estimated from

$$\theta = \sum \theta_i = \sum \frac{T_i l_i}{G_i J_i} \tag{7–19}$$

or, for a constant torque throughout homogeneous material, from

$$\theta = \frac{T}{G} \sum \frac{l_i}{J_i} \tag{7–20}$$

This should be treated only as an estimate, since experimental evidence shows that the actual θ is larger than given by Equations (7–19) and (7–20).[3]

---

[2]C. R. Mischke, "Tabular Method for Transverse Shear Deflection," Section 17.3 in Joseph E. Shigley, Charles R. Mischke, and Thomas H. Brown, Jr. (eds.), *Standard Handbook of Machine Design*, 3rd ed., McGraw Hill, New York, 2004.

[3]R. Bruce Hopkins, *Design Analysis of Shafts and Beams*, McGraw Hill, New York, 1970, pp. 93–99.

If torsional stiffness is defined as $k_i = T_i/\theta_i$ and, since $\theta_i = T_i/k_i$ and $\theta = \Sigma\theta_i = \Sigma(T_i/k_i)$, for constant torque $\theta = T\Sigma(1/k_i)$, it follows that the torsional stiffness of the shaft k in terms of segment stiffnesses is

$$\frac{1}{k} = \sum \frac{1}{k_i} \tag{7–21}$$

7–6 Critical Speeds for Shafts

When a shaft is turning, eccentricity causes a centrifugal force deflection, which is resisted by the shaft's flexural rigidity *EI*. As long as deflections are small, no harm is done. Another potential problem, however, is called *critical speeds:* at certain speeds the shaft is unstable, with deflections increasing without upper bound. It is fortunate that although the dynamic deflection shape is unknown, using a static deflection curve gives an excellent estimate of the lowest critical speed. Such a curve meets the boundary condition of the differential equation (zero moment and deflection at both bearings) and the shaft energy is not particularly sensitive to the exact shape of the deflection curve. Designers seek first critical speeds at least twice the operating speed.

The shaft, because of its own mass, has a critical speed. The ensemble of attachments to a shaft likewise has a critical speed that is much lower than the shaft's intrinsic critical speed. Estimating these critical speeds (and harmonics) is a task of the designer. When geometry is simple, as in a shaft of uniform diameter, simply supported, the task is easy. It can be expressed[4] as

$$\omega_1 = \left(\frac{\pi}{l}\right)^2 \sqrt{\frac{EI}{m}} = \left(\frac{\pi}{l}\right)^2 \sqrt{\frac{gEI}{A\gamma}} \tag{7–22}$$

where m is the mass per unit length, A the cross-sectional area, and γ the specific weight. For an ensemble of attachments, Rayleigh's method for lumped masses gives[5]

$$\omega_1 = \sqrt{\frac{g\Sigma w_i y_i}{\Sigma w_i y_i^2}} \tag{7–23}$$

where w_i is the weight of the ith location and y_i is the deflection at the ith body location. It is possible to use Equation (7–23) for the case of Equation (7–22) by partitioning the shaft into segments and placing its weight force at the segment centroid as seen in Figure 7–12. Computer assistance is often used to lessen the difficulty in finding transverse deflections of a stepped shaft. Rayleigh's equation overestimates the critical speed.

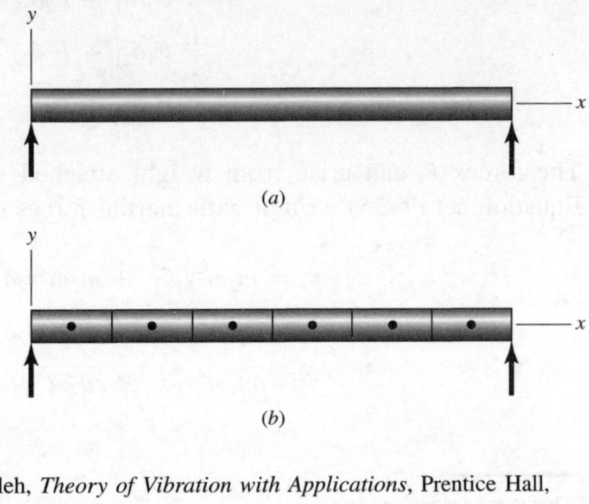

Figure 7–12

(*a*) A uniform-diameter shaft for Equation (7–22). (*b*) A segmented uniform-diameter shaft for Equation (7–23).

[4]William T. Thomson and Marie Dillon Dahleh, *Theory of Vibration with Applications*, Prentice Hall, 5th ed., 1998, p. 273.

[5]Thomson, op. cit., p. 357.

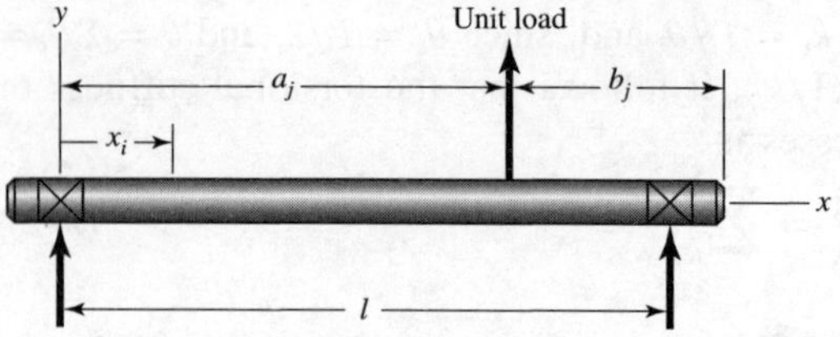

Figure 7–13

The influence coefficient δ_{ij} is the deflection at i due to a unit load at j.

To counter the increasing complexity of detail, we adopt a useful viewpoint. Inasmuch as the shaft is an elastic body, we can use *influence coefficients*. An influence coefficient is the transverse deflection at location i on a shaft due to a unit load at location j on the shaft. From Table A–9–6 we obtain, for a simply supported beam with a single unit load as shown in Figure 7–13,

$$\delta_{ij} = \begin{cases} \dfrac{b_j x_i}{6EIl}\left(l^2 - b_j^2 - x_i^2\right) & x_i \le a_i \\[2ex] \dfrac{a_j(l - x_i)}{6EIl}\left(2lx_i - a_j^2 - x_i^2\right) & x_i > a_i \end{cases} \tag{7–24}$$

For three loads the influence coefficients may be displayed as

	j		
i	**1**	**2**	**3**
1	δ_{11}	δ_{12}	δ_{13}
2	δ_{21}	δ_{22}	δ_{23}
3	δ_{31}	δ_{32}	δ_{33}

Maxwell's reciprocity theorem[6] states that there is a symmetry about the main diagonal, composed of δ_{11}, δ_{22}, and δ_{33}, of the form $\delta_{ij} = \delta_{ji}$. This relation reduces the work of finding the influence coefficients. From the influence coefficients above, one can find the deflections y_1, y_2, and y_3 of Equation (7–23) as follows:

$$\begin{aligned} y_1 &= F_1\delta_{11} + F_2\delta_{12} + F_3\delta_{13} \\ y_2 &= F_1\delta_{21} + F_2\delta_{22} + F_3\delta_{23} \\ y_3 &= F_1\delta_{31} + F_2\delta_{32} + F_3\delta_{33} \end{aligned} \tag{7–25}$$

The forces F_i can arise from weight attached w_i or centrifugal forces $m_i\omega^2 y_i$. The Equation set (7–25) written with inertial forces can be displayed as

$$\begin{aligned} y_1 &= m_1\omega^2 y_1\delta_{11} + m_2\omega^2 y_2\delta_{12} + m_3\omega^2 y_3\delta_{13} \\ y_2 &= m_1\omega^2 y_1\delta_{21} + m_2\omega^2 y_2\delta_{22} + m_3\omega^2 y_3\delta_{23} \\ y_3 &= m_1\omega^2 y_1\delta_{31} + m_2\omega^2 y_2\delta_{32} + m_3\omega^2 y_3\delta_{33} \end{aligned}$$

[6]Thomson, op. cit., p. 167.

which can be rewritten as

$$
\begin{aligned}
(m_1\delta_{11} - 1/\omega^2)y_1 + (m_2\delta_{12})y_2 + (m_3\delta_{13})y_3 &= 0 \\
(m_1\delta_{21})y_1 + (m_2\delta_{22} - 1/\omega^2)y_2 + (m_3\delta_{23})y_3 &= 0 \\
(m_1\delta_{31})y_1 + (m_2\delta_{32})y_2 + (m_3\delta_{33} - 1/\omega^2)y_3 &= 0
\end{aligned} \tag{a}
$$

Equation set (a) is three simultaneous equations in terms of y_1, y_2, and y_3. To avoid the trivial solution $y_1 = y_2 = y_3 = 0$, the determinant of the coefficients of y_1, y_2, and y_3 must be zero (eigenvalue problem). Thus,

$$
\begin{vmatrix}
(m_1\delta_{11} - 1/\omega^2) & m_2\delta_{12} & m_3\delta_{13} \\
m_1\delta_{21} & (m_2\delta_{22} - 1/\omega^2) & m_3\delta_{23} \\
m_1\delta_{31} & m_2\delta_{32} & (m_3\delta_{33} - 1/\omega^2)
\end{vmatrix} = 0 \tag{7–26}
$$

which says that a deflection other than zero exists only at three distinct values of ω, the critical speeds. Expanding the determinant, we obtain

$$
\left(\frac{1}{\omega^2}\right)^3 - (m_1\delta_{11} + m_2\delta_{22} + m_3\delta_{33})\left(\frac{1}{\omega^2}\right)^2 + \cdots = 0 \tag{7–27}
$$

The three roots of Equation (7–27) can be expressed as $1/\omega_1^2$, $1/\omega_2^2$, and $1/\omega_3^2$. Thus Equation (7–27) can be written in the form

$$
\left(\frac{1}{\omega^2} - \frac{1}{\omega_1^2}\right)\left(\frac{1}{\omega^2} - \frac{1}{\omega_2^2}\right)\left(\frac{1}{\omega^2} - \frac{1}{\omega_3^2}\right) = 0
$$

or

$$
\left(\frac{1}{\omega^2}\right)^3 - \left(\frac{1}{\omega_1^2} + \frac{1}{\omega_2^2} + \frac{1}{\omega_3^2}\right)\left(\frac{1}{\omega^2}\right)^2 + \cdots = 0 \tag{7–28}
$$

Comparing Equations (7–27) and (7–28) we see that

$$
\frac{1}{\omega_1^2} + \frac{1}{\omega_2^2} + \frac{1}{\omega_3^2} = m_1\delta_{11} + m_2\delta_{22} + m_3\delta_{33} \tag{7–29}
$$

If we had only a single mass m_1 alone, the critical speed would be given by $1/\omega^2 = m_1\delta_{11}$. Denote this critical speed as ω_{11} (which considers only m_1 acting alone). Likewise for m_2 or m_3 acting alone, we similarly define the terms $1/\omega_{22}^2 = m_2\delta_{22}$ or $1/\omega_{33}^2 = m_3\delta_{33}$, respectively. Thus, Equation (7–29) can be rewritten as

$$
\frac{1}{\omega_1^2} + \frac{1}{\omega_2^2} + \frac{1}{\omega_3^2} = \frac{1}{\omega_{11}^2} + \frac{1}{\omega_{22}^2} + \frac{1}{\omega_{33}^2} \tag{7–30}
$$

If we order the critical speeds such that $\omega_1 < \omega_2 < \omega_3$, then $1/\omega_1^2$ is much greater than $1/\omega_2^2$ and $1/\omega_3^2$. So the first, or fundamental, critical speed ω_1 can be approximated by

$$
\frac{1}{\omega_1^2} \approx \frac{1}{\omega_{11}^2} + \frac{1}{\omega_{22}^2} + \frac{1}{\omega_{33}^2} \tag{7–31}
$$

This idea can be extended to an n-body shaft:

$$
\frac{1}{\omega_1^2} \approx \sum_{i=1}^{n} \frac{1}{\omega_{ii}^2} \tag{7–32}
$$

This is called *Dunkerley's equation.* By ignoring the higher mode term(s), the first critical speed estimate is *lower* than actually is the case.

Since Equation (7–32) has no loads appearing in the equation, it follows that if each load could be placed at some convenient location transformed into an equivalent load, then the critical speed of an array of loads could be found by summing the equivalent loads, all placed at a single convenient location. For the load at station 1, placed at the center of span, denoted with the subscript *c*, the equivalent load is found from

$$\omega_{11}^2 = \frac{1}{m_1\delta_{11}} = \frac{g}{w_1\delta_{11}} = \frac{g}{w_{1c}\delta_{cc}}$$

or

$$w_{1c} = w_1\frac{\delta_{11}}{\delta_{cc}} \tag{7–33}$$

EXAMPLE 7–5

Consider a simply supported steel shaft as depicted in Figure 7–14, with 25 mm diameter and a 790 mm span between bearings, carrying two gears weighing 155 N and 245 N.
(*a*) Find the influence coefficients.
(*b*) Find Σwy and Σwy^2 and the first critical speed using Rayleigh's equation, Equation (7–23).
(*c*) From the influence coefficients, find ω_{11} and ω_{22}.
(*d*) Using Dunkerley's equation, Equation (7–32), estimate the first critical speed.
(*e*) Use superposition to estimate the first critical speed.
(*f*) Estimate the shaft's intrinsic critical speed. Suggest a modification to Dunkerley's equation to include the effect of the shaft's mass on the first critical speed of the attachments.

Solution

(*a*)

$$I = \frac{\pi d^4}{64} = \frac{\pi(25)^4}{64} = 19\ 175\ \text{mm}^4$$

$$6EIl = 6(207\ 000)(19\ 175)790 = 1.88(10^{13})\ \text{N}\cdot\text{mm}^3$$

From Equation set (7–24),

$$\delta_{11} = \frac{610(180)(790^2 - 610^2 - 180^2)}{1.88(10^{13})} = 2.17(10^{-3})\ \text{mm/N}$$

$$\delta_{22} = \frac{280(510)(790^2 - 280^2 - 510^2)}{1.88(10^{-13})} = 1.38(10^{-3})\ \text{mm/N}$$

$$\delta_{12} = \delta_{21} = \frac{280(180)(790^2 - 280^2 - 180^2)}{1.88(10^{13})} = 1.38(10^{-3})\ \text{mm/N}$$

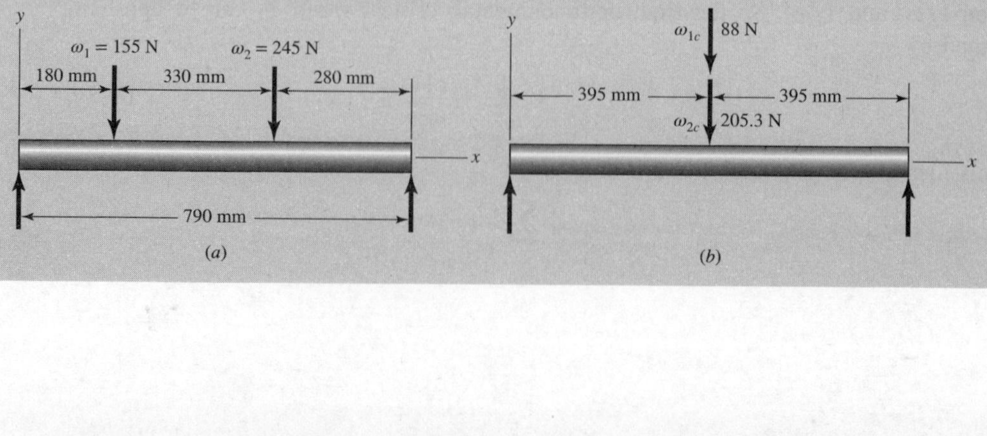

Figure 7–14
(*a*) A 1-in uniform-diameter shaft for Example 7–5.
(*b*) Superposing of equivalent loads at the center of the shaft for the purpose of finding the first critical speed.

Answer

i \ *j*	1	2
1	$1.47(10^{-3})$	$1.38(10^{-3})$
2	$1.38(10^{-3})$	$2.17(10^{-3})$

$$y_1 = w_1\delta_{11} + w_2\delta_{12} = 155(1.47)10^{-3} + 245(1.38)10^{-3} = 0.566 \text{ mm}$$

$$y_2 = w_1\delta_{21} + w_2\delta_{22} = 155(1.38)10^{-3} + 245(2.17)10^{-3} = 0.746 \text{ mm}$$

(*b*)

$$\sum w_i y_i = 155(0.566) + 245(0.746) = 270.5 \text{ N} \cdot \text{mm}$$

Answer

$$\sum w_i y_i^2 = 155(0.566^2) + 245(0.746^2) = 186 \text{ N} \cdot \text{mm}^2$$

Answer

$$\omega = \sqrt{\frac{9810(270.5)}{186}} = 119.4 \text{ rad/s, or } 1141 \text{ rev/min}$$

(*c*)

Answer

$$\frac{1}{\omega_{11}^2} = \frac{w_1}{g}\delta_{11}$$

$$\omega_{11} = \sqrt{\frac{g}{w_1\delta_{11}}} = \sqrt{\frac{9810}{155(1.47)10^{-3}}} = 207.5 \text{ rad/s, or } 1981 \text{ rev/min}$$

Answer

$$\omega_{22} = \sqrt{\frac{g}{w_2\delta_{22}}} = \sqrt{\frac{9810}{245(2.17)10^{-3}}} = 135 \text{ rad/s, or } 1297 \text{ rev/min}$$

(*d*)

$$\frac{1}{\omega_1^2} \approx \sum \frac{1}{\omega_{ii}^2} = \frac{1}{207.5^2} + \frac{1}{135^2} = 7.745(10^{-5}) \qquad (1)$$

Answer

$$\omega_1 \approx \sqrt{\frac{1}{7.745(10^{-5})}} = 113.6 \text{ rad/s, or } 1085 \text{ rev/min}$$

which is less than part *b*, as expected.

(*e*) From Equation (7–24),

$$\delta_{cc} = \frac{b_{cc}x_{cc}(l^2 - b_{cc}^2 - x_{cc}^2)}{6EIl} = \frac{395(395)(790^2 - 395^2 - 395)}{1.88(10^{13})}$$

$$= 2.59(10^{-3})$$

From Equation (7–33),

$$w_{1c} = w_1\frac{\delta_{11}}{\delta_{cc}} = 155\frac{1.47(10^{-3})}{2.59(10^{-3})} = 88 \text{ N}$$

$$w_{2c} = w_2\frac{\delta_{22}}{\delta_{cc}} = 245\frac{2.17(10^{-3})}{2.59(10^{-3})} = 205.3 \text{ N}$$

Answer

$$\omega = \sqrt{\frac{g}{\delta_{cc}\sum w_{ic}}} = \sqrt{\frac{9810}{2.59(10^{-3})(88 + 205.3)}} = 113.6 \text{ rad/s, or } 1085 \text{ rev/min}$$

which, except for rounding, agrees with part *d*, as expected.

(*f*) For the shaft, $E = 207\ 000$ MPa, $\gamma = 7.85(10^{-6})$ Kg/mm^3, and $A = \pi(25^2)/4 = 491$ mm^2. Considering the shaft alone, the critical speed, from Equation (7–22), is

Answer
$$\omega_s = \left(\frac{\pi}{l}\right)^2\sqrt{\frac{gEI}{A\gamma}} = \left(\frac{\pi}{0.79}\right)^2\sqrt{\frac{207\ 000(19\ 175)10^{-12}(10^6)}{491(10^{-6})7850}}$$
$$= 507.5 \text{ rad/s, or } 4846 \text{ rev/min}$$

We can simply add $1/\omega_s^2$ to the right side of Dunkerley's equation, Equation (1), to include the shaft's contribution,

Answer
$$\frac{1}{\omega_1^2} \approx \frac{1}{507.5^2} + 7.745(10^{-5}) = 8.133(10^{-5})$$
$$\omega_1 \approx 110.9 \text{ rad/s, or } 1059 \text{ rev/min}$$

which is slightly less than part *d*, as expected.

The shaft's first critical speed ω_s is just one more single effect to add to Dunkerley's equation. Since it does not fit into the summation, it is usually written up front.

Answer
$$\frac{1}{\omega_1^2} \approx \frac{1}{\omega_s^2} + \sum_{i=1}^{n}\frac{1}{\omega_{ii}^2} \tag{7–34}$$

Common shafts are complicated by the stepped-cylinder geometry, which makes the influence-coefficient determination part of a numerical solution.

7–7 Miscellaneous Shaft Components

Setscrews

Unlike bolts and cap screws, which depend on tension to develop a clamping force, the setscrew depends on compression to develop the clamping force. The resistance to axial motion of the collar or hub relative to the shaft is called *holding power*. This holding power, which is really a force resistance, is due to frictional resistance of the contacting portions of the collar and shaft as well as any slight penetration of the setscrew into the shaft.

Figure 7–15 shows the point types available with socket setscrews. These are also manufactured with screwdriver slots and with square heads.

Table 7–4 lists values of the seating torque and the corresponding holding power for inch-series setscrews. The values listed apply to both axial holding power, for resisting thrust, and the tangential holding power, for resisting torsion. Typical factors of safety are 1.5 to 2.0 for static loads and 4 to 8 for various dynamic loads.

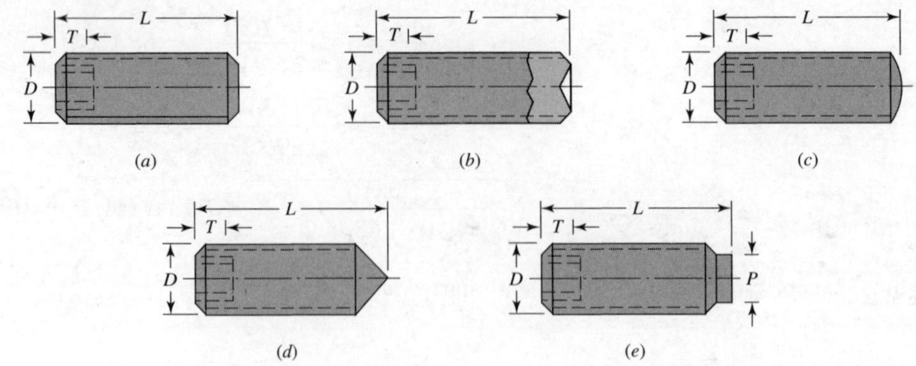

Figure 7–15

Socket setscrews: (*a*) flat point; (*b*) cup point; (*c*) oval point; (*d*) cone point; (*e*) half-dog point.

Table 7–4 Typical Holding Power (Force) for Socket Setscrews*

Size, in	Seating Torque, N · m	Holding Power, N
#0	0.11	222
#1	0.2	289
#2	0.2	378
#3	0.5	534
#4	0.5	712
#5	1.1	890
#6	1.1	1112
#8	2.2	1713
#10	4.0	2403
$\frac{1}{4}$	9.8	4450
$\frac{5}{16}$	18.6	6675
$\frac{3}{8}$	32.8	8900
$\frac{7}{16}$	48.6	11 125
$\frac{1}{2}$	70.0	13 350
$\frac{9}{16}$	70.0	15 575
$\frac{5}{8}$	149.7	17 800
$\frac{3}{4}$	271.2	22 250
$\frac{7}{8}$	587.6	26 700
1	813.6	31 150

Source: Unbrako Division, SPS Technologies, Jenkintown, Pa.
*Based on alloy-steel screw against steel shaft, class 3A coarse or fine threads in class 2B holes, and cup-point socket setscrews.

Setscrews should have a length of about half of the shaft diameter. Note that this practice also provides a rough rule for the radial thickness of a hub or collar.

Keys and Pins

Keys and pins are used on shafts to secure rotating elements, such as gears, pulleys, or other wheels. Keys are used to enable the transmission of torque from the shaft to the shaft-supported element. Pins are used for axial positioning and for the transfer of torque or thrust or both.

Figure 7–16 shows a variety of keys and pins. Pins are useful when the principal loading is shear and when both torsion and thrust are present. Taper pins are sized according to the diameter at the large end. Some of the most useful sizes of these are listed in Table 7–5. The diameter at the small end is

$$d = D - 0.0208L \qquad (7\text{–}35)$$

where d = diameter at small end, mm
D = diameter at large end, mm
L = length, mm

For less important applications, a dowel pin or a drive pin can be used. A large variety of these are listed in manufacturers' catalogs.[7]

[7]See also Joseph E. Shigley, "Unthreaded Fasteners," Chapter 24. In Joseph E. Shigley, Charles R. Mischke, and Thomas H. Brown, Jr. (eds.), *Standard Handbook of Machine Design*, 3rd ed., McGraw Hill, New York, 2004.

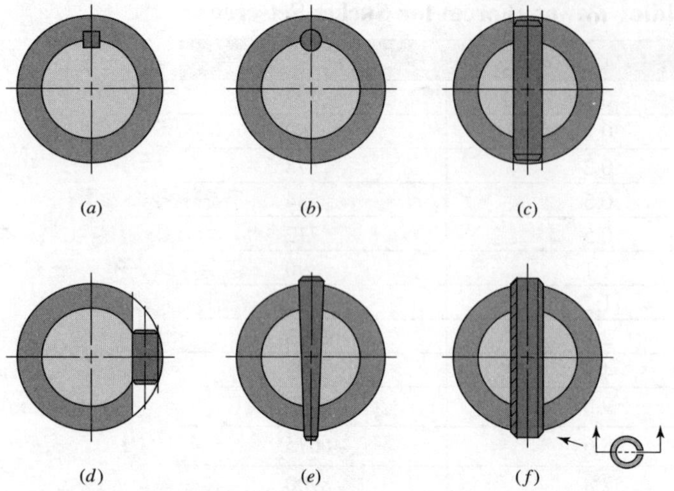

Figure 7–16

(*a*) Square key; (*b*) round key; (*c* and *d*) round pins; (*e*) taper pin; (*f*) split tubular spring pin. The pins in parts (*e*) and (*f*) are shown longer than necessary, to illustrate the chamfer on the ends, but their lengths should be kept smaller than the hub diameters to prevent injuries due to projections on rotating parts.

Table 7–5 Dimensions in mm of some standard taper pins (UNI 7238)

d pin *h* 8	Length		Shaft diameter		Minimum hub length
	from	to	from	to	
1	6	16	3	4	3
2	10	35	6	7	4.5
4	14	55	11	17	6
5	20	60	17	23	7.5
6	25	90	23	30	9
8	25	130	30	45	10

The square key, shown in Figure 7–16*a*, is also available in rectangular sizes. Standard sizes of these, together with the range of applicable shaft diameters, are listed in Table 7–6. The shaft diameter determines standard sizes for width, height, and key depth. The designer chooses an appropriate key length to carry the torsional load. Failure of the key can be by direct shear, or by bearing stress. Example 7–6 demonstrates the process to size the length of a key. The maximum length of a key is limited by the hub length of the attached element, and should generally not exceed about 1.5 times the shaft diameter to avoid excessive twisting with the angular deflection of the shaft. Multiple keys may be used as necessary to carry greater loads, typically oriented at 90° from one another. Excessive safety factors should be avoided in key design, since it is desirable in an overload situation for the key to fail, rather than more costly components.

Stock key material is typically made from low carbon cold-rolled steel, and is manufactured such that its dimensions never exceed the nominal dimension. This allows standard cutter sizes to be used for the keyseats. A setscrew is sometimes used along with a key to hold the hub axially, and to minimize rotational backlash when the shaft rotates in both directions.

The gib-head key, in Figure 7–17*a*, is tapered so that, when firmly driven, it acts to prevent relative axial motion. This also gives the advantage that the hub position

Table 7–6 mm Dimensions for Some Standard Square- and Rectangular-Key Applications

Shaft Diameter		Key Size		
Over	**To (Incl.)**	***w***	***h***	**Keyway Depth**
8	11	2	2	1
11	14	3	2	1
		3	3	1.5
14	22	5	3	1.5
		5	5	2
22	30	6	5	2
		6	6	3
30	36	8	6	3
		8	8	5
36	44	10	6	3
		10	10	5
44	58	12	10	5
		12	12	6
58	70	16	12	5.5
		16	16	8
70	80	20	12	6
		20	20	10

Source: Shigley, E. Joseph. "Unthreaded Fasteners." In *Standard Handbook of Machine Design,* edited by Joseph E. Shigley, Charles R. Mischke, and Thomas H Brown, Jr. 3rd ed. New York: McGraw Hill, 2004.

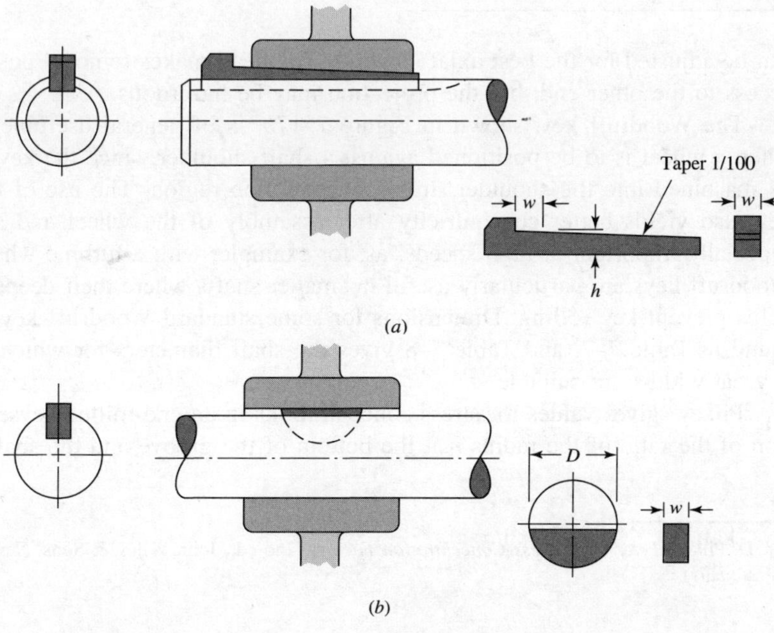

Figure 7–17

(*a*) Gib-head key; (*b*) Woodruff key.

Table 7–7 Dimensions of Woodruff Keys—Millimeter Series

Key Size		Height	Offset	Keyseat Depth	
w	*D*	*b*	*e*	Shaft	Hub
1.5	6	3	0.4	1.8	0.95
1.5	10	4	0.4	3.4	0.95
2	10	4	0.4	3.0	1.3
2	12	5	1	3.8	1.3
2	16	6	1.5	5.0	1.3
3	12	5	1	3.4	1.7
3	16	6	1.5	4.6	1.7
3	20	8	1.5	6.2	1.7
5	16	6	1.5	4.2	2.1
5	20	8	1.5	5.8	2.1
5	22	10	1.5	7.4	2.1
5	20	8	1.5	5.4	2.5
5	22	10	1.5	7.0	2.5
5	25	11	1.5	8.6	2.5
6	22	10	1.5	6.2	3.3
6	25	11	1.5	7.8	3.3
6	30	14	2	10.6	3.3
8	25	11	1.5	7.0	4.1
8	30	14	2	9.8	4.1
8	38	16	3	12.2	4.1
10	30	14	2	9.0	4.9
10	38	16	3	11.4	4.9

Table 7–8 Sizes of Woodruff Keys Suitable for Various Shaft Diameters

Keyseat	Shaft Diameter, mm	
Width, mm	From	To (inclusive)
1.5	8	12
2	10	22
3	10	38
5	12	40
5	14	50
6	18	58
8	20	60
10	25	66

can be adjusted for the best axial location. The head makes removal possible without access to the other end, but the projection may be hazardous.

The Woodruff key, shown in Figure 7–17*b*, is of general usefulness, especially when a wheel is to be positioned against a shaft shoulder, since the keyslot need not be machined into the shoulder stress concentration region. The use of the Woodruff key also yields better concentricity after assembly of the wheel and shaft. This is especially important at high speeds, as, for example, with a turbine wheel and shaft. Woodruff keys are particularly useful in smaller shafts where their deeper penetration helps prevent key rolling. Dimensions for some standard Woodruff key sizes can be found in Table 7–7, and Table 7–8 gives the shaft diameters for which the different keyseat widths are suitable.

Pilkey[8] gives values for stress concentrations in an end-milled keyseat, as a function of the ratio of the radius r at the bottom of the groove and the shaft diameter d.

[8]W. D. Pilkey, *Peterson's Stress-Concentration Factors*, 2nd ed., John Wiley & Sons, New York, 1997, pp. 408–409.

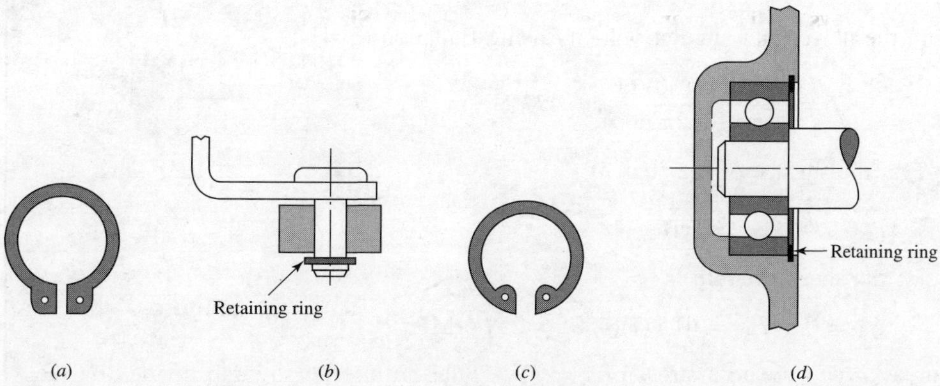

Figure 7–18

Typical uses for retaining rings. (*a*) External ring and (*b*) its application; (*c*) internal ring and (*d*) its application.

For fillets cut by standard milling-machine cutters, with a ratio of $r/d = 0.02$, Peterson's charts give $K_t = 2.14$ for bending and $K_{ts} = 2.62$ for torsion without the key in place, or $K_{ts} = 3.0$ for torsion with the key in place. The stress concentration at the end of the keyseat can be reduced somewhat by using a sled-runner keyseat, eliminating the abrupt end to the keyseat, as shown in Figure 7–17. It does, however, still have the sharp radius in the bottom of the groove on the sides. The sled-runner keyseat can only be used when definite longitudinal key positioning is not necessary. It is also not as suitable near a shoulder. Keeping the end of a keyseat at least a distance of $d/10$ from the start of the shoulder fillet will prevent the two stress concentrations from combining with each other.[9]

Retaining Rings

A retaining ring is frequently used instead of a shaft shoulder or a sleeve to axially position a component on a shaft or in a housing bore. As shown in Figure 7–18, a groove is cut in the shaft or bore to receive the spring retainer. For sizes, dimensions, and axial load ratings, the manufacturers' catalogs should be consulted.

Appendix Tables A–13–16 and A–13–17 give values for stress-concentration factors for flat-bottomed grooves in shafts, suitable for retaining rings. For the rings to seat nicely in the bottom of the groove, and support axial loads against the sides of the groove, the radius in the bottom of the groove must be reasonably sharp, typically about one-tenth of the groove width. This causes comparatively high values for stress-concentration factors, around 5 for bending and axial, and 3 for torsion. Care should be taken in using retaining rings, particularly in locations with high bending stresses.

EXAMPLE 7-6

A UNS G10 350 steel shaft, heat-treated to a minimum yield strength of 515 MPa, has a diameter of 35 mm. The shaft rotates at 600 rev/min and transmits 30 kW through a gear. Select an appropriate key for the gear, with a design factor of 2.8.

Solution

From Table 7–6, a 10 mm square key is selected. Choose a cold-drawn low-carbon mild steel which is generally available for key stock, such as UNS G10 200, with a yield strength of 450 MPa.

[9]W. D. Pilkey, *Peterson's Stress-Concentration Factors*, 2nd ed., John Wiley & Sons, New York, 1997, p. 381.

The torque is obtained from the power and angular velocity using Equation (3–41):

$$T = \frac{P(60)}{2\pi n} = \frac{(30)(60)}{2\pi(600)} = 477\ \text{N} \cdot \text{m}$$

From Figure 7–19, the force F at the surface of the shaft is

$$F = \frac{T}{r} = \frac{477}{(0.95/2)} = 27.25\ \text{kN}$$

By the distortion-energy theory, the shear strength is

$$S_{sy} = 0.577S_y = (0.577)(450) = 259.6\ \text{MPa}$$

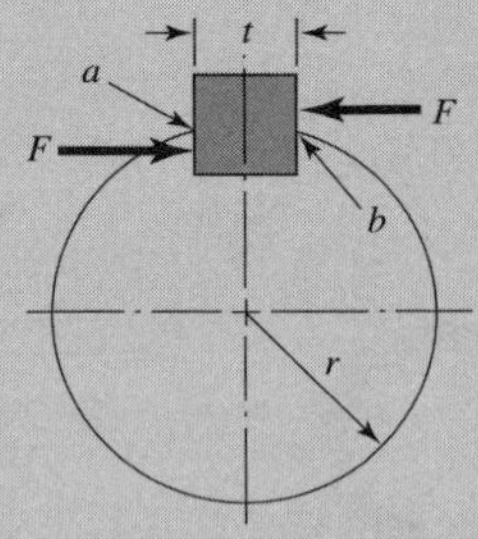

Figure 7–19

Failure by shear across the area ab will create a stress of $\tau = F/tl$. Substituting the strength divided by the design factor for τ gives

$$\frac{S_{sy}}{n} = \frac{F}{tl} \quad \text{or} \quad \frac{256.6}{2.80} = \frac{27\,260}{10(l)}$$

or $l = 29.7$ mm. To resist crushing, the area of one-half the face of the key is used:

$$\frac{S_y}{n} = \frac{F}{tl/2} \quad \text{or} \quad \frac{5}{2.80} = \frac{27\,260}{10(l)/2}$$

and $l = 33.9$ mm. Failure by crushing the key is the dominant failure mode, so it defines the necessary length of the key to be $l = 33.9$ mm.

7–8 Limits and Fits

The designer is free to adopt any geometry of fit for shafts and holes that will ensure the intended function. There is sufficient accumulated experience with commonly recurring situations to make standards useful. There are two standards for limits and fits in the United States, one based on inch units and the other based on metric units.[10] These differ in nomenclature, definitions, and organization. No point would be served by separately studying each of the two systems. The metric version is the newer of the two and is well organized, and so here we present only the metric version but include a set of inch conversions to enable the same system to be used with either system of units.

In using the standard, *capital letters always refer to the hole; lowercase letters are used for the shaft.*

The definitions illustrated in Figure 7–20 are explained as follows:

- ***Basic size*** is the size to which limits or deviations are assigned and is the same for both members of the fit.
- ***Deviation*** is the algebraic difference between a size and the corresponding basic size.
- ***Upper deviation*** is the algebraic difference between the maximum limit and the corresponding basic size.
- ***Lower deviation*** is the algebraic difference between the minimum limit and the corresponding basic size.

[10] *Preferred Limits and Fits for Cylindrical Parts*, ANSI B4.1-1967. *Preferred Metric Limits and Fits*, ANSI B4.2-1978.

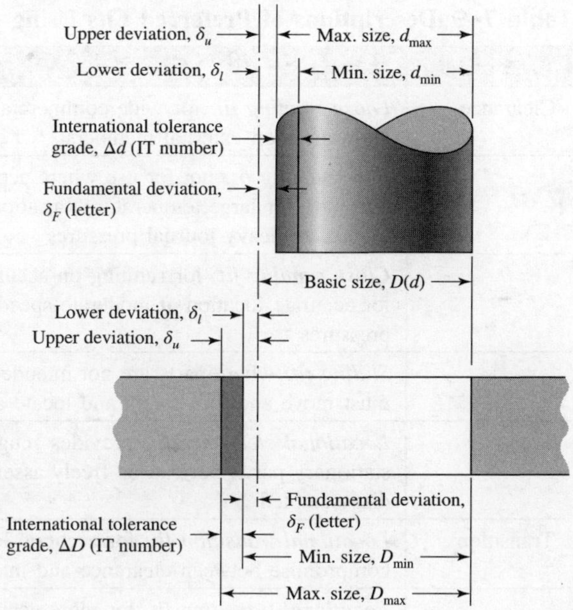

Figure 7–20

Definitions applied to a cylindrical fit.

- ***Fundamental deviation*** is either the upper or the lower deviation, depending on which is closer to the basic size.
- ***Tolerance*** is the difference between the maximum and minimum size limits of a part.
- ***International tolerance grade*** numbers (IT) designate groups of tolerances such that the tolerances for a particular IT number have the same relative level of accuracy but vary depending on the basic size.
- ***Hole basis*** represents a system of fits corresponding to a basic hole size. The fundamental deviation is H.
- ***Shaft basis*** represents a system of fits corresponding to a basic shaft size. The fundamental deviation is h. The shaft-basis system is not included here.

The magnitude of the tolerance zone is the variation in part size and is the same for both the internal and the external dimensions. The tolerance zones are specified in international tolerance grade numbers, called IT numbers. The smaller grade numbers specify a smaller tolerance zone. These range from IT0 to IT16, but only grades IT6 to IT11 are needed for the preferred fits. These are listed in Tables A–11 to A–12 for basic sizes up to 400 mm.

The standard uses *tolerance position letters,* with capital letters for internal dimensions (holes) and lowercase letters for external dimensions (shafts). As shown in Figure 7–20, the fundamental deviation locates the tolerance zone relative to the basic size.

Table 7–9 shows how the letters are combined with the tolerance grades to establish a preferred fit. The ISO symbol for the hole for a sliding fit with a basic size of 32 mm is 32H7. Inch units are not a part of the standard. The capital letter H establishes the fundamental deviation and the number 7 defines a tolerance grade of IT7.

For the sliding fit, the corresponding shaft dimensions are defined by the symbol 32g6.

Table 7–9 Descriptions of Preferred Fits Using the Basic Hole System

Type of Fit	Description	Symbol
Clearance	*Loose running fit:* for wide commercial tolerances or allowances on external members	H11/c11
	Free running fit: not for use where accuracy is essential, but good for large temperature variations, high running speeds, or heavy journal pressures	H9/d9
	Close running fit: for running on accurate machines and for accurate location at moderate speeds and journal pressures	H8/f7
	Sliding fit: where parts are not intended to run freely, but must move and turn freely and locate accurately	H7/g6
	Locational clearance fit: provides snug fit for location of stationary parts, but can be freely assembled and disassembled	H7/h6
Transition	*Locational transition fit:* for accurate location, a compromise between clearance and interference	H7/k6
	Locational transition fit: for more accurate location where greater interference is permissible	H7/n6
Interference	*Locational interference fit:* for parts requiring rigidity and alignment with prime accuracy of location but without special bore pressure requirements	H7/p6
	Medium drive fit: for ordinary steel parts or shrink fits on light sections, the tightest fit usable with cast iron	H7/s6
	Force fit: suitable for parts that can be highly stressed or for shrink fits where the heavy pressing forces required are impractical	H7/u6

Source: Preferred Metric Limits and Fits, ANSI B4.2-1978. See also BS 4500.

The fundamental deviations for shafts are given in Table A-11. For letter codes c, d, f, g, and h,

$$\text{Upper deviation} = \text{fundamental deviation}$$
$$\text{Lower deviation} = \text{upper deviation} - \text{tolerance grade}$$

For letter codes k, n, p, s, and u, the deviations for shafts are

$$\text{Lower deviation} = \text{fundamental deviation}$$
$$\text{Upper deviation} = \text{lower deviation} + \text{tolerance grade}$$

The lower deviation H (for holes) is zero. For these, the upper deviation equals the tolerance grade.

As shown in Figure 7–20, we use the following notation:

$$\begin{aligned} D &= \text{basic size of hole} \\ d &= \text{basic size of shaft} \\ \delta_u &= \text{upper deviation} \\ \delta_l &= \text{lower deviation} \\ \delta_F &= \text{fundamental deviation} \\ \Delta D &= \text{tolerance grade for hole} \\ \Delta d &= \text{tolerance grade for shaft} \end{aligned}$$

Note that these quantities are all deterministic. Thus, for the hole,

$$D_{max} = D + \Delta D \qquad D_{min} = D \tag{7–36}$$

For shafts with clearance fits c, d, f, g, and h,

$$d_{max} = d + \delta_F \qquad d_{min} = d + \delta_F - \Delta d \tag{7–37}$$

For shafts with interference fits k, n, p, s, and u,

$$d_{min} = d + \delta_F \qquad d_{max} = d + \delta_F + \Delta d \tag{7–38}$$

EXAMPLE 7–7

Find the shaft and hole dimensions for a loose running fit with a 34-mm basic size.

Solution

From Table 7–9, the ISO symbol is 34H11/c11. From Table A–11, we find that tolerance grade IT11 is 0.160 mm. The symbol 34H11/c11 therefore says that $\Delta D = \Delta d = 0.160$ mm. Using Equation (7–36) for the hole, we get

Answer $$D_{max} = D + \Delta D = 34 + 0.160 = 34.160 \text{ mm}$$

Answer $$D_{min} = D = 34.000 \text{ mm}$$

The shaft is designated as a 34c11 shaft. From Table A–12, the fundamental deviation is $\delta_F = -0.120$ mm. Using Equation (7–37), we get for the shaft dimensions

Answer $$d_{max} = d + \delta_F = 34 + (-0.120) = 33.880 \text{ mm}$$

Answer $$d_{min} = d + \delta_F - \Delta d = 34 + (-0.120) - 0.160 = 33.720 \text{ mm}$$

EXAMPLE 7–8

Find the hole and shaft limits for a medium drive fit using a basic hole size of 50 mm.

Solution

The symbol for the fit, from Table 7–9, is (50 mm)H7/s6. For the hole, we use Table A–11 and find the IT7 grade to be $\Delta D = 0.025$ mm. Thus, from Equation (7–36),

Answer $$D_{max} = D + \Delta D = 50 + 0.025 = 50.025 \text{ mm}$$

Answer $$D_{min} = D = 50 \text{ mm}$$

The IT6 tolerance for the shaft is $\Delta d = 0.016$ mm. Also, from Table A–12, the fundamental deviation is $\delta_F = 0.026$ mm. Using Equation (7–38), we get for the shaft that

Answer $$d_{min} = d + \delta_F = 50 + 0.026 = 50.026 \text{ mm}$$

Answer $$d_{max} = d + \delta_F + \Delta d = 50 + 0.026 + 0.016 = 50.042 \text{ mm}$$

Stress and Torque Capacity in Interference Fits

Interference fits between a shaft and its components can sometimes be used effectively to minimize the need for shoulders and keyways. The stresses due to an interference fit can be obtained by treating the shaft as a cylinder with a uniform external pressure, and the hub as a hollow cylinder with a uniform internal pressure. Stress equations for these situations were developed in Section 3–16, and will be converted here from radius terms into diameter terms to match the terminology of this section.

The pressure p generated at the interface of the interference fit, from Equation (3–58) converted into terms of diameters, is given by

$$p = \frac{\delta}{\dfrac{d}{E_o}\left(\dfrac{d_o^2 + d^2}{d_o^2 - d^2} + \nu_o\right) + \dfrac{d}{E_i}\left(\dfrac{d^2 + d_i^2}{d^2 - d_i^2} - \nu_i\right)} \tag{7–39}$$

or, in the case where both members are of the same material,

$$p = \frac{E\delta}{2d^3}\left[\frac{(d_o^2 - d^2)(d^2 - d_i^2)}{d_o^2 - d_i^2}\right] \tag{7–40}$$

where d is the nominal shaft diameter, d_i is the inside diameter (if any) of the shaft, d_o is the outside diameter of the hub, E is Young's modulus, and ν is Poisson's ratio, with subscripts o and i for the outer member (hub) and inner member (shaft), respectively. The term δ is the *diametral* interference between the shaft and hub, that is, the difference between the shaft outside diameter and the hub inside diameter.

$$\delta = d_{\text{shaft}} - d_{\text{hub}} \tag{7–41}$$

Since there will be tolerances on both diameters, the maximum and minimum pressures can be found by applying the maximum and minimum interferences. Adopting the notation from Figure 7–20, we write

$$\delta_{\min} = d_{\min} - D_{\max} \tag{7–42}$$

$$\delta_{\max} = d_{\max} - D_{\min} \tag{7–43}$$

where the diameter terms are defined in Equations (7–36) and (7–38). The maximum interference should be used in Equation (7–39) or (7–40) to determine the maximum pressure to check for excessive stress.

From Equations (3–60) and (3–61), with radii converted to diameters, the tangential stresses at the interface of the shaft and hub are

$$\sigma_{t,\,\text{shaft}} = -p\frac{d^2 + d_i^2}{d^2 - d_i^2} \tag{7–44}$$

$$\sigma_{t,\,\text{hub}} = p\frac{d_o^2 + d^2}{d_o^2 - d^2} \tag{7–45}$$

The radial stresses at the interface are simply

$$\sigma_{r,\,\text{shaft}} = -p \tag{7–46}$$

$$\sigma_{r,\,\text{hub}} = -p \tag{7–47}$$

The tangential and radial stresses are orthogonal, and should be combined using a failure theory to compare with the yield strength. If either the shaft or hub yields during assembly, the full pressure will not be achieved, diminishing the torque that can be transmitted. The interaction of the stresses due to the interference fit with the other stresses in the shaft due to shaft loading is not trivial. Finite-element analysis of the interface would be appropriate when warranted. A stress element on the surface of a rotating shaft will experience a completely reversed bending stress in the longitudinal direction, as well as the steady compressive stresses in the tangential and radial directions. This is a three-dimensional stress element. Shear stress due to torsion in the shaft may also be present. Since the stresses due to the press fit are compressive, the fatigue situation is usually actually improved. For this reason, it may be acceptable to simplify the shaft analysis by ignoring the steady compressive stresses due to the press fit. There is, however, a stress concentration effect in the shaft bending stress near the ends of the hub, due to the sudden change from compressed to uncompressed material. The design of the hub geometry, and therefore its uniformity and rigidity, can have a significant effect on the specific value of the stress-concentration factor, making it difficult to report generalized values. For first estimates, values are typically not greater than 2.

The amount of torque that can be transmitted through an interference fit can be estimated with a simple friction analysis at the interface. The friction force is the product of the coefficient of friction f and the normal force acting at the interface. The normal force can be represented by the product of the pressure p and the surface area A of interface. Therefore, the friction force F_f is

$$F_f = fN = f(pA) = f[p2\pi(d/2)l] = \pi f\,pld \tag{7–48}$$

where l is the length of the hub. This friction force is acting with a moment arm of $d/2$ to provide the torque capacity of the joint, so

$$\begin{aligned} T &= F_f d/2 = \pi f\,pld(d/2) \\ T &= (\pi/2) f\,pld^2 \end{aligned} \tag{7–49}$$

The minimum interference, from Equation (7–42), should be used to determine the minimum pressure to check for the maximum amount of torque that the joint should be designed to transmit without slipping.

PROBLEMS

Problems marked with an asterisk (*) are linked to problems in other chapters, as summarized in Table 1–2 of Section 1–17.

7–1 A shaft is loaded in bending and torsion such that $M_a = 70$ N · m, $T_a = 45$ N · m, $M_m = 55$ N · m, and $T_m = 35$ N · m. For the shaft, $S_u = 700$ MPa, $S_y = 560$ MPa, the true fracture strength is 1045 MPa, and a fully corrected endurance limit of $S_e = 210$ MPa is assumed. Let $K_f = 2.2$ and $K_{fs} = 1.8$. With a design factor of 2.0 determine the minimum acceptable diameter of the shaft using the
(*a*) DE-Goodman criterion.
(*b*) DE-Morrow criterion.
(*c*) DE-Gerber criterion.
(*d*) DE-SWT criterion.
Discuss and compare the results.

7–2 A shaft, made of AISI 1050 CD steel, is loaded in bending and torsion such that $M_a = 75$ N · m, $T_a = 45$ N · m, $M_m = 55$ N · m, and $T_m = 35$ N · m. The shaft has a fully corrected endurance limit of $S_e = 207$ MPa and at the critical stress location, $k_f = 2.3$ and $k_{fs} = 1.9$. Estimate the true fracture strength as being 345 MPa greater than the ultimate strength. With a design factor of 2.5 determine the minimum acceptable diameter of the shaft using the
(*a*) DE-Goodman criterion.
(*b*) DE-Morrow criterion.
(*c*) DE-Gerber criterion.
(*d*) DE-SWT criterion.
Discuss and compare the results.

7–3 The section of shaft shown in the figure is to be designed to approximate relative sizes of $d = 0.75D$ and $r = D/20$ with diameter d conforming to that of standard rolling-bearing bore sizes. The shaft is to be made of SAE 2340 steel, heat-treated to obtain minimum strengths in the shoulder area of 1226 MPa ultimate tensile strength and 1130 MPa yield strength with a Brinell hardness not less than 370. At the shoulder the shaft is subjected to a completely reversed bending moment of 70 N · m, accompanied by a steady torsion of 45 N · m. Use a design factor of 2.5 and size the shaft for an infinite life using the DE-Goodman criterion.

Problem 7–3

Section of a shaft containing a grinding-relief groove. Unless otherwise specified, the diameter at the root of the groove $d_r = d - 2r$, and though the section of diameter d is ground, the root of the groove is still a machined surface.

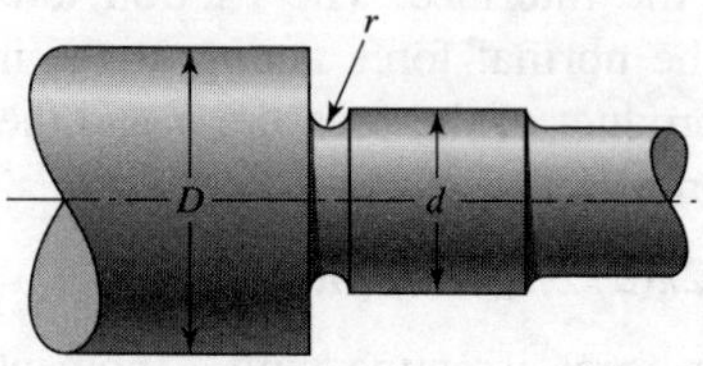

7–4 The rotating solid steel shaft is simply supported by bearings at points B and C and is driven by a gear (not shown) which meshes with the spur gear at D, which has a 150-mm pitch diameter. The force F from the drive gear acts at a pressure angle of 20°. The shaft transmits a torque to point A of $T_A = 340$ N · m. The shaft is machined from steel with $S_y = 420$ MPa and $S_{ut} = 560$ MPa. Using a factor of safety of 2.5, determine the minimum allowable diameter of the 250-mm section of the shaft based on (*a*) a static yield analysis using the distortion energy theory and (*b*) a fatigue-failure analysis. Assume sharp fillet radii at the bearing shoulders for estimating stress-concentration factors.

Problem 7–4

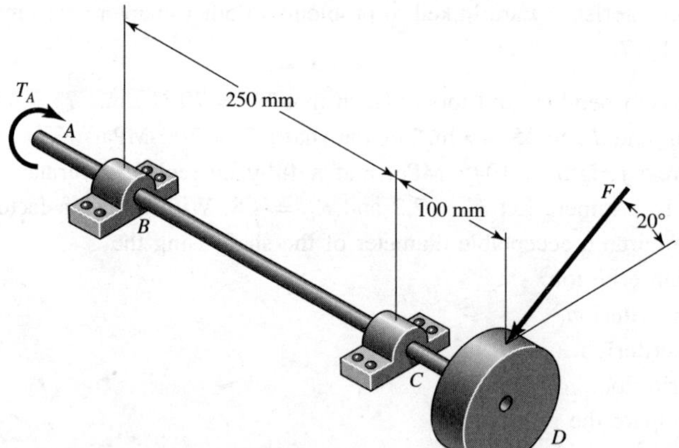

7–5 A rotating step shaft is loaded as shown, where the forces F_A and F_B are constant at 2.6 kN and 1.3 kN, respectively, and the torque T alternates from 0 to 200 N · m. The shaft is to be considered simply supported at points O and C, and is made of AISI 1045 CD steel with a fully corrected endurance limit of $S_e = 276$ MPa. Let $K_f = 2.1$ and $K_{fs} = 1.7$. For a design factor of 2.5 determine the minimal acceptable diameter of section BC using the

(*a*) DE-Gerber criterion.

(*b*) DE-Goodman criterion.

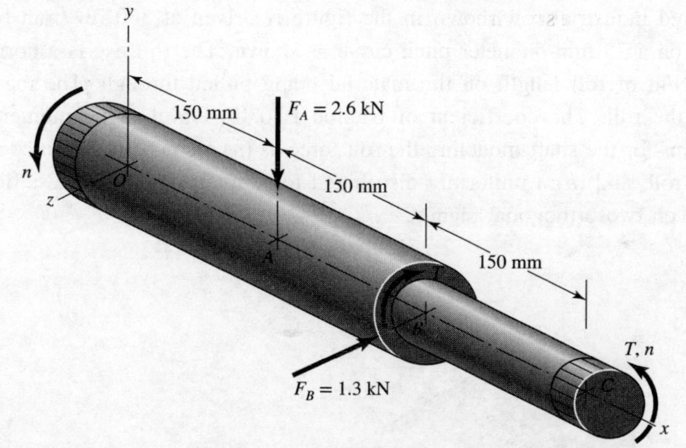

Problem 7–5

7–6 Bevel gears transmit torsional, bending, and axial loads to rotating shafts, such as shown in Figure 13–36. For this problem, the free body diagram for a shaft *ABC* is shown in the figure, where points *A* and *C* are at the bearings. Line *BG* is on the bevel gear extended to the shaft center from point *G*, the contact point with the mating gear. Transmitted from the mating gear are tangential, radial, and axial forces $F_G^y = 2.22$ kN, $F_G^z = 0.26$ kN, and $F_G^x = 0.77$ kN, respectively. The bearing at *C* supports the axial thrust load. For the shaft material, $S_u = 620$ MPa, $S_y = 450$ MPa, and has a fully corrected endurance limit of 275 MPa. Let $(K_f)_a = 2.3$, $(K_f)_b = 2.0$, and $(K_{fs})_t = 1.5$, where the subscripts a, b, and t represent axial, bending, and torsion, respectively. The diameter of shaft *ABC* is 38 mm.

(*a*) For the DE-Goodman criterion, develop a general equation for the design factor using Equations (6–46), (6–66), (6–67), (7–2), and (7–3); and F_m and F_a for the mean and alternating axial forces, respectively.

(*b*) Determine the value of the design factor.

(*c*) Determine the value of the design factor ignoring the axial force. What is the contribution from the axial force?

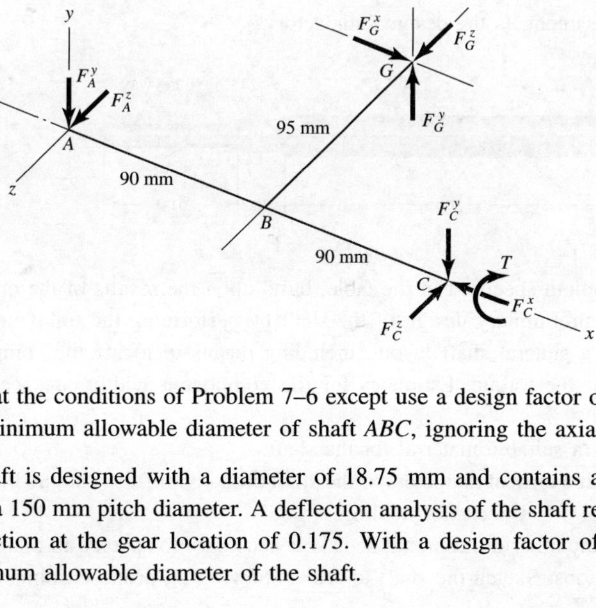

Problem 7–6

7–7 Repeat the conditions of Problem 7–6 except use a design factor of 2.5 and determine the minimum allowable diameter of shaft *ABC*, ignoring the axial force.

7–8 A shaft is designed with a diameter of 18.75 mm and contains a 90-tooth spur gear with a 150 mm pitch diameter. A deflection analysis of the shaft results in a transverse deflection at the gear location of 0.175. With a design factor of 1.5, determine the minimum allowable diameter of the shaft.

7–9 A geared industrial roll shown in the figure is driven at 300 rev/min by a force F acting on a 75 mm-diameter pitch circle as shown. The roll exerts a normal force of 5200 N/m of roll length on the material being pulled through. The material passes under the roll. The coefficient of friction is 0.40. Develop the moment and shear diagrams for the shaft modeling the roll force as (*a*) a concentrated force at the center of the roll, and (*b*) a uniformly distributed force along the roll. These diagrams will appear on two orthogonal planes.

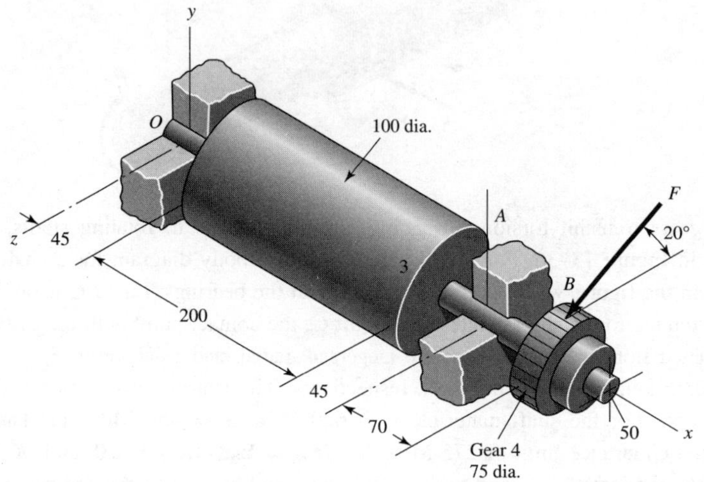

Problem 7–9

Material moves under the roll. Dimensions in millimeters.

7–10 Design a shaft for the situation of the industrial roll of Problem 7–9 with a design factor of 2 and a reliability goal of 0.999 against fatigue failure. Plan for a ball bearing on the left and a cylindrical roller on the right. For deformation use a factor of safety of 2.

7–11 The figure shows a proposed design for the industrial roll shaft of Problem 7–9. Hydrodynamic film bearings are to be used. All surfaces are machined except the journals, which are ground and polished. The material is 1035 HR steel. Perform a design assessment. Is the design satisfactory?

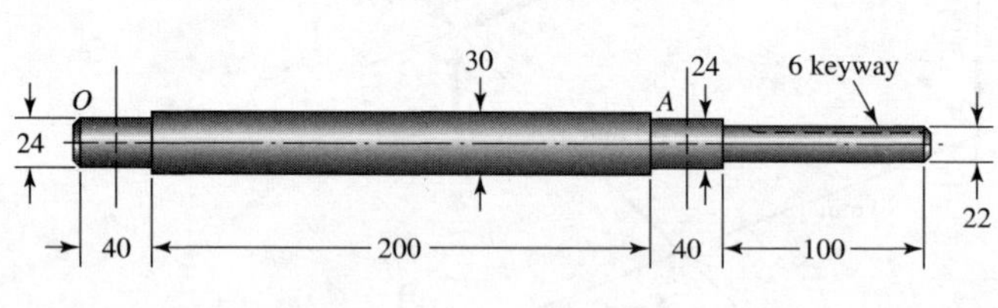

Problem 7–11

Bearing shoulder fillets 0.75 mm, others 1.6 mm. Sled-runner keyway is 90 mm long. Dimensions in millimeters.

7–12* to 7–21* For the problem specified in the table, build upon the results of the original problem to obtain a preliminary design of the shaft by performing the following tasks.

(*a*) Sketch a general shaft layout, including means to locate the components and to transmit the torque. Estimates for the component widths are acceptable at this point.

(*b*) Specify a suitable material for the shaft.

(*c*) Determine critical diameters of the shaft based on infinite fatigue life with a design factor of 1.5. Check for yielding.

(*d*) Make any other dimensional decisions necessary to specify all diameters and axial dimensions. Sketch the shaft to scale, showing all proposed dimensions.

(*e*) Check the deflections at the gears, and the slopes at the gears and the bearings for satisfaction of the recommended limits in Table 7–2. Assume the deflections for any pulleys are not likely to be critical. If any of the deflections exceed the recommended limits, make appropriate changes to bring them all within the limits.

Problem Number	Original Problem Number
7–12*	3–76
7–13*	3–77
7–14*	3–78
7–15*	3–79
7–16*	3–80
7–17*	3–81
7–18*	3–82
7–19*	3–84
7–20*	3–85
7–21*	3–87

7–22 In the double-reduction gear train shown, shaft *a* is driven by a motor attached by a flexible coupling attached to the overhang. The motor provides a torque of 10 N · m at a speed of 1200 rpm. The gears have 20° pressure angles, with diameters shown in the figure. Use an AISI 1020 cold-drawn steel. Design one of the shafts (as specified by the instructor) with a design factor of 1.5 by performing the following tasks.

(*a*) Sketch a general shaft layout, including means to locate the gears and bearings, and to transmit the torque.

(*b*) Perform a force analysis to find the bearing reaction forces, and generate shear and bending moment diagrams.

(*c*) Determine potential critical locations for stress design.

(*d*) Determine critical diameters of the shaft based on fatigue and static stresses at the critical locations.

(*e*) Make any other dimensional decisions necessary to specify all diameters and axial dimensions. Sketch the shaft to scale, showing all proposed dimensions.

(*f*) Check the deflection at the gear, and the slopes at the gear and the bearings for satisfaction of the recommended limits in Table 7–2.

(*g*) If any of the deflections exceed the recommended limits, make appropriate changes to bring them all within the limits.

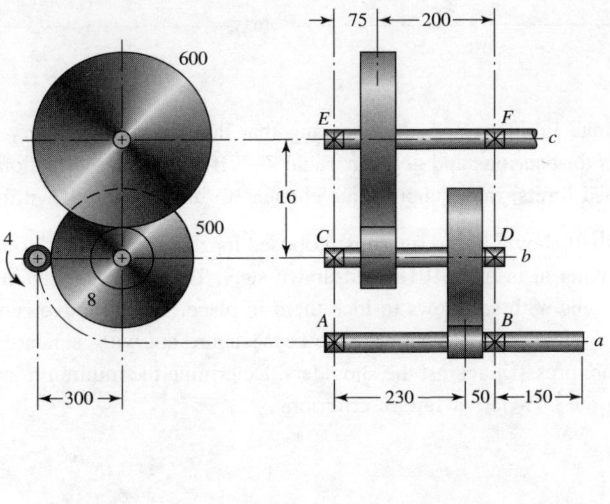

Problem 7–22

Dimensions in millimeters.

7–23 In the figure is a proposed shaft design to be used for the input shaft *a* in Problem 7–22. A ball bearing is planned for the left bearing, and a cylindrical roller bearing for the right.

(*a*) Determine the minimum fatigue factor of safety by evaluating at any critical locations. Use the DE-Goodman fatigue criterion.

(*b*) Check the design for adequacy with respect to deformation, according to the recommendations in Table 7–2.

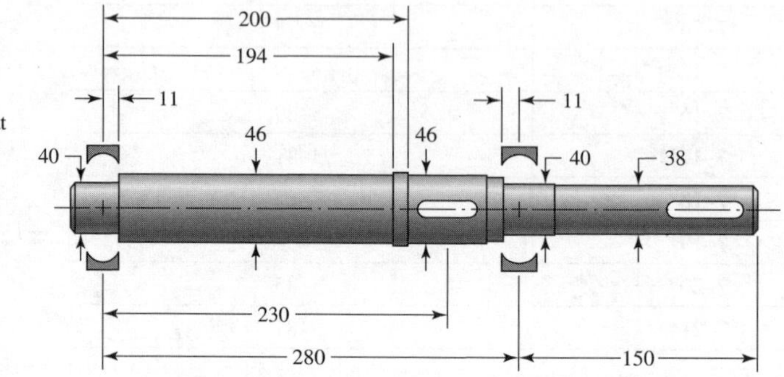

Problem 7–23

Shoulder fillets at bearing seat 0.75 mm radius, others 3 mm radius, except right-hand bearing seat transition, 6 mm. The material is 1030 HR. Keyways 10 mm wide by 5 mm deep. Dimensions in millimeters.

7–24* The shaft shown in the figure is proposed for the application defined in Problem 3–80. The material is AISI 1018 cold-drawn steel. The gears seat against the shoulders, and have hubs with setscrews to lock them in place. The effective centers of the gears for force transmission are shown. The keyseats are cut with standard endmills. The bearings are press-fit against the shoulders. Determine the minimum fatigue factor of safety using the DE-Gerber fatigue criterion.

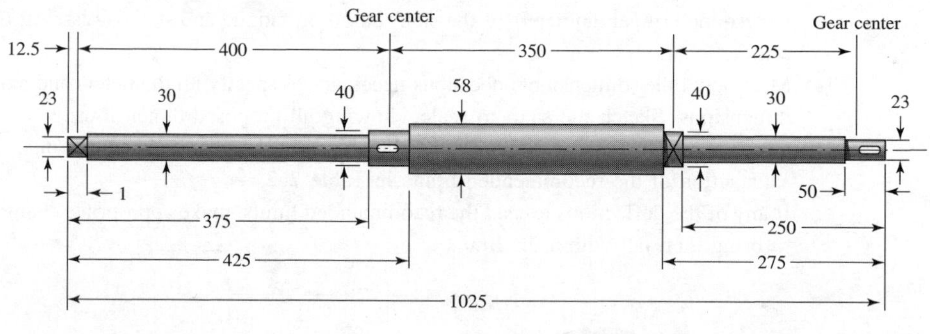

*Problem 7–24**

All fillets 1.6 mm. Dimensions in millimeters.

7–25* Continue Problem 7–24 by checking that the deflections satisfy the suggested minimums for bearings and gears in Table 7–2. If any of the deflections exceed the recommended limits, make appropriate changes to bring them all within the limits.

7–26* The shaft shown in the figure is proposed for the application defined in Problem 3–81. The material is AISI 1018 cold-drawn steel. The gears seat against the shoulders and have hubs with setscrews to lock them in place. The effective centers of the gears for force transmission are shown. The keyseats are cut with standard endmills. The bearings are press-fit against the shoulders. Determine the minimum fatigue factor of safety using the DE-Gerber failure criterion.

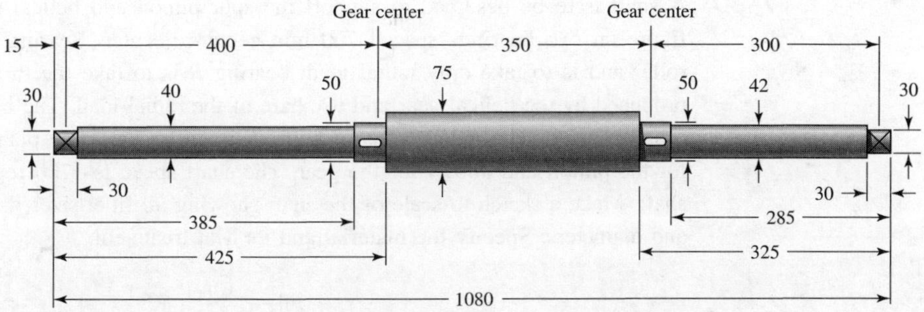

Problem 7–26*
All fillets 2 mm.
Dimensions in millimeters.

7–27* Continue Problem 7–26 by checking that the deflections satisfy the suggested minimums for bearings and gears in Table 7–2. If any of the deflections exceed the recommended limits, make appropriate changes to bring them all within the limits.

7–28 The shaft shown in the figure is driven by a gear at the right keyway, drives a fan at the left keyway, and is supported by two deep-groove ball bearings. The shaft is made from AISI 1020 cold-drawn steel. At steady-state speed, the gear transmits a radial load of 1.1 kN and a tangential load of 3 kN at a pitch diameter of 200 mm.

(*a*) Determine fatigue factors of safety at any potentially critical locations using the DE-Gerber failure criterion.

(*b*) Check that deflections satisfy the suggested minimums for bearings and gears.

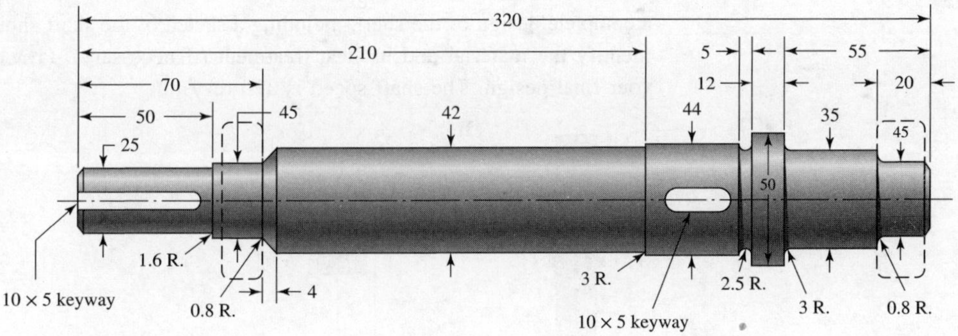

Problem 7–28
Dimensions in millimeters.

7–29 An AISI 1020 cold-drawn steel shaft with the geometry shown in the figure carries a transverse load of 7 kN and a torque of 107 N · m. Examine the shaft for strength and deflection. If the largest allowable slope at the bearings is 0.001 rad and at the gear mesh is 0.0005 rad, what is the factor of safety guarding against damaging distortion? Using the DE-Goodman criterion, what is the factor of safety guarding against a fatigue failure? If the shaft turns out to be unsatisfactory, what would you recommend to correct the problem?

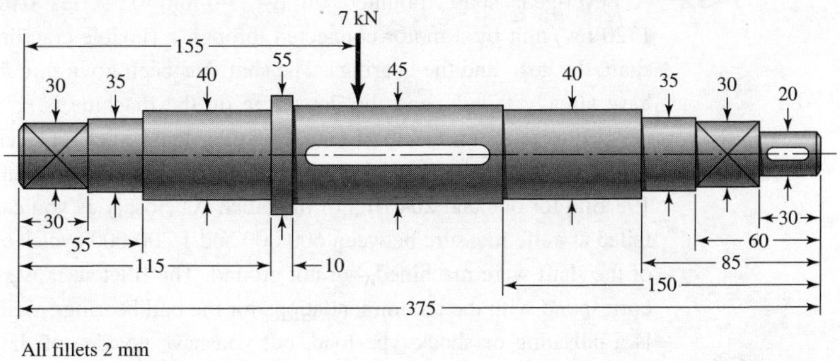

Problem 7–29
Dimensions in millimeters.

All fillets 2 mm

7–30 A shaft is to be designed to support the spur pinion and helical gear shown in the figure on two bearings spaced 700 mm center-to-center. Bearing A is a cylindrical roller and is to take only radial load; bearing B is to take the thrust load of 900 N produced by the helical gear and its share of the radial load. The bearing at B can be a ball bearing. The radial loads of both gears are in the same plane, and are 2.7 kN for the pinion and 900 N for the gear. The shaft speed is 1200 rev/min. Design the shaft. Make a sketch to scale of the shaft showing all fillet sizes, keyways, shoulders, and diameters. Specify the material and its heat treatment.

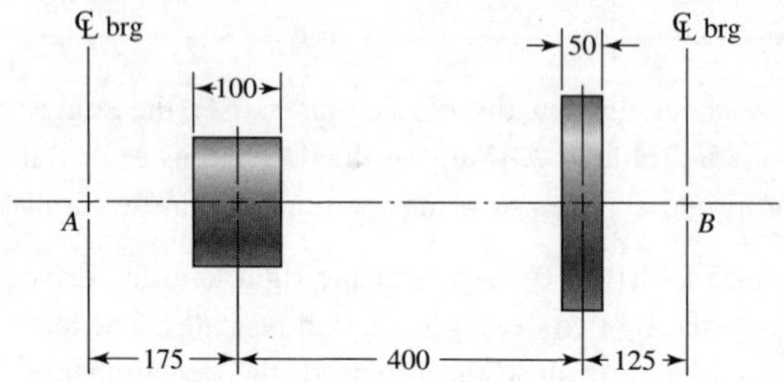

Problem 7–30
Dimensions in millimeters.

7–31 A heat-treated steel shaft is to be designed to support the spur gear and the overhanging worm shown in the figure. A bearing at A takes pure radial load. The bearing at B takes the worm-thrust load for either direction of rotation. The dimensions and the loading are shown in the figure; note that the radial loads are in the same plane. Make a complete design of the shaft, including a sketch of the shaft showing all dimensions. Identify the material and its heat treatment (if necessary). Provide an assessment of your final design. The shaft speed is 310 rev/min.

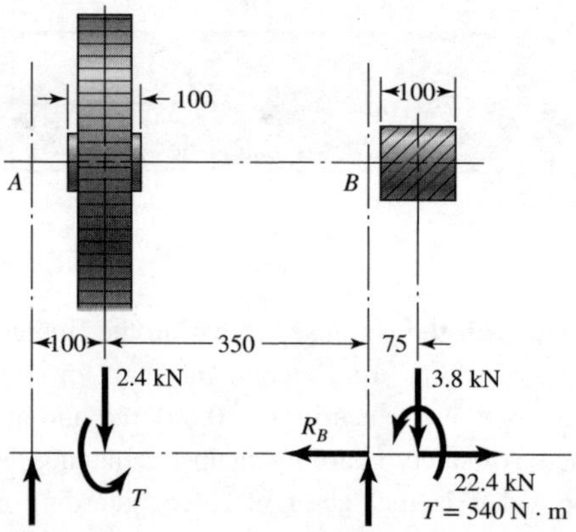

Problem 7–31
Dimensions in millimeters.

7–32 A bevel-gear shaft mounted on two 40-mm 02-series ball bearings is driven at 1720 rev/min by a motor connected through a flexible coupling. The figure shows the shaft, the gear, and the bearings. The shaft has been giving trouble—in fact, two of them have already failed—and the downtime on the machine is so expensive that you have decided to redesign the shaft yourself rather than order replacements. A hardness check of the two shafts in the vicinity of the fracture of the two shafts showed an average of 198 Bhn for one and 204 Bhn of the other. As closely as you can estimate the two shafts failed at a life measure between 600 000 and 1 200 000 cycles of operation. The surfaces of the shaft were machined, but not ground. The fillet sizes were not measured, but they correspond with the recommendations for the ball bearings used. You know that the load is a pulsating or shock-type load, but you have no idea of the magnitude, because the shaft drives an indexing mechanism, and the forces are inertial. The keyways are 10 mm wide by 5 mm deep. The straight-toothed bevel pinion drives a 48-tooth bevel gear. Specify a new shaft in sufficient detail to ensure a long and trouble-free life.

Problem 7–32

Dimensions in millimeters.

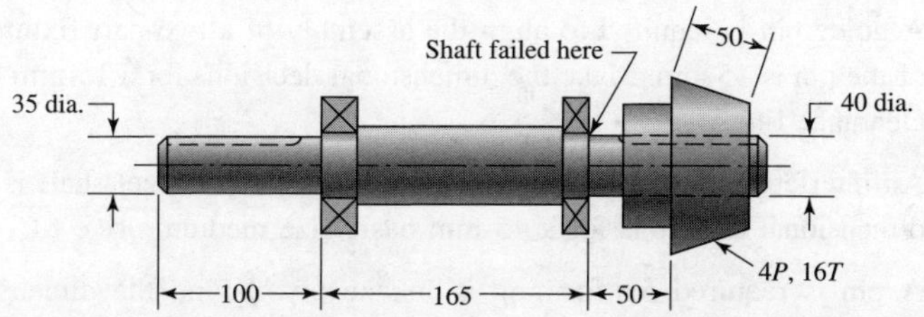

7–33 A 25-mm-diameter uniform steel shaft is 600 mm long between bearings.
(*a*) Find the lowest critical speed of the shaft.
(*b*) If the goal is to double the critical speed, find the new diameter.
(*c*) A half-size model of the original shaft has what critical speed?

7–34 Demonstrate how rapidly Rayleigh's method converges for the uniform-diameter solid shaft of Problem 7–33, by partitioning the shaft into first one, then two, and finally three elements.

7–35 Compare Equation (7–27) for the angular frequency of a two-disk shaft with Equation (7–28), and note that the constants in the two equations are equal.
(*a*) Develop an expression for the *second* critical speed.
(*b*) Estimate the second critical speed of the shaft addressed in Example 7–5, parts *a* and *b*.

7–36 For a uniform-diameter shaft, does hollowing the shaft increase or decrease the critical speed? Determine the ratio of the critical speeds for a solid shaft of diameter d to a hollow shaft of inner diameter $d/2$ and outer diameter d?

7–37 The steel shaft shown in the figure carries an 100 N gear on the left and a 175 N gear on the right. Estimate the first critical speed due to the loads, the shaft's critical speed without the loads, and the critical speed of the combination.

Problem 7–37

Dimensions in millimeters.

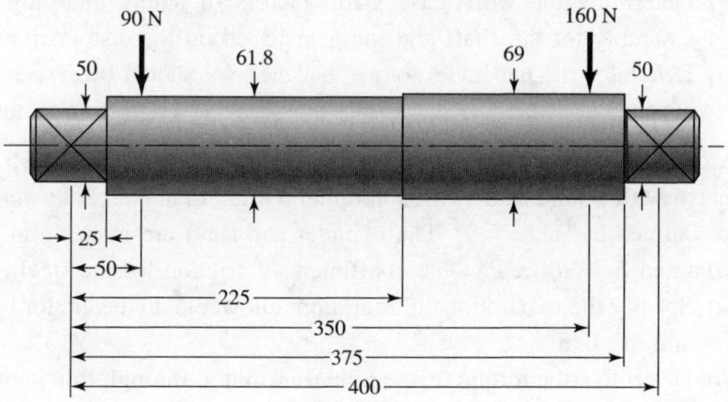

7–38 A transverse drilled and reamed hole can be used in a solid shaft to hold a pin that locates and holds a mechanical element, such as the hub of a gear, in axial position, and allows for the transmission of torque. Since a small-diameter hole introduces high stress concentration, and a larger diameter hole erodes the area resisting bending and torsion, investigate the existence of a pin diameter with minimum adverse affect on the shaft. Specifically, determine the pin diameter, as a percentage of the shaft diameter, that minimizes the peak stress in the shaft. (*Hint:* Use Table A–14.)

7–39* The shaft shown in Problem 7–24 is proposed for the application defined in Problem 3–80. Specify a square key for gear *B*, using a factor of safety of 1.1.

7–40* The shaft shown in Problem 7–26 is proposed for the application defined in Problem 3–81. Specify a square key for gear *B*, using a factor of safety of 1.1.

7–41 A guide pin is required to align the assembly of a two-part fixture. The nominal size of the pin is 15 mm. Make the dimensional decisions for a 15-mm basic size locational clearance fit.

7–42 An interference fit of a cast-iron hub of a gear on a steel shaft is required. Make the dimensional decisions for a 45 mm basic size medium drive fit.

7–43 A pin is required for forming a linkage pivot. Find the dimensions required for a 45-mm basic size pin and clevis with a sliding fit.

7–44 A journal bearing and bushing need to be described. The nominal size is 31.25 mm. What dimensions are needed for a 31.25 mm basic size with a close running fit if this is a lightly loaded journal and bushing assembly?

7–45 A ball bearing has been selected with the bore size specified in the catalog as 35.000 mm to 35.020 mm. Specify appropriate minimum and maximum shaft diameters to provide a locational interference fit.

7–46 A shaft diameter is carefully measured to be 36.05 mm. A bearing is selected with a catalog specification of the bore diameter range from 36 mm to 36.025 mm. Determine if this is an acceptable selection if a locational interference fit is desired.

7–47 A gear and shaft with nominal diameter of 35 mm are to be assembled with a *medium drive fit,* as specified in Table 7–9. The gear has a hub, with an outside diameter of 60 mm, and an overall length of 50 mm. The shaft is made from AISI 1020 CD steel, and the gear is made from steel that has been through hardened to provide $S_u = 700$ MPa and $S_y = 600$ MPa.

(*a*) Specify dimensions with tolerances for the shaft and gear bore to achieve the desired fit.

(*b*) Determine the minimum and maximum pressures that could be experienced at the interface with the specified tolerances.

(*c*) Determine the worst-case static factors of safety guarding against yielding at assembly for the shaft and the gear based on the distortion energy failure theory.

(*d*) Determine the maximum torque that the joint should be expected to transmit without slipping, i.e., when the interference pressure is at a minimum for the specified tolerances.

7–48 A cylinder with a nominal 63 mm ID, a 100-mm OD, and a 75 mm length is to be mated with a solid shaft with a nominal 63 mm diameter. A *medium drive fit* is desired (as defined in Table 7–9). The cylinder and shaft are made from steel, with $S_y = 690$ MPa and $E = 206$ GPa. The coefficient of friction for the steel interface is 0.7.

(*a*) Specify the maximum and minimum allowable diameters for both the cylinder hole and the shaft.

(*b*) Determine the torque that can be transmitted through this joint, assuming the shaft and cylinder are both manufactured within their tolerances such that the minimum interference is achieved.

(*c*) Suppose the shaft and cylinder are both manufactured within their tolerances such that the maximum interference is achieved. Check for yielding of the cylinder at its inner radius by finding the following:

(*i*) The pressure at the interface.

(*ii*) The tangential and radial stresses in the cylinder, at its inner radius.

(*iii*) The factor of safety for static yielding of the cylinder, using the distortion-energy failure theory.

8 Screws, Fasteners, and the Design of Nonpermanent Joints

James Hardy/PhotoAlto

Chapter Outline

The helical-thread screw was undoubtably an extremely important mechanical invention. It is the basis of power screws, which change angular motion to linear motion to transmit power or to develop large forces (presses, jacks, etc.), and threaded fasteners, an important element in nonpermanent joints.

This book presupposes a knowledge of the elementary methods of fastening. Typical methods of fastening or joining parts use such devices as bolts, nuts, cap screws, setscrews, rivets, spring retainers, locking devices, pins, keys, welds, and adhesives. Studies in engineering graphics and in metal processes often include instruction on various joining methods, and the curiosity of any person interested in mechanical engineering naturally results in the acquisition of a good background knowledge of fastening methods. Contrary to first impressions, the subject is one of the most interesting in the entire field of mechanical design.

One of the key targets of current design for manufacture is to reduce the number of fasteners. However, there will always be a need for fasteners to facilitate disassembly for whatever purposes. For example, jumbo jets such as Boeing's 747 require as many as 2.5 million fasteners, some of which cost several dollars apiece. To keep costs down, aircraft manufacturers, and their subcontractors, constantly review new fastener designs, installation techniques, and tooling.

The number of innovations in the fastener field over any period you might care to mention has been tremendous. An overwhelming variety of fasteners are available for the designer's selection. Serious designers generally keep specific notebooks on fasteners alone. Methods of joining parts are extremely important in the engineering of a quality design, and it is necessary to have a thorough understanding of the performance of fasteners and joints under all conditions of use and design.

8–1 Thread Standards and Definitions

The terminology of screw threads, illustrated in Figure 8–1, is explained as follows:

- ***Pitch*** is the distance between adjacent thread forms measured parallel to the thread axis. The pitch in U.S. units is the reciprocal of the number of thread forms per inch N.
- ***Major diameter d*** is the largest diameter of a screw thread.
- ***Minor*** (or root) ***diameter d_r*** is the smallest diameter of a screw thread.
- ***Pitch diameter d_p*** is a theoretical diameter between the major and minor diameters.
- ***Lead l***, not shown, is the distance the nut moves parallel to the screw axis when the nut is given one turn. For a single thread, as in Figure 8–1, the lead is the same as the pitch.

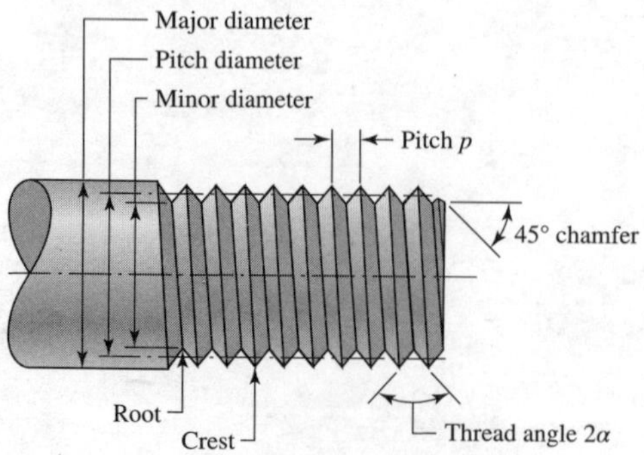

Figure 8–1

Terminology of screw threads. Sharp vee threads shown for clarity; the crests and roots are actually flattened or rounded during the forming operation.

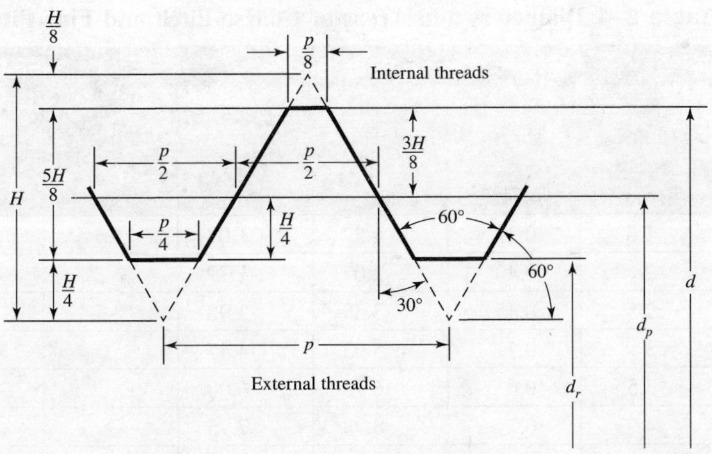

Figure 8–2

Basic profile for metric M and MJ threads.
d = major diameter
d_r = minor diameter
d_p = pitch diameter
p = pitch
$H = \frac{\sqrt{3}}{2}p$

A *multiple-threaded* product is one having two or more threads cut beside each other (imagine two or more strings wound side by side around a pencil). Standardized products such as screws, bolts, and nuts all have single threads; a *double-threaded* screw has a lead equal to twice the pitch, a *triple-threaded* screw has a lead equal to 3 times the pitch, and so on.

All threads are made according to the *right-hand rule* unless otherwise noted. That is, if the bolt is turned clockwise, the bolt advances toward the nut.

The *American National (Unified)* thread standard has been approved in this country and in Great Britain for use on all standard threaded products. The thread angle is 60° and the crests of the thread may be either flat or rounded.

Figure 8–2 shows the thread geometry of the metric M and MJ profiles. The M profile replaces the inch class and is the basic ISO 68 profile with 60° symmetric threads. The MJ profile has a rounded fillet at the root of the external thread and a larger minor diameter of both the internal and external threads. This profile is especially useful where high fatigue strength is required.

Tables 8–1 and 8–2 will be useful in specifying and designing threaded parts. Note that the thread size is specified by giving the pitch p for metric sizes and by giving the number of threads per inch N for the Unified sizes. The screw sizes in Table 8–2 with diameter under $\frac{1}{4}$ in are numbered or gauge sizes. The second column in Table 8–2 shows that a No. 8 screw has a nominal major diameter of 0.1640 in.

A great many tensile tests of threaded rods have shown that an unthreaded rod having a diameter equal to the mean of the pitch diameter and minor diameter will have the same tensile strength as the threaded rod. The area of this unthreaded rod is called the tensile-stress area A_t of the threaded rod; values of A_t are listed in both tables.

Two major Unified thread series are in common use: UN and UNR. The difference between these is simply that a root radius must be used in the UNR series. Because of reduced thread stress-concentration factors, UNR series threads have improved fatigue strengths. Unified threads are specified by stating the nominal major diameter, the number of threads per inch, and the thread series, for example, $\frac{5}{8}$ in-18 UNRF or 0.625 in-18 UNRF.

Metric threads are specified by writing the diameter and pitch in millimeters, in that order. Thus, M12 × 1.75 is a thread having a nominal major diameter of 12 mm and a pitch of 1.75 mm. Note that the letter M, which precedes the diameter, is the clue to the metric designation.

Table 8–1 Diameters and Areas of Coarse-Pitch and Fine-Pitch Metric Threads*

	Coarse-Pitch Series			Fine-Pitch Series		
Nominal Major Diameter d mm	Pitch p mm	Tensile-Stress Area A_t mm^2	Minor-Diameter Area A_r mm^2	Pitch p mm	Tensile-Stress Area A_t mm^2	Minor-Diameter Area A_r mm^2
1.6	0.35	1.27	1.07			
2	0.40	2.07	1.79			
2.5	0.45	3.39	2.98			
3	0.5	5.03	4.47			
3.5	0.6	6.78	6.00			
4	0.7	8.78	7.75			
5	0.8	14.2	12.7			
6	1	20.1	17.9			
8	1.25	36.6	32.8	1	39.2	36.0
10	1.5	58.0	52.3	1.25	61.2	56.3
12	1.75	84.3	76.3	1.25	92.1	86.0
14	2	115	104	1.5	125	116
16	2	157	144	1.5	167	157
20	2.5	245	225	1.5	272	259
24	3	353	324	2	384	365
30	3.5	561	519	2	621	596
36	4	817	759	2	915	884
42	4.5	1120	1050	2	1260	1230
48	5	1470	1380	2	1670	1630
56	5.5	2030	1910	2	2300	2250
64	6	2680	2520	2	3030	2980
72	6	3460	3280	2	3860	3800
80	6	4340	4140	1.5	4850	4800
90	6	5590	5360	2	6100	6020
100	6	6990	6740	2	7560	7470
110				2	9180	9080

*The equations and data used to develop this table have been obtained from ANSI B1.1-1974 and B18.3.1-1978. The minor diameter was found from the equation $d_r = d - 1.226\ 869p$, and the pitch diameter from $d_p = d - 0.649\ 519p$. The mean of the pitch diameter and the minor diameter was used to compute the tensile-stress area.

Source: ANSI/AGMA 2003-B97.

Square and Acme threads, whose profiles are shown in Figures 8–3*a* and *b*, respectively, are used on screws when power is to be transmitted. Table 8–3 lists the preferred pitches for inch-series Acme threads. However, other pitches can be and often are used, since the need for a standard for such threads is not great.

Modifications are frequently made to both Acme and square threads. For instance, the square thread is sometimes modified by cutting the space between the teeth so as

Table 8–2 Diameters and Area of Unified Screw Threads UNC and UNF*

	Coarse Series—UNC				Fine Series—UNF		
Size Designation	Nominal Major Diameter in	Threads per Inch N	Tensile-Stress Area A_t in^2	Minor-Diameter Area A_r in^2	Threads per Inch N	Tensile-Stress Area A_t in^2	Minor-Diameter Area A_r in^2
0	0.0600				80	0.001 80	0.001 51
1	0.0730	64	0.002 63	0.002 18	72	0.002 78	0.002 37
2	0.0860	56	0.003 70	0.003 10	64	0.003 94	0.003 39
3	0.0990	48	0.004 87	0.004 06	56	0.005 23	0.004 51
4	0.1120	40	0.006 04	0.004 96	48	0.006 61	0.005 66
5	0.1250	40	0.007 96	0.006 72	44	0.008 80	0.007 16
6	0.1380	32	0.009 09	0.007 45	40	0.010 15	0.008 74
8	0.1640	32	0.014 0	0.011 96	36	0.014 74	0.012 85
10	0.1900	24	0.017 5	0.014 50	32	0.020 0	0.017 5
12	0.2160	24	0.024 2	0.020 6	28	0.025 8	0.022 6
$\frac{1}{4}$	0.2500	20	0.031 8	0.026 9	28	0.036 4	0.032 6
$\frac{5}{16}$	0.3125	18	0.052 4	0.045 4	24	0.058 0	0.052 4
$\frac{3}{8}$	0.3750	16	0.077 5	0.067 8	24	0.087 8	0.080 9
$\frac{7}{16}$	0.4375	14	0.106 3	0.093 3	20	0.118 7	0.109 0
$\frac{1}{2}$	0.5000	13	0.141 9	0.125 7	20	0.159 9	0.148 6
$\frac{9}{16}$	0.5625	12	0.182	0.162	18	0.203	0.189
$\frac{5}{8}$	0.6250	11	0.226	0.202	18	0.256	0.240
$\frac{3}{4}$	0.7500	10	0.334	0.302	16	0.373	0.351
$\frac{7}{8}$	0.8750	9	0.462	0.419	14	0.509	0.480
1	1.0000	8	0.606	0.551	12	0.663	0.625
$1\frac{1}{4}$	1.2500	7	0.969	0.890	12	1.073	1.024
$1\frac{1}{2}$	1.5000	6	1.405	1.294	12	1.581	1.521

*This table was compiled from ANSI B1.1-1974. The minor diameter was found from the equation $d_r = d - 1.299\,038p$, and the pitch diameter from $d_p = d - 0.649\,519p$. The mean of the pitch diameter and the minor diameter was used to compute the tensile-stress area.
Source: ANSI/AGMA 2003-B97.

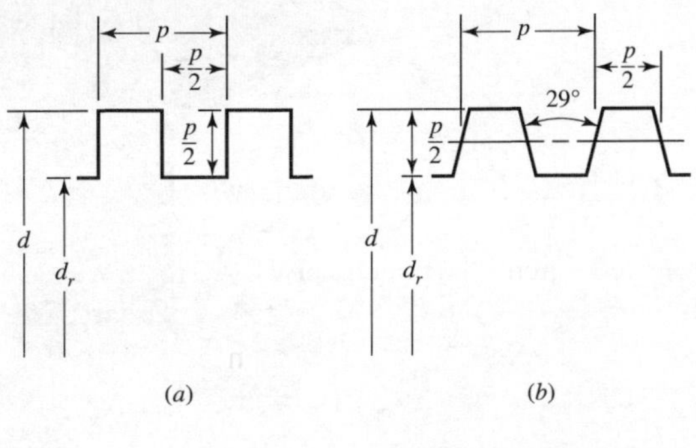

Figure 8–3
(*a*) Square thread; (*b*) Acme thread.

Table 8–3 Preferred Pitches for Acme Threads

d, in	$\frac{1}{4}$	$\frac{5}{16}$	$\frac{3}{8}$	$\frac{1}{2}$	$\frac{5}{8}$	$\frac{3}{4}$	$\frac{7}{8}$	1	$1\frac{1}{4}$	$1\frac{1}{2}$	$1\frac{3}{4}$	2	$2\frac{1}{2}$	3
p, in	$\frac{1}{16}$	$\frac{1}{14}$	$\frac{1}{12}$	$\frac{1}{10}$	$\frac{1}{8}$	$\frac{1}{6}$	$\frac{1}{6}$	$\frac{1}{5}$	$\frac{1}{5}$	$\frac{1}{4}$	$\frac{1}{4}$	$\frac{1}{4}$	$\frac{1}{3}$	$\frac{1}{2}$

to have an included thread angle of 10° to 15°. This is not difficult, since these threads are usually cut with a single-point tool anyhow; the modification retains most of the high efficiency inherent in square threads and makes the cutting simpler. Acme threads are sometimes modified to a stub form by making the teeth shorter. This results in a larger minor diameter and a somewhat stronger screw.

8–2 The Mechanics of Power Screws

A power screw is a device used in machinery to change angular motion into linear motion, and, usually, to transmit power. Familiar applications include the lead screws of lathes, and the screws for vises, presses, and jacks.

An application of power screws to a power-driven jack is shown in Figure 8–4. You should be able to identify the worm, the worm gear, the screw, and the nut. Is the worm gear supported by one bearing or two?

In Figure 8–5 a square-threaded power screw with single thread having a mean diameter d_m, a pitch p, a lead angle λ, and a helix angle ψ is loaded by the axial compressive force F. We wish to find an expression for the torque required to raise this load, and another expression for the torque required to lower the load.

Figure 8–4

The Joyce worm-gear screw jack. *(Courtesy Joyce-Dayton Corp., Dayton, Ohio.)*

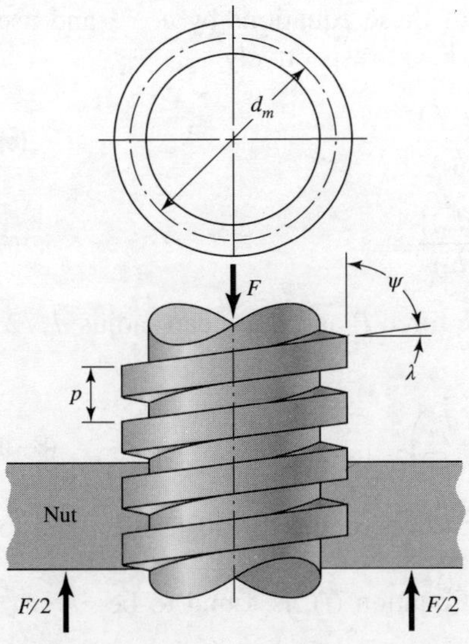

Figure 8–5

Portion of a power screw.

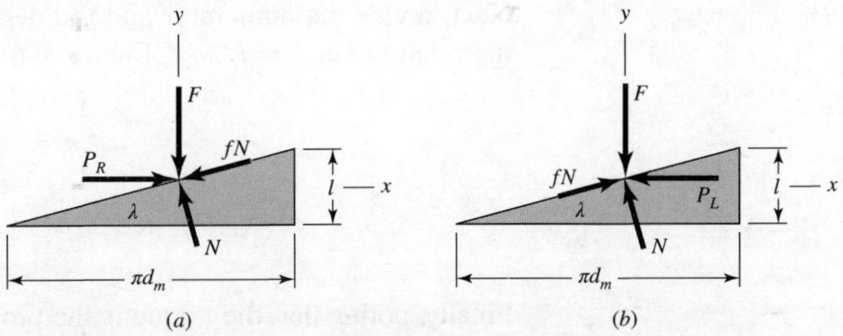

Figure 8–6

Force diagrams: (*a*) lifting the load; (*b*) lowering the load.

First, imagine that a single thread of the screw is unrolled or developed (Figure 8–6) for exactly a single turn. Then one edge of the thread will form the hypotenuse of a right triangle whose base is the circumference of the mean-thread-diameter circle and whose height is the lead. The angle λ, in Figures 8–5 and 8–6, is the lead angle of the thread. We represent the summation of all the axial forces acting upon the normal thread area by F. To raise the load, a force P_R acts to the right (Figure 8–6*a*), and to lower the load, P_L acts to the left (Figure 8–6*b*). The friction force is the product of the coefficient of friction f with the normal force N, and acts to oppose the motion. The system is in equilibrium under the action of these forces, and hence, for raising the load, we have

$$\begin{aligned} \sum F_x &= P_R - N \sin \lambda - fN \cos \lambda = 0 \\ \sum F_y &= -F - fN \sin \lambda + N \cos \lambda = 0 \end{aligned} \tag{a}$$

In a similar manner, for lowering the load, we have

$$\begin{aligned} \sum F_x &= -P_L - N \sin \lambda + fN \cos \lambda = 0 \\ \sum F_y &= -F + fN \sin \lambda + N \cos \lambda = 0 \end{aligned} \tag{b}$$

Since we are not interested in the normal force N, we eliminate it from each of these sets of equations and solve the result for P. For raising the load, this gives

$$P_R = \frac{F(\sin \lambda + f \cos \lambda)}{\cos \lambda - f \sin \lambda} \tag{c}$$

and for lowering the load,

$$P_L = \frac{F(f \cos \lambda - \sin \lambda)}{\cos \lambda + f \sin \lambda} \tag{d}$$

Next, divide the numerator and the denominator of these equations by $\cos\lambda$ and use the relation $\tan\lambda = l/\pi d_m$ (Figure 8–6). We then have, respectively,

$$P_R = \frac{F[(l/\pi d_m) + f]}{1 - (fl/\pi d_m)} \tag{e}$$

$$P_L = \frac{F[f - (l/\pi d_m)]}{1 + (fl/\pi d_m)} \tag{f}$$

Finally, noting that the torque is the product of the force P and the mean radius $d_m/2$, for raising the load we can write

$$T_R = \frac{Fd_m}{2}\left(\frac{l + \pi f d_m}{\pi d_m - fl}\right) \tag{8–1}$$

where T_R is the torque required for two purposes: to overcome thread friction and to raise the load.

The torque required to lower the load, from Equation (*f*), is found to be

$$T_L = \frac{Fd_m}{2}\left(\frac{\pi f d_m - l}{\pi d_m + fl}\right) \tag{8–2}$$

This is the torque required to overcome a part of the friction in lowering the load. It may turn out, in specific instances where the lead is large or the friction is low, that the load will lower itself by causing the screw to spin without any external effort. In such cases, the torque T_L from Equation (8–2) will be negative or zero. When a positive torque is obtained from this equation, the screw is said to be *self-locking*. Thus the condition for self-locking is

$$\pi f d_m > l$$

Now divide both sides of this inequality by πd_m. Recognizing that $l/\pi d_m = \tan\lambda$, we get

$$f > \tan\lambda \tag{8–3}$$

This relation states that self-locking is obtained whenever the coefficient of thread friction is equal to or greater than the tangent of the thread lead angle.

An expression for efficiency is also useful in the evaluation of power screws. If we let $f = 0$ in Equation (8–1), we obtain

$$T_0 = \frac{Fl}{2\pi} \tag{g}$$

which, since thread friction has been eliminated, is the torque required only to raise the load. The thread efficiency is thus defined as

$$e = \frac{T_0}{T_R} = \frac{Fl}{2\pi T_R} \tag{8–4}$$

The preceding equations have been developed for square threads where the normal thread loads are parallel to the axis of the screw. In the case of Acme or other threads, the normal thread load is inclined to the axis because of the thread angle 2α and the lead angle λ. Since lead angles are small, this inclination can be neglected and only the effect of the thread angle (Figure 8–7*a*) considered. The effect of the angle α is to increase the frictional force by the wedging action of the threads. Therefore the

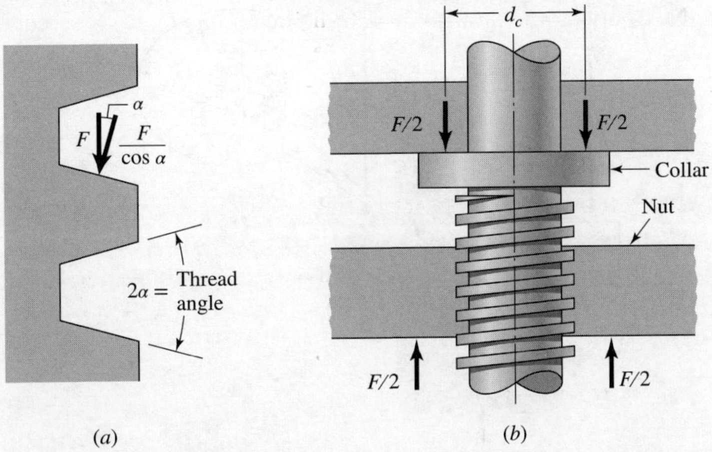

Figure 8–7

(*a*) Normal thread force is increased because of angle α; (*b*) thrust collar has frictional diameter d_c.

frictional terms in Equation (8–1) must be divided by cos α. For raising the load, or for tightening a screw or bolt, this yields

$$T_R = \frac{Fd_m}{2}\left(\frac{l + \pi f d_m \sec\alpha}{\pi d_m - fl \sec\alpha}\right) \tag{8–5}$$

In using Equation (8–5), remember that it is an approximation because the effect of the lead angle has been neglected.

For power screws, the Acme thread is not as efficient as the square thread, because of the additional friction due to the wedging action, but it is often preferred because it is easier to machine and permits the use of a split nut, which can be adjusted to take up for wear.

Usually a third component of torque must be applied in power-screw applications. When the screw is loaded axially, a thrust or collar bearing must be employed between the rotating and stationary members in order to carry the axial component. Figure 8–7*b* shows a typical thrust collar in which the load is assumed to be concentrated at the mean collar diameter d_c. If f_c is the coefficient of collar friction, the torque required is

$$T_c = \frac{F f_c d_c}{2} \tag{8–6}$$

For large collars, the torque should probably be computed in a manner similar to that employed for disk clutches (see Section 16–5).

Nominal body stresses in power screws can be related to thread parameters as follows. The maximum nominal shear stress τ in torsion of the screw body can be expressed as

$$\tau = \frac{16T}{\pi d_r^3} \tag{8–7}$$

The compressive axial stress σ in the body of the screw due to load F is

$$\sigma = -\frac{F}{A} = -\frac{4F}{\pi d_r^2} \tag{8–8}$$

in the absence of column action. For a short column the J. B. Johnson buckling formula is given by Equation (4–48), which is

$$\left(\frac{F}{A}\right)_{\text{crit}} = S_y - \left(\frac{S_y}{2\pi}\frac{l}{k}\right)^2 \frac{1}{CE} \tag{4–48}$$

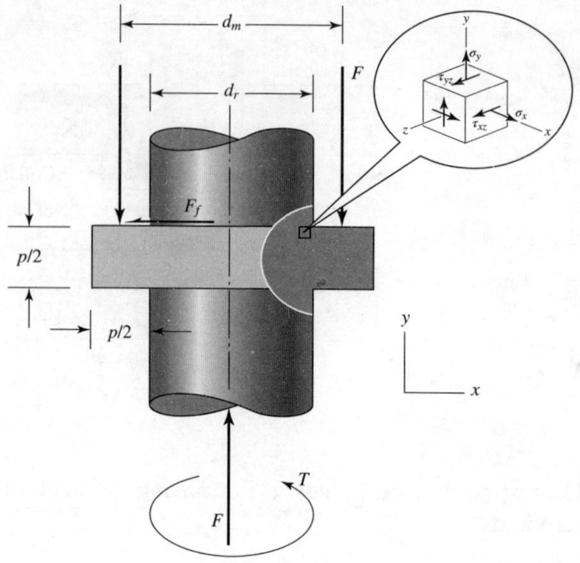

Figure 8–8

Stress in a power screw at the interface of the thread and the screw body.

Nominal thread stresses in power screws can be related to thread parameters as follows. The bearing stress, σ_B, is from the force F pressing into the surface area of the thread, as in Figure 8–8, giving

$$\sigma_B = -\frac{F}{\pi d_m\, n_t p/2} = -\frac{2F}{\pi d_m n_t p} \tag{8–9}$$

where n_t is the number of engaged threads. The bending stress at the root of the thread, in the x direction, is very roughly estimated for such a short and wide cantilever beam (the thread) as

$$Z = \frac{I}{c} = \frac{bh^2}{6} = \frac{(\pi d_r n_t)(p/2)^2}{6} = \frac{\pi}{24} d_r n_t p^2 \qquad M = \frac{Fp}{4}$$

so

$$\sigma_x = \frac{M}{Z} = \frac{Fp}{4}\,\frac{24}{\pi d_r n_t p^2} = \frac{6F}{\pi d_r n_t p} \tag{8–10}$$

Due to the length to height ratio of the thread (see Figure 3–19) the transverse shear stress at the center of the thread root is not critical.

The stress situation where the root of the thread connects to the main screw body is complex and not easily modeled analytically. However, an estimate of the critical stress can be obtained by evaluating a stress element at the outer radius of the screw body adjacent to the root of the thread, as shown in Figure 8–8. The expanded element in Figure 8–8 shows a 3-D representation of the stresses at the root. This stress element experiences all of the screw body stresses, as well as supporting the bending stress from the adjacent stress element in the thread. In addition, isolating the root of the thread of Figure 8–8 from the main screw body also exposes a tangential shear stress that is transferred from the thread to the screw body. This shear is directly related to the torsion that is transmitted through the friction between the nut and the screw thread, and consequently into the screw body. This shear stress is acting on the x face in the positive z direction, hence τ_{xz}. The shear area, A_s, is the circumference of the thread at the root diameter times the total thread thickness. The shear force, $\tau_{xz} A_s$, acting at the root radius, r_r, balances the torsion, T. Thus, the tangential shear stress is

$$\tau_{xz} = \frac{T}{A_s r_r} = \frac{T}{[\pi d_r n_t (p/2)](d_r/2)} = \frac{4T}{\pi d_r^2 n_t p} \tag{8–11}$$

The inset of Figure 8–8 shows the stress element at the outer radius of the screw body, with the stress equations summarized as

$$\left.\begin{array}{ll} \sigma_x = \dfrac{6F}{\pi d_r n_t p} & \tau_{xy} = 0 \\ \sigma_y = -\dfrac{4F}{\pi d_r^2} & |\tau_{yz}| = \dfrac{16T}{\pi d_r^3} \\ \sigma_z = 0 & |\tau_{xz}| = \dfrac{4T}{\pi d_r^2 n_t p} \end{array}\right\} \tag{8–12}$$

The shear stresses are shown as absolute values, as their signs will depend on the direction of the torque and whether the screw is left-hand or right-hand. The magnitude is all that is needed for determining the von Mises stress. The von Mises stress, σ', for this stress element is found by substituting these stresses into Equation (5–14).

The screw-thread form is complicated from an analysis viewpoint. The equations above assume all engaged threads are equally sharing the load, which turns out to be a weak assumption. A power screw lifting a load is in compression and its thread pitch is *shortened* by elastic deformation. Its engaging nut is in tension and its thread pitch is *lengthened*. The engaged threads cannot share the load equally. Some experiments show that the first engaged thread carries 0.38 of the load, the second 0.25, the third 0.18, and the seventh is free of load. In estimating thread stresses by the equations above, substituting $0.38F$ for F and setting n_t to 1 will give the largest level of stresses in the thread-nut combination, assuming there are at least 6 threads engaged.

EXAMPLE 8–1

A square-thread power screw has a major diameter of 32 mm and a pitch of 4 mm with double threads, and it is to be used in an application similar to that in Figure 8–4. The given data include $f = f_c = 0.08$, $d_c = 40$ mm, and $F = 6.4$ kN per screw.

(*a*) Find the thread depth, thread width, pitch diameter, minor diameter, and lead.
(*b*) Find the torque required to raise and lower the load.
(*c*) Find the efficiency during lifting the load.
(*d*) Find the body stresses, torsional and compressive.
(*e*) Find the bearing stress on the first thread.
(*f*) Find the thread bending stress at the root of the first thread.
(*g*) Determine the von Mises stress at the critical stress element where the root of the first thread interfaces with the screw body.

Solution

(*a*) From Figure 8–3*a* the thread depth and width are the same and equal to half the pitch, or 2 mm. Also

$$d_m = d - p/2 = 32 - 4/2 = 30 \text{ mm}$$

Answer

$$d_r = d - p = 32 - 4 = 28 \text{ mm}$$

$$l = np = 2(4) = 8 \text{ mm}$$

(*b*) Using Equations (8–1) and (8–6), the torque required to turn the screw against the load is

$$T_R = \frac{Fd_m}{2}\left(\frac{l + \pi f d_m}{\pi d_m - fl}\right) + \frac{Ff_c d_c}{2}$$

$$= \frac{6.4(30)}{2}\left[\frac{8 + \pi(0.08)(30)}{\pi(30) - 0.08(8)}\right] + \frac{6.4(0.08)40}{2}$$

Answer $$= 15.94 + 10.24 = 26.18 \text{ N} \cdot \text{m}$$

Using Equations (8–2) and (8–6), we find the load-lowering torque is

$$T_L = \frac{Fd_m}{2}\left(\frac{\pi f d_m - l}{\pi d_m + fl}\right) + \frac{Ff_c d_c}{2}$$

$$= \frac{6.4(30)}{2}\left[\frac{\pi(0.08)30 - 8}{\pi(30) + 0.08(8)}\right] + \frac{6.4(0.08)(40)}{2}$$

Answer $$= -0.466 + 10.24 = 9.77 \text{ N} \cdot \text{m}$$

The minus sign in the first term indicates that the screw alone is not self-locking and would rotate under the action of the load except for the fact that the collar friction is present and must be overcome, too. Thus the torque required to rotate the screw "with" the load is less than is necessary to overcome collar friction alone.

(*c*) The overall efficiency in raising the load is

Answer $$e = \frac{Fl}{2\pi T_R} = \frac{6.4(8)}{2\pi(26.18)} = 0.311$$

(*d*) The body shear stress τ given by Equation (8–7) due to torsional moment T_R at the outside of the screw body is

Answer $$\tau = \frac{16T_R}{\pi d_r^3} = \frac{16(26.18)(10^3)}{\pi(28)^3} = 6.07 \text{ MPa}$$

The axial nominal normal stress σ given by Equation (8–8) is

Answer $$\sigma = -\frac{4F}{\pi d_r^2} = -\frac{4(6.4)10^3}{\pi(28)^2} = -10.39 \text{ MPa}$$

(*e*) The bearing stress σ_B given by Equation (8–9) is, with one thread carrying $0.38F$,

Answer $$\sigma_B = -\frac{2(0.38F)}{\pi d_m(1)p} = -\frac{2(0.38)(6.4)10^3}{\pi(30)(1)(4)} = -12.9 \text{ MPa}$$

(*f*) The thread-root bending stress σ_b given by Equation (8–10) with one thread carrying $0.38F$ is

Answer $$\sigma_b = \frac{6(0.38F)}{\pi d_r(1)p} = \frac{6(0.38)(6.4)10^3}{\pi(28)(1)4} = 41.5 \text{ MPa}$$

(*g*) The tangential shear stress given by Equation (8–11) with one thread carrying $0.38\,T$,

Answer $$\tau_{xz} = \frac{4(0.38T)}{\pi d_r^2(1)p} = \frac{4(0.38)26.18(10^3)}{\pi(28^2)4} = 4.04 \text{ MPa}$$

The 3-D stresses for Figure 8–8 from Equation (8–12) are

$$\sigma_x = 41.5 \text{ MPa} \qquad \tau_{xy} = 0$$

$$\sigma_y = -10.39 \text{ MPa} \qquad |\tau_{yz}| = 6.07 \text{ MPa}$$

$$\sigma_z = 0 \qquad |\tau_{xz}| = 4.04 \text{ MPa}$$

For the von Mises stress, Equation (5–14) can be written as

Answer $$\sigma' = \frac{1}{\sqrt{2}} \{[41.5 - (-10.39)]^2 + (-10.39 - 0)^2 + (0 - 41.5)^2 + 6(6.07)^2 + 6(4.04)^2\}$$

$$= 49.2 \text{ MPa}$$

Ham and Ryan[1] showed that the coefficient of friction in screw threads is independent of axial load, practically independent of speed, decreases with heavier lubricants, shows little variation with combinations of materials, and is best for steel on bronze. Sliding coefficients of friction in power screws are about 0.10–0.15.

Table 8–4 shows safe bearing pressures on threads, to protect the moving surfaces from abnormal wear. Table 8–5 shows the coefficients of sliding friction for common material pairs. Table 8–6 shows coefficients of starting and running friction for common material pairs.

Table 8–4 Screw Bearing Pressure p_b

Screw Material	Nut Material	Safe p_b, MPa	Notes
Steel	Bronze	17–24	Low speed
Steel	Bronze	11–17	0.05 m/s
	Cast iron	12–17	0.04 m/s
Steel	Bronze	5.5–9.5	0.1–0.2 m/s
	Cast iron	4–7	0.1–0.2 m/s
Steel	Bronze	1–1.6	0.25 m/s

Source: Data from Rothbart, H. A. and T. H. Brown, Jr. *Mechanical Design Handbook*. 2nd ed. New York: McGraw Hill, 2006.

Table 8–5 Coefficients of Friction f for Threaded Pairs

	Nut Material			
Screw Material	**Steel**	**Bronze**	**Brass**	**Cast Iron**
Steel, dry	0.15–0.25	0.15–0.23	0.15–0.19	0.15–0.25
Steel, machine oil	0.11–0.17	0.10–0.16	0.10–0.15	0.11–0.17
Bronze	0.08–0.12	0.04–0.06	—	0.06–0.09

Source: Data from Rothbart, H. A. and T. H. Brown, Jr. *Mechanical Design Handbook*. 2nd ed. New York: McGraw Hill, 2006.

Table 8–6 Thrust-Collar Friction Coefficients

Combination	Running	Starting
Soft steel on cast iron	0.12	0.17
Hard steel on cast iron	0.09	0.15
Soft steel on bronze	0.08	0.10
Hard steel on bronze	0.06	0.08

Source: Data from Rothbart, H. A. and T. H. Brown, Jr. *Mechanical Design Handbook*. 2nd ed. New York: McGraw Hill, 2006.

[1]Ham and Ryan, *An Experimental Investigation of the Friction of Screw-threads*, Bulletin 247, University of Illinois Experiment Station, Champaign-Urbana, Ill., June 7, 1932.

8–3 Threaded Fasteners

Figure 8–9 is a drawing of a standard hexagon-head bolt. Points of stress concentration are at the fillet, at the start of the threads (runout), and at the thread-root fillet in the plane of the nut when it is present. See Table A–25 for dimensions. The diameter of the washer face is the same as the width across the flats of the hexagon. The thread length for metric bolts is

$$L_T = \begin{cases} 2d + 6 & L \leq 125 \quad d \leq 48 \\ 2d + 12 & 125 < L \leq 200 \\ 2d + 25 & L > 200 \end{cases} \tag{8–13}$$

where the dimensions are in millimeters. The ideal bolt length is one in which only one or two threads project from the nut after it is tightened. Bolt holes may have burrs or sharp edges after drilling. These could bite into the fillet and increase stress concentration. Therefore, washers must always be used under the bolt head to prevent this. They should be of hardened steel and loaded onto the bolt so that the rounded edge of the stamped hole faces the washer face of the bolt. Sometimes it is necessary to use washers under the nut too.

The purpose of a bolt is to clamp two or more parts together. The clamping load stretches or elongates the bolt; the load is obtained by twisting the nut until the bolt has elongated almost to the elastic limit. If the nut does not loosen, this bolt tension remains as the preload or clamping force. When tightening, the mechanic should, if possible, hold the bolt head stationary and twist the nut; in this way the bolt shank will not feel the thread-friction torque.

The head of a hexagon-head cap screw is slightly thinner than that of a hexagon-head bolt. Dimensions of hexagon-head cap screws are listed in Table A–26. Hexagon-head cap screws are used in the same applications as bolts and also in applications in which one of the clamped members is threaded. Three other common cap-screw head styles are shown in Figure 8–10.

A variety of machine-screw head styles are shown in Figure 8–11.

Several styles of hexagonal nuts are illustrated in Figure 8–12; their dimensions are given in Table A–27. The material of the nut must be selected carefully to match that of the bolt. During tightening, the first thread of the nut tends to take the entire load; but yielding occurs, with some strengthening due to the cold work that takes place, and the load is eventually divided over about three nut threads. For this reason you should never reuse nuts; in fact, it can be dangerous to do so.

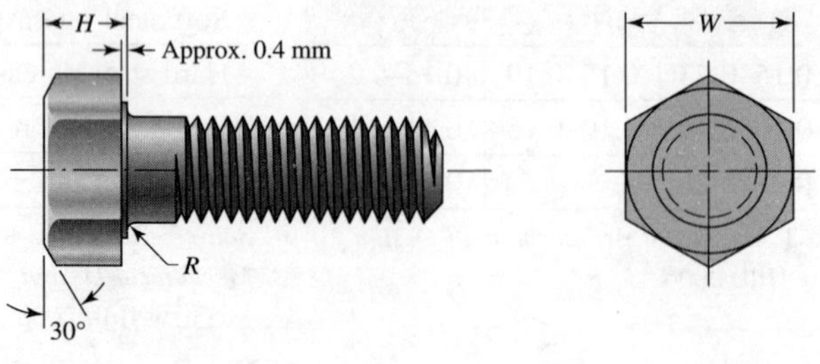

Figure 8–9

Hexagon-head bolt; note the washer face, the fillet under the head, the start of threads, and the chamfer on both ends. Bolt lengths are always measured from below the head.

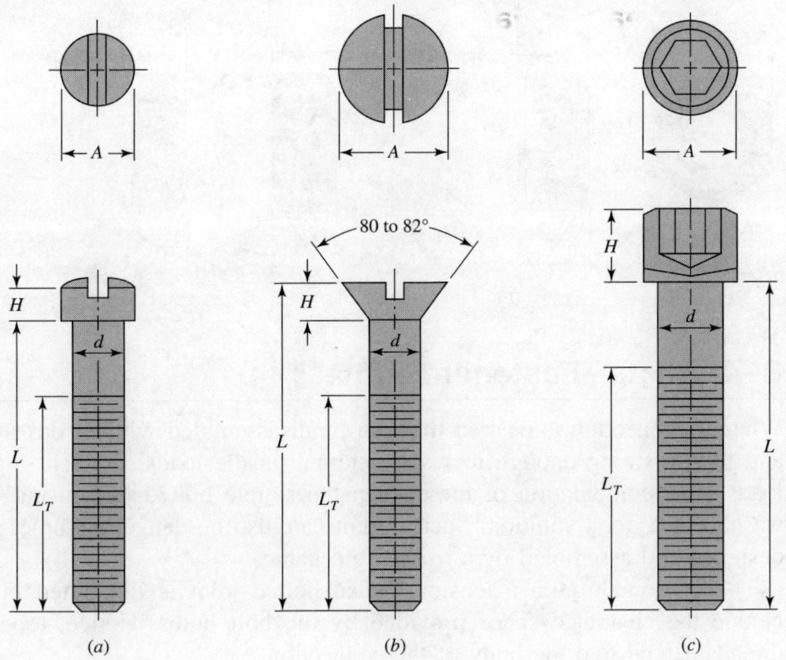

Figure 8–10

Typical cap-screw heads: (*a*) fillister head; (*b*) flat head; (*c*) hexagonal socket head. Cap screws are also manufactured with hexagonal heads similar to the one shown in Figure 8–9, as well as a variety of other head styles. This illustration uses one of the conventional methods of representing threads.

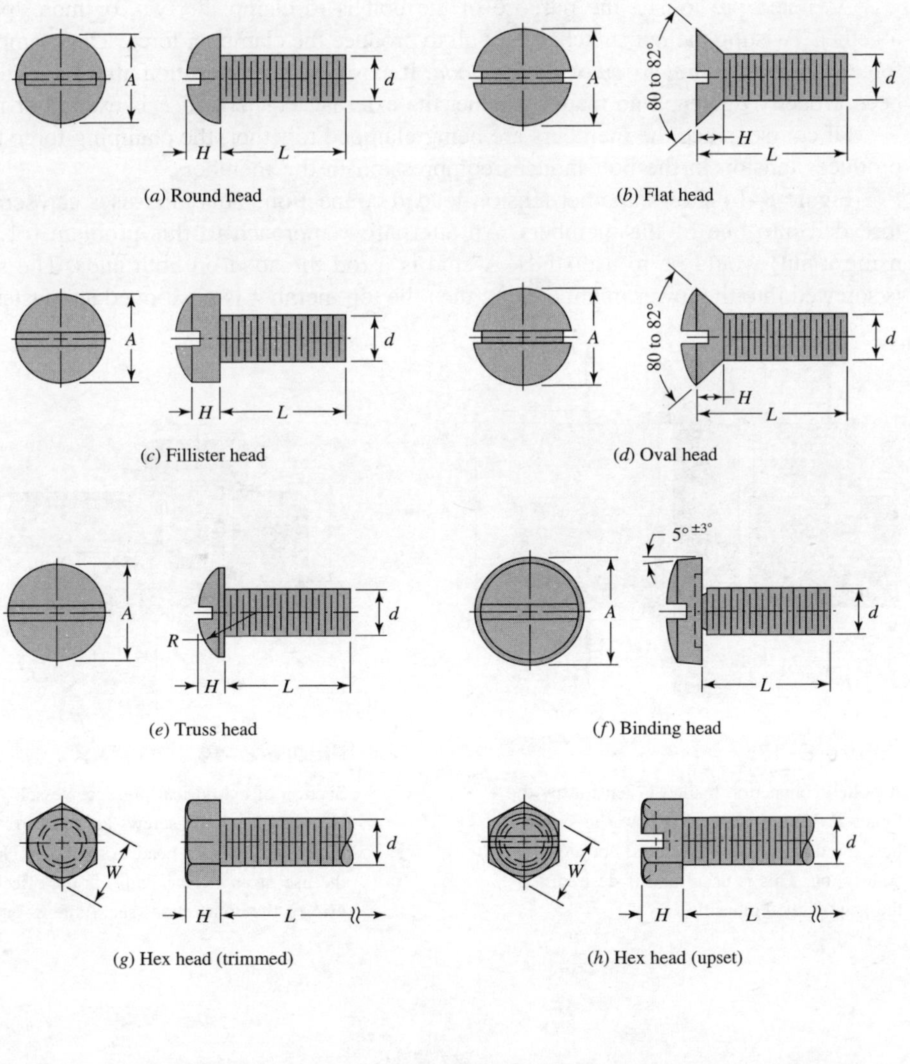

Figure 8–11

Types of heads used on machine screws.

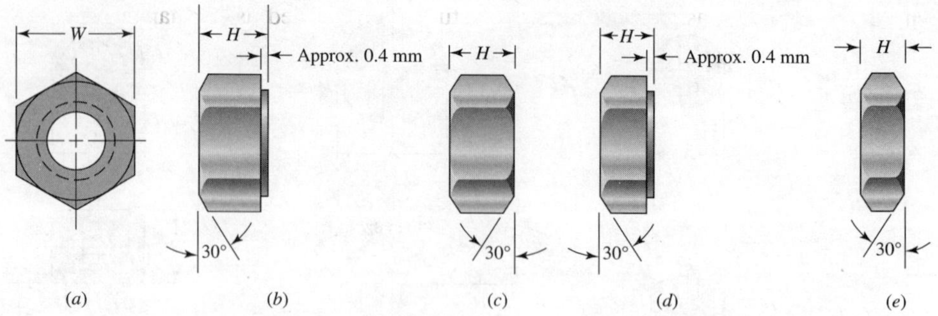

Figure 8–12

Hexagonal nuts: (*a*) end view, general; (*b*) washer-faced regular nut; (*c*) regular nut chamfered on both sides; (*d*) jam nut with washer face; (*e*) jam nut chamfered on both sides.

8–4 Joints—Fastener Stiffness

When a connection is desired that can be disassembled without destructive methods and that is strong enough to resist external tensile loads, moment loads, and shear loads, or a combination of these, then the simple bolted joint using hardened-steel washers is a good solution. Such a joint can also be dangerous unless it is properly designed and assembled by a *trained* mechanic.

A section through a tension-loaded bolted joint is illustrated in Figure 8–13. Notice the clearance space provided by the bolt holes. Notice, too, how the bolt threads extend into the body of the connection.

As noted previously, the purpose of the bolt is to clamp the two, or more, parts together. Twisting the nut stretches the bolt to produce the clamping force. This clamping force is called the *pretension* or *bolt preload*. It exists in the connection after the nut has been properly tightened no matter whether the external tensile load P is exerted or not.

Of course, since the members are being clamped together, the clamping force that produces tension in the bolt induces compression in the members.

Figure 8–14 shows another tension-loaded connection. This joint uses cap screws threaded into one of the members. An alternative approach to this problem (of not using a nut) would be to use studs. A stud is a rod threaded on both ends. The stud is screwed into the lower member first; then the top member is positioned and fastened

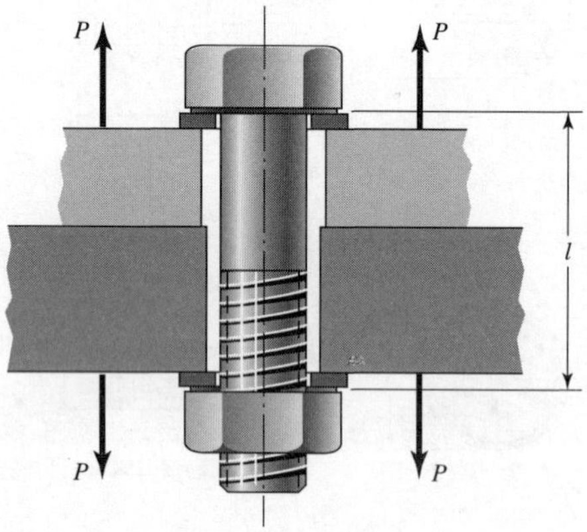

Figure 8–13

A bolted connection loaded in tension by the forces P. Note the use of two washers. Note how the threads extend into the body of the connection. This is usual and is desired. l is the grip of the connection.

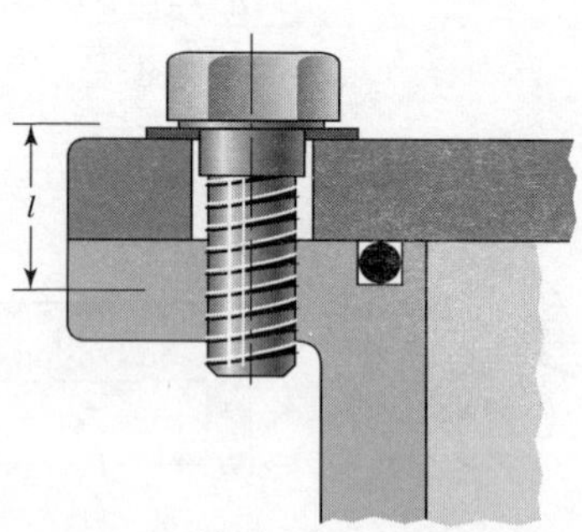

Figure 8–14

Section of cylindrical pressure vessel. Hexagon-head cap screws are used to fasten the cylinder head to the body. Note the use of an O-ring seal. l is the effective grip of the connection (see Table 8–7).

down with hardened washers and nuts. The studs are regarded as permanent, and so the joint can be disassembled merely by removing the nut and washer. Thus the threaded part of the lower member is not damaged by reusing the threads.

The *spring rate* is a limit as expressed in Equation (4–1). For an elastic member such as a bolt, as we learned in Equation (4–2), it is the ratio between the force applied to the member and the deflection produced by that force. We can use Equation (4–4) and the results of Problem 4–1 to find the stiffness constant of a fastener in any bolted connection.

The *grip* l of a connection is the total thickness of the clamped material. In Figure 8–13 the grip is the sum of the thicknesses of both members and both washers. In Figure 8–14 the effective grip is given in Table 8–7.

The stiffness of the portion of a bolt or screw within the clamped zone will generally consist of two parts, that of the unthreaded shank portion and that of the threaded portion. Thus the stiffness constant of the bolt is equivalent to the stiffnesses of two springs in series. Using the results of Problem 4–1, we find

$$\frac{1}{k} = \frac{1}{k_1} + \frac{1}{k_2} \qquad \text{or} \qquad k = \frac{k_1 k_2}{k_1 + k_2} \tag{8–14}$$

for two springs in series. From Equation (4–4), the spring rates of the threaded and unthreaded portions of the bolt in the clamped zone are, respectively,

$$k_t = \frac{A_t E}{l_t} \qquad k_d = \frac{A_d E}{l_d} \tag{8–15}$$

where A_t = tensile-stress area (Tables 8–1, 8–2)
l_t = length of threaded portion of grip
A_d = major-diameter area of fastener
l_d = length of unthreaded portion in grip

Substituting these stiffnesses in Equation (8–14) gives

$$k_b = \frac{A_d A_t E}{A_d l_t + A_t l_d} \tag{8–16}$$

where k_b is the estimated effective stiffness of the bolt or cap screw in the clamped zone. For short fasteners, the one in Figure 8–14, for example, the unthreaded area is small and so the first of the expressions in Equation (8–15) can be used to find k_b. For long fasteners, the threaded area is relatively small, and so the second expression in Equation (8–15) can be used. Table 8–7 is useful.

8–5 Joints—Member Stiffness

In the previous section, we determined the stiffness of the fastener in the clamped zone. In this section, we wish to study the stiffnesses of the members in the clamped zone. Both of these stiffnesses must be known in order to learn what happens when the assembled connection is subjected to an external tensile loading.

There may be more than two members included in the grip of the fastener. All together these act like compressive springs in series, and hence the total spring rate of the members is

$$\frac{1}{k_m} = \frac{1}{k_1} + \frac{1}{k_2} + \frac{1}{k_3} + \cdots + \frac{1}{k_i} \tag{8–17}$$

Table 8–7 Suggested Procedure for Finding Fastener Stiffness

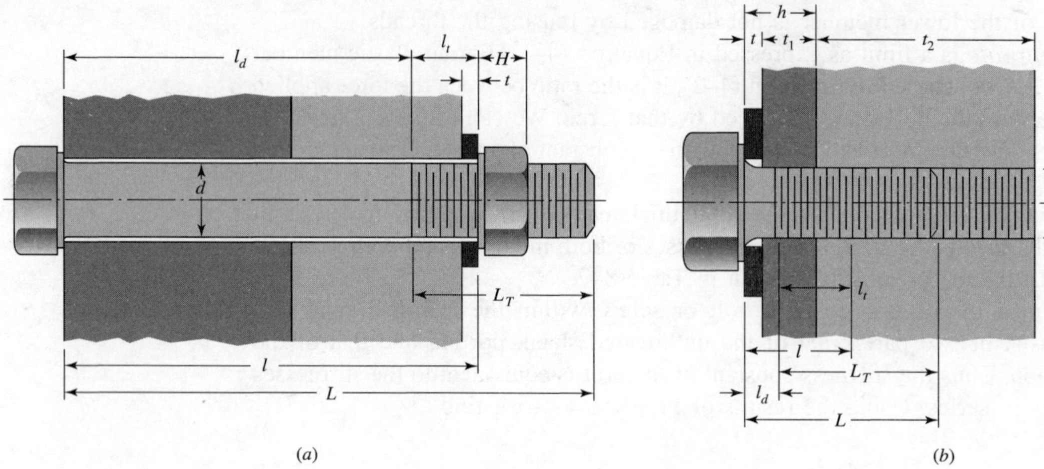

(*a*) (*b*)

Given fastener diameter d and pitch p in mm or number of threads per inch

Washer thickness: t from Table A–28

Nut thickness [Figure (*a*) only]: H from Table A–27

Grip length:

For Figure (*a*): l = thickness of all material squeezed between face of bolt and face of nut

For Figure (*b*): $l = \begin{cases} h + t_2/2, & t_2 < d \\ h + d/2, & t_2 \geq d \end{cases}$

Fastener length (round up using Table A–15*):

For Figure (*a*): $L > l + H$

For Figure (*b*): $L > h + 1.5d$

Threaded length L_T:

$$L_T = \begin{cases} 2d + 6 \text{ mm}, & L \leq 125 \text{ mm}, d \leq 48 \text{ mm} \\ 2d + 12 \text{ mm}, & 125 < L \leq 200 \text{ mm} \\ 2d + 25 \text{ mm}, & L > 200 \text{ mm} \end{cases}$$

Length of unthreaded portion in grip: $l_d = L - L_T$

Length of threaded portion in grip: $l_t = l - l_d$

Area of unthreaded portion: $A_d = \pi d^2/4$

Area of threaded portion: A_t from Table 8–1 or 8–2

Fastener stiffness: $k_b = \dfrac{A_d A_t E}{A_d l_t + A_t l_d}$

*Bolts and cap screws may not be available in all the preferred lengths listed in Table A–15. Large fasteners may not be available in fractional inches or in millimeter lengths ending in a nonzero digit. Check with your bolt supplier for availability.

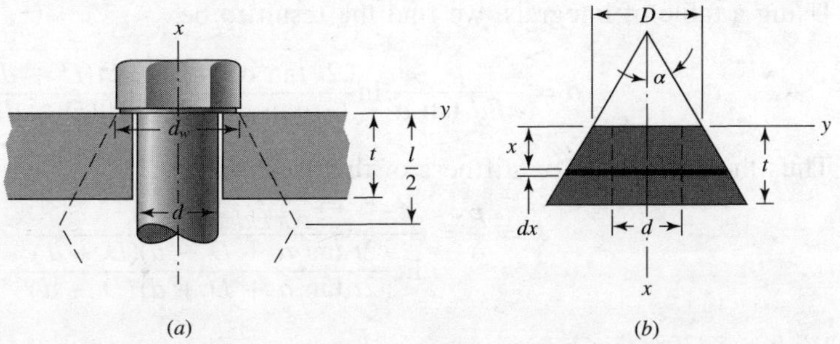

Figure 8–15

Compression of a member with the equivalent elastic properties represented by a frustum of a hollow cone. Here, l represents the grip length.

If one of the members is a soft gasket, its stiffness relative to the other members is usually so small that for all practical purposes the others can be neglected and only the gasket stiffness used.

If there is no gasket, the stiffness of the members is rather difficult to obtain, except by experimentation, because the compression region spreads out between the bolt head and the nut and hence the area is not uniform. There are, however, some cases in which this area can be determined.

Ito[2] has used ultrasonic techniques to determine the pressure distribution at the member interface. The results show that the pressure stays high out to about 1.5 bolt radii. The pressure, however, falls off farther away from the bolt. Thus Ito suggests the use of Rotscher's pressure-cone method for stiffness calculations with a variable cone angle. This method is quite complicated, and so here we choose to use a simpler approach using a fixed cone angle.

Figure 8–15 illustrates the general cone geometry using a half-apex angle α. An angle $\alpha = 45°$ has been used, but Little[3] reports that this overestimates the clamping stiffness. When loading is restricted to a washer-face annulus (hardened steel, cast iron, or aluminum), the proper apex angle is smaller. Osgood[4] reports a range of $25° \le \alpha \le 33°$ for most combinations. In this book we shall use $\alpha = 30°$ except in cases in which the material is insufficient to allow the frusta to exist.

Referring now to Figure 8–15*b*, the contraction of an element of the cone of thickness dx subjected to a compressive force P is, from Equation (4–3),

$$d\delta = \frac{P\,dx}{EA} \tag{a}$$

The area of the element is

$$\begin{aligned} A &= \pi\left(r_o^2 - r_i^2\right) = \pi\left[\left(x \tan\alpha + \frac{D}{2}\right)^2 - \left(\frac{d}{2}\right)^2\right] \\ &= \pi\left(x\tan\alpha + \frac{D+d}{2}\right)\left(x\tan\alpha + \frac{D-d}{2}\right) \end{aligned} \tag{b}$$

Substituting this in Equation (*a*) and integrating gives a total contraction of

$$\delta = \frac{P}{\pi E}\int_0^t \frac{dx}{[x\tan\alpha + (D+d)/2][x\tan a + (D-d)/2]} \tag{c}$$

---

[2] Y. Ito, J. Toyoda, and S. Nagata, "Interface Pressure Distribution in a Bolt-Flange Assembly," ASME paper no. 77-WA/DE-11, 1977.

[3] R. E. Little, "Bolted Joints: How Much Give?" *Machine Design*, November 9, 1967.

[4] C. C. Osgood, "Saving Weight on Bolted Joints," *Machine Design*, October 25, 1979.

Using a table of integrals, we find the result to be

$$\delta = \frac{P}{\pi E d \tan \alpha} \ln \frac{(2t \tan \alpha + D - d)(D + d)}{(2t \tan \alpha + D + d)(D - d)} \tag{d}$$

Thus the spring rate or stiffness of this frustum is

$$k = \frac{P}{\delta} = \frac{\pi E d \tan \alpha}{\ln \dfrac{(2t \tan \alpha + D - d)(D + d)}{(2t \tan \alpha + D + d)(D - d)}} \tag{8–18}$$

With $\alpha = 30°$, this becomes

$$k = \frac{0.5774 \pi E d}{\ln \dfrac{(1.155t + D - d)(D + d)}{(1.155t + D + d)(D - d)}} \tag{8–19}$$

Equation (8–19), or (8–18), must be solved separately for each frustum in the joint. Then individual stiffnesses are assembled to obtain k_m using Equation (8–17).

If the members of the joint have the same Young's modulus E with symmetrical frusta back to back, then they act as two identical springs in series. From Equation (8–17) we learn that $k_m = k/2$. Using the grip as $l = 2t$ and d_w as the diameter of the washer face, from Equation (8–18) we find the spring rate of the members to be

$$k_m = \frac{\pi E d \tan \alpha}{2 \ln \dfrac{(l \tan \alpha + d_w - d)(d_w + d)}{(l \tan \alpha + d_w + d)(d_w - d)}} \tag{8–20}$$

The diameter of the washer face is about 50 percent greater than the fastener diameter for standard hexagon-head bolts and cap screws. Thus we can simplify Equation (8–20) by letting $d_w = 1.5d$. If we also use $\alpha = 30°$, then Equation (8–20) can be written as

$$k_m = \frac{0.5774 \pi E d}{2 \ln \left(5 \dfrac{0.5774 l + 0.5d}{0.5774 l + 2.5d} \right)} \tag{8–21}$$

It is easy to program the numbered equations in this section, and you should do so. The time spent in programming will save many hours of formula plugging.

To see how good Equation (8–20) is, solve it for k_m/E_d:

$$\frac{k_m}{Ed} = \frac{\pi \tan \alpha}{2 \ln \left[\dfrac{(l \tan \alpha + d_w - d)(d_w + d)}{(l \tan \alpha + d_w + d)(d_w - d)} \right]}$$

Earlier in the section use of $\alpha = 30°$ was recommended for hardened steel, cast iron, or aluminum members. Wileman, Choudury, and Green[5] conducted a finite element study of this problem. The results, which are depicted in Figure 8–16, agree with the $\alpha = 30°$ recommendation, coinciding exactly at the aspect ratio $d/l = 0.4$. Additionally, they offered an exponential curve-fit of the form

$$\frac{k_m}{Ed} = A \exp (Bd/l) \tag{8–22}$$

[5] J. Wileman, M. Choudury, and I. Green, "Computation of Member Stiffness in Bolted Connections," *Trans. ASME, J. Mech. Design*, vol. 113, December 1991, pp. 432–437.

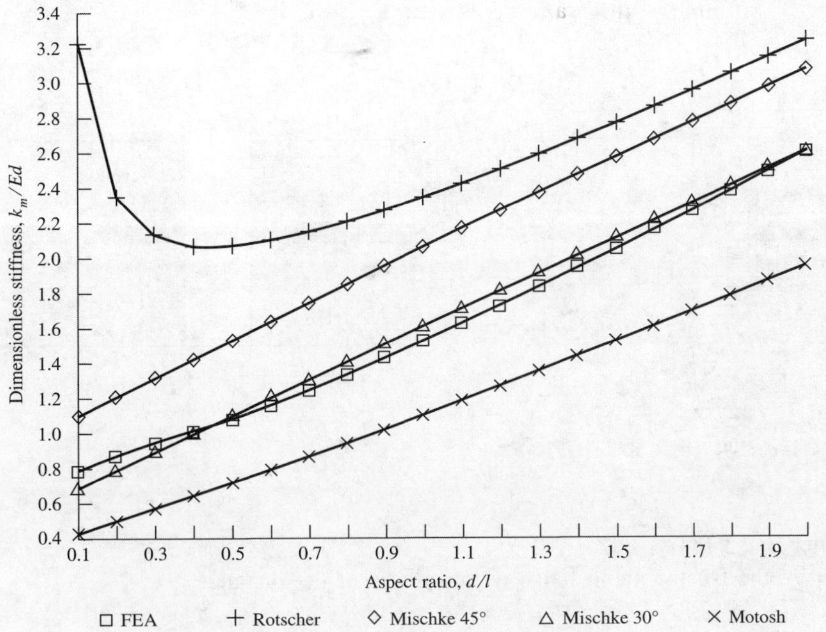

Figure 8–16

The dimensionless plot of stiffness versus aspect ratio of the members of a bolted joint, showing the relative accuracy of methods of Rotscher, Mischke, and Motosh, compared to a finite-element analysis (FEA) conducted by Wileman, Choudury, and Green.

Table 8–8 Stiffness Parameters of Various Member Materials[†]

Material Used	Poisson Ratio	Elastic Modulus GPa	A	B
Steel	0.291	207	0.787 15	0.628 73
Aluminum	0.334	71	0.796 70	0.638 16
Copper	0.326	119	0.795 68	0.635 53
Gray cast iron	0.211	100	0.778 71	0.616 16
General expression			0.789 52	0.629 14

Source: Data from Wileman, J., M. Choudhury, and I. Green. "Computation of Member Stiffness in Bolted Connections." *Transactions of the ASME, Journal of Mechanical Design* 113, no. 4 (December 1991): 432–437.

with constants A and B defined in Table 8–8. Equation (8–22) offers a simple calculation for member stiffness k_m. However, it is very important to note that the *entire joint* must be made up of the *same material*. For departure from these conditions, Equation (8–19) remains the basis for approaching the problem.

EXAMPLE 8–2

As shown in Figure 8–17*a*, two plates are clamped by washer-faced M12 × 1.25 class 8.8, 40 mm length bolts each with a steel plain washer.

(*a*) Determine the member spring rate k_m if the top plate is steel and the bottom plate is gray cast iron.

(*b*) Using the method of conical frusta, determine the member spring rate k_m if both plates are steel.

(*c*) Using Equation (8–22), determine the member spring rate k_m if both plates are steel. Compare the results with part (*b*).

(*d*) Determine the bolt spring rate k_b.

Figure 8–17

Dimensions in millimeters.

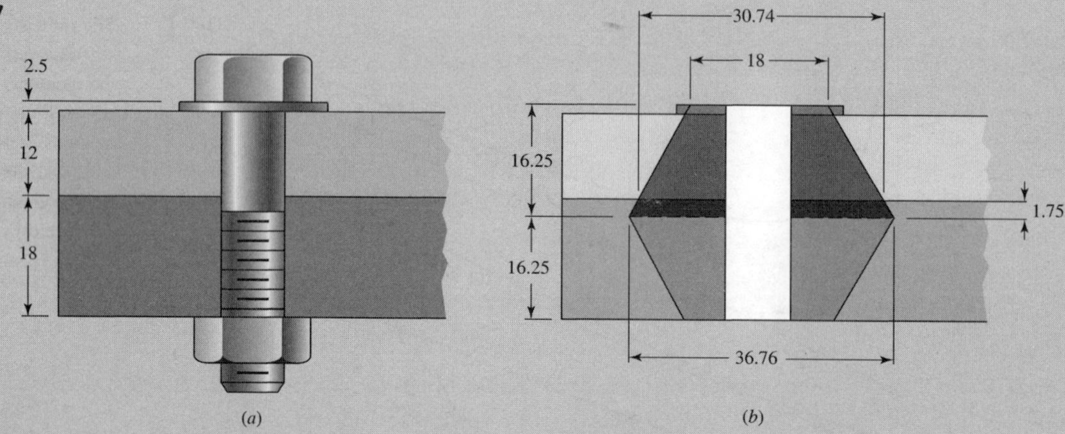

Solution

The thickness of plain washer is 2.5 mm.

(*a*) As shown in Figure 8–17*b*, the frusta extend halfway into the joint the distance

$$\frac{1}{2}(12 + 18 + 2.5) = 16.25 \text{ mm}$$

The distance between the joint line and the dotted frusta line is 16.25 − 12 − 2.5 = 1.75 mm. Thus, the top frusta consist of the steel washer, steel plate, and 1.75 mm of the cast iron. Since the washer and top plate are both steel with E = 207 GPa, they can be considered a single frustum of 14.5 mm thick. The outer diameter of the frustum of the steel member at the joint interface is 18 + 2(14.5) tan 30° = 34.74 mm. The outer diameter at the midpoint of the entire joint is 18 + 2(16.25) tan 30° = 36.76 mm. Using Equation (8–19), the spring rate of the steel is

$$k_1 = \frac{0.5774\pi(207\,000)(12)}{ln\left[\frac{(1.155 \cdot 14.5 + 18 - 12)(18 + 12)}{(1.155 \cdot 14.5 + 18 + 12)(18 - 12)}\right]} = 5.07(10^6) \text{ N/mm}$$

From Table 8–8 or A–5, for gray cast iron, E = 100 GPa. Thus for the upper cast-iron frustum

$$k_2 = \frac{0.5774\pi(100\,000)(12)}{ln\left[\frac{(1.155 \cdot 1.75 + 34.74 - 12)(34.74 + 12)}{(1.155 \cdot 1.75 + 34.74 + 12)(34.74 - 12)}\right]} = 5.08(10^6) \text{ N/mm}$$

For the lower cast-iron frustum

$$k_3 = \frac{0.5774\pi(100\,000)(12)}{ln\left[\frac{(1.155 \cdot 16.25 + 18 - 12)(18 + 12)}{(1.155 \cdot 16.25 + 18 + 12)(18 - 12)}\right]} = 2.34(10^6) \text{ N/mm}$$

The three frusta are in series, so from Equation (8–17)

$$\frac{1}{k_m} = \frac{1}{5.07(10^6)} + \frac{1}{1.21(10^6)} + \frac{1}{2.34(10^6)}$$

Answer

This results in $k_m = 1.22(10^6)$ N/mm.

(*b*) If the entire joint is steel, Equation (8–21) with l = 2(16.25) = 32.5 mm gives

Answer
$$k_m = \frac{0.5774\pi(207\ 000)(12)}{2\ \ln\left(5\dfrac{0.5774 \cdot 32.5 + 0.5 \cdot 12}{0.5774 \cdot 32.5 + 2.5 \cdot 12}\right)} = 2.42(10^6)\ \text{N/mm}.$$

(*c*) From Table 8–8, $A = 0.787\ 15$, $B = 0.628\ 73$. Equation (8–22) gives

Answer
$$k_m = 207\ 000(12)(0.787\ 15)\exp[0.628\ 73(12)/32.5] = 14.92(10^6)\ \text{N/mm}$$

For this case, the difference between the results for Equations (8–21) and (8–22) is less than 2 percent.
(*d*) Following the procedure of Table 8–7, the threaded length of a 12 mm bolt is $L_T = 2(12) + 6 = 30$ mm. The length of the unthreaded portion is $l_d = 40 - 30 = 5$ mm. The length of the unthreaded portion in grip is $l_t = 32.5 - 5 = 27.5$ mm. The major diameter area is $A_d = (\pi/4)(12^2) = 113.10\ \text{mm}^2$. From Table 8–1, the tensile-stress area is $A_t = 92.1\ \text{mm}^2$. From Equation (8–16)

Answer
$$k_b = \frac{113.1(92.1)207\ 000}{113.1(27.5) + 92.1(6)} = 5.88(10^5)\ \text{N/mm}$$

8–6 Bolt Strength

In the specification standards for bolts, the strength is specified by stating SAE or ASTM minimum quantities, the *minimum proof strength,* or *minimum proof load,* and the *minimum tensile strength.* The *proof load* is the maximum load (force) that a bolt can withstand without acquiring a permanent set. The *proof strength* is the quotient of the proof load and the tensile-stress area. The proof strength thus corresponds roughly to the proportional limit and corresponds to 0.0001-in permanent set in the fastener (first measurable deviation from elastic behavior). Tables 8–9, 8–10, and 8–11 provide *minimum* strength specifications for steel bolts. The values of the mean proof strength, the mean tensile strength, and the corresponding standard deviations are not part of the specification codes, so it is the designer's responsibility to obtain these values, perhaps by laboratory testing, if designing to a reliability specification.

The SAE specifications are found in Table 8–9. The bolt grades are numbered according to the tensile strengths, with decimals used for variations at the same strength level. Bolts and screws are available in all grades listed. Studs are available in grades 1, 2, 4, 5, 8, and 8.1. Grade 8.1 is not listed.

ASTM specifications are listed in Table 8–10. ASTM threads are shorter because ASTM deals mostly with structures; structural connections are generally loaded in shear, and the decreased thread length provides more shank area.

Specifications for metric fasteners are given in Table 8–11.

Specifications-grade bolts usually bear a manufacturer's mark or logo, in addition to the grade marking, on the bolt head. Such marks confirm that the bolt meets or exceeds specifications. If such marks are missing, assume the bolt strength is unregulated, or is relatively low and not intended for engineering applications.

Bolts in fatigue axial loading fail at the fillet under the head, at the thread runout, and at the first thread engaged in the nut. If the bolt has a standard shoulder under the head, it has a value of K_f from 2.1 to 2.3, *and* this shoulder fillet is protected from scratching or scoring by a washer. If the thread runout has a 15° or less half-cone angle, the stress is higher at the first engaged thread in the nut. Bolts

Table 8–9 SAE Specifications for Steel Bolts

SAE Grade No.	Size Range Inclusive, in	Minimum Proof Strength,* kpsi (MPa)	Minimum Tensile Strength,* kpsi (MPa)	Minimum Yield Strength,* kpsi (MPa)	Material	Head Marking
1	$\frac{1}{4}$–$1\frac{1}{2}$	33 (230)	60 (410)	36 (250)	Low or medium carbon	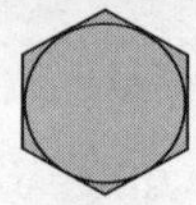
2	$\frac{1}{4}$–$\frac{3}{4}$	55 (380)	74 (510)	57 (390)	Low or medium carbon	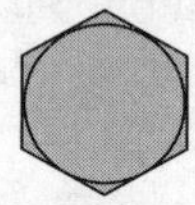
	$\frac{7}{8}$–$1\frac{1}{2}$	33 (230)	60 (410)	36 (250)		
4	$\frac{1}{4}$–$1\frac{1}{2}$	65 (450)	115 (790)	100 (690)	Medium carbon, cold-drawn	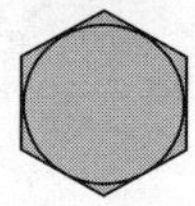
5	$\frac{1}{4}$–1	85 (585)	120 (830)	92 (635)	Medium carbon, Q&T	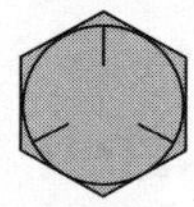
	$1\frac{1}{8}$–$1\frac{1}{2}$	74 (510)	105 (725)	81 (560)		
5.2	$\frac{1}{4}$–1	85 (585)	120 (830)	92 (635)	Low-carbon martensite, Q&T	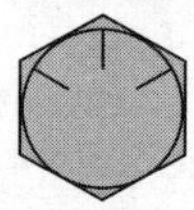
7	$\frac{1}{4}$–$1\frac{1}{2}$	105 (725)	133 (915)	115 (790)	Medium-carbon alloy, Q&T	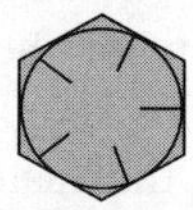
8	$\frac{1}{4}$–$1\frac{1}{2}$	120 (830)	150 (1030)	130 (900)	Medium-carbon alloy, Q&T	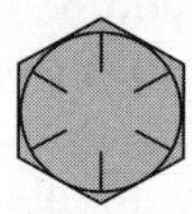
8.2	$\frac{1}{4}$–1	120 (830)	150 (1030)	130 (900)	Low-carbon martensite, Q&T	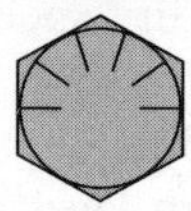

*Minimum strengths are strengths exceeded by 99 percent of fasteners.

are sized by examining the loading at the plane of the washer face of the nut. This is the weakest part of the bolt *if and only if* the conditions above are satisfied (washer protection of the shoulder fillet and thread runout ≤15°). Inattention to this requirement has led to a record of 15 percent fastener fatigue failure under the head, 20 percent at thread runout, and 65 percent where the designer is focusing attention. It does little good to concentrate on the plane of the nut washer face if it is not the weakest location.

Nuts are graded so that they can be mated with their corresponding grade of bolt. The purpose of the nut is to have its threads deflect to distribute the load of the bolt more evenly to the nut. The nut's properties are controlled in order to accomplish this. The grade of the nut should be the grade of the bolt.

Table 8–10 ASTM Specifications for Steel Bolts

ASTM Designation No.	Size Range, Inclusive, in	Minimum Proof Strength,* kpsi (MPa)	Minimum Tensile Strength,* kpsi (MPa)	Minimum Yield Strength,* kpsi (MPa)	Material	Head Marking
A307	$\frac{1}{4}$–$1\frac{1}{2}$	33 (230)	60 (410)	36 (250)	Low carbon	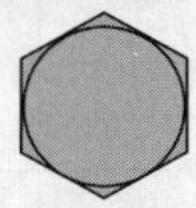
A325, type 1	$\frac{1}{2}$–1	85 (585)	120 (830)	92 (635)	Medium carbon, Q&T	
	$1\frac{1}{8}$–$1\frac{1}{2}$	74 (510)	105 (725)	81 (560)		
A325, type 2	$\frac{1}{2}$–1	85 (585)	120 (830)	92 (635)	Low-carbon, martensite, Q&T	
	$1\frac{1}{8}$–$1\frac{1}{2}$	74 (510)	105 (725)	81 (560)		
A325, type 3	$\frac{1}{2}$–1	85 (585)	120 (830)	92 (635)	Weathering steel, Q&T	
	$1\frac{1}{8}$–$1\frac{1}{2}$	74 (510)	105 (725)	81 (560)		
A354, grade BC	$\frac{1}{4}$–$2\frac{1}{2}$	105 (725)	125 (860)	109 (750)	Alloy steel, Q&T	
	$2\frac{3}{4}$–4	95 (655)	115 (790)	99 (680)		
A354, grade BD	$\frac{1}{4}$–4	120 (830)	150 (1030)	130 (900)	Alloy steel, Q&T	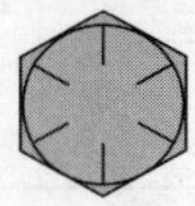
A449	$\frac{1}{4}$–1	85 (585)	120 (830)	92 (635)	Medium-carbon, Q&T	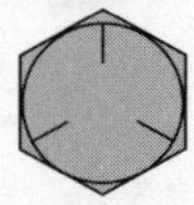
	$1\frac{1}{8}$–$1\frac{1}{2}$	74 (510)	105 (725)	81 (560)		
	$1\frac{3}{4}$–3	55 (380)	90 (625)	58 (400)		
A490, type 1	$\frac{1}{2}$–$1\frac{1}{2}$	120 (830)	150 (1030)	130 (900)	Alloy steel, Q&T	
A490, type 3	$\frac{1}{2}$–$1\frac{1}{2}$	120 (830)	150 (1030)	130 (900)	Weathering steel, Q&T	

*Minimum strengths are strengths exceeded by 99 percent of fasteners.

Table 8–11 Metric Mechanical-Property Classes for Steel Bolts, Screws, and Studs

Property Class	Size Range, Inclusive	Minimum Proof Strength,* MPa	Minimum Tensile Strength,* MPa	Minimum Yield Strength,* MPa	Material	Head Marking
4.6	M5–M36	225	400	240	Low or medium carbon	4.6
4.8	M1.6–M16	310	420	340	Low or medium carbon	4.8
5.8	M5–M24	380	520	420	Low or medium carbon	5.8
8.8	M16–M36	600	830	660	Medium carbon, Q&T	8.8
9.8	M1.6–M16	650	900	720	Medium carbon, Q&T	9.8
10.9	M5–M36	830	1040	940	Low-carbon martensite, Q&T	10.9
12.9	M1.6–M36	970	1220	1100	Alloy, Q&T	12.9

*Minimum strengths are strengths exceeded by 99 percent of fasteners.

8–7 Tension Joints—The External Load

Let us now consider what happens when an external tensile load P is applied to a bolted connection, as shown in Figure 8–18. The nomenclature used is:

F_i = preload

P_{total} = Total external tensile load applied to the joint

P = external tensile load per bolt

P_b = portion of P taken by bolt

P_m = portion of P taken by members

$F_b = P_b + F_i$ = resultant bolt load

$F_m = P_m - F_i$ = resultant load on members

C = fraction of external load P carried by bolt

$1 - C$ = fraction of external load P carried by bolt

N = Number of bolts in the joint

If N bolts equally share the total external load, then

$$P = P_{\text{total}}/N \tag{a}$$

In Figure 8–18*a*, a bolted joint is shown where the nut is in contact with the members, but not yet tightened to introduce a preload. The members and the bolt can be modeled as springs in parallel. The figure schematically represents the members with a large stiff spring and the bolt with a low-stiffness spring, each starting in an undeformed equilibrium condition. In Figure 8–18*b*, the nut is tightened to introduce a preload F_i to the joint. In doing so, the stiff spring of the members is compressed by an amount $\delta_m = F_i/k_m$, while the soft spring of the bolt is stretched by an amount $\delta_b = F_i/k_b$. Figure 8–19 shows a plot of these force-deflection characteristics, where the linear spring stiffness relationships are evident. Note that the ordinate $|F|$ represents the absolute value of the force in each member, where the force in the bolt is positive, and the force in the members is negative. As both the bolt and the members experience the preload F_i, the member loading moves up its stiffness line to point *a*, while the bolt loading moves to point *b*. Since the member stiffness is usually much greater than the bolt stiffness, this results in δ_m being significantly smaller than δ_b.

When an external tensile load P is applied to the members (not directly to the bolt), the bolt elongates further by $\Delta\delta_b$, whereas the members' deflection *decreases* by $\Delta\delta_m$, as shown in Figure 8–18*c*. By inspection of the figure, it can be realized that these deflections must be equal. From Figure 8–19, observe that the external load P is split into two portions, P_b applied to the bolt and P_m applied to the members, apportioned as necessary to ensure that $\Delta\delta_b = \Delta\delta_m$. Thus,

$$\Delta\delta_b = \frac{P_b}{k_b} = \Delta\delta_m = \frac{P_m}{k_m} \tag{b}$$

or,

$$P_m = \frac{k_m}{k_b} P_b \tag{c}$$

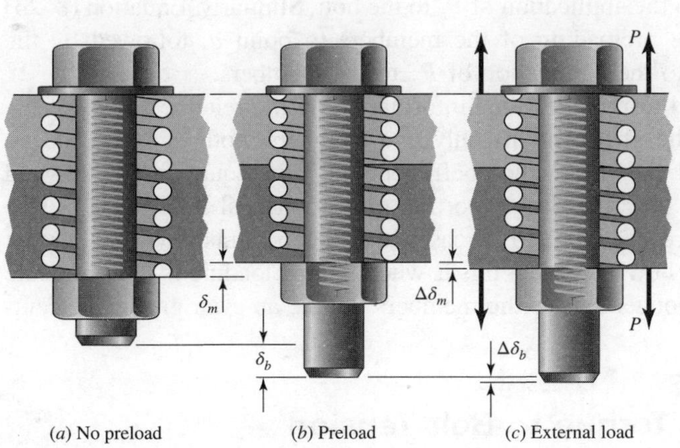

Figure 8–18

Spring model of the effect of preload and external tensile load to a bolted joint.

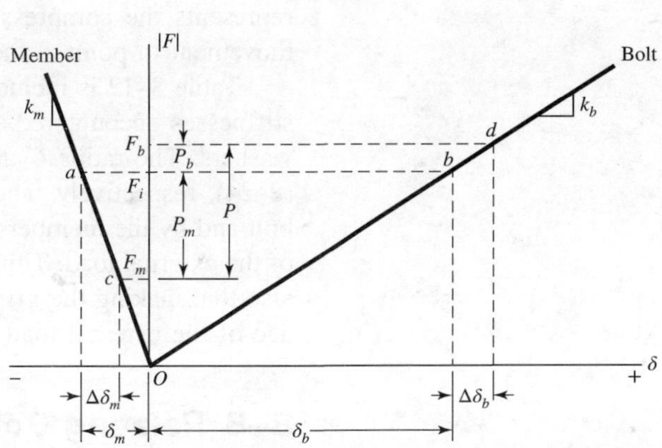

Figure 8–19

Force-deflection characteristics of an externally loaded bolted member with an initial bolt preload force, F_i. The external load is P.

Table 8–12 **An example of Bolt and Member Stiffnesses. Steel members clamped steel bolt.** $C = \dfrac{k_b}{k_b + k_m}$

	Stiffnesses, *M* MN/mm			
Bolt Grip, mm	k_b	k_m	C	$1 - C$
50	0.45	2.22	0.168	0.832
75	0.31	1.98	0.136	0.864
100	0.24	1.86	0.114	0.886

Since $P = P_b + P_m$, then, $P_m = P - P_b$. Substituting this into Equation (*c*) and rearranging we find

$$P_b = \frac{k_b P}{k_b + k_m} = CP \tag{d}$$

and

$$P_m = P - P_b = (1 - C)P \tag{e}$$

where

$$C = \frac{k_b}{k_b + k_m} \tag{f}$$

is called the *stiffness constant of the joint*. The resultant bolt load is

$$F_b = P_b + F_i = CP + F_i \qquad F_m < 0 \tag{8–23}$$

and the resultant load on the connected members is

$$F_m = P_m - F_i = (1 - C)P - F_i \qquad F_m < 0 \tag{8–24}$$

Of course, these results are valid only as long as some clamping load remains in the members; this is indicated by the qualifier in the equations. Returning to Figure 8–19, Equation (8–23) represents the preloading of the bolt to point *b*, followed by the movement to point *d* due to the application of P_b to the bolt. Similarly, Equation (8–24) represents the compressive preloading of the members to point *a*, followed by the movement to point *c* due to the application of P_m to the members.

Table 8–12 is included to provide some information on the relative values of the stiffnesses encountered. The grip contains only two members, both of steel, and no washers. The ratios C and $1 - C$ are the coefficients of P in Equations (8–23) and (8–24), respectively. They describe the proportion of the external load taken by the bolt and by the members, respectively. In all cases, the members take over 80 percent of the external load. Think how important this is when fatigue loading is present. Note also that making the grip longer causes the members to take an even greater percentage of the external load.

8–8 Relating Bolt Torque to Bolt Tension

Having learned that a high preload is very desirable in important bolted connections, we must next consider means of ensuring that the preload is actually developed when the parts are assembled.

If the overall length of the bolt can actually be measured with a micrometer when it is assembled, the bolt elongation due to the preload F_i can be computed using the formula $\delta = F_i l/(AE)$. Then the nut is simply tightened until the bolt elongates through the distance δ. This ensures that the desired preload has been attained.

The elongation of a screw cannot usually be measured, because the threaded end is often in a blind hole. It is also impractical in many cases to measure bolt elongation. In such cases the wrench torque required to develop the specified preload must be estimated. Then torque wrenching, pneumatic-impact wrenching, or the turn-of-the-nut method may be used.

The torque wrench has a built-in dial that indicates the proper torque.

With impact wrenching, the air pressure is adjusted so that the wrench stalls when the proper torque is obtained, or in some wrenches, the air automatically shuts off at the desired torque.

The turn-of-the-nut method requires that we first define the meaning of snug-tight. The *snug-tight* condition is the tightness attained by a few impacts of an impact wrench, or the full effort of a person using an ordinary wrench. When the snug-tight condition is attained, all additional turning develops useful tension in the bolt. The turn-of-the-nut method requires that you compute the fractional number of turns necessary to develop the required preload from the snug-tight condition. For example, for heavy hexagonal structural bolts, the turn-of-the-nut specification states that the nut should be turned a minimum of 180° from the snug-tight condition under optimum conditions. Problems 8–26 to 8–28 illustrate the method further.

Although the coefficients of friction may vary widely, we can obtain a good estimate of the torque required to produce a given preload by combining Equations (8–5) and (8–6):

$$T = \frac{F_i d_m}{2}\left(\frac{l + \pi f d_m \sec\alpha}{\pi d_m - f l \sec\alpha}\right) + \frac{F_i f_c d_c}{2} \tag{a}$$

where d_m is the average of the major and minor diameters. Since $\tan\lambda = l/\pi d_m$, we divide the numerator and denominator of the first term by πd_m and get

$$T = \frac{F_i d_m}{2}\left(\frac{\tan\lambda + f\sec\alpha}{1 - f\tan\lambda\sec\alpha}\right) + \frac{F_i f_c d_c}{2} \tag{b}$$

The diameter of the washer face of a hexagonal nut is the same as the width across flats and equal to $1\frac{1}{2}$ times the nominal size. Therefore the mean collar diameter is $d_c = (d + 1.5d)/2 = 1.25d$. Equation (*b*) can now be arranged to give

$$T = \left[\left(\frac{d_m}{2d}\right)\left(\frac{\tan\lambda + f\sec\alpha}{1 - f\tan\lambda\sec\alpha}\right) + 0.625\, f_c\right] F_i d \tag{c}$$

We now define a *torque coefficient* K as the term in brackets, and so

$$K = \left(\frac{d_m}{2d}\right)\left(\frac{\tan\lambda + f\sec\alpha}{1 - f\tan\lambda\sec\alpha}\right) + 0.625\, f_c \tag{8–25}$$

Equation (*c*) can now be written

$$T = K F_i d \tag{8–26}$$

The coefficient of friction depends upon the surface smoothness, accuracy, and degree of lubrication. On the average, both f and f_c are about 0.15. The interesting fact about Equation (8–25) is that $K \approx 0.20$ for $f = f_c = 0.15$ no matter what size bolts are employed and no matter whether the threads are coarse or fine.

Blake and Kurtz have published results of numerous tests of the torquing of bolts.[6] By subjecting their data to a statistical analysis, we can learn something about the distribution of the torque coefficients and the resulting preload. Blake and Kurtz determined the preload in quantities of unlubricated and lubricated bolts of size $\frac{1}{2}$ in-20 UNF when torqued to 800 lbf · in. This corresponds roughly to an M12 × 1.25 bolt torqued to 90 N · m. The statistical analyses of these two groups of bolts, converted to SI units, are displayed in Tables 8–13 and 8–14.

We first note that both groups have about the same mean preload, 34 kN. The unlubricated bolts have a standard deviation of 4.9 kN. The lubricated bolts have a standard deviation of 3 kN.

The means obtained from the two samples are nearly identical, approximately 34 kN; using Equation (8–26), we find, for both samples, $K = 0.208$.

Bowman Distribution, a large manufacturer of fasteners, recommends the values shown in Table 8–15. In this book we shall use these values and use $K = 0.2$ when the bolt condition is not stated.

Do not rely too much on wrench torque when the range of acceptable preload is narrow. If high reliability is a requirement of the design, then preload should be determined by bolt elongation.

Table 8–13 Distribution of Preload F_i for 20 Tests of Unlubricated Bolts Torqued to 90 N · m

23.6	27.6	28.0	29.4	30.3	30.7	32.9	33.8	33.8	33.8
34.7	35.6	35.6	37.4	37.8	37.8	39.2	40.0	40.5	42.7

Mean value $\overline{F}_i = 34.3$ kN. Standard deviation, $\hat{\sigma} = 4.91$ kN.

Table 8–14 Distribution of Preload F_i for 10 Tests of Lubricated Bolts Torqued to 90 N · m

30.3	32.5	32.5	32.9	32.9	33.8	34.3	34.7	37.4	40.5

Mean value $\overline{F}_i = 34.18$ kN. Standard deviation, $\hat{\sigma} = 2.88$ kN.

Table 8–15 Torque Factors K for Use with Equation (8–26)

Bolt Condition	K
Nonplated, black finish	0.30
Zinc-plated	0.20
Lubricated	0.18
Cadmium-plated	0.16
With Bowman Anti-Seize	0.12
With Bowman-Grip nuts	0.09

[6] J. C. Blake and H. J. Kurtz, "The Uncertainties of Measuring Fastener Preload," *Machine Design*, vol. 37, September 30, 1965, pp. 128–131.

EXAMPLE 8–3

A M20 × 1.5 SAE grade 5 bolt is subjected to a load P of 26.5 kN in a tension joint. The initial bolt tension is F_i = 110 000 N. The bolt and joint stiffnesses are k_b = 1.15 and k_m = 2.4 MN/mm, respectively.
(*a*) Determine the preload and service load stresses in the bolt. Compare these to the SAE minimum proof strength of the bolt.
(*b*) Specify the torque necessary to develop the preload, using Equation (8–26).
(*c*) Specify the torque necessary to develop the preload, using Equation (8–25) with $f = f_c = 0.15$.

Solution
From Table 8–2, A_t = 245 mm^2.
(*a*) The preload stress is

Answer

$$\sigma_i = \frac{F_i}{A_t} = \frac{110\ 000}{245} = 449 \text{ MPa}$$

The stiffness constant is

$$C = \frac{k_b}{k_b + k_m} = \frac{1.14}{1.14 + 2.4} = 0.322$$

From Equation (8–23), the stress under the service load is

$$\sigma_b = \frac{F_b}{A_t} = \frac{CP + F_i}{A_t} = C\frac{P}{A_t} + \sigma_i$$

Answer

$$= 0.322\,\frac{26\ 500}{245} + 449 = 484 \text{ MPa}$$

From Table 8–9, the SAE minimum proof strength of the bolt is S_p = 600 MPa. The preload and service load stresses are respectively 25 and 19 percent less than the proof strength.
(*b*) From Equation (8–26), the torque necessary to achieve the preload is

Answer

$$T = KF_i d = 0.2(110)(10^3)(20) = 440\ 000 \text{ N}\cdot\text{mm}$$

(*c*) The minor diameter can be determined from the minor area in Table 8–2. Thus $d_r = \sqrt{4A_r/\pi} = \sqrt{4(259)/\pi}$ = 18.16 mm. Thus, the mean diameter is d_m = (20 + 18.16)/2 = 19.08 mm. The lead angle is

$$\lambda = \tan^{-1}\frac{l}{\pi d_m} = \tan^{-1}\frac{1}{\pi d_m N} = \tan^{-1}\frac{1.5}{\pi(19.08)} = 1.4342°$$

For α = 30°, Equation (8–25) gives

$$T = \left\{\left[\frac{19.08}{2(20)}\right]\left[\frac{\tan 1.4342° + 0.15(\sec 30°)}{1 - 0.15(\tan 1.4342°)(\sec 30°)}\right] + 0.625(0.15)\right\} 110\ 000(10^3)(20)$$

$$= 428\ 570 \text{ N}\cdot\text{mm}$$

which is 7 percent less than the value found in part (*b*).

8–9 Statically Loaded Tension Joint with Preload

Equations (8–23) and (8–24) represent the forces in a bolted joint with preload. The tensile stress in the bolt can be found as in Example 8–3 as

$$\sigma_b = \frac{F_b}{A_t} = \frac{CP + F_i}{A_t} \tag{a}$$

Thus, the factor of safety guarding against the static stress exceeding the proof strength is

$$n_p = \frac{S_p}{\sigma_b} = \frac{S_p}{(CP + F_i)/A_t} \tag{b}$$

or

$$n_p = \frac{S_p A_t}{CP + F_i} \tag{8–27}$$

This factor of safety is used similarly to a traditional yielding factor of safety, with the proof strength corresponding to the first measurable permanent set. It is sometimes referred to as a yielding factor of safety. Since it is common to load a bolt close to the proof strength, this factor of safety is often not much greater than unity.

Another indicator of yielding that is sometimes used is a *load factor,* which is applied only to the load P as a guard against overloading. Applying such a load factor, n_L, to the load P in Equation (*a*), and equating it to the proof strength gives

$$\frac{Cn_L P + F_i}{A_t} = S_p \tag{c}$$

Solving for the load factor gives

$$n_L = \frac{S_p A_t - F_i}{CP} \tag{8–28}$$

It is also essential for a safe joint that the external load be smaller than that needed to cause the joint to separate. If separation does occur, then the entire external load will be imposed on the bolt. Let P_0 be the value of the external load that would cause joint separation. At separation, $F_m = 0$ in Equation (8–24), and so

$$(1 - C)P_0 - F_i = 0 \tag{d}$$

Let the factor of safety against joint separation be

$$n_0 = \frac{P_0}{P} \tag{e}$$

Substituting $P_0 = n_0 P$ in Equation (*d*), we find

$$n_0 = \frac{F_i}{P(1 - C)} \tag{8–29}$$

as a load factor guarding against joint separation.

Figure 8–20 is the stress-strain diagram of a good-quality bolt material. Notice that there is no clearly defined yield point and that the diagram progresses smoothly up to fracture, which corresponds to the tensile strength. This means that no matter how much preload is given the bolt, it will retain its load-carrying capacity. This is what keeps the bolt tight and determines the joint strength. The pretension is the

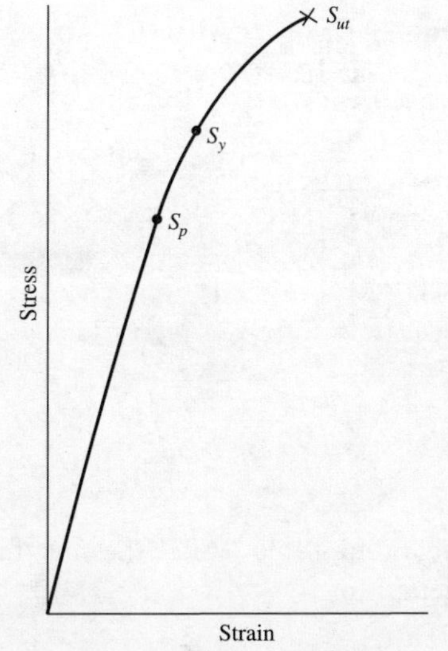

Figure 8–20

Typical stress-strain diagram for bolt materials showing proof strength S_p, yield strength S_y, and ultimate tensile strength S_{ut}.

"muscle" of the joint, and its magnitude is determined by the bolt strength. If the full bolt strength is not used in developing the pretension, then money is wasted and the joint is weaker.

Russell, Burdsall & Ward Inc. (RB&W) recommendations for preload are 60 kpsi for SAE grade 5 bolts for nonpermanent connections, and that A325 bolts (equivalent to SAE grade 5) used in structural applications be tightened to proof load or beyond (85 kpsi up to a diameter of 1 in).[7] Bowman[8] recommends a preload of 75 percent of proof load, which is about the same as the RB&W recommendations for reused bolts. In view of these guidelines, it is recommended for both static and fatigue loading that the following be used for preload:

$$F_i = \begin{cases} 0.75F_p & \text{for nonpermanent connections, reused fasteners} \\ 0.90F_p & \text{for permanent connections} \end{cases} \tag{8–30}$$

where F_p is the proof load, obtained from the equation

$$F_p = A_t S_p \tag{8–31}$$

Here S_p is the proof strength obtained from Tables 8–9 to 8–11. For other materials, an approximate value is $S_p = 0.85S_y$.

EXAMPLE 8–4

Figure 8–21 is a cross section of a grade 25 cast-iron pressure vessel. A total of N bolts are to be used to resist a separating force of 160 kN.

(*a*) Determine k_b, k_m, and C.

(*b*) Find the number of bolts required for a load factor of 2 where the bolts may be reused when the joint is taken apart.

(*c*) With the number of bolts obtained in part (*b*), determine the realized load factor for overload, the yielding factor of safety, and the load factor for joint separation.

[7]Russell, Burdsall & Ward Inc., *Helpful Hints for Fastener Design and Application*, Mentor, Ohio, 1965, p. 42.

[8]Bowman Distribution–Barnes Group, *Fastener Facts*, Cleveland, 1985, p. 90.

Figure 8–21

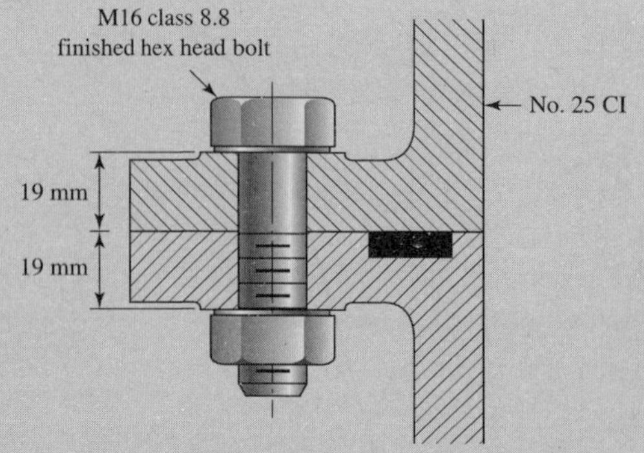

Solution

(*a*) The grip is $l = 38$ mm. From Table A–27, the nut thickness is 13 mm. Adding two threads beyond the nut of 2 mm gives a bolt length of

$$L = 13 + 38 + 4 = 55 \text{ mm}$$

The size bolt is $L = 55$ mm. From Equation (8–13), the thread length is $L_T = 2(16) + 6 = 38$ mm. Thus, the length of the unthreaded portion in the grip is $l_d = 55 - 38 = 17$ mm. The threaded length in the grip is $l_t = l - l_d = 17$ mm. From Table 8–2, $A_t = 157$ mm^2. The major-diameter area is $A_d = \pi(16)^2/4 = 201$ mm^2. The bolt stiffness is then

Answer

$$k_b = \frac{A_d A_t E}{A_d l_t + A_t l_d} = \frac{201\,(157)\,(207\ 000)}{201\,(17) + 157\,(21)}$$

$$= 973 \text{ kN/mm}$$

From Table A–23, for no. 25 cast iron we will use $E = 117\ 700$ MPa. The stiffness of the members, from Equation (8–21), is

Answer

$$k_m = \frac{0.5774\pi E d}{2\ \ln\left(5\,\dfrac{0.5774l + 0.5d}{0.5774l + 2.5d}\right)} = \frac{0.5774\pi(117\ 700)(16)}{2\ \ln\left[5\,\dfrac{0.5774(38) + 0.5(16)}{0.5774(38) + 2.5(16)}\right]}$$

$$= 1935 \text{ kN/mm}$$

If you are using Equation (8–22), from Table 8–8, $A = 0.778\ 71$ and $B = 0.616\ 16$, and

$$\begin{aligned} k_m &= EdA\ \exp\,(Bd/l) \\ &= 117\ 700(16)(0.778\ 71)\ \exp[0.616\ 16(0.625)/38] \\ &= 1901 \text{ kN/mm} \end{aligned}$$

which is only 1.7 percent lower than the previous result.

From the first calculation for k_m, the stiffness constant C is

Answer

$$C = \frac{k_b}{k_b + k_m} = \frac{973}{937 + 1935} = 0.335$$

(*b*) From Table 8–9, $S_p = 600$ MPa. Then, using Equations (8–30) and (8–31), we find the recommended preload to be

$$F_i = 0.75A_tS_p = 0.75(157)(600) = 70\ 650\ \text{N}$$

For N bolts, Equation (8–28) can be written

$$n_L = \frac{S_pA_t - F_i}{C(P_{\text{total}}/N)} \tag{1}$$

or

$$N = \frac{Cn_LP_{\text{total}}}{S_pA_t - F_i} = \frac{0.335(2)(160\ 000)}{600(157) - 70\ 650} = 4.6$$

Answer

Six bolts should be used to provide the specified load factor.

(*c*) With six bolts, the load factor actually realized is

Answer

$$n_L = \frac{600(157) - 70\ 650}{0.335(160\ 000/6)} = 2.6$$

From Equation (8–27), the yielding factor of safety is

Answer

$$n_p = \frac{S_pA_t}{C(P_{\text{total}}/N) + F_i} = \frac{600(157)}{0.335(160\ 000/6) + 70\ 650} = 1.18$$

From Equation (8–29), the load factor guarding against joint separation is

Answer

$$n_0 = \frac{F_i}{(P_{\text{total}}/N)(1 - C)} = \frac{70\ 650}{(160\ 000/6)(1 - 0.335)} = 3.98$$

8–10 Gasketed Joints

If a full gasket of area A_g is present in the joint, the gasket pressure p is found by dividing the force in the member by the gasket area per bolt. Thus, for N bolts,

$$p = -\frac{F_m}{A_g/N} \tag{a}$$

With a load factor n, Equation (8–24) can be written as

$$F_m = (1 - C)nP - F_i \tag{b}$$

Substituting this into Equation (a) gives the gasket pressure as

$$p = [F_i - nP(1 - C)]\frac{N}{A_g} \tag{8–32}$$

In full-gasketed joints uniformity of pressure on the gasket is important. To maintain adequate uniformity of pressure adjacent bolts should not be placed more than six nominal diameters apart on the bolt circle. To maintain wrench clearance, bolts should be placed at least three diameters apart. A rough rule for bolt spacing around a bolt circle is

$$3 \le \frac{\pi D_b}{Nd} \le 6 \tag{8–33}$$

where D_b is the diameter of the bolt circle and N is the number of bolts.

8–11 Fatigue Loading of Tension Joints

Tension-loaded bolted joints subjected to fatigue action can be analyzed directly by the methods of Chapter 6. Table 8–16 lists average fatigue stress-concentration factors for the fillet under the bolt head and also at the beginning of the threads on the bolt shank. These are already corrected for notch sensitivity. Designers should be aware that situations may arise in which it would be advisable to investigate these factors more closely, since they are only average values. Peterson[9] observes that the distribution of typical bolt failures is about 15 percent under the head, 20 percent at the end of the thread, and 65 percent in the thread at the nut face.

Use of rolled threads is the predominant method of thread-forming in screw fasteners. In thread-rolling, the amount of cold work and strain-strengthening is unknown to the designer; therefore, fully corrected (including K_f) axial endurance strength is reported in Table 8–17. Since K_f is included as an endurance strength reducer in Table 8–17, it should not be applied as a stress increaser when using values from this table. For cut threads, the methods of Chapter 6 are useful. Anticipate that the endurance strengths will be considerably lower.

For a general case with a constant preload, and an external load on a per bolt basis fluctuating between P_{min} and P_{max}, a bolt will experience fluctuating forces such that

$$F_{b\min} = CP_{\min} + F_i \tag{a}$$

$$F_{b\max} = CP_{\max} + F_i \tag{b}$$

Table 8–16 Fatigue Stress-Concentration Factors K_f for Threaded Elements

SAE Grade	Metric Grade	Rolled Threads	Cut Threads	Fillet
0 to 2	3.6 to 5.8	2.2	2.8	2.1
4 to 8	6.6 to 10.9	3.0	3.8	2.3

Table 8–17 Fully Corrected Endurance Strengths for Bolts and Screws with Rolled Threads*

Grade or Class	Size Range	Endurance Strength
SAE 5	$\frac{1}{4}$–1 in	18.6 kpsi (128 MPa)
	$1\frac{1}{8}$–$1\frac{1}{2}$ in	16.3 kpsi (112 MPa)
SAE 7	$\frac{1}{4}$–$1\frac{1}{2}$ in	20.6 kpsi (142 MPa)
SAE 8	$\frac{1}{4}$–$1\frac{1}{2}$ in	23.2 kpsi (160 MPa)
ISO 8.8	M16–M36	129 MPa
ISO 9.8	M1.6–M16	140 MPa
ISO 10.9	M5–M36	162 MPa
ISO 12.9	M1.6–M36	190 MPa

*Repeatedly applied, axial loading, fully corrected, including K_f as a strength reducer.

[9] W. D. Pilkey and D. F. Pilkey, *Peterson's Stress-Concentration Factors*, 3rd ed., John Wiley & Sons, New York, 2008, p. 411.

The alternating stress experienced by a bolt is

$$\sigma_a = \frac{(F_{b\max} - F_{b\min})/2}{A_t} = \frac{(CP_{\max} + F_i) - (CP_{\min} + F_i)}{2A_t}$$

$$\sigma_a = \frac{C(P_{\max} - P_{\min})}{2A_t} \tag{8–34}$$

The midrange stress experienced by a bolt is

$$\sigma_m = \frac{(F_{b\max} + F_{b\min})/2}{A_t} = \frac{(CP_{\max} + F_i) + (CP_{\min} + F_i)}{2A_t}$$

$$\sigma_m = \frac{C(P_{\max} + P_{\min})}{2A_t} + \frac{F_i}{A_t} \tag{8–35}$$

A load line typically experienced by a bolt is shown in Figure 8–22, where the stress starts from the preload stress and increases with a constant slope of $\sigma_a/(\sigma_m - \sigma_i)$. The Goodman failure line is also shown in Figure 8–22. The fatigue factor of safety can be found by intersecting the load line and the Goodman line to find the intersection point (S_m, S_a). The load line is given by

Load line: $$S_a = \frac{\sigma_a}{\sigma_m - \sigma_i}(S_m - \sigma_i) \tag{a}$$

The Goodman line, rearranging Equation (6–40), is

Goodman line: $$S_a = S_e - \frac{S_e}{S_{ut}} S_m \tag{b}$$

Equating Equations (*a*) and (*b*), solving for S_m, then substituting S_m back into Equation (*b*) yields

$$S_a = \frac{S_e \sigma_a (S_{ut} - \sigma_i)}{S_{ut}\sigma_a + S_e(\sigma_m - \sigma_i)} \tag{c}$$

The fatigue factor of safety is given by

$$n_f = \frac{S_a}{\sigma_a} \tag{8–36}$$

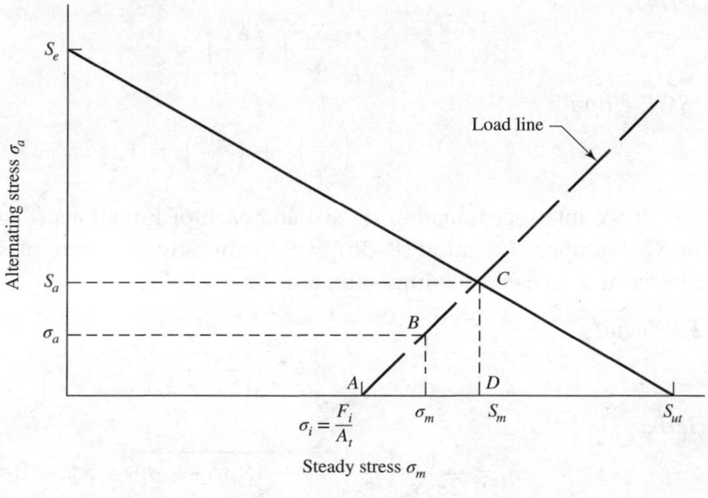

Figure 8–22

Fluctuating-stress diagram showing a Goodman failure line and a commonly used load line for a constant preload and a fluctuating load.

Substituting Equation (*c*) into Equation (8–36) gives

$$n_f = \frac{S_e(S_{ut} - \sigma_i)}{S_{ut}\sigma_a + S_e(\sigma_m - \sigma_i)} \tag{8–37}$$

The same approach can be used for the other failure curves, though for any of the nonlinear failure curves the algebra is a bit more tedious to put in equation form such as Equation (8–37). An easier approach would be to solve in stages numerically, first S_m, then S_a, and finally n_f.

Special Case of Repeated Loading

Often, the type of fatigue loading encountered in the analysis of bolted joints is one in which the externally applied load fluctuates between zero and some maximum force P. This would be the situation in a pressure cylinder, for example, where a pressure either exists or does not exist. For such cases, Equations (8–34) and (8–35) can be simplified by setting $P_{max} = P$ and $P_{min} = 0$, resulting in

$$\sigma_a = \frac{CP}{2A_t} \tag{8–38}$$

$$\sigma_m = \frac{CP}{2A_t} + \frac{F_i}{A_t} \tag{8–39}$$

Note that Equation (8–39) can be viewed as the sum of the alternating stress and the preload stress. If the preload is considered to be constant, the load line relationship between the alternating and midrange stresses can be treated as

$$\sigma_m = \sigma_a + \sigma_i \tag{8–40}$$

This load line has a slope of unity, and is a special case of the load line shown in Figure 8–22. With the simplifications in the algebra, we can now proceed as before to obtain the fatigue factor of safety using a few of the typical failure criteria, duplicated here from Equations (6–40), (6–47), and (6–51), in terms of the intersection point (S_m, S_a).

Goodman:

$$\frac{S_a}{S_e} + \frac{S_m}{S_{ut}} = 1 \tag{8–41}$$

Gerber:

$$\frac{S_a}{S_e} + \left(\frac{S_m}{S_{ut}}\right)^2 = 1 \tag{8–42}$$

ASME-elliptic:

$$\left(\frac{S_a}{S_e}\right)^2 + \left(\frac{S_m}{S_p}\right)^2 = 1 \tag{8–43}$$

Now if we intersect Equation (8–40) and each of Equations (8–41) to (8–43) to solve for S_a, and apply Equation (8–36), we obtain fatigue factors of safety for each failure criteria in a repeated loading situation.

Goodman:

$$n_f = \frac{S_e(S_{ut} - \sigma_i)}{\sigma_a(S_{ut} + S_e)} \tag{8–44}$$

Gerber:

$$n_f = \frac{1}{2\sigma_a S_e}\left[S_{ut}\sqrt{S_{ut}^2 + 4S_e(S_e + \sigma_i)} - S_{ut}^2 - 2\sigma_i S_e\right] \tag{8–45}$$

ASME-elliptic:

$$n_f = \frac{S_e}{\sigma_a\left(S_p^2 + S_e^2\right)}\left(S_p\sqrt{S_p^2 + S_e^2 - \sigma_i^2} - \sigma_i S_e\right) \tag{8–46}$$

Note that Equations (8–44) to (8–46) are only applicable for repeated loads. If K_f is being applied to the stresses, rather than to S_e, be sure to apply it to both σ_a and σ_m. Otherwise, the slope of the load line will not remain 1 to 1.

If desired, σ_a from Equation (8–38) and $\sigma_i = F_i/A_t$ can be directly substituted into any of Equations (8–44) to (8–46). If we do so for the Goodman criteria in Equation (8–44), we obtain

$$n_f = \frac{2S_e\,(S_{ut}A_t - F_i)}{CP(S_{ut} + S_e)} \tag{8–47}$$

when preload F_i is present. With no preload, $C = 1$, $F_i = 0$, and Equation (8–47) becomes

$$n_{f0} = \frac{2S_e S_{ut} A_t}{P(S_{ut} + S_e)} \tag{8–48}$$

Preload is beneficial for resisting fatigue when n_f/n_{f0} is greater than unity. For Goodman, Equations (8–47) and (8–48) with $n_f/n_{f0} \geq 1$ puts an upper bound on the preload F_i of

$$F_i \leq (1 - C)S_{ut}A_t \tag{8–49}$$

If this cannot be achieved, and n_f is unsatisfactory, use the Gerber or ASME-elliptic criterion to obtain a less conservative assessment. If the design is still not satisfactory, additional bolts and/or a different size bolt may be called for.

Preload Considerations

Bolts loosen, as they are friction devices, and cyclic loading and vibration as well as other effects allow the fasteners to lose tension with time. How does one fight loosening? Within strength limitations, the higher the preload the better. A rule of thumb is that preloads of 60 percent of proof load rarely loosen. If more is better, how much more? Well, not enough to create reused fasteners as a future threat. Alternatively, fastener-locking schemes can be employed.

After solving for the fatigue factor of safety, you should also check the possibility of yielding, using the proof strength

$$n_p = \frac{S_p}{\sigma_m + \sigma_a} \tag{8–50}$$

which is equivalent to Equation (8–27).

EXAMPLE 8–5

Figure 8–23 shows a connection using cap screws. The joint is subjected to a fluctuating force whose maximum value is 22 500 N per screw. The required data are: cap screw, M16 class 8.8; hardened-steel washer, $t_w = 2$ mm thick; steel cover plate, $t_1 = 16$ mm, $E_s = 207$ GPa; and cast-iron base, $t_2 = 16$ mm, $E_{ci} = 117.7$ GPa.

(*a*) Find k_b, k_m, and C using the assumptions given in the caption of Figure 8–23.

(*b*) Find all factors of safety and explain what they mean.

Figure 8–23

Pressure-cone frustum member model for a cap screw. For this model the significant sizes are

$$l = \begin{cases} h + t_2/2 & t_2 < d \\ h + d/2 & t_2 \geq d \end{cases}$$

$D_1 = d_w + l \tan \alpha = 1.5d + 0.577l$

$D_2 = d_w = 1.5d$

where $l =$ effective grip. The solutions are for $\alpha = 30°$ and $d_w = 1.5d$.

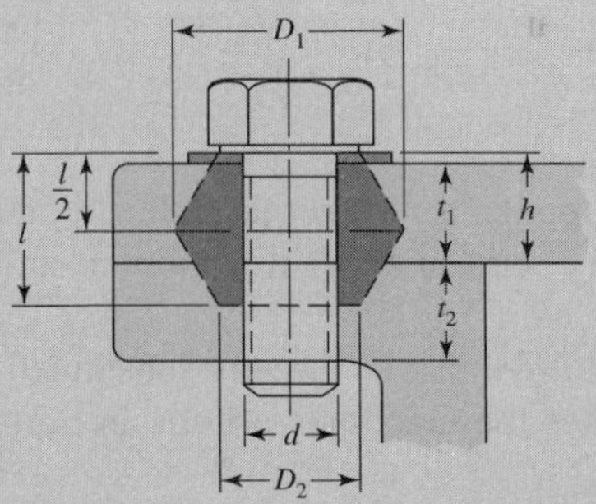

Solution

(*a*) For the symbols of Figures 8–15 and 8–23, $h = t_1 + t_w = 18$ mm, $l = h + d/2 = 26$ mm, and $D_2 = 1.5d = 24$ mm. The joint is composed of three frusta; the upper two frusta are steel and the lower one is cast iron.

For the upper frustum: $t = l/2 = 13$ mm, $D = 24$ mm, and $E = 207\,000$ MPa. Using these values in Equation (8–19) gives $k_1 = 8.14$ MN/mm.

For the middle frustum: $t = h - l/2 = 5$ mm and $D = 24 + 2(l - h) \tan 30° = 33.24$ mm. With these and $E_s = 207\,000$ MPa, Equation (8–19) gives $k_2 = 6.1$ MN/mm.

The lower frustum has $D = 24$ mm, $t = l - h = 8$ mm, and $E_{ci} = 117\,700$ MPa. The same equation yields $k_3 = 6.1$ MN/mm.

Substituting these three stiffnesses into Equation (8–17) gives $k_m = 3.19$ MN/mm. The cap screw is short and threaded all the way. Using $l = 26$ mm for the grip and $A_t = 157$ mm^2 from Table 8–2, we find the stiffness to be $k_b = A_tE/l = 1.25$ MN/mm. Thus the joint constant is

Answer
$$C = \frac{k_b}{k_b + k_m} = \frac{1.25}{1.25 + 3.19} = 0.280$$

(*b*) Equation (8–29) gives the preload as

$$F_i = 0.75F_p = 0.75A_tS_p = 0.75(157)(600) = 70.65 \text{ kN}$$

where from Table 8–9, $S_p = 600$ MPa for an SAE grade 5 cap screw. Using Equation (8–27), we obtain the yielding factor of safety

Answer
$$n_p = \frac{S_pA_t}{CP + F_i} = \frac{600(157)}{0.280(22\,500) + 70\,650} = 1.22$$

This is the traditional factor of safety, which compares the maximum bolt stress to the proof strength.

Using Equation (8–28),

Answer
$$n_L = \frac{S_pA_t - F_i}{CP} = \frac{600(157) - 70\,650}{0.280(22\,500)} = 3.71$$

This factor is an indication of the overload on P that can be applied without exceeding the proof strength.

Next, using Equation (8–29), we have

Answer
$$n_0 = \frac{F_i}{P(1 - C)} = \frac{70\,650}{22\,500(1 - 0.280)} = 4.34$$

If the force P gets too large, the joint will separate and the bolt will take the entire load. This factor guards against that event.

For fatigue, refer to Figure 8–24, where we will show a comparison of the Goodman and Gerber criteria. The intersection of the load line L with the respective failure lines at points C, D, and E defines a set of strengths S_a and S_m at each intersection. Point B represents the stress state σ_m, σ_a. Point A is the preload stress σ_i. Therefore the load line begins at A and makes an angle having a unit slope. This angle is 45° only when both stress axes have the same scale.

The factors of safety are found by dividing the distances AC, AD, and AE by the distance AB. Note that this is the same as dividing S_a for each theory by σ_a.

The quantities shown in the caption of Figure 8–24 are obtained as follows:

Point A

$$\sigma_i = \frac{F_i}{A_t} = \frac{70\ 650}{157} = 450 \text{ MPa}$$

Point B

$$\sigma_a = \frac{CP}{2A_t} = \frac{0.280(22\ 500)}{2(157)} = 20.2 \text{ MPa}$$

$$\sigma_m = \sigma_a + \sigma_i = 20.2 + 450 = 470.2 \text{ MPa}$$

Point C

This is the Goodman criterion. From Table 8–17, we find $S_e = 129$ MPa. Then, using Eq. (8–44), the factor of safety is found to be

Answer
$$n_f = \frac{S_e(S_{ut} - \sigma_i)}{\sigma_a(S_{ut} + S_e)} = \frac{129(830 - 450)}{20.2(830 + 129)} = 2.53$$

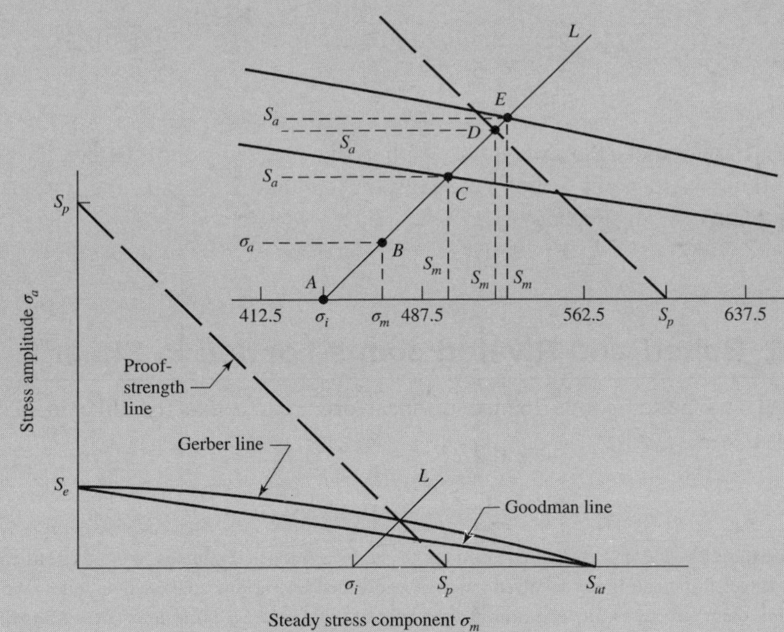

Figure 8–24

Fluctuating-stress diagram for preloaded bolts, drawn to scale, showing the Goodman line, the Gerber curve, and the Langer proof-strength line, with an exploded view of the area of interest. The strengths used are $S_p = 600$ MPa, $S_e = 129$ MPa, and $S_{ut} = 830$ MPa. The coordinates are
A, $\sigma_i = 450$ MPa;
B, $\sigma_a = 20.2$ MPa, $\sigma_m = 470.2$ MPa;
C, $S_a = 51.1$ MPa, $S_m = 501.1$ MPa;
D, $S_a = 75$ MPa, $S_m = 525$ MPa;
E, $S_a = 77$ MPa, $S_m = 523$ MPa.

Point D

This is on the proof-strength line where

$$S_m + S_a = S_p \tag{1}$$

In addition, the horizontal projection of the load line *AD* is

$$S_m = \sigma_i + S_a \tag{2}$$

Solving Equations (1) and (2) simultaneously results in

$$S_a = \frac{S_p - \sigma_i}{2} = \frac{600 - 450}{2} = 75 \text{ MPa}$$

The factor of safety resulting from this is

Answer
$$n_p = \frac{S_a}{\sigma_a} = \frac{75}{20.2} = 3.71$$

which, of course, is identical to the result previously obtained by using Equation (8–28).

A similar analysis of a fatigue diagram could have been done using yield strength instead of proof strength. Though the two strengths are somewhat related, proof strength is a much better and more positive indicator of a fully loaded bolt than is the yield strength. It is also worth remembering that proof-strength values are specified in design codes for bolts; yield strengths are not.

We found $n_f = 2.53$ on the basis of fatigue and the Goodman line, and $n_p = 3.71$ on the basis of proof strength. Thus the danger of failure is by fatigue, not by overproof loading. These two factors should always be compared to determine where the greatest danger lies.

Point E

For the Gerber criterion, from Eq. (8–45), the safety factor is

Answer
$$n_f = \frac{1}{2\sigma_a S_e}\left[S_{ut}\sqrt{S_{ut}^2 + 4S_e(S_e + \sigma_i)} - S_{ut}^2 - 2\sigma_i S_e\right]$$

$$= \frac{1}{2(20.2)(129)}\left[830\sqrt{830^2 + 4(129)(129 + 450)} - 830^2 - 2(450)(129)\right]$$

$$= 3.81$$

which is greater than $n_p = 3.71$ and contradicts the conclusion earlier that the danger of failure is fatigue. Figure 8–24 clearly shows the conflict where point *D* lies between points *C* and *E*. Again, the conservative nature of the Goodman criterion explains the discrepancy.

8–12 Bolted and Riveted Joints Loaded in Shear[10]

Riveted and bolted joints loaded in shear are treated exactly alike in design and analysis.

[10]The design of bolted and riveted connections for boilers, bridges, buildings, and other structures in which danger to human life is involved is strictly governed by various construction codes. When designing these structures, the engineer should refer to the *American Institute of Steel Construction Handbook*, the American Railway Engineering Association specifications, or the Boiler Construction Code of the American Society of Mechanical Engineers.

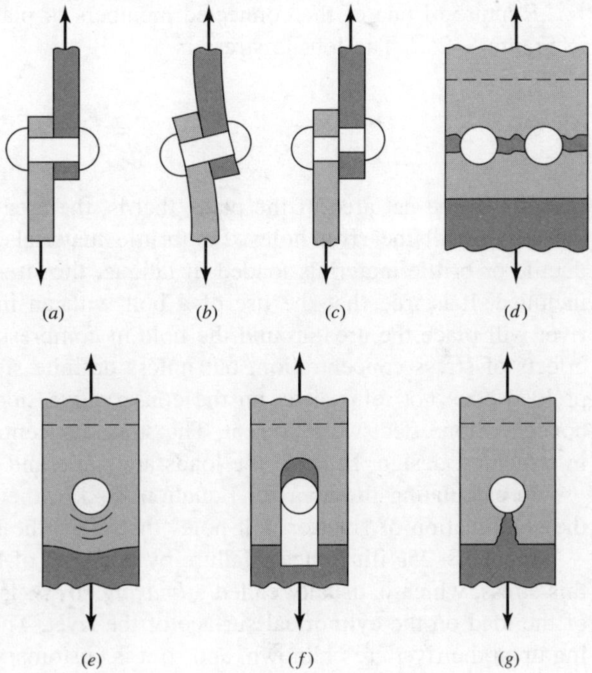

Figure 8–25

Modes of failure in shear loading of a bolted or riveted connection: (*a*) shear loading; (*b*) bending of rivet; (*c*) shear of rivet; (*d*) tensile failure of members; (*e*) bearing of rivet on members or bearing of members on rivet; (*f*) shear tear-out; (*g*) tensile tear-out.

Figure 8–25*a* shows a riveted connection loaded in shear. Let us now study the various means by which this connection might fail.

Figure 8–25*b* shows a failure by bending of the rivet or of the riveted members. The bending moment is approximately $M = Ft/2$, where F is the shearing force and t is the grip of the rivet, that is, the total thickness of the connected parts. The bending stress in the members or in the rivet is, neglecting stress concentration,

$$\sigma = \frac{M}{I/c} \tag{8–51}$$

where I/c is the section modulus for the weakest member or for the rivet or rivets, depending upon which stress is to be found. The calculation of the bending stress in this manner is an assumption, because we do not know exactly how the load is distributed to the rivet or the relative deformations of the rivet and the members. Although this equation can be used to determine the bending stress, it is seldom used in design; instead its effect is compensated for by an increase in the factor of safety.

In Figure 8–25*c* failure of the rivet by pure shear is shown; the stress in the rivet is

$$\tau = \frac{F}{A} \tag{8–52}$$

where A is the cross-sectional area of all the rivets in the group. It may be noted that it is standard practice in structural design to use the nominal diameter of the rivet rather than the diameter of the hole, even though a hot-driven rivet expands and nearly fills up the hole.

Rupture of one of the connected members or plates by pure tension is illustrated in Figure 8–25*d*. The tensile stress is

$$\sigma = \frac{F}{A} \tag{8–53}$$

where A is the net area of the plate, that is, the area reduced by an amount equal to the area of all the rivet holes. For brittle materials and static loads and for either ductile or brittle materials loaded in fatigue, the stress-concentration effects must be included. It is true that the use of a bolt with an initial preload and, sometimes, a rivet will place the area around the hole in compression and thus tend to nullify the effects of stress concentration, but unless definite steps are taken to ensure that the preload does not relax, it is on the conservative side to design as if the full stress-concentration effect were present. The stress-concentration effects are not considered in structural design, because the loads are static and the materials ductile.

In calculating the area for Equation (8–53), the designer should, of course, use the combination of rivet or bolt holes that gives the smallest area.

Figure 8–25*e* illustrates a failure by crushing of the rivet or plate. Calculation of this stress, which is usually called a *bearing stress,* is complicated by the distribution of the load on the cylindrical surface of the rivet. The exact values of the forces acting upon the rivet are unknown, and so it is customary to assume that the components of these forces are uniformly distributed over the projected contact area of the rivet. This gives for the stress

$$\sigma = -\frac{F}{A} \tag{8–54}$$

where the projected area for a single rivet is $A = td$. Here, t is the thickness of the thinnest plate and d is the rivet or bolt diameter.

Edge shearing, or tearing, of the margin is shown in Figure 8–25*f* and *g*, respectively. In structural practice this failure is avoided by spacing the rivets at least $1\frac{1}{2}$ diameters away from the edge. Bolted connections usually are spaced an even greater distance than this for satisfactory appearance, and hence this type of failure may usually be neglected.

In a rivet joint, the rivets all share the load in shear, bearing in the rivet, bearing in the member, and shear in the rivet. Other failures are participated in by only some of the joint. In a bolted joint, shear is taken by clamping friction, and bearing does not exist. When bolt preload is lost, one bolt begins to carry the shear and bearing until yielding slowly brings other fasteners in to share the shear and bearing. Finally, all participate, and this is the basis of most bolted-joint analysis if loss of bolt preload is complete. The usual analysis involves

- Bearing in the bolt (all bolts participate)
- Bearing in members (all holes participate)
- Shear of bolt (all bolts participate eventually)
- Distinguishing between thread and shank shear
- Edge shearing and tearing of member (edge bolts participate)
- Tensile yielding of member across bolt holes
- Checking member capacity

EXAMPLE 8–6

Two 25- by 100 mm 1018 cold-rolled steel bars are butt-spliced with two 12.5 by 100 mm 1018 cold-rolled splice plates using four M20 × 2 class 8.8 bolts as depicted in Figure 8–26. For a design factor of $n_d = 1.5$ estimate the static load F that can be carried if the bolts lose preload.

Solution

From Table A–19, minimum strengths of $S_y = 372$ MPa and $S_{ut} = 440$ MPa are found for the members, and from Table 8–11 minimum strengths of $S_p = 600$ MPa, $S_y = 660$ MPa, and $S_{ut} = 830$ MPa for the bolts are found.

$F/2$ is transmitted by each of the splice plates, but since the areas of the splice plates are half those of the center bars, the stresses associated with the plates are the same. So for stresses associated with the plates, the force and areas used will be those of the center plates.

Bearing in bolts, all bolts loaded:

$$\sigma = \frac{F}{2td} = \frac{S_y}{n_d}$$

$$F = \frac{2td\,S_y}{n_d} = \frac{2(25)(20)660}{1.5} = 440 \text{ kN}$$

Bearing in members, all bolts active:

$$\sigma = \frac{F}{2td} = \frac{(S_y)_{\text{mem}}}{n_d}$$

$$F = \frac{2td(S_y)_{\text{mem}}}{n_d} = \frac{2(25)(20)372}{1.5} = 248 \text{ kN}$$

Shear of bolt, all bolts active: If the bolt threads do not extend into the shear planes for four shanks:

$$\tau = \frac{F}{4\pi d^2/4} = 0.577\frac{S_y}{n_d}$$

$$F = 0.577\pi d^2\frac{S_y}{n_d} = 0.577\pi(20)^2\frac{660}{1.5} = 318 \text{ kN}$$

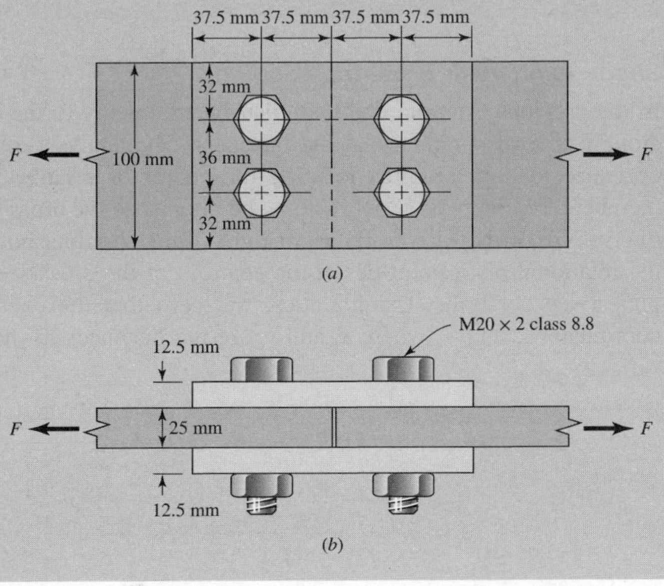

Figure 8–26

If the bolt threads extend into a shear plane:

$$\tau = \frac{F}{4A_r} = 0.577\frac{S_y}{n_d}$$

$$F = \frac{0.577\,(4)A_r S_y}{n_d} = \frac{0.577\,(4)245\,(660)}{1.5} = 226 \text{ kN}$$

Edge shearing of member at two margin bolts: From Figure 8–27,

$$\tau = \frac{F}{4at} = \frac{0.577(S_y)_{\text{mem}}}{n_d}$$

$$F = \frac{4at0.577(S_y)_{\text{mem}}}{n_d} = \frac{4(27.5)(25)(372)}{1.5} = 393.5 \text{ kN}$$

Tensile yielding of members across bolt holes:

$$\sigma = \frac{F}{\left[4 - 2\left(\frac{3}{4}\right)\right]t} = \frac{(S_y)_{\text{mem}}}{n_d}$$

$$F = \frac{\left[4 - 2\left(\frac{3}{4}\right)\right]t(S_y)_{\text{mem}}}{n_d} = \frac{[100 - 2(20)](25)372}{1.5} = 372 \text{ kN}$$

On the basis of bolt shear, the limiting value of the force is 226 kN, assuming the threads extend into a shear plane. However, it would be poor design to allow the threads to extend into a shear plane. So, assuming a *good* design based on bolt shear, the limiting value of the force is 318 kN. For the members, the bearing stress limits the load to 248 kN.

Figure 8–27

Edge shearing of member.

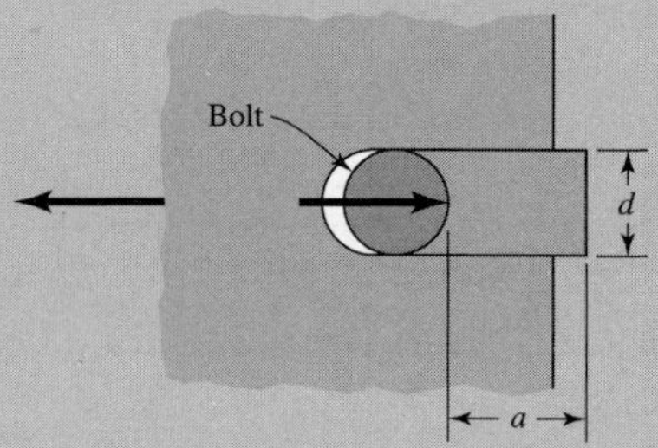

Shear Joints with Eccentric Loading

In the previous example, the load distributed equally to the bolts since the load acted along a line of symmetry of the fasteners. The analysis of a shear joint undergoing eccentric loading requires locating the center of relative motion between the two members. In Figure 8–28 let A_1 to A_5 be the respective cross-sectional areas of a group of five pins, or hot-driven rivets, or tight-fitting shoulder bolts. Under this assumption the rotational pivot point lies at the centroid of the cross-sectional area pattern of the pins, rivets, or bolts. Using statics, we learn that the centroid G is located by the coordinates $\bar{x}$ and $\bar{y}$, where x_i and y_i are the distances to the ith area center:

$$\bar{x} = \frac{A_1x_1 + A_2x_2 + A_3x_3 + A_4x_4 + A_5x_5}{A_1 + A_2 + A_3 + A_4 + A_5} = \frac{\sum_1^n A_i x_i}{\sum_1^n A_i}$$

$$\bar{y} = \frac{A_1y_1 + A_2y_2 + A_3y_3 + A_4y_4 + A_5y_5}{A_1 + A_2 + A_3 + A_4 + A_5} = \frac{\sum_1^n A_i y_i}{\sum_1^n A_i} \qquad (8\text{–}55)$$

In many instances the centroid can be located by symmetry.

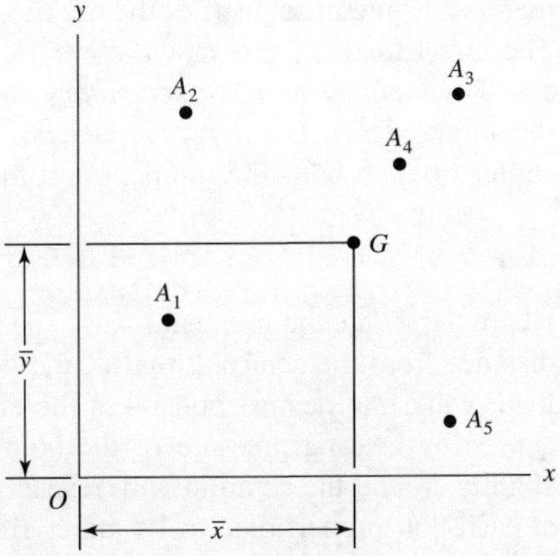

Figure 8–28

Centroid of pins, rivets, or bolts.

An example of eccentric loading of fasteners is shown in Figure 8–29. This is a portion of a machine frame containing a beam subjected to the action of a bending load. In this case, the beam is fastened to vertical members at the ends with specially prepared load-sharing bolts. You will recognize the schematic representation in Figure 8–29*b* as a statically indeterminate beam with both ends fixed and with moment and shear reactions at each end.

For convenience, the centers of the bolts at the left end of the beam are drawn to a larger scale in Figure 8–29*c*. Point O represents the centroid of the group, and it is assumed in this example that all the bolts are of the same diameter. Note that the forces shown in Figure 8–29*c* are the *resultant* forces acting on the pins with a net force and moment equal and opposite to the *reaction* loads V_1 and M_1 acting at O. The total load taken by each bolt will be calculated in three steps. In the first step the shear V_1 is divided equally among the bolts so that each bolt takes $F' = V_1/n$, where n refers to the number of bolts in the group and the force F' is called the *direct load,* or *primary shear*.

It is noted that an equal distribution of the direct load to the bolts assumes an absolutely rigid member. The arrangement of the bolts or the shape and size of the

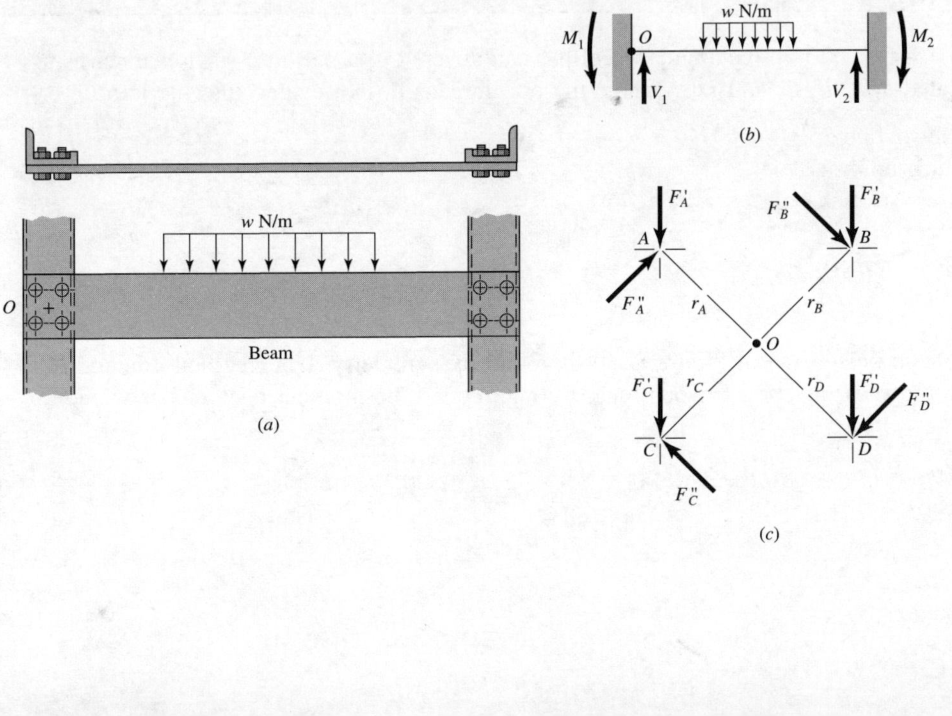

Figure 8–29

(*a*) Beam bolted at both ends with distributed load; (*b*) free-body diagram of beam; (*c*) enlarged view of bolt group centered at O showing primary and secondary resultant shear forces.

members sometimes justifies the use of another assumption as to the division of the load. The direct loads F'_n are shown as vectors on the loading diagram (Figure 8–29*c*).

The *moment load,* or *secondary shear,* is the additional load on each bolt due to the moment M_1. If r_A, r_B, r_C, etc. are the radial distances from the centroid to the center of each bolt, the moment and moment loads are related as follows:

$$M_1 = F''_A r_A + F''_B r_B + F''_C r_C + \cdots \tag{a}$$

where the F'' are the moment loads. The force taken by each bolt depends upon its radial distance from the centroid; that is, the bolt farthest from the centroid takes the greatest load, while the nearest bolt takes the smallest. This is because the moment loads are caused by bearing pressure of the bolt holes against the bolts as the members rotate slightly around the centroid with respect to each other, with the amount of displacement at each bolt proportional to its radius from the centroid. We can therefore write

$$\frac{F''_A}{r_A} = \frac{F''_B}{r_B} = \frac{F''_C}{r_C} \tag{b}$$

where again, the diameters of the bolts are assumed equal. If not, then one replaces F'' in Equation (*b*) with the shear stresses $\tau'' = 4F''/\pi d^2$ for each bolt. Solving Equations (*a*) and (*b*) simultaneously, we obtain

$$F''_n = \frac{M_1 r_n}{r_A^2 + r_B^2 + r_C^2 + \cdots} \tag{8–56}$$

where the subscript n refers to the particular bolt whose load is to be found. Each moment load is a force vector perpendicular to the radial line from the centroid to the bolt center.

In the third step the direct and moment loads are added vectorially to obtain the resultant load on each bolt. Since all the bolts or rivets are usually the same size, only that bolt having the maximum load need be considered. When the maximum load is found, the strength may be determined by using the various methods already described.

EXAMPLE 8–7

Shown in Figure 8–30 is a 15- by 200-mm rectangular steel bar cantilevered to a 250-mm steel channel using four tightly fitted bolts located at *A*, *B*, *C*, and *D*. Assume the bolt threads do not extend into the joint.

For the $F = 16$ kN load shown find

(*a*) The resultant load on each bolt

(*b*) The maximum shear stress in each bolt

(*c*) The maximum bearing stress

(*d*) The critical bending stress in the bar

Solution

(*a*) Point *O*, the centroid of the bolt group in Figure 8–30, is found by symmetry. If a free-body diagram of the beam were constructed, the shear reaction *V* would pass through *O* and the moment reactions *M* would be about *O*. These reactions are

$$V = 16 \text{ kN} \qquad M = 16(300 + 50 + 75) = 6800 \text{ N}\cdot\text{m}$$

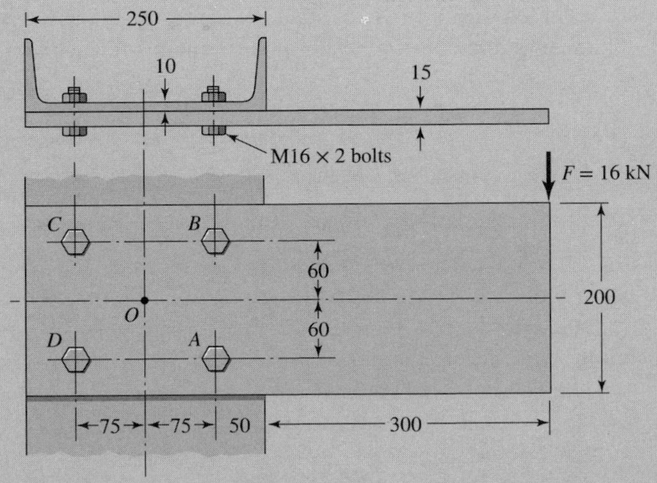

Figure 8–30

Dimensions in millimeters.

In Figure 8–31, the bolt group has been drawn to a larger scale and the reactions and resultants are shown. The distance from the centroid to the center of each bolt is

$$r = \sqrt{(60)^2 + (75)^2} = 96.0 \text{ mm}$$

The resultants are found as follows. The primary shear load per bolt is

$$F' = \frac{V}{n} = \frac{16}{4} = 4 \text{ kN}$$

Since the r_n are equal, the secondary shear forces are equal, and Equation (8–56) becomes

$$F'' = \frac{Mr}{4r^2} = \frac{M}{4r} = \frac{6800}{4(96.0)} = 17.7 \text{ kN}$$

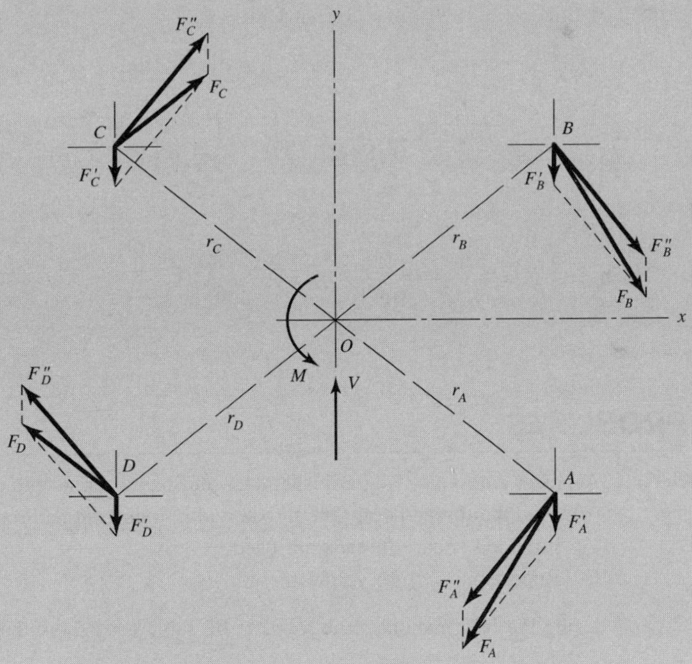

Figure 8–31

The primary and secondary shear forces are plotted to scale in Figure 8–31 and the resultants obtained by using the parallelogram rule. The magnitudes are found by measurement (or analysis) to be

Answer $$F_A = F_B = 21.0 \text{ kN}$$

Answer $$F_C = F_D = 14.8 \text{ kN}$$

(*b*) Bolts *A* and *B* are critical because they carry the largest shear load. The problem stated to assume that the bolt threads are not to extend into the joint. This would require special bolts. If standard nuts and bolts were used, the bolts would need to be 45 mm long with a thread length of $L_T = 38$ mm. Thus the unthreaded portion of the bolt is $45 - 38 = 7$ mm long. This is less than the 15 mm for the plate in Figure 8–30, and the bolts would tend to shear along the minor diameter at a stress of $\tau = F/A_s = 21.0(10)^3/144 = 146$ MPa. Using bolts not extending into the joint, or shoulder bolts, is preferred. For this example, the body area of each bolt is $A = \pi(16)^2/4 = 201.1 \text{ mm}^2$, resulting in a shear stress of

Answer $$\tau = \frac{F}{A} = \frac{21.0(10)^3}{201.1} = 104 \text{ MPa}$$

(*c*) The channel is thinner than the bar, and so the largest bearing stress is due to the pressing of the bolt against the channel web. The bearing area is $A_b = td = 10(16) = 160 \text{ mm}^2$. Thus the bearing stress is

Answer $$\sigma = -\frac{F}{A_b} = -\frac{21.0(10)^3}{160} = -131 \text{ MPa}$$

(*d*) The critical bending stress in the bar is assumed to occur in a section parallel to the *y* axis and through bolts *A* and *B*. At this section the bending moment is

$$M = 16(300 + 50) = 5600 \text{ N} \cdot \text{m}$$

The second moment of area through this section is obtained as follows:

$$I = I_{\text{bar}} - 2(I_{\text{holes}} + \bar{d}^2 A)$$

$$= \frac{15(200)^3}{12} - 2\left[\frac{15(16)^3}{12} + (60)^2(15)(16)\right] = 8.26(10)^6 \text{ mm}^4$$

Then

Answer $$\sigma = \frac{Mc}{I} = \frac{5600(100)}{8.26(10)^6}(10)^3 = 67.8 \text{ MPa}$$

PROBLEMS

8–1 A power screw is 25 mm in diameter and has a thread pitch of 5 mm.

(*a*) Find the thread depth, the thread width, the mean and root diameters, and the lead, provided square threads are used.

(*b*) Repeat part (*a*) for Acme threads.

8–2 Using the information in the footnote of Table 8–1, show that the tensile-stress area is

$$A_t = \frac{\pi}{4}(d - 0.938\ 194p)^2$$

8–3 Show that for zero collar friction the efficiency of a square-thread screw is given by the equation

$$e = \tan\lambda \frac{1 - f\tan\lambda}{\tan\lambda + f}$$

Plot a curve of the efficiency for lead angles up to 45°. Use $f = 0.08$.

8–4 A single-threaded power screw is 25 mm in diameter with a pitch of 5 mm. A vertical load on the screw reaches a maximum of 5 kN. The coefficients of friction are 0.06 for the collar and 0.09 for the threads. The frictional diameter of the collar is 45 mm. Find the overall efficiency and the torque to "raise" and "lower" the load.

8–5 The machine shown in the figure can be used for a tension test but not for a compression test. Why? Can both screws have the same hand?

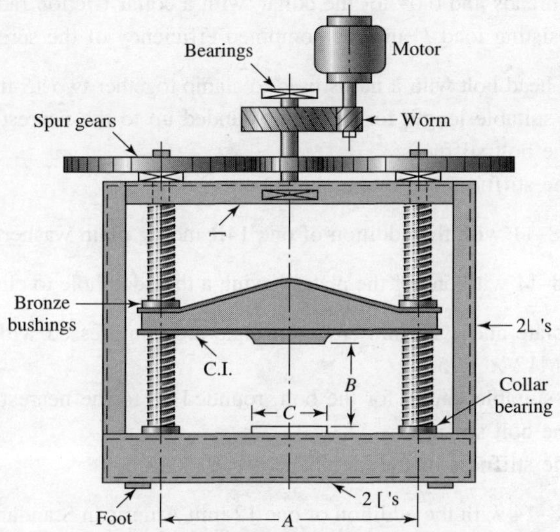

Problem 8–5

8–6 The press shown for Problem 8–5 has a rated load of 22 kN. The twin screws have Acme threads, a diameter of 75 mm, and a pitch of 12 mm. Coefficients of friction are 0.05 for the threads and 0.08 for the collar bearings. Collar diameters are 130 mm. The gears have an efficiency of 95 percent and a speed ratio of 60:1. A slip clutch, on the motor shaft, prevents overloading. The full-load motor speed is 1720 rev/min.

(*a*) When the motor is turned on, how fast will the press head move?

(*b*) What should be the horsepower rating of the motor?

8–7 For the C clamp shown, a force is applied at the end of the 10 mm diameter handle. The screw is a 20 mm Acme thread (see Figure 8–3), and is 250 mm long overall, with a maximum of 200 mm possible in the clamping region. The handle and screw are both made from cold-drawn AISI 1006 steel. The coefficients of friction for the screw and the collar are 0.15. The collar, which in this case is the anvil striker's swivel joint, has a friction diameter of 1 in. It is desired that the handle will yield before the screw will fail. Check this by the following steps.

(*a*) Determine the maximum force that can be applied to the end of the handle to reach the point of yielding of the handle.

(*b*) Using the force from part (*a*), determine the clamping force.

(*c*) Using the force from part (*a*), determine the factor of safety for yielding at the interface of the screw body and the base of the first engaged thread, assuming the first thread carries 38 percent of the total clamping force.

(*d*) Determine a factor of safety for buckling of the screw.

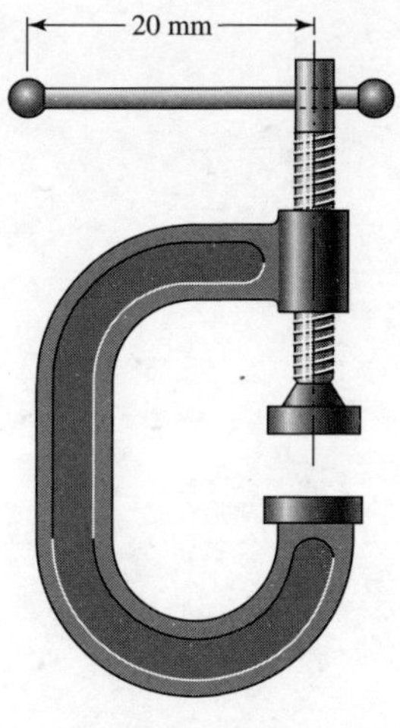

Problem 8–7

8–8 The C clamp shown in the figure for Problem 8–7 uses a 20 mm Acme thread. The frictional coefficients are 0.15 for the threads and for the collar. The collar, which in this case is the anvil striker's swivel joint, has a friction diameter of 25 mm. Calculations are to be based on a maximum force of 35 N applied to the handle at a radius of 90 mm from the screw centerline. Find the clamping force.

8–9 Find the power required to drive a 36 mm power screw having double square threads with a pitch of 6 mm. The nut is to move at a velocity of 50 mm/s and move a load of $F = 10$ kN. The frictional coefficients are 0.10 for the threads and 0.15 for the collar. The frictional diameter of the collar is 55 mm.

8–10 A single square-thread power screw has an input power of 3 kW at a speed of 1 rev/s. The screw has a diameter of 40 mm and a pitch of 8 mm. The frictional coefficients are 0.14 for the threads and 0.09 for the collar, with a collar friction radius of 50 mm. Find the axial resisting load F and the combined efficiency of the screw and collar.

8–11 An M14 × 2 hex-head bolt with a nut is used to clamp together two 15-mm steel plates.
(*a*) Determine a suitable length for the bolt, rounded up to the nearest 5 mm.
(*b*) Determine the bolt stiffness.
(*c*) Determine the stiffness of the members.

8–12 Repeat Problem 8–11 with the addition of one 14R metric plain washer under the nut.

8–13 Repeat Problem 8–11 with one of the plates having a threaded hole to eliminate the nut.

8–14 A 50-mm steel plate and a 25-mm cast-iron plate are compressed with one bolt and nut. The bolt is M12 × 1.75.
(*a*) Determine a suitable length for the bolt, rounded up to the nearest 6 mm.
(*b*) Determine the bolt stiffness.
(*c*) Determine the stiffness of the members.

8–15 Repeat Problem 8–14 with the addition of one 12 mm American Standard plain washer under the head of the bolt, and another identical washer under the nut.

8–16 Repeat Problem 8–14 with the cast-iron plate having a threaded hole to eliminate the nut.

8–17 Two identical aluminum plates are each 50 mm thick, and are compressed with one bolt and nut. Washers are used under the head of the bolt and under the nut.
Washer properties: steel; ID = 13.5 mm; OD = 27 mm; thickness = 1.5 mm
Nut properties: steel; height = 11 mm
Bolt properties: M123 ISO 9.8
Plate properties: aluminum; $E = 71$ GPa; $S_u = 325$ MPa; $S_y = 170$ MPa
(*a*) Determine a suitable length for the bolt, rounded up to the nearest 6 mm.
(*b*) Determine the bolt stiffness.
(*c*) Determine the stiffness of the members.

8–18 Repeat Problem 8–17 with no washer under the head of the bolt, and two washers stacked under the nut.

8–19 A 30-mm thick AISI 1020 steel plate is sandwiched between two 10-mm thick 2024-T3 aluminum plates and compressed with a bolt and nut with no washers. The bolt is M10 × 1.5, property class 5.8.
(*a*) Determine a suitable length for the bolt, rounded up to the nearest 5 mm.
(*b*) Determine the bolt stiffness.
(*c*) Determine the stiffness of the members.

8–20 Repeat Problem 8–19 with the bottom aluminum plate replaced by one that is 20 mm thick.

8–21 Repeat Problem 8–19 with the bottom aluminum plate having a threaded hole to eliminate the nut.

8–22 Two 20-mm steel plates are to be clamped together with a bolt and nut. Specify a coarse thread metric bolt to provide a joint constant C of approximately 0.2.

8–23 A 50-mm steel plate and a 25-mm cast-iron plate are to be compressed with one bolt and nut. Specify a UNC bolt to provide a joint constant C of approximately 0.2.

8–24 An aluminum bracket with a 12 mm thick flange is to be clamped to a steel column with a 20 mm wall thickness. A cap screw passes through a hole in the bracket flange, and threads into a tapped hole through the column wall. Specify a UNC cap screw to provide a joint constant C of approximately 0.25.

8–25 An M14 × 2 hex-head bolt with a nut is used to clamp together two 20-mm steel plates. Compare the results of finding the overall member stiffness by use of Equations (8–19), (8–21), and (8–22).

8–26 A M20 ISO 8.8 bolt has a 20 mm ID steel tube 250 mm long, clamped between washer faces of bolt and nut by turning the nut snug and adding one-third of a turn. The tube OD is the washer-face diameter $d_w = 1.5d = 1.5(0.75) = 28$ mm = OD.

(*a*) Determine the bolt stiffness, the tube stiffness, and the joint constant C.

(*b*) When the one-third turn-of-nut is applied, what is the initial tension F_i in the bolt?

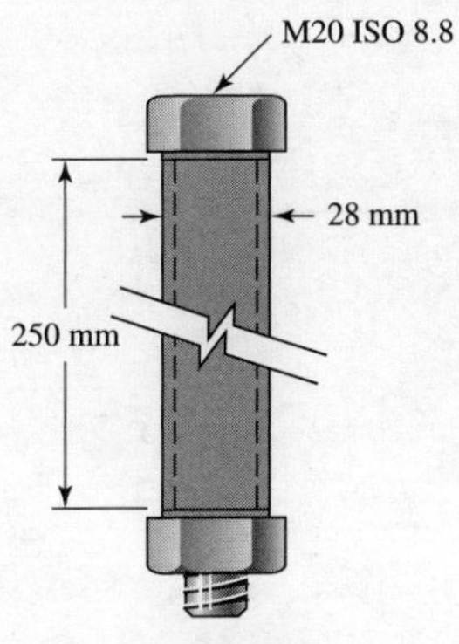

Problem 8–26

8–27 From your experience with Problem 8–26, generalize your solution to develop a turn-of-nut equation

$$N_t = \frac{\theta}{360^\circ} = \left(\frac{k_b + k_m}{k_b k_m}\right) F_i N$$

where N_t = turn of the nut, in rotations, from snug tight
θ = turn of the nut in degrees
N = number of thread/mm ($1/p$ where p is pitch)
F_i = initial preload
k_b, k_m = spring rates of the bolt and members, respectively

Use this equation to find the relation between torque-wrench setting T and turn-of-nut N_t. ("Snug tight" means the joint has been tightened to perhaps half the intended preload to flatten asperities on the washer faces and the members. Then the nut is loosened and retightened finger tight, and the nut is rotated the number of degrees indicated by the equation. Properly done, the result is competitive with torque wrenching.)

8–28 RB&W[11] recommends turn-of-nut from snug fit to preload as follows: $1/3$ turn for bolt grips of 1–4 diameters, $1/2$ turn for bolt grips 4–8 diameters, and $2/3$ turn for grips of 8–12 diameters. These recommendations are for structural steel fabrication (permanent joints), producing preloads of 100 percent of proof strength and beyond. Machinery fabricators with fatigue loadings and possible joint disassembly have much smaller turns-of-nut. The RB&W recommendation enters the nonlinear plastic deformation zone.

[11]Russell, Burdsall & Ward, Inc., Metal Forming Specialists, Mentor, Ohio.

For Example 8–4, use Equation (8–26) with $K = 0.2$ to estimate the torque necessary to establish the desired preload. Then, using the results from Problem 8–27, determine the turn of the nut in degrees. How does this compare with the RB&W recommendations?

8–29 For a bolted assembly with six bolts, the stiffness of each bolt is $k_b = 500$ kN/mm and the stiffness of the members is $k_m = 2000$ kN/mm per bolt. An external load of 80 kips is applied to the entire joint. Assume the load is equally distributed to all the bolts. It has been determined to use M12 ISO 10.9 bolts with rolled threads. Assume the bolts are preloaded to 75 percent of the proof load.
(*a*) Determine the yielding factor of safety.
(*b*) Determine the overload factor of safety.
(*c*) Determine the factor of safety based on joint separation.

8–30 For the bolted assembly of Problem 8–29, it is desired to find the range of torque that a mechanic could apply to initially preload the bolts without expecting failure once the joint is loaded. Assume a torque coefficient of $K = 0.2$.
(*a*) Determine the maximum bolt preload that can be applied without exceeding the proof strength of the bolts.
(*b*) Determine the minimum bolt preload that can be applied while avoiding joint separation.
(*c*) Determine the value of torque in units of N · m that should be specified for preloading the bolts if it is desired to preload to the midpoint of the values found in parts (*a*) and (*b*).

8–31 For a bolted assembly with eight bolts, the stiffness of each bolt is $k_b = 1.0$ MN/mm and the stiffness of the members is $k_m = 2.6$ MN/mm per bolt. The joint is subject to occasional disassembly for maintenance and should be preloaded accordingly. Assume the external load is equally distributed to all the bolts. It has been determined to use M6 × 1 class 5.8 bolts with rolled threads.
(*a*) Determine the maximum external load P_{max} that can be applied to the entire joint without exceeding the proof strength of the bolts.
(*b*) Determine the maximum external load P_{max} that can be applied to the entire joint without causing the members to come out of compression.

8–32 For a bolted assembly, the stiffness of each bolt is $k_b = 700$ kN/mm and the stiffness of the members is $k_m = 2000$ kN/mm per bolt. The joint is subject to occasional disassembly for maintenance and should be preloaded accordingly. A fluctuating external load is applied to the entire joint with $P_{max} = 360$ kN and $P_{min} = 90$ kN. Assume the load is equally distributed to all the bolts. It has been determined to use M12 class 9.8 bolts with rolled threads.
(*a*) Determine the minimum number of bolts necessary to avoid yielding of the bolts.
(*b*) Determine the minimum number of bolts necessary to avoid joint separation.

8–33 to 8–36 The figure illustrates the nonpermanent connection of a steel cylinder head to a grade 30 cast-iron pressure vessel using N bolts. A confined gasket seal has an effective sealing diameter D. The cylinder stores gas at a maximum pressure p_g. For the specifications given in the table for the specific problem assigned, select a suitable bolt length, assuming bolts are available in increments of 5 mm. Then determine the yielding factor of safety n_p, the load factor n_L, and the joint separation factor n_0.

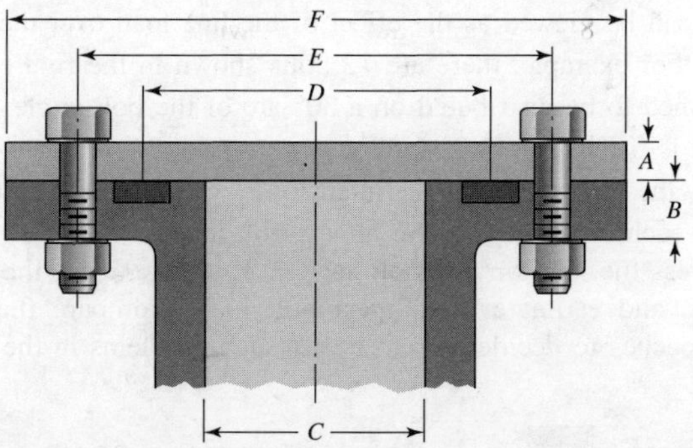

Problems 8–33 to 8–36

Problem Number	8–33	8–34	8–35	8–36
A	20 mm	12 mm	20 mm	10 mm
B	20 mm	16 mm	25 mm	12 mm
C	60 mm	70 mm	0.7 m	70 mm
D	100 mm	90 mm	0.8 m	80 mm
E	200 mm	150 mm	1.0 m	140 mm
F	300 mm	200 mm	1.1 m	180 mm
N	10	10	36	8
p_g	6 MPa	10 MPa	550 kPa	8 MPa
Bolt grade	ISO 9.8	ISO 8.8	ISO 10.9	ISO 10.9
Bolt spec.	M12 × 1.75	M12 × 1.75	M10 × 1.5	M10 × 1.5

8–37 to 8–40 Repeat the requirements for the problem specified in the table if the bolts and nuts are replaced with cap screws that are threaded into tapped holes in the cast-iron cylinder.

Problem Number	Originating Problem Number
8–37	8–33
8–38	8–34
8–39	8–35
8–40	8–36

8–41 to 8–44 For the pressure vessel defined in the problem specified in the table, redesign the bolt specifications to satisfy all of the following requirements.

- Use coarse-thread bolts selecting a class from Table 8–11 for Problems 8–41 and 8–43, or a grade from Table 8–9 for Problems 8–42 and 8–44.
- To ensure adequate gasket sealing around the bolt circle, use enough bolts to provide a maximum center-to-center distance between bolts of four bolt diameters.
- Obtain a joint stiffness constant C between 0.2 and 0.3 to ensure most of the pressure load is carried by the members.
- The bolts may be reused, so the yielding factor of safety should be at least 1.1.
- The overload factor and the joint separation factor should allow for the pressure to exceed the expected pressure by 15 percent.

Problem Number	Originating Problem Number
8–41	8–33
8–42	8–34
8–43	8–35
8–44	8–36

8–45 Bolts distributed about a bolt circle are often called upon to resist an external bending moment as shown in the figure. The external moment is 1400 N · m and the bolt circle has a diameter of 200 mm. The neutral axis for bending is a diameter of the bolt circle. What needs to be determined is the most severe external load seen by a bolt in the assembly.

(*a*) View the effect of the bolts as placing a line load around the bolt circle whose intensity F_b', in newtons per meter, varies linearly with the distance from the neutral axis according to the relation $F_b' = F'_{b,\max} R \sin\theta$. The load on any particular bolt can be viewed as the effect of the line load over the arc associated with the bolt. For example, there are 12 bolts shown in the figure. Thus each bolt load is assumed to be distributed on a 30° arc of the bolt circle. Under these conditions, what is the largest bolt load?

(*b*) View the largest load as the intensity $F'_{b,\max}$ multiplied by the arc length associated with each bolt and find the largest bolt load.

(*c*) Express the load on any bolt as $F = F_{\max} \sin \theta$, sum the moments due to all the bolts, and estimate the largest bolt load. Compare the results of these three approaches to decide how to attack such problems in the future.

Problem 8–45

Bolted connection subjected to bending.

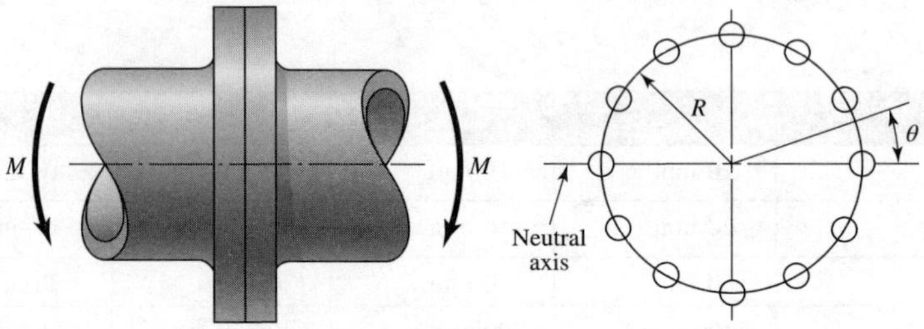

8–46 The figure shows a cast-iron bearing block that is to be bolted to a steel ceiling joist and is to support a gravity load of 18 kN. Bolts used are M24 ISO 8.8 with coarse threads and with 4.6-mm-thick steel washers under the bolt head and nut. The joist flanges are 20 mm in thickness, and the dimension *A*, shown in the figure, is 20 mm. The modulus of elasticity of the bearing block is 135 GPa.

Problem 8–46

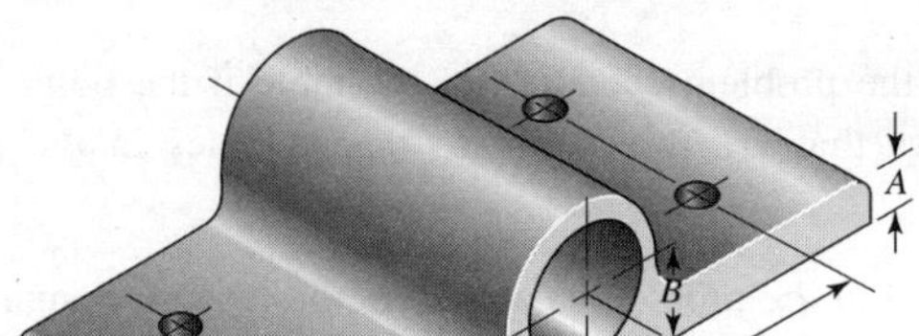

(*a*) Find the wrench torque required if the fasteners are lubricated during assembly and the joint is to be permanent.

(*b*) Determine the factors of safety guarding against yielding, overload, and joint separation.

8–47 The upside-down steel A frame shown in the figure is to be bolted to steel beams on the ceiling of a machine room using ISO grade 8.8 bolts. This frame is to support the 40-kN vertical load as illustrated. The total bolt grip is 48 mm, which includes the thickness of the steel beam, the A-frame feet, and the steel washers used. The bolts are size M20 × 2.5.

(*a*) What tightening torque should be used if the connection is permanent and the fasteners are lubricated?

(*b*) Determine the factors of safety guarding against yielding, overload, and joint separation.

Problem 8–47

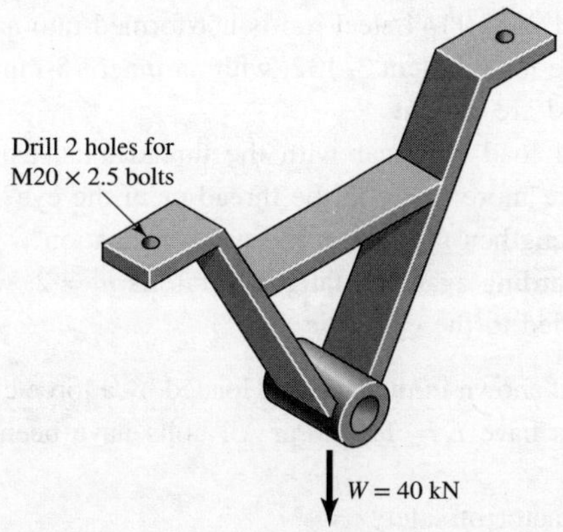

8–48 For the bolted assembly in Problem 8–29, assume the external load is a repeated load. Determine the fatigue factor of safety for the bolts using the following failure criteria:
(*a*) Goodman.
(*b*) Gerber.
(*c*) ASME-elliptic.

8–49 For a bolted assembly with eight bolts, the stiffness of each bolt is $k_b = 1.0$ MN/mm and the stiffness of the members is $k_m = 2.6$ MN/mm per bolt. The bolts are preloaded to 75 percent of proof strength. Assume the external load is equally distributed to all the bolts. The bolts are M6 × 1 class 5.8 with rolled threads. A fluctuating external load is applied to the entire joint with $P_{max} = 60$ kN and $P_{min} = 20$ kN.
(*a*) Determine the yielding factor of safety.
(*b*) Determine the overload factor of safety.
(*c*) Determine the factor of safety based on joint separation.
(*d*) Determine the fatigue factor of safety using the Goodman criterion.

8–50 For the bolted assembly in Problem 8–32, assume 10 bolts are used. Determine the fatigue factor of safety using the Goodman criterion.

8–51 For a bolted assembly with eight M8 × 1.25 class 9.8 bolts with rolled threads, the stiffness of each bolt is $k_b = 1.5$ MN/mm and the stiffness of the members is $k_m =$ 3.9 MN/mm per bolt. The joint is subject to occasional disassembly for maintenance and should be preloaded accordingly. An external load of 50 kN is repeatedly applied to the entire joint. Determine the fatigue factor of safety using the following criteria:
(*a*) Goodman.
(*b*) Gerber.
(*c*) Morrow.

Problem Number	Originating Problem Number
8–52	8–33
8–53	8–34
8–54	8–35
8–55	8–36

8–52 to 8–55 For the pressure cylinder defined in the problem specified in the table, the gas pressure is cycled between zero and p_g. Determine the fatigue factor of safety for the bolts using the following failure criteria:
(*a*) Goodman.
(*b*) Gerber.
(*c*) ASME-elliptic.

Problem Number	Originating Problem Number
8–56	8–33
8–57	8–34
8–58	8–35
8–59	8–36

8–56 to 8–59 For the pressure cylinder defined in the problem specified in the table, the gas pressure is cycled between p_g and $p_g/2$. Determine the fatigue factor of safety for the bolts using the Goodman criterion.

8–60 A 25-mm-diameter hot-rolled AISI 1144 steel rod is hot-formed into an eyebolt similar to that shown in the figure for Problem 3–132, with an inner 75-mm-diameter eye. The threads are M25 × 3 and are die-cut.

(*a*) For a repeatedly applied load collinear with the thread axis, using the Gerber criterion, is fatigue failure more likely in the thread or in the eye?

(*b*) What can be done to strengthen the bolt at the weaker location?

(*c*) If the factor of safety guarding against a fatigue failure is $n_f = 2$, what repeatedly applied load can be applied to the eye?

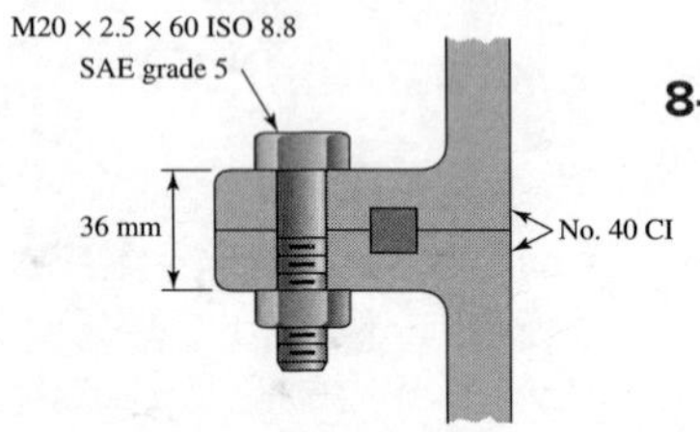

Problem 8–61

8–61 The section of the sealed joint shown in the figure is loaded by a force cycling between 18 and 27 kN. The members have $E = 110$ GPa. All bolts have been carefully preloaded to $F_i = 110$ kN each.

(*a*) Determine the yielding factor of safety.

(*b*) Determine the overload factor of safety.

(*c*) Determine the factor of safety based on joint separation.

(*d*) Determine the fatigue factor of safety using the Goodman criterion.

8–62 Suppose the welded steel bracket shown in the figure is bolted underneath a structural-steel ceiling beam to support a fluctuating vertical load imposed on it by a pin and yoke. The bolts are M12 coarse-thread ISO 9.8, tightened to recommended preload for nonpermanent assembly. The stiffnesses have already been computed and are $k_b = 700$ N/m and $k_m = 28$ N/m.

Problem 8–62

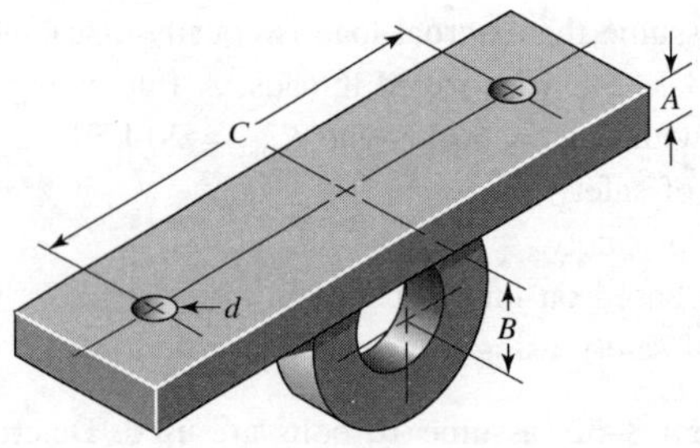

(*a*) Assuming that the bolts, rather than the welds, govern the strength of this design, determine the safe repeated load that can be imposed on this assembly using the Goodman criterion with the load line in Figure 8–22 and a fatigue design factor of 2.

(*b*) Compute the static load factors based on the load found in part (*a*).

8–63 Using the Gerber fatigue criterion and a fatigue-design factor of 2, determine the external repeated load P that a M30 ISO 8.8 coarse-thread bolt can take compared with that for a fine-thread bolt. The joint constants are $C = 0.30$ for coarse- and 0.32 for fine-thread bolts. Assume the bolts are preloaded to 75 percent of the proof load.

8–64 An M30 × 3.5 ISO 8.8 bolt is used in a joint at recommended preload, and the joint is subject to a repeated tensile fatigue load of $P = 65$ kN per bolt. The joint constant is $C = 0.28$. Find the static load factors and the factor of safety guarding against a fatigue failure based on the Gerber fatigue criterion.

8–65 The figure shows a fluid-pressure linear actuator (hydraulic cylinder) in which $D = 100$ mm, $t = 10$ mm, $L = 300$ mm, and $w = 20$ mm. Both brackets as well as the cylinder are of steel. The actuator has been designed for a working pressure of 15 MPa. Six M10 ISO 8.8 coarse-thread bolts are used, tightened to 75 percent of proof load. Assume the bolts are unthreaded within the grip.

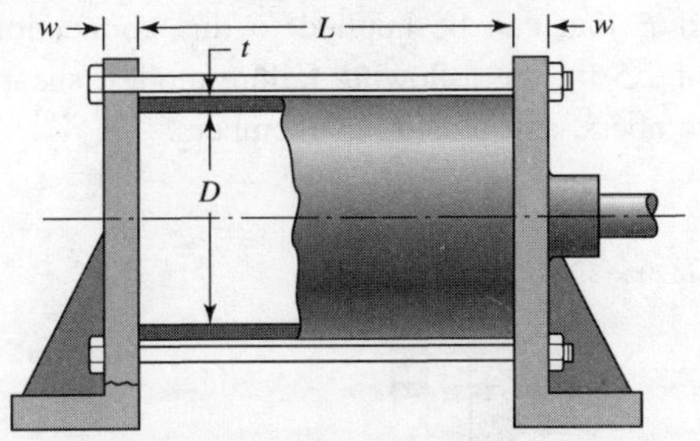

Problem 8–65

(*a*) Find the stiffnesses of the bolts and members, assuming that the entire cylinder is compressed uniformly and that the end brackets are perfectly rigid.

(*b*) Using the Gerber fatigue criterion, find the factor of safety guarding against a fatigue failure.

(*c*) What pressure would be required to cause total joint separation?

8–66 Using the Goodman fatigue criterion, repeat Problem 8–65 with the working pressure cycling between 8 MPa and 14 MPa.

8–67 The figure shows a bolted lap joint that uses ISO 9.8 bolts. The members are made of cold-drawn AISI 1020 steel. Assume the bolt threads do not extend into the joint. Find the safe tensile shear load F that can be applied to this connection to provide a minimum factor of safety of 2 for the following failure modes: shear of bolts, bearing on bolts, bearing on members, and tension of members.

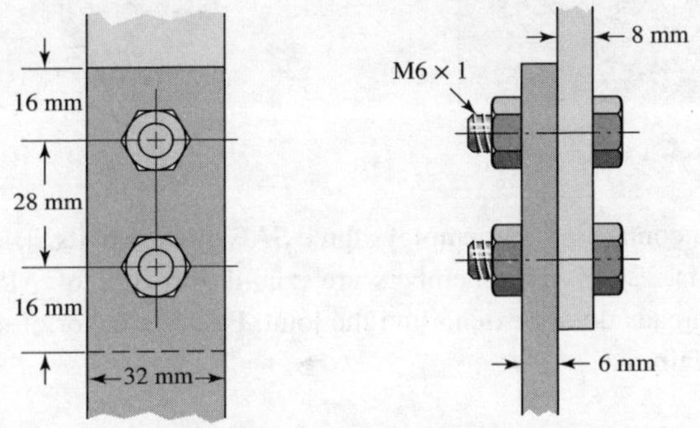

Problem 8–67

8–68 The bolted connection shown in the figure uses ISO class 10.9. The members are hot-rolled AISI 1040 steel. A tensile shear load $F = 22$ kN is applied to the connection. Assume the bolt threads do not extend into the joint. Find the factor of safety for all possible modes of failure.

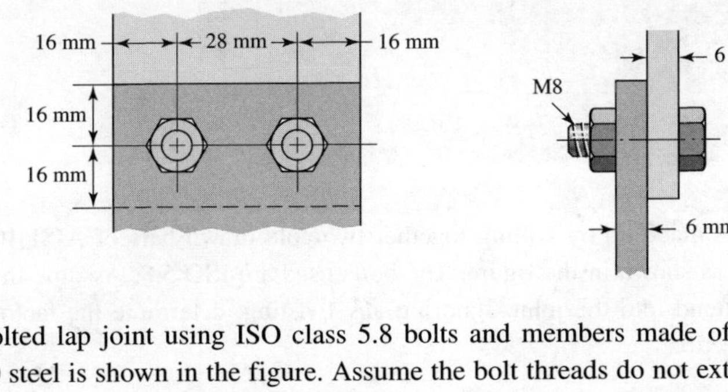

Problem 8–68

8–69 A bolted lap joint using ISO class 5.8 bolts and members made of cold-drawn SAE 1040 steel is shown in the figure. Assume the bolt threads do not extend into the joint.

Find the tensile shear load F that can be applied to this connection to provide a minimum factor of safety of 2.5 for the following failure modes: shear of bolts, bearing on bolts, bearing on members, and tension of members.

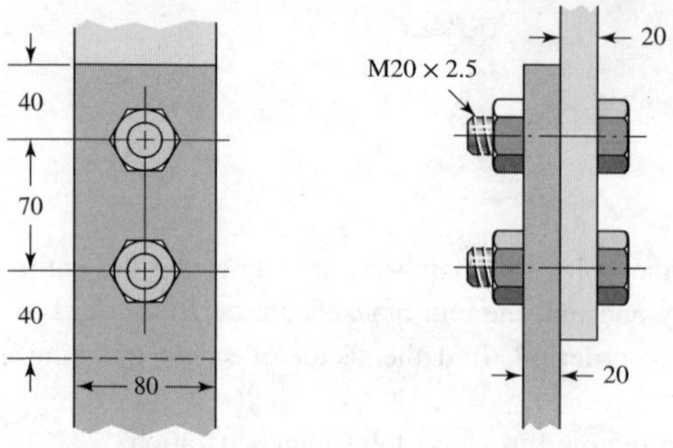

Problem 8–69
Dimensions in millimeters.

8–70 The bolted connection shown in the figure is subjected to a tensile shear load of 90 kN. The bolts are ISO class 5.8 and the material is cold-drawn AISI 1015 steel. Assume the bolt threads do not extend into the joint. Find the factor of safety of the connection for all possible modes of failure.

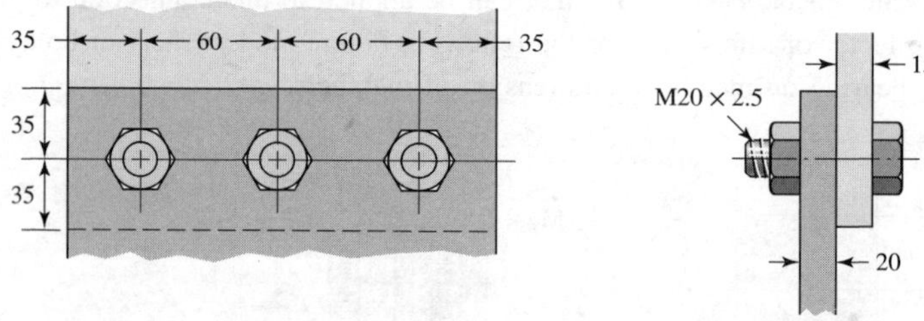

Problem 8–70
Dimensions in millimeters.

8–71 The figure shows a connection that employs three SAE grade 4 bolts. The tensile shear load on the joint is 22 kN. The members are cold-drawn bars of AISI 1020 steel. Assume the bolt threads do not extend into the joint. Find the factor of safety for each possible mode of failure.

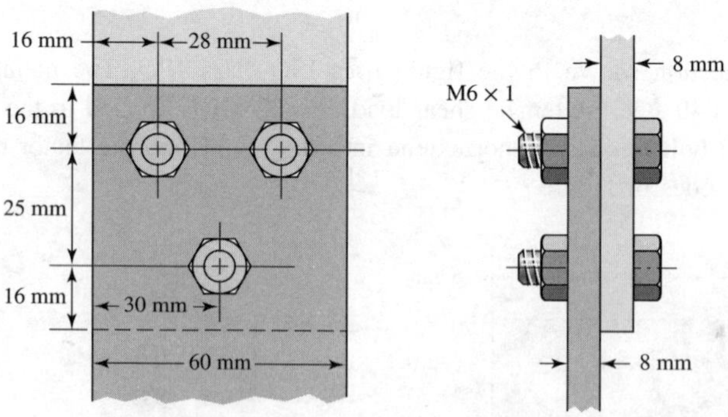

Problem 8–71

8–72 A beam is made up by bolting together two cold-drawn bars of AISI 1018 steel as a lap joint, as shown in the figure. The bolts used are ISO 5.8. Assume the bolt threads do not extend into the joint. Ignoring any twisting, determine the factor of safety of the connection.

Problem 8–72
Dimensions in millimeters.

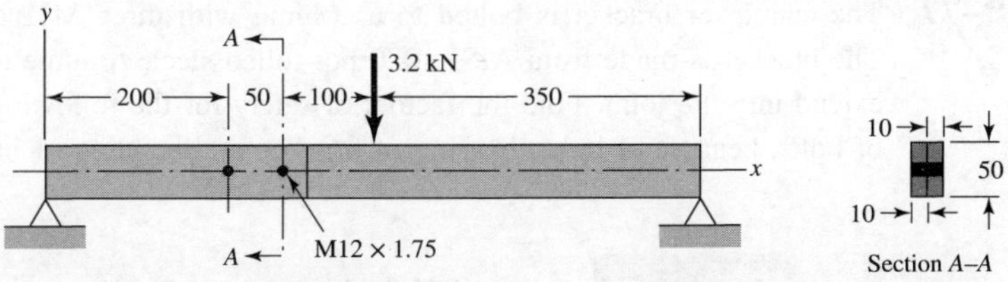

8–73 Standard design practice, as exhibited by the solutions to Problems 8–67 to 8–71, is to assume that the bolts, or rivets, share the shear equally. For many situations, such an assumption may lead to an unsafe design. Consider the yoke bracket of Problem 8–62, for example. Suppose this bracket is bolted to a wide-flange *column* with the centerline through the two bolts in the vertical direction. A vertical load through the yoke-pin hole at distance B from the column flange would place a shear load on the bolts as well as a tensile load. The tensile load comes about because the bracket tends to pry itself about the bottom corner, much like a claw hammer, exerting a large tensile load on the upper bolt. In addition, it is almost certain that both the spacing of the bolt holes and their diameters will be slightly different on the column flange from what they are on the yoke bracket. Thus, unless yielding occurs, only one of the bolts will take the shear load. The designer has no way of knowing which bolt this will be.

In this problem the bracket is 200 mm long, $A = 12$ mm, $B = 75$ mm, $C = 150$ mm, and the column flange is 12 mm thick. The bolts are M12 × 1.75 × 40 mm class 8.8. The nuts are tightened to 75 percent of proof load. The vertical yoke-pin load is 10 kN. If the upper bolt takes all the shear load as well as the tensile load, determine a static factor of safety for the bolt, based on the von Mises stress exceeding the proof strength.

8–74 The bearing of Problem 8–46 is bolted to a vertical surface and supports a horizontal shaft. The bolts used have coarse threads and are M20 ISO 5.8. The joint constant is $C = 0.25$, and the dimensions are $A = 20$ mm, $B = 50$ mm, and $C = 160$ mm. The bearing base is 240 mm long. The bearing load is 14 kN. The bolts are tightened to 75 percent of proof load. Determine a static factor of safety for the bolt, based on the von Mises stress exceeding the proof strength. Use worst-case loading, as discussed in Problem 8–73.

8–75 A split-ring clamp-type shaft collar such as is described in Problem 5–80 must resist an axial load of 5 kN. Using a design factor of $n_d = 3$ and a coefficient of friction of 0.12, specify an ISO 8.8 cap screw using fine threads. What wrench torque should be used if a lubricated screw is used?

8–76 A vertical channel 152 × 76 (see Table A–7) has a cantilever beam bolted to it as shown. The channel is hot-rolled AISI 1006 steel. The bar is of hot-rolled AISI 1015 steel. The shoulder bolts are M10 × 1.5 ISO 5.8. Assume the bolt threads do not extend into the joint. For a design factor of 2.0, find the safe force F that can be applied to the cantilever.

Problem 8–76
Dimensions in millimeters.

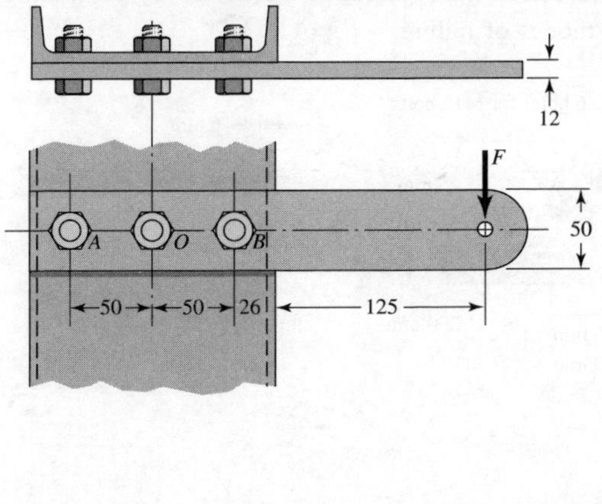

8–77 The cantilever bracket is bolted to a column with three M12 × 1.75 ISO 5.8 bolts. The bracket is made from AISI 1020 hot-rolled steel. Assume the bolt threads do not extend into the joint. Find the factors of safety for the following failure modes: shear of bolts, bearing of bolts, bearing of bracket, and bending of bracket.

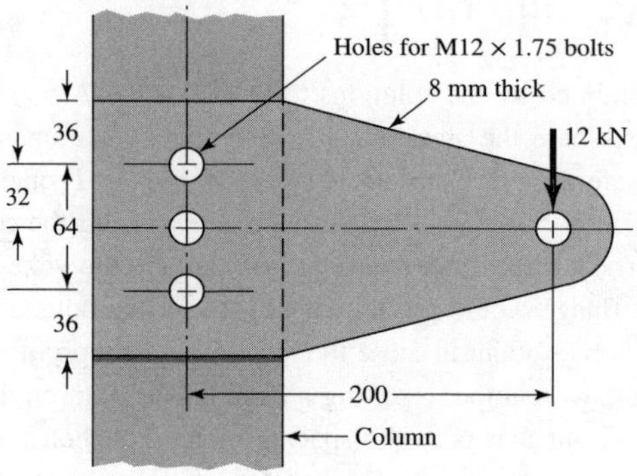

Problem 8–77
Dimensions in millimeters.

8–78 A 10 × 50 AISI 1018 cold-drawn steel bar is cantilevered to support a static load of 1 kN as illustrated. The bar is secured to the support using two M10 ISO 8.8 bolts. Assume the bolt threads do not extend into the joint. Find the factor of safety for the following modes of failure: shear of bolt, bearing on bolt, bearing on member, and strength of member.

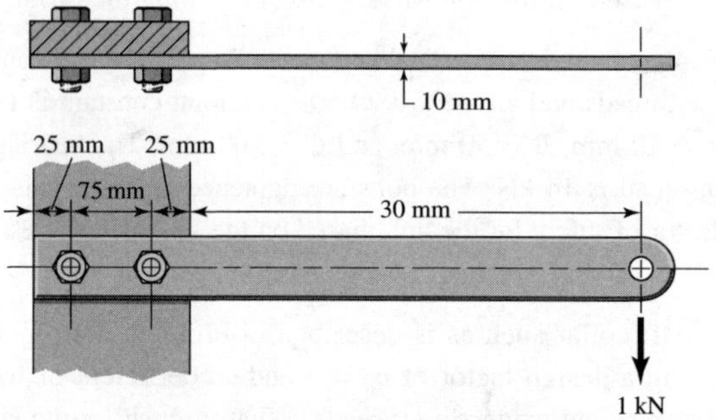

Problem 8–78

8–79 The figure shows a welded fitting which has been tentatively designed to be bolted to a channel so as to transfer the 9 kN load into the channel. The channel and the two fitting plates are of hot-rolled stock having a minimum S_y of 290 MPa. The fitting is to be bolted using six ISO 8.8 shoulder bolts. Assume the bolt threads do not extend into the joint. Check the strength of the design by computing the factor of safety for all possible modes of failure.

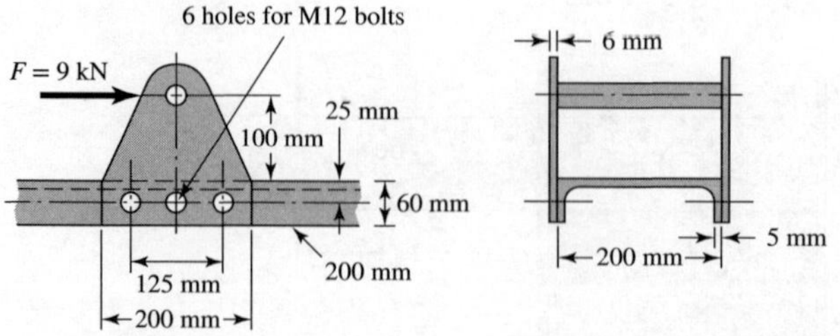

Problem 8–79

8–80 A plate is bolted to two walls as shown. Considering the moment and vertical force reactions at the walls, determine the magnitude of the maximum shear stress in the bolts if they each have a cross-sectional area of 58 mm^2.

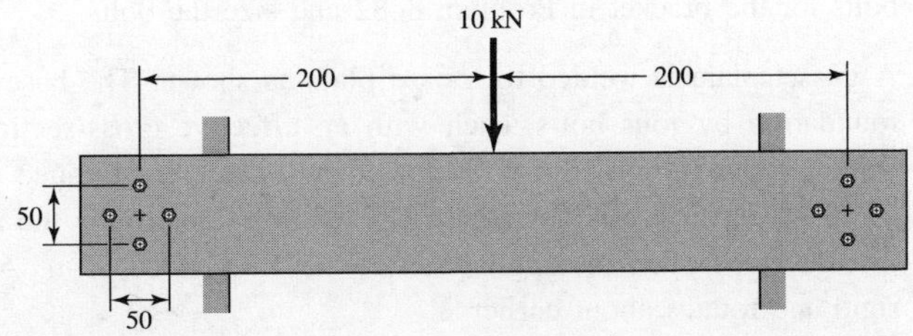

Problem 8–80
Dimensions in millimeters.

8–81 The coupled shafts shown are transmitting 190 kW at 600 rev/min. The shafts are connected by a coupling with eight bolts. The bolt circle diameters are 125 and 250 mm.

(*a*) Determine the minimum nominal shoulder diameter of a bolt if all bolts are to have the same diameter and not exceed a shear stress of 140 MPa. Select a preferred fraction size.

(*b*) Determine the minimum number of bolts of the diameter determined in part (*a*) if the shear stress is not to exceed 140 MPa and the bolts are to be placed only at the outer bolt circle.

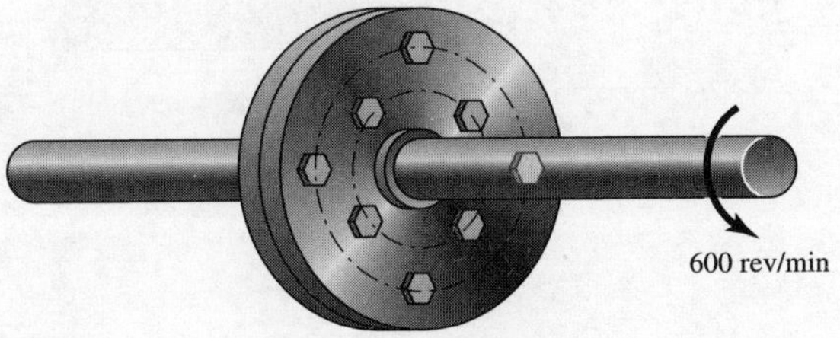

Problem 8–81

8–82 A cantilever is to be attached to the flat side of a 150 mm, 2.3 N/mm channel used as a column. The cantilever is to carry a load as shown in the figure. To a designer the choice of a bolt array is usually an a priori decision. Such decisions are made from a background of knowledge of the effectiveness of various patterns.

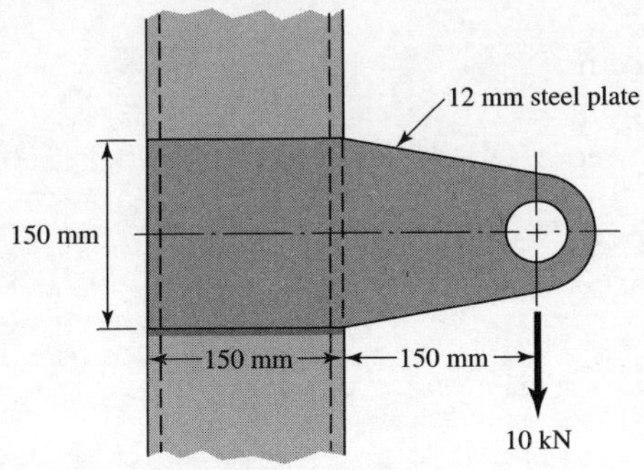

Problem 8–82

(*a*) If two fasteners are used, should the array be arranged vertically, horizontally, or diagonally? How would you decide?

(*b*) If three fasteners are used, should a linear or triangular array be used? For a triangular array, what should be the orientation of the triangle? How would you decide?

8–83 Using your experience with Problem 8–82, specify an optimal bolt pattern for two bolts for the bracket in Problem 8–82 and size the bolts.

8–84 Using your experience with Problem 8–82, specify an optimal bolt pattern for three bolts for the bracket in Problem 8–82 and size the bolts.

8–85 A gusset plate is welded to a base plate as shown. The base plate is secured to the foundation by four bolts, each with an effective cross-sectional area of 130 mm^2. Before application of the 4 kN force, the bolts were torqued down so that a preload of 20 kN tension was developed in each bolt. Determine the maximum tensile stress that develops upon the application of the 4 kN force. *Hint:* Assume that the plate is rigid and rotates about corner *A*.

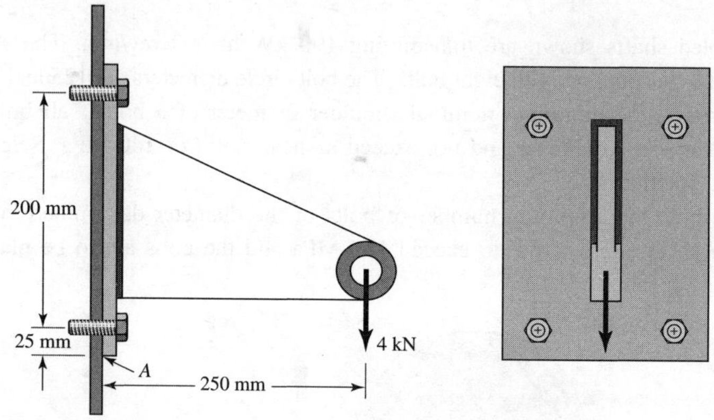

Problem 8–85

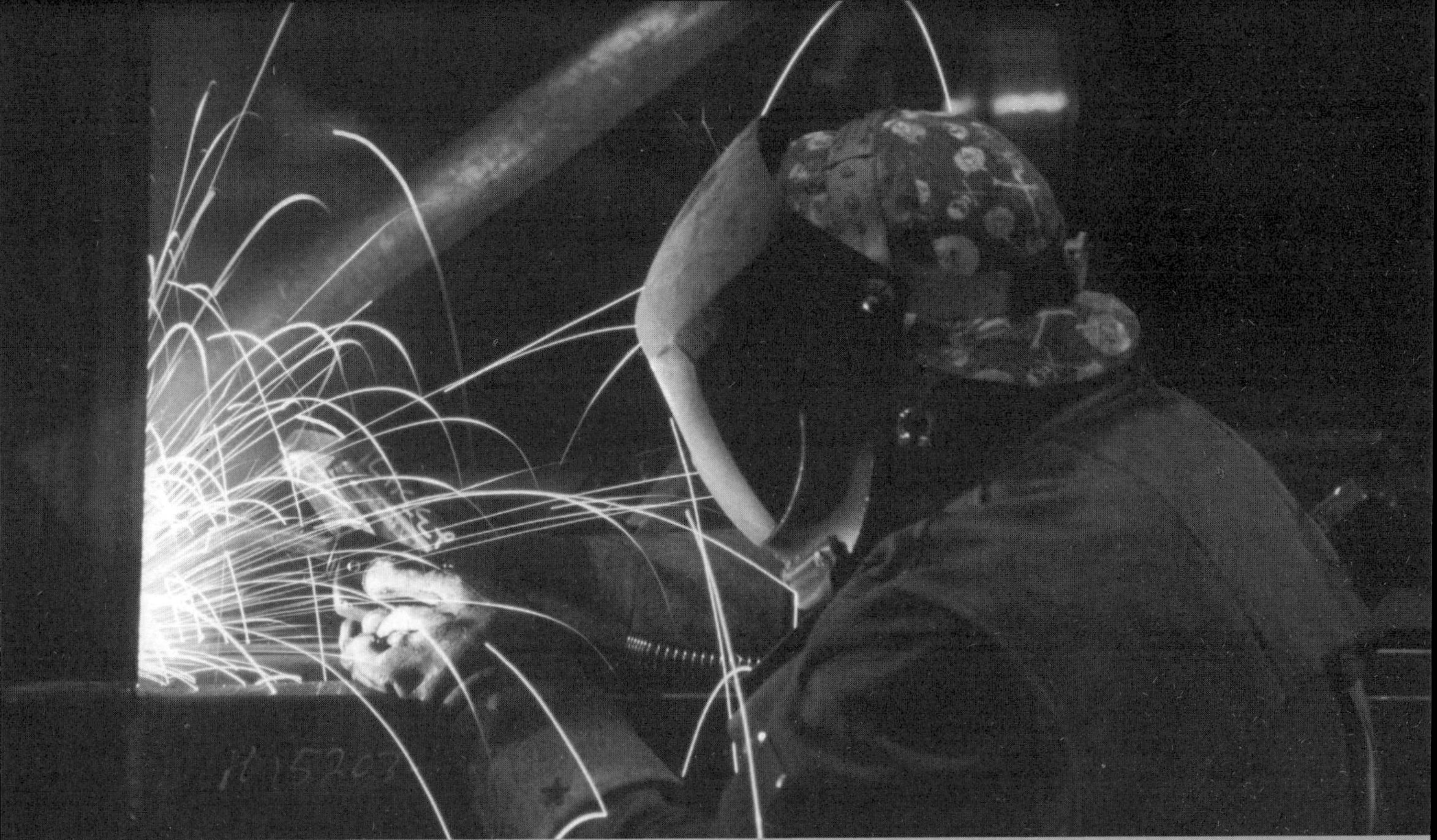

Ingram Publishing/SuperStock

Chapter Outline

Form can more readily pursue function with the help of joining processes such as welding, brazing, soldering, cementing, and gluing—processes that are used extensively in manufacturing today. Whenever parts have to be assembled or fabricated, there is usually good cause for considering one of these processes in preliminary design work. Particularly when sections to be joined are thin, one of these methods may lead to significant savings. The elimination of individual fasteners, with their holes and assembly costs, is an important factor. Also, some of the methods allow rapid machine assembly, furthering their attractiveness.

9–1 Welding Symbols

A weldment is fabricated by welding together a collection of metal shapes, cut to particular configurations. During welding, the several parts are held securely together, often by clamping or jigging. The welds must be precisely specified on working drawings, and this is done by using the welding symbol, shown in Figure 9–1, as standardized by the American Welding Society (AWS). The arrow of this symbol points to the joint to be welded. The body of the symbol contains as many of the following elements as are deemed necessary:

- Reference line
- Arrow
- Basic weld symbols as in Figure 9–2
- Dimensions and other data
- Supplementary symbols
- Finish symbols
- Tail
- Specification or process

Figure 9–1

The AWS standard welding symbol showing the location of the symbol elements.

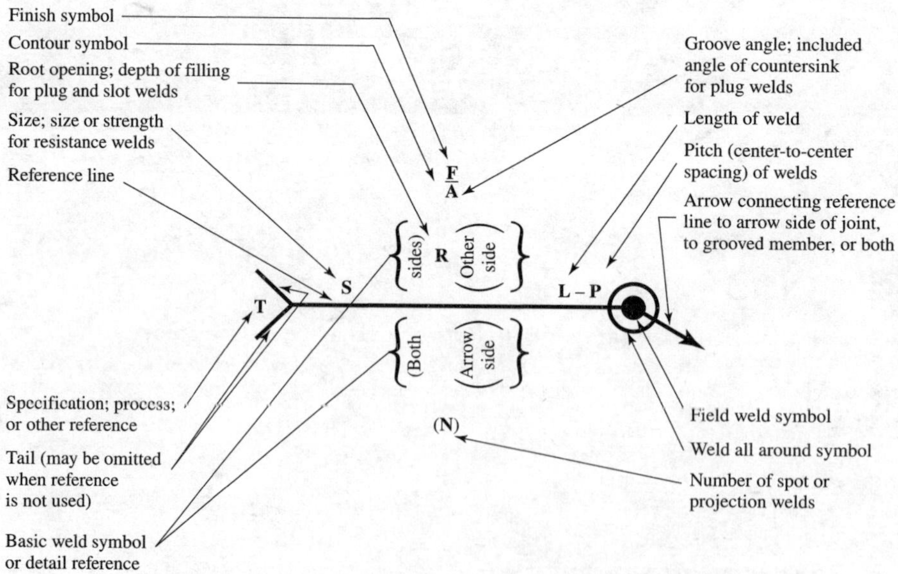

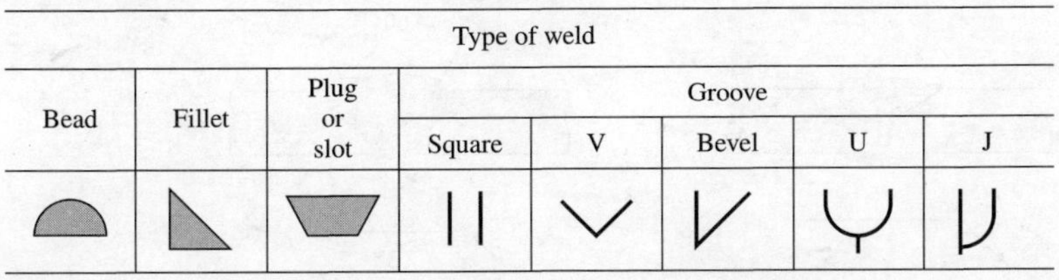

Figure 9–2

Arc- and gas-weld symbols.

The *arrow side* of a joint is the line, side, area, or near member to which the arrow points. The side opposite the arrow side is the *other side*.

Figures 9–3 to 9–6 illustrate the types of welds used most frequently by designers. For general machine elements most welds are fillet welds, though butt welds are used a great deal in designing pressure vessels. Of course, the parts to be joined must be arranged so that there is sufficient clearance for the welding operation. If unusual joints are required because of insufficient clearance or because of the section shape, the design may be a poor one and the designer should begin again and endeavor to synthesize another solution.

Since heat is used in the welding operation, there are metallurgical changes in the parent metal in the vicinity of the weld. Also, residual stresses may be introduced because of clamping or holding or, sometimes, because of the order of welding. Usually these residual stresses are not severe enough to cause concern; in some cases a light heat treatment after welding has been found helpful in relieving them. When the parts to be welded are thick, a preheating will also be of benefit. If the reliability of the component is to be quite high, a testing program should be established to learn what changes or additions to the operations are necessary to ensure the best quality.

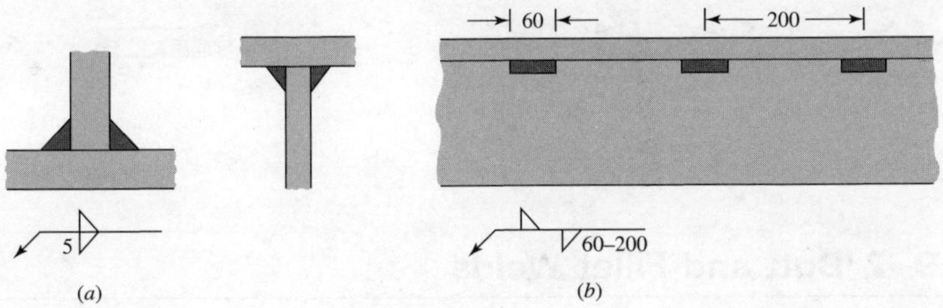

Figure 9–3

Fillet welds. (*a*) The number indicates the leg size; the arrow should point only to one weld when both sides are the same. (*b*) The symbol indicates that the welds are intermittent and staggered 60 mm along on 200-mm centers.

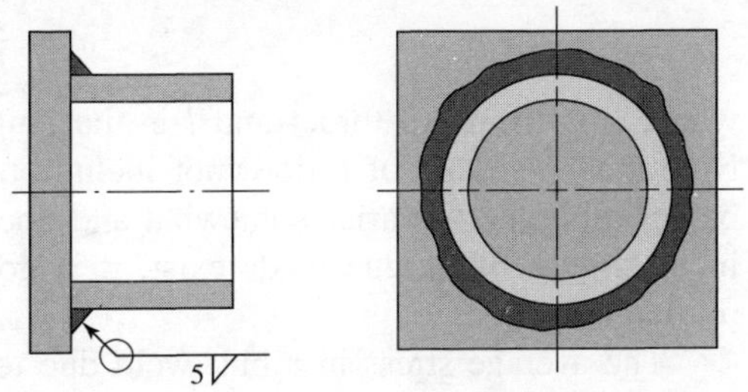

Figure 9–4

The circle on the weld symbol indicates that the welding is to go all around.

Figure 9–5

Butt or groove welds: (*a*) square butt-welded on both sides; (*b*) single V with 60° bevel and root opening of 2 mm; (*c*) double V; (*d*) single bevel.

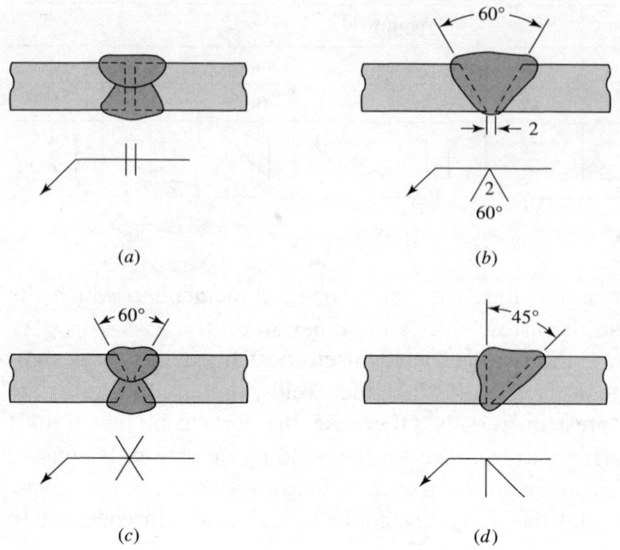

Figure 9–6

Special groove welds: (*a*) T joint for thick plates; (*b*) U and J welds for thick plates; (*c*) corner weld (may also have a bead weld on inside for greater strength but should not be used for heavy loads); (*d*) edge weld for sheet metal and light loads.

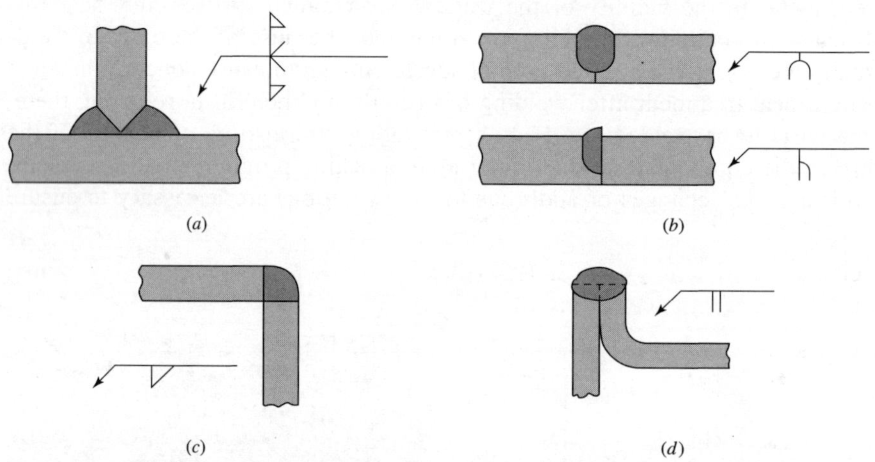

9–2 Butt and Fillet Welds

Figure 9–7*a* shows a single V-groove weld loaded by the tensile force *F*. For either tension or compression loading, the average normal stress is

$$\sigma = \frac{F}{hl} \tag{9–1}$$

where h is the weld throat and l is the length of the weld, as shown in the figure. Note that the value of h does not include the reinforcement. The reinforcement can be desirable, but it varies somewhat and does produce stress concentration at point *A* in the figure. If fatigue loads exist, it is good practice to grind or machine off the reinforcement.

The average stress in a butt weld due to shear loading (Figure 9–7*b*) is

$$\tau = \frac{F}{hl} \tag{9–2}$$

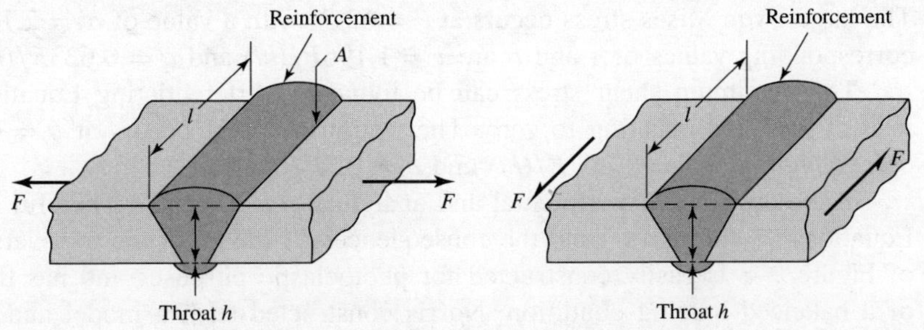

(a) Tensile loading (b) Shear loading

Figure 9–7

A typical butt joint.

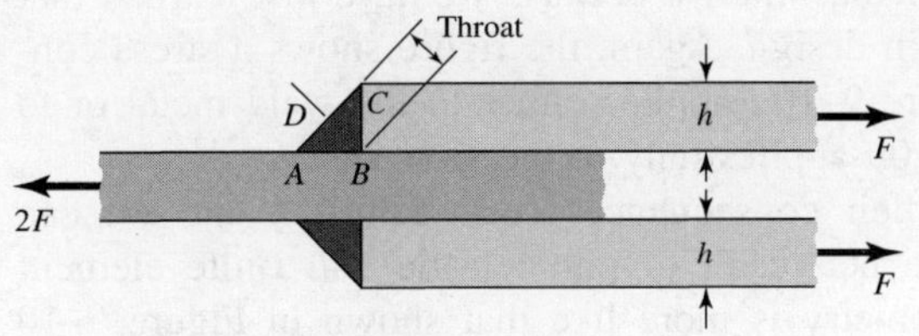

Figure 9–8

A transverse fillet weld.

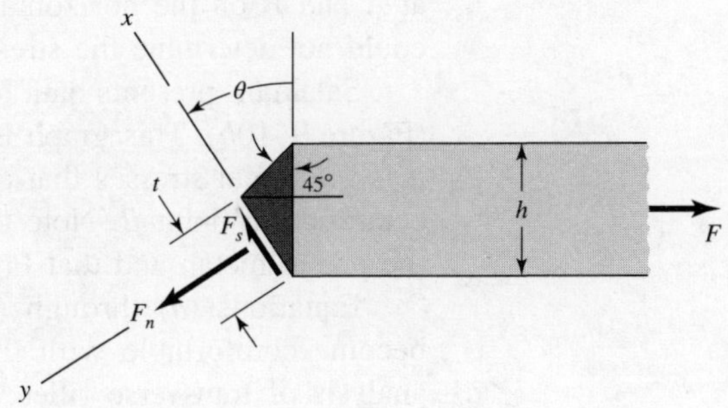

Figure 9–9

Free body from Figure 9–8.

Figure 9–8 illustrates a typical transverse fillet weld. In Figure 9–9 a portion of the welded joint has been isolated from Figure 9–8 as a free body. At angle θ the forces on each weldment consist of a normal force F_n and a shear force F_s. Summing forces in the x and y directions gives

$$F_s = F\sin\theta \tag{a}$$

$$F_n = F\cos\theta \tag{b}$$

Using the law of sines for the triangle in Figure 9–9 yields

$$\frac{t}{\sin 45^\circ} = \frac{h}{\sin(180^\circ - 45^\circ - \theta)} = \frac{h}{\sin(135^\circ - \theta)} = \frac{\sqrt{2}h}{\cos\theta + \sin\theta}$$

Solving for the throat thickness t gives

$$t = \frac{h}{\cos\theta + \sin\theta} \tag{c}$$

The nominal stresses at the angle θ in the weldment, τ and σ, are

$$\tau = \frac{F_s}{A} = \frac{F\sin\theta(\cos\theta + \sin\theta)}{hl} = \frac{F}{hl}(\sin\theta\cos\theta + \sin^2\theta) \tag{d}$$

$$\sigma = \frac{F_n}{A} = \frac{F\cos\theta(\cos\theta + \sin\theta)}{hl} = \frac{F}{hl}(\cos^2\theta + \sin\theta\cos\theta) \tag{e}$$

The von Mises stress σ' at angle θ is

$$\sigma' = (\sigma^2 + 3\tau^2)^{1/2} = \frac{F}{hl}[(\cos^2\theta + \sin\theta\cos\theta)^2 + 3(\sin^2\theta + \sin\theta\cos\theta)^2]^{1/2} \tag{f}$$

The largest von Mises stress occurs at $\theta = 62.5°$ with a value of $\sigma' = 2.16F/(hl)$. The corresponding values of τ and σ are $\tau = 1.196F/(hl)$ and $\sigma = 0.623F/(hl)$.

The maximum shear stress can be found by differentiating Equations (*d*) with respect to θ and equating to zero. The stationary point occurs at $\theta = 67.5°$ with a corresponding $\tau_{max} = 1.207F/(hl)$ and $\sigma = 0.5F/(hl)$.

There are some experimental and analytical results that are helpful in evaluating Equations (*d*) through (*f*) and the consequences. A model of the transverse fillet weld of Figure 9–8 is easily constructed for photoelastic purposes and has the advantage of a balanced loading condition. Norris constructed such a model and reported the stress distribution along the sides *AB* and *BC* of the weld.[1] An approximate graph of the results he obtained is shown as Figure 9–10*a*. Note that stress concentrations exist at *A* and *B* on the horizontal leg and at *B* on the vertical leg. Norris states that he could not determine the stresses at *A* and *B* with any certainty.

Salakian[2] presents data for the stress distribution across the throat of a fillet weld (Figure 9–10*b*). This graph is of particular interest because we have just learned that it is the throat stresses that are used in design. Again, the figure shows a stress concentration at point *B*. Note that Figure 9–10*a* applies either to the weld metal or to the parent metal, and that Figure 9–10*b* applies only to the weld metal.

Equations (*a*) through (*f*) and their consequences seem familiar, and we can become comfortable with them. The net result of photoelastic and finite element analysis of transverse fillet weld geometry is more like that shown in Figure 9–10 than those given by mechanics of materials or elasticity methods. The most important concept here is that we have *no analytical approach that predicts the existing stresses*. The geometry of the fillet is crude by machinery standards, and even if it were ideal, the macrogeometry is too abrupt and complex for our methods. There are also subtle bending stresses due to eccentricities. Still, in the absence of robust analysis,

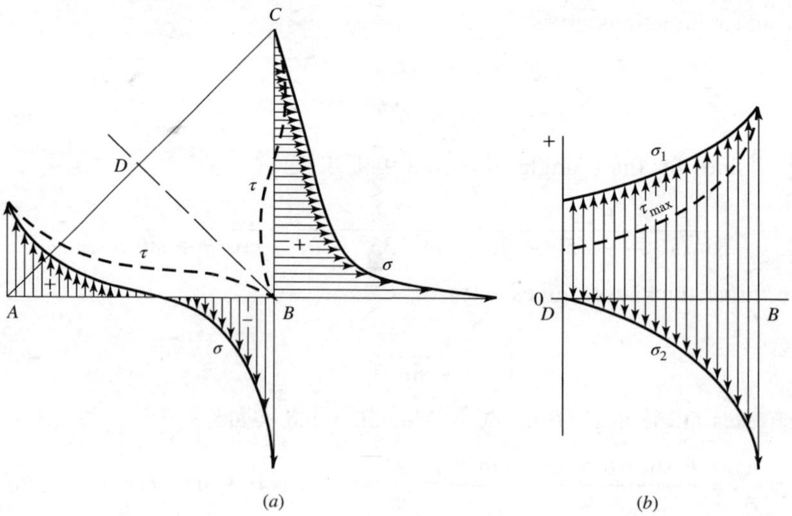

Figure 9–10

Stress distribution in fillet welds: (*a*) stress distribution on the legs as reported by Norris; (*b*) distribution of principal stresses and maximum shear stress as reported by Salakian.

[1]C. H. Norris, "Photoelastic Investigation of Stress Distribution in Transverse Fillet Welds," *Welding J.*, vol. 24, 1945, p. 557s.

[2]A. G. Salakian and G. E. Claussen, "Stress Distribution in Fillet Welds: A Review of the Literature," *Welding J.*, vol. 16, May 1937, pp. 1–24.

weldments must be specified and the resulting joints must be safe. The approach has been to use a simple *and conservative* model, verified by testing as conservative. The approach has been to

- Consider the external loading to be carried by shear forces on the throat area of the weld. By ignoring the normal stress on the throat, the shearing stresses are inflated sufficiently to render the model conservative.
- Use distortion energy for significant stresses.
- Circumscribe typical cases by code.

For this model, the basis for weld analysis or design employs

$$\tau = \frac{F}{0.707hl} = \frac{1.414F}{hl} \tag{9–3}$$

which assumes the entire force F is accounted for by a shear stress in the minimum throat area. Note that this inflates the maximum estimated shear stress by a factor of $1.414/1.207 = 1.17$. Further, consider the parallel fillet welds shown in Figure 9–11 where, as in Figure 9–8, each weld transmits a force F. However, in the case of Figure 9–11, the maximum shear stress *is* at the minimum throat area and corresponds to Equation (9–3).

Under circumstances of combined loading we

- Examine primary shear stresses due to external forces.
- Examine secondary shear stresses due to torsional and bending moments.
- Estimate the strength(s) of the parent metal(s).
- Estimate the strength of deposited weld metal.
- Estimate permissible load(s) for parent metal(s).
- Estimate permissible load for deposited weld metal.

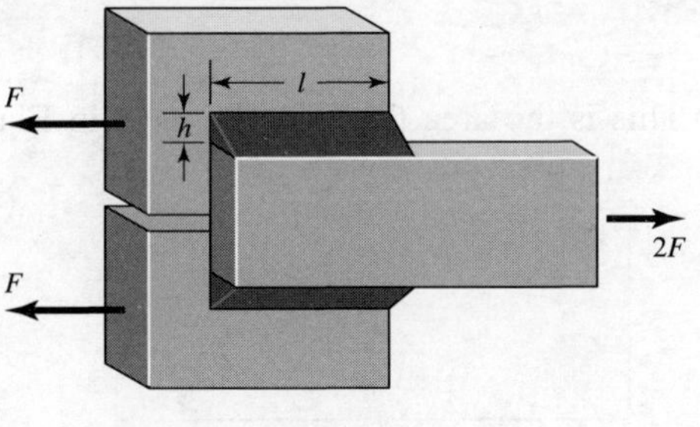

Figure 9–11
Parallel fillet welds.

9–3 Stresses in Welded Joints in Torsion

Figure 9–12 illustrates a cantilever welded to a column by two fillet welds each of length l. The reaction at the support of a cantilever always consists of a shear force V and a moment M. The shear force produces a *primary shear* in the welds of magnitude

$$\tau' = \frac{V}{A} \tag{9–4}$$

where A is the throat area of all the welds.

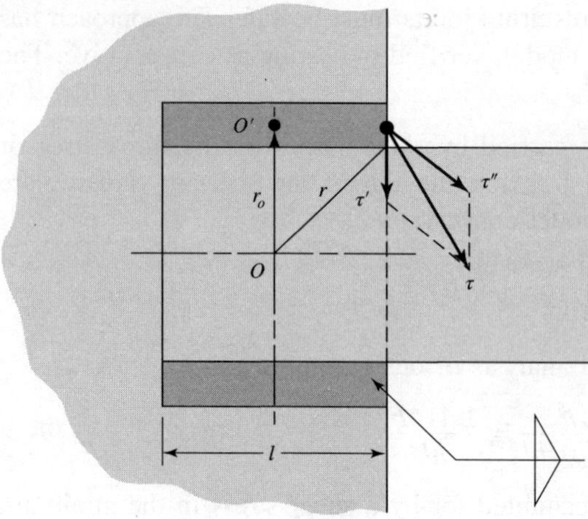

Figure 9–12

This is a *moment connection;* such a connection produces *torsion* in the welds. The shear stresses shown are resultant stresses.

The moment at the support produces *secondary shear* or *torsion* of the welds, and this stress is given by the equation

$$\tau'' = \frac{Mr}{J} \tag{9–5}$$

where r is the distance from the centroid of the weld group to the point in the weld of interest and J is the second polar moment of area of the weld group about the centroid of the group. When the sizes of the welds are known, these equations can be solved and the results combined to obtain the maximum shear stress. Note that r is usually the farthest distance from the centroid of the weld group.

Figure 9–13 shows two welds in a group. The rectangles represent the throat areas of the welds. Weld 1 has a throat thickness $t_1 = 0.707h_1$, and weld 2 has a throat thickness $t_2 = 0.707h_2$. Note that h_1 and h_2 are the respective weld sizes. The throat area of both welds together is

$$A = A_1 + A_2 = t_1 d + t_2 b \tag{a}$$

This is the area that is to be used in Equation (9–4).

Figure 9–13

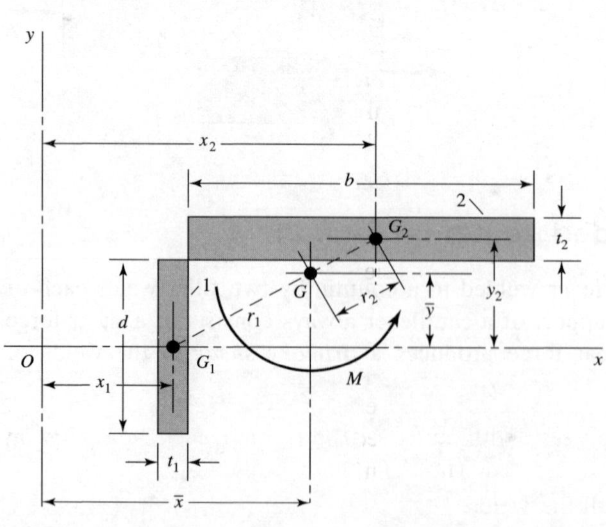

The x axis in Figure 9–13 passes through the centroid G_1 of weld 1. The second moment of area about this axis is

$$I_x = \frac{t_1 d^3}{12}$$

Similarly, the second moment of area about an axis through G_1 parallel to the y axis is

$$I_y = \frac{d t_1^3}{12}$$

Thus the second polar moment of area of weld 1 about its own centroid is

$$J_{G1} = I_x + I_y = \frac{t_1 d^3}{12} + \frac{d t_1^3}{12} \tag{b}$$

In a similar manner, the second polar moment of area of weld 2 about its centroid is

$$J_{G2} = \frac{b t_2^3}{12} + \frac{t_2 b^3}{12} \tag{c}$$

The centroid G of the weld group is located at

$$\bar{x} = \frac{A_1 x_1 + A_2 x_2}{A} \qquad \bar{y} = \frac{A_1 y_1 + A_2 y_2}{A}$$

Using Figure 9–13 again, we see that the distances r_1 and r_2 from G_1 and G_2 to G, respectively, are

$$r_1 = \left[(\bar{x} - x_1)^2 + \bar{y}^2\right]^{1/2} \qquad r_2 = \left[(y_2 - \bar{y})^2 + (x_2 - \bar{x})^2\right]^{1/2}$$

Now, using the parallel-axis theorem, we find the second polar moment of area of the weld group to be

$$J = (J_{G1} + A_1 r_1^2) + (J_{G2} + A_2 r_2^2) \tag{d}$$

This is the quantity to be used in Equation (9–5). The distance r must be measured from G and the moment M computed about G.

The reverse procedure is that in which the allowable shear stress is given and we wish to find the weld size. The usual procedure is to estimate a probable weld size and then to use iteration.

Observe in Equations (b) and (c) the quantities t_1^3 and t_2^3, respectively, which are the cubes of the weld thicknesses. These quantities are small and can be neglected. This leaves the terms $t_1 d^3/12$ and $t_2 b^3/12$, which make J_{G1} and J_{G2} linear in the weld width. Setting the weld thicknesses t_1 and t_2 to unity leads to the idea of treating each fillet weld as a line. The resulting second moment of area is then a *unit second polar moment of area*. The advantage of treating the weld size as a line is that the value of J_u is the same regardless of the weld size. Since the throat width of a fillet weld is $0.707h$, the relationship between J and the unit value is

$$J = 0.707 h J_u \tag{9–6}$$

in which J_u is found by conventional methods for an area having unit width. The transfer formula for J_u must be employed when the welds occur in groups, as in Figure 9–12. Table 9–1 lists the throat areas and the unit second polar moments of area for the most common fillet welds encountered. The example that follows is typical of the calculations normally made.

Table 9–1 Torsional Properties of Fillet Welds*

Weld	Throat Area	Location of *G*	Unit Second Polar Moment of Area
1. 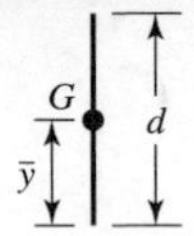	$A = 0.707hd$	$\bar{x} = 0$ $\bar{y} = d/2$	$J_u = d^3/12$
2. 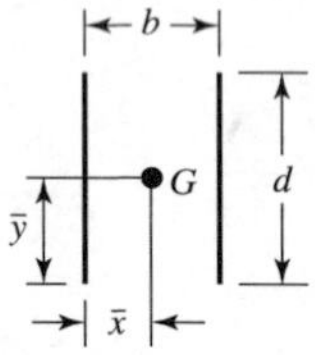	$A = 1.414hd$	$\bar{x} = b/2$ $\bar{y} = d/2$	$J_u = \dfrac{d(3b^2 + d^2)}{6}$
3. 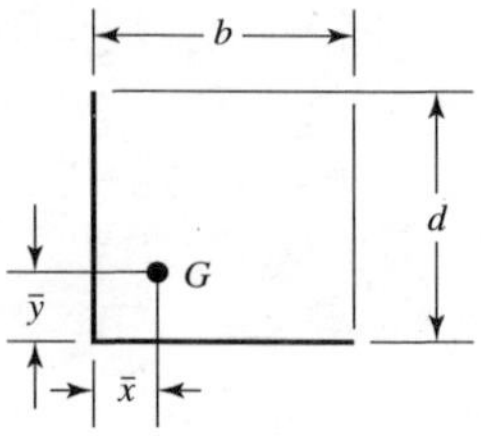	$A = 0.707h(b + d)$	$\bar{x} = \dfrac{b^2}{2(b + d)}$ $\bar{y} = \dfrac{d^2}{2(b + d)}$	$J_u = \dfrac{(b + d)^4 - 6b^2d^2}{12(b + d)}$
4. 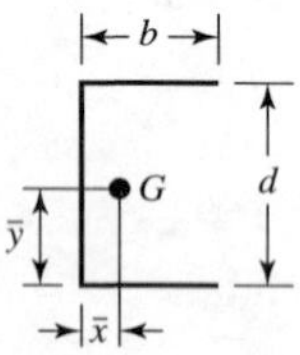	$A = 0.707h(2b + d)$	$\bar{x} = \dfrac{b^2}{2b + d}$ $\bar{y} = d/2$	$J_u = \dfrac{8b^3 + 6bd^2 + d^3}{12} - \dfrac{b^4}{2b + d}$
5. 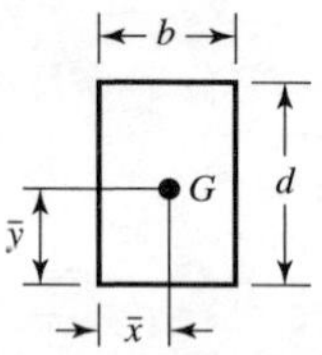	$A = 1.414h(b + d)$	$\bar{x} = b/2$ $\bar{y} = d/2$	$J_u = \dfrac{(b + d)^3}{6}$
6. 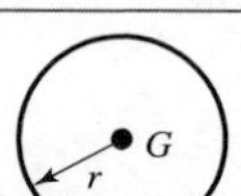	$A = 1.414\pi hr$		$J_u = 2\pi r^3$

**G* is the centroid of weld group; *h* is weld size; plane of torque couple is in the plane of the paper; all welds are of unit width.

EXAMPLE 9–1

A 50-kN load is transferred from a welded fitting into a 200-mm steel channel as illustrated in Figure 9–14. Estimate the maximum stress in the weld.

Solution

(*a*) Label the ends and corners of each weld by letter. See Figure 9–15. Sometimes it is desirable to label each weld of a set by number.

(*b*) Estimate the primary shear stress τ'. As shown in Figure 9–14, each plate is welded to the channel by means of three 6-mm fillet welds. Figure 9–15 shows that we have divided the load in half and are considering only a single plate. From case 4 of Table 9–1 we find the throat area as

$$A = 0.707(6)[2(56) + 190] = 1280 \text{ mm}^2$$

Then the primary shear stress is

$$\tau' = \frac{V}{A} = \frac{25(10)^3}{1280} = 19.5 \text{ MPa}$$

(*c*) Draw the τ' stress, to scale, at each lettered corner or end. See Figure 9–16.
(*d*) Locate the centroid of the weld pattern. Using case 4 of Table 9–1, we find

$$\bar{x} = \frac{(56)^2}{2(56) + 190} = 10.4 \text{ mm}$$

This is shown as point O on Figures 9–15 and 9–16.
(*e*) Find the distances r_i (see Figure 9–16):

$$r_A = r_B = [(190/2)^2 + (56 - 10.4)^2]^{1/2} = 105 \text{ mm}$$

$$r_C = r_D = [(190/2)^2 + (10.4)^2]^{1/2} = 95.6 \text{ mm}$$

These distances can also be scaled from the drawing.
(*f*) Find J. Using case 4 of Table 9–1 again, with Equation (9–6), we get

$$J = 0.707(6)\left[\frac{8(56)^3 + 6(56)(190)^2 + (190)^3}{12} - \frac{(56)^4}{2(56) + 190}\right]$$

$$= 7.07(10)^6 \text{mm}^4$$

(*g*) Find M:

$$M = Fl = 25(100 + 10.4) = 2760 \text{ N} \cdot \text{m}$$

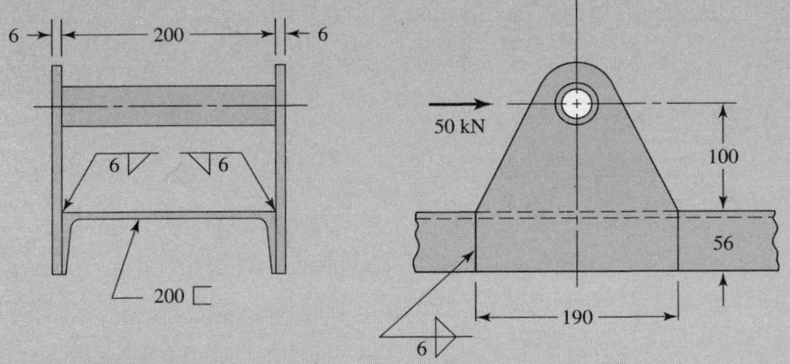

Figure 9–14

Dimensions in millimeters.

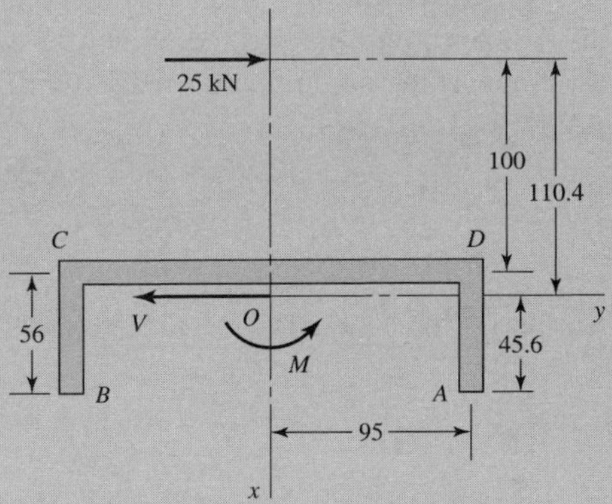

Figure 9–15

Diagram showing the weld geometry on a single plate; all dimensions in millimeters. Note that V and M represent the reaction loads applied by the welds *to the plate*.

Figure 9–16

Free-body diagram of one of the side plates.

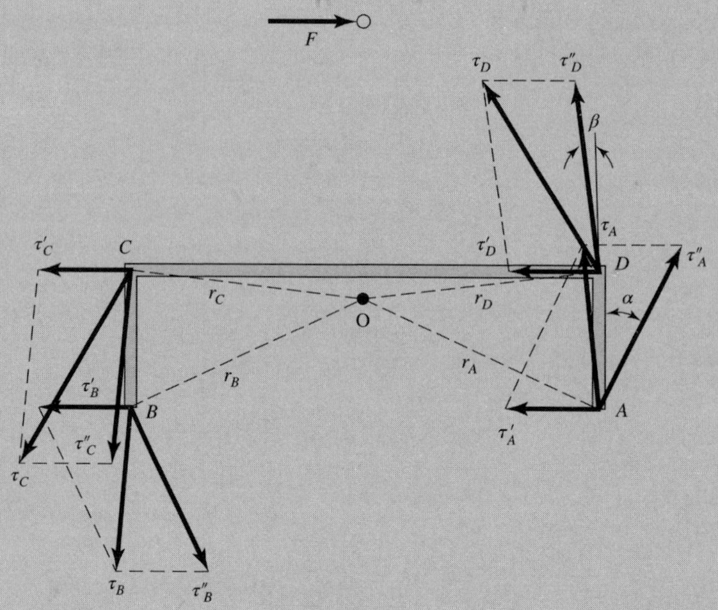

(*h*) Estimate the secondary shear stresses τ'' at each lettered end or corner:

$$\tau''_A = \tau''_B = \frac{Mr}{J} = \frac{2760(10)^3(105)}{7.07(10)^6} = 41.0 \text{ MPa}$$

$$\tau''_C = \tau''_D = \frac{2760(10)^3(95.6)}{7.07(10)^6} = 37.3 \text{ MPa}$$

(*i*) Draw the τ'' stress at each corner and end. See Figure 9–16. Note that this is a free-body diagram of one of the side plates, and therefore the τ' and τ'' stresses represent what the channel is doing to the plate (through the welds) to hold the plate in equilibrium.

(*j*) At each point labeled, combine the two stress components as vectors (since they apply to the same area). At point *A*, the angle that τ''_A makes with the vertical, α, is also the angle r_A makes with the horizontal, which is $\alpha = \tan^{-1}(45.6/95) = 25.64°$. This angle also applies to point *B*. Thus

$$\tau_A = \tau_B = \sqrt{(19.5 - 41.0 \sin 25.64°)^2 + (41.0 \cos 25.64°)^2} = 37.0 \text{ MPa}$$

Similarly, for *C* and *D*, $\beta = \tan^{-1}(10.4/95) = 6.25°$. Thus

$$\tau_C = \tau_D = \sqrt{(19.5 + 37.3 \sin 6.25°)^2 + (37.3 \cos 6.25°)^2} = 43.9 \text{ MPa}$$

(*k*) Identify the most highly stressed point:

Answer $$\tau_{\max} = \tau_C = \tau_D = 43.9 \text{ MPa}$$

9–4 Stresses in Welded Joints in Bending

Figure 9–17*a* shows a cantilever welded to a support by fillet welds at top and bottom. A free-body diagram of the beam would show a shear-force reaction *V* and a moment reaction *M*. The shear force produces a primary shear in the welds of magnitude

$$\tau' = \frac{V}{A} \tag{a}$$

where *A* is the total throat area.

The moment *M* induces a horizontal shear stress component in the welds. Treating the two welds of Figure 9–17*b* as lines we find the unit second moment of area to be

$$I_u = \frac{bd^2}{2} \tag{b}$$

The second moment of area *I*, based on weld throat area, is

$$I = 0.707hI_u = 0.707h\frac{bd^2}{2} \tag{c}$$

The nominal throat shear stress is now found to be

$$\tau'' = \frac{Mc}{I} = \frac{Md/2}{0.707hbd^2/2} = \frac{1.414M}{bdh} \tag{d}$$

The model gives the coefficient of 1.414, in contrast to the predictions of Section 9–2 of 1.197 from distortion energy, or 1.207 from maximum shear. The conservatism of the model's 1.414 is not that it is simply larger than either 1.196 or 1.207, but the tests carried out to validate the model show that it is large enough.

The second moment of area in Equation (*d*) is based on the distance *d* between the two welds. If this moment is found by treating the two welds as having rectangular footprints, the distance between the weld throat centroids is approximately $(d + h)$. This would produce a slightly larger second moment of area, and result in a smaller level of stress. This method of treating welds as a line does not interfere with the conservatism of the model. It also makes Table 9–2 possible with all the conveniences that ensue.

The vertical (primary) shear of Equation (*a*) and the horizontal (secondary) shear of Equation (*d*) are then combined as vectors to give

$$\tau = (\tau'^2 + \tau''^2)^{1/2} \tag{e}$$

For an example problem, see parts (*a*) and (*b*) of Example 9–4.

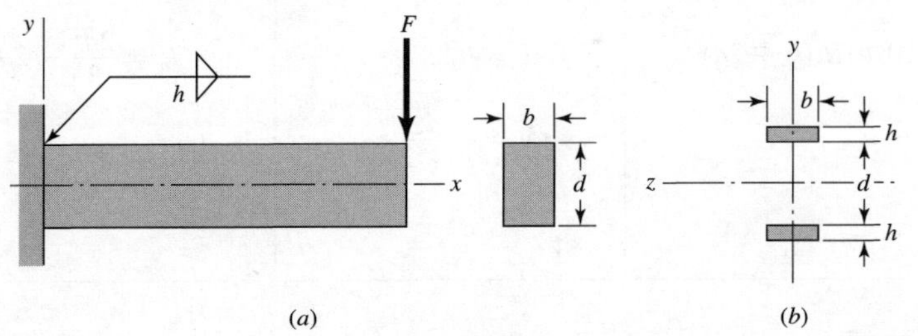

Figure 9–17

A rectangular cross-section cantilever welded to a support at the top and bottom edges.

Table 9–2 Bending Properties of Fillet Welds*

Weld	Throat Area	Location of G	Unit Second Moment of Area
1. 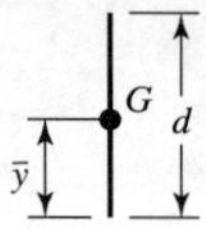	$A = 0.707hd$	$\bar{x} = 0$ $\bar{y} = d/2$	$I_u = \dfrac{d^3}{12}$
2. 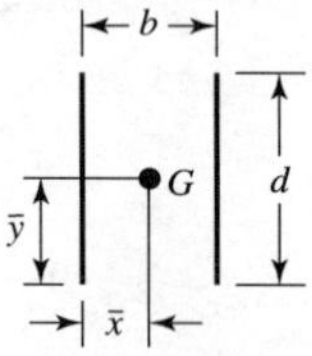	$A = 1.414hd$	$\bar{x} = b/2$ $\bar{y} = d/2$	$I_u = \dfrac{d^3}{6}$
3. 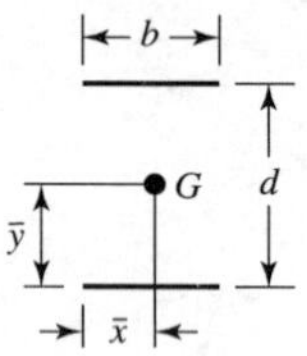	$A = 1.414hb$	$\bar{x} = b/2$ $\bar{y} = d/2$	$I_u = \dfrac{bd^2}{2}$
4. 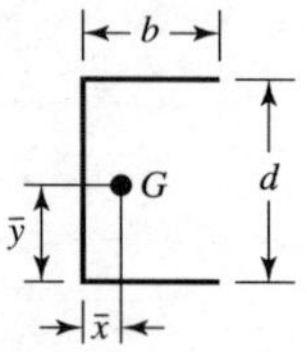	$A = 0.707h(2b + d)$	$\bar{x} = \dfrac{b^2}{2b + d}$ $\bar{y} = d/2$	$I_u = \dfrac{d^2}{12}(6b + d)$
5. 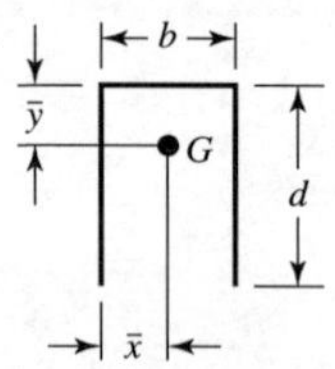	$A = 0.707h(b + 2d)$	$\bar{x} = b/2$ $\bar{y} = \dfrac{d^2}{b + 2d}$	$I_u = \dfrac{2d^3}{3} - 2d^2\bar{y} + (b + 2d)\bar{y}^2$
6. 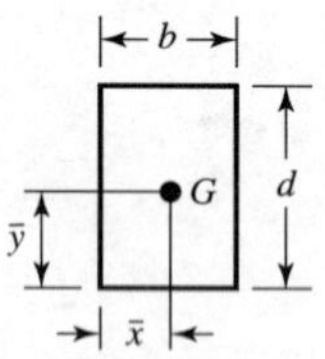	$A = 1.414h(b + d)$	$\bar{x} = b/2$ $\bar{y} = d/2$	$I_u = \dfrac{d^2}{6}(3b + d)$
7. 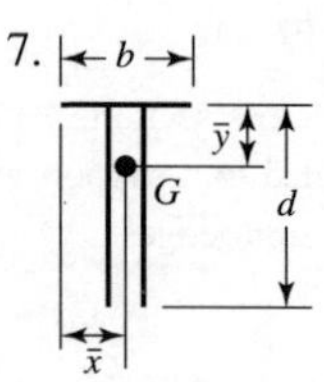	$A = 0.707h(b + 2d)$	$\bar{x} = b/2$ $\bar{y} = \dfrac{d^2}{b + 2d}$	$I_u = \dfrac{2d^3}{3} - 2d^2\bar{y} + (b + 2d)\bar{y}^2$

Table 9–2 ***(Continued)***

Weld	Throat Area	Location of G	Unit Second Moment of Area
8. 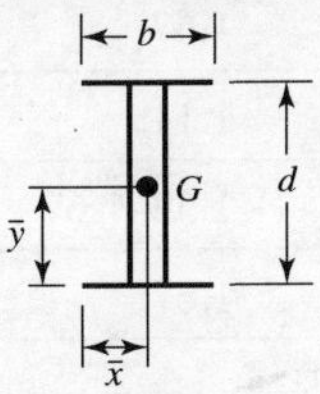	$A = 1.414h(b + d)$	$\bar{x} = b/2$ $\bar{y} = d/2$	$I_u = \dfrac{d^2}{6}(3b + d)$
9. 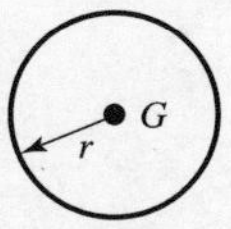	$A = 1.414\pi hr$		$I_u = \pi r^3$

*I_u, unit second moment of area, is taken about a horizontal axis through G, the centroid of the weld group, h is weld size; the plane of the bending couple is normal to the plane of the paper and parallel to the y-axis; all welds are of the same size.

9–5 The Strength of Welded Joints

The matching of the electrode properties with those of the parent metal is usually not so important as speed, operator appeal, and the appearance of the completed joint. The properties of electrodes vary considerably, but Table 9–3 lists the minimum properties for some electrode classes.

It is preferable, in designing welded components, to select a steel that will result in a fast, economical weld even though this may require a sacrifice of other qualities such as machinability. Under the proper conditions, all steels can be welded, but best results will be obtained if steels having a UNS specification between G10 140 and G10 230 are chosen. All these steels have a tensile strength in the hot-rolled condition in the range of 410 to 480 MPa.

The designer can choose factors of safety or permissible working stresses with more confidence by being aware of the values of those used by others. One of the best standards to use is the American Institute of Steel Construction (AISC) code for

Table 9–3 Minimum Weld-Metal Properties

AWS Electrode Number*	Tensile Strength, MPa	Yield Strength, MPa	Percent Elongation
E60xx	427	345	17–25
E70xx	482	393	22
E80xx	551	462	19
E90xx	620	531	14–17
E100xx	689	600	13–16
E120xx	827	737	14

*The American Welding Society (AWS) specification code numbering system for electrodes. This system uses an E prefixed to a four- or five-digit numbering system in which the first two or three digits designate the approximate tensile strength. The last digit includes variables in the welding technique, such as current supply. The next-to-last digit indicates the welding position, as, for example, flat, or vertical, or overhead. The complete set of specifications may be obtained from the AWS upon request.

Table 9–4 Stresses Permitted by the AISC Code for Weld Metal

Type of Loading	Type of Weld	Permissible Stress	n*
Tension	Butt	$0.60S_y$	1.67
Bearing	Butt	$0.90S_y$	1.11
Bending	Butt	$0.60–0.66S_y$	1.52–1.67
Simple compression	Butt	$0.60S_y$	1.67
Shear	Butt or fillet	$0.30S_{ut}^{\dagger}$	

*The factor of safety n has been computed by using the distortion-energy theory.
†Shear stress on base metal should not exceed $0.40S_y$ of base metal.

building construction. The permissible stresses are now based on the yield strength of the material instead of the ultimate strength, and the code permits the use of a variety of ASTM structural steels having yield strengths varying from 230 to 340 MPa. Provided the loading is the same, the code permits the same stress in the weld metal as in the parent metal. Table 9–4 lists the formulas specified by the code for calculating these permissible stresses for various loading conditions. The factors of safety implied by this code are easily calculated. For tension, $n = 1/0.60 = 1.67$. For shear, $n = 0.577/0.40 = 1.44$, using the distortion-energy theory as the criterion of failure.

It is important to observe that the electrode material is often the strongest material present. If a bar of AISI 1010 steel is welded to one of 1018 steel, the weld metal is actually a mixture of the electrode material and the 1010 and 1018 steels. Furthermore, a welded cold-drawn bar has its cold-drawn properties replaced with the hot-rolled properties in the vicinity of the weld. Finally, remembering that the weld metal is usually the strongest, do check the stresses in the parent metals.

The AISC code, as well as the AWS code, for bridges includes permissible stresses when fatigue loading is present. The designer will have no difficulty in using these codes, but their empirical nature tends to obscure the fact that they have been established by means of the same knowledge of fatigue failure already discussed in Chapter 6. Of course, for structures covered by these codes, the actual stresses *cannot* exceed the permissible stresses; otherwise the designer is legally liable. But in general, codes tend to conceal the actual margin of safety involved.

The fatigue stress-concentration factors listed in Table 9–5 are suggested for use. These factors should be used for the parent metal as well as for the weld metal. Table 9–6 gives steady-load information and minimum fillet sizes.

Table 9–5 Fatigue Stress-Concentration Factors, K_{fs}

Type of Weld	K_{fs}
Reinforced butt weld	1.2
Toe of transverse fillet weld	1.5
End of parallel fillet weld	2.7
T-butt joint with sharp corners	2.0

Table 9–6 Allowable Steady Loads and Minimum Fillet Weld Sizes

Schedule A: Allowable Load for Various Sizes of Fillet Welds

	Strength Level of Weld Metal (EXX)						
	60*	70*	80	90*	100	110*	120
	Allowable shear stress on throat, MPa of fillet weld or partial penetration groove weld						
$\tau =$	124	145	165	186	207	228	248
	Allowable Unit Force on Fillet Weld, N/mm						
$^{\dagger}f =$	$87.67h$	$102.52h$	$116.66h$	$131.5h$	$146.35h$	$161.2h$	$175.34h$
Leg Size h, mm	Allowable Unit Force for Various Sizes of Fillet Welds N/mm						
25	2192	2563	2916	3288	3659	4030	4383
22	1929	2255	2566	2893	3220	3546	3857
20	1753	2050	2333	2630	2927	3224	3506
16	1403	1640	1866	2104	2342	2579	2805
12	1052	1230	1400	1578	1756	1934	2104
11	964	1127	1283	1447	1610	1773	1927
10	877	1025	1167	1315	1463	1612	1753
8	701	820	933	1052	1171	1290	1403
6	526	615	700	789	878	967	1052
5	438	513	583	658	732	806	877
3	263	308	350	395	439	484	526
2	175	205	233	263	293	322	351

*Fillet welds actually tested by the joint AISC-AWS Task Committee.

$^{\dagger}f = 0.707h\ \tau_{\text{all}}$.

Schedule B: Minimum Fillet Weld Size, h

Material Thickness of Thicker Part Joined, mm		Weld Size, mm
*To 6 incl.		3
Over 6	To 12	5
Over 12	To 20	6
†Over 20	To 38	8
Over 38	To 58	10
Over 58	To 150	12
Over 150		16

Not to exceed the thickness of the thinner part.

*Minimum size for bridge application does not go below 5 mm.

†For minimum fillet weld size, schedule does not go above 8 mm fillet weld for every 20 mm material.

Source: From Blodgett, W. Omer (ed.). *Stress Allowables Affect Weldment Design, D412*. Cleveland: The James F. Lincoln Arc Welding Foundation, May 1991, p. 3. Reprinted by permission of Lincoln Electric Company.

9–6 Static Loading

Some examples of statically loaded joints are useful in comparing and contrasting the conventional method of analysis and the welding code methodology.

EXAMPLE 9–2

A 12 mm by 50 mm rectangular-cross-section UNS G10 150 HR bar carries a static load of 73 kN. It is welded to a gusset plate with a 10 mm fillet weld 50 mm long on both sides with an E70XX electrode as depicted in Figure 9–18. Use the welding code method.
(*a*) Is the weld metal strength satisfactory?
(*b*) Is the attachment strength satisfactory?

Solution
(*a*) From Table 9–6, allowable force per unit length for a 10 mm E70 electrode metal is 1025 N of weldment; thus

$$F = 1025l = 1025(100) = 102.5 \text{ kN}$$

Since 102.5 > 73, weld metal strength is satisfactory.
(*b*) Check shear in attachment adjacent to the welds. From Table A–19, $S_y = 190$ MPa. Then, from Table 9–4, the allowable attachment shear stress is

$$\tau_{\text{all}} = 0.4S_y = 0.4(190) = 76 \text{ MPa}$$

The shear stress τ on the base metal adjacent to the weld is

$$\tau = \frac{F}{2hl} = \frac{73\,000}{2(10)50} = 73 \text{ MPa}$$

Since $\tau_{\text{all}} \geq \tau$, the attachment is satisfactory near the weld beads. The tensile stress in the shank of the attachment σ is

$$\sigma = \frac{F}{tl} = \frac{73\,000}{(12)50} = 122 \text{ MPa}$$

The allowable tensile stress σ_{all}, from Table 9–4, is $0.6S_y$ and, with welding code safety level preserved,

$$\sigma_{\text{all}} = 0.6S_y = 0.6(190) = 114 \text{ MPa}$$

Since $\sigma \leq \sigma_{\text{all}}$, the shank tensile stress is satisfactory.

Figure 9–18

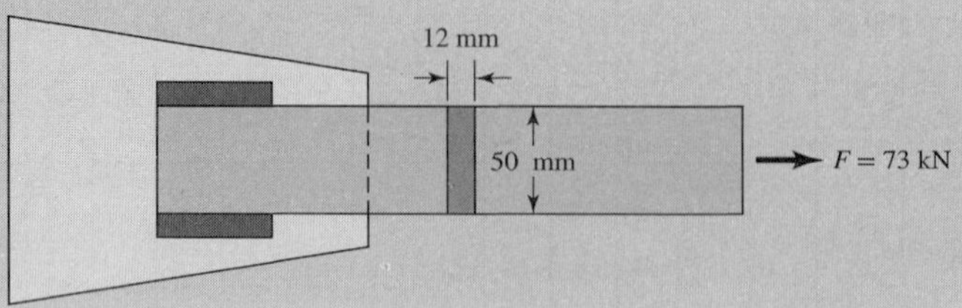

EXAMPLE 9–3

A specially rolled A36 structural steel section for the attachment has a cross section as shown in Figure 9–19 and has yield and ultimate tensile strengths of 250 and 450 MPa, respectively. It is statically loaded through the attachment centroid by a load of $F = 107$ kN. Unsymmetrical weld tracks can compensate for eccentricity such that there is no moment to be resisted by the welds. Specify the weld track lengths l_1 and l_2 for a 8 mm fillet weld using an E70XX electrode. This is part of a design problem in which the design variables include weld lengths and the fillet leg size.

Solution

The y coordinate of the section centroid of the attachment is

$$\bar{y} = \frac{\sum y_i A_i}{\sum A_i} = \frac{25(20)50 + 75(10)50}{20(50) + 10(50)} = 41.6 \text{ mm}$$

Summing moments about point B to zero gives

$$\sum M_B = 0 = -F_1 b + F\bar{y} = -F_1(100) + 107(41.6)$$

from which

$$F_1 = 44.5 \text{ kN}$$

It follows that

$$F_2 = 107 - 44.5 = 62.5 \text{ kN}$$

The weld throat areas have to be in the ratio 62.5/44.5 = 1.4, that is, $l_2 = 1.4l_1$. The weld length design variables are coupled by this relation, so l_1 is the weld length design variable. The other design variable is the fillet weld leg size h, which has been decided by the problem statement. From Table 9–4, the allowable shear stress on the throat τ_{all} is

$$\tau_{\text{all}} = 0.3(482) = 145 \text{ MPa}$$

The shear stress τ on the 45° throat is

$$\tau = \frac{F}{(0.707)h(l_1 + l_2)} = \frac{F}{(0.707)h(l_1 + 1.4l_1)}$$

$$= \frac{F}{(0.707)h(2.4l_1)} = \tau_{\text{all}} = 145 \text{ MPa}$$

from which the weld length l_1 is

$$l_1 = \frac{170\ 000}{145(0.707)8(2.4)} = 54.4 \text{ mm}$$

and

$$l_2 = 1.4l_1 = 1.4(54.4) = 76.1 \text{ mm}$$

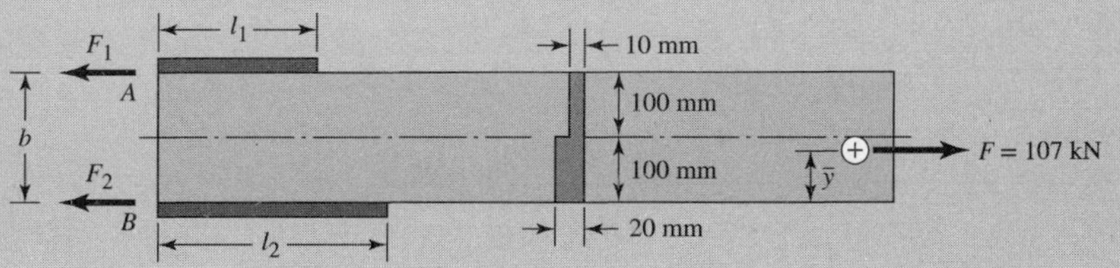

Figure 9–19

These are the weld-bead lengths required by weld metal strength. The attachment shear stress allowable in the base metal, from Table 9–4, is

$$\tau_{\text{all}} = 0.4S_y = 0.4(250) = 100 \text{ MPa}$$

The shear stress τ in the base metal adjacent to the weld is

$$\tau = \frac{F}{h(l_1 + l_2)} = \frac{F}{h(l_1 + 1.4l_1)} = \frac{F}{h(2.4l_1)} = \tau_{\text{all}} = 100 \text{ MPa}$$

from which

$$l_1 = \frac{F}{14.4h(2.4)} = \frac{107\ 000}{100(8)2.4} = 55.73 \text{ mm}$$

$$l_2 = 1.4l_1 = 1.4(55.73) = 78 \text{ mm}$$

These are the weld-bead lengths required by base metal (attachment) strength. The base metal controls the weld lengths. For the allowable tensile stress σ_{all} in the shank of the attachment, the AISC allowable for tension members is $0.6S_y$; therefore,

$$\sigma_{\text{all}} = 0.6S_y = 0.6(250) = 150 \text{ MPa}$$

The nominal tensile stress σ is *uniform* across the attachment cross section because of the load application at the centroid. The stress σ is

$$\sigma = \frac{F}{A} = \frac{107\ 000}{20(50) + 50(10)} = 71.3 \text{ MPa}$$

since $\sigma \le \sigma_{\text{all}}$, the shank section is satisfactory. With l_1 set to a nominal 56 mm, l_2 should be $1.4(56) = 78.4$ mm

Decision

Set $l_1 = 56$ mm, $l_2 = 80$ mm. The small magnitude of the departure from $l_2/l_1 = 1.4$ is not serious. The joint is essentially moment-free.

EXAMPLE 9–4

Perform an adequacy assessment of the statically loaded welded cantilever carrying 2.2 kN depicted in Figure 9–20. The cantilever is made of AISI 1018 HR steel and welded with a 10 mm fillet weld as shown in the figure. An E6010 electrode was used, and the design factor was 3.0.

(*a*) Use the conventional method for the weld metal.

(*b*) Use the conventional method for the attachment (cantilever) metal.

(*c*) Use a welding code for the weld metal.

Figure 9–20

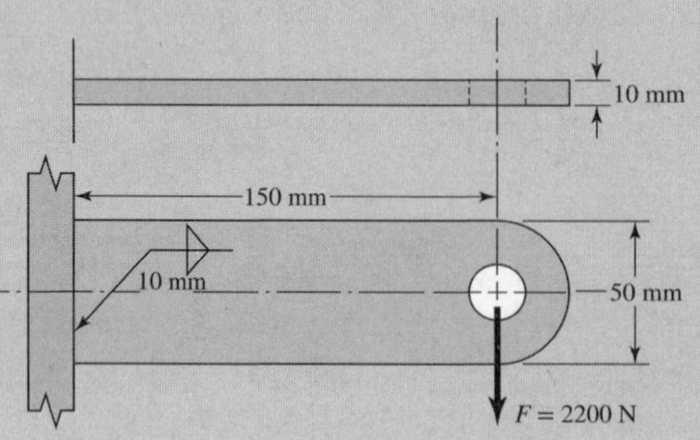

Solution

(*a*) From Table 9–3, $S_y = 345$ MPa, $S_{ut} = 427$ MPa. From Table 9–2, second pattern, $b = 10$ mm, $d = 50$ mm, so

$$A = 1.414hd = 1.414(10)50 = 707 \text{ mm}^2$$

$$I_u = d^3/6 = 50^3/6 = 20\ 833 \text{ mm}^3$$

$$I = 0.707hI_u = 0.707(10)20\ 833 = 147\ 292 \text{ mm}^4$$

Primary shear:

$$\tau' = \frac{F}{A} = \frac{2200}{707} = 3.11 \text{ MPa}$$

Secondary shear:

$$\tau'' = \frac{Mr}{I} = \frac{2200(150)(259)}{147\ 292} = 56 \text{ MPa}$$

The shear magnitude τ is from the vector addition

$$\tau = (\tau'^2 + \tau''^2)^{1/2} = (3.11^2 + 56^2)^{1/2} = 56.1 \text{ MPa}$$

The factor of safety based on a minimum strength and the distortion-energy criterion is

Answer
$$n = \frac{S_{sy}}{\tau} = \frac{0.577(345)}{56.1} = 3.55$$

Since $n \geq n_d$, that is, $3.55 \geq 3.0$, the weld metal has satisfactory strength.

(*b*) From Table A–19, minimum strengths are $S_{ut} = 400$ MPa and $S_y = 220$ MPa. Then

$$\sigma = \frac{M}{I/c} = \frac{M}{bd^2/6} = \frac{2200(150)}{10(50^2)/6} = 79.2 \text{ MPa}$$

Answer
$$n = \frac{S_y}{\sigma} = \frac{220}{79.2} = 2.77$$

Since $n < n_d$, that is, $2.77 < 3.0$, the joint is unsatisfactory as to the attachment strength.

(*c*) From part (*a*), $\tau = 56.1$ MPa. For an E6010 electrode Table 9–6 gives the allowable shear stress τ_{all} as 124 MPa. Since $\tau < \tau_{\text{all}}$, the weld is satisfactory. Since the code already has a design factor of $0.577(345)/124 = 1.6$ included at the equality, the corresponding factor of safety to part (*a*) is

Answer
$$n = 1.6\frac{124}{56.1} = 3.54$$

which is consistent.

9–7 Fatigue Loading

The conventional methods for fatigue are applicable. With only shear loading, we will adapt the fluctuating-stress diagram and a fatigue failure criterion for shear stresses and shear strengths, according to the method outlined in the last subsection of Section 6–13. For the surface factor of Equation (6–18), an as-forged surface should always be assumed for weldments unless a superior finish is specified and obtained.

Some examples of fatigue loading of welded joints follow.

EXAMPLE 9–5

The AISI 1018 HR steel strap of Figure 9–21 has a 4500 N, completely reversed load applied. Determine the factor of safety of the weldment for infinite life.

Solution

From Table A–19 for the 1018 attachment metal the strengths are $S_{ut} = 400$ MPa and $S_y = 220$ MPa. For the E6010 electrode, from Table 9–3 $S_{ut} = 427$ MPa and $S_y = 345$ MPa. The fatigue stress-concentration factor, from Table 9–5, is $K_{fs} = 2.7$. From Table 6–2, $k_a = 12.7(58)^{-0.758} = 0.58$. For case 2 of Table 9–5, the shear area is:

$$A = 1.414(10)(50) = 707 \text{ mm}^2$$

For a uniform shear stress on the throat, $k_b = 1$.

From Equation (6–25), for torsion (shear),

$$k_c = 0.59 \qquad k_d = k_e = 1$$

From Equations (6–10) and (6–17),

$$S_{se} = 0.7(1)0.59(1)(1)(1)0.5(400) = 82.8 \text{ MPa}$$

From Table 9–5, $K_{fs} = 2.7$. Only primary shear is present. So, with $F_a = 4500$ N and $F_m = 0$

$$\tau'_a = \frac{K_{fs}F_a}{A} = \frac{2.7(4500)}{707} = 17.2 \text{ MPa} \qquad \tau'_m = 0 \text{ MPa}$$

In the absence of a mean stress component, the fatigue factor of safety n_f is given by

Answer

$$n_f = \frac{S_{se}}{\tau'_a} = \frac{82.8}{17.2} = 4.81$$

Figure 9–21

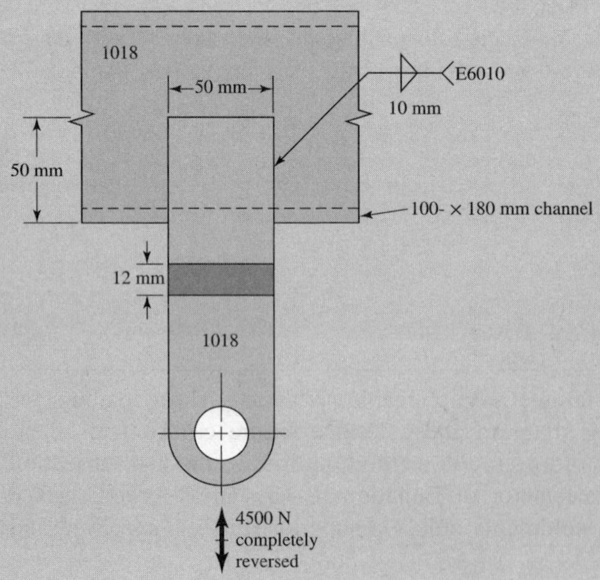

EXAMPLE 9-6

The AISI 1018 HR steel strap of Figure 9–22 has a repeatedly applied load of 9000 N ($F_a = F_m = 4500$ N). Determine the fatigue factor of safety of the weldment.

Solution

From Table 6–2, $k_a = 272(400)^{-0.995} = 0.7$. From case 2 of Table 9–2

$$A = 1.414(10)(10)50 = 707 \text{ mm}^2$$

For uniform shear stress on the throat $k_b = 1$.

Equation (6–25): $k_c = 0.59$

Equations (6–10) and (6–17): $S_{se} = (0.7)(0.59)(0.5)(400) = 82.6$ MPa

From Table 9–5, $K_{fs} = 2$. Only primary shear is present:

$$\tau_a' = \tau_m' = \frac{K_{fs}F_a}{A} = \frac{2(4500)}{707} = 12.7 \text{ MPa}$$

Estimate the shear modulus of rupture with Equation (6–58),

$$S_{su} = 0.67\ S_{ut} = 0.67(400) = 268 \text{ MPa}$$

Choosing the Goodman fatigue failure criterion of Equation (6–41), adapted for shear, gives

Answer
$$n_f = \left(\frac{\tau_a'}{S_{se}} + \frac{\tau_m'}{S_{su}}\right)^{-1} = \left(\frac{12.7}{82.6} + \frac{12.7}{268}\right)^{-1} = 4.97$$

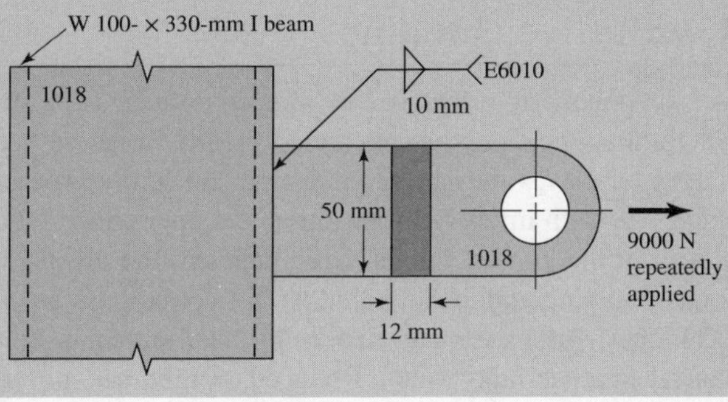

Figure 9–22

9–8 Resistance Welding

The heating and consequent welding that occur when an electric current is passed through several parts that are pressed together is called *resistance welding*. *Spot welding* and *seam welding* are forms of resistance welding most often used. The advantages of resistance welding over other forms are the speed, the accurate regulation of time and heat, the uniformity of the weld, and the mechanical properties that result. In addition the process is easy to automate, and filler metal and fluxes are not needed.

The spot- and seam-welding processes are illustrated schematically in Figure 9–23. Seam welding is actually a series of overlapping spot welds, since the current is applied in pulses as the work moves between the rotating electrodes.

Figure 9–23

(*a*) Spot welding; (*b*) seam welding.

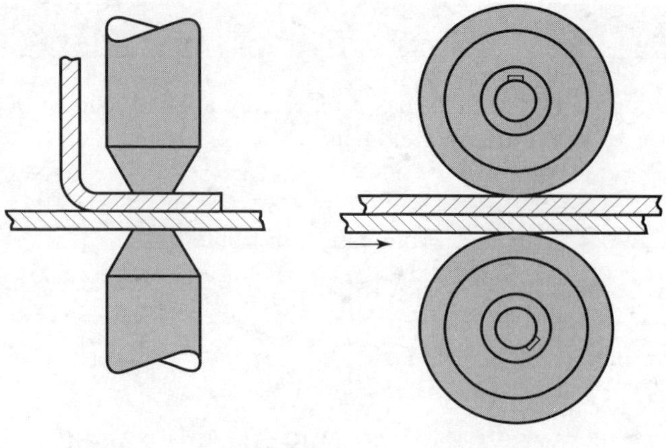

(*a*) (*b*)

Failure of a resistance weld occurs either by shearing of the weld or by tearing of the metal around the weld. Because of the possibility of tearing, it is good practice to avoid loading a resistance-welded joint in tension. Thus, for the most part, design so that the spot or seam is loaded in pure shear. The shear stress is then simply the load divided by the area of the spot. Because the thinner sheet of the pair being welded may tear, the strength of spot welds is often specified by stating the load per spot based on the thickness of the thinnest sheet. Such strengths are best obtained by experiment.

Somewhat larger factors of safety should be used when parts are fastened by spot welding rather than by bolts or rivets, to account for the metallurgical changes in the materials due to the welding.

9–9 Adhesive Bonding[3]

The use of polymeric adhesives to join components for structural, semistructural, and nonstructural applications has expanded greatly in recent years as a result of the unique advantages adhesives may offer for certain assembly processes and the development of new adhesives with improved robustness and environmental acceptability. The increasing complexity of modern assembled structures and the diverse types of materials used have led to many joining applications that would not be possible with more conventional joining techniques. Adhesives are also being used either in conjunction with or to replace mechanical fasteners and welds. Reduced weight, sealing capabilities, and reduced part count and assembly time, as well as improved fatigue and corrosion resistance, all combine to provide the designer with opportunities for customized assembly. Figure 9–24 illustrates the numerous places where adhesives are used on an automobile. Indeed, the fabrication of many modern vehicles, devices, and structures is dependent on adhesives.

In well-designed joints and with proper processing procedures, use of adhesives can result in significant reductions in weight. Eliminating mechanical fasteners eliminates the weight of the fasteners, and also may permit the use of thinner-gauge

[3]For a more extensive discussion of this topic, see J. E. Shigley and C. R. Mischke, *Mechanical Engineering Design*, 6th ed., McGraw-Hill, New York, 2001, Section 9–11. This section was prepared with the assistance of Professor David A. Dillard, Professor of Engineering Science and Mechanics and Director of the Center for Adhesive and Sealant Science, Virginia Polytechnic Institute and State University, Blacksburg, Virginia, and with the encouragement and technical support of the Bonding Systems Division of 3M, Saint Paul, Minnesota.

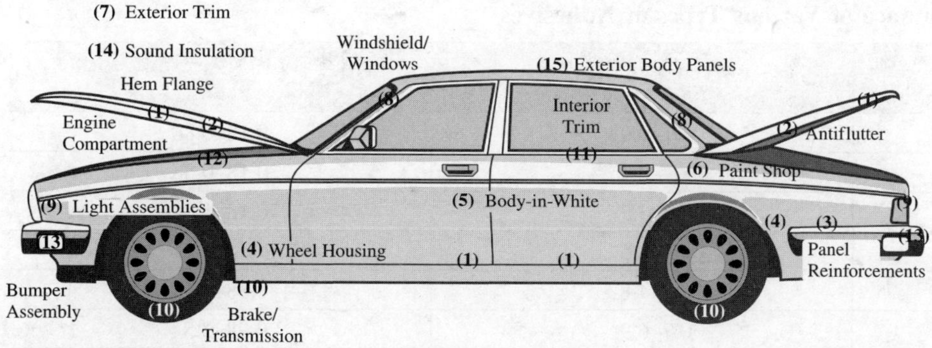

Figure 9–24

Diagram of an automobile body showing at least 15 locations at which adhesives and sealants could be used or are being used. Particular note should be made of the windshield (8), which is considered a load-bearing structure in modern automobiles and is adhesively bonded. Also attention should be paid to hem flange bonding (1), in which adhesives are used to bond and seal. Adhesives are used to bond friction surfaces in brakes and clutches (10). Antiflutter adhesive bonding (2) helps control deformation of hood and trunk lids under wind shear. Thread-sealing adhesives are used in engine applications (12).
(*Source:* Pocius, A. V. *Adhesion and Adhesives Technology.* 3rd ed. Munich: Hanser Publishers, 2012.)

materials because stress concentrations associated with the holes are eliminated. The capability of polymeric adhesives to dissipate energy can significantly reduce noise, vibration, and harshness (NVH), crucial in modern automobile performance. Adhesives can be used to assemble heat-sensitive materials or components that might be damaged by drilling holes for mechanical fasteners. They can be used to join dissimilar materials or thin-gauge stock that cannot be joined through other means.

Types of Adhesive

There are numerous adhesive types for various applications. They may be classified in a variety of ways depending on their chemistry (e.g., epoxies, polyurethanes, polyimides), their form (e.g., paste, liquid, film, pellets, tape), their type (e.g., hot melt, reactive hot melt, thermosetting, pressure sensitive, contact), or their load-carrying capability (structural, semistructural, or nonstructural).

Structural adhesives are relatively strong adhesives that are normally used well below their glass transition temperature; common examples include epoxies and certain acrylics. Such adhesives can carry significant stresses, and they lend themselves to structural applications. For many engineering applications, semistructural applications (where failure would be less critical) and nonstructural applications (of headliners, etc., for aesthetic purposes) are also of significant interest to the design engineer, providing cost-effective means required for assembly of finished products. These include *contact adhesives,* where a solution or emulsion containing an elastomeric adhesive is coated onto both adherends, the solvent is allowed to evaporate, and then the two adherends are brought into contact. Examples include rubber cement and adhesives used to bond laminates to countertops. *Pressure-sensitive adhesives* are very low modulus elastomers that deform easily under small pressures, permitting them to wet surfaces. When the substrate and adhesive are brought into intimate contact, van der Waals forces are sufficient to maintain the contact and provide relatively durable bonds. Pressure-sensitive adhesives are normally purchased as tapes or labels for nonstructural applications, although there are also double-sided foam tapes that can be used in semistructural applications. As the name implies, *hot melts* become liquid when heated, wetting the

Table 9–7 Mechanical Performance of Various Types of Adhesives

Adhesive Chemistry or Type	Room Temperature Lap-Shear Strength, MPa	Peel Strength per Unit Width, kN/m
Pressure-sensitive	0.01–0.07	0.18–0.88
Starch-based	0.07–0.7	0.18–0.88
Cellosics	0.35–3.5	0.18–1.8
Rubber-based	0.35–3.5	1.8–7
Formulated hot melt	0.35–4.8	0.88–3.5
Synthetically designed hot melt	0.7–6.9	0.88–3.5
PVAc emulsion (white glue)	1.4–6.9	0.88–1.8
Cyanoacrylate	6.9–13.8	0.18–3.5
Protein-based	6.9–13.8	0.18–1.8
Anaerobic acrylic	6.9–13.8	0.18–1.8
Urethane	6.9–17.2	1.8–8.8
Rubber-modified acrylic	13.8–24.1	1.8–8.8
Modified phenolic	13.8–27.6	3.6–7
Unmodified epoxy	10.3–27.6	0.35–1.8
Bis-maleimide	13.8–27.6	0.18–3.5
Polyimide	13.8–27.6	0.18–0.88
Rubber-modified epoxy	20.7–41.4	4.4–14

Source: Data from Pocius, A. V. *Adhesion and Adhesives Technology*. 2nd ed. Ohio: Hanser Gardner Publishers, 2002.

surfaces and then cooling into a solid polymer. These materials are increasingly applied in a wide array of engineering applications by more sophisticated versions of the glue guns in popular use. *Anaerobic adhesives* cure within narrow spaces deprived of oxygen; such materials have been widely used in mechanical engineering applications to lock bolts or bearings in place. Cure in other adhesives may be induced by exposure to ultraviolet light or electron beams, or it may be catalyzed by certain materials that are ubiquitous on many surfaces, such as water.

Table 9–7 presents important strength properties of commonly used adhesives.

Stress Distributions

Good design practice normally requires that adhesive joints be constructed in such a manner that the adhesive carries the load in shear rather than tension. Bonds are typically much stronger when loaded in shear rather than in tension across the bond plate. Lap-shear joints represent an important family of joints, both for test specimens to evaluate adhesive properties and for actual incorporation into practical designs. Generic types of lap joints that commonly arise are illustrated in Figure 9–25.

The simplest analysis of lap joints suggests the applied load is uniformly distributed over the bond area. Lap joint test results, such as those obtained following the ASTM D1002 for single-lap joints, report the "apparent shear strength" as the breaking load divided by the bond area. Although this simple analysis can be adequate for stiff adherends bonded with a soft adhesive over a relatively short bond length, significant peaks in shear stress occur except for the most flexible adhesives. In an effort to point out the problems associated with such practice, ASTM D4896 outlines some of the concerns associated with taking this simplistic view of stresses within lap joints.

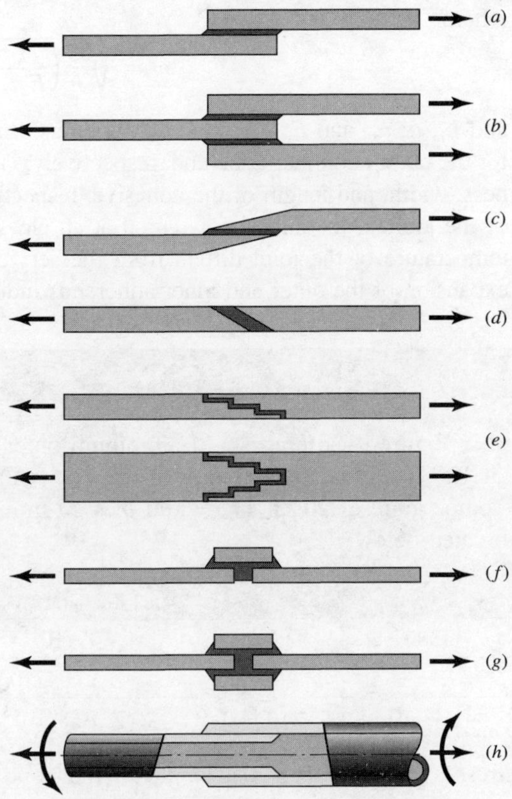

Figure 9–25

Common types of lap joints used in mechanical design: (*a*) single lap; (*b*) double lap; (*c*) scarf; (*d*) bevel; (*e*) step; (*f*) butt strap; (*g*) double butt strap; (*h*) tubular lap. (*Source:* Adapted from Adams, R. D., J. Comyn, and W. C. Wake. *Structural Adhesive Joints in Engineering.* 2nd ed. New York: *Chapman and Hall, 1997.)*

In 1938, O. Volkersen presented an analysis of the lap joint, known as the *shear-lag model*. It provides valuable insights into the shear-stress distributions in a host of lap joints. Bending induced in the single-lap joint due to eccentricity significantly complicates the analysis, so here we will consider a symmetric double-lap joint to illustrate the principles. The shear-stress distribution for the double lap joint of Figure 9–26 is given by

$$\tau(x) = \frac{P\omega}{4b\,\sinh(\omega l/2)}\cosh(\omega x) + \left[\frac{P\omega}{4b\,\cosh(\omega l/2)}\left(\frac{2E_o t_o - E_i t_i}{2E_o t_o + E_i t_i}\right) + \frac{(\alpha_i - \alpha_o)\Delta T\omega}{[1/(E_o t_o) + 2/(E_i t_i)]\,\cosh(\omega l/2)}\right]\sinh(\omega x) \tag{9–7}$$

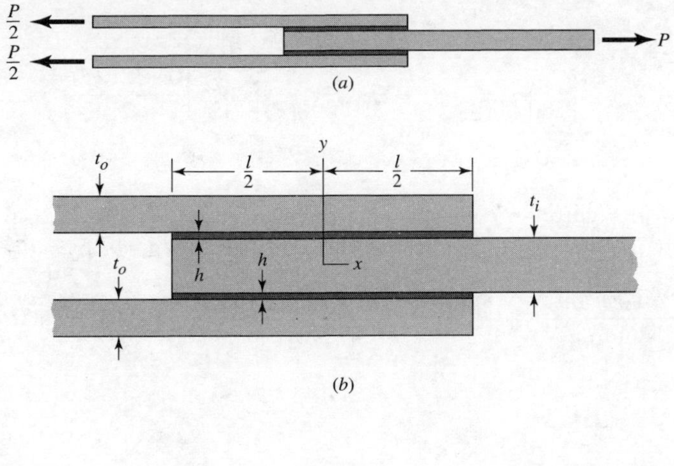

Figure 9–26

Double-lap joint.

where

$$\omega = \sqrt{\frac{G}{h}\left(\frac{1}{E_o t_o} + \frac{2}{E_i t_i}\right)}$$

and E_o, t_o, α_o, and E_i, t_i, α_i, are the modulus, thickness, coefficient of thermal expansion for the outer and inner adherend, respectively; G, h, b, and l are the shear modulus, thickness, width, and length of the adhesive, respectively; and ΔT is a change in temperature of the joint. If the adhesive is cured at an elevated temperature such that the stress-free temperature of the joint differs from the service temperature, the mismatch in thermal expansion of the outer and inner adherends induces a thermal shear across the adhesive.

EXAMPLE 9–7

The double-lap joint depicted in Figure 9–26 consists of aluminum outer adherends and an inner steel adherend. The assembly is cured at 120°C and is stress-free at 90°C. The completed bond is subjected to an axial load of 8900 N at a service temperature of 20°C. The width b is 25 mm, the length of the bond l is 25 mm. Additional information is tabulated below:

	G, MPa	E, MPa	α, mm/(mm · °C)	Thickness, mm
Adhesive	1380		$99(10^{-6})$	0.5
Outer adherend		70 000	$23.94(10^{-6})$	4.5
Inner adherend		210 000	$10.8(10^{-6})$	3

Sketch a plot of the shear stress as a function of the length of the bond due to (*a*) thermal stress, (*b*) load-induced stress, and (*c*) the sum of stresses in *a* and *b*; and (*d*) find where the largest shear stress is maximum.

Solution

In Equation (9–7) the parameter ω is given by

$$\omega = \sqrt{\frac{G}{h}\left(\frac{1}{E_o t_o} + \frac{2}{E_i t_i}\right)}$$

$$= \sqrt{\frac{1380}{0.5}\left[\frac{1}{70\,000(4.5)} + \frac{2}{210\,000(3)}\right]} = 0.132\ \text{mm}^{-1}$$

(*a*) For the thermal stress component, $\alpha_i - \alpha_o = 10.8(10^{-6}) - 23.94(10^{-6}) = -13.14(10^{-6})$ mm/(mm · °C), $\Delta T = 20 - 90 = -70$°C,

$$\tau_{th}(x) = \frac{(\alpha_i - \alpha_o)\Delta T\ \omega \sinh(\omega x)}{[1/(E_o t_o) + 2/(E_i t_i)]\cosh(\omega l/2)}$$

$$\tau_{th}(x) = \frac{-13.4(10^{-6})(-70)\,132\sinh(132x)}{\left[\dfrac{1}{700\,000(4.5)} + \dfrac{2}{210\,000(3)}\right]\cosh\left[\dfrac{0.132(25)}{2}\right]}$$

$$= 7.08\sinh(0.132x)$$

The thermal stress is plotted in Figure 9–27 and tabulated at $x = -12.5$, 0, and 12.5 in the table below.

(*b*) The bond is "balanced" ($E_o t_o = E_i t_i/2$), so the load-induced stress is given by

$$\tau_P(x) = \frac{P\omega\cosh(\omega x)}{4b\sinh(\omega l/2)} = \frac{8900(0.132)\cosh(0.132x)}{4(25)2.82} = 4.165\cosh(0.132x) \qquad (1)$$

The load-induced stress is plotted in Figure 9–27 and tabulated at $x = -12.5$, 0, and 12.5 in the table below.
(*c*) Total stress table (in MPa):

	$\tau(-12.5)$	$\tau(0)$	$\tau(12.5)$
Thermal only	−17.75	0	17.75
Load-induced only	11.24	4.165	11.24
Combined	−6.51	4.165	28.99

(*d*) The maximum shear stress predicted by the shear-lag model will always occur at the ends. See the plot in Figure 9–27. Since the residual stresses are always present, significant shear stresses may already exist prior to application of the load. The large stresses present for the combined-load case could result in local yielding of a ductile adhesive or failure of a more brittle one. The significance of the thermal stresses serves as a caution against joining dissimilar adherends when large temperature changes are involved. Note also that the average shear stress due to the load is $\tau_{\text{avg}} = P/(2bl) = 7.12$ MPa. Equation (1) produced a maximum of 13.7 MPa, almost double the average.

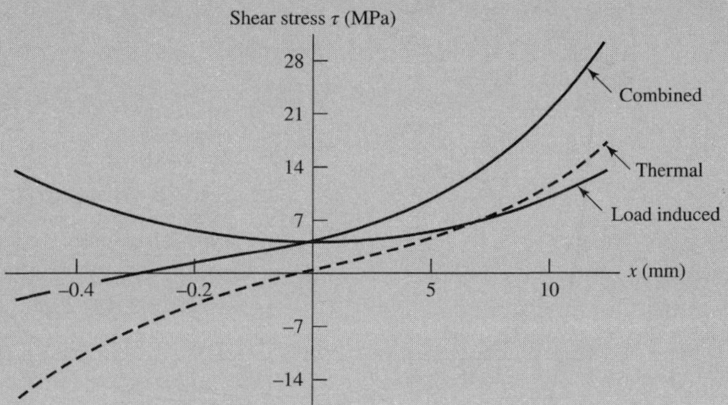

Figure 9–27

Plot for Example 9–7.

Although design considerations for single-lap joints are beyond the scope of this chapter, one should note that the load eccentricity is an important aspect in the stress state of single-lap joints. Adherend bending can result in shear stresses that may be as much as double those given for the double-lap configuration (for a given total bond area). In addition, peel stresses can be quite large and often account for joint failure. Finally, plastic bending of the adherends can lead to high strains, which less ductile adhesives cannot withstand, leading to bond failure as well. Bending stresses in the adherends at the end of the overlap can be four times greater than the average stress within the adherend; thus, they must be considered in the design. Figure 9–28 shows the shear and peel stresses present in a typical single-lap joint that corresponds to the ASTM D1002 test specimen. Note that the shear stresses are significantly larger than predicted by the Volkersen analysis, a result of the increased adhesive strains associated with adherend bending.

Joint Design

Some basic guidelines that should be used in adhesive joint design include:

- Design to place bondline in shear, not peel. Beware of peel stresses focused at bond terminations. When necessary, reduce peel stresses through tapering the adherend ends, increasing bond area where peel stresses occur, or utilizing rivets at bond terminations where peel stresses can initiate failures.

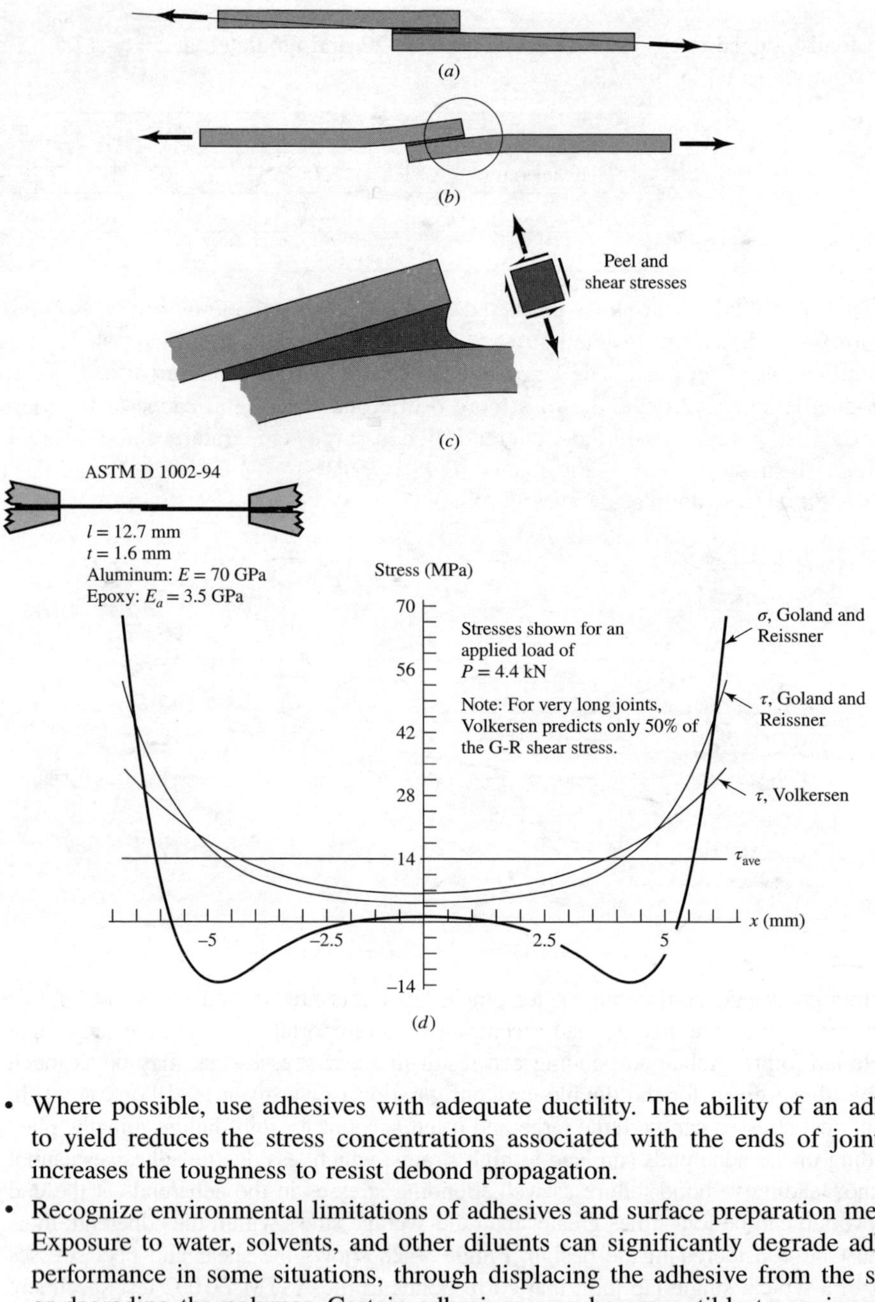

Figure 9–28

Stresses within a single-lap joint. (*a*) Lap-joint tensile forces have a line of action that is not initially parallel to the adherend sides. (*b*) As the load increases the adherends and bond bend. (*c*) In the locality of the end of an adherend peel and shear stresses appear, and the peel stresses often induce joint failure. (*d*) The seminal Goland and Reissner stress predictions. (*Source:* J. Appl Mech., vol 77, 1944 are shown.)

- Where possible, use adhesives with adequate ductility. The ability of an adhesive to yield reduces the stress concentrations associated with the ends of joints and increases the toughness to resist debond propagation.
- Recognize environmental limitations of adhesives and surface preparation methods. Exposure to water, solvents, and other diluents can significantly degrade adhesive performance in some situations, through displacing the adhesive from the surface or degrading the polymer. Certain adhesives may be susceptible to environmental stress cracking in the presence of certain solvents. Exposure to ultraviolet light can also degrade adhesives.
- Design in a way that permits or facilitates inspections of bonds where possible. A missing rivet or bolt is often easy to detect, but debonds or unsatisfactory adhesive bonds are not readily apparent.
- Allow for sufficient bond area so that the joint can tolerate some debonding before going critical. This increases the likelihood that debonds can be detected. Having some regions of the overall bond at relatively low stress levels can significantly improve durability and reliability.

- Where possible, bond to multiple surfaces to offer support to loads in any direction. Bonding an attachment to a single surface can place peel stresses on the bond, whereas bonding to several adjacent planes tends to permit arbitrary loads to be carried predominantly in shear.
- Adhesives can be used in conjunction with spot welding. The process is known as *weld bonding*. The spot welds serve to fixture the bond until it is cured.

Figure 9–29 presents examples of improvements in adhesive bonding.

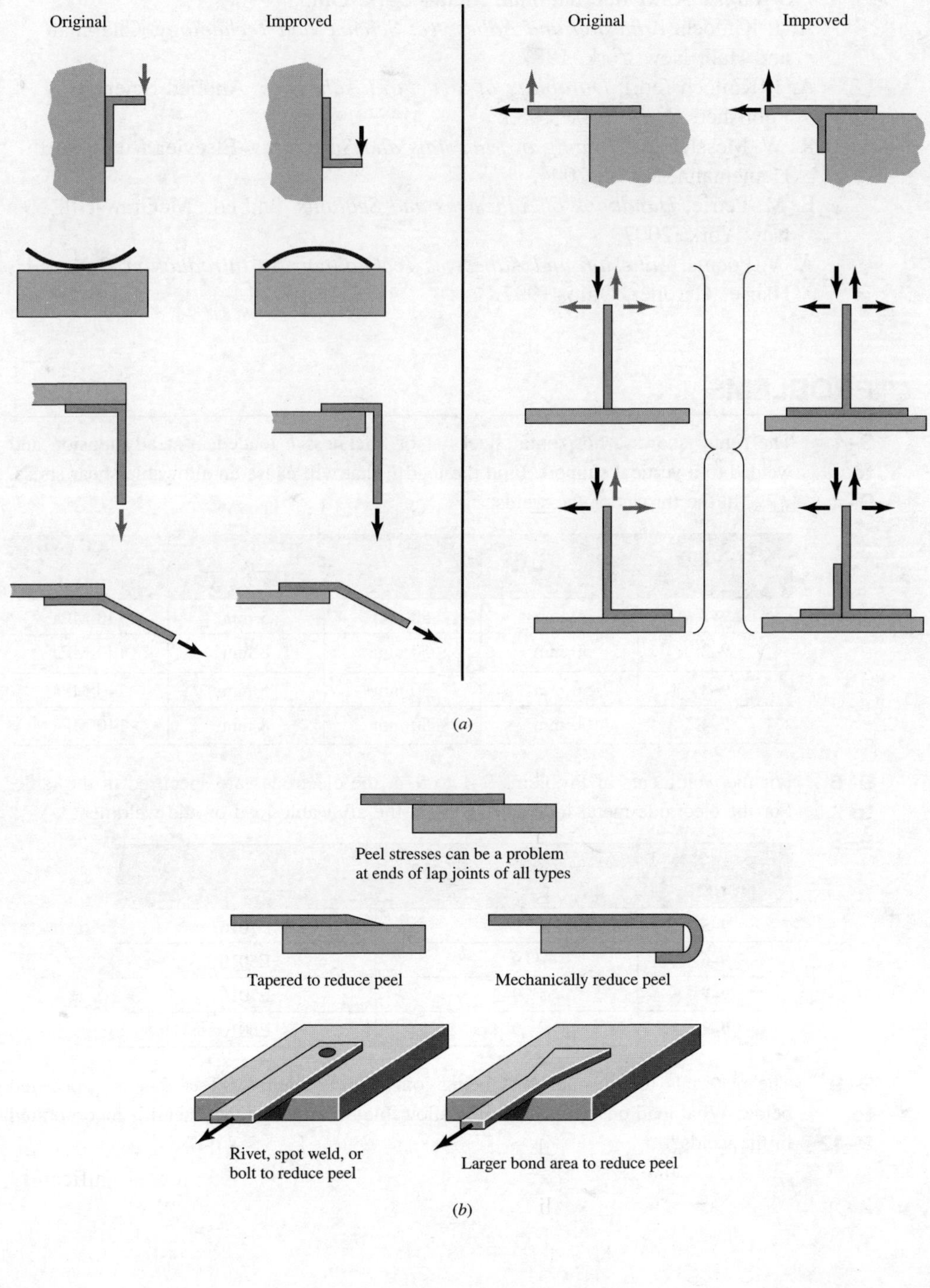

Figure 9–29

Design practices that improve adhesive bonding. (*a*) Gray load vectors are to be avoided as resulting strength is poor. (*b*) Means to reduce peel stresses in lap-type joints.

References

Good references are available for analyzing and designing adhesive bonds, including the following:

R. D. Adams, J. Comyn, and W. C. Wake, *Structural Adhesive Joints in Engineering,* 2nd ed., Chapman and Hall, New York, 1997.

G. P. Anderson, S. J. Bennett, and K. L. DeVries, *Analysis and Testing of Adhesive Bonds,* Academic Press, New York, 1977.

H. F. Brinson (ed.), *Engineered Materials Handbook, vol. 3: Adhesives and Sealants,* ASM International, Metals Park, Ohio, 1990.

A. J. Kinloch, *Adhesion and Adhesives: Science and Technology,* Chapman and Hall, New York, 1987.

A. J. Kinloch (ed.), *Durability of Structural Adhesives,* Applied Science Publishers, New York, 1983.

R. W. Messler, Jr., *Joining of Materials and Structures,* Elsevier Butterworth-Heinemann, Mass., 2004.

E. M. Petrie, *Handbook of Adhesives and Sealants,* 2nd ed., McGraw-Hill, New York, 2007.

A. V. Pocius, *Adhesion and Adhesives Technology: An Introduction,* 2nd ed., Hanser Gardner, Ohio, 1997.

PROBLEMS

9–1 to 9–4 The figure shows a horizontal steel bar of thickness h loaded in steady tension and welded to a vertical support. Find the load F that will cause an allowable shear stress, τ_{allow}, in the throats of the welds.

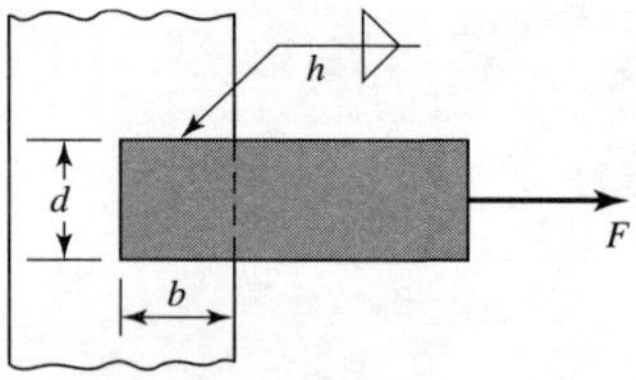

Problems 9–1 to 9–4

Problem Number	b	d	h	τ_{allow}
9–1	50 mm	50 mm	5 mm	140 MPa
9–2	50 mm	50 mm	8 mm	140 MPa
9–3	50 mm	30 mm	5 mm	140 MPa
9–4	100 mm	50 mm	8 mm	140 MPa

9–5 to 9–8 For the weldments of Problems 9–1 to 9–4, the electrodes are specified in the table. For the electrode metal indicated, what is the allowable load on the weldment?

Problem Number	Reference Problem	Electrode
9–5	9–1	E7010
9–6	9–2	E6010
9–7	9–3	E7010
9–8	9–4	E6010

9–9 to 9–12 The materials for the members being joined in Problems 9–1 to 9–4 are specified below. What load on the weldment is allowable because member metal is incorporated in the welds?

Problem Number	Reference Problem	Bar	Vertical Support
9–9	9–1	1018 CD	1018 HR
9–10	9–2	1020 CD	1020 CD
9–11	9–3	1035 HR	1035 CD
9–12	9–4	1035 HR	1020 CD

9–13 to 9–16 A steel bar of thickness h is welded to a vertical support as shown in the figure. What is the shear stress in the throat of the welds due to the force F?

Problem Number	b	d	h	F
9–13	50 mm	50 mm	5 mm	100 kN
9–14	50 mm	50 mm	8 mm	140 kN
9–15	50 mm	30 mm	5 mm	100 kN
9–16	100 mm	50 mm	8 mm	140 kN

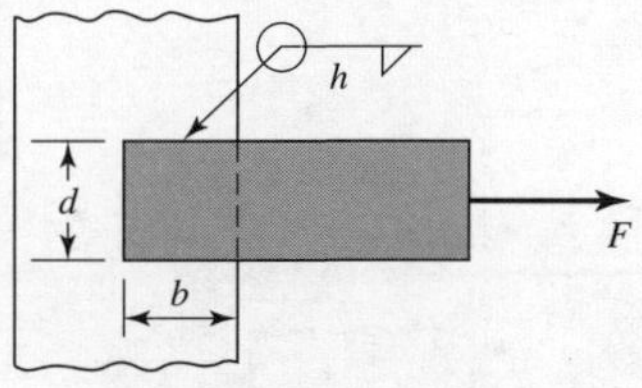

Problems 9–13 to 9–16

9–17 Prove the torsional properties for weld section 3 of Table 9–1.

9–18 Prove the torsional properties for weld section 4 of Table 9–1.

9–19 to 9–22 A steel bar of thickness h, to be used as a beam, is welded to a vertical support by two fillet welds as shown in the figure.

(*a*) Find the safe bending force F if the allowable shear stress in the welds is τ_{allow}.

(*b*) In part *a*, you found a simple expression for F in terms of the allowable shear stress. Find the allowable load if the electrode is E7010, the bar is hot-rolled 1020, and the support is hot-rolled 1015.

Problem Number	b	c	d	h	τ_{allow}
9–19	50 mm	150 mm	50 mm	5 mm	140 MPa
9–20	50 mm	150 mm	50 mm	8 mm	140 MPa
9–21	50 mm	150 mm	30 mm	5 mm	140 MPa
9–22	100 mm	150 mm	50 mm	8 mm	140 MPa

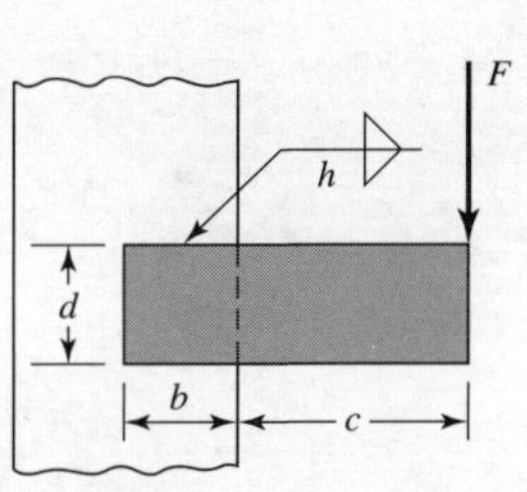

Problems 9–19 to 9–22

9–23 Prove the bending properties for weld section 4 of Table 9–2.

9–24 Prove the bending properties for weld section 5 of Table 9–2.

9–25 to 9–28 The figure shows a weldment just like that for Problems 9–19 to 9–22 except there are four welds instead of two. Find the safe bending force F if the allowable shear stress in the welds is τ_{allow}.

Problem Number	b	c	d	h	τ_{allow}
9–25	50 mm	150 mm	50 mm	5 mm	140 MPa
9–26	50 mm	150 mm	50 mm	8 mm	140 MPa
9–27	50 mm	150 mm	30 mm	5 mm	140 MPa
9–28	100 mm	150 mm	50 mm	8 mm	140 MPa

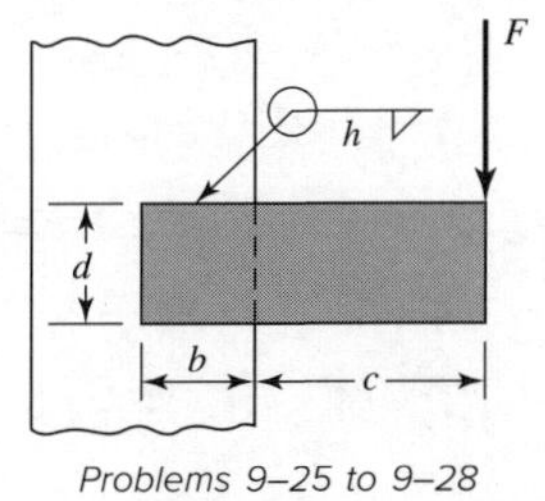

Problems 9–25 to 9–28

9–29 to 9–32 A steel bar of thickness h is welded to a vertical support as shown in the figure. Find the safe force F if the maximum allowable shear stress in the welds is τ_{allow}.

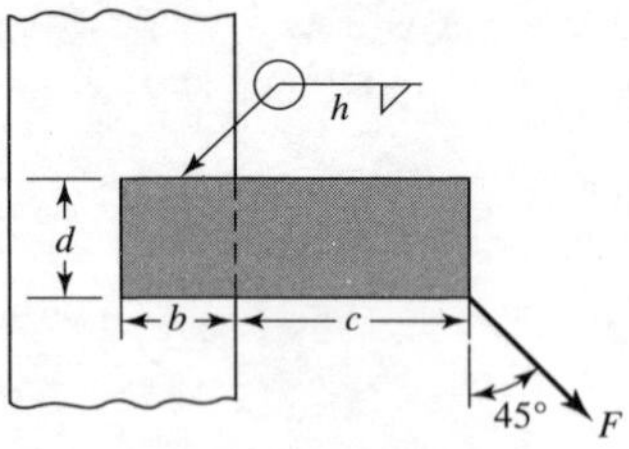

Problems 9–29 to 9–32

Problem Number	b	c	d	h	τ_{allow}
9–29	50 mm	150 mm	50 mm	5 mm	140 MPa
9–30	50 mm	150 mm	50 mm	8 mm	140 MPa
9–31	50 mm	150 mm	30 mm	5 mm	140 MPa
9–32	100 mm	150 mm	50 mm	8 mm	140 MPa

9–33 to 9–36 The weldment shown in the figure is subjected to an alternating force F. The hot-rolled steel bar has a thickness h and is of AISI 1010 steel. The vertical support is likewise AISI 1010 HR steel. The electrode is given in the table below. Estimate the fatigue load F the bar will carry if three fillet welds are used.

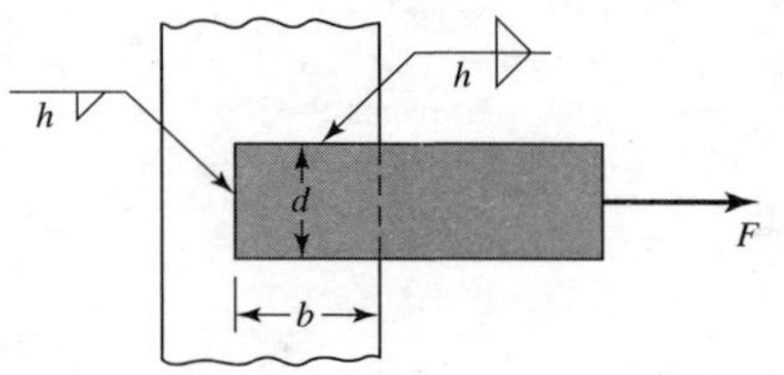

Problems 9–33 to 9–36

Problem Number	b	d	h	Electrode
9–33	50 mm	50 mm	5 mm	E6010
9–34	50 mm	50 mm	8 mm	E6010
9–35	50 mm	30 mm	5 mm	E7010
9–36	100 mm	50 mm	8 mm	E7010

9–37 The permissible shear stress for the weldments shown is 140 MPa. Based on the weldment of the cylinder and plate A, estimate the load, F, that will cause this stress in the weldment throat.

Problem 9–37

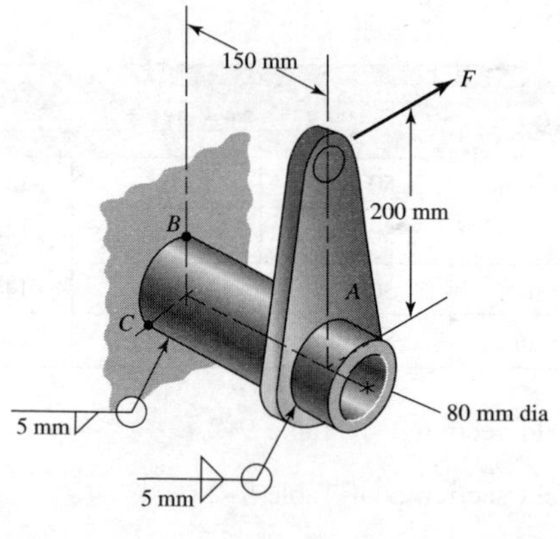

9–38 Repeat the analysis of Problem 9–37 for the weldment at the wall. Base your results on an analysis of points B and C,

9–39 to 9–40 A steel bar of thickness h is subjected to a bending force F. The vertical support is stepped such that the horizontal welds are b_1 and b_2 long. Determine F if the maximum allowable shear stress is τ_{allow}.

Problem Number	b_1	b_2	c	d	h	τ_{allow}
9–39	50 mm	100 mm	150 mm	100 mm	8 mm	140 MPa
9–40	30 mm	50 mm	150 mm	50 mm	5 mm	140 MPa

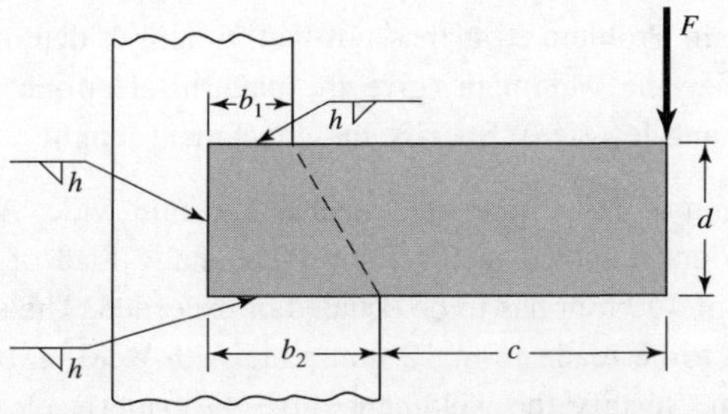

Problems 9–39 to 9–40

9–41 In the design of weldments in torsion it is helpful to have a hierarchical perception of the relative efficiency of common patterns. For example, the weld-bead patterns shown in Table 9–1 can be ranked for desirability. Assume the space available is an $a \times a$ square. Use a formal figure of merit that is directly proportional to J and inversely proportional to the volume of weld metal laid down:

$$\text{fom} = \frac{J}{\text{vol}} = \frac{0.707hJ_u}{(h^2/2)\,l} = 1.414\,\frac{J_u}{hl}$$

A tactical figure of merit could omit the constant, that is, $\text{fom}' = J_u/(hl)$. Rank the six patterns of Table 9–1 from most to least efficient.

9–42 The space available for a weld-bead pattern subject to bending is $a \times a$. Place the patterns of Table 9–2 in hierarchical order of efficiency of weld metal placement to resist bending. A formal figure of merit can be directly proportion to I and inversely proportional to the volume of weld metal laid down:

$$\text{fom} = \frac{I}{\text{vol}} = \frac{0.707h\,I_u}{(h^2/2)l} = 1.414\frac{I_u}{hl}$$

The tactical figure of merit can omit the constant 1.414, that is, $\text{fom}' = I_u/(hl)$. Omit the patterns intended for T beams and I beams. Rank the remaining seven.

9–43 The attachment shown in the figure is made of 1018 HR steel 12 mm thick. The static force is 100 kN. The member is 75 mm wide. Specify the weldment (give the pattern, electrode number, type of weld, length of weld, and leg size).

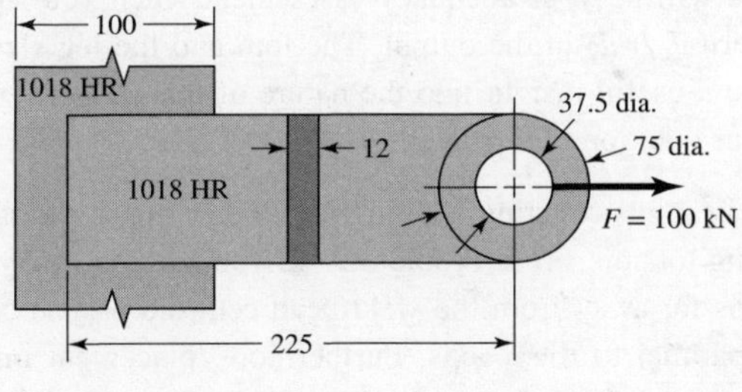

Problem 9–43

Dimensions in millimeters.

9–44 The attachment shown carries a static bending load of 12 kN. The attachment length, l_1, is 225 mm. Specify the weldment (give the pattern, electrode number, type of weld, length of weld, and leg size).

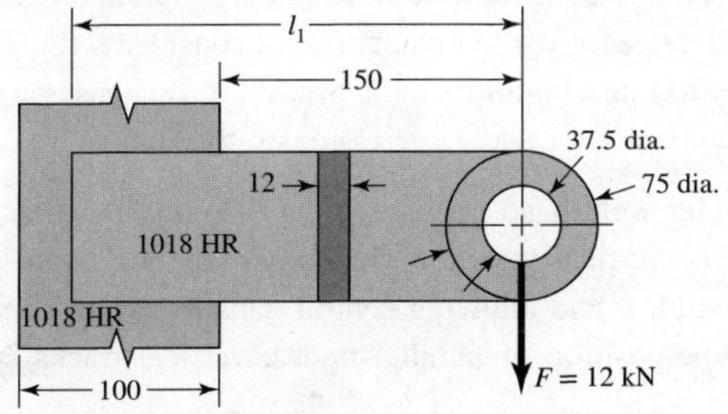

Problem 9–44

Dimensions in millimeters.

9–45 The attachment in Problem 9–44 has not had its length determined. The static force is 12 kN. Specify the weldment (give the pattern, electrode number, type of weld, length of bead, and leg size). Specify the attachment length.

9–46 A vertical column of 1018 hot-rolled steel is 250 mm wide. An attachment has been designed to the point shown in the figure. The static load of 90 kN is applied, and the clearance a of 160 mm has to be equaled or exceeded. The attachment is also 1018 hot-rolled steel, to be made from 12 mm plate with weld-on bosses when all dimensions are known. Specify the weldment (give the pattern, electrode number, type of weld, length of weld bead, and leg size). Specify also the length l_1 for the attachment.

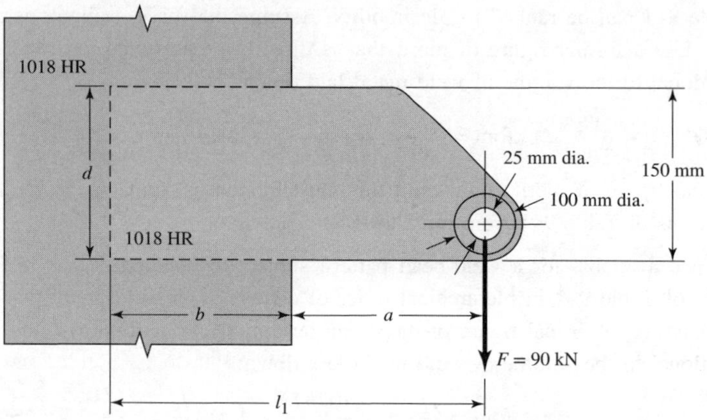

Problem 9–46

Dimensions in millimeters

9–47 Write a computer program to assist with a task such as that of Problem 9–46 with a rectangular weld-bead pattern for a torsional shear joint. In doing so solicit the force F, the clearance a, and the largest allowable shear stress. Then, as part of an iterative loop, solicit the dimensions b and d of the rectangle. These can be your design variables. Output all the parameters after the leg size has been determined by computation. In effect this will be your adequacy assessment when you stop iterating. Include the figure of merit $J_u/(hl)$ in the output. The fom and the leg size h with available width will give you a useful insight into the nature of this class of welds. Use your program to verify your solutions to Problem 9–46.

9–48 Fillet welds in joints resisting bending are interesting in that they can be simpler than those resisting torsion. From Problem 9–42 you learned that your objective is to place weld metal as far away from the weld-bead centroid as you can, but distributed in an orientation parallel to the x axis. Furthermore, placement on the top and bottom of the built-in end of a cantilever with rectangular cross section results in parallel weld beads, each element of which is in the ideal position. The object of this problem is to study the full weld bead and the interrupted weld-bead pattern. Consider the case of Figure 9–17, with $F = 50$ kN, the beam length is 250 mm, $b = 200$ mm, and $d = 200$ mm. For the second case, for the interrupted weld consider a centered gap of $b_1 = 100$ mm existing in the top and bottom welds. Study the two cases with $\tau_{all} = 88$ MPa. What do you notice about τ, σ, and τ_{max}? Compare the fom′.

9–49 For a rectangular weld-bead track resisting bending, develop the necessary equations to treat cases of vertical welds, horizontal welds, and weld-all-around patterns with depth d and width b and allowing central gaps in parallel beads of length b_1 and d_1. Do this by superposition of parallel tracks, vertical tracks subtracting out the gaps.

Then put the two together for a rectangular weld bead with central gaps of length b_1 and d_1. Show that the results are

$$A = 1.414(b - b_1 + d - d_1)h$$

$$I_u = \frac{(b - b_1)d^2}{2} + \frac{d^3 - d_1^3}{6}$$

$$I = 0.707hI_u$$

$$l = 2(b - b_1) + 2(d - d_1)$$

$$\text{fom} = \frac{I_u}{hl}$$

9–50 Write a computer program based on the Problem 9–49 protocol. Solicit the largest allowable shear stress, the force F, and the clearance a, as well as the dimensions b and d. Begin an iterative loop by soliciting b_1 and d_1. Either or both of these can be your design variables. Program to find the leg size corresponding to a shear-stress level at the maximum allowable at a corner. Output all your parameters including the figure of merit. Use the program to check any previous problems to which it is applicable. Play with it in a "what if" mode and learn from the trends in your parameters.

9–51 When comparing two different weldment patterns it is useful to observe the resistance to bending or torsion and the volume of weld metal deposited. The *measure of effectiveness,* defined as second moment of area divided by weld-metal volume, is useful. If a 75 mm by 150 mm section of a cantilever carries a static 50 kN bending load 200 mm from the weldment plane, with an allowable shear stress of 80 MPa realized, compare horizontal weldments with vertical weldments by determining the measure of effectiveness for each weld pattern. The horizontal beads are to be 75 mm long and the vertical beads, 150 mm long.

9–52 to 9–54 A 50 mm dia. steel bar is subjected to the loading indicated. Locate and estimate the maximum shear stress in the weld throat.

Problem Number	F	T
9–52	0	1.6 kN · m
9–53	10 kN	0
9–54	10 kN	1.6 kN · m

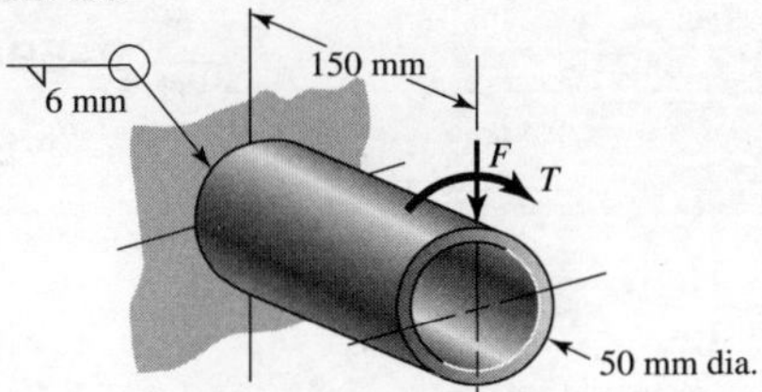

Problems 9–52 to 9–54

9–55 For Problem 9–54, determine the weld size if the maximum allowable shear stress is 140 MPa.

9–56 Find the maximum shear stress in the throat of the weld metal in the figure.

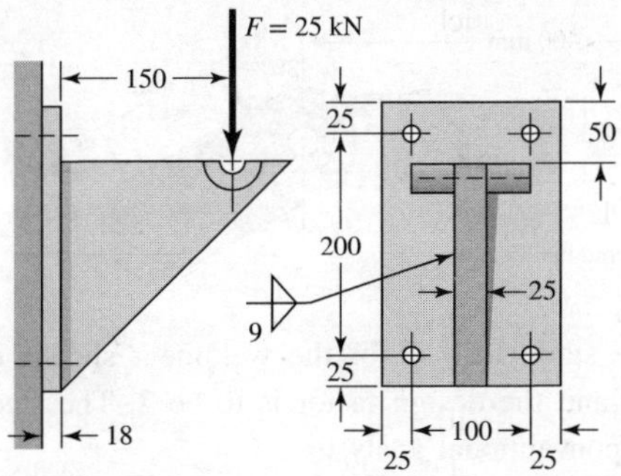

Problem 9–56

Dimensions in millimeters.

9–57 The figure shows a welded steel bracket loaded by a static force F. Estimate the factor of safety if the allowable shear stress in the weld throat is 120 MPa.

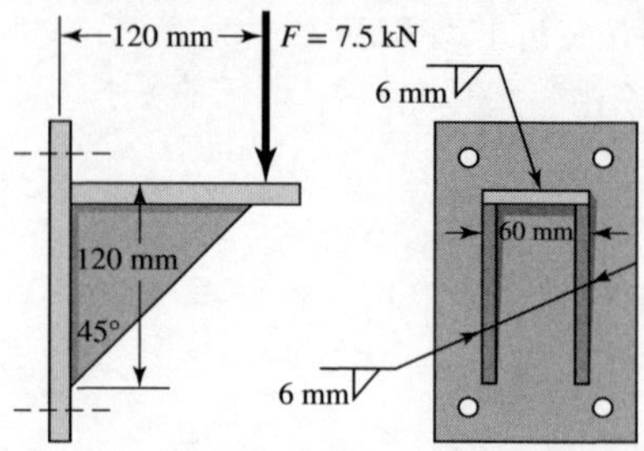

Problem 9–57
Dimensions in millimeters.

9–58 The figure shows a formed sheet-steel bracket. Instead of securing it to the support with machine screws, welding all around the bracket support flange has been proposed. If the combined shear stress in the weld metal is limited to 6 MPa, estimate the total load W the bracket will support. The dimensions of the top flange are the same as the mounting flange.

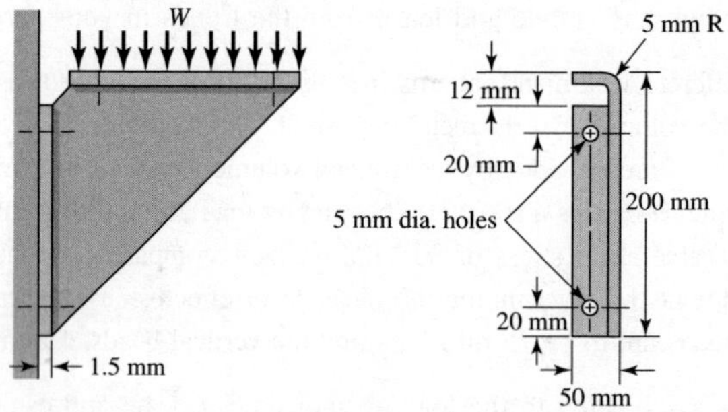

Problem 9–58
Structural support is 1030 HR steel, bracket is 1020 press cold-formed steel. The weld electrode is E6010.

9–59 Without bracing, a machinist can exert only about 400 N on a wrench or tool handle. The lever shown in the figure has t = 12 mm and w = 50 mm. We wish to specify the fillet-weld size to secure the lever to the tubular part at A. Both parts are of steel, and the shear stress in the weld throat should not exceed 20 MPa. Find a safe weld size.

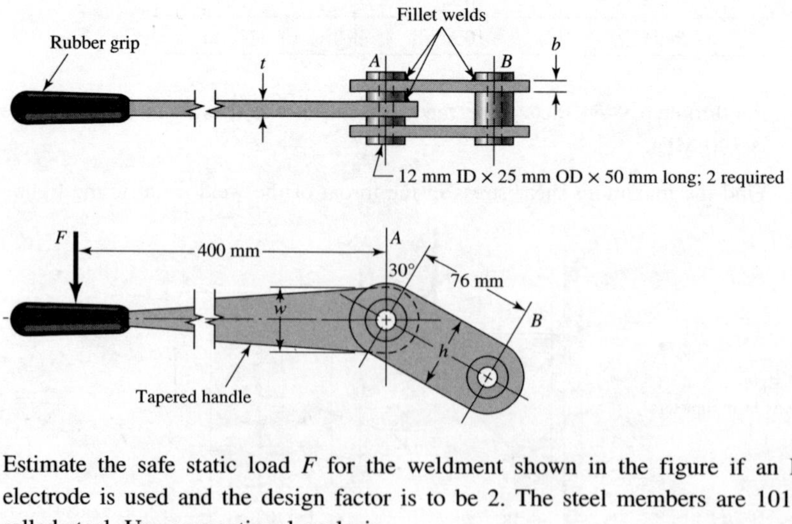

Problem 9–59

9–60 Estimate the safe static load F for the weldment shown in the figure if an E6010 electrode is used and the design factor is to be 2. The steel members are 1015 hot-rolled steel. Use conventional analysis.

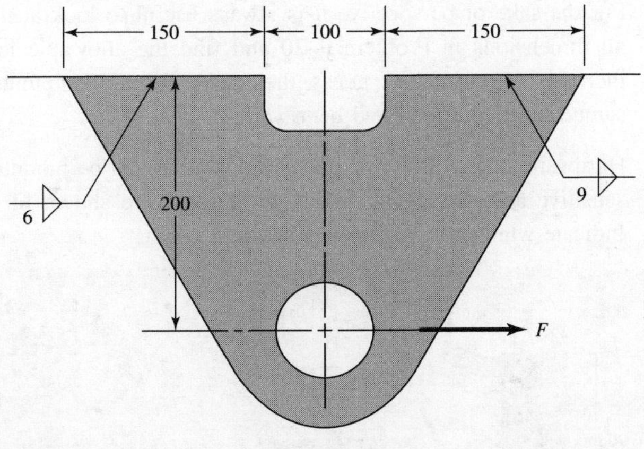

Problem 9–60
Dimensions in millimeters.

9–61 Brackets, such as the one shown, are used in mooring small watercraft. Failure of such brackets is usually caused by bearing pressure of the mooring clip against the side of the hole. Our purpose here is to get an idea of the static and dynamic margins of safety involved. We use a bracket 6 mm thick made of hot-rolled 1018 steel, welded with an E6010 electrode. We then assume wave action on the boat will create force F no greater than 5 kN.

(*a*) Determine the moment M of the force F about the centroid of the weld G. This moment produces a shear stress on the throat resisting bending action with a “tension” at A and “compression” at C.

(*b*) Find the force component F_y that produces a shear stress at the throat resisting a “tension” throughout the weld.

(*c*) Find the force component F_x that produces an in-line shear throughout the weld.

(*d*) Using Table 9–2, determine A, I_u, and I for the bracket.

(*e*) Find the shear stress τ_1 at A due to F_y and M, the shear stress τ_2 due to F_x, and combine to find τ.

(*f*) Find the factor of safety guarding against shear yielding in the weldment. Since the weld material is comprised of a mix of the electrode material and the base material, take the conservative approach of utilizing the strength of the weaker material.

(*g*) Find the factor of safety guarding against a static failure in the parent metal at the weld.

(*h*) Assuming the force F alternates between zero and 5 kN, find the factor of safety guarding against a fatigue failure in the weld metal using a Gerber failure criterion.

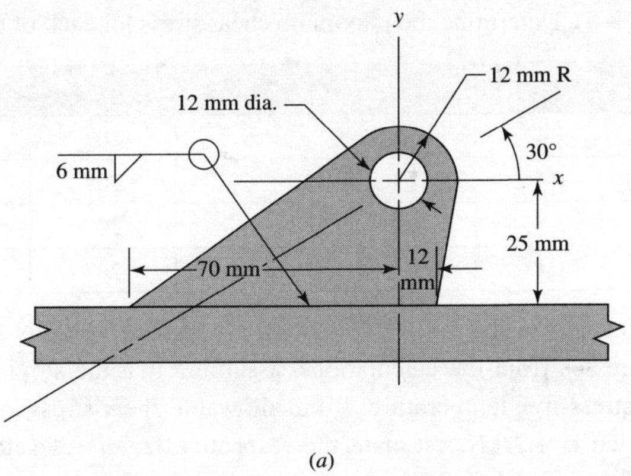

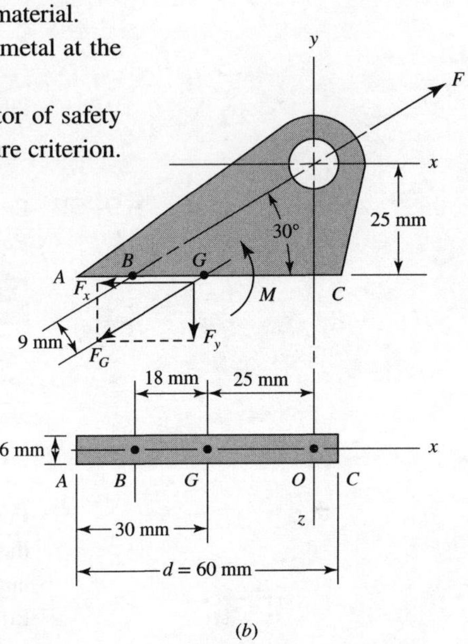

Problem 9–61
Small watercraft mooring bracket.

9–62 For the sake of perspective it is always useful to look at the matter of scale. Double all dimensions in Problem 9–20 and find the allowable load. By what factor has it increased? First make a guess, then carry out the computation. Would you expect the same ratio if the load had been variable?

9–63 Hardware stores often sell plastic hooks that can be mounted on walls with pressure-sensitive adhesive foam tape. Two designs are shown in (*a*) and (*b*) of the figure. Indicate which one you would buy and why.

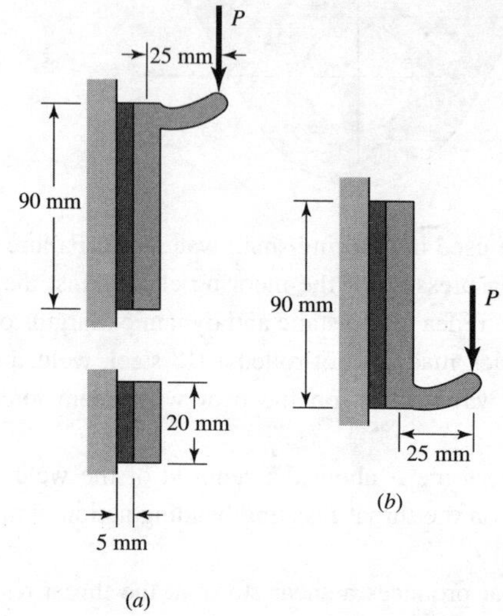

Problem 9–63

9–64 For a balanced double-lap joint cured at room temperature, Volkersen's equation simplifies to

$$\tau(x) = \frac{P\omega \cosh(\omega x)}{4b \sinh(\omega l/2)} = A_1 \cosh(\omega x)$$

(*a*) Show that the average stress $\bar{\tau}$ is $P/(2bl)$.

(*b*) Show that the largest shear stress is $P\omega/[4b \tanh(\omega l/2)]$.

(*c*) Define a stress-augmentation factor K such that

$$\tau(l/2) = K\bar{\tau}$$

and it follows that

$$K = \frac{P\omega}{4b \tanh(\omega l/2)} \frac{2bl}{P} = \frac{\omega l/2}{\tanh(\omega l/2)} = \frac{\omega l}{2} \frac{\exp(\omega l/2) + \exp(-\omega l/2)}{\exp(\omega l/2) - \exp(-\omega l/2)}$$

9–65 Program the shear-lag solution for the shear-stress state into your computer using Equation (9–7). Determine the maximum shear stress for each of the following scenarios:

Part	G, GPa	t_o, mm	t_i, mm	E_o, GPa	E_i, GPa	h, mm
a	1.4	3	6	207	207	0.13
b	1.4	3	6	207	207	0.4
c	1.4	3	3	207	207	0.13
d	1.4	3	6	207	$10(10^6)$	0.13

Provide plots of the actual stress distributions predicted by this analysis. You may omit thermal stresses from the calculations, assuming that the service temperature is similar to the stress-free temperature. If the allowable shear stress is 550 MPa and the load to be carried is 1.2 kN, estimate the respective factors of safety for each geometry. Let l = 30 mm and b = 25 mm.

10 Mechanical Springs

vladimirnenezic/123RF

Chapter Outline

Springs may be used to provide flexibility in a structure, to absorb or store energy, or to provide a relationship between force and deflection. In this chapter we will discuss the more frequently used types of springs, their necessary parametric relationships, and their design.

In general, springs may be classified as wire springs, flat springs, or special-shaped springs, and there are variations within these divisions. Wire springs include helical springs of round or square wire, made to resist and deflect under tensile, compressive, or torsional loads. Flat springs include cantilever and elliptical types, wound motor- or clock-type power springs, and flat spring washers, usually called Belleville springs.

10–1 Stresses in Helical Springs

Figure 10–1*a* shows a round-wire helical compression spring loaded by the axial force F. We designate D as the *mean coil diameter* and d as the *wire diameter*. Isolate a section in the spring, as shown in Figure 10–1*b*. For equilibrium, the isolated section contains a direct shear force F and a torsional moment $T = FD/2$.

The maximum shear stress in the wire may be computed by superposition of the direct shear stress given by Equation (3–26), with $V = F$ and the torsional shear stress given by Equation (3–37). The result is

$$\tau_{\max} = \frac{Tr}{J} + \frac{F}{A} \tag{a}$$

at the *inside* fiber of the spring. Substitution of $\tau_{\max} = \tau$, $T = FD/2$, $r = d/2$, $J = \pi d^4/32$, and $A = \pi d^2/4$ gives

$$\tau = \frac{8FD}{\pi d^3} + \frac{4F}{\pi d^2} \tag{b}$$

Now we define the *spring index*

$$C = \frac{D}{d} \tag{10–1}$$

which is a measure of coil curvature. The preferred value of C ranges from 4 to 12.[1] With this relation, Equation (*b*) can be rearranged to give

$$\tau = K_s \frac{8FD}{\pi d^3} \tag{10–2}$$

where K_s is a *shear stress-correction factor* and is defined by the equation

$$K_s = \frac{2C + 1}{2C} \tag{10–3}$$

The use of square or rectangular wire is not recommended for springs unless space limitations make it necessary. Springs of special wire shapes are not made in large quantities, unlike those of round wire; they have not had the benefit of refining

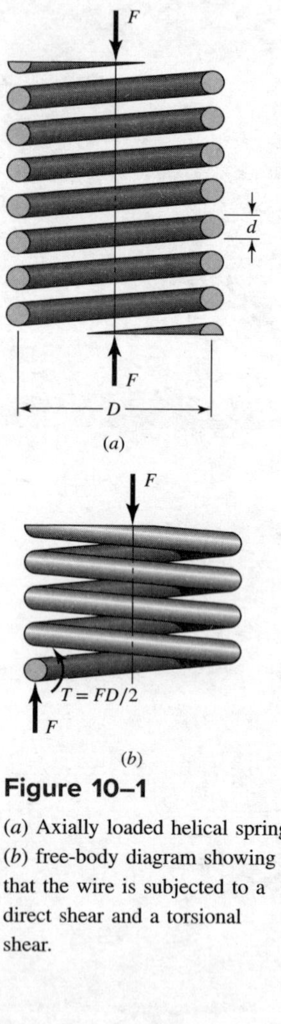

Figure 10–1

(*a*) Axially loaded helical spring; (*b*) free-body diagram showing that the wire is subjected to a direct shear and a torsional shear.

[1] *Design Handbook: Engineering Guide to Spring Design*, Associated Spring-Barnes Group Inc., Bristol, CT, 1987.

development and hence may not be as strong as springs made from round wire. When space is severely limited, the use of nested round-wire springs should always be considered. They may have an economical advantage over the special-section springs, as well as a strength advantage.

10–2 The Curvature Effect

Equation (10–2) is based on the wire being straight. However, the curvature of the wire causes a localized increase in stress on the inner surface of the coil, which can be accounted for with a curvature factor. This factor can be applied in the same way as a stress concentration factor. For static loading, the curvature factor is normally neglected because any localized yielding leads to localized strain strengthening. For fatigue applications, the curvature factor should be included.

Unfortunately, it is necessary to find the curvature factor in a roundabout way. The reason for this is that the published equations also include the effect of the direct shear stress. Suppose K_s in Equation (10–2) is replaced by another K factor, which corrects for both curvature and direct shear. Then this factor is given by either of the equations

$$K_W = \frac{4C - 1}{4C - 4} + \frac{0.615}{C} \tag{10–4}$$

$$K_B = \frac{4C + 2}{4C - 3} \tag{10–5}$$

The first of these is called the *Wahl factor,* and the second, the *Bergsträsser factor.*[2] Since the results of these two equations differ by the order of 1 percent, Equation (10–5) is preferred. The curvature correction factor can now be obtained by canceling out the effect of the direct shear. Thus, using Equation (10–5) with Equation (10–3), the curvature correction factor is found to be

$$K_c = \frac{K_B}{K_s} = \frac{2C(4C + 2)}{(4C - 3)(2C + 1)} \tag{10–6}$$

Now, K_s, K_B or K_W, and K_c are simply stress-correction factors applied multiplicatively to Tr/J at the critical location to estimate a particular stress. There is *no* stress-concentration factor. In this book we will use

$$\tau = K_B \frac{8FD}{\pi d^3} \tag{10–7}$$

to predict the largest shear stress.

10–3 Deflection of Helical Springs

The deflection-force relations are quite easily obtained by using Castigliano's theorem. The total strain energy for a helical spring is composed of a torsional component and a shear component. From Equations (4–22) and (4–23), the strain energy is

$$U = \frac{T^2 l}{2GJ} + \frac{F^2 l}{2AG} \tag{a}$$

[2]Cyril Samónov, "Some Aspects of Design of Helical Compression Springs," *Int. Symp. Design and Synthesis*, Tokyo, 1984.

Substituting $T = FD/2$, $l = \pi DN$, $J = \pi d^4/32$, and $A = \pi d^2/4$ results in

$$U = \frac{4F^2D^3N}{d^4G} + \frac{2F^2DN}{d^2G} \tag{b}$$

where $N = N_a$ = number of active coils. Then using Castigliano's theorem, Equation (4–26), to find total deflection y gives

$$y = \frac{\partial U}{\partial F} = \frac{8FD^3N}{d^4G} + \frac{4FDN}{d^2G} \tag{c}$$

Since $C = D/d$, Equation (c) can be rearranged to yield

$$y = \frac{8FD^3N}{d^4G}\left(1 + \frac{1}{2C^2}\right) \approx \frac{8FD^3N}{d^4G} \tag{10–8}$$

The spring rate, also called the *scale* of the spring, is $k = F/y$, and so

$$k \approx \frac{d^4G}{8D^3N} \tag{10–9}$$

10–4 Compression Springs

The four types of ends generally used for compression springs are illustrated in Figure 10–2. The terminal end of each spring is only shown on the right-end of the spring. A spring with *plain ends* has a noninterrupted helicoid; the ends are the same as if a long spring had been cut into sections. A spring with plain ends that are *squared* or *closed* is obtained by deforming the ends to a zero-degree helix angle. Springs should always be both squared and ground for important applications, because a better transfer of the load is obtained.

Table 10–1 shows how the type of end used affects the number of coils and the spring length.[3] Note that the digits 0, 1, 2, and 3 appearing in Table 10–1 are often used without question. *Some of these need closer scrutiny as they may not be integers.* This depends on how a springmaker forms the ends. Forys[4] pointed out that squared and ground ends give a solid length L_s of

$$L_s = (N_t - a)d$$

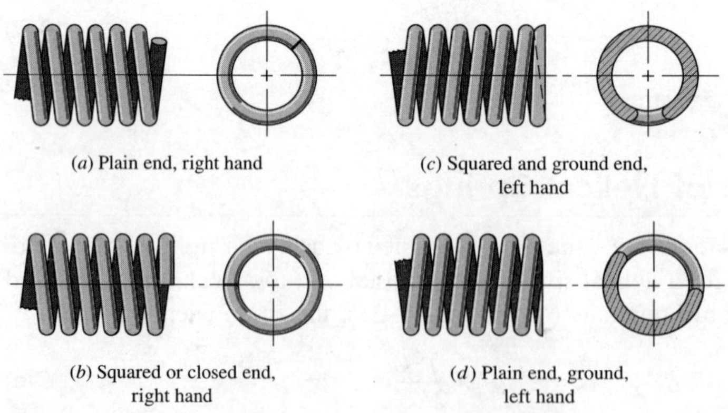

Figure 10–2

Types of ends for compression springs: (a) both ends plain; (b) both ends squared; (c) both ends squared and ground; (d) both ends plain and ground.

[3] For a thorough discussion and development of these relations, see Cyril Samónov, "Computer-Aided Design of Helical Compression Springs," ASME paper No. 80-DET-69, 1980.

[4] Edward L. Forys, "Accurate Spring Heights," *Machine Design*, vol. 56, no. 2, January 26, 1984.

Table 10–1 Formulas for the Dimensional Characteristics of Compression Springs (N_a = Number of Active Coils)

	Type of Spring Ends			
Term	Plain	Plain and Ground	Squared or Closed	Squared and Ground
End coils, N_e	0	1	2	2
Total coils, N_t	N_a	$N_a + 1$	$N_a + 2$	$N_a + 2$
Free length, L_0	$pN_a + d$	$p(N_a + 1)$	$pN_a + 3d$	$pN_a + 2d$
Solid length, L_s	$d(N_t + 1)$	dN_t	$d(N_t + 1)$	dN_t
Pitch, p	$(L_0 - d)/N_a$	$L_0/(N_a + 1)$	$(L_0 - 3d)/N_a$	$(L_0 - 2d)/N_a$

Source: Data from *Design Handbook*, 1987, p. 32.

where a varies, with an average of 0.75, so the entry dN_t in Table 10–1 may be overstated. The way to check these variations is to take springs from a particular springmaker, close them solid, and measure the solid height. Another way is to look at the spring and count the wire diameters in the solid stack.

Set removal or *presetting* or *scragging* is a process used in the manufacture of compression springs to induce useful residual stresses. The terms *set removal* and *presetting* seem counter to each other in meaning. Set removal refers to the original spring forming operation in which the spring wire is wrapped plastically around a mandrel, causing residual stresses in the wire due to differing levels of cold work at different parts of the wire cross-section. This *set* is *removed* by either a normalizing heat treatment, or by inducing a different and more beneficial residual stress. It is done by making the spring longer than needed and then compressing it to a point that partially yields the wire in torsion, usually compressing the spring to its solid height. This operation *sets* the spring to the required final free length. The torsional yield strength has been exceeded, on the outer portion of the wire where the torsional stress is greatest, but usually not in the inner core portion of the wire. When the spring is unloaded, the yielded (permanently deformed) outer diameter does not return to its original dimension, while the inner core elastically attempts to return to its original dimension. This creates residual stress in the wire, which is in the opposite direction to the stress induced during service, and thus beneficial. Springs to be preset should be designed so that 10 to 30 percent of the initial free length is removed during the operation. If the stress at the solid height is greater than 1.3 times the torsional yield strength, distortion may occur. If this stress is much less than 1.1 times, it is difficult to control the resulting free length.

Compression springs are not always preset, as the operation adds an additional cost. Presetting has a particular value when it is desired to increase the load capacity without increasing the size of the spring. Presetting is especially useful when the spring is used for energy-storage purposes. Presetting has less influence on the endurance limit, so it is not as beneficial for fatigue loading.

10–5 Stability

In Chapter 4 we learned that a column will buckle when the load becomes too large. Similarly, compression coil springs may buckle when the deflection becomes too large. The critical deflection is given by the equation

$$y_{cr} = L_0 C_1' \left[1 - \left(1 - \frac{C_2'}{\lambda_{eff}^2} \right)^{1/2} \right] \tag{10–10}$$

Table 10–2 End-Condition Constants α for Helical Compression Springs*

End Condition	Constant α
Spring supported between flat parallel surfaces (fixed ends)	0.5
One end supported by flat surface perpendicular to spring axis (fixed); other end pivoted (hinged)	0.707
Both ends pivoted (hinged)	1
One end clamped; other end free	2

*Ends supported by flat surfaces must be squared and ground.

where y_{cr} is the deflection corresponding to the onset of instability. Samónov[5] states that this equation is cited by Wahl[6] and verified experimentally by Haringx.[7] The quantity λ_{eff} in Equation (10–10) is the *effective slenderness ratio* and is given by the equation

$$\lambda_{eff} = \frac{\alpha L_0}{D} \tag{10–11}$$

C'_1 and C'_2 are dimensionless elastic constants defined by the equations

$$C'_1 = \frac{E}{2(E - G)}$$

$$C'_2 = \frac{2\pi^2(E - G)}{2G + E}$$

Equation (10–11) contains the *end-condition constant* α. This depends upon how the ends of the spring are supported. Table 10–2 gives values of α for usual end conditions. Note how closely these resemble the end conditions for columns.

Absolute stability occurs when, in Equation (10–10), the term C'_2/λ^2_{eff} is greater than unity. This means that the condition for absolute stability is that

$$L_0 < \frac{\pi D}{\alpha}\left[\frac{2(E - G)}{2G + E}\right]^{1/2} \tag{10–12}$$

For steels, this turns out to be

$$L_0 < 2.63\,\frac{D}{\alpha} \tag{10–13}$$

For squared and ground ends supported between flat parallel surfaces, $\alpha = 0.5$ and $L_0 > 5.26D$.

[5]Cyril Samónov "Computer-Aided Design," op. cit.

[6]A. M. Wahl, *Mechanical Springs*, 2nd ed., McGraw Hill, New York, 1963.

[7]J. A. Haringx, "On Highly Compressible Helical Springs and Rubber Rods and Their Application for Vibration-Free Mountings," I and II, *Philips Res. Rep.*, vol. 3, December 1948, pp. 401–449, and vol. 4, February 1949, pp. 49–80.

10–6 Spring Materials

Springs are manufactured either by hot- or cold-working processes, depending upon the size of the material, the spring index, and the properties desired. In general, prehardened wire should not be used if $D/d > 4$ or if $d > 6$ mm. Winding of the spring induces residual stresses through bending, but these are normal to the direction of the torsional working stresses in a coil spring. Quite frequently in spring manufacture, they are relieved, after winding, by a mild thermal treatment.

A great variety of spring materials are available to the designer, including plain carbon steels, alloy steels, and corrosion-resisting steels, as well as nonferrous materials such as phosphor bronze, spring brass, beryllium copper, and various nickel alloys. Descriptions of the most commonly used steels will be found in Table 10–3. The UNS steels listed in Appendix A should be used in designing hot-worked, heavy-coil springs, as well as flat springs, leaf springs, and torsion bars.

Spring materials may be compared by an examination of their tensile strengths; these vary so much with wire size that they cannot be specified until the wire size is known. The material and its processing also, of course, have an effect on tensile strength. It turns out that the graph of tensile strength versus wire diameter is almost

Table 10–3 High-Carbon and Alloy Spring Steels

Name of Material	Similar Specifications	Description
Music wire, 0.80–0.95*C*	UNS G10 850 AISI 1085 ASTM A228-51	This is the best, toughest, and most widely used of all spring materials for small springs. It has the highest tensile strength and can withstand higher stresses under repeated loading than any other spring material. Available in diameters 0.12 to 3 mm. Do not use above 120°C or at subzero temperatures.
Oil-tempered wire, 0.60–0.70*C*	UNS G10 650 AISI 1065 ASTM 229-41	This general-purpose spring steel is used for many types of coil springs where the cost of music wire is prohibitive and in sizes larger than available in music wire. Not for shock or impact loading. Available in diameters 3 to 12 mm, but larger and smaller sizes may be obtained. Not for use above 180°C or at subzero temperatures.
Hard-drawn wire, 0.60–0.70*C*	UNS G10 660 AISI 1066 ASTM A227-47	This is the cheapest general-purpose spring steel and should be used only where life, accuracy, and deflection are not too important. Available in diameters 0.8 to 12 mm. Not for use above 120°C or at subzero temperatures.
Chrome-vanadium	UNS G61 500 AISI 6150 ASTM 231-41	This is the most popular alloy spring steel for conditions involving higher stresses than can be used with the high-carbon steels and for use where fatigue resistance and long endurance are needed. Also good for shock and impact loads. Widely used for aircraft-engine valve springs and for temperatures to 220°C. Available in annealed or pretempered sizes 0.8 to 12 mm in diameter.
Chrome-silicon	UNS G92 540 AISI 9254	This alloy is an excellent material for highly stressed springs that require long life and are subjected to shock loading. Rockwell hardnesses of C50 to C53 are quite common, and the material may be used up to 250°C. Available from 0.8 to 12 mm in diameter.

Source: From Harold C. R. Carlson, "Selection and Application of Spring Materials," *Mechanical Engineering*, vol. 78, 1956, pp. 331–334.

Table 10–4 Constants A and m of $S_{ut} = A/d^m$ for Estimating Minimum Tensile Strength of Common Spring Wires

Material	ASTM No.	Exponent m	Diameter, mm	A, MPa · mm^m	Relative Cost of Wire
Music wire*	A228	0.145	0.10–6.5	2211	2.6
OQ&T wire†	A229	0.187	0.5–12.7	1855	1.3
Hard-drawn wire‡	A227	0.190	0.7–12.7	1783	1.0
Chrome-vanadium wire§	A232	0.168	0.8–11.1	2005	3.1
Chrome-silicon wire‖	A401	0.108	1.6–9.5	1974	4.0
302 Stainless wire#	A313	0.146	0.3–2.5	1867	7.6–11
		0.263	2.5–5	2065	
		0.478	5–10	2911	
Phosphor-bronze wire**	B159	0	0.1–0.6	1000	8.0
		0.028	0.6–2	913	
		0.064	2–7.5	932	

*Surface is smooth, free of defects, and has a bright, lustrous finish.

†Has a slight heat-treating scale which must be removed before plating.

‡Surface is smooth and bright with no visible marks.

§Aircraft-quality tempered wire, can also be obtained annealed.

‖Tempered to Rockwell C49, but may be obtained untempered.

#Type 302 stainless steel.

**Temper CA510.

Source: Data from *Design Handbook,* 1987, p. 19.

a straight line for some materials when plotted on log-log paper. Writing the equation of this line as

$$S_{ut} = \frac{A}{d^m} \tag{10–14}$$

furnishes a good means of estimating minimum tensile strengths when the intercept A and the slope m of the line are known. Values of these constants have been worked out from recent data and are given for strengths in units of MPa in Table 10–4. In Equation (10–14) when d is measured in millimeters, then A is in MPa · mm^m.

Although the torsional yield strength is needed to design the spring and to analyze the performance, spring materials customarily are tested only for tensile strength—perhaps because it is such an easy and economical test to make. A very rough estimate of the torsional yield strength can be obtained by assuming that the tensile yield strength is between 60 and 90 percent of the tensile strength. Then the distortion-energy theory can be employed to obtain the torsional yield strength ($S_{sy} = 0.577S_y$). This approach results in the range

$$0.35S_{ut} \le S_{sy} \le 0.52S_{ut} \tag{10–15}$$

for steels.

For wires listed in Table 10–5, the maximum allowable shear stress in a spring can be seen in column 3. Music wire and hard-drawn steel spring wire have a low

Table 10–5 Mechanical Properties of Some Spring Wires

Material	Elastic Limit, Percent of S_{ut}		Diameter d, mm	E	G
	Tension	Torsion		GPa	GPa
Music wire A228	65–75	45–60	<0.8	203.4	82.7
			0.8–1.6	200	81.7
			1.61–3	196.5	81.0
			>3	193	80.0
HD spring A227	60–70	45–55	<0.8	198.6	80.7
			0.8–1.6	197.9	80.0
			1.61–3	197.2	79.3
			>3	196.5	78.6
Oil tempered A239	85–90	45–50		196.5	77.2
Valve spring A230	85–90	50–60		203.4	77.2
Chrome-vanadium A231	88–93	65–75		203.4	77.2
A232	88–93			203.4	77.2
Chrome-silicon A401	85–93	65–75		203.4	77.2
Stainless steel					
A313*	65–75	45–55		193	69.0
17-7PH	75–80	55–60		208.4	75.8
414	65–70	42–55		200	77.2
420	65–75	45–55		200	77.2
431	72–76	50–55		206	79.3
Phosphor-bronze B159	75–80	45–50		103.4	41.4
Beryllium-copper B197	70	50		117.2	44.8
Inconel alloy X-750	65–70	40–45		213.7	77.2

*Also includes 302, 304, and 316.

Note: See Table 10–6 for allowable torsional stress design values.

end of range $S_{sy} = 0.45S_{ut}$. Valve spring wire, Cr-Va, Cr-Si, and other (not shown) hardened and tempered carbon and low-alloy steel wires as a group have $S_{sy} \geq 0.50S_{ut}$. Many nonferrous materials (not shown) as a group have $S_{sy} \geq 0.35S_{ut}$. In view of this, Joerres[8] uses the maximum allowable torsional stress for static application shown in Table 10–6. For specific materials for which you have torsional yield information use this table as a guide. Joerres provides set-removal information in Table 10–6, that $S_{sy} \geq 0.65S_{ut}$ increases strength through cold work, but at the cost of an additional operation by the springmaker. Sometimes the additional operation can be done by the manufacturer during assembly. Some correlations with carbon steel springs show that the tensile yield strength of spring wire in torsion can be estimated from $0.75S_{ut}$. The corresponding estimate of the yield strength in shear based on distortion energy

[8]Robert E. Joerres, "Springs," Chapter 6 in Joseph E. Shigley, Charles R. Mischke, and Thomas H. Brown, Jr. (eds.), *Standard Handbook of Machine Design*, 3rd ed., McGraw Hill, New York, 2004.

Table 10–6 Maximum Allowable Torsional Stresses for Helical Compression Springs in Static Applications

	Maximum Percent of Tensile Strength	
Material	**Before Set Removed (includes K_W or K_B)**	**After Set Removed (includes K_s)**
Music wire and cold-drawn carbon steel	45	60–70
Hardened and tempered carbon and low-alloy steel	50	65–75
Austenitic stainless steels	35	55–65
Nonferrous alloys	35	55–65

Source: Joerres, E. Robert. "Springs." Chapter 6 in *Standard Handbook of Machine Design,* edited by Joseph E. Shigley, Omrles R. Mischke, and Thomas H. Brown. Jr. 3rd ed. New York: McGraw Hill, 2004.

theory is $S_{sy} = 0.577(0.75)S_{ut} = 0.433S_{ut} \approx 0.45S_{ut}$. Samónov discusses the problem of allowable stress and shows that

$$S_{sy} = \tau_{\text{all}} = 0.56S_{ut} \tag{10–16}$$

for high-tensile spring steels, which is close to the value given by Joerres for hardened alloy steels. He points out that this value of allowable stress is specified by Draft Standard 2089 of the German Federal Republic when Equation (10–2) is used without stress-correction factor.

EXAMPLE 10–1

A helical compression spring is made of no. 16 music wire. The outside coil diameter of the spring is 11 mm. The ends are squared and there are $12\frac{1}{2}$ total turns.

(*a*) Estimate the torsional yield strength of the wire.
(*b*) Estimate the static load corresponding to the yield strength.
(*c*) Estimate the scale of the spring.
(*d*) Estimate the deflection that would be caused by the load in part (*b*).
(*e*) Estimate the solid length of the spring.
(*f*) What length should the spring be to ensure that when it is compressed solid and then released, there will be no permanent change in the free length?
(*g*) Given the length found in part (*f*), is buckling a possibility?
(*h*) What is the pitch of the body coil?

Solution

(*a*) From Table A–25, the wire diameter is $d = 0.94$ mm. From Table 10–4, we find $A = 2211\ \text{MPa} \cdot \text{m}^m$ and $m = 0.145$. Therefore, from Equation (10–14)

$$S_{ut} = \frac{A}{d^m} = \frac{2211}{0.94^{0.145}} = 2231\ \text{MPa}$$

Then, from Table 10–6,

Answer $$S_{sy} = 0.45S_{ut} = 0.45(2231) = 1004 \text{ MPa}$$

(*b*) The mean spring coil diameter is $D = 11 - 0.94 = 10.06$ mm, and so the spring index is $C = 10.06/0.94 = 10.7$. Then, from Equation (10–6),

$$K_B = \frac{4C+2}{4C-3} = \frac{4(10.7)+2}{4(10.7)-3} = 1.126$$

Now rearrange Equation (10–7) replacing τ with S_{sy}, and solve for F resulting in

Answer $$F = \frac{\pi d^3 S_{sy}}{8K_B D} = \frac{\pi(0.94^3)1004}{8(1.126)10.06} = 28.9 \text{ N}$$

(*c*) From Table 10–1, $N_a = 12.5 - 2 = 10.5$ turns. In Table 10–5, $G = 81\ 700$ MPa, and the scale of the spring is found to be, from Equation (10–9),

Answer $$k = \frac{d^4 G}{8D^3 N_a} = \frac{0.94^4\ (81\ 700)}{8(10.06^3)10.5} = 0.75 \text{ N/mm}$$

Answer (*d*) $$y = \frac{F}{k} = \frac{28.9}{0.75} = 38.5 \text{ mm}$$

(*e*) From Table 10–1,

Answer $$L_s = (N_t + 1)d = (12.5 + 1)0.94 = 12.7 \text{ mm}$$

Answer (*f*) $$L_0 = y + L_s = 38.5 + 12.7 = 51.2 \text{ mm.}$$

(*g*) To avoid buckling, Equation (10–13) and Table 10–2 give

$$L_0 < 2.63\frac{D}{\alpha} = 2.63\frac{10.06}{0.5} = 52.9 \text{ mm}$$

Mathematically, a free length of 49.31 mm is less than 52.9 mm, and buckling is unlikely. However, the forming of the ends will control how close α is to 0.5. This has to be investigated and an inside rod or exterior tube or hole may be needed.

(*h*) Finally, from Table 10–1, the pitch of the body coil is

Answer $$p = \frac{L_0 - 3d}{N_a} = \frac{51.2 - 3(0.94)}{10.5} = 4.6 \text{ mm}$$

10–7 Helical Compression Spring Design for Static Service

The preferred range of the spring index is $4 \le C \le 12$, with the lower indexes being more difficult to form (because of the danger of surface cracking) and springs with higher indexes tending to tangle often enough to require individual packing. This can be the first item of the design assessment. The recommended range of active turns is $3 \le N_a \le 15$. To maintain linearity when a spring is about to close, it is necessary to avoid the gradual touching of coils (due to nonperfect pitch). A helical coil spring force-deflection characteristic is ideally linear. Practically, it is nearly so, but not at each end of the force-deflection curve. The spring force is not reproducible for very small deflections, and near closure, nonlinear behavior begins as the number of active turns diminishes as coils begin to touch. The designer confines the spring's operating point to the central 75 percent of the curve between no load, $F = 0$, and closure, $F = F_s$. Thus, the maximum operating force should be limited to $F_{max} \leq \frac{7}{8}F_s$. Defining the fractional overrun to closure as ξ, where

$$F_s = (1 + \xi)F_{max} \tag{10–17}$$

it follows that

$$F_s = (1 + \xi)F_{max} = (1 + \xi)\left(\frac{7}{8}\right)F_s$$

From the outer equality $\xi = 1/7 = 0.143 \approx 0.15$. Thus, it is recommended that $\xi \geq 0.15$.

In addition to the relationships and material properties for springs, we now have some recommended design conditions to follow, namely:

$$4 \leq C \leq 12 \tag{10–18}$$

$$3 \leq N_a \leq 15 \tag{10–19}$$

$$\xi \geq 0.15 \tag{10–20}$$

$$n_s \geq 1.2 \tag{10–21}$$

where n_s is the factor of safety at closure (solid height).

When considering designing a spring for high volume production, the figure of merit can be the cost of the wire from which the spring is wound. The fom would be proportional to the relative material cost, weight density, and volume:

$$\text{fom} = -(\text{relative material cost})\,\frac{\gamma\pi^2 d^2 N_t D}{4} \tag{10–22}$$

For comparisons between steels, the specific weight γ can be omitted.

Spring design is an open-ended process. There are many decisions to be made, and many possible solution paths as well as solutions. In the past, charts, nomographs, and "spring design slide rules" were used by many to simplify the spring design problem. Today, the computer enables the designer to create programs in many different formats—direct programming, spreadsheet, MATLAB, etc. Commercial programs are also available.[9] There are almost as many ways to create a spring-design program as there are programmers. Here, we will suggest one possible design approach.

Design Strategy

Make the a priori decisions, with hard-drawn steel wire the first choice (relative material cost is 1.0). Choose a wire size d. With all decisions made, generate a column of parameters: d, D, C, OD or ID, N_a, L_s, L_0, $(L_0)_{cr}$, n_s, and fom. By incrementing wire sizes available, we can scan the table of parameters and apply the design

[9]For example, see *Advanced Spring Design*, a program developed jointly between the Spring Manufacturers Institute (SMI), www.smihq.org, and Universal Technical Systems (UTS), www.uts.com.

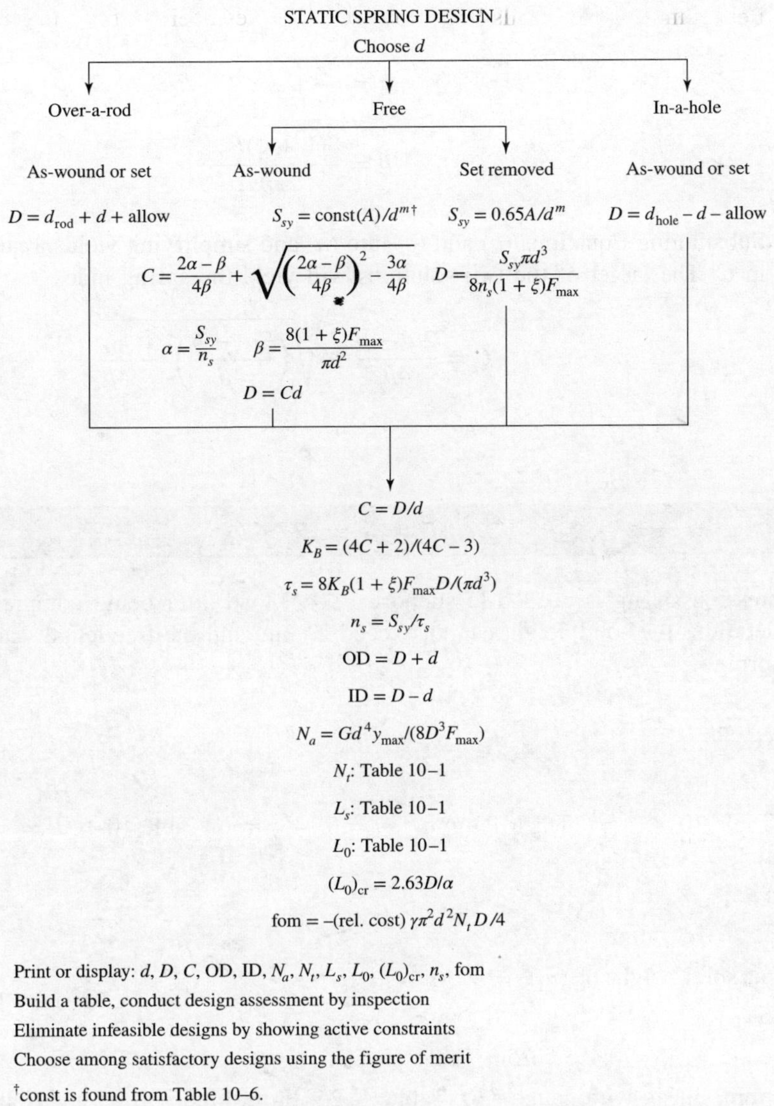

Figure 10–3

Helical coil compression spring design flowchart for static loading.

recommendations by inspection. After wire sizes are eliminated, choose the spring design with the highest figure of merit. This will give the optimal design despite the presence of a discrete design variable d and aggregation of equality and inequality constraints. The column vector of information can be generated by using the flowchart displayed in Figure 10–3. It is general enough to accommodate to the situations of as-wound and set-removed springs, operating over a rod, or in a hole free of rod or hole. In as-wound springs the controlling equation must be solved for the spring index as follows. From Equation (10–7) with $\tau = S_{sy}/n_s$, $C = D/d$, K_B from Equation (10–6), and Equation (10–17),

$$\frac{S_{sy}}{n_s} = K_B \frac{8F_s D}{\pi d^3} = \frac{4C+2}{4C-3}\left[\frac{8(1+\xi)F_{max}C}{\pi d^2}\right] \qquad (a)$$

Let

$$\alpha = \frac{S_{sy}}{n_s} \tag{b}$$

$$\beta = \frac{8(1 + \xi)F_{max}}{\pi d^2} \tag{c}$$

Substituting Equations (*b*) and (*c*) into (*a*) and simplifying yields a quadratic equation in *C*. The larger of the two solutions will yield the spring index

$$C = \frac{2\alpha - \beta}{4\beta} + \sqrt{\left(\frac{2\alpha - \beta}{4\beta}\right)^2 - \frac{3\alpha}{4\beta}} \tag{10–23}$$

EXAMPLE 10–2

A music wire helical compression spring is needed to support a 90 N load after being compressed 50 mm. Because of assembly considerations the solid height cannot exceed 27 mm and the free length cannot be more than 100 mm. Design the spring.

Solution

The a priori decisions are

- Music wire, A228; from Table 10–4, $A = 2211$ MPa-mm^m; $m = 0.145$; from Table 10–5, $E = 196\ 500$ MPa, $G = 81\ 000$ MPa (expecting $d > 1.6$ mm)
- Ends squared and ground
- Function: $F_{max} = 90$ N, $y_{max} = 50$ mm
- Safety: use design factor at solid height of $(n_s)_d = 1.2$
- Robust linearity: $\xi = 0.15$
- Use as-wound spring (cheaper), $S_{sy} = 0.45 S_{ut}$ from Table 10–6
- Decision variable: $d = 2$ mm, music wire gauge #30, Table A–25. From Figure 10–3 and Table 10–6,

$$S_{sy} = 0.45\,\frac{2211}{2^{0.145}} = 900 \text{ MPa}$$

From Figure 10–3 or Equation (10–23)

$$\alpha = \frac{S_{sy}}{n_s} = \frac{900}{1.2} = 750 \text{ MPa}$$

$$\beta = \frac{8(1 + \xi)F_{max}}{\pi d^2} = \frac{8(1 + 0.15)90}{\pi(2^2)} = 66 \text{ MPa}$$

$$C = \frac{2(750) - 66}{4(66)} + \sqrt{\left[\frac{2(750) - 66}{4(66)}\right]^2 - \frac{3(750)}{4(66)}} = 10.03$$

Continuing with Figure 10–3:

$$D = Cd = 10.03(2) = 20.06$$

$$K_B = \frac{4(10.03) + 2}{4(10.03) - 3} = 1.135$$

$$\tau_s = 1.128\,\frac{8(1 + 0.15)90(20.06)}{\pi(2)^3} = 750\text{ MPa}$$

$$n_s = \frac{900}{750} = 1.2$$

$$\text{OD} = 20.1 + 2 = 22.1\text{ mm}$$

$$N_a = \frac{81\,000\,(2)^4(50)}{8(20.1)^3 90} = 11.15\text{ turns}$$

$$N_t = 11.15 + 2 = 13.15\text{ total turns}$$

$$L_s = 2(13.15) = 26.3\text{ mm}$$

$$L_0 = 26.3 + (1 + 0.15)50 = 83.8\text{ mm}$$

$$(L_0)_{\text{cr}} = 2.63(20.1/0.5) = 106\text{ mm}$$

$$\text{fom} = -2.6\pi^2(2)^2 13.15(20.06)/4 = -6770$$

Repeat the above for other wire diameters and form a table (easily accomplished with a spreadsheet program):

d	1.6	1.7	1.8	1.9	2	2.2	2.3	2.4
C	6.10	7.02	7.98	8.98	10.03	12.25	13.42	14.64
D	9.8	11.9	14.4	17.1	20.1	26.9	30.9	35.1
n_s	1.2	1.2	1.2	1.2	1.2	1.2	1.2	1.2
OD	11.4	13.6	16.2	19.0	22.1	29.1	33.2	37.5
N_a	39.70	27.66	19.92	14.74	11.15	6.73	5.35	4.30
L_t	41.70	29.66	21.92	16.74	13.15	8.73	7.35	6.30
L_s	66.7	50.4	39.5	31.8	26.3	19.2	16.9	15.1
L_o	124.2	107.9	97.0	89.3	83.8	76.7	74.4	72.6
L_{cr}	51	63	76	90	106	142	162	185
FOM	−6681	−6561	−6545	−6618	−6770	−7307	−7699	−8181

Now examine the table and perform the adequacy assessment. The shading of the table indicates values outside the range of recommended or specified values. The spring index constraint $4 \leq C \leq 12$ rules out diameters larger than 2.2 mm. The constraint $3 \leq N_a \leq 15$ rules out wire diameters less than 1.9 mm. The $L_s \leq 27$ constraint rules out diameters less than 2 mm. The $L_0 \leq 100$ constraint rules out diameters less than 1.8 mm. The buckling criterion rules out free lengths longer than $(L_0)_{\text{cr}}$, which rules out diameters less than 0.9 mm. The factor of safety n_s is exactly 1.20 because the mathematics forced it. Had the spring been in a hole or over a rod, the helix diameter would be chosen without reference to $(n_s)_d$. The result is that there are only two springs in the feasible domain, one with a wire diameter of 2 mm and the other with a wire diameter of 2.2 mm. The figure of merit decides and the decision is the design with 2 mm wire diameter (−6770 > −7307).

Having designed a spring, will we have it made to our specifications? Not necessarily. There are vendors who stock literally thousands of music wire compression springs. By browsing their catalogs, we will usually find several that are close. Maximum deflection and maximum load are listed in the display of characteristics. Check to see if this allows soliding without damage. Often it does not. Spring rates may only be close. At the very least this situation allows a small number of springs to be ordered "off the shelf" for testing. The decision often hinges on the economics of special order versus the acceptability of a close match.

Spring design is not a closed-form approach and requires iteration. Example 10–2 provided an iterative approach to spring design for static service by first selecting the wire diameter. The diameter selection can be rather arbitrary. In the next example, we will first select a value for the spring index C, which is within the recommended range.

EXAMPLE 10–3

Design a compression spring with plain ends using hard-drawn wire. The deflection is to be 55 mm when the force is 80 N and to close solid when the force is 110 N. Upon closure, use a design factor of 1.2 guarding against yielding. Select the smallest gauge W&M (Washburn & Moen) wire.

Solution

Instead of starting with a trial wire diameter, we will start with an acceptable spring index for C after some preliminaries. From Equation (10–14) and Table 10–6 the shear strength, in MPa, is

$$S_{sy} = 0.45 S_{ut} = 0.45 \left(\frac{A}{d^m}\right) \tag{1}$$

The shear stress given by Equation (10–7) replacing τ and F with $\tau_{\max}$ and $F_{\max}$, respectively, gives

$$\tau_{\max} = K_B \frac{8F_{\max}D}{\pi d^3} = K_B \frac{8F_{\max}C}{\pi d^2} \tag{2}$$

where the Bergsträsser factor, K_B, from Equation (10–5) is

$$K_B = \frac{4C + 2}{4C - 3} \tag{3}$$

Dividing Equation (1) by the design factor n_s and equating this to Equation (2), in MPa, gives

$$\frac{0.45}{n_s}\left(\frac{A}{d^m}\right) = K_B \frac{8F_{\max}C}{\pi d^2} \tag{4}$$

For the problem $F_{\max} = 110$ N and $n_s = 1.2$. Solving for d gives

$$d = \left(163 \frac{K_B C}{A}\right)^{1/(2-m)} \tag{5}$$

Try a trial spring index of $C = 10$. From Equation (3)

$$K_B = \frac{4(10) + 2}{4(10) - 3} = 1.135$$

From Table 10–4, $m = 0.190$ and $A = 1783$ MPa · mm$^{0.190}$. Thus, Equation (5) gives

$$d = \left(163 \frac{1.135(10)}{1783}\right)^{1/(2-0.190)} = 1.02 \text{ mm}$$

From Table A–25, a 12-gauge W&M wire, d = 2.7 mm, is selected. Checking the resulting factor of safety, from Equation (4) with $F_{\max}$ = 110 N

$$n_s = 7.363\,\frac{Ad^{2-m}}{K_B C}$$

$$= 7.363\frac{1783\,(2.7^{2-0.190})}{1.135\,(10)} = 1.55 \qquad (6)$$

which is pretty conservative. If we had selected the 13-gauge wire, d = 2.24 mm, the factor of safety would be n = 1.198, which rounds to 1.2. Taking a little liberty here we will select the W&M 13-gauge wire.

To continue with the design, the spring rate is

$$k = \frac{F}{y} = \frac{80}{55} = 1.45 \text{ N/mm}$$

From Equation (10–9) solving for the active number of coils

$$N_a = \frac{d^4G}{8kD^3} = \frac{dG}{8kC^3} = \frac{2.24\ (81\ 000)}{8(1.45)10^3} = 15.6 \text{ turns}$$

This exceeds the recommended range of $3 \le N_a \le 15$. To decrease N_a, increase C. Repeating the process with C = 12 gives K_B = 1.111 and d = 2.5 mm. Selecting a 12-gauge W&M wire, d = 2.7 mm. From Equation (6), this gives n = 1.32, which is acceptable. The number of active coils is

$$N_a = \frac{dG}{8kC^3} = \frac{2.7\,(81\ 000)}{8\,(1.45)12^3} = 10.91 = 11 \text{ turns}$$

which is acceptable. From Table 10–1, for plain ends, the total number of coils is $N_t = N_a$ = 11 turns. The deflection from free length to solid length of the spring is given by

$$y_s = \frac{F_{\max}}{k} = \frac{110}{1.45} = 75 \text{ mm}$$

From Table 10–1, the solid length is

$$L_s = d(N_t + 1) = 2.7(11 + 1) = 32 \text{ mm}$$

The free length of the spring is then

$$L_0 = L_s + y_s = 32 + 75 = 107 \text{ mm}$$

The mean coil diameter of the spring is

$$D = Cd = 12(2.7) = 32 \text{ mm}$$

and the outside coil diameter of the spring is OD = $D + d$ = 32 + 2.7 = 34.7 mm

To avoid buckling, Equation (10–13) gives

$$\alpha < 2.63\,\frac{D}{L_0} = 2.63\,\frac{32}{107} = 0.780$$

From Table 10–2, the spring is stable provided it is supported between either fixed-fixed or fixed-hinged ends.

The final results are:

Answer

W&M wire size: 12 gauge, d = 2.7 mm

Outside coil diameter: OD = 34.7 mm

Total number of coils: N_t = 11 turns with plain ends

Free length: L_0 = 107 mm

10–8 Critical Frequency of Helical Springs

If a wave is created by a disturbance at one end of a swimming pool, this wave will travel down the length of the pool, be reflected back at the far end, and continue in this back-and-forth motion until it is finally damped out. The same effect occurs in helical springs, and it is called *spring surge*. If one end of a compression spring is held against a flat surface and the other end is disturbed, a compression wave is created that travels back and forth from one end to the other exactly like the swimming-pool wave.

Spring manufacturers have taken slow-motion movies of automotive valve-spring surge. These pictures show a very violent surging, with the spring actually jumping out of contact with the end plates. Figure 10–4 is a photograph of a failure caused by such surging.

When helical springs are used in applications requiring a rapid reciprocating motion, the designer must be certain that the physical dimensions of the spring are not such as to create a natural vibratory frequency close to the frequency of the applied force; otherwise, resonance may occur, resulting in damaging stresses, since the internal damping of spring materials is quite low.

The governing equation for the translational vibration of a spring placed between two flat and parallel plates is the wave equation

$$\frac{\partial^2 u}{\partial x^2} = \frac{W}{kgl^2}\frac{\partial^2 u}{\partial t^2} \tag{10–24}$$

where k = spring rate
g = acceleration due to gravity
l = length of spring between plates
W = weight of spring
x = coordinate along length of spring
u = motion of any particle at distance x

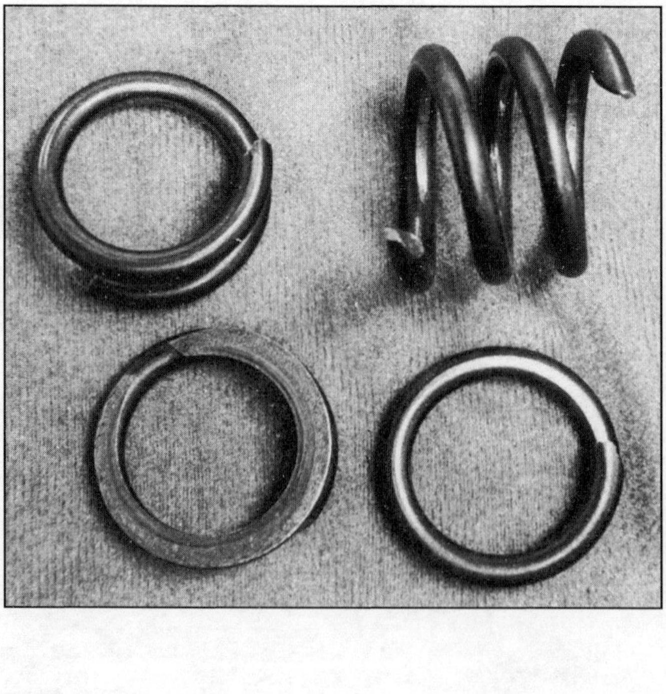

Figure 10–4

Valve-spring failure in an overrevved engine. Fracture is along the 45° line of maximum principal stress associated with pure torsional loading. *(Source: Personal photograph of Larry D. Mitchell, co-author of* Mechanical Engineering Design, *4th ed., McGraw Hill, New York, 1983.)*

The solution to this equation is harmonic and depends on the given physical properties as well as the end conditions of the spring. The harmonic, *natural,* frequencies for a spring placed between two flat and parallel plates, in radians per second, are

$$\omega = m\pi\sqrt{\frac{kg}{W}} \qquad m = 1, 2, 3, \ldots$$

where the fundamental frequency is found for $m = 1$, the second harmonic for $m = 2$, and so on. We are usually interested in the frequency in cycles per second; since $\omega = 2\pi f$, we have, for the fundamental frequency in hertz,

$$f = \frac{1}{2}\sqrt{\frac{kg}{W}} \tag{10–25}$$

assuming the spring ends are always in contact with the plates.

Wolford and Smith[10] show that the frequency is

$$f = \frac{1}{4}\sqrt{\frac{kg}{W}} \tag{10–26}$$

where the spring has one end against a flat plate and the other end free. They also point out that Equation (10–25) applies when one end is against a flat plate and the other end is driven with a sine-wave motion.

The weight of the active part of a helical spring is

$$W = AL\gamma = \frac{\pi d^2}{4}(\pi DN_a)(\gamma) = \frac{\pi^2 d^2 DN_a\gamma}{4} \tag{10–27}$$

where γ is the specific weight.

The fundamental critical frequency should be greater than 15 to 20 times the frequency of the force or motion of the spring in order to avoid resonance with the harmonics. If the frequency is not high enough, the spring should be redesigned to increase k or decrease W.

10–9 Fatigue Loading of Helical Compression Springs

Springs are almost always subject to fatigue loading. In many instances the number of cycles of required life may be small, say, several thousand for a padlock spring or a toggle-switch spring. But the valve spring of an automotive engine must sustain millions of cycles of operation without failure, so it must be designed for infinite life.

To improve the fatigue strength of dynamically loaded springs, shot peening can be used to induce compressive stresses on the surface of the wire. It can increase the torsional fatigue strength by 20 percent or more. Shot size is about 0.4 mm, so spring coil wire diameter and pitch must allow for complete coverage of the spring surface.

For fatigue analysis, we need to determine an endurance limit. The methods from Chapter 6 could be used to estimate the endurance limit, but they are not ideal for helical coil springs. The endurance limit estimates from Chapter 6 are based on

[10] J. C. Wolford and G. M. Smith, "Surge of Helical Springs," *Mech. Eng. News*, vol. 13, no. 1, February 1976, pp. 4–9.

completely reversed bending stresses. The stresses in helical coil springs are torsional shear stresses, and are never completely reversed. Helical springs are loaded from a minimum preload (perhaps zero) to a maximum load, creating a positive mean shear stress that is always equal to or greater than the alternating shear stress, which is far different from completely reversed. Also, as the wire diameter gets smaller, almost always by a cold working process, the relationship between ultimate strength and endurance limit becomes less predictable. Some of the best-known early research results on the torsional endurance limits of spring steels, which are still in common use today, are those reported by Zimmerli.[11] He discovered the surprising fact that for springs cycled between a minimum and maximum load, the size, material, and tensile strength have no effect on the endurance limits (infinite life only) of spring steels in sizes under 10 mm. In other words, he observed that even though the tensile strength is a function of wire diameter and material, the endurance limit was relatively constant. Spring steels that have been cold worked to these smaller diameters generally have high tensile strengths. We have already observed that endurance limits tend to level out at high tensile strengths (Figure 6–22), but the reason for this is not clear. To be clear, Zimmerli's constant endurance limit applies specifically to the loading condition of a spring, and is therefore not the same as the usual endurance limit associated with completely reversed loading. In particular, he preloaded unpeened springs at 138 MPa, and tested to find the repeatedly applied maximum stress that would produce fatigue failure at 10^6 cycles. He found this maximum stress to be 620 MPa, regardless of wire size for any of the standard spring steels. For peened springs, the maximum was found to be 930 MPa. Unpeened springs were tested for infinite life with a minimum torsional preload stress of 138 MPa to a maximum of 620 MPa and peened springs in the range 138 MPa to 930 MPa. The alternating and mean stresses correlating to these tested ranges correspond to a fluctuating-stress condition on the boundary between finite and infinite life, that is, 10^6 cycles for steel spring materials. According to Zimmerli, this fluctuating-stress condition is applicable for any spring steel with diameter under 10 mm. Let us define these fluctuating-stress values as the Zimmerli endurance strength components for infinite life, thus,

Unpeened:

$$S_{sa} = 241 \text{ MPa} \qquad S_{sm} = 379 \text{ MPa} \tag{10–28a}$$

Peened:

$$S_{sa} = 398 \text{ MPa} \qquad S_{sm} = 534 \text{ MPa} \tag{10–28b}$$

From these endurance strength components we can estimate an equivalent completely reversed stress, and consequently an endurance limit, from one of the constant-life curves in Section 6–14. Note that this endurance limit correlates with a theoretical completely reversed loading, which we will never have for a helical spring. It is, however, useful for calculation purposes with the usual fluctuating stress failure criteria. For example, selecting the Goodman criterion, we adapt Equation (6–59) to shear, and apply the endurance strength components as the alternating and mean stresses, resulting in

$$S_{se} = \frac{S_{sa}}{1 - \dfrac{S_{sm}}{S_{su}}} \tag{10–29a}$$

[11]F. P. Zimmerli, "Spring Failures and their Causes," *Metal Progress*, Vol. 62, July 1952, pp. 84–88.

The Gerber criterion is one of the more commonly used for springs. Starting from the Gerber fatigue failure criterion of Equation (6–47), an equation for the completely reversed endurance limit can be developed with the same constant-life approach to be

$$S_{se} = \frac{S_{sa}}{1 - \left(\dfrac{S_{sm}}{S_{su}}\right)^2} \tag{10–29b}$$

Any of the other fatigue criterion could be used in similar fashion.

As an example, given an unpeened spring with $S_{su} = 1460$ MPa, the Gerber criterion of Equation (10–29*b*) predicts the completely reversed endurance limit to be

$$S_{se} = \frac{S_{sa}}{1 - \left(\dfrac{S_{sm}}{S_{su}}\right)^2} = \frac{241}{1 - \left(\dfrac{379}{1460}\right)^2} = 258 \text{ MPa}$$

For the Goodman failure criterion, Equation (10–29*a*) predicts the endurance limit would be 326 MPa. Each possible wire size would change these numbers, since S_{su} would change. An oddity of this approach is that an increased ultimate strength leads to a smaller, completely reversed endurance limit. However, in actual spring loading, the mean stress is always equal to or greater than the alternating stress, so the load line never comes close to the completely reversed condition.

An alternate approach, known as the *Sines failure criterion,* is also available. An extended study[12] of available literature regarding torsional fatigue found that for polished, notch-free, cylindrical specimens subjected to torsional shear stress, the maximum alternating stress that may be imposed without causing failure is *constant* and independent of the mean stress in the cycle provided that the maximum stress range does not equal or exceed the torsional yield strength of the metal. With notches and abrupt section changes this consistency is not found. Springs are free of notches and surfaces are often very smooth.

With only shear stresses present in a helical spring, it is convenient to adapt the fluctuating-stress diagram and a fatigue failure criterion into shear stresses and shear strengths, as described at the end of Section 6–13. Common spring materials have been tested for material properties that are specifically applicable for wire that has been cold drawn to relatively small diameters.[13] From this testing, a recommended relationship for estimating the torsional modulus of rupture S_{su} is

$$S_{su} = 0.67S_{ut} \tag{10–30}$$

In the case of shafts and many other machine members, fatigue loading in the form of completely reversed stresses is quite ordinary. Helical springs, on the other hand, are never used as both compression and extension springs. In fact, they are usually assembled with a preload so that the working load is additional. Thus the stress-time diagram of Figure 6–29*d* expresses the usual condition for helical springs. The worst condition, then, would occur when there is no preload, that is, when $\tau_{\min} = 0$.

[12]Oscar J. Horger (ed.), *Metals Engineering: Design Handbook*, McGraw Hill, New York, 1953, p. 84.

[13]Associated Spring, *Design Handbook: Engineering Guide to Spring Design*, Associated Spring, Barnes Group Inc., Bristol, Conn., 1987.

Now, we define

$$F_a = \frac{F_{\text{max}} - F_{\text{min}}}{2} \tag{10–31a}$$

$$F_m = \frac{F_{\text{max}} + F_{\text{min}}}{2} \tag{10–31b}$$

where the subscripts have the same meaning as those of Figure 6–29*d* when applied to the axial spring force F. Then the shear stress amplitude is

$$\tau_a = K_B \frac{8F_a D}{\pi d^3} \tag{10–32}$$

where K_B is the Bergsträsser factor, obtained from Equation (10–5), and corrects for both direct shear and the curvature effect. As noted in Section 10–2, the Wahl factor K_W can be used instead, if desired.

The mean shear stress is given by the equation

$$\tau_m = K_B \frac{8F_m D}{\pi d^3} \tag{10–33}$$

EXAMPLE 10–4

An as-wound helical compression spring, made of music wire, has a wire size of 2.3 mm, an outside coil diameter of 14 mm, a free length of 110 mm, 21 active coils, and both ends squared and ground. The spring is unpeened. This spring is to be assembled with a preload of 22 N and will operate with a maximum load of 155 N during use.

(*a*) Estimate the factor of safety guarding against fatigue-failure using a torsional Gerber fatigue-failure criterion with Zimmerli data.

(*b*) Repeat part (*a*) using the Sines torsional fatigue criterion (steady stress component has no effect), with Zimmerli data.

(*c*) Repeat using a torsional Goodman failure criterion with Zimmerli data.

(*d*) Estimate the critical frequency of the spring.

Solution

The mean coil diameter is $D = 14 - 2.3 = 11.7$ mm. The spring index is $C = D/d = 11.7/2.3 = 5.09$. Then

$$K_B = \frac{4C + 2}{4C - 3} = \frac{4(5.09) + 2}{4(50.09) - 3} = 1.288$$

From Equations (10–31),

$$F_a = \frac{155 - 22}{2} = 66.5 \text{ N} \qquad F_m = \frac{155 + 22}{2} = 88.5 \text{ N}$$

The alternating shear-stress component is found from Equation (10–32) to be

$$\tau_a = K_B \frac{8F_a D}{\pi d^3} = (1.288)\frac{8(66.5)11.7}{\pi (2.3)^3} = 210 \text{ MPa}$$

Equation (10–33) gives the mean shear-stress component

$$\tau_m = K_B \frac{8F_mD}{\pi d^3} = 1.288 \frac{8(88.5)11.7}{\pi(2.3)^3} = 279 \text{ MPa}$$

From Table 10–4 we find $A = 2211$ MPa · mm^m and $m = 0.145$. The ultimate tensile strength is estimated from Equation (10–14) as

$$S_{ut} = \frac{A}{d^m} = \frac{2211}{2.3^{0.145}} = 1959 \text{ MPa}$$

The shear modulus of rupture is estimated from Equation (10–30)

$$S_{su} = 0.67S_{ut} = 0.67(1959) = 1312 \text{ MPa}$$

(*a*) The completely-reversed endurance limit based on the Gerber criterion, Equation (10–29*b*), and the Zimmerli endurance strength components, Equation (10–28*a*), is

$$S_{se} = \frac{S_{sa}}{1 - (S_{sm}/S_{su})^2} = \frac{241}{1 - (379/1312)^2} = 263 \text{ MPa}$$

The Gerber fatigue criterion from Equation (6–48), adapted for shear, is

$$n_f = \frac{1}{2}\left(\frac{S_{su}}{\tau_m}\right)^2\left(\frac{\tau_a}{S_{se}}\right)\left[-1 + \sqrt{1 + \left(\frac{2\tau_m S_{se}}{S_{su}\tau_a}\right)^2}\,\right]$$

Answer

$$= \frac{1}{2}\left(\frac{1312}{279}\right)^2\left(\frac{210}{263}\right)\left[-1 + \sqrt{1 + \left(\frac{2(279)(263)}{(1312)(210)}\right)^2}\,\right] = 1.17$$

(*b*) The Sines failure criterion ignores S_{sm} so that, for the Zimmerli data of Equation (10–28*a*) with $S_{sa} = 246.5$ MPa,

Answer

$$n_f = \frac{S_{sa}}{\tau_a} = \frac{246.5}{210} = 1.17$$

(*c*) The completely-reversed endurance limit based on the Goodman criterion, Equation (10–29*a*), and the Zimmerli endurance strength components, Equation (10–28*a*), is

$$S_{se} = \frac{S_{sa}}{1 - (S_{sm}/S_{su})} = \frac{246.5}{1 - (379/1312)} = 339 \text{ MPa}$$

The Goodman fatigue criterion from Equation (6–41), adapted for shear, is

Answer

$$n_f = \left(\frac{\tau_a}{S_{se}} + \frac{\tau_m}{S_{su}}\right)^{-1} = \left(\frac{210}{339} + \frac{279}{1312}\right)^{-1} = 1.20$$

(*d*) Using Equation (10–9) and Table 10–5, we estimate the spring rate as

$$k = \frac{d^4G}{8D^3N_a} = \frac{2.3^4[81\ 700]}{8(11.7)^3 21} = 8.5 \text{ N/mm}$$

From Equation (10–27) we estimate the spring weight as

$$W = \frac{\pi^2(2.3^2)\,11.7\,(21)0.0078}{4(1000)} = 0.025 \text{ Kg}$$

and from Equation (10–25) the frequency of the fundamental wave is

Answer

$$f_n = \frac{1}{2}\left[\frac{8500\,(9.81)}{0.025}\right]^{1/2} = 288 \text{ Hz}$$

If the operating or exciting frequency is more than $288/20 = 14.4$ Hz, the spring may have to be redesigned.

We used three approaches to estimate the fatigue factor of safety in Example 10–4. The results, in order of smallest to largest, were 1.18 (Sines), 1.21 (Gerber), and 1.23 (Goodman). Although the results were very close to one another, using the Zimmerli data as we have, the Sines criterion will always be the most conservative and the Goodman the least. If we perform a fatigue analysis using strength properties as was done in Chapter 6, different results would be obtained, but here the Goodman criterion would be more conservative than the Gerber criterion. Be prepared to see designers or design software using any one of these techniques. This is why we cover them. Which criterion is correct? Remember, we are performing *estimates* and only testing will reveal the truth—*statistically*.

10–10 Helical Compression Spring Design for Fatigue Loading

Let us begin with the statement of a problem. In order to compare a static spring to a dynamic spring, we shall design the spring in Example 10–2 for dynamic service.

EXAMPLE 10–5

A music wire helical compression spring with infinite life is needed to resist a dynamic load that varies from 22 to 90 N at 50 Hz while the end deflection varies from 12 to 50 mm. Because of assembly considerations, the solid height cannot exceed 27 mm and the free length cannot be more than 100 mm. The springmaker has the following wire sizes in stock: 1.75, 1.8, 2, 2.15, 2.3, 2.4, 2.7, and 2.8 mm.

Solution

The a priori decisions are:

- Material and condition: for music wire, $A = 2211$ MPa · mm^m, $m = 0.145$, $G = 81\ 000$ MPa; relative cost is 2.6
- Surface treatment: unpeened
- End treatment: squared and ground
- Robust linearity: $\xi = 0.15$
- Set: use in as-wound condition
- Fatigue-safe: $n_f = 1.5$ using the Sines-Zimmerli fatigue-failure criterion
- Function: $F_{min} = 22$ N, $F_{max} = 90$ N, $y_{min} = 12$ mm, $y_{max} = 50$ mm, spring operates free (no rod or hole)
- Decision variable: wire size d

The figure of merit will be the cost of wire to wind the spring, Equation (10–22) without density. The design strategy will be to set wire size d, build a table, inspect the table, and choose the satisfactory spring with the highest figure of merit.

Set $d = 2.8$ mm. Then

$$F_a = \frac{90 - 22}{2} = 34 \text{ N} \qquad F_m = \frac{90 + 22}{2} = 56 \text{ N}$$

$$k = \frac{F_{max}}{y_{max}} = \frac{90}{50} = 1.8 \text{ N/mm}$$

$$S_{ut} = \frac{2211}{2.8^{0.145}} = 1904 \text{ MPa}$$

$$S_{su} = 0.67(1904) = 1276 \text{ MPa}$$

$$S_{sy} = 0.45(1904) = 857 \text{ MPa}$$

From Equation (10–28*a*), with the Sines criterion, $S_{se} = S_{sa} = 241$ MPa. Equation (10–23) can be used to determine C with S_{se}, n_f, and F_a in place of S_{sy}, n_s, and $(1 + \xi)F_{\text{max}}$, respectively. Thus,

$$\alpha = \frac{S_{se}}{n_f} = \frac{241}{1.5} = 161 \text{ MPa}$$

$$\beta = \frac{8F_a}{\pi d^2} = \frac{8(34)}{\pi(2.8^2)} = 11 \text{ MPa}$$

$$C = \frac{2(161) - 11}{4(11)} + \sqrt{\left[\frac{2(161) - 11}{4(11)}\right]^2 - \frac{3(161)}{4(11)}} = 13.22$$

$$D = Cd = 13.22(0.112) = 37 \text{ mm}$$

$$F_s = (1 + \xi)F_{\text{max}} = (1 + 0.15)90 = 103.5 \text{ N}$$

$$N_a = \frac{d^4G}{8D^3k} = \frac{0.112^4(81\ 000)}{8(37)^3 1.8} = 6.81 \text{ turns}$$

$$N_t = N_a + 2 = 6.81 + 2 = 8.81 \text{ turns}$$

$$L_s = dN_t = 0.112(8.81) = 24.7 \text{ mm}$$

$$L_0 = L_s + \frac{F_s}{k} = 24.7 + \frac{103.5}{1.8} = 82.2 \text{ mm}$$

$$\text{ID} = 37 - 2.8 = 34.2 \text{ mm}$$

$$\text{OD} = 37 + 2.8 = 39.8 \text{ mm}$$

$$y_s = L_0 - L_s = 82.2 - 24.7 = 57.5 \text{ mm}$$

$$(L_0)_{\text{cr}} < \frac{2.63D}{\alpha} = 2.63\,\frac{(37)}{0.5} = 195 \text{ mm}$$

$$K_B = \frac{4(13.22) + 2}{4(13.22) - 3} = 1.1$$

$$W = \frac{\pi^2 d^2 D N_a \gamma}{4} = \frac{\pi^2 2.8^2(37)6.81(0.0078)}{4(1000)} = 0.038 \text{ Kg}$$

$$f_n = 0.5\sqrt{\frac{386k}{W}} = 0.5\sqrt{\frac{1800}{0.038}} = 107.7 \text{ Hz}$$

$$\tau_a = K_B\,\frac{8F_aD}{\pi d^3} = 1.094\,\frac{8(34)37}{\pi 2.8^3} = 161 \text{ MPa}$$

$$\tau_m = \tau_a\,\frac{F_m}{F_a} = 161\,\frac{56}{34} = 265 \text{ MPa}$$

$$\tau_s = \tau_a \frac{F_s}{F_a} = 161 \frac{103.5}{34} = 490 \text{ MPa}$$

$$n_f = \frac{S_{sa}}{\tau_a} = \frac{241}{261} = 1.5$$

$$n_s = \frac{S_{sy}}{\tau_s} = \frac{857}{490} = 1.75$$

$$\text{fom} = -(\text{relative material cost})\pi^2 d^2 N_t D/4$$

$$= -2.6\pi^2(2.8^2)(8.81)37/4 = -16\ 409$$

Inspection of the results shows that all conditions are satisfied except for $4 \le C \le 12$. Repeat the process using the other available wire sizes and develop the following table:

d:	1.75	1.8	2	2.15	2.3	2.4	2.7	2.8
D	7.3	8.1	12.0	15.4	19.4	22.4	32.9	37.0
ID	5.5	6.3	10.0	13.3	17.1	20.0	30.2	34.2
OD	9.0	9.9	14.0	17.6	21.7	24.8	35.6	39.8
C	4.16	4.51	5.99	7.18	8.44	9.33	12.20	13.22
N_a	136.93	110.13	52.24	32.64	21.48	16.62	8.37	6.81
L_s	243.1	201.8	108.5	74.5	54.0	44.7	28.0	24.7
L_0	300.6	259.3	166.0	132.0	111.5	102.2	85.5	82.2
$(L_0)_{cr}$	38	43	63	81	102	118	173	195
n_f	1.5	1.5	1.5	1.5	1.5	1.5	1.5	1.5
n_s	1.88	1.87	1.84	1.82	1.80	1.79	1.76	1.75
f_n	86.744	88.999	95.739	99.269	101.992	103.483	106.887	107.761
FOM	−19 862	−18 934	−16 685	−15 864	−15 478	−15 410	−15 973	−16 409

The problem-specific inequality constraints are

$$L_s \le 27 \text{ mm}$$

$$L_0 \le 100 \text{ mm}$$

$$f_n \ge 5(20) = 100 \text{ H}_\text{Z}$$

The general constraints are

$$3 \le N_a \le 15$$

$$4 \le C \le 12$$

$$(L_0)_{cr} > L_0$$

We see that none of the diameters satisfy the given constraints. The 2.7 mm-diameter wire is the closest to satisfying all requirements. The value of $C = 12.14$ is not a serious deviation and can be tolerated. However, the tight constraint on L_s needs to be addressed. If the assembly conditions can be relaxed to accept a solid height of 28 mm, we have a solution. If not, the only other possibility is to use the 2.8 mm diameter and accept a value $C = 14$, individually package the springs, and possibly reconsider supporting the spring in service.

10–11 Extension Springs

Extension springs differ from compression springs in that they carry tensile loading, they require some means of transferring the load from the support to the body of the spring, and the spring body is wound with an initial tension. The load transfer can be done with a threaded plug or a swivel hook; both of these add to the cost of the finished product, and so one of the methods shown in Figure 10–5 is usually employed.

Stresses in the body of the extension spring are handled the same as compression springs. In designing a spring with a hook end, bending and torsion in the hook must be included in the analysis. In Figure 10–6*a* and *b* a commonly used method of designing the end is shown. The maximum tensile stress at *A*, due to bending and axial loading, is given by

$$\sigma_A = F\left[(K)_A \frac{16D}{\pi d^3} + \frac{4}{\pi d^2}\right] \tag{10–34}$$

where $(K)_A$ is a bending stress-correction factor for curvature, given by

$$(K)_A = \frac{4C_1^2 - C_1 - 1}{4C_1(C_1 - 1)} \qquad C_1 = \frac{2r_1}{d} \tag{10–35}$$

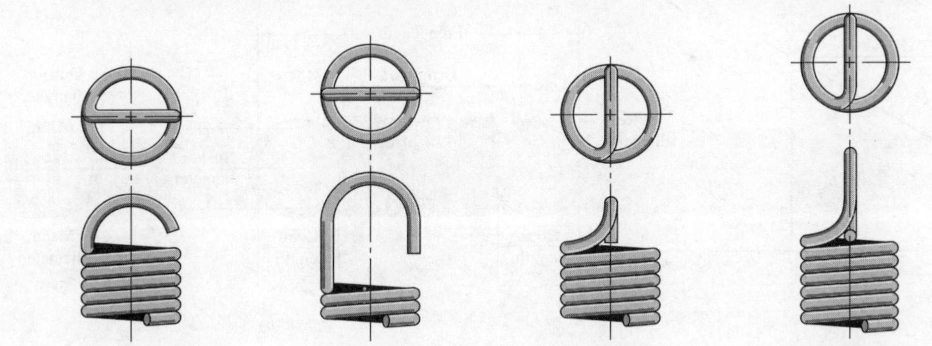

(*a*) Machine half loop–open (*b*) Raised hook (*c*) Short twisted loop (*d*) Full twisted loop

Figure 10–5

Types of ends used on extension springs. *(Source: Courtesy of Associated Spring.)*

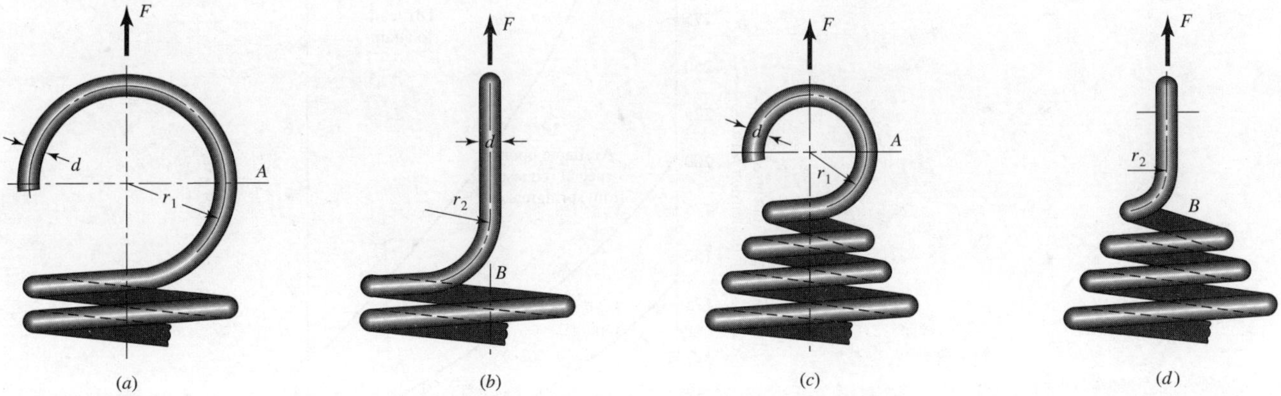

Figure 10–6

Ends for extension springs. (*a*) Usual design; stress at *A* is due to combined axial force and bending moment. (*b*) Side view of part *a*; stress is mostly torsion at *B*. (*c*) Improved design; stress at *A* is due to combined axial force and bending moment. (*d*) Side view of part *c*; stress at *B* is mostly torsion.

Note: Radius r_1 is in the plane of the end coil for curved beam bending stress. Radius r_2 is at a right angle to the end coil for torsional shear stress.

The maximum torsional stress at point B is given by

$$\tau_B = (K)_B \frac{8FD}{\pi d^3} \tag{10–36}$$

where the stress-correction factor for curvature, $(K)_B$, is

$$(K)_B = \frac{4C_2 - 1}{4C_2 - 4} \qquad C_2 = \frac{2r_2}{d} \tag{10–37}$$

Figures 10–6*c* and *d* show an improved design due to a reduced coil diameter.

When extension springs are made with coils in contact with one another, they are said to be *close-wound*. Spring manufacturers prefer some initial tension in close-wound springs in order to hold the free length more accurately. The corresponding load-deflection curve is shown in Figure 10–7*a*, where y is the extension beyond the free length L_0 and F_i is the initial tension in the spring that must be exceeded before the spring deflects. The load-deflection relation is then

$$F = F_i + ky \tag{10–38}$$

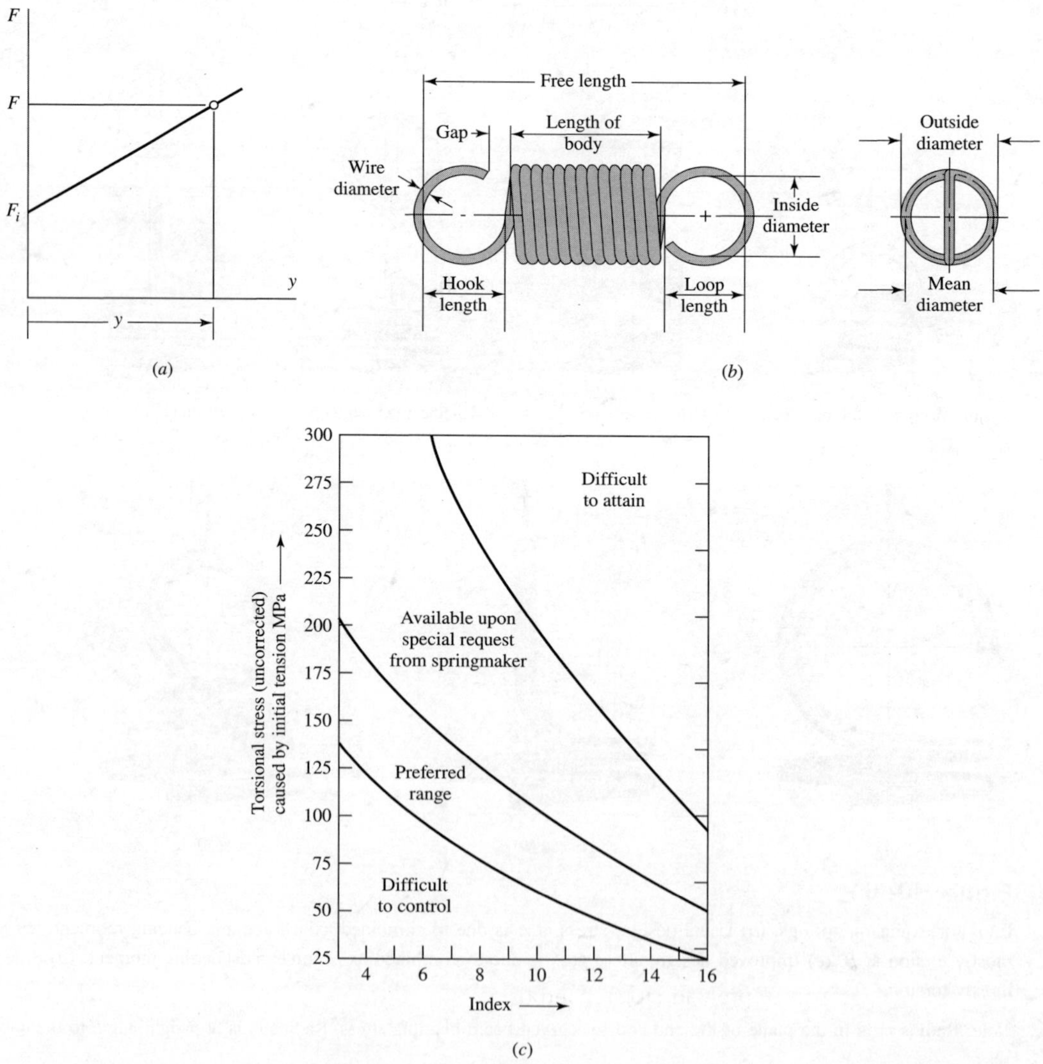

Figure 10–7

(*a*) Geometry of the force F and extension y curve of an extension spring; (*b*) geometry of the extension spring; and (*c*) torsional stresses due to initial tension as a function of spring index C in helical extension springs.

Table 10–7 Maximum Allowable Stresses (K_W or K_B corrected) for Helical Extension Springs in Static Applications

	Percent of Tensile Strength		
	In Torsion		In Bending
Materials	Body	End	End
Patented, cold-drawn or hardened and tempered carbon and low-alloy steels	45–50	40	75
Austenitic stainless steel and nonferrous alloys	35	30	55

This information is based on the following conditions: set not removed and low temperature heat treatment applied. For springs that require high initial tension, use the same percent of tensile strength as for end.

Source: Data from *Design Handbook,* 1987, p. 52.

where k is the spring rate. The free length L_0 of a spring measured inside the end loops or hooks as shown in Figure 10–7*b* can be expressed as

$$L_0 = 2(D - d) + (N_b + 1)d = (2C - 1 + N_b)d \tag{10–39}$$

where D is the mean coil diameter, N_b is the number of body coils, and C is the spring index. With ordinary twisted end loops as shown in Figure 10–7*b*, to account for the deflection of the loops in determining the spring rate k, the equivalent number of active helical turns N_a for use in Equation (10–9) is

$$N_a = N_b + \frac{G}{E} \tag{10–40}$$

where G and E are the shear and tensile moduli of elasticity, respectively (see Problem 10–37).

The initial tension in an extension spring is created in the winding process by twisting the wire as it is wound onto the mandrel. When the spring is completed and removed from the mandrel, the initial tension is locked in because the spring cannot get any shorter. The amount of initial tension that a springmaker can routinely incorporate is as shown in Figure 10–7*c*. The preferred range can be expressed in terms of the initial torsional stress τ_i (uncorrected for curvature) as

$$\tau_i = \frac{231}{\exp(0.105C)} \pm 6.9\left(4 - \frac{C-3}{6.5}\right) \text{MPa} \tag{10–41}$$

where C is the spring index.

Guidelines for the maximum allowable corrected stresses for static applications of extension springs are given in Table 10–7.

EXAMPLE 10–6

A hard-drawn steel wire extension spring has a wire diameter of 0.9 mm, an outside coil diameter of 6.3 mm, hook radii of $r_1 = 2.7$ mm and $r_2 = 2.3$ mm, and an initial tension of 5.5 N. The number of body turns is 12.17. From the given information:

(*a*) Determine the physical parameters of the spring.

(*b*) Check the initial preload stress conditions.

(*c*) Find the factors of safety under a static 24 N load.

Solution

(*a*)
$$D = \text{OD} - d = 6.3 - 0.9 = 5.4 \text{ mm}$$

$$C = \frac{D}{d} = \frac{5.4}{0.9} = 6$$

$$K_B = \frac{4C + 2}{4C - 3} = \frac{4(6) + 2}{4(6) - 3} = 1.238$$

Equation (10–40) and Table 10–5:

$$N_a = N_b + G/E = 12.17 + 80/197.9 = 12.57 \text{ turns}$$

Equation (10–9):
$$k = \frac{d^4 G}{8D^3 N_a} = \frac{0.9^4(80\ 000)}{8(5.4^3)12.57} = 3.31 \text{ N/mm}$$

Equation (10–39):
$$L_0 = (2C - 1 + N_b)d = [2(6) - 1 + 12.17]0.9 = 20.85 \text{ mm}$$

The deflection under the service load is

$$y_{max} = \frac{F_{max} - F_i}{k} = \frac{24 - 5.5}{3.31} = 5.59 \text{ mm}$$

where the spring length becomes $L = L_0 + y = 20.85 + 5.59 = 26.44$ mm.
(*b*) The uncorrected initial stress is given by Equation (10–2) without the correction factor. That is,

$$(\tau_i)_{uncorr} = \frac{8F_i D}{\pi d^3} = \frac{8(5.5)5.4}{\pi(0.9^3)} = 105.6 \text{ MPa}$$

The preferred range is given by Equation (10–41) and for this case is

$$\begin{aligned}(\tau_i)_{pref} &= \frac{231}{\exp(0.105C)} \pm 6.9\left(4 - \frac{C - 3}{6.5}\right) \\ &= \frac{231}{\exp[0.105(6)]} \pm 6.9\left(4 - \frac{6 - 3}{6.5}\right) \\ &= 123 \pm 24.5 = 147.5,\ 98.5 \text{ MPa}\end{aligned}$$

Answer

Thus, the initial tension of 105.6 MPa is in the preferred range.
(*c*) For hard-drawn wire, Table 10–4 gives $m = 0.190$ and $A = 1783 \text{ MPa} \cdot \text{mm}^m$. From Equation (10–14)

$$S_{ut} = \frac{A}{d^m} = \frac{1783}{0.9^{0.190}} = 1819 \text{ MPa}$$

For torsional shear in the main body of the spring, from Table 10–7,

$$S_{sy} = 0.45S_{ut} = 0.45(1819) = 818.5 \text{ MPa}$$

The shear stress under the service load is

$$\tau_{max} = \frac{8K_B F_{max} D}{\pi d^3} = \frac{8(1.238)24(5.4)}{\pi(0.9^3)} = 560 \text{ MPa}$$

Thus, the factor of safety is

Answer
$$n = \frac{S_{sy}}{\tau_{\max}} = \frac{818.5}{560} = 1.45$$

For the end-hook bending at A,

$$C_1 = 2r_1/d = 2(2.7)/0.9 = 6$$

From Equation (10–35)

$$(K)_A = \frac{4C_1^2 - C_1 - 1}{4C_1(C_1 - 1)} = \frac{4(6^2) - 6 - 1}{4(6)(6 - 1)} = 1.14$$

From Equation (10–34)

$$\sigma_A = F_{\max}\left[(K)_A \frac{16D}{\pi d^3} + \frac{4}{\pi d^2}\right]$$
$$= 5.25\left[1.14\frac{16(5.4)}{\pi(0.9^3)} + \frac{4}{\pi(0.9^2)}\right] = 1070 \text{ MPa}$$

The yield strength, from Table 10–7, is given by

$$S_y = 0.75S_{ut} = 0.75(1819) = 1364 \text{ MPa}$$

The factor of safety for end-hook bending at A is then

Answer
$$n_A = \frac{S_y}{\sigma_A} = \frac{1819}{1364} = 1.33$$

For the end-hook in torsion at B, from Equation (10–37)

$$C_2 = 2r_2/d = 2(2.3)/0.9 = 5.11$$

$$(K)_B = \frac{4C_2 - 1}{4C_2 - 4} = \frac{4(5.11) - 1}{4(5.11) - 4} = 1.18$$

and the corresponding stress, given by Equation (10–36), is

$$\tau_B = (K)_B \frac{8F_{\max}D}{\pi d^3} = 1.18\frac{8(24)5.4}{\pi(0.9^3)} = 535 \text{ MPa}$$

Using Table 10–7 for yield strength, the factor of safety for end-hook torsion at B is

Answer
$$n_B = \frac{(S_{sy})_B}{\tau_B} = \frac{0.4(1819)}{535} = 1.35$$

Yield due to bending of the end hook will occur first.

Next, let us consider a fatigue problem.

EXAMPLE 10–7

The helical coil extension spring of Example 10–6 is subjected to a dynamic loading from 7 to 23 N. Estimate the factors of safety using the Gerber failure criterion for (*a*) coil fatigue, (*b*) coil yielding, (*c*) end-hook bending fatigue at point *A* of Figure 10–6*a*, and (*d*) end-hook torsional fatigue at point *B* of Figure 10–6*b*.

Solution

A number of quantities are the same as in Example 10–6: $d = 0.9$ mm, $S_{ut} = 1819$ MPa, $D = 5.4$ mm, $r_1 = 2.7$ mm, $C = 6$, $K_B = 1.238$, $(K)_A = 1.14$, $(K)_B = 1.182$, $N_b = 12.17$ turns, $L_0 = 21.95$ mm, $k = 3.31$ N/mm, $F_i = 5.5$ N, and $(\tau_i)_{\text{uncorr}} = 105.6$ MPa. Then

$$F_a = (F_{\text{max}} - F_{\text{min}})/2 = (23 - 7)/2 = 8 \text{ N}$$

$$F_m = (F_{\text{max}} + F_{\text{min}})/2 = (23 + 7)/2 = 15 \text{ N}$$

The strengths from Example 10–6 include $S_{ut} = 1819$ MPa, $S_y = 1364$ MPa, and $S_{sy} = 818.5$ MPa. The ultimate shear strength is estimated from Equation (10–30) as

$$S_{su} = 0.67 S_{ut} = 0.67(1819) = 1219 \text{ MPa}$$

(*a*) Body-coil fatigue:

$$\tau_a = \frac{8K_B F_a D}{\pi d^3} = \frac{8(1238)8(5.4)}{\pi(0.9^3)} = 163.5 \text{ MPa}$$

$$\tau_m = \frac{F_m}{F_a}\tau_a = \frac{15}{8}163.5 = 306.5 \text{ MPa}$$

Using the Zimmerli data of Equation (10–28*a*) with the Gerber criterion of Equation (10–29*b*) gives

$$S_{se} = \frac{S_{su}}{1 - \left(\dfrac{S_{sm}}{S_{su}}\right)^2} = \frac{241}{1 - \left(\dfrac{379}{1219}\right)^2} = 266.8 \text{ MPa}$$

From Equation (6–48), the Gerber fatigue criterion for shear is

Answer

$$(n_f)_{\text{body}} = \frac{1}{2}\left(\frac{S_{su}}{\tau_m}\right)^2 \frac{\tau_a}{S_{se}}\left[-1 + \sqrt{1 + \left(2\frac{\tau_m}{S_{su}}\frac{S_{se}}{\tau_a}\right)^2}\,\right]$$

$$= \frac{1}{2}\left(\frac{1219}{306.5}\right)^2 \frac{163.5}{266.8}\left[-1 + \sqrt{1 + \left(2\frac{306.5}{1219}\frac{266.8}{163.5}\right)^2}\,\right] = 1.42$$

(*b*) The load-line for the coil body begins at $S_{sm} = \tau_i$ and has a slope $r = \tau_a/(\tau_m - \tau_i)$. It can be shown that the intersection with the yield line is given by $(S_{sa})_y = [r/(r + 1)](S_{sy} - \tau_i)$. Consequently, $\tau_i = (F_i/F_a)\tau_a = (5.5/8)163.5 = 112.4$ MPa, $r = 163.5/(306.5 - 112.4) = 0.850$, and

$$(S_{sa})_y = \frac{0.850}{0.850 + 1}(818.5 - 112.4) = 324.4 \text{ MPa}$$

Thus,

Answer $$(n_y)_{\text{body}} = \frac{(S_{sa})_y}{\tau_a} = \frac{324.4}{163.5} = 1.98$$

(*c*) End-hook bending fatigue: using Equations (10–34) and (10–35) gives

$$\sigma_a = F_a \left[(K)_A \frac{16D}{\pi d^3} + \frac{4}{\pi d^2} \right]$$

$$= 8 \left[1.14 \frac{16(5.4)}{\pi(0.9^3)} + \frac{4}{\pi(0.9^2)} \right] = 349 \text{ MPa}$$

$$\sigma_m = \frac{F_m}{F_a} \sigma_a = \frac{15}{8} 349 = 655 \text{ MPa}$$

To estimate the tensile endurance limit using the distortion-energy theory,

$$S_e = S_{se}/0.577 = 266.8/0.577 = 462.4 \text{ MPa}$$

Using the Gerber criterion for tension from Equation (6–48) gives

Answer $$(n_f)_A = \frac{1}{2}\left(\frac{S_{ut}}{\sigma_m}\right)^2 \frac{\sigma_a}{S_e}\left[-1 + \sqrt{1 + \left(2\frac{\sigma_m}{S_{ut}}\frac{S_e}{\sigma_a}\right)^2}\right]$$

$$= \frac{1}{2}\left(\frac{1819}{655}\right)^2 \frac{349}{462.4}\left[-1 + \sqrt{1 + \left(2\frac{655}{1819}\frac{462.4}{349}\right)^2}\right] = 1.11$$

(*d*) End-hook torsional fatigue: from Equation (10–36)

$$(\tau_a)_B = (K)_B \frac{8F_a D}{\pi d^3} = 1.18 \frac{8(8)5.4}{\pi(0.9^3)} = 178.3 \text{ MPa}$$

$$(\tau_m)_B = \frac{F_m}{F_a}(\tau_a)_B = \frac{15}{8} 178.3 = 335 \text{ MPa}$$

Then, again using the Gerber criterion, we obtain

Answer $$(n_f)_B = \frac{1}{2}\left(\frac{S_{su}}{\tau_m}\right)^2 \frac{\tau_a}{S_{se}}\left[-1 + \sqrt{1 + \left(2\frac{\tau_m}{S_{su}}\frac{S_{se}}{\tau_a}\right)^2}\right]$$

$$= \frac{1}{2}\left(\frac{1219}{335}\right)^2 \frac{178.3}{266.8}\left[-1 + \sqrt{1 + \left(2\frac{335}{1219}\frac{266.8}{178.3}\right)^2}\right] = 1.30$$

The analyses in Examples 10–6 and 10–7 show how extension springs differ from compression springs. The end hooks are usually the weakest part, with bending usually controlling. We should also appreciate that a fatigue failure separates the extension spring under load. Flying fragments, lost load, and machine shutdown are threats to personal safety as well as machine function. For these reasons higher design factors are used in extension-spring design than in the design of compression springs.

In Example 10–7 we estimated the endurance limit for the hook in bending using the Zimmerli data, which are based on torsion in compression springs and the distortion theory. An alternative method is to use Table 10–8, which is based on a stress-ratio of $R = \tau_{\text{min}}/\tau_{\text{max}} = 0$. For this case, $\tau_a = \tau_m = \tau_{\text{max}}/2$. Label the strength values

Table 10–8 Maximum Allowable Stresses for ASTM A228 and Type 302 Stainless Steel Helical Extension Springs in Cyclic Applications

Number of Cycles	Percent of Tensile Strength		
	In Torsion		In Bending
	Body	End	End
10^5	36	34	51
10^6	33	30	47
10^7	30	28	45

This information is based on the following conditions: not shot-peened, no surging and ambient environment with a low temperature heat treatment applied. Stress ratio = 0.

Source: Data from *Design Handbook,* 1987, p. 52.

of Table 10–8 as S_r for bending or S_{sr} for torsion. Then for torsion, for example, $S_{sa} = S_{sm} = S_{sr}/2$. From the Gerber criterion, Equation (10–29*b*), the endurance limit is

$$S_{se} = \frac{S_{sa}}{1 - (S_{sm}/S_{su})^2} = \frac{S_{sr}/2}{1 - \left(\dfrac{S_{sr}/2}{S_{su}}\right)^2} \tag{10–42}$$

So in Example 10–7 an estimate for the bending endurance limit from Table 10–8 would be

$$S_r = 0.45S_{ut} = 0.45(1819) = 818.55 \text{ MPa}$$

and from Equation (10–42)

$$S_e = \frac{S_r/2}{1 - [S_r/(2S_{ut})]^2} = \frac{818.55/2}{1 - \left(\dfrac{818.55/2}{1819}\right)^2} = 431 \text{ MPa}$$

Using this in place of 462.4 MPa in Example 10–7 results in $(n_f)_A = 1.03$, a reduction of 5 percent.

10–12 Helical Coil Torsion Springs

When a helical coil spring is subjected to end torsion, it is called a *torsion spring.* It is usually close-wound, as is a helical coil extension spring, but with negligible initial tension. There are single-bodied and double-bodied types as depicted in Figure 10–8. As shown in the figure, torsion springs have ends configured to apply torsion to the coil body in a convenient manner, with short hook, hinged straight offset, straight torsion, and special ends. The ends ultimately connect a force at a distance from the coil axis to apply a torque. The most frequently encountered (and least expensive) end is the straight torsion end. If intercoil friction is to be avoided completely, the spring can be wound with a pitch that just separates the body coils. Helical coil torsion springs are usually used with a rod or arbor for reactive support when ends cannot be built in, to maintain alignment, and to provide buckling resistance if necessary.

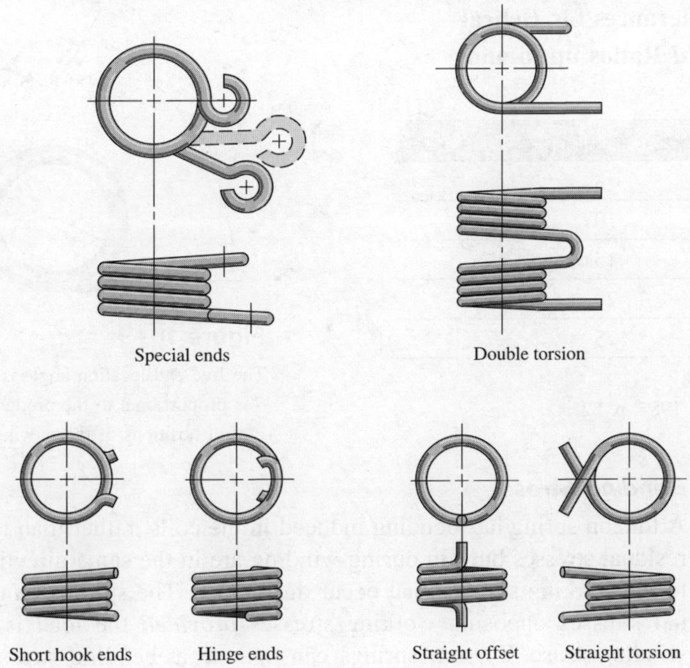

Figure 10–8

Torsion springs. *(Source: Courtesy of Associated Spring.)*

The wire in a torsion spring is in bending, in contrast to the torsion encountered in helical coil compression and extension springs. The springs are designed to wind tighter in service. As the applied torque increases, the inside diameter of the coil decreases. Care must be taken so that the coils do not interfere with the pin, rod, or arbor. The bending mode in the coil might seem to invite square- or rectangular-cross-section wire, but cost, range of materials, and availability discourage its use.

Torsion springs are familiar in clothespins, window shades, and animal traps, where they may be seen around the house, and out of sight in counterbalance mechanisms, ratchets, and a variety of other machine components. There are many stock springs that can be purchased off-the-shelf from a vendor. This selection can add economy of scale to small projects, avoiding the cost of custom design and small-run manufacture.

Describing the End Location

In specifying a torsion spring, the ends must be located relative to each other. Commercial tolerances on these relative positions are listed in Table 10–9. The simplest scheme for expressing the initial unloaded location of one end with respect to the other is in terms of an angle β defining the partial turn present in the coil body as $N_p = \beta/360°$, as shown in Figure 10–9. For analysis purposes the nomenclature of Figure 10–9 can be used. Communication with a springmaker is often in terms of the back-angle α.

The number of body turns N_b is the number of turns in the free spring body by count. The body-turn count is related to the initial position angle β by

$$N_b = \text{integer} + \frac{\beta}{360°} = \text{integer} + N_p$$

where N_p is the number of partial turns. The above equation means that N_b takes on noninteger, discrete values such as 5.3, 6.3, 7.3, . . . , with successive differences of 1 as possibilities in designing a specific spring. This consideration will be discussed later.

Table 10–9 End Position Tolerances for Helical Coil Torsion Springs (for D/d Ratios up to and Including 16)

Total Coils	Tolerance: ±Degrees*
Up to 3	8
Over 3–10	10
Over 10–20	15
Over 20–30	20
Over 30	25

*Closer tolerances available on request.

Source: Data from *Design Handbook,* 1987, p. 52.

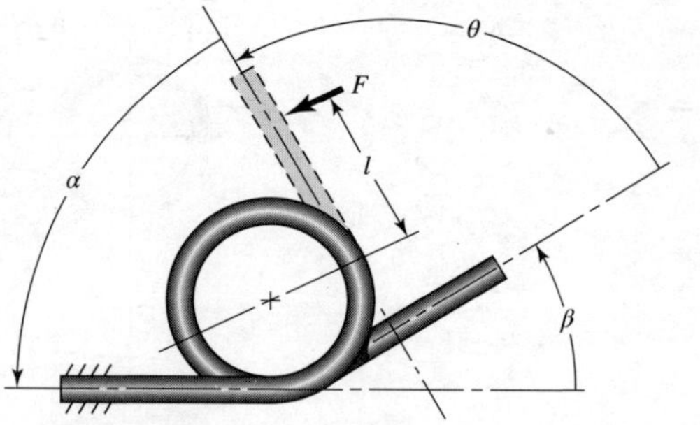

Figure 10–9

The free-end location angle is β. The rotational coordinate θ is proportional to the product Fl. Its back angle is α. For all positions of the moving end $\theta + \alpha = \Sigma =$ constant.

Bending Stress

A torsion spring has bending induced in the coils, rather than torsion. This means that residual stresses built in during winding are in the same direction but of opposite sign to the working stresses that occur during use. The strain-strengthening locks in residual stresses opposing working stresses *provided* the load is always applied in the winding sense. Torsion springs can operate at bending stresses exceeding the yield strength of the wire from which it was wound.

The bending stress can be obtained from curved-beam theory expressed in the form

$$\sigma = K\frac{Mc}{I}$$

where K is a stress-correction factor. The value of K depends on the shape of the wire cross section and whether the stress sought is at the inner or outer fiber. Wahl analytically determined the values of K to be, for round wire,

$$K_i = \frac{4C^2 - C - 1}{4C(C-1)} \qquad K_o = \frac{4C^2 + C - 1}{4C(C+1)} \tag{10–43}$$

where C is the spring index and the subscripts i and o refer to the inner and outer fibers, respectively. In view of the fact that K_o is always less than unity, we shall use K_i to estimate the stresses. When the bending moment is $M = Fr$ and the section modulus $I/c = d^3/32$, we express the bending equation as

$$\sigma = K_i\frac{32Fr}{\pi d^3} \tag{10–44}$$

which gives the bending stress for a round-wire torsion spring.

Deflection and Spring Rate

For torsion springs, angular deflection can be expressed in radians or revolutions (turns). If a term contains revolution units the term will be expressed with a prime sign. The spring rate k' is expressed in units of torque/revolution (N · mm/rev) and moment is proportional to angle θ' expressed in turns rather than radians. The spring rate, if linear, can be expressed as

$$k' = \frac{M_1}{\theta'_1} = \frac{M_2}{\theta'_2} = \frac{M_2 - M_1}{\theta'_2 - \theta'_1} \tag{10–45}$$

where the moment M can be expressed as Fl or Fr.

The angle subtended by the end deflection of a cantilever, when viewed from the built-in ends, is y/l rad. From Table A–9–1,

$$\theta_e = \frac{y}{l} = \frac{Fl^2}{3EI} = \frac{Fl^2}{3E(\pi d^4/64)} = \frac{64Ml}{3\pi d^4 E} \tag{10–46}$$

For a straight torsion end spring, end corrections such as Equation (10–46) must be added to the body-coil deflection. The strain energy in bending is, from Equation (4–24),

$$U = \int \frac{M^2\,dx}{2EI}$$

For a torsion spring, $M = Fl = Fr$, and integration must be accomplished over the length of the body-coil wire. The force F will deflect through a distance $r\theta$ where θ is the angular deflection of the coil body, in radians. Applying Castigliano's theorem gives

$$r\theta = \frac{\partial U}{\partial F} = \int_0^{\pi D N_b} \frac{\partial}{\partial F}\left(\frac{F^2 r^2\,dx}{2EI}\right) = \int_0^{\pi D N_b} \frac{Fr^2\,dx}{EI}$$

Substituting $I = \pi d^4/64$ for round wire and solving for θ gives

$$\theta = \frac{64FrDN_b}{d^4 E} = \frac{64MDN_b}{d^4 E}$$

The total angular deflection in radians is obtained by adding Equation (10–46) for each end of lengths l_1, l_2:

$$\theta_t = \frac{64MDN_b}{d^4 E} + \frac{64Ml_1}{3\pi d^4 E} + \frac{64Ml_2}{3\pi d^4 E} = \frac{64MD}{d^4 E}\left(N_b + \frac{l_1 + l_2}{3\pi D}\right) \tag{10–47}$$

The equivalent number of active turns N_a is expressed as

$$N_a = N_b + \frac{l_1 + l_2}{3\pi D} \tag{10–48}$$

The spring rate k in torque per radian is

$$k = \frac{Fr}{\theta_t} = \frac{M}{\theta_t} = \frac{d^4 E}{64DN_a} \tag{10–49}$$

The spring rate may also be expressed as torque per turn. The expression for this is obtained by multiplying Equation (10–49) by 2π rad/turn. Thus spring rate k' (units torque/turn) is

$$k' = \frac{2\pi d^4 E}{64DN_a} = \frac{d^4 E}{10.2DN_a} \tag{10–50}$$

Tests show that the effect of friction between the coils and arbor is such that the constant 10.2 should be increased to 10.8. The equation above becomes

$$k' = \frac{d^4 E}{10.8DN_a} \tag{10–51}$$

(units torque per turn). Equation (10–51) gives better results. Also Equation (10–47) becomes

$$\theta_t' = \frac{10.8MD}{d^4 E}\left(N_b + \frac{l_1 + l_2}{3\pi D}\right) \tag{10–52}$$

Torsion springs are frequently used over a round bar or pin. When the load is applied to a torsion spring, the spring winds up, causing a decrease in the inside diameter of the coil body. It is necessary to ensure that the inside diameter of the coil never becomes equal to or less than the diameter of the pin, in which case loss of spring function would ensue. The helix diameter of the coil D' becomes

$$D' = \frac{N_b D}{N_b + \theta'_c} \tag{10–53}$$

where θ'_c is the angular deflection of the body of the coil in number of turns, given by

$$\theta'_c = \frac{10.8 M D N_b}{d^4 E} \tag{10–54}$$

The new inside diameter $D'_i = D' - d$ makes the diametral clearance Δ between the body coil and the pin of diameter D_p equal to

$$\Delta = D' - d - D_p = \frac{N_b D}{N_b + \theta'_c} - d - D_p \tag{10–55}$$

Equation (10–55) solved for N_b is

$$N_b = \frac{\theta'_c(\Delta + d + D_p)}{D - \Delta - d - D_p} \tag{10–56}$$

which gives the number of body turns corresponding to a specified diametral clearance of the arbor. This angle may not be in agreement with the necessary partial-turn remainder. Thus the diametral clearance may be exceeded but not equaled.

Static Strength

First column entries in Table 10–6 can be divided by 0.577 (from distortion-energy theory) to give

$$S_y = \begin{cases} 0.78\ S_{ut} & \text{Music wire and cold-drawn carbon steels} \\ 0.87\ S_{ut} & \text{OQ\&T carbon and low-alloy steels} \\ 0.61\ S_{ut} & \text{Austenitic stainless steel and nonferrous alloys} \end{cases} \tag{10–57}$$

Fatigue Strength

Since the spring wire is in bending, the Sines equation is not applicable. The Sines model is in the presence of pure torsion. Since Zimmerli's results were for compression springs (wire in pure torsion), we will use the repeated bending stress ($R = 0$) values provided by Associated Spring in Table 10–10. As in Equation (10–40) we will use the Gerber fatigue-failure criterion incorporating the Associated Spring $R = 0$ fatigue strength S_r:

$$S_e = \frac{S_r/2}{1 - \left(\dfrac{S_r/2}{S_{ut}}\right)^2} \tag{10–58}$$

Table 10–10 Maximum Recommended Bending Stresses (K_B Corrected) for Helical Torsion Springs in Cyclic Applications as Percent of S_{ut}

Fatigue Life, Cycles	ASTM A228 and Type 302 Stainless Steel		ASTM A230 and A232	
	Not Shot-Peened	Shot-Peened*	Not Shot-Peened	Shot-Peened*
10^5	53	62	55	64
10^6	50	60	53	62

This information is based on the following conditions: no surging, springs are in the "as-stress-relieved" condition.

*Not always possible.

Source: Data from Associated Spring.

The value of S_r (and S_e) has been corrected for size, surface condition, and type of loading, but not for temperature or miscellaneous effects.

The Gerber fatigue failure criterion from Equation (6–48) is applicable, and is repeated here:

$$n_f = \frac{1}{2}\frac{\sigma_a}{S_e}\left(\frac{S_{ut}}{\sigma_m}\right)^2\left[-1 + \sqrt{1 + \left(2\frac{\sigma_m}{S_{ut}}\frac{S_e}{\sigma_a}\right)^2}\,\right] \tag{10–59}$$

EXAMPLE 10–8

A stock spring is shown in Figure 10–10. It is made from 1.8 mm-diameter music wire and has 4.25 body turns with straight torsion ends. It works over a pin of 10 mm diameter. The coil outside diameter is 15 mm.
(*a*) Find the maximum operating torque and corresponding rotation for static loading.
(*b*) Estimate the inside coil diameter and pin diametral clearance when the spring is subjected to the torque in part (*a*).
(*c*) Estimate the fatigue factor of safety n_f if the applied moment varies between $M_{min} = 110$ to $M_{max} = 560$ N · mm.

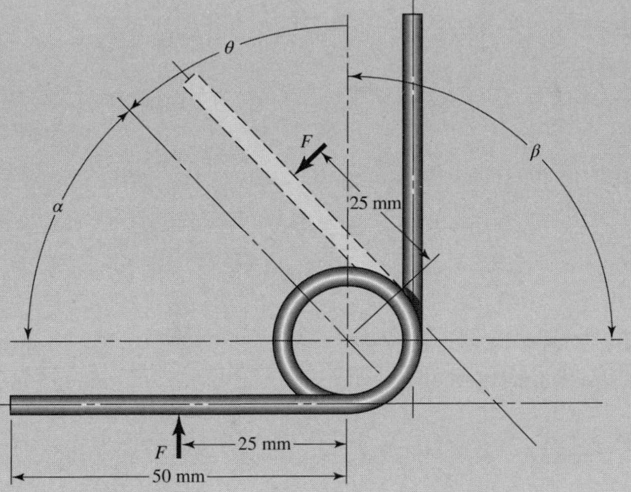

Figure 10–10

Angles α, β, and θ are measured between the straight-end centerline translated to the coil axis. Coil OD is 15 mm.

Solution

(*a*) For music wire, from Table 10–4 we find that $A = 2211$ MPa · mm^m and $m = 0.145$. Therefore,

$$S_{ut} = \frac{A}{d^m} = \frac{2211}{(1.8)^{0.145}} = 2030 \text{ MPa}$$

Using Equation (10–57) gives

$$S_y = 0.78 S_{ut} = 0.78(2030) = 1583 \text{ MPa}$$

The mean coil diameter is $D = 15 - 1.8 = 13.2$ mm. The spring index $C = D/d = 13.2/1.8 = 7.33$. The bending stress-correction factor k_i from Equation (10–43), is

$$K_i = \frac{4(7.33)^2 - 7.33 - 1}{4(7.33)(7.33 - 1)} = 1.113$$

Now rearrange Equation (10–44), substitute S_y for σ, and solve for the maximum torque Fr to obtain

$$M_{\max} = (Fr)_{\max} = \frac{\pi d^3 S_y}{32\,K_i} = \frac{\pi(1.8)^3 1583}{32(1.113)} = 814.3 \text{ N} \cdot \text{mm}$$

Note that no factor of safety has been used. Next, from Equation (10–54) and Table 10–5, the number of turns of the coil body θ'_c is

$$\theta'_c = \frac{10.8 MDN_b}{d^4 E} = \frac{10.8(814.3)13.2(4.25)}{1.8^4(196\ 500)} = 0.239 \text{ turn}$$

Answer

$$(\theta'_c)_{\deg} = 0.239(360°) = 86°$$

The active number of turns N_a, from Equation (10–48), is

$$N_a = N_b + \frac{l_1 + l_2}{3\pi D} = 4.25 + \frac{25 + 25}{3\pi(13.2)} = 4.652 \text{ turns}$$

The spring rate of the complete spring, from Equation (10–51), is

$$k' = \frac{1.8^4(196\ 500)}{10.8(13.2)4.652} = 3110 \text{ N} \cdot \text{mm/turn}$$

The number of turns of the complete spring θ' is

$$\theta' = \frac{M}{k'} = \frac{814}{3110} = 0.262 \text{ turn}$$

Answer

$$(\theta'_s)_{\deg} = 0.262(360°) = 94.22°$$

(*b*) With no load, the mean coil diameter of the spring is 13.2 mm. From Equation (10–53),

$$D' = \frac{N_b D}{N_b + \theta'_c} = \frac{4.25(13.2)}{4.25 + 0.239} = 12.5 \text{ mm}$$

The diametral clearance between the inside of the spring coil and the pin at load is

Answer $$\Delta = D' - d - D_p = 12.5 - 1.8 - 10 = 0.7 \text{ mm}$$

(*c*) Fatigue:

$$M_a = (M_{\text{max}} - M_{\text{min}})/2 = (560 - 110)/2 = 225 \text{ N} \cdot \text{mm}$$

$$M_m = (M_{\text{max}} + M_{\text{min}})/2 = (560 + 110)/2 = 335 \text{ N} \cdot \text{mm}$$

$$r = \frac{M_a}{M_m} = \frac{225}{335}$$

$$\sigma_a = K_i \frac{32M_a}{\pi d^3} = 1.113 \frac{32(225)}{\pi 1.8^3} = 404 \text{ MPa}$$

$$\sigma_m = \frac{M_m}{M_a}\sigma_a = \frac{335}{225}(404) = 601 \text{ MPa}$$

From Table 10–10, $S_r = 0.50S_{ut} = 0.50(2030) = 1015$ MPa. Then

$$S_e = \frac{1015/2}{1 - \left(\dfrac{1015/2}{2030}\right)^2} = 541 \text{ MPa}$$

Applying the Gerber criterion with Equation (10–59),

Answer $$n_f = \frac{1}{2}\left(\frac{S_{ut}}{\sigma_m}\right)^2\left(\frac{\sigma_a}{S_e}\right)\left[-1 + \sqrt{1 + \left(\frac{2\sigma_m S_e}{S_{ut}\sigma_a}\right)^2}\,\right]$$

$$= \frac{1}{2}\left(\frac{2030}{601}\right)^2\left(\frac{404}{541}\right)\left[-1 + \sqrt{1 + \left(\frac{2(601)(541)}{(2030)(404)}\right)^2}\,\right] = 1.165$$

10–13 Belleville Springs

The inset of Figure 10–11 shows the cross-section of a coned-disk spring, commonly called a *Belleville spring*. Although the mathematical treatment is beyond the scope of this book, you should at least become familiar with the remarkable characteristics of these springs.

Aside from the obvious advantage that a Belleville spring occupies only a small space, variation in the h/t ratio will produce a wide variety of load-deflection curve shapes, as illustrated in Figure 10–11. For example, using an h/t ratio of 2.83 or larger gives an S curve that might be useful for snap-acting mechanisms. A reduction of the ratio to a value between 1.41 and 2.1 causes the central portion of the curve to become horizontal, which means that the load is constant over a considerable deflection range.

A higher load for a given deflection may be obtained by nesting, that is, by stacking the springs in parallel. On the other hand, stacking in series provides a larger deflection for the same load, but in this case there is danger of instability.

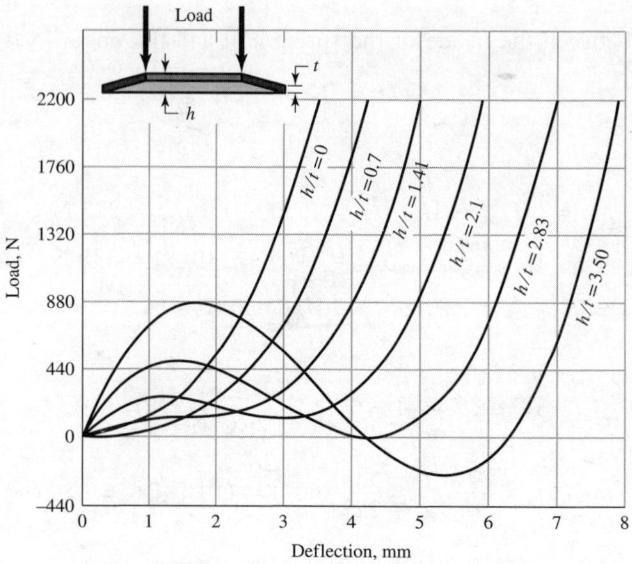

Figure 10–11

Load-deflection curves for Belleville springs. *(Source: Courtesy of Associated Spring.)*

10–14 Miscellaneous Springs

The extension spring shown in Figure 10–12 is made of slightly curved strip steel, not flat, so that the force required to uncoil it remains constant; thus it is called a *constant-force spring*. This is equivalent to a zero spring rate. Such springs can also be manufactured having either a positive or a negative spring rate.

A *volute spring,* shown in Figure 10–13*a*, is a wide, thin strip, or "flat," of material wound on the flat so that the coils fit inside one another. Since the coils do not stack, the solid height of the spring is the width of the strip. A variable-spring scale, in a compression volute spring, is obtained by permitting the coils to contact the support. Thus, as the deflection increases, the number of active coils decreases. The volute spring has another important advantage that cannot be obtained with round-wire springs: if the coils are wound so as to contact or slide on one another during action, the sliding friction will serve to damp out vibrations or other unwanted transient disturbances.

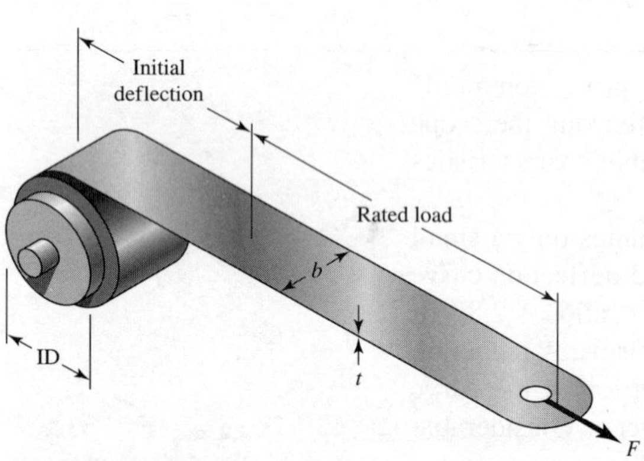

Figure 10–12

Constant-force spring. *(Source: Courtesy of Vulcan Spring & Mfg. Co. Telford, PA, www.vulcanspring.com.)*

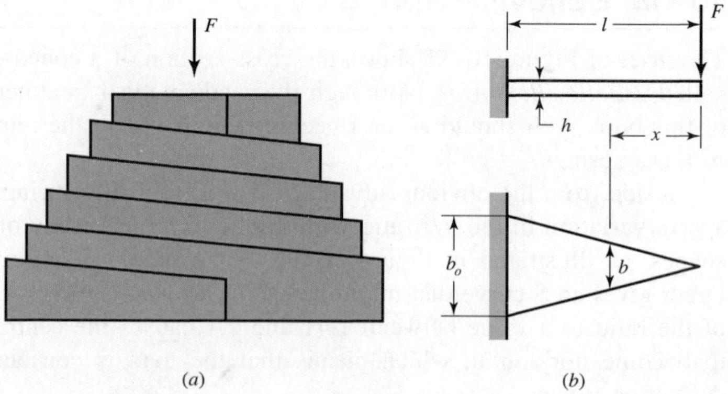

Figure 10–13

(*a*) A volute spring; (*b*) a flat triangular spring.

A *conical spring,* as the name implies, is a coil spring wound in the shape of a cone (see Problem 10–28). Most conical springs are compression springs and are wound with round wire. But a volute spring is a conical spring too. Probably the principal advantage of this type of spring is that it can be wound so that the solid height is only a single wire diameter.

Flat stock is used for a great variety of springs, such as clock springs, power springs, torsion springs, cantilever springs, and hair springs; frequently it is specially shaped to create certain spring actions for fuse clips, relay springs, spring washers, snap rings, and retainers.

In designing many springs of flat stock or strip material, it is often economical and of value to proportion the material so as to obtain a constant stress throughout the spring material. A uniform-section cantilever spring has a stress

$$\sigma = \frac{M}{I/c} = \frac{Fx}{I/c} \tag{a}$$

which is proportional to the distance x if I/c is a constant. But there is no reason why I/c need be a constant. For example, one might design such a spring as that shown in Figure 10–13*b*, in which the thickness h is constant but the width b is permitted to vary. Since, for a rectangular section, $I/c = bh^2/6$, we have, from Equation (*a*),

$$\frac{bh^2}{6} = \frac{Fx}{\sigma}$$

or

$$b = \frac{6Fx}{h^2\sigma} \tag{b}$$

Since b is linearly related to x, the width b_o at the base of the spring is

$$b_o = \frac{6Fl}{h^2\sigma} \tag{10–60}$$

Good approximations for deflections can be found easily by using Castigliano's theorem. To demonstrate this, assume that deflection of the triangular flat spring is primarily due to bending and we can neglect the transverse shear force.[14] The bending moment as a function of x is $M = -Fx$ and the beam width at x can be expressed as $b = b_o x/l$. Thus, the deflection of F is given by Equation (4–31), as

$$y = \int_0^l \frac{M(\partial M/\partial F)}{EI}\,dx = \frac{1}{E}\int_0^l \frac{-Fx(-x)}{\frac{1}{12}(b_o x/l)h^3}\,dx$$

$$= \frac{12Fl}{b_o h^3 E}\int_0^l x\,dx = \frac{6Fl^3}{b_o h^3 E} \tag{10–61}$$

Thus the spring constant, $k = F/y$, is estimated as

$$k = \frac{b_o h^3 E}{6l^3} \tag{10–62}$$

[14]Note that, because of shear, the width of the beam cannot be zero at $x = 0$. So, there is already some simplification in the design model. All of this can be accounted for in a more sophisticated model.

The methods of stress and deflection analysis illustrated in previous sections of this chapter have served to illustrate that springs may be analyzed and designed by using the fundamentals discussed in the earlier chapters of this book. This is also true for most of the miscellaneous springs mentioned in this section, and you should now experience no difficulty in reading and understanding the literature of such springs.

10–15 Summary

In this chapter we have considered helical coil springs in considerable detail in order to show the importance of viewpoint in approaching engineering problems, their analysis, and design. For compression springs undergoing static and fatigue loads, the complete design process was presented. This was not done for extension and torsion springs, as the process is the same, although the governing conditions are not. The governing conditions, however, were provided and extension to the design process from what was provided for the compression spring should be straightforward.

PROBLEMS

10–1 Within the range of recommended values of the spring index, C, determine the maximum and minimum percentage difference between the Bergsträsser factor, K_B, and the Wahl factor, K_W.

10–2 A helical compression spring is wound using 2.5-mm-diameter music wire. The spring has an outside diameter of 31 mm with plain ground ends, and 14 total coils.

(*a*) Estimate the spring rate.

(*b*) What force is needed to compress this spring to closure?

(*c*) What should the free length be to ensure that when the spring is compressed solid the torsional stress does not exceed the yield strength?

(*d*) Is there a possibility that the spring might buckle in service?

10–3 The spring in Problem 10–2 is to be used with a static load of 130 N. Perform a design assessment represented by Equations (10–13) and (10–18) through (10–21) if the spring is closed to solid height.

10–4 A helical compression spring is made with oil-tempered wire with wire diameter of 5 mm, mean coil diameter of 50 mm, a total of 12 coils, a free length of 125 mm, with squared ends.

(*a*) Find the solid length.

(*b*) Find the force necessary to deflect the spring to its solid length.

(*c*) Find the factor of safety guarding against yielding when the spring is compressed to its solid length.

10–5 A helical compression spring is to be made of oil-tempered wire of 4-mm diameter with a spring index of $C = 10$. The spring is to operate inside a hole, so buckling is not a problem and the ends can be left plain. The free length of the spring should be 80 mm. A force of 50 N should deflect the spring 15 mm.

(*a*) Determine the spring rate.

(*b*) Determine the minimum hole diameter for the spring to operate in.

(*c*) Determine the total number of coils needed.

(*d*) Determine the solid length.

(*e*) Determine a static factor of safety based on the yielding of the spring if it is compressed to its solid length.

10–6 A helical compression spring is made of hard-drawn spring steel wire 2 mm in diameter and has an outside diameter of 22 mm. The ends are plain and ground, and there are 8 total coils.

(*a*) The spring is wound to a free length, which is the largest possible with a solid-safe property. Find this free length.

(*b*) What is the pitch of this spring?

(*c*) What force is needed to compress the spring to its solid length?

(*d*) Estimate the spring rate.

(*e*) Will the spring buckle in service?

10–7 The spring of Problem 10–6 is to be used with a static load of 75 N. Perform a design assessment represented by Equations (10–13) and (10–18) through (10–21) if the spring closed to solid height.

10–8 to 10–18 Listed in the tables are six springs described in customary units and five springs described in SI units. Investigate these squared-and-ground-ended helical compression springs to see if they are solid-safe. If not, what is the largest free length to which they can be wound using $n_s = 1.2$?

Problem Number	d, mm	OD, mm	L_0, mm	N_t	Material
10–8	0.15	0.9	16	38	A228 music wire
10–9	0.3	3	20.6	16	B159 phosphor-bronze
10–10	1	6	19.1	11.2	A313 stainless steel
10–11	3.5	50	74.6	5.75	A227 hard-drawn steel
10–12	3.7	24	95.3	12	A229 OQ&T steel
10–13	4.9	76	228.6	8	A232 chrome-vanadium
	d, mm	OD, mm	L_0, mm	N_t	Material
10–14	0.25	0.95	12.1	38	A313 stainless steel
10–15	1.2	6.5	15.7	10.2	A228 music wire
10–16	3.5	50.6	75.5	5.5	A229 OQ&T spring steel
10–17	3.8	31.4	71.4	12.8	B159 phosphor-bronze
10–18	4.5	69.2	215.6	8.2	A232 chrome-vanadium

10–19 Consider the steel spring in the illustration.

(*a*) Find the pitch, solid height, and number of active turns.

(*b*) Find the spring rate. Assume the material is A227 HD steel.

(*c*) Find the force F_s required to close the spring solid.

(*d*) Find the shear stress in the spring due to the force F_s.

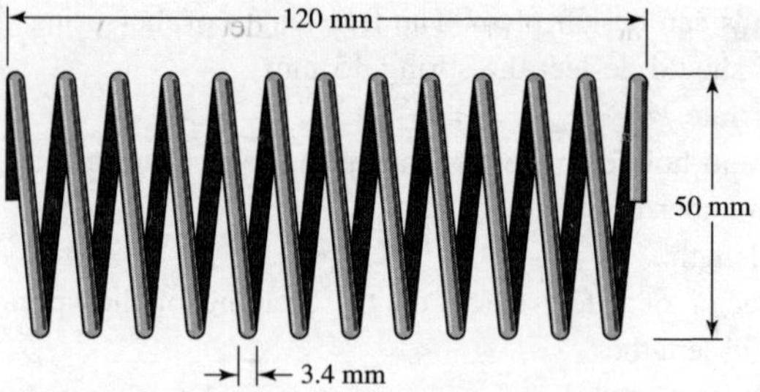

Problem 10–19

10–20 Consider a plain end hard-drawn steel compression helical spring. The spring is made of 12-gauge Washburn & Moen wire, has an outside coil diameter of 20 mm, 20 total coils, and a free length of 95 mm. Determine the

(*a*) spring index, C.
(*b*) pitch of the coil, p.
(*c*) spring deflection when closed solid, y_s.
(*d*) force to close the spring solid, F_s.
(*e*) shear stress at closure, τ_s.
(*f*) factor of safety in yield at closure, n_s.
(*g*) spring rate, k, using the exact and the approximate formulation, and compare the two results.

10–21 Consider a squared and ground oil-tempered compression helical spring. The spring is made of 3-mm wire, has an outer diameter of 30 mm, 32 total coils, and a free length of 240 mm. Determine the

(*a*) spring index, C.
(*b*) pitch of the coil, p.
(*c*) spring deflection when closed solid, y_s.
(*d*) force to close the spring solid, F_s.
(*e*) shear stress at closure, τ_s.
(*f*) factor of safety in yield at closure, n_s.
(*g*) spring rate, k, using the exact and the approximate formulation, and compare the two results.

10–22 Starting with a spring index of $C = 10$, design a compression coil spring of 302 stainless wire with squared ends. The spring is to deflect $y = 50$ mm, for an applied force of $F = 90$ N and to close solid when $y_s = 60$ mm. At closure, use a design factor of 1.2 guarding against yielding. Select the smallest diameter wire where wire diameters are available in 0.2-mm increments between 0.2 and 3.2 mm. For the design, specify the wire diameter, the inside and outside diameters of the coil, the spring rate, the total number of coils, the free length of the spring, the final factor of safety, and stability conditions.

10–23 Starting with a spring index of $C = 10$, design a compression coil spring of phosphor bronze with closed ends. The spring is to deflect $y = 50$ mm, for an applied force of $F = 65$ N and to close solid when $y_s = 75$ mm. At closure, use a design factor of 1.2 guarding against yielding. Determine the smallest-diameter wire based on preferred sizes. For the design, specify the wire diameter, the inside and outside diameters of the coil, the spring rate, the total number of coils, the free length of the spring, the final factor of safety, and stability conditions.

10–24 After determining the wire diameter in Problem 10–23, determine what the spring index should be to maintain a design factor of 1.2 at closure. For this design, specify the wire diameter, the inside and outside diameters of the coil, the spring rate, the total number of coils, the free length of the spring, the final factor of safety, and stability conditions.

10–25 A static service music wire helical compression spring is needed to support a 90 N load after being compressed 50 mm. The solid height of the spring cannot exceed 38 mm. The free length must not exceed 100 mm. The static factor of safety must equal or exceed 1.2. For robust linearity use a fractional overrun to closure ξ of 0.15. There are two springs to be designed. Start with a wire diameter of 1.9 mm.

(*a*) The spring must operate over a 20 mm rod. A 1.3-mm diametral clearance allowance should be adequate to avoid interference between the rod and the spring due to out-of-round coils. Design the spring.

(*b*) The spring must operate in a 25 mm-diameter hole. A 1.2 mm diametral clearance allowance should be adequate to avoid interference between the spring and the hole due to swelling of the spring diameter as the spring is compressed and out-of-round coils. Design the spring.

10–26 Solve Problem 10–25 by iterating with an initial value of $C = 10$. If you have already solved Problem 10–25, compare the steps and the results.

10–27 A holding fixture for a workpiece 37.5 mm thick at the clamp locations is being designed. The detail of one of the clamps is shown in the figure. A spring is required to drive the clamp upward when removing the workpiece with a starting force of 45 N. The clamp screw has an M10 × 1.25 thread. Allow a diametral clearance of 1.25 mm between it and the uncompressed spring. It is further specified that the free length of the spring should be $L_0 \leq 48$ mm, the solid height $L_s \leq 31.5$ mm, and the safety factor when closed solid should be $n_s \geq 1.2$. Starting with $d = 2$ mm, design a suitable helical coil compression spring for this fixture. For A227 HD steel, wire diameters are available in 0.2-mm increments between 0.2 and 3.2 mm.

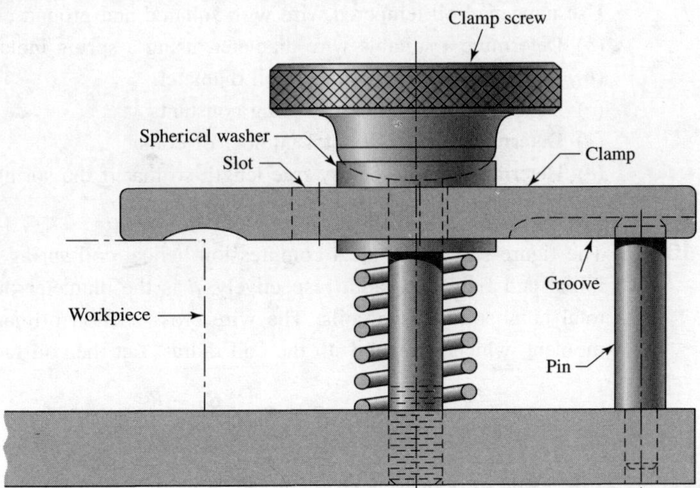

Problem 10–27
Clamping fixture.

10–28 Solve Problem 10–27 by iterating with an initial value of $C = 8$. If you have already solved Problem 10–27, compare the steps and the results.

10–29 Your instructor will provide you with a stock spring supplier's catalog, or pages reproduced from it. Accomplish the task of Problem 10–27 by selecting an available stock spring. (This is design by *selection.*)

10–30 A compression spring is needed to fit over a 12 mm diameter rod. To allow for some clearance, the inside diameter of the spring is to be 15 mm. To ensure a reasonable coil, use a spring index of 10. The spring is to be used in a machine by compressing it from a free length of 125 mm through a stroke of 75 mm to its solid length. The spring should have squared and ground ends, unpeened, and is to be made from cold-drawn wire.

(*a*) Determine a suitable wire diameter.
(*b*) Determine a suitable total number of coils.
(*c*) Determine the spring constant.
(*d*) Determine the static factor of safety when compressed to solid length.
(*e*) Determine the fatigue factor of safety when repeatedly cycled from free length to solid length. Use the Gerber-Zimmerli fatigue-failure criterion.

10–31 A compression spring is needed to fit within a 25 mm diameter hole. To allow for some clearance, the outside diameter of the spring is to be no larger than 22 mm. To ensure a reasonable coil, use a spring index of 8. The spring is to be used in a machine by compressing it from a free length of 75 mm to a solid length of 25 mm. The spring should have squared ends, and is unpeened, and is to be made from music wire.

(*a*) Determine a suitable wire diameter.
(*b*) Determine a suitable total number of coils.
(*c*) Determine the spring constant.
(*d*) Determine the static factor of safety when compressed to solid length.
(*e*) Determine the fatigue factor of safety when repeatedly cycled from free length to solid length. Use the Gerber-Zimmerli fatigue-failure criterion.

10–32 A helical compression spring is to be cycled between 600 N and 12 000 N with a 25 mm stroke. The number of cycles is low, so fatigue is not an issue. The coil must fit in a 52 mm diameter hole with a 2 mm clearance all the way around the spring. Use unpeened oil tempered wire with squared and ground ends.

(*a*) Determine a suitable wire diameter, using a spring index of $C = 7$.
(*b*) Determine a suitable mean coil diameter.
(*c*) Determine the necessary spring constant.
(*d*) Determine a suitable total number of coils.
(*e*) Determine the necessary free length so that if the spring were compressed to its solid length, there would be no yielding.

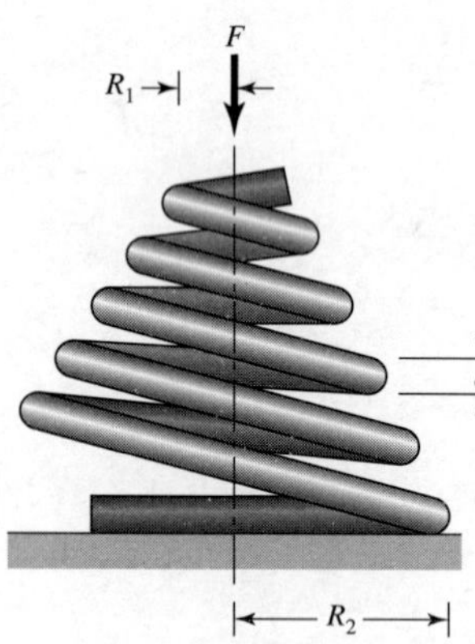

Problem 10–33

10–33 The figure shows a conical compression helical coil spring where R_1 and R_2 are the initial and final coil radii, respectively, d is the diameter of the wire, and N_a is the total number of active coils. The wire cross section primarily transmits a torsional moment, which changes with the coil radius. Let the coil radius be given by

$$R = R_1 + \frac{R_2 - R_1}{2\pi N_a}\theta$$

where θ is in radians. Use Castigliano's method to estimate the spring rate as

$$k = \frac{d^4 G}{16 N_a (R_2 + R_1)\left(R_2^2 + R_1^2\right)}$$

10–34 A helical coil compression spring is needed for food service machinery. The load varies from a minimum of 20 N to a maximum of 90 N. The spring rate k is to be 1660 N/m. The outside diameter of the spring cannot exceed 62 mm. The springmaker has available suitable dies for drawing 2-, 2.3-, 2.6-, and 3-mm-diameter wire. Use squared and ground ends, and an unpeened material. Using a fatigue design factor n_f of 1.5, and the Gerber-Zimmerli fatigue-failure criterion, design a suitable spring.

10–35 Solve Problem 10–34 using the Goodman-Zimmerli fatigue-failure criterion.

10–36 Solve Problem 10–34 using the Sines-Zimmerli fatigue-failure criterion.

10–37 Design the spring of Example 10–5 using the Gerber-Zimmerli fatigue-failure criterion.

10–38 Solve Problem 10–37 using the Goodman-Zimmerli fatigue-failure criterion.

10–39 A hard-drawn spring steel extension spring is to be designed to carry a static load of 90 N with an extension of 12 mm using a design factor of $n_y = 1.5$ in bending. Use full-coil end hooks with the fullest bend radius of $r = D/2$ and $r_2 = 2d$. The free length must be less than 75 mm, and the body turns must be fewer than 30. (Note: Integer and half-integer body turns allow end hooks to be placed in the same plane. However, this adds extra cost and is done only when necessary.)

10–40 The extension spring shown in the figure has full-twisted loop ends. The material is AISI 1065 OQ&T wire. The spring has 84 coils and is close-wound with a preload of 70 N.

(*a*) Find the closed length of the spring.
(*b*) Find the torsional stress in the spring corresponding to the preload.
(*c*) Estimate the spring rate.
(*d*) What load would cause permanent deformation?
(*e*) What is the spring deflection corresponding to the load found in part *d*?

Problem 10–40

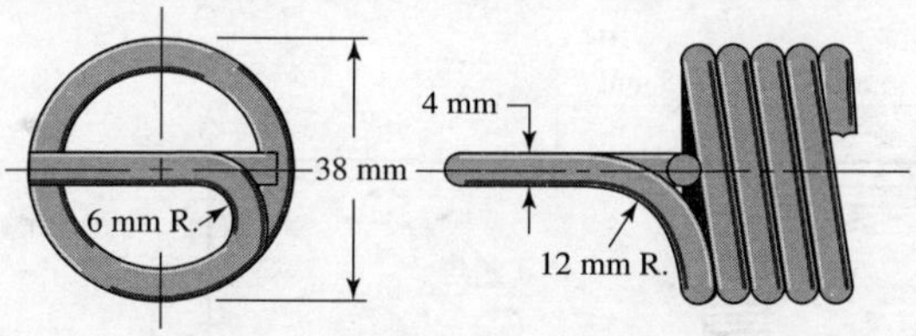

10–41 Design an infinite-life helical coil extension spring with full end loops and generous loop-bend radii for a minimum load of 40 N and a maximum load of 80 N, with an accompanying stretch of 6 mm. The spring is for food-service equipment and must be stainless steel. The outside diameter of the coil cannot exceed 25 mm, and the free length cannot exceed 62 mm. Using a fatigue design factor of $n_f = 2$, complete the design. Use the Gerber criterion with Table 10–8.

10–42 Prove Equation (10–40). *Hint:* Using Castigliano's theorem, determine the deflection due to bending of an end hook alone as if the hook were fixed at the end connecting it to the body of the spring. Consider the wire diameter d small as compared to the mean radius of the hook, $R = D/2$. Add the deflections of the end hooks to the deflection of the main body to determine the final spring constant, then equate it to Equation (10–9).

10–43 The figure shows a finger exerciser used by law-enforcement officers and athletes to strengthen their grip. It is formed by winding A227 hard-drawn steel wire around a mandrel to obtain $2\frac{1}{2}$ turns when the grip is in the closed position. After winding, the wire is cut to leave the two legs as handles. The plastic handles are then molded on, the grip is squeezed together, and a wire clip is placed around the legs to obtain initial "tension" and to space the handles for the best initial gripping position. The clip is formed like a figure 8 to prevent it from coming off. When the grip is in the closed position, the stress in the spring should not exceed the permissible stress.

(*a*) Determine the configuration of the spring before the grip is assembled.

(*b*) Find the force necessary to close the grip.

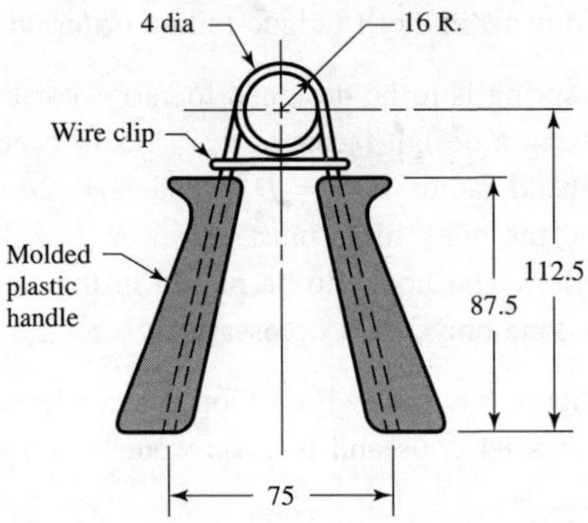

Problem 10–43
Dimensions in millimeters.

10–44 The rat trap shown in the figure uses two opposite-image torsion springs. The wire has a diameter of 2 mm, and the outside diameter of the spring in the position shown is 12 mm. Each spring has 11 turns. Use of a fish scale revealed a force of about 36 N is needed to set the trap.

(*a*) Find the probable configuration of the spring prior to assembly.

(*b*) Find the maximum stress in the spring when the trap is set.

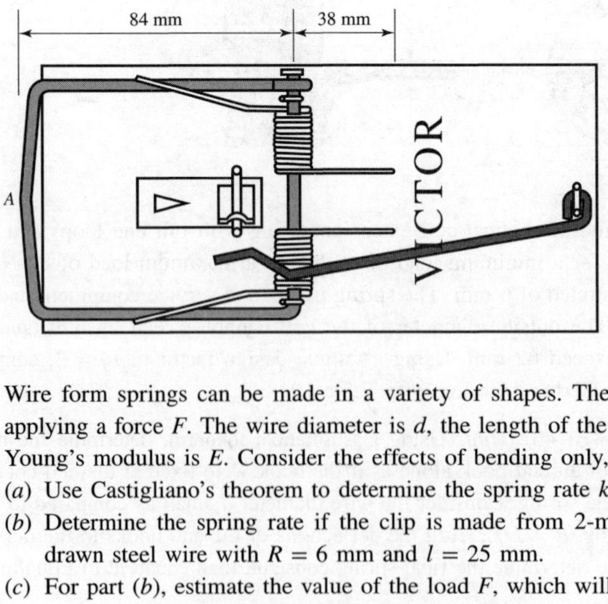

Problem 10–44

10–45 Wire form springs can be made in a variety of shapes. The clip shown operates by applying a force F. The wire diameter is d, the length of the straight section is l, and Young's modulus is E. Consider the effects of bending only, with $d \ll R$.

(*a*) Use Castigliano's theorem to determine the spring rate k.

(*b*) Determine the spring rate if the clip is made from 2-mm diameter A227 hard-drawn steel wire with $R = 6$ mm and $l = 25$ mm.

(*c*) For part (*b*), estimate the value of the load F, which will cause the wire to yield.

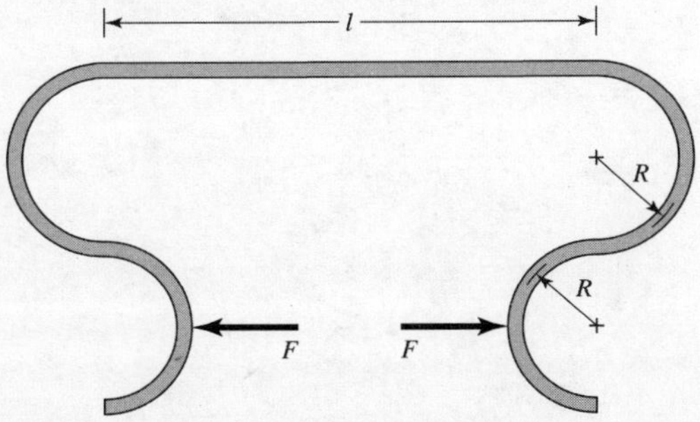

Problem 10–45

10–46 For the wire form shown, the wire diameter is d, the length of the straight section is l, and Young's modulus is E. Consider the effects of bending only, with $d \ll R$.

(a) Use Castigliano's method to determine the spring rate k.

(b) Determine the spring rate if the form is made from 0.7 mm diameter A313 stainless wire with $R = 15$ mm and $l = 12$ mm.

(c) For part (b), estimate the value of the load F, which will cause the wire to yield.

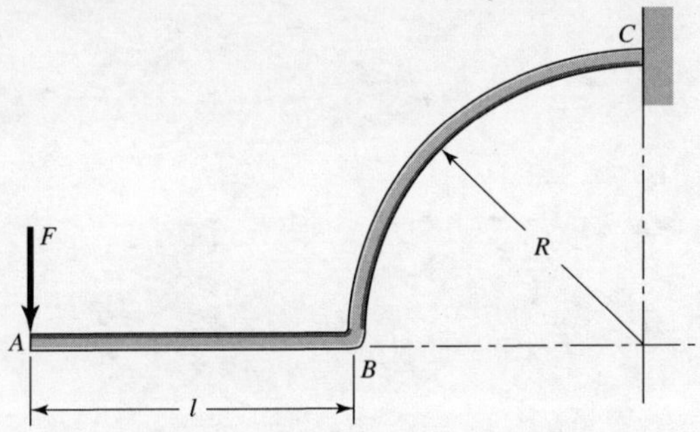

Problem 10–46

10–47 Figure 10–13b shows a spring of constant thickness and constant stress. A constant stress spring can be designed where the width b is constant as shown.

(a) Determine how h varies as a function of x.

(b) Given Young's modulus E, determine the spring rate k in terms of E, l, b, and h_o. Verify the units of k.

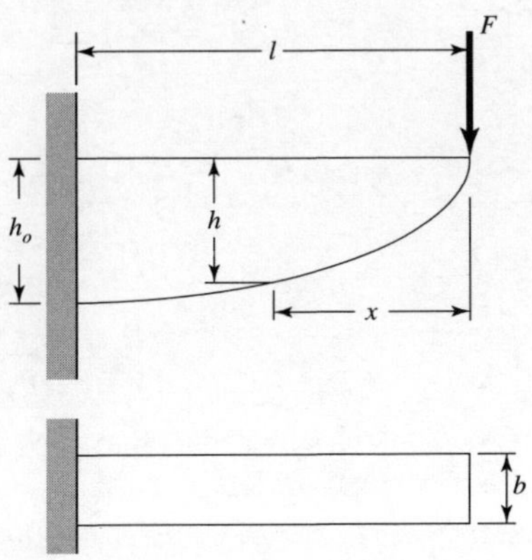

Problem 10–47

10–48 Using the experience gained with Problem 10–34, write a computer program that would help in the design of helical coil compression springs.

10–49 Using the experience gained with Problem 10–41, write a computer program that would help in the design of a helical coil extension spring.

11 Rolling-Contact Bearings

nordroden/123RF

Chapter Outline

The terms *rolling-contact bearing, antifriction bearing,* and *rolling bearing* are all used to describe that class of bearing in which the main load is transferred through elements in rolling contact rather than in sliding contact. In a rolling bearing the starting friction is about twice the running friction, but still it is negligible in comparison with the starting friction of a sleeve bearing. Load, speed, and the operating viscosity of the lubricant do affect the frictional characteristics of a rolling bearing.

From the mechanical designer's standpoint, the study of antifriction bearings differs in several respects when compared with the study of other topics because the bearings they specify have already been designed. The specialist in antifriction-bearing design is confronted with the problem of designing a group of elements that compose a rolling bearing. These elements must be designed to fit into a space whose dimensions are specified; they must be designed to receive a load having certain characteristics; and finally, these elements must be designed to have a satisfactory life when operated under the specified conditions. Bearing specialists must therefore consider such matters as fatigue loading, friction, heat, corrosion resistance, kinematic problems, material properties, lubrication, machining tolerances, assembly, use, and cost. From a consideration of all these factors, bearing specialists arrive at a compromise that, in their judgment, is a good solution to the problem as stated.

We begin with an overview of bearing types; then we note that bearing life cannot be described in deterministic form. We introduce the invariant, the statistical distribution of bearing life, which is described by the Weibull distribution. There are some useful deterministic equations addressing load versus life at constant reliability, and the catalog rating as rating life is introduced. The load-life-reliability relationship combines statistical and deterministic relationships, which gives the designer a way to move from the desired load and life to the catalog rating in *one* equation.

Ball bearings also resist thrust, and a unit of thrust does different damage per revolution than a unit of radial load, so we must find the equivalent pure radial load that does the same damage as the existing radial and thrust loads. Next, variable loading, stepwise and continuous, is approached, and the equivalent pure radial load doing the same damage is quantified. Oscillatory loading is mentioned.

With this preparation we have the tools to consider the selection of ball and cylindrical roller bearings. The question of misalignment is quantitatively approached.

Tapered roller bearings have some unique benefits and design considerations that are covered.

Having the tools to find the proper catalog ratings, we make decisions (selections), we perform a design assessment, and the bearing reliability is quantified. Lubrication and mounting conclude our introduction. Vendors' manuals should be consulted for specific details relating to bearings of their manufacture.

11–1 Bearing Types

Bearings are manufactured to take pure radial loads, pure thrust loads, or a combination of the two kinds of loads. The nomenclature of a ball bearing is illustrated in Figure 11–1, which also shows the four essential parts of a bearing. These are the outer ring, the inner ring, the balls or rolling elements, and the separator. In low-priced bearings, the separator is sometimes omitted, but it has the important function of separating the elements so that rubbing contact will not occur.

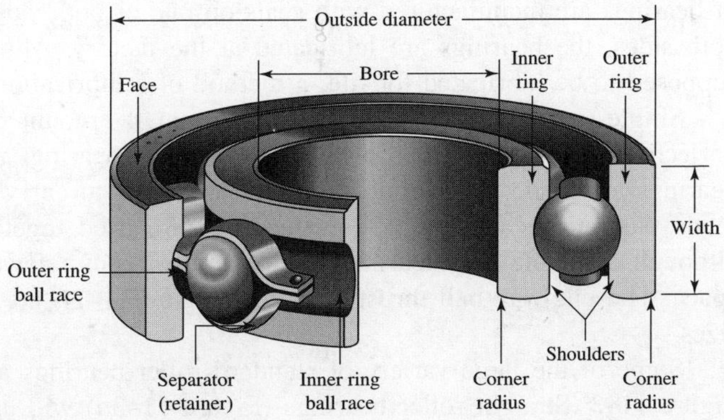

Figure 11–1

Nomenclature of a ball bearing. *(Source: Based on General Motors Corp., GM Media Archives.)*

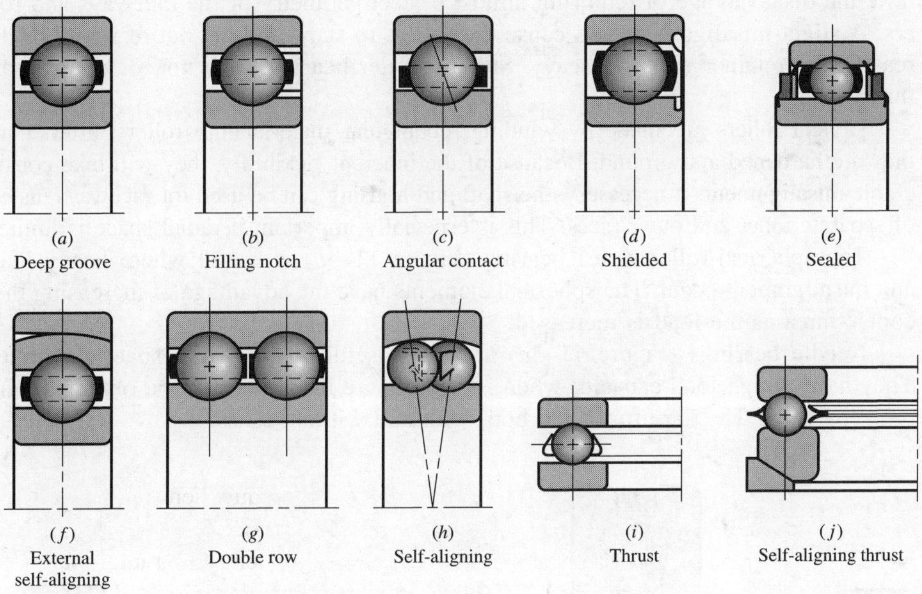

Figure 11–2

Various types of ball bearings.

In this section we include a selection from the many types of standardized bearings that are manufactured. Most bearing manufacturers provide engineering manuals and brochures containing descriptions of the various types available. In the small space available here, only a meager outline of some of the most common types can be given. You should include a survey of bearing manufacturers' literature in your studies of this section.

Some of the various types of standardized bearings that are manufactured are shown in Figure 11–2. The single-row deep-groove bearing will take radial load as well as some thrust load. The balls are inserted into the grooves by moving the inner ring to an eccentric position. The balls are separated after loading, and the separator is then inserted. The use of a filling notch (Figure 11–2*b*) in the inner and outer rings enables a greater number of balls to be inserted, thus increasing the load capacity. The thrust capacity is decreased, however, because of the bumping of the balls against the edge of the notch when thrust loads are present. The angular-contact bearing (Figure 11–2*c*) provides a greater thrust capacity.

All these bearings may be obtained with shields on one or both sides. The shields are not a complete closure but do offer a measure of protection against dirt. A variety of bearings are manufactured with seals on one or both sides. When the seals are on both sides, the bearings are lubricated at the factory. Although a sealed bearing is supposed to be lubricated for life, a method of relubrication is sometimes provided.

Single-row bearings will withstand a small amount of shaft misalignment of deflection, but where this is severe, self-aligning bearings may be used. Double-row bearings are made in a variety of types and sizes to carry heavier radial and thrust loads. Sometimes two single-row bearings are used together for the same reason, although a double-row bearing will generally require fewer parts and occupy less space. The one-way ball thrust bearings (Figure 11–2*i*) are made in many types and sizes.

Some of the large variety of standard roller bearings available are illustrated in Figure 11–3. Straight roller bearings (Figure 11–3*a*) will carry a greater radial load than ball bearings of the same size because of the greater contact area. However, they have the disadvantage of requiring almost perfect geometry of the raceways and rollers. A slight misalignment will cause the rollers to skew and get out of line. For this reason, the retainer must be heavy. Straight roller bearings will not, of course, take thrust loads.

Helical rollers are made by winding rectangular material into rollers, after which they are hardened and ground. Because of the inherent flexibility, they will take considerable misalignment. If necessary, the shaft and housing can be used for raceways instead of separate inner and outer races. This is especially important if radial space is limited.

The spherical-roller thrust bearing (Figure 11–3*b*) is useful where heavy loads and misalignment occur. The spherical elements have the advantage of increasing their contact area as the load is increased.

Needle bearings (Figure 11–3*d*) are very useful where radial space is limited. They have a high load capacity when separators are used, but may be obtained without separators. They are furnished both with and without races.

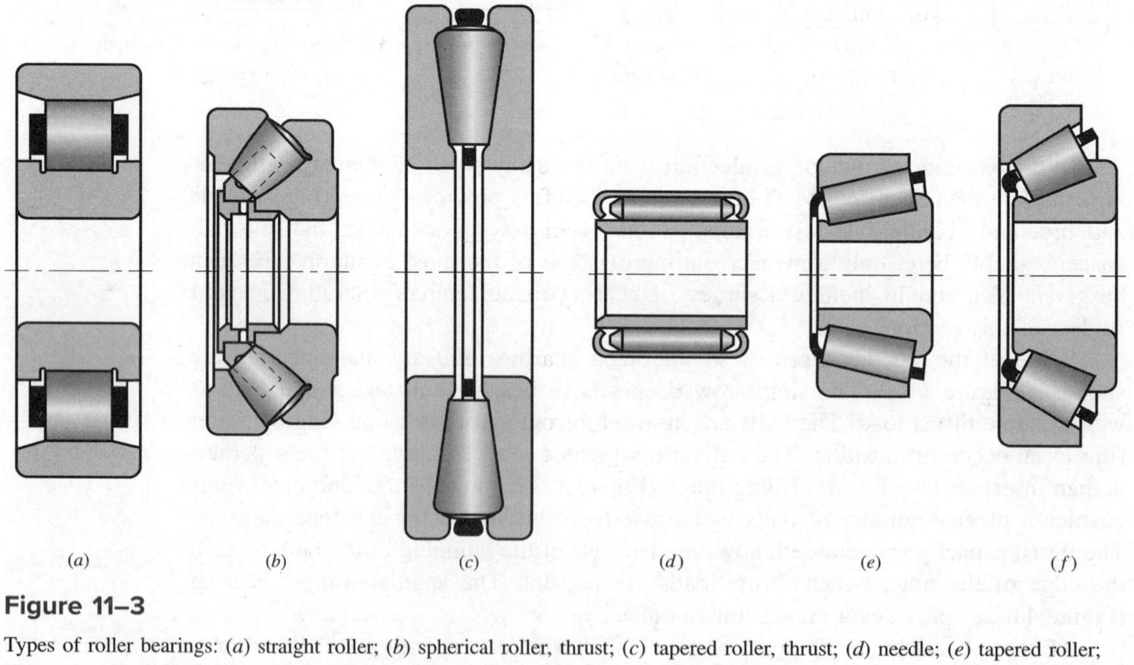

Figure 11–3

Types of roller bearings: (*a*) straight roller; (*b*) spherical roller, thrust; (*c*) tapered roller, thrust; (*d*) needle; (*e*) tapered roller; (*f*) steep-angle tapered roller. *(Source: Redrawn from material Furnished by The Timken Company.)*

Tapered roller bearings (Figure 11–3*e*, *f*) combine the advantages of ball and straight roller bearings, since they can take either radial or thrust loads or any combination of the two, and in addition, they have the high load-carrying capacity of straight roller bearings. The tapered roller bearing is designed so that all elements in the roller surface and the raceways intersect at a common point on the bearing axis.

The bearings described here represent only a small portion of the many available for selection. Many special-purpose bearings are manufactured, and bearings are also made for particular classes of machinery. Typical of these are:

- Instrument bearings, which are high-precision and are available in stainless steel and high-temperature materials
- Nonprecision bearings, usually made with no separator and sometimes having split or stamped sheet-metal races
- Ball bushings, which permit either rotation or sliding motion or both
- Bearings with flexible rollers

11–2 Bearing Life

When the ball or roller of rolling-contact bearings rolls, contact stresses occur on the inner ring, the rolling element, and on the outer ring. Because the curvature of the contacting elements in the axial direction is different from that in the radial direction, the equations for these stresses are more involved than in the Hertz equations presented in Chapter 3. If a bearing is clean and properly lubricated, is mounted and sealed against the entrance of dust and dirt, is maintained in this condition, and is operated at reasonable temperatures, then metal fatigue will be the only cause of failure. Inasmuch as metal fatigue implies many millions of stress applications successfully endured, we need a quantitative life measure. Common life measures are

- Number of revolutions of the inner ring (outer ring stationary) until the first tangible evidence of fatigue
- Number of hours of use at a standard angular speed until the first tangible evidence of fatigue

The commonly used term is *bearing life,* which is applied to either of the measures just mentioned. It is important to realize, as in all fatigue, life as defined above is a stochastic variable and, as such, has both a distribution and associated statistical parameters. The life measure of an individual bearing is defined as the total number of revolutions (or hours at a constant speed) of bearing operation until the failure criterion is developed. Under ideal conditions, the fatigue failure consists of spalling of the load-carrying surfaces. The American Bearing Manufacturers Association (ABMA) standard states that the failure criterion is the first evidence of fatigue. The fatigue criterion used by the Timken Company laboratories is the spalling or pitting of an area of 6.45 mm^2. Timken also observes that the useful life of the bearing may extend considerably beyond this point. This is an operational definition of fatigue failure in rolling bearings.

The *rating life* is a term sanctioned by the ABMA and used by most manufacturers. The rating life of a group of nominally identical ball or roller bearings is defined as the number of revolutions (or hours at a constant speed) that 90 percent of a group of bearings will achieve or exceed before the failure criterion develops. The terms *minimum life*, L_{10} *life,* and B_{10} *life* are also used as synonyms for rating life. The rating life is the 10th percentile location of the bearing group's revolutions-to-failure distribution.

Median life is the 50th percentile life of a group of bearings. The term *average life* has been used as a synonym for median life, contributing to confusion. When many groups of bearings are tested, the median life is between 4 and 5 times the L_{10} life.

Each bearing manufacturer will choose a specific rating life for which load ratings of its bearings are reported. The most commonly used rating life is 10^6 revolutions. The Timken Company is a well-known exception, rating its bearings at 3000 hours at 500 rev/min, which is $90(10^6)$ revolutions. These levels of rating life are actually quite low for today's bearings, but since rating life is an arbitrary reference point, the traditional values have generally been maintained.

11–3 Bearing Load Life at Rated Reliability

When nominally identical groups are tested to the life-failure criterion at different loads, the data are plotted on a graph as depicted in Figure 11–4 using a log-log transformation. To establish a single point, load F_1 and the rating life of group one $(L_{10})_1$ are the coordinates that are logarithmically transformed. The reliability associated with this point, and all other points, is 0.90. Thus we gain a glimpse of the load-life function at 0.90 reliability. Using a regression equation of the form

$$FL^{1/a} = \text{constant} \tag{11–1}$$

the result of many tests for various kinds of bearings result in

- $a = 3$ for ball bearings
- $a = 10/3$ for roller bearings (cylindrical and tapered roller)

A *catalog load rating* is defined as the radial load that causes 10 percent of a group of bearings to fail at the bearing manufacturer's rating life. We shall denote the catalog load rating as C_{10}. The catalog load rating is often referred to as a *Basic Dynamic Load Rating,* or sometimes just Basic Load Rating, if the manufacturer's rating life is 10^6 revolutions. The radial load that would be necessary to cause failure at such a low life would be unrealistically high. Consequently, the Basic Load Rating should be viewed as a reference value, and not as an actual load to be achieved by a bearing.

Figure 11–4

Typical bearing load-life log-log curve.

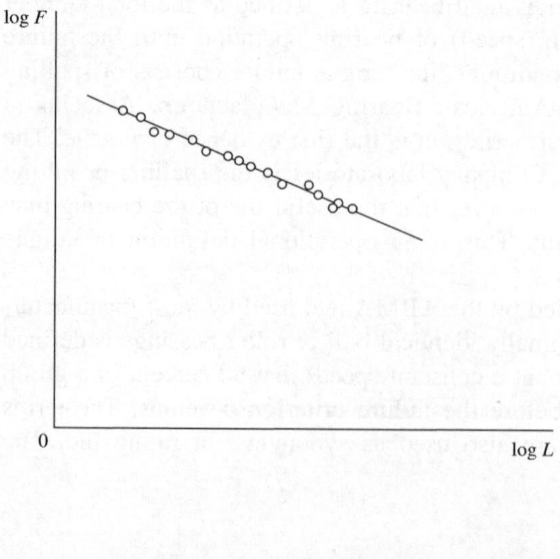

In selecting a bearing for a given application, it is necessary to relate the desired load and life requirements to the published catalog load rating corresponding to the catalog rating life. From Equation (11–1) we can write

$$F_1 L_1^{1/a} = F_2 L_2^{1/a} \tag{11–2}$$

where the subscripts 1 and 2 can refer to any set of load and life conditions. Letting F_1 and L_1 correlate with the catalog load rating and rating life, and F_2 and L_2 correlate with desired load and life for the application, we can express Equation (11–2) as

$$F_R L_R^{1/a} = F_D L_D^{1/a} \tag{a}$$

where the units of L_R and L_D are revolutions, and the subscripts R and D stand for Rated and Desired.

It is sometimes convenient to express the life in hours at a given speed. Accordingly, any life L in revolutions can be expressed as

$$L = 60\,\mathscr{L}n \tag{b}$$

where $\mathscr{L}$ is in hours, n is in rev/min, and 60 min/h is the appropriate conversion factor.

Incorporating Equation (*b*) into Equation (*a*),

$$F_R(\mathscr{L}_R n_R 60)^{1/a} = F_D(\mathscr{L}_D n_D 60)^{1/a} \tag{c}$$

catalog rating, kN (F_R) — desired speed, rev/min (n_D)

rating life in hours ($\mathscr{L}_R$) — desired life, hours ($\mathscr{L}_D$)

rating speed, rev/min (n_R) — desired radial load, kN (F_D)

Solving Equation (*c*) for F_R, and noting that it is simply an alternate notation for the catalog load rating C_{10}, we obtain an expression for a catalog load rating as a function of the desired load, desired life, and catalog rating life.

$$C_{10} = F_R = F_D\left(\frac{L_D}{L_R}\right)^{1/a} = F_D\left(\frac{\mathscr{L}_D n_D 60}{\mathscr{L}_R n_R 60}\right)^{1/a} \tag{11–3}$$

It is sometimes convenient to define $x_D = L_D/L_R$ as a dimensionless *multiple of rating life*.

EXAMPLE 11–1

Consider SKF, which rates its bearings for 1 million revolutions. If you desire a life of 5000 h at 1725 rev/min with a load of 2 kN with a reliability of 90 percent, for which catalog rating would you search in an SKF catalog?

Solution

The rating life is $L_{10} = L_R = \mathscr{L}_R n_R 60 = 10^6$ revolutions. From Equation (11–3),

Answer $$C_{10} = F_D\left(\frac{\mathscr{L}_D n_D 60}{\mathscr{L}_R n_R 60}\right)^{1/a} = 2000\left[\frac{5000(1725)60}{10^6}\right]^{1/3} = 16.1 \text{ kN}$$

11–4 Reliability versus Life—The Weibull Distribution

At constant load, the life measure distribution of rolling-contact bearings is right skewed. Because of its robust ability to adjust to varying amounts of skewness, the *three-parameter Weibull distribution* is used exclusively for expressing the reliability of rolling-contact bearings. Unlike the development of the normal distribution in Section 1–12, we will begin with the definition of the reliability, R, for a Weibull distribution of the life measure, x, as

$$R = \exp\left[-\left(\frac{x - x_0}{\theta - x_0}\right)^b\right] \tag{11–4}$$

where the three parameters are[1]

x_0 = guaranteed, or "minimum," value of x

θ = characteristic parameter. For rolling-contact bearings, this corresponds to the 63.2121 percentile value of x

b = shape parameter that controls the skewness. For rolling-contact bearings, $b \approx 1.5$

The life measure is expressed in dimensionless form as $x = L/L_{10}$.

From Equation (1–8), $R = 1 - p$, where p is the probability of a value of x occurring between $-\infty$ and x, and is the integral of the probability distribution, $f(x)$, between those limits. Accordingly, $f(x) = -dR/dx$. Thus, from the derivative of Equation (11–4), the Weibull probability density function, $f(x)$, is given by

$$f(x) = \begin{cases} \dfrac{b}{\theta - x_0}\left(\dfrac{x - x_0}{\theta - x_0}\right)^{b-1} \exp\left[-\left(\dfrac{x - x_0}{\theta - x_0}\right)^b\right] & x \ge x_0 \ge 0 \\ 0 & x < x_0 \end{cases} \tag{11–5}$$

The mean and standard deviation of $f(x)$ are

$$\mu_x = x_0 + (\theta - x_0)\Gamma(1 + 1/b) \tag{11–6}$$

$$\hat{\sigma}_x = (\theta - x_0)\sqrt{\Gamma(1 + 2/b) - \Gamma^2(1 + 1/b)} \tag{11–7}$$

where Γ is the *gamma function*, and is found tabulated in Table A–29.

Given a specific required reliability, solving Equation (11–4) for x yields

$$x = x_0 + (\theta - x_0)\left(\ln\frac{1}{R}\right)^{1/b} \tag{11–8}$$

EXAMPLE 11–2

Construct the distributional properties of a 02–30 mm deep-groove ball bearing if the Weibull parameters are $x_0 = 0.020$, $\theta = 4.459$, and $b = 1.483$.

Find the mean, median, 10th percentile life, standard deviation, and coefficient of variation.

[1]To estimate the Weibull parameters from data, see J. E. Shigley and C. R. Mischke, *Mechanical Engineering Design*, 5th ed., McGraw Hill, New York, 1989, Sec. 4–12, Ex. 4–10.

Solution

From Equation (11–6) and interpolating Table A–29, the mean dimensionless life is

Answer
$$\mu_x = x_0 + (\theta - x_0)\Gamma(1 + 1/b)$$
$$= 0.020 + (4.459 - 0.020)\Gamma(1 + 1/1.483)$$
$$= 0.020 + 4.439\Gamma(1.67431) = 0.020 + 4.439(0.9040) = 4.033$$

This says that the average bearing life is 4.033 L_{10}.

The median dimensionless life corresponds to $R = 0.50$, or L_{50}, and from Equation (11–8) is

Answer
$$x_{0.50} = x_0 + (\theta - x_0)\left(\ln\frac{1}{0.50}\right)^{1/b}$$
$$= 0.020 + (4.459 - 0.020)\left(\ln\frac{1}{0.50}\right)^{1/1.483} = 3.487$$

or, $L = 3.487\ L_{10}$.

The 10th percentile value of the dimensionless life x is

Answer
$$x_{0.10} = 0.020 + (4.459 - 0.020)\left(\ln\frac{1}{0.90}\right)^{1/1.483} \approx 1 \text{ (as it should be)}$$

The standard deviation of the dimensionless life, given by Equation (11–7), is

Answer
$$\hat{\sigma}_x = (\theta - x_0)\sqrt{\Gamma(1 + 2/b) - \Gamma^2(1 + 1/b)}$$
$$= (4.459 - 0.020)\sqrt{\Gamma(1 + 2/1.483) - \Gamma^2(1 + 1/1.483)}$$
$$= 4.439\sqrt{\Gamma(2.349) - \Gamma^2(1.674)} = 4.439\sqrt{1.2023 - 0.9040^2}$$
$$= 2.755$$

The coefficient of variation of the dimensionless life is

Answer
$$C_x = \frac{\hat{\sigma}_x}{\mu_x} = \frac{2.755}{4.033} = 0.683$$

11–5 Relating Load, Life, and Reliability

This is the designer's problem. The desired load is not the manufacturer's test load or catalog entry. The desired speed is different from the vendor's test speed, and the reliability expectation is typically much higher than the 0.90 accompanying the catalog entry. Figure 11–5 shows the situation. The catalog information is plotted as point A, whose coordinates are (the logs of) C_{10} and $x_{10} = L_{10}/L_{10} = 1$, a point on the 0.90 reliability contour. The design point is at D, with the coordinates (the logs of) F_D and x_D, a point that is on the $R = R_D$ reliability contour. The designer must move from point D to point A via point B as follows. Along a constant reliability contour (BD), Equation (11–2) applies:

$$F_B x_B^{1/a} = F_D x_D^{1/a}$$

from which

$$F_B = F_D\left(\frac{x_D}{x_B}\right)^{1/a} \qquad (a)$$

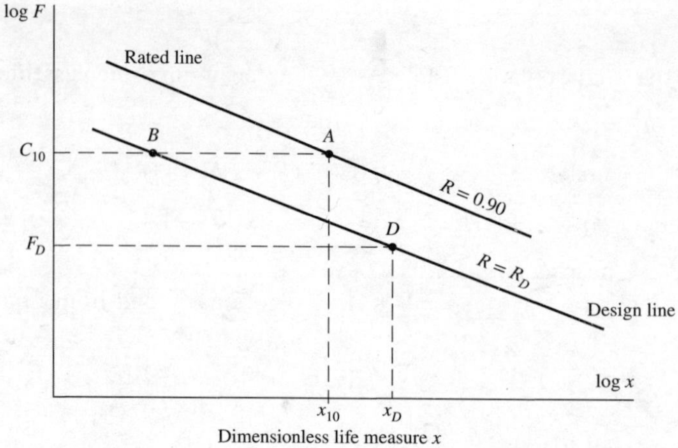

Figure 11–5

Constant reliability contours. Point A represents the catalog rating C_{10} at $x = L/L_{10} = 1$. Point B is on the target reliability design line R_D, with a load of C_{10}. Point D is a point on the desired reliability contour exhibiting the design life $x_D = L_D/L_{10}$ at the design load F_D.

Along a constant load line (AB), Equation (11–4) applies:

$$R_D = \exp\left[-\left(\frac{x_B - x_0}{\theta - x_0}\right)^b\right]$$

Solving for x_B gives

$$x_B = x_0 + (\theta - x_0)\left(\ln\frac{1}{R_D}\right)^{1/b}$$

Now substitute this in Equation (a) to obtain

$$F_B = F_D\left(\frac{x_D}{x_B}\right)^{1/a} = F_D\left[\frac{x_D}{x_0 + (\theta - x_0)[\ln(1/R_D)]^{1/b}}\right]^{1/a}$$

Noting that $F_B = C_{10}$, and including an application factor a_f with the design load,

$$C_{10} = a_f F_D\left[\frac{x_D}{x_0 + (\theta - x_0)[\ln(1/R_D)]^{1/b}}\right]^{1/a} \tag{11–9}$$

The application factor serves as a factor of safety to increase the design load to take into account overload, dynamic loading, and uncertainty. Typical load application factors for certain types of applications will be discussed shortly.

Equation (11–9) can be simplified slightly for calculator entry by noting that

$$\ln\frac{1}{R_D} = \ln\frac{1}{1 - p_f} = \ln(1 + p_f + \cdots) \approx p_f = 1 - R_D$$

where p_f is the probability for failure. Equation (11–9) can be written as

$$C_{10} \approx a_f F_D\left[\frac{x_D}{x_0 + (\theta - x_0)(1 - R_D)^{1/b}}\right]^{1/a} \qquad R \geq 0.90 \tag{11–10}$$

Either Equation (11–9) or Equation (11–10) may be used to convert from a design situation with a desired load, life, and reliability to a catalog load rating based on a rating life at 90 percent reliability. Note that when $R_D = 0.90$, the denominator is equal to one, and the equation reduces to Equation (11–3). The Weibull parameters are usually provided in the manufacturer's catalog. Typical values are given at the beginning of the end-of-chapter problems.

EXAMPLE 11–3

The design load on a ball bearing is 1840 N and an application factor of 1.2 is appropriate. The speed of the shaft is to be 300 rev/min, the life to be 30 kh with a reliability of 0.99. What is the C_{10} catalog entry to be sought (or exceeded) when searching for a deep-groove bearing in a manufacturer's catalog on the basis of 10^6 revolutions for rating life? The Weibull parameters are $x_0 = 0.02$, $(\theta - x_0) = 4.439$, and $b = 1.483$.

Solution

$$x_D = \frac{L_D}{L_R} = \frac{60\mathcal{L}_D n_D}{L_{10}} = \frac{60(30\ 000)300}{10^6} = 540$$

Thus, the design life is 540 times the L_{10} life. For a ball bearing, $a = 3$. Then, from Equation (11–10),

Answer

$$C_{10} = (1.2)(1840)\left[\frac{540}{0.02 + 4.439(1 - 0.99)^{1/1.483}}\right]^{1/3} = 29\ 833\ \text{N}$$

Shafts generally have two bearings. Often these bearings are different. If the bearing reliability of the shaft with its pair of bearings is to be R, then R is related to the individual bearing reliabilities R_A and R_B, using Equation (1–9), as

$$R = R_A R_B$$

First, we observe that if the product $R_A R_B$ equals R, then, in general, R_A and R_B are both greater than R. Since the failure of either or both of the bearings results in the shutdown of the shaft, then A or B or both can create a failure. Second, in sizing bearings one can begin by making R_A and R_B equal to the square root of the reliability goal, $\sqrt{R}$. In Example 11–3, if the bearing was one of a pair, the reliability goal would be $\sqrt{0.99}$, or 0.995. The bearings selected are discrete in their reliability property in your problem, so the selection procedure "rounds up," and the overall reliability exceeds the goal R. Third, it may be possible, if $R_A > \sqrt{R}$, to round down on B yet have the product $R_A R_B$ still exceed the goal R.

11–6 Combined Radial and Thrust Loading

A ball bearing is capable of resisting radial loading and a thrust loading. Furthermore, these can be combined. Consider F_a and F_r to be the axial thrust and radial loads, respectively, and F_e to be the *equivalent radial load* that does the same damage as the combined radial and thrust loads together. A rotation factor V is defined such that $V = 1$ when the inner ring rotates and $V = 1.2$ when the outer ring rotates. Two dimensionless groups can now be formed: $F_e/(VF_r)$ and $F_a/(VF_r)$. When these two dimensionless groups are plotted as in Figure 11–6, the data fall in a gentle curve that is well approximated by two straight-line segments. The abscissa e is defined by the intersection of the two lines. The equations for the two lines shown in Figure 11–6 are

$$\frac{F_e}{VF_r} = 1 \qquad \text{when} \quad \frac{F_a}{VF_r} \le e \tag{11–11a}$$

$$\frac{F_e}{VF_r} = X + Y\frac{F_a}{VF_r} \qquad \text{when} \quad \frac{F_a}{VF_r} > e \tag{11–11b}$$

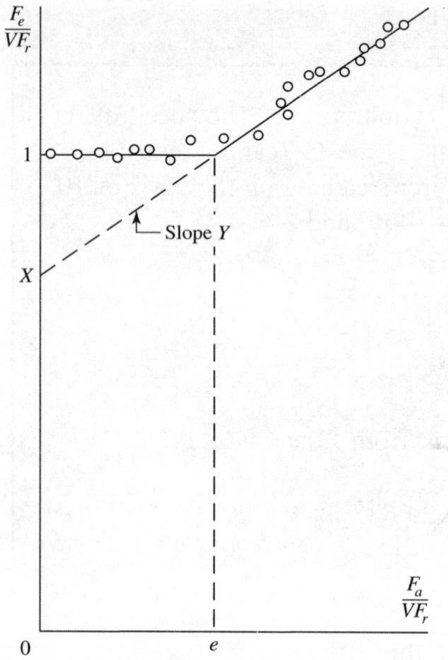

Figure 11–6

The relationship of dimensionless group $F_e/(VF_r)$ and $F_a/(VF_r)$ and the straight-line segments representing the data.

Table 11–1 Equivalent Radial Load Factors for Ball Bearings

		$F_a/(VF_r) \le e$		$F_a/(VF_r) > e$	
F_a/C_0	e	X_1	Y_1	X_2	Y_2
0.014*	0.19	1.00	0	0.56	2.30
0.021	0.21	1.00	0	0.56	2.15
0.028	0.22	1.00	0	0.56	1.99
0.042	0.24	1.00	0	0.56	1.85
0.056	0.26	1.00	0	0.56	1.71
0.070	0.27	1.00	0	0.56	1.63
0.084	0.28	1.00	0	0.56	1.55
0.110	0.30	1.00	0	0.56	1.45
0.17	0.34	1.00	0	0.56	1.31
0.28	0.38	1.00	0	0.56	1.15
0.42	0.42	1.00	0	0.56	1.04
0.56	0.44	1.00	0	0.56	1.00

*Use 0.014 if $F_a/C_0 < 0.014$.

where, as shown, X is the ordinate intercept and Y is the slope of the line for $F_a/(VF_r) > e$. It is common to express Equations (11–11a) and (11–11b) as a single equation,

$$F_e = X_i V F_r + Y_i F_a \tag{11–12}$$

where $i = 1$ when $F_a/(VF_r) \le e$ and $i = 2$ when $F_a/(VF_r) > e$. The X and Y factors depend upon the geometry and construction of the specific bearing. Table 11–1 lists representative values of X_1, Y_1, X_2, and Y_2 as a function of e, which in turn is a function of F_a/C_0, where C_0 is the basic static load rating. The *basic static load rating* is the load that will produce a total permanent deformation in the raceway and rolling element at any contact point of 0.0001 times the diameter of the rolling element. The basic static load rating is typically tabulated, along with the basic dynamic load rating C_{10}, in bearing manufacturers' publications. See Table 11–2, for example.

In these equations, the rotation factor V is intended to correct for the rotating-ring conditions. The factor of 1.2 for outer-ring rotation is simply an acknowledgment that the fatigue life is reduced under these conditions. Self-aligning bearings are an exception: they have $V = 1$ for rotation of either ring.

Since straight or cylindrical roller bearings will take no axial load, or very little, the Y factor is always zero.

The ABMA has established standard boundary dimensions for bearings, which define the bearing bore, the outside diameter (OD), the width, and the fillet sizes on the shaft and housing shoulders. The basic plan covers all ball and straight roller bearings in the metric sizes. The plan is quite flexible in that, for a given bore, there is an assortment of widths and outside diameters. Furthermore, the outside diameters selected are such that, for a particular outside diameter, one can usually find a variety of bearings having different bores and widths.

Table 11–2 Dimensions and Load Ratings for Single-Row 02-Series Deep-Groove and Angular-Contact Ball Bearings

Bore, mm	OD, mm	Width, mm	Fillet Radius, mm	Shoulder Diameter, mm		Load Ratings, kN Deep Groove		Load Ratings, kN Angular Contact	
				d_S	d_H	C_{10}	C_0	C_{10}	C_0
10	30	9	0.6	12.5	27	5.07	2.24	4.94	2.12
12	32	10	0.6	14.5	28	6.89	3.10	7.02	3.05
15	35	11	0.6	17.5	31	7.80	3.55	8.06	3.65
17	40	12	0.6	19.5	34	9.56	4.50	9.95	4.75
20	47	14	1.0	25	41	12.7	6.20	13.3	6.55
25	52	15	1.0	30	47	14.0	6.95	14.8	7.65
30	62	16	1.0	35	55	19.5	10.0	20.3	11.0
35	72	17	1.0	41	65	25.5	13.7	27.0	15.0
40	80	18	1.0	46	72	30.7	16.6	31.9	18.6
45	85	19	1.0	52	77	33.2	18.6	35.8	21.2
50	90	20	1.0	56	82	35.1	19.6	37.7	22.8
55	100	21	1.5	63	90	43.6	25.0	46.2	28.5
60	110	22	1.5	70	99	47.5	28.0	55.9	35.5
65	120	23	1.5	74	109	55.9	34.0	63.7	41.5
70	125	24	1.5	79	114	61.8	37.5	68.9	45.5
75	130	25	1.5	86	119	66.3	40.5	71.5	49.0
80	140	26	2.0	93	127	70.2	45.0	80.6	55.0
85	150	28	2.0	99	136	83.2	53.0	90.4	63.0
90	160	30	2.0	104	146	95.6	62.0	106	73.5
95	170	32	2.0	110	156	108	69.5	121	85.0

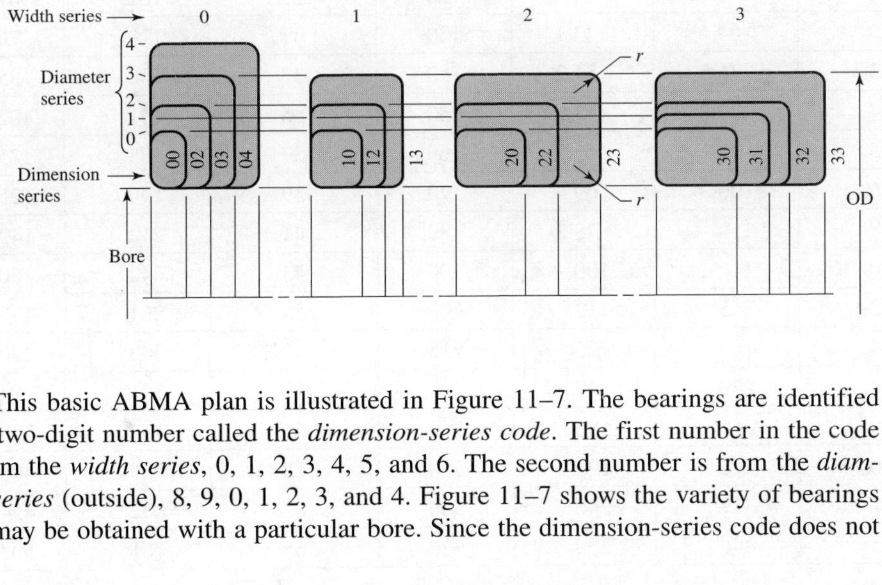

Figure 11–7
The basic ABMA plan for boundary dimensions. These apply to ball bearings, straight roller bearings, and spherical roller bearings, but not to inchseries ball bearings or tapered roller bearings. The contour of the corner is not specified. It may be rounded or chamfered, but it must be small enough to clear the fillet radius specified in the standards.

This basic ABMA plan is illustrated in Figure 11–7. The bearings are identified by a two-digit number called the *dimension-series code*. The first number in the code is from the *width series*, 0, 1, 2, 3, 4, 5, and 6. The second number is from the *diameter series* (outside), 8, 9, 0, 1, 2, 3, and 4. Figure 11–7 shows the variety of bearings that may be obtained with a particular bore. Since the dimension-series code does not

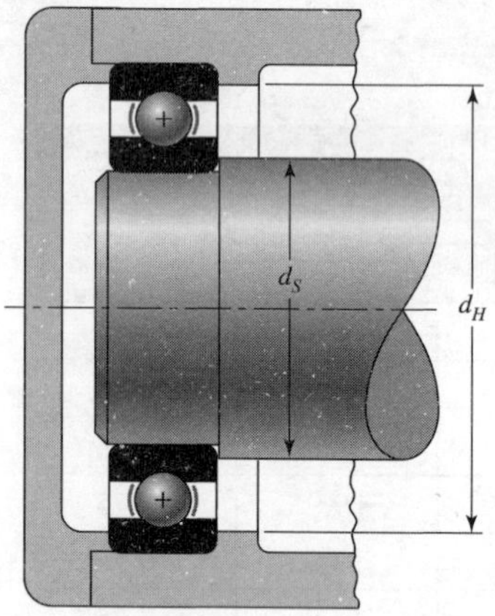

Figure 11–8

Shaft and housing shoulder diameters d_S and d_H should be adequate to ensure good bearing support.

reveal the dimensions directly, it is necessary to resort to tabulations. The 02 series is used here as an example of what is available. See Table 11–2.

The housing and shaft shoulder diameters listed in the tables should be used whenever possible to secure adequate support for the bearing and to resist the maximum thrust loads (Figure 11–8). Table 11–3 lists the dimensions and load ratings of some straight roller bearings.

To assist the designer in the selection of bearings, most of the manufacturers' handbooks contain data on bearing life for many classes of machinery, as well as information on load-application factors. Such information has been accumulated the hard way, that is, by experience, and the beginner designer should utilize this information until enough experience is gained to know when deviations are possible. Table 11–4 contains recommendations on bearing life for some classes of machinery. The load-application factors in Table 11–5 serve the same purpose as factors of safety; use them to increase the equivalent load before selecting a bearing.

Table 11–3 Dimensions and Basic Load Ratings for Cylindrical Roller Bearings

	02-Series				03-Series			
			Load Rating, kN				Load Rating, kN	
Bore, mm	OD, mm	Width, mm	C_{10}	C_0	OD, mm	Width, mm	C_{10}	C_0
25	52	15	16.8	8.8	62	17	28.6	15.0
30	62	16	22.4	12.0	72	19	36.9	20.0
35	72	17	31.9	17.6	80	21	44.6	27.1
40	80	18	41.8	24.0	90	23	56.1	32.5
45	85	19	44.0	25.5	100	25	72.1	45.4
50	90	20	45.7	27.5	110	27	88.0	52.0
55	100	21	56.1	34.0	120	29	102	67.2
60	110	22	64.4	43.1	130	31	123	76.5
65	120	23	76.5	51.2	140	33	138	85.0
70	125	24	79.2	51.2	150	35	151	102
75	130	25	93.1	63.2	160	37	183	125
80	140	26	106	69.4	170	39	190	125
85	150	28	119	78.3	180	41	212	149
90	160	30	142	100	190	43	242	160
95	170	32	165	112	200	45	264	189
100	180	34	183	125	215	47	303	220
110	200	38	229	167	240	50	391	304
120	215	40	260	183	260	55	457	340
130	230	40	270	193	280	58	539	408
140	250	42	319	240	300	62	682	454
150	270	45	446	260	320	65	781	502

Table 11–4 Bearing-Life Recommendations for Various Classes of Machinery

Type of Application	Life, kh
Instruments and apparatus for infrequent use	Up to 0.5
Aircraft engines	0.5–2
Machines for short or intermittent operation where service interruption is of minor importance	4–8
Machines for intermittent service where reliable operation is of great importance	8–14
Machines for 8-h service that are not always fully utilized	14–20
Machines for 8-h service that are fully utilized	20–30
Machines for continuous 24-h service	50–60
Machines for continuous 24-h service where reliability is of extreme importance	100–200

Table 11–5 Load-Application Factors

Type of Application	Load Factor
Precision gearing	1.0–1.1
Commercial gearing	1.1–1.3
Applications with poor bearing seals	1.2
Machinery with no impact	1.0–1.2
Machinery with light impact	1.2–1.5
Machinery with moderate impact	1.5–3.0

EXAMPLE 11–4

An SKF 6210 angular-contact ball bearing has an axial load F_a of 1780 N and a radial load F_r of 2225 N applied with the outer ring stationary. The basic static load rating C_0 is 19 795 N and the basic load rating C_{10} is 35 140 N. Estimate the $\mathscr{L}_{10}$ life at a speed of 720 rev/min.

Solution

$V = 1$ and $F_a/C_0 = 1780/19\ 795 = 0.090$. Interpolate for e in Table 11–1:

F_a/C_0	e	
0.084	0.28	
0.090	e	from which $e = 0.285$
0.110	0.30	

$F_a/(VF_r) = 1780(1)2225 = 0.8 > 0.285$. Thus, interpolate for Y_2:

F_a/C_0	Y_2	
0.084	1.55	
0.090	Y_2	from which $Y_2 = 1.527$
0.110	1.45	

From Equation (11–12),

$$F_e = X_2VF_r + Y_2F_a = 0.56(1)2225 + 1.527(1780) = 3972 \text{ N}$$

With $\mathscr{L}_D = \mathscr{L}_{10}$ and $F_D = F_e$, solving Equation (11–3) for $\mathscr{L}_{10}$ gives

Answer
$$\mathscr{L}_{10} = \frac{60\mathscr{L}_R n_R}{60 n_D}\left(\frac{C_{10}}{F_e}\right)^a = \frac{10^6}{60(720)}\left(\frac{35\ 140}{3927}\right)^3 = 16\ 028 \text{ h}$$

We now know how to combine a steady radial load and a steady thrust load into an equivalent steady radial load F_e that inflicts the same damage per revolution as the radial-thrust combination.

11–7 Variable Loading

Bearing loads are frequently variable and occur in some identifiable patterns:

- Piecewise constant loading in a cyclic pattern
- Continuously variable loading in a repeatable cyclic pattern
- Random variation

Equation (11–1) can be written as

$$F^a L = \text{constant} = K \tag{a}$$

Note that F may already be an equivalent steady radial load for a radial–thrust load combination. Figure 11–9 is a plot of F^a as ordinate and L as abscissa for Equation (a). If a load level of F_1 is selected and run to the failure criterion, then the area under the F_1^a-L_1 trace is numerically equal to K. The same is true for a load level F_2; that is, the area under the F_2^a-L_2 trace is numerically equal to K. The linear damage theory says that in the case of load level F_1, the area from $L = 0$ to $L = L_A$ does damage measured by $F_1^a L_A = D$.

Consider the piecewise continuous cycle depicted in Figure 11–10. The loads F_{ei} are equivalent steady radial loads for combined radial–thrust loads. The damage done by loads F_{e1}, F_{e2}, and F_{e3} is

$$D = F_{e1}^a l_1 + F_{e2}^a l_2 + F_{e3}^a l_3 \tag{b}$$

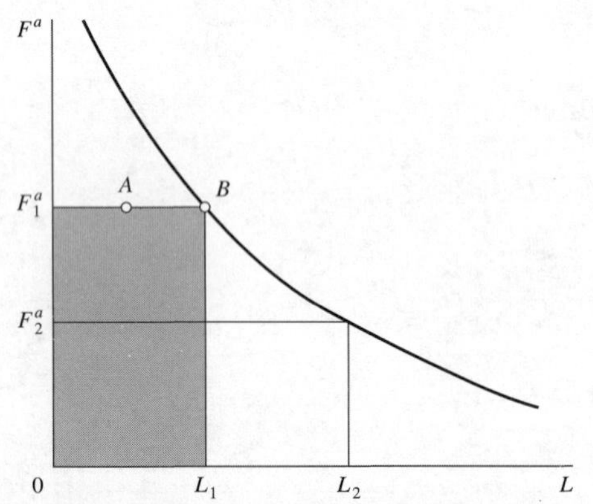

Figure 11–9

Plot of F^a as ordinate and L as abscissa for F^aL = constant. The linear damage hypothesis says that in the case of load F_1, the area under the curve from $L = 0$ to $L = L_A$ is a measure of the damage $D = F_1^a L_A$. The complete damage to failure is measured by $C_{10}^a L_B$.

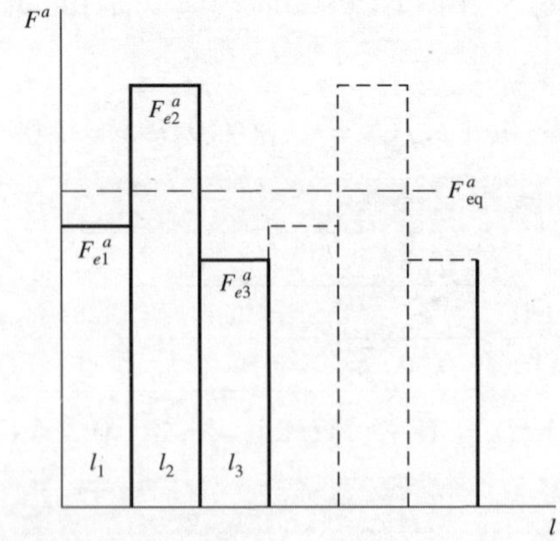

Figure 11–10

A three-part piecewise-continuous periodic loading cycle involving loads F_{e1}, F_{e2}, and F_{e3}. F_{eq} is the equivalent steady load inflicting the same damage when run for $l_1 + l_2 + l_3$ revolutions, doing the same damage D per period.

where l_i is the number of revolutions at life L_i. The equivalent steady load F_{eq} when run for $l_1 + l_2 + l_3$ revolutions does the same damage D. Thus

$$D = F_{eq}^a(l_1 + l_2 + l_3) \tag{c}$$

Equating Equations (*b*) and (*c*), and solving for F_{eq}, we get

$$F_{eq} = \left[\frac{F_{e1}^a l_1 + F_{e2}^a l_2 + F_{e3}^a l_3}{l_1 + l_2 + l_3}\right]^{1/a} = \left[\sum f_i F_{ei}^a\right]^{1/a} \tag{11–13}$$

where f_i is the fraction of revolution run up under load F_{ei}. Since l_i can be expressed as $n_i t_i$, where n_i is the rotational speed at load F_{ei} and t_i is the duration of that speed, then it follows that

$$F_{eq} = \left[\frac{\sum n_i t_i F_{ei}^a}{\sum n_i t_i}\right]^{1/a} \tag{11–14}$$

The character of the individual loads can change, so an application factor (a_f) can be prefixed to each F_{ei} as $(a_{fi}F_{ei})^a$; then Equation (11–13) can be written

$$F_{eq} = \left[\sum f_i(a_{fi}F_{ei})^a\right]^{1/a} \qquad L_{eq} = \frac{K}{F_{eq}^a} \tag{11–15}$$

EXAMPLE 11–5

A ball bearing is run at four piecewise continuous steady loads as shown in the following table. Columns (1), (2), and (5) to (8) are given.

(1) Time Fraction	(2) Speed, rev/min	(3) Product, Column (1) × (2)	(4) Turns Fraction, (3)/Σ(3)	(5) F_{ri}, N	(6) F_{ai}, N	(7) F_{ei}, N	(8) a_{fi}	(9) $a_{fi}F_{ei}$, N
0.1	2000	200	0.077	2668	1334	3532	1.10	3883
0.1	3000	300	0.115	1334	1334	2784	1.25	3536
0.3	3000	900	0.346	3336	1334	3905	1.10	4297
0.5	2400	1200	0.462	1668	1334	2971	1.25	3714
		2600	1.000					

Columns 1 and 2 are multiplied to obtain column 3. The column 3 entry is divided by the sum of column 3, 2600, to give column 4. Columns 5, 6, and 7 are the radial, axial, and equivalent loads, respectively. Column 8 is the appropriate application factor. Column 9 is the product of columns 7 and 8.

Solution

From Equation (11–13), with $a = 3$, the equivalent radial load F_e is

Answer $F_e = [0.077(3883)^3 + 0.115(3536)^3 + 0.346(4297)^3 + 0.462(3714)^3]^{1/3} = 3930$ N

Sometimes the question after several levels of loading is: How much life is left if the next level of stress is held until failure? Failure occurs under the linear damage hypothesis when the damage D equals the constant $K = F^a L$. Taking the first form of Equation (11–13), we write

$$F_{\text{eq}}^a L_{\text{eq}} = F_{e1}^a l_1 + F_{e2}^a l_2 + F_{e3}^a l_3$$

and note that

$$K = F_{e1}^a L_1 = F_{e2}^a L_2 = F_{e3}^a L_3$$

and K also equals

$$K = F_{e1}^a l_1 + F_{e2}^a l_2 + F_{e3}^a l_3 = \frac{K}{L_1} l_1 + \frac{K}{L_2} l_2 + \frac{K}{L_3} l_3 = K \sum \frac{l_i}{L_i}$$

From the outer parts of the preceding equation we obtain

$$\sum \frac{l_i}{L_i} = 1 \qquad \text{(11–16)}$$

This equation was advanced by Palmgren in 1924, and again by Miner in 1945. See Equation (6–69).

The second kind of load variation mentioned is continuous, periodic variation, depicted by Figure 11–11. The differential damage done by F^a during rotation through the angle $d\theta$ is

$$dD = F^a d\theta$$

An example of this would be a cam whose bearings rotate with the cam through the angle $d\theta$. The total damage during a complete cam rotation is given by

$$D = \int dD = \int_0^{\phi} F^a d\theta = F_{\text{eq}}^a \phi$$

from which, solving for the equivalent load, we obtain

$$F_{\text{eq}} = \left[\frac{1}{\phi}\int_0^{\phi} F^a d\theta\right]^{1/a} \qquad L_{\text{eq}} = \frac{K}{F_{\text{eq}}^a} \qquad \text{(11–17)}$$

The value of ϕ is often 2π, although other values occur. Numerical integration is often useful to carry out the indicated integration, particularly when a is not an integer and trigonometric functions are involved. We have now learned how to find the steady equivalent load that does the same damage as a continuously varying cyclic load.

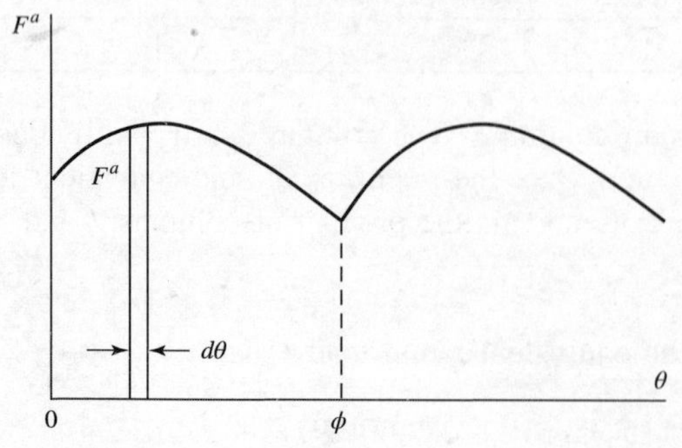

Figure 11–11

A continuous load variation of a cyclic nature whose period is ϕ.

EXAMPLE 11–6

The operation of a particular rotary pump involves a power demand of $P = \overline{P} + A' \sin\theta$ where $\overline{P}$ is the average power. The bearings feel the same variation as $F = \overline{F} + A \sin\theta$. Develop an application factor a_f for this application of ball bearings.

Solution

From Equation (11–17), with $a = 3$,

$$F_{\text{eq}} = \left(\frac{1}{2\pi}\int_0^{2\pi} F^a d\theta\right)^{1/a} = \left(\frac{1}{2\pi}\int_0^{2\pi} (\overline{F} + A \sin\theta)^3 d\theta\right)^{1/3}$$

$$= \left[\frac{1}{2\pi}\left(\int_0^{2\pi} \overline{F}^3 d\theta + 3\overline{F}^2 A \int_0^{2\pi} \sin\theta\, d\theta + 3\overline{F}A^2 \int_0^{2\pi} \sin^2\theta\, d\theta \right.\right.$$

$$\left.\left. + A^3 \int_0^{2\pi} \sin^3\theta\, d\theta\right)\right]^{1/3}$$

$$F_{\text{eq}} = \left[\frac{1}{2\pi}(2\pi\overline{F}^3 + 0 + 3\pi\overline{F}A^2 + 0)\right]^{1/3} = \overline{F}\left[1 + \frac{3}{2}\left(\frac{A}{\overline{F}}\right)^2\right]^{1/3}$$

In terms of $\overline{F}$, the application factor is

Answer

$$a_f = \left[1 + \frac{3}{2}\left(\frac{A}{\overline{F}}\right)^2\right]^{1/3}$$

We can present the result in tabular form:

$A/\overline{F}$	a_f
0	1
0.2	1.02
0.4	1.07
0.6	1.15
0.8	1.25
1.0	1.36

11–8 Selection of Ball and Cylindrical Roller Bearings

We have enough information concerning the loading of rolling-contact ball and roller bearings to develop the steady equivalent radial load that will do as much damage to the bearing as the existing loading. Now let's put it to work.

EXAMPLE 11–7

The second shaft on a parallel-shaft 18.6 kW foundry crane speed reducer contains a helical gear with a pitch diameter of 205.25 mm. Helical gears transmit components of force in the tangential, radial, and axial directions (see Chapter 13). The components of the gear force transmitted to the second shaft are shown in Figure 11–12, at point A. The bearing reactions at C and D, assuming simple-supports, are also shown. A ball bearing

Figure 11–12

Forces in pounds applied to the second shaft of the helical gear speed reducer of Example 11–7.

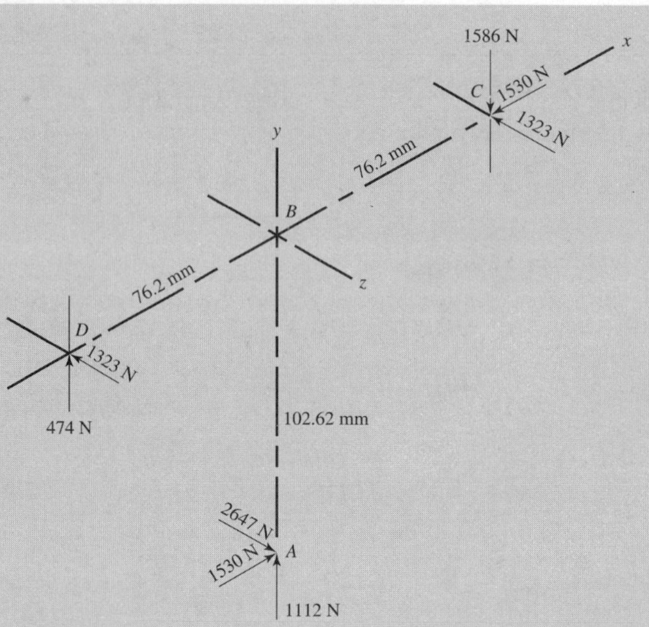

is to be selected for location C to accept the thrust, and a cylindrical roller bearing is to be utilized at location D. The life goal of the speed reducer is 10 kh, with a reliability factor for the ensemble of all four bearings (both shafts) to equal or exceed 0.96 for the Weibull parameters of Example 11–3. The application factor is to be 1.2.

(*a*) Select the roller bearing for location D.

(*b*) Select the ball bearing (angular contact) for location C, assuming the inner ring rotates.

Solution

The torque transmitted is $T = 2647$ N(102.62 mm) $= 271.6$ N · m. The speed at the rated horsepower, given by Equation (3–42), is

$$n_D = \frac{60H}{2(\pi)T} = \frac{60(18.65)}{2(\pi)271.6} = 655.4 \text{ rev/min}$$

The radial load at D is $\sqrt{474^2 + 1323^2} = 1405$ N, and the radial load at C is $\sqrt{1586^2 + 1323^2} = 2065$ N. The individual bearing reliabilities, if equal, must be at least $\sqrt[4]{0.96} = 0.98985 \approx 0.99$. The dimensionless design life for both bearings is

$$x_D = \frac{L_D}{L_{10}} = \frac{60\mathcal{L}_D n_D}{L_{10}} = \frac{60(10\ 000)655.4}{10^6} = 393.2$$

(*a*) From Equation (11–10), the Weibull parameters of Example 11–3, an application factor of 1.2, and $a = 10/3$ for the roller bearing at D, the catalog rating should be equal to or greater than

$$C_{10} = a_f F_D \left[\frac{x_D}{x_0 + (\theta - x_0)(1 - R_D)^{1/b}}\right]^{1/a}$$

$$= 1.2(316.0)\left[\frac{393.2}{0.02 + 4.439(1 - 0.99)^{1/1.483}}\right]^{3/10} = 16.0 \text{ kN}$$

Answer

The absence of a thrust component makes the selection procedure simple. Choose a 02-25 mm series, or a 03-25 mm series cylindrical roller bearing from Table 11–3.

(*b*) The ball bearing at C involves a thrust component. This selection procedure requires an iterative procedure. Assuming $F_a/(VF_r) > e$,

1. Choose Y_2 from Table 11–1.
2. Find C_{10}.
3. Tentatively identify a suitable bearing from Table 11–2, note C_0.
4. Using F_a/C_0 enter Table 11–1 to obtain a new value of Y_2.
5. Find C_{10}.
6. If the same bearing is obtained, stop.
7. If not, take next bearing and go to step 4.

As a first approximation, take the middle entry from Table 11–1:

$$X_2 = 0.56 \qquad Y_2 = 1.63.$$

From Equation (11–12), with $V = 1$,

$$F_e = XVF_r + YF_a = 0.56(1)(464.4) + 1.63(344) = 3.65 \text{ kN}$$

From Equation (11–10), with $a = 3$,

$$C_{10} = 1.2(3.65)\left[\frac{393.2}{0.02 + 4.439(1 - 0.99)^{1/1.483}}\right]^{1/3} = 53.2 \text{ kN}$$

From Table 11–2, angular-contact bearing 02-60 mm has $C_{10} = 55.9$ kN. C_0 is 35.5 kN. Step 4 becomes, with F_a in kN,

$$\frac{F_a}{C_0} = \frac{344(4.45)10^{-3}}{35.5} = 0.0431$$

which makes e from Table 11–1 approximately 0.24. Now $F_a/(VF_r) = 344/[(1)464.4] = 0.74$, which is greater than 0.24, so we find Y_2 by interpolation:

F_a/C_0	Y_2	
0.042	1.85	
0.043	Y_2	from which $Y_2 = 1.84$
0.056	1.71	

From Equation (11–12),

$$F_e = 0.56(1)(464.4) + 1.84(344) = 3.97 \text{ kN}$$

The prior calculation for C_{10} changes only in F_e, so

$$C_{10} = \frac{3.97}{3.65}53.2 = 57.9 \text{ kN}$$

From Table 11–2 an angular contact bearing 02-65 mm has $C_{10} = 63.7$ kN and C_0 of 41.5 kN. Again,

$$\frac{F_a}{C_0} = \frac{344(4.45)10^{-3}}{41.5} = 0.0369$$

making e approximately 0.23. Now from before, $F_a/(VF_r) = 0.74$, which is greater than 0.23. We find Y_2 again by interpolation:

F_a/C_0	Y_2	
0.028	1.99	
0.0369	Y_2	from which $Y_2 = 1.90$
0.042	1.85	

From Equation (11–12),

$$F_e = 0.56(1)(464.4) + 1.90(344) = 4.07 \text{ kN}$$

The prior calculation for C_{10} changes only in F_e, so

$$C_{10} = \frac{4.07}{3.65} 53.2 = 59.3 \text{ kN}$$

Answer

From Table 11–2 an angular-contact 02-65 mm is still selected, so the iteration is complete.

11–9 Selection of Tapered Roller Bearings

Tapered roller bearings have a number of features that make them complicated. As we address the differences between tapered roller and ball and cylindrical roller bearings, note that the underlying fundamentals are the same, but that there are differences in detail. Moreover, bearing and cup combinations are not necessarily priced in proportion to capacity. Any catalog displays a mix of high-production, low-production, and successful special-order designs. Bearing suppliers have computer programs that will take your problem descriptions, give intermediate design assessment information, and list a number of satisfactory cup-and-cone combinations in order of decreasing cost. Company sales offices provide access to comprehensive engineering services to help designers select and apply their bearings. At a large original equipment manufacturer's plant, there may be a resident bearing company representative.

Bearing suppliers provide a wealth of engineering information and detail in their catalogs and engineering guides, both in print and online. It is strongly recommended that the designer become familiar with the specifics of the supplier. It will usually utilize a similar approach as presented here, but may include various modifying factors for such things as temperature and lubrication. Many of the suppliers will provide online software tools to aid in bearing selection. The engineer will always benefit from a general understanding of the theory utilized in such software tools. Our goal here is to introduce the vocabulary, show congruence to fundamentals that were learned earlier, offer examples, and develop confidence. Finally, problems should reinforce the learning experience.

The four components of a tapered roller bearing assembly are the

1 Cone (inner ring)
2 Cup (outer ring)
3 Tapered rollers
4 Cage (spacer-retainer)

The assembled bearing consists of two separable parts: (1) the cone assembly: the cone, the rollers, and the cage; and (2) the cup. Bearings can be made as single-row, two-row, four-row, and thrust-bearing assemblies. Additionally, auxiliary components such as spacers and closures can be used. Figure 11–13 shows the nomenclature of a

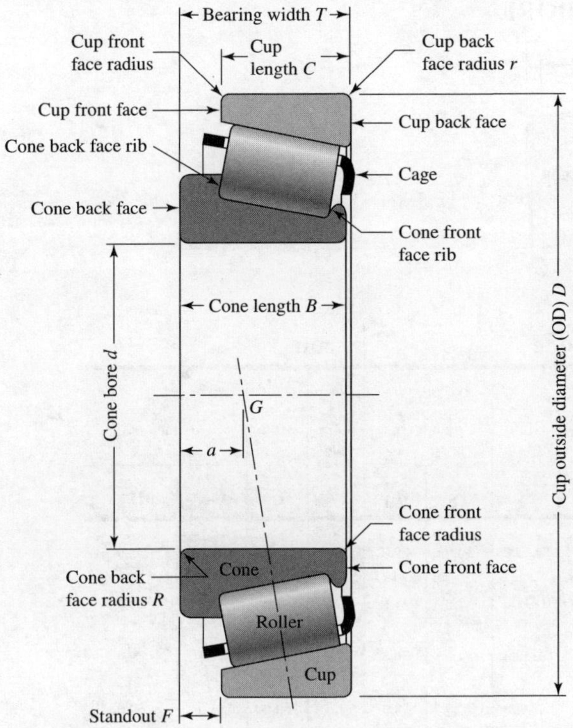

Figure 11–13

Nomenclature of a tapered roller bearing. Point G is the location of the effective load center; use this point to estimate the radial bearing load. *(Source: Redrawn from material Furnished by The Timken Company.)*

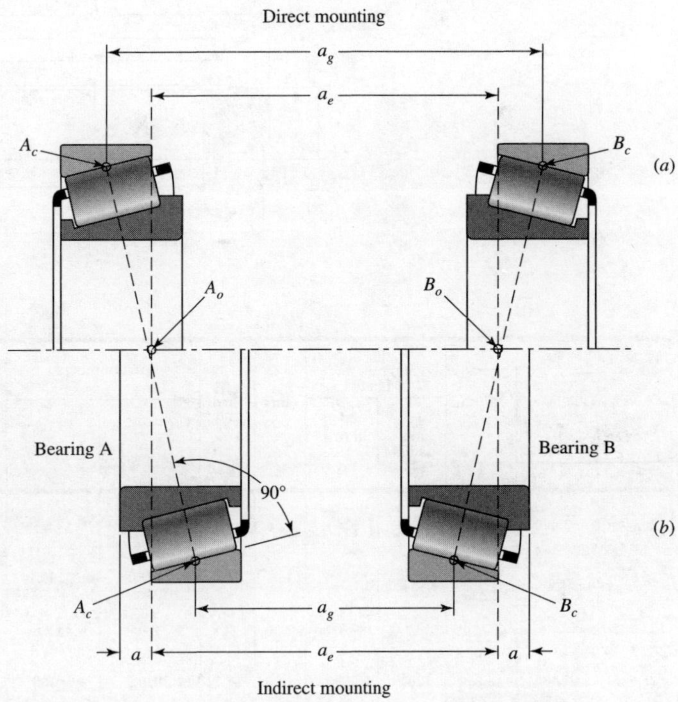

Figure 11–14

Comparison of mounting stability between indirect and direct mountings. *(Source: Redrawn from material Furnished by The Timken Company.)*

tapered roller bearing, and the point G through which radial and axial components of load act.

A tapered roller bearing can carry both radial and thrust (axial) loads, or any combination of the two. However, even when an external thrust load is not present, the radial load will induce a thrust reaction within the bearing because of the taper. To avoid the separation of the races and the rollers, this thrust must be resisted by an equal and opposite force. One way of generating this force is to always use at least two tapered roller bearings on a shaft. Two bearings can be mounted with the cone backs facing each other, in a configuration called *direct mounting,* or with the cone fronts facing each other, in what is called *indirect mounting*.

Figure 11–14 shows a pair of tapered roller bearings mounted directly (*a*) and indirectly (*b*) with the bearing reaction locations A_0 and B_0 shown for the shaft. For the shaft as a beam, the span is a_e, the effective spread. It is through points A_0 and B_0 that the radial loads act perpendicular to the shaft axis, and the thrust loads act along the shaft axis. The geometric spread a_g for the direct mounting is greater than for the indirect mounting. With indirect mounting the bearings are closer together compared to the direct mounting; however, the system stability is the same (a_e is the same in both cases). Thus direct and indirect mounting involve space and compactness needed or desired, but with the same system stability.

In addition to the usual ratings and geometry information, catalog data for tapered roller bearings will include the location of the effective force center. Two sample pages from a Timken catalog are shown in Figure 11–15.

SINGLE-ROW STRAIGHT BORE

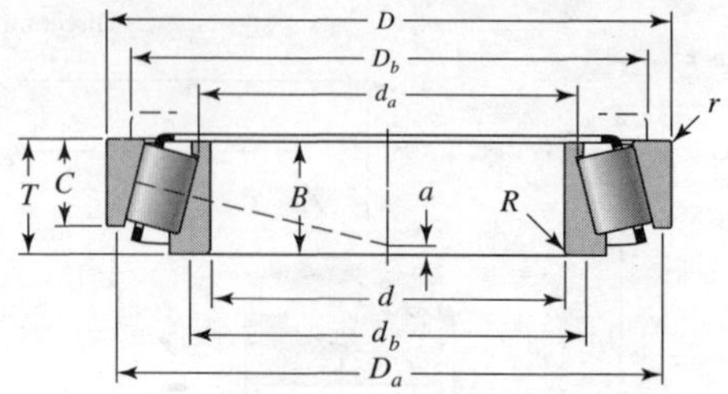

bore	outside diameter	width	rating at 500 rpm for 3000 hours L_{10}		factor	eff. load center	part numbers		cone				cup			
			one-row radial	thrust			cone	cup	max shaft fillet radius	width	backing shoulder diameters		max housing fillet radius	width	backing shoulder diameters	
d	D	T	N	N	K	a②			R①	B	d_b	d_a	r①	C	D_b	D_a
25.000	52.000	16.250	8190	5260	1.56	−3.6	◆30205	◆30205	1.0	15.000	30.5	29.0	1.0	13.000	46.0	48.5
25.000	52.000	19.250	9520	9510	1.00	−3.0	◆32205-B	◆32205-B	1.0	18.000	34.0	31.0	1.0	15.000	43.5	49.5
25.000	52.000	22.000	13200	7960	1.66	−7.6	◆33205	◆33205	1.0	22.000	34.0	30.5	1.0	18.000	44.5	49.0
25.000	62.000	18.250	13000	6680	1.95	−5.1	◆30305	◆30305	1.5	17.000	32.5	30.0	1.5	15.000	55.0	57.0
25.000	62.000	25.250	17400	8930	1.95	−9.7	◆32305	◆32305	1.5	24.000	35.0	31.5	1.5	20.000	54.0	57.0
25.159	50.005	13.495	6990	4810	1.45	−2.8	07096	07196	1.5	14.260	31.5	29.5	1.0	9.525	44.5	47.0
25.400	50.005	13.495	6990	4810	1.45	−2.8	07100	07196	1.0	14.260	30.5	29.5	1.0	9.525	44.5	47.0
25.400	50.005	13.495	6990	4810	1.45	−2.8	07100-S	07196	1.5	14.260	31.5	29.5	1.0	9.525	44.5	47.0
25.400	50.292	14.224	7210	4620	1.56	−3.3	L44642	L44610	3.5	14.732	36.0	29.5	1.3	10.668	44.5	47.0
25.400	50.292	14.224	7210	4620	1.56	−3.3	L44643	L44610	1.3	14.732	31.5	29.5	1.3	10.668	44.5	47.0
25.400	51.994	15.011	6990	4810	1.45	−2.8	07100	07204	1.0	14.260	30.5	29.5	1.3	12.700	45.0	48.0
25.400	56.896	19.368	10900	5740	1.90	−6.9	1780	1729	0.8	19.837	30.5	30.0	1.3	15.875	49.0	51.0
25.400	57.150	19.431	11700	10900	1.07	−3.0	M84548	M84510	1.5	19.431	36.0	33.0	1.5	14.732	48.5	54.0
25.400	58.738	19.050	11600	6560	1.77	−5.8	1986	1932	1.3	19.355	32.5	30.5	1.3	15.080	52.0	54.0
25.400	59.530	23.368	13900	13000	1.07	−5.1	M84249	M84210	0.8	23.114	36.0	32.5	1.5	18.288	49.5	56.0
25.400	60.325	19.842	11000	6550	1.69	−5.1	15578	15523	1.3	17.462	32.5	30.5	1.5	15.875	51.0	54.0
25.400	61.912	19.050	12100	7280	1.67	−5.8	15101	15243	0.8	20.638	32.5	31.5	2.0	14.288	54.0	58.0
25.400	62.000	19.050	12100	7280	1.67	−5.8	15100	15245	3.5	20.638	38.0	31.5	1.3	14.288	55.0	58.0
25.400	62.000	19.050	12100	7280	1.67	−5.8	15101	15245	0.8	20.638	32.5	31.5	1.3	14.288	55.0	58.0

Figure 11–15 *(Continued on next page)*

Catalog entry of single-row straight-bore Timken roller bearings, in part. *(Courtesy of The Timken Company.)*

SINGLE-ROW STRAIGHT BORE

bore	outside diameter	width	rating at 500 rpm for 3000 hours L_{10}		factor	eff. load center	part numbers		cone				cup			
			one-row radial	thrust			cone	cup	max shaft fillet radius	width	backing shoulder diameters		max housing fillet radius	width	backing shoulder diameters	
d	D	T	N	N	K	a②			R①	B	d_b	d_a	r①	C	D_b	D_a
25.400	62.000	19.050	12100	7280	1.67	−5.8	15102	15245	1.5	20.638	34.0	31.5	1.3	14.288	55.0	58.0
25.400	62.000	20.638	12100	7280	1.67	−5.8	15101	15244	0.8	20.638	32.5	31.5	1.3	15.875	55.0	58.0
25.400	63.500	20.638	12100	7280	1.67	−5.8	15101	15250	0.8	20.638	32.5	31.5	1.3	15.875	56.0	59.0
25.400	63.500	20.638	12100	7280	1.67	−5.8	15101	15250X	0.8	20.638	32.5	31.5	1.5	15.875	55.0	59.0
25.400	64.292	21.433	14500	13500	1.07	−3.3	M86643	M86610	1.5	21.433	38.0	36.5	1.5	16.670	54.0	61.0
25.400	65.088	22.225	13100	16400	0.80	−2.3	23100	23256	1.5	21.463	39.0	34.5	1.5	15.875	53.0	63.0
25.400	66.421	23.812	18400	8000	2.30	−9.4	2687	2631	1.3	25.433	33.5	31.5	1.3	19.050	58.0	60.0
25.400	68.262	22.225	15300	10900	1.40	−5.1	02473	02420	0.8	22.225	34.5	33.5	1.5	17.462	59.0	63.0
25.400	72.233	25.400	18400	17200	1.07	−4.6	HM88630	HM88610	0.8	25.400	39.5	39.5	2.3	19.842	60.0	69.0
25.400	72.626	30.162	22700	13000	1.76	−10.2	3189	3120	0.8	29.997	35.5	35.0	3.3	23.812	61.0	67.0
26.157	62.000	19.050	12100	7280	1.67	−5.8	15103	15245	0.8	20.638	33.0	32.5	1.3	14.288	55.0	58.0
26.162	63.100	23.812	18400	8000	2.30	−9.4	2682	2630	1.5	25.433	34.5	32.0	0.8	19.050	57.0	59.0
26.162	66.421	23.812	18400	8000	2.30	−9.4	2682	2631	1.5	25.433	34.5	32.0	1.3	19.050	58.0	60.0
26.975	58.738	19.050	11600	6560	1.77	−5.8	1987	1932	0.8	19.355	32.5	31.5	1.3	15.080	52.0	54.0
† 26.988	50.292	14.224	7210	4620	1.56	−3.3	L44649	L44610	3.5	14.732	37.5	31.0	1.3	10.668	44.5	47.0
† 26.988	60.325	19.842	11000	6550	1.69	−5.1	15580	15523	3.5	17.462	38.5	32.0	1.5	15.875	51.0	54.0
† 26.988	62.000	19.050	12100	7280	1.67	−5.8	15106	15245	0.8	20.638	33.5	33.0	1.3	14.288	55.0	58.0
† 26.988	66.421	23.812	18400	8000	2.30	−9.4	2688	2631	1.5	25.433	35.0	33.0	1.3	19.050	58.0	60.0
28.575	56.896	19.845	11600	6560	1.77	−5.8	1985	1930	0.8	19.355	34.0	33.5	0.8	15.875	51.0	54.0
28.575	57.150	17.462	11000	6550	1.69	−5.1	15590	15520	3.5	17.462	39.5	33.5	1.5	13.495	51.0	53.0
28.575	58.738	19.050	11600	6560	1.77	−5.8	1985	1932	0.8	19.355	34.0	33.5	1.3	15.080	52.0	54.0
28.575	58.738	19.050	11600	6560	1.77	−5.8	1988	1932	3.5	19.355	39.5	33.5	1.3	15.080	52.0	54.0
28.575	60.325	19.842	11000	6550	1.69	−5.1	15590	15523	3.5	17.462	39.5	33.5	1.5	15.875	51.0	54.0
28.575	60.325	19.845	11600	6560	1.77	−5.8	1985	1931	0.5	19.355	34.0	33.5	1.3	15.875	52.0	55.0

① These maximum fillet radii will be cleared by the bearing corners.
② Minus value indicates center is inside cone backface.
† For standard class **ONLY**, the maximum metric size is a whole mm value.
* For "J" part tolerances—see metric tolerances, page 73, and fitting practice, page 65.
◆ ISO cone and cup combinations are designated with a common part number and should be purchased as an assembly.
For ISO bearing tolerances—see metric tolerances, page 73, and fitting practice, page 65.

Figure 11–15 *(Continued)*

A radial load on a tapered roller bearing will induce a thrust reaction. The *load zone* includes about half the rollers and subtends an angle of approximately 180°. Using the symbol F_i for the induced thrust load from a radial load with a 180° load zone, Timken provides the equation

$$F_i = \frac{0.47F_r}{K} \tag{11–18}$$

where the K factor is geometry-specific, and is the ratio of the radial load rating to the thrust load rating. The K factor can be first approximated with 1.5 for a radial bearing and 0.75 for a steep angle bearing in the preliminary selection process. After a possible bearing is identified, the exact value of K for each bearing can be found in the bearing catalog.

A shaft supported by a pair of direct-mounted tapered roller bearings is shown in Figure 11–16. Force vectors are shown as applied to the shaft. F_{rA} and F_{rB} are the radial loads carried by the bearings, applied at the effective force centers G_A and G_B. The induced loads F_{iA} and F_{iB} due to the effect of the radial loads on the tapered bearings are also shown. Additionally, there may be an externally applied thrust load F_{ae} on the shaft from some other source, such as the axial load on a helical gear. Since the bearings experience both radial and thrust loads, it is necessary to determine equivalent radial loads. Following the form of Equation (11–12), where $F_e = XVF_r + YF_a$, Timken recommends using $X = 0.4$ and $V = 1$ for all cases, and using the K factor for the specific bearing for Y. This gives an equation of the form

$$F_e = 0.4F_r + KF_a \tag{a}$$

The axial load F_a is the net axial load carried by the bearing due to the combination of the induced axial load from the other bearing and the external axial load. However, only one of the bearings will carry the net axial load, and which one it is depends on the direction the bearings are mounted, the relative magnitudes of the induced loads, the direction of the external load, and whether the shaft or the housing is the moving part. Timken handles it with a table containing each of the configurations and a sign convention on the external loads. It further requires the application to be oriented horizontally with left and right bearings that must match the left and right

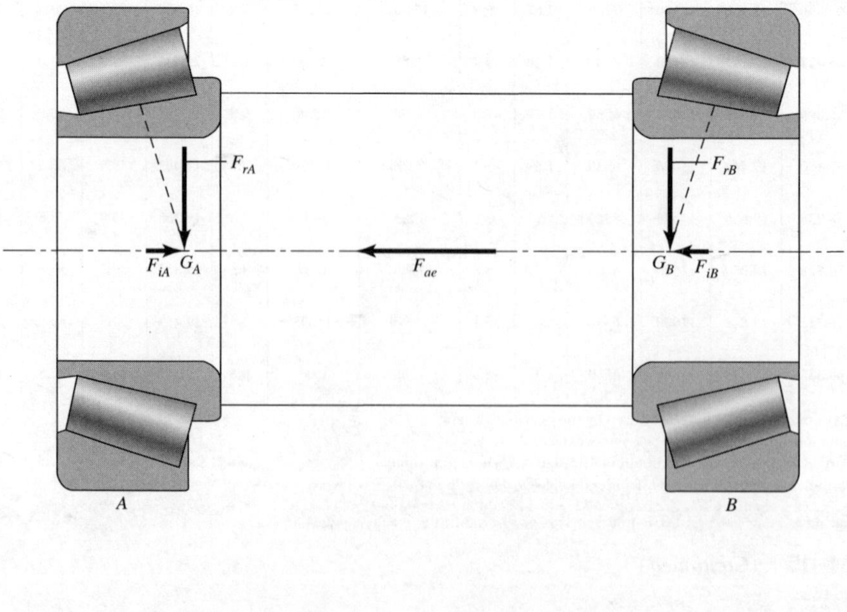

Figure 11–16

Direct-mounted tapered roller bearings, showing radial, induced thrust, and external thrust loads.

sign conventions. Here, we will present a method that gives equivalent results, but that is perhaps more conducive to visualizing and understanding the logic behind it.

First, determine visually which bearing is being "squeezed" by the external thrust load, and label it as bearing *A*. Label the other bearing as bearing *B*. For example, in Figure 11–16, the external thrust F_{ae} causes the shaft to push to the left against the cone of the left bearing, squeezing it against the rollers and the cup. On the other hand, it tends to pull apart the cup from the right bearing. The left bearing is therefore labeled as bearing *A*. If the direction of F_{ae} were reversed, then the right bearing would be labeled as bearing *A*. This approach to labeling the bearing being squeezed by the external thrust is applied similarly regardless of whether the bearings are mounted directly or indirectly, regardless of whether the shaft or the housing carries the external thrust, and regardless of the orientation of the assembly. To clarify by example, consider the vertical shaft and cylinder in Figure 11–17 with direct-mounted bearings. In Figure 11–17*a*, an external load is applied in the upward direction to a rotating shaft, compressing the top bearing, which should be labeled as bearing *A*. On the other hand, in Figure 11–17*b*, an upward external load is applied to a rotating outer cylinder with a stationary shaft. In this case, the lower bearing is being squeezed and should be labeled as bearing *A*. If there is no external thrust, then either bearing can arbitrarily be labeled as bearing *A*.

Second, determine which bearing actually carries the net axial load. Generally, it would be expected that bearing *A* would carry the axial load, since the external thrust F_{ae} is directed toward *A*, along with the induced thrust F_{iB} from bearing *B*. However, if the induced thrust F_{iA} from bearing *A* happens to be larger than the combination of the external thrust and the thrust induced by bearing *B*, then bearing *B* will carry the net thrust load. We will use Equation (*a*) for the bearing carrying the thrust load. Timken recommends leaving the other bearing at its original radial load, rather than reducing it due to the negative net thrust load. The results are presented in equation form below, where the induced thrusts are defined by Equation (11–18).

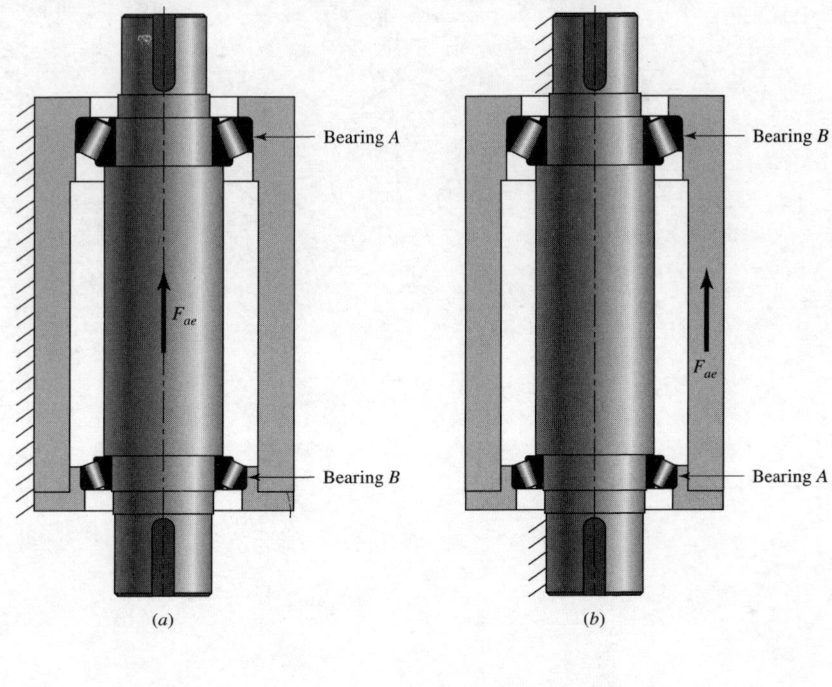

Figure 11–17

Examples of determining which bearing carries the external thrust load. In each case, the compressed bearing is labeled as bearing *A*. (*a*) External thrust applied to rotating shaft; (*b*) External thrust applied to rotating cylinder.

$$\text{If} \quad F_{iA} \le (F_{iB} + F_{ae}) \qquad \begin{cases} F_{eA} = 0.4F_{rA} + K_A(F_{iB} + F_{ae}) & \text{(11–19a)} \\ F_{eB} = F_{rB} & \text{(11–19b)} \end{cases}$$

$$\text{If} \quad F_{iA} > (F_{iB} + F_{ae}) \qquad \begin{cases} F_{eB} = 0.4F_{rB} + K_B(F_{iA} + F_{ae}) & \text{(11–20a)} \\ F_{eA} = F_{rA} & \text{(11–20b)} \end{cases}$$

In any case, if the equivalent radial load is ever less than the original radial load, then the original radial load should be used.

Once the equivalent radial loads are determined, they should be used to find the catalog rating load using any of Equations (11–3), (11–9), or (11–10) as before. Timken uses a Weibull model with $x_0 = 0$, $\theta = 4.48$, and $b = 3/2$. Note that since K_A and K_B are dependent on the specific bearing chosen, it may be necessary to iterate the process.

EXAMPLE 11–8

The shaft depicted in Figure 11–18*a* carries a helical gear with a tangential force of 3980 N, a radial force of 1770 N, and a thrust force of 1690 N at the pitch cylinder with directions shown. The pitch diameter of the gear is 200 mm. The shaft runs at a speed of 800 rev/min, and the span (effective spread) between the direct-mount bearings is 150 mm. The design life is to be 5000 h and an application factor of 1 is appropriate. If the reliability of the bearing set is to be 0.99, select suitable single-row tapered-roller Timken bearings.

Solution

The reactions in the *xz* plane from Figure 11–18*b* are

$$R_{zA} = \frac{3980(50)}{150} = 1327 \text{ N}$$

$$R_{zB} = \frac{3980(100)}{150} = 2653 \text{ N}$$

Figure 11–18

Essential geometry of helical gear and shaft. Length dimensions in mm, loads in N, couple in N · mm. (*a*) Sketch (not to scale) showing thrust, radial, and tangential forces. (*b*) Forces in *xz* plane. (*c*) Forces in *xy* plane.

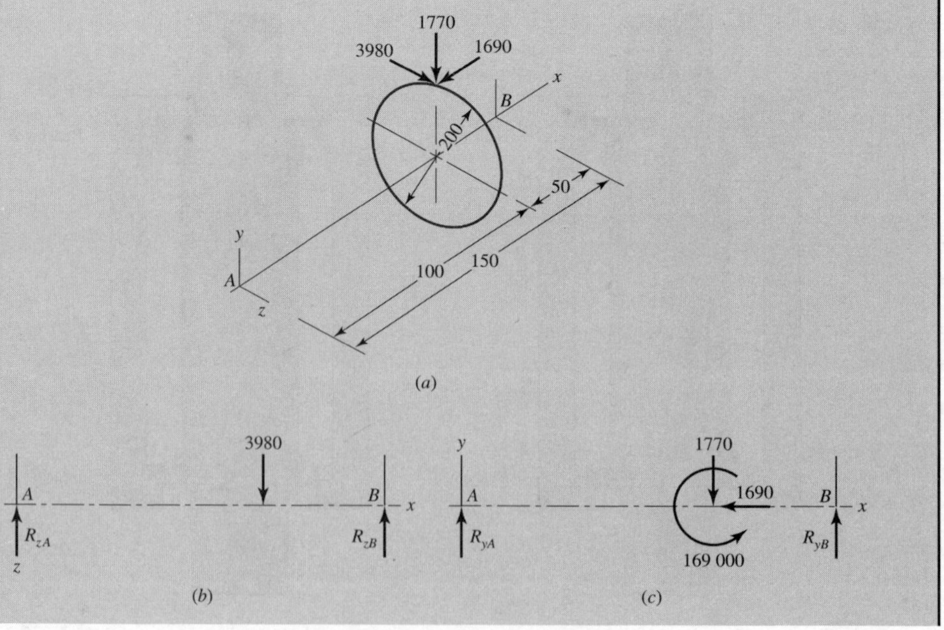

The reactions in the xy plane from Figure 11–18c are

$$R_{yA} = \frac{1770(50)}{150} + \frac{169\ 000}{150} = 1716.7 = 1717 \text{ N}$$

$$R_{yB} = \frac{1770(100)}{150} - \frac{169\ 000}{150} = 53.3 \text{ N}$$

The radial loads F_{rA} and F_{rB} are the vector additions of R_{yA} and R_{zA}, and R_{yB} and R_{zB}, respectively:

$$F_{rA} = \left(R_{zA}^2 + R_{yA}^2\right)^{1/2} = (1327^2 + 1717^2)^{1/2} = 2170 \text{ N}$$

$$F_{rB} = \left(R_{zB}^2 + R_{yB}^2\right)^{1/2} = (2653^2 + 53.3^2)^{1/2} = 2654 \text{ N}$$

Trial 1: With direct mounting of the bearings and application of the external thrust to the shaft, the squeezed bearing is bearing A as labeled in Figure 11–18a. Using K of 1.5 as the initial guess for each bearing, the induced loads from the bearings are

$$F_{iA} = \frac{0.47F_{rA}}{K_A} = \frac{0.47(2170)}{1.5} = 680 \text{ N}$$

$$F_{iB} = \frac{0.47F_{rB}}{K_B} = \frac{0.47(2654)}{1.5} = 832 \text{ N}$$

Since F_{iA} is clearly less than $F_{iB} + F_{ae}$, bearing A carries the net thrust load, and Equation (11–19) is applicable. Therefore, the dynamic equivalent loads are

$$F_{eA} = 0.4F_{rA} + K_A(F_{iB} + F_{ae}) = 0.4(2170) + 1.5(832 + 1690) = 4651 \text{ N}$$

$$F_{eB} = F_{rB} = 2654 \text{ N}$$

The multiple of rating life is

$$x_D = \frac{L_D}{L_R} = \frac{\mathcal{L}_D n_D 60}{L_R} = \frac{(5000)(800)(60)}{90(10^6)} = 2.67$$

Estimate R_D as $\sqrt{0.99} = 0.995$ for each bearing. For bearing A, from Equation (11–10) the catalog entry C_{10} should equal or exceed

$$C_{10} = (1)(4651)\left[\frac{2.67}{(4.48)(1 - 0.995)^{2/3}}\right]^{3/10} = 11\ 486 \text{ N}$$

From Figure 11–15, tentatively select type TS 15 100 cone and 15 245 cup, which will work: $K_A = 1.67$, $C_{10} = 12\ 100$ N.

For bearing B, from Equation (11–10), the catalog entry C_{10} should equal or exceed

$$C_{10} = (1)2654\left[\frac{2.67}{(4.48)(1 - 0.995)^{2/3}}\right]^{3/10} = 6554 \text{ N}$$

Tentatively select the bearing identical to bearing A, which will work: $K_B = 1.67$, $C_{10} = 12\ 100$ N.

Trial 2: Repeat the process with $K_A = K_B = 1.67$ from tentative bearing selection.

$$F_{iA} = \frac{0.47F_{rA}}{K_A} = \frac{0.47(2170)}{1.67} = 611 \text{ N}$$

$$F_{iB} = \frac{0.47F_{rB}}{K_B} = \frac{0.47(2654)}{1.67} = 747 \text{ N}$$

Since F_{iA} is still less than $F_{iB} + F_{ae}$, Equation (11–19) is still applicable.

$$F_{eA} = 0.4F_{rA} + K_A(F_{iB} + F_{ae}) = 0.4(2170) + 1.67(747 + 1690) = 4938 \text{ N}$$

$$F_{eB} = F_{rB} = 2654 \text{ N}$$

For bearing A, from Equation (11–10) the corrected catalog entry C_{10} should equal or exceed

$$C_{10} = (1)(4938)\left[\frac{2.67}{(4.48)(1 - 0.995)^{2/3}}\right]^{3/10} = 12\,195 \text{ N}$$

Although this catalog entry exceeds slightly the tentative selection for bearing A, we will keep it since the reliability of bearing B exceeds 0.995. In the next section we will quantitatively show that the combined reliability of bearing A and B will exceed the reliability goal of 0.99.

For bearing B, $F_{eB} = F_{rB} = 2654$ N. From Equation (11–10),

$$C_{10} = (1)2654\left[\frac{2.67}{(4.48)(1 - 0.995)^{2/3}}\right]^{3/10} = 6554 \text{ N}$$

Select cone and cup 15 100 and 15 245, respectively, for both bearing A and B. Note from Figure 11–14 the effective load center is located at $a = -5.8$ mm, that is, 5.8 mm into the cup from the back. Thus the shoulder-to-shoulder dimension should be $150 - 2(5.8) = 138.4$ mm. Note that in each iteration of Equation (11–10) to find the catalog load rating, the bracketed portion of the equation is identical and need not be re-entered on a calculator each time.

11–10 Design Assessment for Selected Rolling-Contact Bearings

In textbooks, machine elements typically are treated singly. This can lead the reader to the presumption that a design assessment involves only that element, in this case a rolling-contact bearing. The immediately adjacent elements (the shaft journal and the housing bore) have immediate influence on the performance. Other elements, further removed (gears producing the bearing load), also have influence. Just as some say, "If you pull on something in the environment, you find that it is attached to everything else." This should be intuitively obvious to those involved with machinery. How, then, can one check shaft attributes that aren't mentioned in a problem statement? Possibly, because the bearing hasn't been designed yet (in fine detail). All this points out the necessary iterative nature of designing, say, a speed reducer. If power, speed, and reduction are stipulated, then gear sets can be roughed in, their sizes, geometry, and location estimated, shaft forces and moments identified, bearings tentatively selected, seals identified; the bulk is beginning to make itself evident, the housing and lubricating scheme as well as the cooling considerations become clearer, shaft overhangs and coupling accommodations appear. It is time to iterate, now addressing each element again, knowing much more about all of the others. When you have completed the necessary iterations, you will know what you need for the design assessment for the bearings. In the meantime you do as much of the design assessment as you can, avoiding bad selections, even if tentative. Always keep in mind that you eventually have to do it all in order to pronounce your completed design satisfactory.

An outline of a design assessment for a rolling contact bearing includes, at a minimum,

- Bearing reliability for the load imposed and life expected
- Shouldering on shaft and housing satisfactory
- Journal finish, diameter and tolerance compatible
- Housing finish, diameter and tolerance compatible
- Lubricant type according to manufacturer's recommendations; lubricant paths and volume supplied to keep operating temperature satisfactory
- Preloads, if required, are supplied

Since we are focusing on rolling-contact bearings, we can address bearing reliability quantitatively, as well as shouldering. Other quantitative treatment will have to wait until the materials for shaft and housing, surface quality, and diameters and tolerances are known.

Bearing Reliability

Equation (11–9) can be solved for the reliability R_D in terms of C_{10}, the basic load rating of the selected bearing:

$$R = \exp\left(-\left\{\frac{x_D\left(\frac{a_f F_D}{C_{10}}\right)^a - x_0}{\theta - x_0}\right\}^b\right) \tag{11–21}$$

Equation (11–10) can likewise be solved for R_D:

$$R \approx 1 - \left\{\frac{x_D\left(\frac{a_f F_D}{C_{10}}\right)^a - x_0}{\theta - x_0}\right\}^b \qquad R \geq 0.90 \tag{11–22}$$

EXAMPLE 11–9

In Example 11–3, the minimum required load rating for 99 percent reliability, at $x_D = L_D/L_{10} = 540$, is $C_{10} =$ 29.8 kN. From Table 11–2 a 02-40 mm deep-groove ball bearing would satisfy the requirement. If the bore in the application had to be 70 mm or larger (selecting a 02-70 mm deep-groove ball bearing), what is the resulting reliability?

Solution

From Table 11–2, for a 02-70 mm deep-groove ball bearing, $C_{10} = 61.8$ kN. Using Equation (11–22), recalling from Example 11–3 that $a_f = 1.2$, $F_D = 1840$ N, $x_0 = 0.02$, $(\theta - x_0) = 4.439$, and $b = 1.483$, we can write

Answer
$$R \approx 1 - \left\{\frac{\left[540\left[\frac{1.2(1840)}{61\ 800}\right]^3 - 0.02\right]}{4.439}\right\}^{1.483} = 0.999\ 965$$

which, as expected, is much higher than 0.99 from Example 11–3.

In tapered roller bearings, or other bearings for a two-parameter Weibull distribution, Equation (11–21) becomes, for $x_0 = 0$, $\theta = 4.48$, $b = \frac{3}{2}$,

$$R = \exp\left\{ -\left[\frac{x_D}{\theta[C_{10}/(a_f F_D)]^a} \right]^b \right\}$$

$$= \exp\left\{ -\left[\frac{x_D}{4.48[C_{10}/(a_f F_D)]^{10/3}} \right]^{3/2} \right\} \qquad \text{(11–23)}$$

and Equation (11–22) becomes

$$R \approx 1 - \left\{ \frac{x_D}{\theta[C_{10}/(a_f F_D)]^a} \right\}^b = 1 - \left\{ \frac{x_D}{4.48[C_{10}/(a_f F_D)]^{10/3}} \right\}^{3/2} \qquad \text{(11–24)}$$

EXAMPLE 11–10

In Example 11–8 bearings A and B (cone 15 100 and cup 15 245) have $C_{10} = 12\ 100$ N. What is the reliability of the pair of bearings A and B?

Solution

The desired life x_D was $5000(800)60/[90(10^6)] = 2.67$ rating lives. Using Equation (11–24) for bearing A, where from Example 11–8, $F_D = F_{eA} = 4938$ N, and $a_f = 1$, gives

$$R_A \approx 1 - \left\{ \frac{2.67}{4.48[12\ 100/(1 \times 4938)]^{10/3}} \right\}^{3/2} = 0.994\ 791$$

which is less than 0.995, as expected. Using Equation (11–24) for bearing B with $F_D = F_{eB} = 2654$ N gives

$$R_B \approx 1 - \left\{ \frac{2.67}{4.48[12\ 100/(1 \times 2654)]^{10/3}} \right\}^{3/2} = 0.999\ 766$$

Answer

The reliability of the bearing pair is

$$R = R_A R_B = 0.994\ 791(0.999\ 766) = 0.994\ 558$$

which is greater than the overall reliability goal of 0.99. When two bearings are made identical for simplicity, or reducing the number of spares, or other stipulation, and the loading is not the same, both can be made smaller and still meet a reliability goal. If the loading is disparate, then the more heavily loaded bearing can be chosen for a reliability goal just slightly larger than the overall goal.

An additional example is useful to show what happens in cases of pure thrust loading.

EXAMPLE 11–11

Consider a constrained housing as depicted in Figure 11–19 with two direct-mount tapered roller bearings resisting an external thrust F_{ae} of 8000 N. The shaft speed is 950 rev/min, the desired life is 10 000 h, the expected shaft diameter is approximately 1 in. The reliability goal is 0.95. The application factor is appropriately $a_f = 1$.

(*a*) Choose a suitable tapered roller bearing for A.

(*b*) Choose a suitable tapered roller bearing for B.

(*c*) Find the reliabilities R_A, R_B, and R.

Solution

(*a*) By inspection, note that the left bearing carries the axial load and is properly labeled as bearing *A*. The bearing reactions at *A* are

$$F_{rA} = F_{rB} = 0$$

$$F_{aA} = F_{ae} = 8000 \text{ N}$$

Since bearing *B* is unloaded, we will start with $R = R_A = 0.95$.

With no radial loads, there are no induced thrust loads. Equation (11–19) is applicable.

$$F_{eA} = 0.4F_{rA} + K_A(F_{iB} + F_{ae}) = K_A F_{ae}$$

If we set $K_A = 1$, we can find C_{10} in the thrust column and avoid iteration:

$$F_{eA} = (1)8000 = 8000 \text{ N}$$

$$F_{eB} = F_{rB} = 0$$

The multiple of rating life is

$$x_D = \frac{L_D}{L_R} = \frac{\mathscr{L}_D n_D 60}{L_R} = \frac{(10\ 000)(950)(60)}{90(10^6)} = 6.333$$

Then, from Equation (11–10), for bearing *A*

$$C_{10} = a_f F_{eA}\left[\frac{x_D}{4.48(1 - R_D)^{2/3}}\right]^{3/10}$$

$$= (1)8000\left[\frac{6.33}{4.48(1 - 0.95)^{2/3}}\right]^{3/10} = 16\ 159 \text{ N}$$

Answer

Figure 11–15 presents one possibility in the 1-in bore (25.4-mm) size: cone, HM88 630, cup HM88 610 with a thrust rating $(C_{10})_a = 17\ 200$ N.

Answer

(*b*) Bearing *B* experiences no load, and the cheapest bearing of this bore size will do, including a ball or roller bearing.

(*c*) The actual reliability of bearing *A*, from Equation (11–24), is

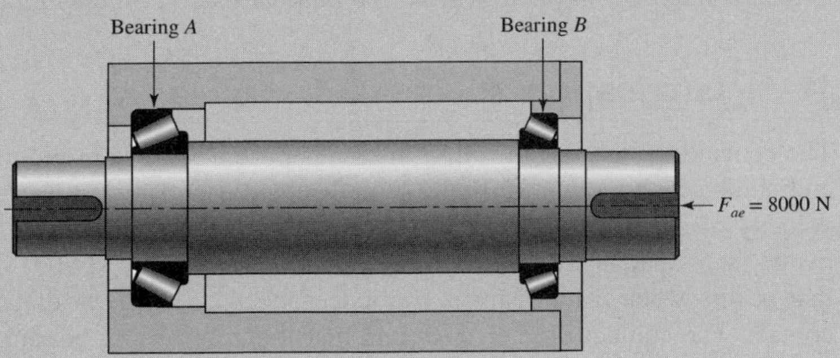

Figure 11–19

The constrained housing of Example 11–11.

Answer
$$R_A \approx 1 - \left\{ \frac{x_D}{4.48[C_{10}/(a_f F_D)]^{10/3}} \right\}^{3/2}$$

$$\approx 1 - \left\{ \frac{6.333}{4.48[17\ 200/(1 \times 8000)]^{10/3}} \right\}^{3/2} = 0.963$$

which is greater than 0.95, as one would expect. For bearing B,

$$F_D = F_{eB} = 0$$

Answer
$$R_B \approx 1 - \left[\frac{6.333}{0.85(17\ 200/0)^{10/3}} \right]^{3/2} = 1 - 0 = 1$$

as one would expect. The combined reliability of bearings A and B as a pair is

Answer
$$R = R_A R_B = 0.963(1) = 0.963$$

which is greater than the reliability goal of 0.95, as one would expect.

Matters of Fit

Table 11–2 (and Figure 11–8), which shows the rating of single-row, 02-series, deep-groove and angular-contact ball bearings, includes shoulder diameters recommended for the shaft seat of the inner ring and the shoulder diameter of the outer ring, denoted d_S and d_H, respectively. The shaft shoulder can be greater than d_S but not enough to obstruct the annulus. It is important to maintain concentricity and perpendicularity with the shaft centerline, and to that end the shoulder diameter should equal or exceed d_S. The housing shoulder diameter d_H is to be equal to or less than d_H to maintain concentricity and perpendicularity with the housing bore axis. Neither the shaft shoulder nor the housing shoulder features should allow interference with the free movement of lubricant through the bearing annulus.

In a tapered roller bearing (Figure 11–15), the cup housing shoulder diameter should be equal to or less than D_b. The shaft shoulder for the cone should be equal to or greater than d_b. Additionally, free lubricant flow is not to be impeded by obstructing any of the annulus. In splash lubrication, common in speed reducers, the lubricant is thrown to the housing cover (ceiling) and is directed in its draining by ribs to a bearing. In direct mounting, a tapered roller bearing pumps oil from outboard to inboard. An oil passageway to the outboard side of the bearing needs to be provided. The oil returns to the sump as a consequence of bearing pump action. With an indirect mount, the oil is directed to the inboard annulus, the bearing pumping it to the outboard side. An oil passage from the outboard side to the sump has to be provided.

11–11 Lubrication

The contacting surfaces in rolling bearings have a relative motion that is both rolling and sliding, and so it is difficult to understand exactly what happens. If the relative velocity of the sliding surfaces is high enough, then the lubricant action is hydrodynamic (see Chapter 12). *Elastohydrodynamic lubrication* (EHD) is the phenomenon that occurs when a lubricant is introduced between surfaces that are in pure rolling contact. The contact of gear teeth and that found in rolling bearings and in cam-and-follower surfaces are typical examples. When a lubricant is trapped between two surfaces in rolling contact, a tremendous increase in the pressure within the lubricant film occurs. But viscosity is exponentially related to pressure, and so a very large increase in viscosity occurs in the lubricant that is trapped between the surfaces. Leibensperger[2] observes that the change in viscosity in and out of contact pressure is equivalent to the difference between cold asphalt and light sewing machine oil.

The purposes of an antifriction-bearing lubricant may be summarized as follows:

- To provide a film of lubricant between the sliding and rolling surfaces
- To help distribute and dissipate heat
- To prevent corrosion of the bearing surfaces
- To protect the parts from the entrance of foreign matter

Either oil or grease may be employed as a lubricant. The following rules may help in deciding between them.

Use Grease When	Use Oil When
• The temperature is not over 200°F. • The speed is low. • Unusual protection is required from the entrance of foreign matter. • Simple bearing enclosures are desired. • Operation for long periods without attention is desired.	• Speeds are high. • Temperatures are high. • Oiltight seals are readily employed. • Bearing type is not suitable for grease lubrication. • The bearing is lubricated from a central supply which is also used for other machine parts.

11–12 Mounting and Enclosure

There are so many methods of mounting antifriction bearings that each new design is a real challenge to the ingenuity of the designer. The housing bore and shaft outside diameter must be held to very close limits, which of course is expensive. There are usually one or more counterboring operations, several facing operations and drilling, tapping, and threading operations, all of which must be performed on the shaft, housing, or cover plate. Each of these operations contributes to the cost of production, so that the designer, in ferreting out a trouble-free and low-cost mounting, is faced with a difficult and important problem. The various bearing manufacturers' handbooks give many mounting details in almost every design area. In a text of this nature, however, it is possible to give only the barest details.

The most frequently encountered mounting problem is that which requires one bearing at each end of a shaft. Such a design might use one ball bearing at each end, one tapered roller bearing at each end, or a ball bearing at one end and a straight roller bearing at the other. One of the bearings usually has the added function of positioning or axially locating the shaft. Figure 11–20 shows a very common solution to this problem. The inner rings are backed up against the shaft shoulders and are held in position by round nuts threaded onto the shaft. The outer ring of the left-hand bearing is backed up against a housing shoulder and is held in position by a device that is not shown. The outer ring of the right-hand bearing floats in the housing.

[2]R. L. Leibensperger, "When Selecting a Bearing," *Machine Design*, vol. 47, no. 8, April 3, 1975, pp. 142–147.

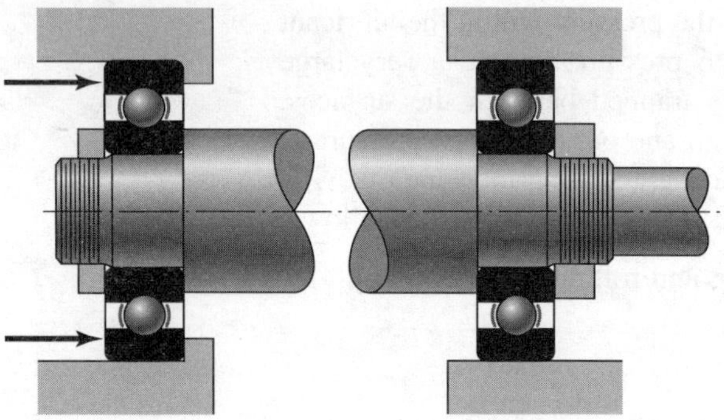

Figure 11–20

A common bearing mounting.

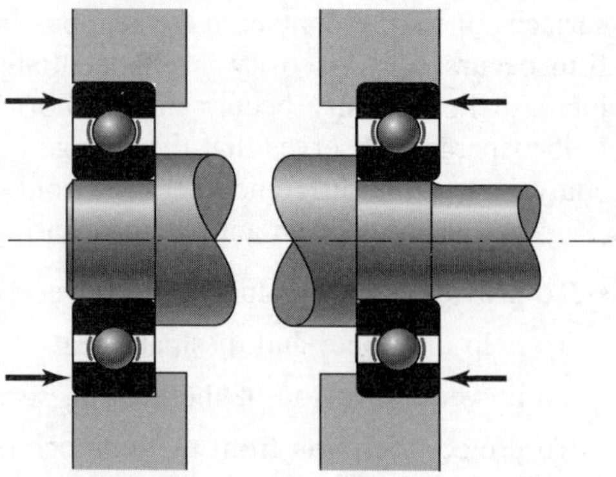

Figure 11–21

An alternative bearing mounting to that in Figure 11–20.

There are many variations possible on the method shown in Figure 11–20. For example, the function of the shaft shoulder may be performed by retaining rings, by the hub of a gear or pulley, or by spacing tubes or rings. The round nuts may be replaced by retaining rings or by washers locked in position by screws, cotters, or taper pins. The housing shoulder may be replaced by a retaining ring; the outer ring of the bearing may be grooved for a retaining ring, or a flanged outer ring may be used. The force against the outer ring of the left-hand bearing is usually applied by the cover plate, but if no thrust is present, the ring may be held in place by retaining rings.

Figure 11–21 shows an alternative method of mounting in which the inner races are backed up against the shaft shoulders as before but no retaining devices are required. With this method the outer races are completely retained. This eliminates the grooves or threads, which cause stress concentration on the overhanging end, but it requires accurate dimensions in an axial direction or the employment of adjusting means. This method has the disadvantage that if the distance between the bearings is great, the temperature rise during operation may expand the shaft enough to destroy the bearings.

It is frequently necessary to use two or more bearings at one end of a shaft. For example, two bearings could be used to obtain additional rigidity or increased load capacity or to cantilever a shaft. Several two-bearing mountings are shown in Figure 11–22. These may be used with tapered roller bearings, as shown, or with ball bearings. In either case it should be noted that the effect of the mounting is to preload the bearings in an axial direction.

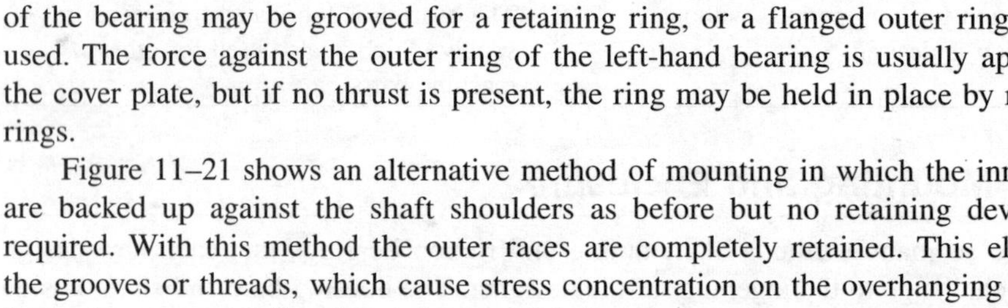

(a) (b)

Figure 11–22

Two-bearing mountings. *(Source: Redrawn from material Furnished by The Timken Company.)*

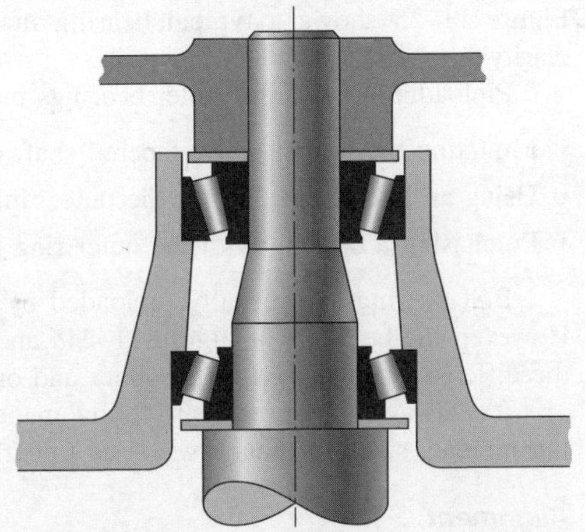

Figure 11–23

Mounting for a washing-machine spindle. *(Source: Redrawn from material Furnished by The Timken Company.)*

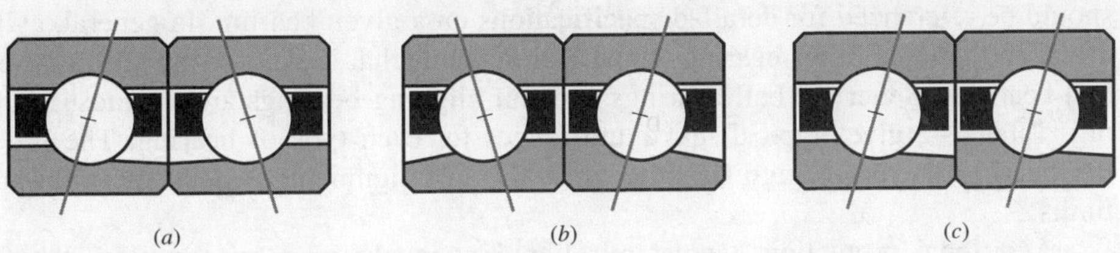

Figure 11–24

Arrangements of angular ball bearings. (*a*) DF mounting; (*b*) DB mounting; (*c*) DT mounting. *(Source: Redrawn from material Furnished by The Timken Company.)*

Figure 11–23 shows another two-bearing mounting. Note the use of washers against the cone backs.

When maximum stiffness and resistance to shaft misalignment is desired, pairs of angular-contact ball bearings (Figure 11–2) are often used in an arrangement called *duplexing*. Bearings manufactured for duplex mounting have their rings ground with an offset, so that when a pair of bearings is tightly clamped together, a preload is automatically established. As shown in Figure 11–24, three mounting arrangements are used. The face-to-face mounting, called DF, will take heavy radial loads and thrust loads from either direction. The DB mounting (back to back) has the greatest aligning stiffness and is also good for heavy radial loads and thrust loads from either direction. The tandem arrangement, called the DT mounting, is used where the thrust is always in the same direction; since the two bearings have their thrust functions in the same direction, a preload, if required, must be obtained in some other manner.

Bearings are usually mounted with the rotating ring a press fit, whether it be the inner or outer ring. The stationary ring is then mounted with a push fit. This permits the stationary ring to creep in its mounting slightly, bringing new portions of the ring into the load-bearing zone to equalize wear.

Preloading

The object of preloading is to:

- Remove the internal clearance usually found in bearings
- Increase the fatigue life
- Decrease the shaft slope at the bearing

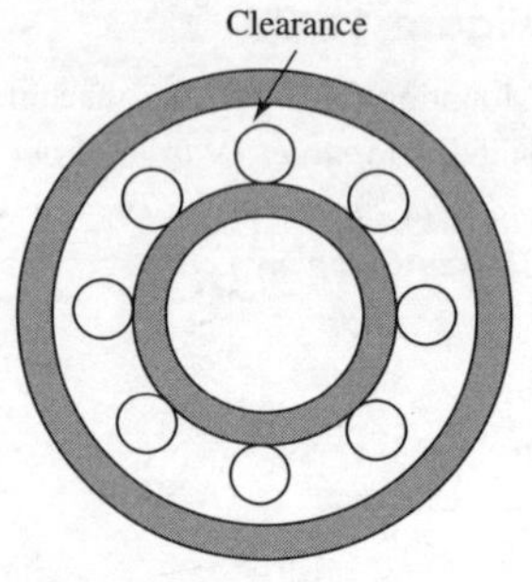

Figure 11–25

Clearance in an off-the-shelf bearing, exaggerated for clarity.

Figure 11–25 shows a typical bearing in which the clearance is exaggerated for clarity.

Preloading of straight roller bearings may be obtained by:

- Mounting the bearing on a tapered shaft or sleeve to expand the inner ring
- Using an interference fit for the outer ring
- Purchasing a bearing with the outer ring preshrunk over the rollers

Ball bearings are usually preloaded by the axial load built in during assembly. However, the bearings of Figure 11–24*a* and *b* are preloaded in assembly because of the differences in widths of the inner and outer rings.

It is always good practice to follow manufacturers' recommendations in determining preload, since too much will lead to early failure.

Alignment

The permissible misalignment in bearings depends on the type of bearing and the geometric and material properties of the specific bearing. Manufacturers' catalogs should be referenced for detailed specifications on a given bearing. In general, cylindrical and tapered roller bearings require alignments that are closer than deep-groove ball bearings. Spherical ball bearings and self-aligning bearings are the most forgiving. Table 7–2 gives typical maximum ranges for each type of bearing. The life of the bearing decreases significantly when the misalignment exceeds the allowable limits.

Additional protection against misalignment is obtained by providing the full shoulders (see Figure 11–8) recommended by the manufacturer. Also, if there is any misalignment at all, it is good practice to provide a safety factor of around 2 to account for possible increases during assembly.

Enclosures

To exclude dirt and foreign matter and to retain the lubricant, the bearing mountings must include a seal. The three principal methods of sealings are the felt seal, the commercial seal, and the labyrinth seal (Figure 11–26).

1. ***Felt seals*** may be used with grease lubrication when the speeds are low. The rubbing surfaces should have a high polish. Felt seals should be protected from dirt by placing them in machined grooves or by using metal stampings as shields.
2. The ***commercial seal*** is an assembly consisting of the rubbing element and, generally, a spring backing, which are retained in a sheet-metal jacket. These seals are usually made by press fitting them into a counterbored hole in the bearing cover. Since they obtain the sealing action by rubbing, they should not be used for high speeds.

Figure 11–26

Typical sealing methods. *(Source: General Motors Corp. Used with permission, GM Media Archives.)*

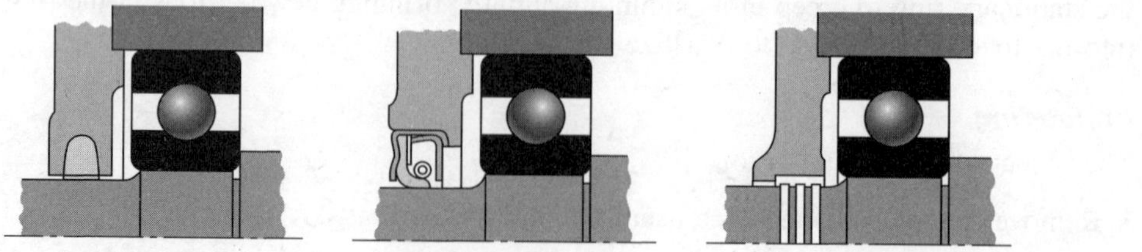

(*a*) Felt seal (*b*) Commercial seal (*c*) Labyrinth seal

3 The *labyrinth seal* is especially effective for high-speed installations and may be used with either oil or grease. It is sometimes used with flingers. At least three grooves should be used, and they may be cut on either the bore or the outside diameter. The clearance may vary from 0.25 to 1 mm, depending upon the speed and temperature.

PROBLEMS

Problems marked with an asterisk (*) are linked to problems in other chapters, as summarized in Table 1–2 of Section 1–17.

Since each bearing manufacturer makes individual decisions with respect to materials, treatments, and manufacturing processes, manufacturers' experiences with bearing life distribution differ. In solving the following problems, we will use the experience of two manufacturers, tabulated in Table 11–6.

Table 11–6 Typical Weibull Parameters for Two Manufacturers

Manufacturer	Rating Life, Revolutions	Weibull Parameters Rating Lives		
		x_0	θ	b
1	$90(10^6)$	0	4.48	1.5
2	$1(10^6)$	0.02	4.459	1.483

Tables 11–2 and 11–3 are based on manufacturer 2.

11–1 Timken rates its bearings for 3000 hours at 500 rev/min. Determine the catalog rating for a ball bearing running for 10 000 hours at 1800 rev/min with a load of 2.75 kN with a reliability of 90 percent.

11–2 A certain application requires a ball bearing with the inner ring rotating, with a design life of 25 kh at a speed of 350 rev/min. The radial load is 2.5 kN and an application factor of 1.2 is appropriate. The reliability goal is 0.90. Find the multiple of rating life required, x_D, and the catalog rating C_{10} with which to enter a bearing table. Choose a 02-series deep-groove ball bearing from Table 11–2, and estimate the reliability in use.

11–3 An angular-contact, inner ring rotating, 02-series ball bearing is required for an application in which the life requirement is 40 kh at 480 rev/min. The design radial load is 2745. The application factor is 1.4. The reliability goal is 0.90. Find the multiple of rating life x_D required and the catalog rating C_{10} with which to enter Table 11–2. Choose a bearing and estimate the existing reliability in service.

11–4 The other bearing on the shaft of Problem 11–3 is to be a 03-series cylindrical roller bearing with inner ring rotating. For a 10 kN radial load, find the catalog rating C_{10} with which to enter Table 11–3. The reliability goal is 0.90. Choose a bearing and estimate its reliability in use.

11–5 Problems 11–3 and 11–4 raise the question of the reliability of the bearing pair on the shaft. Since the combined reliabilities R is R_1R_2, what is the reliability of the two bearings (probability that either or both will not fail) as a result of your decisions in Problems 11–3 and 11–4? What does this mean in setting reliability goals for each of the bearings of the pair on the shaft?

11–6 Combine Problems 11–3 and 11–4 for an overall reliability of $R = 0.90$. Reconsider your selections, and meet this overall reliability goal.

11–7 For Problem 11–1, determine the catalog rating for a ball bearing with a reliability of 96 percent with an application factor of 1.2. The Weibull parameters for Timken bearings are $x_0 = 0$, $\theta = 4.48$, and $b = 1.5$.

11–8 A straight (cylindrical) roller bearing is subjected to a radial load of 20 kN. The life is to be 8000 h at a speed of 950 rev/min and exhibit a reliability of 0.95. What basic load rating should be used in selecting the bearing from a catalog of manufacturer 2 in Table 11–6?

11–9 Two ball bearings from different manufacturers are being considered for a certain application. Bearing A has a catalog rating of 2.0 kN based on a catalog rating system of 3000 hours at 500 rev/min. Bearing B has a catalog rating of 7.0 kN based on a catalog that rates at 10^6 cycles. For a given application, determine which bearing can carry the larger load.

11–10 to 11–15 For the bearing application specifications given in the table for the assigned problem, determine the Basic Load Rating for a ball bearing with which to enter a bearing catalog of manufacturer 2 in Table 11–6. Assume an application factor of one.

Problem Number	Radial Load	Design Life	Desired Reliability
11–10	2 kN	10^9 rev	90%
11–11	8 kN	5 kh, 900 rev/min	90%
11–12	4 kN	8 kh, 500 rev/min	90%
11–13	3 kN	5 yrs, 40 h/week, 400 rev/min	95%
11–14	9 kN	10^8 rev	99%
11–15	8 kN	5 kh, 900 rev/min	99%

11–16* to 11–19* For the problem specified in the table, build upon the results of the original problem to obtain a Basic Load Rating for a ball bearing at C with a 95 percent reliability, assuming distribution data from manufacturer 2 in Table 11–6. The shaft rotates at 1200 rev/min, and the desired bearing life is 15 kh. Use an application factor of 1.2.

Problem Number	Original Problem Number
11–16*	3–76
11–17*	3–77
11–18*	3–78
11–19*	3–79

11–20* For the shaft application defined in Problem 3–85, the input shaft *EG* is driven at a constant speed of 191 rev/min. Obtain a Basic Load Rating for a ball bearing at *A* for a life of 12 kh with a 95 percent reliability, assuming distribution data from manufacturer 2 in Table 11–6.

11–21* For the shaft application defined in Problem 3–87, the input shaft *EG* is driven at a constant speed of 280 rev/min. Obtain a Basic Load Rating for a cylindrical roller bearing at *A* for a life of 14 kh with a 98 percent reliability, assuming distribution data from manufacturer 2 in Table 11–6.

11–22 An 02-series single-row deep-groove ball bearing with a 65-mm bore (see Tables 11–1 and 11–2 for specifications) is loaded with a 3-kN axial load and a 7-kN radial load. The outer ring rotates at 500 rev/min.

(*a*) Determine the equivalent radial load that will be experienced by this particular bearing.

(*b*) Determine whether this bearing should be expected to carry this load with a 95 percent reliability for 10 kh.

11–23 An 02-series single-row deep-groove ball bearing with a 30-mm bore (see Tables 11–1 and 11–2 for specifications) is loaded with a 2-kN axial load and a 5-kN radial load. The inner ring rotates at 400 rev/min.

(*a*) Determine the equivalent radial load that will be experienced by this particular bearing.

(*b*) Determine the predicted life (in revolutions) that this bearing could be expected to give in this application with a 99 percent reliability.

11–24 to 11–28 An 02-series single-row deep-groove ball bearing is to be selected from Table 11–2 for the application conditions specified in the table. Assume Table 11–1 is applicable if needed. Specify the smallest bore size from Table 11–2 that can satisfy these conditions.

Problem Number	Radial Load	Axial Load	Design Life	Ring Rotating	Desired Reliability
11–24	8 kN	0 kN	10^9 rev	Inner	90%
11–25	8 kN	2 kN	10 kh, 400 rev/min	Inner	99%
11–26	8 kN	3 kN	10^8 rev	Outer	90%
11–27	10 kN	5 kN	12 kh, 300 rev/min	Inner	95%
11–28	9 kN	3 kN	10^8 rev	Outer	99%

11–29* The shaft shown in the figure is proposed as a preliminary design for the application defined in Problem 3–80. The effective centers of the gears for force transmission are shown. The dimensions for the bearing surfaces (indicated with cross markings) have been estimated. The shaft rotates at 1200 rev/min, and the desired bearing life is 15 kh with a 95 percent reliability in each bearing, assuming distribution data from manufacturer 2 in Table 11–6. Use an application factor of 1.2.

(*a*) Obtain a Basic Load Rating for a ball bearing at the right end.

(*b*) Use an online bearing catalog to find a specific bearing that satisfies the needed Basic Load Rating and the geometry requirements. If necessary, indicate appropriate adjustments to the dimensions of the bearing surface.

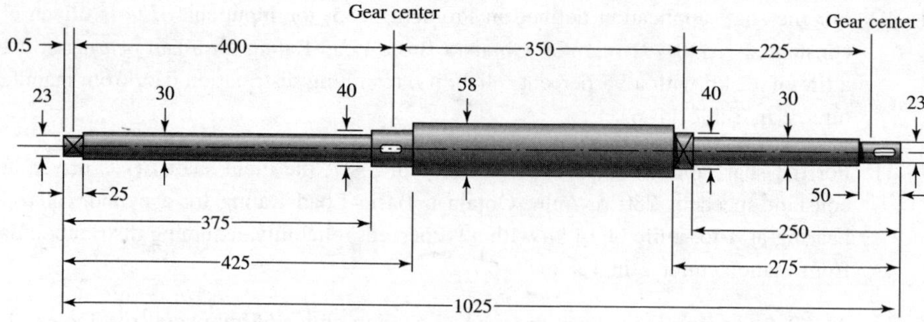

Problem 11–29*

All fillets 2 mm. Dimensions in millimeters.

11–30* Repeat the requirements of Problem 11–29 for the bearing at the left end of the shaft.

11–31* The shaft shown in the figure is proposed as a preliminary design for the application defined in Problem 3–81. The effective centers of the gears for force transmission are shown. The dimensions for the bearing surfaces (indicated with cross markings) have been estimated. The shaft rotates at 900 rev/min, and the desired bearing life is 12 kh with a 98 percent reliability in each bearing, assuming distribution data from manufacturer 2 in Table 11–6. Use an application factor of 1.2.

(*a*) Obtain a Basic Load Rating for a ball bearing at the right end.

(*b*) Use an online bearing catalog to find a specific bearing that satisfies the needed Basic Load Rating and the geometry requirements. If necessary, indicate appropriate adjustments to the dimensions of the bearing surface.

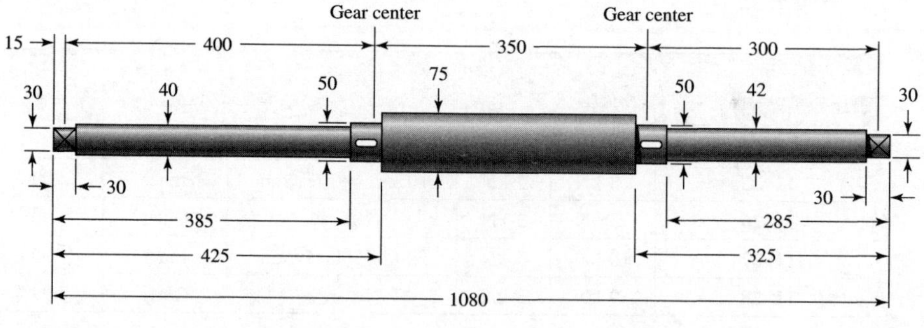

Problem 11–31*

All fillets 2 mm. Dimensions in millimeters.

11–32* Repeat the requirements of Problem 11–31 for the bearing at the left end of the shaft.

11–33 Shown in the figure is a gear-driven squeeze roll that mates with an idler roll. The roll is designed to exert a normal force of 5.25 N/mm of roll length and a pull of 4.2 N/mm on the material being processed. The roll speed is 300 rev/min, and a design life of 30 kh is desired. Use an application factor of 1.2, and select a pair of angular-contact 02-series ball bearings from Table 11–2 to be mounted at 0 and *A*. Use the same size bearings at both locations and a combined reliability of at least 0.92, assuming distribution data from manufacturer 2 in Table 11–6.

Problem 11–33
Dimensions in millimeters.

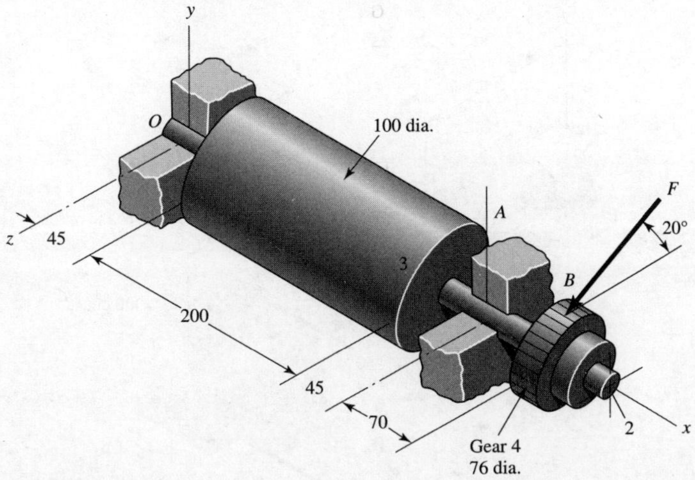

11–34 The figure shown is a geared countershaft with an overhanging pinion at C. Select an angular-contact ball bearing from Table 11–2 for mounting at O and an 02-series cylindrical roller bearing from Table 11–3 for mounting at B. The force on gear A is $F_A = 2670$ N, and the shaft is to run at a speed of 480 rev/min. Solution of the statics problem gives force of bearings against the shaft at O as $\mathbf{R}_O = 1722\mathbf{j} + 2078\mathbf{k}$ N, and at B as $\mathbf{R}_B = 1406\mathbf{j} - 7187\mathbf{k}$ N. Specify the bearings required, using an application factor of 1.4, a desired life of 50 kh, and a combined reliability goal of 0.9, assuming distribution data from manufacturer 2 in Table 11–6.

Problem 11–34
Dimensions in millimeters.

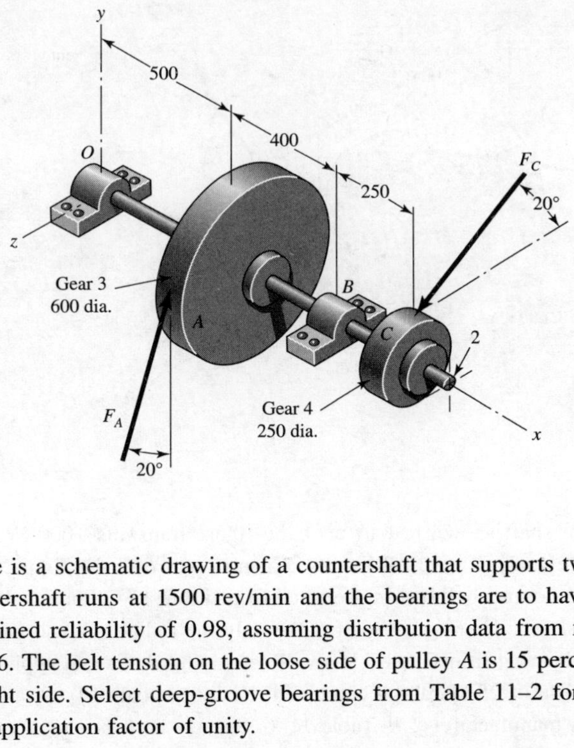

11–35 The figure is a schematic drawing of a countershaft that supports two V-belt pulleys. The countershaft runs at 1500 rev/min and the bearings are to have a life of 60 kh at a combined reliability of 0.98, assuming distribution data from manufacturer 2 in Table 11–6. The belt tension on the loose side of pulley A is 15 percent of the tension on the tight side. Select deep-groove bearings from Table 11–2 for use at O and E, using an application factor of unity.

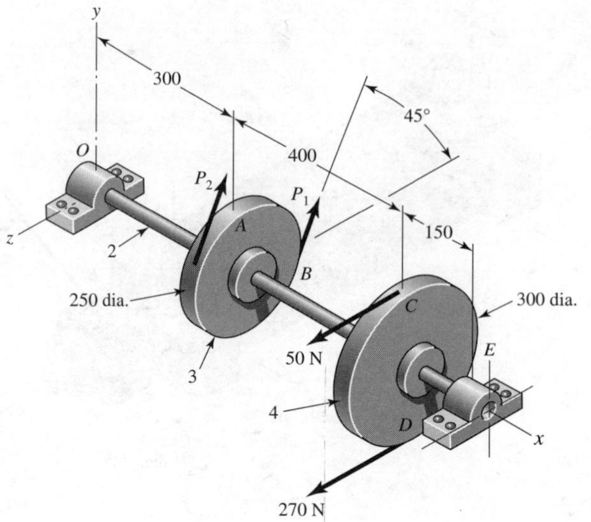

Problem 11–35
Dimensions in millimeters.

11–36 A gear-reduction unit uses the countershaft depicted in the figure. Find the two bearing reactions. The bearings are to be angular-contact ball bearings, having a desired life of 40 kh when used at 200 rev/min. Use 1.2 for the application factor and a reliability goal for the bearing pair of 0.95, assuming distribution data from manufacturer 2 in Table 11–6. Select the bearings from Table 11–2.

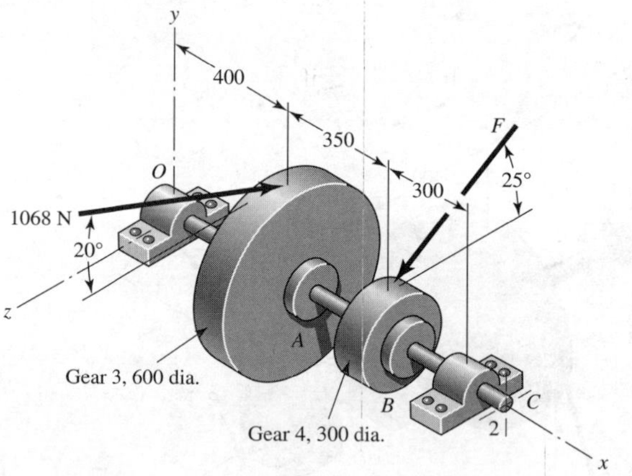

Problem 11–36
Dimensions in millimeters.

11–37 The worm shaft shown in part *a* of the figure transmits 1000 W at 600 rev/min. A static force analysis gave the results shown in part *b* of the figure. Bearing *A* is to be an angular-contact ball bearing selected from Table 11–2, mounted to take the 2470-N thrust load. The bearing at *B* is to take only the radial load, so an 02-series cylindrical roller bearing from Table 11–3 will be employed. Use an application factor of 1.3, a desired life of 25 kh, and a combined reliability goal of 0.99, assuming distribution data from manufacturer 2 in Table 11–6. Specify each bearing.

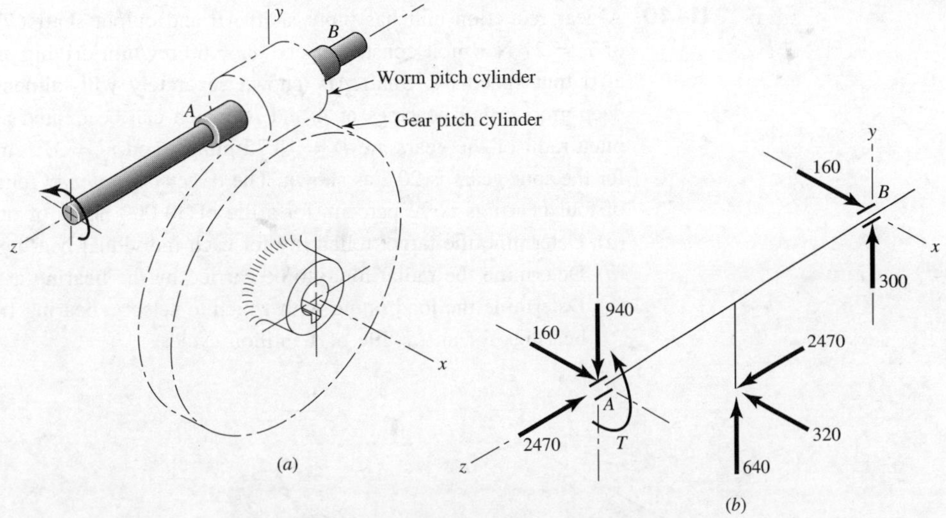

Problem 11–37

(*a*) Worm and worm gear; (*b*) force analysis of worm shaft, forces in Newtons.

11–38 In bearings tested at 2000 rev/min with a steady radial load of 18 kN, a set of bearings showed an L_{10} life of 115 h and an L_{80} life of 600 h. The basic load rating of this bearing is 39.6 kN. Estimate the Weibull shape factor b and the characteristic life θ for a two-parameter model. This manufacturer rates ball bearings at 1 million revolutions.

11–39 A 16-tooth pinion drives the double-reduction spur-gear train in the figure. All gears have 25° pressure angles. The pinion rotates ccw at 1200 rev/min and transmits power to the gear train. The shaft has not yet been designed, but the free bodies have been generated. The shaft speeds are 1200 rev/min, 240 rev/min, and 80 rev/min. A bearing study is commencing with a 10-kh life and a gearbox bearing ensemble reliability of 0.99, assuming distribution data from manufacturer 2 in Table 11–6. An application factor of 1.2 is appropriate. For each shaft, specify a matched pair of 02-series cylindrical roller bearings from Table 11–3.

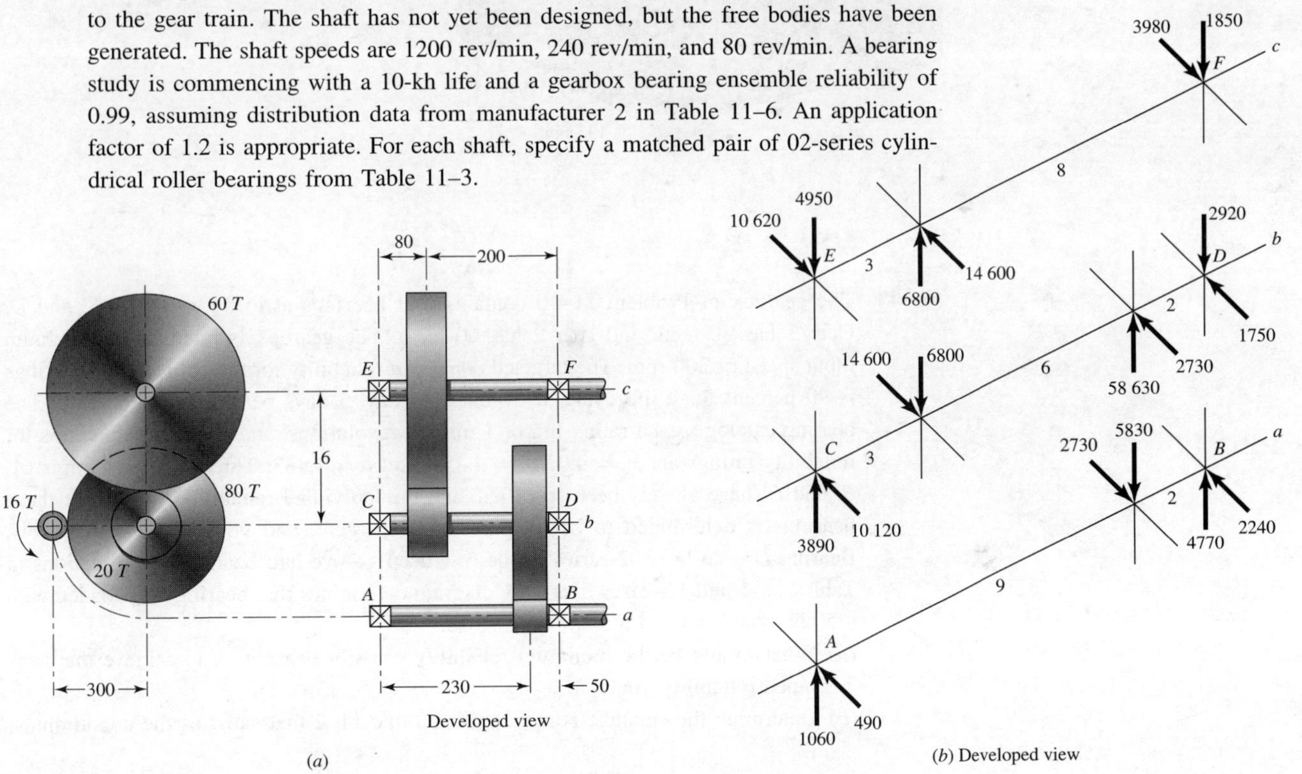

Problem 11–39

(*a*) Drive detail; (*b*) force analysis on shafts. Forces in Newtons; linear dimensions in millimeters.

11–40 A gear reduction unit has input shaft AB and output shaft CD, with an input torque of $T_i = 27$ N · m at constant speed $\omega_i = 60$ rev/min driving an output load torque T_o at output speed ω_o. Shaft AB (shown separately with dimensions) is supported by deep-groove ball bearings at A and B, which can be treated as simple supports. The pitch radii of the gears are $r_1 = 18.75$ mm in and $r_2 = 37.5$ mm. The pressure angle for the spur gears is 20°, as shown. The targeted *combined* reliability for the entire set of four bearings is 92 percent, for a life of 30 000 hours of operation.

(*a*) Determine the target reliability for each individual bearing.

(*b*) Determine the radial force to be carried by the bearing at A.

(*c*) Determine the load rating with which to select a bearing from a catalog that rates bearings for an L_{10} life of 1 million cycles.

Problem 11–40

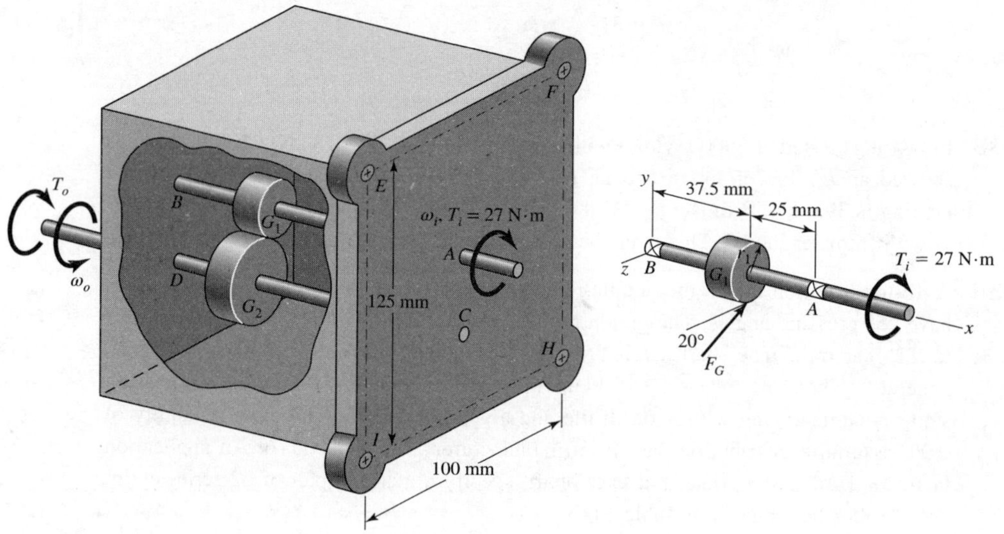

11–41 The gearbox of Problem 11–40 contains four bearings at locations A, B, C, and D. Gear 1 has 30 teeth and gear 2 has 60 teeth. The gearbox is rated for a maximum input speed of 200 rpm. The targeted *combined* reliability for the entire set of bearings is 90 percent for a life cycle of 8 hours per day, 5 days per week, for 6 years. The bearing catalog uses a rating life of 1 million revolutions, and Weibull parameters for reliability ratings are $x_0 = 0.02$, $\theta = 4.459$, and $b = 1.483$. Three of the bearings (A, B, and C) have already been specified, and the individual reliability for each of these bearings is determined to be 97 percent, 96 percent, and 98 percent, respectively. Bearing D is to be a 02-series single-row deep groove ball bearing (specifications in Tables 11–1 and 11–2). A free body diagram determines that bearing D is loaded with a 5-kN axial load and a 9-kN radial load.

(*a*) What should be the minimum reliability goal for bearing D, to achieve the combined reliability goal?

(*b*) Determine the smallest bore size from Table 11–2 that satisfies the conditions.

11–42 Given a 03-series 30-mm cylindrical roller bearing, which was subjected to 300 000 revolutions with a radial load of 22 kN. Estimate the remaining life of the bearing if subsequently subjected to 35 kN.

11–43 A shaft cycles such that each bearing undergoes the following radial loads: 30 percent of the time at 20 kN, 50 percent of the time at 25 kN, and 20 percent of the time at 30 kN. If the design life of a bearing is to be at least 5 (10^6) revolutions, select the smallest 03-series cylindrical roller bearing that will accomplish this.

11–44 Estimate the remaining life in revolutions of an 02-30 mm angular-contact ball bearing already subjected to 200 000 revolutions with a radial load of 18 kN, if it is now to be subjected to a change in load to 30 kN.

11–45 The same 02-30 angular-contact ball bearing as in Problem 11–44 is to be subjected to a two-step loading cycle of 4 min with a loading of 18 kN, and one of 6 min with a loading of 30 kN. This cycle is to be repeated until failure. Estimate the total life in revolutions, hours, and loading cycles.

11–46 A countershaft is supported by two tapered roller bearings using an indirect mounting. The radial bearing loads are 2240 N for the left-hand bearing and 1095 for the right-hand bearing. An axial load of 800 N is carried by the left bearing. The shaft rotates at 400 rev/min and is to have a desired life of 40 kh. Use an application factor of 1.4 and a combined reliability goal of 0.90, assuming distribution data from manufacturer 1 in Table 11–6. Using an initial $K = 1.5$, find the required radial rating for each bearing. Select the bearings from Figure 11–15.

11–47* For the shaft application defined in Problem 3–82, perform a preliminary specification for tapered roller bearings at C and D. A bearing life of 10^8 revolutions is desired with a 90 percent combined reliability for the bearing set, assuming distribution data from manufacturer 1 in Table 11–6. Should the bearings be oriented with direct mounting or indirect mounting for the axial thrust to be carried by the bearing at C? Assuming bearings are available with $K = 1.5$, find the required radial rating for each bearing. For this preliminary design, assume an application factor of one.

11–48* For the shaft application defined in Problem 3–84, perform a preliminary specification for tapered roller bearings at A and B. A bearing life of 500 million revolutions is desired with a 90 percent combined reliability for the bearing set, assuming distribution data from manufacturer 1 in Table 11–6. Should the bearings be oriented with direct mounting or indirect mounting for the axial thrust to be carried by the bearing at A? Assuming bearings are available with $K = 1.5$, find the required radial rating for each bearing. For this preliminary design, assume an application factor of one.

11–49 An outer hub rotates around a stationary shaft, supported by two tapered roller bearings as shown in Figure 11–23. The device is to operate at 250 rev/min, 8 hours per day, 5 days per week, for 5 years, before bearing replacement is necessary. A reliability of 90 percent on each bearing is acceptable. A free body analysis determines the radial force carried by the upper bearing to be 12 kN and the radial force at the lower bearing to be 25 kN. In addition, the outer hub applies a downward force of 5 kN. Assuming bearings are available from manufacturer 1 in Table 11–6 with $K = 1.5$, find the required radial rating for each bearing. Assume an application factor of 1.2.

11–50 The gear-reduction unit shown has a gear that is press fit onto a cylindrical sleeve that rotates around a stationary shaft. The helical gear transmits an axial thrust load T of 1 kN as shown in the figure. Tangential and radial loads (not shown) are also transmitted through the gear, producing radial ground reaction forces at the bearings of 3.5 kN for bearing A and 2.5 kN for bearing B. The desired life for each bearing is 90 kh at a speed of 150 rev/min with a 90 percent reliability. The first iteration of the shaft design indicates approximate diameters of 28 mm at A and 25 mm at B. Assuming distribution data from manufacturer 1 in Table 11–6, select suitable tapered roller bearings from Figure 11–15.

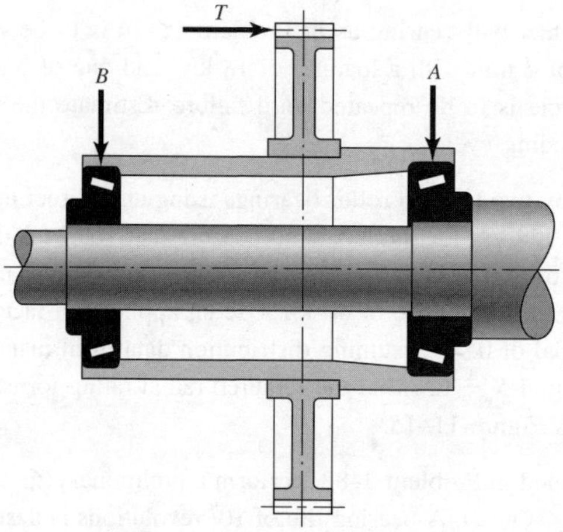

Problem 11–50

(Source: Redrawn from material Furnished by The Timken Company.)

12 Lubrication and Journal Bearings

Chase Somero/Shutterstock

Chapter Outline

The object of lubrication is to reduce friction, wear, and heating of machine parts that move relative to each other. A lubricant is any substance that, when inserted between the moving surfaces, accomplishes these purposes. In a sleeve bearing, a shaft, or *journal,* rotates or oscillates within a sleeve, or *bushing,* and the relative motion is sliding. In an antifriction bearing, the main relative motion is rolling. A follower may either roll or slide on the cam. Gear teeth mate with each other by a combination of rolling and sliding. Pistons slide within their cylinders. All these applications require lubrication to reduce friction, wear, and heating.

The field of application for journal bearings is immense. The crankshaft and connecting-rod bearings of an automotive engine must operate for thousands of miles at high temperatures and under varying load conditions. The journal bearings used in the steam turbines of power-generating stations are said to have reliabilities approaching 100 percent. At the other extreme there are thousands of applications in which the loads are light and the service relatively unimportant; a simple, easily installed bearing is required, using little or no lubrication. In such cases an antifriction bearing might be a poor answer because of the cost, the elaborate enclosures, the close tolerances, the radial space required, the high speeds, or the increased inertial effects. Instead, a nylon bearing requiring no lubrication, a powder-metallurgy bearing with the lubrication "built in," or a bronze bearing with ring oiling, wick feeding, or solid-lubricant film or grease lubrication might be a very satisfactory solution. Recent metallurgy developments in bearing materials, combined with increased knowledge of the lubrication process, now make it possible to design journal bearings with satisfactory lives and very good reliabilities.

Much of the material we have studied thus far in this book has been based on fundamental engineering studies, such as statics, dynamics, the mechanics of solids, metal processing, mathematics, and metallurgy. In the study of lubrication and journal bearings, additional fundamental studies, such as chemistry, fluid mechanics, thermodynamics, and heat transfer, must be utilized in developing the material. While we shall not utilize all of them in the material to be included here, you can now begin to appreciate better how the study of mechanical engineering design is really an integration of most of your previous studies and a directing of this total background toward the resolution of a single objective.

12–1 Types of Lubrication

Five distinct forms of lubrication may be identified:

1. Hydrodynamic
2. Hydrostatic
3. Elastohydrodynamic
4. Boundary
5. Solid film

Hydrodynamic lubrication means that the load-carrying surfaces of the bearing are separated by a relatively thick film of lubricant, so as to prevent metal-to-metal contact. Hydrodynamic lubrication does not depend upon the introduction of the lubricant under pressure, though that may occur; but it does require the existence of an adequate supply at all times. The film pressure is created by the moving surface itself pulling the lubricant into a wedge-shaped zone at a velocity sufficiently high to create the pressure necessary to separate the surfaces against the load on the bearing. Hydrodynamic lubrication is also called *full-film,* or *fluid lubrication.*

Hydrostatic lubrication is obtained by introducing the lubricant, which is sometimes air or water, into the load-bearing area at a pressure high enough to separate the surfaces with a relatively thick film of lubricant. So, unlike hydrodynamic lubrication, this kind of lubrication does not require motion of one surface relative to another. We shall not deal with hydrostatic lubrication in this book, but the subject should be considered in designing bearings where the velocities are small or zero and where the frictional resistance is to be an absolute minimum.

Elastohydrodynamic lubrication (EHL) is the phenomenon where surface deformations play a key role in the development of a lubricant film. *Hard EHL* occurs when a lubricant is introduced between surfaces that are in rolling contact, such as mating gears, cam-tappet contacts, or roller bearings. The size of lubricant film region between the surfaces is of the order of that predicted by Hertzian contact mechanics and is thus much smaller than the dimensions of the bodies themselves. On the other hand, *soft EHL* occurs where the size of the contact region can extend over a substantial portion of the mating journal and sleeve surfaces, for example, in big-end connecting rod bearings.

Insufficient surface area, a drop in the velocity of the moving surface, a lessening in the quantity of lubricant delivered to a bearing, an increase in the bearing load, or an increase in lubricant temperature resulting in a decrease in viscosity—any one of these—may prevent the buildup of a film thick enough for full-film lubrication. When this happens, the highest asperities may be separated by lubricant films only several molecular dimensions in thickness. This is called ***boundary lubrication.*** The change from hydrodynamic to boundary lubrication is not at all a sudden or abrupt one. It is probable that a mixed hydrodynamic- and boundary-type lubrication occurs first, and as the surfaces move closer together, the boundary-type lubrication becomes predominant. The viscosity of the lubricant is not of as much importance with boundary lubrication as is the chemical composition.

When bearings must be operated at extreme temperatures, a ***solid-film lubricant*** such as graphite or molybdenum disulfide must be used because the ordinary mineral oils are not satisfactory. Much research is currently being carried out in an effort to find composite bearing materials with low wear rates as well as small frictional coefficients.

12–2 Viscosity

In Figure 12–1 consider a plate with surface area A moving with a velocity U on a film of lubricant of thickness h. We imagine the film as composed of a series of horizontal layers and the force F causing these layers to deform or slide on one another just like a deck of cards. The layers in contact with the moving plate are assumed to

Figure 12–1

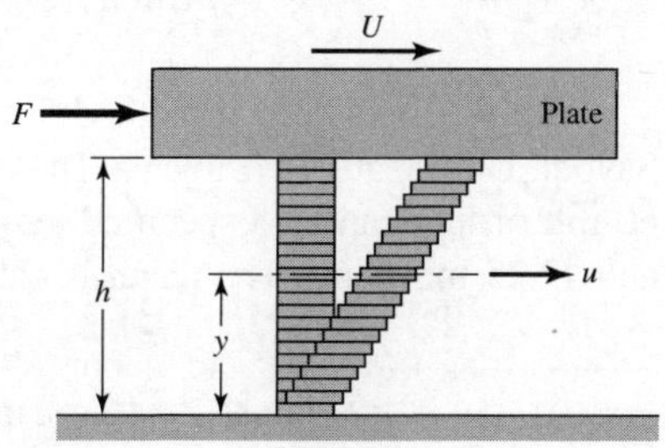

have a velocity U; those in contact with the stationary surface are assumed to have a zero velocity. Intermediate layers have velocities that depend upon their distances y from the stationary surface. The fluid is said to be Newtonian when the shear stress in the fluid is proportional to the rate of change of velocity with respect to y. Thus

$$\tau = \frac{F}{A} = \mu \frac{du}{dy} \tag{12–1}$$

where μ is the constant of proportionality and defines *dynamic viscosity,* also called *absolute viscosity*. The derivative du/dy is the rate of change of velocity with distance and may be called the rate of shear, or the velocity gradient. The viscosity μ is thus a measure of the internal frictional resistance of the fluid. For most lubricating fluids, the rate of shear is constant, and $du/dy = U/h$. Thus, from Equation (12–1),

$$\tau = \frac{F}{A} = \mu \frac{U}{h} \tag{12–2}$$

The unit of dynamic viscosity in the ips system is seen to be the pound-force-second per square inch; this is the same as stress or pressure multiplied by time. The ips unit is called the *reyn,* in honor of Sir Osborne Reynolds. When using ips units, the microreyn (μreyn) is often more convenient. The symbol μ' will be used to designate viscosity in μreyn such that $\mu = \mu'/(10^6)$.

The dynamic viscosity in SI units is measured by the pascal-second (Pa · s); this is the same as a Newton-second per square meter. The conversion from ips units to SI is the same as for stress. For example, multiply the dynamic viscosity in reyns by 6894 to convert to units of Pa · s.

The American Society of Mechanical Engineers (ASME) has published a list of cgs units that are not to be used in ASME documents.[1] This list results from a recommendation by the International Committee of Weights and Measures (CIPM) that the use of cgs units with special names be discouraged. Included in this list is a unit of force called the *dyne* (dyn), a unit of dynamic viscosity called the *poise* (P), and a unit of kinematic viscosity called the *stoke* (St). All of these units have been, and still are, used extensively in lubrication studies.

The poise is the cgs unit of dynamic or absolute viscosity, and its unit is the dyne-second per square centimeter (dyn · s/cm^2). It has been customary to use the centipoise (cP) in analysis, because its value is more convenient. When the viscosity is expressed in centipoises, it is often designated by Z. The conversion from cgs units to SI and ips units is as follows:

$$\mu(\text{Pa} \cdot \text{s}) = (10)^{-3} Z \text{ (cP)}$$

$$\mu(\text{reyn}) = \frac{Z \text{ (cP)}}{6.89(10)^6}$$

$$\mu(\text{mPa} \cdot \text{s}) = 6.89\ \mu' \ (\mu\text{reyn})$$

The ASTM standard method for determining viscosity uses an instrument called the Saybolt Universal Viscosimeter. The method consists of measuring the time in seconds for 60 mL of lubricant at a specified temperature to run through a tube 1.76 mm in diameter and 12.25 mm long. The result is called the *kinematic viscosity,* and in the past the unit

[1]*ASME Orientation and Guide for Use of Metric Units,* 2nd ed., American Society of Mechanical Engineers, 1972, p. 13.

of the square centimeter per second has been used. One square centimeter per second is defined as a *stoke*. By the use of the *Hagen-Poiseuille law,* the kinematic viscosity based upon seconds Saybolt, also called *Saybolt Universal viscosity* (SUV) in seconds, is

$$Z_k = \left(0.22t - \frac{180}{t}\right) \tag{12–3}$$

where Z_k is in centistokes (cSt) and t is the number of seconds Saybolt.

In SI, the kinematic viscosity ν has the unit of the square meter per second (m^2/s), and the conversion is

$$\nu(\mathrm{m}^2/\mathrm{s}) = 10^{-6} Z_k(\mathrm{cSt})$$

Thus, Equation (12–3) becomes

$$\nu = \left(0.22t - \frac{180}{t}\right)(10^{-6}) \tag{12–4}$$

To convert to dynamic viscosity, we multiply ν by the density in SI units. Designating the density as ρ with the unit of the kilogram per cubic meter, we have

$$\mu = \rho\left(0.22t - \frac{180}{t}\right)(10^{-6}) \tag{12–5}$$

where μ is in pascal-seconds.

Viscosity can vary considerably with temperature in a nonlinear fashion. Figure 12–2 show the dynamic viscosity in SI units for common grades of lubricating oils employed in machinery. The ordinates in Figure 12–2 are not logarithmic, as the decades are of differing vertical length. Figure 12–3 shows the dynamic viscosity in the ips system for a number of other fluids often used for lubrication purposes and their variation in temperature.

The viscosity-temperature trends are provided here only in graphical form. However, various curve fits have been developed. See Table 12–1 for one particular example. Booker[2] provides general viscosity-temperature functional forms based on sets of two or three measured temperature-viscosity data points.

12–3 Petroff's Equation

The phenomenon of bearing friction was first explained by Petroff on the assumption that the shaft is concentric with its bushing. Though we shall seldom make use of Petroff's method of analysis in the material to follow, it is important because it defines groups of dimensionless parameters and because the coefficient of friction predicted by this law turns out to be quite good even when the shaft is not concentric.

Let us now consider a vertical shaft rotating in a guide bearing. It is assumed that the bearing carries a very small load, that the clearance space is uniform and completely filled with oil, and that leakage is negligible (Figure 12–4). We denote the radius of the shaft by r, the radial clearance by c, and the length of the bearing by l, all dimensions being in meters. Note, in tribology literature, the clearance may be denoted as C. If the shaft rotates at N rev/s, then its surface velocity is $U = 2\pi rN$ m/s. Since the shearing stress in the lubricant is equal to the velocity gradient times the viscosity, from Equation (12–2) we have

$$\tau = \mu\frac{U}{h} = \frac{2\pi r\mu N}{c} \tag{a}$$

[2]J. F. Booker, "Dynamically-Loaded Journal Bearings: Numerical Application of the Mobility Method," *ASME Journal of Lubrication Technology*, vol. 93, 1971, pp. 168–176, 315.

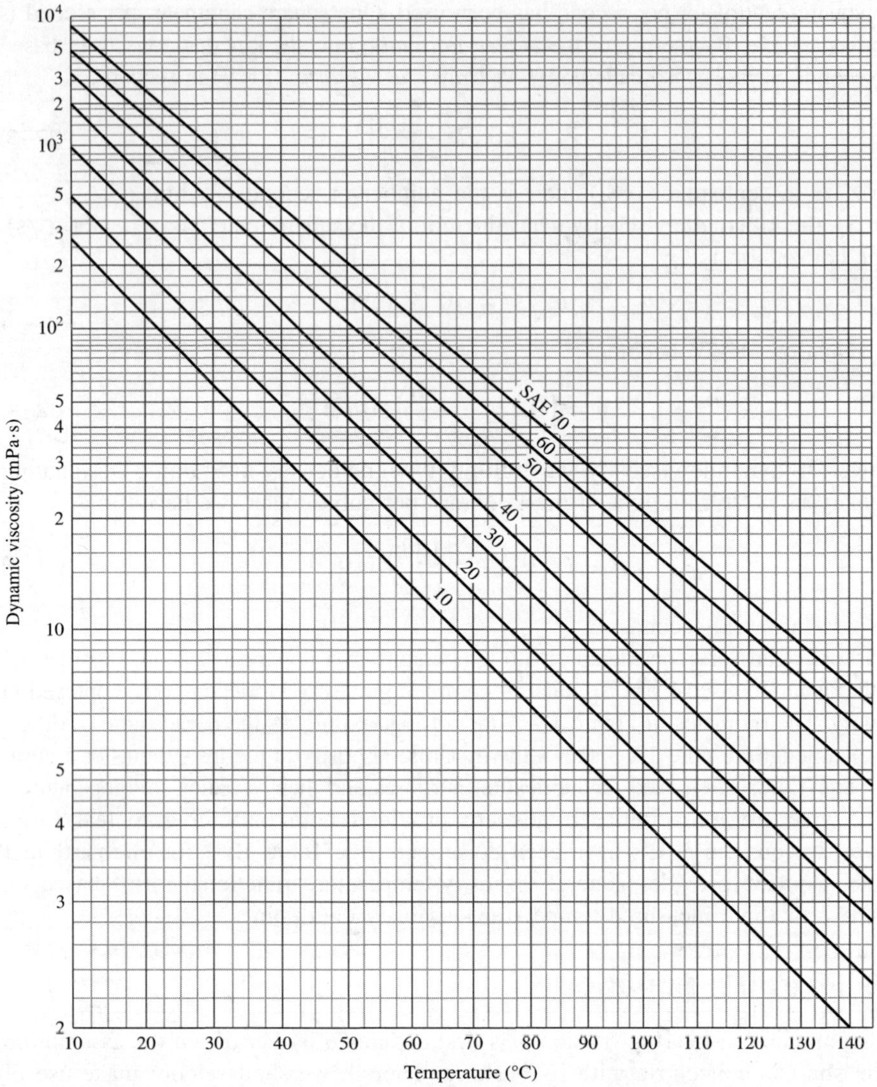

Figure 12–2

Viscosity–temperature chart in SI units. *(Source: Adapted from Raimondi and Boyd.)*

where the radial clearance c has been substituted for the distance h. The force required to shear the film is the stress times the area. The torque is the force times the lever arm r. Thus

$$T = (\tau A)(r) = \left(\frac{2\pi r\mu\, N}{c}\right)(2\pi rl)(r) = \frac{4\pi^2 r^3 l\mu\, N}{c} \tag{12–6}$$

If we now designate a small force on the bearing by W, in pounds-force, then the bearing pressure P acting upon the projected area of the bearing is

$$P = \frac{W}{2rl} \tag{12–7}$$

The frictional force is fW, where f is the coefficient of friction, and so the frictional torque is

$$T = fWr = (f)(2rlP)(r) = 2r^2 flP \tag{12–8}$$

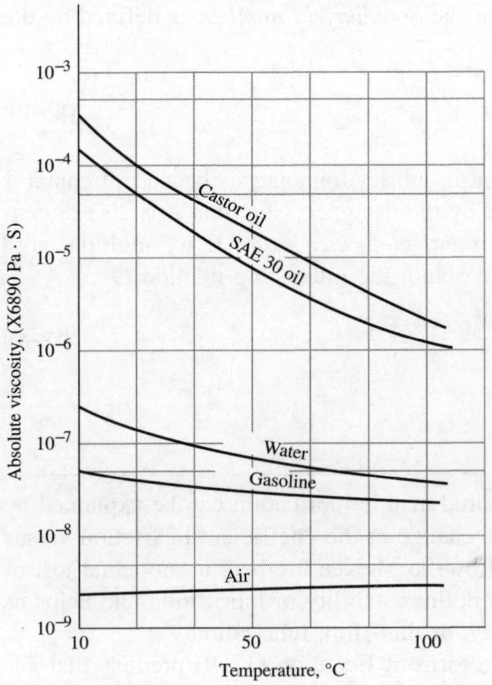

Figure 12–3

A comparison of the viscosities of various fluids.

Table 12–1 Curve Fits* to Approximate the Viscosity versus Temperature Functions for SAE Grades 10 to 60

Oil Grade, SAE	Viscosity μ_0, mPa · s	Constant b, °C
10	0.109	635.3
20	0.094	688.7
30	0.097	737.8
40	0.083	801.3
50	0.117	820.9
60	0.129	851.1

*$\mu = \mu_0 \cdot \exp[9/5\ b + 32)/(9/5\ T + 127)]$, T in °C.

Source: A. S. Seireg and S. Dandage. "Empirical Design Procedure for the Thermodynamic Behavior of Journal Bearings." *Journal of Lubrication Technology* 104 (April 1982): 135–148.

Substituting the value of the torque from Equation (12–8) in Equation (12–6) and solving for the coefficient of friction, we find

$$f = 2\pi^2 \frac{\mu N}{P} \frac{r}{c} \tag{12–9}$$

Equation (12–9) is called *Petroff's equation* and was first published in 1883. The two quantities $\mu N/P$ and r/c are very important parameters in lubrication. Substitution of the appropriate dimensions in each parameter will show that they are dimensionless.

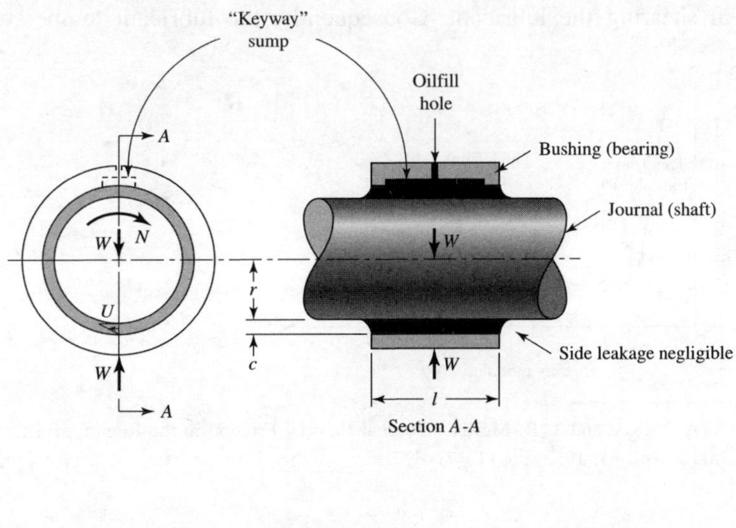

Figure 12–4

Petroff's lightly loaded journal bearing consisting of a shaft journal and a bushing with an axial-groove internal lubricant reservoir. The linear velocity gradient is shown in the end view. The radial clearance c is several thousandths of an millimeter and is grossly exaggerated for presentation purposes.

The *bearing characteristic number,* or the *Sommerfeld number,* is defined by the equation

$$S = \left(\frac{r}{c}\right)^2 \frac{\mu N}{P} \tag{12–10}$$

The Sommerfeld number is very important in lubrication analysis because it contains many of the parameters that are specified by the designer. Note that it is also dimensionless. The quantity r/c is called the *radial clearance ratio.* If we multiply both sides of Equation (12–9) by this ratio, we obtain the interesting relation

$$f\frac{r}{c} = 2\pi^2 \frac{\mu N}{P}\left(\frac{r}{c}\right)^2 = 2\pi^2 S \tag{12–11}$$

12–4 Stable Lubrication

The difference between boundary and hydrodynamic lubrication can be explained by reference to Figure 12–5. This plot of the change in the coefficient of friction versus the *bearing modulus* $\mu N/P$ was obtained by the McKee brothers in an actual test of friction.[3] The plot is important because it defines stability of lubrication and helps us to understand hydrodynamic and boundary, or thin-film, lubrication.

Recall Petroff's bearing model in the form of Equation (12–9) predicts that f is proportional to $\mu N/P$, that is, a straight line from the origin in the first quadrant. On the coordinates of Figure 12–5 the locus to the right of point C is an example. Petroff's model presumes thick-film lubrication, that is, no metal-to-metal contact, the surfaces being completely separated by a lubricant film.

The McKee abscissa was ZN/P (Pa × rpm/s/Pa, that is, dimensionless) and the value of abscissa B in Figure 12–5 was $0.33(10^{-6})$. Designers keep $\mu N/P \geq 1.7(10^{-6})$, which corresponds to $ZN/P \geq 150$. A design constraint to keep thick-film lubrication is to be sure that

$$\frac{\mu N}{P} \geq 1.7(10^{-6}) \tag{a}$$

Suppose we are operating to the right of line BA and something happens, say, an increase in lubricant temperature. This results in a lower viscosity and hence a smaller value of $\mu N/P$. The coefficient of friction decreases and not as much heat is generated in shearing the lubricant. Consequently, the lubricant temperature drops, and the

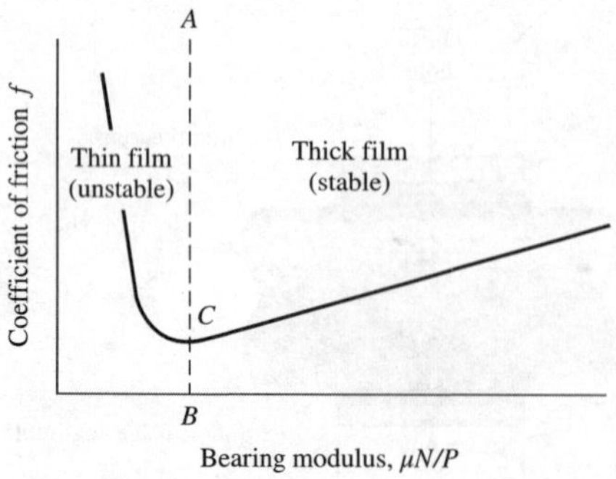

Figure 12–5
The variation of the coefficient of friction f with $\mu N/P$.

[3]S. A. McKee and T. R. McKee, "Journal Bearing Friction in the Region of Thin Film Lubrication," *SAE J.*, vol. 31, 1932, pp. (T)371–377.

viscosity increases. Thus the region to the right of line *BA* defines *stable lubrication* because variations are self-correcting.

To the left of line *BA*, a decrease in viscosity would increase the friction. A temperature rise would ensue, and the viscosity would be reduced still more. The result would be compounded. Thus the region to the left of line *BA* represents *unstable lubrication*.

It is also helpful to see that a small viscosity, and hence a small $\mu N/P$, means that the lubricant film is very thin and that there will be a greater possibility of some metal-to-metal contact, and hence of more friction. Thus, point *C* is a reasonable estimation of the beginning of metal-to-metal contact as $\mu N/P$ becomes smaller.

12–5 Thick-Film Lubrication

The nomenclature of a *complete journal bearing* is shown in Figure 12–6. The bearing comprises a *journal* and *bushing* (or *sleeve*) separated by a thin lubricant film. Both journal and bushing surfaces are assumed perfectly cylindrical, and their respective axes of rotation are assumed parallel. The dimension c is the radial clearance and is the difference in the radii of the journal and bushing. The radial clearance as drawn in Figure 12–6 is exaggerated for clarity. The radial clearance is typically very small compared to the journal radius (typically $c/r \approx 0.001$, where r is journal radius). The displacement of the journal center O relative to the sleeve center O' in the clearance space is called the *eccentricity* and is denoted by e. The film thickness at any point on the bearing is denoted as h and to a high approximation can be written as

$$h = c + e \cos \theta \tag{12–12}$$

where bearing angle θ is measured from the line of action of the journal and sleeve centers as shown. For design purposes, it is convenient to define the eccentricity ratio, ε, as

$$\varepsilon = \frac{e}{c} \tag{12–13}$$

where the kinematic limit $\varepsilon = 1$ denotes journal to bushing contact.

Suppose a steady load W is applied to a journal rotating clockwise at a *steady* speed N. The *attitude angle* ϕ of the bearing is defined as the included angle between the eccentricity and load vector. When a lubricant is introduced into the top of the bearing, the action of the rotating journal is to pump the lubricant around the bearing in a clockwise direction. The lubricant is drawn into a convergent wedge-shaped space and forces the journal center to a steady position over to the side. The *minimum film thickness* h_0 occurs, not along the load line, but at a point displaced clockwise from the load. The magnitude of h_0 is $c\ (1 - \varepsilon)$ and is found by setting $\theta = 180°$ in Equation (12–12).

A complete lubricant film is established from the inlet region to a point just beyond the minimum film thickness value. It is in this fluid-converging region that film pressure is generated in the lubricant film to support the applied load. As the fluid continues to flow clockwise, the journal and bushing surfaces diverge, which results in a significant pressure drop. The film pressure in this region drops to sub-ambient levels, whereupon the fluid film *ruptures* or *cavitates* into liquid streamers separated by a liquid–gas mixture. The cavitation region typically extends over approximately half of the bearing surface.

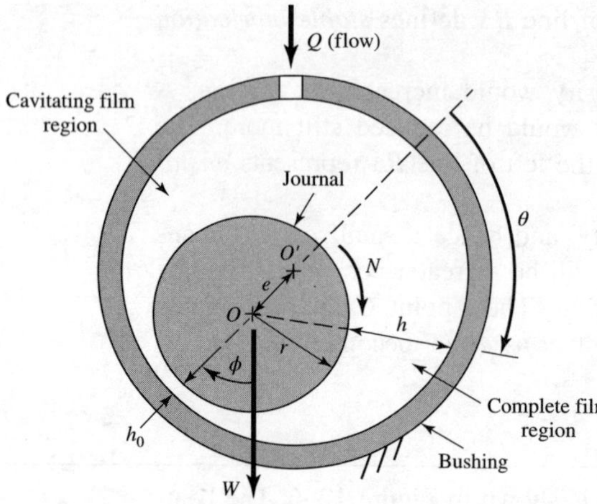

Figure 12–6

Nomenclature of a complete journal bearing.

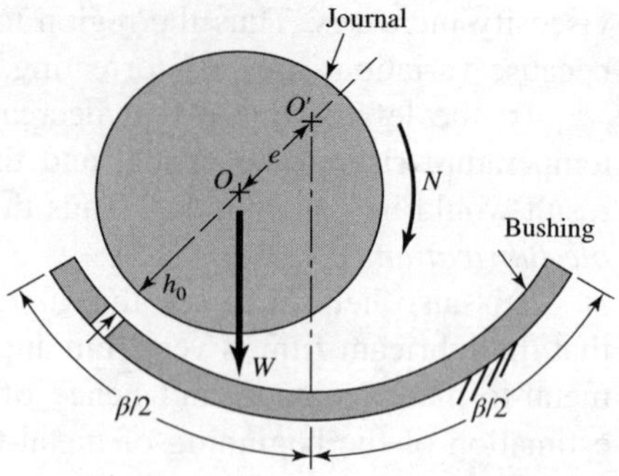

Figure 12–7

Nomenclature of a partial journal bearing.

A *partial journal bearing* is shown in Figure 12–7. If the radius of the bushing is the same as that of the journal ($c = 0$), it is known as a *fitted bearing*. The angle β defines the angular extent of the partial bearing, in this case measured from the load line. If the angular extent is small and if h_0 is near the edge, cavitation is unlikely to occur in the lubricant film.

12–6 Hydrodynamic Theory

The present theory of hydrodynamic lubrication originated in the laboratory of Beauchamp Tower in the early 1880s in England. Tower had been employed to study the friction in railroad journal bearings and learn the best methods of lubricating them. It was an accident or error, during the course of this investigation, that prompted Tower to look at the problem in more detail and that resulted in a discovery that eventually led to the development of the theory.

Figure 12–8 is a schematic drawing of the journal bearing that Tower investigated. It is a partial bearing, having a diameter of 4 in, a length of 6 in, a bearing arc of 157°, and bath-type lubrication, as shown. The coefficients of friction obtained by Tower in his investigations on this bearing were quite low, which is now not surprising. After testing this bearing, Tower later drilled a 12.5-mm-diameter lubricator hole

Figure 12–8

Schematic representation of the partial bearing used by Tower.

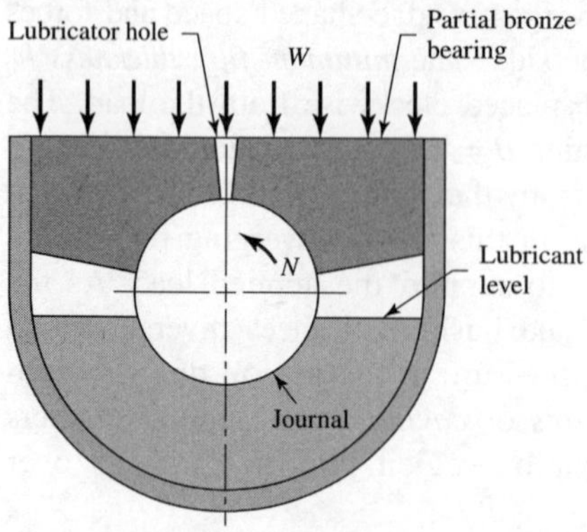

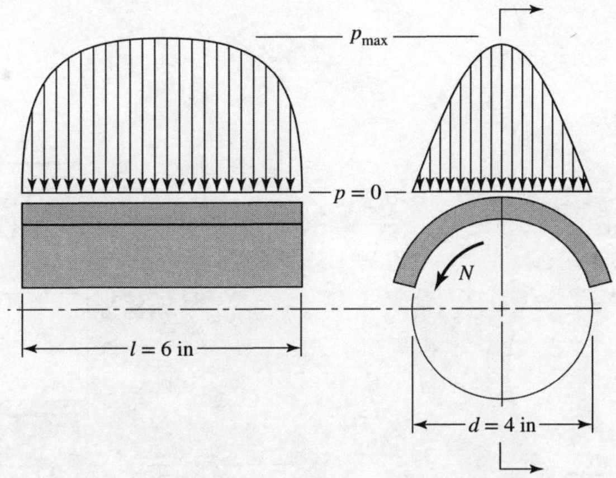

Figure 12–9

Approximate pressure-distribution curves obtained by Tower. *(Source: Tower, Beauchamp. "First Report on Friction Experiments."* Proceedings of the Institution of Mechanical Engineers 35, no. 1 *(November 1883): 632–666; "Second Report." ibid. (1885): 58–70; "Third Report." ibid. (1888): 173–205; "Fourth Report." ibid. (1891): 111–140.)*

through the top. But when the apparatus was set in motion, oil flowed out of this hole. In an effort to prevent this, a cork stopper was used, but this popped out, and so it was necessary to drive a wooden plug into the hole. When the wooden plug was pushed out too, Tower, at this point, undoubtedly realized that he was on the verge of discovery. A pressure gauge connected to the hole indicated a pressure of more than twice the unit bearing load. He further investigated the bearing film pressures in detail throughout the bearing width and length and reported a distribution similar to that of Figure 12–9.[4]

The results obtained by Tower had such regularity that Osborne Reynolds concluded that there must be a definite equation relating the friction, the pressure, and the surface velocity. The present mathematical theory of lubrication is based upon Reynolds's work following the experiment by Tower.[5] The solution is a challenging problem that has interested many investigators ever since then, and it is still the starting point for lubrication studies.

Reynolds pictured the lubricant as adhering to both surfaces and being pulled by the moving surface into a narrowing, wedge-shaped space so as to create a fluid or film pressure of sufficient intensity to support the bearing load. One of the important simplifying assumptions resulted from Reynolds's realization that the fluid films were so thin in comparison with the bearing radius that the curvature could be neglected. This enabled him to replace the cylindrical bearing with a flat "unwrapped" bearing, called a *plane slider bearing*. Other assumptions made were:

1 The lubricant is Newtonian, Equation (12–1).

2 The forces due to the inertia of the lubricant are neglected.

3 The lubricant is assumed to be incompressible.

4 The viscosity is assumed to be constant throughout the film.

5 The pressure does not vary in the axial direction.

Figure 12–10*a* shows a journal rotating in the clockwise direction at a constant angular velocity ω_j, supported by a film of lubricant of variable thickness h on a

---

[4]Beauchamp Tower, "First Report on Friction Experiments," *Proc. Inst. Mech. Eng.,* November 1883, pp. 632–666; "Second Report," ibid., 1885, pp. 58–70; "Third Report," ibid., 1888, pp. 173–205; "Fourth Report," ibid., 1891, pp. 111–140.

[5]Osborne Reynolds, "Theory of Lubrication, Part I," *Phil. Trans. Roy. Soc. London,* 1886.

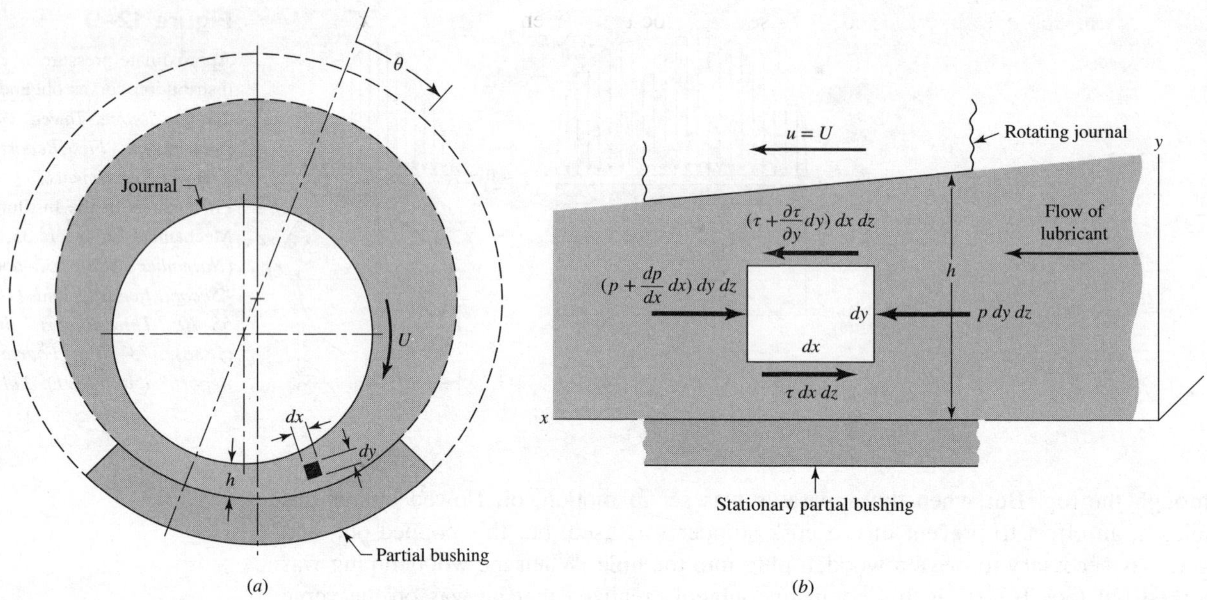

Figure 12–10

partial bearing, which is fixed. We specify that the journal has a constant surface velocity, $U = r\omega_j$. Using Reynolds's assumption that curvature can be neglected, we fix a right-handed xyz reference system to the stationary bushing, where $x = r\theta$. We now make the following additional assumptions:

6 The bushing and journal extend infinitely in the z direction; this means there can be no lubricant flow in the z direction.

7 The film pressure is constant in the y direction. Thus the pressure depends only on the coordinate x.

8 The velocity of any particle of lubricant in the film depends only on the coordinates x and y.

We now select an element of lubricant in the film (Figure 12–10*a*) of dimensions dx, dy, and dz, and compute the forces that act on the sides of this element. As shown in Figure 12–10*b*, normal forces, due to the pressure, act upon the right and left sides of the element, and shear forces, due to the viscosity and to the velocity, act upon the top and bottom sides. Summing the forces in the x direction gives

$$\sum F_x = p\, dy\, dz - \left(p + \frac{dp}{dx} dx\right) dy\, dz - \tau\, dx\, dz + \left(\tau + \frac{\partial \tau}{\partial y} dy\right) dx\, dz = 0 \tag{a}$$

This reduces to

$$\frac{dp}{dx} = \frac{\partial \tau}{\partial y} \tag{b}$$

From Equation (12–1), we have

$$\tau = \mu \frac{\partial u}{\partial y} \tag{c}$$

where the partial derivative is used because the velocity u depends upon both x and y. Substituting Equation (c) in Equation (b), we obtain

$$\frac{dp}{dx} = \mu \frac{\partial^2 u}{\partial y^2} \tag{d}$$

Holding x constant, we now integrate this expression twice with respect to y. This gives

$$\frac{\partial u}{\partial y} = \frac{1}{\mu}\frac{dp}{dx}y + C_1$$

$$u = \frac{1}{2\mu}\frac{dp}{dx}y^2 + C_1 y + C_2 \tag{e}$$

Note that the act of holding x constant means that C_1 and C_2 can be functions of x. We now assume that there is no slip between the lubricant and the boundary surfaces. This gives two sets of boundary conditions for evaluating C_1 and C_2:

$$\text{At} \quad y = 0,\ u = 0$$

$$\text{At} \quad y = h,\ u = U \tag{f}$$

Notice, in the second condition, that h is a function of x. Substituting these conditions in Equation (e) and solving for C_1 and C_2 gives

$$C_1 = \frac{U}{h} - \frac{h}{2\mu}\frac{dp}{dx} \qquad C_2 = 0$$

or

$$u = \frac{1}{2\mu}\frac{dp}{dx}(y^2 - hy) + \frac{U}{h}y \tag{12–14}$$

This equation gives the velocity distribution of the lubricant in the film as a function of the coordinate y and the pressure gradient dp/dx. The equation shows that the velocity distribution across the film (from $y = 0$ to $y = h$) is obtained by superposing a parabolic distribution onto a linear distribution. Figure 12–11 shows the superposition of these distributions to obtain the velocity for particular values of x and dp/dx. In general, the parabolic term may be additive or subtractive to the linear term, depending upon the sign of the pressure gradient. When the pressure is maximum, $dp/dx = 0$ and the velocity is

$$u = \frac{U}{h}y \tag{g}$$

which is a linear relation.

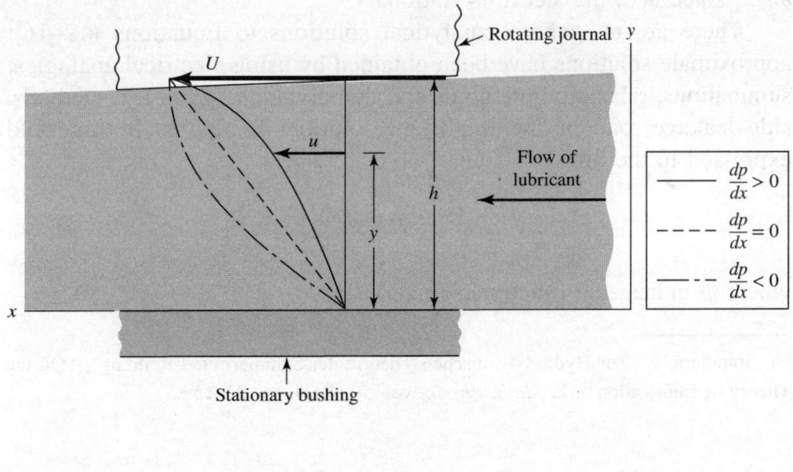

Figure 12–11
Velocity of the lubricant.

We next define Q as the volume of lubricant flowing in the x direction per unit time. By using a width of unity in the z direction, the volume may be obtained by the expression

$$Q = \int_0^h u\,dy \tag{h}$$

Substituting the value of u from Equation (12–14) and integrating gives

$$Q = \frac{Uh}{2} - \frac{h^3}{12\mu}\frac{dp}{dx} \tag{i}$$

The next step uses the assumption of an incompressible lubricant and states that the flow is the same for any cross section. Thus

$$\frac{dQ}{dx} = 0$$

From Equation (i),

$$\frac{dQ}{dx} = \frac{U}{2}\frac{dh}{dx} - \frac{d}{dx}\left(\frac{h^3}{12\mu}\frac{dp}{dx}\right) = 0$$

or

$$\frac{d}{dx}\left(\frac{h^3}{\mu}\frac{dp}{dx}\right) = 6U\frac{dh}{dx} \tag{12–15}$$

which is the classical Reynolds equation for one-dimensional flow. It neglects side leakage, that is, flow in the z direction. A similar development is used when side leakage is not neglected. The resulting Reynolds equation for general two-dimensional flow is

$$\frac{\partial}{\partial x}\left(\frac{h^3}{\mu}\frac{\partial p}{\partial x}\right) + \frac{\partial}{\partial z}\left(\frac{h^3}{\mu}\frac{\partial p}{\partial z}\right) = 6U\frac{\partial h}{\partial x} \tag{12–16}$$

Returning to the curved partial bearing, with $x = r\theta$ and $U = r\omega_j$, Equation (12–16) can be rewritten as

$$\frac{1}{r^2}\frac{\partial}{\partial \theta}\left(\frac{h^3}{12\mu}\frac{\partial p}{\partial \theta}\right) + \frac{\partial}{\partial z}\left(\frac{h^3}{12\mu}\frac{\partial p}{\partial z}\right) = \frac{\omega_j}{2}\frac{\partial h}{\partial \theta} \tag{12–17}$$

The wedge effect depends on the average velocity of the journal and sleeve, which is $\omega_j/2$, since here the sleeve is stationary.

There are no general analytical solutions to Equations (12–16) and (12–17); approximate solutions have been obtained by using electrical analogies, mathematical summations, relaxation methods, and numerical and graphical methods. Assuming no side leakage, one of the important solutions is due to Sommerfeld[6] and can be expressed in the form

$$\frac{r}{c}f = \phi\left[\left(\frac{r}{c}\right)^2\frac{\mu N}{P}\right] \tag{12–18}$$

where ϕ indicates a functional relationship.

[6]A. Sommerfeld, "Zur Hydrodynamischen Theorie der Schmiermittel-Reibung" ("On the Hydrodynamic Theory of Lubrication"), *Z. Math. Physik,* vol. 50, 1904, pp. 97–155.

12–7 Design Variables

We may distinguish between two groups of variables in the design of sliding bearings. In the first group are those whose values either are given or are under the control of the designer. These are:

1 The viscosity μ

2 The load W or the nominal bearing pressure P

3 The speed N

4 The bearing dimensions r, c, β, and l

Of these four variables, the designer usually has no control over the speed, because it is specified by the overall design of the machine. Sometimes the viscosity is specified in advance, as, for example, when the oil is stored in a sump and is used for lubricating and cooling a variety of bearings. The remaining variables, and sometimes the viscosity, may be controlled by the designer and are therefore the *decisions* the designer makes. In other words, when these four decisions are made, the design is complete.

In the second group are the dependent variables. The designer cannot control these except indirectly by changing one or more of the first group. These are:

1 The coefficient of friction f

2 The temperature rise ΔT

3 The volume flow rate of oil Q

4 The minimum film thickness h_0

This group of variables tells us how well the bearing is performing, and hence we may regard them as *performance factors*. Certain limitations on their values must be imposed by the designer to ensure satisfactory performance. These limitations are specified by the characteristics of the bearing materials and of the lubricant. The fundamental problem in bearing design, therefore, is to define satisfactory limits for the second group of variables and then to decide upon values for the first group such that these limitations are not exceeded.

Trumpler's Design Criteria for Journal Bearings

Because the bearing assembly creates the lubricant pressure to carry a load, it reacts to loading by changing its eccentricity, which reduces the minimum film thickness h_0 until the load is carried. What is the limit of smallness of h_0? Close examination reveals that the moving adjacent surfaces of the journal and bushing are not smooth but consist of a series of asperities that pass one another, separated by a lubricant film. In starting a bearing under load from rest there is metal-to-metal contact and surface asperities are broken off, free to move and circulate with the oil. Unless a filter is provided, this debris accumulates. Such particles have to be free to tumble at the section containing the minimum film thickness without snagging in a togglelike configuration, creating additional damage and debris. Trumpler, an accomplished bearing designer, provides a throat of at least 200 μ in to pass particles from ground surfaces.[7] He also provides for the influence of size (tolerances tend to increase with size) by stipulating

$$h_0 \geq 0.00508 + 0.000\,04d \text{ mm} \qquad (a)$$

where d is the journal diameter in millimeters.

[7]P. R. Trumpler, *Design of Film Bearings,* Macmillan, New York, 1966, pp. 192–194.

A lubricant is a mixture of hydrocarbons that reacts to increasing temperature by vaporizing the lighter components, leaving behind the heavier. This process (bearings have lots of time) slowly increases the viscosity of the remaining lubricant, which increases heat generation rate and elevates lubricant temperatures. This sets the stage for future failure. For light oils, Trumpler limits the maximum film temperature $T_{\max}$ to

$$T_{\max} \leq 121°\text{C} \tag{b}$$

Some oils can operate at slightly higher temperatures. Always check with the lubricant manufacturer.

A journal bearing often consists of a ground steel journal working against a softer, usually nonferrous, bushing. In starting under load there is metal-to-metal contact, abrasion, and the generation of wear particles, which, over time, can change the geometry of the bushing. The starting load divided by the projected area is limited to

$$\frac{W_{st}}{lD} \leq 2068 \text{ kPA} \tag{c}$$

If the load on a journal bearing is suddenly increased, the increase in film temperature in the annulus is immediate. Since ground vibration due to passing trucks, trains, and earth tremors is often present, Trumpler used a design factor of 2 or more on the running load, but not on the starting load of Equation (c):

$$n_d \geq 2 \tag{d}$$

Many of Trumpler's designs are operating today, long after his consulting career was over; clearly they constitute good advice to the beginning designer.

12–8 The Relations of the Variables

Before proceeding to the problem of design, it is necessary to establish the relationships between the variables. Albert A. Raimondi and John Boyd, of Westinghouse Research Laboratories, used an iteration technique to solve Reynolds's equation on the digital computer.[8] This is the first time such extensive data have been available for use by designers, and consequently we shall employ them in this book.[9]

The Raimondi and Boyd papers were published in three parts and contain 45 detailed charts and six tables of numerical information. In all three parts, charts are used to define the variables for length-diameter (l/d) ratios of 1:4, 1:2, and 1 and for beta angles of 60° to 360°. Part III of the Raimondi-Boyd papers assumes that the oil film is ruptured or cavitated when the film pressure becomes zero. Part III also contains data for the infinitely long bearing; since it has no ends, this means that there is no side leakage. The charts appearing in this book are from Part III of the papers, and are for full journal bearings ($\beta = 360°$) only.

[8]A. A. Raimondi and John Boyd, "A Solution for the Finite Journal Bearing and Its Application to Analysis and Design, Parts I, II, and III," *Trans. ASLE,* vol. 1, no. 1, in *Lubrication Science and Technology,* Pergamon, New York, 1958, pp. 159–209.

[9]See also the earlier companion paper, John Boyd and Albert A. Raimondi, "Applying Bearing Theory to the Analysis and Design of Journal Bearings, Parts I and II," *J. Appl. Mechanics,* vol. 73, 1951, pp. 298–316.

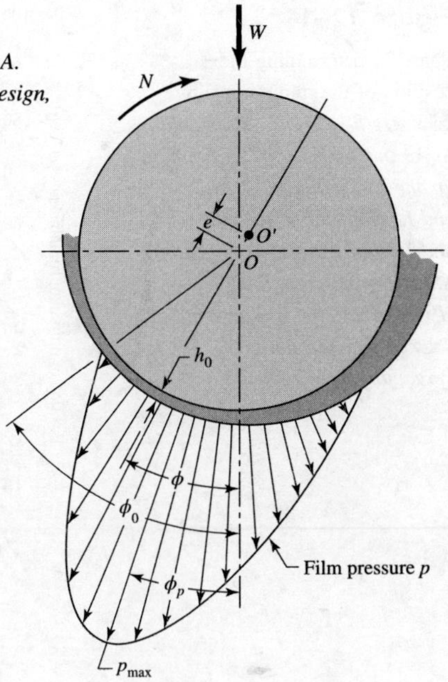

Figure 12–12

Polar diagram of the film–pressure distribution showing the notation used. *(Source: Raimondi, A. A. and John Boyd. "A Solution for the Finite Journal Bearing and Its Application to Analysis and Design, Parts I, II, and III." ASLE Transactions 1, no. 1 (1958): 159–209, in Lubrication Science and Technology, Pergamon, New York.)*

The charts from Raimondi and Boyd relate several variables to the Sommerfeld number. Figure 12–12 shows the notation used for the variables. These variables are

Minimum film thickness and attitude angle (Figures 12–13 and 12–14)
Coefficient of friction (Figure 12–15)
Lubricant flow (Figures 12–16 and 12–17)
Film pressure (Figures 12–18 and 12–19)

We will describe the use of these curves in a series of four examples using the same set of given parameters.

Minimum Film Thickness

In Figure 12–13, *the minimum film thickness* variable h_0/c and *eccentricity ratio* $\varepsilon = e/c$ are plotted against the Sommerfeld number S with contours for various values of l/d. The corresponding attitude angle or angular position of the minimum film thickness relative to the load is found in Figure 12–14.

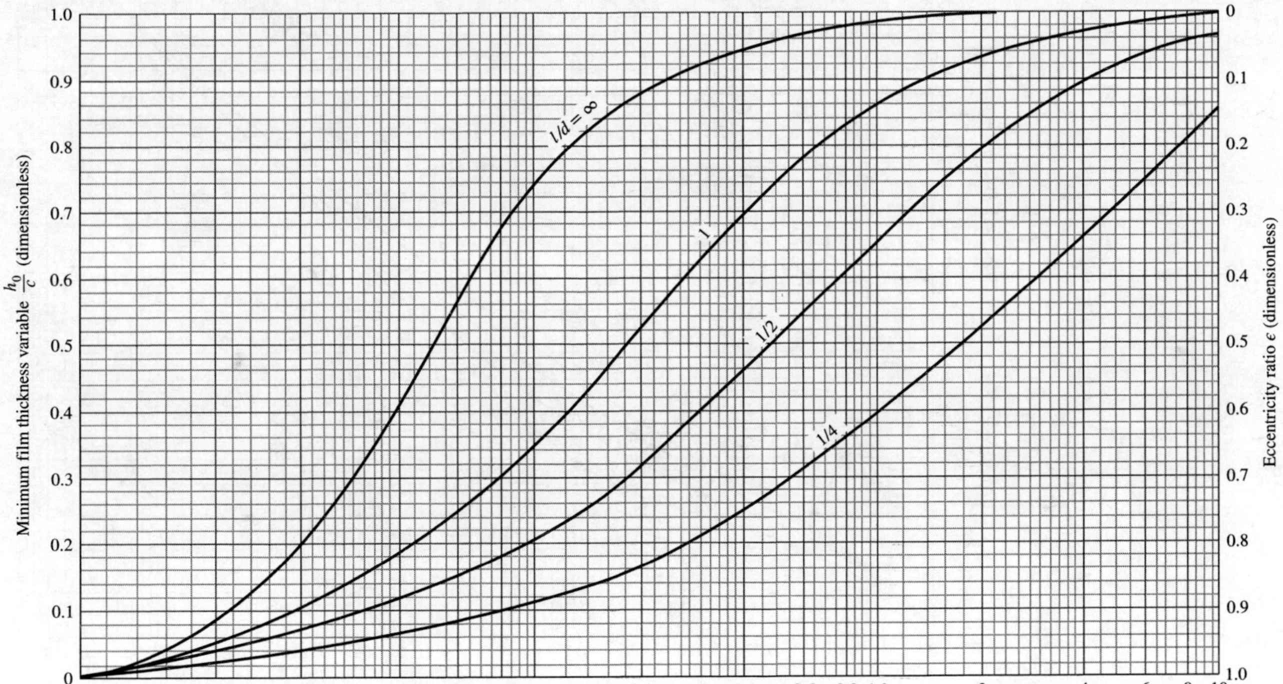

Figure 12–13

Chart for minimum film thickness variable and eccentricity ratio. *(Source: Raimondi, A. A. and John Boyd. "A Solution for the Finite Journal Bearing and Its Application to Analysis and Design, Parts I, II, and III." ASLE Transactions 1, no. 1 (1958): 159–209, in Lubrication Science and Technology, Pergamon, New York.)*

Figure 12–14

Chart for determining the position of the minimum film thickness h_0. *(Source: Raimondi, A. A. and John Boyd. "A Solution for the Finite Journal Bearing and Its Application to Analysis and Design, Parts I, II, and III." ASLE Transactions 1, no. 1 (1958): 159–209, in Lubrication Science and Technology, Pergamon, New York.)*

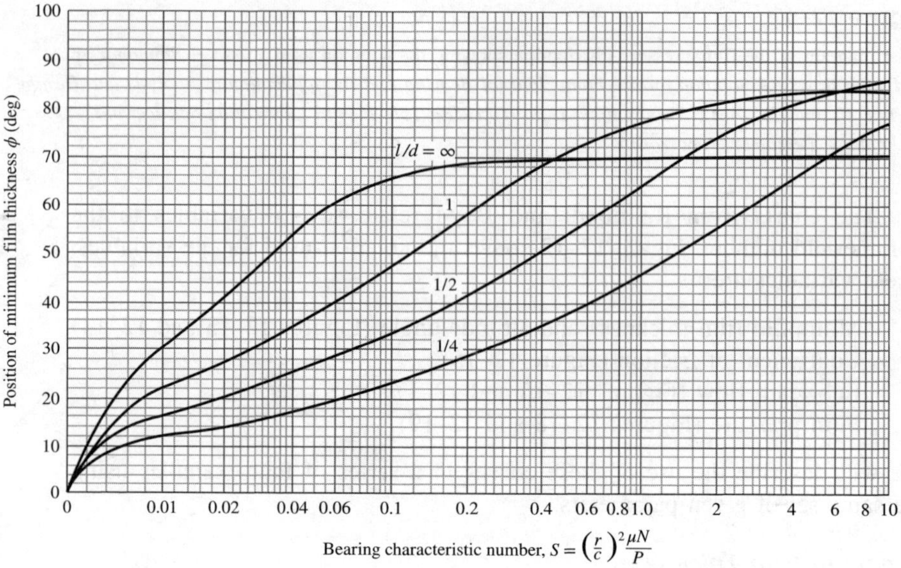

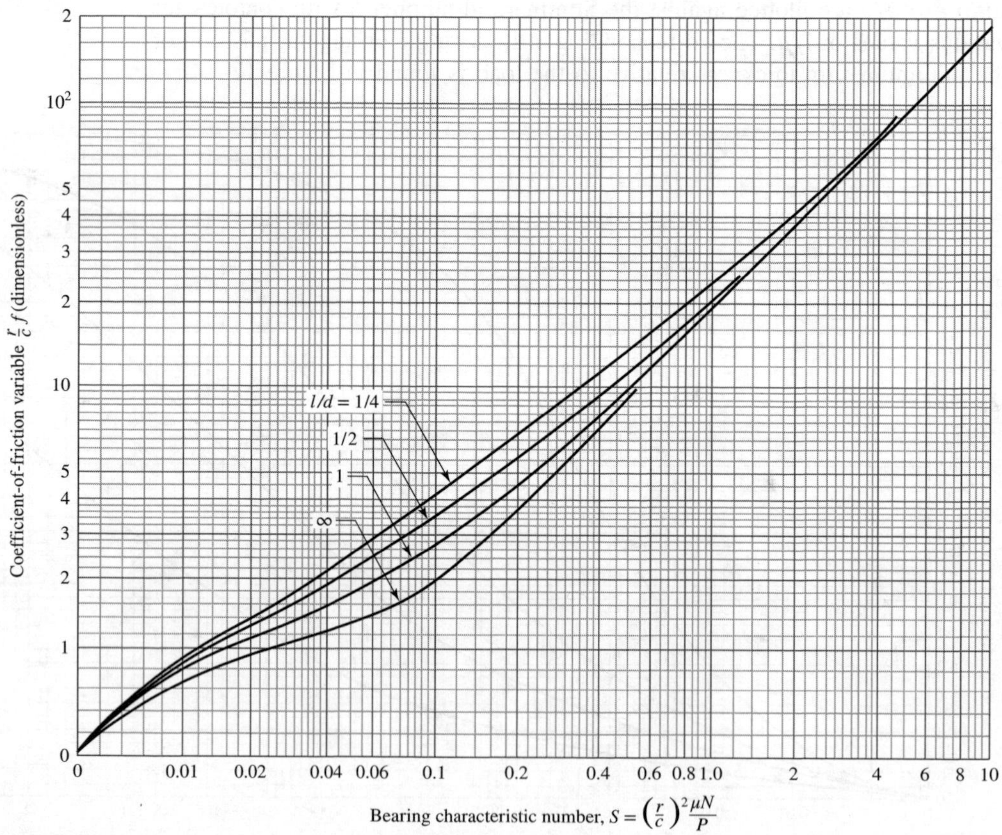

Figure 12–15

Chart for coefficient-of-friction variable; note that Petroff's equation is the asymptote. *(Source: Raimondi, A. A. and John Boyd. "A Solution for the Finite Journal Bearing and Its Application to Analysis and Design, Parts I, II, and III." ASLE Transactions 1, no. 1 (1958): 159–209, in Lubrication Science and Technology, Pergamon, New York.)*

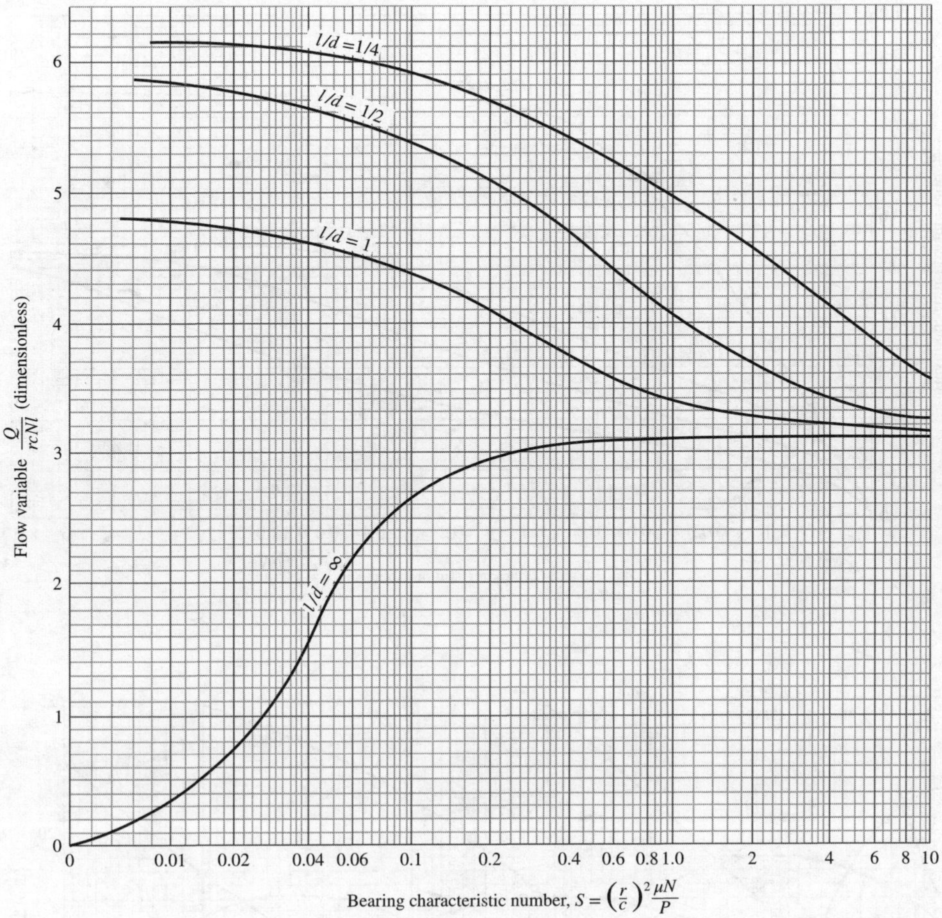

Figure 12–16

Chart for total flow variable. *Note:* Not for pressure-fed bearings. *(Source: Raimondi, A. A. and John Boyd. "A Solution for the Finite Journal Bearing and Its Application to Analysis and Design, Parts I, II, and III." ASLE Transactions 1, no. 1 (1958): 159–209, in Lubrication Science and Technology, Pergamon, New York.)*

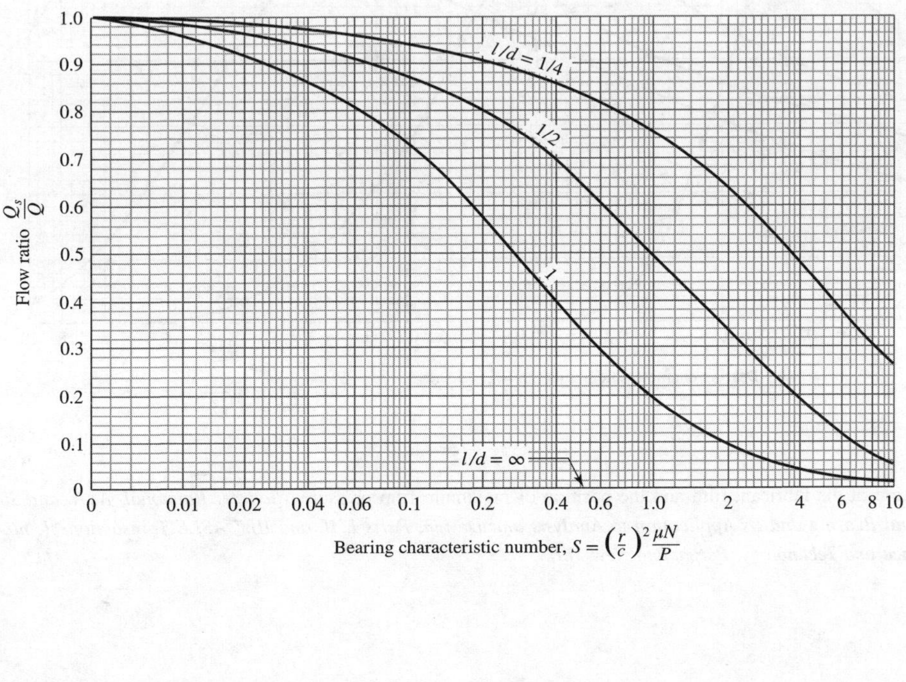

Figure 12–17

Chart for determining the ratio of side flow to total flow. *(Source: Raimondi, A. A. and John Boyd. "A Solution for the Finite Journal Bearing and Its Application to Analysis and Design, Parts I, II, and III." ASLE Transactions 1, no. 1 (1958): 159–209, in Lubrication Science and Technology, Pergamon, New York.)*

Figure 12–18

Chart for determining the maximum film pressure. *Note:* Not for pressure-fed bearings. *(Source: Raimondi, A. A. and John Boyd. "A Solution for the Finite Journal Bearing and Its Application to Analysis and Design, Parts I, II, and III." ASLE Transactions 1, no. 1 (1958): 159–209, in Lubrication Science and Technology, Pergamon, New York.)*

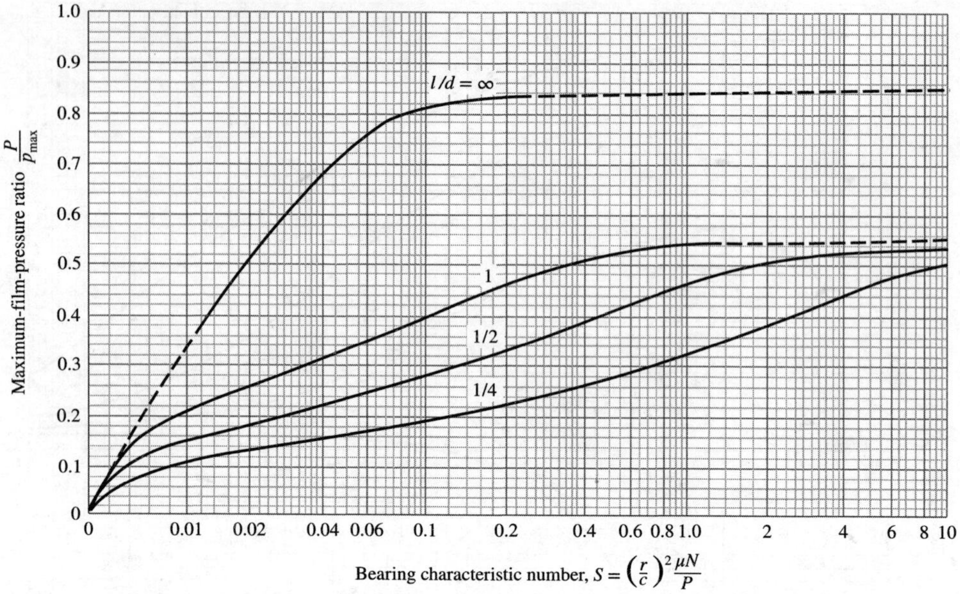

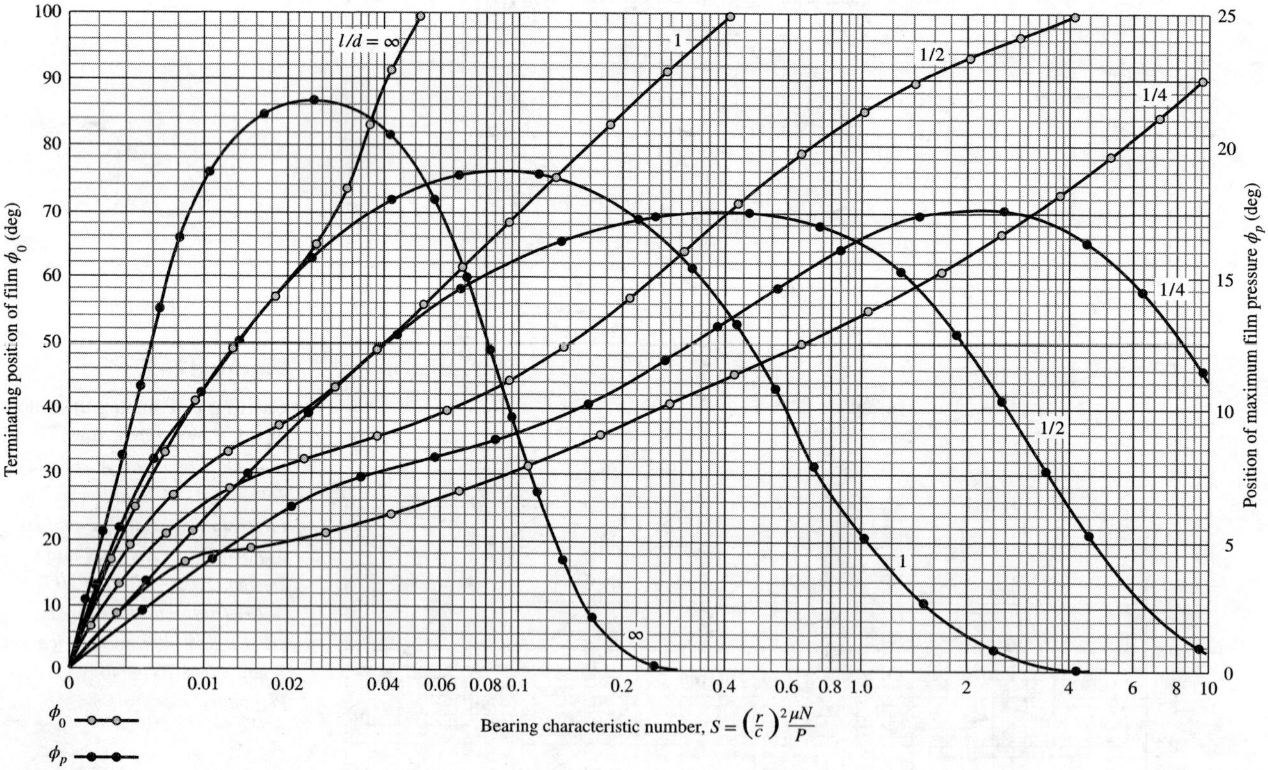

Figure 12–19

Chart for finding the terminating position of the lubricant film and the position of maximum film pressure. *(Source: Raimondi, A. A. and John Boyd. "A Solution for the Finite Journal Bearing and Its Application to Analysis and Design, Parts I, II, and III." ASLE Transactions 1, no. 1 (1958): 159–209, in Lubrication Science and Technology, Pergamon, New York.)*

EXAMPLE 12–1

Determine h_0 and e using the following given parameters: $\mu = 27.5 \cdot 10^{-3}$ Pa · s, $N = 30$ rev/s, $W = 2225$ N, (bearing load), $r = 19$ mm, $c = 0.038$ mm, and $l = 38$ mm.

Solution

From Equation (12–7), the nominal bearing pressure (in projected area of the journal) is

$$P = \frac{W}{2rl} = \frac{2225}{2(19)38} = 1.54 \text{ MPa}$$

The Sommerfeld number is, from Equation (12–10), where $N = N_j = 30$ rev/s,

$$S = \left(\frac{r}{c}\right)^2 \left(\frac{\mu N}{P}\right) = \left(\frac{19}{0.038}\right)^2 \left[\frac{27.5(10^{-3})30}{1.54(10^6)}\right] = 0.134$$

Also, $l/d = 38/[2(19)] = 1$. Entering Figure 12–13 with $S = 0.135$ and $l/d = 1$ gives $h_0/c = 0.42$ and $\varepsilon = 0.58$. Since $c = 0.038$ mm, the minimum film thickness h_0 is

$$h_0 = 0.42(0.038) = 0.016 \text{ mm}$$

We can find the angular location ϕ of the minimum film thickness from the chart of Figure 12–14. Entering with $S = 0.134$ and $l/d = 1$ gives $\phi = 53°$.

The eccentricity ratio is $\varepsilon = e/c = 0.58$. This means the eccentricity e is

$$e = 0.58(0.038) = 0.022 \text{ mm}$$

Coefficient of Friction

The friction chart, Figure 12–15, has the *friction variable* $(r/c)f$ plotted against Sommerfeld number S with contours for various values of the l/d ratio.

EXAMPLE 12–2

Using the parameters given in Example 12–1, determine the coefficient of friction, the torque to overcome friction, and the power loss to friction.

Solution

We enter Figure 12–15 with $S = 0.134$ and $l/d = 1$ and find $(r/c)f = 3.50$. The coefficient of friction f is

$$f = 3.50\, c/r = 3.50(0.038/19) = 0.0070$$

The friction torque on the journal is

$$T = f\,Wr = 0.007(2225)19 = 0.30 \text{ N} \cdot \text{m}$$

The power loss in Watt is

$$(hp)_{\text{loss}} = 2\pi TN = 2\pi(0.30)30 = 57 \text{ W}$$

Lubricant Flow

Figures 12–16 and 12–17 are used to determine the lubricant total flow and side flow.

EXAMPLE 12–3

Continuing with the parameters of Example 12–1, determine the total volumetric flow rate Q and the side flow rate Q_s.

Solution

To estimate the lubricant flow, enter Figure 12–16 with $S = 0.134$ and $l/d = 1$ to obtain $Q/(rcNl) = 4.28$. The total volumetric flow rate is

$$Q = 4.58rcNl = 4.28(19(10^{-3})0.038(10^{-3})(30)38(10^{-3}) = 3.52(10^{-6})\ \text{m}^3/\text{s}$$

From Figure 12–17 we find the *flow ratio* $Q_s/Q = 0.655$ and Q_s is

$$Q_s = 0.655Q = 0.655(3.52(10^{-6}) = 2.31(10^{-6})\ \text{m}^3/\text{s}$$

The side leakage Q_s is from the lower part of the bearing, where the internal pressure is above atmospheric pressure. The leakage forms a fillet at the journal-bushing external junction, and it is carried by journal motion to the top of the bushing, where the internal pressure is below atmospheric pressure and the gap is much larger, to be "sucked in" and returned to the lubricant sump. That portion of side leakage that leaks away from the bearing has to be made up by adding oil to the bearing sump periodically by maintenance personnel.

Film Pressure

The maximum pressure developed in the film can be estimated by finding the pressure ratio P/p_{max} from the chart in Figure 12–18. The locations where the terminating and maximum pressures occur, as defined in Figure 12–12, are determined from Figure 12–19.

EXAMPLE 12–4

Using the parameters given in Example 12–1, determine the maximum film pressure and the locations of the maximum and terminating pressures.

Solution

Entering Figure 12–18 with $S = 0.138$ and $l/d = 1$, we find $P/p_{\text{max}} = 0.42$. The maximum pressure p_{max} is therefore

$$p_{\text{max}} = \frac{P}{0.42} = \frac{1.54}{0.42} = 3.67\ \text{MPa}$$

With $S = 0.134$ and $l/d = 1$, from Figure 12–19, $\phi_p = 18.5°$ and the terminating position ϕ_0 is 75°.

Examples 12–1 to 12–4 demonstrate how the Raimondi and Boyd charts are used. It should be clear that we do not have journal-bearing parametric relations as equations, but in the form of charts. Moreover, the examples were simple because the steady-state equivalent viscosity was given. We will now show how the average film temperature (and the corresponding viscosity) is found from energy considerations.

Lubricant Temperature Rise

The temperature of the lubricant rises until the rate at which work is done by the journal on the film through fluid shear is the same as the rate at which heat is transferred to the greater surroundings. The specific arrangement of the bearing plumbing affects the quantitative relationships. Figure 12–20 shows a lubricant sump (internal or external to the bearing housing) which supplies lubricant at sump temperature T_s to the bearing annulus at temperature $T_s = T_1$. The lubricant passes once around the bushing and is delivered at a higher lubricant temperature $T_1 + \Delta T$ to the sump. Some of the lubricant leaks out of the bearing at a mixing-cup temperature of $T_1 + \Delta T/2$ and is returned to the sump. The sump may be a keyway-like groove in the bearing cap or a larger chamber up to half the bearing circumference. It can occupy "all" of the bearing cap of a split bearing. In such a bearing the side leakage occurs from the lower portion and is sucked back in, into the ruptured film arc. The sump could be well removed from the journal-bushing interface.

Let

Q = volumetric oil-flow rate into the bearing, m³/s

Q_s = volumetric side-flow leakage rate out of the bearing and to the sump, m³/s

$Q - Q_s$ = volumetric oil-flow discharge from annulus to sump, m³/s

T_1 = oil inlet temperature (equal to sump temperature T_s), °C

ΔT = temperature rise in oil between inlet and outlet, °C

ρ = lubricant density, kg/m³

C_p = specific heat capacity of lubricant, J/(kg · °C)

J = Joulean heat equivalent, m · N/J

H = heat rate, J/s

Using the sump as a control region, we can write an enthalpy balance. Using T_1 as the datum temperature gives

$$H_{\text{loss}} = \rho C_p Q_s \Delta T/2 + \rho C_p (Q - Q_s)\Delta T = \rho C_p Q \Delta T\left(1 - 0.5\frac{Q_s}{Q}\right) \tag{a}$$

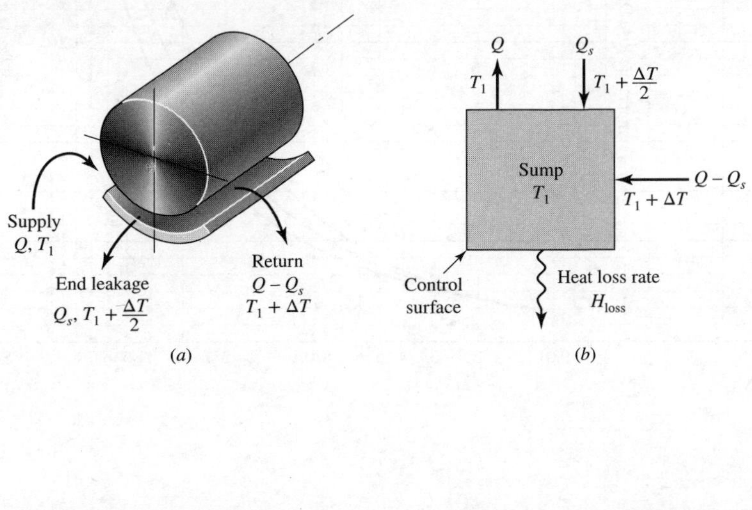

Figure 12–20

Schematic of a journal bearing with an external sump with cooling; lubricant makes one pass before returning to the sump.

The thermal energy loss at steady state H_{loss} is equal to the rate the journal does work on the film, that is, $H_{\text{loss}} = \dot{\mathscr{W}} = 2\pi TN/J$. The torque $T = fWr$, the load in terms of pressure is $W = 2Prl$, and multiplying numerator and denominator by the clearance c gives

$$H_{\text{loss}} = \frac{4\pi\, PrlNc}{J}\frac{rf}{c} \tag{b}$$

Equating Equations (*a*) and (*b*) and rearranging results in

$$\frac{J\rho C_p\,\Delta T}{4\pi P} = \frac{rf/c}{(1 - 0.5Q_s/Q)[Q/(rcNl)]} \tag{c}$$

For common petroleum lubricants $\rho = 862$ kg/m^3, $C_p = 1757$ J/(kg · °C), and $J = 0.99566$ m · N/J; therefore the left term of Equation (*c*) is

$$\frac{J\rho C_p\,\Delta T}{4\pi\,P} = \frac{0.99566(862)1757\Delta T_C}{4\pi\,P_{\text{MPa}}} = 0.12\,\frac{\Delta T_C}{P_{\text{MPa}}}$$

thus

$$\frac{0.12\Delta T_C}{P_{\text{MPa}}} = \frac{rf/c}{(1 - 0.5Q_s/Q)[Q/(rcN_j l)]} \tag{12–19}$$

where ΔT_C is the temperature rise in °C and P_{MPa} is the bearing pressure in MPa. The right side of Equation (12–17) can be evaluated from Figures 12–15, 12–16, and 12–17 for various Sommerfeld numbers and l/d ratios to give Figure 12–21. The ordinate in Figure 12–21 is $0.120\Delta T_C/P_{\text{MPa}}$, which is not surprising since both are

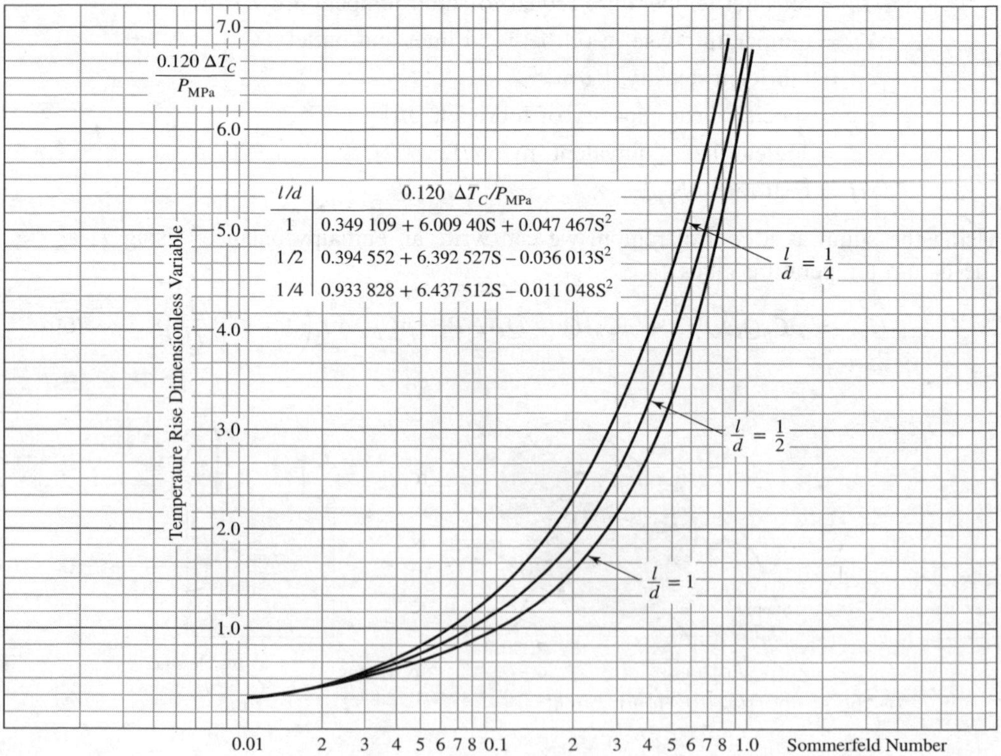

Figure 12–21

Figures 12–15, 12–16, and 12–17 combined to reduce iterative table look-up. *(Source: Chart based on work of Raimondi and Boyd boundary condition (2), i.e., no negative lubricant pressure developed. Chart is for full journal bearing using single lubricant pass, side flow emerges with temperature rise ΔT/2, thru flow emerges with temperature rise ΔT, and entire flow is supplied at datum sump temperature.)*

dimensionless in proper units and *identical in magnitude*. Since solutions to bearing problems involve iteration and reading many graphs can introduce errors, Figure 12–21 reduces three graphs to one, a step in the proper direction.

Interpolation

For l/d ratios other than the ones given in the charts, Raimondi and Boyd have provided the following interpolation equation

$$y = \frac{1}{(l/d)^3}\left[-\frac{1}{8}\left(1-\frac{l}{d}\right)\left(1-2\frac{l}{d}\right)\left(1-4\frac{l}{d}\right)y_\infty + \frac{1}{3}\left(1-2\frac{l}{d}\right)\left(1-4\frac{l}{d}\right)y_1 \right.$$
$$\left. -\frac{1}{4}\left(1-\frac{l}{d}\right)\left(1-4\frac{l}{d}\right)y_{1/2} + \frac{1}{24}\left(1-\frac{l}{d}\right)\left(1-2\frac{l}{d}\right)y_{1/4}\right] \tag{12–20}$$

where y is the desired variable within the interval $\infty > l/d > \frac{1}{4}$ and y_∞, y_1, $y_{1/2}$, and $y_{1/4}$ are the variables corresponding to l/d ratios of ∞, 1, $\frac{1}{2}$, and $\frac{1}{4}$, respectively.

Oil Outlet Temperature and Effective Viscosity

The determination of oil outlet temperature is a trial-and-error process when the oil type and its inlet temperature are specified. In an analysis, the temperature rise will first be estimated. This allows for the viscosity to be determined from the chart. With the value of the viscosity, the analysis is performed where the temperature rise is then computed. With this, a new estimate of the temperature rise is established. This process is continued until the estimated and computed temperatures agree.

As discussed previously, some of the lubricant that enters the bearing emerges as a side flow, which carries away some of the heat. The balance of the lubricant flows through the load-bearing zone and carries away the balance of the heat generated. In determining the viscosity to be used, we shall employ a temperature that is the average of the inlet and outlet temperatures, or

$$T_{av} = T_1 + \Delta T/2 \tag{12–21}$$

where T_1 is the inlet temperature and ΔT is the temperature rise of the lubricant from inlet to outlet. Of course, the viscosity used in the analysis must correspond to T_{av}.

12–9 Steady-State Conditions in Self-Contained Bearings

The case in which the lubricant carries away all of the enthalpy increase from the journal-bushing pair has already been discussed. Bearings in which the warm lubricant stays within the bearing housing will now be addressed. These bearings are called *self-contained* bearings because the lubricant sump is within the bearing housing and the lubricant is cooled within the housing. These bearings are described as *pillow-block* or *pedestal* bearings. They find use on fans, blowers, pumps, and motors, for example. Integral to design considerations for these bearings is dissipating heat from the bearing housing to the surroundings at the same rate that enthalpy is being generated within the fluid film.

In a self-contained bearing the sump can be positioned as a keywaylike cavity in the bushing, the ends of the cavity not penetrating the end planes of the bushing. Film oil exits the annulus at about one-half of the relative peripheral speeds of the journal and bushing and slowly tumbles the sump lubricant, mixing with the sump contents. Since the film in the top "half" of the cap has cavitated, it contributes essentially nothing to the support of the load, but it does contribute friction. Bearing caps are in use in which the "keyway" sump is expanded peripherally to encompass the top half of the bearing. This reduces friction for the same load, but the included angle β of the bearing has been reduced to 180°. Charts for this case were included in the Raimondi and Boyd paper.

The heat given up by the bearing housing may be estimated from the equation

$$H_{\text{loss}} = \hbar_{\text{CR}} A(T_b - T_\infty) \tag{12–22}$$

where H_{loss} = heat dissipated, J/s

$\hbar_{\text{CR}}$ = combined overall coefficient of radiation and convection heat transfer, J/(s · m² · °C)

A = surface area of bearing housing, m²

T_b = surface temperature of the housing, °C

T_∞ = ambient temperature, °C

The overall coefficient $\hbar_{\text{CR}}$ depends on the material, surface coating, geometry, even the roughness, the temperature difference between the housing and surrounding objects, and air velocity. After Karelitz,[10] and others, in ordinary industrial environments, the overall coefficient $\hbar_{\text{CR}}$ can be treated as a constant. Some representative values are

$$\hbar_{\text{CR}} = \begin{cases} 11.3 \text{ J/(s} \cdot \text{m}^2 \cdot {}^\circ\text{C)} & \text{for still air} \\ 15.3 \text{ J/(s} \cdot \text{m}^2 \cdot {}^\circ\text{C)} & \text{for shaft-stirred air} \\ 33.5 \text{ J/(s} \cdot \text{m}^2 \cdot {}^\circ\text{C)} & \text{for air moving at 2.5 m/s} \end{cases} \tag{12–23}$$

An expression similar to Equation (12–22) can be written for the temperature difference $T_f - T_b$ between the lubricant film and the housing surface. This is possible because the bushing and housing are metal and very nearly isothermal. If one defines $\bar{T}_f$ as the *average* film temperature (halfway between the lubricant inlet temperature T_s and the outlet temperature $T_s + \Delta T$), then the following proportionality has been observed between $\bar{T}_f - T_b$ and the difference between the housing surface temperature and the ambient temperature, $T_b - T_\infty$:

$$\bar{T}_f - T_b = \alpha(T_b - T_\infty) \tag{a}$$

where $\bar{T}_f$ is the average film temperature and α is a constant depending on the lubrication scheme and the bearing housing geometry. Equation (*a*) may be used to estimate the bearing housing temperature. Table 12–2 provides some guidance concerning suitable values of α. The work of Karelitz allows the broadening of the application of the charts of Raimondi and Boyd, to be applied to a variety of bearings beyond the natural circulation pillow-block bearing.

Table 12–2

Lubrication System	Conditions	Range of α
Oil ring	Moving air	1–2
	Still air	$\frac{1}{2}$–1
Oil bath	Moving air	$\frac{1}{2}$–1
	Still air	$\frac{1}{5}$–$\frac{2}{5}$

[10]G. B. Karelitz, "Heat Dissipation in Self-Contained Bearings," *Trans. ASME,* vol. 64, 1942, p. 463; D. C. Lemmon and E. R. Booser, "Bearing Oil-Ring Performance," *Trans. ASME,* J. Bas. Engin., vol. 88, 1960, p. 327.

Solving Equation (*a*) for T_b and substituting into Equation (12–22) gives the bearing heat loss rate to the surroundings as

$$H_{\text{loss}} = \frac{\hbar_{\text{CR}} A}{1+\alpha}(\overline{T}_f - T_\infty) \tag{12–24}$$

and rewriting Equation (*a*) gives

$$T_b = \frac{\overline{T}_f + \alpha T_\infty}{1+\alpha} \tag{12–25}$$

In beginning a steady-state analysis the average film temperature is unknown, hence the viscosity of the lubricant in a self-contained bearing is unknown. Finding the equilibrium temperatures is an iterative process wherein a trial average film temperature (and the corresponding viscosity) is used to compare the heat generation rate and the heat loss rate. An adjustment is made to bring these two heat rates into agreement. This can be done on paper with a tabular array to help adjust $\overline{T}_f$ to achieve equality between heat generation and loss rates. A root-finding algorithm can be used.

Because of the shearing action there is a uniformly distributed energy release in the lubricant that heats the lubricant as it works its way around the bearing. The temperature is uniform in the radial direction but increases from the sump temperature T_s by an amount ΔT during the lubricant pass. The exiting lubricant mixes with the sump contents, being cooled to sump temperature. The lubricant in the sump is cooled because the bushing and housing metal are at a nearly uniform lower temperature because of heat losses by convection and radiation to the surroundings at ambient temperature T_∞. In the usual configurations of such bearings, the bushing and housing metal temperature is approximately midway between the average film temperature $\overline{T}_f = T_s + \Delta T/2$ and the ambient temperature T_∞. The heat generation rate H_{gen}, at steady state, is equal to the work rate from the frictional torque T. Expressing this in J/s results in $H_{\text{gen}} = 2\pi\, TN$. Then from Equation (12–6) the torque is $T = 4\pi^2 r^3 l\mu N/c$, resulting in

$$H_{\text{gen}} = \frac{248\mu N^2 l r^3}{c} \tag{b}$$

Equating this to Equation (12–24) and solving for $\overline{T}_f$ gives

$$\overline{T}_f = T_\infty + 248(1+\alpha)\frac{\mu N^2 l r^3}{\hbar_{\text{CR}} A c} \tag{12–26}$$

EXAMPLE 12–5

Consider a pillow-block bearing with a keyway sump, whose journal rotates at 900 rev/min in shaft-stirred air at 21°C with $\alpha = 1$. The lateral area of the bearing is $2.58{\cdot}10^{-2}$ m^2. The lubricant is SAE grade 20 oil. The gravity radial load is 445 N and the l/d ratio is unity. The bearing has a journal diameter of $50.8 + 0.000/{-0.05}$ in, a bushing bore of $50.85 + 0.10/{-0.000}$ mm. For a minimum clearance assembly estimate the steady-state temperatures as well as the minimum film thickness and coefficient of friction.

Solution

The minimum radial clearance, $c_{\min}$, is

$$c_{\min} = \frac{50.85 - 50.8}{2} = 0.025 \text{ mm}$$

$$P = \frac{W}{ld} = \frac{455}{50.8(50.8)} = 0.172 \text{ N/mm}^2 \tag{1}$$

$$S = \left(\frac{r}{c}\right)^2 \frac{\mu N}{P} = \left(\frac{25.4}{0.025}\right)^2 \frac{\mu'(15)}{0.172(10^6)} = 90\ \mu'$$

where μ' is viscosity in Pa(s). The friction power (watt) loss, $(\text{H})_f$, is found as follows:

$$(\text{H})_f = (fWr)(2\pi N/60) = W\,\frac{2\pi N}{60}\,c\left(\frac{fr}{c}\right) = 445\,\frac{2\pi\cdot 900}{60}\,0.025\cdot 10^{-3}\frac{fr}{c} = 1.049\,\frac{fr}{c}\ \text{J/s} \tag{2}$$

The heat generation rate Hgen at the regime equals the heat friction power (H)f.

From Equation (12–24) with $\hbar_{\text{CR}} = 15.3$ J/(s(m^2)°C), the rate of heat loss to the environment H_{loss} is

$$H_{\text{loss}} = \frac{\hbar_{\text{CR}} A}{\alpha + 1}(\overline{T}_f - 21) = \frac{15.3(40/144)}{(1+1)}(\overline{T}_f - 21) = 0.198(\overline{T}_f - 21)\ \text{J/s} \tag{3}$$

Build a table as follows for trial values of $\overline{T}_f$ of 88° and 91°C:

1. With $\overline{T}_f$, calculate μ' from Table 12–1 for SAE grade 20 oil.
2. Evaluate S from Equation (1).
3. Determine fr/c from Figure 12–15.
4. Determine $H_{gen} = Hf$ from Equation (2).
5. Determine H_{loss} from Equation (3).

Trial $\overline{T}_f$	μ [Pa · s]	S	fr/c	H_{gen} [J/s]	H_{loss} [J/s]
88	$7.18\cdot 10^{-3}$	0.65	12.8	13.4	13.1
91	$6.42\cdot 10^{-3}$	0.58	11.6	12.2	13.9

The temperature at which $H_{\text{gen}} = H_{\text{loss}} = 13.2$ J/s is 87.7°C. Rounding $\overline{T}_f$ to 90°C we find $\mu' = 6.6(10^{-3}$ Pa) s and $S = 0.59$. From Figure 12–16 and 12–17, $Q_s/Q = 0.3$ and $Q/rcNl = 3.6$, by Equation 12–19 ΔT_c (0.120)/P_{Mpa} = 3.9°C/MPa and thus

$$\Delta T_C = 3.9P/0.120 = 3.9(0.172)/0.120 = 5.6°C$$

$$T_1 = T_s = \overline{T}_f - \Delta T_C/2 = 90 - 5.6/2 = 87.2°C$$

$$T_{\max} = T_1 - \Delta T_C = 87.2 + 5.6 = 92.8°C$$

From Equation (12–25)

$$T_b = \frac{T_f + \alpha T_\infty}{1 + \alpha} = \frac{90 + (1)21}{1 + 1} = 55.5°\text{C}$$

with $S = 0.59$, the minimum film thickness from Figure 12–13 is

$$h_0 = \frac{h_0}{c}c = 0.78(0.025) = 0.0195 \text{ mm}$$

The coefficient of friction from Figure 12–15 is

$$f = \frac{fr}{c}\frac{c}{r} = 11.9\frac{0.025}{25.4} = 0.012$$

The parasitic friction torque T is

$$T = fWr = 0.012(445)(25.4(10^{-3})) = 0.136 \text{ N} \cdot \text{m}$$

12–10 Clearance

In designing a journal bearing for thick-film lubrication, the engineer must select the grade of oil to be used, together with suitable values for P, N, r, c, and l. A poor selection of these or inadequate control of them during manufacture or in use may result in a film that is too thin, so that the oil flow is insufficient, causing the bearing to overheat and, eventually, fail. Furthermore, the radial clearance c is difficult to hold accurate during manufacture, and it may increase because of wear. What is the effect of an entire range of radial clearances, expected in manufacture, and what will happen to the bearing performance if c increases because of wear? Most of these questions can be answered and the design optimized by plotting curves of the performance as functions of the quantities over which the designer has control.

Figure 12–22 shows the results obtained when the performance of a particular bearing is calculated for a whole range of radial clearances and is plotted with clearance as the independent variable. The bearing used for this graph is the one of Examples 12–1 to 12–4 with SAE 20 oil at an inlet temperature of 37.8°C. The graph shows that if the clearance is too tight, the temperature will be too high and the minimum film thickness too low. High temperatures may cause the bearing to fail by fatigue. If the oil film is too thin, dirt particles may be unable to pass without scoring or may embed themselves in the bearing. In either event, there will be excessive wear and friction, resulting in high temperatures and possible seizing.

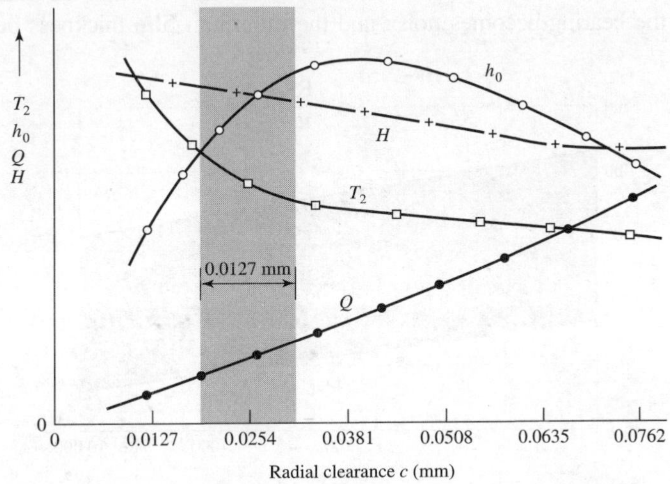

Figure 12–22

A plot of some performance characteristics of the bearing of Examples 12–1 to 12–4 for radial clearances of 0.0127 to 0.0762 mm. The bearing outlet temperature is designated T_2. New bearings should be designed for the shaded zone, because wear will move the operating point to the right.

Table 12–3 **Maximum, Minimum, and Average Clearances for 38 mm Diameter Journal Bearings Based on Type of Fit**

Type of Fit	Symbol	Clearance c, mm Maximum	Average	Minimum
Close-running	H8/f7	0.04445	0.02858	0.01270
Free-running	H9/d9	0.10033	0.06985	0.03937

Table 12–4 **Performance of 1.5-in-Diameter Journal Bearing with Various Clearances. (SAE 20 Lubricant, T_1 = 38°C, N = 30 rev/s, W = 2225 N, L = 38 mm)**

c, mm	T_2, °C	h_0, mm	f	Q, mm³/s,	H, J/s
0.01270	108	0.00965	0.0113	999	0.0903
0.02858	61	0.01651	0.0090	2507	0.0714
0.03937	56	0.01956	0.0087	3571	0.0693
0.04445	53	0.01930	0.0084	4129	0.0672
0.06985	48	0.01854	0.0079	6864	0.063
0.10033	45	0.01753	0.0077	10 108	0.062

To investigate the problem in more detail, Table 12–3 was prepared using the two types of preferred running fits that seem to be most useful for journal-bearing design (see Table 7–9). The results shown in Table 12–3 were obtained by using Equations (7–36) and (7–37) of Section 7–8. Notice that there is a slight overlap, but the range of clearances for the free-running fit is about twice that of the close-running fit.

The six clearances of Table 12–3 were used in a computer program to obtain the numerical results shown in Table 12–4. These conform to the results of Figure 12–22, too. Both the table and the figure show that a tight clearance results in a high temperature. Figure 12–23 can be used to estimate an upper temperature limit when the characteristics of the application are known.

It would seem that a large clearance will permit the dirt particles to pass through and also will permit a large flow of oil, as indicated in Table 12–4. This lowers the temperature and increases the life of the bearing. However, if the clearance becomes too large, the bearing becomes noisy and the minimum film thickness begins to decrease again.

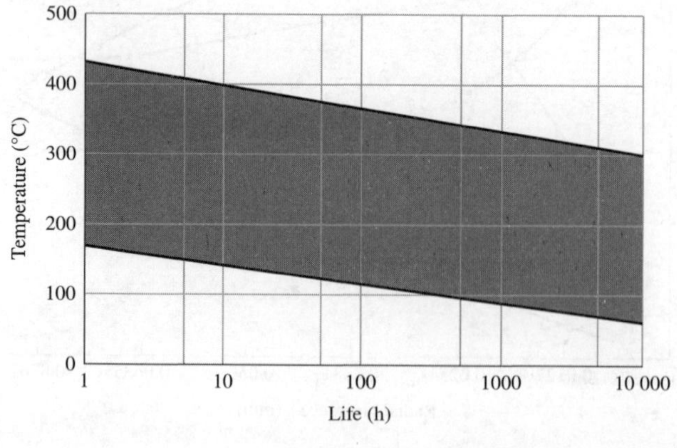

Figure 12–23

Temperature limits for mineral oils. The lower limit is for oils containing antioxidants and applies when oxygen supply is unlimited. The upper limit applies when insignificant oxygen is present. The life in the shaded zone depends on the amount of oxygen and catalysts present. *(Source: Neale, M. J. (ed.). Tribology Handbook. Section B1. London: Newnes-Butterworth, 1975.)*

In between these two limitations there exists a rather large range of clearances that will result in satisfactory bearing performance.

When both the production tolerance and the future wear on the bearing are considered, it is seen, from Figure 12–22, that the best compromise is a clearance range slightly to the left of the top of the minimum-film-thickness curve. In this way, future wear will move the operating point to the right and increase the film thickness and decrease the operating temperature.

12–11 Pressure-Fed Bearings

The load-carrying capacity of self-contained natural-circulating journal bearings is quite restricted. The factor limiting better performance is the heat-dissipation capability of the bearing. A first thought of a way to increase heat dissipation is to cool the sump with an external fluid such as water. The high-temperature problem is in the film where the heat is generated but cooling is not possible in the film until later. This does not protect against exceeding the maximum allowable temperature of the lubricant. A second alternative is to reduce the *temperature rise* in the film by dramatically increasing the rate of lubricant flow. The lubricant itself is reducing the temperature rise. A water-cooled sump may still be in the picture. To increase lubricant flow, an external pump must be used with lubricant supplied at pressures on the order of a few bar. Because the lubricant is supplied to the bearing under pressure, such bearings are called *pressure-fed bearings.*

To force a greater flow through the bearing and thus obtain an increased cooling effect, a common practice is to use a circumferential groove at the center of the bearing, with an oil-supply hole located opposite the load-bearing zone. Such a bearing is shown in Figure 12–24. The effect of the groove is to create two half-bearings, each having a smaller l/d ratio than the original. The groove divides the pressure-distribution curve into two lobes and reduces the minimum film thickness, but it has wide acceptance among lubrication engineers because such bearings carry more load without overheating.

To set up a method of solution for oil flow, we shall assume a groove ample enough that the pressure drop in the groove itself is small. Initially we will neglect eccentricity and then apply a correction factor for this condition. The oil flow, then, is the amount that flows out of the two halves of the bearing in the direction of the concentric shaft. If we neglect the rotation of the shaft, the flow of the lubricant is caused by the supply pressure p_s, shown in Figure 12–25. Laminar flow is assumed, with the pressure varying linearly from $p = p_s$ at $x = 0$, to $p = 0$ at $x = l'$. Consider the static equilibrium of an element of thickness dx, height $2y$, and unit depth. Note particularly that the origin of the reference system has been chosen at the midpoint of the clearance space and symmetry about the x axis is implied with the shear stresses τ being equal on the top and bottom surfaces. The equilibrium equation in the x direction is

$$-2y(p + dp) + 2yp + 2\tau\, dx = 0 \qquad (a)$$

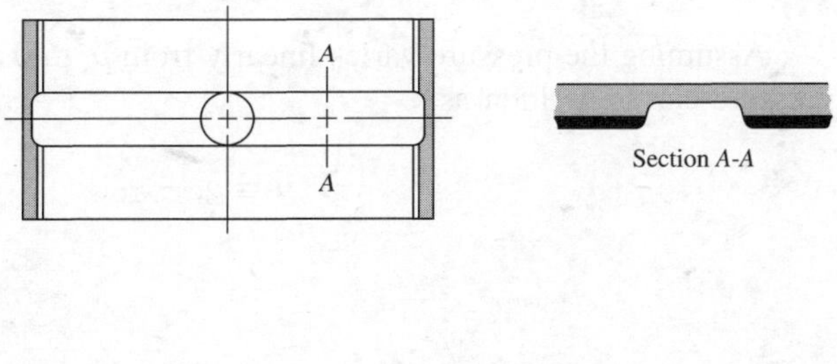

Figure 12–24

Centrally located full annular groove. *(Source: Based on Cleveland Graphite Bronze Company. Division of Clevite Corporation.)*

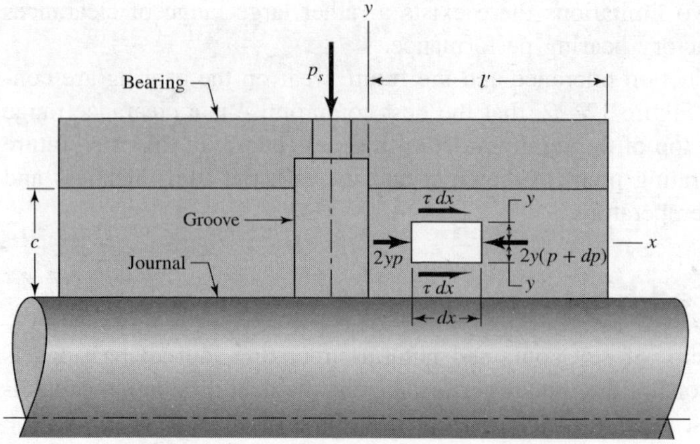

Figure 12–25

Flow of lubricant from a pressure-fed bearing having a central annular groove.

Expanding and canceling terms, we find that

$$\tau = y\frac{dp}{dx} \tag{b}$$

Newton's equation for viscous flow, Equation (12–1), is

$$\tau = \mu\frac{du}{dy} \tag{c}$$

Now eliminating τ from Equations (*b*) and (*c*) gives

$$\frac{du}{dy} = \frac{1}{\mu}\frac{dp}{dx}y \tag{d}$$

Treating dp/dx as a constant and integrating with respect to y gives

$$u = \frac{1}{2\mu}\frac{dp}{dx}y^2 + C_1 \tag{e}$$

At the boundaries, where $y = \pm c/2$, the velocity u is zero. Using one of these conditions in Equation (*e*) gives

$$0 = \frac{1}{2\mu}\frac{dp}{dx}\left(\frac{c}{2}\right)^2 + C_1$$

or

$$C_1 = -\frac{c^2}{8\mu}\frac{dp}{dx}$$

Substituting C_1 in Equation (*e*) yields

$$u = \frac{1}{8\mu}\frac{dp}{dx}(4y^2 - c^2) \tag{f}$$

Assuming the pressure varies linearly from p_s to 0 at $x = 0$ to l', respectively, the pressure can be written as

$$p = p_s - \frac{p_s}{l'}x \tag{g}$$

and therefore the pressure gradient is given by

$$\frac{dp}{dx} = -\frac{p_s}{l'} \tag{h}$$

We can now substitute Equation (*h*) in Equation (*f*) and the relationship between the oil velocity and the coordinate *y* is

$$u = \frac{p_s}{8\mu l'}(c^2 - 4y^2) \tag{12–27}$$

Figure 12–26 shows a graph of this relation fitted into the clearance space *c* so that you can see how the velocity of the lubricant varies from the journal surface to the bearing surface. The distribution is parabolic, as shown, with the maximum velocity occurring at the center, where $y = 0$. The magnitude is, from Equation (12–27),

$$u_{\max} = \frac{p_s c^2}{8\mu l'} \tag{i}$$

To consider eccentricity, as shown in Figure 12–27, the film thickness is $h = c - e\cos\theta$. Substituting *h* for *c* in Equation (*i*), with the average ordinate of a parabola being two-thirds the maximum, the average velocity at any angular position θ is

$$u_{av} = \frac{2}{3}\frac{p_s h^2}{8\mu l'} = \frac{p_s}{12\mu l'}(c - e\cos\theta)^2 \tag{j}$$

Now that we have an expression for the lubricant velocity, we can compute the amount of lubricant that flows out both ends. The elemental side flow at any position θ (Figure 12–27) is

$$dQ_s = 2u_{av}\,dA = 2u_{av}(rh\,d\theta) \tag{k}$$

where *dA* is the elemental area. Substituting u_{av} from Equations (*j*) and (*h*) from Figure 12–27 gives

$$dQ_s = \frac{p_s r}{6\mu l'}(c - e\cos\theta)^3\,d\theta \tag{l}$$

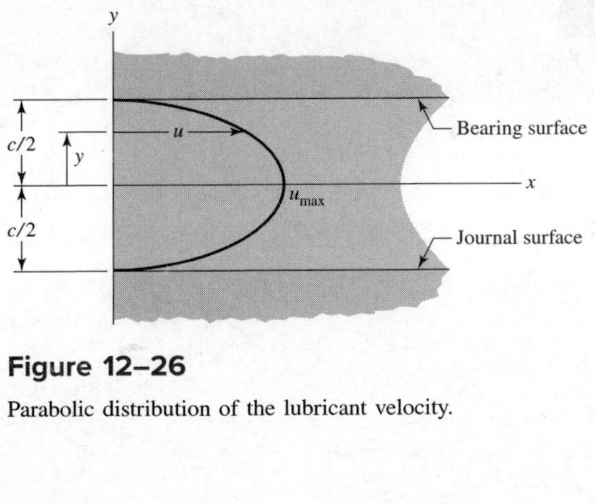

Figure 12–26

Parabolic distribution of the lubricant velocity.

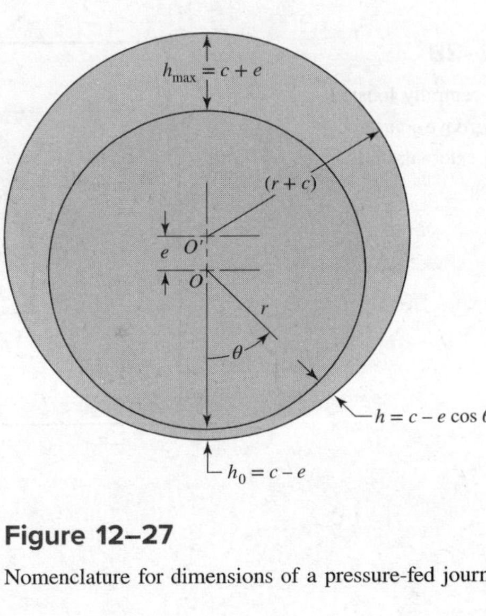

Figure 12–27

Nomenclature for dimensions of a pressure-fed journal bearing.

Integrating around the bearing gives the total side flow as

$$Q_s = \int dQ_s = \frac{p_s r}{6\mu l'} \int_0^{2\pi} (c - e \cos \theta)^3 d\theta = \frac{p_s r}{6\mu l'} (2\pi c^3 + 3\pi ce^2)$$

Rearranging, with $\varepsilon = e/c$, gives

$$Q_s = \frac{\pi p_s r c^3}{3\mu l'} (1 + 1.5\varepsilon^2) \tag{12–28}$$

In analyzing the performance of pressure-fed bearings, the bearing length should be taken as l', as defined in Figure 12–25. The characteristic pressure in each of the two bearings that constitute the pressure-fed bearing assembly P is given by

$$P = \frac{W/2}{2rl'} = \frac{W}{4rl'} \tag{12–29}$$

The charts for flow variable and flow ratio (Figures 12–16 and 12–17) do not apply to pressure-fed bearings. Also, the maximum film pressure given by Figure 12–18 must be increased by the oil supply pressure p_s to obtain the total film pressure.

Since the oil flow has been increased by forced feed, Equation (12–21) will give a temperature rise that is too high because the side flow carries away all the heat generated. The plumbing in a pressure-fed bearing is depicted schematically in Figure 12–28. The oil leaves the sump at the externally maintained temperature T_s at the volumetric rate Q_s. The heat gain of the fluid passing through the bearing is

$$H_{\text{gain}} = 2\rho C_p (Q_s/2)\Delta T = \rho C_p Q_s \Delta T \tag{12–30}$$

At steady state, the rate at which the journal does frictional work on the fluid film is

$$H_f = \frac{2\pi TN}{J} = \frac{2\pi f\, WrN}{J} = \frac{2\pi\, WNc}{J} \frac{fr}{c} \tag{m}$$

Equating the heat gain to the frictional work and solving for ΔT gives

$$\Delta T = \frac{2\pi WNc}{J\rho C_p Q_s} \frac{fr}{c} \tag{n}$$

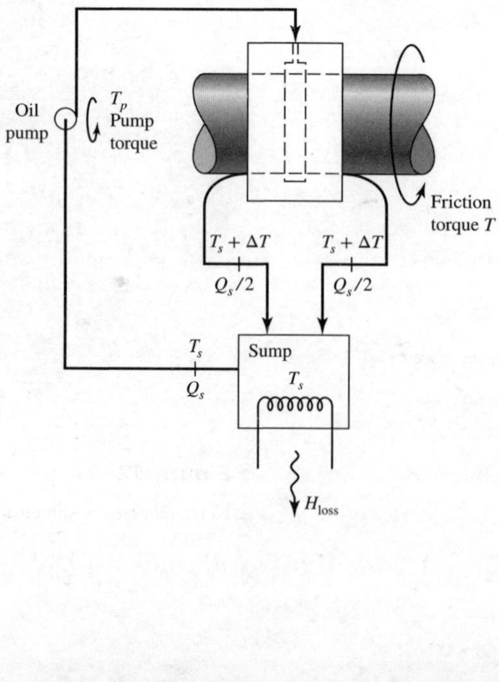

Figure 12–28

Pressure-fed centrally located full annular-groove journal bearing with external, coiled lubricant sump.

Substituting Equation (12–28) for Q_s in the equation for ΔT gives

$$\Delta T = \frac{2\pi}{J\rho C_p} WNc \frac{fr}{c} \frac{3\mu l'}{(1 + 1.5\varepsilon^2)\pi p_s r c^3}$$

The Sommerfeld number may be expressed as

$$S = \left(\frac{r}{c}\right)^2 \frac{\mu N}{P} = \left(\frac{r}{c}\right)^2 \frac{4rl'\mu N}{W}$$

Solving for $\mu Nl'$ in the Sommerfeld expression; substituting in the ΔT expression; and using $J = 0.995\ 66$ N · m/J, $\rho = 862$ kg/m^3, and $C_p = 1757$ J/(kg · °C), we find

$$\Delta T_C = \frac{9.92(10^{-7})}{1 + 1.5\varepsilon^2} \frac{(fr/c)SW^2}{p_s r^4} \tag{12–31}$$

where the bearing load W in kN, lubricant supply pressure ps in Pa, and the journal radius r in m. An analysis example of a pressure-fed bearing will be useful.

EXAMPLE 12–6

A circumferential-groove pressure-fed bearing is lubricated with SAE grade 20 oil supplied at a gauge pressure of 2 bar. The journal diameter d_j is 44.5 mm, with a unilateral tolerance of −0.05 mm. The central circumferential bushing has a diameter d_b of 44.58 mm, with a unilateral tolerance of 0.1 mm. The l'/d ratio of the two "half-bearings" that constitute the complete pressure-fed bearing is 1/2. The journal angular speed is 3000 rev/min, or 50 rev/s, and the radial steady load is 4000 N. The external sump is maintained at 50°C as long as the necessary heat transfer does not exceed 235 J/s.

(*a*) Find the steady-state average film temperature.

(*b*) Estimate the volumetric side flow Q_s, the heat loss rate H_{loss}, and the parasitic friction torque.

Solution (*a*)

$$r = \frac{d_j}{2} = \frac{44.5}{2} = 22.25 \text{ mm}$$

$$c_{\min} = \frac{(d_b)_{\min} - (d_j)_{\max}}{2} = \frac{44.58 - 44.5}{2} = 0.04 \text{ mm}$$

Since $l'/d = 1/2$, $l' = d/2 = r = 22.25$ mm. Then the pressure due to the load is

$$P = \frac{W}{4rl'} = \frac{4000}{4(22.25)22.25} = 2 \text{ MPa}$$

The Sommerfeld number S can be expressed as

$$S = \left(\frac{r}{c}\right)^2 \frac{\mu N}{P} = \left(\frac{22.25}{0.04}\right)^2 \frac{\mu'}{(10^6)} \frac{50}{2} = 7.735\mu' \tag{1}$$

We will use a tabulation method to find the average film temperature. The first trial average film temperature $\overline{T}_f$ will be 75°C. Using the Seireg curve fit of Table 12–1, we obtain

$$\mu' = 0.094 \exp\left[\frac{688.7\cdot(9/5) + 32}{(9/5)\cdot 75 + 127}\right] = 12.05 \text{ MPa} \cdot \text{s}$$

From Equation (1)

$$S = 7.735\mu' = 7.735(12.5(10^{-3}) = 0.0955$$

From Figure 12–15, $fr/c = 3.349$, and from Figure 12–13, $\varepsilon = 0.803$. From Equation (12–31),

$$\Delta T_F = 9.92\cdot 10^{-7}\ \frac{3.349\cdot 0.0932\cdot 4000^2}{(1 + 1.5\cdot 0.803^2)2\cdot 10^5\cdot 0.02225^4} = 51.4°\text{C}$$

$$T_{\text{av}} = T_s + \frac{\Delta T}{2} = 50 + \frac{51.4}{2} = 75.7°\text{C}$$

We form a table, adding a second line with $\overline{T}_f = 75.7$°C:

Trial $\overline{T}_f$	μ'	S	fr/c	ε	ΔT_F	T_{av}
75.0	12.05	0.0932	3.349	0.803	51.4	75.7
75.7	11.93	0.0923	3.326	0.804	50.5	75.3

For the next iteration, plot trial $\overline{T}_f$ against resulting T_{av} and draw a straight line between them, the intersection with a $\overline{T}_f = T_{\text{av}}$ line defining the new trial $\overline{T}_f$.

Answer

The result of the final iteration is $\overline{T}_f = 75.3$°C and $\Delta T_F = 50.5$°C. Furthermore, $T_{\text{max}} = 50 + 50.5 = 100.5$°C.

(*b*) From Equation (12–28),

Answer

$$Q_s = \frac{\pi(2(10^5)22.25(10^{-3})(0.04)^3}{3(11.93)10^{-3}(22.25)(10^{-3})}\ [1 + 1.5(0.803)^2] = 2.21(10^{-6})\ \text{m}^3/\text{s}$$

Assuming steady-state conditions, $H_{\text{loss}} = H_{\text{gain}}$, and Equation (12–30) gives

$$H_{\text{loss}} = \rho C_p Q_s \Delta T = 860(1758)2.21(10^{-6})(50.5) = 168.7 \text{ J/s}$$

The parasitic friction torque T is

Answer

$$T = fWr = \frac{fr}{c}\ Wc = 3.326(4000)40(10^{-6}) = 0.53 \text{ N} \cdot \text{m}$$

12–12 Loads and Materials

Some help in choosing unit loads and bearing materials is afforded by Tables 12–5 and 12–6. Since the diameter and length of a bearing depend upon the unit load, these tables will help the designer to establish a starting point in the design.

The length-diameter ratio l/d of a bearing depends upon whether it is expected to run under thin-film-lubrication conditions. A long bearing (large l/d ratio) reduces the coefficient of friction and the side flow of oil and therefore is desirable where thin-film or boundary-value lubrication is present. On the other hand, where forced-feed or positive lubrication is present, the l/d ratio should be relatively small. The short bearing length results in a greater flow of oil out of the ends, thus keeping the bearing cooler. Current practice is to use an l/d ratio of about unity, in general, and then to increase this ratio if thin-film lubrication is likely to occur and to decrease it for thick-film lubrication or high temperatures. If shaft deflection is likely to be severe, a short bearing should be used to prevent metal-to-metal contact at the ends of the bearings.

You should always consider the use of a partial bearing if high temperatures are a problem, because relieving the non-load-bearing area of a bearing can very substantially reduce the heat generated.

The two conflicting requirements of a good bearing material are that it must have a satisfactory compressive and fatigue strength to resist the externally applied loads and that it must be soft and have a low melting point and a low modulus of elasticity. The second set of requirements is necessary to permit the material to wear or break in, since the material can then conform to slight irregularities and absorb and release foreign particles. The resistance to wear and the coefficient of friction are also important because all bearings must operate, at least for part of the time, with thin-film or boundary lubrication.

Additional considerations in the selection of a good bearing material are its ability to resist corrosion and, of course, the cost of producing the bearing. Some of the commonly used materials are listed in Table 12–6, together with their composition and characteristics.

Table 12–5 Range of Unit Loads in Current Use for Sleeve Bearings

Application	Unit Load MPa
Diesel engines:	
Main bearings	6–12
Crankpin	8–15
Wristpin	14–15
Electric motors	0.8–1.5
Steam turbines	0.8–1.5
Gear reducers	0.8–1.5
Automotive engines:	
Main bearings	4–5
Crankpin	10–15
Air compressors:	
Main bearings	1–2
Crankpin	2–4
Centrifugal pumps	0.6–1.2

Table 12–6 Some Characteristics of Bearing Alloys

Alloy Name	Thickness, mm	SAE Number	Clearance Ratio r/c	Load Capacity	Corrosion Resistance
Tin-base babbitt	0.6	12	600–1000	1.0	Excellent
Lead-base babbitt	0.6	15	600–1000	1.2	Very good
Tin-base babbitt	0.1	12	600–1000	1.5	Excellent
Lead-base babbitt	0.1	15	600–1000	1.5	Very good
Leaded bronze	Solid	792	500–1000	3.3	Very good
Copper-lead	0.6	480	500–1000	1.9	Good
Aluminum alloy	Solid		400–500	3.0	Excellent
Silver plus overlay	0.3	17P	600–1000	4.1	Excellent
Cadmium (1.5% Ni)	0.6	18	400–500	1.3	Good
Trimetal 88*				4.1	Excellent
Trimetal 77†				4.1	Very good

*This is a 0.2 mm layer of copper-lead on a steel back plus 0.025 mm of tin-base babbitt.
†This is a 0.3 mm layer of copper-lead on a steel back plus 0.025 mm of lead-base babbitt.

Bearing life can be increased very substantially by depositing a layer of babbitt, or other white metal, in thicknesses from 0.025 to 0.35 mm over steel backup material. In fact, a copper-lead layer on steel to provide strength, combined with a babbitt overlay to enhance surface conformability and corrosion resistance, makes an excellent bearing.

Small bushings and thrust collars are often expected to run with thin-film or boundary lubrication. When this is the case, improvements over a solid bearing material can be made to add significantly to the life. A powder-metallurgy bushing is porous and permits the oil to penetrate into the bushing material. Sometimes such a bushing may be enclosed by oil-soaked material to provide additional storage space. Bearings are frequently ball-indented to provide small basins for the storage of lubricant while the journal is at rest. This supplies some lubrication during starting. Another method of reducing friction is to indent the bearing wall and to fill the indentations with graphite.

With all these tentative decisions made, a lubricant can be selected and the hydrodynamic analysis made as already presented. The values of the various performance parameters, if plotted as in Figure 12–22, for example, will then indicate whether a satisfactory design has been achieved or additional iterations are necessary.

12–13 Bearing Types

A bearing may be as simple as a hole machined into a cast-iron machine member. It may still be simple yet require detailed design procedures, as, for example, the two-piece grooved pressure-fed connecting-rod bearing in an automotive engine. Or it may be as elaborate as the large water-cooled, ring-oiled bearings with built-in reservoirs used on heavy machinery.

Figure 12–29 shows two types of bushings. The solid bushing is made by casting, by drawing and machining, or by using a powder-metallurgy process. The lined bushing is usually a split type. In one method of manufacture the molten lining material is cast continuously on thin strip steel. The babbitted strip is then processed through

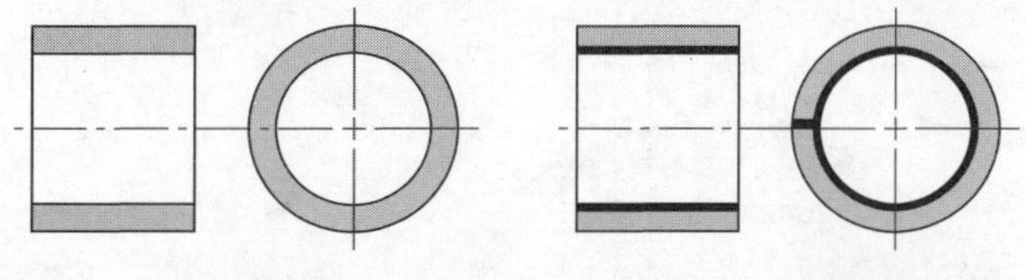

Figure 12–29
Sleeve bushings.

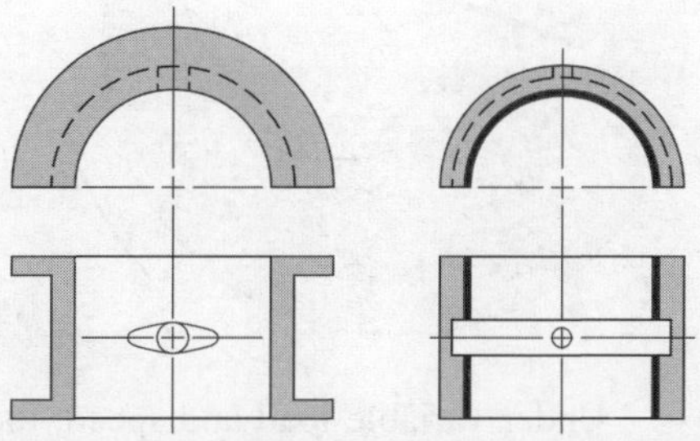

Figure 12–30
Two-piece bushings.

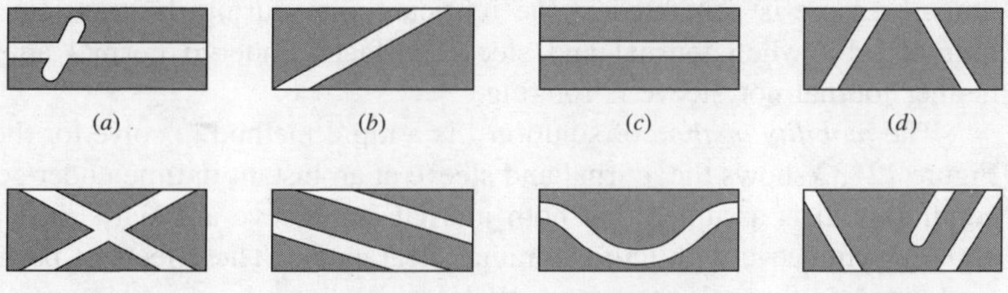

Figure 12–31
Developed views of typical groove patterns. *(Source: Cleveland Graphite Bronze Company, Division of Clevite Corporation.)*

presses, shavers, and broaches, resulting in a lined bushing. Any type of grooving may be cut into the bushings. Bushings are assembled as a press fit and finished by boring, reaming, or burnishing.

Flanged and straight two-piece bearings are shown in Figure 12–30. These are available in many sizes in both thick- and thin-wall types, with or without lining material. A locking lug positions the bearing and effectively prevents axial or rotational movement of the bearing in the housing.

Some typical groove patterns are shown in Figure 12–31. In general, the lubricant may be brought in from the end of the bushing, through the shaft, or through the bushing. The flow may be intermittent or continuous. The preferred practice is to bring the oil in at the center of the bushing so that it will flow out both ends, thus increasing the flow and cooling action.

12–14 Dynamically Loaded Journal Bearings

So far, this chapter has focused on journal bearings that are subjected to a steady load at a steady journal speed. Many journal bearing systems, however, are subjected to loads and speeds that can vary considerably in magnitude and direction over time. Examples are engine bearings employed in automotive and off-highway applications, reciprocating gas compressor bearings, actuator bearings in reciprocating machinery, and bearings in intermittent motion mechanisms.

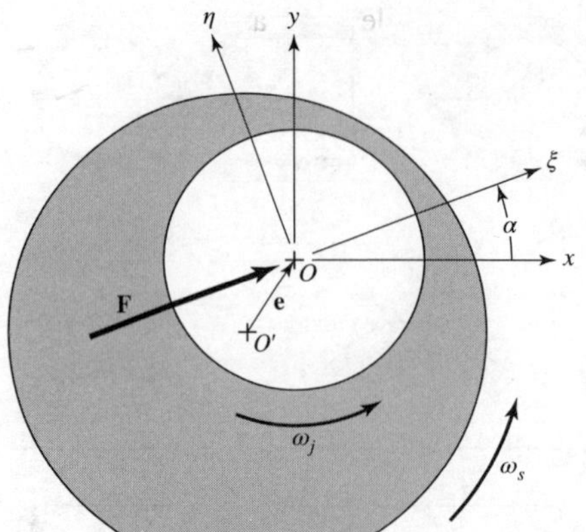

Figure 12–32

Schematic of a dynamically loaded journal bearing.

Under variable load and speed, the journal center can move over the entire clearance space, its path typically referred to as the *journal orbit*. A new load-carrying mechanism, *squeeze-film action*, can occur in the lubricant film under these conditions. Pressure is generated in the lubricant film during the transient portion of the journal orbit when journal and sleeve surfaces undergo normal approach, even if neither journal nor sleeve is rotating.

The *mobility method* of solution[11] is a rapid method to solve for the journal orbit. Figure 12–32 shows the journal and sleeve at an instant in time under general dynamic conditions. It is assumed that both journal and sleeve are cylindrical and circumferentially symmetric with uniform radial clearance c. The effects of partial grooving or feed holes on journal motion are neglected.

The mobility method employs an *xyz computation frame* with its origin at the sleeve center as shown in Figure 12–32. Both the journal and sleeve can rotate about the z axis relative to this frame with non-steady angular velocities ω_j and ω_s, respectively. The journal eccentricity is represented by a two-dimensional vector **e**. The load vector **F**, which can vary in both magnitude and direction in the xy plane, is transmitted from journal to sleeve through the lubricant film.

Employing the nomenclature of Figure 12–6, the Reynolds equation in this case becomes

$$\frac{1}{r^2}\frac{\partial}{\partial\theta}\left(\frac{h^3}{12\mu}\frac{\partial p}{\partial\theta}\right)+\frac{\partial}{\partial z}\left(\frac{h^3}{12\mu}\frac{\partial p}{\partial z}\right)=\bar{\omega}\frac{\partial h}{\partial\theta}+\frac{\partial h}{\partial t} \tag{12–32}$$

where

$$\bar{\omega}=\frac{\omega_j+\omega_s}{2}$$

and Equation (12–32) includes the additional squeeze-film term $\partial h/\partial t$ on the right-hand side.

At a given instant in time, the instantaneous time rate of change of the journal center relative to the sleeve, also referred as *journal translational velocity* or eccentricity rate, can be found from the equation

$$\frac{d\mathbf{e}}{dt}=\frac{F(c/r)^2}{\mu l d/c}\mathbf{M}(\boldsymbol{\varepsilon},\, l/d)+\bar{\boldsymbol{\omega}}\times\mathbf{e} \tag{12–33}$$

[11]J. F. Booker, "Dynamically Loaded Journal Bearings: Numerical Application of the Mobility Method," *ASME Journal of Lubrication Technology*, vol. 93, 1971, pp. 168–176, 315.

where $\overline{\boldsymbol{\omega}} = \overline{\omega}\mathbf{k}$ and the journal length and diameter are denoted as l and d, respectively, with $d = 2r$. The mobility vector $\mathbf{M}$ is a function of the journal eccentricity ratio vector $\boldsymbol{\varepsilon} = \mathbf{e}/c$ and l/d ratio only and is determined from the solution of Equation (12–32) (for details, see footnote 11). The new journal position at time $t + \Delta t$ is computed using a variety of standard explicit numerical solution methods, such as the Euler or Runge-Kutta methods.

Mobility data is typically collected and curve-fitted relative to a *ξη map frame* of reference attached to the load as shown in Figure 12–32. One common map is developed from the solution of the *short-bearing approximation* of the Reynolds equation, which ignores the first term on the left-hand side of Equation (12–32). This neglected term represents flow in the circumferential direction which, for bearings with small l/d ratio, can be considered small compared to the second term, which represents flow in the axial direction. The curve fit of the mobility vector data for this short-bearing film model has components[12]

$$M_\xi(\varepsilon_\xi, \varepsilon_\eta) = \frac{M_0(\varepsilon_\xi)}{(l/d)^2} \tag{12–34a}$$

$$M_\eta(\varepsilon_\xi, \varepsilon_\eta) = \frac{-4\varepsilon_\eta M_0(\varepsilon_\xi)}{\pi(1 - \varepsilon_\xi)(l/d)^2} \tag{12–34b}$$

where
$$M_0(\varepsilon_\xi) = \frac{\left(1 - \varepsilon_\xi^2\right)^{5/2}}{6\varepsilon_\xi\left(1 - \varepsilon_\xi^2\right)^{1/2} + 2\left(1 + 2\varepsilon_\xi^2\right)\cos^{-1}(-\varepsilon_\xi)} \tag{a}$$

Figure 12–33 shows the mobility vector components computed from Equations (12–34) plotted in the map frame of reference over the clearance space for $l/d = 1$. The dashed and solid lines on the *mobility map* indicate mobility vector magnitude and direction, respectively, at a given point in the clearance space. The physical interpretation of the map is that the direction lines indicate *pure-squeeze* motion of the journal center under a steady load and zero average surface velocity starting from an initial position in the clearance space.

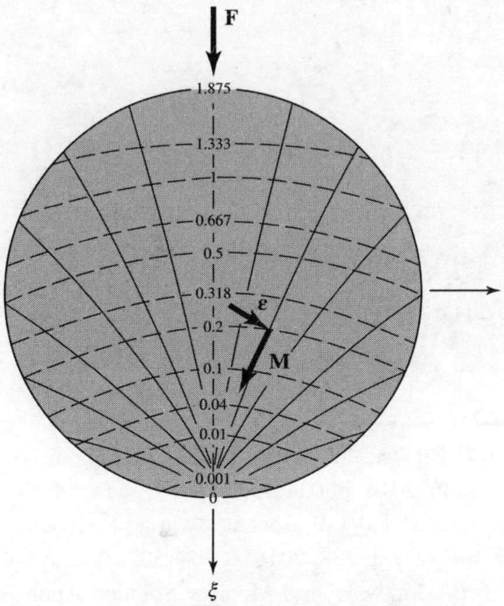

Figure 12–33
Short bearing mobility map for $l/d = 1$.

[12]J. F. Booker, "Dynamically Loaded Journal Bearings: Maximum Film Pressure," *ASME Journal of Lubrication Technology*, vol. 91, 1969, pp. 534–537.

It is a relatively straightforward matter of transforming the mobility vector from the map frame to the computation frame to find the instantaneous journal velocity. Example 12–7 illustrates the method.

A wide variety of mobility maps based on different methods of solution of the Reynolds equation are available to the bearing designer.[13,14] While these maps have been developed assuming perfect journal and sleeve axial alignment, it is possible to employ them on a limited basis for prescribed journal angular misalignment.[15] Maps to determine instantaneous maximum film pressure and its location have also been constructed.[12,16]

EXAMPLE 12–7

A main engine bearing has journal diameter 40 mm, length 20 mm, and radial clearance of 25 μm. Fluid dynamic viscosity is 5 mPa-s. At the instant $t = 0$, the journal is rotating at a constant angular velocity of 400 rad/s counterclockwise (positive) about the z axis of Figure 12–32, and the sleeve is fixed. Journal eccentricity ratio components in the computing frame at this instant are $\varepsilon_x = 0.7$, $\varepsilon_y = 0.3$. At this same instant, bearing load **F** with components $F_x = 3000$ N, $F_y = 6000$ N is transmitted from journal to sleeve. Using Euler's method, find the journal eccentricity ratio at time $t = 40$ μs.

Solution

At time $t = 0$ s, the load magnitude F and its direction cosines (see Figure 12–32) are given by

$$F = \sqrt{F_x^2 + F_y^2} = \sqrt{(3000)^2 + (6000)^2} = 6708.2 \text{ N}$$

$$\cos\alpha = F_x/F = 3000/6708.2 = 0.4472$$

$$\sin\alpha = F_y/F = 6000/6708.2 = 0.8944$$

Transforming the eccentricity ratio components from the xy computing frame to the $\xi\eta$ map frame we obtain

$$\varepsilon_\xi = \varepsilon_x \cos\alpha + \varepsilon_y \sin\alpha = (0.7)(0.4472) + (0.3)(0.8944) = 0.5814$$

$$\varepsilon_\eta = -\varepsilon_x \sin\alpha + \varepsilon_y \cos\alpha = -(0.7)(0.8944) + (0.3)(0.4472) = -0.4919$$

From Equation (a)

$$M_0 = \frac{[1 - (0.5814)^2]^{5/2}}{6(0.5814)[1 - (0.5814)^2]^{1/2} + 2[1 + 2(0.5814)^2]\cos^{-1}(-0.5814)} = 0.0350$$

The mobility components in the map frame are then found from Equations (12–34) to be

$$M_\xi = (0.0350)/(0.02/0.04)^2 = 0.1400$$

$$M_\eta = -4(-0.4919)(0.0350)/[\pi(1 - 0.5814)(0.02/0.04)^2] = 0.2095$$

[13] J. F. Booker, "Mobility/Impedance Methods: A Guide for Application," *ASME Journal of Tribology*, vol. 136, 2014, p. 024 501.

[14] S. Boedo and J. F. Booker, "Warner-Sommerfeld Impedance and Mobility Maps," *ASME Journal of Tribology*, vol. 138, 2016, p. 034 502. Errata: vol. 140, 2018, p. 017 001.

[15] S. Boedo, "A Hybrid Mobility Solution Approach for Dynamically Loaded Misaligned Journal Bearings," *ASME Journal of Tribology*, vol. 135, 2013, p. 024 501.

[16] P. K. Goenka, "Analytical Curve Fits for Solution Parameters of Dynamically Loaded Journal Bearings," *ASME Journal of Tribology*, vol. 106, 1984, pp. 421–428.

The mobility components in the computing frame are given by

$$M_x = M_\xi \cos\alpha - M_\eta \sin\alpha = (0.1400)(0.4472) - (0.2905)(0.8944) = -0.1248$$

$$M_y = M_\xi \sin\alpha + M_\eta \cos\alpha = (0.1400)(0.8944) + (0.2095)(0.4472) = 0.2189$$

The coefficient of the mobility vector in Equation (12–33) is

$$\frac{F(c/r)^2}{\mu l d/c} = \frac{6708.2[(25)(10^{-6})/0.02]^2}{0.005(0.02)(0.04)/[(25)(10^{-6})]} = 0.06551 \text{ m/s}$$

and the average angular velocity of the journal and sleeve is

$$\bar{\omega} = \frac{\omega_j + \omega_s}{2} = \frac{400 + 0}{2} = 200 \text{ rad/s}$$

The journal center velocity components, v_x and v_y, at time $t = 0$ are obtained from Equation (12–33).

$$v_x = \frac{de_x}{dt} = \frac{F(c/r)^2}{\mu l d/c} M_x - \bar{\omega} e_y$$

$$= (0.06551)(-0.1248) - (200)(0.3)(25)(10^{-6}) = -9.676(10^{-3}) \text{ m/s}$$

$$v_y = \frac{de_y}{dt} = \frac{F(c/r)^2}{\mu l d/c} M_y + \bar{\omega} e_x$$

$$= (0.06551)(0.2189) + (200)(0.7)(25)(10^{-6}) = 1.784(10^{-2}) \text{ m/s}$$

Using Euler's method, and $\varepsilon = e/c$, the journal eccentricity ratio components at $t = 40$ μs are given by

Answer

$$\varepsilon_x|_{t=40\,\mu s} = \varepsilon_x|_{t=0} + (v_x/c)\Delta t$$

$$= 0.7 + (-9.676)(10^{-3})(40)(10^{-6})/[(25)(10^{-6})] = 0.6845$$

Answer

$$\varepsilon_y|_{t=40\,\mu s} = \varepsilon_y|_{t=0} + (v_y/c)\Delta t$$

$$= 0.3 + (1.784)(10^{-2})(40)(10^{-6})/[(25)(10^{-6})] = 0.3285$$

Design of Big-End Connecting Rod Bearings

In reciprocating machinery, the connecting rod is a mechanical linkage that converts reciprocating piston motion into rotary crankshaft motion. Four-stroke diesel and gasoline engines found in automotive and off-highway applications form a very large market share of such mechanisms. Figure 19–1 in Chapter 19 shows the representative geometry of a connecting rod. The big end and small end of the connecting rod are connected to the crankshaft and piston assemblies, respectively, using fluid-film journal bearings.

The design of the big-end connecting rod bearing is of particular interest as it is subjected to extreme variations in loading coupled with non-steady sleeve rotation over its periodic duty cycle. Figure 12–34 shows a representative periodic polar load diagram of bearing load transmitted from the crankpin to the big-end sleeve over a four-stroke periodic duty cycle. The degree markings indicate time in degrees of crankshaft rotation. The *xyz* computation frame in this case is attached to the moving connecting rod with its origin at the big-end center and with the *x* axis situated along the bearing line of centers.

Figure 12–35 shows the periodic journal orbit resulting from the periodic duty cycle loading. Note the large excursion of journal translational motion over the entire clearance space. The *cyclic-minimum film thickness* h_{min} is the minimum value of the minimum thickness obtained from the journal orbit, and this value is used as a performance variable in the design process.

Figure 12–36 shows a design chart developed from running hundreds of short bearing mobility orbit simulations that can be used to predict the cyclic minimum film thickness

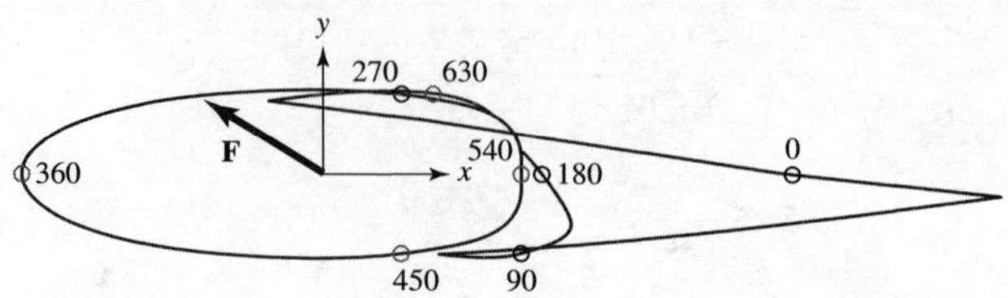

Figure 12–34

Typical polar load diagram for big-end connecting rod bearing: four-stroke engine application.

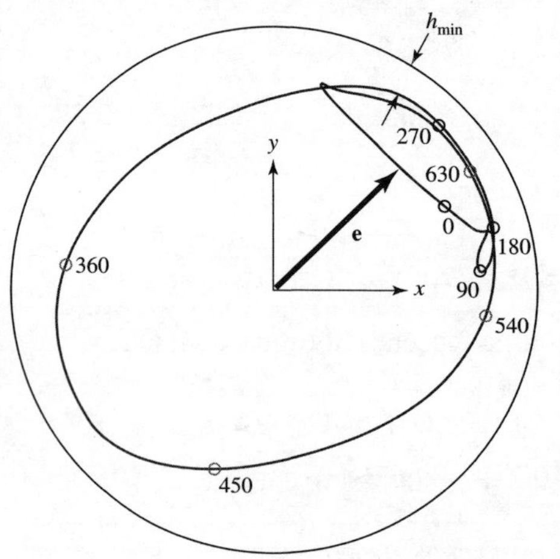

Figure 12–35

Periodic journal orbit for big-end connecting rod bearing: four-stroke engine application.

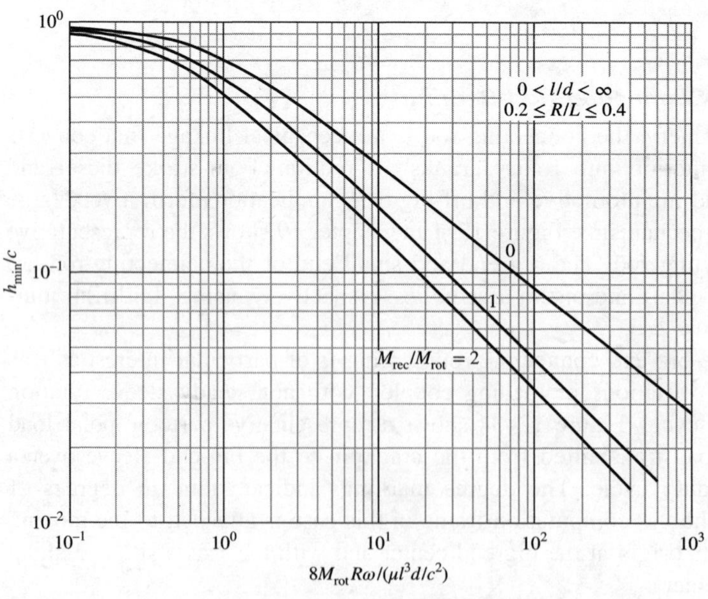

Figure 12–36

Cyclic minimum film thickness prediction for big-end connecting rod bearing: four-stroke application. *(Source: Based on data in Fig. 10 of Booker, J.F. "Squeeze Films and Bearing Dynamics." In CRC Handbook of Lubrication (Theory and Practice of Tribology), Volume II: Theory and Design, edited by E.R. Booser, 121–137. Boca Raton, Florida: CRC Press, 1984.)*

in a big-end connecting rod bearing.[17] The connecting rod is modeled as two point masses situated at the bearing centers. The point masses are usually selected so that the center of mass and total mass of the model are identical to that in the actual connecting rod.

The following design variables employed in the chart can be defined in any set of consistent units:

M_{rec}	Combined small-end (reciprocating) connecting rod mass and piston mass
M_{rot}	Big-end (rotating) connecting mass
R	Crankshaft radius (1/2 of engine stroke)
L	Connecting rod center-to-center length
l	Bearing length
d	Bearing diameter
P_{cyl}	Maximum value of periodic cylinder gas pressure
A_{cyl}	Piston bore cross-sectional area
ω	Engine crankshaft angular velocity in rad/time
c	Radial clearance
μ	Fluid dynamic viscosity

The chart is applicable when

$$P_{cyl}A_{cyl} < 7F_{rot} \tag{12–35}$$

where

$$F_{rot} = M_{rot}R\omega^2 \tag{12–36}$$

is the inertia load contribution from the big-end mass alone. Comparing computed cyclic-minimum film thickness using the short bearing theory with field experience, Table 12–7 lists recommended danger levels for automotive and industrial applications.[18] Example 12–8 illustrates the use of the design chart.

Table 12–7 Film Thickness Danger Levels for Big-End Connecting Rod Bearings

	d (typical) mm	h_{min} from Figure 12–36 μm
Automotive (Otto)	50	1.0
Automotive (Diesel)	75–100	1.75
Industrial (Diesel)	250	2.5

Booker, J.F. "Squeeze Films and Bearing Dynamics." In *CRC Handbook of Lubrication (Theory and Practice of Tribology), Volume II: Theory and Design,* edited by E.R. Booser, 121–137. Boca Raton, Florida: CRC Press, 1984.

[17]Based on data in Figure 10 of J. F. Booker, "Squeeze Films and Bearing Dynamics," *CRC Handbook of Lubrication (Theory and Practice of Tribology), Volume II: Theory and Design*, E. R. Booser, ed., CRC Press, 1984, pp. 121–137.

[18]Based on data in Table 1 of footnote 17.

EXAMPLE 12–8

A four-stroke automotive gasoline engine has a stroke of 86.4 mm and is running at 6000 rev/min. The connecting rod has a mass of 0.55 kg, center-to-center length of 144 mm, and the center of mass is located 37 mm from the big-end center. The piston has a bore diameter of 74 mm, mass of 0.5 kg, and the peak cylinder pressure over the four-stroke cycle is 7 MPa. The big-end rod bearing has a diameter of 50.6 mm, length 25.3 mm, radial clearance 20 μm, and the dynamic viscosity of the oil is 4.332 mPa · s. Determine whether the bearing specifications are acceptable for this application.

Solution

The first step is to represent the connecting rod as two point masses. Assuming that the actual connecting rod and the model have the same mass and center of mass, the small end mass M_s and big-end mass M_b have values

$$M_s = \frac{M\bar{x}}{L} = \frac{(0.55)(0.037)}{(0.144)} = 0.14 \text{ kg}$$

$$M_b = M - M_s = 0.55 - 0.14 = 0.41 \text{ kg}$$

The reciprocating and rotating masses and their ratio are thus

$$M_{\text{rec}} = M_s + M_{\text{piston}} = 0.14 + 0.5 = 0.64 \text{ kg}$$

$$M_{\text{rot}} = M_b = 0.41 \text{ kg}$$

$$M_{\text{rec}}/M_{\text{rot}} = 0.64/0.41 = 1.57$$

Next, determine whether the chart, Figure 12–36, is applicable.

$$A_{\text{cyl}} = \frac{\pi d_{\text{bore}}^2}{4} = \frac{\pi(0.074^2)}{4} = 0.0043 \text{ m}^2$$

$$\omega = \frac{2\pi N}{60} = \frac{2\pi(6000)}{60} = 628.32 \text{ rad/s}$$

$$\frac{P_{\text{cyl}}A_{\text{cyl}}}{F_{\text{rot}}} = \frac{P_{\text{cyl}}A_{\text{cyl}}}{M_{\text{rot}}R\omega^2} = \frac{7(10^6)(0.0043)}{(0.41)(0.0432)(628.32)^2} = 4.3 < 7 \qquad \text{OK}$$

$$0 \le M_{\text{rec}}/M_{\text{rot}} = 1.57 \le 2 \qquad \text{OK}$$

$$0.2 \le R/L = 0.0432/0.144 = 0.3 \le 0.4 \qquad \text{OK}$$

Now,

$$\frac{8M_{\text{rot}}R\omega c^2}{\mu l^3 d} = \frac{8(0.41)(0.0432)(628.32)[(20)(10^{-6})]^2}{(0.004332)(0.0253^3)(0.0506)} = 10.0$$

From Figure 12–36, $h_{\text{min}}/c \approx 0.16$, so $h_{\text{min}} \approx (0.16)\,[20(10^{-6})] = 3.2(10^{-6})$ m, or 3.2 μm. From Table 12–7, $h_{\text{min}} > 1\ \mu$m, so the bearing specifications are acceptable for this application.

12–15 Boundary-Lubricated Bearings

When two surfaces slide relative to each other with only a partial lubricant film between them, *boundary lubrication* is said to exist. Boundary- or thin-film lubrication occurs in hydrodynamically lubricated bearings when they are starting or stopping, when the load increases, when the supply of lubricant decreases, or whenever other operating changes happen to occur. There are, of course, a very large number of cases in design in which boundary-lubricated bearings must be used because of the type of application or the competitive situation.

The coefficient of friction for boundary-lubricated surfaces may be greatly decreased by the use of animal or vegetable oils mixed with the mineral oil or grease. Fatty acids, such as stearic acid, palmitic acid, or oleic acid, or several of these, which occur in animal and vegetable fats, are called *oiliness agents*. These acids appear to reduce friction, either because of their strong affinity for certain metallic surfaces or because they form a soap film that binds itself to the metallic surfaces by a chemical reaction. Thus the fatty-acid molecules bind themselves to the journal and bearing surfaces with such great strength that the metallic asperities of the rubbing metals do not weld or shear.

Fatty acids will break down at temperatures of 120°C or more, causing increased friction and wear in thin-film-lubricated bearings. In such cases the *extreme-pressure*, or EP, lubricants may be mixed with the fatty-acid lubricant. These are composed of chemicals such as chlorinated esters or tricresyl phosphate, which form an organic film between the rubbing surfaces. Though the EP lubricants make it possible to operate at higher temperatures, there is the added possibility of excessive chemical corrosion of the sliding surfaces.

When a bearing operates partly under hydrodynamic conditions and partly under dry or thin-film conditions, a *mixed-film lubrication* exists. If the lubricant is supplied by hand oiling, by drop or mechanical feed, or by wick feed, for example, the bearing is operating under mixed-film conditions. In addition to occurring with a scarcity of lubricant, mixed-film conditions may be present when

- The viscosity is too low.
- The bearing speed is too low.
- The bearing is overloaded.
- The clearance is too tight.
- Journal and bearing are not properly aligned.

Relative motion between surfaces in contact in the presence of a lubricant is called *boundary lubrication*. This condition is present in hydrodynamic film bearings during starting, stopping, overloading, or lubricant deficiency. Some bearings are boundary lubricated (or dry) at all times. To signal this an adjective is placed before the word "bearing." Commonly applied adjectives (to name a few) are thin-film, boundary friction, Oilite, Oiles, and bushed-pin. The applications include situations in which thick film will not develop and there are low journal speed, oscillating journal, padded slides, light loads, and lifetime lubrication. The characteristics include considerable friction, ability to tolerate expected wear without losing function, and light loading. Such bearings are limited by lubricant temperature, speed, pressure, galling, and cumulative wear. Table 12–8 gives some properties of a range of bushing materials.

Linear Sliding Wear

Consider the sliding block depicted in Figure 12–37, moving along a plate with contact pressure P acting over area A, in the presence of a coefficient of sliding friction f_s. The linear measure of wear w is expressed in millimeters. The work done by force f_sPA during displacement S is f_sPAS or f_sPAVt, where V is the sliding velocity and t is time. The material volume removed due to wear is wA and is proportional to the work done, that is, $wA \propto f_sPAVt$, or

$$wA = KPAVt$$

Table 12–8 Some Materials for Boundary-Lubricated Bearings and Their Operating Limits

Material	Maximum Load, MPa	Maximum Temperature, °C	Maximum Speed, m/s	Maximum *PV* Value*
Cast bronze	30	160	7.5	$175 \cdot 10^6$
Porous bronze	30	65	7.5	$175 \cdot 10^6$
Porous iron	55	65	4	$175 \cdot 10^6$
Phenolics	40	90	12.5	$0.53 \cdot 10^6$
Nylon	7	90	5	$0.11 \cdot 10^6$
Teflon	3.5	260	0.5	$0.04 \cdot 10^6$
Reinforced Teflon	17	260	5	$0.35 \cdot 10^6$
Teflon fabric	410	260	0.3	$0.88 \cdot 10^6$
Delrin	7	80	5	$0.11 \cdot 10^6$
Carbon-graphite	4	400	12.5	$0.53 \cdot 10^6$
Rubber	0.3	65	20	
Wood	14	65	10	$0.53 \cdot 10^6$

*P = load, Pa; V = speed, m/s.

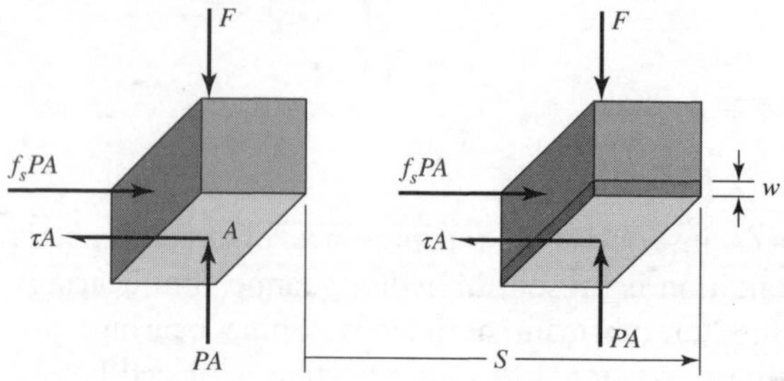

Figure 12–37

Sliding block subjected to wear.

where K is the proportionality factor, which includes f_s, and is determined from laboratory testing. The linear wear is then expressed as

$$w = KPVt \tag{12–37}$$

SI units commonly used for K are $\text{cm}^3 \cdot \text{min}/(\text{kgf} \cdot \text{m} \cdot \text{h})$, where 1 kgf = 9.806 N. Tables 12–9 and 12–10 give some wear factors and coefficients of friction from one manufacturer. Some care in the selection of the wear coefficient K is warranted here, as K can be dependent on a variety of factors, such as surface finish, application of coatings, and surface speeds.

Bushing Wear

Consider a pin of diameter D, rotating at speed N, in a bushing of length L, and supporting a stationary radial load F. The nominal pressure P is given by

$$P = \frac{F}{DL} \tag{12–38}$$

Table 12–9 Wear Factors in SI Units*

Bushing Material	Wear Factor K	Limiting PV
Oiles 800	$60.5\cdot10^{-18}$	$6.3\cdot10^5$
Oiles 500	$12.0\cdot10^{-18}$	$16.41\cdot10^5$
Polyactal copolymer	$1006.0\cdot10^{-18}$	$1.76\cdot10^5$
Polyactal homopolymer	$1207.5\cdot10^{-18}$	$1.05\cdot10^5$
66 nylon	$4024.5\cdot10^{-18}$	$0.70\cdot10^5$
66 nylon + 15% PTFE	$261.5\cdot10^{-18}$	$2.46\cdot10^5$
+ 15% PTFE + 30% glass	$322.0\cdot10^{-18}$	$3.51\cdot10^5$
+ 2.5% MoS_2	$4024.5\cdot10^{-18}$	$0.70\cdot10^5$
6 nylon	$4024.5\cdot10^{-18}$	$0.70\cdot10^5$
Polycarbonate + 15% PTFE	$1509.0\cdot10^{-18}$	$2.46\cdot10^5$
Sintered bronze	$2052.5\cdot10^{-18}$	$2.99\cdot10^5$
Phenol + 25% glass fiber	$161.0\cdot10^{-18}$	$4.04\cdot10^5$

*dim[K] = m^3/(W · s), dim [PV] = Pa · m/s.
Source: Data from Oiles America Corp., Plymouth, MI 48 170.

Table 12–10 Coefficients of Friction

Type	Bearing	f_s
Placetic	Oiles 80	0.05
Composite	Drymet ST	0.03
	Toughmet	0.05
Met	Cermet M	0.05
	Oiles 2000	0.03
	Oiles 300	0.03
	Oiles 500SP	0.03

Source: Data from Oiles America Corp., Plymouth, MI 48 170.

and if N is in rev/min and D is in meters, velocity in m/s is given by

$$V = \pi DN \tag{12–39}$$

Thus PV, in Pa · m/s, is

$$PV = \pi \frac{FN}{L} \tag{12–40}$$

Note the independence of PV from the journal diameter D.

A time-wear equation similar to Equation (12–37) can be written. However, before doing so, it is important to note that Equation (12–38) provides the nominal value of P. Figure 12–38 provides a more accurate representation of the pressure distribution, which can be written as

$$p = P_{\max} \cos \theta \qquad -\frac{\pi}{2} \le \theta \le \frac{\pi}{2}$$

The vertical component of $p\, dA$ is $p\, dA \cos \theta = [pL(D/2)\, d\theta]\cos \theta = P_{\max}\, (DL/2)$ $\cos^2 \theta\, d\theta$. Integrating this from $\theta = -\pi/2$ to $\pi/2$ yields F. Thus,

$$\int_{-\pi/2}^{\pi/2} P_{\max} \left(\frac{DL}{2}\right) \cos^2 \theta\, d\theta = \frac{\pi}{4} P_{\max}\, DL = F$$

or

$$P_{\max} = \frac{4}{\pi} \frac{F}{DL} \tag{12–41}$$

Substituting V from Equation (12–39) and $P_{\max}$ for P from Equation (12–41) into Equation (12–36) gives

$$w = K \frac{4}{\pi} \frac{F}{DL} \pi DNt = \frac{4KFNt}{L} \tag{12–42}$$

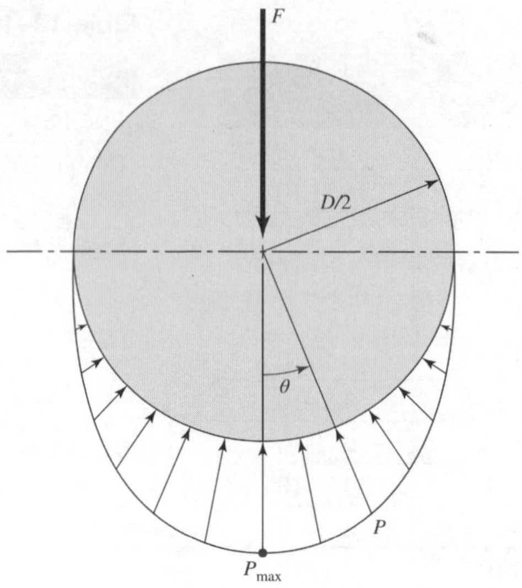

Figure 12–38
Pressure distribution on a boundary-lubricated bushing.

In designing a bushing, because of various trade-offs it is recommended that the length/diameter ratio be in the range

$$0.5 \le L/D \le 2 \tag{12–43}$$

EXAMPLE 12–9

An Oiles SP 500 alloy brass bushing is 25 mm long with a 25 mm bore and operates in a clean environment at 20°C. The allowable wear without loss of function is 0.13 mm. The radial load is 3000 N. The peripheral velocity is 0.17 m/s. Estimate the number of revolutions for radial wear to be 0.13 mm. See Figure 12–39 and Table 12–11 from the manufacturer.

Solution

From Table 12–9, $K = 12(10^{-18})$ m³/(W · s); Table 12–11, $PV = 1.641{\cdot}10^6$ Pa · m/s, $P_{max} = 25.5{\cdot}10^6$ Pa, $V_{max} = 0.51$ m/s. From Equations (12–41), (12–39), and (12–40),

$$P_{max} = \frac{4}{\pi}\frac{F}{DL} = \frac{4}{\pi}\frac{3000}{(25)(25)} = 6.11 \text{ MPa} < 25.5 \text{ MPa} \qquad \text{(OK)}$$

$$P = \frac{F}{DL} = \frac{3000}{(25)(25)} = 4.8 \text{ MPa}$$

$$V = 0.17 \text{ m/s} < 0.51 \text{ m/s} \qquad \text{(OK)}$$

$$PV = 4.8{\cdot}10^6(0.17) = 8.16{\cdot}10^5 \text{ Ps(m/s)} < 1.641{\cdot}10^6 \text{ Pa} \cdot \text{m/s} \qquad \text{(OK)}$$

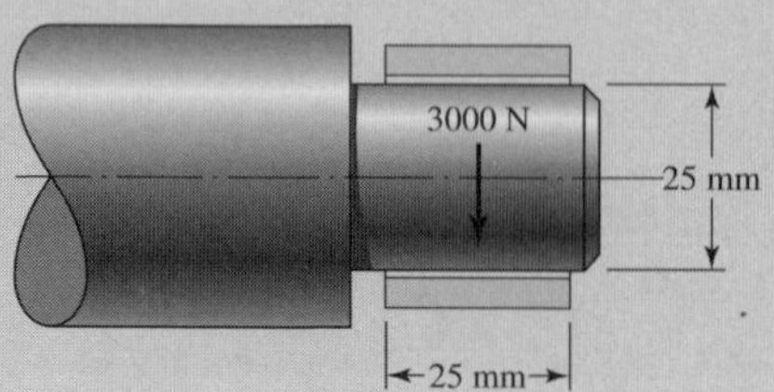

Figure 12–39
Journal/bushing for Example 12–9.

Table 12–11 **Oiles 500 SP (SPBN·SPWN) Service Range and Properties**

Service Range	Units	Allowable
Characteristic pressure P_{max}	Pa	$<25.5 \cdot 10^6$
Velocity V_{max}	m/s	<0.51
PV product	Pa(m/s)	<1 641 000
Temperature T	°C	<149
Properties	**Test Method, Units**	**Value**
Tensile strength	Pa	$>758.4 \cdot 10^6$
Elongation	%	>12
Compressive strength	Pa	$343.2 \cdot 10^6$
Brinell hardness	HB	>210
Coefficient of thermal expansion	$(10^{-5})°C$	>1.6
Specific gravity		8.2

Source: Data from Oiles America Corp., Plymouth, MI 48 170.

Equation (12–42) with Equation (12–39) is

$$w = K \frac{4}{\pi} \frac{F}{DL} \frac{\pi DNt}{12} = K \frac{4}{\pi} \frac{F}{DL} Vt$$

Solving for t gives

$$t = \frac{\pi DLw}{4\ KVF} = \frac{\pi \cdot 25 \cdot 10^{-3} \cdot 25 \cdot 10^{-3} \cdot 0.13 \cdot 10^{-3}}{4 \cdot 1.3 \cdot 1 \cdot 12 \cdot 10^{-18} \cdot 0.17 \cdot 3000} = 1.04 \cdot 10^6 \text{ s}$$

The rotational speed is

$$N = \frac{V}{\pi D} = \frac{(0.17)}{\pi(25 \cdot 10^{-3})} = 2.16 \text{ rev/s}$$

Answer

$$\text{Cycles} = Nt = 2.16(1.04 \cdot 10^6) = 22.5(10^6) \text{ rev}$$

Temperature Rise

At steady state, the rate at which work is done against bearing friction equals the rate at which heat is transferred from the bearing housing to the surroundings by convection and radiation. The rate of heat generation in J/s is given by $f_s FV$, or

$$H_{gen} = \frac{\pi f_s FDN}{J} \tag{12–44}$$

where N is journal speed in rev/min. The rate at which heat is transferred to the surroundings, in J/s, is

$$H_{loss} = \hbar_{CR} A \Delta T = \hbar_{CR} A(T_b - T_\infty) = \frac{\hbar_{CR} A}{2}(T_f - T_\infty) \tag{12–45}$$

where A = housing surface area, m^2

$\hbar_{CR}$ = overall combined coefficient of heat transfer, $W/(m^2 \cdot °C)$

T_b = housing metal temperature, °C

T_f = lubricant temperature, °C

The empirical observation that T_b is about midway between T_f and T_∞ has been incorporated in Equation (12–45). Equating Equations (12–44) and (12–45) gives

$$T_f = T_\infty + \frac{2\pi f_s FDN}{\hbar_{CR} A} \tag{12–46}$$

Although this equation seems to indicate the temperature rise $T_f - T_\infty$ is independent of length L, the housing surface area generally is a function of L. The housing surface area can be initially estimated, and as tuning of the design proceeds, improved results will converge. If the bushing is to be housed in a pillow block, the surface area can be roughly estimated from

$$A \approx 2\pi DL \tag{12–47}$$

Substituting Equation (12–47) into Equation (12–46) gives

$$T_f \approx = T_\infty + \frac{f_s FN}{\hbar_{CR} L} \tag{12–48}$$

EXAMPLE 12–10

Choose an Oiles 500 bushing to give a maximum wear of 0.025 mm for 800 h of use with a 5 rev/s journal and 220 N radial load. Use $\hbar_{CR} = 13.3\ \text{W}/(\text{m}^2 \cdot {}^\circ\text{C})$, $T_{max} = 149°\text{C}$, $f_s = 0.03$ and a design factor $n_d = 2$. Table 12–12 lists the available bushing sizes from the manufacturer.

Solution

Using Equation (12–48) with $n_d F$ for F, $f_s = 0.03$ from Table 12–10, and $\hbar_{CR} = 13.3\ \text{W/(m} \cdot {}^\circ\text{C)}$, gives

$$L \geq \frac{f_s n_d FN}{\hbar_{CR}(T_f - T_\infty)} = \frac{(0.03)2(220)5}{(13.3)(149 - 21)} = 0.0388 \text{ m} = 38.8 \text{ mm}$$

From Table 12–12, the smallest available bushing has an ID = 16 mm, OD = 22 mm, and L = 38 mm. However, for this case $L/D = 38/16 = 2.375$, and is outside of the recommendations of Equation (12–43). Thus, for the first trial, try the bushing with ID = 20 mm, OD = 28 mm, and L = 38 mm ($L/D = 38/20 = 1.9$). Thus,

Equation (12–41): $$P_{max} = \frac{4}{\pi}\frac{n_d F}{DL} = \frac{4}{\pi}\frac{2(220)}{20(38)} = 0.74 \text{ MPa} < 25.5 \text{ MPa} \qquad \text{(OK)}$$

$$P = \frac{n_d F}{DL} = \frac{2(220)}{20(38)} = 0.58 \text{ MPa}$$

Equation (12–39): $$V = \pi DN = \pi(20)5 = 0.31 \text{ m/s} < 0.51 \text{ m/s} \qquad \text{(OK)}$$

$$PV = 0.58(0.31) = 0.18 \text{ MPa(m/s)} < 1.64 \text{ MPa(m/s)} \qquad \text{(OK)}$$

Equation (12–42), with Table 12–9:

$$w = \frac{4Kn_d FNt}{L} = \frac{(4)12(10^{-18})2(220)5(2\ 880\ 000)10^3}{0.038} = 0.008 \text{ mm} < 0.025 \text{ mm} \qquad \text{(OK)}$$

Answer

Select ID = 20 mm, OD = 28 mm, and L = 38 mm.

Table 12–12 Available Bushing Sizes (in millimeters) of One Manufacturer*

		L													
ID	OD	12	16	20	22	25	30	38	44	50	62	75	90	100	125
12	20	•	•	•	•	•									
16	22		•	•		•		•							
20	28		•	•		•		•							
22	30			•		•	•	•							
25	35			•		•	•	•	•	•					
25	38			•		•		•		•					
30	40					•	•	•	•	•					
38	50					•	•	•	•	•					
44	58						•	•	•	•	•	•	•	•	
50	60							•		•	•	•			
58	70							•		•	•	•			
62	75							•		•		•			
70	84							•		•	•	•			
75	90									•	•	•	•		
90	103									•		•		•	
100	120									•		•		•	
115	135											•		•	•
125	150											•		•	•

*In a display such as this a manufacturer is likely to show catalog numbers where the • appears.

PROBLEMS

12–1 A full journal bearing has a journal diameter of 25 mm, with a unilateral tolerance of −0.03 mm. The bushing bore has a diameter of 25.03 mm and a unilateral tolerance of 0.04 mm. The l/d ratio is 1/2. The load is 1.2 kN and the journal runs at 1100 rev/min. If the average viscosity is 55 mPa · s, find the minimum film thickness, the power loss, and the side flow for the minimum clearance assembly.

12–2 A full journal bearing has a journal diameter of 32 mm, with a unilateral tolerance of −0.012 mm. The bushing bore has a diameter of 32.05 mm and a unilateral tolerance of 0.032 mm. The bearing is 64 mm long. The journal load is 1.75 kN and it runs at a speed of 900 rev/min. Using an average viscosity of 55 mPa · s find the minimum film thickness, the maximum film pressure, and the total oil-flow rate for the minimum clearance assembly.

12–3 A journal bearing has a journal diameter of 75 mm, with a unilateral tolerance of –0.025 mm. The bushing bore has a diameter of 75.125 mm and a unilateral tolerance of 0.1 mm. The bushing is 37.5 mm long. The journal speed is 600 rev/min and the load is 3450 N. For both SAE 10 and SAE 40, lubricants, find the minimum film thickness and the maximum film pressure for an operating temperature of 62°C for the minimum clearance assembly.

12–4 A journal bearing has a journal diameter of 75 mm with a unilateral tolerance of –0.075 mm. The bushing bore has a diameter of 75.15 mm and a unilateral tolerance of 0.1 mm. The bushing is 75 mm long and supports an 2585 N load. The journal speed is 750 rev/min. Find the minimum oil film thickness and the maximum film pressure for both SAE 20 and SAE 20W-40 lubricants, for the tightest assembly if the operating film temperature is 65°C.

12–5 A full journal bearing has a journal with a diameter of 50 mm and a unilateral tolerance of –0.03 mm. The bushing has a bore with a diameter of 50.06 mm and a unilateral tolerance of 0.05 mm. The bushing is 25 mm long and supports a load of 2590 N at a speed of 800 rev/min. Find the minimum film thickness, the power loss, and the total lubricant flow if the average film temperature is 54°C and SAE 20 lubricant is used. The tightest assembly is to be analyzed.

12–6 A full journal bearing has a shaft journal diameter of 25 mm with a unilateral tolerance of –0.01 mm. The bushing bore has a diameter of 25.04 mm with a unilateral tolerance of 0.03 mm. The l/d ratio is unity. The bushing load is 1.25 kN, and the journal rotates at 1200 rev/min. Analyze the minimum clearance assembly if the average viscosity is 50 mPa · s to find the minimum oil film thickness, the power loss, and the percentage of side flow.

12–7 A full journal bearing has a shaft journal with a diameter of 30 mm and a unilateral tolerance of –0.015 mm. The bushing bore has a diameter of 30.05 mm with a unilateral tolerance of 0.035. The bushing bore is 50 mm in length. The bearing load is 2.75 kN and the journal rotates at 1120 rev/min. Analyze the minimum clearance assembly and find the minimum film thickness, the coefficient of friction, and the total oil flow if the average viscosity is 60 mPa · s.

12–8 A journal bearing has a shaft diameter of 75.00 mm with a unilateral tolerance of –0.02 mm. The bushing bore has a diameter of 75.10 mm with a unilateral tolerance of 0.06 mm. The bushing is 36 mm long and supports a load of 2 kN. The journal speed is 720 rev/min. For the minimum clearance assembly find the minimum film thickness, the heat loss rate, and the maximum lubricant pressure for SAE 20 and SAE 40 lubricants operating at an average film temperature of 60°C.

12–9 A full journal bearing is 28 mm long. The shaft journal has a diameter of 56 mm with a unilateral tolerance of –0.012 mm. The bushing bore has a diameter of 56.05 mm with a unilateral tolerance of 0.012 mm. The load is 2.4 kN and the journal speed is 900 rev/min. For the minimum clearance assembly find the minimum oil-film thickness, the power loss, and the side flow if the operating temperature is 65°C and SAE 40 lubricating oil is used.

12–10 A full journal bearing has a shaft diameter of 80 mm with a unilateral tolerance of –0.01 mm. The l/d ratio is unity. The bushing has a bore diameter of 80.08 mm with a unilateral tolerance of 0.03 mm. The SAE 40 oil supply is in an axial-groove sump with a steady-state temperature of 60°C. The radial load is 3 kN. Estimate the average film temperature, the minimum film thickness, the heat loss rate, and the lubricant side-flow rate for the minimum clearance assembly, if the journal speed is 8 rev/s.

12–11 A 64 × 64 mm sleeve bearing uses grade 20 lubricant. The axial-groove sump has a steady-state temperature of 43°C. The shaft journal has a diameter of 63.5 mm with a unilateral tolerance of −0.025 mm. The bushing bore has a diameter of 63.6 mm with a unilateral tolerance of 0.025 mm. The journal speed is 1120 rev/min and the radial load is 5.34 kN. Estimate

(*a*) The magnitude and location of the minimum oil-film thickness.
(*b*) The eccentricity.
(*c*) The coefficient of friction.
(*d*) The power loss rate.
(*e*) Both the total and side oil-flow rates.
(*f*) The maximum oil-film pressure and its angular location.
(*g*) The terminating position of the oil film.
(*h*) The average temperature of the side flow.
(*i*) The oil temperature at the terminating position of the oil film.

12–12 A set of sleeve bearings has a specification of shaft journal diameter of 31.25 mm with a unilateral tolerance of −0.025 mm. The bushing bore has a diameter of 31.3 mm with a unilateral tolerance of 0.075 mm. The bushing is 31.25 mm long. The radial load is 1113 N and the shaft rotational speed is 1750 rev/min. The lubricant is SAE 10 oil and the axial-groove sump temperature at steady state T_s is 49°C. For the $c_{\min}$, c_{median}, and $c_{\max}$ assemblies analyze the bearings and observe the changes in S, ε, f, Q, Q_s, ΔT, $T_{\max}$, $\bar{T}_f$, and h_0.

12–13 An interpolation equation was given by Raimondi and Boyd, and it is displayed as Equation (12–20). This equation is a good candidate for a computer program. Write such a program for interactive use. Once ready for service it can save time and reduce errors. Another version of this program can be used with a subprogram that contains curve fits to Raimondi and Boyd charts for computer use.

12–14 A natural-circulation pillow-block bearing with $l/d = 1$ has a journal diameter D of 62.5 mm with a unilateral tolerance of −0.025 mm. The bushing bore diameter B is 62.6 mm with a unilateral tolerance of 0.1 mm. The shaft runs at an angular speed of 1120 rev/min; the bearing uses SAE grade 20 oil and carries a steady load of 1350 N in shaft-stirred air at 21°C with $\alpha = 1$. The lateral area of the pillow-block housing is 38 700 mm^2. Perform a design assessment using minimum radial clearance for a load of 2700 N and 1350 N. Use Trumpler's criteria.

12–15 An eight-cylinder diesel engine has a front main bearing with a journal diameter of 87.5 mm and a unilateral tolerance of −0.025 mm. The bushing bore diameter is 87.625 mm with a unilateral tolerance of 0.125 mm. The bushing length is 50 mm. The pressure-fed bearing has a central annular groove 6.25 mm wide. The SAE 30 oil comes from a sump at 49°C using a supply pressure of 345 kPa. The sump's heat-dissipation capacity is 1.46 kW per bearing. For a minimum radial clearance, a speed of 2000 rev/min, and a radial load of 20.7 kN, find the average film temperature and apply Trumpler's criteria in your design assessment.

12–16 A pressure-fed bearing has a journal diameter of 50.00 mm with a unilateral tolerance of −0.05 mm. The bushing bore diameter is 50.084 mm with a unilateral tolerance of 0.10 mm. The length of the bushing is 55 mm. Its central annular groove is 5 mm wide and is fed by SAE 30 oil is 55°C at 200 kPa supply gauge pressure. The journal speed is 2880 rev/min carrying a load of 10 kN. The sump can dissipate 300 watts per bearing if necessary. For minimum radial clearances, perform a design assessment using Trumpler's criteria.

12–17 Design a central annular-groove pressure-fed bearing with an l'/d ratio of 0.5, using SAE grade 20 oil, the lubricant supplied at 207 kPa. The exterior oil cooler can maintain the sump temperature at 49°C for heat dissipation rates up to 438 W. The load to be carried is 3860 N at 3000 rev/min. The groove width is 6.25 mm. Use nominal journal diameter d as one design variable and c as the other. Use Trumpler's criteria for your adequacy assessment.

12–18 Repeat design problem Problem 12–17 using the nominal bushing bore B as one decision variable and the radial clearance c as the other. Again, Trumpler's criteria to be used.

12–19 Table 12–1 gives the Seireg and Dandage curve fit approximation for the absolute viscosity in customary U.S. engineering units. Show that in SI units of mPa · s and a temperature of C degrees Celsius, the viscosity can be expressed as

$$\mu = 6.89(10^6)\mu_0 \exp[(b/(1.8C + 127))]$$

where μ_0 and b are from Table 12–1. If the viscosity μ_0' is expressed in μreyn, then

$$\mu = 6.89\mu_0' \exp[(b/(1.8C + 127))]$$

What is the viscosity of a grade 50 oil at 70°C? Compare your results with Figure 12–2.

12–20 For Problem 12–17 a satisfactory design is

$$d = 50^{+0}_{-0.025}\ \text{mm} \qquad b = 50.125^{+0.075}_{-0}\ \text{mm}$$

Double the size of the bearing dimensions and quadruple the load to 16 kN.

(*a*) Analyze the scaled-up bearing for median assembly.

(*b*) Compare the results of a similar analysis for the 50 mm bearing, median assembly.

12–21 Consider a journal bearing of length 25 mm, diameter 50 mm, and radial clearance 25 μm. The dynamic viscosity of the lubricant is 7 mPa · s. In the computing reference frame of Figure 12–32, the journal is rotating at a constant angular velocity of 200 rad/s, and the sleeve is fixed. At time $t = 0$, the journal is initially positioned with journal eccentricity ratio components $\varepsilon_x = -0.8$, $\varepsilon_y = -0.3$. At time $t = 0$, a steady load with components $F_x = 3000$ N, $F_y = 0$ N is applied to the journal.

(*a*) Calculate and plot the journal orbit in the clearance space from $t = 0$ to the time it takes the journal to rotate one revolution (360°). Use Euler integration and a time step equal to 0.5° of journal rotation. Indicate the location of the journal eccentricity ratio at 90° intervals on the plot.

(*b*) In part (*a*), the journal has essentially reached an equilibrium position. Compare the minimum film thickness and attitude angle obtained at journal rotation of 360° with that predicted by Figures 12–13 and 12–14.

12–22 With the same bearing dimensional specifications and fluid viscosity used in Problem 12–21, the load now rotates with the journal; i.e., $F_x = 3000 \cos \omega t$, $F_y = 3000 \sin \omega t$, where $\omega = 200$ rad/s is the steady journal angular velocity.

(*a*) Starting the journal at the same initial position as in Problem 12–21, calculate and plot the journal orbit in the clearance space from $t = 0$ to the time it takes the journal to rotate one revolution (360°). Use Euler integration and a time step equal to 0.5° of journal rotation. Indicate the location of the journal eccentricity ratio at 90° intervals on the plot.

(*b*) Describe the resulting periodic orbit shape and estimate the cyclic minimum film thickness. Compare your answer with that obtained from Problem 12–21.

12–23 With the same bearing dimensional specifications and fluid viscosity used in Problem 12–21, the load now rotates at the half speed of the journal; i.e., $F_x = 3000 \cos \omega t/2$, $F_y = 3000 \sin \omega t/2$, where $\omega = 200$ rad/s is the steady journal angular velocity.

(*a*) Starting the journal at the same initial position as in Problem 12–21, calculate and plot the journal orbit in the clearance space from $t = 0$ to the time it takes the journal to rotate four revolutions (1440°). Use Euler integration and a time step equal to 0.5° of journal rotation. Indicate the location of the journal eccentricity ratio at 360° intervals on the plot.

(*b*) Describe and explain the resulting journal orbit.

12–24 In Problem 12–21, assume the journal has reached its equilibrium position under the specified steady load and journal speed. If the load is suddenly removed entirely, describe and explain the resulting journal orbit.

12–25 Using the bearing specifications in Example 12–8, prepare a plot of cyclic minimum film thickness in a big-end connecting bearing as a function of engine speed over the range 500–8000 rev/min. Assume the peak cylinder pressure is constant over the specified speed range. Indicate on the plot the speed range where the predicted film thicknesses from the chart are applicable, and indicate the range of speeds where the bearing performance is acceptable for this application.

12–26 An Oiles SP 500 alloy brass bushing is 19 mm long with a 19 mm dia bore and operates in a clean environment at 20°C. The allowable wear without loss of function is 0.1 mm. The radial load is 1800 N. The shaft speed is 250 rev/min. Estimate the number of revolutions for radial wear to be 0.1 mm.

12–27 Choose an Oiles SP 500 alloy brass bushing to give a maximum wear of 0.05 mm for 1000 h of use with a 200 rev/min journal and 450 N radial load. Use $\hbar_{CR} = 15.3$ W/(m$^2 \cdot$ °C), $T_{max} = 149$°C, $f_s = 0.03$, and a design factor $n_d = 2$. The bearing is to operate in a clean environment at 20°C. Table 12–12 lists the bushing sizes available from the manufacturer.

13 Gears—General

Baloncici/iStock/Getty Images

Chapter Outline

This chapter addresses gear geometry, the kinematic relations, and the forces transmitted by the four principal types of gears: spur, helical, bevel, and worm gears. The forces transmitted between meshing gears supply torsional moments to shafts for motion and power transmission and create forces and moments that affect the shaft and its bearings. The next two chapters will address stress, strength, safety, and reliability of the four types of gears.

13–1 Types of Gears

Spur gears, illustrated in Figure 13–1, have teeth parallel to the axis of rotation and are used to transmit motion from one shaft to another, parallel, shaft. Of all types, the spur gear is the simplest and, for this reason, will be used to develop the primary kinematic relationships of the tooth form.

Helical gears, shown in Figure 13–2, have teeth inclined to the axis of rotation. Helical gears can be used for the same applications as spur gears and, when so used, are not as noisy, because of the more gradual engagement of the teeth during meshing. The inclined tooth also develops thrust loads and bending couples, which are not present with spur gearing. Sometimes helical gears are used to transmit motion between nonparallel shafts.

Bevel gears, shown in Figure 13–3, have teeth formed on conical surfaces and are used mostly for transmitting motion between intersecting shafts. The figure actually illustrates *straight-tooth bevel gears*. *Spiral bevel gears* are cut so the tooth is no longer straight, but forms a circular arc. *Hypoid gears* are quite similar to spiral bevel gears except that the shafts are offset and nonintersecting.

Worms and worm gears, shown in Figure 13–4, represent the fourth basic gear type. As shown, the worm resembles a screw. The direction of rotation of the worm gear, also called the worm wheel, depends upon the direction of rotation of the worm and upon whether the worm teeth are cut right-hand or left-hand. Worm gearsets are also made so that the teeth of one or both wrap partly around the other. Such sets are called *single-enveloping* and *double-enveloping* worm gearsets. Worm gearsets are mostly used when the speed ratios of the two shafts are quite high, say, 3 or more.

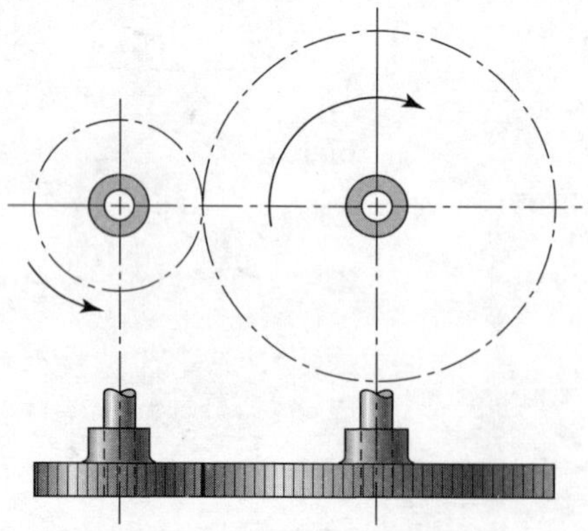

Figure 13–1

Spur gears are used to transmit rotary motion between parallel shafts.

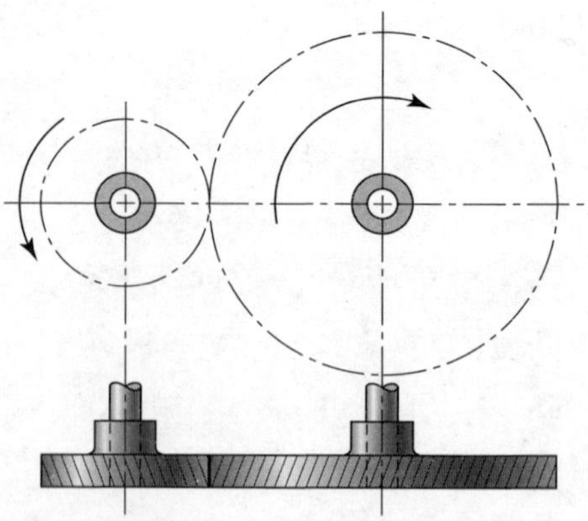

Figure 13–2

Helical gears are used to transmit motion between parallel or nonparallel shafts.

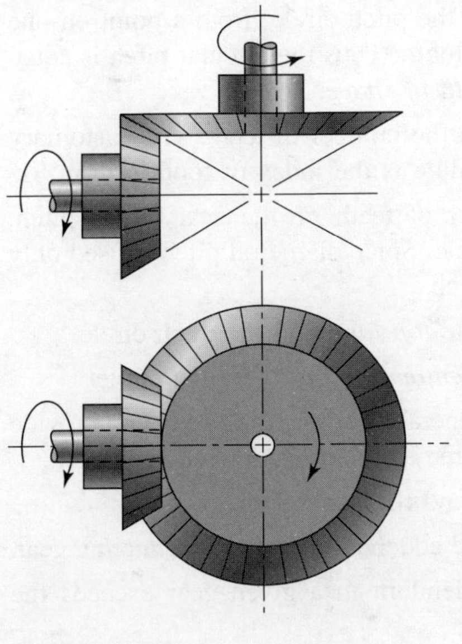

Figure 13–3

Bevel gears are used to transmit rotary motion between intersecting shafts.

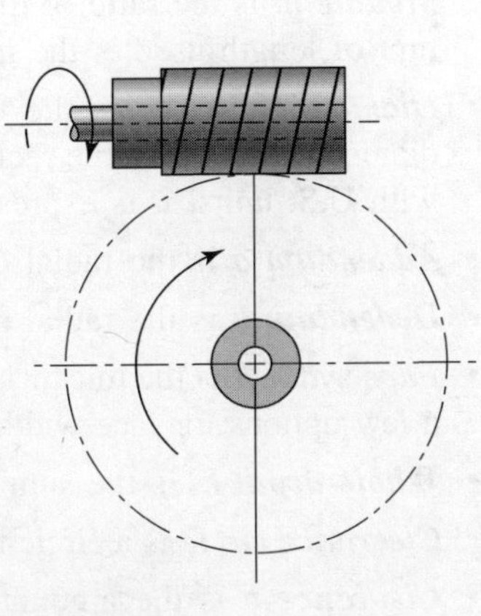

Figure 13–4

Worm gearsets are used to transmit rotary motion between nonparallel and nonintersecting shafts.

13–2 Nomenclature

The terminology of spur-gear teeth is illustrated in Figure 13–5.

- ***Pitch circle*** is a theoretical circle upon which all calculations are usually based. The pitch circles of a pair of mating gears are tangent to each other.
- ***Pitch diameter*** is the diameter of the pitch circle.
- ***Pinion*** is the smaller of two mating gears. The larger is often called the *gear*.

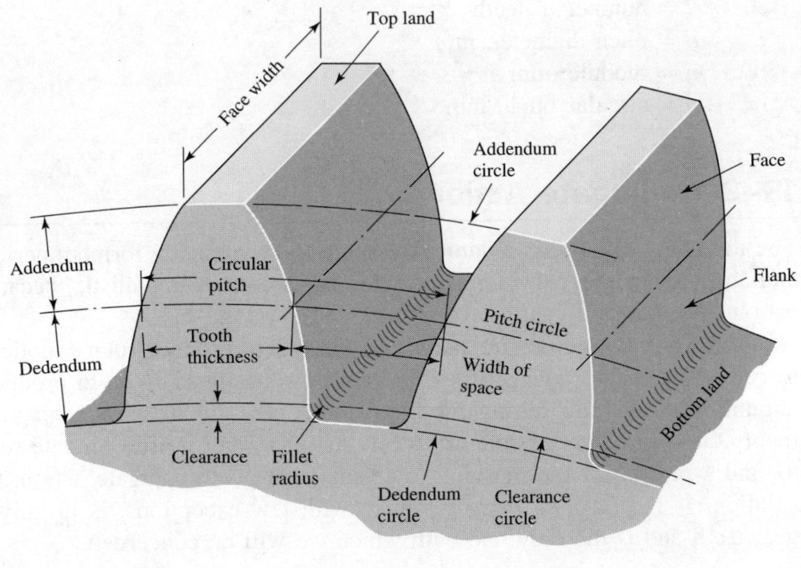

Figure 13–5

Nomenclature of spur-gear teeth.

- ***Circular pitch p*** is the distance, measured on the pitch circle, from a point on one tooth to a corresponding point on an adjacent tooth. Thus the circular pitch is equal to the sum of the *tooth thickness* and the *width of space*.
- ***Module m*** is the ratio of the pitch diameter to the number of teeth. The customary unit of length used is the millimeter. The module is the index of tooth size in SI.
- ***Diametral pitch P*** is the ratio of the number of teeth on the gear to the pitch. diameter. Thus, it is the reciprocal of the module. Since diametral pitch is used only with U.S. units, it is expressed as teeth per inch.
- ***Addendum a*** is the radial distance between the *top land* and the pitch circle.
- ***Dedendum b*** is the radial distance from the *bottom land* to the pitch circle.
- ***Face width F*** is the thickness of the tooth. In general, gear manufacturers will provide a few options for face width in the range of three to five times the circular pitch.
- ***Whole depth*** h_t is the sum of the addendum and the dedendum.
- ***Clearance circle*** is a circle that is tangent to the addendum circle of the mating gear.
- ***Clearance c*** is the amount by which the dedendum in a given gear exceeds the addendum of its mating gear.
- ***Backlash*** is the amount by which the width of a tooth space exceeds the thickness of the engaging tooth measured on the pitch circles.

You should prove for yourself the validity of the following useful relations:

$$P = \frac{N}{d} \tag{13–1}$$

$$m = \frac{d}{N} \tag{13–2}$$

$$p = \frac{\pi d}{N} = \pi m \tag{13–3}$$

$$pP = \pi \tag{13–4}$$

where P = diametral pitch, teeth per m
N = number of teeth
d = pitch diameter, mm
m = module, mm
p = circular pitch, mm

13–3 Conjugate Action

The following discussion assumes the teeth to be perfectly formed, perfectly smooth, and absolutely rigid. Such an assumption is, of course, unrealistic, because the application of forces will cause deflections.

Mating gear teeth acting against each other to produce rotary motion are similar to cams. When the tooth profiles, or cams, are designed so as to produce a constant angular-velocity ratio during meshing, these are said to have *conjugate action*. In theory, at least, it is possible arbitrarily to select any profile for one tooth and then to find a profile for the meshing tooth that will give conjugate action. One of these solutions is the *involute profile*, which, with few exceptions, is in universal use for gear teeth and is the only one with which we will be concerned.

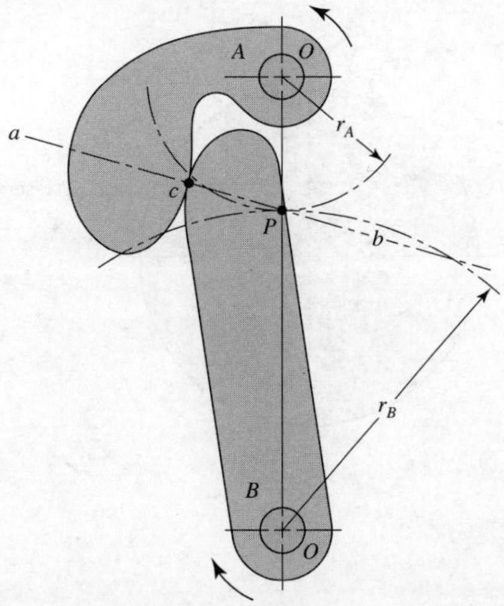

Figure 13–6

Cam *A* and follower *B* in contact. When the contacting surfaces are involute profiles, the ensuing conjugate action produces a constant angular-velocity ratio.

When one curved surface pushes against another (Figure 13–6), the point of contact occurs where the two surfaces are tangent to each other (point *c*), and the forces at any instant are directed along the common normal *ab* to the two curves. The line *ab*, representing the direction of action of the forces, is called the *line of action*. The line of action will intersect the line of centers *O-O* at some point *P*. The angular-velocity ratio between the two arms is inversely proportional to their radii to the point *P*. Circles drawn through point *P* from each center are called *pitch circles,* and the radius of each circle is called the *pitch radius*. Point *P* is called the *pitch point.*

Figure 13–6 is useful in making another observation. A pair of gears is really a pair of cams that act through a small arc and, before running off the involute contour, are replaced by another identical pair of cams. The cams can run in either direction and are configured to transmit a constant angular-velocity ratio. If involute curves are used, the gears tolerate changes in center-to-center distance with *no* variation in constant angular-velocity ratio. Furthermore, the rack profiles are straight-flanked, making primary tooling simpler.

To transmit motion at a constant angular-velocity ratio, the pitch point must remain fixed; that is, all the lines of action for every instantaneous point of contact must pass through the same point *P*. In the case of the involute profile, it will be shown that all points of contact occur on the same straight line *ab*, that all normals to the tooth profiles at the point of contact coincide with the line *ab*, and, thus, that these profiles transmit uniform rotary motion.

13–4 Involute Properties

An involute curve may be generated as shown in Figure 13–7*a*. Cord *def*, held tight, is wrapped around cylinder *A*. Point *b* on the cord represents the tracing point, and as the cord is wrapped and unwrapped about the cylinder, point *b* will trace out the involute curve *ac*. The radius of the curvature of the involute varies continuously, being zero at point *a* and a maximum at point *c*. At point *b* the radius is equal to the distance *be*, since point *b* is instantaneously rotating about point *e*. Thus the generating line *de* is normal to the involute at all points of intersection and, at the same time, is always tangent to the cylinder *A*. The circle on which the involute is generated is called the *base circle*.

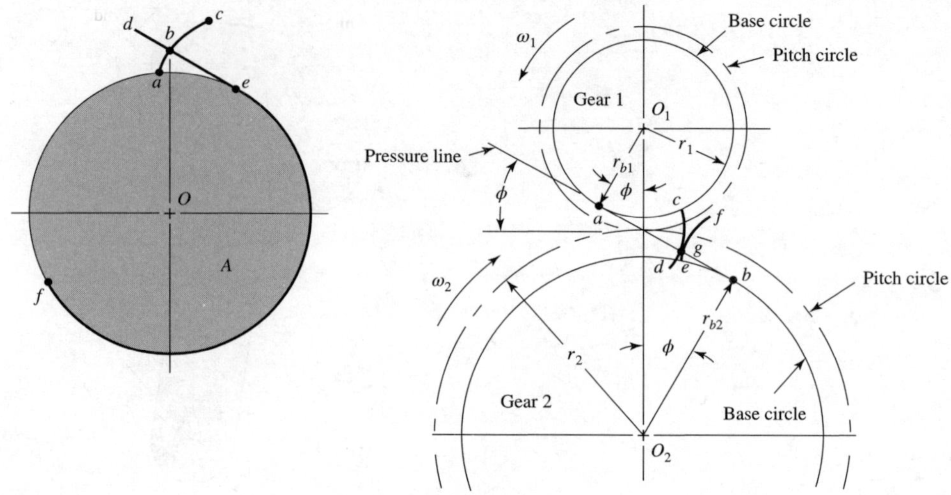

Figure 13–7

(*a*) Generation of an involute; (*b*) involute action.

Let us now examine the involute profile to see how it satisfies the requirement for the transmission of uniform motion. In Figure 13–7*b*, two gear blanks with fixed centers at O_1 and O_2 are shown having base circles whose respective radii are O_1a and O_2b. We now imagine that a cord is wound counterclockwise around the base circle of gear 1, pulled tight between points a and b, and wound clockwise around the base circle of gear 2. If, now, the base circles are rotated in different directions so as to keep the cord tight, a point g on the cord will trace out the involutes cd on gear 1 and ef on gear 2. The involutes are thus generated simultaneously by the tracing point. The tracing point, therefore, represents the point of contact, while the portion of the cord ab is the generating line. The point of contact moves along the generating line; the generating line does not change position, because it is always tangent to the base circles; and since the generating line is always normal to the involutes at the point of contact, the requirement for uniform motion is satisfied.

13–5 Fundamentals

When two gears are in mesh, their pitch circles roll on one another without slipping. As shown in Figure 13–7*b*, designate the pitch radii as r_1 and r_2 and the angular velocities as ω_1 and ω_2, respectively. Then the pitch-line velocity is

$$V = |r_1\omega_1| = |r_2\omega_2|$$

Thus the relation between the radii on the angular velocities is

$$\left|\frac{\omega_1}{\omega_2}\right| = \frac{r_2}{r_1} \tag{13–5}$$

Suppose now we wish to design a speed reducer such that the input speed is 1800 rev/min and the output speed is 1200 rev/min. This is a ratio of 3:2; the gear pitch diameters would be in the same ratio, for example, a 100 mm pinion driving a 150 mm gear. The various dimensions found in gearing are always based on the pitch circles.

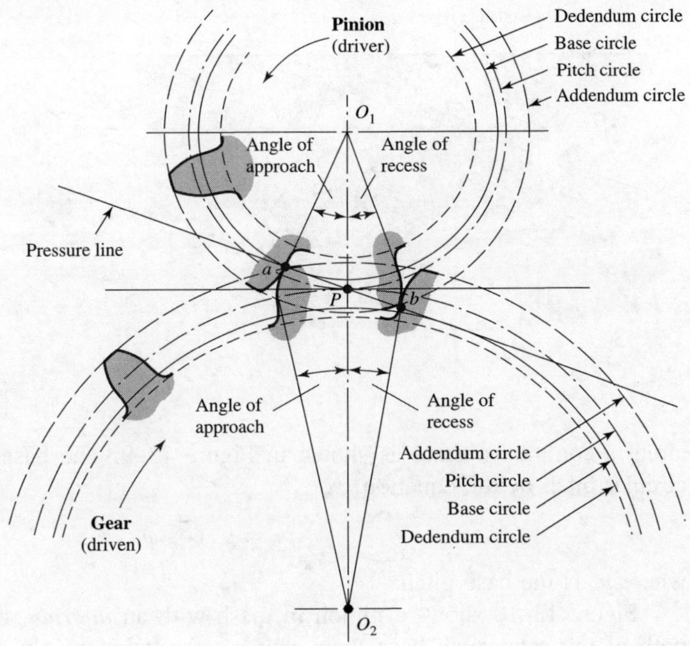

Figure 13–8

Tooth action.

The line *ab* in Figure 13–7*b* has three names, all of which are in general use. It is called the *pressure line*, the *generating line*, and the *line of action*. The angle between the pressure line and the perpendicular to the line of the gear centers, O_1O_2, is called the *pressure angle, ϕ*. The pressure angle usually has values of 20° or 25°, although $14\frac{1}{2}°$ was once used. The pressure line is tangent to each gear at their respective base circle, where the base radius of a gear is given by

$$r_b = r\cos\phi \tag{13–6}$$

where r is the pitch radius of the gear.

Figure 13–8 contains further definitions of the gears and depicts the meshing process. The initial contact will take place when the flank of the driver comes into contact with the tip of the driven tooth. This occurs at point *a* where the addendum circle of the driven gear crosses the pressure line. This defines the *angle of approach* for each gear. As the mesh proceeds, the point of contact will slide up the side of the driving tooth so that the tip of the driver will be in contact just before contact ends. The final point of contact will therefore be where the addendum circle of the driver crosses the pressure line. This is point *b* in Figure 13–8. This defines the *angle of recess* for each gear in a manner similar to that of the angle of approach. The sum of the angle of approach and the angle of recess for either gear is called the *angle of action*.

We may imagine a *rack* as a spur gear having an infinitely large pitch diameter. Therefore, the rack has an infinite number of teeth and a base circle that is an infinite distance from the pitch point. The sides of involute teeth on a rack are straight lines making an angle to the line of centers equal to the pressure angle. Figure 13–9 shows an involute rack in mesh with a pinion. Corresponding sides on involute teeth are parallel curves; the *base pitch* is the constant and fundamental distance between them

Figure 13–9

Involute-toothed pinion and rack.

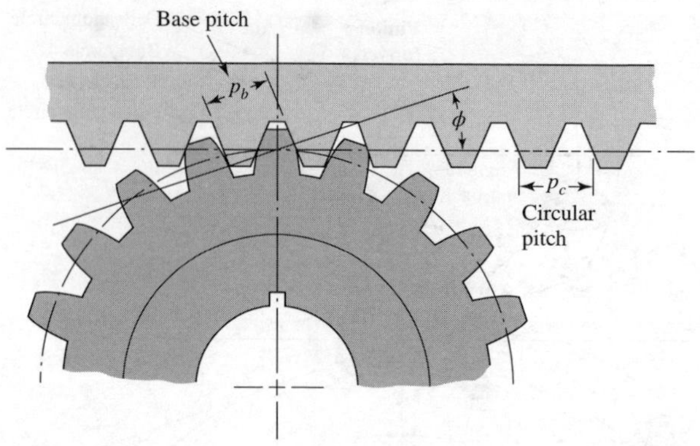

along a common normal as shown in Figure 13–9. The base pitch is related to the circular pitch by the equation

$$p_b = p_c \cos\phi \tag{13–7}$$

where p_b is the base pitch.

Figure 13–10 shows a pinion in mesh with an *internal,* or *ring, gear.* Note that both of the gears now have their centers of rotation on the same side of the pitch point. Thus the positions of the addendum and dedendum circles with respect to the pitch circle are reversed; the addendum circle of the internal gear lies *inside* the pitch circle. Note, too, from Figure 13–10, that the base circle of the internal gear lies inside the pitch circle near the addendum circle.

Another interesting observation concerns the fact that the operating diameters of the pitch circles of a pair of meshing gears need not be the same as the respective design pitch diameters of the gears. If we increase the center distance, we create two new operating pitch circles having larger diameters because they must be tangent to each other at the pitch point. Thus the pitch circles of gears really do not come into existence until a pair of gears are brought into mesh.

Changing the center distance has no effect on the base circles, because these were used to generate the tooth profiles. Thus the base circle is basic to a gear. Increasing the center distance increases the pressure angle and decreases the length of the line of action, but the teeth are still conjugate, the requirement for uniform motion transmission is still satisfied, and the angular-velocity ratio has not changed.

Figure 13–10

Internal gear and pinion.

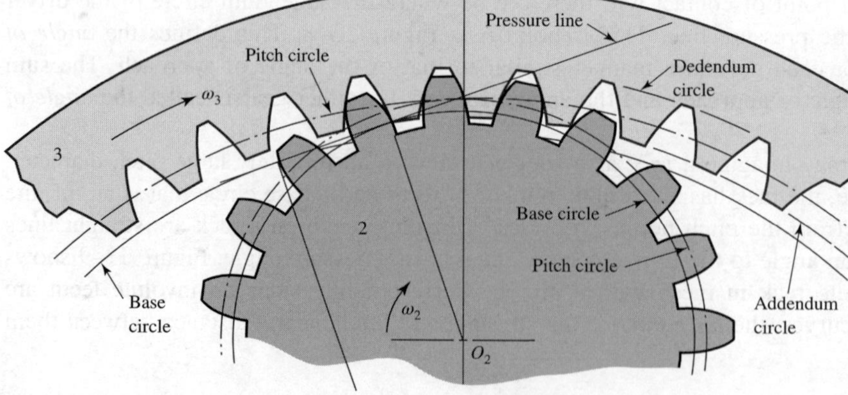

EXAMPLE 13–1

A gearset consists of a 16-tooth pinion driving a 40-tooth gear. The module, is 12 mm, and the addendum and dedendum are 1 m and 1.25 m, respectively. The gears are cut using a pressure angle of 20°.
(*a*) Compute the circular pitch, the center distance, and the radii of the base circles.
(*b*) In mounting these gears, the center distance was incorrectly made 6 mm larger. Compute the new values of the pressure angle and the pitch-circle diameters.

Solution

Answer (*a*)

$$p = \pi m = \frac{\pi}{2} = 37.7 \text{ mm}$$

The pitch diameters of the pinion and gear are, respectively,

$$d_P = N_{Pm} = 16(12) = 192 \text{ mm} \qquad d_G = N_G \cdot m = 40(12) = 480 \text{ mm}$$

Therefore the center distance is

Answer

$$\frac{d_P + d_G}{2} = \frac{192 + 480}{2} = 336 \text{ mm}$$

From Equation (13–6), with a 20° pressure angle, the base radii are

Answer

$$(r_b)_{\text{pinion}} = \frac{192}{2} \cos 20° = 90.2 \text{ mm}$$

Answer

$$(r_b)_{\text{gear}} = \frac{480}{2} \cos 20° = 225.5 \text{ mm}$$

(*b*) Designating d'_P and d'_G as the new pitch-circle diameters, the 6 mm increase in the center distance requires that

$$\frac{d'_P + d'_G}{2} = 342 \tag{1}$$

Also, the velocity ratio does not change, and hence

$$\frac{d'_P}{d'_G} = \frac{16}{40} \tag{2}$$

Solving Equations (1) and (2) simultaneously yields

Answer

$$d'_P = 195.43 \text{ mm} \qquad d'_G = 488.57 \text{ mm}$$

Since $r_b = r \cos \phi$, using either the pinion or gear, the new pressure angle is

Answer

$$\phi' = \cos^{-1}\frac{(r_b)_{\text{pinion}}}{d'_P/2} = \cos^{-1}\frac{90.2}{194.43/2} = 22.62°$$

13–6 Contact Ratio

The zone of action of meshing gear teeth is shown in Figure 13–11. We recall that tooth contact begins and ends at the intersections of the two addendum circles with the pressure line. In Figure 13–11 initial contact occurs at a and final contact at b. Tooth profiles drawn through these points intersect the pitch circle at A and B, respectively. As shown, the distance AP is called the *arc of approach* q_a, and the distance PB, the *arc of recess* q_r. The sum of these is the *arc of action* q_t.

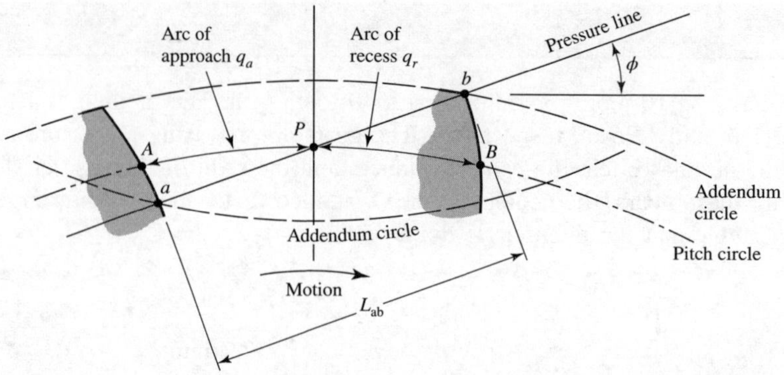

Figure 13–11

Definition of contact ratio.

Now, consider a situation in which the arc of action is exactly equal to the circular pitch, that is, $q_t = p$. This means that one tooth and its space will occupy the entire arc AB. In other words, when a tooth is just beginning contact at a, the previous tooth is simultaneously ending its contact at b. Therefore, during the tooth action from a to b, there will be exactly one pair of teeth in contact.

Next, consider a situation in which the arc of action is greater than the circular pitch, but not very much greater, say, $q_t \approx 1.2p$. This means that when one pair of teeth is just entering contact at a, another pair, already in contact, will not yet have reached b. Thus, for a short period of time, there will be two teeth in contact, one in the vicinity of A and another near B. As the meshing proceeds, the pair near B must cease contact, leaving only a single pair of contacting teeth, until the procedure repeats itself.

Because of the nature of this tooth action, either one or two pairs of teeth in contact, it is convenient to define the term *contact ratio* m_c as

$$m_c = \frac{q_t}{p} \tag{13–8}$$

a number that indicates the average number of pairs of teeth in contact. Note that this ratio is also equal to the length of the path of contact divided by the base pitch. Gears should not generally be designed having contact ratios less than about 1.20, because inaccuracies in mounting might reduce the contact ratio even more, increasing the possibility of impact between the teeth as well as an increase in the noise level.

An easier way to obtain the contact ratio is to measure the line of action ab instead of the arc distance AB. Since ab in Figure 13–11 is tangent to the base circle when extended, the base pitch p_b must be used to calculate m_c instead of the circular pitch as in Equation (13–8). If the length of the line of action is L_{ab}, the contact ratio is

$$m_c = \frac{L_{ab}}{p \cos \phi} \tag{13–9}$$

in which Equation (13–7) was used for the base pitch.

13–7 Interference

The contact of portions of tooth profiles that are not conjugate is called *interference*. Consider Figure 13–12. Illustrated are two 16-tooth gears that have been cut to the now obsolete $14\frac{1}{2}^\circ$ pressure angle. The driver, gear 2, turns clockwise. The initial and final points of contact are designated A and B, respectively, and are located on the pressure line. Now notice that the points of tangency of the pressure line with the base circles C and D are located *inside* of points A and B. Interference is present.

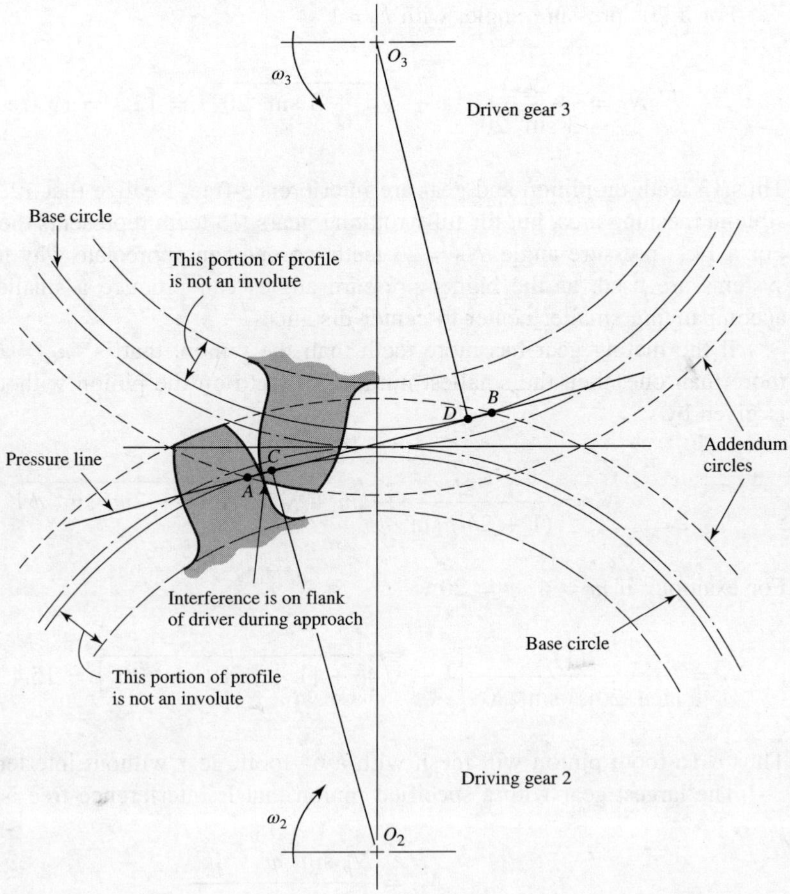

Figure 13–12

Interference in the action of gear teeth.

The interference is explained as follows. Contact begins when the tip of the driven tooth contacts the flank of the driving tooth. In this case the flank of the driving tooth first makes contact with the driven tooth at point A, and this occurs *before* the involute portion of the driving tooth comes within range. In other words, contact is occurring below the base circle of gear 2 on the *noninvolute* portion of the flank. The actual effect is that the involute tip or face of the driven gear tends to dig out the noninvolute flank of the driver.

In this example the same effect occurs again as the teeth leave contact. Contact should end at point D or before. Since it does not end until point B, the effect is for the tip of the driving tooth to dig out, or interfere with, the flank of the driven tooth.

When gear teeth are produced by a generation process, interference is automatically eliminated because the cutting tool removes the interfering portion of the flank. This effect is called *undercutting;* if undercutting is at all pronounced, the undercut tooth is considerably weakened. Thus the effect of eliminating interference by a generation process is merely to substitute another problem for the original one.

The smallest number of teeth on a spur pinion and gear,[1] one-to-one gear ratio, which can exist without interference is N_P. This number of teeth for spur gears is given by

$$N_P = \frac{2k}{3\sin^2\phi}\left(1 + \sqrt{1 + 3\sin^2\phi}\right) \tag{13–10}$$

where $k = 1$ for full-depth teeth, 0.8 for stub teeth and $\phi =$ pressure angle.

[1]Robert Lipp, "Avoiding Tooth Interference in Gears," *Machine Design*, vol. 54, no. 1, 1982, pp. 122–124.

For a 20° pressure angle, with $k = 1$,

$$N_P = \frac{2(1)}{3\sin^2 20°}\left(1 + \sqrt{1 + 3\sin^2 20°}\right) = 12.3 = 13 \text{ teeth}$$

Thus 13 teeth on pinion and gear are interference-free. Realize that 12.3 teeth is possible in meshing arcs, but for fully rotating gears, 13 teeth represents the least number. For a $14\frac{1}{2}°$ pressure angle, $N_P = 23$ teeth, so one can appreciate why few $14\frac{1}{2}°$-tooth systems are used, as the higher pressure angles can produce a smaller pinion with accompanying smaller center-to-center distances.

If the mating gear has more teeth than the pinion, that is, $m_G = N_G/N_P = m$ is more than one, then the smallest number of teeth on the pinion without interference is given by

$$N_P = \frac{2k}{(1 + 2m)\sin^2\phi}\left(m + \sqrt{m^2 + (1 + 2m)\sin^2\phi}\right) \tag{13–11}$$

For example, if $m = 4$, $\phi = 20°$,

$$N_P = \frac{2(1)}{[1 + 2(4)]\sin^2 20°}\left[4 + \sqrt{4^2 + [1 + 2(4)]\sin^2 20°}\right] = 15.4 = 16 \text{ teeth}$$

Thus a 16-tooth pinion will mesh with a 64-tooth gear without interference.

The largest gear with a specified pinion that is interference-free is

$$N_G = \frac{N_P^2\sin^2\phi - 4k^2}{4k - 2N_P\sin^2\phi} \tag{13–12}$$

For example, for a 13-tooth pinion with a pressure angle ϕ of 20°,

$$N_G = \frac{13^2\sin^2 20° - 4(1)^2}{4(1) - 2(13)\sin^2 20°} = 16.45 = 16 \text{ teeth}$$

For a 13-tooth spur pinion, the maximum number of gear teeth possible without interference is 16.

Table 13–1 provides a convenient summary of Equations (13–11) and (13–12) for 20° pressure angle. The first column gives the useful range of minimum pinion teeth numbers. The second column shows the associated maximum teeth numbers calculated from Equation (13–12). The third column rounds the maximum teeth numbers down to the nearest integer. The fourth column calculates the maximum gear ratio that can be obtained, interference free. If the first column is extended to 18 teeth and beyond, the maximum number of teeth becomes undefined (infinite). Thus, 17 is the largest value that is practically needed, as it is generally recommended to limit the gear ratio for a pair of gears to no more than 10 to 1. If fewer than 13 teeth are attempted for the minimum, the maximum number of teeth calculated from Equation (13–12) is less than the minimum, which is practically meaningless. Thus, 13 is the lowest practical number of teeth for a pinion to avoid interference. To use the table to determine the minimum number of teeth to avoid interference, use column four to find the gear ratio

Table 13–1 Gear Teeth Limits to Avoid Interference. Spur gears with 20° pressure angle.

Minimum N_P	Max N_G	Integer Max N_G	Max Gear Ratio $m_G = N_G/N_P$
13	16.45	16	1.23
14	26.12	26	1.86
15	45.49	45	3
16	101.07	101	6.31
17	1309.86	1309	77

Table 13–2 Gear Teeth Limits to Avoid Interference. Spur gears with 25° pressure angle.

Minimum N_P	Max N_G	Integer Max N_G	Max Gear Ratio $m_G = N_G/N_P$
9	13.33	13	1.44
10	32.39	32	3.2
11	249.23	249	22.64

that just exceeds the desired gear ratio. From that row, in column one obtain the minimum number of teeth for the pinion. For example, for a desired gear ratio of 4 to 1, from column four select 6.31 as the first ratio greater than the desired ratio. From this row, determine that 16 is the minimum number of teeth for the pinion. A 16-tooth pinion driving a 64-tooth gear will provide the 4:1 ratio without interference.

Table 13–2 gives a similar summary for a 25° pressure angle. The table shows the primary motivation for using a larger pressure angle; the minimum numbers of teeth are significantly less than for the 20° pressure angle. For the same diametral pitch (tooth size), the smaller numbers of teeth correlate with smaller pitch diameters, and thus a smaller packaging size for the gear train. To understand why the larger pressure angle reduces the interference problem, consider Figure 13–12. Increasing the pressure angle results in smaller base circles, which are tangent to the pressure line, with no change in the addendum circles, dedendum circles, or pitch circles. This moves points *A* and *B*, the first and last points of contact, toward points *C* and *D*, the tangent points to the base circle, where the involute profile is effective. Another way of seeing it is that since the involute profile is generated from the base circle, reducing the base circle makes more of the tooth profile to be involute. There is also a negative aspect to increasing the pressure angle. The load is transmitted between the gear teeth along the normal to the contacting teeth, which is the pressure line. Separating this load into tangential and radial components (see Figure 13–29), we see that a higher-pressure angle produces a higher radial component. The radial component serves no useful purpose in transmitting torque, but leads to larger reaction forces at the bearing which must be carried by the shaft. So, the practical advantage of increasing the pressure angle is smaller package size for the gear train. The practical tradeoff is the need for larger bearings and larger shafts. The standard pressure angle of 20° is the accepted compromise for most designs. The 25° pressure angle is used when the design warrants more emphasis on the small size.

The smallest spur pinion that will operate with a rack without interference is

$$N_P = \frac{2(k)}{\sin^2 \phi} \tag{13–13}$$

For a 20° pressure angle full-depth tooth the smallest number of pinion teeth to mesh with a rack is

$$N_P = \frac{2(1)}{\sin^2 20°} = 17.1 = 18 \text{ teeth}$$

Since gear-shaping tools amount to contact with a rack, and the gear-hobbing process is similar, the minimum number of teeth to prevent interference to prevent undercutting by the hobbing process is equal to the value of N_P when N_G is infinite.

The importance of the problem of teeth that have been weakened by undercutting cannot be overemphasized. Of course, interference can be eliminated by using more teeth on the pinion. However, if the pinion is to transmit a given amount of power, more teeth can be used only by increasing the pitch diameter.

13–8 The Forming of Gear Teeth

There are a large number of ways of forming the teeth of gears, such as *sand casting, shell molding, investment casting, permanent-mold casting, die casting,* and *centrifugal casting*. Teeth can also be formed by using the *powder-metallurgy process;* or, by using *extrusion,* a single bar of aluminum may be formed and then sliced into gears. Gears that carry large loads in comparison with their size are usually made of steel and are cut with either *form cutters* or *generating cutters.* In form cutting, the tooth space takes the exact form of the cutter. In generating, a tool having a shape different from the tooth profile is moved relative to the gear blank so as to obtain the proper tooth shape. One of the newest and most promising of the methods of forming teeth is called *cold forming,* or *cold rolling,* in which dies are rolled against steel blanks to form the teeth. The mechanical properties of the metal are greatly improved by the rolling process, and a high-quality generated profile is obtained at the same time.

Gear teeth may be machined by milling, shaping, or hobbing. They may be finished by shaving, burnishing, grinding, or lapping.

Gears made of thermoplastics such as nylon, polycarbonate, and acetal are quite popular and are easily manufactured by *injection molding*. These gears are of low to moderate precision, low in cost for high production quantities, and capable of light loads, and can run without lubrication.

Milling

Gear teeth may be cut with a form milling cutter shaped to conform to the tooth space. With this method it is theoretically necessary to use a different cutter for each gear, because a gear having 25 teeth, for example, will have a different-shaped tooth space from one having, say, 24 teeth. Actually, the change in space is not too great, and it has been found that eight cutters may be used to cut with reasonable accuracy any gear in the range of 12 teeth to a rack. A separate set of cutters is, of course, required for each pitch.

Shaping

Teeth may be generated with either a pinion cutter or a rack cutter. The pinion cutter (Figure 13–13) reciprocates along the vertical axis and is slowly fed into the gear blank to the required depth. When the pitch circles are tangent, both the cutter and the blank rotate slightly after each cutting stroke. Since each tooth of the cutter is a cutting tool, the teeth are all cut after the blank has completed one rotation. The sides of an involute rack tooth are straight. For this reason, a rack-generating tool provides an accurate method of cutting gear teeth. This is also a shaping operation and is illustrated by the drawing of Figure 13–14. In operation, the cutter reciprocates and is first fed into the gear blank until the pitch circles are tangent. Then, after each cutting stroke, the gear blank and cutter roll slightly on their pitch circles. When the blank and cutter have rolled a distance equal to the circular pitch, the cutter is returned to the starting point, and the process is continued until all the teeth have been cut.

Figure 13–13

Generating a spur gear with a pinion cutter. *(Courtesy of Boston Gear–Altra Industrial Motion)*

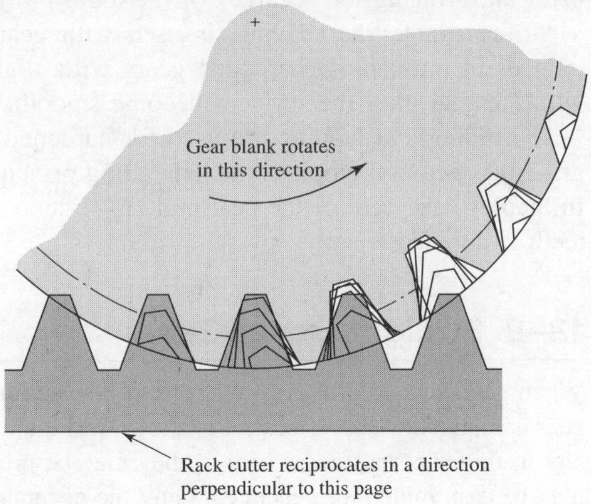

Figure 13–14

Shaping teeth with a rack.

Hobbing

The hobbing process is illustrated in Figure 13–15. The hob is simply a cutting tool that is shaped like a worm. The teeth have straight sides, as in a rack, but the hob axis must be turned through the lead angle in order to cut spur-gear teeth. For this reason, the teeth generated by a hob have a slightly different shape from those generated by a rack cutter. Both the hob and the blank must be rotated at the proper angular-velocity ratio. The hob is then fed slowly across the face of the blank until all the teeth have been cut.

Figure 13–15

Hobbing a spur gear. *(Dmitry Kalinovsky/Shutterstock)*

Finishing

Gears that run at high speeds and transmit large forces may be subjected to additional dynamic forces if there are errors in tooth profiles. Errors may be diminished somewhat by finishing the tooth profiles. The teeth may be finished, after cutting, by either shaving or burnishing. Several shaving machines are available that cut off a minute amount of metal, bringing the accuracy of the tooth profile within the limits of 250 μin.

Burnishing, like shaving, is used with gears that have been cut but not heat-treated. In burnishing, hardened gears with slightly oversize teeth are run in mesh with the gear until the surfaces become smooth.

Grinding and lapping are used for hardened gear teeth after heat treatment. The grinding operation employs the generating principle and produces very accurate teeth. In lapping, the teeth of the gear and lap slide axially so that the whole surface of the teeth is abraded equally.

13–9 Straight Bevel Gears

When gears are used to transmit motion between intersecting shafts, some form of bevel gear is required. A bevel gearset is shown in Figure 13–16. Although bevel gears are usually made for a shaft angle of 90°, they may be produced for almost any angle. The teeth may be cast, milled, or generated. Only the generated teeth may be classed as accurate.

The terminology of bevel gears is illustrated in Figure 13–16. The pitch of bevel gears is measured at the large end of the tooth, and both the circular pitch and the pitch diameter are calculated in the same manner as for spur gears. It should be noted that the clearance is uniform. The pitch angles are defined by the pitch cones meeting at the apex, as shown in the figure. They are related to the tooth numbers as follows:

$$\tan \gamma = \frac{N_P}{N_G} \qquad \tan \Gamma = \frac{N_G}{N_P} \tag{13–14}$$

where the subscripts P and G refer to the pinion and gear, respectively, and where γ and Γ are, respectively, the pitch angles of the pinion and gear.

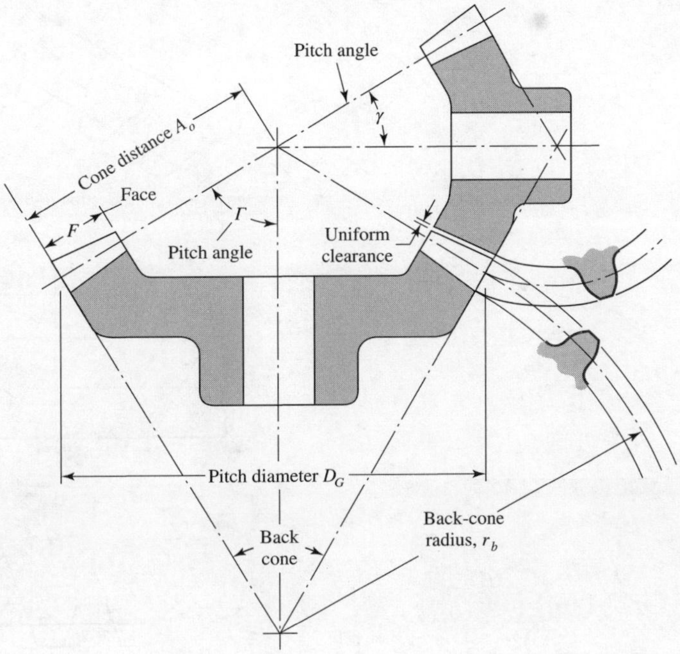

Figure 13–16

Terminology of bevel gears.

Figure 13–16 shows that the shape of the teeth, when projected on the back cone, is the same as in a spur gear having a radius equal to the back-cone distance r_b. This is called Tredgold's approximation. The number of teeth in this imaginary gear is

$$N' = \frac{2\pi r_b}{p} \tag{13–15}$$

where N' is the *virtual number of teeth* and p is the circular pitch measured at the large end of the teeth. Standard straight-tooth bevel gears are cut by using a 20° pressure angle, unequal addenda and dedenda, and full-depth teeth. This increases the contact ratio, avoids undercut, and increases the strength of the pinion.

13–10 Parallel Helical Gears

Helical gears, used to transmit motion between parallel shafts, are shown in Figure 13–2. The helix angle is the same on each gear, but one gear must have a right-hand helix and the other a left-hand helix. The shape of the tooth is an involute helicoid and is illustrated in Figure 13–17. If a piece of paper cut in the shape of a parallelogram is wrapped around a cylinder, the angular edge of the paper becomes a helix. If we unwind this paper, each point on the angular edge generates an involute curve. This surface obtained when every point on the edge generates an involute is called an *involute helicoid.*

The initial contact of spur-gear teeth is a line extending all the way across the face of the tooth. The initial contact of helical-gear teeth is a point that extends into a line as the teeth come into more engagement. In spur gears the line of contact is parallel to the axis of rotation; in helical gears the line is diagonal across the face of the tooth. It is this gradual engagement of the teeth and the smooth transfer of load from one tooth to another that gives helical gears the ability to transmit heavy loads at high speeds. Because of the nature of contact between helical gears, the contact

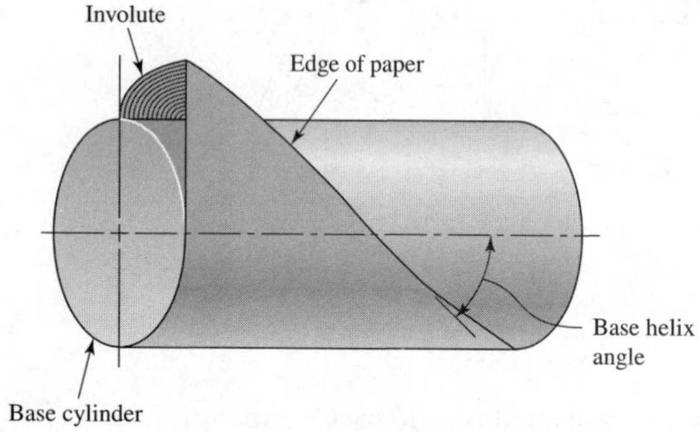

Figure 13–17

An involute helicoid.

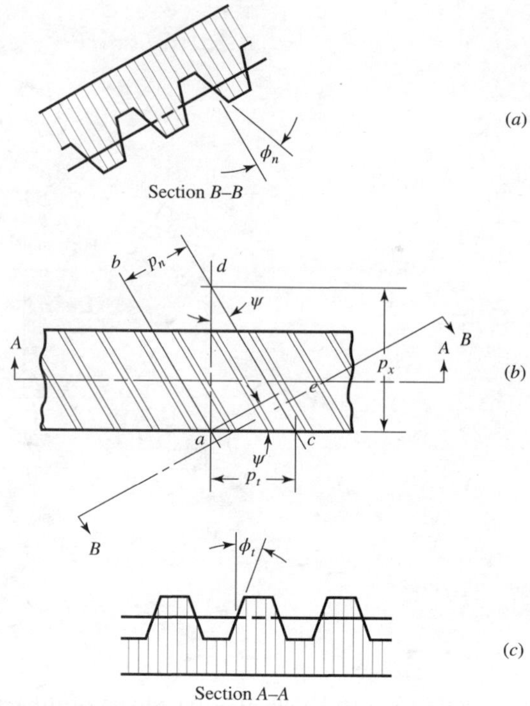

Figure 13–18

Nomenclature of helical gears.

ratio is of only minor importance, and it is the contact area, which is proportional to the face width of the gear, that becomes significant.

Helical gears subject the shaft bearings to both radial and thrust loads. When the thrust loads become high or are objectionable for other reasons, it may be desirable to use double helical gears. A double helical gear (herringbone) is equivalent to two helical gears of opposite hand, mounted side by side on the same shaft. They develop opposite thrust reactions and thus cancel out the thrust load.

When two or more single helical gears are mounted on the same shaft, the hand of the gears should be selected so as to produce the minimum thrust load.

Figure 13–18 represents a portion of the top view of a helical rack. Lines *ab* and *cd* are the centerlines of two adjacent helical teeth taken on the same pitch plane. The angle ψ is the *helix angle.* The distance *ac* is the *transverse circular pitch* p_t in the plane of rotation (usually called the *circular pitch*). The distance *ae* is the *normal circular pitch* p_n and is related to the transverse circular pitch as follows:

$$p_n = p_t \cos \psi \tag{13–16}$$

The distance *ad* is called the *axial pitch* p_x and is related by the expression

$$p_x = \frac{p_t}{\tan \psi} \tag{13–17}$$

Since $p_n = \pi m_n$, the *normal module* is

$$m_n = m_t \cos \psi \tag{13–18}$$

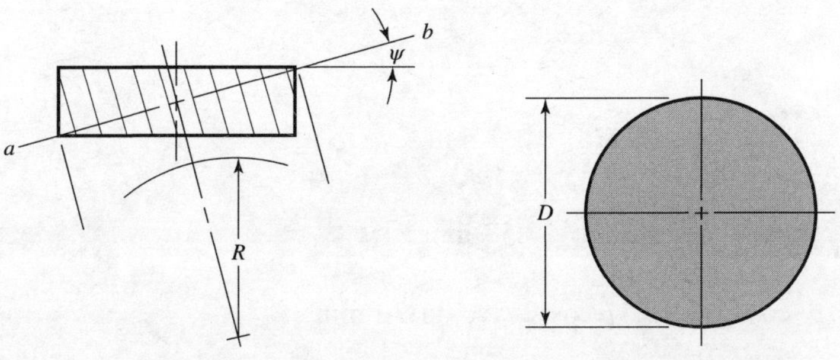

Figure 13–19

A cylinder cut by an oblique plane.

The pressure angle ϕ_n in the normal direction is different from the pressure angle ϕ_t in the direction of rotation, because of the angularity of the teeth. These angles are related by the equation

$$\cos\psi = \frac{\tan\phi_n}{\tan\phi_t} \tag{13–19}$$

Figure 13–19 illustrates a cylinder cut by an oblique plane *ab* at an angle ψ to a right section. The oblique plane cuts out an arc having a radius of curvature of R. For the condition that $\psi = 0$, the radius of curvature is $R = D/2$. If we imagine the angle ψ to be slowly increased from zero to 90°, we see that R begins at a value of $D/2$ and increases until, when $\psi = 90°$, $R = \infty$. The radius R is the apparent pitch radius of a helical-gear tooth when viewed in the direction of the tooth elements. A gear of the same pitch and with the radius R will have a greater number of teeth, because of the increased radius. In helical-gear terminology this is called the *virtual number of teeth.* It can be shown by analytical geometry that the virtual number of teeth is related to the actual number by the equation

$$N' = \frac{N}{\cos^3\psi} \tag{13–20}$$

where N' is the virtual number of teeth and N is the actual number of teeth. It is necessary to know the virtual number of teeth in design for strength and also, sometimes, in cutting helical teeth. This apparently larger radius of curvature means that few teeth may be used on helical gears, because there will be less undercutting.

EXAMPLE 13–2

A stock helical gear has a normal pressure angle of 20°, a helix angle of 25°, and a transverse module of 4 mm, and has 18 teeth. Find:

(*a*) The pitch diameter
(*b*) The transverse, the normal, and the axial pitches
(*c*) The normal diametral pitch
(*d*) The transverse pressure angle

Solution

Answer (a) $$d = Nm = 18(4) = 72 \text{ mm}$$

Answer (b) $$p_t = m_t\pi = \pi 4 = 12.57 \text{ mm}$$

Answer $$p_n = p_t \cos\psi = 12.57 \cos 25° = 11.39 \text{ mm}$$

Answer $$p_x = \frac{p_t}{\tan\psi} = \frac{12.57}{\tan 25°} = 26.95 \text{ mm}$$

Answer (c) $$m_n = m_t\cos(\psi) = 4\cos(25°) = 3.96 \text{ mm}$$

Answer (d) $$\phi_t = \tan^{-1}\left(\frac{\tan\phi_n}{\cos\psi}\right) = \tan^{-1}\left(\frac{\tan 20°}{\cos 25°}\right) = 21.88°$$

Just like teeth on spur gears, helical-gear teeth can interfere. Equation (13–19) can be solved for the pressure angle ϕ_t in the tangential (rotation) direction to give

$$\phi_t = \tan^{-1}\left(\frac{\tan\phi_n}{\cos\psi}\right)$$

The smallest tooth number N_P of a helical-spur pinion that will run without interference[2] with a gear with the same number of teeth is

$$N_P = \frac{2k\cos\psi}{3\sin^2\phi_t}\left(1 + \sqrt{1 + 3\sin^2\phi_t}\right) \tag{13–21}$$

For example, if the normal pressure angle ϕ_n is 20°, the helix angle ψ is 30°, then ϕ_t is

$$\phi_t = \tan^{-1}\left(\frac{\tan 20°}{\cos 30°}\right) = 22.80°$$

$$N_P = \frac{2(1)\cos 30°}{3\sin^2 22.80°}\left(1 + \sqrt{1 + 3\sin^2 22.80°}\right) = 8.48 = 9 \text{ teeth}$$

For a given gear ratio $m_G = N_G/N_P = m$, the smallest pinion tooth count is

$$N_P = \frac{2k\cos\psi}{(1 + 2m)\sin^2\phi_t}\left[m + \sqrt{m^2 + (1 + 2m)\sin^2\phi_t}\right] \tag{13–22}$$

The largest gear with a specified pinion is given by

$$N_G = \frac{N_P^2\sin^2\phi_t - 4k^2\cos^2\psi}{4k\cos\psi - 2N_P\sin^2\phi_t} \tag{13–23}$$

For example, for a nine-tooth pinion with a pressure angle ϕ_n of 20°, a helix angle ψ of 30°, and recalling that the tangential pressure angle ϕ_t is 22.80°,

$$N_G = \frac{9^2\sin^2 22.80° - 4(1)^2\cos^2 30°}{4(1)\cos 30° - 2(9)\sin^2 22.80°} = 12.02 = 12$$

[2]Op. cit., Robert Lipp, *Machine Design*, pp. 122–124.

The smallest pinion that can be run with a rack is

$$N_P = \frac{2k \cos \psi}{\sin^2 \phi_t} \tag{13–24}$$

For a normal pressure angle ϕ_n of 20° and a helix angle ψ of 30°, and $\phi_t = 22.80°$,

$$N_P = \frac{2(1) \cos 30°}{\sin^2 22.80°} = 11.5 = 12 \text{ teeth}$$

For helical-gear teeth the number of teeth in mesh across the width of the gear will be greater than unity and a term called *face-contact ratio* is used to describe it. This increase of contact ratio, and the gradual sliding engagement of each tooth, results in quieter gears.

13–11 Worm Gears

The nomenclature of a worm gearset is shown in Figure 13–20. The worm and worm gear of a set have the same hand of helix as for crossed helical gears, but the helix angles are usually quite different. The helix angle on the worm is generally quite large, and that on the gear very small. Because of this, it is usual to specify the lead angle λ on the worm and helix angle ψ_G on the gear; the two angles are equal for a 90° shaft angle. The worm lead angle is the complement of the worm helix angle, as shown in Figure 13–20.

In specifying the pitch of worm gearsets, it is customary to state the *axial pitch* p_x of the worm and the *transverse circular pitch* p_t, often simply called the circular pitch, of the mating gear. These are equal if the shaft angle is 90°. The pitch

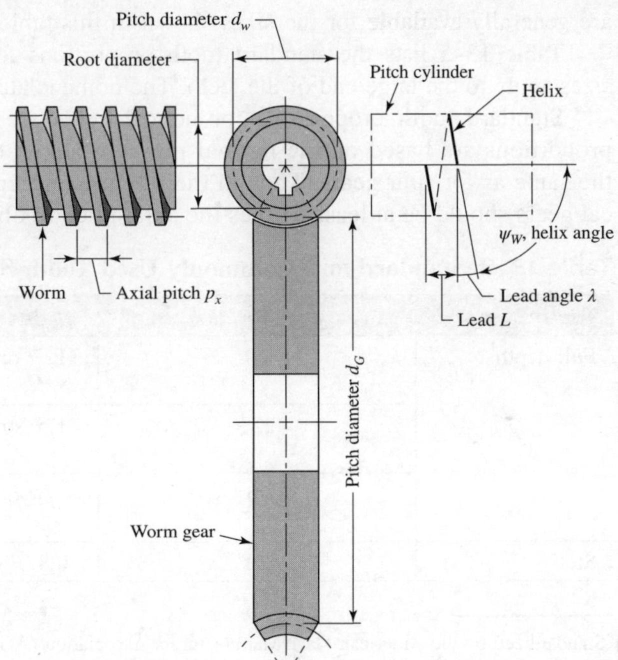

Figure 13–20

Nomenclature of a single-enveloping worm gearset.

diameter of the gear is the diameter measured on a plane containing the worm axis, as shown in Figure 13–20; it is the same as for spur gears and is

$$d_G = \frac{N_G p_t}{\pi} \tag{13–25}$$

Since it is not related to the number of teeth, the worm may have any pitch diameter; this diameter should, however, be the same as the pitch diameter of the hob used to cut the worm-gear teeth. Generally, the pitch diameter of the worm should be selected so as to fall into the range

$$\frac{C^{0.875}}{3.0} \le d_W \le \frac{C^{0.875}}{1.7} \tag{13–26}$$

where C is the center distance. These proportions appear to result in optimum horsepower capacity of the gearset.

The *lead L* and the *lead angle* λ of the worm have the following relations:

$$L = p_x N_W \tag{13–27}$$

$$\tan \lambda = \frac{L}{\pi d_W} \tag{13–28}$$

13–12 Tooth Systems[3]

A *tooth system* is a standard that specifies the relationships involving addendum, dedendum, working depth, tooth thickness, and pressure angle. The standards were originally planned to attain interchangeability of gears of all tooth numbers, but of the same pressure angle and pitch.

Table 13–3 contains the standards most used for spur gears. A $14\frac{1}{2}°$ pressure angle was once used for these but is now obsolete; the resulting gears had to be comparatively larger to avoid interference problems.

Table 13–4 is particularly useful in selecting the pitch or module of a gear. Cutters are generally available for the sizes shown in this table.

Table 13–5 lists the standard tooth proportions for straight bevel gears. These sizes apply to the large end of the teeth. The nomenclature is defined in Figure 13–16.

Standard tooth proportions for helical gears are listed in Table 13–6. Tooth proportions are based on the normal pressure angle; these angles are standardized the same as for spur gears. Though there will be exceptions, the face width of helical gears should be at least 2 times the axial pitch to obtain good helical-gear action.

Table 13–3 Standard and Commonly Used Tooth Systems for Spur Gears

Tooth System	Pressure Angle ϕ, deg	Addendum a	Dedendum b
Full depth	20	$1/P$ or m	$1.25/P$ or $1.25m$ $1.35/P$ or $1.35m$
	$22\frac{1}{2}$	$1/P$ or m	$1.25/P$ or $1.25m$ $1.35/P$ or $1.35m$
	25	$1/P$ or m	$1.25/P$ or $1.25m$ $1.35/P$ or $1.35m$
Stub	20	$0.8/P$ or $0.8m$	$1/P$ or m

[3]Standardized by the American Gear Manufacturers Association (AGMA). Write AGMA for a complete list of standards, because changes are made from time to time. The address is: 1001 N. Fairfax Street, Suite 500, Alexandria, VA 22 314–1587; or, www.agma.org.

Table 13–4 Tooth Sizes in General Uses

Module m (mm)	
Preferred	1, 1.25, 1.5, 2, 2.5, 3, 4, 5, 6, 8, 10, 12, 16, 20, 25, 32, 40, 50
Next Choice	1.125, 1.375, 1.75, 2.25, 2.75, 3.5, 4.5, 5.5, 7, 9, 11, 14, 18, 22, 28, 36, 45

Table 13–5 Tooth Proportions for 20° Straight Bevel-Gear Teeth

Item	Formula
Working depth	$h_k = 2.0$ m
Clearance	$c = 0.188$ m $+ 0.05$ mm
Addendum of gear	$a_G = 0.54 \text{ m} + \dfrac{0.46 \text{ m}}{P(m_{90})^2}$
Gear ratio	$m_G = N_G/N_P$
Equivalent 90° ratio	$m_{90} = m_G$ when $\Gamma = 90°$ $m_{90} = \sqrt{m_G \dfrac{\cos\gamma}{\cos\Gamma}}$ when $\Gamma \neq 90°$
Face width	$F = \dfrac{A_0}{3}$ or F = 10 m, whichever is smaller
Minimum number of teeth	Pinion: 16, 15, 14, 13 Gear: 16, 17, 20, 30

Table 13–6 Standard Tooth Proportions for Helical Gears

Quantity*	Formula	Quantity*	Formula
Addendum	$1.0\ m_n$	External gears:	
Dedendum	$1.25\ m_n$	Standard center distance	$\dfrac{D+d}{2}$
Pinion pitch diameter	N_P	Gear outside diameter	$D + 2a$
Gear pitch diameter	N_G	Pinion outside diameter	$d + 2a$
Normal arc tooth thickness†	$\pi m_n - \dfrac{B_n}{2}$	Gear root diameter	$D - 2b$
Pinion base diameter	$d \cos \phi_t$	Pinion root diameter	$d - 2b$
		Internal gears:	
Gear base diameter	$D \cos \phi_t$	Center distance	$\dfrac{D-d}{2}$
Base helix angle	$\tan^{-1}(\tan\psi \cos\phi_t)$	Inside diameter	$D - 2a$
		Root diameter	$D + 2b$

*All dimensions are in millimeters.

†B_n is the normal backlash.

Table 13–7 Recommended Pressure Angles and Tooth Depths for Worm Gearing

Lead Angle λ, deg	Pressure Angle ϕ_n, deg	Addendum a	Dedendum b_G
0–15	$14\frac{1}{2}$	$0.3683p_x$	$0.3683p_x$
15–30	20	$0.3683p_x$	$0.3683p_x$
30–35	25	$0.2865p_x$	$0.3314p_x$
35–40	25	$0.2546p_x$	$0.2947p_x$
40–45	30	$0.2228p_x$	$0.2578p_x$

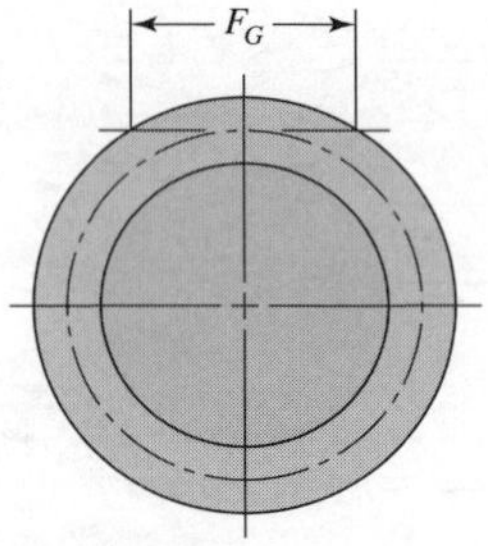

Figure 13–21

A graphical depiction of the face width of the worm of a worm gearset.

Tooth forms for worm gearing have not been highly standardized, perhaps because there has been less need for it. The pressure angles used depend upon the lead angles and must be large enough to avoid undercutting of the worm-gear tooth on the side at which contact ends. A satisfactory tooth depth, which remains in about the right proportion to the lead angle, may be obtained by making the depth a proportion of the axial circular pitch. Table 13–7 summarizes what may be regarded as good practice for pressure angle and tooth depth.

The *face width* F_G of the worm gear should be made equal to the length of a tangent to the worm pitch circle between its points of intersection with the addendum circle, as shown in Figure 13–21.

13–13 Gear Trains

Consider a pinion 2 driving a gear 3. The speed of the driven gear is

$$n_3 = \left|\frac{N_2}{N_3}n_2\right| = \left|\frac{d_2}{d_3}n_2\right| \tag{13–29}$$

where n = revolutions or rev/min

N = number of teeth

d = pitch diameter

Equation (13–29) applies to any gearset no matter whether the gears are spur, helical, bevel, or worm. The absolute-value signs are used to permit complete freedom in choosing positive and negative directions. In the case of spur and parallel helical gears, the directions in the viewing plane ordinarily correspond to the right-hand rule—positive for counterclockwise rotation and negative for clockwise rotation.

Rotational directions are somewhat more difficult to deduce for worm and crossed helical gearsets. Figure 13–22 will be of help in these situations.

The gear train shown in Figure 13–23 is made up of five gears. Considering gear 2 to be the primary driving gear, the speed of gear 6 is

$$n_6 = -\frac{N_2}{N_3}\frac{N_3}{N_4}\frac{N_5}{N_6}n_2 \tag{a}$$

Hence we notice that gear 3 is an idler, that its tooth numbers cancel in Equation (a), and hence that it affects only the direction of rotation of gear 6. We notice,

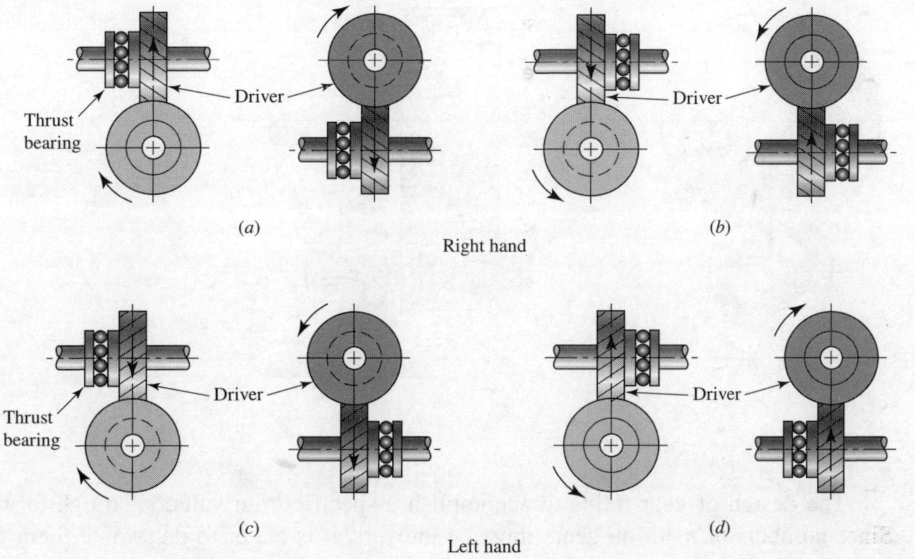

Figure 13–22

Thrust, rotation, and hand relations for crossed helical gears. Note that each pair of drawings refers to a single gearset. These relations also apply to worm gearsets. *(Source: Boston Gear Division, Colfax Corp.)*

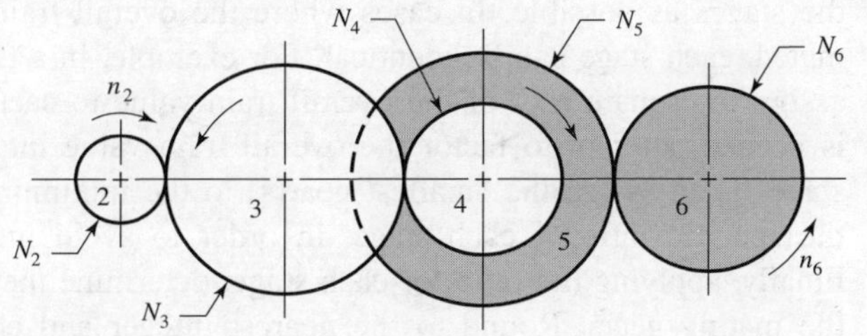

Figure 13–23

A gear train.

furthermore, that gears 2, 3, and 5 are drivers, while 3, 4, and 6 are driven members. We define the *train value e* as

$$e = \pm \frac{\text{product of driving tooth numbers}}{\text{product of driven tooth numbers}} \tag{13–30}$$

Note that pitch diameters can be used in Equation (13–30) as well. When Equation (13–30) is used for spur gears, e is positive if the last gear rotates in the same sense as the first, and negative if the last rotates in the opposite sense. Another way of establishing the sign is to count the number of meshes. If the number is odd, then e is negative. If the number is even, then e is positive. Do not count the mesh when a mesh involves an internal gear such as the ring gear in Figure 13–26.

Now we can write

$$n_L = e n_F \tag{13–31}$$

where n_L is the speed of the last gear in the train and n_F is the speed of the first.

As a rough guideline, a train value of up to 10 to 1 can be obtained with one pair of gears. Greater ratios can be obtained in less space and with fewer dynamic problems by compounding additional pairs of gears. A two-stage compound gear train, such as shown in Figure 13–24, can obtain a train value of up to 100 to 1.

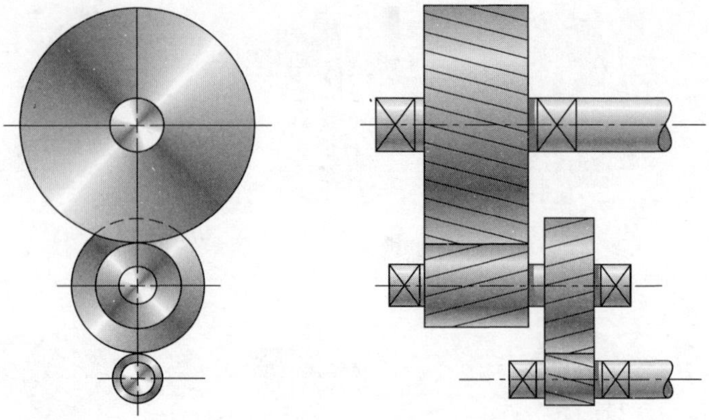

Figure 13–24

A two-stage compound gear train.

The design of gear trains to accomplish a specific train value is straightforward. Since numbers of teeth on gears must be integers, it is better to determine them first, and then obtain pitch diameters second. Determine the number of stages necessary to obtain the overall ratio, then divide the overall ratio into portions to be accomplished in each stage. To minimize package size, keep the portions as evenly divided between the stages as possible. In cases where the overall train value need only be approximated, each stage can be identical. For example, in a two-stage compound gear train, assign the square root of the overall train value to each stage. If an exact train value is needed, attempt to factor the overall train value into integer components for each stage. Then assign the smallest gear(s) to the minimum number of teeth allowed for the specific ratio of each stage, in order to avoid interference (see Section 13–7). Finally, applying the ratio for each stage, determine the necessary number of teeth for the mating gears. Round to the nearest integer and check that the resulting overall ratio is within acceptable tolerance.

EXAMPLE 13–3

A gearbox is needed to provide a 30:1 (± 1 percent) increase in speed, while minimizing the overall gearbox size. Specify appropriate tooth numbers.

Solution

Since the ratio is greater than 10:1, but less than 100:1, a two-stage compound gear train, such as in Figure 13–24, is needed. The portion to be accomplished in each stage is $\sqrt{30} = 5.4772$. For this ratio, assuming a typical 20° pressure angle, the minimum number of teeth to avoid interference is 16, according to Equation (13–11). The number of teeth necessary for the mating gears is

Answer

$$16\sqrt{30} = 87.64 \approx 88$$

From Equation (13–30), the overall train value is

$$e = (88/16)(88/16) = 30.25$$

This is within the 1 percent tolerance. If a closer tolerance is desired, then increase the pinion size to the next integer and try again.

EXAMPLE 13–4

A gearbox is needed to provide an *exact* 30:1 increase in speed, while minimizing the overall gearbox size. Specify appropriate teeth numbers.

Solution

The previous example demonstrated the difficulty with finding integer numbers of teeth to provide an exact ratio. In order to obtain integers, factor the overall ratio into two integer stages.

$$e = 30 = (6)(5)$$

$$N_2/N_3 = 6 \qquad \text{and} \qquad N_4/N_5 = 5$$

With two equations and four unknown numbers of teeth, two free choices are available. Choose N_3 and N_5 to be as small as possible without interference. Assuming a 20° pressure angle, Equation (13–11) gives the minimum as 16.

Then

$$N_2 = 6N_3 = 6(16) = 96$$

$$N_4 = 5N_5 = 5(16) = 80$$

The overall train value is then exact.

$$e = (96/16)(80/16) = (6)(5) = 30$$

It is sometimes desirable for the input shaft and the output shaft of a two-stage compound gear train to be in-line, as shown in Figure 13–25. This configuration is called a *compound reverted gear train*. This requires the distances between the shafts to be the same for both stages of the train, which adds to the complexity of the design task. The distance constraint is

$$d_2/2 + d_3/2 = d_4/2 + d_5/2$$

The diametral pitch relates the diameters and the numbers of teeth, $P = N/d$. Replacing all the diameters gives

$$N_2/(2P) + N_3/(2P) = N_4/(2P) + N_5/(2P)$$

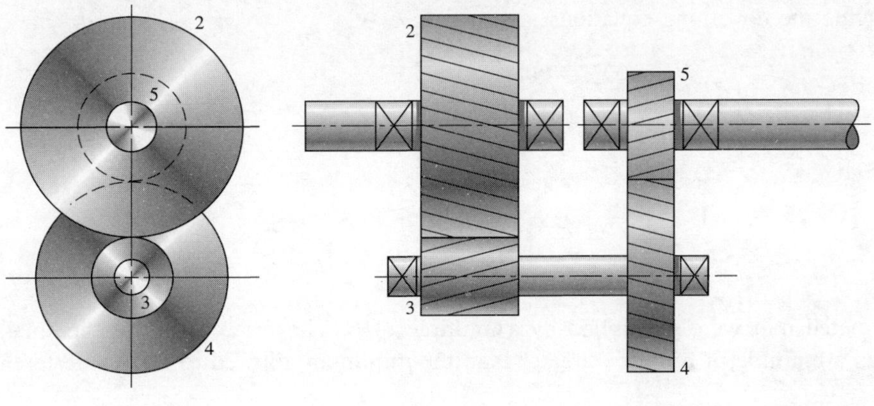

Figure 13–25
A compound reverted gear train.

Assuming a constant diametral pitch in both stages, we have the geometry condition stated in terms of numbers of teeth:

$$N_2 + N_3 = N_4 + N_5$$

This condition must be exactly satisfied, in addition to the previous ratio equations, to provide for the in-line condition on the input and output shafts.

EXAMPLE 13–5

A gearbox is needed to provide an exact 30:1 increase in speed, while minimizing the overall gearbox size. The input and output shafts should be in-line. Specify appropriate teeth numbers.

Solution

The governing equations are

$$N_2/N_3 = 6$$

$$N_4/N_5 = 5$$

$$N_2 + N_3 = N_4 + N_5$$

With three equations and four unknown numbers of teeth, only one free choice is available. Of the two smaller gears, N_3 and N_5, the free choice should be used to minimize N_3 since a greater gear ratio is to be achieved in this stage. To avoid interference, the minimum for N_3 is 16.

Applying the governing equations yields

$$N_2 = 6N_3 = 6(16) = 96$$

$$N_2 + N_3 = 96 + 16 = 112 = N_4 + N_5$$

Substituting $N_4 = 5N_5$ gives

$$112 = 5N_5 + N_5 = 6N_5$$

$$N_5 = 112/6 = 18.67$$

If the train value need only be approximated, then this can be rounded to the nearest integer. But for an exact solution, it is necessary to choose the initial free choice for N_3 such that solution of the rest of the teeth numbers results exactly in integers. This can be done by trial and error, letting $N_3 = 17$, then 18, etc., until it works. Or, the problem can be normalized to quickly determine the minimum free choice. Beginning again, let the free choice be $N_3 = 1$. Applying the governing equations gives

$$N_2 = 6N_3 = 6(1) = 6$$

$$N_2 + N_3 = 6 + 1 = 7 = N_4 + N_5$$

Substituting $N_4 = 5N_5$, we find

$$7 = 5N_5 + N_5 = 6N_5$$

$$N_5 = 7/6$$

This fraction could be eliminated if it were multiplied by a multiple of 6. The free choice for the smallest gear N_3 should be selected as a multiple of 6 that is greater than the minimum allowed to avoid interference.

This would indicate that $N_3 = 18$. Repeating the application of the governing equations for the final time yields

$$N_2 = 6N_3 = 6(18) = 108$$

$$N_2 + N_3 = 108 + 18 = 126 = N_4 + N_5$$

$$126 = 5N_5 + N_5 = 6N_5$$

$$N_5 = 126/6 = 21$$

$$N_4 = 5N_5 = 5(21) = 105$$

Thus,

Answer

$$N_2 = 108$$

$$N_3 = 18$$

$$N_4 = 105$$

$$N_5 = 21$$

Checking, we calculate $e = (108/18)(105/21) = (6)(5) = 30$.

And checking the geometry constraint for the in-line requirement, we calculate

$$N_2 + N_3 = N_4 + N_5$$

$$108 + 18 = 105 + 21$$

$$126 = 126$$

Unusual effects can be obtained in a gear train by permitting some of the gear axes to rotate about others. Such trains are called *planetary,* or *epicyclic, gear trains.* Planetary trains always include a *sun gear,* a *planet carrier* or *arm,* and one or more *planet gears,* as shown in Figure 13–26. Planetary gear trains are unusual mechanisms because they have two degrees of freedom; that is, for constrained motion, a planetary train must have two inputs. For example, in Figure 13–26 these two inputs could be the motion of any two of the elements of the train. We might, say, specify that the sun gear rotates at 100 rev/min clockwise and that the ring gear rotates at 50 rev/min counterclockwise; these are the inputs. The output would be the motion of the arm. In most planetary trains one of the elements is attached to the frame and has no motion. Figure 13–27 shows a planetary train composed of a sun gear 2, an arm or carrier 3, and planet gears 4 and 5. The angular velocity of gear 2 relative to the arm in rev/min is

$$n_{23} = n_2 - n_3 \tag{b}$$

Also, the velocity of gear 5 relative to the arm is

$$n_{53} = n_5 - n_3 \tag{c}$$

Dividing Equation (*c*) by Equation (*b*) gives

$$\frac{n_{53}}{n_{23}} = \frac{n_5 - n_3}{n_2 - n_3} \tag{d}$$

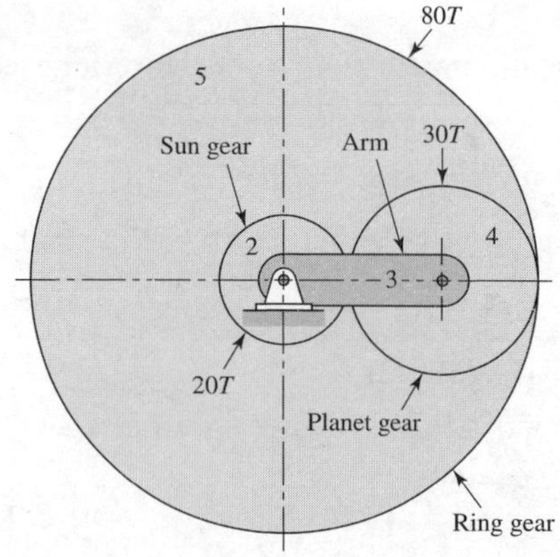

Figure 13–26

A planetary gear train.

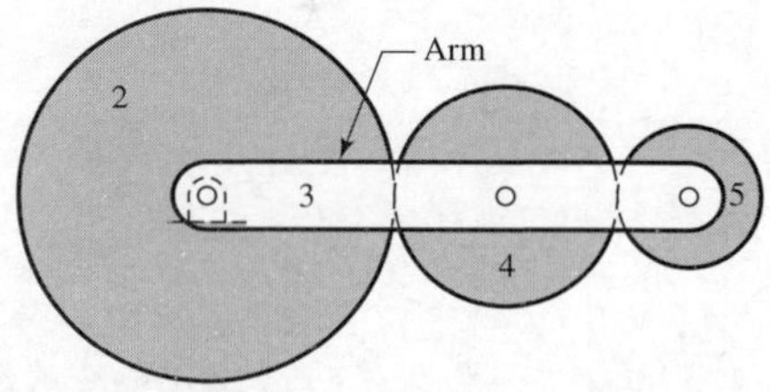

Figure 13–27

A gear train on the arm of a planetary gear train.

Equation (d) expresses the ratio of gear 5 to that of gear 2, and both velocities are taken relative to the arm. Now this ratio is the same and is proportional to the tooth numbers, whether the arm is rotating or not. It is the train value. Therefore, we may write

$$e = \frac{n_5 - n_3}{n_2 - n_3} \tag{e}$$

This equation can be used to solve for the output motion of any planetary train. It is more conveniently written in the form

$$e = \frac{n_L - n_A}{n_F - n_A} \tag{13–32}$$

where n_F = rev/min of first gear in planetary train
n_L = rev/min of last gear in planetary train
n_A = rev/min of arm

EXAMPLE 13–6

In Figure 13–26 the sun gear is the input, and it is driven clockwise at 100 rev/min. The ring gear is held stationary by being fastened to the frame. Find the rev/min and direction of rotation of the arm and gear 4.

Solution

Let $n_F = n_2 = -100$ rev/min, and $n_L = n_5 = 0$. For e, unlock gear 5 and fix the arm. Then, planet gear 4 and ring gear 5 rotate in the *same* direction, *opposite* of sun gear 2. Thus, e is negative. Alternatively, the number of meshes, *not* counting the one internal mesh is one, again making e negative and

$$e = -\left(\frac{N_2}{N_4}\right)\left(\frac{N_4}{N_5}\right) = -\left(\frac{20}{30}\right)\left(\frac{30}{80}\right) = -0.25$$

Substituting this value in Equation (13–32) gives

$$-0.25 = \frac{0 - n_A}{(-100) - n_A}$$

or

Answer $$n_A = -20 \text{ rev/min} = 20 \text{ rev/min clockwise}$$

To obtain the speed of gear 4, we follow the procedure outlined by Equations (*b*), (*c*), and (*d*). Thus

$$n_{43} = n_4 - n_3 \qquad n_{23} = n_2 - n_3$$

and so

$$\frac{n_{43}}{n_{23}} = \frac{n_4 - n_3}{n_2 - n_3} \tag{1}$$

But

$$\frac{n_{43}}{n_{23}} = -\frac{20}{30} = -\frac{2}{3} \tag{2}$$

Substituting the known values in Equation (1) gives

$$-\frac{2}{3} = \frac{n_4 - (-20)}{(-100) - (-20)}$$

Solving gives

Answer $$n_4 = +33\tfrac{1}{3} \text{ rev/min} = 33\tfrac{1}{3} \text{ rev/min counterclockwise}$$

13–14 Force Analysis—Spur Gearing

Before beginning the force analysis of gear trains, let us agree on the notation to be used. Beginning with the numeral 1 for the frame of the machine, we shall designate the input gear as gear 2, and then number the gears successively 3, 4, etc., until we arrive at the last gear in the train. Next, there may be several shafts involved, and usually one or two gears are mounted on each shaft as well as other elements. We shall designate the shafts, using lowercase letters of the alphabet, *a*, *b*, *c*, etc.

With this notation we can now speak of the force exerted by gear 2 against gear 3 as F_{23}. The force of gear 2 against shaft *a* is F_{2a}. We can also write F_{a2} to mean the force of shaft *a* against gear 2. Unfortunately, it is also necessary to use superscripts to indicate directions. The coordinate directions will usually be indicated by the *x*, *y*, and *z* coordinates, and the radial and tangential directions by superscripts *r* and *t*. With this notation, F^t_{43} is the tangential component of the force of gear 4 acting against gear 3.

Figure 13–28*a* shows a pinion mounted on shaft a rotating clockwise at n_2 rev/min and driving a gear on shaft *b* at n_3 rev/min. The reactions between the mating teeth occur along the pressure line. In Figure 13–28*b* the pinion has been separated from the gear and the shaft, and their effects have been replaced by forces. F_{a2} and T_{a2} are the force and torque, respectively, exerted by shaft *a* against pinion 2. F_{32} is the force exerted by gear 3 against the pinion. Using a similar approach, we obtain the free-body diagram of the gear shown in Figure 13–28*c*.

Figure 13–28

Free-body diagrams of the forces and moments acting upon two gears of a simple gear train.

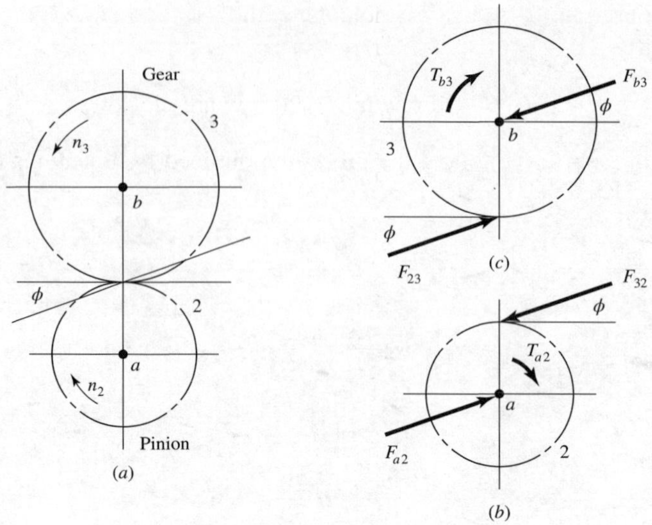

Figure 13–29

Resolution of gear forces.

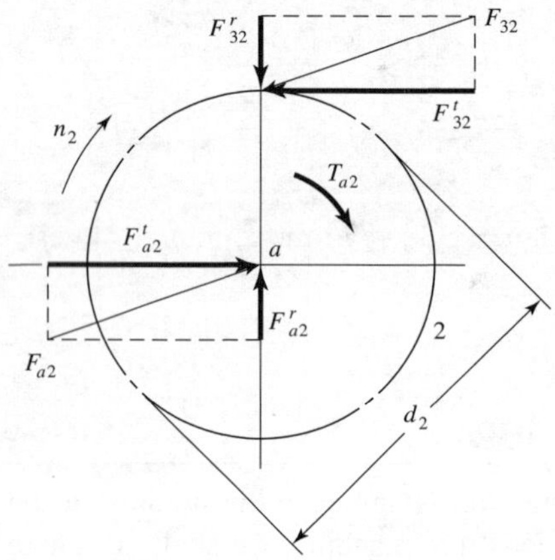

In Figure 13–29, the free-body diagram of the pinion has been redrawn and the forces have been resolved into tangential and radial components. We now define

$$W_t = F^t_{32} \tag{a}$$

as the *transmitted load*. This tangential load is really the useful component, because the radial component F^r_{32} serves no useful purpose. It does not transmit power. The applied torque and the transmitted load are seen to be related by the equation

$$T = \frac{d}{2} W_t \tag{b}$$

where we have used $T = T_{a2}$ and $d = d_2$ to obtain a general relation.

The power H transmitted through a rotating gear can be obtained from the standard relationship of the product of torque T and angular velocity ω.

$$H = T\omega = (W_t d/2)\omega \tag{13–33}$$

While any units can be used in this equation, the units of the resulting power will obviously be dependent on the units of the other parameters. It will often be desirable to work with the power in either horsepower or kilowatts, and appropriate conversion factors should be used.

Since meshed gears are reasonably efficient, with losses of less than 2 percent, the power is generally treated as constant through the mesh. Consequently, with a pair of meshed gears, Equation (13–33) will give the same power regardless of which gear is used for d and ω.

Gear data is often tabulated using *pitch-line velocity*, which is the linear velocity of a point on the gear at the radius of the pitch circle; thus $V = (d/2)\omega$. Converting this to customary units gives

$$V = \pi dn/60\ 000 \tag{13–34}$$

where V = pitch-line velocity, m/s
d = gear diameter, mm
n = gear speed, rev/min

Many gear design problems will specify the power and speed, so it is convenient to solve Equation (13–33) for W_t. With the pitch-line velocity and appropriate conversion factors incorporated, Equation (13–33) can be rearranged and expressed in SI units as

$$W_t = \frac{60\ 000\ H}{\pi dn} \tag{13–35}$$

where W_t = transmitted load, kN
H = power, kW
d = gear diameter, mm
n = speed, rev/min

EXAMPLE 13–7

Pinion 2 in Figure 13–30*a* runs at 1750 rev/min and transmits 2.5 kW to idler gear 3. The teeth are cut on the 20° full-depth system and have a module of m = 2.5 mm. Draw a free-body diagram of gear 3 and show all the forces that act upon it.

Solution

The pitch diameters of gears 2 and 3 are

$$d_2 = N_2 m = 20(2.5) = 50 \text{ mm}$$

$$d_3 = N_3 m = 50(2.5) = 125 \text{ mm}$$

From Equation (13–35) we find the transmitted load to be

$$W_t = \frac{60\ 000 H}{\pi d_2 n} = \frac{60\ 000(2.5)}{\pi(50)(1750)} = 0.546 \text{ kN}$$

Figure 13–30

A gear train containing an idler gear. (*a*) The gear train. (*b*) Free-body of the idler gear.

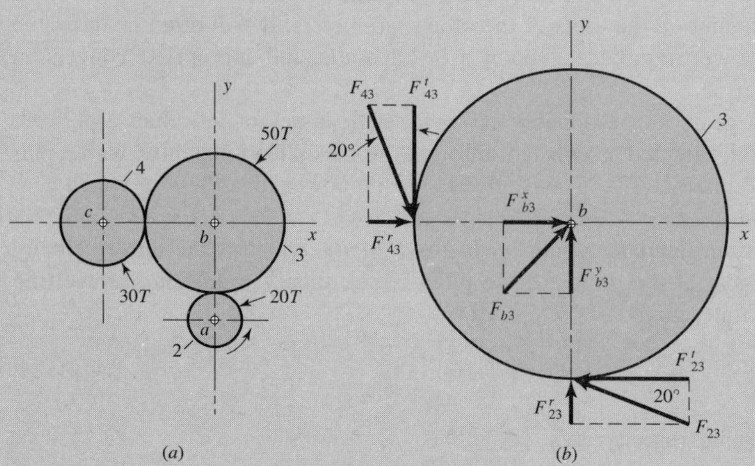

Thus, the tangential force of gear 2 on gear 3 is $F^t_{23} = 0.546$ kN, as shown in Figure 13–30*b*. Therefore

$$F^r_{23} = F^t_{23} \tan 20° = (0.546) \tan 20° = 0.199 \text{ kN}$$

and so

$$F_{23} = \frac{F^t_{23}}{\cos 20°} = \frac{0.546}{\cos 20°} = 0.581 \text{ kN}$$

Since gear 3 is an idler, it transmits no power (torque) to its shaft, and so the tangential reaction of gear 4 on gear 3 is also equal to W_t. Therefore

$$F^t_{43} = 0.546 \text{ kN} \qquad F^r_{43} = 0.199 \text{ kN} \qquad F_{43} = 0.581 \text{ kN}$$

and the directions are shown in Figure 13–30*b*.

The shaft reactions in the x and y directions are

$$F^x_{b3} = -(F^t_{23} + F^r_{43}) = -(-0.546 + 0.199) = 0.347 \text{ kN}$$

$$F^y_{b3} = -(F^r_{23} + F^t_{43}) = -(0.199 - 0.546) = 0.347 \text{ kN}$$

The resultant shaft reaction is

$$F_{b3} = \sqrt{(0.347)^2 + (0.347)^2} = 0.491 \text{ kN}$$

These are shown on the figure.

13–15 Force Analysis—Bevel Gearing

In determining shaft and bearing loads for bevel-gear applications, the usual practice is to use the tangential or transmitted load that would occur if all the forces were concentrated at the midpoint of the tooth. While the actual resultant occurs somewhere between the midpoint and the large end of the tooth, there is only a small error in making this assumption. For the transmitted load, this gives

$$W_t = \frac{T}{r_{\text{av}}} \tag{13–36}$$

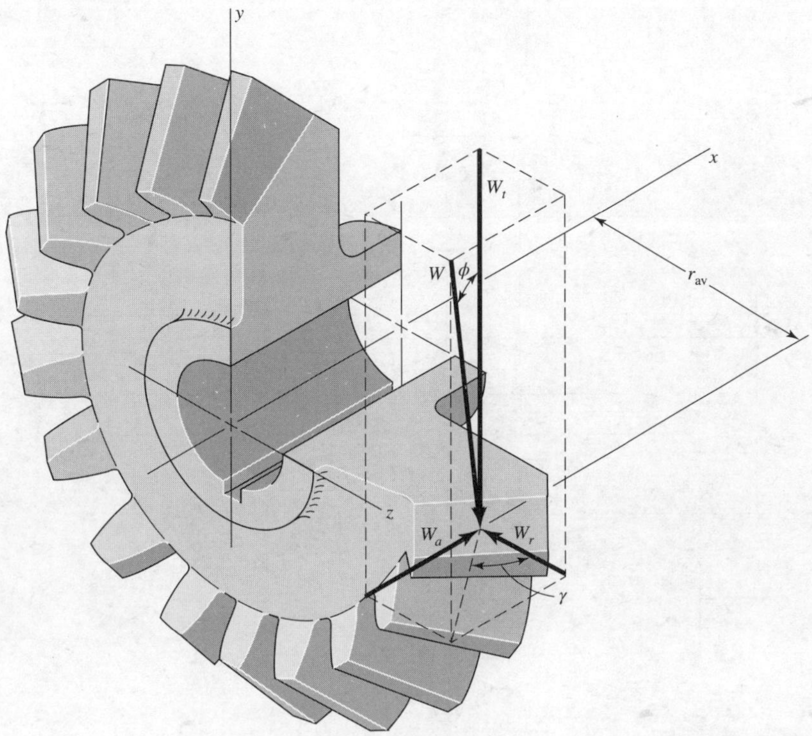

Figure 13–31

Bevel-gear tooth forces.

where T is the torque and r_{av} is the pitch radius at the midpoint of the tooth for the gear under consideration.

The forces acting at the center of the tooth are shown in Figure 13–31. The resultant force W has three components: a tangential force W_t, a radial force W_r, and an axial force W_a. From the trigonometry of the figure,

$$W_r = W_t \tan \phi \cos \gamma$$
$$W_a = W_t \tan \phi \sin \gamma \tag{13–37}$$

The three forces W_t, W_r, and W_a are at right angles to each other and can be used to determine the bearing loads by using the methods of statics.

EXAMPLE 13–8

The bevel pinion in Figure 13–32*a* rotates at 600 rev/min in the direction shown and transmits 3.5 kW to the gear. The mounting distances, the location of all bearings, and the average pitch radii of the pinion and gear are shown in the figure. For simplicity, the teeth have been replaced by pitch cones. Bearings A and C should take the thrust loads. Find the bearing forces on the gearshaft.

Solution

The pitch angles are

$$\gamma = \tan^{-1}\left(\frac{3}{9}\right) = 18.4^\circ \qquad \Gamma = \tan^{-1}\left(\frac{9}{3}\right) = 71.6^\circ$$

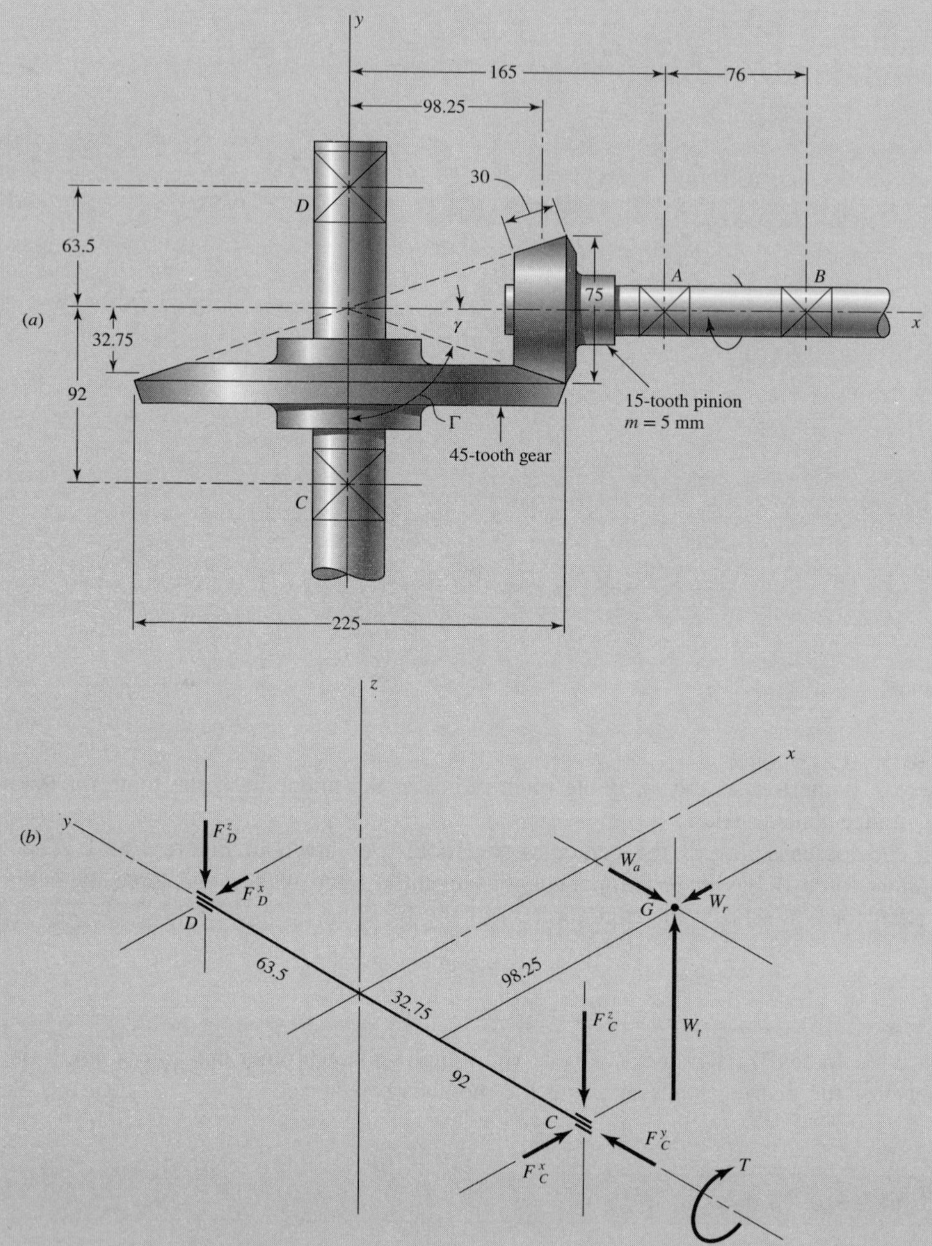

Figure 13–32

(*a*) Bevel gearset of Example 13–8.
(*b*) Free-body diagram of shaft *CD*. Dimensions in millimeters.

The pitch-line velocity corresponding to the average pitch radius is

$$V = \frac{2\pi r_P n}{60\ 000} = \frac{2\pi(32.75)(600)}{60\ 000} = 2.06 \text{ m/s}$$

Therefore the transmitted load is

$$W_t = \frac{H}{V} = \frac{3500}{2.06} = 1.7 \text{ kN}$$

and from Equation (13–37), with Γ replacing γ, we have

$$W_r = W_t \tan \phi \cos \Gamma = 1.7 \tan 20^\circ \cos 71.6^\circ = 0.20 \text{ kN}$$

$$W_a = W_t \tan \phi \sin \Gamma = 1.7 \tan 20^\circ \sin 71.6^\circ = 0.59 \text{ kN}$$

where W_t acts in the positive z direction, W_r in the $-x$ direction, and W_a in the $-y$ direction, as illustrated in the isometric sketch of Figure 13–32*b*.

In preparing to take a sum of the moments about bearing D, define the position vector from D to G as

$$\mathbf{R}_G = 98.25\mathbf{i} - (63.5 + 32.75)\mathbf{j} = 98.25\mathbf{i} - 96.25\mathbf{j}$$

We shall also require a vector from D to C:

$$\mathbf{R}_C = -(63.5 + 92)\mathbf{j} = -155.5\mathbf{j}$$

Then, summing moments about D gives

$$\mathbf{R}_G \times \mathbf{W} + \mathbf{R}_C \times \mathbf{F}_C + \mathbf{T} = \mathbf{0} \tag{1}$$

When we place the details in Equation (1), we get

$$(98.25\mathbf{i} - 3.793\mathbf{j}) \times (-0.20\mathbf{i} - 0.59\mathbf{j} + 1.7\mathbf{k}) + (-155.5\mathbf{j}) \times \left(F_C^x\mathbf{i} + F_C^y\mathbf{j} + F_C^z\mathbf{k}\right) + T\mathbf{j} = \mathbf{0} \tag{2}$$

After the two cross products are taken, the equation becomes

$$(-163.6\mathbf{i} - 167.0\mathbf{j} - 77.2\mathbf{k}) + \left(-155.5F_C^z\mathbf{i} + 155.5F_C^x\mathbf{k}\right) + T\mathbf{j} = \mathbf{0}$$

from which

$$\mathbf{T} = 167\mathbf{j} \text{ N} \cdot \text{m} \qquad F_C^x = 0.49 \text{ kN} \qquad F_C^z = -1.05 \text{ kN} \tag{3}$$

Now sum the forces to zero. Thus

$$\mathbf{F}_D + \mathbf{F}_C + \mathbf{W} = \mathbf{0} \tag{4}$$

When the details are inserted, Equation (4) becomes

$$\left(F_D^x\mathbf{i} + F_D^z\mathbf{k}\right) + \left(118\mathbf{i} + F_C^y\mathbf{j} - 251\mathbf{k}\right) + (-46.6\mathbf{i} - 140\mathbf{j} + 406\mathbf{k}) = \mathbf{0} \tag{5}$$

First we see that $F_C^y = 0.59$ kN, and so

Answer
$$\mathbf{F}_C = 0.49\mathbf{i} + 0.59\mathbf{j} - 1.5\mathbf{k} \text{ kN}$$

Then, from Equation (5),

Answer
$$\mathbf{F}_D = -0.29\mathbf{i} - 0.65\mathbf{k} \text{ kN}$$

These are all shown in Figure 13–32*b* in the proper directions. The analysis for the pinion shaft is quite similar.

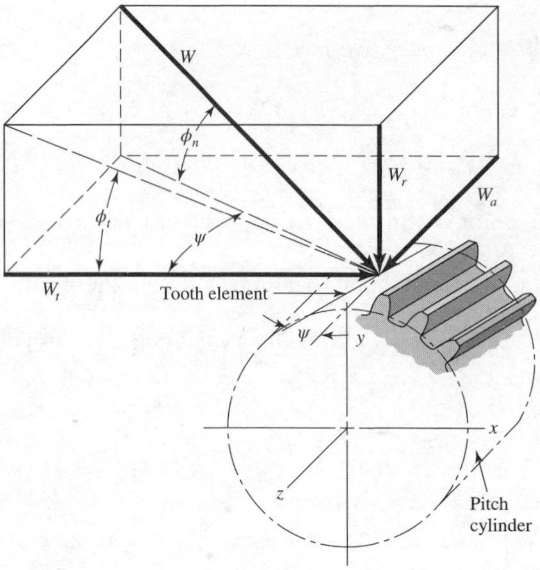

Figure 13–33

Tooth forces acting on a right-hand helical gear.

13–16 Force Analysis—Helical Gearing

Figure 13–33 is a three-dimensional view of the forces acting against a helical-gear tooth. The point of application of the forces is in the pitch plane and in the center of the gear face. From the geometry of the figure, the three components of the total (normal) tooth force W are

$$\begin{aligned} W_r &= W \sin \phi_n \\ W_t &= W \cos \phi_n \cos \psi \\ W_a &= W \cos \phi_n \sin \psi \end{aligned} \tag{13–38}$$

where W = total force

W_r = radial component

W_t = tangential component, also called the transmitted load

W_a = axial component, also called the thrust load

Usually W_t is given and the other forces are desired. In this case, it is not difficult to discover that

$$\begin{aligned} W_r &= W_t \tan \phi_t \\ W_a &= W_t \tan \psi \\ W &= \frac{W_t}{\cos \phi_n \cos \psi} \end{aligned} \tag{13–39}$$

EXAMPLE 13–9

In Figure 13–34 an electric motor transmits 0.75 kW at 1800 rev/min in the clockwise direction, as viewed from the positive x axis. Keyed to the motor shaft is an 18-tooth helical pinion having a normal pressure angle of 20°, a helix angle of 30°, and a normal module of 2 mm. The hand of the helix is shown in the figure. Make a three-dimensional sketch of the motor shaft and pinion, and show the forces acting on the pinion and the bearing reactions at A and B. The thrust should be taken out at A.

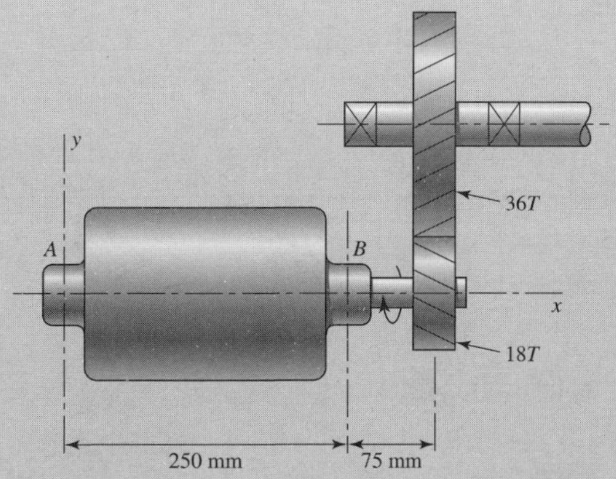

Figure 13–34

The motor and gear train of Example 13–9.

Solution

From Equation (13–19) we find

$$\phi_t = \tan^{-1}\frac{\tan\phi_n}{\cos\psi} = \tan^{-1}\frac{\tan 20^\circ}{\cos 30^\circ} = 22.8^\circ$$

Also, $m_t = m_n/\cos\psi = 2/\cos 30^\circ = 2.31$ mm. Therefore the pitch diameter of the pinion is $d_p = 18(2.31) = 41.6$ mm. The pitch-line velocity is

$$V = \frac{\pi dn}{60\,000} = \frac{\pi(41.6)(1800)}{60\,000} = 3.9 \text{ m/s}$$

The transmitted load is

$$W_t = \frac{H}{V} = \frac{750}{3.9} = 0.19 \text{ kN}$$

From Equation (13–39) we find

$$W_r = W_t \tan\phi_t = (0.19)\tan 22.8^\circ = 0.08 \text{ kN}$$

$$W_a = W_t \tan\psi = (0.19)\tan 30^\circ = 0.11 \text{ kN}$$

$$W = \frac{W_t}{\cos\phi_n \cos\psi} = \frac{0.19}{\cos 20^\circ \cos 30^\circ} = 0.23 \text{ kN}$$

These three forces, $W_r = 0.08$ kN in the $-y$ direction, $W_a = 0.11$ kN in the $-x$ direction, and $W_t = 0.19$ kN in the $+z$ direction, are shown acting at point C in Figure 13–35. We assume bearing reactions at A and B as shown. Then $F_A^x = W_a = 0.11$ kN. Taking moments about the z axis,

$$-(0.08)(325) + (0.11)(20.8) + 250F_B^y = 0$$

or $F_B^y = 0.095$ kN. Summing forces in the y direction then gives $F_A^y = 0.01$ kN. Taking moments about the y axis, next

$$250F_B^z - (0.19)(325) = 0$$

or $F_B^z = 0.25$ kN. Summing forces in the z direction and solving gives $F_A^z = 0.06$ kN. Also, the torque is $T = W_t d_p/2 = (0.19)(41.6/2) = 3.95$ N · m.

For comparison, solve the problem again using vectors. The force at C is

$$\mathbf{W} = -0.11\mathbf{i} - 0.08\mathbf{j} + 0.19\mathbf{k} \text{ kN}$$

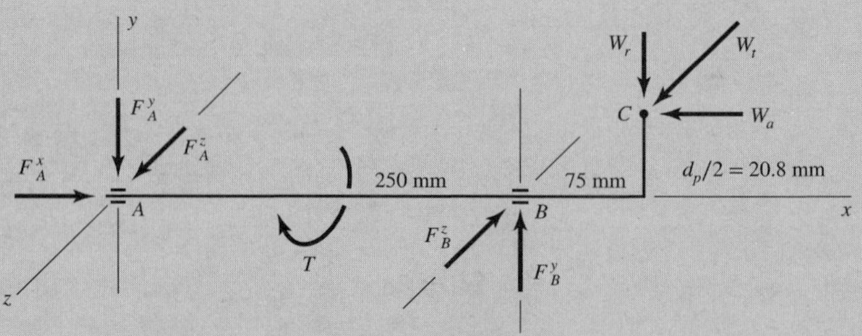

Figure 13–35
Free-body diagram of motor shaft of Example 13–9. Forces in kN.

Position vectors to B and C from origin A are

$$\mathbf{R}_B = 250\mathbf{i} \qquad \mathbf{R}_C = 325\mathbf{i} + 20.8\mathbf{j}$$

Taking moments about A, we have

$$\mathbf{R}_B \times \mathbf{F}_B + \mathbf{T} + \mathbf{R}_C \times \mathbf{W} = \mathbf{0}$$

Using the directions assumed in Figure 13–35 and substituting values gives

$$250\mathbf{i} \times \left(F_B^y\mathbf{j} - F_B^z\mathbf{k}\right) - T\mathbf{i} + (325\mathbf{i} + 20.8\mathbf{j}) \times (-0.11\mathbf{i} - 0.08\mathbf{j} + 0.19\mathbf{k}) = \mathbf{0}$$

When the cross products are evaluated we get

$$\left(250F_B^y\mathbf{k} + 250F_B^z\mathbf{j}\right) - T\mathbf{i} + (3.95\mathbf{i} - 61.8\mathbf{j} - 23.7\mathbf{k}) = \mathbf{0}$$

obtaining $T = 3.95$ N · m, $F_B^y = 0.09$ kN, and $F_B^z = 0.25$ kN.

Next,

$$\mathbf{F}_A = -\mathbf{F}_B - \mathbf{W}, \text{ and so } \mathbf{F}_A = 0.11\mathbf{i} - 0.01\mathbf{j} + 0.06\mathbf{k} \text{ kN.}$$

13–17 Force Analysis—Worm Gearing

If friction is neglected, then the only force exerted by the gear will be the force W, shown in Figure 13–36, having the three orthogonal components W^x, W^y, and W^z. From the geometry of the figure, we see that

$$\begin{aligned} W^x &= W \cos \phi_n \sin \lambda \\ W^y &= W \sin \phi_n \\ W^z &= W \cos \phi_n \cos \lambda \end{aligned} \tag{13–40}$$

We now use the subscripts W and G to indicate forces acting against the worm and gear, respectively. We note that W^y is the separating, or radial, force for both the worm and the gear. The tangential force on the worm is W^x and is W^z on the gear, assuming a 90° shaft angle. The axial force on the worm is W^z, and on the gear, W^x. Since the gear forces are opposite to the worm forces, we can summarize these relations by writing

$$\begin{aligned} W_{Wt} &= -W_{Ga} = W^x \\ W_{Wr} &= -W_{Gr} = W^y \\ W_{Wa} &= -W_{Gt} = W^z \end{aligned} \tag{13–41}$$

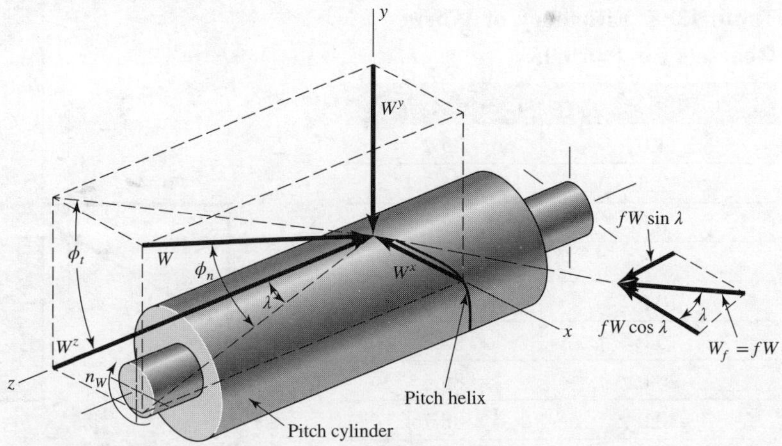

Figure 13–36

Drawing of the pitch cylinder of a worm, showing the forces exerted upon it by the worm gear.

It is helpful in using Equation (13–40) and also Equation (13–41) to observe that *the gear axis is parallel to the x direction and the worm axis is parallel to the z direction* and that we are employing a right-handed coordinate system.

In our study of spur-gear teeth we have learned that the motion of one tooth relative to the mating tooth is primarily a rolling motion; in fact, when contact occurs at the pitch point, the motion is pure rolling. In contrast, the relative motion between worm and worm-gear teeth is pure sliding, and so we must expect that friction plays an important role in the performance of worm gearing. By introducing a coefficient of friction f, we can develop another set of relations similar to those of Equation (13–40). In Figure 13–36 we see that the force W acting normal to the worm-tooth profile produces a frictional force $W_f = fW$, having a component $fW \cos \lambda$ in the negative x direction and another component $fW \sin \lambda$ in the positive z direction. Equation (13–40) therefore becomes

$$
\begin{aligned}
W^x &= W(\cos \phi_n \sin \lambda + f \cos \lambda) \\
W^y &= W \sin \phi_n \\
W^z &= W(\cos \phi_n \cos \lambda - f \sin \lambda)
\end{aligned}
\tag{13–42}
$$

Equation (13–41), of course, still applies.

Inserting $-W_{Gt}$ from Equation (13–41) for W^z in Equation (13–42) and multiplying both sides by f, we find the frictional force W_f to be

$$
W_f = fW = \frac{fW_{Gt}}{f \sin \lambda - \cos \phi_n \cos \lambda} \tag{13–43}
$$

A useful relation between the two tangential forces, W_{Wt} and W_{Gt}, can be obtained by equating the first and third parts of Equations (13–41) and (13–42) and eliminating W. The result is

$$
W_{Wt} = W_{Gt} \frac{\cos \phi_n \sin \lambda + f \cos \lambda}{f \sin \lambda - \cos \phi_n \cos \lambda} \tag{13–44}
$$

Efficiency η can be defined by using the equation

$$
\eta = \frac{W_{Wt} \text{ (without friction)}}{W_{Wt} \text{ (with friction)}} \tag{a}
$$

Table 13–8 Efficiency of Worm Gearsets for $f = 0.05$

Lead Angle λ, deg	Efficiency η, %
1.0	25.2
2.5	45.7
5.0	62.6
7.5	71.3
10.0	76.6
15.0	82.7
20.0	85.6
30.0	88.7

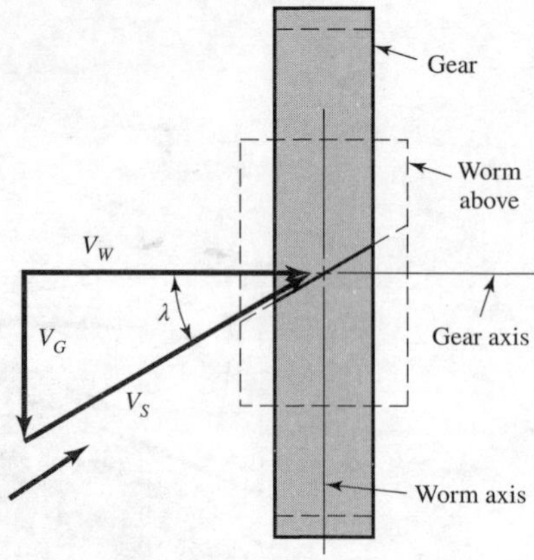

Figure 13–37

Velocity components in worm gearing.

Substitute Equation (13–44) with $f = 0$ in the numerator of Equation (a) and the same equation in the denominator. After some rearranging, you will find the efficiency to be

$$\eta = \frac{\cos \phi_n - f \tan \lambda}{\cos \phi_n + f \cot \lambda} \tag{13–45}$$

Selecting a typical value of the coefficient of friction, say $f = 0.05$, and the pressure angles shown in Table 13–7, we can use Equation (13–45) to get some useful design information. Solving this equation for lead angles from 1° to 30° gives the interesting results shown in Table 13–8.

Many experiments have shown that the coefficient of friction is dependent on the relative or sliding velocity. In Figure 13–37, V_G is the pitch-line velocity of the gear and V_W the pitch-line velocity of the worm. Vectorially, $\mathbf{V}_W = \mathbf{V}_G + \mathbf{V}_S$; consequently, the sliding velocity is

$$V_S = \frac{V_W}{\cos \lambda} \tag{13–46}$$

Published values of the coefficient of friction vary as much as 20 percent, undoubtedly because of the differences in surface finish, materials, and lubrication. The values on the chart of Figure 13–38 are representative and indicate the general trend.

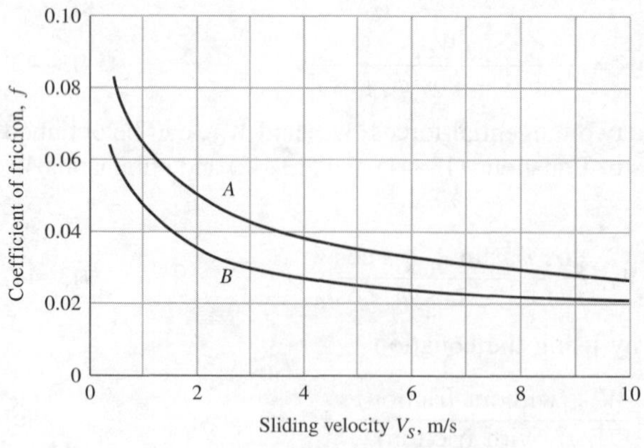

Figure 13–38

Representative values of the coefficient of friction for worm gearing. These values are based on good lubrication. Use curve B for high-quality materials, such as a case-hardened steel worm mating with a phosphor-bronze gear. Use curve A when more friction is expected, as with a cast-iron worm mating with a cast-iron worm gear.

EXAMPLE 13–10

A 2-tooth right-hand worm transmits 0.75 kW at 1200 rev/min to a 30-tooth worm gear. The gear has a transverse module of 4 mm and a face width of 25 mm. The worm has a pitch diameter of 50 mm and a face width of 62.5. The normal pressure angle is 14.5°. The materials and quality of the gearing to be used are such that curve *B* of Figure 13–38 should be used to obtain the coefficient of friction.
(*a*) Find the axial pitch, the center distance, the lead, and the lead angle.
(*b*) Figure 13–39 is a drawing of the worm gear oriented with respect to the coordinate system described earlier in this section; the gear is supported by bearings *A* and *B*. Find the forces exerted by the bearings against the worm-gear shaft, and the output torque.

Solution
(*a*) The axial pitch is the same as the transverse circular pitch of the gear, which is

Answer
$$p_x = p_t = m\pi = 4\pi = 12.57 \text{ mm}$$

The pitch diameter of the gear is $d_G = N_G m = 30(4) = 120$ mm. Therefore, the center distance is

Answer
$$C = \frac{d_W + d_G}{2} = \frac{50 + 120}{2} = 85 \text{ mm}$$

From Equation (13–27), the lead is

$$L = p_x N_W = (12.57)(2) = 25.14 \text{ mm}$$

Answer
Also using Equation (13–28), we find

Answer
$$\lambda = \tan^{-1}\frac{L}{\pi d_W} = \tan^{-1}\frac{25.14}{\pi(50)} = 9.09°$$

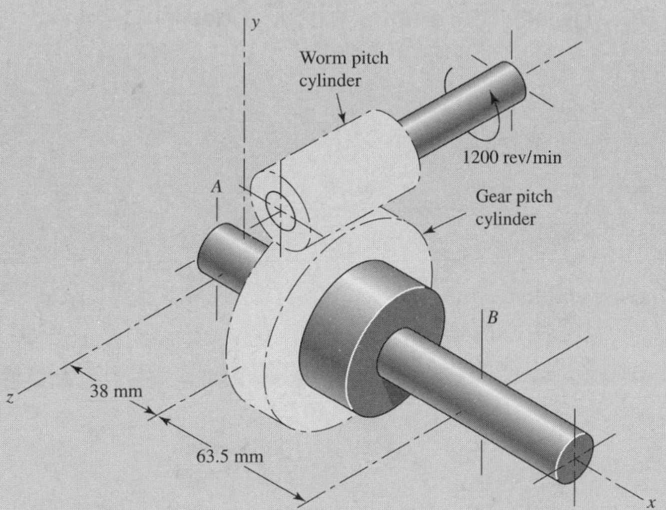

Figure 13–39
The pitch cylinders of the worm gear train of Example 13–10.

(*b*) Using the right-hand rule for the rotation of the worm, you will see that your thumb points in the positive z direction. Now use the bolt-and-nut analogy (the worm is right-handed, as is the screw thread of a bolt), and turn the bolt clockwise with the right hand while preventing nut rotation with the left. The nut will move axially along the bolt toward your right hand. Therefore the surface of the gear (Figure 13–39) in contact with the worm will move in the negative z direction. Thus, viewing the gear in the negative x direction, the gear rotates clockwise about the x axis

The pitch-line velocity of the worm is

$$V_W = \frac{\pi d_W n_W}{60\ 000} = \frac{\pi(50)(1200)}{60\ 000} = 3.14 \text{ m/s}$$

The speed of the gear is $n_G = \left(\frac{2}{30}\right)(1200) = 80$ rev/min. Therefore the pitch-line velocity of the gear is

$$V_G = \frac{\pi d_G n_G}{60\ 000} = \frac{\pi(120)(80)}{60\ 000} = 0.503 \text{ m/s}$$

Then, from Equation (13–46), the sliding velocity V_S is found to be

$$V_S = \frac{V_W}{\cos \lambda} = \frac{3.14}{\cos 9.09^\circ} = 3.18 \text{ m/s}$$

Getting to the forces now, we begin with the horsepower formula

$$W_{Wt} = \frac{H}{V_W} = \frac{750}{3.18} = 238.85 \text{ N}$$

This force acts in the negative x direction, the same as in Figure 13–36. Using Figure 13–38, we find $f = 0.03$. Then, the first equation of Equation (13–42) gives

$$W = \frac{W^x}{\cos \phi_n \sin \lambda + f \cos \lambda}$$

$$= \frac{238.85}{\cos 14.5^\circ \sin 9.09^\circ + 0.03 \cos 9.09^\circ} = 1308.2 \text{ N}$$

Also, from Equation (13–42),

$$W^y = W \sin \phi_n = 1308 \sin 14.5^\circ = 327.5 \text{ N}$$

$$W^z = W(\cos \phi_n \cos \lambda - f \sin \lambda)$$

$$= 1308.2(\cos 14.5^\circ \cos 9.09^\circ - 0.03 \sin 9.09^\circ) = 1244.4 \text{ N}$$

We now identify the components acting on the gear as

$$W_{Ga} = -W^x = 238.85 \text{ N}$$

$$W_{Gr} = -W^y = -327.5 \text{ N}$$

$$W_{Gt} = -W^z = -1244.4 \text{ N}$$

A free-body diagram showing the forces and torsion acting on the gearshaft is shown in Figure 13–40.

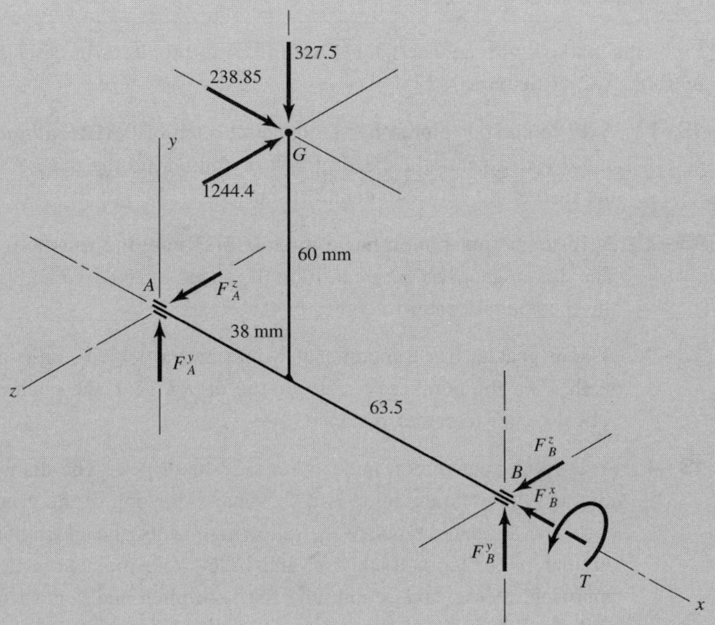

Figure 13–40

Free-body diagram for Example 13–10. Forces are given in N.

We shall make B a thrust bearing in order to place the gearshaft in compression. Thus, summing forces in the x direction gives

Answer $$F_B^x = -238.85 \text{ N}$$

Taking moments about the z axis, we have

Answer $$-(238.85)(60) - (327.5)(38) + 101.5F_B^y = 0 \qquad F_B^y = 263.8 \text{ N}$$

Taking moments about the y axis,

Answer $$(1244.4)(38) - 101.5F_B^z = 0 \qquad F_B^z = 465.9 \text{ N}$$

Summing forces in the y direction,

Answer $$-327.5 + 263.8 + F_A^y = 0 \qquad F_A^y = 63.7 \text{ N}$$

Similarly, summing forces in the z direction,

Answer $$-1244.4 + 465.9 + F_A^z = 0 \qquad F_A^z = 778.5 \text{ N}$$

We still have one more equation to write. Summing moments about x,

Answer $$-(1244.4)(60) + T = 0 \qquad T = 74.66 \text{ N} \cdot \text{m}$$

It is because of the frictional loss that this output torque is less than the product of the gear ratio and the input torque.

PROBLEMS

Problems marked with an asterisk (*) are linked to problems in other chapters, as summarized in Table 1–2 of Section 1–17.

13–1 A 17-tooth spur pinion has a diametral pitch of 0.31 teeth/mm, runs at 1120 rev/min, and drives a gear at a speed of 544 rev/min. Find the number of teeth on the gear and the theoretical center-to-center distance.

13–2 A 15-tooth spur pinion has a module of 3 mm and runs at a speed of 1600 rev/min. The driven gear has 60 teeth. Find the speed of the driven gear, the circular pitch, and the theoretical center-to-center distance.

13–3 A spur gearset has a module of 6 mm and a velocity ratio of 4. The pinion has 16 teeth. Find the number of teeth on the driven gear, the pitch diameters, and the theoretical center-to-center distance.

13–4 A 21-tooth spur pinion mates with a 28-tooth gear. The diametral pitch is 0.12 teeth/mm and the pressure angle is 20°. Make a drawing of the gears showing one tooth on each gear. Find and tabulate the following results: the addendum, dedendum, clearance, circular pitch, tooth thickness, and base-circle diameters; the lengths of the arc of approach, recess, and action; and the base pitch and contact ratio.

13–5 A 20° straight-tooth bevel pinion having 14 teeth and module of 4 mm drives a 32-tooth gear. The two shafts are at right angles and in the same plane. Find:
(*a*) The cone distance
(*b*) The pitch angles
(*c*) The pitch diameters
(*d*) The face width

13–6 A parallel helical gearset uses a 20-tooth pinion driving a 36-tooth gear. The pinion has a right-hand helix angle of 30°, a normal pressure angle of 25°, and a normal module of 6 mm. Find:
(*a*) The normal, transverse, and axial circular pitches
(*b*) The normal base circular pitch
(*c*) The transverse diametral pitch and the transverse pressure angle
(*d*) The addendum, dedendum, and pitch diameter of each gear

13–7 A parallel helical gearset consists of a 19-tooth pinion driving a 57-tooth gear. The pinion has a left-hand helix angle of 30°, a normal pressure angle of 20°, and a normal module of 2.5 mm. Find:
(*a*) The normal, transverse, and axial circular pitches
(*b*) The transverse diametral pitch and the transverse pressure angle
(*c*) The addendum, dedendum, and pitch diameter of each gear

13–8 To avoid the problem of interference in a pair of spur gears using a 20° pressure angle, specify the minimum number of teeth allowed on the pinion for each of the following gear ratios.
(*a*) 2 to 1
(*b*) 3 to 1
(*c*) 4 to 1
(*d*) 5 to 1

13–9 Repeat Problem 13–8 with a 25° pressure angle.

13–10 For a spur gearset with $\phi = 20°$, while avoiding interference, find:
(*a*) The smallest pinion tooth count that will run with itself
(*b*) The smallest pinion tooth count at a ratio $m_G = 2.5$, and the largest gear tooth count possible with this pinion
(*c*) The smallest pinion that will run with a rack

13–11 Repeat Problem 13–10 for a helical gearset with $\phi_n = 20°$ and $\psi = 30°$.

13–12 The decision has been made to use $\phi_n = 20°$, $P_t = 0.2362$ teeth/mm, and $\psi = 30°$ for a 2:1 reduction. Choose the smallest acceptable full-depth pinion and gear tooth count to avoid interference.

13–13 Repeat Problem 13–12 with $\psi = 45°$.

13–14 By employing a pressure angle larger than standard, it is possible to use fewer pinion teeth, and hence obtain smaller gears without undercutting during machining. If the gears are full-depth spur gears, what is the smallest possible pressure angle ϕ that can be obtained without undercutting for a 9-tooth pinion to mesh with a rack?

13–15 A parallel-shaft gearset consists of an 18-tooth helical pinion driving a 32-tooth gear. The pinion has a left-hand helix angle of 25°, a normal pressure angle of 20°, and a normal module of 3 mm. Find:
(*a*) The normal, transverse, and axial circular pitches
(*b*) The transverse module and the transverse pressure angle
(*c*) The pitch diameters of the two gears

13–16 The double-reduction helical gearset shown in the figure is driven through shaft a at a speed of 700 rev/min. Gears 2 and 3 have a normal module of 2 mm, a 30° helix angle, and a normal pressure angle of 20°. The second pair of gears in the train, gears

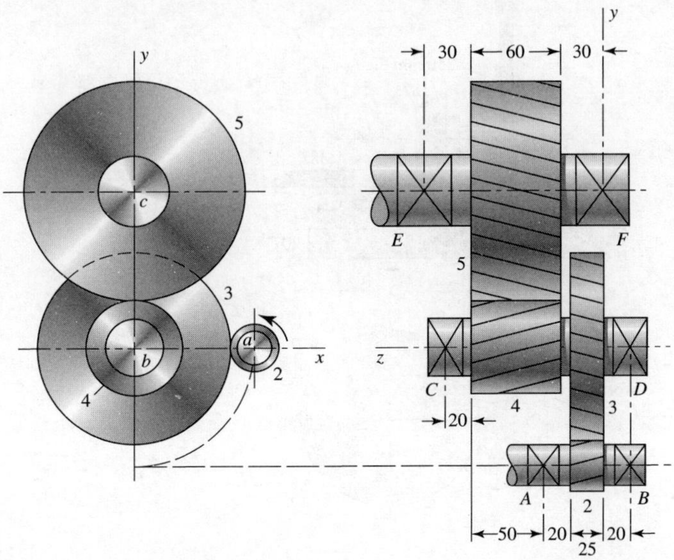

Problem 13–16
Dimensions in millimeters.

4 and 5, have a normal module of 4 mm, a 25° helix angle, and a normal pressure angle of 20°. The tooth numbers are: $N_2 = 12$, $N_3 = 48$, $N_4 = 16$, $N_5 = 36$. Find:

(*a*) The directions of the thrust force exerted by each gear upon its shaft

(*b*) The speed and direction of shaft *c*

(*c*) The center distance between shafts

13–17 Shaft *a* in the figure rotates at 600 rev/min in the direction shown. Find the speed and direction of rotation of shaft *d*.

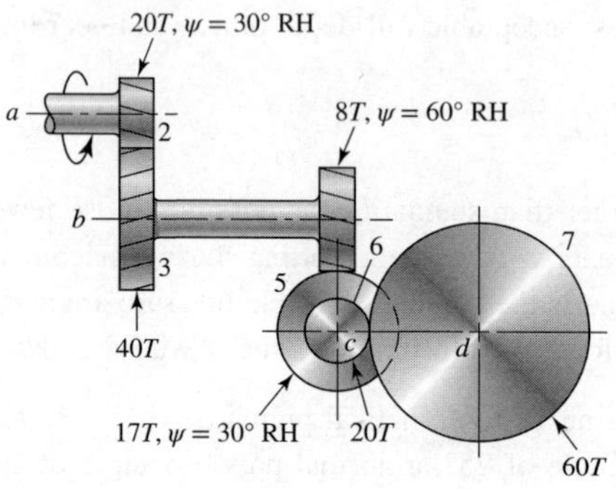

Problem 13–17

13–18 The mechanism train shown consists of an assortment of gears and pulleys to drive gear 9. Pulley 2 rotates at 1200 rev/min in the direction shown. Determine the speed and direction of rotation of gear 9.

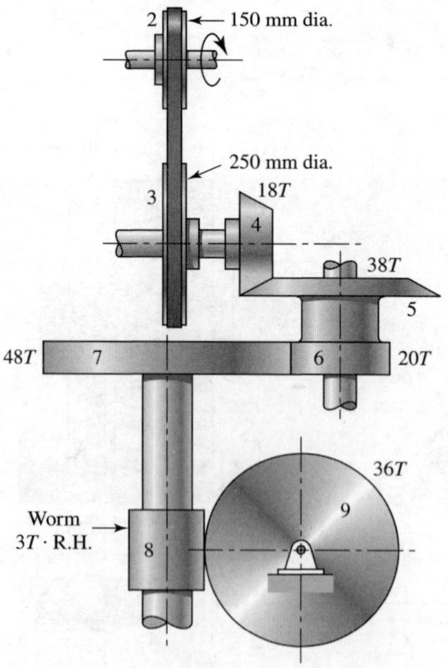

Problem 13–18

13–19 The figure shows a gear train consisting of a pair of helical gears and a pair of miter gears. The helical gears have a $17\frac{1}{2}°$ normal pressure angle and a helix angle as shown. Find:

(*a*) The speed of shaft *c*

(*b*) The distance between shafts *a* and *b*

(*c*) The pitch diameter of the miter gears

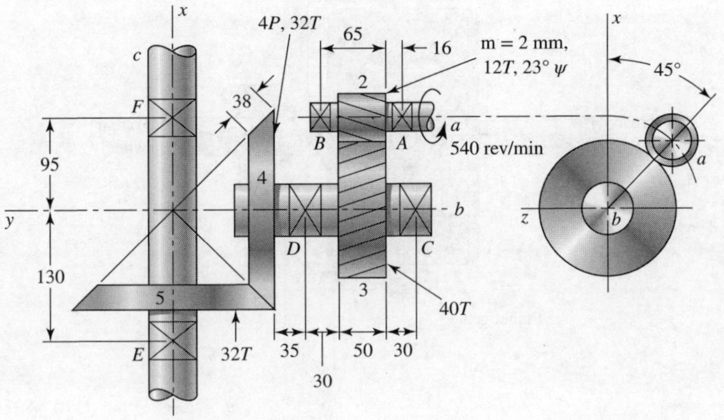

Problem 13–19

Dimensions in millimeters.

13–20 A compound reverted gear train is to be designed as a speed increaser to provide a total increase of speed of exactly 45 to 1. With a 20° pressure angle, specify appropriate numbers of teeth to minimize the gearbox size while avoiding the interference problem in the teeth. Assume all gears will have the same diametral pitch.

13–21 Repeat Problem 13–20 with a 25° pressure angle.

13–22 Repeat Problem 13–20 for a gear ratio of exactly 30 to 1.

13–23 Repeat Problem 13–20 for a gear ratio of *approximately* 45 to 1.

13–24 A gearbox is to be designed with a compound reverted gear train that transmits 18 kW with an input speed of 2500 rev/min. The output should deliver the power at a rotational speed in the range of 280 to 300 rev/min. Spur gears with 20° pressure angle are to be used. Determine suitable numbers of teeth for each gear, to minimize the gearbox size while providing an output speed within the specified range. Be sure to avoid an interference problem in the teeth.

13–25 The tooth numbers for the automotive differential shown in the figure are $N_2 = 16$, $N_3 = 48$, $N_4 = 14$, $N_5 = N_6 = 20$. The drive shaft turns at 900 rev/min.

(*a*) What are the wheel speeds if the car is traveling in a straight line on a good road surface?

(*b*) Suppose the right wheel is jacked up and the left wheel resting on a good road surface. What is the speed of the right wheel?

(*c*) Suppose, with a rear-wheel drive vehicle, the auto is parked with the right wheel resting on a wet icy surface. Does the answer to part (*b*) give you any hint as to what would happen if you started the car and attempted to drive on?

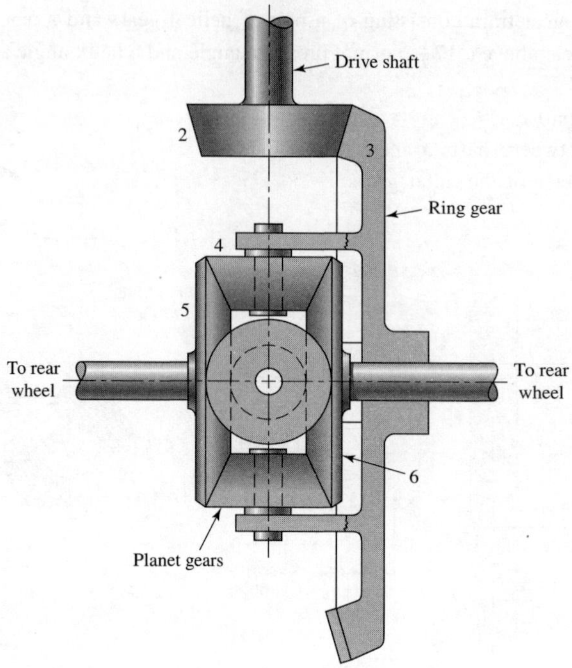

Problem 13–25

13–26 The figure illustrates an all-wheel drive concept using three differentials, one for the front axle, another for the rear, and the third connected to the drive shaft.

(*a*) Explain why this concept may allow greater acceleration.

(*b*) Suppose either the center or the rear differential, or both, can be locked for certain road conditions. Would either or both of these actions provide greater traction? Why?

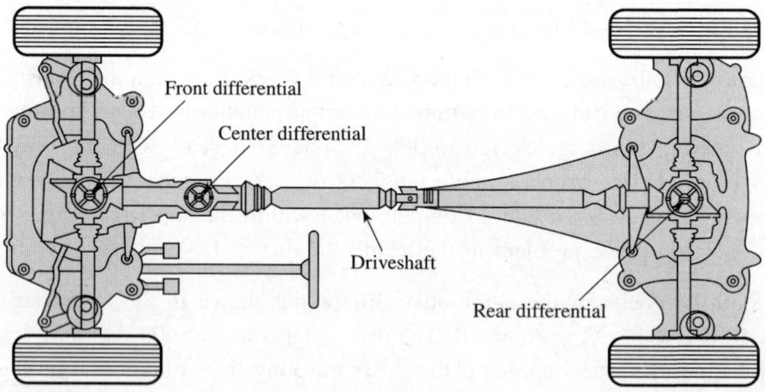

Problem 13–26
The Audi "Quattro concept," showing the three differentials that provide permanent all-wheel drive.
(Source: Audi of America, Inc.)

13–27 In the reverted planetary train illustrated, find the speed and direction of rotation of the arm if gear 2 is unable to rotate and gear 6 is driven at 12 rev/min in the clockwise direction as viewed from the bottom of the figure.

Problem 13–27

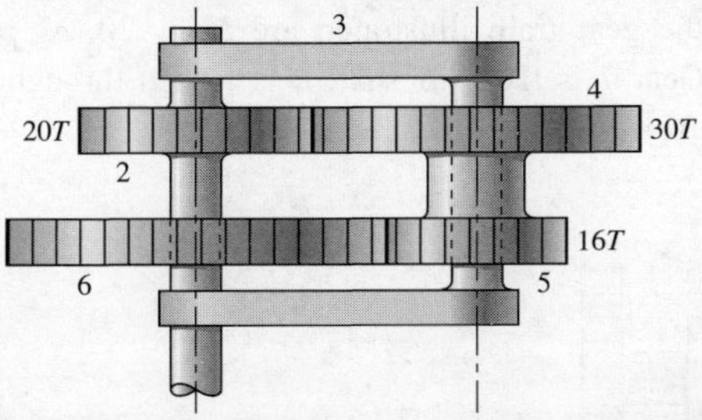

13–28 In the gear train of Problem 13–27, let gear 6 be driven at 85 rev/min counterclockwise (as viewed from the bottom of the figure) while gear 2 is held stationary. What is the speed and direction of rotation of the arm?

13–29 A compound reverted gear train is to be designed to provide the appropriate ratio between the minute hand and the hour hand of a clock. The hour hand is attached to the shaft of gear 5, such that the rotational speed of the hour hand will equal the rotational speed of gear 5. The shaft of gear 2 passes through the center of the hollow shaft of gear 5, and has the minute hand attached to it. Use spur gears with a 20° pressure angle. Specify appropriate numbers of teeth, while minimizing the gearbox size and avoiding the interference problem in the teeth.

Problem 13–29

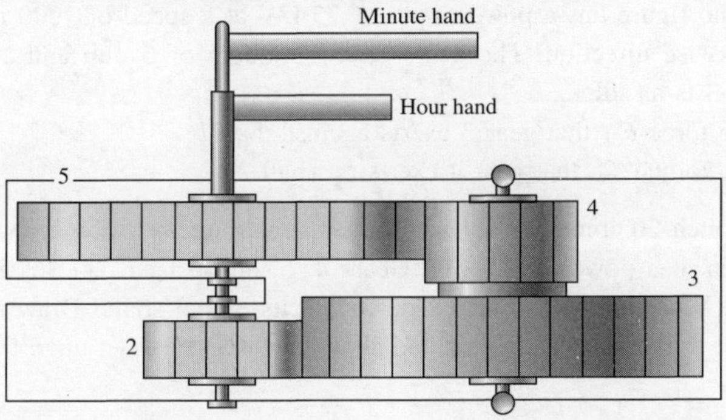

13–30 Tooth numbers for the gear train shown in the figure are $N_2 = 12$, $N_3 = 16$, and $N_4 = 12$. How many teeth must internal gear 5 have? Suppose gear 5 is fixed. What is the speed of the arm if shaft a rotates at 320 rev/min counterclockwise as viewed from the left side of the figure?

Problem 13–30

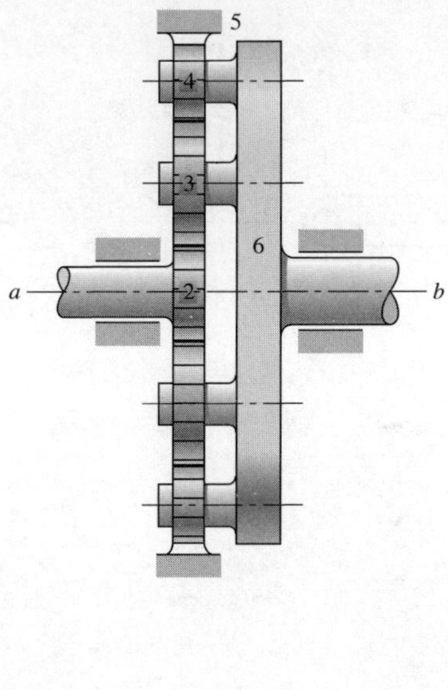

13–31 The tooth numbers for the gear train illustrated are $N_2 = 20$, $N_3 = 16$, $N_4 = 30$, $N_6 = 36$, and $N_7 = 46$. Gear 7 is fixed. If shaft a is turned through 10 revolutions, how many turns will shaft b make?

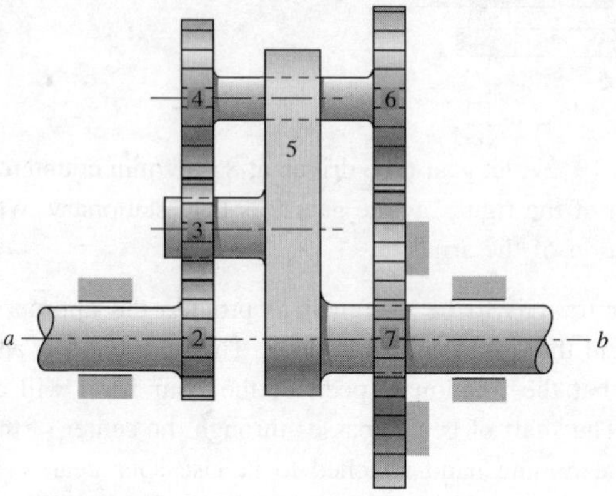

Problem 13–31

13–32 Shaft a in the figure has a power input of 75 kW at a speed of 1000 rev/min in the counterclockwise direction. The gears have a module of 5 mm and a 20° pressure angle. Gear 3 is an idler.

(*a*) Find the force F_{3b} that gear 3 exerts against shaft b.

(*b*) Find the torque T_{4c} that gear 4 exerts on shaft c.

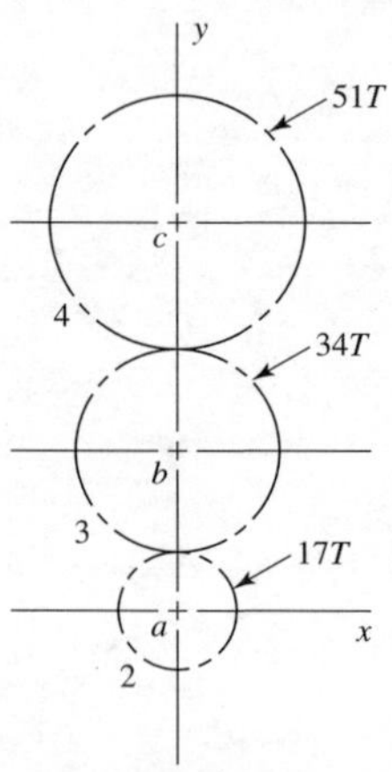

Problem 13–32

13–33 The 24T 6-pitch 20° pinion 2 shown in the figure rotates clockwise at 1000 rev/min and is driven at a power of 20 kW. Gears 4, 5, and 6 have 24, 36, and 144 teeth, respectively. What torque can arm 3 deliver to its output shaft? Draw free-body diagrams of the arm and of each gear and show all forces that act upon them.

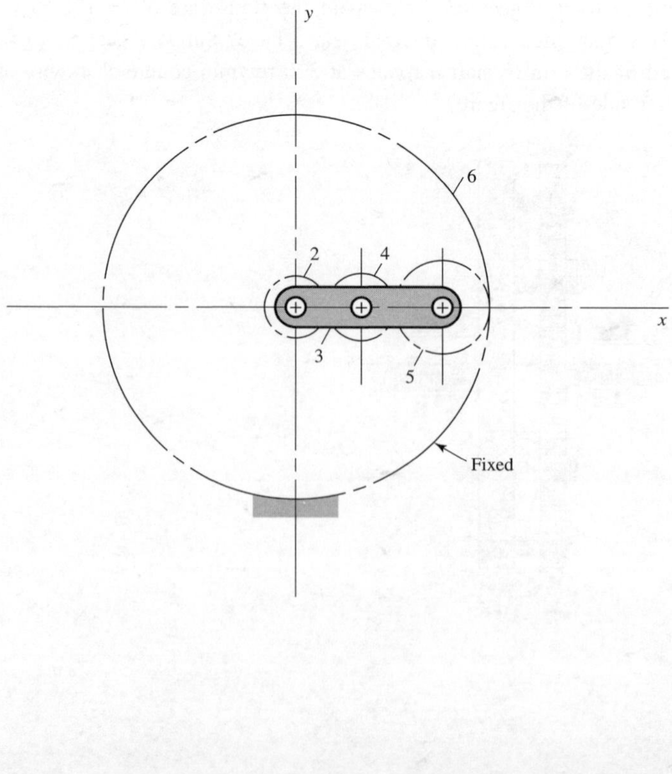

Problem 13–33

13–34 A compound reverted gear train is to be designed as a speed increaser to provide a total increase of speed of exactly 40 to 1. Use spur gears with a 20° pressure angle and a module of 4 mm.

(*a*) Specify appropriate numbers of teeth to minimize the gearbox size while avoiding the interference problem in the teeth.

(*b*) Determine the gearbox outside dimension *Y*, assuming the wall thickness of the box is 20 mm, and allowing 15 mm clearances between the tips of the gear teeth and the walls.

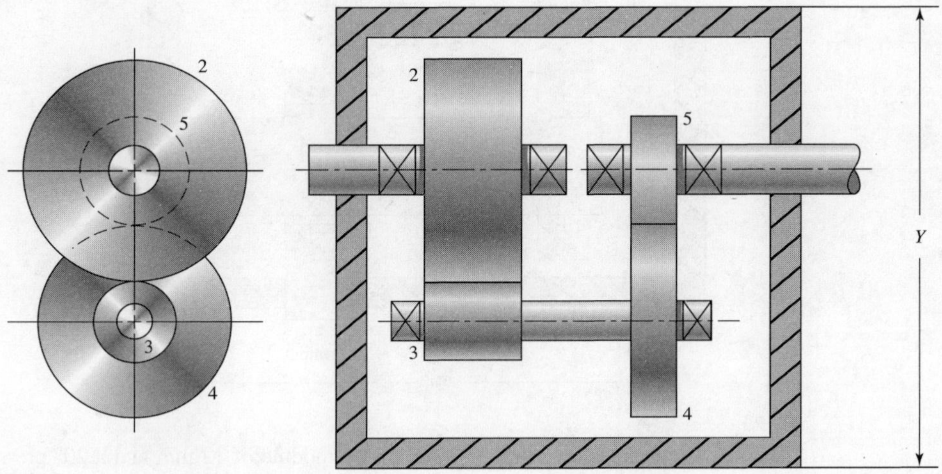

Problem 13–34

13–35 Repeat Problem 13–34, using a 25° pressure angle. Take note of the effect of this variation on the overall size of the gearbox.

13–36 Repeat Problem 13–34, using helical gears with a 20° pressure angle and a 45° helix angle. Take note of the effect of this variation on the overall size of the gearbox.

13–37 Repeat Problem 13–34, except instead of an exact speed ratio of 40 to 1, the speed ratio may be anywhere between 38 and 42. Take note of the effect of this variation on the overall size of the gearbox.

13–38 In the gearbox shown, gears 4 and 5 are compounded to the same shaft. The gearbox receives an input power of 3 kW at a speed of 300 rpm. The gears have a module of 4 mm, a 20° pressure angle, and the following numbers of teeth: N_7 = 80 teeth, N_5 = N_3 = 20 teeth, N_4 = 60 teeth, N_2 = 30 teeth. Assume all shafts lie in the same plane. Assume the gearbox is reasonably efficient, so losses can be neglected for this analysis. Determine

(*a*) the number of teeth needed on gear 6 to make the input and output shafts be in-line.

(*b*) the minimum inside dimension of the gearbox *Y* before adding any clearance.

(*c*) the pitch line velocity for gear 2, in units of m/s.

(*d*) the tangential force transmitted between gears 2 and 3.

(*e*) the radial force transmitted between gears 2 and 3.

(*f*) the input torque, in units of N · m.

(*g*) the output torque, in units of N · m.

(*h*) the output speed, in units of rev/min.
(*i*) the output power, in units of kW.

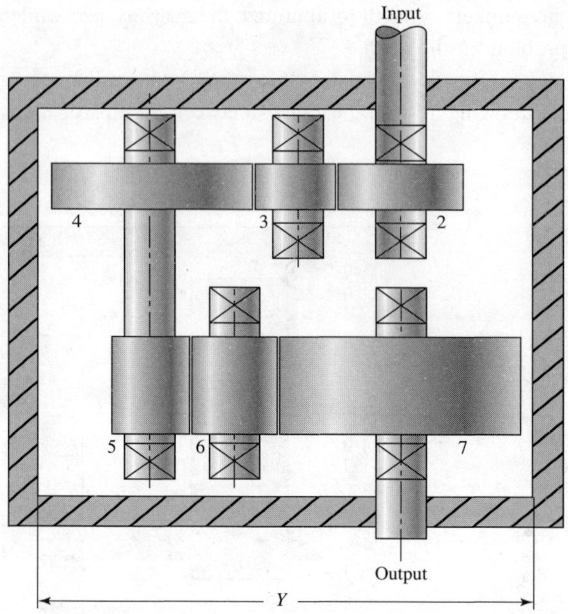

Problem 13–38

13–39 The gears shown in the figure have a module of 12 mm and a 20° pressure angle. The pinion rotates at 1800 rev/min clockwise and transmits 150 kW through the idler pair to gear 5 on shaft *c*. What forces do gears 3 and 4 transmit to the idler shaft?

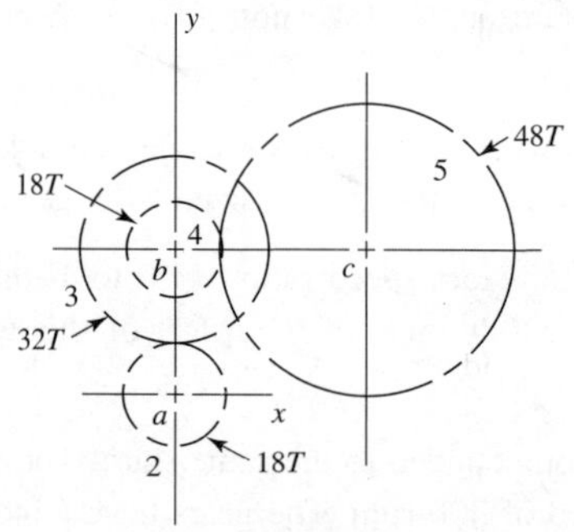

Problem 13–39

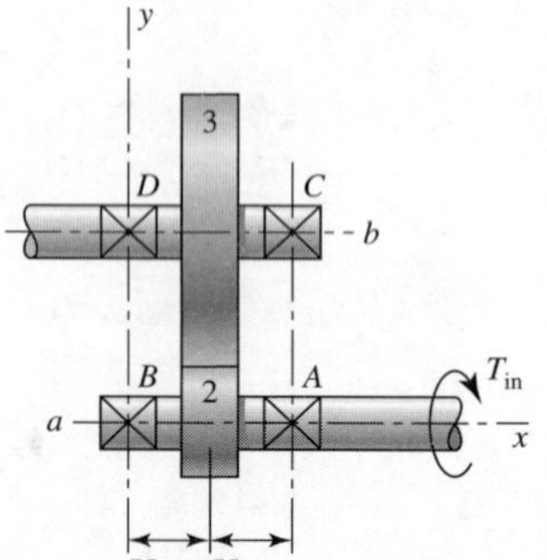

Problem 13–40

13–40 The figure shows a pair of shaft-mounted spur gears having a module of 5 mm with an 18-tooth 20° pinion driving a 45-tooth gear. The power input is 23 kW at 1800 rev/min. Find the direction and magnitude of the forces acting on bearings *A*, *B*, *C*, and *D*.

13–41 The figure shows the electric-motor frame dimensions for a 20 kW 900 rev/min motor. The frame is bolted to its support using four 20 mm bolts spaced 285 mm apart in the view shown and 356 apart when viewed from the end of the motor. A module m = 6 mm 20° spur pinion having 20 teeth and a face width of 50 mm is keyed to and flush with the end of the motor shaft. This pinion drives another gear whose axis is in the same xz plane and directly behind the motor shaft. Determine the maximum shear and tensile forces on the mounting bolts based on 200 percent overload torque. Does the direction of rotation matter?

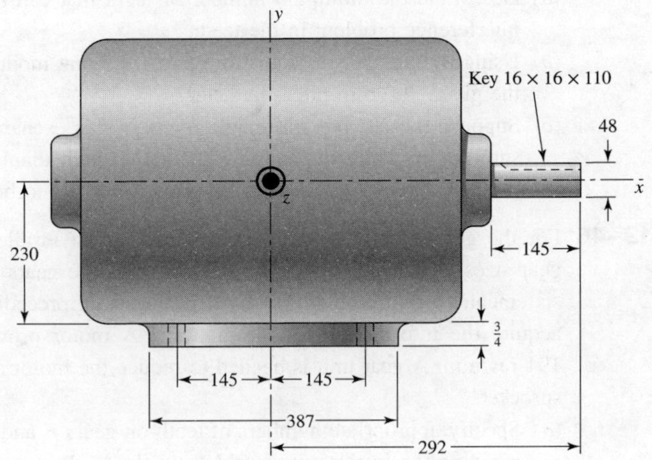

Problem 13–41

NEMA No. 364 frame; dimensions in millimeters. The z axis is directed out of the paper.

13–42 Continue Problem 13–24 by finding the following information, assuming a module of 4 mm.

(*a*) Determine pitch diameters for each of the gears.
(*b*) Determine the pitch line velocities (in m/s) for each set of gears.
(*c*) Determine the magnitudes of the tangential, radial, and total forces transmitted between each set of gears.
(*d*) Determine the input torque.
(*e*) Determine the output torque, neglecting frictional losses.

13–43 A speed-reducer gearbox containing a compound reverted gear train transmits 26.1 kW with an input speed of 1200 rev/min. Spur gears with 20° pressure angle are used, with 16 teeth on each of the small gears and 48 teeth on each of the larger gears. A module of 2.5 mm is proposed.

(*a*) Determine the speeds of the intermediate and output shafts.
(*b*) Determine the pitch line velocities (in m/s) for each set of gears.
(*c*) Determine the magnitudes of the tangential, radial, and total forces transmitted between each set of gears.
(*d*) Determine the input torque.
(*e*) Determine the output torque, neglecting frictional losses.

13–44* For the countershaft in Problem 3–80 assume the gear ratio from gear B to its mating gear is 2 to 1.

(*a*) Determine the minimum number of teeth that can be used on gear B without an interference problem in the teeth.
(*b*) Using the number of teeth from part (*a*), what diametral pitch is required to also achieve the given module of 3 mm?

(*c*) Suppose the 20° pressure angle gears are exchanged for gears with 25° pressure angle, while maintaining the same pitch diameters and module. Determine the new forces F_A and F_B if the same power is to be transmitted.

13–45* For the countershaft in Problem 3–81, assume the gear ratio from gear *B* to its mating gear is 5 to 1.

(*a*) Determine the minimum number of teeth that can be used on gear *B* without an interference problem in the teeth.

(*b*) Using the number of teeth from part (*a*), what module is required to also achieve the given 300-mm pitch diameter?

(*c*) Suppose the 20° pressure angle for gear *A* is exchanged for a gear with 25° pressure angle, while maintaining the same pitch diameters and module. Determine the new forces F_A and F_B if the same power is to be transmitted.

13–46* For the gear and sprocket assembly analyzed in Problem 3–85, information for the gear sizes and the forces transmitted through the gears was provided in the problem statement. In this problem, we will perform the preceding design steps necessary to acquire the information for the analysis. A motor providing 2 kW is to operate at 191 rev/min. A gear unit is needed to reduce the motor speed by half to drive a chain sprocket.

(*a*) Specify appropriate numbers of teeth on gears *F* and *C* to minimize the size while avoiding the interference problem in the teeth.

(*b*) Assuming an initial guess of 125-mm pitch diameter for gear *F*, what is the module that should be used for the stress analysis of the gear teeth?

(*c*) Calculate the input torque applied to shaft *EFG*.

(*d*) Calculate the magnitudes of the radial, tangential, and total forces transmitted between gears *F* and *C*.

13–47* For the gear and sprocket assembly analyzed in Problem 3–87, information for the gear sizes and the forces transmitted through the gears was provided in the problem statement. In this problem, we will perform the preceding design steps necessary to acquire the information for the analysis. A motor providing 0.75 kW is to operate at 70 rev/min. A gear unit is needed to double the motor speed to drive a chain sprocket.

(*a*) Specify appropriate numbers of teeth on gears *F* and *C* to minimize the size while avoiding the interference problem in the teeth.

(*b*) Assuming an initial guess of 250 mm pitch diameter for gear *F*, what is the diametral pitch that should be used for the stress analysis of the gear teeth?

(*c*) Calculate the input torque applied to shaft *EFG*.

(*d*) Calculate the magnitudes of the radial, tangential, and total forces transmitted between gears *F* and *C*.

13–48* For the bevel gearset in Problems 3–82 and 3–84, shaft *AB* is rotating at 600 rev/min and transmits 7.5 Kw. The gears have a 20° pressure angle.

(*a*) Determine the bevel angle γ for the gear on shaft *AB*.

(*b*) Determine the pitch-line velocity.

(*c*) Determine the tangential, radial, and axial forces acting on the pinion. Were the forces given in Problem 3–82 correct?

13–49 The figure shows a 16*T* 20° straight bevel pinion driving a 32*T* gear, and the location of the bearing centerlines. Pinion shaft *a* receives 2 kW at 240 rev/min. Determine the bearing reactions at *A* and *B* if *A* is to take both radial and thrust loads.

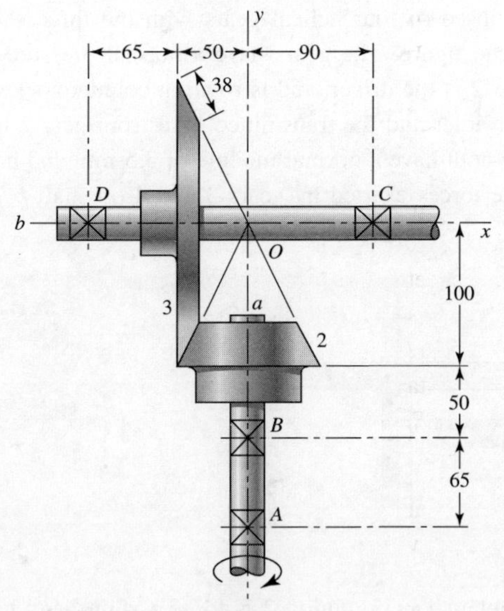

Problem 13–49

Dimensions in millimeters.

13–50 The figure shows a 250 mm diametral pitch 18-tooth 20° straight bevel pinion driving a 30-tooth gear. The transmitted load is 110 N. Find the bearing reactions at *C* and *D* on the output shaft if *D* is to take both radial and thrust loads.

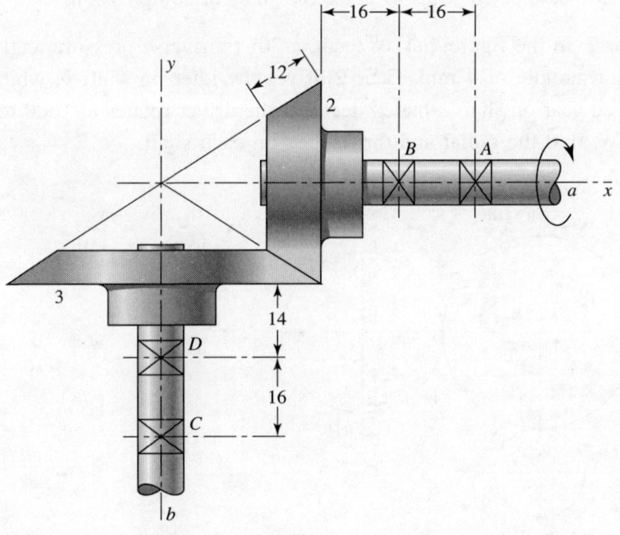

Problem 13–50

Dimensions in millimeters.

13–51 The gears shown in the figure have a normal module of 5 mm, a normal pressure angle of 20°, and a 30° helix angle. The transmitted load is 3500 N. The pinion rotates counterclockwise about the *y* axis, as viewed from the positive *y* axis. Find the force exerted by each gear on its shaft.

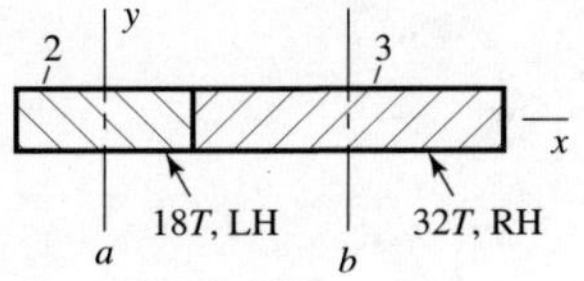

Problem 13–51

13–52 The gears shown in the figure have a normal module of 5 mm, a normal pressure angle of 20°, and a 30° helix angle. The transmitted load is 3500 N. Gear 2 rotates clockwise about the *y* axis, as viewed from the positive *y* axis. Gear 3 is an idler. Find the forces exerted by gears 2 and 3 on their shafts.

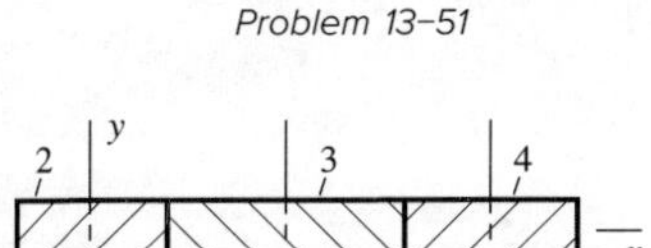

Problem 13–52

13–53 A gear train is composed of four helical gears with the three shaft axes in a single plane, as shown in the figure. The gears have a normal pressure angle of 20° and a 30° helix angle. Gear 2 is the driver, and is rotating counterclockwise as viewed from the top. Shaft b is an idler and the transmitted load from gear 2 to gear 3 is 2200 N. The gears on shaft b both have a normal module of 3.5 mm and have 54 and 14 teeth, respectively. Find the forces exerted by gears 3 and 4 on shaft b.

Problem 13–53

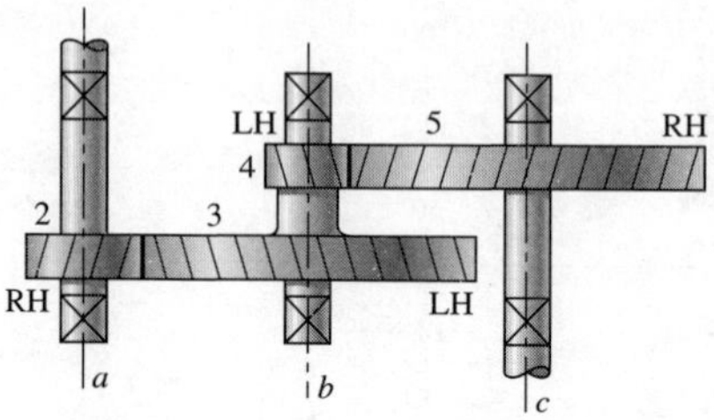

13–54 In the figure for Problem 13–35, pinion 2 is to be a right-hand helical gear having a helix angle of 30°, a normal pressure angle of 20°, 16 teeth, and a normal module of 4 mm. A motor delivers 18 kW to shaft a at a speed of 1720 rev/min clockwise about the x axis. Gear 3 has 42 teeth. Find the reaction exerted by bearings C and D on shaft b. One of these bearings is to take both radial and thrust loads. This bearing should be selected so as to place the shaft in compression.

13–55 Gear 2, in the figure, has 16 teeth, a 20° transverse pressure angle, a 15° helix angle, and a module of 4 mm. Gear 2 drives the idler on shaft b, which has 36 teeth. The driven gear on shaft c has 28 teeth. If the driver rotates at 1600 rev/min and transmits 6 kW, find the radial and thrust load on each shaft.

Problem 13–55

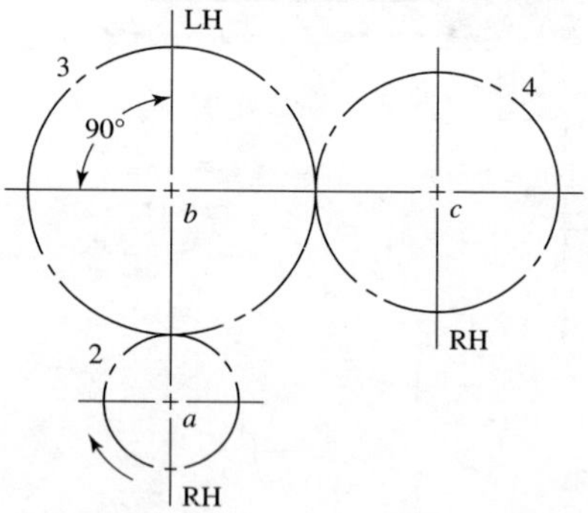

13–56 The figure shows a double-reduction helical gearset. Pinion 2 is the driver, and it receives a torque of 135 N · m from its shaft in the direction shown. Pinion 2 has a normal module of 3 mm, 14 teeth, and a normal pressure angle of 20° and is cut right-handed with a helix angle of 30°. The mating gear 3 on shaft b has 36 teeth. Gear 4, which is the driver for the second pair of gears in the train, has a normal module of 5 mm, 15 teeth, and a normal pressure angle of 20° and is cut left-handed with a helix angle of 15°. Mating gear 5 has 45 teeth. Find the magnitude and direction of the force exerted by the bearings C and D on shaft b if bearing C can take only a radial load while bearing D is mounted to take both radial and thrust loads.

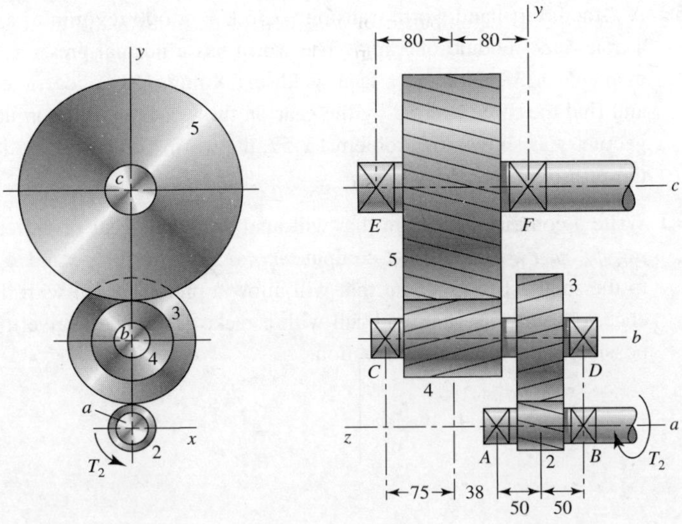

Problem 13–56

Dimensions in millimeters.

13–57 A right-hand single-tooth hardened-steel (hardness not specified) worm has a catalog rating of 2000 W at 600 rev/min when meshed with a 48-tooth cast-iron gear. The axial pitch of the worm is 25 mm, the normal pressure angle is 14.5°, the pitch diameter of the worm is 100 mm, and the face widths of the worm and gear are, respectively, 100 mm and 50 mm. Bearings are centered at locations A and B on the worm shaft. Determine which should be the thrust bearing (so that the axial load in the shaft is in compression), and find the magnitudes and directions of the forces exerted by both bearings.

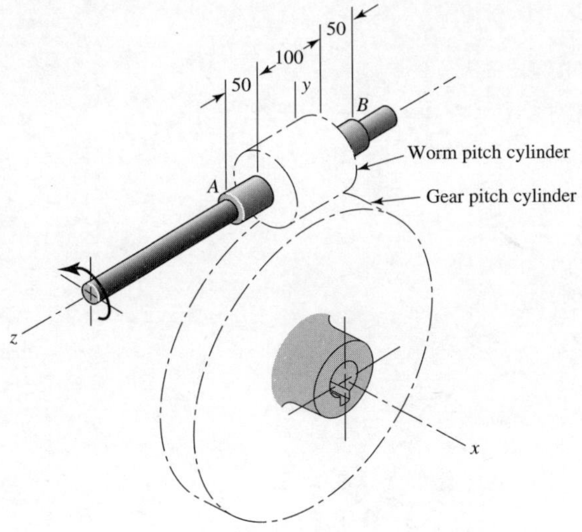

Problem 13–57

Dimensions in millimeters.

13–58 The hub diameter and projection for the gear of Problem 13–57 are 100 and 37.5 mm, respectively. The face width of the gear is 50 mm. Locate bearings C and D on opposite sides, spacing C 10 mm from the gear on the hidden face (see figure) and D 10 mm from the hub face. Choose one as the thrust bearing, so that the axial load in the shaft is in compression. Find the output torque and the magnitudes and directions of the forces exerted by the bearings on the gearshaft.

13–59 A 2-tooth left-hand worm transmits 0.56 kW at 600 rev/min to a 36-tooth gear having a transverse module of 3 mm. The worm has a normal pressure angle of 20°, a pitch diameter of 38 mm, and a face width of 38 mm. Use a coefficient of friction of 0.05 and find the force exerted by the gear on the worm and the torque input. For the same geometry as shown for Problem 13–57, the worm velocity is clockwise as viewed from the positive z axis.

13–60 Write a computer program that will analyze a spur gear or helical-mesh gear, accepting ϕ_n, ψ, P_t, N_p, and N_G; compute m_G, d_p, d_G, p_t, p_n, p_x, and ϕ_t; and give advice as to the smallest tooth count that will allow a pinion to run with itself without interference, run with its gear, and run with a rack. Also have it give the largest tooth count possible with the intended pinion.

14 Spur and Helical Gears

koi88/Alamy Stock Photo

Chapter Outline

This chapter is devoted to analysis and design of spur and helical gears to resist bending failure of the teeth as well as pitting failure of tooth surfaces. Failure by bending will occur when the significant tooth stress equals or exceeds either the yield strength or the bending endurance strength. A surface failure occurs when the significant contact stress equals or exceeds the surface endurance strength. The first two sections present a little of the history of the analyses from which current methodology developed.

The American Gear Manufacturers Association[1] (AGMA) has for many years been the responsible authority for the dissemination of knowledge pertaining to the design and analysis of gearing. The methods this organization presents are in general use in the United States when strength and wear are primary considerations. In view of this fact it is important that the AGMA approach to the subject be presented here.

The general AGMA approach requires a great many charts and graphs—too many for a single chapter in this book. We have omitted many of these here by choosing a single pressure angle and by using only full-depth teeth. This simplification reduces the complexity but does not prevent the development of a basic understanding of the approach. Furthermore, the simplification makes possible a better development of the fundamentals and hence should constitute an ideal introduction to the use of the general AGMA method.[2] Sections 14–1 and 14–2 are elementary and serve as an examination of the foundations of the AGMA method. Table 14–1 is largely AGMA nomenclature.

14–1 The Lewis Bending Equation

Wilfred Lewis introduced an equation for estimating the bending stress in gear teeth in which the tooth form entered into the formulation. The equation, announced in 1892, still remains the basis for most gear design today.

To derive the basic Lewis equation, refer to Figure 14–1*a*, which shows a rectangular cantilever beam of cross-sectional dimensions F and t, having a length l and a load W^t, uniformly distributed across the face width F. The section modulus I/c is $Ft^2/6$, and therefore the bending stress is

$$\sigma = \frac{M}{I/c} = \frac{6W^t l}{Ft^2} \tag{a}$$

Gear designers denote the components of gear-tooth forces as W_t, W_r, W_a or W^t, W^r, W^a interchangeably. The latter notation leaves room for post-subscripts essential to free-body diagrams. For instance, for gears 2 and 3 in mesh, W^t_{23} is the transmitted force

[1]1001 N. Fairfax Street, Suite 500, Alexandria, VA 22 314-1587.

[2]The standards ANSI/AGMA 2001-D04 (revised AGMA 2001-C95) and ANSI/AGMA 2101-D04 (metric edition of ANSI/AGMA 2001-D04), *Fundamental Rating Factors and Calculation Methods for Involute Spur and Helical Gear Teeth*, are used in this chapter. The use of American National Standards is completely voluntary; their existence does not in any respect preclude people from manufacturing, marketing, purchasing, or using products, processes, or procedures not conforming to the standards.

The American National Standards Institute does not develop standards and will in no circumstances give an interpretation of any American National Standard. Requests for interpretation of these standards should be addressed to the American Gear Manufacturers Association. [Tables or other self-supporting sections may be quoted or extracted in their entirety. Credit line should read: "Extracted from ANSI/AGMA Standard 2001-D04 or 2101-D04 *Fundamental Rating Factors and Calculation Methods for Involute Spur and Helical Gear Teeth*" with the permission of the publisher, American Gear Manufacturers Association, 1001 N. Fairfax Street, Suite 500, Alexandria, Virginia 22 314-1587.] The foregoing is adapted in part from the ANSI foreword to these standards.

Table 14–1 Symbols, Their Names, and Locations

Symbol*	Name	Where Found
C_e	Mesh alignment correction factor	Eq. (14–33)
$C_f\ (Z_R)$	Surface condition factor	Eq. (14–14)
$C_H\ (Z_W)$	Hardness-ratio factor	Eq. (14–16)
C_{ma}	Mesh alignment factor	Eq. (14–32)
C_{mc}	Load correction factor	Eq. (14–29)
C_{mf}	Face load-distribution factor	Eq. (14–28)
$C_p\ (Z_E)$	Elastic coefficient	Eq. (14–11)
C_{pf}	Pinion proportion factor	Eq. (14–30)
C_{pm}	Pinion proportion modifier	Eq. (14–31)
d	Pitch diameter	Ex. 14–1
d_p	Pitch diameter, pinion	Eq. (14–20)
d_G	Pitch diameter, gear	Eq. (14–20)
$F\ (b)$	Net face width of narrowest member	Eq. (14–13)
f_P	Pinion surface finish	Fig. 14–13
H	Power	Fig. 14–17
H_B	Brinell hardness	Ex. 14–3
H_{BG}	Brinell hardness of gear	Sec. 14–12
H_{BP}	Brinell hardness of pinion	Sec. 14–12
hp	Horsepower	Ex. 14–1
h_t	Gear-tooth whole depth	Sec. 14–16
$I\ (Z_I)$	Geometry factor of pitting resistance	Eq. (14–14)
$J\ (Y_J)$	Geometry factor for bending strength	Eq. (14–13)
K_B	Rim-thickness factor	Eq. (14–38)
K_f	Fatigue stress-concentration factor	Eq. (14–7)
$K_m\ (K_H)$	Load-distribution factor	Eq. (14–28)
K_o	Overload factor	Eq. (14–13)
$K_R\ (Y_Z)$	Reliability factor	Eq. (14–15)
K_s	Size factor	Sec. 14–10
$K_T\ (Y_\theta)$	Temperature factor	Eq. (14–15)
K_v	Dynamic factor	Eq. (14–25)
m	Module	Eq. (14–13)
m_B	Backup ratio	Eq. (14–37)
m_F	Face-contact ratio	Eq. (14–17)
m_G	Gear ratio (never less than 1)	Eq. (14–20)
m_N	Load-sharing ratio	Eq. (14–19)
m_t	Transverse module	Eq. (14–13)
N	Number of stress cycles	Fig. 14–14
N_G	Number of teeth on gear	Eq. (14–20)
N_P	Number of teeth on pinion	Eq. (14–20)

(Continued)

Table 14–1 Symbols, Their Names, and Locations *(Continued)*

Symbol*	Name	Where Found
n	Speed, in rev/min	Eq. (13–34)
n_P	Pinion speed, in rev/min	Ex. 14–4
n_b	Bending fatigue factor of safety	Eq. (14–39)
n_c	Contact stress factor of safety	Eq. (14–40)
$n_{c,\text{linear}}$	n_c linearized with W^t	Eq. (14–41)
P	Transverse diametral pitch	Eq. (14–13)
P_n	Normal diametral pitch	Ex. 14–5
P_N	Normal base pitch	Eq. (14–22)
p_n	Normal circular pitch	Eq. (14–22)
p_x	Axial pitch	Eq. (14–17)
Q_v	Quality number	Eq. (14–26)
R	Reliability	Eq. (14–36)
R_a	Root-mean-squared roughness	Fig. 14–13
r_f	Tooth fillet radius	Fig. 14–1
r_G	Pitch-circle radius, gear	In standard
r_P	Pitch-circle radius, pinion	In standard
r_{bP}	Pinion base-circle radius	Eq. (14–23)
r_{bG}	Gear base-circle radius	Eq. (14–23)
S_C	Buckingham surface endurance strength	Ex. 14–3
S_c	AGMA surface endurance strength	Eq. (14–16)
S_t	AGMA bending strength	Eq. (14–15)
S	Bearing span	Fig. 14–10
S_1	Pinion offset from center span	Fig. 14–10
S_F	AGMA safety factor—bending	Sec. 14–17
S_H	AGMA safety factor—pitting	Sec. 14–17
W^t or W_t	Transmitted load	Fig. 14–1
Y_N	Stress-cycle factor for bending strength	Fig. 14–14
Z_N	Stress-cycle factor for pitting resistance	Fig. 14–15
β	Exponent	Eq. (14–42)
σ_b	Bending stress, AGMA	Eq. (14–13)
σ_C	Contact stress from Hertzian relationships	Eq. (14–12)
σ_c	Contact stress from AGMA relationships	Eq. (14–14)
$\sigma_{b,\text{all}}$	Allowable bending stress, AGMA	Eq. (14–15)
$\sigma_{c,\text{all}}$	Allowable contact stress, AGMA	Eq. (14–16)
ϕ	Pressure angle	Eq. (14–10)
ϕ_n	Normal pressure angle	Eq. (14–22)
ϕ_t	Transverse pressure angle	Eq. (14–21)
ψ	Helix angle	Ex. 14–5

*Where applicable, the alternate symbol for the metric standard is shown in parenthesis.

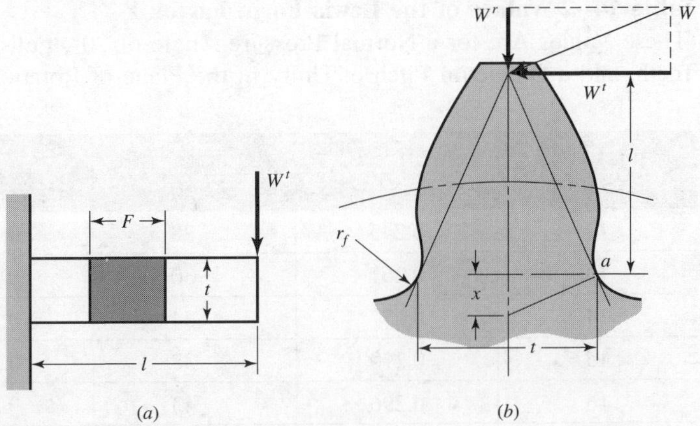

Figure 14–1

of body 2 on body 3, and W_{32}^t is the transmitted force of body 3 on body 2. Since gear-force components rarely take exponents, this causes no complication.

Referring to Figure 14–1*b*, we assume that the maximum stress in a gear tooth occurs at point *a*. By similar triangles, you can write

$$\frac{t/2}{x} = \frac{l}{t/2} \qquad \text{or} \qquad x = \frac{t^2}{4l} \qquad \text{or} \qquad l = \frac{t^2}{4x} \tag{b}$$

By rearranging Equation (*a*),

$$\sigma = \frac{6W^t l}{Ft^2} = \frac{W^t}{F}\frac{1}{t^2/6l} = \frac{W^t}{F}\frac{1}{t^2/4l}\frac{1}{\frac{4}{6}} \tag{c}$$

If we now substitute the value of *l* from Equation (*b*) in Equation (*c*) and multiply the numerator and denominator by the circular pitch *p*, we find

$$\sigma = \frac{W^t p}{F\left(\frac{2}{3}\right) xp} \tag{d}$$

Letting $y = 2x/(3p)$, we have

$$\sigma = \frac{W^t}{Fpy} \tag{14–1}$$

This completes the development of the original Lewis equation. The factor *y* is called the *Lewis form factor,* and it may be obtained by a graphical layout of the gear tooth or by digital computation.

In using this equation, most engineers prefer to employ the diametral pitch in determining the stresses. This is done by substituting $p = \pi/P$ and $y = Y/\pi$ in Equation (14–1). This gives

$$\sigma = \frac{W^t P}{FY} \tag{14–2}$$

where

$$Y = \frac{2xP}{3} \tag{14–3}$$

The use of this equation for *Y* means that only the bending of the tooth is considered and that the compression due to the radial component of the force is neglected. Values of *Y* obtained from this equation are tabulated in Table 14–2.

Table 14–2 **Values of the Lewis Form Factor *Y***
(These Values Are for a Normal Pressure Angle of 20°, Full-Depth Teeth, and a Diametral Pitch of Unity in the Plane of Rotation)

Number of Teeth	Y	Number of Teeth	Y
12	0.245	28	0.353
13	0.261	30	0.359
14	0.277	34	0.371
15	0.290	38	0.384
16	0.296	43	0.397
17	0.303	50	0.409
18	0.309	60	0.422
19	0.314	75	0.435
20	0.322	100	0.447
21	0.328	150	0.460
22	0.331	300	0.472
24	0.337	400	0.480
26	0.346	Rack	0.485

The use of Equation (14–3) also implies that the teeth do not share the load and that the greatest force is exerted at the tip of the tooth. But the contact ratio should be somewhat greater than unity, say about 1.5, to achieve a quality gearset. If, in fact, the gears are cut with sufficient accuracy, the tip-load condition is not the worst, because another pair of teeth will be in contact when this condition occurs. Therefore the maximum stress probably occurs while a single pair of teeth is carrying the full load near the middle of the teeth, when another pair of teeth is just on the verge of coming into contact.

Dynamic Effects

When a pair of gears is driven at moderate or high speed and noise is generated, it is certain that dynamic effects are present. One of the earliest efforts to account for an increase in the load due to velocity employed a number of gears of the same size, material, and strength. Several of these gears were tested to destruction by meshing and loading them at zero velocity. The remaining gears were tested to destruction at various pitch-line velocities. For example, if a pair of gears failed at 2000 N tangential load at zero velocity and at 1000 N at velocity V_1, then a *velocity factor,* designated K_v, of 2 was specified for the gears at velocity V_1. Then another, identical, pair of gears running at a pitch-line velocity V_1 could be assumed to have a load equal to twice the tangential or transmitted load.

Note that the definition of dynamic factor K_v has been altered. AGMA standards ANSI/AGMA 2001-D04 and 2101-D04 contain this caution:

> Dynamic factor K_v has been redefined as the reciprocal of that used in previous AGMA standards. It is now greater than 1.0. In earlier AGMA standards it was less than 1.0.

Care must be taken in referring to work done prior to this change in the standards.

In the nineteenth century, Carl G. Barth first expressed the velocity factor, and in SI units, AGMA suggests the following expressions for the dynamic factor Kv they are represented as

$$K_v = \frac{3.05 + V}{3.05} \quad \text{(cast iron, cast profile)} \tag{14–4a}$$

$$K_v = \frac{6.1 + V}{6.1} \quad \text{(cut or milled profile)} \tag{14–4b}$$

$$K_v = \frac{3.56 + \sqrt{V}}{3.56} \quad \text{(hobbed or shaped profile)} \tag{14–4c}$$

$$K_v = \sqrt{\frac{5.56 + \sqrt{V}}{5.56}} \quad \text{(shaved or ground profile)} \tag{14–4d}$$

where V is in meters per second (m/s).

Introducing the velocity factor into Equation (14–2) gives

$$\sigma = \frac{K_v W^t P}{FY} \tag{14–5}$$

The metric version of this equation is

$$\sigma = \frac{K_v W^t}{FmY} \tag{14–6}$$

where the face width F and the module m are both in millimeters (mm). Expressing the tangential component of load W^t in newtons (N) then results in stress units of megapascals (MPa).

As a general rule, spur gears should have a face width F from three to five times the circular pitch p.

Equations (14–5) and (14–6) are important because they form the basis for the AGMA approach to the bending strength of gear teeth. They are in general use for estimating the capacity of gear drives when life and reliability are not important considerations. The equations can be useful in obtaining a preliminary estimate of gear sizes needed for various applications.

EXAMPLE 14–1

A stock spur gear is available having a module of 3 mm, a 38 mm face, 16 teeth, and a pressure angle of 20° with full-depth teeth. The material is AISI 1020 steel in as-rolled condition. Use a design factor of $n_d = 3$ to rate the horsepower output of the gear corresponding to a speed of 1200 rev/m and moderate applications.

Solution

The term *moderate applications* seems to imply that the gear can be rated by using the yield strength as a criterion of failure. From Table A–19, we find $S_{ut} = 380$ MPa and $S_y = 210$ MPa. A design factor of 3 means that the allowable bending stress is $210/3 = 70$ MPa. The pitch diameter is $d = N \cdot m = 16 \cdot 3 = 48$ mm, so the pitch-line velocity is

$$V = \frac{\pi dn}{60} = \frac{\pi 48(10^{-3})1200}{60} = 3.014 \text{ m/s}$$

The velocity factor from Equation (14–4*b*) is found to be

$$K_v = \frac{(6.1 + V)}{6.1} = \frac{6.1 + 3.014}{6.1} = 1.5$$

Table 14–2 gives the form factor as $Y = 0.296$ for 16 teeth. We now arrange and substitute in Equation (14–5) as follows:

$$W^t = \frac{FY\sigma_{\text{all}}}{K_v P} = \frac{(38(3)0.296(70))}{1.5} = 1574.7 \text{ N}$$

The horsepower that can be transmitted is

Answer $$hp = W^t V = 1574.7(3.014) = 4746.2 \text{ W}$$

It is important to emphasize that this is a rough estimate, and that this approach must not be used for important applications. The example is intended to help you understand some of the fundamentals that will be involved in the AGMA approach.

EXAMPLE 14–2

Estimate the horsepower rating of the gear in the previous example based on obtaining an infinite life in bending.

Solution

The rotating-beam endurance limit is estimated from Eq. (6–10),

$$S_e' = 0.5S_{ut} = 0.5(380) = 190 \text{ MPa}$$

To obtain the surface finish Marin factor k_a we refer to Table 6–2 for machined surface, finding $a = 4.50$ and $b = -0.265$. Then Eq. (6–18) gives the surface finish Marin factor k_a as

$$k_a = aS_{ut}^b = 4.50(380)^{-0.265} = 0.93$$

The next step is to estimate the size factor k_b. From Table 13–1, the sum of the addendum and dedendum is

$$l = \text{m} + 1.25 \text{ m} = 3 + 1.25(3) = 6.75 \text{ mm}$$

The tooth thickness t in Fig. 14–1b is given in Sec. 14–1 [Eq. (b)] as $t = (4lx)^{1/2}$ when $x = 3\text{ m}Y/2$. Therefore, since from Ex. 14–1 $Y = 0.296$ and m = 3,

$$x = \frac{3\text{ m}Y}{2} = \frac{(3)(3)(0.296)}{2} = 1.332\text{ mm}$$

then

$$t = (4lx)^{1/2} = [4(6.75)1.332]^{1/2} = 6\text{ mm}$$

We have recognized the tooth as a cantilever beam of rectangular cross section, so the equivalent rotating-beam diameter must be obtained from Eq. (6–24):

$$d_e = 0.808(hb)^{1/2} = 0.808(Ft)^{1/2} = 0.808[38(6)]^{1/2} = 12.2\text{ mm}$$

Then, Eq. (6–19) gives k_b as

$$k_b = \left(\frac{d_e}{7.62}\right)^{-0.107} = \left(\frac{12.2}{7.62}\right)^{-0.107} = 0.95$$

The load factor k_c from Eq. (6–25) is unity. With no information given concerning temperature and reliability we will set $k_d = k_e = 1$.

In general, a gear tooth is subjected only to one-way bending. Exceptions include idler gears and gears used in reversing mechanisms. We will account for one-way bending by establishing a miscellaneous-effects Marin factor k_f.

For one-way bending the steady and alternating stress components are $\sigma_a = \sigma_m = \sigma/2$ where σ is the largest repeatedly applied bending stress as given in Eq. (14–5). If a material exhibited a Goodman failure locus,

$$\frac{\sigma_a}{S'_e} + \frac{\sigma_m}{S_{ut}} = 1$$

Since σ_a and σ_m are equal for one-way bending, we substitute σ_a for σ_m and solve the preceding equation for σ_a, giving

$$\sigma_a = \frac{S'_e S_{ut}}{S'_e + S_{ut}}$$

Now replace σ_a with $\sigma/2$, and in the denominator replace S'_e with $0.5S_{ut}$ to obtain

$$\sigma = \frac{2S'_e S_{ut}}{0.5S_{ut} + S_{ut}} = \frac{2S'_e}{0.5 + 1} = 1.33S'_e$$

Now defining a miscellaneous Marin factor $k_f = \sigma/S'_e = 1.33S'_e/S'_e = 1.33$. Similarly, if we were to use a Gerber fatigue locus,

$$\frac{\sigma_a}{S'_e} + \left(\frac{\sigma_m}{S_{ut}}\right)^2 = 1$$

Setting $\sigma_a = \sigma_m$ and solving the quadratic in σ_a gives

$$\sigma_a = \frac{S_{ut}^2}{2S'_e}\left(-1 + \sqrt{1 + \frac{4S'^2_e}{S_{ut}^2}}\right)$$

Setting $\sigma_a = \sigma/2$, $S_{ut} = S'_e/0.5$ gives

$$\sigma = \frac{S'_e}{0.5^2}\left[-1 + \sqrt{1 + 4(0.5)^2}\right] = 1.66S'_e$$

and $k_f = \sigma/S'_e = 1.66$. Since a Gerber locus runs in and among fatigue data and Goodman does not, we will use $k_f = 1.66$. The Marin equation for the fully corrected endurance strength is

$$S_e = k_a k_b k_c k_d k_e k_f S'_e$$
$$= 0.93(0.95)(1)(1)(1)1.66(190) = 279\text{ MPa}$$

For stress, we will first determine the fatigue stress-concentration factor K_f. For a 20° full-depth tooth the radius of the root fillet is denoted r_f, with a typically proportioned value of

$$r_f = 0.300 \text{ m} = 0.300(3) = 0.9 \text{ mm}$$

From Fig. A–13–6

$$\frac{r}{d} = \frac{r_f}{t} = \frac{0.9}{6.35} = 0.14$$

Since $D/d = \infty$, we approximate with $D/d = 3$, giving $K_t = 1.68$. From Fig. 6–26, $q = 0.62$. From Eq. (6–32),

$$K_f = 1 + (0.62)(1.68 - 1) = 1.42$$

For a design factor of $n_d = 3$, as used in Ex. 14–1, applied to the load or strength, the maximum bending stress is

$$\sigma_{\max} = K_f \sigma_{\text{all}} = \frac{S_e}{n_d}$$

$$\sigma_{\text{all}} = \frac{S_e}{K_f n_d} = \frac{279}{1.42(3)} = 65.5 \text{ MPa}$$

The transmitted load W^t is

$$W^t = \frac{F(m)Y\sigma_{\text{all}}}{K_v} = \frac{(38)(3)(0.296)(65.5)}{1.5} = 1473.5 \text{ N}$$

and the power is, with $V = 3.014$ m/s from Ex. 14–1,

$$hp = W^t V = 1473.5(3.014) = 4441 \text{ W}$$

Again, it should be emphasized that these results should be accepted *only* as preliminary estimates to alert you to the nature of bending in gear teeth.

In Example 14–2 our resources (Figure A–13–6) did not directly address stress concentration in gear teeth. A photoelastic investigation by Dolan and Broghamer reported in 1942 constitutes a primary source of information on stress concentration.[3] Mitchiner and Mabie[4] interpret the results in term of fatigue stress-concentration factor K_f as

$$K_f = H + \left(\frac{t}{r}\right)^L \left(\frac{t}{l}\right)^M \tag{14–7}$$

where

$$H = 0.34 - 0.458\,366\,2\phi$$

$$L = 0.316 - 0.458\,366\,2\phi$$

$$M = 0.290 + 0.458\,366\,2\phi$$

$$r = \frac{(b - r_f)^2}{(d/2) + b - r_f}$$

[3]T. J. Dolan and E. I. Broghamer, *A Photoelastic Study of the Stresses in Gear Tooth Fillets*, Bulletin 335, Univ. Ill. Exp. Sta., March 1942. See also W. D. Pilkey and D. F. Pilkey, *Peterson's Stress-Concentration Factors*, 3rd ed., John Wiley & Sons, Hoboken, NJ, 2008, pp. 407–409, 434–437.

[4]R. G. Mitchiner and H. H. Mabie, "Determination of the Lewis Form Factor and the AGMA Geometry Factor J of External Spur Gear Teeth," *J. Mech. Des.*, Vol. 104, No. 1, Jan. 1982, pp. 148–158.

In these equations l and t are from the layout in Figure 14–1, ϕ is the pressure angle, r_f is the fillet radius, b is the dedendum, and d is the pitch diameter. It is left as an exercise for the reader to compare K_f from Equation (14–7) with the results of using the approximation of Figure A–13–6 in Example 14–2.

14–2 Surface Durability

In this section we are interested in the failure of the surfaces of gear teeth, which is generally called *wear. Pitting,* as explained in Section 6–18, is a surface fatigue failure due to many repetitions of high contact stresses. Other surface failures are *scoring,* which is a lubrication failure, and *abrasion,* which is wear due to the presence of foreign material.

To obtain an expression for the surface-contact stress, we shall employ the Hertz theory. In Equation (3–76), it was shown that the contact stress between two cylinders may be computed from the equation

$$p_{max} = \frac{2F}{\pi b l} \tag{a}$$

where p_{max} = largest surface pressure

F = force pressing the two cylinders together

l = length of cylinders

and half-width b is obtained from Equation (3–75), given by

$$b = \left[\frac{2F}{\pi l} \frac{(1-\nu_1^2)/E_1 + (1-\nu_2^2)/E_2}{1/d_1 + 1/d_2}\right]^{1/2} \tag{14–8}$$

where ν_1, ν_2, E_1, and E_2 are the elastic constants and d_1 and d_2 are the diameters, respectively, of the two contacting cylinders.

To adapt these relations to the notation used in gearing, we replace F by $W^t/\cos\phi$, d by $2r$, and l by the face width F. With these changes, we can substitute the value of b as given by Equation (14–8) in Equation (*a*). Replacing p_{max} by σ_C, the *surface-contact stress (Hertzian stress)* is found from the equation

$$\sigma_C = -\left[\frac{W^t}{\pi F \cos\phi} \frac{1/r_1 + 1/r_2}{(1-\nu_1^2)/E_1 + (1-\nu_2^2)/E_2}\right]^{1/2} \tag{14–9}$$

where r_1 and r_2 are the instantaneous values of the radii of curvature on the pinion- and gear-tooth profiles, respectively, at the point of contact. By accounting for load sharing in the value of W^t used, Equation (14–9) can be solved for the Hertzian stress for any or all points from the beginning to the end of tooth contact. Of course, pure rolling exists only at the pitch point. Elsewhere the motion is a mixture of rolling and sliding. Equation (14–9) does not account for any sliding action in the evaluation of stress. We note that AGMA uses μ for Poisson's ratio instead of ν as is used here.

We have already noted that the first evidence of wear occurs near the pitch line. The radii of curvature of the tooth profiles at the pitch point are

$$r_1 = \frac{d_P \sin\phi}{2} \qquad r_2 = \frac{d_G \sin\phi}{2} \tag{14–10}$$

where ϕ is the pressure angle and d_P and d_G are the pitch diameters of the pinion and gear, respectively.

Note, in Equation (14–9), that the denominator of the second group of terms contains four elastic constants, two for the pinion and two for the gear. As a simple means of combining and tabulating the results for various combinations of pinion and gear materials, AGMA defines an *elastic coefficient* C_P by the equation

$$C_P = \left[\frac{1}{\pi \left(\dfrac{1 - \nu_P^2}{E_P} + \dfrac{1 - \nu_G^2}{E_G} \right)} \right]^{1/2} \tag{14–11}$$

With this simplification, and the addition of a velocity factor K_v, Equation (14–9) can be written as

$$\sigma_C = -C_p \left[\frac{K_v W^t}{F \cos \phi} \left(\frac{1}{r_1} + \frac{1}{r_2} \right) \right]^{1/2} \tag{14–12}$$

where the sign is negative because σ_C is a compressive stress.

EXAMPLE 14–3

The pinion of Exs. 14–1 and 14–2 is to be mated with a 50-tooth gear manufactured of ASTM No. 50 cast iron. Using the tangential load of 1700 N, estimate the factor of safety of the drive based on the possibility of a surface fatigue failure. The surface endurance strength of cast iron can be estimated from $S_c = 2.21\ H_B$ MPa for 10^8 cycles.

Solution

From Table A–5 we find the elastic constants to be E_P = 207 000 MPa, ν_P = 0.292, E_G = 99 900 MPa, ν_G = 0.211. We substitute these in Eq. (14–11) to get the elastic coefficient as

$$C_P = \left\{ \pi \left[\frac{1 - (0.292)^2}{207\ 000} + \frac{1 - (0.211)^2}{999\ 000} \right] \right\}^{-(1/2)} = 150.91$$

From Ex. 14–1, the pinion pitch diameter is d_P = 48 mm. The value for the gear is d_G = 50(3) = 150 mm. Then Eq. (14–12) is used to obtain the radii of curvature at the pitch points. Thus

$$r_1 = \frac{48 \sin 20^\circ}{2} = 8.208 \text{ mm} \qquad r_2 = \frac{150 \sin 20^\circ}{2} = 25.652 \text{ mm}$$

The face width is given as F = 38 mm. Use K_v = 1.50 from Ex. 14–1. Substituting all these values in Eq. (14–12) with $\phi = 20^\circ$ gives the contact stress as

$$\sigma_C = -150.91 \left[\frac{(1.50)(1552.2)}{(38)(\cos 20^\circ)} \left(\frac{1}{8.208} + \frac{1}{25.652} \right) \right]^{1/2} = -489 \text{ MPa}$$

Table A–23 gives H_B = 262 for ASTM No. 50 cast iron. Therefore S_C = 2.21 (262) = 579 MPa. Contact stress is not linear with respect to the transmitted load [see Eq. (14–12)]. If the factor of safety is defined as the loss-of-function load divided by the imposed load, then the ratio of loads is the ratio of stresses squared. In other words,

$$n = \frac{\text{loss-of-function load}}{\text{imposed load}} = \frac{S_C^2}{\sigma_C^2} = \left(\frac{579}{489} \right)^2 = 1.40$$

One is free to define the factor of safety as S_C/σ_C. Awkwardness comes when one compares the factor of safety in bending fatigue with the factor of safety in surface fatigue for a particular gear. Suppose the factor of safety of this gear in bending fatigue is 1.20 and the factor of safety in surface fatigue is 1.40 as above. The threat, since 1.40 is greater than 1.20, is in bending fatigue since both numbers are based on load ratios. If the factor of safety in surface fatigue is based on $S_C/\sigma_C = \sqrt{1.40} = 1.18$, then 1.20 is greater than 1.18, but the threat is not from surface fatigue. The surface fatigue factor of safety can be defined either way. One way has the burden of requiring a squared number before numbers that instinctively seem comparable can be compared.

14–3 AGMA Stress Equations

The AGMA method considers two modes of failure—fracture of a tooth due to bending fatigue and pitting of the tooth surface due to contact-stress fatigue. Alternate terms are used interchangeably for surface failure, including pitting, contact stress, surface fatigue, and wear. Both failure modes are due to fatigue; therefore, the AGMA method is applicable to situations in which the gears are expected to endure many cycles at stress levels significantly below the yield strength. The AGMA method embeds the details of fatigue analysis into equations for stress and allowable stress that are compared directly with one another. In this section, the stress equations are presented. In AGMA terminology, the stresses are called *stress numbers,* as contrasted with actual applied stresses, and are designated by a lowercase letter s instead of the Greek lowercase σ we have used in this book (and shall continue to use). The fundamental stress equations for bending are

$$\sigma_b = W^t K_o K_v K_s \frac{1}{bm_t} \frac{K_H K_B}{Y_J} \tag{14–13}$$

where,

W^t is the tangential transmitted load, N
K_o is the overload factor
K_v is the dynamic factor
K_s is the size factor
P is the transverse diametral pitch
b is the face width of the narrower member, mm
K_H is the load-distribution factor
K_B is the rim-thickness factor
Y_J is the geometry factor for bending strength (which includes root fillet stress-concentration factor K_f)
(m_t) is the transverse metric module

The fundamental equations for pitting resistance (contact stress) are

$$\sigma_c = Z_E \sqrt{W^t K_o K_v K_s \frac{K_H}{d_{w1} b} \frac{Z_R}{Z_I}} \tag{14–14}$$

where W^t, K_o, K_v, K_s, K_m, F, and b are the same terms as defined for Equation (14–13). For SI units, the additional terms are

Z_E is an elastic coefficient, $\sqrt{\text{N/mm}^2}$
Z_R is the surface condition factor
d_{w1} is the pitch diameter of the *pinion,* mm
Z_I is the geometry factor for pitting resistance

The evaluation of all these factors is explained in the sections that follow.

14–4 AGMA Strength Equations

Instead of using the term *strength,* AGMA uses data termed *allowable stress numbers* and designates these by the symbols s_{at} and s_{ac}. It will be less confusing here if we continue the practice in this book of using the uppercase letter S to designate strength and the lowercase Greek letters σ and τ for stress. To make it perfectly clear we shall use the term *gear strength* as a replacement for the phrase *allowable stress numbers* as used by AGMA.

Following this convention, values for *gear bending strength,* designated here as S_t, are to be found in Figures 14–2, 14–3, and 14–4, and in Tables 14–3 and 14–4. These gear strengths are determined or estimated from testing and accumulated field experiences. They are based on 10 million stress cycles, unidirectional loading, and 99 percent reliability of material strength. These values are suitable for determining achievable strengths for design purposes, but manufacturer data should be obtained for the specific gears to be purchased. Since gear strengths are not identified with other strengths such as S_{ut}, S_e, or S_y as used elsewhere in this book, their use should be restricted to gear problems.

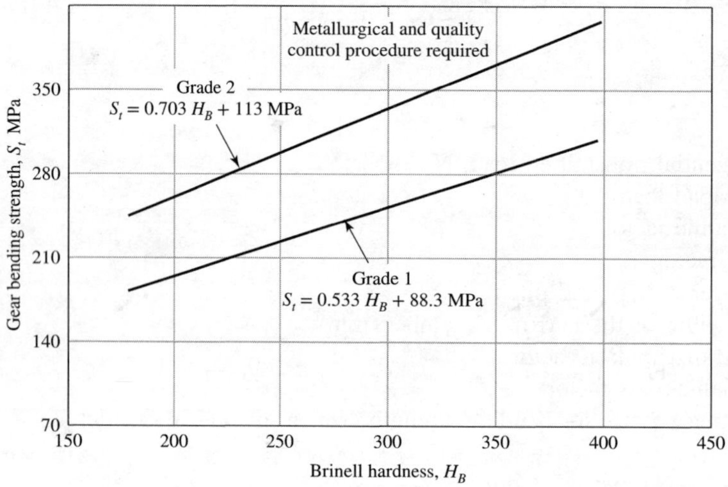

Figure 14–2

Gear bending strength for through-hardened steels, S_t. *(Source: Data from ANSI/AGMA 2001-D04 and 2101-D04.)*

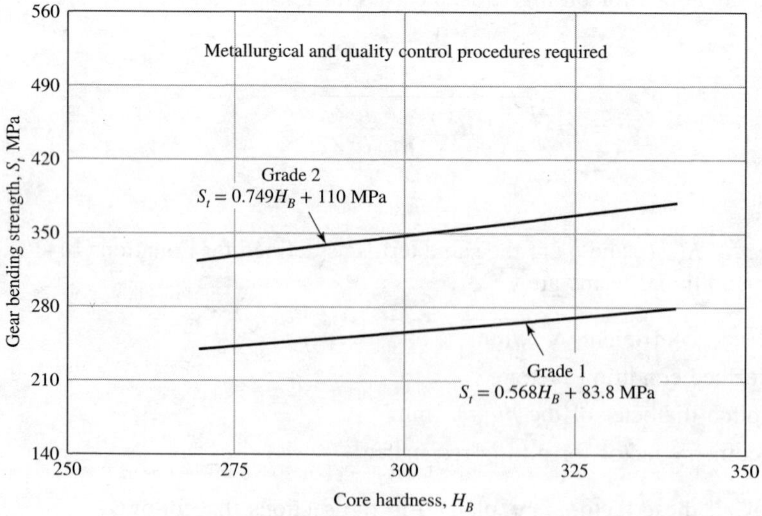

Figure 14–3

Gear bending strength for nitrided through-hardened steel gears (i.e., AISI 4140, 4340), S_t. *(Source: Data from ANSI/AGMA 2001-D04 and 2101-D04.)*

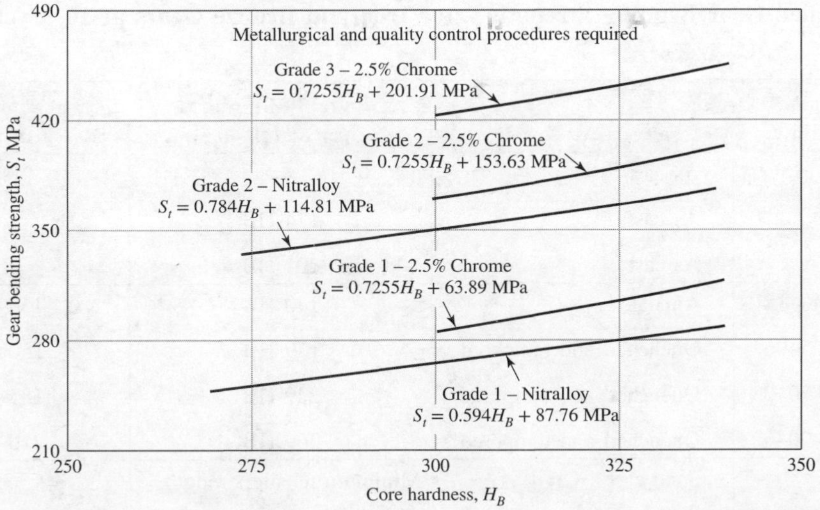

Figure 14–4

Gear bending strength for nitriding steel gears, S_t. *(Source: Data from ANSI/AGMA 2001-D04, 2101-D04.)*

Table 14–3 Repeatedly Applied Gear Bending Strength S_t at 10^7 Cycles and 0.99 Reliability for Steel Gears

Material Designation	Heat Treatment	Minimum Surface Hardness[1]	Gear Bending Strength S_t,[2] MPa		
			Grade 1	Grade 2	Grade 3
Steel[3]	Through-hardened Flame[4] or induction hardened[4] with type A pattern[5]	See Fig. 14–2 See Table 8*	See Fig. 14–2 310	See Fig. 14–2 380	— —
	Flame[4] or induction hardened[4] with type B pattern[5]	See Table 8*	150	150	—
	Carburized and hardened	See Table 9*	380	440 or 483[6]	517
	Nitrided[4,7] (through-hardened steels)	83.5 HR15N	See Fig. 14–3	See Fig. 14–3	—
Nitralloy 135M, Nitralloy N, and 2.5% chrome (no aluminum)	Nitrided[4,7]	87.5 HR15N	See Fig. 14–4	See Fig. 14–4	See Fig. 14–4

Notes: See ANSI/AGMA 2001-D04 for references cited in notes 1–7.

[1]Hardness to be equivalent to that at the root diameter in the center of the tooth space and face width.

[2]See tables 7 through 10 for major metallurgical factors for each stress grade of steel gears.

[3]The steel selected must be compatible with the heat treatment process selected and hardness required.

[4]The gear bending strength indicated may be used with the case depths prescribed in 16.1.

[5]See figure 12 for type A and type B hardness patterns.

[6]If bainite and microcracks are limited to grade 3 levels, 483 MPa may be used.

[7]The overload capacity of nitrided gears is low. Since the shape of the effective S-N curve is flat, the sensitivity to shock should be investigated before proceeding with the design. [7]

*Tables 8 and 9 of ANSI/AGMA 2001-D04 are comprehensive tabulations of the major metallurgical factors affecting S_t and S_c of flame-hardened and induction-hardened (Table 8) and carburized and hardened (Table 9) steel gears.

Source: Data from ANSI/AGMA 2001-D04.

Table 14–4 Repeatedly Applied Gear Bending Strength S_t for Iron and Bronze Gears at 10^7 Cycles and 0.99 Reliability

Material	Material Designation[1]	Heat Treatment	Typical Minimum Surface Hardness[2]	Gear Bending Strength, S_t,[3] MPa
ASTM A48 gray cast iron	Class 20	As cast	—	34
	Class 30	As cast	174 HB	58
	Class 40	As cast	201 HB	89
ASTM A536 ductile (nodular) Iron	Grade 60–40–18	Annealed	140 HB	150–227
	Grade 80–55–06	Quenched and tempered	179 HB	150–227
	Grade 100–70–03	Quenched and tempered	229 HB	185–275
	Grade 120–90–02	Quenched and tempered	269 HB	210–300
Bronze		Sand cast	Minimum tensile strength 275 MPa	39
	ASTM B–148 Alloy 954	Heat treated	Minimum tensile strength 620 MPa	160

Notes:

[1]See ANSI/AGMA 2004-B89, *Gear Materials and Heat Treatment Manual.*

[2]Measured hardness to be equivalent to that which would be measured at the root diameter in the center of the tooth space and face width.

[3]The lower values should be used for general design purposes. The upper values may be used when:

High quality material is used.

Section size and design allow maximum response to heat treatment.

Proper quality control is effected by adequate inspection.

Operating experience justifies their use.

Source: Data from ANSI/AGMA 2001-D04.

When two-way (reversed) loading occurs, as with idler gears, AGMA recommends using 70 percent of S_t values. This is equivalent to $1/0.70 = 1.43$ as a value of k_f in Example 14–2. The recommendation falls between the value of $k_f = 1.33$ for a Goodman failure locus and $k_f = 1.66$ for a Gerber failure locus.

The strengths are modified by various factors that produce limiting values of the bending stress and the contact stress.

The equation for the allowable bending stress is

$$\sigma_{b,\text{all}} = \frac{S_t}{S_F}\frac{Y_N}{Y_\theta Y_Z} \tag{14–15}$$

where for SI units,

S_t is the gear bending strength, N/mm^2
Y_N is the stress-cycle factor for bending stress
Y_θ are the temperature factors
Y_Z are the reliability factors
S_F is the AGMA safety factor

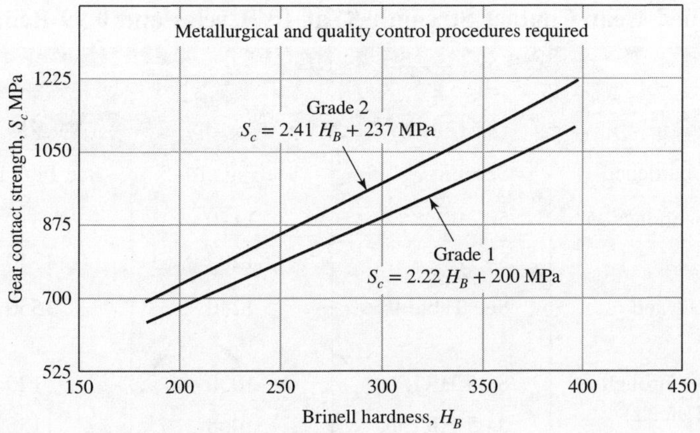

Figure 14–5

Gear contact strength S_c at 10^7 cycles and 0.99 reliability for through-hardened steel gears. *(Source: Data from ANSI/AGMA 2001-D04 and 2101-D04.)*

The equation for the allowable contact stress $\sigma_{c,\text{all}}$ is

$$\sigma_{c,\text{all}} = \frac{S_c}{S_H}\frac{Z_N Z_W}{Y_\theta Y_Z} \tag{14–16}$$

where

S_c is the gear contact strength, N/mm^2
Z_N is the stress-cycle factor
Z_W are the hardness ratio factors for pitting resistance
Y_θ are the temperature factors
Y_Z are the reliability factors
S_H is the AGMA safety factor

The values for the gear contact strength, designated here as S_c, are to be found in Figure 14–5 and Tables 14–5, 14–6, and 14–7.

Table 14–5 Nominal Temperature Used in Nitriding and Hardnesses Obtained

Steel	Temperature Before Nitriding, °C	Nitriding, °C	Hardness, Rockwell *C* Scale	
			Case	Core
Nitralloy 135*	620	524	62–65	30–35
Nitralloy 135M	620	524	62–65	32–36
Nitralloy N	538	524	62–65	40–44
AISI 4340	593	524	48–53	27–35
AISI 4140	593	524	49–54	27–35
31 Cr Mo V 9	593	524	58–62	27–33

*Nitralloy is a trademark of the Nitralloy Corp., New York.

Source: Dudley, Darle W. *Handbook of Practical Gear Design*. Rev. ed. New York: McGraw Hill, 1984.

Table 14–6 Repeatedly Applied Gear Contact Strength S_c at 10^7 Cycles and 0.99 Reliability for Steel Gears

Material Designation	Heat Treatment	Minimum Surface Hardness[1]	Gear Contact Strength,[2] S_c, MPa		
			Grade 1	Grade 2	Grade 3
Steel[3]	Through hardened[4]	See Fig. 14–5	See Fig. 14–5	See Fig. 14–5	—
	Flame[5] or induction hardened[5]	50 HRC	1170	1310	—
		54 HRC	1205	1340	—
	Carburized and hardened[5]	See Table 9*	1240	1550	1896
	Nitrided[5] (through hardened steels)	83.5 HR15N	1030	1120	1205
		84.5 HR15N	1068	1158	1240
2.5% chrome (no aluminum)	Nitrided[5]	87.5 HR15N	1068	1185	1300
Nitralloy 135M	Nitrided[5]	90.0 HR15N	1170	1260	1340
Nitralloy N	Nitrided[5]	90.0 HR15N	1185	1295	1410
2.5% chrome (no aluminum)	Nitrided[5]	90.0 HR15N	1210	1350	1490

Notes: See ANSI/AGMA 2001-D04 for references cited in notes 1–5.

[1]Hardness to be equivalent to that at the start of active profile in the center of the face width.

[2]See Tables 7 through 10 for major metallurgical factors for each stress grade of steel gears.

[3]The steel selected must be compatible with the heat treatment process selected and hardness required.

[4]These materials must be annealed or normalized as a minimum.

[5]The gear contact strengths indicated may be used with the case depths prescribed in 16.1.

*Table 9 of ANSI/AGMA 2001-D04 is a comprehensive tabulation of the major metallurgical factors affecting S_t and S_c of carburized and hardened steel gears.

Source: Data from ANSI/AGMA 2001-D04.

14–5 Geometry Factors Z_I and Y_J

We have seen how the factor Y is used in the Lewis equation to introduce the effect of tooth form into the stress equation. The AGMA factors[5] I and J are intended to accomplish the same purpose in a more involved manner.

The determination of I and J depends upon the *face-contact ratio* m_F. This is defined as

$$m_F = \frac{F}{p_x} \tag{14–17}$$

where p_x is the axial pitch and F is the face width. For spur gears, $m_F = 0$.

Low-contact-ratio (LCR) helical gears having a small helix angle or a thin face width, or both, have face-contact ratios less than unity ($m_F \leq 1$), and will not be considered here. Such gears have a noise level not too different from that for spur gears. Consequently we shall consider here only spur gears with $m_F = 0$ and conventional helical gears with $m_F > 1$.

[5]A useful reference is AGMA 908-B89, *Geometry Factors for Determining Pitting Resistance and Bending Strength of Spur, Helical and Herringbone Gear Teeth.*

Table 14–7 Repeatedly Applied Gear Contact Strength S_c 10^7 Cycles and 0.99 Reliability for Iron and Bronze Gears

Material	Material Designation[1]	Heat Treatment	Typical Minimum Surface Hardness[2]	Gear Contact Strength,[3] S_c, MPa
ASTM A48 gray cast iron	Class 20	As cast	—	344–413
	Class 30	As cast	174 HB	448–517
	Class 40	As cast	201 HB	517–585
ASTM A536 ductile (nodular) iron	Grade 60–40–18	Annealed	140 HB	530–634
	Grade 80–55–06	Quenched and tempered	179 HB	530–634
	Grade 100–70–03	Quenched and tempered	229 HB	634–772
	Grade 120–90–02	Quenched and tempered	269 HB	1710–868
Bronze	—	Sand cast	Minimum tensile strength 275 MPa	206
	ASTM B-148 Alloy 954	Heat treated	Minimum tensile strength 620 MPa	448

Notes:

[1]See ANSI/AGMA 2004-B89, *Gear Materials and Heat Treatment Manual.*

[2]Hardness to be equivalent to that at the start of active profile in the center of the face width.

[3]The lower values should be used for general design purposes. The upper values may be used when:

High-quality material is used.

Section size and design allow maximum response to heat treatment.

Proper quality control is effected by adequate inspection.

Operating experience justifies their use.

Source: Data from ANSI/AGMA 2001-D04.

Bending-Strength Geometry Factor Y_J

The AGMA factor Y_J employs a modified value of the Lewis form factor, also denoted by Y; a *fatigue stress-concentration factor* K_f; and a tooth *load-sharing ratio* m_N. The resulting equation for Y_J for spur and helical gears is

$$Y_J = \frac{Y}{K_f m_N} \tag{14–18}$$

It is important to note that the form factor Y in Equation (14–18) is *not* the Lewis factor at all. The value of Y here is obtained from calculations within AGMA 908-B89, and is often based on the highest point of single-tooth contact.

The factor K_f in Equation (14–18) is called a *stress-correction factor* by AGMA. It is based on a formula deduced from a photoelastic investigation of stress concentration in gear teeth over 50 years ago.

The load-sharing ratio m_N is equal to the face width divided by the minimum total length of the lines of contact. This factor depends on the transverse contact ratio m_p, the face-contact ratio m_F, the effects of any profile modifications, and the tooth deflection. For spur gears, $m_N = 1.0$. For helical gears having a face-contact ratio $m_F > 2.0$, a conservative approximation is given by the equation

$$m_N = \frac{p_N}{0.95Z} \tag{14–19}$$

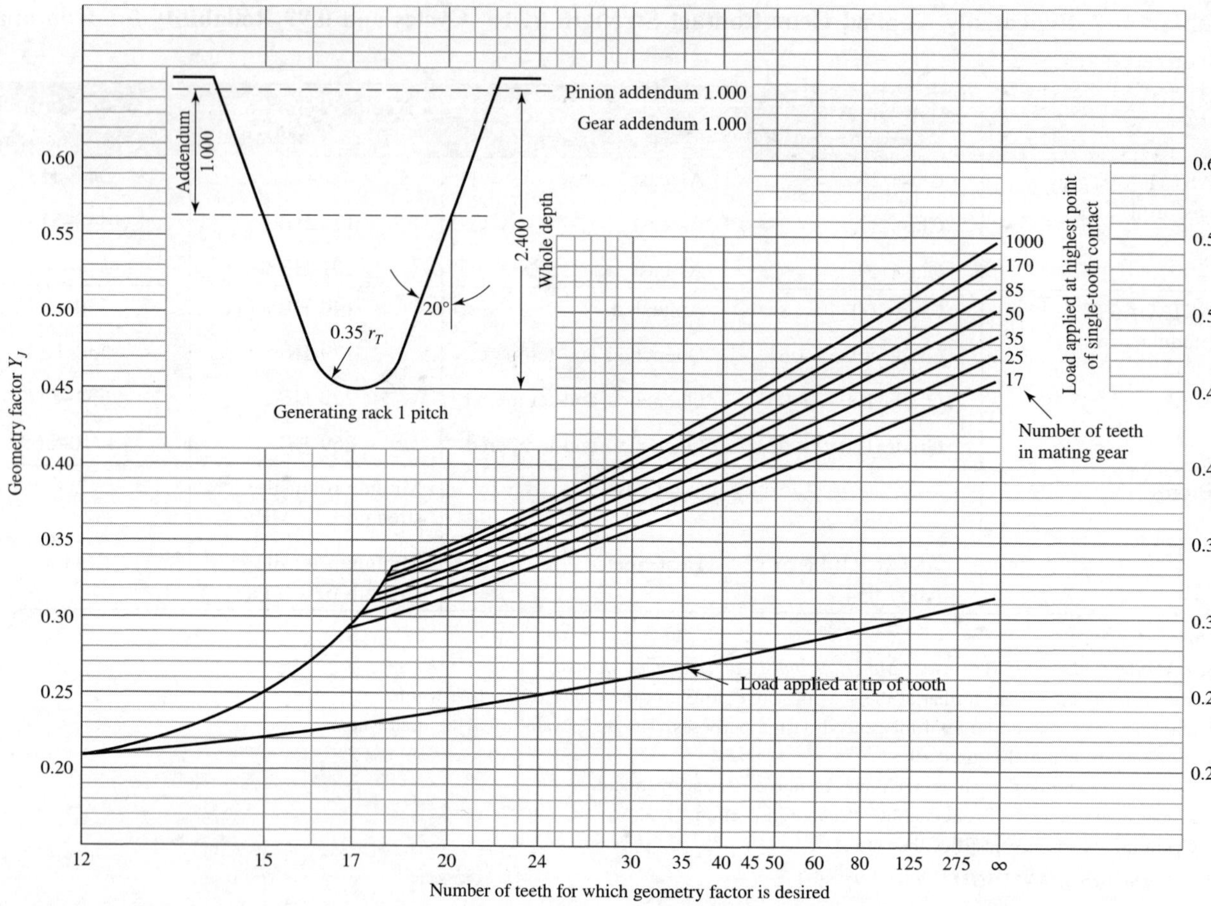

Figure 14–6

Spur-gear geometry factors Y_J. *(Source: The graph is from AGMA 218.01, which is consistent with tabular data from the current AGMA 908-B89. The graph is convenient for design purposes.)*

where p_N is the normal base pitch and Z is the length of the line of action in the transverse plane (distance L_{ab} in Figure 13–11).

Use Figure 14–6 to obtain the geometry factor Y_J for spur gears having a 20° pressure angle and full-depth teeth. Use Figures 14–7 and 14–8 for helical gears having a 20° normal pressure angle and face-contact ratios of $m_F = 2$ or greater. For other gears, consult the AGMA standard.

Surface-Strength Geometry Factor Z_I

The factor Z_I is also called the *pitting-resistance geometry factor* by AGMA. We will develop an expression for I by noting that the sum of the reciprocals of Equation (14–12), from Equation (14–10), can be expressed as

$$\frac{1}{r_1} + \frac{1}{r_2} = \frac{2}{\sin \phi_t}\left(\frac{1}{d_P} + \frac{1}{d_G}\right) \tag{a}$$

where we have replaced ϕ by ϕ_t, the transverse pressure angle, so that the relation will apply to helical gears too. Now define *speed ratio* m_G as

$$m_G = \frac{N_G}{N_P} = \frac{d_G}{d_P} \tag{14–20}$$

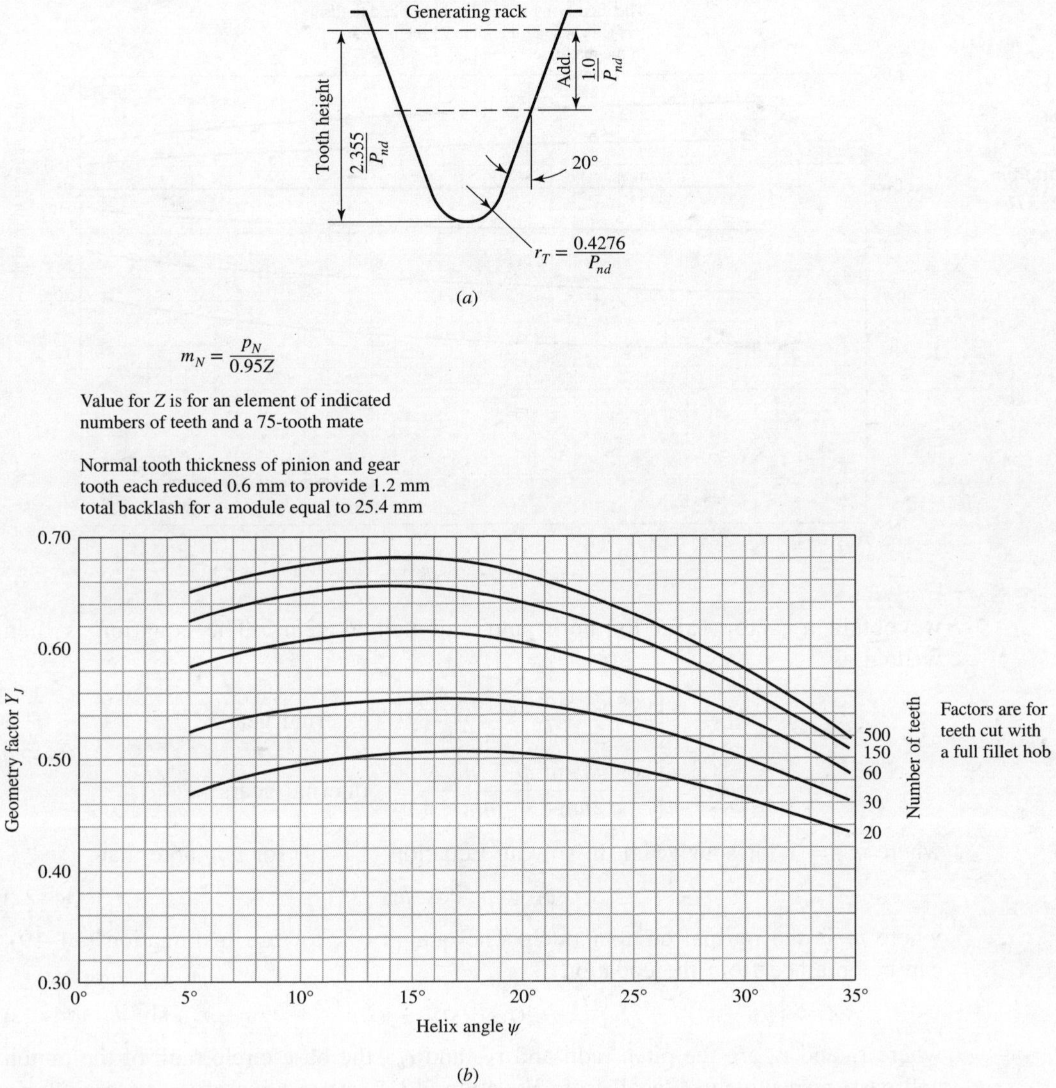

Figure 14–7

Helical-gear geometry factors Z_I'. *(Source: The graph is from AGMA 218.01, which is consistent with tabular data from the current AGMA 908-B89. The graph is convenient for design purposes.)*

Equation (*a*) can now be written

$$\frac{1}{r_1} + \frac{1}{r_2} = \frac{2}{d_P \sin \phi_t} \frac{m_G + 1}{m_G} \tag{b}$$

Now substitute Equation (*b*) for the sum of the reciprocals in Equation (14–12). The result is found to be

$$\sigma_c = |\sigma_C| = C_p \left[\frac{K_v W^t}{d_P F} \frac{1}{\dfrac{\cos \phi_t \sin \phi_t}{2} \dfrac{m_G}{m_G + 1}} \right]^{1/2} \tag{c}$$

The geometry factor I for external spur and helical gears is the denominator of the second term in the brackets in Equation (*c*). By adding the load-sharing ratio m_N,

Figure 14–8

Z_I'-factor multipliers for use with Figure 14–7 to find Y_J. *(Source: The graph is from AGMA 218.01, which is consistent with tabular data from the current AGMA 908-B89. The graph is convenient for design purposes.)*

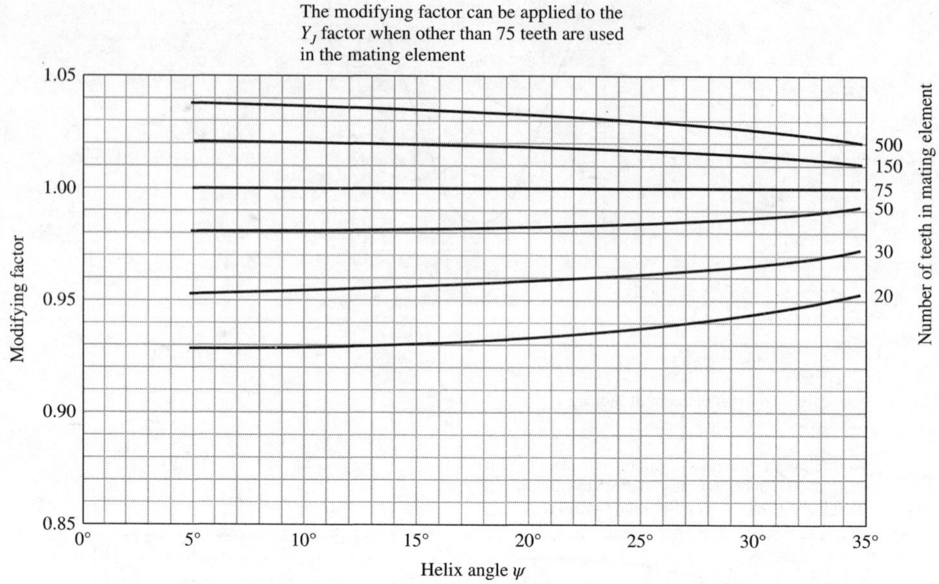

we obtain a factor valid for both spur and helical gears. The equation is then written as

$$Y_J = \begin{cases} \dfrac{\cos \phi_t \sin \phi_t}{2m_N} \dfrac{m_G}{m_G + 1} & \text{external gears} \\ \dfrac{\cos \phi_t \sin \phi_t}{2m_N} \dfrac{m_G}{m_G - 1} & \text{internal gears} \end{cases} \tag{14–21}$$

where $m_N = 1$ for spur gears. In solving Equation (14–19) for m_N, note that

$$p_N = p_n \cos \phi_n \tag{14–22}$$

where p_n is the normal circular pitch. The quantity Z, for use in Equation (14–19), can be obtained from the equation

$$Z = \left[(r_P + a)^2 - r_{bP}^2\right]^{1/2} + \left[(r_G + a)^2 - r_{bG}^2\right]^{1/2} - (r_P + r_G) \sin \phi_t \tag{14–23}$$

where r_P and r_G are the pitch radii and r_{bP} and r_{bG} the base-circle radii of the pinion and gear, respectively.[6] Recall from Equation (13–6), the radius of the base circle is

$$r_b = r \cos \phi_t \tag{14–24}$$

Certain precautions must be taken in using Equation (14–23). The tooth profiles are not conjugate below the base circle, and consequently, if either one or the other of the first two terms in brackets is larger than the third term, then it should be replaced by the third term. In addition, the effective outside radius is sometimes less than $r + a$, owing to removal of burrs or rounding of the tips of the teeth. When this is the case, always use the effective outside radius instead of $r + a$.

14–6 The Elastic Coefficient Z_E

Values of Z_E may be computed directly from Equation (14–11) or obtained from Table 14–8.

[6]For a development, see Joseph E. Shigley and John J. Uicker Jr., *Theory of Machines and Mechanisms*, McGraw Hill, New York, 1980, p. 262.

Table 14–8 Elastic Coefficient Z_E, $\sqrt{\text{MPa}}$

Pinion Material	Pinion Modulus of Elasticity E_P MPa*	Gear Material and Modulus of Elasticity E_G, MPa*					
		Steel 200 000	Malleable Iron 175 000	Nodular Iron 170 000	Cast Iron 150 000	Aluminum Bronze 120 000	Tin Bronze 110 000
Steel	200 000	191	181	179	174	162	158
Malleable iron	175 000	181	174	172	168	158	154
Nodular iron	170 000	179	172	170	166	156	152
Cast iron	150 000	174	168	166	163	154	149
Aluminum bronze	120 000	162	158	156	154	145	141
Tin bronze	110 000	158	154	152	149	141	137

Poisson's ratio = 0.30.

*When more exact values for modulus of elasticity are obtained from roller contact tests, they may be used.

Source: Data from AGMA 218.01.

14–7 Dynamic Factor K_v

As noted earlier, dynamic factors are used to account for inaccuracies in the manufacture and meshing of gear teeth in action. *Transmission error* is defined as the departure from uniform angular velocity of the gear pair. Some of the effects that produce transmission error are:

- Inaccuracies produced in the generation of the tooth profile; these include errors in tooth spacing, profile lead, and runout
- Vibration of the tooth during meshing due to the tooth stiffness
- Magnitude of the pitch-line velocity
- Dynamic unbalance of the rotating members
- Wear and permanent deformation of contacting portions of the teeth
- Gearshaft misalignment and the linear and angular deflection of the shaft
- Tooth friction

In an attempt to account for these effects, AGMA has defined a set of *quality numbers*, Q_v.[7] These numbers define the tolerances for gears of various sizes manufactured to a specified accuracy. Quality numbers 3 to 7 will include most commercial-quality gears. Quality numbers 8 to 12 are of precision quality. The following equations for the dynamic factor are based on these Q_v numbers:

$$K_v = \left(\frac{A + \sqrt{200V}}{A}\right)^B \qquad V \text{ in m/s} \tag{14–25a}$$

where

$$\begin{aligned} A &= 50 + 56(1 - B) \\ B &= 0.25(12 - Q_v)^{2/3} \end{aligned} \tag{14–25b}$$

Figure 14–9 graphically represents Equation (14–25). The maximum recommended pitch-line velocity for a given quality number is represented by the end point of each Q_v curve, and is given by

$$(V_t)_{\max} = \frac{[A + (Q_v - 3)]^2}{200} \quad \text{m/s} \tag{14–26}$$

14–8 Overload Factor K_o

The overload factor K_o is intended to make allowance for loading conditions that may lead to increases in the nominal tangential load W^t in a particular application. Examples include variations in torque from the mean value due to firing of cylinders in an

[7]AGMA 2000-A88. ANSI/AGMA 2001-D04, adopted in 2004, replaced the quality number Q_v with the transmission accuracy level number A_v and incorporated ANSI/AGMA 2015-1-A01. A_v ranges from 6 to 12, with lower numbers representing greater accuracy. The Q_v approach was maintained as an alternate approach, and resulting K_v values are comparable.

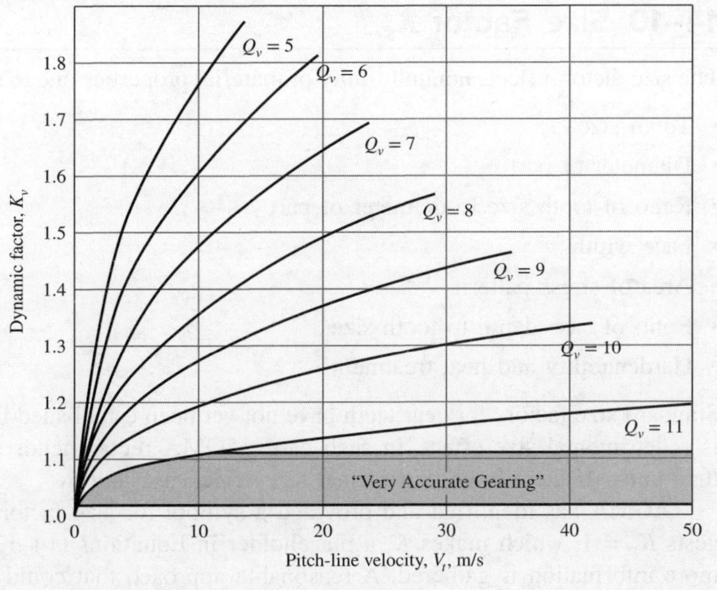

Figure 14–9

Dynamic factor K_v. The equations to these curves are given by Equation (14–25) and the end points by Equation (14–26).

(Source: Data from ANSI/AGMA 2001-D04, Annex A.)

Table 14–9 Suggested Values for Overload Factor, K_o

	Driven Machine		
Power source	**Uniform**	**Moderate shock**	**Heavy shock**
Uniform	1.00	1.25	1.75
Light shock	1.25	1.50	2.00
Medium shock	1.50	1.75	2.25

internal combustion engine or reaction to torque variations in a piston pump drive. Table 14–9 provides suggested values based on the nature of the load and the power source. There are other similar factors such as application factor or service factor. These factors are established after considerable field experience in a particular application.[8]

14–9 Surface Condition Factor Z_R

The surface condition factor Z_R is to account for surface conditions that may detrimentally affect pitting resistance. It depends on

- Surface finish as affected by, but not limited to, cutting, shaving, lapping, grinding, shotpeening
- Residual stress
- Plastic effects (work hardening)

Standard surface conditions for gear teeth have not yet been established. When a detrimental surface finish effect is known to exist, AGMA specifies a value of Z_R greater than unity.

[8]An extensive list of service factors appears in Howard B. Schwerdlin, "Couplings," Chap. 16 in Joseph E. Shigley, Charles R. Mischke, and Thomas H. Brown, Jr. (eds.), *Standard Handbook of Machine Design*, 3rd ed., McGraw Hill, New York, 2004.

14–10 Size Factor K_s

The size factor reflects nonuniformity of material properties due to size. It depends upon

- Tooth size
- Diameter of part
- Ratio of tooth size to diameter of part
- Face width
- Area of stress pattern
- Ratio of case depth to tooth size
- Hardenability and heat treatment

Standard size factors for gear teeth have not yet been established for cases where there is a detrimental size effect. In such cases AGMA recommends a size factor greater than unity. If there is no detrimental size effect, use unity.

AGMA has identified and provided a symbol for size factor. Also, AGMA suggests $K_s = 1$, which makes K_s a placeholder in Equations (14–13) and (14–14) until more information is gathered. A reasonable approach that could be applied, particularly for larger gear teeth, incorporates the method for the size factor used in Equation (6–19) for fatigue of a non-rotating rectangular section. From Table 13–1, $l = a + b = 2.25/P$. The tooth thickness t in Figure 14–6 is given in Section 14–1, Equation (*b*), as $t = \sqrt{4lx}$ where $x = 3Y/(2P)$ from Equation (14–3). From Equation (6–24), the equivalent diameter d_e of a rectangular section in bending is $d_e = 0.808\sqrt{Ft}$. From Equation (6–19), $k_b = (d_e/0.3)^{-0.107}$. Noting that K_s is the reciprocal of k_b, we find the result of all the algebraic substitution is

$$K_s = \frac{1}{k_b} = 1.192\left(\frac{F\sqrt{Y}}{P}\right)^{0.0535} \tag{14–27}$$

where Y is the Lewis form factor from Table 14–2. K_s can be viewed as Lewis's geometry incorporated into the Marin size factor in fatigue. You may set $K_s = 1$, when the size of the teeth is not expected to be a detrimental factor. For larger teeth, if more specific data are not available, Equation (14–27) provides some compensation for the detrimental effect on bending fatigue. There is no rationale to use Equation (14–27) for contact stress.

14–11 Load-Distribution Factor K_H

The load-distribution factor modifies the stress equations to reflect nonuniform distribution of load across the line of contact. The ideal is to locate the gear "midspan" between two bearings at the zero slope place when the load is applied. However, this is not always possible. The following procedure is applicable to

- Net face width to pinion pitch diameter ratio $b/d_P \leq 2$
- Gear elements mounted between the bearings
- Face widths up to 1000 mm
- Contact, when loaded, across the full width of the narrowest member

The load-distribution factor under these conditions is currently given by the *face load distribution factor*, C_{mf}, where

$$K_H = C_{mf} = 1 + C_{mc}(C_{pf}C_{pm} + C_{ma}C_e) \tag{14–28}$$

Table 14–10 Empirical Constants *A*, *B*, and *C* for Equation (14–32), Face Width *F* in mm

Condition	*A*	*B*	*C*
Open gearing	0.247	$6.57 \cdot 10^{-4}$	$-1.19 \cdot 10^{-7}$
Commercial, enclosed units	0.127	$6.22 \cdot 10^{-4}$	$-1.44 \cdot 10^{-7}$
Precision, enclosed units	0.0675	$5.04 \cdot 10^{-4}$	$-1.44 \cdot 10^{-7}$
Extraprecision enclosed gear units	0.00360	$4.02 \cdot 10^{-4}$	$-1.27 \cdot 10^{-7}$

Source: ANSI/AGMA 2101-D04.

The parameters that contribute to the load-distribution factor are:

C_{mc} – lead (crown) correction factor
C_{pf} – pinion proportion factor
C_{pm} – pinion proportion modifier
C_{ma} – mesh alignment factor
C_e – mesh alignment correction factor

These factors are defined as follows:

$$C_{mc} = \begin{cases} 1 & \text{for uncrowned teeth} \\ 0.8 & \text{for crowned teeth} \end{cases} \tag{14–29}$$

$$C_{pf} = \begin{cases} \dfrac{b}{10d_P} - 0.025 & b \le 25 \text{ mm} \\ \dfrac{b}{10d_P} - 0.0375 + 0.0125b & 25 < b \le 430 \text{ mm} \\ \dfrac{b}{10d_P} - 0.1109 + 0.0207b - 0.000\,228b^2 & 430 < b \le 1000 \text{ mm} \end{cases} \tag{14–30}$$

Note that for values of $F/(10d_P) < 0.05$, $b/(10d_P) = 0.05$ is used.

$$C_{pm} = \begin{cases} 1 & \text{for straddle-mounted pinion with } S_1/S < 0.175 \\ 1.1 & \text{for straddle-mounted pinion with } S_1/S \ge 0.175 \end{cases} \tag{14–31}$$

$$C_{ma} = A + Bb + Cb^2 \qquad \text{(see Table 14–10 for values of } A, B, \text{ and } C) \tag{14–32}$$

$$C_e = \begin{cases} 0.8 & \text{for gearing adjusted at assembly, or compatibility is improved by lapping, or both} \\ 1 & \text{for all other conditions} \end{cases} \tag{14–33}$$

See Figure 14–10 for definitions of S and S_1 for use with Equation (14–31), and see Figure 14–11 for graph of C_{ma}.

14–12 Hardness-Ratio Factor Z_W

The pinion generally has a smaller number of teeth than the gear and consequently is subjected to more cycles of contact stress. If both the pinion and the gear are through-hardened, then a uniform surface strength can be obtained by making the pinion harder than the gear. A similar effect can be obtained when a surface-hardened pinion is mated with a through-hardened gear. The hardness-ratio factor Z_W is used *only for the gear*.

Figure 14–10

Definition of distances S and S_1 used in evaluating C_{pm}, Equation (14–31).
(Source: Data from ANSI/AGMA 2001-D04.)

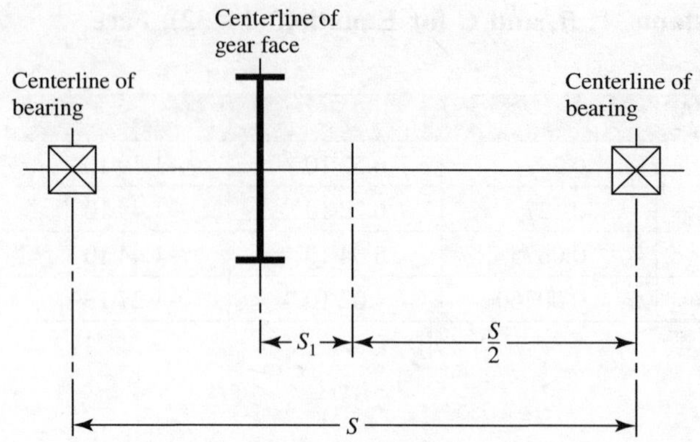

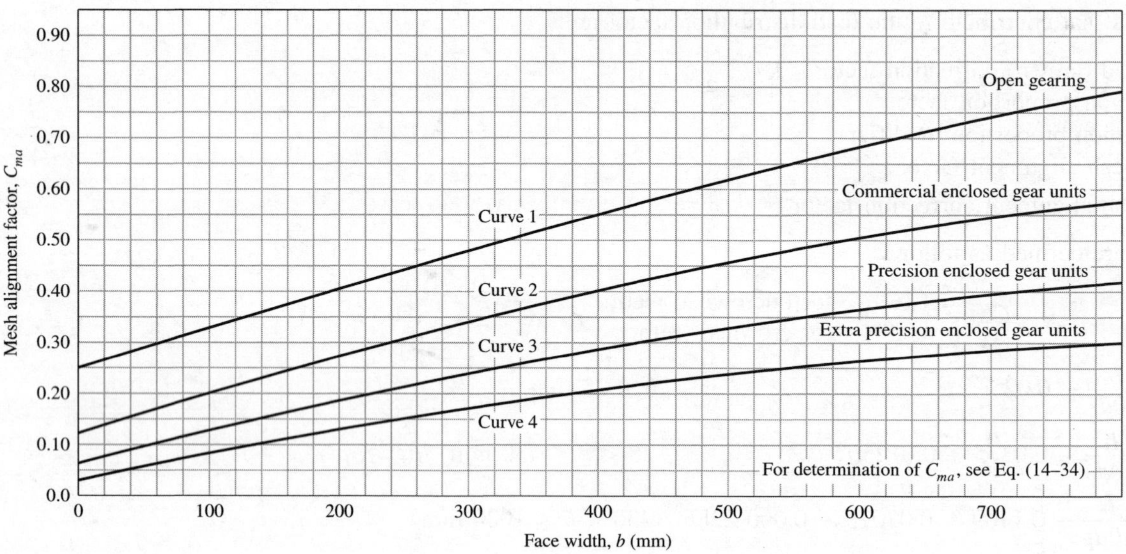

Figure 14–11

Mesh alignment factor C_{ma}. Curve-fit equations in Table 14–10.
(Source: Data from ANSI/AGMA 2001-D04.)

Its purpose is to adjust the surface strengths for this effect. For the pinion, $C_H = 1$. For the gear, C_H is obtained from the equation

$$Z_W = 1.0 + A'(m_G - 1.0) \tag{14–34}$$

where

$$A' = 8.98(10^{-3})\left(\frac{H_{BP}}{H_{BG}}\right) - 8.29(10^{-3}) \qquad 1.2 \le \frac{H_{BP}}{H_{BG}} \le 1.7$$

The terms H_{BP} and H_{BG} are the Brinell hardness (10-mm ball at 3000-kg load) of the pinion and gear, respectively. The term m_G is the speed ratio and is given by Equation (14–20). See Figure 14–12 for a graph of Equation (14–34). For

$$\frac{H_{BP}}{H_{BG}} < 1.2, \qquad A' = 0$$

$$\frac{H_{BP}}{H_{BG}} > 1.7, \qquad A' = 0.006\,98$$

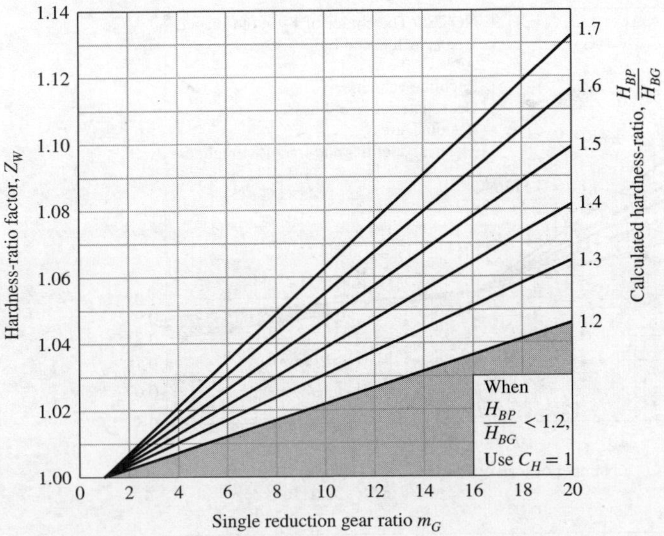

Figure 14–12

Hardness-ratio factor Z_W (through-hardened steel). *(Source: Data from ANSI/AGMA 2001-D04.)*

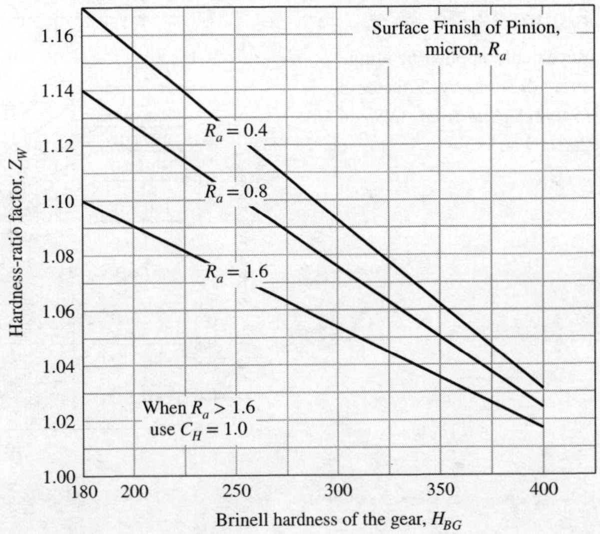

Figure 14–13

Hardness-ratio factor Z_W (surface-hardened steel pinion). *(Source: Data from ANSI/AGMA 2001-D04.)*

When surface-hardened pinions with hardnesses of 48 Rockwell C scale (Rockwell C48) or harder are run with through-hardened gears (180–400 Brinell), a work hardening occurs. The C_H factor is a function of pinion surface finish f_P and the mating gear hardness. Figure 14–13 displays the relationships:

$$C_H = 1 + B'(450 - H_{BG}) \tag{14–35}$$

where $B' = 0.000\,75 \exp[-0.0112 f_P]$ and f_P is the surface finish of the pinion expressed as root-mean-square roughness R_a in μ in.

14–13 Stress-Cycle Factors Y_N and Z_N

The AGMA strengths as given in Figures 14–2 through 14–4, in Tables 14–3 and 14–4 for bending fatigue, and in Figure 14–5 and Tables 14–5 and 14–6 for contact-stress fatigue are based on 10^7 load cycles applied. The purpose of the stress-cycle factors Y_N and Z_N is to modify the gear strength for lives other than 10^7 cycles. Values for these factors are given in Figures 14–14 and 14–15. Note that for 10^7 cycles $Y_N = Z_N = 1$ on each graph. Note also that the equations for Y_N and Z_N change on either side of 10^7 cycles. For life goals slightly higher than 10^7 cycles, the mating gear may be experiencing fewer than 10^7 cycles and the equations for $(Y_N)_P$ and $(Y_N)_G$ can be different. The same comment applies to $(Z_N)_P$ and $(Z_N)_G$.

14–14 Reliability Factor Y_Z

The reliability factor accounts for the effect of the statistical distributions of material fatigue failures. This factor is not addressing the overall reliability of a gear within an application, nor does it address load variation. It should be thought of as a material strength reliability factor. The gear strengths S_t and S_c are based on a reliability

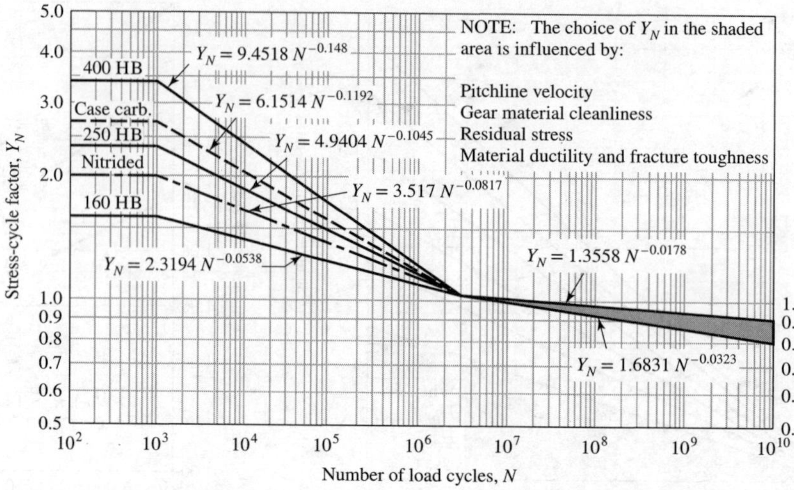

Figure 14–14

Repeatedly applied bending strength stress-cycle factor Y_N. *(Source: Data from ANSI/AGMA 2001-D04.)*

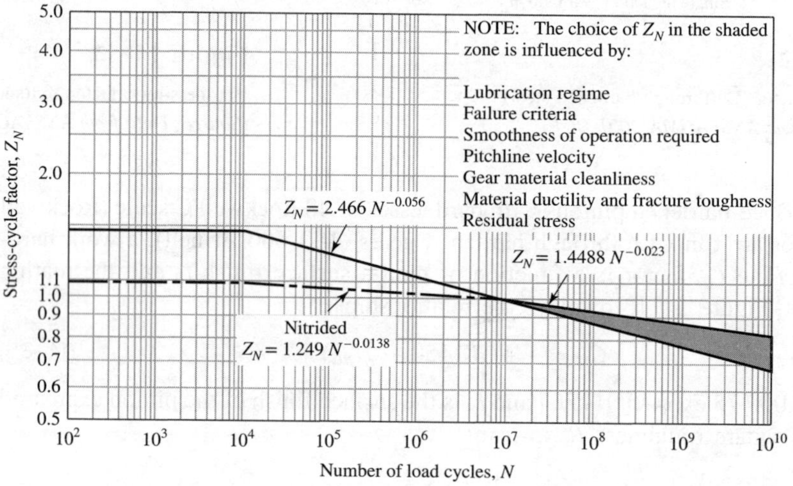

Figure 14–15

Pitting resistance stress-cycle factor Z_N. *(Source: Data from ANSI/AGMA 2001-D04.)*

Table 14–11 Reliability Factors Y_Z

Reliability	Y_Z
0.9999	1.50
0.999	1.25
0.99	1.00
0.90	0.85
0.50	0.70

Source: Data from ANSI/AGMA 2001-D04.

of 99 percent. Table 14–11 is based on data developed by the U.S. Navy for bending and contact-stress fatigue failures.

The functional relationship between Y_Z and reliability is highly nonlinear. When interpolation is required, linear interpolation is too crude. A log transformation to each quantity produces a linear string. A least-squares regression fit is

$$Y_Z = \begin{cases} 0.658 - 0.0759 \ln (1 - R) & 0.5 < R < 0.99 \\ 0.50 - 0.109 \ln (1 - R) & 0.99 \le R \le 0.9999 \end{cases} \tag{14–36}$$

For cardinal values of R, take Y_Z from the table. Otherwise use the logarithmic interpolation afforded by Equations (14–36).

14–15 Temperature Factor Y_θ

For oil or gear-blank temperatures up to 120°C, use $Y_\theta = 1.0$. For higher temperatures, the factor should be greater than unity. Heat exchangers may be used to ensure that operating temperatures are considerably below this value, as is desirable for the lubricant.

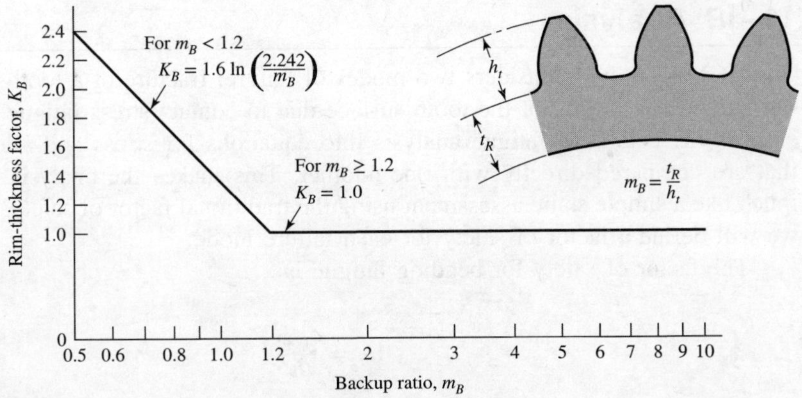

Figure 14–16

Rim-thickness factor K_B. *(Source: Data from ANSI/AGMA 2001-D04.)*

14–16 Rim-Thickness Factor K_B

When the rim thickness is not sufficient to provide full support for the tooth root, the location of bending fatigue failure may be through the gear rim rather than at the tooth fillet. In such cases, the use of a stress-modifying factor K_B is recommended. This factor, the *rim-thickness factor* K_B, adjusts the estimated bending stress for the thin-rimmed gear. It is a function of the backup ratio m_B,

$$m_B = \frac{t_R}{h_t} \tag{14–37}$$

where t_R = rim thickness below the tooth, and h_t = the tooth height. The geometry is depicted in Figure 14–16. The rim-thickness factor K_B is given by

$$K_B = \begin{cases} 1.6 \ln \dfrac{2.242}{m_B} & m_B < 1.2 \\ 1 & m_B \geq 1.2 \end{cases} \tag{14–48}$$

Figure 14–16 also gives the value of K_B graphically.

14–17 Safety Factors S_F and S_H

The AGMA standard contains a safety factor S_F guarding against bending fatigue failure and a safety factor S_H guarding against pitting failure. These factors are for a different purpose than the overload factor K_o, which is targeted specifically to include predictable excursions of load beyond the nominal transmitted load. The safety factors are to account for unquantifiable aspects of the application, such as economic risk, variations in manufacturing, or safety risks. The greater the uncertainties or consequences of such considerations, the higher the safety factor should be. Appropriate values for these safety factors are largely determined based on experience within the industry, and sometimes are influenced by industry standards or codes. When used in this manner as part of the design process, the AGMA safety factors are design factors. Sometimes these factors are also used to document the analysis of traditional strength compared to stress. This is traditionally called a factor of safety (rather than a design factor or safety factor). For clarity, we will choose to limit the safety factors to AGMA's definition and define the overall factors of safety in the next section.

14–18 Analysis

The AGMA method considers two modes of failure: fracture of a tooth due to bending fatigue and pitting of the tooth surface due to contact stress fatigue. The method embeds the details of fatigue analysis into equations for stress and allowable stress that are compared directly with one another. This makes the analysis process look much like a simple static assessment using the traditional factor of safety. Accordingly, we will define a factor of safety for each failure mode.

The factor of safety for bending fatigue is

$$n_b = \frac{\sigma_{b,\text{all}}}{\sigma_b} \tag{14–39}$$

where $\sigma_{b,\text{all}}$ is defined in Equation (14–15) and σ_b is defined in Equation (14–13).

The factor of safety for surface fatigue from contact stress is

$$n_c = \frac{\sigma_{c,\text{all}}}{\sigma_c} \tag{14–40}$$

where $\sigma_{c,\text{all}}$ is defined in Equation (14–16) and σ_c is defined in Equation (14–14).

There is an overlap in the use of the AGMA safety factors S_F and S_H with these factors of safety n_b and n_c. If S_F and S_H are used as design factors to incorporate the desired margin of safety into the design, then n_b and n_c could be expected to be near unity. Alternatively, if S_F and S_H are set to unity, then n_b and n_c should be sufficiently greater than unity to account for the desired margin of safety.

Caution is required when comparing n_b with n_c in an analysis in order to ascertain the nature and severity of the threat to loss of function. In Equation (14–39), the bending stress is linear with the transmitted load W^t. But in Equation (14–40), the contact stress is not linear with W^t. To render n_c linear with W^t, it can be defined as

$$n_{c,\text{linear}} = \left(\frac{\sigma_{c,\text{all}}}{\sigma_c}\right)^i \tag{14–41}$$

where i is 2 for linear or helical contact and 3 for spherical contact (crowned teeth). Accordingly, to identify whether bending fatigue or contact fatigue is closer to failure due to an increase of W^t, it is reasonable to compare n_b with $n_{c,\text{linear}}$.

Restating for clarity, note that the actual factors of safety for the failure modes are given by n_b and n_c. These are perfectly suitable for use in the traditional sense of comparing stress to allowable stress, especially when they are used to encapsulate broad aspects to design safety, for example, economic issues, acceptability of risk, manufacturing variabilities, material uncertainties, etc. But if it is desired to compare these two factors to one another with the assumption that they are primarily guarding against overload of W_t, it may be desirable to compare n_b to $n_{c,\text{linear}}$.

The AGMA method has many details. A summary is provided in Figure 14–17 for bending fatigue and in Figure 14–18 for contact-stress fatigue. The equations in these figures are in terms of U.S. customary units. Similar road maps can readily be generated in terms of SI units.

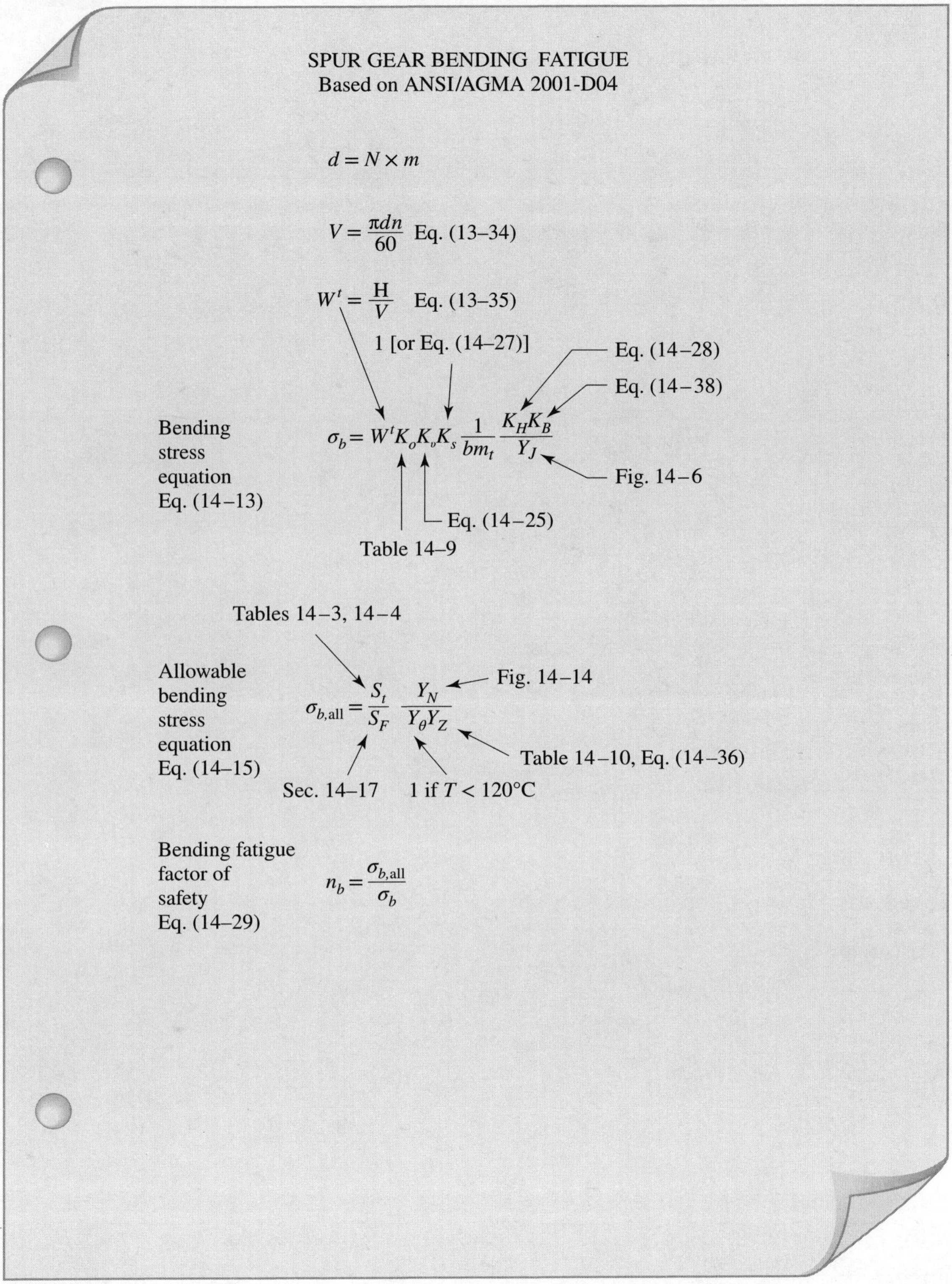

Figure 14–17

Road map of gear bending equations based on AGMA standards. *(Source: Data from ANSI/AGMA 2001-D04.)*

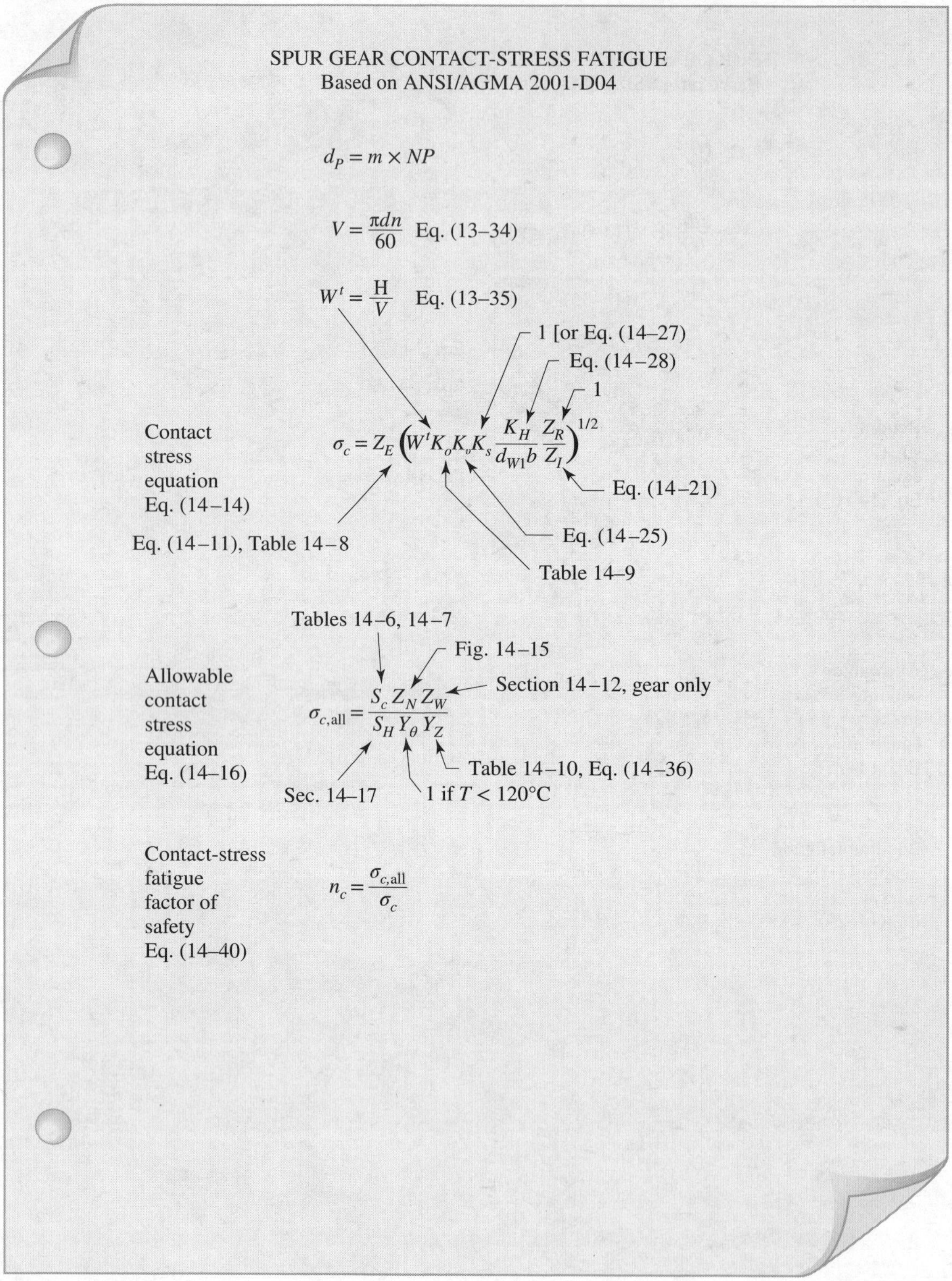

Figure 14–18

Road map of gear contact stress equations based on AGMA standards. *(Source: Data from ANSI/AGMA 2001-D04.)*

EXAMPLE 14–4

A 17-tooth 20° pressure angle spur pinion rotates at 1800 rev/min and transmits 3000 W to a 52-tooth disk gear. The module is 2.5 mm, the face width 38 mm, and the quality standard is No. 6. The gears are straddle-mounted with bearings immediately adjacent. The pinion is a grade 1 steel with a hardness of 240 Brinell tooth surface and through-hardened core. The gear is steel, grade 1 material, with a Brinell hardness of 240, tooth surface and through-hardened core. Poisson's ratio is 0.30, $(Y_J)_p = 0.30$, $(Y_J)_G = 0.40$, and Young's modulus is 200 GPa. The loading is smooth because of motor and load. Assume a pinion life of 10^8 cycles and a material strength reliability of 0.90, and use $Y_N = 1.3558N^{-0.0178}$, $Z_N = 1.4488N^{-0.023}$. The tooth profile is uncrowned. This is a commercial enclosed gear unit.
(*a*) Find the factor of safety of the gears in bending fatigue.
(*b*) Find the factor of safety of the gears in surface fatigue.
(*c*) By examining the factors of safety, identify the threat to each gear and to the mesh.

Solution

There will be many terms to obtain so use Figs. 14–17 and 14–18 as guides to what is needed.

$$d_P = N_P(m_t) = 17 \times 2.5 = 42.5 \text{ mm} \qquad d_G = 52(2.5) = 130 \text{ mm}$$

$$V = \frac{\pi d_P n_P}{60} = \frac{\pi(42.5)\,1800}{12} = 4 \text{ m/s}$$

$$W^t = \frac{H}{V} = \frac{3000}{4} = 750 \text{ N}$$

Assuming uniform loading, $K_o = 1$. To evaluate K_v, from Eq. (14–25*b*) with a quality number $Q_v = 6$,

$$B = 0.25(12 - 6)^{2/3} = 0.8255$$
$$A = 50 + 56(1 - 0.8255) = 59.77$$

Then from Eq. (14–25*a*) the dynamic factor is

$$K_v = \left(\frac{59.77 + \sqrt{200(4)}}{59.77}\right)^{0.8255} = 1.377$$

To determine the size factor, K_s, the Lewis form factor is needed. From Table 14–2, with $N_P = 17$ teeth, $Y_P = 0.303$. Interpolation for the gear with $N_G = 52$ teeth yields $Y_G = 0.412$. Thus from Eq. (14–27), with $F = 38$ mm,

$$(K_s)_P = 0.843\left(38(2.5)\sqrt{0.303}\right)^{0.0535} = 1.043$$

$$(K_s)_G = 0.843\left(38(2.5)\sqrt{0.412}\right)^{0.0535} = 1.052$$

Note that for this tooth size, the AGMA recommendation of a size factor of unity is satisfactory. The load distribution factor K_m is determined from Eq. (14–28), where five terms are needed. They are, where $F = 38$ mm when needed:

Uncrowned, Eq. (14–28): $C_{mc} = 1$,
Eq. (14–30): $C_{pf} = 38/[10(42.5)] - 0.0375 + 4.92(10^{-4})(38) = 0.0706$
Bearings immediately adjacent, Eq. (14–31): $C_{pm} = 1$
Commercial enclosed gear units (Fig. 14–11): $C_{ma} = 0.15$
Eq. (14–33): $C_e = 1$

Thus,

$$K_m = 1 + C_{mc}(C_{pf}C_{pm} + C_{ma}C_e) = 1 + (1)[0.0706(1) + 0.15(1)] = 1.22$$

Assuming constant thickness gears, the rim-thickness factor $K_B = 1$. The speed ratio is $m_G = N_G/N_P = 52/17 = 3.059$. The load cycle factors given in the problem statement, with $N(\text{pinion}) = 10^8$ cycles and $N(\text{gear}) = 10^8/m_G = 10^8/3.059$ cycles, are

$$(Y_N)_P = 1.3558(10^8)^{-0.0178} = 0.977$$

$$(Y_N)_G = 1.3558(10^8/3.059)^{-0.0178} = 0.996$$

From Table 14–11, with a reliability of 0.9, $K_R = 0.85$. From Fig. 14–18, the temperature and surface condition factors are $K_T = 1$ and $C_f = 1$. From Eq. (14–21), with $m_N = 1$ for spur gears,

$$Z_I = \frac{\cos 20° \sin 20°}{2} \frac{3.059}{3.059 + 1} = 0.121$$

From Table 14–8, $Z_E = 191\ \sqrt{\text{MPa}}$.

Next, we need the terms for the gear allowable stress equations. From Table 14–3, for grade 1 steel with $H_{BP} = 240$ and $H_{BG} = 200$, we use Fig. 14–2, which gives

$$(S_t)_P = 0.533(240) + 88.3 = 216 \text{ MPa}$$

$$(S_t)_G = 0.533(200) + 88.3 = 195 \text{ MPa}$$

Similarly, from Table 14–6, we use Fig. 14–5, which gives

$$(S_c)_P = 2.2(240) + 200 = 728 \text{ MPa}$$

$$(S_c)_G = 2.2(200) + 200 = 640 \text{ MPa}$$

From Fig. 14–15,

$$(Z_N)_P = 1.4488(10^8)^{-0.023} = 0.948$$

$$(Z_N)_G = 1.4488(10^8/3.059)^{-0.023} = 0.973$$

For the hardness ratio factor Z_W, the hardness ratio is $H_{BP}/H_{BG} = 240/200 = 1.2$. Then, from Sec. 14–12,

$$A' = 8.98(10^{-3})(H_{BP}/H_{BG}) - 8.29(10^{-3})$$

$$= 8.98(10^{-3})(1.2) - 8.29(10^{-3}) = 0.002\,49$$

Thus, from Eq. (14–34),

$$Z_W = 1 + 0.002\,49(3.059 - 1) = 1.005$$

Design factors should be determined as appropriate for the application. Lacking sufficient background information in this problem statement, design factors of unity will be used. The final factors of safety should therefore be viewed with the awareness that no design factor has been incorporated.

(*a*) **Pinion tooth bending.** Substituting the appropriate terms for the pinion into Eq. (14–13) gives

$$(\sigma_b)_P = \left(W^t K_o K_v K_s \frac{1}{(bm_t)} \frac{K_H K_B}{Y_j}\right)_P = 750(1)1.377(1.043)\frac{1}{38(2.5)}\frac{1.22(1)}{0.30}$$

$$= 46 \text{ MPa}$$

Substituting the appropriate terms for the pinion into Eq. (14–15) gives

$$(\sigma_{b,\text{all}})_P = \left(\frac{S_t Y_N}{S_F Y_{teta} Y_B}\right)_P = \frac{216(0.977)}{1(1)0.85} = 248 \text{ MPa}$$

The factor of safety for the pinion bending fatigue is found from comparison of the bending stress to the allowable bending stress, as given in Eq. (14–39),

Answer
$$(n_b)_P = \frac{(\sigma_{b,\text{all}})_P}{(\sigma_b)_P} = \frac{248}{46} = 5.39$$

Gear tooth bending. Substituting the appropriate terms for the gear into Eq. (14–13) gives

$$(\sigma_b)_G = 750(1)1(1.377)1.052\frac{1}{38(2.5)}\frac{1.22(1)}{0.40} = 35 \text{ MPa}$$

Substituting the appropriate terms for the gear into Eq. (14–15) gives

$$(\sigma_{b,\text{all}})_G = \left(\frac{S_t Y_N}{S_F Y_{teta} Y_z}\right)_G = \frac{(195)(0.966)}{(1)(1)(0.85)} = 221.6 \text{ MPa}$$

The factor of safety for the gear bending fatigue is found from comparison of the bending stress to the allowable bending stress, as given in Eq. (14–39),

Answer
$$(n_b)_G = \frac{(\sigma_{b,\text{all}})_G}{(\sigma_b)_G} = \frac{221.6}{35} = 6.33$$

(*b*) **Pinion tooth wear.** Substituting the appropriate terms for the pinion into Eq. (14–14) gives

$$(\sigma_c)_P = Z_E\left(W^t K_o K_v K_s \frac{\text{H}}{\text{b}d_{w1}}\frac{Z_R}{Z_I}\right)_P^{1/2}$$

$$= 191\left[750(1)(1)(3.77)1.052\frac{1.22}{38(42.5)}\frac{1}{0.121}\right]^{1/2} = 497.4 \text{ MPa}$$

Substituting the appropriate terms for the pinion into Eq. (14–16) gives

$$(\sigma_{c,\text{all}})_P = \left(\frac{S_c Z_N C_H}{S_H K_T K_R}\right)_P = \frac{(728)(0.948)(1)}{(1)(1)(0.85)} = 812 \text{ MPa}$$

The factor of safety for the pinion contact-stress fatigue is found from comparison of the contact stress to the allowable contact stress, as given in Eq. (14–40),

Answer
$$(n_c)_P = \frac{(\sigma_{c,\text{all}})_P}{(\sigma_c)_P} = \frac{812}{497.4} = 1.63$$

Gear tooth wear. The only term in Eq. (14–14) that changes for the gear is K_s. Thus,

$$(\sigma_c)_G = \left[\frac{(K_s)_G}{(K_s)_P}\right]^{1/2}(\sigma_c)_P = \left(\frac{1.052}{1.043}\right)^{1/2} 497.4 = 500 \text{ MPa}$$

Substituting the appropriate terms for the gear into Eq. (14–16) gives

$$(\sigma_{c,\text{all}})_G = \left(\frac{S_c Z_N C_H}{S_H K_T K_R}\right)_G = \frac{(640)(0.973)(1.005)}{(1)(1)(0.85)} = 777.7 \text{ MPa}$$

From Eq. (14–40),

Answer
$$(n_c)_G = \frac{(\sigma_{c,\text{all}})_G}{(\sigma_c)_G} = \frac{777.7}{500} = 1.55$$

(*c*) To compare the susceptibility of failure by bending fatigue versus surface fatigue, we want to compare n_b with $n_{c,\text{linear}}$.

For the pinion,

$$(n_b)_P = 5.39$$

$$(n_{c,\text{linear}})_P = (n_c)_P^2 = 1.63^2 = 2.66$$

Thus, pinion failure is driven by surface fatigue.

For the gear,

$$(n_b)_G = 6.33$$

$$(n_{c,\text{linear}})_G = (n_c)_G^2 = 1.55^2 = 2.4$$

Thus, gear failure is also driven by surface fatigue. Also, comparing the gear to the pinion, we see that the overall minimum factor of safety is with the surface fatigue of the gear.

There are perspectives to be gained from Example 14–4. First, the pinion is overly strong in bending compared to wear. The performance in wear can be improved by surface-hardening techniques, such as flame or induction hardening, nitriding, or carburizing and case hardening, as well as shot peening. Second, in bending, the gear is stronger than the pinion, indicating that both the gear core hardness and tooth size could be reduced; that is, we may increase P and reduce the diameters of the gears, or perhaps allow a cheaper material.

Having followed a spur-gear analysis in detail in Example 14–4, it is timely to analyze a helical gearset under similar circumstances to observe similarities and differences.

EXAMPLE 14–5

A 17-tooth 20° normal pitch-angle helical pinion with a right-hand helix angle of 30° rotates at 1800 rev/min when transmitting 3000 W to a 52-tooth helical gear. The normal module is 2.5 mm, the face width is 38 mm, and the set has a quality number of 6. The gears are straddle-mounted with bearings immediately adjacent. The pinion and gear are made from a through-hardened steel with surface and core hardnesses of 240 Brinell on the pinion and surface and core hardnesses of 200 Brinell on the gear. The transmission is smooth, connecting an electric motor and a centrifugal pump. Assume a pinion life of 10^8 cycles and a material reliability of 0.9 and use the upper curves in Figs. 14–14 and 14–15.

(*a*) Find the factors of safety of the gears in bending fatigue.

(*b*) Find the factors of safety of the gears in surface fatigue.

(*c*) By examining the factors of safety identify the threat to each gear and to the mesh.

Solution

All of the parameters in this example are the same as in Ex. 14–4 with the exception that we are using helical gears. Thus, several terms will be the same as Ex. 14–4. The following terms remain unchanged: $K_o = 1$, $Y_P = 0.303$, $Y_G = 0.412$, $m_G = 3.059$, $(K_s)_P = 1.043$, $(K_s)_G = 1.052$, $(Y_N)_P = 0.977$, $(Y_N)_G = 0.996$, $K_R = 0.85$, $K_T = 1$, $C_f = 1$, $Z_E = 191\ \sqrt{\text{MPa}}$, $(S_t)_P = 216$ MPa, $(S_t)_G = 195$ MPa, $(S_c)_P = 728$ MPa, $(S_c)_G = 640$ MPa, $(Z_N)_P = 0.948$, $(Z_N)_G = 0.973$, and $C_H = 1.005$.

For helical gears, the transverse module, given by Eq. (13–18), is

$$m_t = m_n/\cos\psi = 2.5/\cos 30° = 2.887 \text{ mm}$$

Thus, the pitch diameters are $d_P = N_P(m_t) = 17(2.887) = 49.079$ mm and $d_G = 52(2.887) = 150.124$ mm. The pitch-line velocity and transmitted force are

$$V = \frac{\pi d_P n_P}{60} = \frac{\pi(49.079)1800}{60} = 4.626 \text{ m/s}$$

$$W^t = \frac{H}{V} = \frac{300}{4.626} = 648.51 \text{ N}$$

As in Ex. 14–4, for the dynamic factor, $B = 0.8255$ and $A = 59.77$. Thus, Eq. (14–25) gives

$$K_v = \left(\frac{59.77 + \sqrt{200(4.626)}}{59.77}\right)^{0.8255} = 1.404$$

The geometry factor I for helical gears requires a little work. First, the transverse pressure angle is given by Eq. (13–19),

$$\phi_t = \tan^{-1}\left(\frac{\tan \phi_n}{\cos \psi}\right) = \tan^{-1}\left(\frac{\tan 20^\circ}{\cos 30^\circ}\right) = 22.80^\circ$$

The radii of the pinion and gear are $r_P = 49.079/2 = 24.54$ mm and $r_G = 150.124/2 = 75.06$ mm, respectively. The addendum is $a = m_n = 2.5$ mm, and the base-circle radii of the pinion and gear are given by Eq. (13–6), with $\phi = \phi_t$:

$$(r_b)_P = r_P \cos \phi_t = 24.54 \cos 22.80^\circ = 22.62 \text{ mm}$$

$$(r_b)_G = 75.06 \cos 22.80^\circ = 69.20 \text{ mm}$$

From Eq. (14–23), the surface strength geometry factor

$$\begin{aligned} Z &= \sqrt{(24.54 + 2.5)^2 - 22.64^2} + \sqrt{(75.06 + 2.5)^2 - 69^2} \\ &\quad -(24.54 + 75.06) \sin 22.80^\circ \\ &= 14.785 + 35.42 - 38.59 = 11.6 \text{ mm} \end{aligned}$$

Since the first two terms are less than 38.59, the equation for Z stands. From Eq. (14–22) the normal base pitch p_N is

$$p_N = p_n \cos \phi_n = \pi(m_n) \cos 20^\circ = \pi(2.5) \cos 20^\circ = 7.38 \text{ mm}$$

From Eq. (14–19), the load sharing ratio

$$m_N = \frac{p_N}{0.95Z} = \frac{7.38}{(0.95)(11.215)} = 0.6927$$

Substituting in Eq. (14–21), the geometry factor Z_I is

$$Z_I = \frac{\sin 22.80^\circ \cos 22.80^\circ}{2(0.6927)} \frac{3.06}{3.06 + 1} = 0.195$$

From Fig. 14–7, geometry factors $J'_P = 0.45$ and $J'_G = 0.54$. Also from Fig. 14–8 the J-factor multipliers are 0.94 and 0.98, correcting J'_P and J'_G to

$$J_P = 0.45(0.94) = 0.423$$

$$J_G = 0.54(0.98) = 0.529$$

The load-distribution factor K_m is estimated from Eq. (14–30):

$$C_{pf} = \frac{38}{(10)(49.079)} - 0.0375 + 4.92(10–4)(38) = 0.0586$$

with $C_{mc} = 1$, $C_{pm} = 1$, $C_{ma} = 0.15$ from Fig. 14–11, and $C_e = 1$. Therefore, from Eq. (14–28),

$$K_m = 1 + (1)[0.0586(1) + 0.15(1)] = 1.208$$

(*a*) **Pinion tooth bending.** Substituting the appropriate terms into Eq. (14–13), noting that P is the transverse diametral pitch, gives

$$(\sigma_b)_P = \left(W^t K_o K_v K_s \frac{1}{(b)(m_t)} \frac{K_H K_B}{Y_j}\right)_P = 648.51(1)(1.404)1.403 \frac{1}{(38)(2.887)} \frac{1.208(1)}{0.423}$$
$$= 24.7 \text{ MPa}$$

Substituting the appropriate terms for the pinion into Eq. (14–15) gives

$$(\sigma_{b,\text{all}})_P = \left(\frac{S_t Y_N}{S_F Y_{teta} Y_Z}\right)_P = \frac{(216.2)(0.977)}{(1)(1)(0.85)} = 248.5 \text{ MPa}$$

From Eq. (14–39),

Answer $$(n_b)_P = \frac{(\sigma_{b,\text{all}})_P}{(\sigma_b)_P} = \frac{248.5}{24.7} = 10.06$$

Gear tooth bending. Substituting the appropriate terms for the gear into Eq. (14–13) gives

$$(\sigma_b)_G = 648.51(1)1.404(1.052) \frac{1}{(38)(2.887)} \frac{1.208(1)}{0.529} = 19.9 \text{ MPa}$$

Substituting the appropriate terms for the gear into Eq. (14–15) gives

$$(\sigma_{b,\text{all}})_G = \left(\frac{S_t Y_N}{S_F Y_{teta} Y_Z}\right)_G = \frac{(194.9)(0.996)}{(1)(1)(0.85)} = 228.38 \text{ MPa}$$

From Eq. (14–39),

Answer $$(n_b)_G = \frac{(\sigma_{b,\text{all}})_G}{(\sigma_b)_G} = \frac{228.38}{19.9} = 11.48$$

(*b*) **Pinion tooth wear.** Substituting the appropriate terms for the pinion into Eq. (14–14) gives

$$(\sigma_c)_P = Z_E \left(W^t K_o K_v K_s \frac{K_H}{\text{b}(d_{w1})} \frac{Z_R}{Z_I}\right)_P^{1/2}$$
$$= 191 \left[648.51(1)1.404(1.043) \frac{1.208}{(38)(49.079)} \frac{1}{0.195}\right]^{1/2} = 340 \text{ MPa}$$

Substituting the appropriate terms for the pinion into Eq. (14–16) gives

$$(\sigma_{c,\text{all}})_P = \left(\frac{S_c Z_N C_H}{S_H Y_{teta} Y_Z}\right)_P = \frac{(728)(0.948)(1)}{(1)(1)(0.85)} = 812 \text{ MPa}$$

From Eq. (14–40),

Answer $$(n_c)_P = \frac{(\sigma_{c,\text{all}})_P}{(\sigma_c)_P} = \frac{812}{340} = 2.39$$

Gear tooth wear. The only term in Eq. (14–14) that changes for the gear is K_s. Thus,

$$(\sigma_c)_G = \left[\frac{(K_s)_G}{(K_s)_P}\right]^{1/2} (\sigma_c)_P = \left(\frac{1.052}{1.043}\right)^{1/2} 340 = 341.5 \text{ MPa}$$

Substituting the appropriate terms for the gear into Eq. (14–40)

$$(\sigma_{c,\text{all}})_G = \left(\frac{S_c Z_N C_H}{S_H Y_{teta} Y_Z}\right)_G = 640\frac{(0.973)(1.005)}{(1)(0.85)} = 736.3\ \text{MPa}$$

From Eq. (14–40),

Answer
$$(n_c)_G = \frac{(\sigma_{c,\text{all}})_G}{(\sigma_c)_G} = \frac{736.3}{341.5} = 2.16$$

(*c*) To compare the susceptibility of failure by bending fatigue versus surface fatigue, we want to compare n_b with $n_{c,\text{linear}}$.

For the pinion,

$$(n_b)_P = 10.06$$

$$(n_{c,\text{linear}})_P = (n_c)_P^2 = 2.39^2 = 5.71$$

Thus, pinion failure is driven by surface fatigue.

For the gear,

$$(n_b)_G = 11.48$$

$$(n_{c,\text{linear}})_G = (n_c)_G^2 = 2.16^2 = 4.67$$

Thus, gear failure is also driven by surface fatigue. Also, for the entire mesh, we see that the overall minimum factor of safety is with the surface fatigue of the gear.

It is worthwhile to compare Example 14–4 with Example 14–5. The spur and helical gearsets were placed in nearly identical circumstances. The helical gear teeth are of greater length because of the helix and identical face widths. The pitch diameters of the helical gears are larger. The J factors and the I factor are larger, thereby reducing stresses. The result is larger factors of safety. In the design phase the gearsets in Example 14–4 and Example 14–5 can be made smaller with control of materials and relative hardnesses.

Now that examples have given the AGMA parameters substance, it is time to examine some desirable (and necessary) relationships between material properties of spur gears in mesh. We shall consider the relationship between the material strength needed for the pinion versus the gear for bending fatigue. The stress and allowable stress equations contain many terms that are identical for the pinion and the gear. If we set the bending factors of safety for the pinion and the gear equal to each other, and cancel all identical terms, and solve for $(S_t)_G$, we obtain a relationship between the strengths of the pinion and the gear.

$$(S_t)_G = (S_t)_P \frac{(Y_N)_P}{(Y_N)_G} \frac{(Y_J)_P}{(Y_J)_G} \tag{a}$$

The stress-cycle factor Y_N comes from Figure 14–14, where for a particular hardness, $Y_N = \alpha N^\beta$. For the pinion, $(Y_N)_P = \alpha N_P^\beta$, and for the gear, $(Y_N)_G = \alpha(N_P/m_G)^\beta$. Substituting these into Equation (*a*) and simplifying gives

$$(S_t)_G = (S_t)_P m_G^\beta \frac{(Y_J)_P}{(Y_J)_G} \tag{14–42}$$

Normally, $m_G > 1$ and $(Y_J)_G > (Y_J)_P > J_P$, so Equation (14–42) shows that the gear can be less strong (lower Brinell hardness) than the pinion for the same safety factor.

EXAMPLE 14–6

In a set of spur gears, a 300-Brinell 18-tooth 1.55 mm-module 20° full-depth pinion meshes with a 64-tooth gear. Both gear and pinion are of grade 1 through-hardened steel. Using $\beta = -0.023$, what hardness can the gear have for the same factor of safety?

Solution

For through-hardened grade 1 steel the pinion strength $(S_t)_P$ is given in Figure 14–2:

$$(S_t)_P = 0.533\,(300) + 88.3 = 248.2 \text{ MPa}$$

From Figure 14–6 the form factors are $(Y_J)_P = 0.32$ and $(Y_J)_G = 0.41$. Equation (14–42) gives

$$(S_t)_G = 248.2\left(\frac{64}{18}\right)^{-0.023}\frac{0.32}{0.41} = 188.1 \text{ MPa}$$

Use the equation in Figure 14–2 again.

Answer

$$(H_B)_G = \frac{188.1 - 88.3}{0.533} = 187 \text{ Brinell}$$

Similarly, we compare the material strength needed for the pinion versus the gear for surface fatigue. Setting the surface fatigue factors of safety for the pinion and the gear equal to each other, and canceling all identical terms, and solving for $(S_c)_G$, we obtain

$$(S_c)_G = (S_c)_P\frac{(Z_N)_P}{(Z_N)_G}\left(\frac{1}{Z_W}\right)_G = (S_C)_P m_G^{\beta}\left(\frac{1}{Z_W}\right)_G$$

where, as in the development of Equation (14–42), $(Z_N)_P/(Z_N)_G = m_G^{\beta}$ and the value of β for wear comes from Figure 14–15. Since C_H is so close to unity, it is usually neglected; therefore

$$(S_c)_G = (S_c)_P m_G^{\beta} \tag{14–43}$$

EXAMPLE 14–7

For $\beta = -0.056$ for a through-hardened steel, grade 1, continue Example 14–6 for wear.

Solution

From Figure 14–5,

$$(S_c)_P = 2.22\,(300) + 200 = 866 \text{ MPa}$$

From Equation (14–43),

$$(S_c)_G = (S_c)_P\left(\frac{64}{18}\right)^{-0.056} = 866\left(\frac{64}{18}\right)^{-0.056} = 806.6\text{MPa}$$

Answer

$$(H_B)_G = \frac{806.6 - 200}{2.22} = 273 \text{ Brinell}$$

which is slightly less than the pinion hardness of 300 Brinell.

Equations (14–42) and (14–43) apply as well to helical gears.

14–19 Design of a Gear Mesh

A useful decision set for spur and helical gears includes

a priori decisions:

- Function: load, speed, reliability, life, K_o
- Unquantifiable risk: design factors or safety factors
- Tooth system: ϕ, ψ, addendum, dedendum, root fillet radius
- Gear ratio m_G, N_p, N_G
- Quality number Q_v

design decisions:

- module m
- Face width b
- Pinion material, core hardness, case hardness
- Gear material, core hardness, case hardness

The first item to notice is the dimensionality of the decision set. There are four design decision categories, eight different decisions if you count them separately. This is a larger number than we often encounter. It is important to use a design strategy that is convenient in either longhand execution or computer implementation. The design decisions have been placed in order of importance (impact on the amount of work to be redone in iterations). The steps, after the a priori decisions have been made are

1 Choose a diametral pitch.
2 Examine implications on face width, pitch diameters, and material properties. If not satisfactory, return to pitch decision for change.
3 Choose a pinion material and examine core and case hardness requirements. If not satisfactory, return to pitch decision and iterate until no decisions are changed.
4 Choose a gear material and examine core and case hardness requirements. If not satisfactory, return to pitch decision and iterate until no decisions are changed.

With these plan steps in mind, we can consider them in more detail.

First select a trial diametral pitch.

Pinion bending:

- Select a median face width for this pitch, $F = 4\pi m$
- Find the range of necessary ultimate strengths
- Choose a material and a core hardness
- Find face width to meet factor of safety in bending
- If face width is not in the readily available range, select a new diametral pitch and iterate
- Check factor of safety in bending

Gear bending:

- Find necessary companion core hardness
- Choose a material and core hardness
- Check factor of safety in bending

Pinion wear:

- Find necessary S_c and attendant case hardness
- Choose a case hardness
- Check factor of safety in wear

Gear wear:

- Find companion case hardness
- Choose a case hardness
- Check factor of safety in wear

Completing this set of steps will yield a satisfactory design. Additional designs with diametral pitches adjacent to the first satisfactory design will produce several among which to choose. A figure of merit is necessary in order to choose the best. Unfortunately, a figure of merit in gear design is complex in an academic environment because material and processing costs vary. The possibility of using a process depends on the manufacturing facility if gears are made in house.

After examining Examples 14–4 and 14–5 and seeing the wide range of factors of safety, one might entertain the notion of setting all factors of safety equal. In steel gears, wear is usually controlling and $(n_c)_P$ and $(n_c)_G$ can be brought close to equality. The use of softer cores can bring down $(n_b)_P$ and $(n_b)_G$, but there is value in keeping them higher. A tooth broken by bending fatigue not only can destroy the gearset, but can bend shafts, damage bearings, and produce inertial stresses up- and downstream in the power train, causing damage elsewhere if the gear box locks.

EXAMPLE 14–8

Design a 4:1 spur-gear reduction for a 75 kW, three-phase squirrel-cage induction motor running at 1120 rev/min for 10^9 revolutions of the pinion. The load is smooth. A material strength reliability of 0.95 is acceptable. Gearing space is meager. Use Nitralloy 135M, grade 1 material to keep the gear size small. The gears are heat-treated first then nitrided. The industry standard for this design is a factor of safety of 2 guarding against complete failure (e.g., tooth fracture from bending fatigue) and 1.4 guarding against a decline of performance (e.g., pitting of tooth surfaces).

Solution

Make the a priori decisions:

- Function: 75 kW, 1120 rev/min, $R = 0.95$, $N = 10^9$ cycles
- Tooth system: $\phi_n = 20°$
- Tooth count: $N_P = 18$ teeth, $N_G = 72$ teeth (no interference, Sec. 13–7)
- Quality number: $Q_v = 6$, use grade 1 material
- Assume $m_B \geq 1.2$ in Eq. (14–38), $K_B = 1$

Design factors: With a uniform load and a uniform power source, the expectation of overload is minimal. From Table 14–9, we select $K_o = 1$.

We choose to set AGMA design factors to unity and handle the overall unquantifiable exigencies (including company policy regarding risk and consequences of failure) in the overall factors of safety. $n_b = 2$; $n_c = 1.4$

Pitch: Select a trial module of m = 6 mm. Thus, $d_P = 18(6) = 108$ mm and $d_G = 72(6) = 432$ mm. From Table 14–2, $Y_P = 0.309$, $Y_G = 0.4324$ (interpolated). From Fig. 14–6, $(Y_j)_P = 0.32$, $(Y_j)_G = 0.415$.

$$V = \frac{\pi d_P n_P}{60} = \frac{\pi(108)\,1120}{12} = 6.33 \text{ m/s}$$

$$W^t = \frac{33\ 000H}{V} = \frac{75\ 000}{6.33} = 11\ 848 \text{ N}$$

From Eq. (14–25),

$$B = 0.25(12 - Q_v)^{2/3} = 0.25(12 - 6)^{2/3} = 0.8255$$

$$A = 50 + 56(1 - 0.8255) = 59.77$$

$$K_v = \left(\frac{59.77 + \sqrt{200(6.33)}}{59.77}\right)^{0.8255} = 1.47$$

From Eq. (14–36), $K_R = 0.658 - 0.0759 \ln(1 - 0.95) = 0.885$. From Fig. 14–14,

$$(Y_N)_P = 1.3558(10^9)^{-0.0178} = 0.938$$

$$(Y_N)_G = 1.3558(10^9/4)^{-0.0178} = 0.961$$

From Fig. 14–15,

$$(Z_N)_P = 1.4488(10^9)^{-0.023} = 0.900$$

$$(Z_N)_G = 1.4488(10^9/4)^{-0.023} = 0.929$$

From the recommendation for usual commercial availability, $3p \le F \le 5p$. The midrange is $F = 4p = 4\pi\text{m} = 75.36$ mm. Try $F = 75$ mm. From Eq. (14–27)

$$K_s = 0.843\left(F(m)\sqrt{Y}\right)^{0.0535} = 0.843\left(75(6)\sqrt{0.309}\right)^{0.0535} = 1.133$$

From Eqs. (14–29), (14–31), and (14–32), $C_{mc} = C_{pm} = C_e = 1$. From Fig. 14–11, $C_{ma} = 0.175$ for commercial enclosed gear units. From Eq. (14–30), $F/(10d_P) = 75/[10(108)] = 0.069$. Thus,

$$C_{pf} = 0.069 - 0.0375 + 4.92(10^{-4})(75) = 0.068$$

From Eq. (14–28),

$$K_H = 1 + (1)[0.068(1) + 0.175(1)] = 1.243$$

From Table 14–8, for steel gears, $Z_E = 191\sqrt{\text{MPa}}$. From Eq. (14–21), with $m_G = 4$ and $m_N = 1$,

$$Z_I = \frac{\cos 20^\circ \sin 20^\circ}{2}\frac{4}{4+1} = 0.1286$$

Pinion tooth bending. The bending stress in the pinion, from Eq. (14–13) is

$$(\sigma_b)_P = \left(W^t K_o K_v K_s \frac{1}{(b)(m_t)}\frac{K_H K_B}{Y_J}\right)_P = 11\ 848(1)(1.48)(1.133)\frac{1}{(75)(6)}\frac{4}{3}\frac{(1.242)(1)}{0.32} = 171.5 \text{ MPa}$$

Applying the factor of safety goal with Eq. (14–39),

$$n_b = 2 = \frac{(\sigma_{b,\text{all}})_P}{(\sigma_b)_P}$$

$$(\sigma_{b,\text{all}})_P = 1.2(171.5) = 206 \text{ MPa}$$

To estimate the material strength needed, use Eq. (14–16), solving for S_t,

$$S_t = (\sigma_{b,\text{all}})_P \left(\frac{S_F Y_{teta} Y_Z}{Y_N}\right)_P = (206\text{ MPa})\frac{(1)(1)(0.885)}{0.938} = 195\text{ MPa}$$

From Table 14–5 the hardness range of Nitralloy 135M is Rockwell C32–36 (302–335 Brinell). Choosing a midrange hardness as attainable, use 320 Brinell. From Fig. 14–4,

$$S_t = 0.594(320) + 87.76 = 278\text{ MPa}$$

This is a little low. We could increase the hardness to obtain higher strength. But another option is to go back and increase the face width, staying within the recommended range. Select $F = 95$ mm. Correct K_s and K_m:

$$K_s = 0.843\left(95(6)\sqrt{0.309}\right)^{0.0535} = 1.147$$

$$\frac{F}{10d_P} = \frac{95}{(10)(108)} = 0.088$$

$$C_{pf} = 0.088 - 0.0375 + 4.92(10^{-4})(95) = 0.0972$$

$$K_H = 1 + (1)[0.0972(1) + 0.175(1)] = 1.272$$

The bending stress from Eq. (14–13) is

$$(\sigma_b)_P = 11\ 848(1)1.48(1.147)\frac{1}{(96)(6)}\frac{(1.272)(1)}{0.32} = 135\text{ MPa}$$

The allowable stress in bending of the pinion, from Eq. (14–15), is

$$(\sigma_{b,\text{all}})_P = \left(\frac{S_t Y_N}{S_F Y_{teta} Y_Z}\right)_P = \frac{(278)(0.938)}{(1)(1)(0.885)} = 295\text{ MPa}$$

The factor of safety for the pinion bending fatigue, from Eq. (14–39),

$$(n_b)_P = \frac{(\sigma_{b,\text{all}})_P}{(\sigma_b)_P} = \frac{295}{135} = 2.18$$

This is moderately greater than the goal of 2.0, so we will continue.

Decision

Gear tooth bending. Use cast gear blank because of the 432 mm pitch diameter. Use the same material, heat treatment, and nitriding. The only change is in Y_J, so the load-induced bending stress is in the ratio of Y_{jP}/Y_{jG}. Then

$$(\sigma_b)_G = 135\frac{0.32}{0.415} = 104.1\text{ MPa}$$

The allowable stress in bending of the gear, from Eq. (14–15), is

$$(\sigma_{b,\text{all}})_G = \left(\frac{S_t Y_N}{S_F Y_{teta} Y_Z}\right)_G = \frac{(278)(0.961)}{(1)(1)(0.885)} = 301.9\text{ MPa}$$

The factor of safety for the gear bending fatigue, from Eq. (14–39),

$$(n_b)_G = \frac{(\sigma_{b,\text{all}})_G}{(\sigma_b)_G} = \frac{301.9}{104.1} = 2.9$$

This is more than needed. We have the option of reducing the strength of the gear material. But we will proceed first to check the suitability of the current material choice for surface failure.

Pinion tooth wear. The contact stress, given by Eq. (14–14), is

$$(\sigma_c)_P = 191\left[11\ 848(1)\,1.48\,(1.147)\frac{1.272}{(108)(95)}\,\frac{1}{0.129}\right]^{1/2} = 838 \text{ MPa}$$

From Table 14–6, for the material choice of Nitralloy 135M, S_c = 1170 MPa. The allowable stress for contact stress of the pinion, from Eq. (14–16), is

$$(\sigma_{c,\text{all}})_P = \left(\frac{S_c Z_N Z_W}{S_H Y_{teta} Y_Z}\right)_P = \frac{(1170)(0.900)(1)}{(1)(1)(0.885)} = 1190 \text{ MPa}$$

From Eq. (14–40),

$$(n_c)_P = \frac{(\sigma_{c,\text{all}})_P}{(\sigma_c)_P} = \frac{1190}{838} = 1.42$$

Gear tooth wear. The hardness of the gear and pinion are the same. Thus, from Fig. 14–12, Z_W = 1, the contact stress on the gear is the same as the pinion, $(\sigma_c)_G$ = 838 MPa. The wear strength is also the same, S_c = 11 790 MPa.

The allowable stress for contact stress of the gear is

$$(\sigma_{c,\text{all}})_G = \left(\frac{S_c Z_N Z_W}{S_H Y_{teta} Y_Z}\right)_G = \frac{(1170)(0.929)(1)}{(1)(1)(0.885)} = 1228 \text{ MPa}$$

From Eq. (14–40),

$$(n_c)_G = \frac{(\sigma_{c,\text{all}})_G}{(\sigma_c)_G} = \frac{1228}{838} = 1.47$$

The factors of safety all exceed the goals. The factors for wear, particularly for the pinion, are only marginally above the goal of 1.4. If it had been too low, we have several choices for adjustment, including increasing the surface hardness, choosing a different material, or increasing the tooth size and face width. Since the bending factors of safety are satisfactory, increasing the surface hardness is potentially a good choice since it primarily helps the surface strength.

We have chosen not to use the version of the factor of safety for contact stress that is linear with respect to the load, that is $n_{c,\text{linear}}$. This choice is in recognition that the goals for the factors of safety were stated relevant to the uncertainties within the entirety of the design rather than with respect to the load. In fact, the load itself was deemed to be rather predictable. Also, the factors of safety were stated with respect to the severity of the consequences of failure, and specific goals for each mode of failure were readily interpreted.

Rim. Keep $m_B \geq 1.2$. The whole depth is h_t = addendum + dedendum = m_t + 1.25(mt) = 2.25(mt) = 2.25(6) = 13.5 mm. The rim thickness t_R is

$$t_R \geq m_B h_t = 1.2(13.5 \text{ mm}) = 16.2 \text{ mm}$$

In the design of the gear blank, be sure the rim thickness exceeds 16.2 mm; if it does not, review and modify this mesh design.

PROBLEMS

Problems marked with an asterisk (*) are linked to problems in other chapters, as summarized in Table 1–2 of Section 1–17.

Because the results will vary depending on the method used, the problems are presented by section.

Section 14–1

14–1 A steel spur pinion has a module of 4 mm, 22 full-depth teeth, and a 20° pressure angle. The pinion runs at a speed of 1200 rev/min and transmits 11 kW to a 60-tooth gear. If the face width is 50 mm, estimate the bending stress.

14–2 A steel spur pinion has a module of 2.5 mm, 18 teeth cut full-depth with a 20° pressure angle, and a face width of 25 mm. This pinion is expected to transmit 1.5 kW at a speed of 600 rev/min. Determine the bending stress.

14–3 A steel spur pinion has a module of 1.25 mm, 18 teeth cut on the 20° full-depth system, and a face width of 12 mm. At a speed of 1800 rev/min, this pinion is expected to carry a steady load of 0.5 kW. Determine the bending stress.

14–4 A steel spur pinion has 16 teeth cut on the 20° full-depth system with a module of 8 mm and a face width of 90 mm. The pinion rotates at 150 rev/min and transmits 6 kW to the mating steel gear. What is the bending stress?

14–5 A steel spur pinion has a module of 1 mm and 16 teeth cut on the 20° full-depth system and is to carry 0.15 kW at 400 rev/min. Determine a suitable face width based on an allowable bending stress of 150 MPa.

14–6 A 20° full-depth steel spur pinion has 20 teeth and a module of 2 mm and is to transmit 0.5 kW at a speed of 200 rev/min. Find an appropriate face width if the bending stress is not to exceed 75 MPa.

14–7 A 20° full-depth steel spur pinion has a module of 5 mm and 24 teeth and transmits 4.5 kW at a speed of 50 rev/min. Find an appropriate face width if the allowable bending stress is 140 MPa.

14–8 A steel spur pinion is to transmit 15 kW at a speed of 400 rev/min. The pinion is cut on the 20° full-depth system and has a module of 6 mm and 16 teeth. Find a suitable face width based on an allowable stress of 80 MPa.

14–9 A 20° full-depth steel spur pinion with 18 teeth is to transmit 2 kW at a speed of 600 rev/min. Determine appropriate values for the face width and diametral pitch based on an allowable bending stress of 70 MPa.

14–10 A 20° full-depth steel spur pinion is to transmit 1.5 kW hp at a speed of 900 rev/min. If the pinion has 18 teeth, determine suitable values for the module and face width. The bending stress should not exceed 75 MPa.

Section 14–2

14–11 The steel pinion of Problem 14–4 is to mesh with a steel gear with a gear ratio of 4:1. The Brinell hardness of the teeth is 200, and the tangential load transmitted by the gears is 6 kN. If the contact fatigue strength of the steel can be estimated from the AGMA formula of $S_c = 2.22\ H_B + 200$ MPa, estimate the factor of safety of the drive based on a surface fatigue failure.

14–12 A speed reducer has 20° full-depth teeth and consists of a 20-tooth steel spur pinion driving a 50-tooth cast-iron gear. The power transmitted is 9 kW at a pinion speed of 1200 rev/min. For a module of 3 mm and a face width of 38 mm, find the contact stress.

14–13 A gear drive consists of a 16-tooth 20° steel spur pinion and a 48-tooth cast-iron gear having a module of 2 mm. For a power input of 1 kW at a pinion speed of 700 rev/min, select a face width based on an allowable contact stress of 690 MPa.

14–14 A gearset has a module of 5 mm, a 20° pressure angle, and a 24-tooth cast-iron spur pinion driving a 48-tooth cast-iron gear. The pinion is to rotate at 50 rev/min. What power input can be used with this gearset if the contact stress is limited to 690 MPa and $b = 60$ mm?

14–15 A 20° 20-tooth cast-iron spur pinion having a module of 4 mm drives a 32-tooth cast-iron gear. Find the contact stress if the pinion speed is 1000 rev/min, the face width is 50 mm, and 10 kW of power is transmitted.

14–16 A steel spur pinion and gear have a module of 2 mm, milled teeth, 17 and 30 teeth, respectively, a 20° pressure angle, a face width of 22 mm, and a pinion speed of 525 rev/min. The tooth properties are $S_{ut} = 524$ MPa, $S_y = 290$ MPa and the Brinell hardness is 149. Use the Gerber criteria to compensate for one-way bending. For a design factor of 2.25, what is the power rating of the gearset?

14–17 A milled-teeth steel pinion and gear pair have $S_{ut} = 780$ MPa, $S_y = 590$ MPa and a hardness at the involute surface of 262 Brinell. The module is 8 mm, the face width is 50 mm, and the pinion speed is 870 rev/min. The tooth counts are 20 and 100. Use the Gerber criteria to compensate for one-way bending. For a design factor of 1.5, rate the gearset for power considering both bending and wear.

14–18 A 20° full-depth steel spur pinion rotates at 1145 rev/min. It has a module of 6 mm, a face width of 75 mm, and 16 milled teeth. The ultimate tensile strength at the involute is 900 MPa exhibiting a Brinell hardness of 260. The gear is steel with 30 teeth and has identical material strengths. Use the Gerber criteria to compensate for one-way bending. For a design factor of 1.3 find the power rating of the gearset based on the pinion and the gear resisting bending and wear fatigue.

14–19 A steel spur pinion has a module of 4 mm, 17 full-depth milled teeth, and a pressure angle of 20°. The pinion has an ultimate tensile strength at the involute surface of 800 MPa, a Brinell hardness of 232, and a yield strength of 620 MPa. Its shaft speed is 1120 rev/min, its face width is 2 in, and its mating gear has 51 teeth. Use a design factor of 2.

(*a*) Pinion bending fatigue imposes what power limitation? Use the Gerber criteria to compensate for one-way bending.

(*b*) Pinion surface fatigue imposes what power limitation? The gear has identical strengths to the pinion with regard to material properties.

(*c*) Determine power limitations due to gear bending and wear.

(*d*) Specify the power rating for the gearset.

Section 14–3 to 14–19

14–20 A commercial enclosed gear drive is to be designed. Initially, the following is specified. The drive is to transmit 3.7 kW with an input pinion speed of 300 rev/min. The drive is to consist of two spur gears, with a 20° pressure angle, and the pinion is to have 16 teeth driving a 48-tooth gear. The gears are to be grade 1 steel, through-hardened at 200 Brinell, made to No. 6 quality standards. The gears will be uncrowned, centered on their shafts between bearings. A pinion life of 10^8 cycles is desired, with a 90 percent reliability.

(*a*) Use a trial value of module of 4 mm and a face width of 50 mm. For the pinion,
  (*i*) determine the bending stress, allowable bending stress, and bending factor of safety.
  (*ii*) determine the contact stress, allowable contact stress, and contact factor of safety.

(*b*) Assume gears are readily available in face-width increments of 12 mm, with face widths in the range of three to five times the circular pitch. Specify a diametral pitch and face width such that the minimum factor of safety for the pinion is equal to 2.

14–21 A 20° spur pinion with 20 teeth and a module of 2.5 mm transmits 120 W to a 36-tooth gear. The pinion speed is 100 rev/min, and the gears are grade 1, 18-mm face width, through-hardened steel at 200 Brinell, uncrowned, manufactured to a No. 6 quality standard, and considered to be of open gearing quality installation. Find the AGMA bending and contact stresses and the corresponding factors of safety for a pinion life of 10^8 cycles and a reliability of 0.95.

14–22 Repeat Problem 14–20, part (*a*), using helical gears each with a 20° normal pitch angle and a helix angle of 30° and a normal module of 4 mm.

14–23 A spur gearset has 17 teeth on the pinion and 51 teeth on the gear. The pressure angle is 20° and the overload factor $K_o = 1$. The module is 4 mm and the face width is 50 mm. The pinion speed is 1120 rev/min and its cycle life is to be 10^8 revolutions at a reliability $R = 0.99$. The quality number is 5. The material is a through-hardened steel, grade 1, with Brinell hardnesses of 232 core and case of both gears. For a load-based design factor of 2, rate the gearset for these conditions using the AGMA method.

14–24 In Section 14–10, Equation (*a*) is given for K_s based on the procedure in Example 14–2. Derive this equation.

14–25 A speed-reducer has 20° full-depth teeth, and the single-reduction spur-gear gearset has 22 and 60 teeth. The module is 6 mm and the face width is 80 mm. The pinion shaft speed is 1145 rev/min. The life goal of 5-year 24-hour-per-day service is about $3(10^9)$ pinion revolutions. The absolute value of the pitch variation is such that the quality number is 6. The materials are 4340 through-hardened grade 1 steels, heat-treated to 250 Brinell, core and case, both gears. The load is moderate shock and the power is smooth. For a reliability of 0.99, rate the speed reducer for power.

14–26 The speed reducer of Problem 14–25 is to be used for an application requiring 30 kW at 1145 rev/min. For the gear and the pinion, estimate the factors of safety for bending and wear. By examining the factors of safety based on load and stress, identify the threat to each gear and to the mesh.

14–27 The gearset of Problem 14–25 needs improvement of wear capacity. Toward this end the gears are nitrided so that the grade 1 materials have hardnesses as follows: The pinion core is 250 and the pinion case hardness is 390 Brinell, and the gear core hardness is 250 core and 390 case. Estimate the power rating for the new gearset.

14–28 The gearset of Problem 14–25 has had its gear specification changed to 9310 for carburizing and surface hardening with the result that the pinion Brinell hardnesses are 285 core and 580–600 case, and the gear hardnesses are 285 core and 580–600 case. Estimate the power rating for the new gearset.

14–29 The gearset of Problem 14–28 is going to be upgraded in material to a quality of grade 2 (9310) steel. Estimate the power rating for the new gearset.

14–30 Matters of scale always improve insight and perspective. Reduce the physical size of the gearset in Problem 14–25 by one-half and note the result on the estimates of transmitted load W^t and power.

14–31 AGMA procedures with cast-iron gear pairs differ from those with steels because life predictions are difficult; consequently $(Y_N)_P$, $(Y_N)_G$, $(Z_N)_P$, and $(Z_N)_G$ are set to unity. The consequence of this is that the fatigue strengths of the pinion and gear materials are the same. The reliability is 0.99 and the life is 10^7 revolution of the pinion ($K_R = 1$). For longer lives the reducer is derated in power. For the pinion and gearset of Problem 14–25, use grade 40 cast iron for both gears (H_B = 201 Brinell). Rate the reducer for power with S_F and S_H equal to unity.

14–32 Spur-gear teeth have rolling and slipping contact (often about 8 percent slip). Spur gears tested to wear failure are reported at 10^8 cycles as Buckingham's surface fatigue load-stress factor K. This factor is related to Hertzian contact strength S_C by

$$S_C = \sqrt{\frac{1.4K}{(1/E_1 + 1/E_2)\sin\phi}}$$

where ϕ is the normal pressure angle. Cast iron grade 20 gears with $\phi = 14\frac{1}{2}^\circ$ and 20° pressure angle exhibit a minimum K of 0.56 and 0.77 MPa, respectively. How does this compare with $S_C = 2.2H_B$ MPa?

14–33 You've probably noticed that although the AGMA method is based on two equations, the details of assembling all the factors is computationally intensive. To reduce error and omissions, a computer program would be useful. Write a program to perform a power rating of an existing gearset, then use Problems 14–25, 14–27, 14–28, 14–29, and 14–30 to test your program by comparing the results to your longhand solutions.

14–34 In Example 14–5 use nitrided grade 1 steel (4140) which produces Brinell hardnesses of 250 core and 500 at the surface (case). Use the upper fatigue curves on Figures 14–14 and 14–15. Estimate the power capacity of the mesh with safety factors of $S_F = S_H = 1$.

14–35 In Example 14–5 use carburized and case-hardened gears of grade 1. Carburizing and case-hardening can produce a 550 Brinell case. The core hardnesses are 200 Brinell. Estimate the power capacity of the mesh with safety factors of $S_F = S_H = 1$, using the lower fatigue curves in Figures 14–14 and 14–15.

14–36 In Example 14–5, use carburized and case-hardened gears of grade 2 steel. The core hardnesses are 200, and surface hardnesses are 600 Brinell. Use the lower fatigue curves of Figures 14–14 and 14–15. Estimate the power capacity of the mesh using $S_F = S_H = 1$. Compare the power capacity with the results of Problem 14–35.

14–37* The countershaft in Problem 3–83 is part of a speed reducing compound gear train using 20° spur gears. A gear on the input shaft drives gear A. Gear B drives a gear on the output shaft. The input shaft runs at 2400 rev/min. Each gear reduces the speed (and thus increases the torque) by a 2 to 1 ratio. All gears are to be of the same material. Since gear B is the smallest gear, transmitting the largest load, it will likely be critical, so a preliminary analysis is to be performed on it. Use a module of 12 mm, a face width of 4 times the circular pitch, a Grade 2 steel through-hardened to a Brinell hardness of 300, and a desired life of 15 kh with a 95 percent reliability. Determine factors of safety for bending and wear.

14–38* The countershaft in Problem 3–84 is part of a speed reducing compound gear train using 20° spur gears. A gear on the input shaft drives gear *A* with a 2 to 1 speed reduction. Gear *B* drives a gear on the output shaft with a 5 to 1 speed reduction. The input shaft runs at 1800 rev/min. All gears are to be of the same material. Since gear *B* is the smallest gear, transmitting the largest load, it will likely be critical, so a preliminary analysis is to be performed on it. Use a module of 18.75 mm, a face width of 4 times the circular pitch, a Grade 2 steel through-hardened to a Brinell hardness of 300, and a desired life of 12 kh with a 98 percent reliability. Determine factors of safety for bending and wear.

14–39* Build on the results of Problem 13–46 to find factors of safety for bending and wear for gear *F*. Both gears are made from Grade 2 carburized and hardened steel. Use a face width of 4 times the circular pitch. The desired life is 12 kh with a 95 percent reliability.

14–40* Build on the results of Problem 13–47 to find factors of safety for bending and wear for gear *C*. Both gears are made from Grade 2 carburized and hardened steel. Use a face width of 4 times the circular pitch. The desired life is 14 kh with a 98 percent reliability.

15 Bevel and Worm Gears

kadmy/123RF

Chapter Outline

The American Gear Manufacturers Association (AGMA) has established standards for the analysis and design of the various kinds of bevel and worm gears. Chapter 14 was an introduction to the AGMA methods for spur and helical gears and contains many of the definitions of terms used in this chapter. AGMA has established similar methods for other types of gearing, which all follow the same general approach.

15–1 Bevel Gearing—General

Bevel gears may be classified as follows:

- Straight bevel gears
- Spiral bevel gears
- Zerol bevel gears
- Hypoid gears
- Spiroid gears

A ***straight bevel gear*** was illustrated in Figure 13–31. These gears are usually used for pitch-line velocities up to 5 m/s when the noise level is not an important consideration. They are available in many stock sizes and are less expensive to produce than other bevel gears, especially in small quantities.

A ***spiral bevel gear*** is shown in Figure 15–1; the definition of the *spiral angle* is illustrated in Figure 15–2. These gears are recommended for higher speeds and where the noise level is an important consideration. Spiral bevel gears are the bevel counterpart of the helical gear; it can be seen in Figure 15–1 that the pitch surfaces and the nature of contact are the same as for straight bevel gears except for the differences brought about by the spiral-shaped teeth.

The ***Zerol bevel gear*** is a patented gear having curved teeth but with a zero spiral angle. The axial thrust loads permissible for Zerol bevel gears are not as large as those for the spiral bevel gear, and so they are often used instead of straight bevel gears. The Zerol bevel gear is generated by the same tool used for regular spiral bevel

Figure 15–1

Spiral bevel gears. *(Jim Francis/ Shutterstock)*

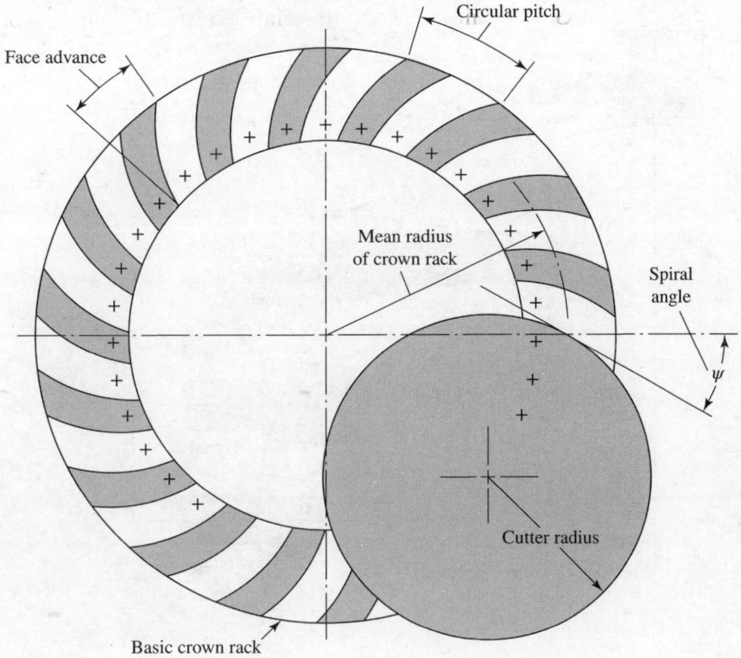

Figure 15–2

Cutting spiral-gear teeth on the basic crown rack.

gears. For design purposes, use the same procedure as for straight bevel gears and then simply substitute a Zerol bevel gear.

It is frequently desirable, as in the case of automotive differential applications, to have gearing similar to bevel gears but with the shafts offset. Such gears are called ***hypoid gears,*** because their pitch surfaces are hyperboloids of revolution. The tooth action between such gears is a combination of rolling and sliding along a straight line and has much in common with that of worm gears. Figure 15–3 shows a pair of hypoid gears providing input to a differential.

Figure 15–4 is included to assist in the classification of spiral bevel gearing. It is seen that the hypoid gear has a relatively small shaft offset. For larger offsets, the pinion begins to resemble a tapered worm and the set is then called ***spiroid gearing.***

Figure 15–3

Hypoid gear used to drive a differential. *(charistoone-images/ Alamy Stock Photo)*

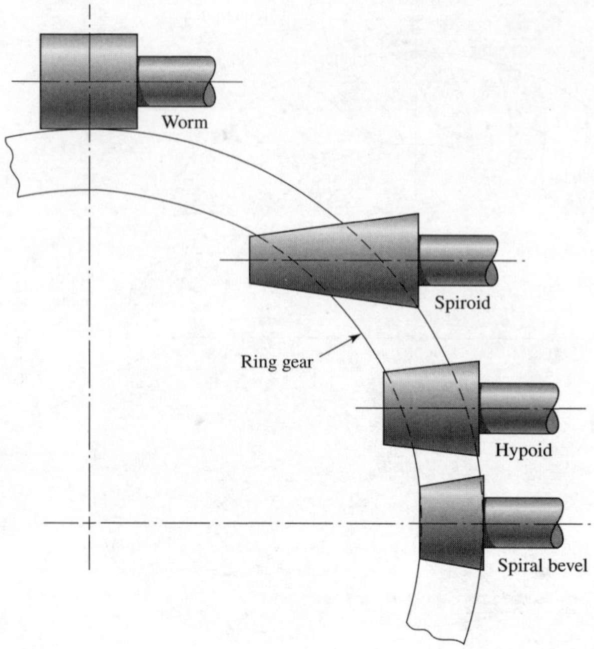

Figure 15–4

Comparison of intersecting- and offset-shaft bevel-type gearings. *(Source: From* Gear Handbook *by Darle W. Dudley, 1962, pp. 2–24.)*

15–2 Bevel-Gear Stresses and Strengths

In a typical bevel-gear mounting, Figure 13–32, for example, one of the gears is often mounted outboard of the bearings. This means that the shaft deflections can be more pronounced and can have a greater effect on the nature of the tooth contact. Another difficulty that occurs in predicting the stress in bevel-gear teeth is the fact that the teeth are tapered. Thus, to achieve perfect line contact passing through the cone center, the teeth should bend more at the large end than at the small end. To obtain this condition requires that the load be proportionately greater at the large end. Because of this varying load across the face of the tooth, it is desirable to have a fairly short face width.

Because of the complexity of bevel, spiral bevel, Zerol bevel, hypoid, and spiroid gears, as well as the limitations of space, only a portion of the applicable standards that refer to straight-bevel gears is presented here.[1] Table 15–1 gives the symbols used in ANSI/AGMA 2003-B97.

Fundamental Contact Stress Equation

$$\sigma_H = Z_E \left(\frac{W^t}{bdZ_I} K_A K_v K_{H\beta} Z_x Z_{xc} \right)^{1/2} \tag{15–1}$$

[1]Figures 15–5 to 15–13 and Tables 15–1 to 15–7 have been extracted from ANSI/AGMA 2003-B97, *Rating the Pitting Resistance and Bending Strength of Generated Straight Bevel, Zerol Bevel and Spiral Bevel Gear Teeth* with the permission of the publisher, the American Gear Manufacturers Association, 1001 N. Fairfax Street, Suite 500, Alexandria, VA, 22314-1587.

Table 15–1 Symbols Used in Bevel Gear Rating Equations, ANSI/AGMA 2003-B97 Standard

ISO Symbol	Description	Units
R_m	Mean cone distance	mm
R_e	Outer cone distance	mm
Z_W	Hardness ratio factor for pitting resistance	
Z_i	Inertia factor for pitting resistance	
Z_{NT}	Stress cycle factor for pitting resistance	
Z_E	Elastic coefficient	$[\mathrm{N/mm^2}]^{0.5}$
Z_Z	Reliability factor for pitting	
	Service factor for pitting resistance	
Z_x	Size factor for pitting resistance	
Z_{xc}	Crowning factor for pitting resistance	
d_{e2}, d_{e1}	Outer pitch diameters of gear and pinion, respectively	mm
E_2, E_1	Young's modulus of elasticity for materials of gear and pinion, respectively	$\mathrm{N/mm^2}$
e	Base of natural (Napierian) logarithms	
b	Net face width	mm
b'_2, b'_1	Effective face widths of gear and pinion, respectively	mm
R_{a1}	Pinion surface roughness	μm
H_{B2}	Minimum Brinell hardness number for gear material	
H_{B1}	Minimum Brinell hardness number for pinion material	
$E_{ht\,\min}$	Minimum total case depth at tooth middepth	mm
h'_c	Minimum effective case depth	mm
$h'_{c\,\lim}$	Suggested maximum effective case depth limit at tooth middepth	mm
Z_I	Geometry factor for pitting resistance	
Y_J	Geometry factor for bending strength	
Y_{J2}, Y_{J1}	Geometry factor for bending strength for gear and pinion, respectively	
Y_F	Stress correction and concentration factor	
Y_i	Inertia factor for bending strength	
Y_{NT}	Stress cycle factor for bending strength	
$K_{H\beta}$	Load distribution factor	
K_A	Overload factor	
Y_z	Reliability factor for bending strength	
Y_X	Size factor for bending strength	
	Service factor for bending strength	
K_θ	Temperature factor	
K_v	Dynamic factor	
Y_β	Lengthwise curvature factor for bending strength	
m_{et}	Outer transverse module	mm
m_{mt}	Mean transverse module	mm
m_{mn}	Mean normal module	mm
ε_{NI}	Load sharing ratio, pitting	
ε_{NJ}	Load sharing ratio, bending	

(*Continued*)

Table 15–1 Symbols Used in Bevel Gear Rating Equations, ANSI/AGMA 2003-B97 Standard (*Continued*)

ISO Symbol	Description	Units
z_2	Number of gear teeth	
n_L	Number of load cycles	
z_1	Number of pinion teeth	
n_1	Pinion speed	
P	Design power through gear pair	kW
P_a	Allowable transmitted power	kW
P_{az}	Allowable transmitted power for pitting resistance	kW
P_{azu}	Allowable transmitted power for pitting resistance at unity service factor	kW
P_{ay}	Allowable transmitted power for bending strength	kW
P_{ayu}	Allowable transmitted power for bending strength at unity service factor	kW
	Outer transverse diametral pitch	
	Mean transverse diametral pitch	
	Mean normal diametral pitch	
Q_v	Transmission accuracy number	
q	Exponent used in formula for lengthwise curvature factor	
r_{mpt2}, r_{mpt1}	Mean transverse pitch radii for gear and pinion, respectively	mm
r_{myo2}, r_{myo1}	Mean transverse radii to point of load application for gear and pinion, respectively	mm
r_{c0}	Cutter radius used for producing Zerol bevel and spiral bevel gears	mm
g_c	Length of the instantaneous line of contact between mating tooth surfaces	mm
$\sigma_{H\ \lim}$	Allowable contact stress number	N/mm^2
$\sigma_{F\ \lim}$	Bending stress number (allowable)	N/mm^2
σ_H	Calculated contact stress number	N/mm^2
S_F	Bending safety factor	
S_H	Contact safety factor	
σ_F	Calculated bending stress number	N/mm^2
σ_{HP}	Permissible contact stress number	N/mm^2
σ_{FP}	Permissible bending stress number	N/mm^2
T_1	Operating pinion torque	N · m
θ_T	Operating gear blank temperature	°C
s_{ai}	Normal tooth top land thickness at narrowest point	mm
U_c	Core hardness coefficient for nitrided gear	N/mm^2
U_H	Hardening process factor for steel	N/mm^2
v_{et}	Pitch-line velocity at outer pitch circle	m/s
Y_{K2}, Y_{K1}	Tooth form factors including stress-concentration factor for gear and pinion, respectively	
ν_2, ν_1	Poisson's ratio for materials of gear and pinion, respectively	
ρy_o	Relative radius of profile curvature at point of maximum contact stress between mating tooth surfaces	mm
α_n	Normal pressure angle at pitch surface	
α_{wt}	Transverse pressure angle at pitch point	
β_m	Mean spiral angle at pitch surface	
β_{mb}	Mean base spiral angle	

Source: ANSI/AGMA 2003-B97.

The first term in each equation is the AGMA symbol, whereas σ_c, our normal notation, is directly equivalent.

Permissible Contact Stress Number (Strength) Equation

$$\sigma_{HP} = \frac{\sigma_{H\,\lim} Z_{NT} Z_W}{S_H K_\theta Z_Z} \tag{15–2}$$

Bending Stress

$$\sigma_F = \frac{W^t}{b} \frac{K_A K_v}{m_{et}} \frac{Y_x K_{H\beta}}{Y_\beta Y_J} \tag{15–3}$$

Permissible Bending Stress Equation

$$\sigma_{FP} = \frac{\sigma_{F\,\lim} Y_{NT}}{S_F K_\theta Y_z} \tag{15–4}$$

15–3 AGMA Equation Factors

Overload Factor K_A

The overload factor makes allowance for any externally applied loads in excess of the nominal transmitted load. Table 15–2, from Appendix A of 2003-B97, is included for your guidance.

Safety Factors S_H and S_F

The factors of safety S_H and S_F as defined in 2003-B97 are adjustments to strength, not load, and consequently cannot be used as is to assess (by comparison) whether the threat is from wear fatigue or bending fatigue. Since W^t is the same for the pinion and gear, the comparison of $\sqrt{S_H}$ to S_F allows direct comparison.

Table 15–2 Overload Factors K_A

Character of Prime Mover	Character of Load on Driven Machine			
	Uniform	Light Shock	Medium Shock	Heavy Shock
Uniform	1.00	1.25	1.50	1.75 or higher
Light shock	1.10	1.35	1.60	1.85 or higher
Medium shock	1.25	1.50	1.75	2.00 or higher
Heavy shock	1.50	1.75	2.00	2.25 or higher

Note: This table is for speed-decreasing drives. For speed-increasing drives, add $0.01(N/n)^2$ or $0.01(z_2/z_1)^2$ to the above factors.

Source: ANSI/AGMA 2003-B97.

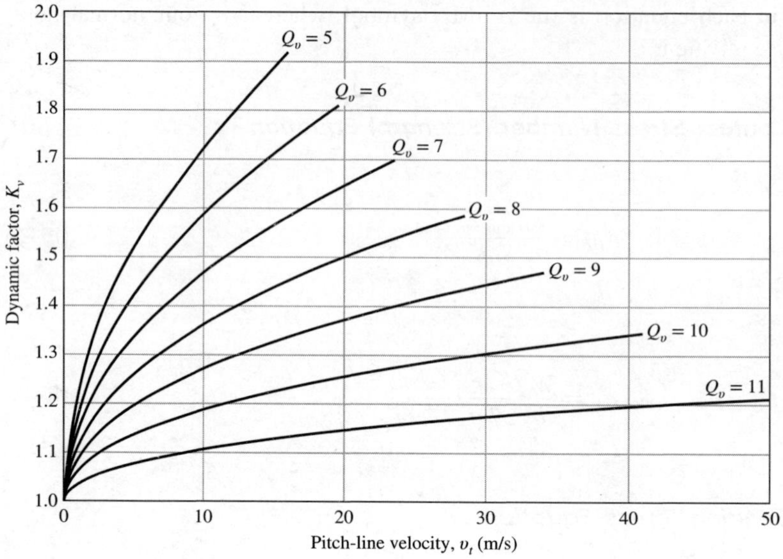

Figure 15–5

Dynamic factor K_v.
(Source: ANSI/AGMA 2003-B97.)

Dynamic Factor K_v

In 2003-C87 AGMA changed the definition of K_v to its reciprocal but used the same symbol. Other standards have yet to follow this move. The dynamic factor K_v makes allowance for the effect of gear-tooth quality related to speed and load, and the increase in stress that follows. AGMA uses a *transmission accuracy number* Q_v to describe the precision with which tooth profiles are spaced along the pitch circle. Figure 15–5 shows graphically how pitch-line velocity and transmission accuracy number are related to the dynamic factor K_v. Curve fits are

$$K_v = \left(\frac{A + \sqrt{200v_{et}}}{A}\right)^B \tag{15–5}$$

where

$$\begin{aligned} A &= 50 + 56(1 - B) \\ B &= 0.25(12 - Q_v)^{2/3} \end{aligned} \tag{15–6}$$

and v_{et} is the pitch-line velocity at outside pitch diameter, expressed in (m/s):

$$v_{et} = 5.236(10^{-5})d_1 n_1 \tag{15–7}$$

The maximum recommended pitch-line velocity is associated with the abscissa of the terminal points of the curve in Figure 15–5:

$$v_{et\ \max} = \frac{[A + (Q_v - 3)]^2}{200} \tag{15–8}$$

where $v_{et\ \max}$ is in m/s, respectively.

Size Factor for Pitting Resistance Z_x

$$Z_x = \begin{cases} 0.5 & b < 12.7 \text{ mm} \\ 0.004\,92b + 0.4375 & 12.7 \le b \le 114.3 \text{ mm} \\ 1 & b > 114.3 \text{ mm} \end{cases} \tag{15–9}$$

Size Factor for Bending Y_x

$$Y_x = \begin{cases} 0.5 & m_{et} < 1.6 \text{ mm} \\ 0.4867 + 0.008\,339 m_{et} & 1.6 \le m_{et} \le 50 \text{ mm} \end{cases} \tag{15–10}$$

Load-Distribution Factor $K_{H\beta}$

$$K_{H\beta} = K_{mb} + 5.6(10^{-6})b^2 \tag{15–11}$$

where

$$K_{mb} = \begin{cases} 1.00 & \text{both members straddle-mounted} \\ 1.10 & \text{one member straddle-mounted} \\ 1.25 & \text{neither member straddle-mounted} \end{cases}$$

Crowning Factor for Pitting Z_{xc}

The teeth of most bevel gears are crowned in the lengthwise direction during manufacture to accommodate the deflection of the mountings.

$$Z_{xc} = \begin{cases} 1.5 & \text{properly crowned teeth} \\ 2.0 & \text{or larger uncrowned teeth} \end{cases} \tag{15–12}$$

Lengthwise Curvature Factor for Bending Strength Y_β

For straight-bevel gears,

$$Y_\beta = 1 \tag{15–13}$$

Pitting Resistance Geometry Factor Z_I

Figure 15–6 shows the geometry factor Z_I for straight-bevel gears with a 20° pressure angle and 90° shaft angle. Enter the figure ordinate with the number of pinion teeth, move to the number of gear-teeth contour, and read from the abscissa.

Bending Strength Geometry Factor Y_J

Figure 15–7 shows the geometry factor J for straight-bevel gears with a 20° pressure angle and 90° shaft angle.

Figure 15–6

Contact geometry factor Z_I for coniflex straight-bevel gears with a 20° normal pressure angle and a 90° shaft angle. *(Source: ANSI/AGMA 2003-B97.)*

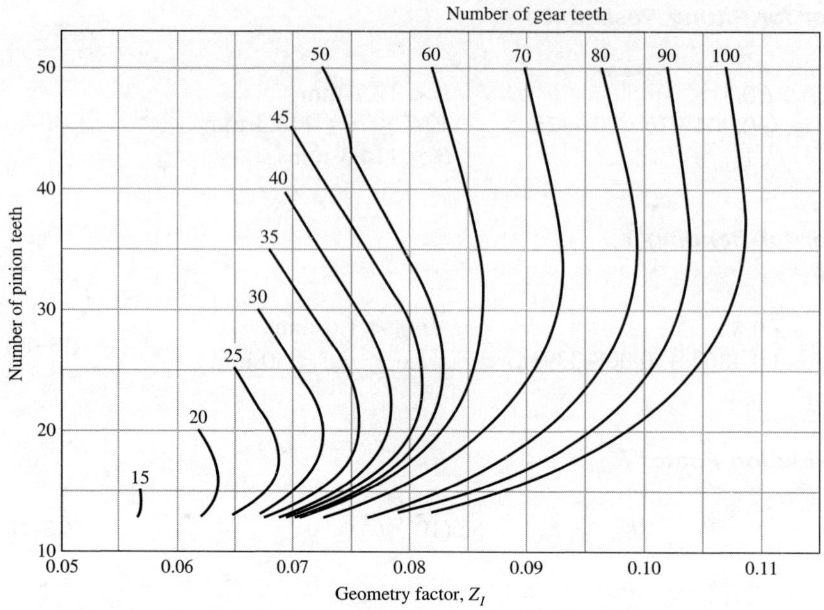

Figure 15–7

Bending factor Y_J for coniflex straight-bevel gears with a 20° normal pressure angle and 90° shaft angle. *(Source: ANSI/AGMA 2003-B97.)*

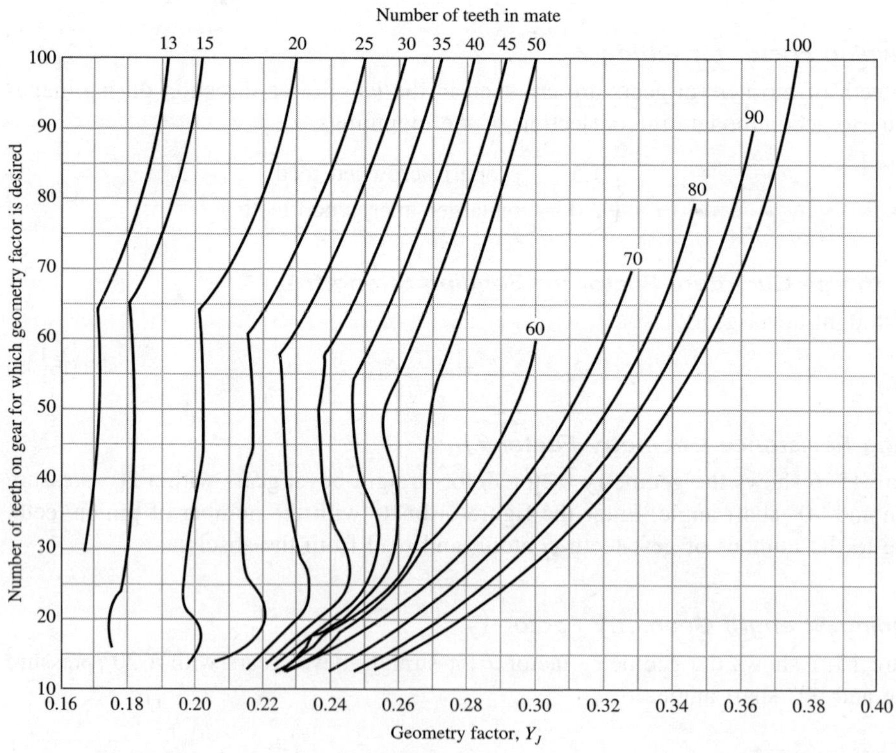

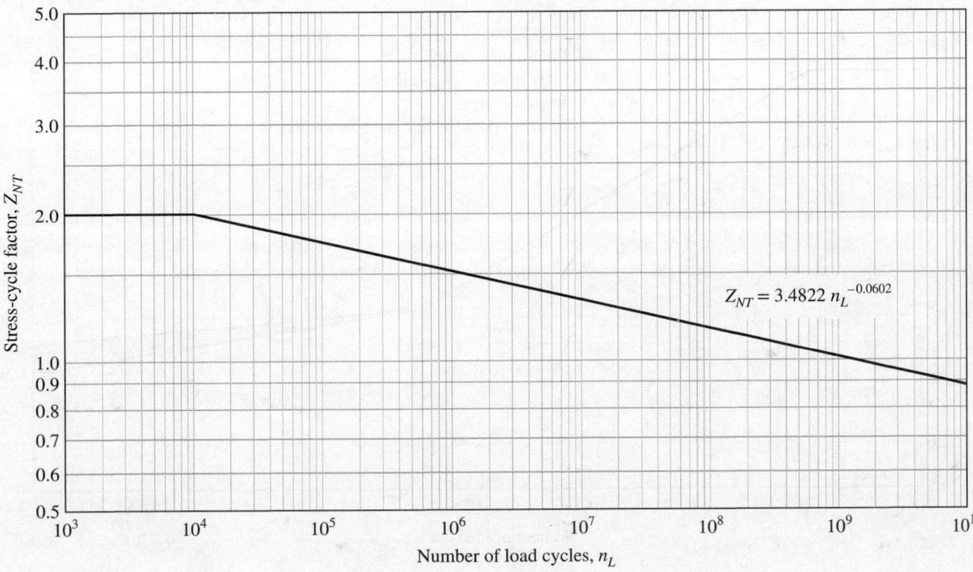

Figure 15–8

Contact stress-cycle factor for pitting resistance Z_{NT} for carburized case-hardened steel bevel gears.
(Source: ANSI/AGMA 2003-B97.)

Stress-Cycle Factor for Pitting Resistance Z_{NT}

$$Z_{NT} = \begin{cases} 2 & 10^3 \le n_L < 10^4 \\ 3.4822 n_L^{-0.0602} & 10^4 \le n_L \le 10^{10} \end{cases} \tag{15–14}$$

See Figure 15–8 for a graphical presentation of Equations (15–14).

Stress-Cycle Factor for Bending Strength Y_{NT}

$$Y_{NT} = \begin{cases} 2.7 & 10^2 \le n_L < 10^3 \\ 6.1514 n_L^{-0.1192} & 10^3 \le n_L < 3(10^6) \\ 1.683 n_L^{-0.0323} & 3(10^6) \le n_L \le 10^{10} \quad \text{critical} \\ 1.3558 n_L^{-0.0178} & 3(10^6) \le n_L \le 10^{10} \quad \text{general} \end{cases} \tag{15–15}$$

See Figure 15–9 for a plot of Equations (15–15).

Hardness-Ratio Factor Z_W

$$Z_W = 1 + B_1(z_2/z_1 - 1) \qquad B_1 = 0.008\,98(H_{B1}/H_{B2}) - 0.008\,29 \tag{15–16}$$

The preceding equations are valid when $1.2 \le H_{B1}/H_{B2} \le 1.7$. Figure 15–10 graphically displays Equations (15–16). When a surface-hardened pinion (48 HRC or harder) is run with a through-hardened gear ($180 \le H_B \le 400$), a work-hardening effect

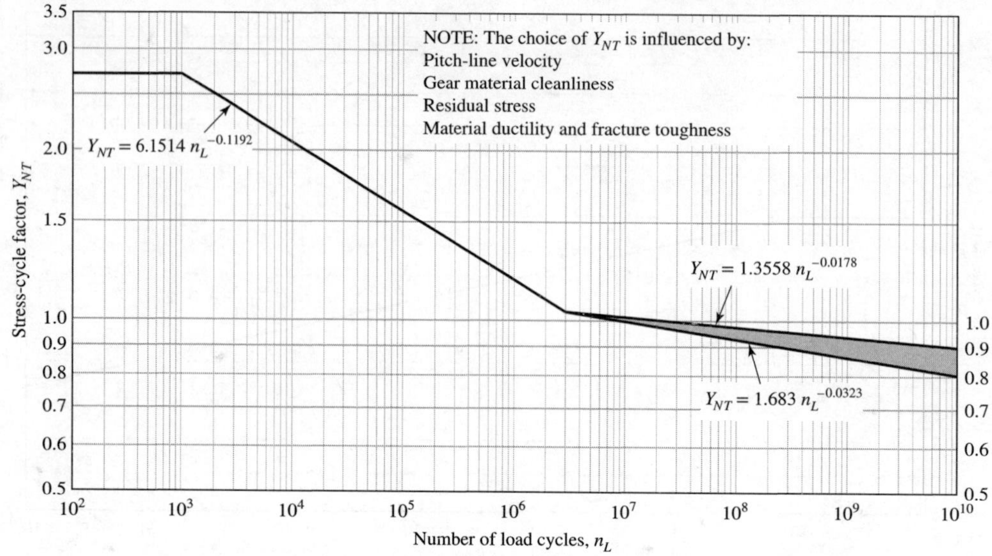

Figure 15–9

Stress-cycle factor for bending strength Y_{NT} for carburized case-hardened steel bevel gears.
(Source: ANSI/AGMA 2003-B97.)

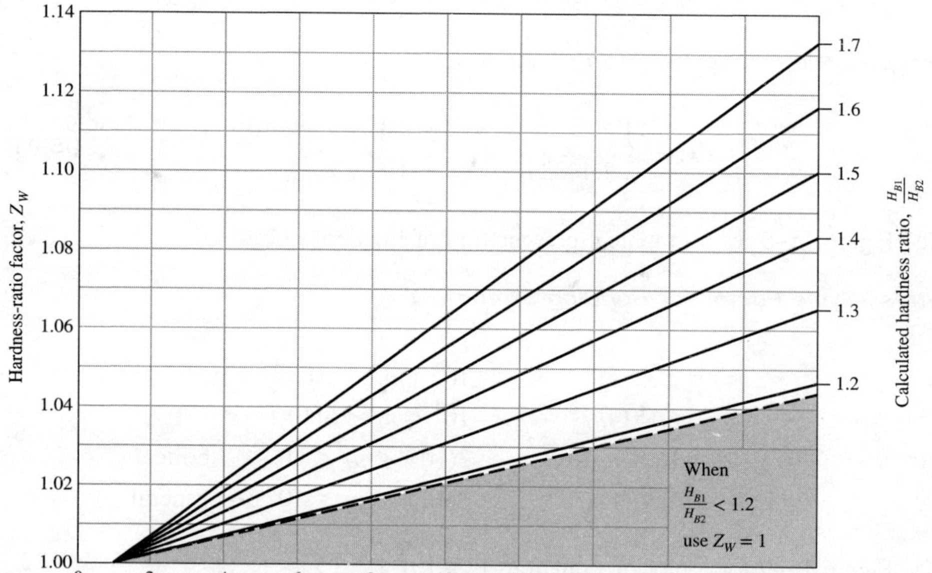

Figure 15–10

Hardness-ratio factor Z_W for through-hardened pinion and gear.
(Source: ANSI/AGMA 2003-B97.)

occurs. The Z_W factor varies with pinion surface roughness R_{a1} and the mating-gear hardness:

$$Z_W = 1 + B_2(450 - H_{B2}) \qquad B_2 = 0.000\,75 \exp(-0.52\, R_{a1}) \tag{15–17}$$

where R_{a1} = pinion surface hardness μm

H_{B2} = minimum Brinell hardness of the gear

See Figure 15–11 for carburized steel gear pairs of approximately equal hardness $Z_W = 1$.

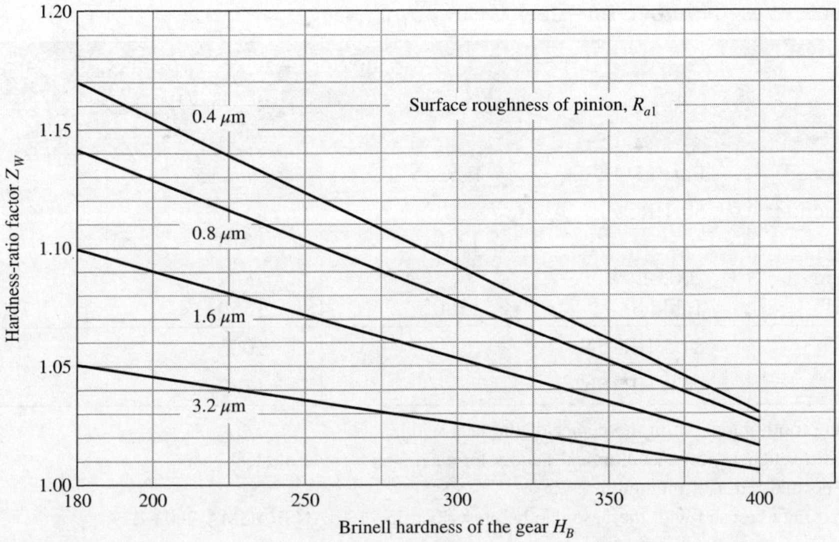

Figure 15–11

Hardness-ratio factor Z_W for surface-hardened pinions. *(Source: ANSI/AGMA 2003-B97.)*

Temperature Factor K_θ

$$K_\theta = \begin{cases} 1 & 0^\circ\text{C} \le \theta \le 120^\circ\text{C} \\ (273 + \theta)/393 & \theta > 120^\circ\text{C} \end{cases} \tag{15–18}$$

Reliability Factors Z_Z and Y_Z

Table 15–3 displays the reliability factors. Note that $Z_Z = \sqrt{Y_Z}$. Logarithmic interpolation equations are

$$Y_Z = \begin{cases} 0.50 - 0.25 \log(1 - R) & 0.99 \le R \le 0.999 \quad (15\text{–}19) \\ 0.70 - 0.15 \log(1 - R) & 0.90 \le R < 0.99 \quad (15\text{–}20) \end{cases}$$

Table 15–3 Reliability Factors

	Reliability Factors for Steel*	
Requirements of Application	Z_Z	Y_Z†
Fewer than one failure in 10 000	1.22	1.50
Fewer than one failure in 1000	1.12	1.25
Fewer than one failure in 100	1.00	1.00
Fewer than one failure in 10	0.92	0.85‡
Fewer than one failure in 2	0.84	0.70§

*At the present time there are insufficient data concerning the reliability of bevel gears made from other materials.

†Tooth breakage is sometimes considered a greater hazard than pitting. In such cases a greater value of Y_Z is selected for bending.

‡At this value plastic flow might occur rather than pitting.

§From test data extrapolation.

Source: ANSI/AGMA 2003-B97.

Table 15–4 Allowable Contact Stress Number for Steel Gears, $\sigma_{H\,\lim}$

Material Designation	Heat Treatment	Minimum Surface* Hardness	Allowable Contact Stress Number, $\sigma_{H\,\lim}$ N/mm²		
			Grade 1†	Grade 2†	Grade 3†
Steel	Through-hardened‡	Fig. 15–12	Fig. 15–12	Fig. 15–12	
	Flame or induction hardened§	50 HRC	1210	1310	
	Carburized and case hardened§	2003-B97 Table 8	1380	1550	1720
AISI 4140	Nitrided§	84.5 HR15N		1000	
Nitralloy 135M	Nitrided§	90.0 HR15N		1100	

*Hardness to be equivalent to that at the tooth middepth in the center of the face width.

†See ANSI/AGMA 2003-B97, Tables 8 through 11, for metallurgical factors for each stress grade of steel gears.

‡These materials must be annealed or normalized as a minimum.

§The allowable stress numbers indicated may be used with the case depths prescribed in 21.1, ANSI/AGMA 2003-B97.

Source: ANSI/AGMA 2003-B97.

Table 15–5 Allowable Contact Stress Number for Iron Gears, $\sigma_{H\,\lim}$

Material Designation			Heat Treatment	Typical Minimum Surface Hardness	Allowable Contact Stress Number, $\sigma_{H\,\lim}$ N/mm²
Material	ASTM	ISO			
Cast iron	ASTM A48	ISO/DR 185			
	Class 30	Grade 200	As cast	175 HB	345
	Class 40	Grade 300	As cast	200 HB	450
Ductile (nodular) iron	ASTM A536	ISO/DIS 1083			
	Grade 80-55-06	Grade 600-370-03	Quenched and tempered	180 HB	650
	Grade 120-90-02	Grade 800-480-02		300 HB	930

Source: ANSI/AGMA 2003-B97.

The reliability of the stress (fatigue) numbers allowable in Tables 15–4, 15–5, 15–6, and 15–7 is 0.99.

Elastic Coefficient for Pitting Resistance Z_E

$$Z_E = \sqrt{\frac{1}{\pi[(1-\nu_1^2)/E_1 + (1-\nu_2^2)/E_2]}} \tag{15–21}$$

where Z_E = elastic coefficient, 190 $\sqrt{\text{N/mm}^2}$ for steel

E_1 and E_2 = Young's moduli for pinion and gear respectively, N/mm²

Table 15–6 Allowable Bending Stress Numbers for Steel Gears, $\sigma_{F\ \text{lim}}$

Material Designation	Heat Treatment	Minimum Surface Hardness	Bending Stress Number (Allowable), $\sigma_{F\ \text{lim}}$ N/mm^2		
			Grade 1*	Grade 2*	Grade 3*
Steel	Through-hardened	Fig. 15–13	Fig. 15–13	Fig. 15–13	
	Flame or induction hardened				
	Unhardened roots	50 HRC	85	95	
	Hardened roots		154		
	Carburized and case hardened†	2003-B97 Table 8	205	240	275
AISI 4140	Nitrided†,‡	84.5 HR15N		150	
Nitralloy 135M	Nitrided†,‡	90.0 HR15N		165	

*See ANSI/AGMA 2003-B97, Tables 8–11, for metallurgical factors for each stress grade of steel gears.

†The allowable stress numbers indicated may be used with the case depths prescribed in 21.1, ANSI/AGMA 2003-B97.

‡The overload capacity of nitrided gears is low. Since the shape of the effective S-N curve is flat, the sensitivity to shock should be investigated before proceeding with the design.

Source: ANSI/AGMA 2003-B97.

Table 15–7 Allowable Bending Stress Number for Iron Gears, $\sigma_{F\ \text{lim}}$

Material Designation			Heat Treatment	Typical Minimum Surface Hardness	Bending Stress Number (Allowable), $\sigma_{F\ \text{lim}}$ N/mm^2
Material	ASTM	ISO			
Cast iron	ASTM A48	ISO/DR 185			
	Class 30	Grade 200	As cast	175 HB	30
	Class 40	Grade 300	As cast	200 HB	45
Ductile (nodular) iron	ASTM A536	ISO/DIS 1083			
	Grade 80-55-06	Grade 600-370-03	Quenched and tempered	180 HB	70
	Grade 120-90-02	Grade 800-480-02		300 HB	95

Source: ANSI/AGMA 2003-B97.

Allowable Contact Stress

Tables 15–4 and 15–5 provide values of σ_H for steel gears and for iron gears, respectively. Figure 15–12 graphically displays allowable stress for grade 1 and 2 materials.

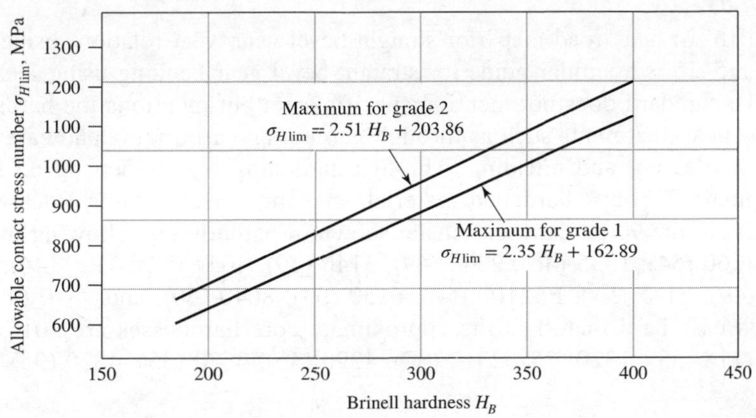

Figure 15–12
Allowable contact stress number for through-hardened steel gears, $\sigma_{H\ \text{lim}}$. *(Source: ANSI/AGMA 2003-B97.)*

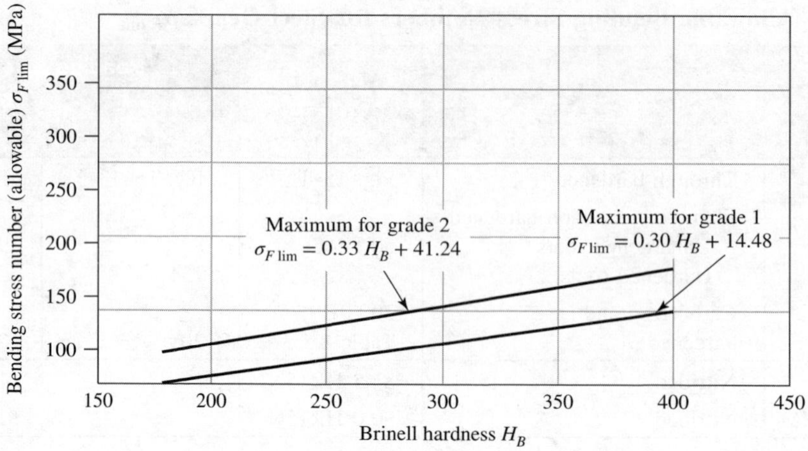

Figure 15–13

Allowable bending stress number for through-hardened steel gears, $\sigma_{F\,\text{lim}}$.
(Source: ANSI/AGMA 2003-B97.)

The equations are

$$\begin{aligned} \sigma_{H\,\text{lim}} &= 2.35H_B + 162.89 \text{ MPa} \qquad \text{grade 1} \\ \sigma_{H\,\text{lim}} &= 2.51H_B + 203.86 \text{ MPa} \qquad \text{grade 2} \end{aligned} \tag{15–22}$$

Allowable Bending Stress Numbers

Tables 15–6 and 15–7 provide $\sigma_{F\,\text{lim}}$ for steel gears and for iron gears, respectively. Figure 15–13 shows graphically allowable bending stress $\sigma_{F\,\text{lim}}$ for through-hardened steels. The equations are

$$\begin{aligned} \sigma_{F\,\text{lim}} &= 0.30H_B + 14.48 \text{ MPa} \qquad \text{grade 1} \\ \sigma_{F\,\text{lim}} &= 0.33H_B + 41.24 \text{ MPa} \qquad \text{grade 2} \end{aligned} \tag{15–23}$$

Reversed Loading

AGMA recommends use of 70 percent of allowable strength in cases where tooth load is completely reversed, as in idler gears and reversing mechanisms.

Summary

Figure 15–14 is a "road map" for straight-bevel gear wear relations using 2003-B97. Figure 15–15 is a similar guide for straight-bevel gear bending using 2003-B97.

The standard does not mention specific steel but mentions the hardness attainable by heat treatments such as through-hardening, carburizing and case-hardening, flame-hardening, and nitriding. Through-hardening results depend on size (diametral pitch). Through-hardened materials and the corresponding Rockwell C-scale hardness at the 90 percent martensite shown in parentheses following include 1045 (50), 1060 (54), 1335 (46), 2340 (49), 3140 (49), 4047 (52), 4130 (44), 4140 (49), 4340 (49), 5145 (51), E52100 (60), 6150 (53), 8640 (50), and 9840 (49). For carburized case-hard materials the approximate core hardnesses are 1015 (22), 1025 (37), 1118 (33), 1320 (35), 2317 (30), 4320 (35), 4620 (35), 4820 (35), 6120 (35),

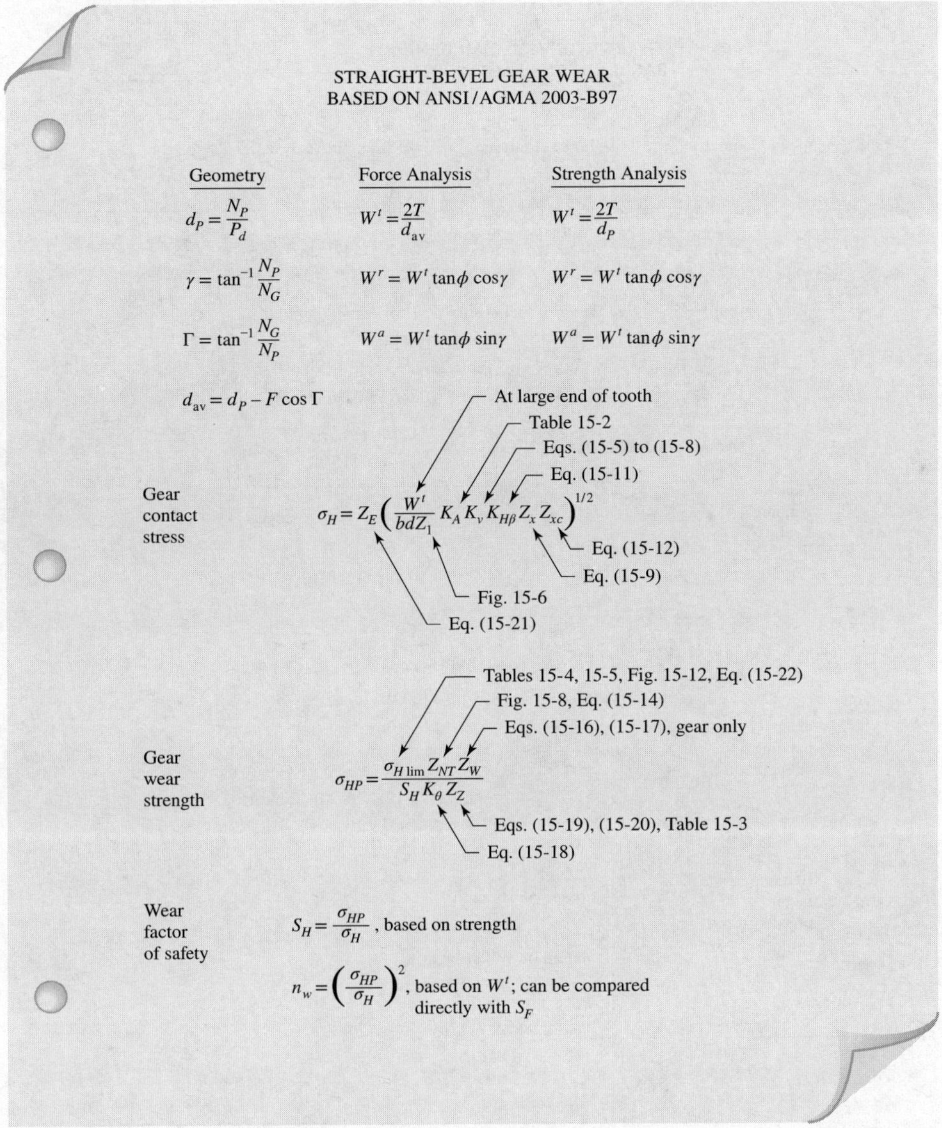

Figure 15–14

"Road map" summary of principal straight-bevel gear wear equations and their parameters.

8620 (35), and E9310 (30). The conversion from HRC to H_B (300-kg load, 10-mm ball) is

HRC	42	40	38	36	34	32	30	28	26	24	22	20	18	16	14	12	10
H_B	388	375	352	331	321	301	285	269	259	248	235	223	217	207	199	192	187

Most bevel-gear sets are made from carburized case-hardened steel, and the factors incorporated in 2003-B97 largely address these high-performance gears. For through-hardened gears, 2003-B97 is silent on K_L and C_L, and Figures 15–8 and 15–9 should prudently be considered as approximate.

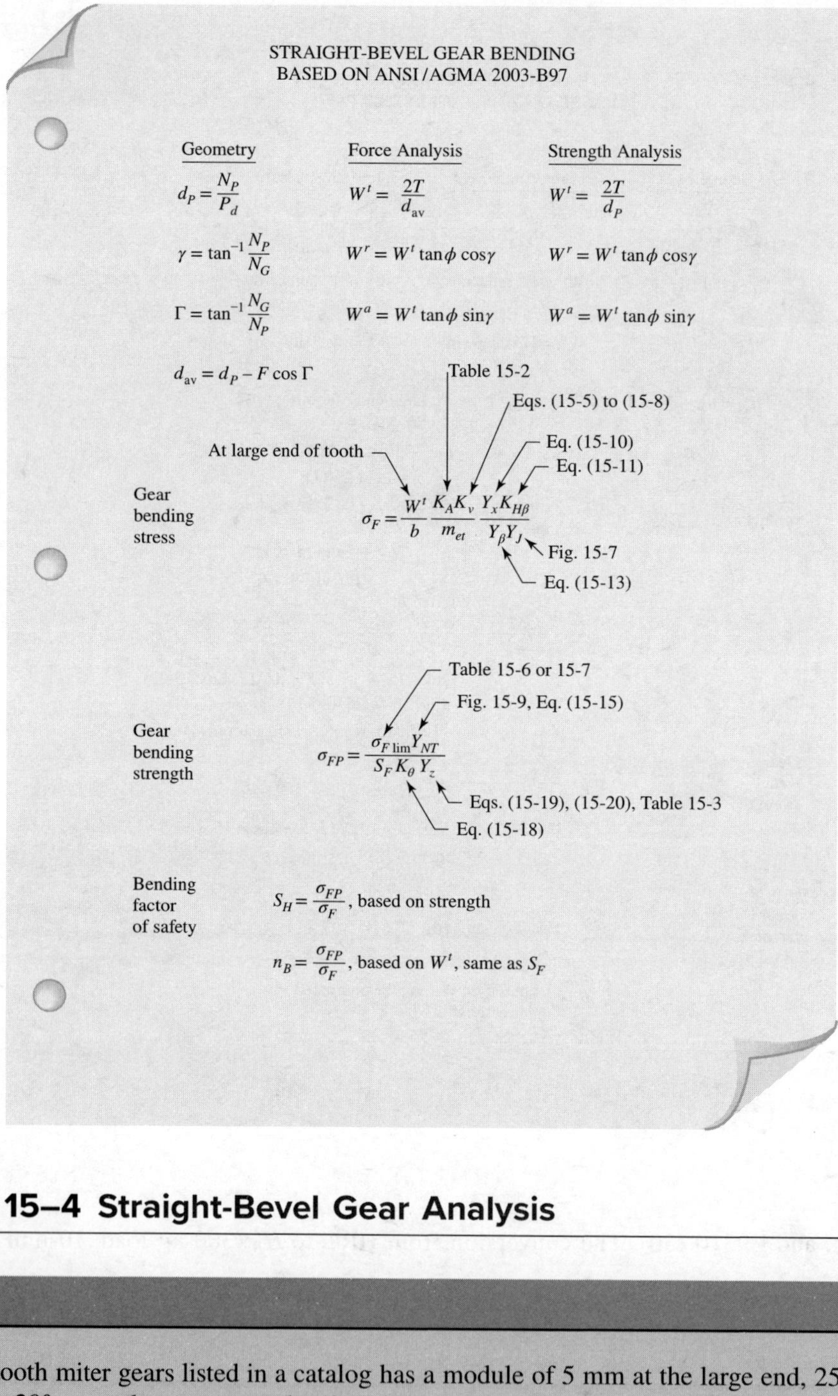

Figure 15–15

"Road map" summary of principal straight-bevel gear bending equations and their parameters.

15–4 Straight-Bevel Gear Analysis

EXAMPLE 15–1

A pair of identical straight-tooth miter gears listed in a catalog has a module of 5 mm at the large end, 25 teeth, a 28 mm face width, and a 20° normal pressure angle; the gears are grade 1 steel through-hardened with a core and case hardness of 180 Brinell. The gears are uncrowned and intended for general industrial use. They have a quality number of $Q_v = 7$. It is likely that the application intended will require outboard mounting of the gears. Use a safety factor of 1, a 10^7 cycle life, and a 0.99 reliability.

(*a*) For a speed of 600 rev/min find the power rating of this gearset based on AGMA bending strength.

(*b*) For the same conditions as in part (*a*) find the power rating of this gearset based on AGMA wear strength.

(*c*) For a reliability of 0.995, a gear life of 10^9 revolutions, and a safety factor of $S_F = S_H = 1.5$, find the power rating for this gearset using AGMA strengths.

Solution

From Figs. 15–14 and 15–15,

$$d_P = zm_{et} = 25(5) = 125 \text{ mm}$$

$$v_{et} = \pi d_P n_P/60 = \pi(0.125)\,600/60 = 3.93 \text{ m/s}$$

Overload factor: uniform-uniform loading, Table 15–2, $K_A = 1.00$.
Safety factor: $S_F = 1$, $S_H = 1$.
Dynamic factor K_v: from Eq. (15–6),

$$B = 0.25(12 - 7)^{2/3} = 0.731$$

$$A = 50 + 56(1 - 0.731) = 65.06$$

$$K_v = \left(\frac{65.06 + \sqrt{785.4}}{65.06}\right)^{0.731} = 1.299$$

From Eq. (15–8),

$$v_{et\ \max} = [65.06 + (7 - 3)]^2/200 = 23.8 \text{ m/s}$$

$v_t < v_{t\ \max}$, that is, $3.93 < 23.8$ m/s, therefore K_v is valid. From Eq. (15–10),

$$Y_s = 0.4867 + 0.008339(5) = 0.528$$

From Eq. (15–11),

$$K_{mb} = 1.25 \qquad \text{and} \qquad K_m = 1.25 + 5.6(10^{-6})28^2 = 1.254$$

From Eq. (15–13), $Y_\beta = 1$. From Fig. 15–6, $Z_I = 0.065$; from Fig. 15–7, $Y_{J1} = 0.216$, $Y_{J2} = 0.216$. From Eq. (15–15),

$$Y_{NT} = 1.3558(10^7)^{-0.0178} = 1.0176 \approx 1$$

From Eq. (15–14),

$$Z_{NT} = 3.4822(10^7)^{-0.0602} = 1.32$$

Since $H_{BP}/H_{BG} = 1$, then from Fig. 15–10, $Z_W = 1$. From Eqs. (15–13) and (15–18), $Y_\beta = 1$ and $K_\theta = 1$, respectively. From Eq. (15–20),

$$Y_Z = 0.70 - 0.15 \log(1 - 0.99) = 1, \qquad Z_Z = \sqrt{Y_Z} = \sqrt{1} = 1$$

(*a*) *Bending:* From Eq. (15–23),

$$\sigma_{F\lim} = 0.30(180) + 14.48 = 68.48 \text{ MPa}$$

From Eq. (15–3),

$$\sigma_F = \frac{W^t}{b}\frac{K_A K_v}{m_{et}}\frac{Y_x K_{H\beta}}{Y_\beta Y_J} = \frac{W^t}{28}\frac{1{\cdot}1.299}{5}\frac{0.528{\cdot}1.254}{1{\cdot}0.216}$$

$$= 0.02844W^t$$

From Eq. (15–4),

$$\sigma_{FP} = \frac{\sigma_{F\lim} Y_{NT}}{S_F K_\theta Y_Z} = \frac{68.48(1)}{(1)(1)(1)} = 68.48 \text{ MPa}$$

Equating σ_F and σ_{FP},

$$0.02844W^t = 68.48 \qquad W^t = 2408 \text{ N}$$

Answer $$H = W^t v_t = 2408(3.93) = 9.46 \text{ kW}$$

(*b*) *Wear:* From Fig. 15–12 or Eq. (15–22),

$$\sigma_{H\lim} = 2.35(180) + 162.89 = 586 \text{ MPa}$$

From Eq. (15–2),

$$\sigma_{HP} = \frac{\sigma_{H\lim} Z_{NT} Z_W}{S_H K_\theta Z_Z} = \frac{586(1.32)(1)}{(1)(1)(1)} = 773 \text{ MPa}$$

Now $Z_E = 190\backslash\sqrt{\text{N/mm}^2}$ from definitions following Eq. (15–21). From Eq. (15–9),

$$Z_X = 0.00492(28) + 0.4375 = 0.575$$

From Eq. (15–12), $Z_{XC} = 2$. Substituting in Eq. (15–1) gives

$$\sigma_H = Z_E\left(\frac{W^t}{bdZ_I} K_A K_v K_{H\beta} Z_x Z_{xc}\right)^{1/2}$$

$$= 190\left[\frac{W^t}{28(125)0.065}(1)1.299(1.254)0.575(2)\right]^{1/2} = 17.24\sqrt{W^t}$$

Equating σ_H and σ_{HP} gives

$$17.24\sqrt{W^t} = 773, \qquad W^t = 2010 \text{ N}$$

$$H = 2010(3.93) = 7.9 \text{ kW}$$

Rated power for the gearset is

Answer $$H = \min(9.4,\ 7.9) = 7.9 \text{ kW}$$

(*c*) Life goal 10^9 cycles, $R = 0.995$, $S_F = S_H = 1.5$, and from Eq. (15–15),

$$Y_{NT} = 1.3558(10^9)^{-0.0178} = 0.9375$$

From Eq. (15–19),

$$Y_Z = 0.50 - 0.25 \log(1 - 0.995) = 1.075, \qquad Z_Z = \sqrt{Y_Z} = \sqrt{1.075} = 1.037$$

From Eq. (15–14),

$$Z_{NT} = 3.4822(10^9)^{-0.0602} = 1$$

Bending: From Eq. (15–23) and part (*a*), $\sigma_{F\lim} = 68.48$ MPa. From Eq. (15–3),

$$\sigma = \frac{W^t}{28(5)}(1)1.299\frac{0.528(1.254)}{(1)0.216} = 0.02844W^t$$

From Eq. (15–4),

$$\sigma_{FP} = \frac{\sigma_{F\,\text{lim}} Y_{NT}}{S_F K_\theta Y_Z} = \frac{68.48(0.9375)}{1.5(1)(1.075)} = 39.81 \text{ MPa}$$

Equating σ_F to σ_{FP} gives

$$0.02844W^t = 39.81 \qquad W^t = 1400 \text{ N}$$

$$H = 1400(3.93) = 5.5 \text{ kW}$$

Wear: From Eq. (15–22), and part (*b*), $\sigma_{H\,\text{lim}} = 585$ MPa.
Substituting into Eq. (15–2) gives

$$\sigma_{HP} = \frac{\sigma_{H\,\text{lim}} Z_{NT} Z_W}{S_H K_\theta Z_Z} = \frac{585(1)(1)}{1.5(1)1.037} = 376 \text{ MPa}$$

From part (*b*), $\sigma_H = 17.24\sqrt{W^t}$ and equating σ_H to σ_{HP} gives

$$\sigma_H = \sigma_{HP} = 376 = 17.24\sqrt{W^t} \qquad W^t = 476 \text{ N}$$

The wear power is

$$H = 476(3.93) = 1.87 \text{ kW}$$

Answer
The mesh rated power is $H = \min(5.5,\ 1.87) = 1.87$ kW.

15–5 Design of a Straight-Bevel Gear Mesh

A useful decision set for straight-bevel gear design is

A priori decisions:

- Function: power, speed, m_G, R,
- Design factor: n_d
- Tooth system
- Tooth count: N_P, N_G

Design decisions:

- Pitch and face width: P_d, b
- Quality number: Q_v
- Gear material, core and case hardness
- Pinion material, core and case hardness

In bevel gears the quality number is linked to the wear strength. The J factor for the gear can be smaller than for the pinion. Bending strength is not linear with face width, because added material is placed at the small end of the teeth. Consequently, face width is roughly prescribed as

$$b = \min(0.3A_0,\ 10m_{et}) \tag{15–24}$$

where A_0 is the cone distance (see Figure 13–16), given by

$$A_0 = \frac{d_P}{2 \sin \gamma} = \frac{d_G}{2 \sin \Gamma} \tag{15–25}$$

EXAMPLE 15–2

Design a straight-bevel gear mesh for shaft centerlines that intersect perpendicularly, to deliver 5 kW at 900 rev/min with a gear ratio of 3:1, temperature of 150°C, normal pressure angle of 20°, using a design factor of 2. The load is uniform-uniform. Use a pinion of 20 teeth. The material is to be AGMA grade 1 and the teeth are to be crowned. The reliability goal is 0.995 with a pinion life of 10^9 revolutions.

Solution

First we list the a priori decisions and their immediate consequences.

Function: 5 kW at 900 rev/min, gear ratio $m_G = 3$, 150°C environment, neither gear straddle-mounted, $K_{mb} = 1.25$ [Eq. (15–11)], $R = 0.995$ at 10^9 revolutions of the pinion,

Eq. (15–14): $$(Z_{NT})_G = 3.4822(10^9/3)^{-0.0602} = 1.068$$

$$(Z_{NT})_P = 3.4822(10^9)^{-0.0602} = 1$$

Eq. (15–15): $$(Y_{NT})_G = 1.683(10^9/3)^{-0.0323} = 0.8929$$

$$(Y_{NT})_P = 1.683(10^9)^{-0.0323} = 0.8618$$

Eq. (15–19): $$Y_Z = 0.50 - 0.25\log(1 - 0.995) = 1.075$$

$$Z_Z = \sqrt{Y_Z} = \sqrt{1.075} = 1.037$$

Eq. (15–18): $$K_\theta = (273 + 150)/393 = 1.076$$

Design factor: $n_d = 2$, $S_F = 2$, $S_H = \sqrt{2} = 1.414$.

Tooth system: crowned, straight-bevel gears, normal pressure angle 20°,

Eq. (15–13): $$Y_\beta = 1$$

Eq. (15–12): $$Z_{xc} = 1.5.$$

With $Z_1 = 20$ teeth, $Z_2 = (3)20 = 60$ teeth and from Fig. 15–14,

$$\gamma = \tan^{-1}(Z_1/Z_2) = \tan^{-1}(20/60) = 18.43^\circ \qquad \Gamma = \tan^{-1}(60/20) = 71.57^\circ$$

From Figs. 15–6 and 15–7, $Z_I = 0.0825$, $Y_{J1} = 0.248$, and $Y_{J2} = 0.202$. Note that $Y_{J1} > Y_{J2}$.

Decision 1: Trial diametral pitch, $m_{et} = 3$ mm.

Eq. (15–10): $$K_s = 0.4867 + 0.008339(3) = 0.5117$$

$$d_P = m_{et}\,Z_1 = 3(20) = 60 \text{ mm}$$

$$d_G = 60(3) = 180 \text{ mm}$$

$$v_t = \pi d_P n_P/12 = \pi(2.5)900/12 = 2.83 \text{ m/s}$$

$$W^t = \text{hp}/v_t = 5000/2.83 = 1767 \text{ W}$$

Eq. (15–25): $$A_0 = d_p/(2\sin\gamma) = 60/(2\sin 18.43^\circ) = 94.89 \text{ mm}$$

Eq. (15–24): $$b = \min(0.3A_0, 10m_{et}) = \min[0.3(94.89), 10(3)] = \min(28.5, 30) = 28.5 \text{ mm}$$

Decision 2: Let $b = 30$ mm. Then,

Eq. (15–9): $$Z_X = 0.00492(30) + 0.4375 = 0.5851$$

Eq. (15–11): $$K_{H\beta} = 1.25 + 5.6(10^{-6})(30)^2 = 1.255$$

Decision 3: Let the transmission accuracy number be 6. Then, from Eq. (15–6),

$$B = 0.25(12 - 6)^{2/3} = 0.8255$$

$$A = 50 + 56(1 - 0.8255) = 59.77$$

Eq. (15–5):

$$K_v = \left(\frac{59.77 + \sqrt{200(2.83)}}{59.77}\right)^{0.8255} = 1.319$$

Decision 4: Pinion and gear material and treatment. Carburize and case-harden grade ASTM 1320 to

Core 21 HRC (H_B is 229 Brinell)
Case 55-64 HRC (H_B is 515 Brinell)

From Table 15–4, $\sigma_{H\,\text{lim}}$ = 1380 MPa and from Table 15–6, $\sigma_{F\,\text{lim}}$ = 205 MPa.

Gear bending: From Eq. (15–3), the bending stress is

$$(\sigma_F)_G = \frac{W^t}{b}\frac{K_A K_v}{m_{et}}\frac{Y_x K_{H\beta}}{Y_\beta Y_J} = \frac{1767}{30}\frac{1(1.319)}{3}\frac{0.5117(1.255)}{(1)0.202}$$

$$= 82 \text{ MPa}$$

The bending strength, from Eq. (15–4), is given by

$$(\sigma_{FP})_G = \left(\frac{\sigma_{F\,\text{lim}} Y_{NT}}{S_F K_\theta Y_Z}\right)_G = \frac{205(0.8929)}{2(1.076)1.075} = 79 \text{ MPa}$$

The strength exceeds the stress by a factor of 79/82 = 0.96, giving an actual factor of safety of $(S_F)_G = 2(0.96)$ = 1.92.

Pinion bending: The bending stress can be found from

$$(\sigma_F)_P = (\sigma_F)_G \frac{Y_{J2}}{Y_{J1}} = 82\frac{0.202}{0.248} = 67 \text{ MPa}$$

The bending strength, again from Eq. (15–4), is given by

$$(\sigma_{FP})_P = \left(\frac{\sigma_{F\,\text{lim}} Y_{NT}}{S_F K_\theta Y_Z}\right)_P = \frac{205(0.8618)}{2(1.076)1.075} = 79 \text{ MPa}$$

The strength exceeds the stress by a factor of 79/67 = 1.18, giving an actual factor of safety of $(S_F)_P = 2(1.18)$ = 2.36.

Gear wear: The load-induced contact stress for the pinion and gear, from Eq. (15–1), is

$$\sigma_H = Z_E\left(\frac{W^t}{bdZ_I} K_A K_v K_{H\beta} Z_x Z_{xc}\right)^{1/2}$$

$$= 190\left[\frac{1767}{30(60)0.0825}(1)1.319(1.255)0.5851(1.5)\right]^{1/2}$$

$$= 790 \text{ MPa}$$

From Eq. (15–2) the contact strength of the gear is

$$(\sigma_{HP})_G = \left(\frac{\sigma_{H\,\text{lim}} Z_{NT} Z_W}{S_H K_\theta Z_Z}\right)_G = \frac{1380(1.068)(1)}{\sqrt{2}(1.076)1.037} = 934 \text{ MPa}$$

The strength exceeds the stress by a factor of 934/790 = 1.18, giving an actual factor of safety of $(S_H)_G^2 = 1.18^2(2) = 2.78$.

Pinion wear: From Eq. (15–2) the contact strength of the pinion is

$$(\sigma_{HP})_P = \left(\frac{\sigma_{H\lim} Z_{NT} Z_W}{S_H K_\theta Z_Z}\right)_G = \frac{1380(1)(1)}{\sqrt{2}(1.076)1.037} = 875 \text{ MPa}$$

The strength exceeds the stress by a factor of 875/790 = 1.1, giving an actual factor of safety of $(S_H)_P^2 = 1.1^2(2) = 2.42$.

The actual factors of safety are 1.92, 2.36, 2.78, and 2.42. Making a direct comparison of the factors, we note that the primary threat is from gear bending. We also note that the other three factors of safety are considerably higher than the target design factor. If optimization is desired, our goal would be to make changes in the design decisions that drive the factors closer to 2.

15–6 Worm Gearing—AGMA Equation[2]

Sections 13–11 and 13–17 introduced worm gearing and its force analysis and efficiency. Here, we will present a condensed version of the AGMA recommendations for cylindrical (single-enveloping) worm gearing.[3] For brevity, the equations will be shown for U.S. customary units only. Similar equations for SI units are available in the AGMA standards.

Since they are essentially nonenveloping worm gears, the *crossed helical* gears, shown in Figure 15–16, can be considered with other worm gearing. Crossed helical gears, and worm gears too, usually have a 90° shaft angle, though this need not be so. The relation between the shaft and helix angles is

$$\Sigma = \psi_P \pm \psi_G \tag{15–26}$$

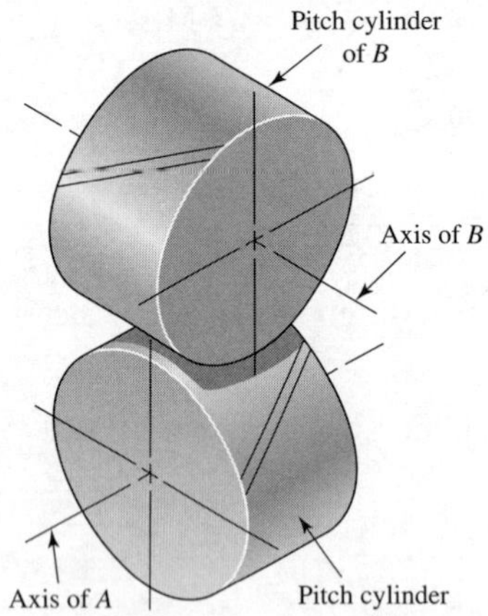

Figure 15–16

View of the pitch cylinders of a pair of crossed helical gears.

[2]This paragraph and the following ones report values in U.S. customary units, as specified by AGMA standards; for values in SI units, please refer to the BS standards.

[3]ANSI/AGMA 6034-B92, February 1992, *Practice for Enclosed Cylindrical Wormgear Speed-Reducers and Gear Motors*; and ANSI/AGMA 6022-C93, Dec. 1993, *Design Manual for Cylindrical Wormgearing*. Note: Equations (15–32) to (15–38) are contained in Annex C of 6034-B92 for informational purposes only. To comply with ANSI/AGMA 6034-B92, use the tabulations of these rating factors provided in the standard.

Table 15–8 Cylindrical Worm Dimensions Common to Both Worm and Gear*

		ϕ_n		
		14.5°	20°	25°
Quantity	**Symbol**	$N_W \le 2$	$N_W \le 2$	$N_W > 2$
Addendum	a	$0.3183p_x$	$0.3183p_x$	$0.286p_x$
Dedendum	b	$0.3683p_x$	$0.3683p_x$	$0.349p_x$
Whole depth	h_t	$0.6866p_x$	$0.6866p_x$	$0.635p_x$

*The table entries are for a tangential diametral pitch of the gear of $P_t = 1$.

where Σ is the shaft angle. The plus sign is used when both helix angles are of the same hand, and the minus sign when they are of opposite hand. The subscript P in Equation (15–26) refers to the pinion (worm); the subscript W is used for this same purpose. The subscript G refers to the gear, also called *gear wheel, worm wheel,* or simply the *wheel.* Table 15–8 gives cylindrical worm dimensions common to worm and gear.

Good proportions indicate the worm pitch diameter d falls in the range

$$\frac{C^{0.875}}{3} \le d \le \frac{C^{0.875}}{1.6} \tag{15–27}$$

where C is the center-to-center distance. AGMA relates the allowable tangential force on the worm-gear tooth $(W^t)_{\text{all}}$ to other parameters by

$$(W^t)_{\text{all}} = C_s D_m^{0.8} F_e C_m C_v \tag{15–28}$$

where C_s = materials factor

D_m = mean gear diameter, in

F_e = effective face width of the gear (actual face width, but not to exceed $0.67d_m$, the mean worm diameter), in

C_m = ratio correction factor

C_v = velocity factor

The friction force W_f is given by

$$W_f = \frac{fW^t}{\cos\lambda\cos\phi_n} \tag{15–29}$$

where f = coefficient of friction

λ = lead angle at mean worm diameter

ϕ_n = normal pressure angle

The sliding velocity V_s at the mean worm diameter, in feet per minute, is

$$V_s = \frac{\pi n_W d_m}{12\cos\lambda} \tag{15–30}$$

where n_W = rotative speed of the worm and d_m = mean worm diameter. The torque at the worm gear is

$$T_G = \frac{W^t D_m}{2} \tag{15–31}$$

where D_m is the mean gear diameter.

The parameters in Equation (15–28) are, quantitatively,

$$C_s = 720 + 10.37C^3 \qquad C \le 3 \text{ in} \tag{15–32}$$

For sand-cast gears,

$$C_s = \begin{cases} 1000 & C > 3 \quad D_m \le 2.5 \text{ in} \\ 1190 - 477 \log D_m & C > 3 \quad D_m > 2.5 \text{ in} \end{cases} \tag{15–33}$$

For chilled-cast gears,

$$C_s = \begin{cases} 1000 & C > 3 \quad D_m \le 8 \text{ in} \\ 1412 - 456 \log D_m & C > 3 \quad D_m > 8 \text{ in} \end{cases} \tag{15–34}$$

For centrifugally cast gears,

$$C_s = \begin{cases} 1000 & C > 3 \quad D_m \le 25 \text{ in} \\ 1251 - 180 \log D_m & C > 3 \quad D_m > 25 \text{ in} \end{cases} \tag{15–35}$$

The ratio correction factor C_m for gear ratio m_G is given by

$$C_m = \begin{cases} 0.02\sqrt{-m_G^2 + 40m_G - 76} + 0.46 & 3 < m_G \le 20 \\ 0.0107\sqrt{-m_G^2 + 56m_G + 5145} & 20 < m_G \le 76 \\ 1.1483 - 0.006\,58m_G & m_G > 76 \end{cases} \tag{15–36}$$

The velocity factor C_v is given by

$$C_v = \begin{cases} 0.659 \exp(-0.0011V_s) & V_s < 700 \text{ ft/min} \\ 13.31\, V_s^{-0.571} & 700 \le V_s < 3000 \text{ ft/min} \\ 65.52\, V_s^{-0.774} & V_s > 3000 \text{ ft/min} \end{cases} \tag{15–37}$$

AGMA reports the coefficient of friction f as

$$f = \begin{cases} 0.15 & V_s = 0 \\ 0.124 \exp\left(-0.074V_s^{0.645}\right) & 0 < V_s \le 10 \text{ ft/min} \\ 0.103 \exp\left(-0.110V_s^{0.450}\right) + 0.012 & V_s > 10 \text{ ft/min} \end{cases} \tag{15–38}$$

Now we examine some worm-gear mesh geometry. The addendum a and dedendum b are

$$a = \frac{p_x}{\pi} = 0.3183p_x \tag{15–39}$$

$$b = \frac{1.157p_x}{\pi} = 0.3683p_x \tag{15–40}$$

The full depth h_t is

$$h_t = \begin{cases} \dfrac{2.157p_x}{\pi} = 0.6866p_x & p_x \ge 0.16 \text{ in} \\ \dfrac{2.200p_x}{\pi} + 0.002 = 0.7003p_x + 0.002 & p_x < 0.16 \text{ in} \end{cases} \tag{15–41}$$

The worm outside diameter d_o is

$$d_o = d + 2a \tag{15–42}$$

The worm root diameter d_r is

$$d_r = d - 2b \tag{15–43}$$

The worm-gear throat diameter D_t is

$$D_t = D + 2a \tag{15–44}$$

where D is the worm-gear pitch diameter. The worm-gear root diameter D_r is

$$D_r = D - 2b \tag{15–45}$$

The clearance c is

$$c = b - a \tag{15–46}$$

The worm face width (maximum) $(F_W)_{\max}$ is

$$(F_W)_{\max} = 2\sqrt{\left(\frac{D_t}{2}\right)^2 - \left(\frac{D}{2} - a\right)^2} = 2\sqrt{2Da} \tag{15–47}$$

which was simplified using Equation (15–44). The worm-gear face width F_G is

$$F_G = \begin{cases} 2d_m/3 & p_x > 0.16 \text{ in} \\ 1.125\sqrt{(d_o + 2c)^2 - (d_o - 4a)^2} & p_x \le 0.16 \text{ in} \end{cases} \tag{15–48}$$

The heat loss rate H_{loss} from the worm-gear case in ft · lbf/min is

$$H_{\text{loss}} = 33\,000(1 - e)H_{\text{in}} \tag{15–49}$$

where e is efficiency, given by Equation (13–45), and H_{in} is the input horsepower from the worm. The overall coefficient $\hbar_{\text{CR}}$ for combined convective and radiative heat transfer from the worm-gear case in ft · lbf/(min · in^2 · °F) is

$$\hbar_{\text{CR}} = \begin{cases} \dfrac{n_W}{6494} + 0.13 & \text{no fan on worm shaft} \\[2ex] \dfrac{n_W}{3939} + 0.13 & \text{fan on worm shaft} \end{cases} \tag{15–50}$$

The temperature of the oil sump t_s is given by

$$t_s = t_a + \frac{H_{\text{loss}}}{\hbar_{\text{CR}}A} = \frac{33\,000(1 - e)(H)_{\text{in}}}{\hbar_{\text{CR}}A} + t_a \tag{15–51}$$

where A is the case lateral area in in^2, and t_a is the ambient temperature in °F. Bypassing Equations (15–49), (15–50), and (15–51) one can apply the AGMA recommendation for minimum lateral area $A_{\min}$ in in^2 using

$$A_{\min} = 43.20C^{1.7} \tag{15–52}$$

Because worm teeth are inherently much stronger than worm-gear teeth, they are not considered. The teeth in worm gears are short and thick on the edges of the face; midplane they are thinner as well as curved. Buckingham[4] adapted the Lewis equation for this case:

$$\sigma_a = \frac{W_G^t}{p_n F_e y} \tag{15–53}$$

where $p_n = p_x \cos\lambda$ and y is the Lewis form factor related to circular pitch. For $\phi_n = 14.5°$, $y = 0.100$; $\phi_n = 20°$, $y = 0.125$; $\phi_n = 25°$, $y = 0.150$; $\phi_n = 30°$, $y = 0.175$.

[4]Earle Buckingham, *Analytical Mechanics of Gears*, McGraw Hill, New York, 1949, p. 495.

15–7 Worm-Gear Analysis

Compared to other gearing systems worm-gear meshes have a much lower mechanical efficiency. Cooling, for the benefit of the lubricant, becomes a design constraint sometimes resulting in what appears to be an oversize gear case in light of its contents. If the heat can be dissipated by natural cooling, or simply with a fan on the wormshaft, simplicity persists. Water coils within the gear case or lubricant outpumping to an external cooler is the next level of complexity. For this reason, gear-case area is a design decision.

To reduce cooling load, use multiple-thread worms. Also keep the worm pitch diameter as small as possible.

Multiple-thread worms can remove the self-locking feature of many worm-gear drives. When the worm drives the gearset, the mechanical efficiency e_W is given by

$$e_W = \frac{\cos\phi_n - f\tan\lambda}{\cos\phi_n + f\cot\lambda} \tag{15–54}$$

With the gear driving the gearset, the mechanical efficiency e_G is given by

$$e_G = \frac{\cos\phi_n - f\cot\lambda}{\cos\phi_n + f\tan\lambda} \tag{15–55}$$

To ensure that the worm gear will drive the worm,

$$f_{\text{stat}} < \cos\phi_n \tan\lambda \tag{15–56}$$

where values of f_{stat} can be found in ANSI/AGMA 6034-B92. To prevent the worm gear from driving the worm, refer to clause 9 of 6034-B92 for a discussion of self-locking in the static condition.

It is important to have a way to relate the tangential component of the gear force W^t_G to the tangential component of the worm force W^t_W, which includes the role of friction and the angularities of ϕ_n and λ. Refer to Equation (13–44), solved for W^t_W:

$$W^t_W = W^t_G \frac{\cos\phi_n \sin\lambda + f\cos\lambda}{\cos\phi_n \cos\lambda - f\sin\lambda} \tag{15–57}$$

In the absence of friction

$$W^t_W = W^t_G \tan\lambda$$

The mechanical efficiency of most gearing is very high, which allows power in and power out to be used almost interchangeably. Worm gearsets have such poor efficiencies that we work with, and speak of, output power. The magnitude of the gear transmitted force W^t_G can be related to the output horsepower H_0, the application factor K_a, the efficiency e, and design factor n_d by

$$W^t_G = \frac{33\,000 n_d H_0 K_a}{V_G e} \tag{15–58}$$

We use Equation (15–57) to obtain the corresponding worm force W^t_W. It follows that the worm and gear transmitted powers in hp are

$$H_W = \frac{W^t_W V_W}{33\,000} = \frac{\pi d_W n_W W^t_W}{12(33\,000)} \tag{15–59}$$

$$H_G = \frac{W^t_G V_G}{33\,000} = \frac{\pi d_G n_G W^t_G}{12(33\,000)} \tag{15–60}$$

From Equation (13–43),

$$W_f = \frac{f W_G^t}{f \sin \lambda - \cos \phi_n \cos \lambda} \tag{15–61}$$

The sliding velocity of the worm at the pitch cylinder V_s is

$$V_s = \frac{\pi d n_W}{12 \cos \lambda} \tag{15–62}$$

and the friction power H_f is given by

$$H_f = \frac{|W_f|\, V_s}{33\ 000} \text{ hp} \tag{15–63}$$

Table 15–9 gives the largest lead angle λ_{max} associated with normal pressure angle ϕ_n.

Table 15–9 Largest Lead Angle Associated with a Normal Pressure Angle ϕ_n for Worm Gearing

ϕ_n	Maximum Lead Angle λ_{max}
14.5°	16°
20°	25°
25°	35°
30°	45°

EXAMPLE 15–3

A single-thread steel worm rotates at 1800 rev/min, meshing with a 24-tooth worm gear transmitting 3 hp to the output shaft. The worm pitch diameter is 3 in and the tangential diametral pitch of the gear is 4 teeth/in. The normal pressure angle is 14.5°. The ambient temperature is 70°F. The application factor is 1.25 and the design factor is 1; gear face width is 2 in, lateral case area 600 in^2, and the gear is chill-cast bronze.

(*a*) Find the gear geometry.
(*b*) Find the transmitted gear forces and the mesh efficiency.
(*c*) Is the mesh sufficient to handle the loading?
(*d*) Estimate the lubricant sump temperature.

Solution

(*a*) $m_G = N_G/N_W = 24/1 = 24$, gear: $D = N_G/P_t = 24/4 = 6.000$ in, worm: $d = 3.000$ in. The axial circular pitch p_x is $p_x = \pi/P_t = \pi/4 = 0.7854$ in. $C = (3 + 6)/2 = 4.5$ in.

Eq. (15–39): $a = p_x/\pi = 0.7854/\pi = 0.250$ in

Eq. (15–40): $b = 0.3683p_x = 0.3683(0.7854) = 0.289$ in

Eq. (15–41): $h_t = 0.6866p_x = 0.6866(0.7854) = 0.539$ in

Eq. (15–42): $d_o = 3 + 2(0.250) = 3.500$ in

Eq. (15–43): $d_r = 3 - 2(0.289) = 2.422$ in

Eq. (15–44): $D_t = 6 + 2(0.250) = 6.500$ in

Eq. (15–45): $D_r = 6 - 2(0.289) = 5.422$ in

Eq. (15–46): $c = 0.289 - 0.250 = 0.039$ in

Eq. (15–47): $(F_W)_{max} = 2\sqrt{2(6)0.250} = 3.464$ in

The tangential speeds of the worm, V_W, and gear, V_G, are, respectively,

$$V_W = \pi(3)1800/12 = 1414 \text{ ft/min} \qquad V_G = \frac{\pi(6)1800/24}{12} = 117.8 \text{ ft/min}$$

The lead of the worm, from Eq. (13–27), is $L = p_x N_W = 0.7854(1) = 0.7854$ in. The lead angle λ, from Eq. (13–28), is

$$\lambda = \tan^{-1}\frac{L}{\pi d} = \tan^{-1}\frac{0.7854}{\pi(3)} = 4.764^\circ$$

The normal diametral pitch for a worm gear is the same as for a helical gear, which from Eq. (13–18), with $\psi = \lambda$ is

$$P_n = \frac{P_t}{\cos\lambda} = \frac{4}{\cos 4.764^\circ} = 4.014$$

$$p_n = \frac{\pi}{P_n} = \frac{\pi}{4.014} = 0.7827 \text{ in}$$

The sliding velocity, from Eq. (15–62), is

$$V_s = \frac{\pi d n_W}{12\cos\lambda} = \frac{\pi(3)1800}{12\cos 4.764^\circ} = 1419 \text{ ft/min}$$

(*b*) The coefficient of friction, from Eq. (15–38), is

$$f = 0.103\exp[-0.110(1419)^{0.450}] + 0.012 = 0.0178$$

The efficiency e, from Eq. (13–45), is

Answer
$$e = \frac{\cos\phi_n - f\tan\lambda}{\cos\phi_n + f\cot\lambda} = \frac{\cos 14.5^\circ - 0.0178\tan 4.764^\circ}{\cos 14.5^\circ + 0.0178\cot 4.764^\circ} = 0.818$$

The designer used $n_d = 1$, $K_a = 1.25$ and an output horsepower of $H_0 = 3$ hp. The gear tangential force component W_G^t, from Eq. (15–58), is

Answer
$$W_G^t = \frac{33\,000 n_d H_0 K_a}{V_G e} = \frac{33\,000(1)3(1.25)}{117.8(0.818)} = 1284 \text{ lbf}$$

Answer
The tangential force on the worm is given by Eq. (15–57):

$$W_W^t = W_G^t\frac{\cos\phi_n\sin\lambda + f\cos\lambda}{\cos\phi_n\cos\lambda - f\sin\lambda}$$

$$= 1284\frac{\cos 14.5^\circ\sin 4.764^\circ + 0.0178\cos 4.764^\circ}{\cos 14.5^\circ\cos 4.764^\circ - 0.0178\sin 4.764^\circ} = 131 \text{ lbf}$$

(*c*)

Eq. (15–34): $C_s = 1000$

Eq. (15–36): $C_m = 0.0107\sqrt{-24^2 + 56(24) + 5145} = 0.823$

Eq. (15–37): $C_v = 13.31(1419)^{-0.571} = 0.211$

Eq. (15–28):
$$(W^t)_{\text{all}} = C_s D^{0.8}(F_e)_G C_m C_v$$
$$= 1000(6)^{0.8}(2)0.823(0.211) = 1456 \text{ lbf}$$

The friction force W_f is given by Eq. (15–61):

$$W_f = \frac{f W_G^t}{f \sin \lambda - \cos \phi_n \cos \lambda} = \frac{0.0178(1284)}{0.0178 \sin 4.764° - \cos 14.5° \cos 4.764°}$$

$$= -23.7 \text{ lbf}$$

The power dissipated in frictional work H_f is given by Eq. (15–63):

$$H_f = \frac{|W_f| V_s}{33\,000} = \frac{|-23.7| 1419}{33\,000} = 1.02 \text{ hp}$$

The worm and gear transmitted powers, H_W and H_G, are given by

$$H_W = \frac{W_W^t V_W}{33\,000} = \frac{131(1414)}{33\,000} = 5.61 \text{ hp} \qquad H_G = \frac{W_G^t V_G}{33\,000} = \frac{1284(117.8)}{33\,000} = 4.58 \text{ hp}$$

Answer

Gear power is satisfactory. Now,

$$P_n = P_t/\cos \lambda = 4/\cos 4.764° = 4.014$$

$$p_n = \pi/P_n = \pi/4.014 = 0.7827 \text{ in}$$

The bending stress in a gear tooth is given by Buckingham's adaptation of the Lewis equation, Eq. (15–53), as

$$(\sigma)_G = \frac{W_G^t}{p_n F_G y} = \frac{1284}{0.7827(2)(0.1)} = 8200 \text{ psi}$$

Answer

Stress in gear is satisfactory.

(*d*)

Eq. (15–52): $$A_{\min} = 43.2C^{1.7} = 43.2(4.5)^{1.7} = 557 \text{ in}^2$$

The gear case has a lateral area of 600 in^2.

Eq. (15–49): $$H_{\text{loss}} = 33\,000\,(1 - e)H_{\text{in}} = 33\,000(1 - 0.818)5.61$$

$$= 33\,690 \text{ ft} \cdot \text{lbf/min}$$

Eq. (15–50): $$\hbar_{\text{CR}} = \frac{n_W}{3939} + 0.13 = \frac{1800}{3939} + 0.13 = 0.587 \text{ ft} \cdot \text{lbf/(min} \cdot \text{in}^2 \cdot °\text{F)}$$

Answer Eq. (15–51): $$t_s = t_a + \frac{H_{\text{loss}}}{\hbar_{\text{CR}} A} = 70 + \frac{33\,690}{0.587(600)} = 166°\text{F}$$

15–8 Designing a Worm-Gear Mesh

A usable decision set for a worm-gear mesh includes

A priori decisions:

- Function: power, speed, m_G, K_a
- Design factor: n_d
- Tooth system
- Materials and processes
- Number of threads on the worm: N_W

Design decisions:

- Axial pitch of worm: p_x
- Pitch diameter of the worm: d_W
- Face width of gear: F_G
- Lateral area of case: A

Reliability information for worm gearing is not well developed at this time. The use of Equation (15–28) together with the factors C_s, C_m, and C_v, with an alloy steel case-hardened worm together with customary nonferrous worm-wheel materials, will result in lives in excess of 25 000 h. The worm-gear materials in the experience base are principally bronzes:

- Tin- and nickel-bronzes (chilled-casting produces hardest surfaces)
- Lead-bronze (high-speed applications)
- Aluminum- and silicon-bronze (heavy load, slow-speed application)

The factor C_s for bronze in the spectrum sand-cast, chilled-cast, and centrifugally cast increases in the same order.

Standardization of tooth systems is not as far along as it is in other types of gearing. For the designer this represents freedom of action, but acquisition of tooling for tooth-forming is more of a problem for in-house manufacturing. When using a subcontractor the designer must be aware of what the supplier is capable of providing with on-hand tooling.

Axial pitches for the worm are usually integers, and quotients of integers are common. Typical pitches are $\frac{1}{4}$, $\frac{5}{16}$, $\frac{3}{8}$, $\frac{1}{2}$, $\frac{3}{4}$, 1, $\frac{5}{4}$, $\frac{6}{4}$, $\frac{7}{4}$, and 2, but others are possible. Table 15–8 shows dimensions common to both worm gear and cylindrical worm for proportions often used. Teeth frequently are stubbed when lead angles are 30° or larger.

Worm-gear design is constrained by available tooling, space restrictions, shaft center-to-center distances, gear ratios needed, and the designer's experience. ANSI/AGMA 6022-C93, *Design Manual for Cylindrical Wormgearing* offers the following guidance. Normal pressure angles are chosen from 14.5°, 17.5°, 20°, 22.5°, 25°, 27.5°, and 30°. The recommended minimum number of gear teeth is given in Table 15–10. The normal range of the number of threads on the worm is 1 through 10. Mean worm pitch diameter is usually chosen in the range given by Equation (15–27).

Table 15–10 **Minimum Number of Gear Teeth for Normal Pressure Angle ϕ_n**

ϕ_n	$(N_G)_{min}$
14.5	40
17.5	27
20	21
22.5	17
25	14
27.5	12
30	10

A design decision is the axial pitch of the worm. Since acceptable proportions are couched in terms of the center-to-center distance, which is not yet known, one chooses a trial axial pitch p_x. Having N_W and a trial worm diameter d,

$$N_G = m_G N_W \qquad P_t = \frac{\pi}{p_x} \qquad D = \frac{N_G}{P_t}$$

Then

$$(d)_{\text{lo}} = C^{0.875}/3 \qquad (d)_{\text{hi}} = C^{0.875}/1.6$$

Examine $(d)_{\text{lo}} \le d \le (d)_{\text{hi}}$, and refine the selection of mean worm-pitch diameter to d_1 if necessary. Recompute the center-to-center distance as $C = (d_1 + D)/2$. There is even an opportunity to make C a round number. Choose C and set

$$d_2 = 2C - D$$

Equations (15–39) through (15–48) apply to one usual set of proportions.

EXAMPLE 15–4

Design a 10-hp 11:1 worm-gear speed-reducer mesh for a lumber mill planer feed drive for 3- to 10-h daily use. A 1720-rev/min squirrel-cage induction motor drives the planer feed ($K_a = 1.25$), and the ambient temperature is 70°F.

Solution

Function: $H_0 = 10$ hp, $m_G = 11$, $n_W = 1720$ rev/min.
Design factor: $n_d = 1.2$.
Materials and processes: case-hardened alloy steel worm, sand-cast bronze gear.
Worm threads: double, $N_W = 2$, $N_G = m_G N_W = 11(2) = 22$ gear teeth acceptable for $\phi_n = 20°$, according to Table 15–10.

Decision 1: Choose an axial pitch of worm $p_x = 1.5$ in. Then,

$$P_t = \pi/p_x = \pi/1.5 = 2.0944$$

$$D = N_G/P_t = 22/2.0944 = 10.504 \text{ in}$$

Eq. (15–39): $a = 0.3183p_x = 0.3183(1.5) = 0.4775$ in (addendum)

Eq. (15–40): $b = 0.3683(1.5) = 0.5525$ in (dedendum)

Eq. (15–41): $h_t = 0.6866(1.5) = 1.030$ in

Decision 2: Choose a mean worm diameter $d = 2.000$ in. Then

$$C = (d + D)/2 = (2.000 + 10.504)/2 = 6.252 \text{ in}$$

$$(d)_{\text{lo}} = 6.252^{0.875}/3 = 1.657 \text{ in}$$

$$(d)_{\text{hi}} = 6.252^{0.875}/1.6 = 3.107 \text{ in}$$

The range, given by Eq. (15–27), is $1.657 \le d \le 3.107$ in, which is satisfactory. Try $d = 2.500$ in. Recompute C:

$$C = (2.5 + 10.504)/2 = 6.502 \text{ in}$$

The range is now $1.715 \le d \le 3.216$ in, which is still satisfactory. Decision: $d = 2.500$ in. Then

Eq. (13–27): $L = p_x N_W = 1.5(2) = 3.000$ in

Eq. (13–28): $\lambda = \tan^{-1}[L/(\pi d)] = \tan^{-1}[3/(\pi 2.5)] = 20.905°$ (from Table 15–9 lead angle OK)

Eq. (15–62): $$V_s = \frac{\pi d n_W}{12 \cos \lambda} = \frac{\pi(2.5)1720}{12 \cos 20.905^\circ} = 1205.1 \text{ ft/min}$$

$$V_W = \frac{\pi d n_W}{12} = \frac{\pi(2.5)1720}{12} = 1125.7 \text{ ft/min}$$

$$V_G = \frac{\pi D n_G}{12} = \frac{\pi(10.504)1720/11}{12} = 430.0 \text{ ft/min}$$

Eq. (15–33): $$C_s = 1190 - 477 \log 10.504 = 702.8$$

Eq. (15–36): $$C_m = 0.02\sqrt{-11^2 + 40(11) - 76} + 0.46 = 0.772$$

Eq. (15–37): $$C_v = 13.31(1205.1)^{-0.571} = 0.232$$

Eq. (15–38): $$f = 0.103 \exp[-0.11(1205.1)^{0.45}] + 0.012 = 0.0191$$

Eq. (15–54): $$e_W = \frac{\cos 20^\circ - 0.0191 \tan 20.905^\circ}{\cos 20^\circ + 0.0191 \cot 20.905^\circ} = 0.942$$

(If the worm gear drives, $e_G = 0.939$.) To ensure nominal 10-hp output, with adjustments for K_a, n_d, and e,

Eq. (15–57): $$W_W^t = 1222 \frac{\cos 20^\circ \sin 20.905^\circ + 0.0191 \cos 20.905^\circ}{\cos 20^\circ \cos 20.905^\circ - 0.0191 \sin 20.905^\circ} = 495.4 \text{ lbf}$$

Eq. (15–58): $$W_G^t = \frac{33\,000(1.2)10(1.25)}{430(0.942)} = 1222 \text{ lbf}$$

Eq. (15–59): $$H_W = \frac{\pi(2.5)1720(495.4)}{12(33\,000)} = 16.9 \text{ hp}$$

Eq. (15–60): $$H_G = \frac{\pi(10.504)1720/11(1222)}{12(33\,000)} = 15.92 \text{ hp}$$

Eq. (15–61): $$W_f = \frac{0.0191(1222)}{0.0191 \sin 20.905^\circ - \cos 20^\circ \cos 20.905^\circ} = -26.8 \text{ lbf}$$

Eq. (15–63): $$H_f = \frac{|-26.8|\, 1205.1}{33\,000} = 0.979 \text{ hp}$$

From Eq. (15–28), with $C_s = 702.8$, $C_m = 0.772$, and $C_v = 0.232$,

$$(F_e)_{\text{req}} = \frac{W_G^t}{C_s D^{0.8} C_m C_v} = \frac{1222}{702.8(10.504)^{0.8} 0.772(0.232)} = 1.479 \text{ in}$$

Decision 3: The available range of $(F_e)_G$ is $1.479 \le (F_e)_G \le 2d/3$ or $1.479 \le (F_e)_G \le 1.667$ in. Set $(F_e)_G = 1.5$ in.

Eq. (15–28): $$W_{\text{all}}^t = 702.8(10.504)^{0.8} 1.5(0.772)0.232 = 1239 \text{ lbf}$$

This is greater than 1222 lbf. There is a little excess capacity. The force analysis stands.

Decision 4:

Eq. (15–50): $$\hbar_{\text{CR}} = \frac{n_W}{6494} + 0.13 = \frac{1720}{6494} + 0.13 = 0.395 \text{ ft} \cdot \text{lbf/(min} \cdot \text{in}^2 \cdot {}^\circ\text{F)}$$

Eq. (15–49): $$H_{\text{loss}} = 33\,000(1 - e)H_W = 33\,000(1 - 0.942)16.9 = 32\,347 \text{ ft} \cdot \text{lbf/min}$$

The AGMA area, from Eq. (15–52), is $A_{\min} = 43.2C^{1.7} = 43.2(6.502)^{1.7} = 1041.5\ \text{in}^2$. A rough estimate of the lateral area for 6-in clearances:

Vertical:	$d + D + 6 = 2.5 + 10.5 + 6 = 19$ in
Width:	$D + 6 = 10.5 + 6 = 16.5$ in
Thickness:	$d + 6 = 2.5 + 6 = 8.5$ in
Area:	$2(19)16.5 + 2(8.5)19 + 16.5(8.5) \approx 1090\ \text{in}^2$

Expect an area of 1100 in^2. Choose: Air-cooled, no fan on worm, with an ambient temperature of 70°F.

$$t_s = t_a + \frac{H_{\text{loss}}}{\hbar_{\text{CR}}A} = 70 + \frac{32\ 350}{0.395(1100)} = 70 + 74.5 = 144.5°\text{F}$$

Lubricant is safe with some margin for smaller area.

Eq. (13–18):
$$P_n = \frac{P_t}{\cos\lambda} = \frac{2.094}{\cos 20.905°} = 2.242$$

$$p_n = \frac{\pi}{P_n} = \frac{\pi}{2.242} = 1.401\ \text{in}$$

Gear bending stress, for reference, is

Eq. (15–53):
$$\sigma = \frac{W_G^t}{p_n F_e y} = \frac{1222}{1.401(1.5)0.125} = 4652\ \text{psi}$$

The risk is from wear, which is addressed by the AGMA method that provides $\left(W_G^t\right)_{\text{all}}$.

15–9 Buckingham Wear Load

A precursor to the AGMA method was the method of Buckingham, which identified an allowable wear load in worm gearing. Buckingham showed that the allowable gear-tooth loading for wear can be estimated from

$$\left(W_G^t\right)_{\text{all}} = K_w d_G F_e \qquad (15\text{–}64)$$

where K_w = worm-gear load factor
d_G = gear-pitch diameter
F_e = worm-gear effective face width

Table 15–11 gives values for K_w for worm gearsets as a function of the material pairing and the normal pressure angle.

EXAMPLE 15–5

Estimate the allowable gear wear load $\left(W_G^t\right)_{\text{all}}$ for the gearset of Example 15–4 using Buckingham's wear equation.

Solution

From Table 15–11 for a hardened steel worm and a bronze bear, K_w is given as 80 for $\phi_n = 20°$. Equation (15–64) gives

$$\left(W_G^t\right)_{\text{all}} = 80(10.504)1.5 = 1260\ \text{lbf}$$

which is larger than the 1239 lbf of the AGMA method. The method of Buckingham does not have refinements of the AGMA method. [Is $\left(W_G^t\right)_{\text{all}}$ linear with gear diameter?]

Table 15–11 Wear Factor K_w for Worm Gearing

Material		Thread Angle ϕ_n			
Worm	Gear	$14\frac{1}{2}°$	20°	25°	30°
Hardened steel*	Chilled bronze	90	125	150	180
Hardened steel*	Bronze	60	80	100	120
Steel, 250 BHN (min.)	Bronze	36	50	60	72
High-test cast iron	Bronze	80	115	140	165
Gray iron†	Aluminum	10	12	15	18
High-test cast iron	Gray iron	90	125	150	180
High-test cast iron	Cast steel	22	31	37	45
High-test cast iron	High-test cast iron	135	185	225	270
Steel 250 BHN (min.)	Laminated phenolic	47	64	80	95
Gray iron	Laminated phenolic	70	96	120	140

*Over 500 BHN surface.

†For steel worms, multiply given values by 0.6.

Source: Buckingham, Earle. *Design of Worm and Spiral Gears*. New York: Industrial Press, 1981.

For material combinations not addressed by AGMA, Buckingham's method allows quantitative treatment.

PROBLEMS

15–1 An uncrowned straight-bevel pinion has 20 teeth, a module of 4 mm, and a transmission accuracy number of 6. Both the pinion and gear are made of through-hardened steel with a Brinell hardness of 300. The driven gear has 60 teeth. The gearset has a life goal of 10^9 revolutions of the pinion with a reliability of 0.999. The shaft angle is 90°, and the pinion speed is 900 rev/min. The face width is 32 mm, and the normal pressure angle is 20°. The pinion is mounted outboard of its bearings, and the gear is straddle-mounted. Based on the AGMA bending strength, what is the power rating of the gearset? Use $K_0 = 1$ and $S_F = S_H = 1$.

15–2 For the gearset and conditions of Problem 15–1, find the power rating based on the AGMA surface durability.

15–3 An uncrowned straight-bevel pinion has 30 teeth, a module of 4 mm, and a transmission accuracy number of 6. The driven gear has 60 teeth. Both are made of No. 30 cast iron. The shaft angle is 90°. The face width is 32 mm, the pinion speed is 900 rev/min, and the normal pressure angle is 20°. The pinion is mounted outboard of its bearings and the bearings of the gear straddle it. What is the power rating based on AGMA bending strength? Note: For cast iron gearsets reliability information has not yet been developed. We say that if the life is greater than 10^7 revolutions, then set $Y_{NT} = 1$, $Z_{NT} = 1$, $Z_Z = 1$, $Y_Z = 1$, and apply a factor of safety. Use $S_F = 2$ and $S_H = \sqrt{2}$.

15–4 For the gearset and conditions of Problem 15–3, find the power rating based on AGMA surface durability.

15–5 An uncrowned straight-bevel pinion has 22 teeth, a module of 4 mm, and a transmission accuracy number of 5. The pinion and the gear are made of through-hardened steel, both having core and case hardnesses of 180 Brinell. The pinion drives the

24-tooth bevel gear. The shaft angle is 90°, the pinion speed is 1800 rev/min, the face width is 25 mm, and the normal pressure angle is 20°. Both gears have an outboard mounting. Find the power rating based on AGMA pitting resistance if the life goal is 10^9 revolutions of the pinion at 0.999 reliability.

15–6 For the gearset and conditions of Problem 15–5, find the power rating for AGMA bending strength.

15–7 In straight-bevel gearing, there are some analogs to Equations (14–42) and (14–43), respectively. If we have a pinion core with a hardness of $(H_B)_{11}$ and we try equal power ratings, the transmitted load W^t can be made equal in all four cases. It is possible to find these relations:

	Core	Case
Pinion	$(H_B)_{11}$	$(H_B)_{12}$
Gear	$(H_B)_{21}$	$(H_B)_{22}$

(*a*) For carburized case-hardened gear steel with core AGMA bending strength $(\sigma_{F\lim})_G$ and pinion core strength $(\sigma_{F\lim})_P$, show that the relationship is

$$(\sigma_{F\lim})_G = (\sigma_{F\lim})_{P)} \frac{Y_{JP}}{Y_{JG}} m - 0.0323_G m\ G{-}0.0323$$

This allows $(H_B)_{21}$ to be related to $(H_B)_{11}$.

(*b*) Show that the AGMA contact strength of the gear case $(\sigma_{H\lim})_G$ can be related to the AGMA core bending strength of the pinion core $(\sigma_{H\lim})_G$ by

$$(\sigma_{H\lim})_G = \frac{Z_E}{(Z_{NT})_G Z_W} \sqrt{\frac{S_H^2}{S_F} \frac{(\sigma_{H\lim})_G (Y_{NT})_P Y_\beta (Y_J)_P K_\theta Z_X Z_{xc}}{N_P Z_I Y_x}}$$

If factors of safety are applied to the transmitted load W^t, then $S_H = \sqrt{S_F}$ and S_H^2/S_F is unity. The result allows $(H_B)_{22}$ to be related to $(H_B)_{11}$.

(*c*) Show that the AGMA contact strength of the gear $(\sigma_{H\lim})_G$ is related to the contact strength of the pinion $(\sigma_{H\lim})_P$ by

$$(\sigma_{H\lim})_P = (\sigma_{H\lim})_G\, m_G^{0.0602}\, Z_W$$

15–8 Refer to your solution to Problems 15–1 and 15–2. If the pinion core hardness is 300 Brinell, use the relations from Problem 15–7 to determine the required hardness of the gear core and the case hardnesses of both gears to ensure equal power ratings.

15–9 Repeat Problems 15–1 and 15–2 with the hardness protocol

	Core	Case
Pinion	300	373
Gear	339	345

which can be established by the relations in Problem 15–7, and see if the result matches transmitted loads W^t in all four cases.

15–10 A catalog of stock bevel gears lists a power rating of 3.9 kW at 1200 rev/min pinion speed for a straight-bevel gearset consisting of a 20-tooth pinion driving a 40-tooth gear. This gear pair has a 20° normal pressure angle, a face width of 18 mm, a module of 2.5 mm, and is through-hardened to 300 BHN. Assume the gears are for general industrial use, are generated to a transmission accuracy number of 5, and are uncrowned. Also assume the gears are rated for a life of 3×10^6 revolutions with a 99 percent reliability. Given these data, what do you think about the stated catalog power rating?

15–11 Apply the relations of Problem 15–7 to Example 15–1 and find the Brinell case hardness of the gears for equal allowable load W^t in bending and wear. Check your work by reworking Example 15–1 to see if you are correct. How would you go about the heat treatment of the gears?

15–12 Your experience with Example 15–1 and problems based on it will enable you to write an interactive computer program for power rating of through-hardened steel gears. Test your understanding of bevel-gear analysis by noting the ease with which the coding develops. The hardness protocol developed in Problem 15–7 can be incorporated at the end of your code, first to display it, then as an option to loop back and see the consequences of it.

15–13 Use your experience with Problem 15–11 and Example 15–2 to design an interactive computer-aided design program for straight-steel bevel gears, implementing the ANSI/AGMA 2003-B97 standard. It will be helpful to follow the decision set in Section 15–5, allowing the return to earlier decisions for revision as the consequences of earlier decisions develop.

15–14 A single-threaded steel worm rotates at 1725 rev/min, meshing with a 56-tooth worm gear transmitting 1 hp to the output shaft. The pitch diameter of the worm is 1.50. The tangential module of the gear is 4 mm and the normal pressure angle is 20°. The ambient temperature is 21°C, the application factor is 1.25, the design factor is 1, the gear face is 12 mm, the lateral case area is 550 mm^2, and the gear is sand-cast bronze.
(*a*) Determine and evaluate the geometric properties of the gears.
(*b*) Determine the transmitted gear forces and the mesh efficiency.
(*c*) Is the mesh sufficient to handle the loading?
(*d*) Estimate the lubricant sump temperature, assuming fan-stirred air.

15–15 to 15–22 As in Example 15–4, design a cylindrical worm-gear mesh to connect a squirrel-cage induction motor to a liquid agitator. The motor speed is 1125 rev/min, and the velocity ratio is to be 10:1. The output power requirement is 19 kW. The shaft axes are 90° to each other. An overload factor K_o (see Table 15–2) makes allowance for external dynamic excursions of load from the nominal or average load W^t. For this service $K_A = 1.25$ is appropriate. Additionally, a design factor n_d of 1.1 is to be included to address other unquantifiable risks. For Problems 15–15 to 15–17 use the AGMA method for $(W_G^t)_{\text{all}}$ whereas for Problems 15–18 to 15–22, use the Buckingham method. See Table 15–12.

Table 15–12 Table Supporting Problems 15–15 to 15–22

Problem No.	Method	Materials	
		Worm	Gear
15–15	AGMA	Steel, HRC 58	Sand-cast bronze
15–16	AGMA	Steel, HRC 58	Chilled-cast bronze
15–17	AGMA	Steel, HRC 58	Centrifugal-cast bronze
15–18	Buckingham	Steel, 500 BHN	Chilled-cast bronze
15–19	Buckingham	Steel, 500 BHN	Cast bronze
15–20	Buckingham	Steel, 250 BHN	Cast bronze
15–21	Buckingham	High-test cast iron	Cast bronze
15–22	Buckingham	High-test cast iron	High-test cast iron

16 Clutches, Brakes, Couplings, and Flywheels

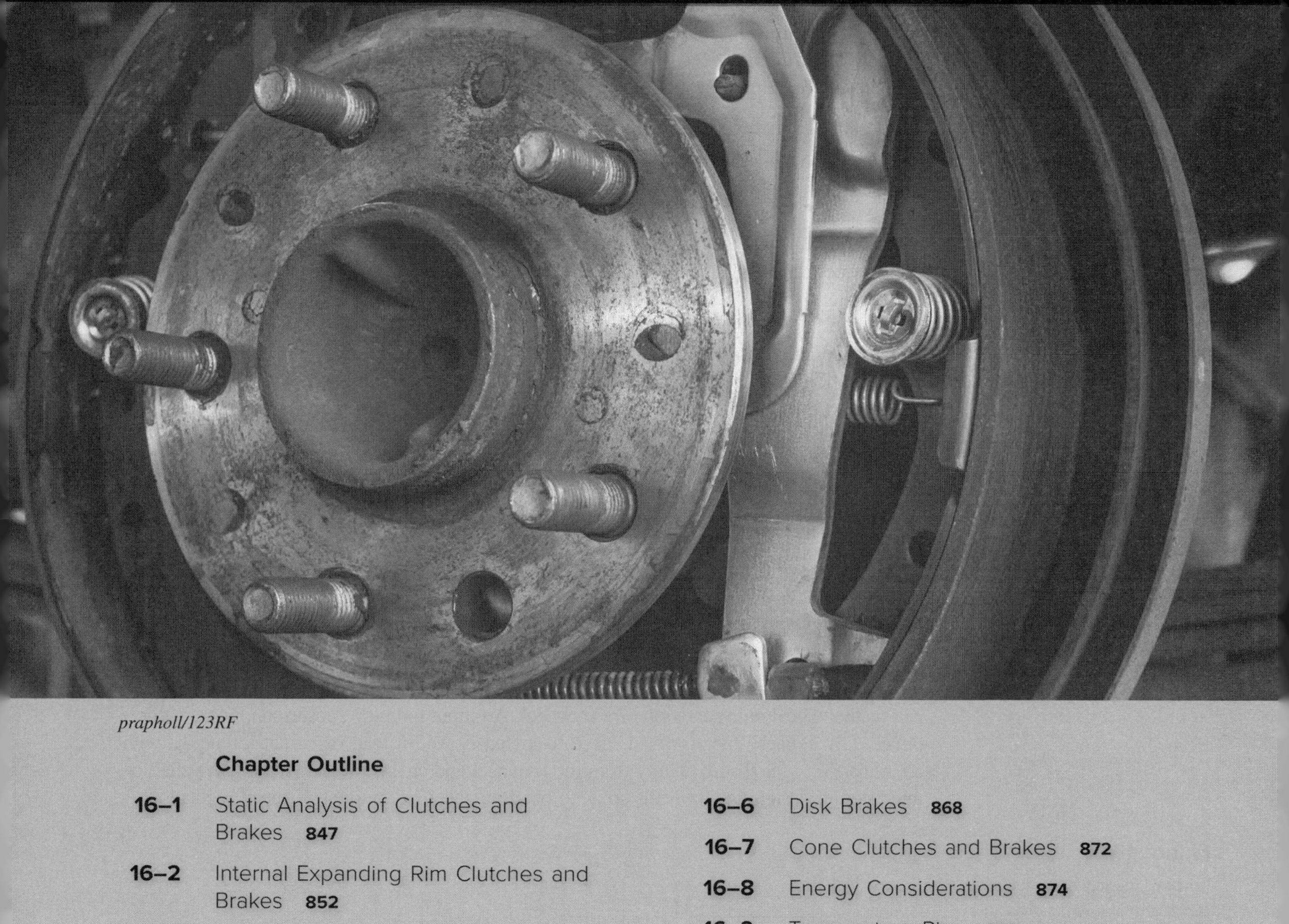

prapholl/123RF

Chapter Outline

This chapter is concerned with a group of elements usually associated with rotation that have in common the function of storing and/or transferring rotational energy. Because of this similarity of function, clutches, brakes, couplings, and flywheels are treated together in this chapter.

A simplified dynamic representation of a friction clutch or brake is shown in Figure 16–1*a*. Two inertias, I_1 and I_2, traveling at the respective angular velocities ω_1 and ω_2, one of which may be zero in the case of brakes, are to be brought to the same speed by engaging the clutch or brake. Slippage occurs because the two elements are running at different speeds and energy is dissipated during actuation, resulting in a temperature rise. In analyzing the performance of these devices we shall be interested in:

- The actuating force
- The torque transmitted
- The energy loss
- The temperature rise

The torque transmitted is related to the actuating force, the coefficient of friction, and the geometry of the clutch or brake. This is a problem in statics, which will have to be studied separately for each geometric configuration. However, temperature rise is related to energy loss and can be studied without regard to the type of brake or clutch, because the geometry of interest is that of the heat-dissipating surfaces.

The various types of devices to be studied may be classified as follows:

- Rim types with internal expanding shoes
- Rim types with external contracting shoes
- Band types
- Disk or axial types
- Cone types
- Miscellaneous types

A flywheel is an inertial energy-storage device. It absorbs mechanical energy by increasing its angular velocity and delivers energy by decreasing its velocity. Figure 16–1*b* is a mathematical representation of a flywheel. An input torque T_i, corresponding to a coordinate θ_i, will cause the flywheel speed to increase. And a load or output torque T_o, with coordinate θ_o, will absorb energy from the flywheel and cause it to slow down. We shall be interested in designing flywheels so as to obtain a specified amount of speed regulation.

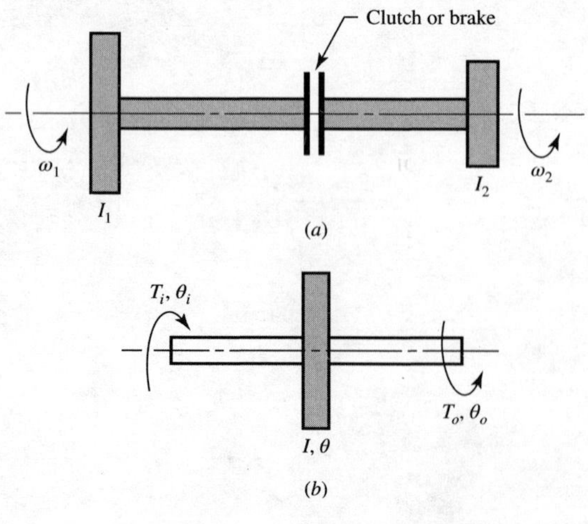

Figure 16–1

(*a*) Dynamic representation of a clutch or brake; (*b*) mathematical representation of a flywheel.

16–1 Static Analysis of Clutches and Brakes

Many types of clutches and brakes can be analyzed by following a general procedure. The procedure entails the following tasks:

- Estimate, model, or measure the pressure distribution on the friction surfaces.
- Find a relationship between the largest pressure and the pressure at any point.
- Use the conditions of static equilibrium to find the braking force or torque and the support reactions.

Let us apply these tasks to the doorstop depicted in Figure 16–2*a*. The stop is hinged at pin *A*. A normal pressure distribution $p(u)$ is shown under the friction pad as a function of position u, taken from the right edge of the pad. A similar distribution of shearing frictional traction is on the surface, of intensity $fp(u)$, in the direction of the motion of the floor relative to the pad, where f is the coefficient of friction. The width of the pad into the page is w_2. The net force in the y direction and moment about C from the pressure are respectively,

$$N = w_2 \int_0^{w_1} p(u)\, du = p_{\text{av}} w_1 w_2 \tag{a}$$

$$w_2 \int_0^{w_1} p(u) u\, du = \bar{u}\, w_2 \int_0^{w_1} p(u)\, du = p_{\text{av}} w_1 w_2 \bar{u} \tag{b}$$

We sum the forces in the x-direction to obtain

$$\sum F_x = R_x \mp w_2 \int_0^{w_1} fp(u)\, du = 0$$

where − or + is for rightward or leftward relative motion of the floor, respectively. Assuming f constant, solving for R_x gives

$$R_x = \pm w_2 \int_0^{w_1} fp(u)\, du = \pm\, f w_1 w_2 p_{\text{av}} \tag{c}$$

Summing the forces in the y direction gives

$$\sum F_y = -F + w_2 \int_0^{w_1} p(u)\, du + R_y = 0$$

from which

$$R_y = F - w_2 \int_0^{w_1} p(u)\, du = F - p_{\text{av}} w_1 w_2 \tag{d}$$

for either direction. Summing moments about the pin located at A we have

$$\sum M_A = Fb - w_2 \int_0^{w_1} p(u)(c + u)\, du \mp afw_2 \int_0^{w_1} p(u)\, du = 0$$

A brake shoe is *self-energizing* if its moment sense helps set the brake, *self-deenergizing* if the moment resists setting the brake. Continuing,

$$F = \frac{w_2}{b}\left[\int_0^{w_1} p(u)(c + u)\, du \pm af \int_0^{w_1} p(u)\, du\right] \tag{e}$$

Figure 16–2

A common doorstop. (*a*) Free body of the doorstop. (*b*) Trapezoidal pressure distribution on the foot pad based on linear deformation of pad. (*c*) Free-body diagram for leftward movement of the floor, uniform pressure, Example 16–1. (*d*) Free-body diagram for rightward movement of the floor, uniform pressure, Example 16–1. (*e*) Free-body diagram for leftward movement of the floor, trapezoidal pressure, Example 16–1. For (*c*), (*d*), and (*e*), units of forces and dimensions are N and mm, respectively.

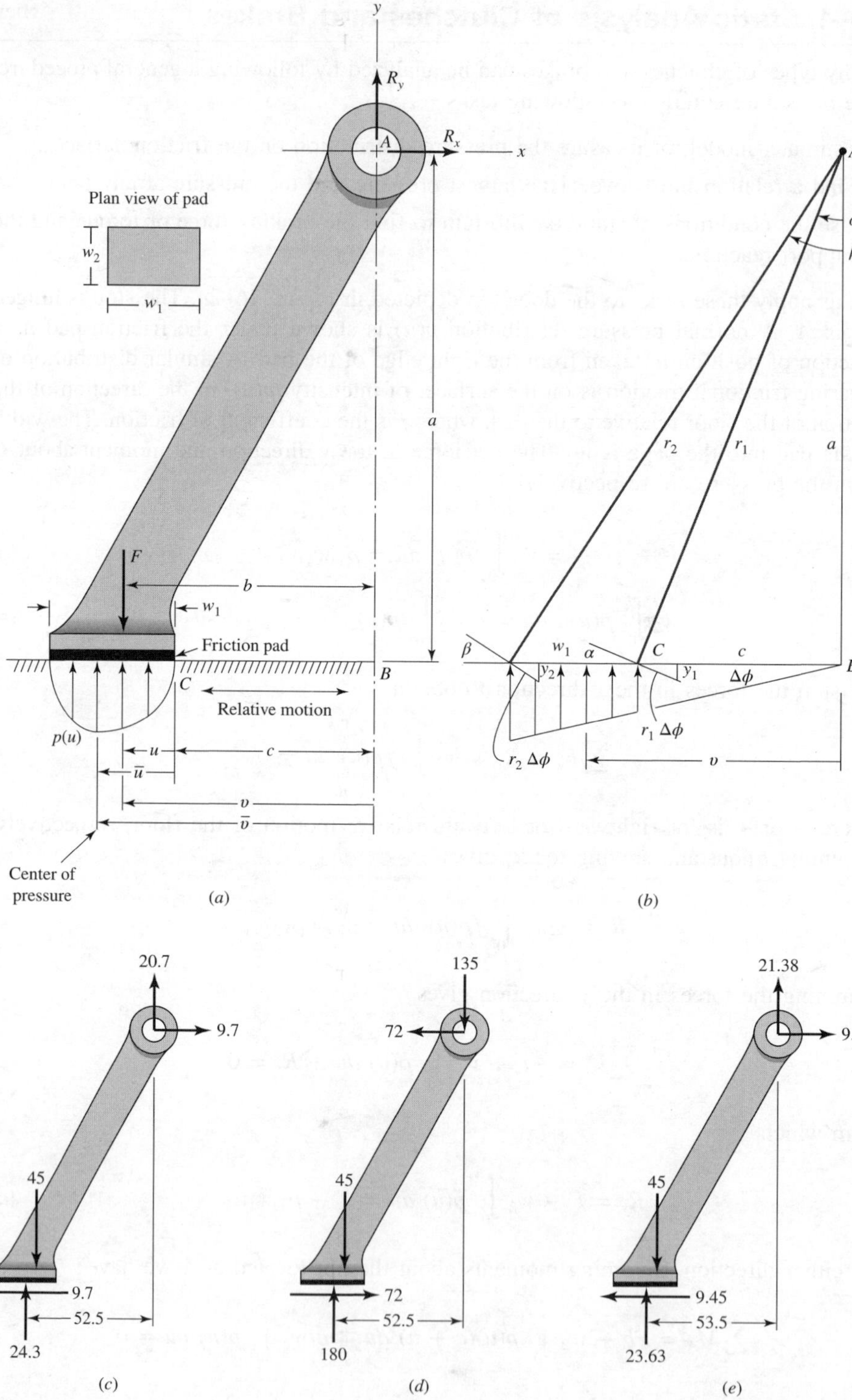

Can F be equal to or less than zero? Only during rightward motion of the floor when the expression in brackets in Equation (e) is equal to or less than zero. We set the brackets to zero or less:

$$\int_0^{w_1} p(u)(c + u)\, du - af \int_0^{w_1} p(u)\, du \leq 0$$

from which

$$f_{cr} \geq \frac{1}{a} \frac{\int_0^{w_1} p(u)(c + u)\, du}{\int_0^{w_1} p(u)\, du} = \frac{1}{a} \frac{c\int_0^{w_1} p(u)\, du + \int_0^{w_1} p(u)u\, du}{\int_0^{w_1} p(u)\, du} \qquad (f)$$

$$f_{cr} \geq \frac{c + \bar{u}}{a}$$

where $\bar{u}$ is the distance of the center of pressure from the right edge of the pad. The conclusion that a *self-acting* or *self-locking* phenomenon is present is independent of our knowledge of the normal pressure distribution $p(u)$. Our ability to *find* the critical value of the coefficient of friction f_{cr} is dependent on our knowledge of $p(u)$, from which we derive $\bar{u}$.

EXAMPLE 16–1

The doorstop depicted in Figure 16–2a has the following dimensions: $a = 100$ mm, $b = 50$ mm, $c = 40$ mm, $w_1 = 25$ mm, $w_2 = 18$ mm, where w_2 is the depth of the pad into the plane of the paper.

(a) For a leftward relative movement of the floor, an actuating force F of 45 N, a coefficient of friction of 0.4, use a uniform pressure distribution p_{av}, find R_x, R_y, p_{av}, and the largest pressure p_a.

(b) Repeat part (a) for rightward relative movement of the floor.

(c) Model the normal pressure to be the "crush" of the pad, much as if it were composed of many small helical coil springs. Find R_x, R_y, p_{av}, and p_a for leftward relative movement of the floor and other conditions as in part (a).

(d) For rightward relative movement of the floor, is the doorstop a self-acting brake?

Solution (a)

Eq. (c): $R_x = f\, p_{av} w_1 w_2 = 0.4(25)(18)p_{av} = 180p_{av}$

Eq. (d): $R_y = F - p_{av} w_1 w_2 = 45 - p_{av}(25)(18) = 45 - 450p_{av}$

$$\text{Eq. }(e)\text{:}\quad F = \frac{w_2}{b}\left[\int_0^{25} p_{av}(c + u)\, du + af \int_0^{25} p_{av}\, du\right]$$

$$= \frac{w_2}{b}\left(p_{av} c \int_0^{25} du + p_{av} \int_0^{25} u\, du + af\, p_{av} \int_0^{25} du\right)$$

$$= (25)\frac{w_2 p_{av}}{b}(c + 25(0.5) + af) = \frac{(25)18}{50}[40 + 25(0.5) + 100(0.4)]p_{av}$$

$$= 832.5p_{av}$$

Solving for p_{av} gives

$$p_{av} = \frac{F}{1.3875} = \frac{45}{724.5} = 0.054 \text{ MPa}$$

We evaluate R_x and R_y as

Answer $$R_x = 180(0.054) = 9.7 \text{ N}$$

Answer $$R_y = 45 - 450(0.054) = 20.7 \text{ N}$$

The normal force N on the pad is $F - R_y = 45 - 20.7 = 24.3$ N, upward. The line of action is through the center of pressure, which is at the center of the pad. The friction force is $fN = 0.4(24.3) = 9.7$ N directed to the left. A check of the moments about A gives

$$\sum M_A = Fb - fNa - N(w_1/2 + c)$$

$$= 45(50) - 0.4(24.3)100 - 24.3(25/2 + 40) \approx 0$$

Answer
The maximum pressure $p_a = p_{av} = 0.054$ MPa
(*b*)

Eq. (*c*): $$R_x = -f\, p_{av} w_1 w_2 = -0.4(25)(18)p_{av} = -180p_{av}$$

Eq. (*d*): $$R_y = F - p_{av} w_1 w_2 = 45 - p_{av}(25)(18) = 45 - 450p_{av}$$

Eq. (*e*): $$F = \frac{w_2}{b}\left[\int_0^{25} p_{av}(c + u)\, du + af \int_0^{25} p_{av}\, du\right]$$

$$= \frac{w_2}{b}\left(p_{av} c \int_0^{25} du + p_{av} \int_0^{25} u\, du + af\, p_{av} \int_0^{25} du\right)$$

$$= \frac{(25)18}{50} p_{av}[40 + 25(0.5) - 100(0.4)] = 112.5p_{av}$$

from which

$$p_{av} = \frac{F}{112.5} = \frac{45}{112.5} = 0.4 \text{ MPa}$$

which makes

Answer $$R_x = -180(0.4) = -72 \text{ N}$$

Answer $$R_y = 45 - 450(0.4) = -135 \text{ N}$$

The normal force N on the pad is $45 + 135 = 180$ N upward. The friction shearing force is $fN = 0.4(180) = 72$ N to the right. We now check the moments about A:

$$M_A = fNa + Fb - N(c + 0.5) = 72(100) + 45(50) - 180(40 + 25/2) = 0$$

Note the change in average pressure from 0.054 MPa in part (*a*) to 0.4 MPa. Also note how directions of forces have changed. The maximum pressure p_a is the same as p_{av}, which has changed from 0.054 MPa to 0.4 MPa.
(*c*) We will model the deformation of the pad as follows. If the doorstop rotates $\Delta\phi$ counterclockwise, the right and left edges of the pad will deform down y_1 and y_2, respectively (Figure 16–2*b*). From similar triangles,

$y_1/(r_1\ \Delta\phi) = c/r_1$ and $y_2/(r_2\ \Delta\phi) = (c + w_1)/r_2$. Thus, $y_1 = c\ \Delta\phi$ and $y_2 = (c + w_1)\Delta\phi$. This means that y is directly proportional to the horizontal distance from the pivot point A. That is, $y = C_1 v$, where C_1 is a constant (see Figure 16–2*b*). Assuming the pressure is directly proportional to deformation, then $p(v) = C_2 v$, where C_2 is a constant. In terms of u, the pressure is $p(u) = C_2(c + u) = C_2(40 + u)$.

Eq. (*e*):

$$F = \frac{w_2}{b}\left[\int_0^{w_1} p(u)c\, du + \int_0^{w_1} p(u)u\, du + af\int_0^{w_1} p(u)\, du\right]$$

$$= \frac{18}{50}\left[\int_0^{25} C_2(40 + u)40du + \int_0^{25} C_2(40 + u)\, u\, du + af\int_0^{25} C_2(40 + u)\, du\right]$$

$$= 0.36C_2[(40(25) + 25^2(0.5))40 + (40(25^2)/2 + 25^3/3) + 100(0.4)(40(25) + 25^2(0.5))] = 44\ 175C_2$$

Since $F = 45$ N, then $C_2 = 45/44\ 175 = 0.001$ N/mm, and $p(u) = 0.001(40 + u)$. The average pressure is given by

Answer $$p_{av} = \frac{1}{w_1}\int_0^{w_1} p(u)\, du = \frac{1}{25}\int_0^{25} 0.001(40 + u)\, du = \frac{1}{(25)(0.001)}(40(25) + \frac{25^2}{2}) = 0.0525 \text{ MPa}$$

The maximum pressure occurs at $u = 25$ in, and is

Answer $$p_a = 0.001(40 + 25) = 0.065 \text{ MPa}$$

Equations (*c*) and (*d*) of Sec. 16–1 are still valid. Thus,

Answer $$R_x = 180p_{av} = 180(0.0525) = 9.45 \text{ N}$$
$$R_y = 45 - 450p_{av} = 45 - 450(0.0525) = 21.38 \text{ N}$$

From statics, the floor normal force is 23.63 N at a horizontal distance from A of 53.5 mm. The average pressure is $p_{av} = 0.0525$ MPa and the maximum pressure is $p_a = 0.065$ MPa, which is approximately 24 percent higher than the average pressure. The presumption that the pressure was uniform in part (*a*) (because the pad was small, or because the arithmetic would be easier?) underestimated the peak pressure. Modeling the pad as a one-dimensional springset is better, but the pad is really a three-dimensional continuum. A theory of elasticity approach or a finite element modeling may be overkill, given uncertainties inherent in this problem, but it still represents better modeling.

(*d*) To evaluate $\bar{u}$ we need to evaluate two integrations

$$\int_0^{wl} p(u)u\, du = \int_0^{25} 0.001(40 + u)u\, du = 0.001(40(25^2)0.5 + 25^3/3) = 17.71 \text{ N}$$

$$\int_0^{wl} p(u)\, du = \int_0^{25} 0.001(40 + u)du = 0.001(40(25) + 25^2(0.5)) = 1.31 \text{ N/mm}$$

Thus $\bar{u} = 17.71/1.31 = 13.5$ mm. Then, from Eq. (*f*) of Sec. 16–1, the critical coefficient of friction is

Answer $$f_{cr} \geq \frac{c + \bar{u}}{a} = \frac{40 + 13.5}{100} = 0.535$$

The doorstop friction pad does not have a high enough coefficient of friction to make the doorstop a self-acting brake. The configuration must change and/or the pad material specification must be changed to sustain the function of a doorstop.

16–2 Internal Expanding Rim Clutches and Brakes

The internal-shoe rim clutch shown in Figure 16–3 consists essentially of three elements:

1 the mating frictional surface
2 the means of transmitting the torque to and from the surfaces
3 the actuating mechanism

Depending upon the operating mechanism, such clutches are further classified as

- expanding-ring
- centrifugal
- magnetic
- hydraulic
- pneumatic

The ***expanding-ring clutch*** is often used in textile machinery, excavators, and machine tools where the clutch may be located within the driving pulley. Expanding-ring clutches benefit from centrifugal effects; transmit high torque, even at low speeds; and require both positive engagement and ample release force.

The ***centrifugal clutch*** is used mostly for automatic operation. If no spring is used, the torque transmitted is proportional to the square of the speed. This is particularly useful for electric-motor drives where, during starting, the driven machine comes up to speed without shock. Springs can also be used to prevent engagement until a certain motor speed is reached, but some shock may occur.

Magnetic clutches are particularly useful for automatic and remote-control systems. Such clutches are also useful in drives subject to complex load cycles (see Section 11–7).

Hydraulic and pneumatic clutches are also useful in drives having complex loading cycles and in automatic machinery, or in robots. Here the fluid flow can be controlled remotely using solenoid valves. These clutches are also available as disk, cone, and multiple-plate clutches.

In braking systems, the *internal-shoe* or *drum* brake is used mostly for automotive applications.

Figure 16–3

An internal expanding centrifugal-acting rim clutch.

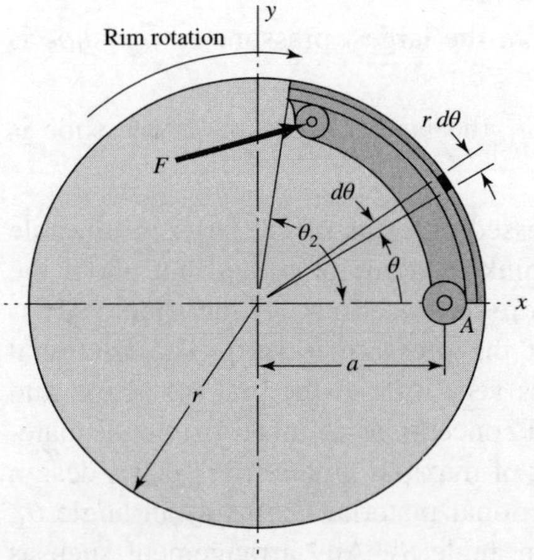

Figure 16–4

Internal friction shoe geometry.

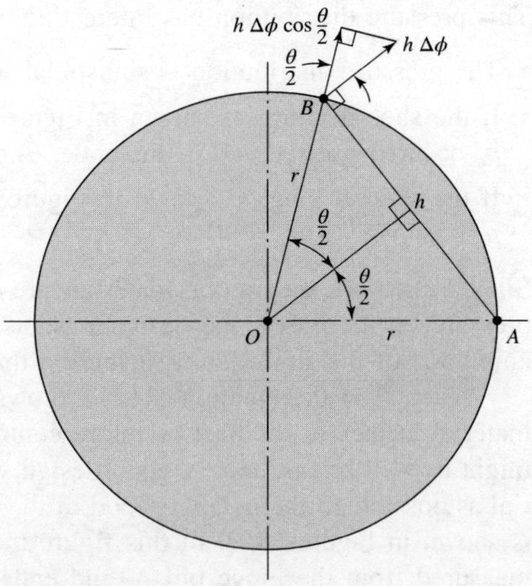

Figure 16–5

The geometry associated with an arbitrary point on the shoe.

To analyze an internal-shoe device, refer to Figure 16–4, which shows a shoe pivoted at point A, with the actuating force acting at the other end of the shoe. Since the shoe is long, we cannot make the assumption that the distribution of normal forces is uniform. The mechanical arrangement permits no pressure to be applied at the heel, and we will therefore assume the pressure at this point to be zero.

It is the usual practice to omit the friction material for a short distance away from the heel (point A). This eliminates interference, and the material would contribute little to the performance anyway, as will be shown. In some designs the hinge pin is made movable to provide additional heel pressure. This gives the effect of a floating shoe. (Floating shoes will not be treated in this book, although their design follows the same general principles.)

Let us consider the pressure p acting upon an element of area of the frictional material located at an angle θ from the hinge pin (Figure 16–4). We designate the maximum pressure p_a located at an angle θ_a from the hinge pin. To find the pressure distribution on the periphery of the internal shoe, consider point B on the shoe (Figure 16–5). As in Example 16–1, if the shoe deforms by an infinitesimal rotation $\Delta\phi$ about the pivot point A, deformation perpendicular to AB is $h\,\Delta\phi$. From the isosceles triangle AOB, $h = 2r\sin(\theta/2)$, so

$$h\,\Delta\phi = 2r\,\Delta\phi\,\sin(\theta/2)$$

The deformation perpendicular to the rim is $h\,\Delta\phi\cos(\theta/2)$, which is

$$h\,\Delta\phi\cos(\theta/2) = 2r\,\Delta\phi\sin(\theta/2)\cos(\theta/2) = r\,\Delta\phi\sin\theta$$

Thus, the deformation, and consequently the pressure, is proportional to $\sin\theta$. In terms of the pressure at B and where the pressure is a maximum, this means

$$\frac{p}{\sin\theta} = \frac{p_a}{\sin\theta_a} \tag{a}$$

Rearranging gives

$$p = \frac{p_a}{\sin\theta_a}\sin\theta \tag{16–1}$$

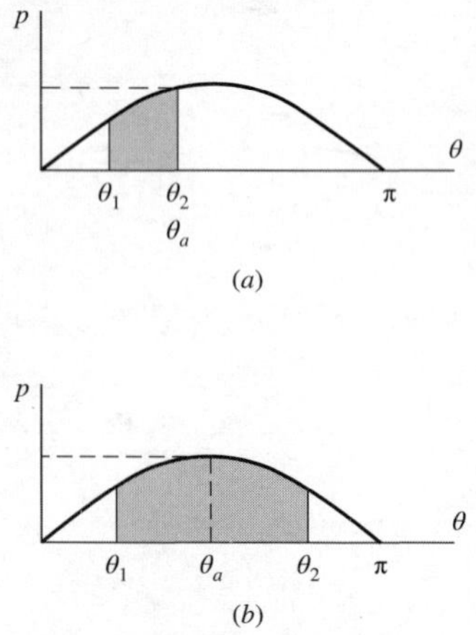

Figure 16–6

Defining the angle θ_a at which the maximum pressure p_a occurs when (*a*) shoe exists in zone $\theta_1 \leq \theta_2 \leq \pi/2$ and (*b*) shoe exists in zone $\theta_1 \leq \pi/2 \leq \theta_2$.

This pressure distribution has interesting and useful characteristics:

- The pressure distribution is sinusoidal with respect to the angle θ.
- If the shoe is short, as shown in Figure 16–6*a*, the largest pressure *on the shoe* is p_a occurring at the end of the shoe, θ_2.
- If the shoe is long, as shown in Figure 16–6*b*, the largest pressure on the shoe is p_a occurring at $\theta_a = 90°$.

Since limitations on friction materials are expressed in terms of the largest allowable pressure on the lining, the designer wants to think in terms of p_a and not about the amplitude of the sinusoidal distribution that addresses locations off the shoe.

When $\theta = 0$, Equation (16–1) shows that the pressure is zero. The frictional material located at the heel therefore contributes very little to the braking action and might as well be omitted. A good design would concentrate as much frictional material as possible in the neighborhood of the point of maximum pressure. Such a design is shown in Figure 16–7. In this figure the frictional material begins at an angle θ_1, measured from the hinge pin A, and ends at an angle θ_2. Any arrangement such as this will give a good distribution of the frictional material.

Proceeding now (Figure 16–7), the hinge-pin reactions are R_x and R_y. The actuating force F has components F_x and F_y and operates at distance c from the hinge pin. At any angle θ from the hinge pin there acts a differential normal force dN whose magnitude is

$$dN = pbr\ d\theta \tag{b}$$

where b is the face width (perpendicular to the page) of the friction material. Substituting the value of the pressure from Equation (16–1), the normal force is

$$dN = \frac{p_a br \sin\theta\ d\theta}{\sin\theta_a} \tag{c}$$

The normal force dN has horizontal and vertical components $dN \cos\theta$ and $dN \sin\theta$, as shown in the figure. The frictional force fdN has horizontal and vertical components

Figure 16–7

Forces on the shoe.

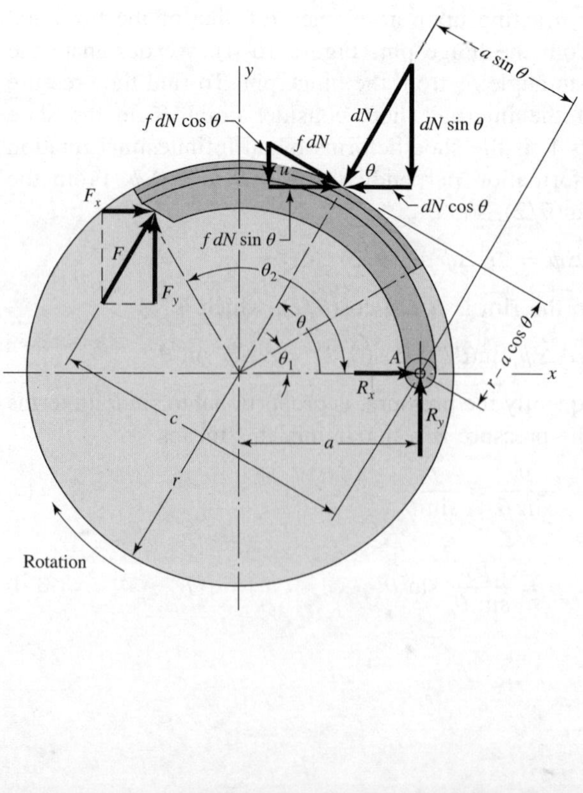

whose magnitudes are $f dN \sin\theta$ and $f dN \cos\theta$, respectively. By applying the conditions of static equilibrium, we may find the actuating force F, the torque T, and the pin reactions R_x and R_y.

We shall find the actuating force F, using the condition that the summation of the moments about the hinge pin is zero. The frictional forces have a moment arm about the pin of $r - a\cos\theta$. The moment M_f of these frictional forces is

$$M_f = \int f dN(r - a\cos\theta) = \frac{f p_a b r}{\sin\theta_a}\int_{\theta_1}^{\theta_2} \sin\theta(r - a\cos\theta)\,d\theta \tag{16–2}$$

which is obtained by substituting the value of dN from Equation (*c*). It is convenient to integrate Equation (16–2) for each problem, and we shall therefore retain it in this form. The moment arm of the normal force dN about the pin is $a\sin\theta$. Designating the moment of the normal forces by M_N and summing these about the hinge pin give

$$M_N = \int dN(a\sin\theta) = \frac{p_a b r a}{\sin\theta_a}\int_{\theta_1}^{\theta_2} \sin^2\theta\,d\theta \tag{16–3}$$

The actuating force F must balance these moments. Thus

$$F = \frac{M_N - M_f}{c} \tag{16–4}$$

We see here that a condition for zero actuating force exists. In other words, if we make $M_N = M_f$, self-locking is obtained, and no actuating force is required. This furnishes us with a method for obtaining the dimensions for some self-energizing action. So to *avoid self-locking*, the dimension a in Figure 16–7 must be such that

$$M_N > M_f \tag{16–5}$$

The torque T applied to the drum by the brake shoe is the sum of the frictional forces $f dN$ times the radius of the drum:

$$\begin{aligned} T &= \int f r\, dN = \frac{f p_a b r^2}{\sin\theta_a}\int_{\theta_1}^{\theta_2} \sin\theta\,d\theta \\ &= \frac{f p_a b r^2(\cos\theta_1 - \cos\theta_2)}{\sin\theta_a} \end{aligned} \tag{16–6}$$

The hinge-pin reactions are found by taking a summation of the horizontal and vertical forces. Thus, for R_x, we have

$$\begin{aligned} R_x &= \int dN\cos\theta - \int f\,dN\sin\theta - F_x \\ &= \frac{p_a b r}{\sin\theta_a}\left(\int_{\theta_1}^{\theta_2}\sin\theta\cos\theta\,d\theta - f\int_{\theta_1}^{\theta_2}\sin^2\theta\,d\theta\right) - F_x \end{aligned} \tag{d}$$

The vertical reaction is found in the same way:

$$\begin{aligned} R_y &= \int dN\sin\theta + \int f\,dN\cos\theta - F_y \\ &= \frac{p_a b r}{\sin\theta_a}\left(\int_{\theta_1}^{\theta_2}\sin^2\theta\,d\theta + f\int_{\theta_1}^{\theta_2}\sin\theta\cos\theta\,d\theta\right) - F_y \end{aligned} \tag{e}$$

The direction of the frictional forces is reversed if the rotation is reversed. Thus, for counterclockwise rotation the actuating force is

$$F = \frac{M_N + M_f}{c} \tag{16–7}$$

and since both moments have the same sense, the self-energizing effect is lost. Also, for counterclockwise rotation the signs of the frictional terms in the equations for the pin reactions change, and Equations (*d*) and (*e*) become

$$R_x = \frac{p_a br}{\sin \theta_a} \left(\int_{\theta_1}^{\theta_2} \sin \theta \cos \theta \, d\theta + f \int_{\theta_1}^{\theta_2} \sin^2 \theta \, d\theta \right) - F_x \tag{f}$$

$$R_y = \frac{p_a br}{\sin \theta_a} \left(\int_{\theta_1}^{\theta_2} \sin^2 \theta \, d\theta - f \int_{\theta_1}^{\theta_2} \sin \theta \cos \theta \, d\theta \right) - F_y \tag{g}$$

Equations (*d*), (*e*), (*f*), and (*g*) can be simplified to ease computations. Thus, let

$$\begin{aligned} A &= \int_{\theta_1}^{\theta_2} \sin \theta \cos \theta \, d\theta = \left(\frac{1}{2} \sin^2 \theta \right)_{\theta_1}^{\theta_2} \\ B &= \int_{\theta_1}^{\theta_2} \sin^2 \theta \, d\theta = \left(\frac{\theta}{2} - \frac{1}{4} \sin 2\theta \right)_{\theta_1}^{\theta_2} \end{aligned} \tag{16–8}$$

Then, for clockwise rotation as shown in Figure 16–7, the hinge-pin reactions are

$$\begin{aligned} R_x &= \frac{p_a br}{\sin \theta_a}(A - fB) - F_x \\ R_y &= \frac{p_a br}{\sin \theta_a}(B + fA) - F_y \end{aligned} \tag{16–9}$$

For counterclockwise rotation, Equations (*f*) and (*g*) become

$$\begin{aligned} R_x &= \frac{p_a br}{\sin \theta_a}(A + fB) - F_x \\ R_y &= \frac{p_a br}{\sin \theta_a}(B - fA) - F_y \end{aligned} \tag{16–10}$$

In using these equations, the reference system always has its origin at the center of the drum. The positive x axis is taken through the hinge pin. The positive y axis is always in the direction of the shoe, even if this should result in a left-handed system.

The following assumptions are implied by the preceding analysis:

1. The pressure at any point on the shoe is assumed to be proportional to the distance from the hinge pin, being zero at the heel. This should be considered from the standpoint that pressures specified by manufacturers are averages rather than maxima.
2. The effect of centrifugal force has been neglected. In the case of brakes, the shoes are not rotating, and no centrifugal force exists. In clutch design, the effect of this force must be considered in writing the equations of static equilibrium.

3 The shoe is assumed to be rigid. Since this cannot be true, some deflection will occur, depending upon the load, pressure, and stiffness of the shoe. The resulting pressure distribution may be different from that which has been assumed.

4 The entire analysis has been based upon a coefficient of friction that does not vary with pressure. Actually, the coefficient may vary with a number of conditions, including temperature, wear, and environment.

EXAMPLE 16–2

The brake shown in Figure 16–8 is 300 mm in diameter and is actuated by a mechanism that exerts the same force F on each shoe. The shoes are identical and have a face width of 32 mm. The lining is a molded asbestos having a coefficient of friction of 0.32 and a pressure limitation of 1000 kPa. Estimate the maximum
(*a*) Actuating force F.
(*b*) Braking capacity.
(*c*) Hinge-pin reactions.

Solution

(*a*) The right-hand shoe is self-energizing, and so the force F is found on the basis that the maximum pressure will occur on this shoe. Here $\theta_1 = 0°$, $\theta_2 = 126°$, $\theta_a = 90°$, and $\sin\theta_a = 1$. Also,

$$a = \sqrt{(112)^2 + (50)^2} = 122.7 \text{ mm}$$

Integrating Equation (16–2) from 0 to θ_2 yields

$$M_f = \frac{fp_abr}{\sin\theta_a}\left[\left(-r\cos\theta\right)_0^{\theta_2} - a\left(\frac{1}{2}\sin^2\theta\right)_0^{\theta_2}\right]$$

$$= \frac{fp_abr}{\sin\theta_a}\left(r - r\cos\theta_2 - \frac{a}{2}\sin^2\theta_2\right)$$

Changing all lengths to meters, we have

$$M_f = (0.32)[1000(10)^3](0.032)(0.150) \times \left[0.150 - 0.150\cos 126° - \left(\frac{0.1227}{2}\right)\sin^2 126°\right]$$

$$= 304 \text{ N} \cdot \text{m}$$

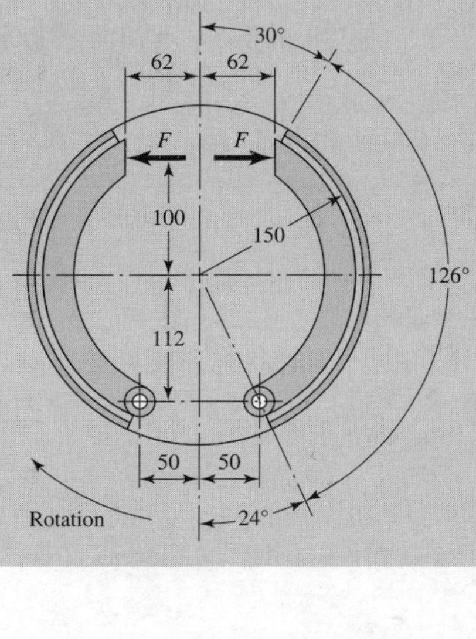

Figure 16–8

Brake with internal expanding shoes; dimensions in millimeters.

The moment of the normal forces is obtained from Equation (16–3). Integrating from 0 to θ_2 gives

$$
\begin{aligned}
M_N &= \frac{p_a bra}{\sin\theta_a}\left(\frac{\theta}{2} - \frac{1}{4}\sin 2\theta\right)_0^{\theta_2} \\
&= \frac{p_a bra}{\sin\theta_a}\left(\frac{\theta_2}{2} - \frac{1}{4}\sin 2\theta_2\right) \\
&= [1000(10)^3](0.032)(0.150)(0.1227)\left\{\frac{\pi}{2}\frac{126}{180} - \frac{1}{4}\sin[(2)(126°)]\right\} \\
&= 788 \text{ N}\cdot\text{m}
\end{aligned}
$$

From Equation (16–4), the actuating force is

Answer $$F = \frac{M_N - M_f}{c} = \frac{788 - 304}{100 + 112} = 2.28 \text{ kN}$$

(*b*) From Equation (16–6), the torque applied by the right-hand shoe is

$$
\begin{aligned}
T_R &= \frac{f p_a b r^2(\cos\theta_1 - \cos\theta_2)}{\sin\theta_a} \\
&= \frac{0.32[1000(10)^3](0.032)(0.150)^2(\cos 0° - \cos 126°)}{\sin 90°} = 366 \text{ N}\cdot\text{m}
\end{aligned}
$$

The torque contributed by the left-hand shoe cannot be obtained until we learn its maximum operating pressure. Equations (16–2) and (16–3) indicate that the frictional and normal moments are proportional to this pressure. Thus, for the left-hand shoe,

$$M_N = \frac{788 p_a}{1000} \qquad M_f = \frac{304 p_a}{1000}$$

Then, from Equation (16–7),

$$F = \frac{M_N + M_f}{c}$$

or

$$2.28 = \frac{(788/1000)p_a + (304/1000)p_a}{100 + 112}$$

Solving gives $p_a = 443$ kPa. Then, from Equation (16–6), the torque on the left-hand shoe is

$$T_L = \frac{f p_a b r^2(\cos\theta_1 - \cos\theta_2)}{\sin\theta_a}$$

Since $\sin\theta_a = \sin 90° = 1$, we have

$$T_L = 0.32[443(10)^3](0.032)(0.150)^2(\cos 0° - \cos 126°) = 162 \text{ N}\cdot\text{m}$$

The braking capacity is the total torque:

Answer $$T = T_R + T_L = 366 + 162 = 528 \text{ N}\cdot\text{m}$$

(*c*) In order to find the hinge-pin reactions, we note that $\sin\theta_a = 1$ and $\theta_1 = 0$. Then Equation (16–8) gives

$$A = \frac{1}{2}\sin^2\theta_2 = \frac{1}{2}\sin^2 126° = 0.3273$$

$$B = \frac{\theta_2}{2} - \frac{1}{4}\sin 2\theta_2 = \frac{\pi(126)}{2(180)} - \frac{1}{4}\sin\,[(2)(126°)] = 1.3373$$

Also, let

$$D = \frac{p_a br}{\sin\theta_a} = \frac{1000(0.032)(0.150)}{1} = 4.8 \text{ kN}$$

where $p_a = 1000$ kPa for the right-hand shoe. Then, using Equation (16–9), we have

$$R_x = D(A - fB) - F_x = 4.8[0.3273 - 0.32(1.3373)] - 2.28 \sin 24° = -1.410 \text{ kN}$$

$$R_y = D(B + fA) - F_y = 4.8[1.3373 + 0.32(0.3273)] - 2.28 \cos 24° = 4.839 \text{ kN}$$

The resultant on this hinge pin is

Answer $$R = \sqrt{(-1.410)^2 + (4.839)^2} = 5.04 \text{ kN}$$

The reactions at the hinge pin of the left-hand shoe are found using Equations (16–10) for a pressure of 443 kPa. They are found to be $R_x = 0.678$ kN and $R_y = 0.538$ kN. The resultant is

Answer $$R = \sqrt{(0.678)^2 + (0.538)^2} = 0.866 \text{ kN}$$

The reactions for both hinge pins, together with their directions, are shown in Figure 16–9.

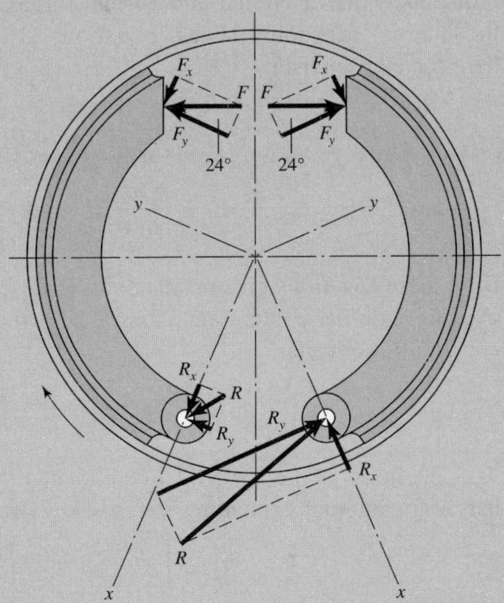

Figure 16–9

This example dramatically shows the benefit to be gained by arranging the shoes to be self-energizing. If the left-hand shoe were turned over so as to place the hinge pin at the top, it could apply the same torque as the right-hand shoe. This would make the capacity of the brake (2)(366) = 732 N · m instead of the present 528 N · m, a 30 percent improvement. In addition, some of the friction material at the heel could be eliminated without seriously affecting the capacity, because of the low pressure in this area. This change might actually improve the overall design because the additional rim exposure would improve the heat-dissipation capacity.

16–3 External Contracting Rim Clutches and Brakes

The patented clutch-brake of Figure 16–10 has external contracting friction elements, but the actuating mechanism is pneumatic. Here we shall study only pivoted external shoe brakes and clutches, though the methods presented can easily be adapted to the clutch-brake of Figure 16–10.

Operating mechanisms can be classified as:

- Solenoids
- Levers, linkages, or toggle devices
- Linkages with spring loading
- Hydraulic and pneumatic devices

The static analysis required for these devices has already been covered in Section 3–1. The methods there apply to any mechanism system, including all those used in brakes and clutches. It is not necessary to repeat the material in Chapter 3 that applies directly to such mechanisms. Omitting the operating mechanisms from consideration allows us to concentrate on brake and clutch performance without the extraneous influences introduced by the need to analyze the statics of the control mechanisms.

The notation for external contracting shoes is shown in Figure 16–11. The moments of the frictional and normal forces about the hinge pin are the same as for the internal expanding shoes. Equations (16–2) and (16–3) apply and are repeated here for convenience:

$$M_f = \frac{fp_a br}{\sin \theta_a} \int_{\theta_1}^{\theta_2} \sin \theta (r - a \cos \theta)\, d\theta \tag{16–2}$$

$$M_N = \frac{p_a bra}{\sin \theta_a} \int_{\theta_1}^{\theta_2} \sin^2 \theta \, d\theta \tag{16–3}$$

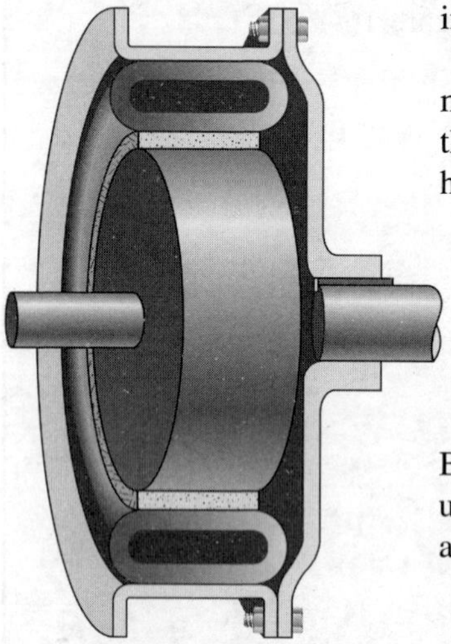

Figure 16–10

An external contracting clutch-brake that is engaged by expanding the flexible tube with compressed air. *(Source: Twin Disc Clutch Company.)*

Both these equations give positive values for clockwise moments (Figure 16–11) when used for external contracting shoes. The actuating force must be large enough to balance both moments:

$$F = \frac{M_N + M_f}{c} \tag{16–11}$$

The horizontal and vertical reactions at the hinge pin are found in the same manner as for internal expanding shoes. They are

$$R_x = \int dN \cos \theta + \int f\, dN \sin \theta - F_x \tag{a}$$

$$R_y = \int f\, dN \cos \theta - \int dN \sin \theta + F_y \tag{b}$$

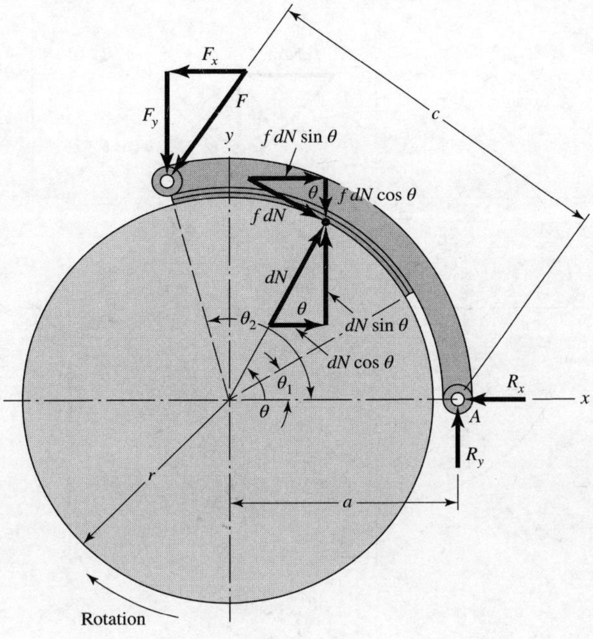

Figure 16–11

Notation of external contracting shoes.

By using Equation (16–8) and Equation (c) from Section 16–2, we have

$$R_x = \frac{p_a br}{\sin\theta_a}(A + fB) - F_x$$

$$R_y = \frac{p_a br}{\sin\theta_a}(fA - B) + F_y \tag{16–12}$$

If the rotation is counterclockwise, the sign of the frictional term in each equation is reversed. Thus Equation (16–11) for the actuating force becomes

$$F = \frac{M_N - M_f}{c} \tag{16–13}$$

and self-energization exists for counterclockwise rotation. The horizontal and vertical reactions are found, in the same manner as before, to be

$$R_x = \frac{p_a br}{\sin\theta_a}(A - fB) - F_x$$

$$R_y = \frac{p_a br}{\sin\theta_a}(-fA - B) + F_y \tag{16–14}$$

It should be noted that, when external contracting designs are used as clutches, the effect of centrifugal force is to decrease the normal force. Thus, as the speed increases, a larger value of the actuating force F is required.

A special case arises when the pivot is symmetrically located and also placed so that the moment of the friction forces about the pivot is zero. The geometry of such a brake will be similar to that of Figure 16–12a. To get a pressure-distribution relation, we note that lining wear is such as to retain the cylindrical shape, much as a

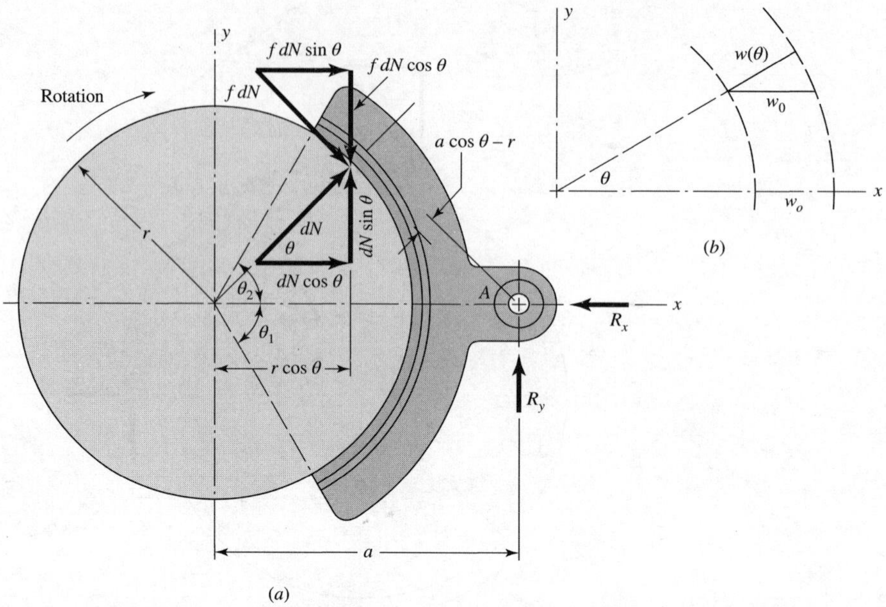

Figure 16–12

(*a*) Brake with symmetrical pivoted shoe; (*b*) wear of brake lining.

milling machine cutter feeding in the x direction would do to the shoe held in a vise. See Figure 16–12*b*. This means the abscissa component of wear is w_0 for all positions θ. If wear in the radial direction is expressed as $w(\theta)$, then

$$w(\theta) = w_0 \cos \theta$$

Using Equation (12–37), to express radial wear $w(\theta)$ as

$$w(\theta) = KPVt$$

where K is a material constant, P is pressure, V is rim velocity, and t is time. Then, denoting P as $p(\theta)$ above and solving for $p(\theta)$ gives

$$p(\theta) = \frac{w(\theta)}{KVt} = \frac{w_0 \cos \theta}{KVt}$$

Since all elemental surface areas of the friction material see the same rubbing speed for the same duration, $w_0/(KVt)$ is a constant and

$$p(\theta) = (\text{constant}) \cos \theta = p_a \cos \theta \tag{c}$$

where p_a is the maximum value of $p(\theta)$.

Proceeding to the force analysis, we observe from Figure 16–12*a* that

$$dN = pbr\, d\theta \tag{d}$$

or

$$dN = p_a br \cos \theta\, d\theta \tag{e}$$

The distance a to the pivot is chosen by finding where the moment of the frictional forces M_f is zero. First, this ensures that reaction R_y is at the correct location to establish symmetrical wear. Second, a cosinusoidal pressure distribution is sustained, preserving our predictive ability. Symmetry means $\theta_1 = \theta_2$, so

$$M_f = 2\int_0^{\theta_2} (f dN)(a \cos \theta - r) = 0$$

Substituting Equation (*e*) gives

$$2fp_abr\int_0^{\theta_2}(a\cos^2\theta - r\cos\theta)\,d\theta = 0$$

from which

$$a = \frac{4r\sin\theta_2}{2\theta_2 + \sin 2\theta_2} \tag{16–15}$$

The distance a depends on the pressure distribution. Mislocating the pivot makes M_f zero about a different location, so the brake lining adjusts its local contact pressure, through wear, to compensate. The result is unsymmetrical wear, retiring the shoe lining, hence the shoe, sooner.

With the pivot located according to Equation (16–15), the moment about the pin is zero, and the horizontal and vertical reactions are

$$R_x = 2\int_0^{\theta_2} dN\cos\theta = 2\int_0^{\theta_2}(p_abr\cos\theta\,d\theta)\cos\theta = \frac{p_abr}{2}(2\theta_2 + \sin 2\theta_2) \tag{16–16}$$

where, because of symmetry,

$$\int f\,dN\sin\theta = 0$$

Also,

$$R_y = 2\int_0^{\theta_2} f\,dN\cos\theta = 2\int_0^{\theta_2} f(p_abr\cos\theta\,d\theta)\cos\theta = \frac{p_abrf}{2}(2\theta_2 + \sin 2\theta_2) \tag{16–17}$$

where

$$\int dN\sin\theta = 0$$

also because of symmetry. Note, too, that $R_x = N$ and $R_y = fN$, as might be expected for the particular choice of the dimension a. Also, it can be shown that the torque is

$$T = afN \tag{16–18}$$

16–4 Band-Type Clutches and Brakes

Flexible clutch and brake bands are used in power excavators and in hoisting and other machinery. The analysis follows the notation of Figure 16–13.

Because of friction and the rotation of the drum, the actuating force P_2 is less than the pin reaction P_1. Any element of the band, of angular length $d\theta$, will be in equilibrium under the action of the forces shown in the figure. Summing these forces in the vertical direction, we have

$$dN - P\sin\frac{d\theta}{2} - (P + dP)\sin\frac{d\theta}{2} = 0 \tag{a}$$

For small (infinitesimal) $d\theta$ and dP, $\sin d\theta/2 \approx d\theta/2$ and $dP\sin d\theta/2 \approx 0$. Thus, Equation (*a*) reduces to

$$dN = P\,d\theta \tag{b}$$

Figure 16–13

Forces on a brake band.

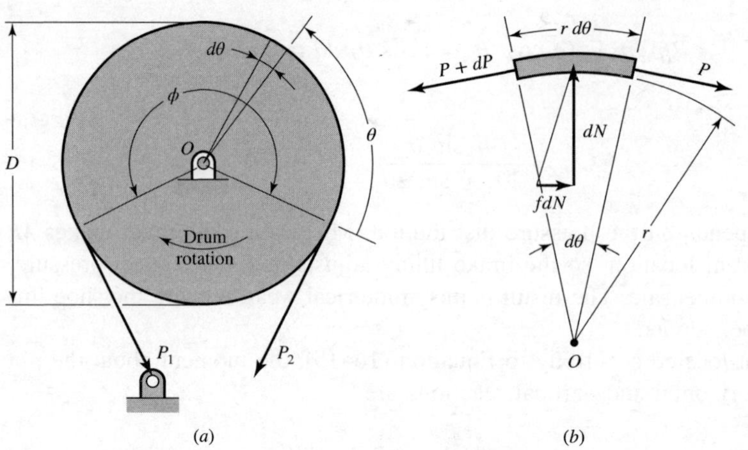

Summing the forces in the horizontal direction gives

$$f\,dN + P\cos\frac{d\theta}{2} - (P + dP)\cos\frac{d\theta}{2} = 0 \tag{c}$$

For small (infinitesimal) $d\theta$, $\cos(d\theta/2) \approx 1$, and Equation (*c*) simplifies to

$$dP = f\,dN \tag{d}$$

Substituting the value of dN from Equation (*b*) in (*d*) and integrating gives

$$\int_{P_2}^{P_1}\frac{dP}{P} = f\int_0^{\phi} d\theta \qquad \text{or} \qquad \ln\frac{P_1}{P_2} = f\phi$$

and thus

$$\frac{P_1}{P_2} = e^{f\phi} \tag{16–19}$$

The torque may be obtained from the equation

$$T = (P_1 - P_2)\frac{D}{2} \tag{16–20}$$

The normal force dN acting on an element of area of width b and length $r\,d\theta$ is

$$dN = pbr\,d\theta \tag{e}$$

where p is the pressure. Substitution of the value of dN from Equation (*b*) gives

$$P\,d\theta = pbr\,d\theta$$

Therefore

$$p = \frac{P}{br} = \frac{2P}{bD} \tag{16–21}$$

The pressure is therefore proportional to the tension in the band. The maximum pressure p_a will occur at the toe and has the value

$$p_a = \frac{2P_1}{bD} \tag{16–22}$$

16–5 Frictional-Contact Axial Clutches

An axial clutch is one in which the mating frictional members are moved in a direction parallel to the shaft. One of the earliest of these is the cone clutch, which is simple in construction and quite powerful. However, except for relatively simple installations, it has been largely displaced by the disk clutch employing one or more disks as the operating members. Advantages of the disk clutch include the freedom from centrifugal effects, the large frictional area that can be installed in a small space, the more effective heat-dissipation surfaces, and the favorable pressure distribution. Figure 16–14 shows a single-plate disk clutch; a multiple-disk clutch-brake is shown in Figure 16–15. Let us now determine the capacity of such a clutch or brake in terms of the material and geometry.

Figure 16–16 shows a friction disk having an outside diameter D and an inside diameter d. We are interested in obtaining the axial force F necessary to produce a certain torque T and pressure p. Two methods of solving the problem, depending upon the construction of the clutch, are in general use. If the disks are rigid, then the greatest amount of wear will at first occur in the outer areas, since the work of friction is greater in those areas. After a certain amount of wear has taken place, the pressure distribution will change so as to permit the wear to be uniform. This is the basis of the first method of solution.

Another method of construction employs springs to obtain a uniform pressure over the area. It is this assumption of uniform pressure that is used in the second method of solution.

Uniform Wear

After initial wear has taken place and the disks have worn down to a point where uniform wear is established, the axial wear can be expressed by Equation (12–37) as

$$w = KPVt$$

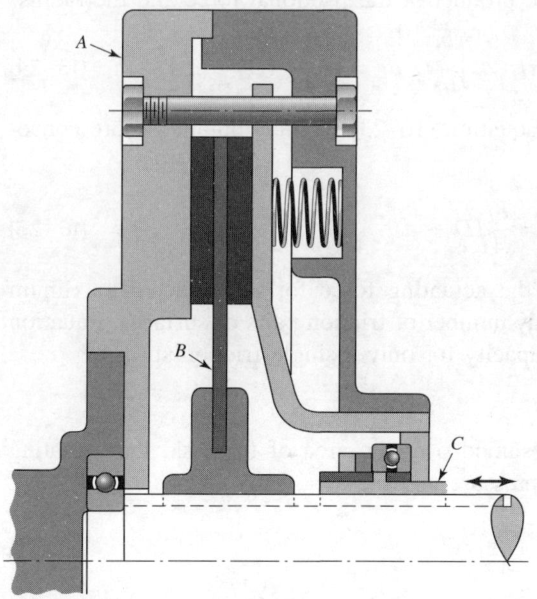

Figure 16–14

Cross-sectional view of a single-plate clutch; *A*, driver; *B*, driven plate (keyed to driven shaft); *C*, actuator.

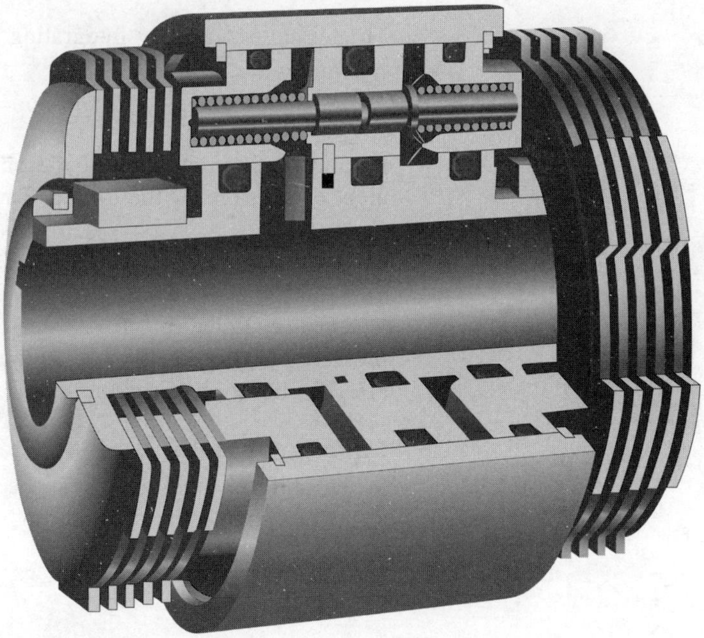

Figure 16–15

An oil-actuated multiple-disk clutch-brake for operation in an oil bath or spray. It is especially useful for rapid cycling. *(Source: Twin Disc Clutch Company.)*

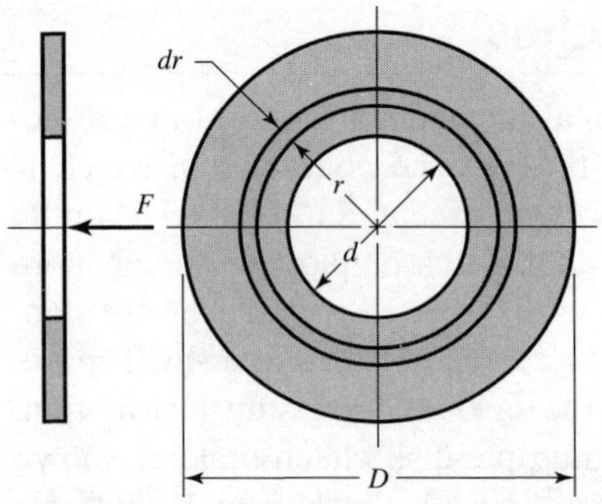

Figure 16–16

Disk friction member.

where the pressure P and velocity V can vary in the wear area. For uniform wear, w is constant, therefore PV is constant. Setting $p = P$, and $V = r\omega$, where ω is the angular velocity of the rotating member, we find that in the wear area, pr is constant. The maximum pressure p_a occurs where r is minimum, $r = d/2$, and thus

$$pr = p_a \frac{d}{2} \tag{a}$$

We can take an expression from Equation (a), which is the condition for having the same amount of work done at radius r as is done at radius $d/2$. Referring to Figure 16–16, we have an element of area of radius r and thickness dr. The area of this element is $2\pi r\, dr$, so that the normal force acting upon this element is $dF = 2\pi pr\, dr$. We can find the total normal force by letting r vary from $d/2$ to $D/2$ and integrating. Thus, with Equation (a),

$$F = \int_{d/2}^{D/2} 2\pi pr\, dr = \pi p_a d \int_{d/2}^{D/2} dr = \frac{\pi p_a d}{2}(D - d) \tag{16–23}$$

The torque is found by integrating the product of the frictional force and the radius:

$$T = \int_{d/2}^{D/2} 2\pi f pr^2\, dr = \pi f p_a d \int_{d/2}^{D/2} r\, dr = \frac{\pi f p_a d}{8}(D^2 - d^2) \tag{16–24}$$

By substituting the value of F from Equation (16–23) we may obtain a more convenient expression for the torque. Thus

$$T = \frac{Ff}{4}(D + d) \tag{16–25}$$

In use, Equation (16–23) gives the actuating force for the selected maximum pressure p_a. This equation holds for any number of friction pairs or surfaces. Equation (16–25), however, gives the torque capacity for only a single friction surface.

Uniform Pressure

When uniform pressure p_a can be assumed over the area of the disk, the actuating force F is simply the product of the pressure and the area. This gives

$$F = \frac{\pi p_a}{4}(D^2 - d^2) \tag{16–26}$$

As before, the torque is found by integrating the product of the frictional force and the radius:

$$T = 2\pi f p_a \int_{d/2}^{D/2} r^2\, dr = \frac{\pi f p_a}{12}(D^3 - d^3) \tag{16–27}$$

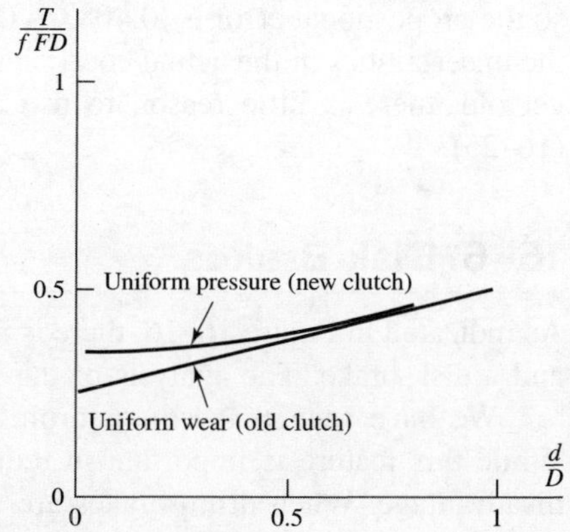

Figure 16–17

Dimensionless plot of Equations (*b*) and (*c*).

From Equation (16–26) we can rewrite Equation (16–27) as

$$T = \frac{Ff}{3}\frac{D^3 - d^3}{D^2 - d^2} \tag{16–28}$$

It should be noted for both equations that the torque is for a single pair of mating surfaces. This value must therefore be multiplied by the number of pairs of surfaces in contact.

Let us express Equation (16–25) for torque during uniform wear (old clutch) as

$$\frac{T}{f\,FD} = \frac{1 + d/D}{4} \tag{b}$$

and Equation (16–28) for torque during uniform pressure (new clutch) as

$$\frac{T}{f\,FD} = \frac{1}{3}\frac{1 - (d/D)^3}{1 - (d/D)^2} \tag{c}$$

and plot these in Figure 16–17. What we see is a dimensionless presentation of Equations (*b*) and (*c*) that reduces the number of variables from five (T, f, F, D, and d) to three (T/FD, f, and d/D), which are dimensionless. This is the method of Buckingham. The dimensionless groups (called pi terms) are

$$\pi_1 = \frac{T}{FD} \qquad \pi_2 = f \qquad \pi_3 = \frac{d}{D}$$

This allows a five-dimensional space to be reduced to a three-dimensional space. Further, because of the "multiplicative" relation between f and T in Equations (*b*) and (*c*), it is possible to plot π_1/π_2 versus π_3 in a two-dimensional space (the plane of a sheet of paper) to view all cases over the domain of existence of Equations (*b*) and (*c*) and to compare, without risk of oversight! By examining Figure 16–17 we can conclude that a new clutch, Equation (*c*), always transmits more torque than an old clutch, Equation (*b*). Furthermore, since clutches of this type are typically proportioned to make the diameter ratio d/D fall in the range $0.6 \le d/D \le 1$, the largest discrepancy between Equation (*b*) and Equation (*c*) will be

$$\frac{T}{f\,FD} = \frac{1 + 0.6}{4} = 0.400 \qquad \text{uniform wear (old clutch)}$$

$$\frac{T}{f\,FD} = \frac{1}{3}\frac{1 - 0.6^3}{1 - 0.6^2} = 0.4083 \qquad \text{uniform pressure (new clutch)}$$

so the proportional error is (0.4083 – 0.400)/0.400 = 0.021, or about 2 percent. Given the uncertainties in the actual coefficient of friction and the certainty that new clutches get old, there is little reason to use anything but Equations (16–23), (16–24), and (16–25).

16–6 Disk Brakes

As indicated in Figure 16–16, there is no fundamental difference between a disk clutch and a disk brake. The analysis of the preceding section applies to disk brakes too.

We have seen that rim or drum brakes can be designed for self-energization. While this feature is important in reducing the braking effort required, it also has a disadvantage. When drum brakes are used as vehicle brakes, only a slight change in the coefficient of friction will cause a large change in the pedal force required for braking. A not unusual 30 percent reduction in the coefficient of friction due to a temperature change or moisture, for example, can result in a 50 percent change in the pedal force required to obtain the same braking torque obtainable prior to the change. The disk brake has no self-energization, and hence is not so susceptible to changes in the coefficient of friction.

Another type of disk brake is the *floating caliper brake,* shown in Figure 16–18. The caliper supports a single floating piston actuated by hydraulic pressure. The

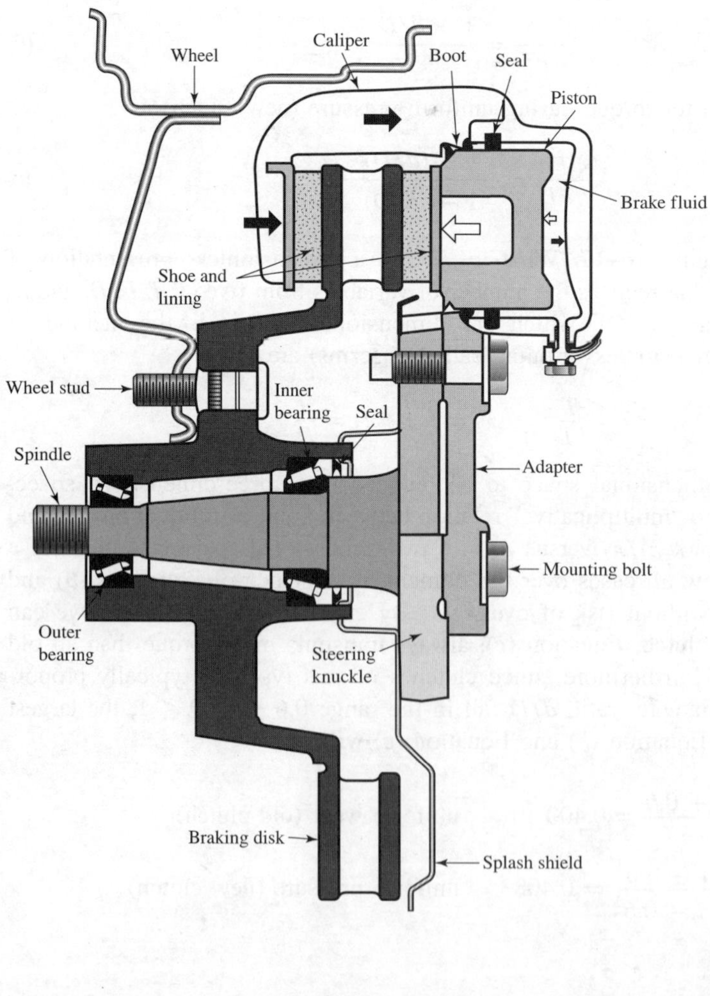

Figure 16–18

An automotive disk brake. *(Source: DaimlerChrysler Corporation.)*

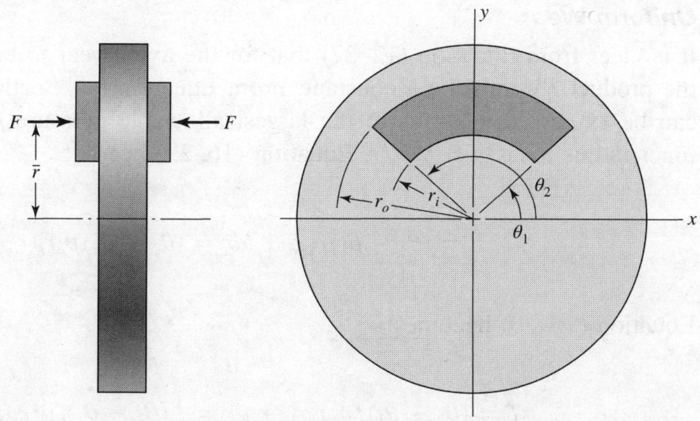

Figure 16–19

Geometry of contact area of an annular-pad segment of a caliper brake.

action is much like that of a screw clamp, with the piston replacing the function of the screw. The floating action also compensates for wear and ensures a fairly constant pressure over the area of the friction pads. The seal and boot of Figure 16–18 are designed to obtain clearance by backing off from the piston when the piston is released.

Caliper brakes (named for the nature of the actuating linkage) and disk brakes (named for the shape of the unlined surface) press friction material against the face(s) of a rotating disk. Depicted in Figure 16–19 is the geometry of an annular-pad brake contact area. The governing axial wear equation is Equation (12–37)

$$w = KPVt$$

The coordinate $\bar{r}$ locates the line of action of force F that intersects the y axis. Of interest also is the effective radius r_e, which is the radius of an equivalent shoe of infinitesimal radial thickness. If p is the local contact pressure, the actuating force F and the friction torque T are given by

$$F = \int_{\theta_1}^{\theta_2} \int_{r_i}^{r_o} pr\, dr\, d\theta = (\theta_2 - \theta_1) \int_{r_i}^{r_o} pr\, dr \tag{16–29}$$

$$T = \int_{\theta_1}^{\theta_2} \int_{r_i}^{r_o} f\, pr^2\, dr\, d\theta = (\theta_2 - \theta_1) f \int_{r_i}^{r_o} pr^2\, dr \tag{16–30}$$

The equivalent radius r_e can be found from $f\,Fr_e = T$, or

$$r_e = \frac{T}{fF} = \frac{\int_{r_i}^{r_o} pr^2\, dr}{\int_{r_i}^{r_o} pr\, dr} \tag{16–31}$$

The locating coordinate $\bar{r}$ of the activating force is found by taking moments about the x axis:

$$M_x = F\bar{r} = \int_{\theta_1}^{\theta_2} \int_{r_i}^{r_o} pr(r \sin\theta)\, dr\, d\theta = (\cos\theta_1 - \cos\theta_2) \int_{r_i}^{r_o} pr^2\, dr$$

$$\bar{r} = \frac{M_x}{F} = \frac{(\cos\theta_1 - \cos\theta_2)}{\theta_2 - \theta_1} r_e \tag{16–32}$$

Uniform Wear

It is clear from Equation (12–37) that for the axial wear to be the same everywhere, the product *PV* must be a constant. From Equation (*a*), Section 16–5, the pressure p can be expressed in terms of the largest allowable pressure p_a (which occurs at the inner radius r_i) as $p = p_a r_i / r$. Equation (16–29) becomes

$$F = (\theta_2 - \theta_1) p_a r_i \int_{r_i}^{r_o} dr = (\theta_2 - \theta_1) p_a r_i (r_o - r_i) \tag{16–33}$$

Equation (16–30) becomes

$$T = (\theta_2 - \theta_1) f p_a r_i \int_{r_i}^{r_o} r\, dr = \frac{1}{2}(\theta_2 - \theta_1) f p_a r_i (r_o^2 - r_i^2) \tag{16–34}$$

Equation (16–31) becomes

$$r_e = \frac{p_a r_i \int_{r_i}^{r_o} r\, dr}{p_a r_i \int_{r_i}^{r_o} dr} = \frac{r_o^2 - r_i^2}{2} \frac{1}{r_o - r_i} = \frac{r_o + r_i}{2} \tag{16–35}$$

Equation (16–32) becomes

$$\bar{r} = \frac{\cos \theta_1 - \cos \theta_2}{\theta_2 - \theta_1} \frac{r_o + r_i}{2} \tag{16–36}$$

Uniform Pressure

In this situation, approximated by a new brake, $p = p_a$. Equation (16–29) becomes

$$F = (\theta_2 - \theta_1) p_a \int_{r_i}^{r_o} r\, dr = \frac{1}{2}(\theta_2 - \theta_1) p_a (r_o^2 - r_i^2) \tag{16–37}$$

Equation (16–30) becomes

$$T = (\theta_2 - \theta_1) f p_a \int_{r_i}^{r_o} r^2\, dr = \frac{1}{3}(\theta_2 - \theta_1) f p_a (r_o^3 - r_i^3) \tag{16–38}$$

Equation (16–31) becomes

$$r_e = \frac{p_a \int_{r_i}^{r_o} r^2\, dr}{p_a \int_{r_i}^{r_o} r\, dr} = \frac{r_o^3 - r_i^3}{3} \frac{2}{r_o^2 - r_i^2} = \frac{2}{3} \frac{r_o^3 - r_i^3}{r_o^2 - r_i^2} \tag{16–39}$$

Equation (16–32) becomes

$$\bar{r} = \frac{\cos \theta_1 - \cos \theta_2}{\theta_2 - \theta_1} \frac{2}{3} \frac{r_o^3 - r_i^3}{r_o^2 - r_i^2} = \frac{2}{3} \frac{r_o^3 - r_i^3}{r_o^2 - r_i^2} \frac{\cos \theta_1 - \cos \theta_2}{\theta_2 - \theta_1} \tag{16–40}$$

EXAMPLE 16–3

Two annular pads, $r_i = 98.4$ mm, $r_o = 139.7$ mm, subtend an angle of 108°, have a coefficient of friction of 0.37, and are actuated by a pair of hydraulic cylinders 38 mm in diameter. The torque requirement is 1470 N · m. For uniform wear

(*a*) Find the largest normal pressure p_a.

(*b*) Estimate the actuating force F.

(*c*) Find the equivalent radius r_e and force location $\bar{r}$.

(*d*) Estimate the required hydraulic pressure.

Solution

(*a*) From Eq. (16–34), with $T = 1470/2 = 735$ N · m for each pad,

Answer

$$p_a = \frac{2T}{(\theta_2 - \theta_1) f r_i (r_o^2 - r_i^2)}$$

$$= \frac{2(735)}{(144° - 36°)(\pi/180)0.37(98.4)(139.7^2 - 98.4^2)} = 2.178 \text{ MPa}$$

(*b*) From Eq. (16–33),

Answer

$$F = (\theta_2 - \theta_1) p_a r_i (r_o - r_i) = (144° - 36°)(\pi/180)2.18(98.4)\ (139.7 - 98.4)$$
$$= 16\ 684 \text{ N}$$

(*c*) From Eq. (16–35),

Answer

$$r_e = \frac{r_o + r_i}{2} = \frac{139.7 + 98.4}{2} = 118 \text{ mm}$$

From Eq. (16–36),

Answer

$$\bar{r} = \frac{\cos\theta_1 - \cos\theta_2}{\theta_2 - \theta_1}\frac{r_o + r_i}{2} = \frac{\cos 36° - \cos 144°}{(144° - 36°)(\pi/180)}\frac{139.7 + 98.4}{2}$$
$$= 102.2 \text{ mm}$$

(*d*) Each cylinder supplies the actuating force, 16 684 N.

Answer

$$p_{\text{hydraulic}} = \frac{F}{A_P} = \frac{16\ 684}{\pi(38^2/4)} = 14.71 \text{ MPa}$$

Circular (Button or Puck) Pad Caliper Brake

Figure 16–20 displays the pad geometry. Numerical integration is necessary to analyze this brake since the boundaries are difficult to handle in closed form. Table 16–1 gives the parameters for this brake as determined by Fazekas. The effective radius is given by

$$r_e = \delta e \tag{16–41}$$

The actuating force is given by

$$F = \pi R^2 p_{\text{av}} \tag{16–42}$$

and the torque is given by

$$T = f F r_e \tag{16–43}$$

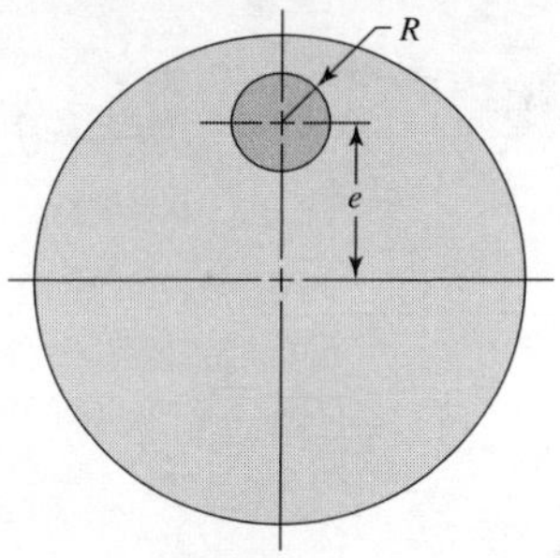

Figure 16–20

Geometry of circular pad of a caliper brake.

Table 16–1 Parameters for a Circular-Pad Caliper Brake

$\frac{R}{e}$	$\delta = \frac{r_e}{e}$	$\frac{p_{max}}{p_{av}}$
0.0	1.000	1.000
0.1	0.983	1.093
0.2	0.969	1.212
0.3	0.957	1.367
0.4	0.947	1.578
0.5	0.938	1.875

Source: Fazekas, G. A. "On Circular Spot Brakes." *Translated by ASME, J. Engineering for Industry* 94 series B, no. 3 (August 1972): 859–863.

EXAMPLE 16–4

A button-pad disk brake uses dry sintered metal pads. The pad radius is 12.7 and its center is 50.8 mm from the axis of rotation of the 90 mm diameter disk. Using half of the largest allowable pressure, $p_{max} = 2400$ kPa, find the actuating force and the brake torque. The coefficient of friction is 0.31.

Solution

Since the pad radius $R = 12.7$ mm and eccentricity $e = 50.8$ mm,

$$\frac{R}{e} = \frac{12.7}{50.8} = 0.25$$

From Table 16–1, by interpolation, $\delta = 0.963$ and $p_{max}/p_{av} = 1.290$. It follows that the effective radius e is found from Equation (16–41):

$$r_e = \delta e = 0.963(50.8) = 48.9 \text{ mm}$$

and the average pressure is

$$p_{av} = \frac{p_{max}/2}{1.290} = \frac{2.4/2}{1.290} = 0.930 \text{ MPa}$$

The actuating force F is found from Equation (16–42) to be

Answer $$F = \pi R^2 p_{av} = \pi(12.7)^2 0.930 = 471 \text{ N} \qquad \text{(one side)}$$

The brake torque T is

Answer $$T = fFr_e = 0.31(471)48.9 = 7140 \text{ N} \cdot \text{mm} \qquad \text{(one side)}$$

16–7 Cone Clutches and Brakes

The drawing of a *cone clutch* in Figure 16–21 shows that it consists of a *cup* keyed or splined to one of the shafts, a *cone* that must slide axially on splines or keys on the mating shaft, and a helical *spring* to hold the clutch in engagement. The clutch is disengaged by means of a fork that fits into the shifting groove on the friction cone. The important geometric design parameters are the *cone angle α* and the diameter and face width of the cone. If the cone angle is too small, say, less than about 8°, then

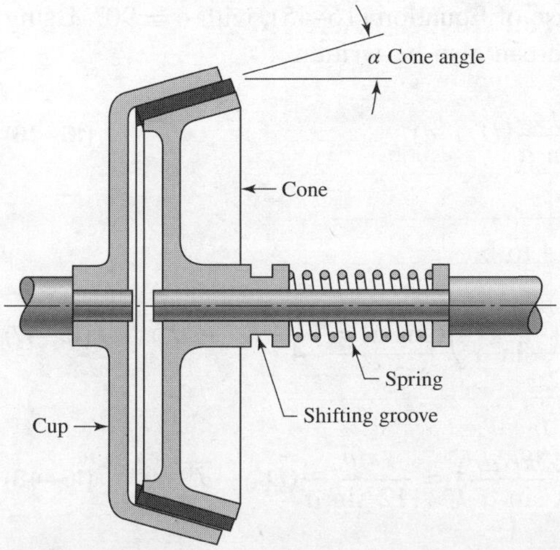

Figure 16–21

Cross section of a cone clutch.

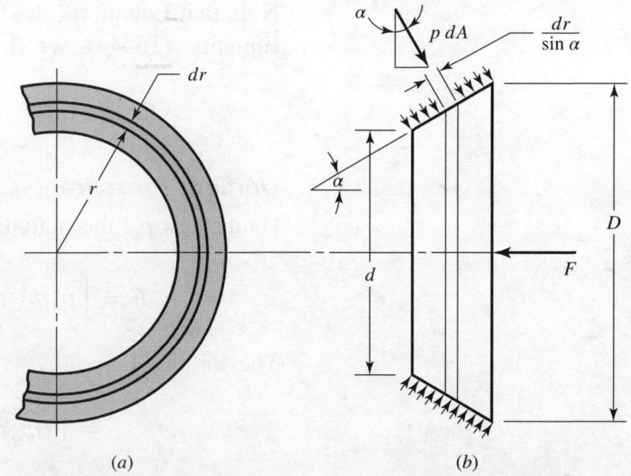

Figure 16–22

Contact area of a cone clutch.

the force required to disengage the clutch may be quite large. The wedging effect lessens rapidly when larger cone angles are used. Depending upon the characteristics of the friction materials, a good compromise can usually be found using cone angles between 10° and 15°.

To find a relation between the operating force F and the torque transmitted, designate the dimensions of the friction cone as shown in Figure 16–22. As in the case of the axial clutch, we can obtain one set of relations for a uniform-wear and another set for a uniform-pressure assumption.

Uniform Wear

The pressure relation is the same as for the axial clutch:

$$p = p_a \frac{d}{2r} \tag{a}$$

Next, referring to Figure 16–22, we see that we have an element of area dA of radius r and width $dr/\sin\alpha$. Thus $dA = (2\pi r dr)/\sin\alpha$. As shown in Figure 16–22, the operating force will be the integral of the axial component of the differential force pdA. Thus

$$\begin{aligned} F &= \int pdA \sin\alpha = \int_{d/2}^{D/2} \left(p_a \frac{d}{2r}\right)\left(\frac{2\pi r\, dr}{\sin\alpha}\right)(\sin\alpha) \\ &= \pi p_a d \int_{d/2}^{D/2} dr = \frac{\pi p_a d}{2}(D - d) \end{aligned} \tag{16–44}$$

which is the same result as in Equation (16–23).

The differential friction force is $fpdA$, and the torque is the integral of the product of this force with the radius. Thus,

$$\begin{aligned} T &= \int rfpdA = \int_{d/2}^{D/2} (rf)\left(p_a \frac{d}{2r}\right)\left(\frac{2\pi r\, dr}{\sin\alpha}\right) \\ &= \frac{\pi f p_a d}{\sin\alpha} \int_{d/2}^{D/2} r\, dr = \frac{\pi f p_a d}{8 \sin\alpha}(D^2 - d^2) \end{aligned} \tag{16–45}$$

Note that Equation (16–24) is a special case of Equation (16–45), with $\alpha = 90°$. Using Equation (16–44), we find that the torque can also be written

$$T = \frac{Ff}{4 \sin \alpha}(D + d) \tag{16–46}$$

Uniform Pressure

Using $p = p_a$, the actuating force is found to be

$$F = \int p_a dA \sin \alpha = \int_{d/2}^{D/2} (p_a)\left(\frac{2\pi r\, dr}{\sin \alpha}\right)(\sin \alpha) = \frac{\pi p_a}{4}(D^2 - d^2) \tag{16–47}$$

The torque is

$$T = \int r f p_a dA = \int_{d/2}^{D/2} (rfp_a)\left(\frac{2\pi r\, dr}{\sin \alpha}\right) = \frac{\pi f p_a}{12 \sin \alpha}(D^3 - d^3) \tag{16–48}$$

Using Equation (16–47) in Equation (16–48) gives

$$T = \frac{Ff}{3 \sin \alpha} \frac{D^3 - d^3}{D^2 - d^2} \tag{16–49}$$

As in the case of the axial clutch, we can write Equation (16–46) dimensionlessly as

$$\frac{T \sin \alpha}{f FD} = \frac{1 + d/D}{4} \tag{b}$$

and write Equation (16–49) as

$$\frac{T \sin \alpha}{f FD} = \frac{1}{3} \frac{1 - (d/D)^3}{1 - (d/D)^2} \tag{c}$$

This time there are six (T, α, f, F, D, and d) parameters and four pi terms:

$$\pi_1 = \frac{T}{FD} \qquad \pi_2 = f \qquad \pi_3 = \sin \alpha \qquad \pi_4 = \frac{d}{D}$$

As in Figure 16–17, we plot $T \sin \alpha/(fFD)$ as ordinate and d/D as abscissa. The plots and conclusions are the same. There is little reason for using equations other than Equations (16–44), (16–45), and (16–46).

16–8 Energy Considerations

When the rotating members of a machine are caused to stop by means of a brake, the kinetic energy of rotation must be absorbed by the brake. This energy appears in the brake in the form of heat. In the same way, when the members of a machine that are initially at rest are brought up to speed, slipping must occur in the clutch until the driven members have the same speed as the driver. Kinetic energy is absorbed during slippage of either a clutch or a brake, and this energy appears as heat.

We have seen how the torque capacity of a clutch or brake depends upon the coefficient of friction of the material and upon a safe normal pressure. However, the character of the load may be such that, if this torque value is permitted, the clutch or brake may be destroyed by its own generated heat. The capacity of a clutch is therefore limited by two factors, the characteristics of the material and the ability of the clutch to dissipate heat. In this section we shall consider the amount of heat generated by a clutching or braking operation. If the heat is generated faster than it is dissipated, we have a temperature-rise problem; that is the subject of the next section.

To get a clear picture of what happens during a simple clutching or braking operation, refer to Figure 16–1*a*, which is a mathematical model of a two-inertia system connected by a clutch. As shown, inertias I_1 and I_2 have initial angular velocities of ω_1 and ω_2, respectively. During the clutch operation both angular velocities change and eventually become equal. We assume that the two shafts are rigid and that the clutch torque is constant.

Writing the equation of motion for inertia 1 gives

$$I_1\ddot{\theta}_1 = -T \qquad (a)$$

where $\ddot{\theta}_1$ is the angular acceleration of I_1 and T is the clutch torque. A similar equation for I_2 is

$$I_2\ddot{\theta}_2 = T \qquad (b)$$

We can determine the instantaneous angular velocities $\dot{\theta}_1$ and $\dot{\theta}_2$ of I_1 and I_2 after any period of time t has elapsed by integrating Equations (a) and (b). The results are

$$\dot{\theta}_1 = -\frac{T}{I_1}t + \omega_1 \qquad (c)$$

$$\dot{\theta}_2 = \frac{T}{I_2}t + \omega_2 \qquad (d)$$

where $\dot{\theta}_1 = \omega_1$ and $\dot{\theta}_2 = \omega_2$ at $t = 0$. The difference in the velocities, sometimes called the relative velocity, is

$$\dot{\theta} = \dot{\theta}_1 - \dot{\theta}_2 = -\frac{T}{I_1}t + \omega_1 - \left(\frac{T}{I_2}t + \omega_2\right) = \omega_1 - \omega_2 - T\left(\frac{I_1 + I_2}{I_1 I_2}\right)t \qquad (e)$$

It is convenient to manipulate the inertial terms in Equation (e) into a form that mathematically works well for clutches (where both inertias are needed) and for brakes (where the inertia of the grounded body, I_2, can be set to infinity). This gives

$$\dot{\theta} = \omega_1 - \omega_2 - \left(\frac{I_1/I_2 + 1}{I_1}\right)Tt \qquad (16\text{–}50)$$

The clutching operation is completed at the instant in which the two angular velocities $\dot{\theta}_1$ and $\dot{\theta}_2$ become equal. Let the time required for the entire operation be t_1. Then $\dot{\theta} = 0$ when $\dot{\theta}_1 = \dot{\theta}_2$, and so Equation (16–50) gives the time as

$$t_1 = \left(\frac{I_1}{I_1/I_2 + 1}\right)\frac{\omega_1 - \omega_2}{T} \qquad (16\text{–}51)$$

This equation shows that the time required for the engagement operation is directly proportional to the velocity difference and inversely proportional to the torque.

We have assumed the clutch torque to be constant. Therefore, using Equation (16–50), we find the rate of energy-dissipation during the clutching operation to be

$$u = T\dot{\theta} = T\left[\omega_1 - \omega_2 - \left(\frac{I_1/I_2 + 1}{I_1}\right)Tt\right] \tag{f}$$

This equation shows that the energy-dissipation rate is greatest at the start, when $t = 0$.

The total energy dissipated during the clutching operation or braking cycle is obtained by integrating Equation (*e*) from $t = 0$ to $t = t_1$. The result is found to be

$$\begin{aligned} E &= \int_0^{t_1} u\, dt = T\int_0^{t_1}\left[\omega_1 - \omega_2 - \left(\frac{I_1/I_2 + 1}{I_1}\right)Tt\right] dt \\ &= \left(\frac{I_1}{I_1/I_2 + 1}\right)\frac{(\omega_1 - \omega_2)^2}{2} \end{aligned} \tag{16–52}$$

where Equation (16–51) was employed. Note that the energy dissipated is proportional to the velocity difference squared and is independent of the clutch torque.

Note that E in Equation (16–52) is the energy lost or dissipated and is the energy that is absorbed by the clutch or brake. In SI, the inertias are expressed in kilogram-meter^2 units, and the energy dissipated is expressed in joules.

$$H = E \tag{16–53}$$

16–9 Temperature Rise

The temperature rise of the clutch or brake assembly can be approximated by the classic expression

$$\Delta T = \frac{E}{C_p m} \tag{16–54}$$

where ΔT = temperature rise, °C

C_p = specific heat capacity, use 500 J/kg · °C for steel or cast iron

m = mass of clutch or brake parts, kg

The temperature-rise equations above can be used to explain what happens when a clutch or brake is operated. However, there are so many variables involved that it would be most unlikely that such an analysis would even approximate experimental results. For this reason such analyses are most useful, for repetitive cycling, in pinpointing those design parameters that have the greatest effect on performance.

If an object is at initial temperature T_1 in an environment of temperature T_∞, then Newton's cooling model is expressed as

$$\frac{T - T_\infty}{T_1 - T_\infty} = \exp\left(-\frac{\hbar_{CR}A}{WC_p}t\right) \tag{16–55}$$

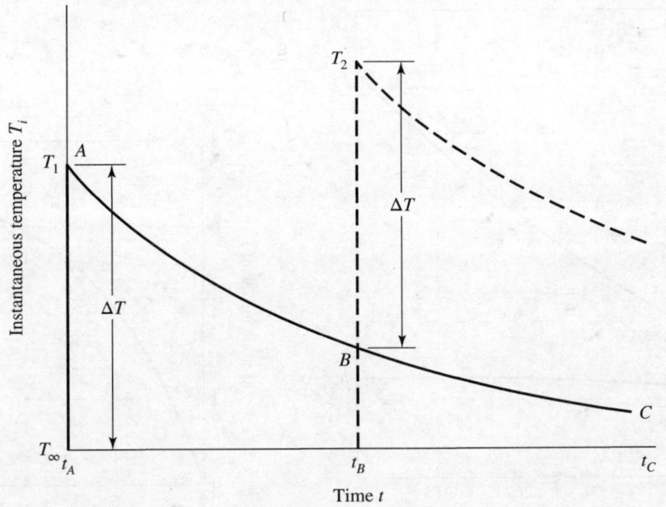

Figure 16–23
The effect of clutching or braking operations on temperature. T_∞ is the ambient temperature. Note that the temperature rise ΔT may be different for each operation.

where T = temperature at time t, °C
T_1 = initial temperature, °C
T_∞ = environmental temperature, °C
$\hbar_{CR}$ = overall coefficient of heat transfer, J/(mm^2 · s · °C)
A = lateral surface area, mm^2
W = mass of the object, kg
C_p = specific heat capacity of the object, J/(kg · °C)

Figure 16–23 shows an application of Equation (16–55). The curve ABC is the exponential decline of temperature given by Equation (16–55). At time t_B a second application of the brake occurs. The temperature quickly rises to temperature T_2, and a new cooling curve is started. For repetitive brake applications, subsequent temperature peaks $T_3, T_4, \ldots,$ may be higher than the previous peaks if insufficient cooling has occurred between applications. If this is a production situation with brake applications every t_1 seconds, then a steady state develops in which all the peaks T_{max} and all the valleys T_{min} are repetitive.

The heat-dissipation capacity of disk brakes has to be planned to avoid reaching disk and pad temperatures that are detrimental to the parts. When a disk brake has a rhythm such as discussed above, then the rate of heat transfer is described by another Newtonian equation:

$$H_{loss} = \hbar_{CR}A(T - T_\infty) = (h_r + f_v h_c)A(T - T_\infty) \tag{16–56}$$

where H_{loss} = rate of energy loss, J/s
$\hbar_{CR}$ = overall coefficient of heat transfer, J/(mm^2 · s · °C)
h_r = radiation component of $\hbar_{CR}$, J/(mm^2 · s · °C), Figure 16–24*a*
h_c = convective component of $\hbar_{CR}$, J/(mm^2 · s · °C), Figure 16–24*a*
f_v = ventilation factor, Figure 16–24*b*
T = disk temperature, °C
T_∞ = ambient temperature, °C

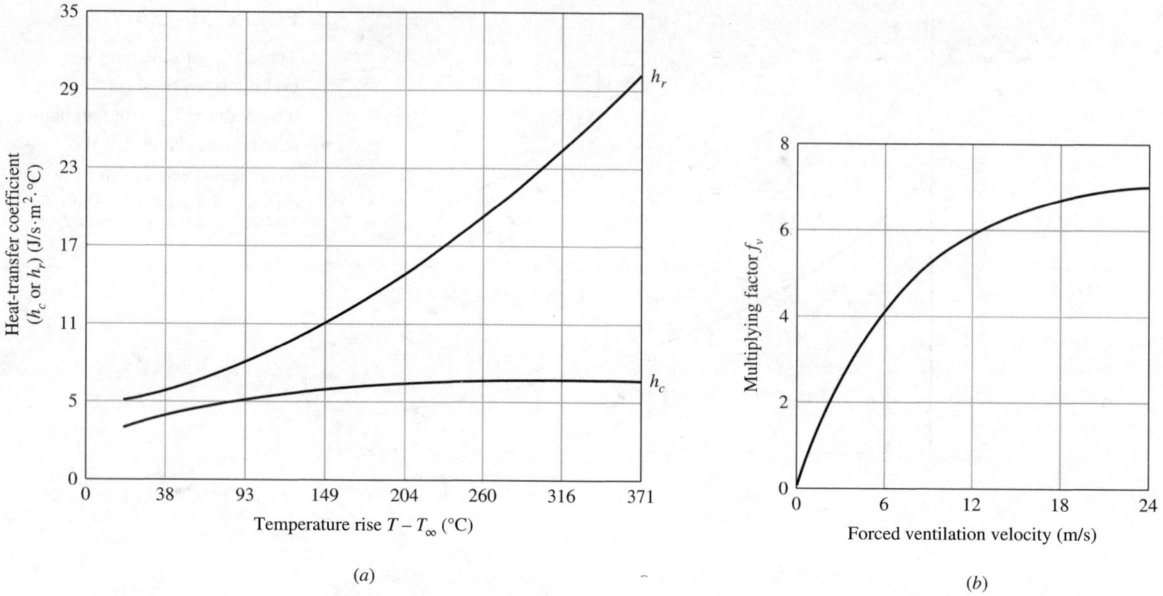

Figure 16–24

(*a*) Heat-transfer coefficient in still air. (*b*) Ventilation factors. *(Source: Data from Tolo-o-matic.)*

The energy E absorbed by the brake stopping a rotary inertia I in terms of original and final angular velocities ω_o and ω_f is the change in kinetic energy, $I(\omega_o^2 - \omega_f^2)/2$. Expressing this in J,

$$E = \frac{1}{2} I \left(\omega_o^2 - \omega_f^2\right) \tag{16–57}$$

The temperature rise ΔT due to a single stop is

$$\Delta T = \frac{E}{mC} \tag{16–58}$$

$T_{\max}$ has to be high enough to transfer E in t_1 seconds. For steady state, rearrange Equation (16–55) as

$$\frac{T_{\min} - T_\infty}{T_{\max} - T_\infty} = \exp(-\beta t_1)$$

where $\beta = \hbar_{\text{CR}} A/(WC_p)$. Cross-multiply, multiply the equation by -1, add $T_{\max}$ to both sides, set $T_{\max} - T_{\min} = \Delta T$, and rearrange, obtaining

$$T_{\max} = T_\infty + \frac{\Delta T}{1 - \exp(-\beta t_1)} \tag{16–59}$$

EXAMPLE 16–5

A caliper brake is used 24 times per hour to arrest a machine shaft from a speed of 250 rev/min to rest. The ventilation of the brake provides a mean air speed of 7.6 m/s. The equivalent rotary inertia of the machine as seen from the brake shaft is 32.65 kg · m^2. The disk is steel with a density ρ = 7811 kg/m^3, a specific heat capacity of 450 J/(kg · °C), a diameter of 152 mm, a thickness of 6.35 mm. The pads are dry sintered metal. The lateral area of the brake surface is 32 260 mm^2. Find $T_{\max}$ and $T_{\min}$ for the steady-state operation if T_∞ = 21°C.

Solution $t_1 = 60^2/24 = 150\ \text{s}$

Assuming a temperature rise of $T_{max} - T_\infty = 93°\text{C}$, from Fig. 16–24*a*,

$$h_r = 8\ \text{J/(m}^2 \cdot \text{s} \cdot °\text{C)}$$

$$h_c = 5\ \text{J/(m}^2 \cdot \text{s} \cdot °\text{C)}$$

Fig. 16–24*b* $f_v = 4.8$

$$\hbar_{CR} = h_r + f_v h_c = 8 + 4.8(5) = 32\ \text{J/(m}^2 \cdot \text{s} \cdot °\text{C)}$$

The mass of the disk is

$$W = \frac{\pi \rho D^2 h}{4} = \frac{\pi(7811)0.152^2(6.35(10)^{-3})}{4} = 0.9\ \text{kg}$$

Eq. (16–57): $$E = \frac{1}{2} I\left(\omega_o^2 - \omega_f^2\right) = \frac{32.65}{2}\left(\frac{2\pi}{60}250\right)^2 = 11\ 189\ \text{J}$$

$$\beta = \frac{\hbar_{CR} A}{W C_p} = \frac{32(32260)10^{-6}}{0.9(450)} = 2.55(10^{-3})\ \text{s}^{-1}$$

Eq. (16–58): $$\Delta T = \frac{E}{W C_p} = \frac{11\ 189}{0.9(450)} = 27.6°\text{C}$$

Answer Eq. (16–59): $$T_{max} = 21 + \frac{27.6}{1 - \exp[-2.55(10^{-3})150]} = 107.8°\text{C}$$

Answer $$T_{min} = 107.8 - 27.6 = 80.2°\text{C}$$

The predicted temperature rise here is $T_{max} - T_\infty = 107.8 - 21 = 86.8°\text{C}$.

Table 16–3 for dry sintered metal pads gives a continuous operating maximum temperature of 300–350°C. There is no danger of overheating.

16–10 Friction Materials

A brake or friction clutch should have the following lining material characteristics to a degree that is dependent on the severity of service:

- High and reproducible coefficient of friction
- Imperviousness to environmental conditions, such as moisture
- The ability to withstand high temperatures, together with good thermal conductivity and diffusivity, as well as high specific heat capacity
- Good resiliency
- High resistance to wear, scoring, and galling
- Compatible with the environment
- Flexibility

Table 16–2 **Area of Friction Material Required for a Given Average Braking Power**

Duty Cycle	Typical Applications	Ratio of Area to Average Braking Power, $m^2/(J/s)$		
		Band and Drum Brakes	Plate Disk Brakes	Caliper Disk Brakes
Infrequent	Emergency brakes	$1.61 \cdot 10^{-4}$	$5.29 \cdot 10^{-4}$	$5.29 \cdot 10^{-5}$
Intermittent	Elevators, cranes, and winches	$5.29 \cdot 10^{-4}$	$1.34 \cdot 10^{-3}$	$1.32 \cdot 10^{-4}$
Heavy-duty	Excavators, presses	$(1.06–1.30) \cdot 10^{-3}$	$2.57 \cdot 10^{-3}$	$2.66 \cdot 10^{-4}$

Sources: Neale, M. J. *The Tribology Handbook.* London: Butterworth, 1973; Ferodo Ltd. *Friction Materials for Engineers*. Chapel-en-le-frith, England: Feredo, 1968.

Table 16–2 gives area of friction surface required for several braking powers. Table 16–3 gives important characteristics of some friction materials for brakes and clutches.

The manufacture of friction materials is a highly specialized process, and it is advisable to consult manufacturers' catalogs and handbooks, as well as manufacturers directly, in selecting friction materials for specific applications. Selection involves a consideration of the many characteristics as well as the standard sizes available.

The *woven-cotton lining* is produced as a fabric belt that is impregnated with resins and polymerized. It is used mostly in heavy machinery and is usually supplied in rolls up to 15 m in length. Thicknesses available range from 3 to 25 mm, in widths up to about 300 mm.

A *woven-asbestos lining* is made in a similar manner to the cotton lining and may also contain metal particles. It is not quite as flexible as the cotton lining and comes in a smaller range of sizes. Along with the cotton lining, the asbestos lining was widely used as a brake material in heavy machinery.

Molded-asbestos linings contain asbestos fiber and friction modifiers; a thermoset polymer is used, with heat, to form a rigid or semirigid molding. The principal use was in drum brakes.

Molded-asbestos pads are similar to molded linings but have no flexibility; they were used for both clutches and brakes.

Sintered-metal pads are made of a mixture of copper and/or iron particles with friction modifiers, molded under high pressure and then heated to a high temperature to fuse the material. These pads are used in both brakes and clutches for heavy-duty applications.

Cermet pads are similar to the sintered-metal pads and have a substantial ceramic content.

Table 16–4 lists properties of typical brake linings. The linings may consist of a mixture of fibers to provide strength and ability to withstand high temperatures, various friction particles to obtain a degree of wear resistance as well as a higher coefficient of friction, and bonding materials.

Table 16–5 includes a wider variety of clutch friction materials, together with some of their properties. Some of these materials may be run wet by allowing them to dip in oil or to be sprayed by oil. This reduces the coefficient of friction somewhat but carries away more heat and permits higher pressures to be used.

Table 16–3 Characteristics of Friction Materials for Brakes and Clutches

Material	Friction Coefficient f	Maximum Pressure p_{max}, MPa	Maximum Temperature Instantaneous, °C	Maximum Temperature Continuous, °C	Maximum Velocity V_{max}, m/s	Applications
Cermet	0.32	1.0	815	400		Brakes and clutches
Sintered metal (dry)	0.29–0.33	2.1–2.8	500–550	300–350	18	Clutches and caliper disk brakes
Sintered metal (wet)	0.06–0.08	3.4	500	300	18	Clutches
Rigid molded asbestos (dry)	0.35–0.41	0.7	350–400	180	18	Drum brakes and clutches
Rigid molded asbestos (wet)	0.06	2.1	350	180	18	Industrial clutches
Rigid molded asbestos pads	0.31–0.49	5.2	500–750	230–350	24	Disk brakes
Rigid molded nonasbestos	0.33–0.63	0.7–1.0		260–400	24–38	Clutches and brakes
Semirigid molded asbestos	0.37–0.41	0.7	350	150	18	Clutches and brakes
Flexible molded asbestos	0.39–0.45	0.7	350–400	150–180	18	Clutches and brakes
Wound asbestos yarn and wire	0.38	0.7	350	150	18	Vehicle clutches
Woven asbestos yarn and wire	0.38	0.7	260	130	18	Industrial clutches and brakes
Woven cotton	0.47	0.7	110	75	18	Industrial clutches and brakes
Resilient paper (wet)	0.09–0.15	2.8	150		$PV < 18$ MPa · m/s	Clutches and transmission bands

Sources: Ferodo Ltd., Chapel-en-le-frith, England; Scan-pac, Mequon, Wisc; Raybestos, New York, N.Y. and Stratford, Conn.; Gatke Corp., Chicago, Ill.; General Metals Powder Co., Akron, Ohio; D. A. B. Industries, Troy, Mich; Friction Products Co., Medina, Ohio.

Table 16–4 Some Properties of Brake Linings

	Woven Lining	Molded Lining	Rigid Block
Compressive strength, MPa	70–100	70–125	70–100
Tensile strength, MPa	17–21	27–35	21–27
Max. temperature, °C	200–260	260	400
Max. speed, m/s	38	25	38
Max. pressure, kPa	340–690	690	1000
Frictional coefficient, mean	0.45	0.47	0.40–45

Table 16–5 Friction Materials for Clutches

	Friction Coefficient		Max. Temperature	Max. Pressure
Material	Wet	Dry	°C	kPa
Cast iron on cast iron	0.05	0.15–0.20	320	1000–1750
Powdered metal* on cast iron	0.05–0.1	0.1–0.4	540	1000
Powdered metal* on hard steel	0.05–0.1	0.1–0.3	540	2100
Wood on steel or cast iron	0.16	0.2–0.35	150	400–620
Leather on steel or cast iron	0.12	0.3–0.5	100	70–280
Cork on steel or cast iron	0.15–0.25	0.3–0.5	100	50–100
Felt on steel or cast iron	0.18	0.22	140	35–70
Woven asbestos* on steel or cast iron	0.1–0.2	0.3–0.6	175–260	350–700
Molded asbestos* on steel or cast iron	0.08–0.12	0.2–0.5	260	350–1000
Impregnated asbestos* on steel or cast iron	0.12	0.32	260–400	1000
Carbon graphite on steel	0.05–0.1	0.25	370–540	2100

*The friction coefficient can be maintained with ±5 percent for specific materials in this group.

16–11 Miscellaneous Clutches and Couplings

The square-jaw clutch shown in Figure 16–25*a* is one form of positive-contact clutch. These clutches have the following characteristics:

- They do not slip.
- No heat is generated.
- They cannot be engaged at high speeds.
- Sometimes they cannot be engaged when both shafts are at rest.
- Engagement at any speed is accompanied by shock.

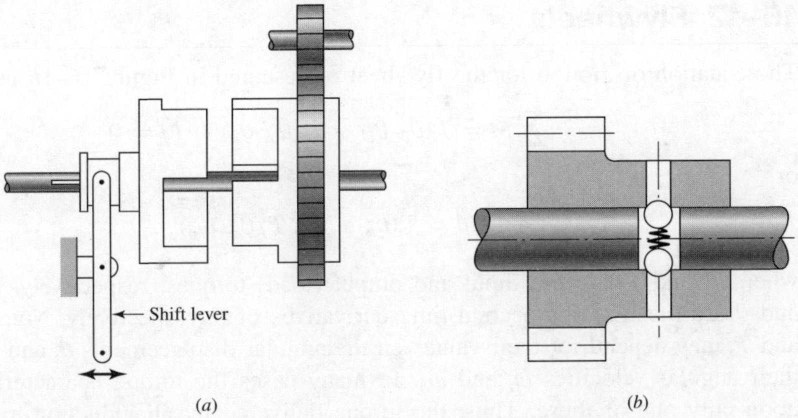

Figure 16–25

(*a*) Square-jaw clutch; (*b*) overload release clutch using a detent.

The greatest differences among the various types of positive clutches are concerned with the design of the jaws. To provide a longer period of time for shift action during engagement, the jaws may be ratchet-shaped, spiral-shaped, or gear-tooth-shaped. Sometimes a great many teeth or jaws are used, and they may be cut either circumferentially, so that they engage by cylindrical mating, or on the faces of the mating elements.

Although positive clutches are not used to the extent of the frictional-contact types, they do have important applications where synchronous operation is required, as, for example, in power presses or rolling-mill screw-downs.

Devices such as linear drives or motor-operated screwdrivers must run to a definite limit and then come to a stop. An overload-release type of clutch is required for these applications. Figure 16–25*b* is a schematic drawing illustrating the principle of operation of such a clutch. These clutches are usually spring-loaded so as to release at a predetermined torque. The clicking sound that is heard when the overload point is reached is considered to be a desirable signal.

Both fatigue and shock loads must be considered in obtaining the stresses and deflections of the various portions of positive clutches. In addition, wear must generally be considered. The application of the fundamentals discussed in Parts 1 and 2 of this book is usually sufficient for the complete design of these devices.

An overrunning clutch or coupling permits the driven member of a machine to "freewheel" or "overrun" because the driver is stopped or because another source of power increases the speed of the driven mechanism. The construction uses rollers or balls mounted between an outer sleeve and an inner member having cam flats machined around the periphery. Driving action is obtained by wedging the rollers between the sleeve and the cam flats. This clutch is therefore equivalent to a pawl and ratchet with an infinite number of teeth.

There are many varieties of overrunning clutches available, and they are built in capacities up to hundreds of horsepower. Since no slippage is involved, the only power loss is that due to bearing friction and windage.

The shaft couplings shown in Figure 16–26 are representative of the selection available in catalogs.

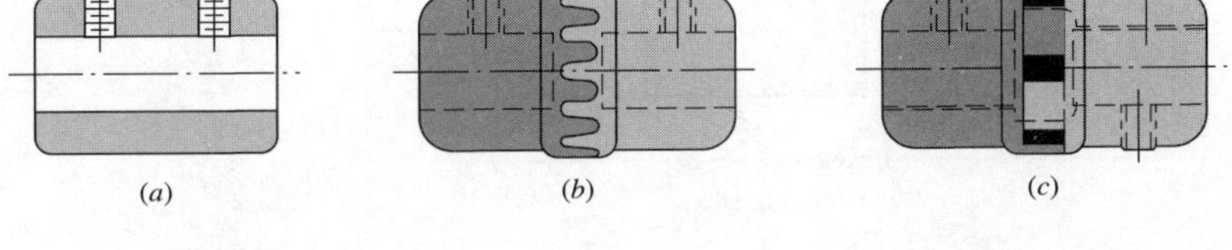

Figure 16–26

Shaft couplings. (*a*) Plain. (*b*) Light-duty toothed coupling. (*c*) Three-jaw coupling available with bronze, rubber, or polyurethane insert to minimize vibration. *(Source: Boston Gear Division, Colfax Corp.)*

16–12 Flywheels

The equation of motion for the flywheel represented in Figure 16–1*b* is

$$\sum M = T_i(\theta_i,\ \dot{\theta}_i) - T_o(\theta_o,\ \dot{\theta}_o) - I\ddot{\theta} = 0$$

or

$$I\ddot{\theta} = T_i(\theta_i,\ \omega_i) - T_o(\theta_o,\ \omega_o) \tag{a}$$

where T_i and T_o are the input and output (load) torques, respectively; and where $\dot{\theta}$ and $\ddot{\theta}$ are the first and second time derivatives of θ, respectively. Note that both T_i and T_o may depend for their values on the angular displacements θ_i and θ_o as well as their angular velocities ω_i and ω_o. In many cases the torque characteristic depends upon only one of these. Thus, the torque delivered by an induction motor depends upon the speed of the motor. In fact, motor manufacturers publish charts detailing the torque-speed characteristics of their various motors.

When the input and output torque functions are given, Equation (*a*) can be solved for the motion of the flywheel using well-known techniques for solving linear and nonlinear differential equations. We can dispense with this here by assuming a rigid shaft, giving $\theta_i = \theta_o = \theta$ and $\omega_i = \omega_o = \omega$. Thus, Equation (*a*) becomes

$$I\ddot{\theta} = T_i(\theta,\ \omega) - T_o(\theta,\ \omega) \tag{b}$$

When the two torque functions are known and the starting values of the displacement θ and velocity ω are given, Equation (*b*) can be solved for θ, ω, and $\ddot{\theta}$ as functions of time. However, we are not really interested in the instantaneous values of these terms at all. Primarily we want to know the overall performance of the flywheel. What should its moment of inertia be? How do we match the power source to the load? And what are the resulting performance characteristics of the system that we have selected?

To gain insight into the problem, a hypothetical situation is diagrammed in Figure 16–27. An input power source subjects a flywheel to a constant torque T_i while the shaft rotates from θ_1 to θ_2. This is a positive torque and is plotted upward. Equation (*b*) indicates that a positive acceleration $\ddot{\theta}$ will be the result, and so the shaft velocity increases from ω_1 to ω_2. As shown, the shaft now rotates from θ_2 to θ_3 with zero torque and hence, from Equation (*b*), with zero acceleration. Therefore $\omega_3 = \omega_2$. From θ_3 to θ_4 a load, or output torque, of constant magnitude is applied, causing the shaft to slow down from ω_3 to ω_4. Note that the output torque is plotted in the negative direction in accordance with Equation (*b*).

The work input to the flywheel is the area of the rectangle between θ_1 and θ_2, or

$$U_i = T_i(\theta_2 - \theta_1) \tag{c}$$

Figure 16–27

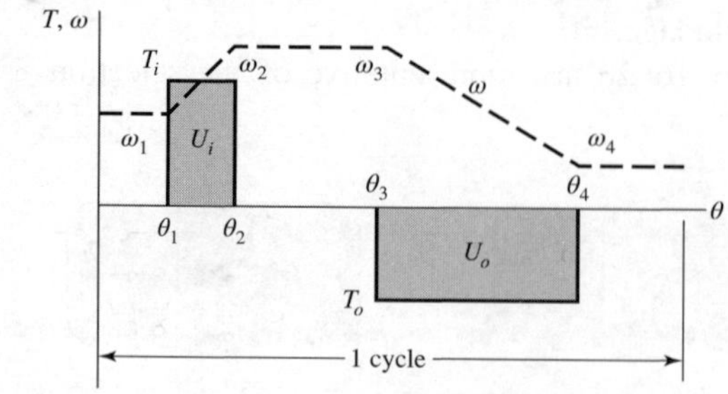

The work output of the flywheel is the area of the rectangle from θ_3 to θ_4, or

$$U_o = T_o(\theta_4 - \theta_3) \tag{d}$$

If U_o is greater than U_i, the load uses more energy than has been delivered to the flywheel and so ω_4 will be less than ω_1. If $U_o = U_i$, assuming no friction losses, ω_4 will be equal to ω_1 because the gains and losses are equal. And finally, ω_4 will be greater than ω_1 if $U_i > U_o$.

We can also write these relations in terms of kinetic energy. At $\theta = \theta_1$ the flywheel has a velocity of ω_1 rad/s, and so its kinetic energy is

$$E_1 = \frac{1}{2} I \omega_1^2 \tag{e}$$

At $\theta = \theta_2$ the velocity is ω_2, and so

$$E_2 = \frac{1}{2} I \omega_2^2 \tag{f}$$

Thus the change in kinetic energy is

$$E_2 - E_1 = \frac{1}{2} I\left(\omega_2^2 - \omega_1^2\right) \tag{16–61}$$

Many of the torque displacement functions encountered in practical engineering situations are so complicated that they must be integrated by numerical methods. Figure 16–28, for example, is a typical plot of the engine torque for one cycle of motion of a single-cylinder internal combustion engine. Since a part of the torque curve is negative, the flywheel must return part of the energy back to the engine. Integrating this curve from $\theta = 0$ to 4π and dividing the result by 4π yields the mean torque T_m available to drive a load during the cycle.

It is convenient to define a *coefficient of speed fluctuation* as

$$C_s = \frac{\omega_2 - \omega_1}{\omega} \tag{16–62}$$

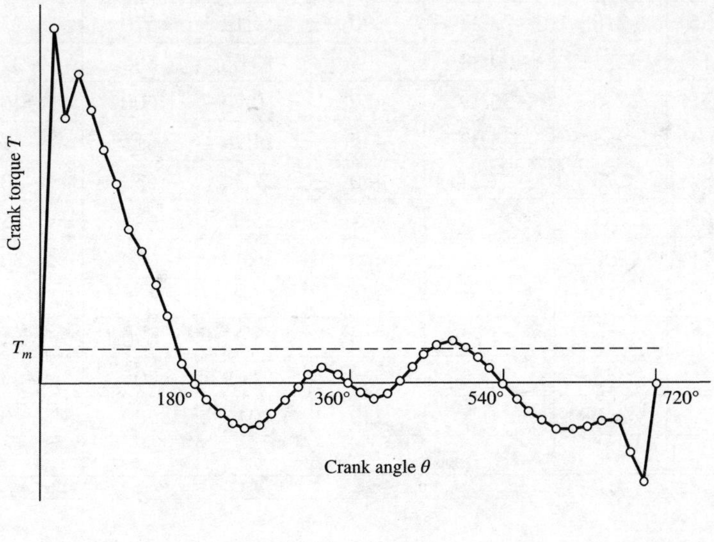

Figure 16–28

Relation between torque and crank angle for a one-cylinder, four-stroke–cycle internal combustion engine.

where ω is the nominal angular velocity, given by

$$\omega = \frac{\omega_2 + \omega_1}{2} \tag{16–63}$$

Equation (16–61) can be factored to give

$$E_2 - E_1 = \frac{I}{2}(\omega_2 - \omega_1)(\omega_2 + \omega_1)$$

Since $\omega_2 - \omega_1 = C_s\omega$ and $\omega_2 + \omega_1 = 2\omega$, we have

$$E_2 - E_1 = C_s I\omega^2 \tag{16–64}$$

Equation (16–64) can be used to obtain an appropriate flywheel inertia corresponding to the energy change $E_2 - E_1$.

EXAMPLE 16–6

Table 16–6 lists values of the torque used to plot Figure 16–28. The nominal speed of the engine is to be 250 rad/s.
(*a*) Integrate the torque-displacement function for one cycle and find the energy that can be delivered to a load during the cycle.
(*b*) Determine the mean torque T_m (see Figure 16–28).
(*c*) The greatest energy fluctuation is approximately between $\theta = 15°$ and $\theta = 150°$ on the torque diagram; see Figure 16–28 and note that $T_o = -T_m$. Using a coefficient of speed fluctuation $C_s = 0.1$, find a suitable value for the flywheel inertia.
(*d*) Find ω_2 and ω_1.

Solution

(*a*) Using $n = 48$ intervals of $\Delta\theta = 4\pi/48$, numerical integration of the data of Table 16–6 yields

Answer $$E = 380.5 \text{ J}$$

Table 16–6 Plotting Data for Figure 16–28

θ, deg	T, N · m	θ, deg	T, N · m	θ, deg	T, N · m	θ, deg	T, N · m
0	0	195	−12.09	375	−9.60	555	−12.09
15	316.36	210	−23.27	390	−14.12	570	−23.27
30	236.14	225	−29.38	405	−10.06	585	−32.99
45	274.55	240	−36.49	420	0.90	600	−40.11
60	244.05	255	−35.03	435	14.24	615	−41.92
75	207.89	270	−27.34	450	27.34	630	−40.90
90	179.65	285	−14.24	465	35.03	645	−35.25
105	136.71	300	−0.90	480	36.49	660	−30.73
120	120.44	315	10.06	495	31.64	675	−30.96
135	90.73	330	14.12	510	23.27	690	−61.92
150	60.11	345	9.60	525	12.09	705	−85.87
165	20.79	360	0	540	0	720	0
180	0						

This is the energy that can be delivered to the load.

Answer (*b*)

$$T_m = \frac{380.5}{4(\pi)} = 30.3 \text{ N} \cdot \text{m}$$

(*c*) The largest positive loop on the torque-displacement diagram occurs between $\theta = 0°$ and $\theta = 180°$. We select this loop as yielding the largest speed change. Subtracting 30.3 N · m from the values in Table 16–6 for this loop gives, respectively, −30.30, 286.06, 205.84, 244.25, 213.75, 177.59, 149.35, 106.41, 90.14, 60.43, 29.81, −9.51, and −30.3 N · m. Numerically integrating $T - T_m$ with respect to θ yields $E_2 - E_1 = 400$ N · m. We now solve Equation (16–64) for I. This gives

Answer

$$I = \frac{E_2 - E_1}{C_s\omega^2} = \frac{400}{0.1(250)^2} = 0.064 \text{ N} \cdot \text{m} \cdot \text{s}^2$$

(*d*) Equations (16–62) and (16–63) can be solved simultaneously for ω_2 and ω_1. Substituting appropriate values in these two equations yields

Answer

$$\omega_2 = \frac{\omega}{2}(2 + C_s) = \frac{250}{2}(2 + 0.1) = 262.5 \text{ rad/s}$$

Answer

$$\omega_1 = 2\omega - \omega_2 = 2(250) - 262.5 = 237.5 \text{ rad/s}$$

These two speeds occur at $\theta = 180°$ and $\theta = 0°$, respectively.

Punch-press torque demand often takes the form of a severe impulse and the running friction of the drive train. The motor overcomes the minor task of overcoming friction while attending to the major task of restoring the flywheel's angular speed. The situation can be idealized as shown in Figure 16–29. Neglecting the running friction, Euler's equation can be written as

$$T(\theta_1 - 0) = \frac{1}{2}I\left(\omega_1^2 - \omega_2^2\right) = E_2 - E_1$$

where the only significant inertia is that of the flywheel. Punch presses can have the motor and flywheel on one shaft, then, through a gear reduction, drive a slider-crank mechanism that carries the punching tool. The motor can be connected to the punch

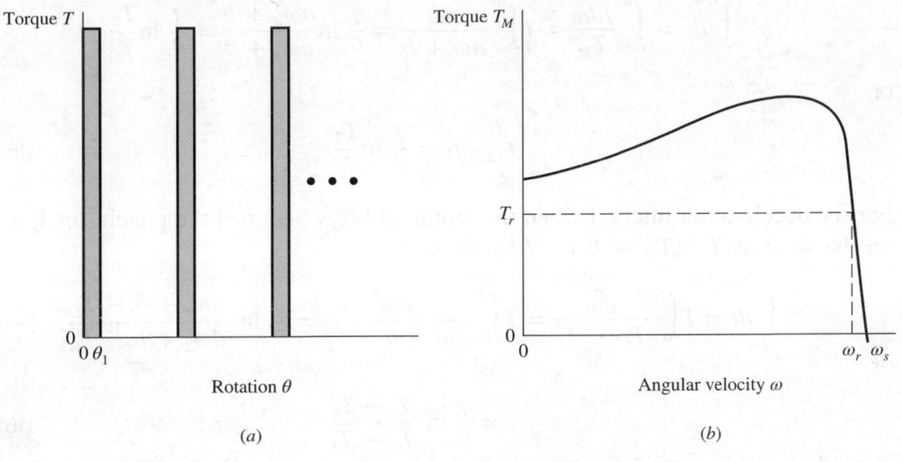

Figure 16–29

(*a*) Punch-press torque demand during punching. (*b*) Squirrel-cage electric motor torque-speed characteristic.

continuously, creating a punching rhythm, or it can be connected on command through a clutch that allows one punch and a disconnect. The motor and flywheel must be sized for the most demanding service, which is steady punching. The work done is given by

$$W = \int_{\theta_1}^{\theta_2} [T(\theta) - \bar{T}]d\theta = \frac{1}{2}I(\omega_{\max}^2 - \omega_{\min}^2)$$

This equation can be arranged to include the coefficient of speed fluctuation C_s as follows:

$$W = \frac{1}{2}I(\omega_{\max}^2 - \omega_{\min}^2) = \frac{I}{2}(\omega_{\max} - \omega_{\min})(\omega_{\max} + \omega_{\min})$$
$$= \frac{I}{2}(C_s\bar{\omega})(2\omega_0) = IC_s\bar{\omega}\omega_0$$

When the speed fluctuation is low, $\omega_0 \approx \bar{\omega}$, and

$$I = \frac{W}{C_s\bar{\omega}^2}$$

An induction motor has a linear torque characteristic $T_M = a\omega + b$ in the range of operation. The constants a and b can be found from the nameplate speed ω_r and the synchronous speed ω_s:

$$a = \frac{T_r - T_s}{\omega_r - \omega_s} = \frac{T_r}{\omega_r - \omega_s} = -\frac{T_r}{\omega_s - \omega_r}$$
$$b = \frac{T_r\omega_s - T_s\omega_r}{\omega_s - \omega_r} = \frac{T_r\omega_s}{\omega_s - \omega_r} \tag{16–65}$$

For example, a 2.2 kW three-phase squirrel-cage ac motor with a synchronous speed of 1200 rev/min is rated at 1125 rev/min and has a torque of $T_r = 1000(30)2.2\ /1125\pi = 18.7$ N · m. The rated angular velocity is $\omega_r = 2\pi n_r/60 = 2\pi(1125)/60 = 117.81$ rad/s, and the synchronous angular velocity $\omega_s = 2\pi(1200)/60 = 125.66$ rad/s. Thus $a = -18.7/(125.66 - 117.81) = -2.38$ N · m/rad, and $b = 299$ N · m, and we can express $T(\omega)$ as $a\omega + b$. During the interval from t_1 to t_2 the motor accelerates the flywheel according to $I\ddot{\theta} = T_M$(i.e., $Id\omega/dt = T_M$). Separating the equation $T_M = Id\omega/dt$ we have

$$\int_{t_1}^{t_2} dt = \int_{\omega_r}^{\omega_2} \frac{Id\omega}{T_M} = I\int_{\omega_r}^{\omega_2} \frac{d\omega}{a\omega + b} = \frac{I}{a}\ln\frac{a\omega_2 + b}{a\omega_r + b} = \frac{I}{a}\ln\frac{T_2}{T_r}$$

or

$$t_2 - t_1 = \frac{I}{a}\ln\frac{T_2}{T_r} \tag{16–66}$$

For the deceleration interval when the motor and flywheel feel the punch (load) torque on the shaft as T_L, $(T_M - T_L) = Id\omega/dt$, or

$$\int_0^{t_1} dt = I\int_{\omega_2}^{\omega_r} \frac{d\omega}{T_M - T_L} = I\int_{\omega_2}^{\omega_r} \frac{d\omega}{a\omega + b - T_L} = \frac{I}{a}\ln\frac{a\omega_r + b - T_L}{a\omega_2 + b - T_L}$$

or

$$t_1 = \frac{I}{a}\ln\frac{T_r - T_L}{T_2 - T_L} \tag{16–67}$$

We can divide Equation (16–66) by Equation (16–67) to obtain

$$\frac{T_2}{T_r} = \left(\frac{T_L - T_r}{T_L - T_2}\right)^{(t_2-t_1)/t_1} \tag{16–68}$$

Equation (16–68) can be solved for T_2 numerically. Having T_2 the flywheel inertia is, from Equation (16–66),

$$I = \frac{a(t_2 - t_1)}{\ln (T_2/T_r)} \tag{16–69}$$

It is important that a be in units of N · m · s/rad so that I has proper units. The constant a should not be in N · m per rev/min or N · m per rev/s.

PROBLEMS

16–1 The figure shows an internal rim-type brake having an inside rim diameter of 300 mm and a dimension R = 125 mm. The shoes have a face width of 40 mm and are both actuated by a force of 2.2 kN. The drum rotates clockwise. The mean coefficient of friction is 0.28.

(*a*) Find the maximum pressure and indicate the shoe on which it occurs.

(*b*) Estimate the braking torque effected by each shoe, and find the total braking torque.

(*c*) Estimate the resulting hinge-pin reactions.

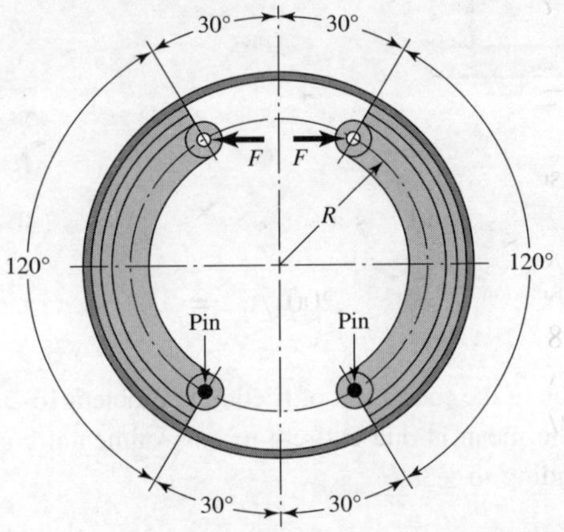

Problem 16–1

16–2 For the brake in Problem 16–1, consider the pin and actuator locations to be the same. However, instead of 120°, let the friction surface of the brake shoes be 90° and centrally located. Find the maximum pressure and the total braking torque.

16–3 In the figure for Problem 16–1, the inside rim diameter is 280 mm and the dimension R is 90 mm. The shoes have a face width of 32 mm. Find the braking torque and the maximum pressure for each shoe if the actuating force is 1 kN, the drum rotation is counterclockwise, and f = 0.30.

16–4 The figure shows a 400-mm-diameter brake drum with four internally expanding shoes. Each of the hinge pins A and B supports a pair of shoes. The actuating mechanism is to be arranged to produce the same force F on each shoe. The face width of the shoes is 75 mm. The material used permits a coefficient of friction of 0.24 and a maximum pressure of 1000 kPa.

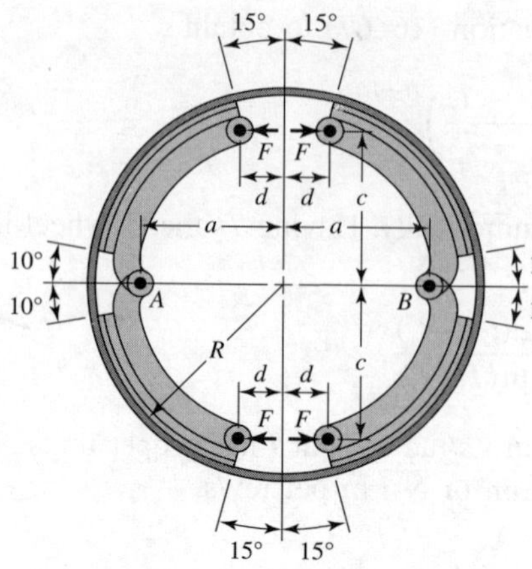

Problem 16–4

The dimensions in millimeters are $a = 150$, $c = 165$, $R = 200$, and $d = 50$.

(*a*) Determine the maximum actuating force.

(*b*) Estimate the brake capacity.

(*c*) Noting that rotation may be in either direction, estimate the hinge-pin reactions.

16–5 The block-type hand brake shown in the figure has a face width of 30 mm and a mean coefficient of friction of 0.25. For an estimated actuating force of 400 N, find the maximum pressure on the shoe and find the braking torque.

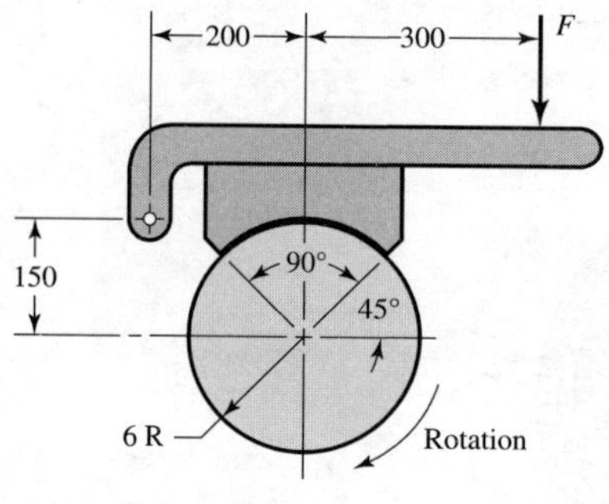

Problem 16–5

Dimensions in millimeters.

16–6 Suppose the standard deviation of the coefficient of friction in Problem 16–5 is $\hat{\sigma}_f = 0.025$, where the deviation from the mean is due entirely to environmental conditions. Find the brake torques corresponding to $\pm 3\hat{\sigma}_f$.

16–7 The brake shown in the figure has a coefficient of friction of 0.30, a face width of 50 mm, and a limiting shoe lining pressure of 1000 kPa. Find the limiting actuating force F and the torque capacity.

16–8 Refer to the symmetrical pivoted external brake shoe of Figure 16–12 and Equation (16–15). Suppose the pressure distribution was uniform, that is, the pressure p is independent of θ. What would the pivot distance a' be? If $\theta_1 = \theta_2 = 60°$, compare a with a'.

16–9 The shoes on the brake depicted in the figure subtend a 90° arc on the drum of this external pivoted-shoe brake. The actuation force P is applied to the lever. The rotation direction of the drum is counterclockwise, and the coefficient of friction is 0.30.

(*a*) What should the dimension e be, in order to eliminate frictional moments on each shoe?

(*b*) Draw the free-body diagrams of the handle lever and both shoe levers, with forces expressed in terms of the actuation force P.

(*c*) Does the direction of rotation of the drum affect the braking torque?

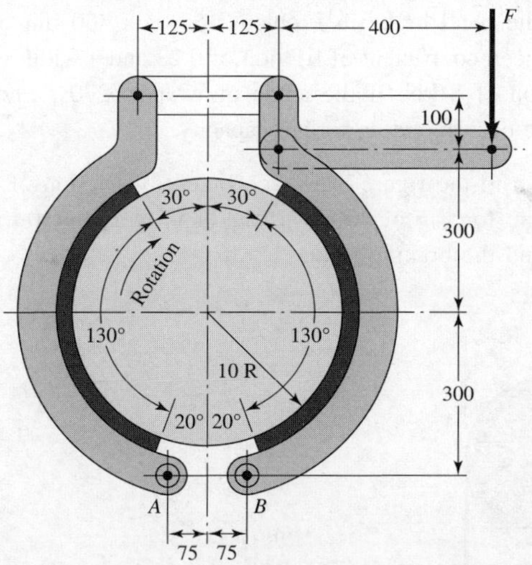

Problem 16–7
Dimensions in millimeters.

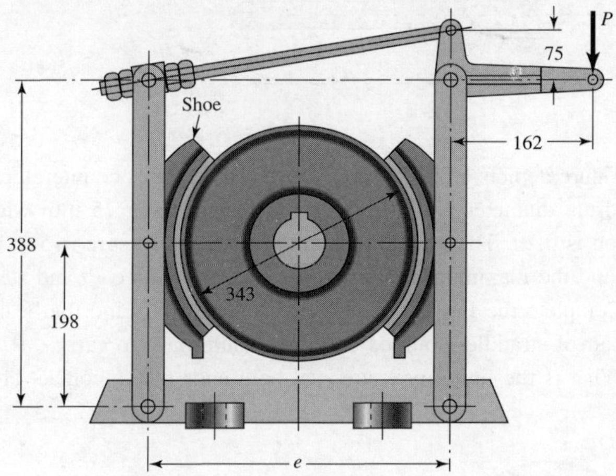

Problem 16–9
Dimensions in millimeters.

16–10 Problem 16–9 is preliminary to analyzing the brake. A rigid molded non-asbestos lining is used dry in the brake of Problem 16–9 on a cast-iron drum. The shoes are 6 in wide and subtend a 90° arc. Conservatively estimate the maximum allowable actuation force and the braking torque.

16–11 The maximum band interface pressure on the brake shown in the figure is 620 kPa. Use a 350 mm-diameter drum, a band width of 25 mm, a coefficient of friction of 0.30, and an angle-of-wrap of 270°. Find the maximum band tensions and the torque capacity.

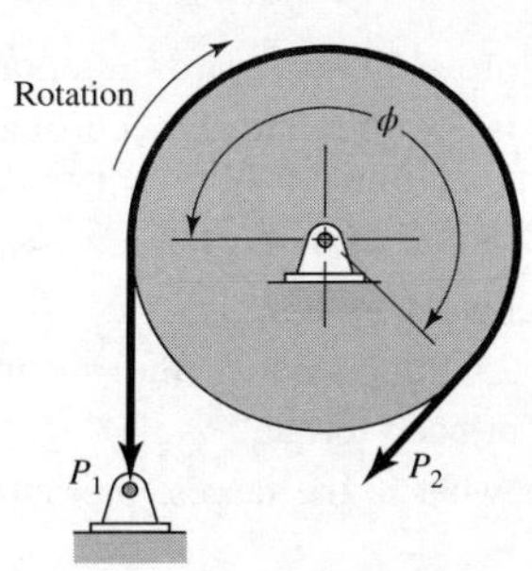

Problem 16–11

16–12 The drum for the band brake in Problem 16–11 is 300 mm in diameter. The band selected has a mean coefficient of friction of 0.28 and a width of 80 mm. It can safely support a tension of 8 kN. If the angle of wrap is 270°, find the maximum lining pressure and the corresponding torque capacity.

16–13 The brake shown in the figure has a coefficient of friction of 0.30 and is to operate using a maximum force F of 400 N. If the band width is 50 mm, find the maximum band tensions and the braking torque.

Problem 16–13

Dimensions in millimeters.

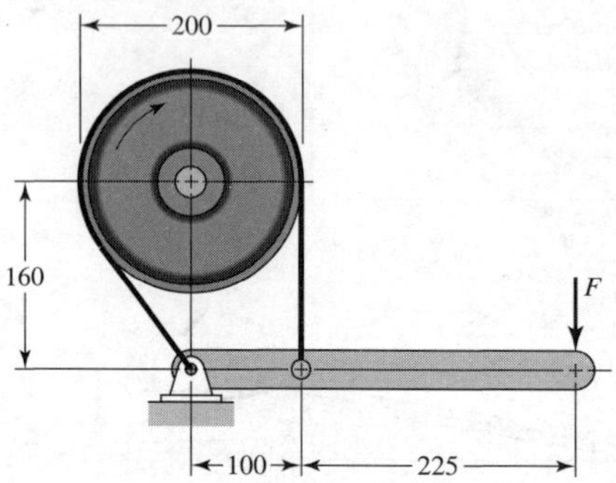

16–14 The figure depicts a band brake whose drum rotates counterclockwise at 200 rev/min. The drum diameter is 400 mm and the band lining 75 mm wide. The coefficient of friction is 0.20. The maximum lining interface pressure is 500 kPa.

(*a*) Find the maximum brake torque, necessary force P, and steady-state power.

(*b*) Complete the free-body diagram of the drum. Find the bearing radial load that a pair of straddle-mounted bearings would have to carry.

(*c*) What is the lining pressure p at both ends of the contact arc?

Problem 16–14

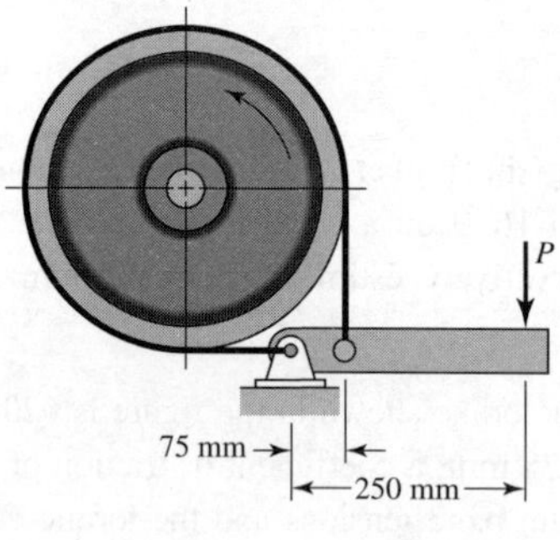

16–15 The figure shows a band brake designed to prevent "backward" rotation of the shaft. The angle of wrap is 270°, the band width is 53 mm, and the coefficient of friction is 0.20. The torque to be resisted by the brake is 200 N · m. The diameter of the pulley is 210 mm.

(*a*) What dimension c_1 will just prevent backward motion?

(*b*) If the rocker was designed with $c_1 = 25$ mm, what is the maximum pressure between the band and drum at 200 N · m back torque?

(*c*) If the back-torque demand is 10 N · m, what is the largest pressure between the band and drum?

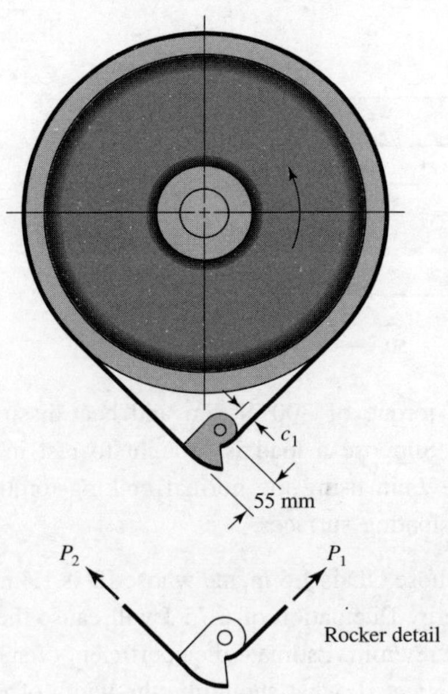

Problem 16–15

16–16 A plate clutch has a single pair of mating friction surfaces 250-mm OD by 175-mm ID. The mean value of the coefficient of friction is 0.30, and the actuating force is 4 kN.
(*a*) Find the maximum pressure and the torque capacity using the uniform-wear model.
(*b*) Find the maximum pressure and the torque capacity using the uniform-pressure model.

16–17 A hydraulically operated multidisk plate clutch has an effective disk outer diameter of 165 mm and an inner diameter of 100 mm. The coefficient of friction is 0.24, and the limiting pressure is 830 kPa. There are six planes of sliding present.
(*a*) Using the uniform wear model, estimate the limiting axial force F and the torque T.
(*b*) Let the inner diameter of the friction pairs d be a variable. Complete the following table:

d, mm	50	75	100	125	150
T, N · m					

(*c*) What does the table show?

16–18 Look again at Problem 16–17.
(*a*) Show how the optimal diameter d^* is related to the outside diameter D.
(*b*) What is the optimal inner diameter?
(*c*) What does the tabulation show about maxima?
(*d*) Common proportions for such plate clutches lie in the range $0.45 \leq d/D \leq 0.80$. Is the result in part (*a*) useful?

16–19 A cone clutch has $D = 300$ mm, $d = 275$ mm, a cone length of 55 mm, and a coefficient of friction of 0.28. A torque of 200 N · m is to be transmitted. For this requirement, estimate the actuating force and maximum pressure by both models.

16–20 Show that for the caliper brake the $T/(fFD)$ versus d/D plots are the same as Equations (*b*) and (*c*) of Section 16–5.

16–21 A two-jaw clutch has the dimensions shown in the figure and is made of ductile steel. The clutch has been designed to transmit 2 kW at 500 rev/min. Find the bearing and shear stresses in the key and the jaws.

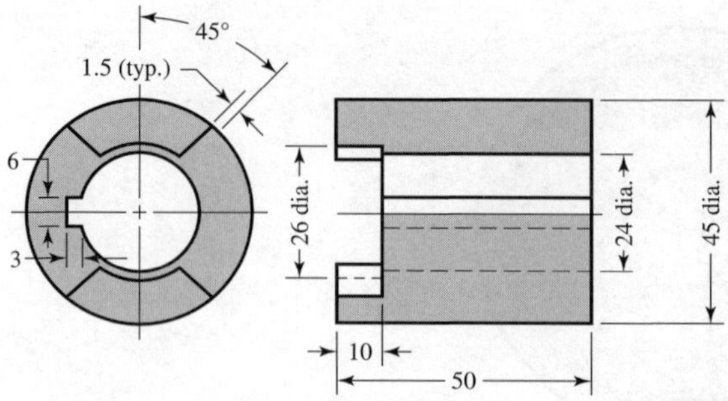

Problem 16–21

Dimensions in millimeters.

16–22 A brake has a normal braking torque of 300 N · m and heat-dissipating cast-iron surfaces whose mass is 20 kg. Suppose a load is brought to rest in 8.0 s from an initial angular speed of 1600 rev/min using the normal braking torque; estimate the temperature rise of the heat-dissipating surfaces.

16–23 A cast-iron flywheel has a rim whose OD is 1.5 m and whose ID is 1.4 m. The flywheel weight is to be such that an energy fluctuation of 6.75 J will cause the angular speed to vary no more than 240 to 260 rev/min. Estimate the coefficient of speed fluctuation. If the weight of the spokes is neglected, what should be the width of the rim?

16–24 A single-geared blanking press has a stroke of 200 mm and a rated capacity of 320 kN. A cam-driven ram is assumed to be capable of delivering the full press load at constant force during the last 15 percent of a constant-velocity stroke. The camshaft has an average speed of 90 rev/min and is geared to the flywheel shaft at a 6:1 ratio. The total work done is to include an allowance of 20 percent for friction.

(*a*) Estimate the maximum energy fluctuation.

(*b*) Find the rim weight for an effective diameter of 1.2 m and a coefficient of speed fluctuation of 0.10.

16–25 Using the data of Table 16–6, find the mean output torque and flywheel inertia required for a three-cylinder in-line engine corresponding to a nominal speed of 2400 rev/min. Use $C_s = 0.30$.

16–26 When a motor armature inertia, a pinion inertia, and a motor torque reside on a motor shaft, and a gear inertia, a load inertia, and a load torque exist on a second shaft, it is useful to reflect all the torques and inertias to one shaft, say, the armature shaft. We need some rules to make such reflection easy. Consider the pinion and gear as disks of pitch radius.

- A torque on a second shaft is reflected to the motor shaft as the load torque divided by the negative of the stepdown ratio.
- An inertia on a second shaft is reflected to the motor shaft as its inertia divided by the stepdown ratio squared.
- The inertia of a disk gear on a second shaft in mesh with a disk pinion on the motor shaft is reflected to the pinion shaft as the *pinion* inertia multiplied by the stepdown ratio squared.

(*a*) Verify the three rules.

(*b*) Using the rules, reduce the two-shaft system in the figure to a motor-shaft shish kebab equivalent. Correctly done, the dynamic response of the shish kebab and the real system are identical.

(*c*) For a stepdown ratio of $n = 10$ compare the shish kebab inertias.

Problem 16–26

Dimensions in millimeters.

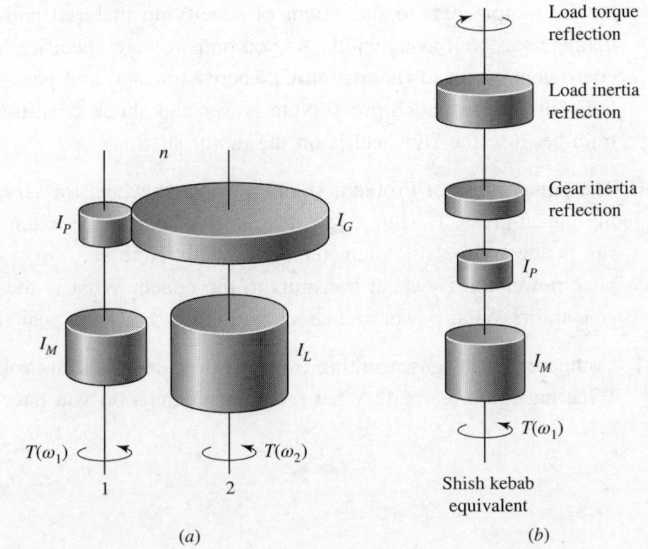

16–27 Apply the rules of Problem 16–26 to the three-shaft system shown in the figure to create a motor shaft shish kebab.

(*a*) Show that the equivalent inertia I_e is given by

$$I_e = I_M + I_P + n^2 I_P + \frac{I_P}{n^2} + \frac{m^2 I_P}{n^2} + \frac{I_L}{m^2 n^2}$$

(*b*) If the overall gear reduction R is a constant nm, show that the equivalent inertia becomes

$$I_e = I_M + I_P + n^2 I_P + \frac{I_P}{n^2} + \frac{R^2 I_P}{n^4} + \frac{I_L}{R^2}$$

(*c*) If the problem is to minimize the gear-train inertia, find the ratios n and m for the values of $I_P = 1$, $I_M = 10$, $I_L = 100$, and $R = 10$.

Problem 16–27

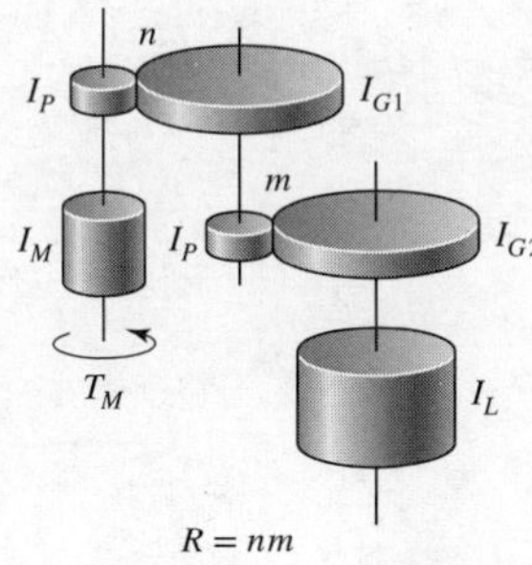

16–28 For the conditions of Problem 16–27, make a plot of the equivalent inertia I_e as ordinate and the stepdown ratio n as abscissa in the range $1 \le n \le 10$. How does the minimum inertia compare to the single-step inertia?

16–29 A punch-press geared 10:1 is to make six punches per minute under circumstances where the torque on the crankshaft is 1800 N · m for $\frac{1}{2}$ s. The motor's nameplate reads 2.5 kW at 1125 rev/min for continuous duty. Design a satisfactory flywheel for use on the motor shaft to the extent of specifying material and rim inside and outside diameters as well as its width. As you prepare your specifications, note $\omega_{\max}$, $\omega_{\min}$, the coefficient of speed fluctuation C_s, energy transfer, and peak power that the flywheel transmits to the punch-press. Note power and shock conditions imposed on the gear train because the flywheel is on the motor shaft.

16–30 The punch-press of Problem 16–29 needs a flywheel for service on the crankshaft of the punch-press. Design a satisfactory flywheel to the extent of specifying material, rim inside and outside diameters, and width. Note $\omega_{\max}$, $\omega_{\min}$, C_s, energy transfer, and peak power the flywheel transmits to the punch. What is the peak power seen in the gear train? What power and shock conditions must the gear train transmit?

16–31 Compare the designs resulting from the tasks assigned in Problems 16–29 and 16–30. What have you learned? What recommendations do you have?

17 Flexible Mechanical Elements

Stason4ik/Shutterstock

Chapter Outline

Belts, ropes, chains, and other similar elastic or flexible machine elements are used in conveying systems and in the transmission of power over comparatively long distances. It often happens that these elements can be used as a replacement for gears, shafts, bearings, and other relatively rigid power-transmission devices. In many cases their use simplifies the design of a machine and substantially reduces the cost.

In addition, since these elements are elastic and usually quite long, they play an important part in absorbing shock loads and in damping out and isolating the effects of vibration. This is an important advantage as far as machine life is concerned.

Most flexible elements do not have an infinite life. When they are used, it is important to establish an inspection schedule to guard against wear, aging, and loss of elasticity. The elements should be replaced at the first sign of deterioration.

17–1 Belts

The four principal types of belts are shown, with some of their characteristics, in Table 17–1. *Crowned pulleys* are used for flat belts, and *grooved pulleys,* or *sheaves,* for round and V belts. Timing belts require *toothed wheels,* or *sprockets*. In all cases, the pulley axes must be separated by a certain minimum distance, depending upon the belt type and size, to operate properly. Other characteristics of belts are:

- They may be used for long center distances.
- Except for timing belts, there is some slip and creep, and so the angular-velocity ratio between the driving and driven shafts is neither constant nor exactly equal to the ratio of the pulley diameters.
- In some cases an idler or tension pulley can be used to avoid adjustments in center distance that are ordinarily necessitated by age or the installation of new belts.

Figure 17–1 illustrates the geometry of open and crossed flat-belt drives. For a flat belt with this drive the belt tension is such that the sag or droop is visible in Figure 17–2*a*, when the belt is running. Although the top is preferred for the loose side of the belt, for other belt types either the top or the bottom may be used, because their installed tension is usually greater.

Table 17–1 Characteristics of Some Common Belt Types

Figures are cross sections except for the timing belt, which is a side view.

Belt Type	Figure	Joint	Size Range	Center Distance
Flat	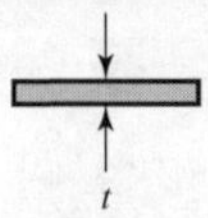	Yes	$t = 0.75$ to 5 mm	No upper limit
Round	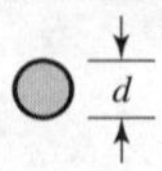	Yes	$d = 3$ to 19 mm	No upper limit
V	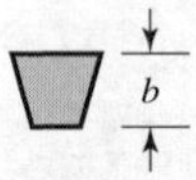	None	$b = 8$ to 19 mm	Limited
Timing	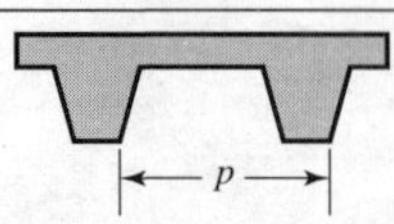	None	$p = 2$ mm and up	Limited

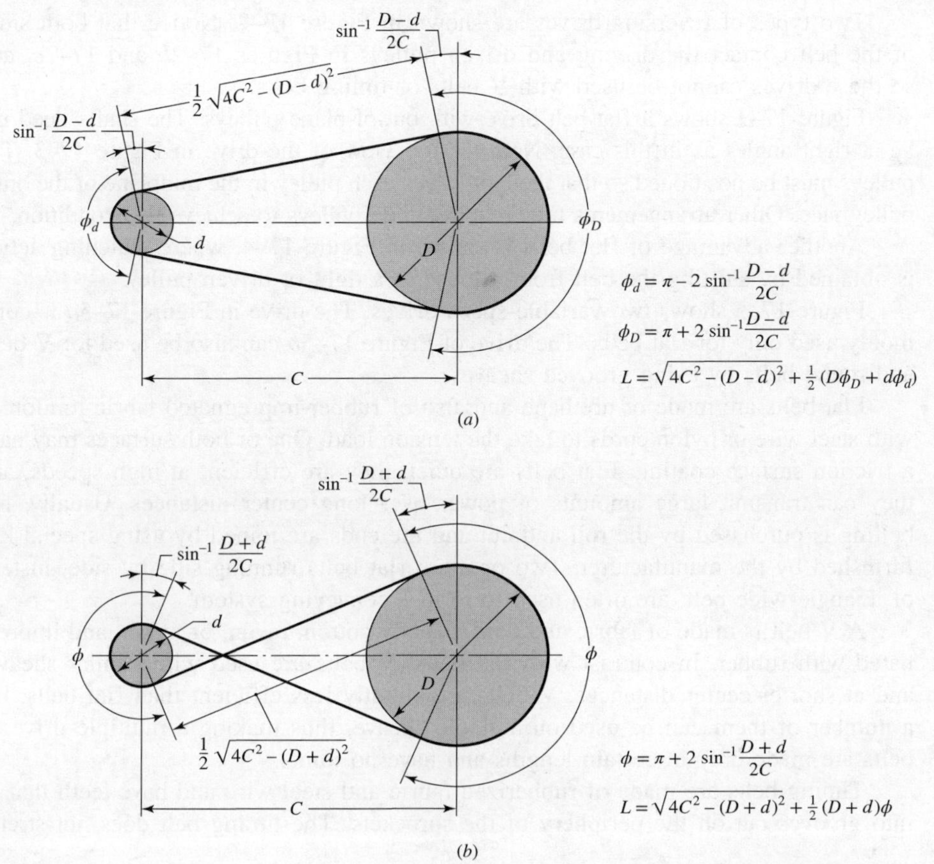

Figure 17–1

Flat-belt geometry. (*a*) Open belt. (*b*) Crossed belt.

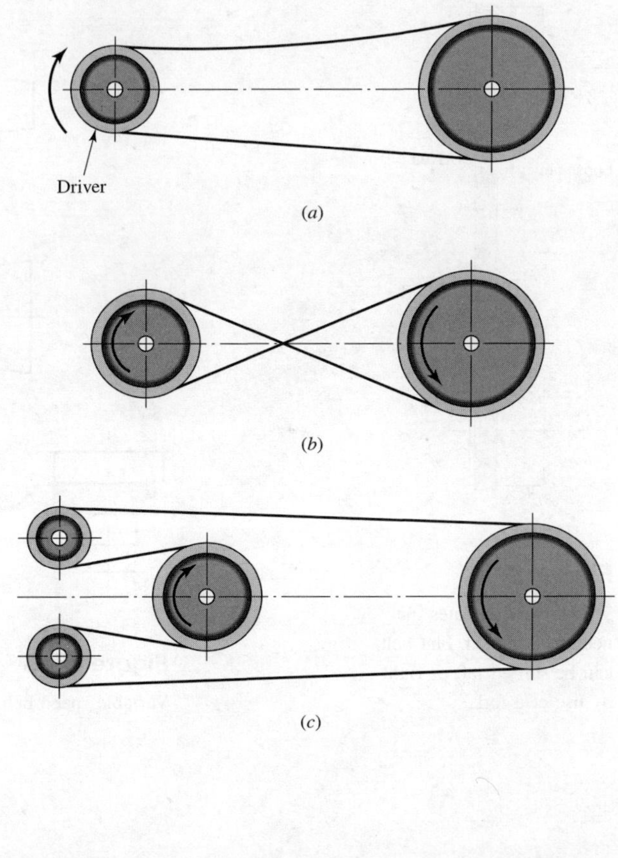

Figure 17–2

Nonreversing and reversing belt drives. (*a*) Nonreversing open belt. (*b*) Reversing crossed belt. Crossed belts must be separated to prevent rubbing if high-friction materials are used. (*c*) Reversing open-belt drive.

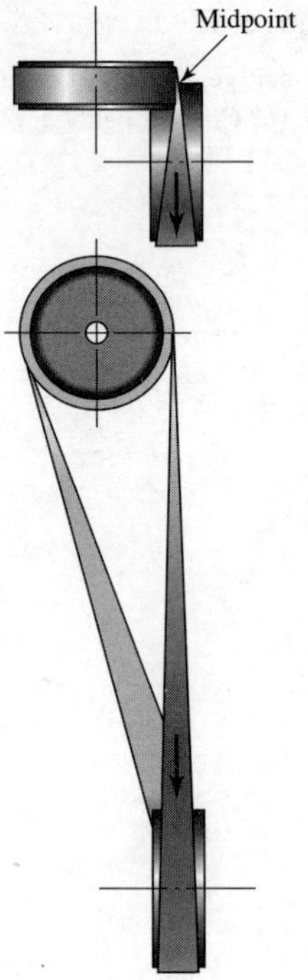

Figure 17–3

Quarter-twist belt drive; an idler guide pulley must be used if motion is to be in both directions.

Two types of reversing drives are shown in Figure 17–2. Notice that both sides of the belt contact the driving and driven pulleys in Figures 17–2*b* and 17–2*c*, and so these drives cannot be used with V belts or timing belts.

Figure 17–3 shows a flat-belt drive with out-of-plane pulleys. The shafts need not be at right angles as in this case. Note the top view of the drive in Figure 17–3. The pulleys must be positioned so that the belt leaves each pulley in the midplane of the other pulley face. Other arrangements may require guide pulleys to achieve this condition.

Another advantage of flat belts is shown in Figure 17–4, where clutching action is obtained by shifting the belt from a loose to a tight or driven pulley.

Figure 17–5 shows two variable-speed drives. The drive in Figure 17–5*a* is commonly used only for flat belts. The drive of Figure 17–5*b* can also be used for V belts and round belts by using grooved sheaves.

Flat belts are made of urethane and also of rubber-impregnated fabric reinforced with steel wire or nylon cords to take the tension load. One or both surfaces may have a friction surface coating. Flat belts are quiet, they are efficient at high speeds, and they can transmit large amounts of power over long center distances. Usually, flat belting is purchased by the roll and cut and the ends are joined by using special kits furnished by the manufacturer. Two or more flat belts running side by side, instead of a single wide belt, are often used to form a conveying system.

A V belt is made of fabric and cord, usually cotton, rayon, or nylon, and impregnated with rubber. In contrast with flat belts, V belts are used with similar sheaves and at shorter center distances. V belts are slightly less efficient than flat belts, but a number of them can be used on a single sheave, thus making a multiple drive. V belts are made only in certain lengths and have no joints.

Timing belts are made of rubberized fabric and steel wire and have teeth that fit into grooves cut on the periphery of the sprockets. The timing belt does not stretch

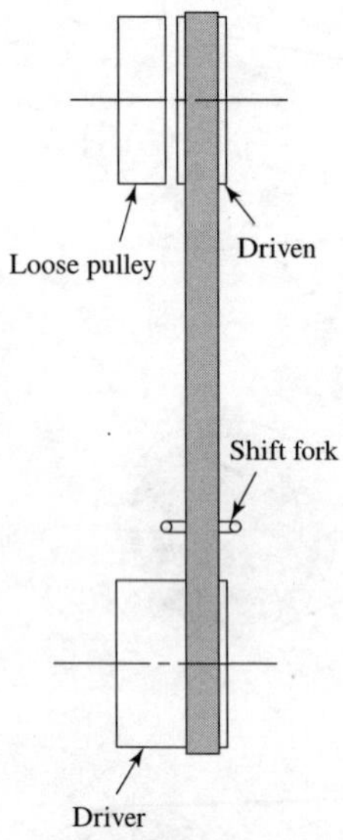

Figure 17–4

This drive eliminates the need for a clutch. Flat belt can be shifted left or right by use of a fork.

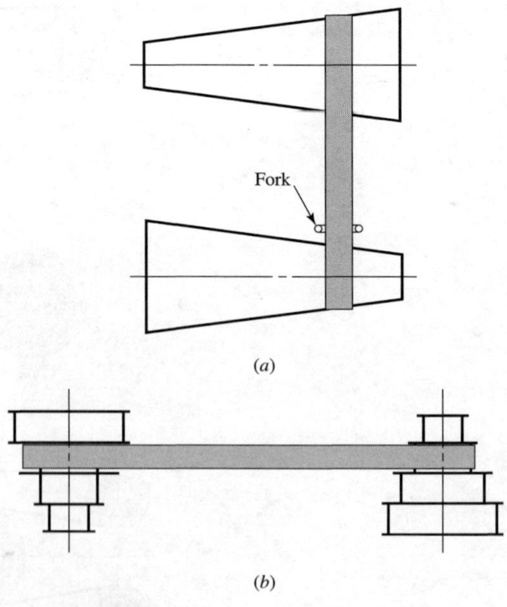

Figure 17–5

Variable-speed belt drives.

or slip and consequently transmits power at a constant angular-velocity ratio. The fact that the belt is toothed provides several advantages over ordinary belting. One of these is that no initial tension is necessary, so that fixed-center drives may be used. Another is the elimination of the restriction on speeds; the teeth make it possible to run at nearly any speed, slow or fast. Disadvantages are the first cost of the belt, the necessity of grooving the sprockets, and the attendant dynamic fluctuations caused at the belt-tooth meshing frequency.

17–2 Flat- and Round-Belt Drives

Modern flat-belt drives consist of a strong elastic core surrounded by an elastomer. These drives have distinct advantages over gear drives or V-belt drives. A flat-belt drive has an efficiency of about 98 percent, which is about the same as for a gear drive. On the other hand, the efficiency of a V-belt drive ranges from about 70 to 96 percent.[1] Flat-belt drives produce very little noise and absorb more torsional vibration from the system than either V-belt or gear drives.

When an open-belt drive (Figure 17–1*a*) is used, the wrap angles are found to be

$$\begin{aligned} \phi_d &= \pi - 2\sin^{-1}\frac{D-d}{2C} \\ \phi_D &= \pi + 2\sin^{-1}\frac{D-d}{2C} \end{aligned} \tag{17–1}$$

where D = diameter of large pulley
d = diameter of small pulley
C = center distance
ϕ = wrap angle

The length of the belt is found by summing the two arc lengths with twice the distance between the beginning and end of contact. The result is

$$L = [4C^2 - (D-d)^2]^{1/2} + \frac{1}{2}(D\phi_D + d\phi_d) \tag{17–2}$$

A similar set of equations can be derived for the crossed belt of Figure 17–2*b*. For this belt, the angle of wrap is the same for both pulleys and is

$$\phi = \pi + 2\sin^{-1}\frac{D+d}{2C} \tag{17–3}$$

The belt length for crossed belts is found to be

$$L = [4C^2 - (D+d)^2]^{1/2} + \frac{1}{2}(D+d)\phi \tag{17–4}$$

Firbank[2] explains flat-belt-drive theory in the following way. A change in belt tension due to friction forces between the belt and pulley will cause the belt to elongate or contract and move relative to the surface of the pulley. This motion is caused by *elastic creep* and is associated with sliding friction as opposed to static friction. The action at the driving pulley, through that portion of the angle of wrap that is actually transmitting power, is such that the belt moves more slowly than the surface

[1]A. W. Wallin, "Efficiency of Synchronous Belts and V-Belts," *Proc. Nat. Conf. Power Transmission*, vol. 5, Illinois Institute of Technology, Chicago, Nov. 7–9, 1978, pp. 265–271.

[2]T. C. Firbank, *Mechanics of the Flat Belt Drive*, ASME paper no. 72-PTG-21.

speed of the pulley because of the elastic creep. The angle of wrap is made up of the *effective arc,* through which power is transmitted, and the *idle arc*. For the driving pulley the belt first contacts the pulley with a *tight-side tension* F_1 and a velocity V_1, which is the same as the surface velocity of the pulley. The belt then passes through the idle arc with no change in F_1 or V_1. Then creep or sliding contact begins, and the belt tension changes in accordance with the friction forces. At the end of the effective arc the belt leaves the pulley with a *loose-side tension* F_2 and a reduced speed V_2.

Firbank has used this theory to express the mechanics of flat-belt drives in mathematical form and has verified the results by experiment. His observations include the finding that substantially more power is transmitted by static friction than sliding friction. He also found that the coefficient of friction for a belt having a nylon core and leather surface was typically 0.7, but that it could be raised to 0.9 by employing special surface finishes.

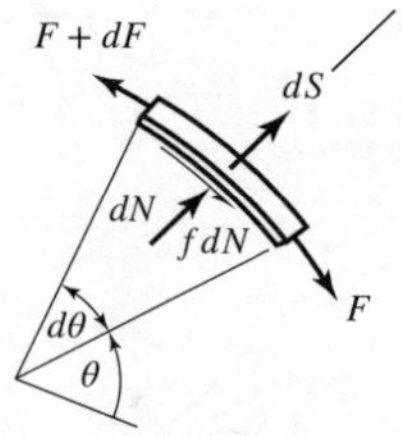

Figure 17–6

Free body of an infinitesimal element of a flat belt in contact with a pulley.

Our model will assume that the friction force on the belt is proportional to the normal pressure along the arc of contact. We seek first a relationship between the tight side tension and slack side tension, similar to that of band brakes but incorporating the consequences of movement, that is, centrifugal tension in the belt. In Figure 17–6 we see a free body of a small segment of the belt. The differential force dS is due to centrifugal force, dN is the normal force between the belt and pulley, and $f\,dN$ is the shearing traction due to friction at the point of slip. The belt width is b and the thickness is t. The belt mass per unit length is m. The centrifugal force dS can be expressed as

$$dS = (mr\,d\theta)r\omega^2 = mr^2\omega^2\,d\theta = mV^2\,d\theta = F_c\,d\theta \tag{a}$$

where V is the belt speed and F_c is the hoop tension due to the centrifugal force. Summing forces radially, with $\sin(d\theta/2) \approx d\theta/2$, gives

$$\sum F_r = -(F + dF)\frac{d\theta}{2} - F\frac{d\theta}{2} + dN + dS = 0$$

Ignoring the higher-order term, we have

$$dN = F\,d\theta - dS \tag{b}$$

Summing forces tangentially gives

$$\sum F_t = -f dN - F + (F + dF) = 0$$

from which, incorporating Equations (*a*) and (*b*), we obtain

$$dF = f dN = fF\,d\theta - f dS = fF\,d\theta - fmr^2\omega^2\,d\theta$$

or

$$\frac{dF}{d\theta} - fF = -fmr^2\omega^2 \tag{c}$$

The solution to this nonhomogeneous first-order linear differential equation is

$$F = A\exp(f\theta) + mr^2\omega^2 \tag{d}$$

where A is an arbitrary constant. As shown in Figure 17–7, θ starts at the loose side, and the boundary condition that F at $\theta = 0$ equals F_2 gives $A = F_2 - mr^2\omega^2$. The solution is

$$F = (F_2 - mr^2\omega^2)\exp(f\theta) + mr^2\omega^2 \tag{17–5}$$

At the end of the angle of wrap ϕ, the tight side,

$$F|_{\theta=\phi} = F_1 = (F_2 - mr^2\omega^2)\exp(f\phi) + mr^2\omega^2 \tag{17–6}$$

Now we can write

$$\frac{F_1 - mr^2\omega^2}{F_2 - mr^2\omega^2} = \frac{F_1 - F_c}{F_2 - F_c} = \exp(f\phi) \tag{17–7}$$

where, from Equation (*a*), $F_c = mr^2\omega^2$. Equation (17–7) is referred to as *the belting equation* and can be written as

$$F_1 - F_2 = (F_1 - F_c)\,\frac{\exp(f\phi) - 1}{\exp(f\phi)} \tag{17–8}$$

Now F_c is found as follows: with n being the rotational speed, in rev/min, of the pulley of diameter d, in millimeters, the belt speed is

$$V = \pi\, dn/1000 \text{ m/min}$$

The weight w of a meter of belt is given in terms of the density ρ in kg/m^3 as $w = \rho bt$ kg/m where b and t are in meters. F_c is written as

$$F_c = w\left(\frac{V}{60}\right)^2 \tag{e}$$

Figure 17–7 shows a free body diagram of a pulley and part of the belt. The tight side tension F_1 and the loose side tension F_2 have the following additive components:

$$F_1 = F_i + F_c + \Delta F/2 = F_i + F_c + T/d \tag{f}$$

$$F_2 = F_i + F_c - \Delta F/2 = F_i + F_c - T/d \tag{g}$$

where F_i = initial tension
F_c = hoop tension due to centrifugal force
$\Delta F/2$ = tension due to the transmitted torque T
d = diameter of the pulley

The difference between F_1 and F_2 is related to the pulley torque. Subtracting Equation (*g*) from Equation (*f*) gives

$$F_1 - F_2 = \frac{2T}{d} \tag{h}$$

Adding Equations (*f*) and (*g*) gives

$$F_1 + F_2 = 2F_i + 2F_c$$

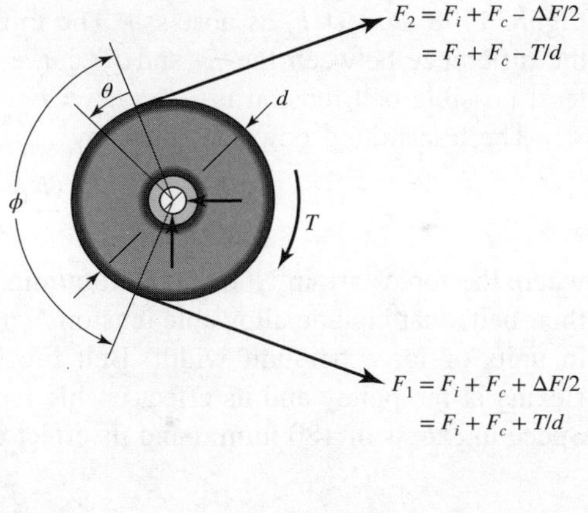

Figure 17–7

Forces and torques on a pulley.

from which

$$F_i = \frac{F_1 + F_2}{2} - F_c \tag{i}$$

Dividing Equation (*i*) by Equation (*h*), manipulating, and using Equation (17–7) gives

$$\frac{F_i}{T/d} = \frac{(F_1 + F_2)/2 - F_c}{(F_1 - F_2)/2} = \frac{F_1 + F_2 - 2F_c}{F_1 - F_2} = \frac{(F_1 - F_c) + (F_2 - F_c)}{(F_1 - F_c) - (F_2 - F_c)}$$

$$= \frac{(F_1 - F_c)/(F_2 - F_c) + 1}{(F_1 - F_c)/(F_2 - F_c) - 1} = \frac{\exp(f\phi) + 1}{\exp(f\phi) - 1}$$

from which

$$F_i = \frac{T}{d}\frac{\exp(f\phi) + 1}{\exp(f\phi) - 1} \tag{17–9}$$

Equation (17–9) give us a fundamental insight into flat belting. If F_i equals zero, then T equals zero: no initial tension, no torque transmitted. The torque is in proportion to the initial tension. This means that if there is to be a satisfactory flat-belt drive, the initial tension must be (1) provided, (2) sustained, (3) in the proper amount, and (4) maintained by routine inspection.

From Equation (*f*), incorporating Equation (17–9) gives

$$F_1 = F_i + F_c + \frac{T}{d} = F_c + F_i + F_i\frac{\exp(f\phi) - 1}{\exp(f\phi) + 1}$$

$$= F_c + \frac{F_i[\exp(f\phi) + 1] + F_i[\exp(f\phi) - 1]}{\exp(f\phi) + 1}$$

$$F_1 = F_c + F_i\frac{2\exp(f\phi)}{\exp(f\phi) + 1} \tag{17–10}$$

From Equation (*g*), incorporating Equation (17–9) gives

$$F_2 = F_i + F_c - \frac{T}{d} = F_c + F_i - F_i\frac{\exp(f\phi) - 1}{\exp(f\phi) + 1}$$

$$= F_c + \frac{F_i[\exp(f\phi) + 1] - F_i[\exp(f\phi) - 1]}{\exp(f\phi) + 1}$$

$$F_2 = F_c + F_i\frac{2}{\exp(f\phi) + 1} \tag{17–11}$$

Equation (17–7) is the belting equation, but Equations (17–9), (17–10), and (17–11) reveal how belting works. We plot Equations (17–10) and (17–11) as shown in Figure 17–8 against F_i as abscissa. The initial tension needs to be sufficient so that the difference between the F_1 and F_2 curve is $2T/d$. With no torque transmitted, the least possible belt tension is $F_1 = F_2 = F_c$.

The transmitted power is given by

$$H = \frac{(F_1 - F_2)V}{60} \tag{j}$$

where the forces are in N and V is in m/min. Manufacturers provide specifications for their belts that include allowable tension F_a (or stress σ_{all}), the tension being expressed in units of force per unit width. Belt life is usually several years. The severity of flexing at the pulley and its effect on life is reflected in a pulley correction factor C_p. Speed in excess of 180 m/min and its effect on life is reflected in a velocity correction

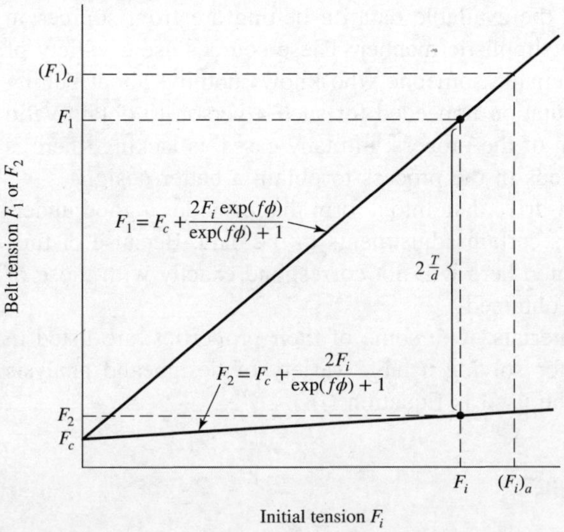

Figure 17–8

Plot of initial tension F_i against belt tension F_1 or F_2, showing the intercept F_c, the equations of the curves, and where $2T/d$ is to be found.

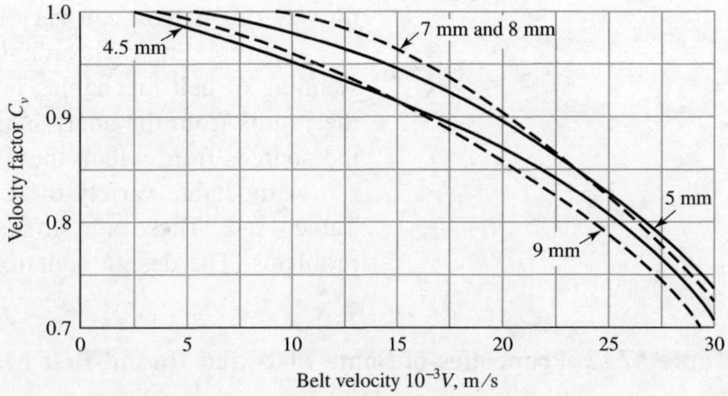

Figure 17–9

Velocity correction factor C_v for leather belts for various thicknesses. *(Source: Data from* Machinery's Handbook. 20th ed. *New York: Industrial Press, 1976, 1047.)*

factor C_v. For polyamide and urethane belts use $C_v = 1$. For leather belts see Figure 17–9. A service factor K_s is used for excursions of load from nominal, applied to the nominal power as $H_d = H_{\text{nom}}K_s n_d$, where n_d is the design factor for exigencies. These effects are incorporated as follows:

$$(F_1)_a = bF_aC_pC_v \tag{17–12}$$

where $(F_1)_a$ = allowable largest tension, N
b = belt width, mm
F_a = manufacturer's allowed tension, N/mm
C_p = pulley correction factor (Table 17–4)
C_v = velocity correction factor

The steps in analyzing a flat-belt drive can include (see Example 17–1)

1 Find $\exp(f\phi)$ from belt-drive geometry and friction from Table 17–2
2 From belt geometry, material (Table 17–2), and speed, find F_c from Equation (*e*)
3 Using Equation (3–42), with $H = H_d = H_{\text{nom}}K_s n_d$, find the necessary torque from $T = 30H_{\text{nom}}K_s n_d/\pi n$
4 From torque T find the necessary $(F_1)_a - F_2 = 2T/d$
5 From Tables 17–2 and 17–4, and Equation (17–12) determine $(F_1)_a$
6 Find F_2 from $(F_1)_a - [(F_1)_a - F_2]$
7 From Equation (*i*) find the necessary initial tension F_i
8 Check the friction development, $f' < f$. Use Equation (17–7) solved for f':

$$f' = \frac{1}{\phi}\ln\frac{(F_1)_a - F_c}{F_2 - F_c}$$

9 Find the factor of safety from $n_{fs} = H_a/(H_{\text{nom}}K_s)$

It is unfortunate that many of the available data on belting are from sources in which they are presented in a very simplistic manner. These sources use a variety of charts, nomographs, and tables to enable someone who knows nothing about belting to apply them. Little, if any, computation is needed for such a person to obtain valid results. Since a basic understanding of the process, in many cases, is lacking, there is no way this person can vary the steps in the process to obtain a better design.

Incorporating the available belt-drive data into a form that provides a good understanding of belt mechanics involves certain adjustments in the data. Because of this, the results from the analysis presented here will not correspond exactly with those of the sources from which they were obtained.

A moderate variety of belt materials, with some of their properties, are listed in Table 17–2. These are sufficient for solving a large variety of design and analysis problems. The design equation to be used is Equation (*j*).

Table 17–2 Properties of Some Flat- and Round-Belt Materials

Diameter = d, thickness = t, width = w.

Material	Specification	Size, mm	Minimum Pulley Diameter, mm	Allowable Tension per Unit Width at 3 m/s, (10^3) N/m	Specific Weight, kN/m^3	Coefficient of Friction
Leather	1 ply	$t = 4.5$	75	5	9.5–12.2	0.4
		$t = 5$	90	6	9.5–12.2	0.4
	2 ply	$t = 7$	115	7	9.5–12.2	0.4
		$t = 8$	150	9	9.5–12.2	0.4
		$t = 9$	230	10	9.5–12.2	0.4
Polyamide[b]	F–0[c]	$t = 0.8$	15	1.8	9.5	0.5
	F–1[c]	$t = 1.3$	25	6	9.5	0.5
	F–2[c]	$t = 1.8$	60	10	13.8	0.5
	A–2[c]	$t = 2.8$	60	10	10.0	0.8
	A–3[c]	$t = 3.3$	110	18	11.4	0.8
	A–4[c]	$t = 5.0$	240	30	10.6	0.8
	A–5[c]	$t = 6.4$	340	48	10.6	0.8
Urethane[d]	$w = 12.7$	$t = 1.6$	See Table 17–3	1.0[e]	10.3–12.2	0.7
	$w = 19$	$t = 2.0$		1.7[e]	10.3–12.2	0.7
	$w = 32$	$t = 2.3$		3.3[e]	10.3–12.2	0.7
	Round	$d = 6$	See Table 17–3	1.4[e]	10.3–12.2	0.7
		$d = 10$		3.3[e]	10.3–12.2	0.7
		$d = 12$		5.8[e]	10.3–12.2	0.7
		$d = 20$		13[e]	10.3–12.2	0.7

[a]Add 50 mm to pulley size for belts 200 mm wide or more.

[b]*Source: Habasit Engineering Manual*, Habasit Belting, Inc., Chamblee (Atlanta), Ga.

[c]Friction cover of acrylonitrile-butadiene rubber on both sides.

[d]*Source*: Eagle Belting Co., Des Plaines, Ill.

[e]At 6% elongation; 12% is maximum allowable value.

Table 17–3 Minimum Pulley Sizes for Flat and Round Urethane Belts

Listed are the pulley diameters in millimeters.

Belt Style	Belt Size, mm	Ratio of Pulley Speed to Belt Length, rev/(m · s)		
		Up to 14	14 to 27	28 — 55
Flat	12.7 × 1.6	9.7	11.2	12.7
	19 × 2.0	12.7	16	19
	32 × 2.3	12.7	16	19
Round	6	38.1	44.5	50.8
	10	57.1	66.5	76.2
	12	76.2	88.9	101.6
	20	127	152	177.8

Source: Eagle Belting Co., Des Plaines, Ill.

The values given in Table 17–2 for the allowable belt tension are based on a belt speed of 3 m/s. For higher speeds, use Figure 17–9 to obtain C_v values for leather belts. For polyamide and urethane belts, use $C_v = 1.0$.

The service factors K_s for V-belt drives, given in Table 17–15 in Section 17–3, are also recommended here for flat- and round-belt drives.

Minimum pulley sizes for the various belts are listed in Tables 17–2 and 17–3. The pulley correction factor accounts for the amount of bending or flexing of the belt and how this affects the life of the belt. For this reason it is dependent on the size and material of the belt used. See Table 17–4. Use $C_p = 1.0$ for urethane belts.

Flat-belt pulleys should be crowned to keep belts from running off the pulleys. If only one pulley is crowned, it should be the larger one. Both pulleys must be crowned whenever the pulley axes are not in a horizontal position. Use Table 17–5 for the crown height.

Table 17–4 Pulley Correction Factor C_p for Flat Belts

Material	Small-Pulley Diameter, mm					
	40–100	115–200	220–310	355–405	460–800	Over 800
Leather	0.5	0.6	0.7	0.8	0.9	1.0
Polyamide, F–0	0.95	1.0	1.0	1.0	1.0	1.0
F–1	0.70	0.92	0.95	1.0	1.0	1.0
F–2	0.73	0.86	0.96	1.0	1.0	1.0
A–2	0.73	0.86	0.96	1.0	1.0	1.0
A–3	—	0.70	0.87	0.94	0.96	1.0
A–4	—	—	0.71	0.80	0.85	0.92
A–5	—	—	—	0.72	0.77	0.91

Source: Average values of C_p for the given ranges were approximated from curves in the *Habasit Engineering Mannual*, Habasit Belting, Inc., Chamblee (Atlanta), Ga.

Table 17–5 Crown Height and ISO Pulley Diameters for Flat Belts*

ISO Crown Pulley Diameter, mm	ISO Height, mm	Pulley Diameter, mm	Crown Height, mm	
			$w \leq 250$ mm	$w > 250$ mm
40, 50, 62	0.3	315, 355	0.75	0.75
70, 80	0.3	315, 355	1.0	1.0
90, 100, 115	0.3	570, 635, 710	1.3	1.3
125, 142	0.4	800, 900	1.3	1.5
160, 180	0.5	1015	1.3	1.5
200, 230	0.6	1140, 1270, 1420	1.5	2.0
250, 285	0.75	1600, 1800, 2030	1.8	2.5

*Crown should be rounded, not angled; maximum roughness is R_a = AA 1500 μmm.

EXAMPLE 17–1

A polyamide A-3 flat belt 150 mm wide is used to transmit 11 kW under light shock conditions where $K_s = 1.25$, and a factor of safety equal to or greater than 1.1 is appropriate. The pulley rotational axes are parallel and in the horizontal plane. The shafts are 2400 mm apart. The 150 mm driving pulley rotates at 1750 rev/min in such a way that the loose side is on top. The driven pulley is 450 mm in diameter. See Figure 17–10. The factor of safety is for unquantifiable exigencies (see the steps outlined earlier).

(*a*) Estimate the centrifugal tension F_c and the torque T.

(*b*) Estimate the allowable F_1, F_2, F_i and allowable power H_a.

(*c*) Estimate the factor of safety. Is it satisfactory?

Solution

(*a*) Eq. (17–1):

$$\phi_d = \pi - 2\sin^{-1}\left[\frac{18 - 6}{2(8)12}\right] = 3.0165 \text{ rad}$$

Table 17–2:

$$\rho = 1140 \text{ kg/m}^3 \qquad f = 0.8 \qquad F_a = 18\ 000 \text{ N/m}$$

Step 1:

$$\exp(f\phi) = \exp[0.8(3.0165)] = 11.17$$

$$V = \pi(150)1750/1000 = 824.7 \text{ m/min}$$

$$w = \gamma bt = (1140)0.15(0.003) = 0.513 \text{ Kg/m}$$

Answer Step 2:

$$F_c = w\left(\frac{V}{60}\right)^2 = 0.513\left(\frac{824.7}{60}\right)^2 = 96.92 \text{ N}$$

Step 3:

$$T = \frac{60 H_{\text{nom}} K_s n_d}{2(\varphi)n} = \frac{60(11\ 000)\,1.25(1.1)}{2(\varphi)1750}$$

Answer

$$= 82.53 \text{ N} \cdot \text{m}$$

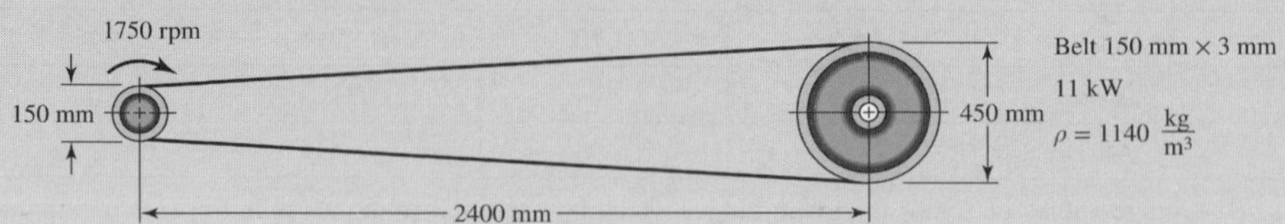

Figure 17–10

The flat-belt drive of Ex. 17–1. (Drawing is not to scale)

(*b*) From step 4, the necessary $(F_1)_a - F_2$ to transmit the torque T, from Eq. (*h*), is

Step 4: $$(F_1)_a - F_2 = \frac{2T}{d} = \frac{2(82.53)}{0.150} = 1100 \text{ N}$$

For polyamide belts $C_v = 1$, and from Table 17–4 $C_p = 0.70$. From Eq. (17–12) the allowable largest belt tension $(F_1)_a$ is

Answer Step 5: $$(F_1)_a = bF_aC_pC_v = 150(18)0.70(1) = 1890 \text{ N}$$

then

Answer Step 6: $$F_2 = (F_1)_a - [(F_1)_a - F_2] = 1890 - 1100 = 790 \text{ N}$$

and from Eq. (*i*)

Answer Step 7: $$F_i = \frac{(F_1)_a + F_2}{2} - F_c = \frac{1890 + 790}{2} - 96.92 = 1437 \text{ N}$$

Answer
The combination $(F_1)_a$, F_2, and F_i will transmit the design power of $H_a = H_{\text{nom}}K_sn_d = 11(1.25)(1.1) = 15.13$ kW and protect the belt. From step 8, we check the friction development by solving Eq. (17–7) for f':

Step 8: $$f' = \frac{1}{\phi} \ln \frac{(F_1)_a - F_c}{F_2 - F_c} = \frac{1}{3.0165} \ln \frac{420 - 25.6}{172.4 - 25.6} = 0.315$$

As determined earlier, $f = 0.8$. Since $f' < f$, there is no danger of slipping.
(*c*) From step 9,

Answer Step 9: $$n_{fs} = \frac{H_a}{H_{\text{nom}}K_s} = \frac{15.13}{11(1.25)} = 1.1 \qquad \text{(as expected)}$$

Answer
The belt is satisfactory and the maximum allowable belt tension exists. If the initial tension is maintained, the capacity is the design power of 15.13 kN.

Initial tension is the key to the functioning of the flat belt as intended. There are ways of controlling initial tension. One way is to place the motor and drive pulley on a pivoted mounting plate so that the weight of the motor, pulley, and mounting plate and a share of the belt weight induces the correct initial tension and maintains it. A second way is the use of a spring-loaded idler pulley, adjusted to the same task. Both of these methods accommodate to temporary or permanent belt stretch. See Figure 17–11.

Because flat belts were used for long center-to-center distances, the weight of the belt itself can provide the initial tension. The static belt deflects to an approximate catenary curve, and the dip from a straight belt can be measured against a stretched music wire. This provides a way of measuring and adjusting the dip. From catenary theory the dip is related to the initial tension by

$$dip = \frac{L^2w}{8F_i} g(10^{-3}) \tag{17–13}$$

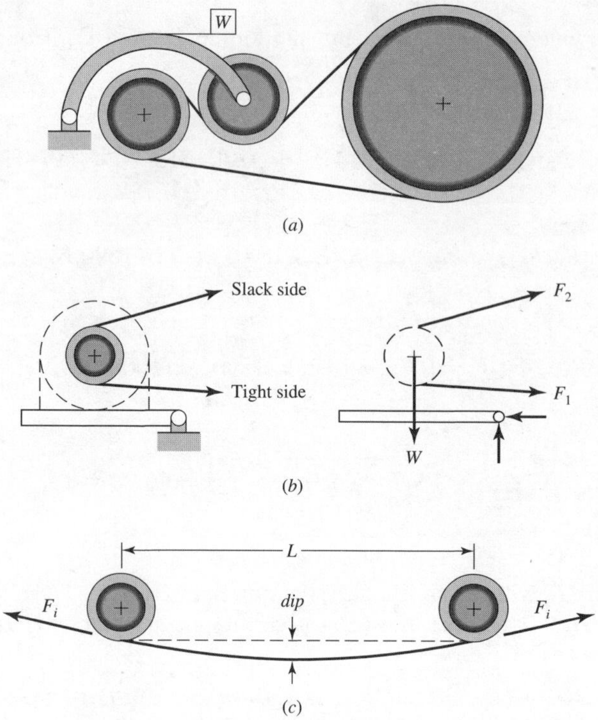

Figure 17–11

Belt-tensioning schemes.
(*a*) Weighted idler pulley.
(*b*) Pivoted motor mount.
(*c*) Catenary-induced tension.

where dip = dip, mm
L = center-to-center distance, mm
w = density per meter of the belt, kg/m
F_i = initial tension, N
g = gravitational acceleration, m/s^2

In Example 17–1 the dip corresponding to a 1800 N initial tension is

$$dip = \frac{2400^2(0.576)}{8(1180)} 9.81(10^{-3}) = 3.45 \text{ mm}$$

A decision set for a flat belt can be

- Function: power, speed, durability, reduction, service factor, center distance
- Design factor: n_d
- Initial tension maintenance
- Belt material
- Drive geometry, d, D
- Belt thickness: t
- Belt width: b

Depending on the problem, some or all of the last four could be design variables. Belt cross-sectional area is really the design decision, but available belt thicknesses and widths are discrete choices. Available dimensions are found in suppliers' catalogs.

EXAMPLE 17–2

Design a flat-belt drive to connect horizontal shafts on 4900 mm centers. The velocity ratio is to be 2.25:1. The angular speed of the small driving pulley is 860 rev/min, and the nominal power transmission is to be 45 kW under very light shock.

Solution

- Function: $H_{\text{nom}} = 45$ kW, 860 rev/min, 2.25:1 ratio, $K_s = 1.15$, $L = 4900$ mm
- Design factor: $n_d = 1.05$
- Initial tension maintenance: catenary
- Belt material: polyamide
- Drive geometry, d, D
- Belt thickness: t
- Belt width: b

The last four could be design variables. Let's make a few more a priori decisions.

Decision $\quad d = 400$ mm, $D = 2.25d = 2.25(16) = 900$ mm.

Decision

Use polyamide A-3 belt; therefore $t = 3.3$ mm and $C_v = 1$.

Now there is one design decision remaining to be made, the belt width b.

Table 17–2: $\quad \rho = 1140\ \text{kg/m}^3 \qquad f = 0.8 \qquad F_a = 18$ N/mm at 600 rev/min

Table 17–4: $\quad C_p = 0.94$

Eq. (17–12):

$$(F_1)_a = b(18)0.94(1) = 16.9b\ \text{N} \tag{1}$$

$$H_d = H_{\text{nom}} K_s n_d = 45(1.15)1.05 = 54.34\ \text{kW}$$

$$T = \frac{60H_d}{2(\pi)n} = \frac{60(54\ 340)}{2(\pi)860} = 603\ \text{N} \cdot \text{m}$$

Estimate $\exp(f\phi)$ for full friction development:

Eq. (17–1):

$$\phi_d = \pi - 2\sin^{-1}\frac{(900-400)}{2(4900)} = 3.039\ \text{rad}$$

$$\exp(f\phi) = \exp[0.80(3.039)] = 11.37$$

Estimate centrifugal tension F_c in terms of belt width b:

$$w = \gamma bt = (1140)b10^{-3}\ (0.0033) = 3.76(10^{-3})b\ \text{kg/m}$$

$$V = \pi dn/1000 = \pi(400)860/1000 = 1081\ \text{m/min}$$

Eq. (e):

$$F_c = w\left(\frac{V}{60}\right)^2 = 3.76(10^{-3})b\left(\frac{1081}{60}\right)^2 = 1.22b\ \text{N} \tag{2}$$

For design conditions, that is, at H_d power level, using Eq. (h) gives

$$(F_1)_a - F_2 = 2T/d = 2(603)/0.4 = 3015\ \text{N} \tag{3}$$

$$F_2 = (F_1)_a - [(F_1)_a - F_2] = 16.9b - 3015\ \text{N} \tag{4}$$

Using Eq. (*i*) gives

$$F_i = \frac{(F_1)_a + F_2}{2} - F_c = \frac{16.9b + 16.9b - 3015}{2} - 1.22b = 15.68b - 1508 \text{ N} \tag{5}$$

Place friction development at its highest level, using Eq. (17–7):

$$f\phi = \ln\frac{(F_1)_a - F_c}{F_2 - F_c} = \ln\frac{16.9b - 1.22b}{16.9b - 3015 - 1.22b} = \ln\frac{15.68b}{15.68b - 3015}$$

Solving the preceding equation for belt width b at which friction is fully developed gives

$$b = \frac{3015}{15.68}\frac{\exp(f\phi)}{\exp(f\phi) - 1} = \frac{3015}{15.68}\frac{11.37}{11.37 - 1} = 210.8 \text{ mm}$$

A belt width greater than 210.8 mm will develop friction less than $f = 0.80$. The manufacturer's data indicate that the next available larger width is 250 mm.

Decision

Use 250-mm-wide belt.
It follows that for a 250-mm-wide belt

Eq. (2): $$F_c = 1.22(250) = 305 \text{ N}$$

Eq. (1): $$(F_1)_a = 16.9(250) = 4225 \text{ N}$$

Eq. (4): $$F_2 = 16.9(250) - 3015 = 1210 \text{ N}$$

Eq. (5): $$F_i = 15.68(250) - 1508 = 2412 \text{ N}$$

The transmitted power, from Eq. (j), is

$$H_t = \frac{[(F_1)_a - F_2]V}{60} = \frac{3015(1081)}{60} = 54.3 \text{ kW}$$

and the level of friction development f', from Eq. (17–7) is

$$f' = \frac{1}{\phi}\ln\frac{(F_1)_a - F_c}{F_2 - F_c} = \frac{1}{3.039}\ln\frac{(4225-305)}{(1210-305)} = 0.482$$

which is less than $f = 0.8$, and thus is satisfactory. Had a 225 mm belt width been available, the analysis would show $(F_1)_a = 3803$ N, $F_2 = 788$ N, $F_i = 2020$ N, and $f' = 0.634$. With a figure of merit available reflecting cost, thicker belts (A-4 or A-5) could be examined to ascertain which of the satisfactory alternatives is best. From Eq. (17–13) the catenary dip is

$$dip = \frac{L^2 w}{8F_i}\, g = \frac{4.9^2\,(0.793)}{8(2412)} = 9.81 = 0.009\,68 \text{ m} = 9.68 \text{ mm}$$

Figure 17–12 illustrates the variation of flexible flat-belt tensions at some cardinal points during a belt pass.

Flat Metal Belts

Thin flat metal belts with their attendant strength and geometric stability could not be fabricated until laser welding and thin rolling technology made possible belts as

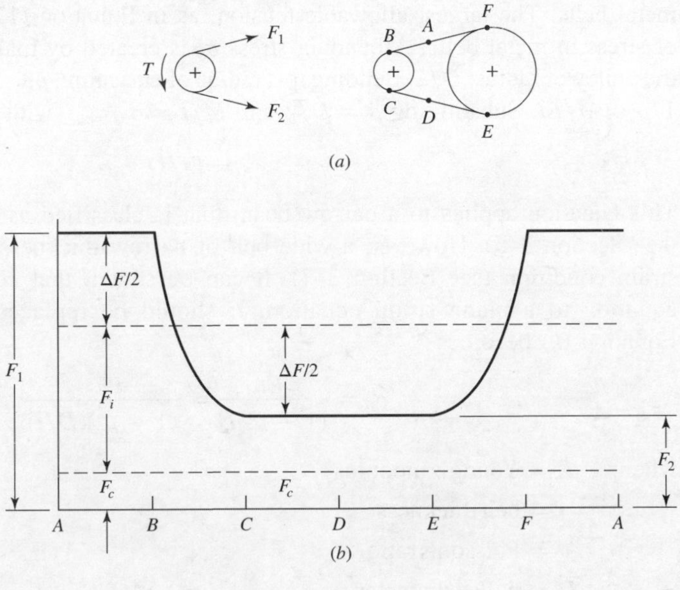

Figure 17–12

Flat-belt tensions.

thin as 0.05 mm and as narrow as 0.65 mm. The introduction of perforations allows no-slip applications. Thin metal belts exhibit

- High strength-to-weight ratio
- Dimensional stability
- Accurate timing
- Usefulness to temperatures up to 370°C
- Good electrical and thermal conduction properties

In addition, stainless steel alloys offer "inert," nonabsorbent belts suitable to hostile (corrosive) environments, and can be made sterile for food and pharmaceutical applications.

Thin metal belts can be classified as friction drives, timing or positioning drives, or tape drives. Among friction drives are plain, metal-coated, and perforated belts. Crowned pulleys are used to compensate for tracking errors.

Figure 17–13 shows a thin flat metal belt with the tight tension F_1 and the slack side tension F_2 revealed. The relationship between F_1 and F_2 and the driving torque T is the same as in Equation (h). Equations (17–9), (17–10), and (17–11) also apply, with the hoop tension due to centrifugal force typically neglected for the very thin

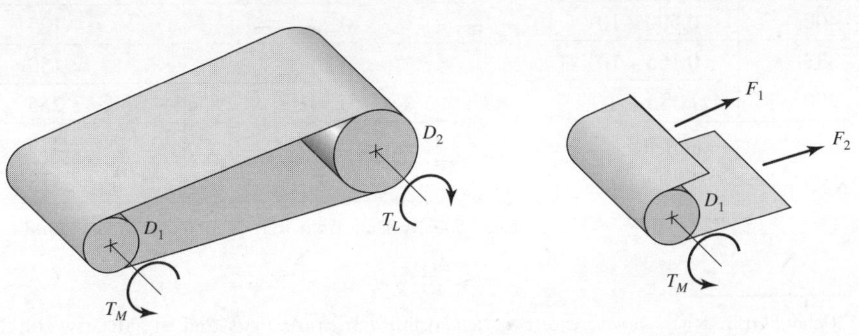

Figure 17–13

Metal-belt tensions and torques.

metal belts. The largest allowable tension, as in Equation (17–12), is posed in terms of stress in metal belts. A bending stress σ_b is created by making the belt conform to the pulley of radius, $D/2$. Bending to a radius of curvature, ρ, is given by Equation (4–7), $1/\rho = M/EI$. Substituting $\rho = D/2$ and $M/I = \sigma_b/y_{max}$ with $y_{max} = t/2$ gives

$$\sigma_b = Et/D \tag{k}$$

This equation applies to a narrow beam that is classified as a plane stress condition (see Section 3–5). However, a wide belt of narrow thickness is classified as a plane strain condition (see Section 3–7). It can be shown that converting a plane stress equation to a plane strain equation, E should be replaced by $E/(1-\nu^2)$.[3] Thus, Equation (k) becomes

$$\sigma_b = \frac{Et}{(1-\nu^2)D} = \frac{E}{(1-\nu^2)(D/t)} \tag{17–14}$$

where E = Young's modulus
t = belt thickness
ν = Poisson's ratio
D = pulley diameter

The tensile stresses $(\sigma)_1$ and $(\sigma)_2$ imposed by the belt tensions F_1 and F_2 are

$$(\sigma)_1 = F_1/(bt) \qquad \text{and} \qquad (\sigma)_2 = F_2/(bt)$$

The largest tensile stress is $(\sigma_b)_1 + F_1/(bt)$ and the smallest is $(\sigma_b)_2 + F_2/(bt)$. During a belt pass both levels of stress appear.

Although the belts are of simple geometry, the method of Marin is not used because the condition of the butt weldment (to form the loop) is not accurately known, and the testing of weld coupons is difficult. The belts are run to failure on two equal-sized pulleys. Information concerning fatigue life, as shown in Table 17–6, is obtainable. Tables 17–7 and 17–8 give additional information.

Table 17–6 Belt Life for Stainless Steel Friction Drives

$\frac{D}{t}$	Belt Passes
625	$\geq 10^6$
400	$0.500 \cdot 10^6$
333	$0.165 \cdot 10^6$
200	$0.085 \cdot 10^6$

Source: Data from Belt Technologies, Agawam, Mass.

Table 17–7 Minimum Pulley Diameter

Belt Thickness, mm	Minimum Pulley Diameter, mm
0.05	30
0.08	45
0.13	75
0.20	125
0.25	150
0.38	255
0.50	315
1.00	635

Source: Data from Belt Technologies, Agawam, Mass.

[3]Richard G. Budynas, *Advanced Strength and Applied Stress Analysis*, 2nd ed., McGraw Hill, New York, 1999, Table 4.1.1, p. 224.

Table 17–8 Typical Material Properties, Metal Belts

Alloy	Yield Strength, MPa	Young's Modulus, GPa	Poisson's Ratio
301 or 302 stainless steel	1206	193	0.285
BeCu	1170	117	0.220
1075 or 1095 carbon steel	1585	207	0.287
Titanium	1034	103	—
Inconel	1103	207	0.284

Source: Data courtesy of Belt Technologies, Agawam, Mass.

Table 17–6 shows metal belt life expectancies for a stainless steel belt. From Equation (17–14) with $E = 200$ GPa and $\nu = 0.29$, the bending stresses corresponding to the four entries of the table are 350, 546, 656, and 1092 MPa. Using a natural log transformation on stress and passes shows that the regression line ($r^2 = 0.9303$) is

$$\sigma = 97\ 702 N_p^{-0.407} \tag{17–15}$$

where N_p is the number of belt passes.

The selection of a metal flat belt can consist of the following steps:

1 Find $\exp(f\phi)$ from geometry and friction

2 Find endurance strength

$$S_f = 97\ 702 N_p^{-0.407} \qquad \text{301, 302 stainless}$$

$$S_f = S_y/3 \qquad \text{others}$$

3 Allowable tension

$$(F_1)_a = \left[S_f - \frac{Et}{(1-\nu^2)D}\right] tb = ab$$

4 $\Delta F = 2T/D$

5 $F_2 = (F_1)_a - \Delta F = ab - \Delta F$

6 $F_i = \dfrac{(F_1)_a + F_2}{2} = \dfrac{ab + ab - \Delta F}{2} = ab - \dfrac{\Delta F}{2}$

7 $b_{\min} = \dfrac{\Delta F}{a} \dfrac{\exp(f\phi)}{\exp(f\phi) - 1}$

8 Choose $b > b_{\min}$, $(F_1)_a = ab$, $F_2 = ab - \Delta F$, $F_i = ab - \Delta F/2$, $T = \Delta F D/2$

9 Check frictional development f':

$$f' = \frac{1}{\phi} \ln \frac{(F_1)_a}{F_2} \qquad f' < f$$

EXAMPLE 17–3

A friction-drive stainless steel metal belt runs over two 100 mm metal pulleys ($f = 0.35$). The belt thickness is to be 0.075 mm. For a life exceeding 10^6 belt passes with smooth torque ($K_s = 1$), (*a*) select the belt if the torque is to be 3500 N · mm, and (*b*) find the initial tension F_i.

Solution

(*a*) From step 1, $\phi_d = \pi$, therefore $\exp(0.35\pi) = 3.00$. From step 2,

$$(S_f)_{10^6} = 97\ 702(10^6)^{-0.407} = 353 \text{ MPa}$$

From steps 3, 4, 5, and 6,

$$(F_1)_a = \left[353 - \frac{(200\ 000)\ 0.075}{(1 - 0.285^2)\ 100}\right] 0.075b = 14.23b \qquad (1)$$

$$\Delta F = 2T/D = 2(3500)/100 = 70 \text{ N}$$

$$F_2 = (F_1)_a - \Delta F = 14.23b - 70 \qquad (2)$$

$$F_i = \frac{(F_1)_a + F_2}{2} = 14.23b - \frac{70}{2} \qquad (3)$$

From step 7,

$$b_{\min} = \frac{\Delta F}{a} \frac{\exp(f\phi)}{\exp(f\phi) - 1} = \frac{70}{14.23} \frac{3.00}{3.00 - 1} = 7.38 \text{ mm}$$

Decision

From step 8, select an available 19 mm wide belt 0.075 mm thick.

Eq. (1): $$(F_1)_a = 14.23(19) = 270 \text{ N}$$

Eq. (2): $$F_2 = 14.23(19) - 70 = 200 \text{ N}$$

Eq. (3): $$F_i = 270 - 70/2 = 235 \text{ N}$$

From step 9, $$f' = \frac{1}{\phi} \ln \frac{(F_1)_a}{F_2} = \frac{1}{\pi} \ln \frac{270}{200} = 0.0955$$

Note $f' < f$, that is, $0.0955 < 0.35$.

17–3 V Belts

The cross-sectional dimensions of V-belts have been standardized by manufacturers, with each section designated by a number for metric sizes and by a letter of the alphabet for inch-based sizes. Though these have not been included here, the procedure for analyzing and designing them is the same as presented here. Dimensions, minimum sheave diameters, and the horsepower range for each of the lettered sections are listed in Table 17–9.

To specify a V belt, give the belt-section letter, followed by the inside circumference in millimeters (standard circumferences are listed in Table 17–10). For example, B1950 is a B-section belt having an inside circumference of 1950 mm.

Table 17–9 Standard V-Belt Sections

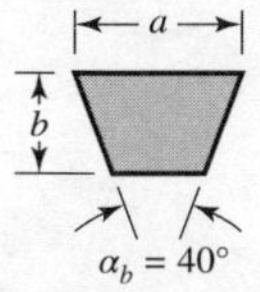

Belt Section	Width a, mm	Thickness b, mm	Minimum Sheave Diameter, mm	kW Range, One or More Belts
A	12	8.5	75	0.2–7.5
B	16	11	135	0.7–18.5
C	22	13	230	11–75
D	30	19	325	37–186
E	38	25	540	75 and up

Table 17–10 Inside Circumferences of Standard V Belts

Section	Circumference, mm
A	650, 775, 825, 875, 950, 1050, 1150, 1200, 1275, 1325, 1375, 1425, 1500, 1550, 1600, 1650, 1700, 1775, 1875, 1950, 2000, 2125, 2250, 2400, 2625, 2800, 3000, 3200
B	875, 950, 1050, 1150, 1200, 1275, 1325, 1375, 1425, 1500, 1550, 1600, 1650, 1700, 1775, 1875, 1950, 2000, 2125, 2250, 2400, 2625, 2800, 3000, 3200, 3275, 3400, 3450, 3950, 4325, 4500, 4875, 5250, 6000, 6750, 7500
C	1275, 1500, 1700, 1875, 2025, 2125, 2250, 2400, 2625, 2800, 3000, 3200, 3400, 3600, 3950, 4050, 4350, 4500, 4875, 5250, 6000, 6750, 7500, 8250, 9000, 9750, 10 500
D	3000, 3200, 3600, 3950, 4050, 4350, 4500, 4875, 5250, 6000, 6750, 7500, 8250, 9000, 9750, 10 500, 12 000, 13 500, 15 000, 16 500
E	4500, 4875, 5250, 6000, 6750, 7500, 8250, 9000, 9750, 10 500, 12 000, 13 500, 15 000, 16 500

Table 17–11 Length Conversion Dimensions

Add the listed quantity to the inside circumference to obtain the pitch length in millimeters.

Belt section	A	B	C	D	E
Quantity to be added	32	46	72	82	112

Calculations involving the belt length are usually based on the pitch length. For any given belt section, the pitch length is obtained by adding a quantity to the inside circumference (Tables 17–10 and 17–11). For example, a B1950 belt has a pitch length of 1996. Similarly, calculations of the velocity ratios are made using the pitch diameters of the sheaves, and for this reason the stated diameters are usually understood to be the pitch diameters even though they are not always so specified.

The groove angle of a sheave, α_s, is made somewhat smaller than the belt-section angle, α_b. This causes the belt to wedge itself into the groove, thus increasing friction. The exact value of this angle depends on the belt section, the sheave diameter, and the angle of wrap. If it is made too much smaller than the belt, the force required to pull the belt out of the groove as the belt leaves the pulley will be excessive. Optimum values are given in the commercial literature.

The minimum sheave diameters have been listed in Table 17–9. For best results, a V belt should be run quite fast: 1200 m/min is a good speed. Trouble may be encountered if the belt runs much faster than 1500 m/min or much slower than 300 m/min.

The *pitch length* L_p and the center-to-center distance C are

$$L_p = 2C + \pi(D + d)/2 + (D - d)^2/(4C) \tag{17–16a}$$

$$C = 0.25\left\{\left[L_p - \frac{\pi}{2}(D + d)\right] + \sqrt{\left[L_p - \frac{\pi}{2}(D + d)\right]^2 - 2(D - d)^2}\right\} \tag{17–16b}$$

where D = pitch diameter of the large sheave and d = pitch diameter of the small sheave.

In the case of flat belts, there is virtually no limit to the center-to-center distance. Long center-to-center distances are not recommended for V belts because the excessive vibration of the slack side will shorten the belt life materially. In general, the center-to-center distance should be no greater than three times the sum of the sheave diameters and no less than the diameter of the larger sheave. Link-type V belts have less vibration, because of better balance, and hence may be used with longer center-to-center distances.

The basis for power ratings of V belts depends somewhat on the manufacturer; it is not often mentioned quantitatively in vendors' literature but is available from vendors. The basis may be a number of hours, 24 000 h, for example, or a life of 10^8 or 10^9 belt passes. Since the number of belts must be an integer, an undersized belt set that is augmented by one belt can be substantially oversized. Table 17–12 gives power ratings of standard V belts.

The rating, whether in terms of hours or belt passes, is for a belt running on equal-diameter sheaves (180° of wrap), of moderate length, and transmitting a steady load. Deviations from these laboratory test conditions are acknowledged by multiplicative adjustments. If the tabulated power of a belt for a C-section belt is 7.06 kW for a 300 mm diameter sheave at a peripheral speed of 15 m/s (Table 17–12), then, when the belt is used under other conditions, the tabulated value H_{tab} is adjusted as follows:

$$H_a = K_1 K_2 H_{\text{tab}} \tag{17–17}$$

where H_a = allowable power, per belt

K_1 = angle-of-wrap (ϕ) correction factor, Table 17–13

K_2 = belt length correction factor, Table 17–14

The allowable power can be near to H_{tab}, depending upon circumstances.

In a V belt the effective coefficient of friction f' is $f/\sin(\alpha_s/2)$, which amounts to an augmentation by a factor of about 3 due to the grooves. The effective coefficient of friction f' is sometimes tabulated against *sheave* groove angles of α_s = 30°, 34°, and 38°. The corresponding tabulated values are f' = 0.50, 0.45, and 0.40, respectively, revealing a belt material-on-metal coefficient of friction of $f = 0.13$ for each case. The Gates Rubber Company declares its effective coefficient of friction to be 0.5123 for grooves. Thus,

$$\frac{F_1 - F_c}{F_2 - F_c} = \exp(0.5123\phi) \tag{17–18}$$

The design power is given by

$$H_d = H_{\text{nom}} K_s n_d \tag{17–19}$$

Table 17–12 Horsepower Ratings of Standard V Belts

Belt Section	Sheave Pitch Diameter, mm	Belt Speed, m/s				
		5	10	15	20	25
A	65	0.35	0.46	0.40	0.11	
	75	0.49	0.75	0.84	0.69	0.28
	85	0.60	0.98	1.17	1.64	0.84
	95	0.69	1.16	1.43	1.49	1.28
	105	0.77	1.30	1.64	1.78	1.63
	115	0.83	1.41	1.82	2.01	1.93
	125 and up	0.87	1.51	1.97	2.21	2.16
B	105	0.80	1.18	1.25	0.94	0.16
	115	0.95	1.48	1.71	1.55	0.92
	125	1.07	1.74	2.09	2.06	1.57
	135	1.19	1.95	2.42	2.49	2.10
	145	1.28	2.14	2.69	2.87	2.57
	155	1.36	2.31	2.94	3.19	2.98
	165	1.43	2.45	3.16	3.48	3.34
	175 and up	1.50	2.58	3.35	3.74	3.66
C	150	1.37	1.98	2.03	1.40	
	175	1.85	2.94	3.46	3.31	2.33
	200	2.21	3.66	4.54	4.74	4.12
	225	2.49	4.21	5.38	5.86	5.51
	250	2.72	4.66	6.05	7.16	6.63
	275	2.89	5.03	6.59	7.46	7.53
	300 and up	3.05	5.33	7.06	8.13	8.28
D	250	3.09	4.57	4.89	3.80	1.01
	275	3.73	5.84	6.80	6.34	4.19
	300	4.26	6.91	8.36	8.50	6.85
	325	4.71	7.83	9.70	10.30	9.10
	350	5.09	8.58	10.89	11.79	11.04
	375	5.42	9.25	11.86	13.13	12.68
	400	5.71	9.85	12.76	14.32	14.17
	425 and up	5.98	10.37	13.50	15.37	15.44
E	400	6.48	10.44	13.06	13.50	11.41
	450	7.40	12.46	15.82	17.16	16.04
	500	8.13	13.95	18.05	20.07	19.69
	550	8.73	15.14	19.84	22.53	22.75
	600	9.25	16.11	21.34	24.54	25.22
	650	9.70	17.01	22.60	26.19	27.38
	700 and up	10.00	17.68	23.72	27.68	29.17

where H_{nom} is the nominal power, K_s is the service factor given in Table 17–15, and n_d is the design factor. The number of belts, N_b, is usually the next higher integer to H_d/H_a.

That is,

$$N_b \geq \frac{H_d}{H_a} \qquad N_b = 1, 2, 3, \ldots \tag{17–20}$$

Designers work on a per-belt basis.

Table 17–13 Angle of Wrap Correction Factor K_1 for VV and V-Flat Drives[†]

$\frac{D-d}{C}$	ϕ, deg	K_1	
		VV*	V-Flat
0.00	180	1.00	0.75
0.10	174.3	0.99	0.76
0.20	166.5	0.97	0.78
0.30	162.7	0.96	0.79
0.40	156.9	0.94	0.80
0.50	151.0	0.93	0.81
0.60	145.1	0.91	0.83
0.70	139.0	0.89	0.84
0.80	132.8	0.87	0.85
0.90	126.5	0.85	0.85
1.00	120.0	0.82	0.82
1.10	113.3	0.80	0.80
1.20	106.3	0.77	0.77
1.30	98.9	0.73	0.73
1.40	91.1	0.70	0.70
1.50	82.8	0.65	0.65

*A curve fit for the VV column in terms of ϕ is
$K_1 = 0.143\ 543 + 0.007\ 468\ \phi - 0.000\ 015\ 052\ \phi^2$ in the range $90° \leq \phi \leq 180°$.
[†]VV drives are when the driver and driven pulleys are both V pulleys. A V-flat drive uses a V pulley, typically the smaller on the driver shaft, and a flat pulley on the driven shaft.

Table 17–14 Belt-Length Correction Factor K_2^*

Length Factor	Nominal Belt Length, m				
	A Belts	B Belts	C Belts	D Belts	E Belts
0.85	Up to 0.88	Up to 1.15	Up to 1.88	Up to 3.2	
0.90	0.95–1.15	1.2–1.5	2.03–2.4	3.6–4.05	Up to 4.88
0.95	1.2–1.38	1.55–1.88	2.63–3.0	4.33–5.25	Up to 4.88
1.00	1.5–1.88	1.95–2.43	3.2–3.95	6.0	6.75–7.5
1.05	1.95–2.25	2.63–3.0	4.05–4.88	6.75–8.25	8.25–9.75
1.10	2.4–2.8	3.2–3.6	5.25–6.0	9.0–10.5	10.5–12.0
1.15	3.0 and up	3.95–4.5	6.75–7.5	12.0	13.5–15.0
1.20		4.88 and up	8.25 and up	13.5 and up	16.5

*Multiply the rated horsepower per belt by this factor to obtain the corrected horsepower.

Table 17–15 Suggested Service Factors K_S for V-Belt Drives

Driven Machinery	Source of Power	
	Normal Torque Characteristic	High or Nonuniform Torque
Uniform	1.0 to 1.2	1.1 to 1.3
Light shock	1.1 to 1.3	1.2 to 1.4
Medium shock	1.2 to 1.4	1.4 to 1.6
Heavy shock	1.3 to 1.5	1.5 to 1.8

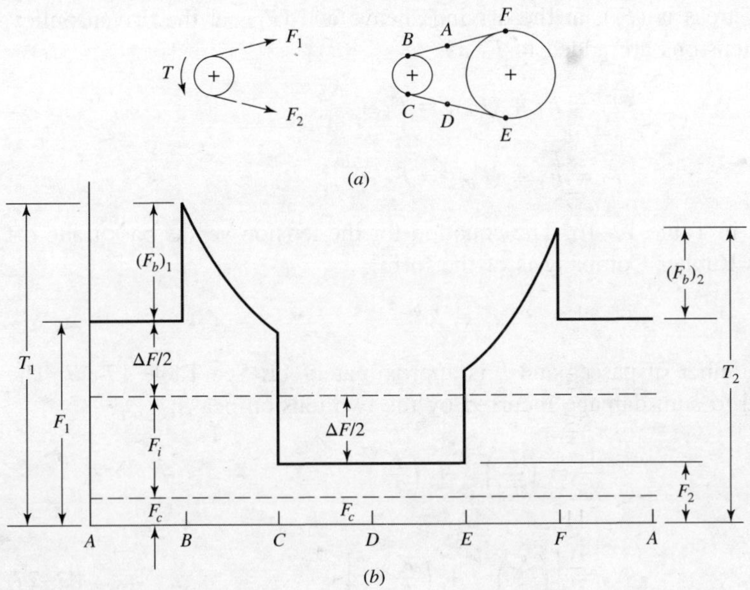

Figure 17–14

V-belt tensions.

Table 17–16 Some V-Belt Parameters

Belt Section	K_b	K_c
A	24 855	26.86
B	65 076	46.21
C	180 767	82.17
D	641 722	167.49
E	1 225 824	241.38
3V	25 985	20.35
5V	124 051	58.27
8V	545 690	157.44

Source: Data from Gates Rubber Co., Denver, Colo.

The flat-belt tensions shown in Figure 17–12 ignored the tension induced by bending the belt about the pulleys. This is more pronounced with V belts, as shown in Figure 17–14.

The centrifugal tension F_c is given by

$$F_c = K_c\left(\frac{V}{1000}\right)^2 \tag{17–21}$$

where K_c is from Table 17–16.

The power that is transmitted per belt is based on $\Delta F = F_1 - F_2$, where

$$\Delta F = \frac{\left(\frac{30}{\pi}\right)(10^{-6})\frac{H_d}{N_b}}{n\left(\frac{d}{2}\right)} \tag{17–22}$$

then from Equation (17–8) the larger tension F_1 is given by

$$F_1 = F_c + \frac{\Delta F \exp(f'\phi)}{\exp(f'\phi) - 1} \tag{17–23}$$

From the definition of ΔF, the smaller tension F_2 is

$$F_2 = F_1 - \Delta F \tag{17–24}$$

From Equation (*j*) in Section 17–2

$$F_i = \frac{F_1 + F_2}{2} - F_c \tag{17–25}$$

The factor of safety is

$$n_{fs} = \frac{H_a N_b}{H_{\text{nom}} K_s} \tag{17–26}$$

Durability (life) correlations are complicated by the fact that the bending induces flexural stresses in the belt; the corresponding belt tension that induces the same maximum tensile stress is $(F_b)_1$ at the driving sheave and $(F_b)_2$ at the driven pulley. These equivalent tensions are added to F_1 as

$$T_1 = F_1 + (F_b)_1 = F_1 + \frac{K_b}{d}$$

$$T_2 = F_1 + (F_b)_2 = F_1 + \frac{K_b}{D}$$

where K_b is given in Table 17–16. The equation for the tension versus pass trade-off used by the Gates Rubber Company is of the form

$$T^b N_P = K^b$$

where N_P is the number of passes and b is approximately 11. See Table 17–17. The Miner rule is used to sum damage incurred by the two tension peaks:

$$\frac{1}{N_P} = \left(\frac{K}{T_1}\right)^{-b} + \left(\frac{K}{T_2}\right)^{-b}$$

or

$$N_P = \left[\left(\frac{K}{T_1}\right)^{-b} + \left(\frac{K}{T_2}\right)^{-b}\right]^{-1} \tag{17–27}$$

The lifetime t in hours is given by

$$t = \frac{N_P L_p}{60V} \tag{17–28}$$

The constants K and b have their ranges of validity. If $N_P > 10^9$, report that $N_P = 10^9$ and $t > N_P L_p/(60V)$ without placing confidence in numerical values beyond the validity interval. See the statement about N_P and t near the conclusion of Example 17–4.

The analysis of a V-belt drive can consist of the following steps:

1 Find V, L_p, C, ϕ, and $\exp(0.5123\phi)$

2 Find H_d, H_a, and N_b from H_d/H_a and round up

3 Find F_c, ΔF, F_1, F_2, and F_i, and n_{fs}

4 Find belt life in number of passes, or hours, if possible

Table 17–17 Durability Parameters for Some V-Belt Sections

Belt Section	10^8 to 10^9 Force Peaks		10^9 to 10^{10} Force Peaks		Minimum Sheave Diameter, mm
	K	b	K	b	
A	2999	11.089			75
B	5309	10.926			125
C	9069	11.173			215
D	18 726	11.105			325
E	26 791	11.100			540
3V	3240	12.464	4726	10.153	66
5V	7360	12.593	10 653	10.283	177
8V	16 189	12.629	23 376	10.319	312

Source: Data from Spotts, M. E. *Design of Machine Elements*. 6th ed. Englewood Cliffs, New Jersey: Prentice Hall, 1985.

EXAMPLE 17–4

A 7.5 kW split-phase motor running at 1750 rev/min is used to drive a rotary pump, which operates 24 hours per day. An engineer has specified a 188 mm small sheave, a 280 mm large sheave, and three B2845 belts. The service factor of 1.2 was augmented by 0.1 because of the continuous-duty requirement. Analyze the drive and estimate the belt life in passes and hours.

Solution

For step 1, the peripheral speed V of the belt is

$$V = \pi\, dn/1000 = \pi(188)1750/1000 = 1034 \text{ m/min}$$

Table 17–11: $L_p = L + L_c = 2845 + 46 = 2891$ mm

Eq. (17–16*b*):

$$C = 0.25\left\{\left[2891 - \frac{\pi}{2}(280 + 188)\right] + \sqrt{\left[2891 - \frac{\pi}{2}(280 + 188)\right]^2 - 2(280 - 188)^2}\right\}$$
$$= 1077 \text{ mm}$$

Eq. (17–1):

$$\phi_d = \pi - 2\sin^{-1}(280 - 188)/[2(1077)] = 3.057 \text{ rad}$$

$$\exp[0.5123(3.057)] = 4.788$$

For step 2, interpolating in Table 17–12 for $V = 1034$ m/min gives $H_{\text{tab}} = 3.5$ kW. The wrap angle in degrees is $3.057(180)/\pi = 175°$. From Table 17–13, $K_1 = 0.99$. From Table 17–14, $K_2 = 1.05$. Thus, from Eq. (17–17),

$$H_a = K_1K_2H_{\text{tab}} = 0.99(1.05)3.5 = 3.64 \text{ kW}$$

Eq. (17–19):

$$H_d = H_{\text{nom}}K_s n_d = 7.5(1.2 + 0.1)(1) = 9.75 \text{ kW}$$

Eq. (17–20):

$$N_b \geq H_d/H_a = 9.75/3.64 = 2.67 \rightarrow 3$$

For step 3, from Table 17–16, $K_c = 46.21$. Thus, from Eq. (17–21),

$$F_c = 46.21(1034/1000)^2 = 49.40 \text{ N}$$

Eq. (17–22):

$$\Delta F = \frac{30/\pi\,(10^6)(9.75)/3}{1750(188/2)} = 188.7 \text{ N}$$

Eq. (17–23):

$$F_1 = 49.40 + \frac{188.7(4.788)}{4.788 - 1} = 287.9 \text{ N}$$

Eq. (17–24):

$$F_2 = F_1 - \Delta F = 287.9 - 188.7 = 99.2 \text{ N}$$

Eq. (17–25):

$$F_i = \frac{287.9 + 99.2}{2} - 49.40 = 144.2 \text{ N}$$

Eq. (17–26):

$$n_{fs} = \frac{H_aN_b}{H_{\text{nom}}K_s} = \frac{3.64(3)}{7.5(1.3)} = 1.12$$

Life: For step 4, from Table 17–16, $K_b = 65\ 076$.

$$(F_b)_1 = \frac{K_b}{d} = \frac{65\ 076}{188} = 346\ \text{N}$$

$$(F_b)_2 = \frac{65\ 075}{280} = 232\ \text{N}$$

$$T_1 = F_1 + (F_b)_1 = 287.9 + 346 = 634\ \text{N}$$

$$T_2 = F_1 + (F_b)_2 = 287.9 + 232 = 520\ \text{N}$$

From Table 17–17, $K = 5309$ and $b = 10.926$.

Eq. (17–27): $$N_P = \left[\left(\frac{5309}{634}\right)^{-10.926} + \left(\frac{5309}{520}\right)^{-10.926}\right]^{-1} = 11(10^9)\ \text{passes}$$

Answer

Since N_P is out of the validity range of Eq. (17–27), life is reported as greater than 10^9 passes. Then

Answer Eq. (17–28): $$t > \frac{10^9(2891)}{60(1034)} = 46\ 600\ \text{h}$$

17–4 Timing Belts

A timing belt is made of a rubberized fabric coated with a nylon fabric, and has steel wire within to take the tension load. It has teeth that fit into grooves cut on the periphery of the pulleys (Figure 17–15). A timing belt does not stretch appreciably or slip and consequently transmits power at a constant angular-velocity ratio. No initial tension is needed. Such belts can operate over a very wide range of speeds, have efficiencies in the range of 97 to 99 percent, require no lubrication, and are quieter than chain drives. There is no chordal-speed variation, as in chain drives (see Section 17–5), and so they are an attractive solution for precision-drive requirements.

The steel wire, the tension member of a timing belt, is located at the belt pitch line (Figure 17–15). Thus the pitch length is the same regardless of the thickness of the backing.

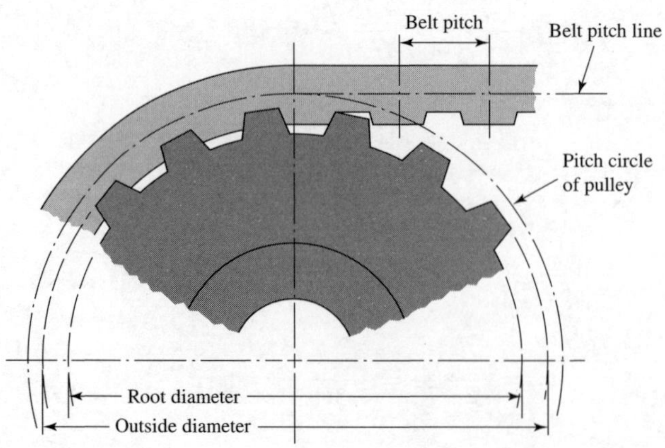

Figure 17–15

Timing-belt drive showing portions of the pulley and belt. Note that the pitch diameter of the pulley is greater than the diametral distance across the top lands of the teeth.

Table 17–18 Standard Pitches of Timing Belts

Service	Designation	Pitch p, mm
Extra light	XL	5
Light	L	10
Heavy	H	12
Extra heavy	XH	22
Double extra heavy	XXH	30

The five standard series pitches available are listed in Table 17–18 with their letter designations. Standard pitch lengths are available in sizes from 150 to 4600 mm. Pulleys come in sizes from 15 mm pitch diameter up to 900 mm and with groove numbers from 10 to 120.

The design and selection process for timing belts is so similar to that for V belts that the process will not be presented here. As in the case of other belt drives, the manufacturers will provide an ample supply of information and details on sizes and strengths.

17–5 Roller Chain

Basic features of chain drives include a constant ratio, since no slippage or creep is involved; long life; and the ability to drive a number of shafts from a single source of power.

Roller chains have been standardized as to sizes by the ANSI. Figure 17–16 shows the nomenclature. The pitch is the linear distance between the centers of the rollers. The width is the space between the inner link plates. These chains are manufactured in single, double, triple, and quadruple strands. The dimensions of standard sizes are listed in Table 17–19.

Figure 17–17 shows a sprocket driving a chain and rotating in a counterclockwise direction. Denoting the chain pitch by p, the pitch angle by γ, and the pitch diameter of the sprocket by D, from the trigonometry of the figure we see

$$\sin\frac{\gamma}{2} = \frac{p/2}{D/2} \quad \text{or} \quad D = \frac{p}{\sin(\gamma/2)} \tag{a}$$

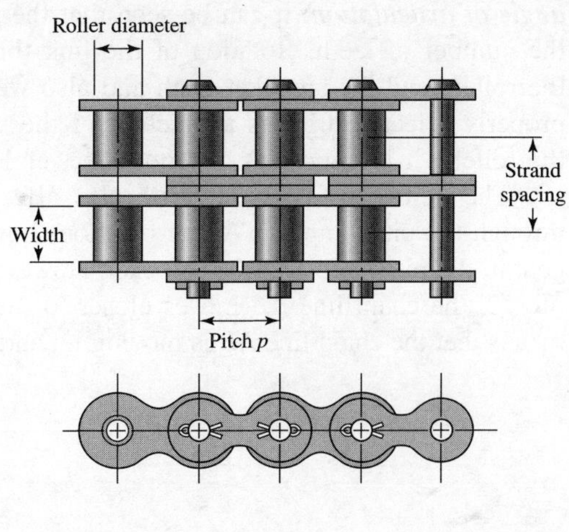

Figure 17–16

Portion of a double-strand roller chain.

Table 17–19 Dimensions of American Standard Roller Chains—Single Strand

ANSI Chain Number	Pitch, mm	Width, mm	Minimum Tensile Strength, N	Average Weight, N/m	Roller Diameter, mm	Multiple-Strand Spacing, mm
25	6.35	3.18	3 470	1.31	3.30	6.40
35	9.52	4.76	7 830	3.06	5.08	10.13
41	12.70	6.35	6 670	3.65	7.77	—
40	12.70	7.94	13 920	6.13	7.92	14.38
50	15.88	9.52	21 700	10.1	10.16	18.11
60	19.05	12.7	31 300	14.6	11.91	22.78
80	25.40	15.88	55 600	25.0	15.87	29.29
100	31.75	19.05	86 700	37.7	19.05	35.76
120	38.10	25.40	124 500	56.5	22.22	45.44
140	44.45	25.40	169 000	72.2	25.40	48.87
160	50.80	31.75	222 000	96.5	28.57	58.55
180	57.15	35.71	280 000	132.2	35.71	65.84
200	63.50	38.10	347 000	159.9	39.67	71.55
240	76.70	47.63	498 000	239	47.62	87.83

Source: Compiled from ANSI B29.1-1975.

Since $\gamma = 360°/N$, where N is the number of sprocket teeth, Equation (*a*) can be written

$$D = \frac{p}{\sin(180°/N)} \tag{17–29}$$

The angle $\gamma/2$, through which the link swings as it enters contact, is called the *angle of articulation*. It can be seen that the magnitude of this angle is a function of the number of teeth. Rotation of the link through this angle causes impact between the rollers and the sprocket teeth and also wear in the chain joint. Since the life of a properly selected drive is a function of the wear and the surface fatigue strength of the rollers, it is important to reduce the angle of articulation as much as possible.

The number of sprocket teeth also affects the velocity ratio during the rotation through the pitch angle γ. At the position shown in Figure 17–17, the chain AB is tangent to the pitch circle of the sprocket. However, when the sprocket has turned an angle of $\gamma/2$, the chain line AB moves closer to the center of rotation of the sprocket. This means that the chain line AB is moving up and down, and that the lever arm varies with

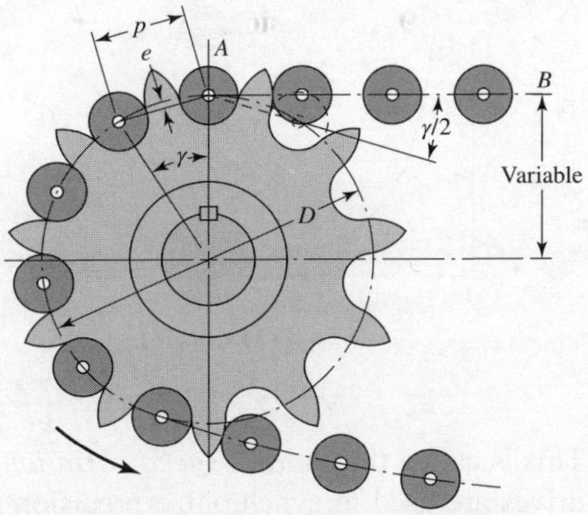

Figure 17–17

Engagement of a chain and sprocket.

rotation through the pitch angle, all resulting in an uneven chain exit velocity. You can think of the sprocket as a polygon in which the exit velocity of the chain depends upon whether the exit is from a corner, or from a flat of the polygon. Of course, the same effect occurs when the chain first enters into engagement with the sprocket.

The chain velocity V is defined as the number of meters coming off the sprocket per unit time. Thus the chain velocity in meters per minute is

$$V = \frac{Npn}{1000} \tag{17–30}$$

where N = number of sprocket teeth
p = chain pitch, mm
n = sprocket speed, rev/min

The maximum exit velocity of the chain is

$$v_{\max} = \frac{\pi Dn}{1000} = \frac{\pi np}{1000 \sin(\gamma/2)} \tag{b}$$

where Equation (*a*) has been substituted for the pitch diameter D. The minimum exit velocity occurs at a diameter d, smaller than D. Using the geometry of Figure 17–17, we find

$$d = D \cos\frac{\gamma}{2} \tag{c}$$

Thus the minimum exit velocity is

$$v_{\min} = \frac{\pi dn}{1000} = \frac{\pi np}{1000} \frac{\cos(\gamma/2)}{\sin(\gamma/2)} \tag{d}$$

Now substituting $\gamma/2 = 180°/N$ and employing Equations (17–30), (*b*), and (*d*), we find the speed variation to be

$$\frac{\Delta V}{V} = \frac{v_{\max} - v_{\min}}{V} = \frac{\pi}{N}\left[\frac{1}{\sin(180°/N)} - \frac{1}{\tan(180°/N)}\right] \tag{17–31}$$

Figure 17–18

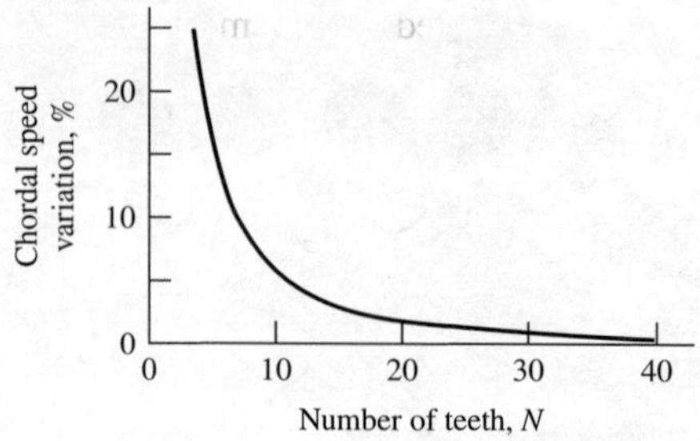

This is called the *chordal speed variation* and is plotted in Figure 17–18. When chain drives are used to synchronize precision components or processes, due consideration must be given to these variations. For example, if a chain drive synchronized the cutting of photographic film with the forward drive of the film, the lengths of the cut sheets of film might vary too much because of this chordal speed variation. Such variations can also cause vibrations within the system.

Although a large number of teeth is considered desirable for the driving sprocket, in the usual case it is advantageous to obtain as small a sprocket as possible, and this requires one with a small number of teeth. For smooth operation at moderate and high speeds it is considered good practice to use a driving sprocket with at least 17 teeth; 19 or 21 will, of course, give a better life expectancy with less chain noise. Where space limitations are severe or for very slow speeds, smaller tooth numbers may be used by sacrificing the life expectancy of the chain.

Driven sprockets are not made in standard sizes over 120 teeth, because the pitch elongation will eventually cause the chain to “ride” high long before the chain is worn out. The most successful drives have velocity ratios up to 6:1, but higher ratios may be used at the sacrifice of chain life.

Roller chains seldom fail because they lack tensile strength; they more often fail because they have been subjected to a great many hours of service. Actual failure may be due either to wear of the rollers on the pins or to fatigue of the surfaces of the rollers. Roller-chain manufacturers have compiled tables that give the horsepower capacity corresponding to a life expectancy of 15 kh for various sprocket speeds. These capacities are tabulated in Table 17–20 for 17-tooth sprockets. Table 17–21 displays available tooth counts on sprockets of one supplier. Table 17–22 lists the tooth correction factors for other than 17 teeth. Table 17–23 shows the multiple-strand factors K_2.

The capacities of chains are based on the following:

- 15 000 h at full load
- Single strand
- ANSI proportions
- Service factor of unity
- 100 pitches in length
- Recommended lubrication
- Elongation maximum of 3 percent
- Horizontal shafts
- Two 17-tooth sprockets

The fatigue strength of link plates governs capacity at lower speeds. The American Chain Association (ACA) publication *Chains for Power Transmission and Materials Handling* (1982) gives, for single-strand chain, the nominal power H_1, link-plate limited, as

$$H_1 = 2.98{\cdot}10^{-3}{\cdot}N_1^{1.08}n_1^{0.9}\left(\frac{p}{25.4}\right)^{(3-2.76{\cdot}10^{-3}p)} \qquad \text{kW} \tag{17–32}$$

and the nominal power H_2, roller-limited, as

$$H_2 = \frac{745.7K_rN_1^{1.5}(p/25.4)^{0.8}}{n_1^{1.5}} \qquad \text{kW} \tag{17–33}$$

where N_1 = number of teeth in the smaller sprocket

n_1 = sprocket speed, rev/min

p = pitch of the chain, mm

K_r = 29 for chain numbers 25, 35; 3.4 for chain 41; and 17 for chains 40–240

Table 17–20 Rated Capacity in kW of Single-Strand Single-Pitch Roller Chain for a 17-Tooth Sprocket

Sprocket Speed, rev/min	ANSI Chain Number					
	25	35	40	41	50	60
50	0.037	0.12	0.28	0.15	0.54	0.93
100	0.067	0.21	0.51	0.28	0.99	1.72
150	0.097*	0.30*	0.74*	0.42*	1.43*	2.48
200	0.12*	0.40*	0.96	0.53	1.87	3.20
300	0.17	0.58	1.38	0.75	2.69	4.63
400	0.22*	0.75*	1.80	0.98	3.50	6.00
500	0.28	0.93	2.20	1.20	4.25	7.32
600	0.33*	1.10*	2.60*	1.42*	5.01*	8.65
700	0.37	1.25	2.96	1.63	5.77	9.92
800	0.42*	1.40*	3.34*	1.84*	6.5*	11.20
900	0.46	1.56	3.72	2.04	7.23	12.50
1000	0.51*	1.72*	4.1	2.25	7.98	13.65
1200	0.60	2.04	4.81	2.45	9.40	16.11
1400	0.69*	2.33*	5.53	1.95	10.74	13.50
1600	0.78*	2.63*	6.24	1.60	9.55	11.00
1800	0.86	2.93	6.68	1.33	7.98	9.25
2000	0.95*	3.22*	5.76*	1.13*	6.89*	7.90
2500	1.16	3.94	4.11*	0.82*	4.90*	5.64
3000	1.37	4.2	3.11	0.62	3.72	4.30
Type A		Type B			Type C	

*Estimated from ANSI tables by linear interpolation.

Note: Type A—manual or drip lubrication; type B—bath or disk lubrication; type C—oil-stream lubrication.

Source: Compiled from ANSI B29.1-1975 information only section, and from B29.9-1958.

(*Continued*)

Table 17–20 Rated Capacity in kW of Single-Strand Single-Pitch Roller Chain for a 17-Tooth Sprocket (*Continued*)

Sprocket Speed, rev/min		ANSI Chain Number							
		80	100	120	140	160	180	200	240
50	Type A	2.15	4.11	7	10.7	15.6	21.6	28.6	46.1
100		4.01	7.7	13	20	29.2	40.3	53.4	85.8
150		5.78	11	18.7	29	42	58	76.8	123.8
200		7.46	14.3	24.2	37.5	54.4	75.3	100	160.4
300		10.82	20.7	35	54.4	78.3	108	144	231.3
400		14	26.8	45.2	70	101.5	140.2	185.8	268.0
500	Type B	17	32.7	55.3	85.8	123.8	152.1	165.6	0
600		20.1	38.6	65.1	94.7	105.2	115.6	126	
700		23.1	44.3	66.4	75.3	83.6	91.8	0	
800		26.1	47	54.3	61.5	68.4	75.3		
900		29.8	39.4	45.5	51.5	57.3	63		
1000		28.1	33.6	38.8	44	49	53.8		
1200		21.4	25.6	29.5	33.5	37.2	0		
1400		16.9	20.3	23.5	26.5	0			
1600		13.9	18.6	19.3	0				
1800		11.6	14	16.1					
2000		9.92	11.9	0					
2500		7.13	0.3						
3000		5.4	0						
Type C				Type C′					

Note: Type A—manual or drip lubrication; type B—bath or disk lubrication; type C—oil-stream lubrication; type C′—type C, but this is a galling region; submit design to manufacturer for evaluation.

Table 17–21 Single-Strand Sprocket Tooth Counts Available from One Supplier

No.	Available Sprocket Tooth Counts
25	8-30, 32, 34, 35, 36, 40, 42, 45, 48, 54, 60, 64, 65, 70, 72, 76, 80, 84, 90, 95, 96, 102, 112, 120
35	4-45, 48, 52, 54, 60, 64, 65, 68, 70, 72, 76, 80, 84, 90, 95, 96, 102, 112, 120
41	6-60, 64, 65, 68, 70, 72, 76, 80, 84, 90, 95, 96, 102, 112, 120
40	8-60, 64, 65, 68, 70, 72, 76, 80, 84, 90, 95, 96, 102, 112, 120
50	8-60, 64, 65, 68, 70, 72, 76, 80, 84, 90, 95, 96, 102, 112, 120
60	8-60, 62, 63, 64, 65, 66, 67, 68, 70, 72, 76, 80, 84, 90, 95, 96, 102, 112, 120
80	8-60, 64, 65, 68, 70, 72, 76, 78, 80, 84, 90, 95, 96, 102, 112, 120
100	8-60, 64, 65, 67, 68, 70, 72, 74, 76, 80, 84, 90, 95, 96, 102, 112, 120
120	9-45, 46, 48, 50, 52, 54, 55, 57, 60, 64, 65, 67, 68, 70, 72, 76, 80, 84, 90, 96, 102, 112, 120
140	9-28, 30, 31, 32, 33, 34, 35, 36, 37, 39, 40, 42, 43, 45, 48, 54, 60, 64, 65, 68, 70, 72, 76, 80, 84, 96
160	8-30, 32-36, 38, 40, 45, 46, 50, 52, 53, 54, 56, 57, 60, 62, 63, 64, 65, 66, 68, 70, 72, 73, 80, 84, 96
180	13-25, 28, 35, 39, 40, 45, 54, 60
200	9-30, 32, 33, 35, 36, 39, 40, 42, 44, 45, 48, 50, 51, 54, 56, 58, 59, 60, 63, 64, 65, 68, 70, 72
240	9-30, 32, 35, 36, 40, 44, 45, 48, 52, 54, 60

Source: Data from Morse chain company, Ithaca, NY, Type B hub sprockets.

Table 17–22 Tooth Correction Factors, K_1

Number of Teeth on Driving Sprocket	K_1 Pre-extreme power	K_1 Post-extreme power
11	0.62	0.52
12	0.69	0.59
13	0.75	0.67
14	0.81	0.75
15	0.87	0.83
16	0.94	0.91
17	1.00	1.00
18	1.06	1.09
19	1.13	1.18
20	1.19	1.28
N	$(N_1/17)^{1.08}$	$(N_1/17)^{1.5}$

Table 17–23 Multiple-Strand Factors, K_2

Number of Strands	K_2
1	1.0
2	1.7
3	2.5
4	3.3
5	3.9
6	4.6
8	6.0

The constant $2.98{\cdot}10^{-3}$ in Equation (17–32) becomes $1.64{\cdot}10^{-3}$ for no. 41 light-weight chain. The nominal horsepower in Table 17–20 is $H_{nom} = \min(H_1, H_2)$. For example, for $N_1 = 17$, $n_1 = 1000$ rev/min, no. 40 chain with $p = 12.7$ mm, from Equation (17–32),

$$H_1 = 2.98{\cdot}10^{-3}{\cdot}17^{1.08}{\cdot}1000^{0.9}\left(\frac{12.7}{25.4}\right)^{(13-2.76\cdot10^{-3}\cdot12.7)} = 4.079 \text{ kW}$$

From Equation (17–33),

$$H_2 = \frac{745.7{\cdot}17{\cdot}17^{1.5}(12.7/25.4)^{0.8}}{1000^{1.5}} = 16.14 \text{ kW}$$

The tabulated value in Table 17–20 is $H_{tab} = \min(4.08, 16.14) = 4.08$ kW.

It is preferable to have an odd number of teeth on the driving sprocket (17, 19, . . .) and an even number of pitches in the chain to avoid a special link. The approximate length of the chain L in pitches is

$$\frac{L}{P} \approx \frac{2C}{p} + \frac{N_1 + N_2}{2} + \frac{(N_2 - N_1)^2}{4\pi^2 C/p} \tag{17–34}$$

The center-to-center distance C is given by

$$C = \frac{p}{4}\left[-A + \sqrt{A^2 - 8\left(\frac{N_2 - N_1}{2\pi}\right)^2}\right] \tag{17–35}$$

where

$$A = \frac{N_1 + N_2}{2} - \frac{L}{p} \tag{17–36}$$

The allowable power H_a is given by

$$H_a = K_1 K_2 H_{\text{tab}} \tag{17–37}$$

where K_1 = correction factor for tooth number other than 17 (Table 17–22)
K_2 = strand correction (Table 17–23)

The horsepower that must be transmitted H_d is given by

$$H_d = H_{\text{nom}} K_s n_d \tag{17–38}$$

where K_s is a service factor to account for nonuniform loads, and n_d is a design factor.

Equation (17–32) is the basis of the pre-extreme power entries (vertical entries) of Table 17–20, and the chain power is limited by link-plate fatigue. Equation (17–33) is the basis for the post-extreme power entries of these tables, and the chain power performance is limited by impact fatigue. The entries are for chains of 100 pitch length and 17-tooth sprocket. For a deviation from this

$$H_2 = 1000\left[K_r\left(\frac{N_1}{n_1}\right)^{1.5}\left(\frac{p}{25.4}\right)^{0.8}\left(\frac{L_p}{745.7}\right)^{0.4}\left(\frac{15\ 000}{h}\right)^{0.4}\right] \tag{17–39}$$

where L_p is the chain length in pitches and h is the chain life in hours. Viewed from a deviation viewpoint, Equation (17–39) can be written as a trade-off equation in the following form:

$$\frac{H_2^{2.5} h}{N_1^{3.75} L_p} = \text{constant} \tag{17–40}$$

If tooth-correction factor K_1 is used, then omit the term $N_1^{3.75}$.

In Equation (17–40) one would expect the h/L_p term because doubling the hours can require doubling the chain length, other conditions constant, for the same number of cycles. Our experience with contact stresses leads us to expect a load (tension) life relation of the form $F^a L$ = constant. In the more complex circumstance of roller-bushing impact, the Diamond Chain Company has identified $a = 2.5$.

The maximum speed (rev/min) for a chain drive is limited by galling between the pin and the bushing. Tests suggest

$$n_1 \le 1000\left[\frac{82.5}{7.95^{p/25.4}(1.0278)^{N_1}(1.323)^{F/4448}}\right]^{1/\left(1.59\log\left(\frac{p}{25.4}\right)+1.873\right)} \quad \text{rev/min}$$

where F is the chain tension in N.

EXAMPLE 17–5

Select drive components for a 2:1 reduction, 67 kW input at 300 rev/min, moderate shock, an abnormally long 18-hour day, poor lubrication, cold temperatures, dirty surroundings, short drive $C/p = 25$.

Solution

Function: $H_{\text{nom}} = 67$ kW, $n_1 = 300$ rev/min, $C/p = 25$
Design factor: Choose $n_d = 1.5$
Service factor: Choose $K_s = 1.3$ for moderate shock
Sprocket teeth: $N_1 = 17$ teeth, $N_2 = 34$ teeth, $K_1 = 1$, $K_2 = 1, 1.7, 2.5, 3.3$
Chain number of strands: From Equations (17–37) and (17–38), with $H_a = H_d$,

$$H_{\text{tab}} = \frac{n_d K_s H_{\text{nom}}}{K_1 K_2} = \frac{1.5(1.3)67}{(1)K_2} = \frac{131}{K_2}$$

Form a table:

Number of Strands	$131/K_2$ (Table 17–23)	Chain Number (Table 17–20)	Lubrication Type
1	$131/1 = 176$	200	C′
2	$131/1.7 = 104$	160	C
3	$131/2.5 = 70.4$	140	B
4	$131/3.3 = 53.3$	140	B

Decision

3 strands of number 140 chain (from Table 17–20, H_{tab} is 53.99 kW).

Number of pitches in the chain:

$$\frac{L}{p} = \frac{2C}{p} + \frac{N_1 + N_2}{2} + \frac{(N_2 - N_1)^2}{4\pi^2 C/p}$$

$$= 2(25) + \frac{17 + 34}{2} + \frac{(34 - 17)^2}{4\pi^2(25)} = 75.79 \text{ pitches}$$

Decision

Use 76 pitches. Then $L/p = 76$.

Identify the center-to-center distance: From Equations (17–35) and (17–36),

$$A = \frac{N_1 + N_2}{2} - \frac{L}{p} = \frac{17 + 34}{2} - 76 = -50.5$$

$$C = \frac{p}{4}\left[-A + \sqrt{A^2 - 8\left(\frac{N_2 - N_1}{2\pi}\right)^2}\right]$$

$$= \frac{p}{4}\left[50.5 + \sqrt{50.5^2 - 8\left(\frac{34 - 17}{2\pi}\right)^2}\right] = 25.104p$$

For a 140 chain, $p = 44.45$ mm. Thus,

$$C = 25.104p = 25.104(44.45) = 1115.9 \text{ mm}$$

Lubrication: Type B

Comment: This is operating on the pre-extreme portion of the power, so durability estimates other than 15 000 h are not available. Given the poor operating conditions, life will be much shorter.

Lubrication of roller chains is essential in order to obtain a long and trouble-free life. Either a drip feed or a shallow bath in the lubricant is satisfactory. A medium or light mineral oil, without additives, should be used. Except for unusual conditions, heavy oils and greases are not recommended, because they are too viscous to enter the small clearances in the chain parts.

17–6 Wire Rope

Wire rope is made with two types of winding, as shown in Figure 17–19. The *regular lay,* which is the accepted standard, has the wire twisted in one direction to form the strands, and the strands twisted in the opposite direction to form the rope. In the completed rope the visible wires are approximately parallel to the axis of the rope. Regular-lay ropes do not kink or untwist and are easy to handle.

Figure 17–19

Types of wire rope; both lays are available in either right or left hand.

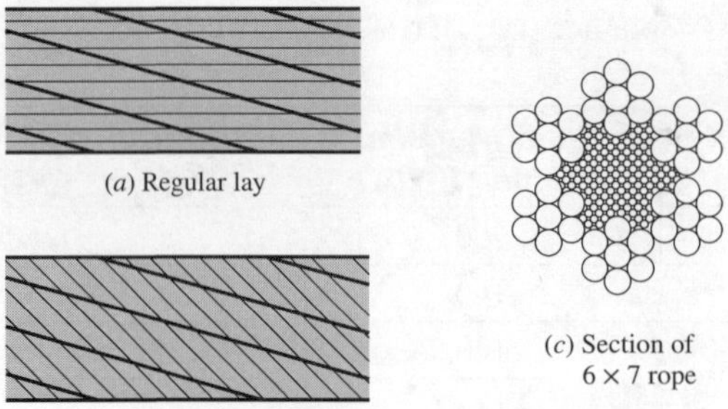

(*a*) Regular lay

(*b*) Lang lay

(*c*) Section of 6 × 7 rope

Lang-lay ropes have the wires in the strand and the strands in the rope twisted in the same direction, and hence the outer wires run diagonally across the axis of the rope. Lang-lay ropes are more resistant to abrasive wear and failure due to fatigue than are regular-lay ropes, but they are more likely to kink and untwist.

Standard ropes are made with a hemp core, which supports and lubricates the strands. When the rope is subjected to heat, either a steel center or a wire-strand center must be used.

Wire rope is designated as, for example, a 38-mm 6 × 7 haulage rope. The first figure is the diameter of the rope (Figure 17–19*c*). The second and third figures are the number of strands and the number of wires in each strand, respectively. Table 17–24

Table 17–24 Wire-Rope Data

Rope	Weight per meter, (10^{-3})N	Minimum Sheave Diameter, mm	Standard Sizes *d*, mm	Material	Size of Outer Wires	Modulus of Elasticity,* GPa	Strength,† MPa
6 × 7 haulage	$33.92d^2$	$42d$	6 – 38	Monitor steel	$d/9$	96	690
				Plow steel	$d/9$	96	608
				Mild plow steel	$d/9$	96	524
6 × 19 standard hoisting	$36.18d^2$	$26d$–$34d$	6 – 70	Monitor steel	$d/13$–$d/16$	83	730
				Plow steel	$d/13$–$d/16$	83	640
				Mild plow steel	$d/13$–$d/16$	83	550
6 × 37 special flexible	$35.05d^2$	$18d$	6 – 90	Monitor steel	$d/22$	76	690
				Plow steel	$d/22$	76	608
8 × 19 extra flexible	$32.79d^2$	$21d$–$26d$	6 – 38	Monitor steel	$d/15$–$d/19$	69	634
				Plow steel	$d/15$–$d/19$	69	550
7 × 7 aircraft	$38.45d^2$	—	1.6 – 10	Corrosion-resistant steel	—	—	850
				Carbon steel	—	—	850
7 × 9 aircraft	$39.58d^2$	—	3 – 36	Corrosion-resistant steel	—	—	930
				Carbon steel	—	—	986
19-wire aircraft	$48.62d^2$	—	0.8 – 8	Corrosion-resistant steel	—	—	1137
				Carbon steel	—	—	1137

*The modulus of elasticity is only approximate; it is affected by the loads on the rope and, in general, increases with the life of the rope.

†The strength is based on the nominal area of the rope. The figures given are only approximate and are based on 25.4 mm rope sizes and 6.35 mm aircraft-cable sizes.

Source: Compiled from *American Steel and Wire Company Handbook.*

lists some of the various ropes that are available, together with their characteristics and properties. The area of the metal in standard hoisting and haulage rope is about $A_m = 0.38d^2$.

When a wire rope passes around a sheave, there is a certain amount of readjustment of the elements. Each of the wires and strands must slide on several others, and presumably some individual bending takes place. It is probable that in this complex action there exists some stress concentration. The stress in one of the wires of a rope passing around a sheave may be calculated as follows. From solid mechanics, we have

$$M = \frac{EI}{\rho} \qquad \text{and} \qquad M = \frac{\sigma I}{c} \tag{a}$$

where the quantities have their usual meaning. Eliminating M and solving for the stress gives

$$\sigma = \frac{Ec}{\rho} \tag{b}$$

For the radius of curvature ρ, we can substitute the sheave radius $D/2$. Also, $c = d_w/2$, where d_w is the wire diameter. These substitutions give

$$\sigma = E_r \frac{d_w}{D} \tag{c}$$

where E_r is the *modulus of elasticity of the rope,* not the wire. To understand this equation, observe that the individual wire makes a corkscrew figure in space and if you pull on it to determine E it will stretch or give more than its native E would suggest. Therefore E is still the modulus of elasticity of the *wire,* but in its peculiar configuration as part of the rope, its modulus is smaller. For this reason we say that E_r in Equation (*c*) is the modulus of elasticity of the rope, not the wire, recognizing that one can quibble over the name used.

Equation (*c*) gives the tensile stress σ in the outer wires. The sheave diameter is represented by D. This equation reveals the importance of using a large-diameter sheave. The suggested minimum sheave diameters in Table 17–24 are based on a D/d_w ratio of 400. If possible, the sheaves should be designed for a larger ratio. For elevators and mine hoists, D/d_w is usually taken from 800 to 1000. If the ratio is less than 200, heavy loads will often cause a permanent set in the rope.

A wire rope tension giving the same tensile stress as the sheave bending is called the *equivalent bending load* F_b, given by

$$F_b = \sigma A_m = \frac{E_r d_w A_m}{D} \tag{17–41}$$

A wire rope may fail because the static load exceeds the ultimate strength of the rope. Failure of this nature is generally not the fault of the designer, but rather that of the operator in permitting the rope to be subjected to loads for which it was not designed.

The first consideration in selecting a wire rope is to determine the static load. This load is composed of the following items:

- The known or dead weight
- Additional loads caused by sudden stops or starts
- Shock loads
- Sheave-bearing friction

When these loads are summed, the total can be compared with the ultimate strength of the rope to find a factor of safety. However, the ultimate strength used in this determination must be reduced by the strength loss that occurs when the rope passes over a curved surface such as a stationary sheave or a pin; see Figure 17–20.

For an average operation, use a factor of safety of 5. Factors of safety up to 8 or 9 are used if there is danger to human life and for very critical situations. Table 17–25 lists minimum factors of safety for a variety of design situations. Here, the factor of safety is defined as

$$n = \frac{F_u}{F_t}$$

where F_u is the ultimate wire load and F_t is the largest working tension.

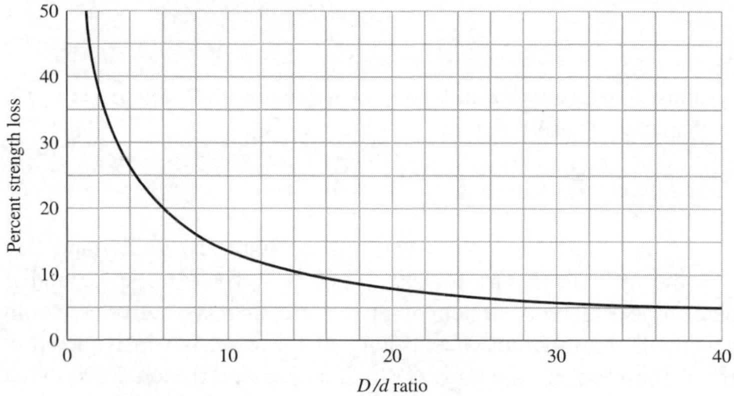

Figure 17–20

Percent strength loss due to different D/d ratios; derived from standard test data for 6 × 19 and 6 × 17 class ropes. *(Source: Data from the Wire Rope Technical Board (WRTB),* Wire Rope Users Manual Third Edition, *Second printing.)*

Table 17–25 Minimum Factors of Safety for Wire Rope*

Track cables	3.2	Passenger elevators, m/s:	
Guys	3.5	0.25	7.60
Mine shafts, ft:		1.52	9.20
Up to 152.5	8.0	4.06	11.25
305–610	7.0	6.10	11.80
610–915	6.0	7.62	11.90
Over 915	5.0	Freight elevators, m/s:	
Hoisting	5.0	0.25	6.65
Haulage	6.0	1.52	8.20
Cranes and derricks	6.0	4.06	10.00
Electric hoists	7.0	6.10	10.50
Hand elevators	5.0	7.62	10.55
Private elevators	7.5	Powered dumbwaiters, m/s:	
Hand dumbwaiter	4.5	0.25	4.8
Grain elevators	7.5	1.52	6.6
		4.06	8.0

*Use of these factors does not preclude a fatigue failure.

Source: Compiled from a variety of sources, including ANSI A17.1-1978.

Once you have made a tentative selection of a rope based upon static strength, the next consideration is to ensure that the wear life of the rope and the sheave or sheaves meets certain requirements. When a loaded rope is bent over a sheave, the rope stretches like a spring, rubs against the sheave, and causes wear of both the rope and the sheave. The amount of wear that occurs depends upon the pressure of the rope in the sheave groove. This pressure is called the *bearing pressure;* a good estimate of its magnitude is given by

$$P = \frac{2F}{dD} \tag{17–42}$$

where F = tensile force on rope

d = rope diameter

D = sheave diameter

The allowable pressures given in Table 17–26 are to be used only as a rough guide; they may not prevent a fatigue failure or severe wear. They are presented here because they represent past practice and furnish a starting point in design.

A fatigue diagram not unlike an *S-N* diagram can be obtained for wire rope. Such a diagram is shown in Figure 17–21. Here the ordinate is the pressure-strength ratio p/S_u, and S_u is the ultimate tensile strength of the *wire*. The abscissa is the number of bends that occur in the total life of the rope. The curve implies that a wire rope has a fatigue limit; but this is not true at all. A wire rope that is used over sheaves will eventually fail in fatigue or in wear. However, the graph does show that the rope will have a long life if the ratio p/S_u is less than 0.001. Substitution of this ratio in Equation (17–42) gives

$$S_u = \frac{2000F}{dD} \tag{17–43}$$

where S_u is the ultimate strength of the *wire,* not the rope, and the units of S_u are related to the units of F. This interesting equation contains the wire strength, the load,

Table 17–26 Maximum Allowable Bearing Pressures of Ropes on Sheaves (in MPa)

	Sheave Material				
Rope	Wood[a]	Cast Iron[b]	Cast Steel[c]	Chilled Cast Irons[d]	Manganese Steel[e]
Regular lay:					
6 × 7	1.0	2.1	3.8	4.5	10.1
6 × 19	1.7	3.3	6.2	7.6	16.6
6 × 37	2.1	4.0	7.4	9.1	20.7
8 × 19	2.4	4.7	8.7	10.7	24.1
Lang lay:					
6 × 7	1.1	2.4	4.1	4.9	11.4
6 × 19	1.9	3.8	6.9	8.3	19.0
6 × 37	2.3	4.6	8.1	10.0	22.8

[a]On end grain of beech, hickory, or gum.

[b]For H_B (min.) = 125.

[c]30–40 carbon; H_B (min.) = 160.

[d]Use only with uniform surface hardness.

[e]For high speeds with balanced sheaves having ground surfaces.

Source: Data from Wire Rope Users Manual, AISI, 1979.

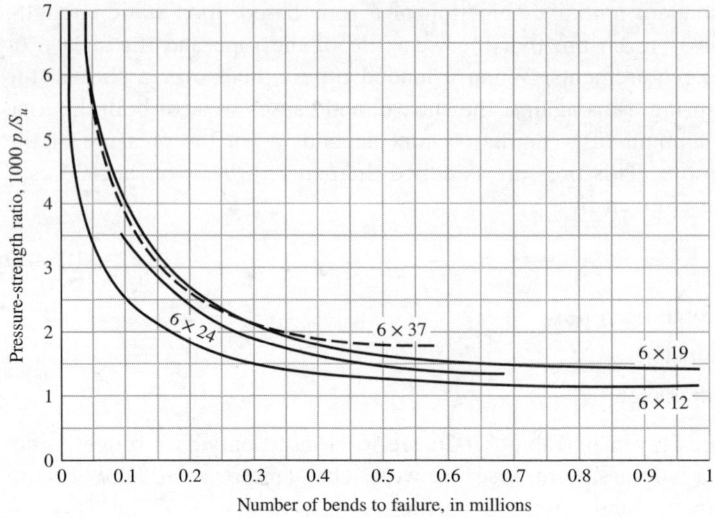

Figure 17–21

Experimentally determined relation between the fatigue life of wire rope and the sheave pressure.

the rope diameter, and the sheave diameter—all four variables in a single equation! Dividing both sides of Equation (17–42) by the ultimate strength of the wires S_u and solving for F gives

$$F_f = \frac{(p/S_u)S_u dD}{2} \tag{17–44}$$

where F_f is interpreted as the allowable fatigue tension as the wire is flexed a number of times corresponding to p/S_u selected from Figure 17–21 for a particular rope and life expectancy. The factor of safety can be defined in fatigue as

$$n_f = \frac{F_f - F_b}{F_t} \tag{17–45}$$

where F_f is the rope tension strength under flexing and F_t is the tension at the place where the rope is flexing. Unfortunately, the designer often has vendor information that tabulates ultimate rope tension and gives no ultimate-strength S_u information concerning the wires from which the rope is made. Some guidance in strength of individual wires is

Improved plow steel (monitor)	$1655 < S_u < 1930$ MPa
Plow steel	$1448 < S_u < 1655$ MPa
Mild plow steel	$1241 < S_u < 1448$ MPa

In wire-rope usage, the factor of safety is defined for static loading as

$$n_s = \frac{F_u - F_b}{F_t} \tag{17–46}$$

where F_b is the rope tension that would induce the same outer-wire stress as that given by Equation (c). Be careful when comparing recommended static factors of safety to Equation (17–46), as n_s is sometimes defined as F_u/F_t. The factor of safety in fatigue loading can be defined as in Equation (17–45), or by using a static analysis and compensating with a large factor of safety applicable to static loading, as in Table 17–25.

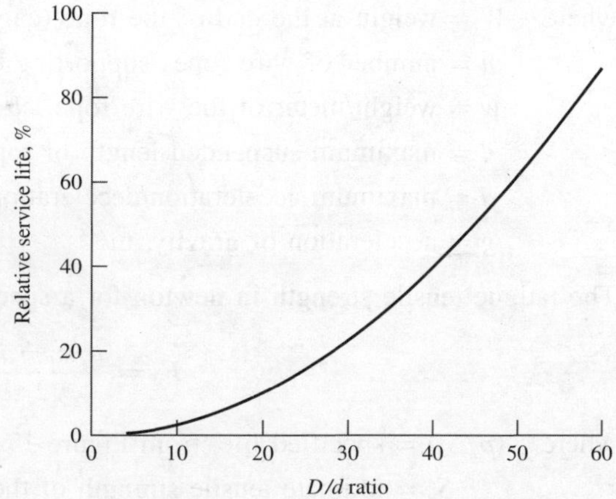

Figure 17–22

Service-life curve based on bending and tensile stresses only. This curve shows that the life corresponding to $D/d = 48$ is twice that of $D/d = 33$. *(Source: Wire Rope Technical Board (WRTB),* Wire Rope Users Manual Third Edition, *Second printing.)*

When using factors of safety expressed in codes, standards, corporate design manuals, or wire-rope manufacturers' recommendations or from the literature, be sure to ascertain upon which basis the factor of safety is to be evaluated, and proceed accordingly.

If the rope is made of plow steel, the wires are probably hard-drawn AISI 1070 or 1080 carbon steel. Referring to Table 10–3, we see that this lies somewhere between hard-drawn spring wire and music wire. But the constants m and A needed to solve Equation (10–14), for S_u are lacking.

Practicing engineers who desire to solve Equation (17–43) should determine the wire strength S_u for the rope under consideration by unraveling enough wire to test for the Brinell hardness. Then S_u can be found using Equation (2–21). Fatigue failure in wire rope is not sudden, as in solid bodies, but progressive, and shows as the breaking of an outside wire. This means that the beginning of fatigue can be detected by periodic routine inspection.

Figure 17–22 is another graph showing the gain in life to be obtained by using large D/d ratios. In view of the fact that the life of wire rope used over sheaves is only finite, it is extremely important that the designer specify and insist that periodic inspection, lubrication, and maintenance procedures be carried out during the life of the rope. Table 17–27 gives useful properties of some wire ropes.

For a mine-hoist problem we present here the working equations, some repeated from the preceding presentation. The wire rope tension F_t due to load and acceleration/deceleration is

$$F_t = \left(\frac{W}{m} + wl\right)\left(1 + \frac{a}{g}\right) \tag{17–47}$$

Table 17–27 Some Useful Properties of 6 × 7, 6 × 19, and 6 × 37 Wire Ropes

Wire Rope	Weight per Foot w, N/m	Weight per Foot Including Core w, N/m	Minimum Sheave Diameter D, mm	Better Sheave Diameter D, mm	Diameter of Wires d_w, mm	Area of Metal A_m, mm^2	Rope Young's Modulus E_r, GPa
6 × 7	$0.0344d^2$		$42d$	$72d$	$0.111d$	$0.38d^2$	89.6
6 × 19	$0.0362d^2$	$0.0398d^2$	$30d$	$45d$	$0.067d$	$0.40d^2$	82.7
6 × 37	$0.0351d^2$	$0.0387d^2$	$18d$	$27d$	$0.048d$	$0.40d^2$	82.7

where W = weight at the end of the rope (cage and load), N
m = number of wire ropes supporting the load
w = weight/meter of the wire rope, N/m
l = maximum suspended length of rope, m
a = maximum acceleration/deceleration experienced, m/s^2
g = acceleration of gravity, m/s^2

The fatigue tensile strength in newton for a specified life is

$$F_f = \frac{(p/S_u)S_uDd}{2} \tag{17–44}$$

where (p/S_u) = specified life, from Figure 17–21
S_u = ultimate tensile strength of the wires, Pa
D = sheave or winch drum diameter, m
d = nominal wire rope size, m

The *equivalent bending load* F_b is

$$F_b = \frac{E_r d_w A_m}{D} \tag{17–41}$$

where E_r = Young's modulus for the wire rope, Table 17–24 or 17–27, Pa
d_w = diameter of the wires, m
A_m = metal cross-sectional area, Table 17–27, m^2
D = sheave or winch drum diameter, m

The static factor of safety n_s is

$$n_s = \frac{F_u - F_b}{F_t} \tag{17–46}$$

sometimes defined as F_u/F_t. The fatigue factor of safety n_f is

$$n_f = \frac{F_f - F_b}{F_t} \tag{17–45}$$

EXAMPLE 17–6

Given a 6 × 19 monitor steel (S_u = 1650 MPa) wire rope.
(*a*) Develop the expressions for rope tension F_t, fatigue tension F_f, equivalent bending tensions F_b, and fatigue factor of safety n_f for a 162 m, 8900 N cage-and-load mine hoist with a starting acceleration of 0.6 m/s^2 as depicted in Fig. 17–23. The sheave diameter is 1830 mm.
(*b*) Using the expressions developed in part (*a*), examine the variation in factor of safety n_f for various wire rope diameters d and number of supporting ropes m.

Solution

(*a*) Rope tension F_t from Eq. (17–47), using Table 17–24 for w, is given by

Answer

$$F_t = \left(\frac{W}{m} + wl\right)\left(1 + \frac{a}{g}\right) = \left[\frac{8900}{m} + 0.0362d^2(162)\right]\left(1 + \frac{0.6}{9.81}\right)$$

$$= \frac{9444}{m} + 6.22d^2$$

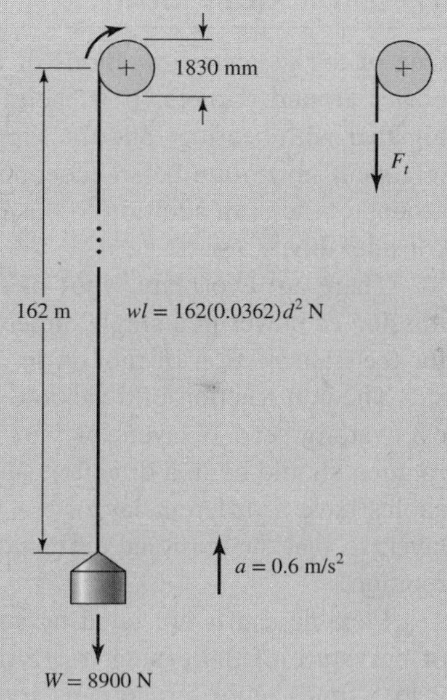

Figure 17–23

Geometry of the mine hoist of Example 17–6.

From Fig. 17–21, use $p/S_u = 0.0014$. Fatigue tension F_f from Eq. (17–44) is given by

Answer
$$F_f = \frac{(p/S_u)S_uDd}{2} = \frac{0.0014(1650)1830d}{2} = 2114d$$

Equivalent bending tension F_b from Eq. (17–41) and Table 17–27 is given by

Answer
$$F_b = \frac{E_r d_w A_m}{D} = \frac{82.7(10^3)0.067d(0.40d^2)}{1830} = 1.211d^3$$

Factor of safety n_f in fatigue from Eq. (17–45) is given by

Answer
$$n_f = \frac{F_f - F_b}{F_t} = \frac{2114d - 1.211d^3}{9444/m + 6.22d^2}$$

(*b*) Using a spreadsheet program, form a table as follows:

	n_f			
d	$m = 1$	$m = 2$	$m = 3$	$m = 4$
6.5	1.381	2.690	3.931	5.110
10	1.980	3.729	5.286	6.681
13	2.365	4.299	5.911	7.274
16	2.615	4.571	6.089	7.301
19	2.726	4.573	5.907	6.916
23	2.661	4.229	5.263	5.996
25	2.545	3.941	4.822	5.430

Wire rope sizes are discrete, as is the number of supporting ropes. Note that for each *m* the factor of safety exhibits a maximum. Predictably the largest factor of safety increases with *m*. If the required factor of safety were to be 6, only three or four ropes could meet the requirement. The sizes are different: $\frac{5}{8}$-in ropes with three ropes or $\frac{3}{8}$-in ropes with four ropes. The costs include not only the wires, but the grooved winch drums.

17–7 Flexible Shafts

One of the greatest limitations of the solid shaft is that it cannot transmit motion or power around corners. It is therefore necessary to resort to belts, chains, or gears, together with bearings and the supporting framework associated with them. The flexible shaft may often be an economical solution to the problem of transmitting motion around corners. In addition to the elimination of costly parts, its use may reduce noise considerably.

There are two main types of flexible shafts: the power-drive shaft for the transmission of power in a single direction, and the remote-control or manual-control shaft for the transmission of motion in either direction.

The construction of a flexible shaft is shown in Figure 17–24. The cable is made by winding several layers of wire around a central core. For the power-drive shaft, rotation should be in a direction such that the outer layer is wound up. Remote-control cables have a different lay of the wires forming the cable, with more wires in each layer, so that the torsional deflection is approximately the same for either direction of rotation.

Flexible shafts are rated by specifying the torque corresponding to various radii of curvature of the casing. A 15-in radius of curvature, for example, will give from 2 to 5 times more torque capacity than a 7-in radius. When flexible shafts are used in a drive in which gears are also used, the gears should be placed so that the flexible shaft runs at as high a speed as possible. This permits the transmission of the maximum amount of horsepower.

Figure 17–24

Flexible shaft: (*a*) construction details; (*b*) a variety of configurations. *(Courtesy of S. S. White Technologies, Inc.)*

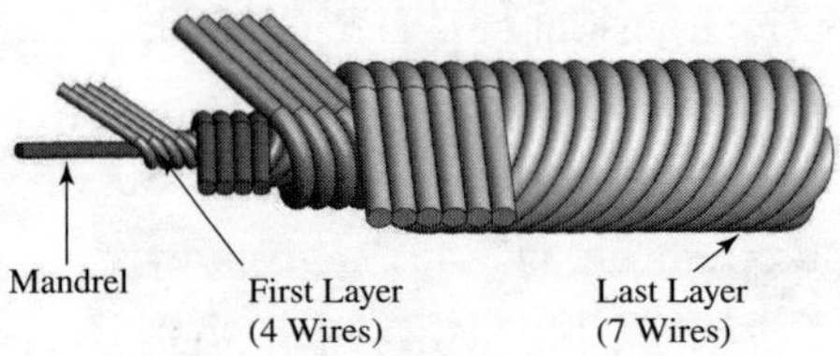

(*a*)

(*b*)

PROBLEMS

17–1 Example 17–2 resulted in selection of a 250 mm wide A-3 polyamide flat belt. Show that the value of F_1 restoring f to 0.80 is

$$F_1 = \frac{(\Delta F + F_c)\exp(f\phi) - F_c}{\exp(f\phi) - 1}$$

and compare the initial tensions.

17–2 A 150-mm wide polyamide F-1 flat belt is used to connect a 50 mm diameter pulley to drive a larger pulley with an angular velocity ratio of 0.5. The center-to-center distance is 2.7 m. The angular speed of the small pulley is 1750 rev/min as it delivers 1.5 kW. The service is such that a service factor K_s of 1.25 is appropriate.
(*a*) Find F_c, F_i, $(F_1)_a$, and F_2, assuming operation at the maximum tension limit.
(*b*) Find H_a, n_{fs}, and belt length.
(*c*) Find the dip.

17–3 Perspective and insight can be gained by doubling all geometric dimensions and observing the effect on problem parameters. Take the drive of Problem 17–2, double the dimensions, and compare.

17–4 A flat-belt drive is to consist of two 1.2 m diameter cast-iron pulleys spaced 4.8 m apart. Select a belt type to transmit 45 kW at a pulley speed of 380 rev/min. Use a service factor of 1.1 and a design factor of 1.0.

17–5 In solving problems and examining examples, you probably have noticed some recurring forms:

$$w = \gamma bt = (\gamma t)b = a_1 b,$$

$$(F_1)_a = F_a b C_p C_v = (F_a C_p C_v)b = a_0 b$$

$$F_c = \frac{wV^2}{g} = a_1 b\ V^2 = a_2 b$$

$$(F_1)_a - F_2 = 2T/d = H_d/V = H_{\text{nom}} K_s n_d / V$$

$$F_2 = (F_1)_a - [(F_1)_a - F_2] = a_0 b - 2T/d$$

$$f\phi = \ln\frac{(F_1)_a - F_c}{F_2 - F_c} = \ln\frac{(a_0 - a_2)b}{(a_0 - a_2)b - 2T/d}$$

Show that

$$b = \frac{1}{a_0 - a_2}\,\frac{H_d}{V}\,\frac{\exp(f\phi)}{\exp(f\phi) - 1}$$

17–6 Return to Example 17–1 and complete the following.
(*a*) What is the minimum initial tension, F_i, that would put the drive as built at the point of slip?
(*b*) With the tension from part (*a*), find the belt width b that exhibits $n_{fs} = n_d = 1.1$.
(*c*) For the belt width from part (*b*) find the corresponding $(F_1)_a$, F_c, F_i, F_2, power, and n_{fs}, assuming operation at the tension limit.
(*d*) What have you learned?

17–7 Take the drive of Example 17–1 and double the belt width. Compare F_c, F_i, $(F_1)_a$, F_2, H_a, n_{fs}, and dip, assuming operation at the tension limit.

17–8 Belted pulleys place loads on shafts, inducing bending and loads on bearings. Examine Figure 17–7 and develop an expression for the load the belt places on the pulley, and then apply it to Example 17–2.

17–9 The line shaft illustrated in the figure is used to transmit power from an electric motor by means of flat-belt drives to various machines. Pulley *A* is driven by a vertical belt from the motor pulley. A belt from pulley *B* drives a machine tool at an angle of 70° from the vertical and at a center-to-center distance of 2.7 m. Another belt from pulley *C* drives a grinder at a center-to-center distance of 3.4 m. Pulley *C* has a double width to permit belt shifting as shown in Figure 17–4. The belt from pulley *D* drives a dust-extractor fan whose axis is located horizontally 2.4 m from the axis of the lineshaft. Additional data are

Machine	Speed, rev/min	Power, kW	Lineshaft Pulley	Diameter, mm
Machine tool	400	9.3	*B*	400
Grinder	300	3.4	*C*	350
Dust extractor	500	6.0	*D*	450

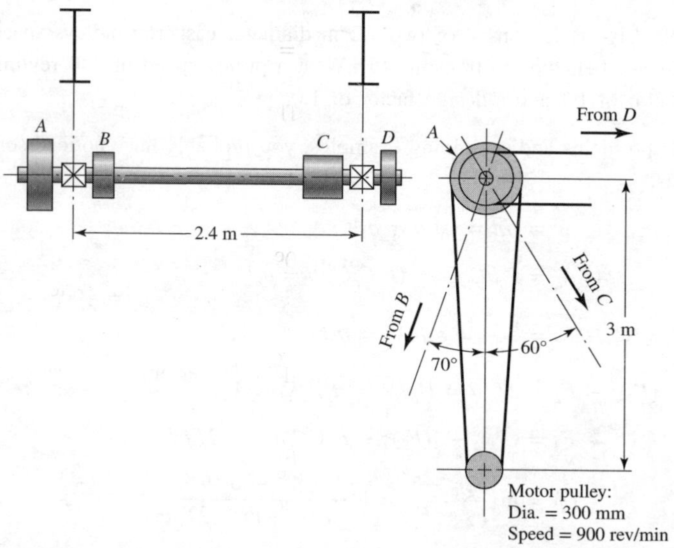

Problem 17–9

(Source: Dr. Ahmed F. Abdel Azim, Zagazig University, Cairo.)

The power requirements, listed above, account for the overall efficiencies of the equipment. The two line-shaft bearings are mounted on hangers suspended from two overhead wide-flange beams. Select the belt types and sizes for each of the four drives. Make provision for replacing belts from time to time because of wear or permanent stretch.

17–10 Two shafts 6 m apart, with axes in the same horizontal plane, are to be connected with a flat belt in which the driving pulley, powered by a six-pole squirrel-cage induction motor with a 75 kW rating at 1140 rev/min, drives the second shaft at half its angular speed. The driven shaft drives light-shock machinery loads. Select a flat belt.

17–11 The mechanical efficiency of a flat-belt drive is approximately 98 percent. Because of its high value, the efficiency is often neglected. If a designer should choose to include it, where should it be inserted in the flat-belt protocol?

17–12 In metal belts, the centrifugal tension F_c is ignored as negligible. Convince yourself that this is a reasonable problem simplification.

17–13 A designer has to select a metal-belt drive to transmit a power of H_{nom} under circumstances where a service factor of K_s and a design factor of n_d are appropriate. The design goal becomes $H_d = H_{\text{nom}} K_s n_d$. Use Equation (17–8) with negligible centrifugal force to show that the minimum belt width is given by

$$b_{\min} = \frac{1}{a}\left(\frac{H_d}{V}\right)\frac{\exp(f\phi)}{\exp(f\phi) - 1}$$

where a is the constant from $(F_1)_a = ab$.

17–14 Design a friction metal flat-belt drive to connect a 1 kW, four-pole squirrel-cage motor turning at 1750 rev/min to a shaft 380 mm away, running at half speed. The circumstances are such that a service factor of 1.2 and a design factor of 1.05 are appropriate. The life goal is 10^6 belt passes, $f = 0.35$, and the environmental considerations require a stainless steel belt.

17–15 A beryllium-copper metal flat belt with $S_f = 390$ MPa is to transmit 3.7 kW at 1125 rev/min with a life goal of 10^6 belt passes between two shafts 500 mm apart whose centerlines are in a horizontal plane. The coefficient of friction between belt and pulley is 0.32. The conditions are such that a service factor of 1.25 and a design factor of 1.1 are appropriate. The driven shaft rotates at one-third the motor-pulley speed. Specify your belt, pulley sizes, and initial tension at installation.

17–16 For the conditions of Problem 17–15 use a 1095 plain carbon-steel heat-treated belt. Conditions at the driving pulley hub require a pulley outside diameter of 75 mm or more. Specify your belt, pulley sizes, and initial tension at installation.

17–17 A single Gates Rubber V belt is to be selected to deliver engine power to the wheel-drive transmission of a riding tractor. A 2.2 kW single-cylinder engine is used. At most, 60 percent of this power is transmitted to the belt. The driving sheave has a diameter of 155 mm, the driven, 300 mm. The belt selected should be as close to a 2.3 m pitch length as possible. The engine speed is governor-controlled to a maximum of 3100 rev/min. Select a satisfactory belt and assess the factor of safety and the belt life in passes.

17–18 Two B2125 V belts are used in a drive composed of a 135 mm driving sheave, rotating at 1200 rev/min, and a 400 mm driven sheave. Find the power capacity of the drive based on a service factor of 1.25, and find the center-to-center distance.

17–19 A 45 kW four-cylinder internal combustion engine is used to drive a medium-shock brick-making machine under a schedule of two shifts per day. The drive consists of two 650 mm sheaves spaced about 3.6 m apart, with a sheave speed of 400 rev/min. Select a V-belt arrangement. Find the factor of safety, and estimate the life in passes and hours.

17–20 A reciprocating air compressor has a 1.5-m diameter flywheel 350 mm wide, and it operates at 170 rev/min. An eight-pole squirrel-cage induction motor has nameplate data 37 kW at 875 rev/min.

(*a*) Design a V-belt drive to transmit power from the motor to the compressor flywheel.

(*b*) Can cutting the V-belt grooves in the flywheel be avoided by using a V-flat drive?

17–21 The geometric implications of a V-flat drive are interesting.

(*a*) If the earth's equator was an inextensible string, snug to the spherical earth, and you spliced 1.8 m of string into the equatorial cord and arranged it to be concentric to the equator, how far off the ground is the string?

(*b*) Using the solution to part (*a*), formulate the modifications to the expressions for m_G, ϕ_d and ϕ_D, L_p, and C for a V-flat drive.

(*c*) As a result of this exercise, how would you revise your solution to part (*b*) of Problem 17–20?

17–22 A 1.5-kW electric motor running at 1720 rev/min is to drive a blower at a speed of 240 rev/min. Select a V-belt drive for this application and specify standard V belts, sheave sizes, and the resulting center-to-center distance. The motor size limits the center distance to at least 550 mm.

17–23 The standard roller-chain number indicates the chain pitch in inches, construction proportions, series, and number of strands as follows:

254 0 H-2

- 2 — two strands
- H — heavy series
- 0 — standard proportions
- 254 — pitch is 254/8 mm

This convention makes the pitch directly readable from the chain number. In Example 17–5 ascertain the pitch from the selected chain number and confirm from Table 17–19.

17–24 Equate Equations (17–32) and (17–33) to find the rotating speed n_1 at which the power equates and marks the division between the premaximum and the postmaximum power domains.

(*a*) Show that

$$n_1 = \left[\frac{0.25(10^6)K_r N_1^{0.42}}{p/25.4^{(2.2-0.07p/25.4)}}\right]^{1/2.4}$$

(*b*) Find the speed n_1 for a no. 60 chain, $p = 19$ mm, $N_1 = 17$, $K_r = 17$, and confirm from Table 17–20.

(*c*) At which speeds is Equation (17–40) applicable?

17–25 A double-strand no. 60 roller chain is used to transmit power between a 13-tooth driving sprocket rotating at 300 rev/min and a 52-tooth driven sprocket.

(*a*) What is the allowable horsepower of this drive?

(*b*) Estimate the center-to-center distance if the chain length is 82 pitches.

(*c*) Estimate the torque and bending force on the driving shaft by the chain if the actual horsepower transmitted is 30 percent less than the corrected (allowable) power.

17–26 A four-strand no. 40 roller chain transmits power from a 21-tooth driving sprocket to an 84-tooth driven sprocket. The angular speed of the driving sprocket is 2000 rev/min. A life of 20 000 hours is desired.

(*a*) Estimate the chain length if the center-to-center distance is to be about 508 mm.

(*b*) Estimate the allowable horsepower from Table 17–20, adjusting with K_1 and K_2, and neglecting the deviations from the table's life and length assumptions.

(*c*) Estimate the allowable horsepower from Equations (17–32) and (17–33) (which are the source of Table 17–20). Compare answers with part (*b*).

(*d*) Estimate the allowable horsepower H_2 from Equation (17–39), which is a broader version of Equation (17–33) that allows for adjustment for life and length. Compare the answer with part (*c*).

(*e*) Estimate the average chain velocity in meters per minute.

(*f*) Estimate the torque that is transmitted from the driving sprocket when operating at the maximum allowable power from part (*d*).

(*g*) Estimate the tension in the chain when operating at the maximum allowable power from part (*d*).

17–27 A 700 rev/min 18.65 kW squirrel-cage induction motor is to drive a two-cylinder reciprocating pump, out-of-doors under a shed. A service factor K_s of 1.5 and a design factor of 1.1 are appropriate. The pump speed is 140 rev/min. Select a suitable chain and sprocket sizes.

17–28 A centrifugal pump is driven by a 37.3 kW synchronous motor at a speed of 1800 rev/min. The pump is to operate at 900 rev/min. Despite the speed, the load is smooth ($K_s = 1.2$). For a design factor of 1.1 specify a chain and sprockets that will realize a 50 000-h life goal. Let the sprockets be 19T and 38T.

17–29 A mine hoist uses a 50 mm 6 × 19 monitor-steel wire rope. The rope is used to haul loads of 36 kN from the shaft 145 m deep. The drum has a diameter of 1.8 m, the sheaves are of good-quality cast steel, and the smallest is 1 m in diameter.

(*a*) Using a maximum hoisting speed of 6 m/s and a maximum acceleration of 0.6 m/s^2, estimate the stresses in the rope.

(*b*) Estimate the various factors of safety.

17–30 A temporary construction elevator is to be designed to carry workers and materials to a height of 30 m. The maximum estimated load to be hoisted is 25 kN at a velocity not to exceed 0.6 m/s. For minimum sheave diameters and acceleration of 1.2 m/s^2, specify the number of ropes required if the 12 mm plow-steel 6 × 19 hoisting strand is used.

17–31 A 600 m mine hoist operates with a 1.8 m drum using 6 × 19 monitor-steel wire rope. The cage and load weigh 36 kN, and the cage is subjected to an acceleration of 0.6 m/s^2 when starting.

(*a*) For a single-strand hoist how does the factor of safety $n_f = F_f/F_t$, neglecting bending, vary with the choice of rope diameter?

(*b*) For four supporting strands of wire rope attached to the cage, how does the factor of safety vary with the choice of rope diameter?

17–32 Generalize the results of Problem 17–31 by representing the factor of safety n as

$$n_f = \frac{ad}{(b/m) + cd^2}$$

where m is the number of ropes supporting the cage, and a, b, and c are constants. Show that the optimal diameter is $d^* = [b/(mc)]^{1/2}$ and the corresponding maximum attainable factor of safety is $n_f^* = a[m/(bc)]^{1/2}/2$.

17–33 From your results in Problem 17–32, show that to meet a fatigue factor of safety n_1 the optimal solution is

$$m = \frac{4bcn_1}{a^2} \text{ ropes}$$

having a diameter of

$$d = \frac{a}{2cn_1}$$

Solve Problem 17–31 if a factor of safety of 2 is required. Show what to do in order to accommodate to the necessary discreteness in the rope diameter d and the number of ropes m.

17–34 For Problem 17–29 estimate the elongation of the rope if a 31 kN loaded mine cart is placed on the cage which weighs 4.5 kN. The results of Problem 4–7 may be useful.

Computer Programs

In approaching the ensuing computer problems, the following suggestions may be helpful:

- Decide whether an analysis program or a design program would be more useful. In problems as simple as these, you will find the programs similar. For maximum instructional benefit, try the design problem.
- Creating a design program without a figure of merit precludes ranking alternative designs but does not hinder the attainment of satisfactory designs. Your instructor can provide the class design library with commercial catalogs, which not only have price information but define available sizes.
- Quantitative understanding and logic of interrelations are required for programming. Difficulty in programming is a signal to you and your instructor to increase your understanding. The following programs can be accomplished in 100 to 500 lines of code.
- Make programs interactive and user-friendly.
- Let the computer do what it can do best; the user should do what a human can do best.
- Assume the user has a copy of the text and can respond to prompts for information.
- If interpolating in a table is in order, solicit table entries in the neighborhood, and let the computer crunch the numbers.

- In decision steps, allow the user to make the necessary decision, even if it is undesirable. This allows learning of consequences and the use of the program for analysis.
- Display a lot of information in the summary. Show the decision set used up-front for user perspective.
- When a summary is complete, adequacy assessment can be accomplished with ease, so consider adding this feature.

17–35 Your experience with Problems 17–1 through 17–11 has placed you in a position to write an interactive computer program to design/select flat-belt drive components. A possible decision set is

A Priori Decisions

- Function: H_{nom}, rev/min, velocity ratio, approximate C
- Design factor: n_d
- Initial tension maintenance: catenary
- Belt material: t, d_{min}, allowable tension, density, f
- Drive geometry: d, D
- Belt thickness: t (in material decision)

Design Decisions

- Belt width: b

17–36 Problems 17–12 through 17–16 have given you some experience with flat metal friction belts, indicating that a computer program could be helpful in the design/selection process. A possible decision set is

A Priori Decisions

- Function: H_{nom}, rev/min, velocity ratio approximate C
- Design factor: n_d
- Belt material: S_y, E, ν, d_{min}
- Drive geometry: d, D
- Belt thickness: t

Design Decisions

- Belt width: b
- Length of belt (often standard loop periphery)

17–37 Problems 17–17 through 17–22 have given you enough experience with V belts to convince you that a computer program would be helpful in the design/selection of V-belt drive components. Write such a program.

17–38 Experience with Problems 17–23 through 17–28 can suggest an interactive computer program to help in the design/selection process of roller-chain elements. A possible decision set is

A Priori Decisions

- Function: power, speed, space, K_s, life goal
- Design factor: n_d
- Sprocket tooth counts: N_1, N_2, K_1, K_2

Design Decisions

- Chain number
- Strand count
- Lubrication system
- Chain length in pitches

(center-to-center distance for reference)

18 Power Transmission Case Study

Hanoi Photography/123RF

Chapter Outline

Transmission of power from a source, such as an engine or motor, through a machine to an output actuation is one of the most common machine tasks. An efficient means of transmitting power is through rotary motion of a shaft that is supported by bearings. Gears, belt pulleys, or chain sprockets may be incorporated to provide for torque and speed changes between shafts. Most shafts are cylindrical (solid or hollow), and include stepped diameters with shoulders to accommodate the positioning and support of bearings, gears, etc.

The design of a system to transmit power requires attention to the design and selection of individual components (gears, bearings, shafts, etc.). However, as is often the case in design, these components are not independent. For example, in order to design the shafts for stress and deflection, it is necessary to know the applied forces. If the forces are transmitted through gears, it is necessary to know the gear specifications in order to determine the forces that will be transmitted to the shafts. But stock gears come with certain bore sizes, requiring knowledge of the necessary shaft diameters. It is no surprise that the design process is interdependent and iterative, but where should a designer start?

The nature of machine design textbooks is to focus on each component separately. This chapter will focus on an overview of a power transmission system design, demonstrating how to incorporate the details of each component into an overall design process. A typical two-stage gear reduction such as shown in Figure 18–1 will be assumed for this discussion. The design sequence is similar for variations of this particular transmission system.

The following outline will help clarify a logical design sequence. Discussion of how each part of the outline affects the overall design process will be given in sequence in this chapter. Details on the specifics for designing and selecting major components are covered in separate chapters, particularly Chapter 7 on shaft design, Chapter 11 on bearing selection, and Chapters 13 and 14 on gear specification. A complete case study is presented as a specific vehicle to demonstrate the process.

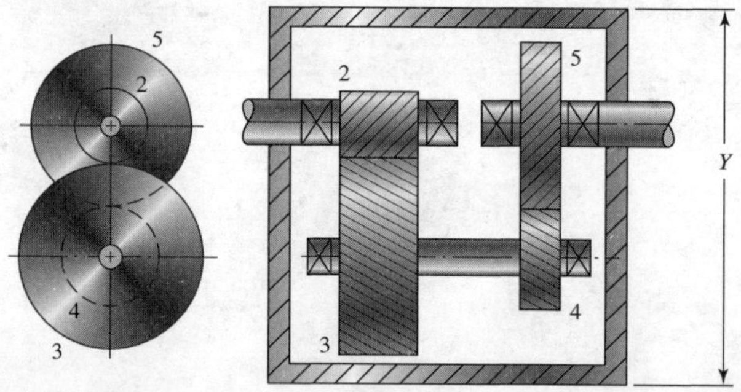

Figure 18–1

A compound reverted gear train.

CASE STUDY EXAMPLE 18–1, PART 1 PROBLEM SPECIFICATION

Section 1-18 presents the background for this case study involving a speed reducer. A two-stage, compound reverted gear train such as shown in Figure 18-1 will be designed. In this chapter, the design of the intermediate shaft and its components is presented, taking into account the other shafts as necessary. A subset of the pertinent design specifications that will be needed for this part of the design are given here.

Power to be delivered: 15 kW
Input speed: 1750 rpm
Output speed: 82–88 rev/min
Usually low shock levels, occasional moderate shock
Input and output shafts extend 100 mm outside gearbox
Maximum gearbox size: 350 mm × 350 mm base, 550 mm height
Output shaft and input shaft in-line
Gear and bearing life > 12 000 hours; infinite shaft life

18–1 Design Sequence for Power Transmission

There is not a precise sequence of steps for any design process. By nature, design is an iterative process in which it is necessary to make some tentative choices, and to build a skeleton of a design, and to determine which parts of the design are critical. However, much time can be saved by understanding the dependencies between the parts of the problem, allowing the designer to know what parts will be affected by any given change. In this section, only an outline is presented, with a short explanation of each step. Further details will be discussed in the following sections.

- ***Power and torque requirements.*** Power considerations should be addressed first, as this will determine the overall sizing needs for the entire system. Any necessary speed or torque ratio from input to output must be determined before addressing gear/pulley sizing.
- ***Gear specification.*** Necessary gear ratios and torque transmission issues can now be addressed with selection of appropriate gears. Note that a full force analysis of the shafts is not yet needed, as only the transmitted loads are required to specify the gears.
- ***Shaft layout.*** The general layout of the shafts, including axial location of gears and bearings, must now be specified. Decisions on how to transmit the torque from the gears to each shaft need to be made (keys, splines, etc.), as well as how to hold gears and bearings in place (retaining rings, press fits, nuts, etc.). However, it is not necessary at this point to size these elements, since their standard sizes allow estimation of stress-concentration factors.
- ***Force analysis.*** Once the gear/pulley diameters are known, and the axial locations of the gears and bearings are known, the free-body, shear force, and bending moment diagrams for the shafts can be produced. Forces at the bearings can be determined.
- ***Shaft material selection.*** Since fatigue design depends so heavily on the material choice, it is usually easier to make a reasonable material selection first, then check for satisfactory results.
- ***Shaft design for stress (fatigue and static).*** At this point, a stress design of each shaft should look very similar to a typical design problem from the shaft chapter (Chapter 7). Shear force and bending moment diagrams are known, critical locations can be predicted, approximate stress concentrations can be used, and estimates for shaft diameters can be determined.

- ***Shaft design for deflection.*** Since deflection analysis is dependent on the entire shaft geometry, it is saved until this point. With all shaft geometry now estimated, the critical deflections at the bearing and gear locations can be checked by analysis.
- ***Bearing selection.*** Specific bearings from a catalog may now be chosen to match the estimated shaft diameters. The diameters can be adjusted slightly as necessary to match the catalog specifications.
- ***Key and retaining ring selection.*** With shaft diameters settling in to stable values, appropriate keys and retaining rings can be specified in standard sizes. This should make little change in the overall design if reasonable stress-concentration factors were assumed in previous steps.
- ***Final analysis.*** Once everything has been specified, iterated, and adjusted as necessary for any specific part of the task, a complete analysis from start to finish will provide a final check and specific safety factors for the actual system.

18–2 Power and Torque Requirements

Power transmission systems will typically be specified by a power capacity, for example, a 40-horsepower gearbox. This rating specifies the combination of torque and speed that the unit can endure. Remember that, in the ideal case, *power in* equals *power out,* so that we can refer to the power being the same throughout the system. In reality, there are small losses due to factors like friction in the bearings and gears. In many transmission systems, the losses in the rolling bearings will be negligible. Gears have a reasonably high efficiency, with about 1 to 2 percent power loss in a pair of meshed gears. Thus, in the double-reduction gearbox in Figure 18–1, with two pairs of meshed gears the output power is likely to be about 2 to 4 percent less than the input power. Since this is a small loss, it is common to speak of simply the power of the system, rather than input power and output power. Flat belts and timing belts have efficiencies typically in the mid to upper 90 percent range. V belts and worm gears have efficiencies that may dip much lower, requiring a distinction between the necessary input power to obtain a desired output power.

Torque, on the other hand, is typically not constant throughout a transmission system. Remember that power equals the product of torque and speed. Since *power in* = *power out*, we know that for a gear train

$$H = T_i\omega_i = T_o\omega_o \tag{18–1}$$

With a constant power, a gear ratio to decrease the angular velocity will simultaneously increase torque. The gear ratio, or train value, for the gear train is

$$e = \omega_o/\omega_i = T_i/T_o \tag{18–2}$$

A typical power transmission design problem will specify the desired power capacity, along with either the input and output angular velocities, or the input and output torques. There will usually be a tolerance specified for the output values. After the specific gears are specified, the actual output values can be determined.

18–3 Gear Specification

With the gear train value known, the next step is to determine appropriate gears. As a rough guideline, a train value of up to 10 to 1 can be obtained with one pair of gears. Greater ratios can be obtained by compounding additional pairs of gears (see

Section 13–13). The compound reverted gear train in Figure 18–1 can obtain a train value of up to 100 to 1.

Since numbers of teeth on gears must be integers, it is best to design with teeth numbers rather than diameters. See Examples 13–3, 13–4, and 13–5 for details on designing appropriate numbers of teeth to satisfy the gear train value and any necessary geometry condition, such as in-line condition of input and output shaft. Care should be taken at this point to find the best combination of teeth numbers to minimize the overall package size. If the train value only needs to be approximate, use this flexibility to try different options of tooth numbers to minimize the package size. A difference of one tooth on the smallest gear can result in a significant increase in size of the overall package.

If designing for large production quantities, gears can be purchased in large enough quantities that it is not necessary to worry about preferred sizes. For small lot production, consideration should be given to the trade-offs between smaller gearbox size and extra cost for odd gear sizes that are difficult to purchase off the shelf. If stock gears are to be used, their availability in prescribed numbers of teeth with anticipated diametral pitch should be checked at this time. If necessary, iterate the design for numbers of teeth that are available.

CASE STUDY EXAMPLE 18–1, PART 2 SPEED, TORQUE, AND GEAR RATIOS

Continue the case study by determining appropriate tooth counts to reduce the input speed of $\omega_i = 1750$ rev/min to an output speed within the range

$$82 \text{ rev/min} < \omega_o < 88 \text{ rev/min}$$

Once final tooth counts are specified, determine values of

(a) Speeds for the intermediate and output shafts

(b) Torques for the input, intermediate and output shafts, to transmit 15 kW.

Solution

Use the notation for gear numbers from Figure 18-1. Choose mean value for initial design, $\omega_5 = 85$ rev/min.

$$e = \frac{\omega_5}{\omega_2} = \frac{85}{1750} = \frac{1}{20.59} \qquad \text{Equation (18–2)}$$

For a compound reverted gear train,

$$e = \frac{1}{20.59} = \frac{N_2 N_4}{N_3 N_5} \qquad \text{Equation (13–30)}$$

For smallest package size, let both stages be the same reduction. Also, by making the two stages identical, the in-line condition on the input and output shaft will automatically be satisfied.

$$\frac{N_2}{N_3} = \frac{N_4}{N_5} = \sqrt{\frac{1}{20.59}} = \frac{1}{4.54}$$

For this ratio, assuming a pressure angle of $\phi = 20°$, the minimum number of teeth from Equation (13-11) is

$$N = \frac{2(1)}{[1 + 2(4.54)]\sin^2 20°}\left[4.54 + \sqrt{4.54^2 + [1 + 2(4.54)]\sin^2 20°}\right] = 15.6 = 16 \text{ teeth}$$

Thus,

$$N_2 = N_4 = 16 \text{ teeth}$$
$$N_3 = 4.54(N_2) = 72.64$$

Try rounding down and check if ω_5 is within limits.

$$\omega_5 = \left(\frac{16}{72}\right)\left(\frac{16}{72}\right)(1750) = 86.42 \text{ rev/min} \qquad \text{Acceptable}$$

Proceed with

$$\boxed{\begin{aligned} N_2 &= N_4 = 16 \text{ teeth} \\ N_3 &= N_5 = 72 \text{ teeth} \end{aligned}}$$

$$\boxed{e = \left(\frac{16}{72}\right)\left(\frac{16}{72}\right) = \frac{1}{20.25}}$$

$$\boxed{\omega_5 = 86.42 \text{ rev/min}}$$

$$\boxed{\omega_3 = \omega_4 = \left(\frac{16}{72}\right)(1750) = 388.9 \text{ rev/min}}$$

To determine the torques, return to the power relationship,

$$H = T_2\omega_2 = T_5\omega_5 \qquad \text{Equation (18–1)}$$

$$T_2 = H/\omega_2 = \left(\frac{15000 \text{ W}}{1750 \text{ rev/min}}\right)\left(\frac{1 \text{ rev}}{2\pi \text{ rad}}\right)\left(60\frac{\text{s}}{\text{min}}\right)$$

$$\boxed{T_2 = 81.9 \text{ N} \cdot \text{m}}$$

$$\boxed{T_3 = T_2\frac{\omega_2}{\omega_3} = 81.9\frac{1750}{388.9} = 368.5 \text{ N} \cdot \text{m}}$$

$$\boxed{T_5 = T_2\frac{\omega_2}{\omega_5} = 81.9\frac{1750}{86.42} = 1658.5 \text{ N} \cdot \text{m}}$$

If a maximum size for the gearbox has been specified in the problem specification, a minimum diametral pitch (maximum tooth size) can be estimated at this point by writing an expression for the gearbox size in terms of gear diameters, and converting to numbers of teeth through the diametral pitch. For example, from Figure 18–1, the overall height of the gearbox is

$$Y = d_3 + d_2/2 + d_5/2 + 2m + \text{clearances} + \text{wall thicknesses}$$

where the $2m$ term accounts for the addendum height of the teeth on gears 3 and 5 that extend beyond the pitch diameters. Substituting $d_i = mN_i$ gives

$$Y = mN_3 + mN_2/2 + mN_5/2 + 2m + \text{clearances} + \text{wall thicknesses}$$

Solving this for m, we find

$$m = (Y - \text{clearances} - \text{wall thicknesses})/(N_3 + N_2/2 + N_5/2 + 2) \quad \text{(18–3)}$$

This is the minimum value that can be used for diametral pitch, and therefore the maximum tooth size, to stay within the overall gearbox constraint. It should be rounded *up* to the next standard diametral pitch, which reduces the maximum tooth size.

The AGMA approach, as described in Chapter 14, for both bending and contact stress should be applied next to determine suitable gear parameters. The primary design parameters to be specified by the designer include material, diametral pitch, and face width. A recommended procedure is to start with an estimated diametral pitch. This allows determination of gear diameters ($d = mN$), pitch-line velocities [Equation (13–34)], and transmitted loads [Equation (13–35)]. Typical spur gears are available with face widths from 3 to 5 times the circular pitch p. Using an average of 4, a first estimate can be made for face width $F = 4p = 4\pi m$. Alternatively, the designer can simply perform a quick search of online gear catalogs to find available face widths for the diametral pitch and number of teeth.

Next, the AGMA equations in Chapter 14 can be used to determine appropriate material choices to provide desired safety factors. It is generally most efficient to attempt to analyze the most critical gear first, as it will determine the limiting values of diametral pitch and material strength. Usually, the critical gear will be the smaller gear, on the high-torque (low-speed) end of the gearbox.

If the required material strengths are too high, such that they are either too expensive or not available, iteration with a smaller diametral pitch (larger tooth) will help. Of course, this will increase the overall gearbox size. Often the excessive stress will be in one of the small gears. Rather than increase the tooth size for all gears, it is sometimes better to reconsider the design of tooth counts, shifting more of the gear ratio to the pair of gears with less stress, and less ratio to the pair of gears with the excessive stress. This will allow the offending gear to have more teeth and therefore larger diameter, decreasing its stress.

If contact stress turns out to be more limiting than bending stress, consider gear materials that have been heat treated or case hardened to increase the surface strength. Adjustments can be made to the diametral pitch if necessary to achieve a good balance of size, material, and cost. If the stresses are all much lower than the material strengths, a larger diametral pitch is in order, which will reduce the size of the gears and the gearbox.

Everything up to this point should be iterated until acceptable results are obtained, as this portion of the design process can usually be accomplished independently from the next stages of the process. The designer should be satisfied with the gear selection before proceeding to the shaft. Selection of specific gears from catalogs at this point will be helpful in later stages, particularly in knowing overall width, bore size, recommended shoulder support, and maximum fillet radius.

CASE STUDY EXAMPLE 18–1, PART 3 GEAR SPECIFICATION

Continue the case study by specifying appropriate gears, including pitch diameter, module, face width, and material. Achieve factors of safety of at least 1.2 for wear and bending.

Solution

Estimate the minimum diametral pitch for overall gearbox height = 550 mm.
From Equation (18-3) and Figure 18-1,

$$m_{max} = \frac{(Y - \text{clearances} - \text{wall thickness})}{\left(N_3 + \dfrac{N_3}{2} + \dfrac{N_5}{2 + 2}\right)}$$

Allow 40 mm for clearances and wall thickness.

$$m_{max} = \frac{(550 - 40)}{\left(72 + \dfrac{16}{2} + \dfrac{72}{2 + 2}\right)} = 4.32 \text{ mm}$$

Start with $m = \underline{4 \text{ mm}}$

$$\boxed{\begin{aligned} d_2 = d_4 &= N_2 m = 16(4) = 64 \text{ mm} \\ d_3 = d_5 &= 72(4) = 288 \text{ mm} \end{aligned}}$$

Shaft speeds were previously determined to be

$$\omega_2 = 1750 \text{ rev/min} \qquad \omega_3 = \omega_4 = 388.9 \text{ rev/min} \qquad \omega_5 = 86.4 \text{ rev/min}$$

Get pitch-line velocities and transmitted loads for later use.

$$V_{23} = \frac{\pi d_2 \omega_2}{60\,000} = \frac{\pi(64)(1750)}{60\,000} = \underline{5.86 \text{ m/s}} \qquad \text{Equation (13–34)}$$

$$V_{45} = \frac{\pi d_5 \omega_5}{60\,000} = \underline{1.30 \text{ m/s}}$$

$$W^t_{23} = \frac{H}{V_{23}} = \left(\frac{15}{5.86}\right) = \underline{2.56 \text{ kN}} \qquad \text{Equation (13–35)}$$

$$W^t_{45} = \frac{H}{V_{45}} = \left(\frac{15}{1.3}\right) = \underline{11.54 \text{ kN}}$$

Start with gear 4, since it is the smallest gear, transmitting the largest load. It will likely be critical.
Start with wear by contact stress, since it is often the limiting factor.

Gear 4 Wear

$$Z_I = \frac{\cos 20° \sin 20°}{2(1)} \left(\frac{4.5}{4.5 + 1}\right) = 0.1315 \qquad \text{Equation (14–21)}$$

For K_v, assume $Q_v = 7$. $B = 0.731$, $A = 65.1$ Equation (14–26)

$$K_v = \left(\frac{65.1 + \sqrt{200(1.30)}}{65.1}\right)^{0.731} = 1.18 \qquad \text{Equation (14–25a)}$$

Face width b is typically from 3 to 5 times circular pitch. Try

$$b = 4 \cdot \pi m = 50.3 \text{ mm.}$$

Choose $b = \underline{50 \text{ mm}}$.

For K_H, $C_{pf} = 0.0624$ Equation (14–30)

$C_{mc} = 1$ uncrowned teeth Equation (14–29)

$C_{pm} = 1$ straddle-mounted Equation (14–31)

$C_{ma} = 0.15$ commercial enclosed unit Equation (14–32)

$C_e = 1$ Equation (14–33)

$K_H = 1.21$ Equation (14–28)

$Z_E = 191$ Table 14–8

$K_o = K_s = Z_R = 1$

$$\sigma_c = 191\sqrt{\frac{11\ 540(1.18)(1.21)}{64(50)(0.1315)}} = \underline{1195\ \text{MPa}}$$ Equation (14–14)

Get factors for $\sigma_{c,\text{all}}$. For life factor Z_N, get number of cycles for specified life of 12 000 h.

$$L_4 = (12\ 000\ \text{h})\left(60\frac{\text{min}}{\text{h}}\right)\left(389\frac{\text{rev}}{\text{min}}\right) = 2.8 \times 10^8\ \text{rev}$$

$Z_N = 0.9$ Figure 14–15

$Y_Z = Y_\theta = Z_W = 1$

We will choose to set the AGMA design factors to unity and handle the design factor of 1.2 with the overall factors of safety. From Equation (14-40),

$$n_c = 1.2 = \frac{\sigma_{c,\text{all}}}{\sigma_c}$$

$$\sigma_{c,\text{all}} = 1.2(1195) = 1434\ \text{MPa}$$

From Equation (14-16), solving for S_c,

$$S_c = \frac{\sigma_{c,\text{all}}}{Z_N} = \frac{1434}{0.9} = 1593\ \text{MPa}$$

From Table 14-6 this strength is achievable with Grade 3 carburized and hardened with $S_c = \underline{1896\ \text{MPa}}$. To find the achieved factor of safety, $n_c = \sigma_{c,\text{all}}/\sigma_c$. The factor of safety for wear of gear 4 is

$$n_c = \frac{\sigma_{c,\text{all}}}{\sigma_c} = \frac{S_c Z_N}{\sigma_c} = \frac{1896(0.9)}{1195} = \underline{1.43}$$

Gear 4 Bending

$J = 0.27$ Figure 14–6

$K_B = 1$

Everything else is the same as before.

$$\sigma_b = W_t K_v \frac{1}{bm_t}\frac{K_H}{Y_J} = (11\ 540)(1.18)\left(\frac{1}{50}(4)\right)\left(\frac{1.21}{0.27}\right) \qquad \text{Equation (14–13)}$$

$$\underline{\sigma_b = 305 \text{ MPa}}$$

$$Y_N = 0.9 \qquad \text{Figure 14–14}$$

Using Grade 2 carburized and hardened, find $S_t = 440$ MPa (Table 14-3).

$$\sigma_{\text{all}} = S_t Y_N = 396 \text{ MPa}$$

The factor of safety for bending of gear 4 is

$$\boxed{n_b = \frac{\sigma_{b,\text{all}}}{\sigma_b} = \frac{396}{305} = 1.3}$$

Gear 5 Bending and Wear

Everything is the same as for gear 4, except Y_J, Y_N, and Z_N.

$$Y_J = 0.41 \qquad \text{Figure 14–6}$$

$$L_5 = (12\ 000 \text{ h})(60 \text{ min/h})(86.4 \text{ rev/min}) = 6.2 \times 10^7 \text{ rev}$$

$$Y_N = 0.97 \qquad \text{Figure 14–14}$$

$$Z_N = 1.0 \qquad \text{Figure 14–15}$$

$$\sigma_c = 191\sqrt{\frac{11\ 540(1.18)(1.21)}{64(50)(0.1315)}} = 1195 \text{ MPa}$$

$$\sigma_b = (11\ 540)(1.18)\left(\frac{1}{50}(4)\right)\left(\frac{1.21}{0.41}\right) = 201 \text{ MPa}$$

Choose Grade 2 carburized and hardened

$$\boxed{\begin{aligned} n_c &= \frac{\sigma_{c,\text{all}}}{\sigma_c} = \frac{1550}{1195} = 1.30 \\ n_b &= \frac{\sigma_{b,\text{all}}}{\sigma_b} = \frac{440(0.97)}{201} = 2.12 \end{aligned}}$$

Gear 2 Wear

Gears 2 and 3 are evaluated similarly. Only selected results are shown.

$$K_v = 1.37$$

Try $b = 40$ mm, since the loading is less on gears 2 and 3.

$$K_H = 1.19$$

All other factors are the same as those for gear 4.

$$\sigma_c = 191\sqrt{\frac{(2560)(1.37)(1.19)}{64(40)(0.1315)}} = 673 \text{ MPa}$$

$$L_2 = (12\ 000 \text{ h})(60 \text{ min/h})(1750 \text{ rev/min}) = 1.26 \times 10^9 \text{ rev} \qquad Z_N = 0.8$$

Try grade 1 flame-hardened, $S_c = 1170$ MPa

$$\boxed{n_c = \frac{\sigma_{c,\text{all}}}{\sigma_c} = \frac{1170(0.8)}{673} = 1.40}$$

Gear 2 Bending

$$Y_J = 0.27 \qquad Y_N = 0.88$$

$$\sigma_b = 2560(1.37)\frac{1.19}{40(4)(0.41)} = 63.6 \text{ MPa}$$

$$n_b = \frac{\sigma_{b,\text{all}}}{\sigma_b} = \frac{310(0.88)}{96.6} = 2.82$$

Gear 3 Wear and Bending

$$Y_J = 0.41 \qquad Y_N = 0.9 \qquad Z_N = 0.9$$

$$\sigma_c = 191\sqrt{\frac{(2560)(1.37)(1.19)}{64(40)(0.1315)}} = 673 \text{ MPa}$$

$$\sigma_b = 2560(1.37)\frac{1}{40(4)}\frac{(1.19)}{(0.41)} = 63.6 \text{ MPa}$$

Try Grade 1 steel, through-hardened to 300 H_B. From Figure 14-2, $S_t = 248$ MPa and from Figure 14-5, $S_c = 866$ MPa.

$$n_c = \frac{866(0.9)}{673} = 1.16$$

$$n_b = \frac{\sigma_{b,\text{all}}}{\sigma_b} = \frac{248(0.9)}{63.6} = 3.51$$

In summary, the resulting gear specifications are:

All gears, $m = 4$ mm
Gear 2, Grade 1 flame-hardened, $S_c = 1170$ MPa and $S_t = 310$ MPa
$d_2 = 64$ mm, face width $= 40$ mm
Gear 3, Grade 1 through-hardened to 300 H_B, $S_c = 866$ MPa and $S_t = 248$ MPa
$d_3 = 288$ mm, face width $= 40$ mm
Gear 4, Grade 3 carburized and hardened, $S_c = 1896$ MPa and $S_t = 440$ MPa
$d_4 = 64$ mm, face width $= 50$ mm
Gear 5, Grade 2 carburized and hardened, $S_c = 1550$ MPa and $S_t = 440$ MPa
$d_5 = 288$ mm, face width $= 50$ mm

18–4 Shaft Layout

The general layout of the shafts, including axial location of gears and bearings, must now be specified in order to perform a free-body force analysis and to obtain shear force and bending moment diagrams. If there is no existing design to use as a starter, then the determination of the shaft layout may have many solutions. Section 7–3 discusses the issues involved in shaft layout. In this section the focus will be on how the decisions relate to the overall process.

A free-body force analysis can be performed without knowing shaft diameters, but can not be performed without knowing axial distances between gears and bearings. It is extremely important to keep axial distances small. Even small forces can create large bending moments if the moment arms are large. Also, recall that beam deflection equations typically include length terms raised to the third power.

It is worth examining the entirety of the gearbox at this time, to determine what factors drive the length of the shaft and the placement of the components. A rough sketch, such as shown in Figure 18–2, is sufficient for this purpose.

CASE STUDY EXAMPLE 18–1, PART 4 SHAFT LAYOUT

Continue the case study by preparing a sketch of the gearbox sufficient to determine the axial dimensions. In particular, estimate the overall length, and the distance between the gears of the intermediate shaft, in order to fit with the mounting requirements of the other shafts.

Solution

Figure 18-2 shows the rough sketch. It includes all three shafts, with consideration of how the bearings are to mount in the case. The gear widths are known at this point. Bearing widths are guessed, allowing a little more space for larger bearings on the intermediate shaft where bending moments will be greater. Small changes in bearing widths will have minimal effect on the force analysis, since the location of the ground reaction force will change very little. The 4-in distance between the two gears on the countershaft is dictated by the requirements of the input and output shafts, including the space for the case to mount the bearings. Small allotments are given for the retaining rings, and for space behind the bearings. Adding it all up gives the intermediate shaft length as 292 mm.

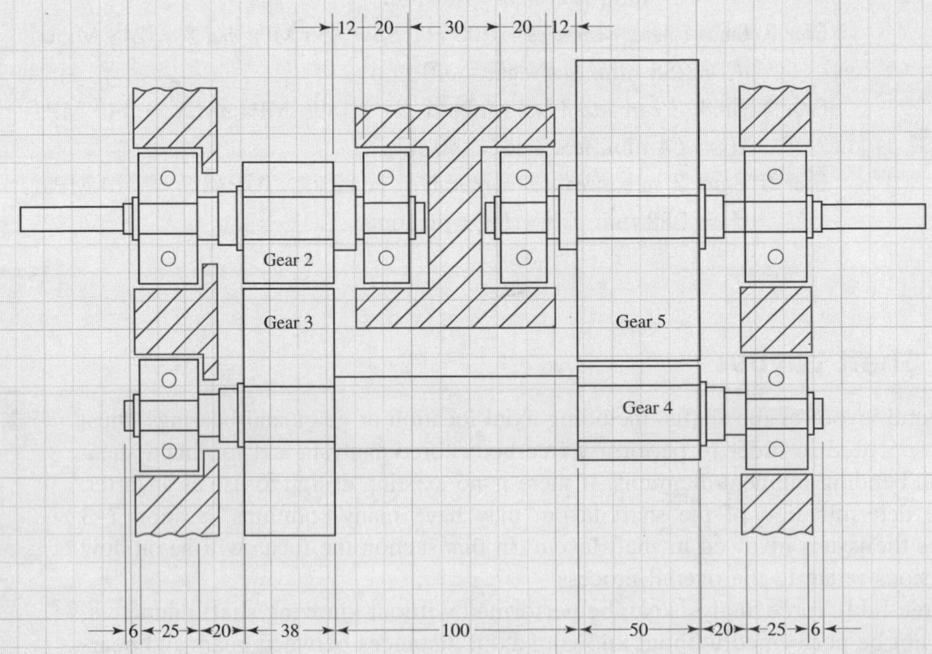

Figure 18–2

Sketch for shaft layout. Dimensions are in millimeters.

Wider face widths on gears require more shaft length. Originally, gears with hubs were considered for this design to allow the use of set screws instead of high-stress-concentration retaining rings. However, the extra hub lengths added several millimeters to the shaft lengths and the gearbox housing.

Several points are worth noting in the layout in Figure 18–2. The gears and bearings are positioned against shoulders, with retaining rings to hold them in position. While it is desirable to place gears near the bearings, a little extra space is provided between them to accommodate any housing that extends behind the bearing, and to allow for a bearing puller to have space to access the back of the bearing. The extra change in diameter between the bearings and the gears allows the shoulder height for the bearing and the bore size for the gear to be different. This diameter can have loose tolerances and a large fillet radius.

Each bearing is restrained axially on its shaft, but only one bearing on each shaft is axially fixed in the housing, allowing for slight axial thermal expansion of the shafts.

18–5 Force Analysis

Once the gear diameters are known, and the axial locations of the components are set, the free-body diagrams and shear force and bending moment diagrams for the shafts can be produced. With the known transmitted loads, determine the radial and axial loads transmitted through the gears (see Sections 13–14 through 13–17). From summation of forces and moments on each shaft, ground reaction forces at the bearings can be determined. For shafts with gears and pulleys, the forces and moments will usually have components in two planes along the shaft. For rotating shafts, usually only the resultant magnitude is needed, so force components at bearings are summed as vectors. Shear force and bending moment diagrams are usually obtained in two planes, then summed as vectors at any point of interest. A torque diagram should also be generated to clearly visualize the transfer of torque from an input component, through the shaft, and to an output component.

See the beginning of Example 7–2 for the force analysis portion of the case study for the intermediate shaft. The bending moment is largest at gear 4. This is predictable, since gear 4 is smaller, and must transmit the same torque that entered the shaft through the much larger gear 3.

While the force analysis is not difficult to perform manually, if beam software is to be used for the deflection analysis, it will necessarily calculate reaction forces, along with shear force and bending moment diagrams in the process of calculating deflections. The designer can enter guessed values for diameters into the software at this point, just to get the force information, and later enter actual diameters to the same model to determine deflections.

18–6 Shaft Material Selection

A trial material for the shaft can be selected at any point before the stress design of the shaft, and can be modified as necessary during the stress design process. Section 7–2 provides details for decisions regarding material selection. For the case study, Example 7–2 provides the stress analysis. Initially, an inexpensive steel, 1020 CD, is selected. After the stress analysis, a slightly higher strength 1050 CD is chosen to avoid increasing the shaft diameters.

18–7 Shaft Design for Stress

The critical shaft diameters are to be determined by stress analysis at critical locations. Section 7–4 provides a detailed examination of the issues involved in shaft design for stress.

CASE STUDY EXAMPLE 18–1, PART 5 DESIGN FOR STRESS

Proceed with the next phase of the case study design, in which appropriate diameters for each section of the shaft are estimated, based on providing sufficient fatigue and static stress capacity for infinite life of the shaft, with minimum safety factor of 1.5.

Solution

The solution to this phase of the design is presented in Example 7-2.

Since the bending moment is highest at gear 4, potentially critical stress points are at its shoulder, keyway, and retaining ring groove. It turns out that the keyway is the critical location. It seems that shoulders often get the most attention. This example demonstrates the danger of neglecting other stress concentration sources, such as keyways.

The material choice was changed in the course of this phase, choosing to pay for a higher strength to limit the shaft diameter to 2 in. If the shaft were to get much bigger, the small gear would not be able to provide an adequate bore size. If it becomes necessary to increase the shaft diameter any more, the gearing specification will need to be redesigned.

18–8 Shaft Design for Deflection

Section 7–5 provides a detailed discussion of deflection considerations for shafts. Typically, a deflection problem in a shaft will not cause catastrophic failure of the shaft, but will lead to excess noise and vibration, and premature failure of the gears or bearings.

CASE STUDY EXAMPLE 18–1, PART 6 DEFLECTION CHECK

Proceed with the next phase of the case study by checking that deflections and slopes at the gears and bearings on the intermediate shaft are within acceptable ranges.

Solution

The solution to this phase of the design is presented in Example 7-3.

It turns out that in this problem all the deflections are within recommended limits for bearings and gears. This is not always the case, and it would be a poor choice to neglect the deflection analysis. In a first iteration of this case study, with longer shafts due to using gears with hubs, the deflections were more critical than the stresses.

18–9 Bearing Selection

Bearing selection is straightforward now that the bearing reaction forces and the approximate bore diameters are known. See Chapter 11 for general details on bearing selection. Rolling-contact bearings are available with a wide range of load capacities and dimensions, so it is usually not a problem to find a suitable bearing that is close to the estimated bore diameter and width.

CASE STUDY EXAMPLE 18–1, PART 7 BEARING SELECTION

Continue the case study by selecting appropriate bearings for the intermediate shaft, with a reliability of 99 percent. The problem specifies a design life of 12 000 h. The intermediate shaft speed is 389 rev/min. The estimated bore size is 30 mm, and the estimated bearing width is 25 mm.

Solution

From the free-body diagram (see Example 7-2),

$$R_{Az} = 547 \text{ N} \qquad R_{Ay} = 1856 \text{ N} \qquad R_A = 1935 \text{ N}$$

$$R_{Bz} = 10\ 233 \text{ N} \qquad R_{By} = 4304 \text{ N} \qquad R_B = 11\ 101 \text{ N}$$

At the shaft speed of 389 rev/min, the design life of 12 000 h correlates to a bearing life of $L_D = (12\ 000 \text{ h})(60 \text{ min/h})(389 \text{ rev/min}) = 2.8 \times 10^8$ rev.

Start with bearing B since it has the higher loads and will likely raise any lurking problems. From Equation (11-10), assuming a ball bearing with $a = 3$ and $L = 2.8 \times 10^8$ rev,

$$C_{10} = (1)11\ 101 \left[\frac{2.8 \times 10^8/10^6}{0.02 + (4.459 - 0.02)(1 - 0.99)^{1/1.483}} \right]^{1/3} = 120.5 \text{ kN}$$

A check on the Internet for available bearings (www.globalspec.com is one good starting place) shows that this load is relatively high for a ball bearing with bore size in the neighborhood of 30 mm. Try a cylindrical roller bearing. Recalculating C_{10} with the exponent $a = 10/3$ for roller bearings, we obtain

$$C_{10} = 95 \text{ kN}$$

Cylindrical roller bearings are available from several sources in this range. A specific one is chosen from a common supplier of bearings, with the following specifications:

Cylindrical roller bearing at right end of shaft
$C = 100$ kN, ID $= 30$ mm, OD $= 72$ mm, $W = 27$ mm
Shoulder diameter range $= 37$ mm to 39 mm, and maximum fillet radius $= 1.1$ mm

For bearing A, again assuming a ball bearing,

$$C_{10} = 1935\left[\frac{2.8 \times 10^8/10^6}{0.02 + 4.439(1 - 0.99)^{1/1.483}}\right]^{1/3} = 21 \text{ kN}$$

A specific ball bearing is chosen from the SKF Internet catalog.

Deep-groove ball bearing at left end of shaft
$C = 29$ kN, ID $= 30$ mm, OD $= 72$ mm, $W = 27$ mm
Shoulder diameter range $= 37$ mm *to* 39 mm, *and maximum fillet radius* $= 1.1$ mm

At this point, the actual bearing dimensions can be checked against the initial assumptions. For bearing B the bore diameter of 30 mm is slightly larger than the original 25 mm. There is no reason for this to be a problem as long as there is room for the shoulder diameter. The original estimate for shoulder support diameters was 35 mm. As long as this diameter is less than 42 mm, the next step of the shaft, there should not be any problem. In the case study, the recommended shoulder support diameters are within the acceptable range. The original estimates for stress concentration at the bearing shoulder assumed a fillet radius such that $r/d = 0.02$. The actual bearings selected have ratios of 0.036 and 0.080. This allows the fillet radii to be increased from the original design, decreasing the stress-concentration factors.

The bearing widths are close to the original estimates. Slight adjustments should be made to the shaft dimensions to match the bearings. No redesign should be necessary.

18–10 Key and Retaining Ring Selection

The sizing and selection of keys is discussed in Section 7–7, with an example in Example 7–6. The cross-sectional size of the key will be dictated to correlate with the shaft size (see Tables 7–6 and 7–8), and must certainly match an integral keyway in the gear bore. The design decision includes the length of the key, and if necessary an upgrade in material choice.

The key could fail by shearing across the key, or by crushing due to bearing stress. For a square key, it turns out that checking only the crushing failure is adequate, since the shearing failure will be less critical according to the distortion energy failure theory, and equal according to the maximum shear stress failure theory. Check Example 7–6 to investigate why.

CASE STUDY EXAMPLE 18–1, PART 8 KEY DESIGN

Continue the case study by specifying appropriate keys for the two gears on the intermediate shaft to provide a factor of safety of 2. The gears are to be custom bored and keyed to the required specifications. Previously obtained information includes the following:

Transmitted torque: $T = 360$ N · m
Bore diameters: $d_3 = d_4 = 42$ mm
Gear hub lengths: $l_3 = 38$ mm, $l_4 = 50$ mm

Solution

From Table 7-6, for a shaft diameter of 42 mm, choose a square key with side dimension $t = 10$ mm. Choose 1020 CD material, with $S_y = 390$ MPa. The force on the key at the surface of the shaft is

$$F = \frac{T}{r} = \frac{360}{0.042/2} = 17.14 \text{ kN}$$

Checking for failure by crushing, we find the area of one-half the face of the key is used.

$$n = \frac{S_y}{\sigma} = \frac{S_y}{F/(tl/2)}$$

Solving for l gives

$$l = \frac{2Fn}{tS_y} = \frac{2(17\ 140)(2)}{10(390)} = 18 \text{ mm}$$

Since both gears have the same bore diameter and transmit the same torque, the same key specification can be used for both.

Retaining ring selection is simply a matter of checking catalog specifications. The retaining rings are listed for nominal shaft diameter, and are available with different axial load capacities. Once selected, the designer should make note of the depth of the groove, the width of the groove, and the fillet radius in the bottom of the groove. The catalog specification for the retaining ring also includes an edge margin, which is the minimum distance to the next smaller diameter change. This is to ensure support for the axial load carried by the ring. It is important to check stress-concentration factors with actual dimensions, as these factors can be rather large. In the case study, a specific retaining ring was already chosen during the stress analysis (see Example 7–2) at the potentially critical location of gear 4. The other locations for retaining rings were not at points of high stress, so it is not necessary to worry about the stress concentration due to the retaining rings in these locations. Specific retaining rings should be selected at this time to complete the dimensional specifications of the shaft.

For the case study, retaining rings specifications are entered into globalspec, and specific rings are selected from Truarc Co., with the following specifications:

	Both Gears	Left Bearing	Right Bearing
Nominal shaft diameter	42 mm	25 mm	30 mm
Groove diameter	38 ± 0.125 mm	24 ± 0.1mm	28 ± 0.1 mm
Groove width	$1.7^{+0.1}_{-0.0}$ mm	$1.2^{+0.1}_{-0.0}$ mm	$1.4^{+0.1}_{-0.0}$ mm
Nominal groove depth	1.2 mm	0.8 mm	0.9 mm
Max groove fillet radius	0.25 mm	0.25 mm	0.25 mm
Minimum edge margin	3.6 mm	2.6 mm	2.6 mm
Allowable axial thrust	52.7 kN	26.7 kN	31.2 kN

These are within the estimates used for the initial shaft layout, and should not require any redesign. The final shaft should be updated with these dimensions.

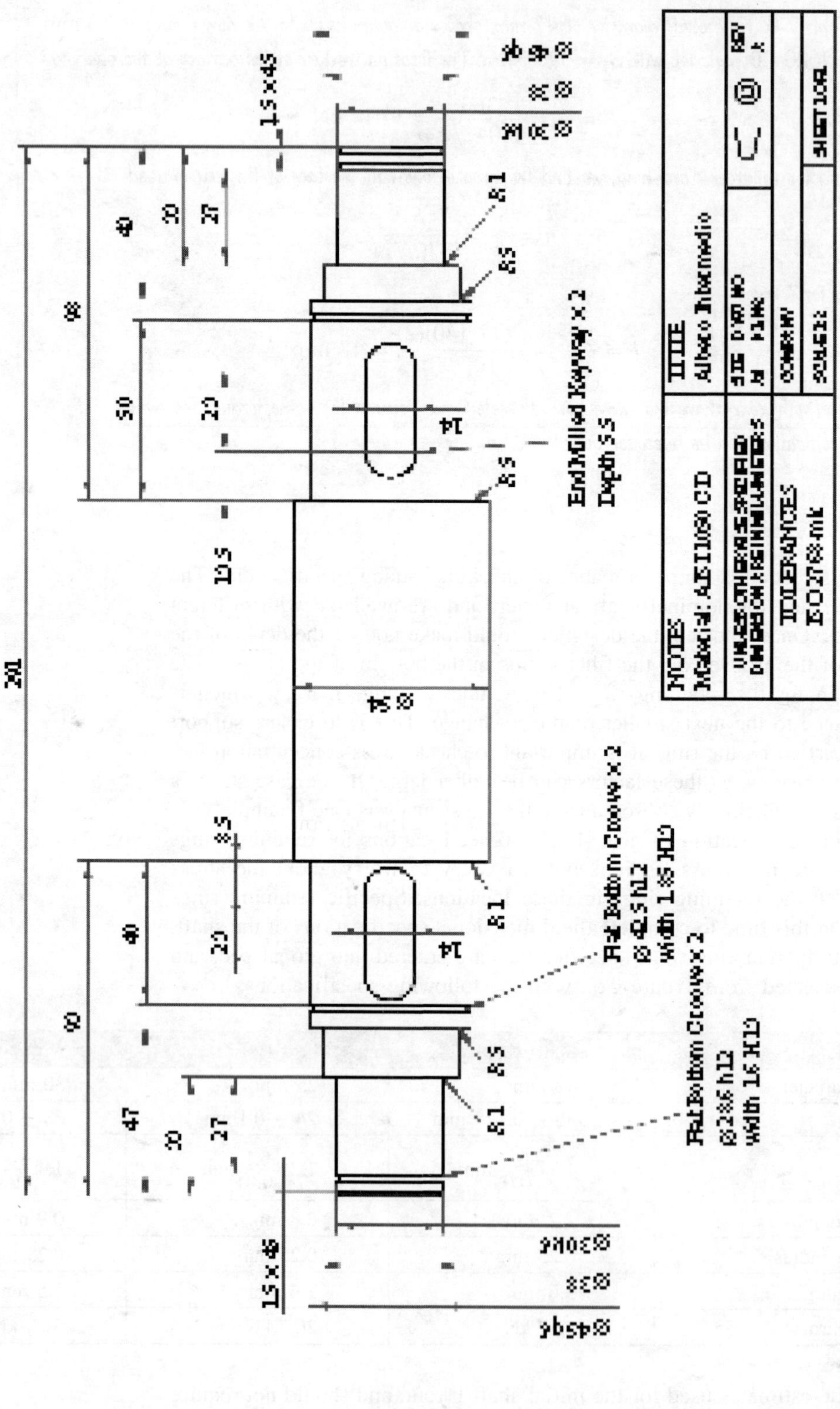
201
End Milled Keyway x 2
Depth 5.5
Flat Bottom Groove x 2
NOTES
TITLE
TOLERANCES

Figure 18–3

18–11 Final Analysis

At this point in the design, everything seems to check out. Final details include determining dimensions and tolerances for appropriate fits with the gears and bearings. See Section 7–8 for details on obtaining specific fits. Any small changes from the nominal diameters already specified will have negligible effect on the stress and deflection analysis. However, for manufacturing and assembly purposes, the designer should not overlook the tolerance specification. Improper fits can lead to failure of the design. Lack of attention to tolerance specification can make the part nonfunctional or overly expensive to manufacture. Further insight on tolerance specification is given in Section 1–14. The final drawing for the intermediate shaft is shown in Figure 18–3. This drawing shows the important dimensions and dimensional tolerances in a form that is generally considered satisfactory for small production quantities where direct attention is given to manufacturing methods. A more robust method of part specification that also addresses the allowed variations from perfect form (e.g., straightness or concentricity) is known as Geometric Dimensioning and Tolerancing and is introduced in Chapter 20.

For documentation purposes, and for a check on the design work, the design process should conclude with a complete analysis of the final design. Remember that analysis is much more straightforward than design, so the investment of time for the final analysis will be relatively small.

PROBLEMS

18–1 For the case study problem, design the input shaft, including complete specification of the gear, bearings, key, retaining rings, and shaft.

18–2 For the case study problem, design the output shaft, including complete specification of the gear, bearings, key, retaining rings, and shaft.

18–3 For the case study problem, use helical gears and design the intermediate shaft. Compare your results with the spur gear design presented in this chapter.

18–4 Perform a final analysis for the resulting design of the intermediate shaft of the case study problem presented in this chapter. Produce a final drawing with dimensions and tolerances for the shaft. Does the final design satisfy all the requirements? Identify the critical aspects of the design with the lowest factor of safety.

18–5 For the case study problem, change the power requirement to 30 kW. Design the intermediate shaft, including complete specification of the gears, bearings, keys, retaining rings, and shaft.

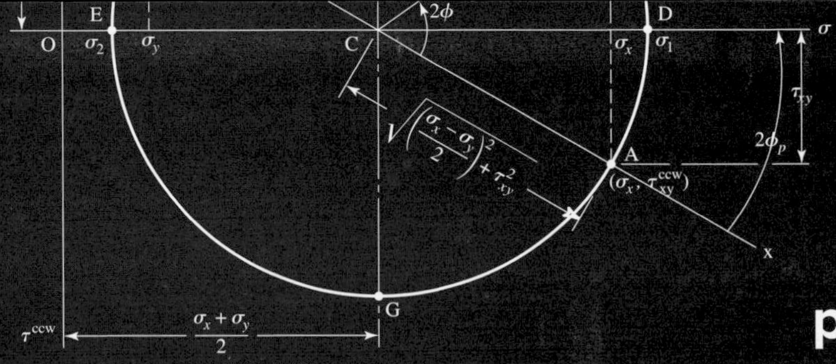

part 4

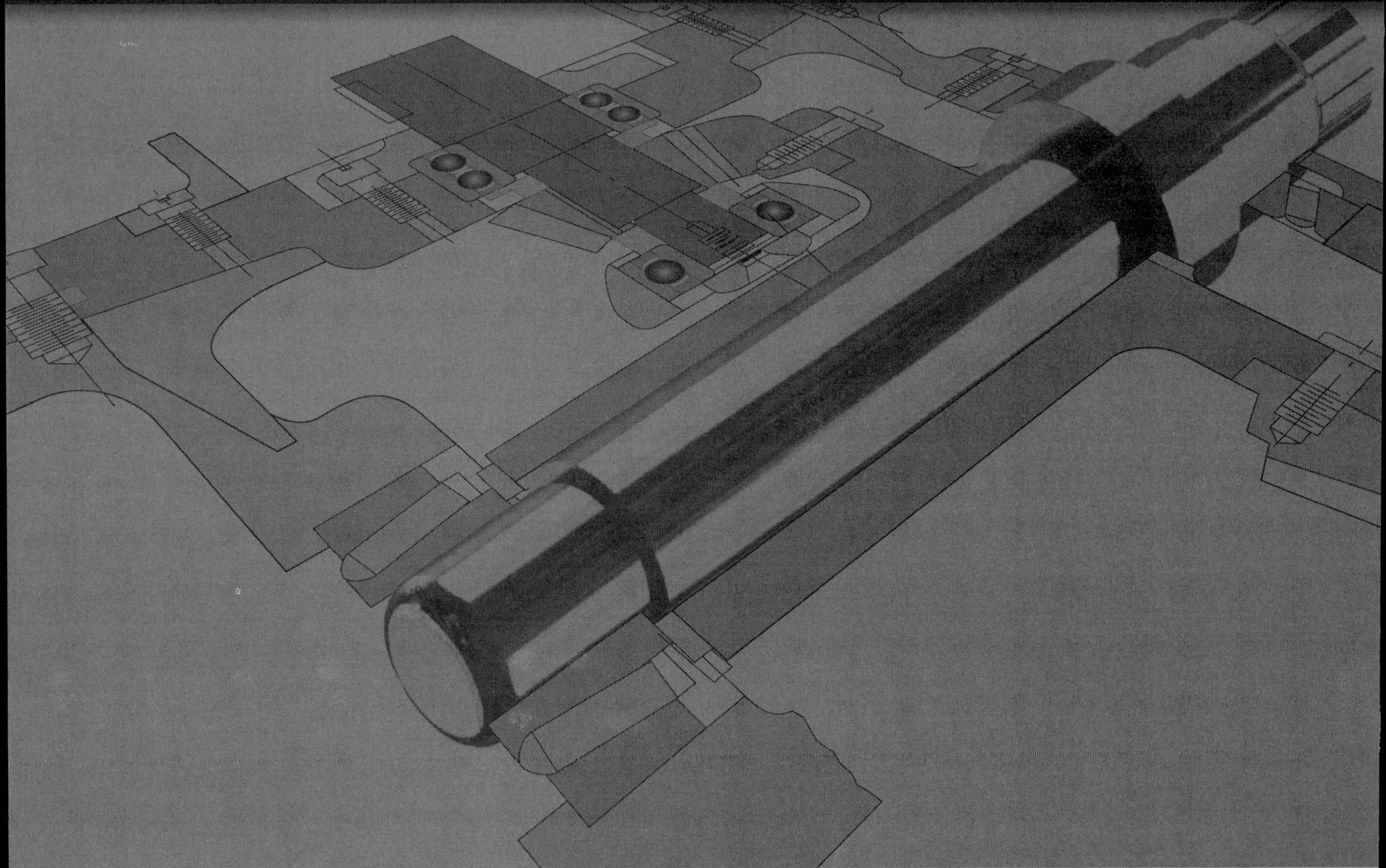

Courtesy of Dee Dehokenanan

Special Topics

19 Finite-Element Analysis

Mathew Alexander/Shutterstock

Chapter Outline

Mechanical components in the form of simple bars, beams, etc., can be analyzed quite easily by basic methods of mechanics that provide closed-form solutions. Actual components, however, are rarely so simple, and the designer is forced to less effective approximations of closed-form solutions, experimentation, or numerical methods. There are a great many numerical techniques used in engineering applications for which the digital computer is very useful. In mechanical design, where computer-aided design (CAD) software is heavily employed, the analysis method that integrates well with CAD is *finite-element analysis (FEA)*. The mathematical theory and applications of the method are vast. There are also a number of commercial FEA software packages that are available, such as ANSYS, MSC/NASTRAN, ALGOR, etc.

The purpose of this chapter is only to expose the reader to some of the fundamental aspects of FEA, and therefore the coverage is extremely introductory in nature. For further detail, the reader is urged to consult the many references cited at the end of this chapter. Figure 19–1 shows a finite-element model of a connecting rod that was developed to study the effects of dynamic elastohydrodynamic lubrication on bearing and structural performance.[1]

There are a multitude of FEA applications such as static and dynamic, linear and nonlinear, stress and deflection analysis; free and forced vibrations; heat transfer (which can be combined with stress and deflection analysis to provide thermally

Figure 19–1

Model of a connecting rod using ANSYS finite-element software. (*a*) Meshed model; (*b*) stress contours. *(Courtesy of Dr. Stephen Boedo)*

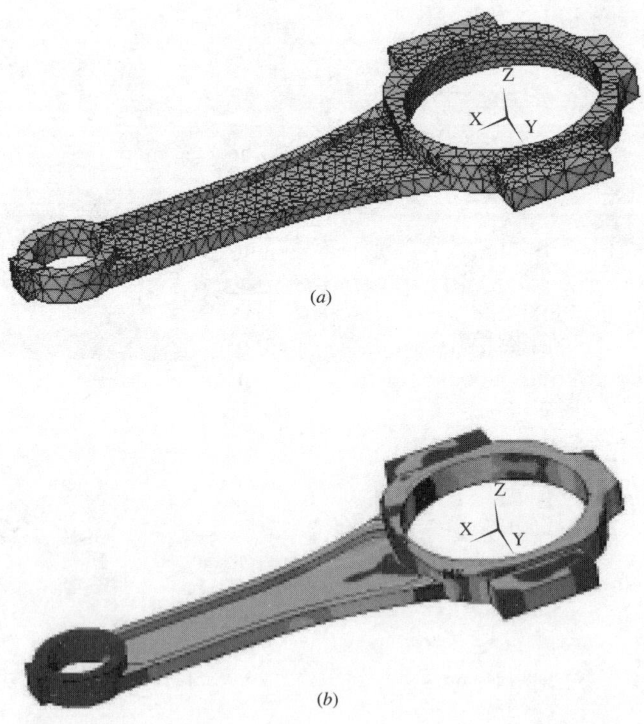

[1]S. Boedo, "Elastohydrodynamic Lubrication of Conformal Bearing Systems," *Proceedings of 2002 ANSYS Users Conference,* Pittsburgh, PA, April 22–24, 2002.

induced stresses and deflections); elastic instability (buckling); acoustics; electrostatics and magnetics (which can be combined with heat transfer); fluid dynamics; piping analysis; and multiphysics. For the purposes of this chapter, we will limit ourselves to basic mechanics analyses.

An actual mechanical component is a continuous elastic structure (continuum). FEA divides (discretizes) the structure into small but finite, well-defined, elastic substructures (elements). By using polynomial functions, together with matrix operations, the continuous elastic behavior of each element is developed in terms of the element's material and geometric properties. Loads can be applied within the element (gravity, dynamic, thermal, etc.), on the surface of the element, or at the *nodes* of the element. The element's nodes are the fundamental governing entities of the element, as it is the node where the element connects to other elements, where elastic properties of the element are eventually established, where boundary conditions are assigned, and where forces (contact or body) are ultimately applied. A node possesses *degrees of freedom* (dof's). Degrees of freedom are the independent translational and rotational motions that can exist at a node. At most, a node can possess three translational and three rotational degrees of freedom. Once each element within a structure is defined *locally* in matrix form, the elements are then *globally* assembled (attached) through their common nodes (dof's) into an overall system matrix. Applied loads and boundary conditions are then specified and through matrix operations the values of all unknown displacement degrees of freedom are determined. Once this is done, it is a simple matter to use these displacements to determine strains and stresses through the constitutive equations of elasticity.

19–1 The Finite-Element Method

The modern development of the finite-element method began in the 1940s in the field of structural mechanics with the work of Hrennikoff,[2] McHenry,[3] and Newmark,[4] who used a lattice of line elements (rods and beams) for the solution of stresses in continuous solids. In 1943, from a 1941 lecture, Courant[5] suggested piecewise polynomial interpolation over triangular subregions as a method to model torsional problems. With the advent of digital computers in the 1950s it became practical for engineers to write and solve the stiffness equations in matrix form.[6,7,8] A classic paper by Turner, Clough, Martin, and Topp published in 1956 presented the matrix stiffness equations

[2]A. Hrennikoff, "Solution of Problems in Elasticity by the Frame Work Method," *Journal of Applied Mechanics,* Vol. 8, No. 4, pp. 169–175, December 1941.

[3]D. McHenry, "A Lattice Analogy for the Solution of Plane Stress Problems," *Journal of Institution of Civil Engineers*, Vol. 21, pp. 59–82, December 1943.

[4]N. M. Newmark, "Numerical Methods of Analysis in Bars, Plates, and Elastic Bodies," *Numerical Methods in Analysis in Engineering* (ed. L. E. Grinter), Macmillan, 1949.

[5]R. Courant, "Variational Methods for the Solution of Problems of Equilibrium and Vibrations," *Bulletin of the American Mathematical Society,* Vol. 49, pp. 1–23, 1943.

[6]S. Levy, "Structural Analysis and Influence Coefficients for Delta Wings," *Journal of Aeronautical Sciences,* Vol. 20, No. 7, pp. 449–454, July 1953.

[7]J. H. Argyris, "Energy Theorems and Structural Analysis," *Aircraft Engineering,* October, November, December 1954 and February, March, April, May 1955.

[8]J. H. Argyris and S. Kelsey, *Energy Theorems and Structural Analysis,* Butterworths, London, 1960 (reprinted from *Aircraft Engineering,* 1954–55).

for the truss, beam, and other elements.[9] The expression *finite element* is first attributed to Clough.[10] Since these early beginnings, a great deal of effort has been expended in the development of the finite element method in the areas of element formulations and computer implementation of the entire solution process. The major advances in computer technology include the rapidly expanding computer hardware capabilities, efficient and accurate matrix solver routines, and computer graphics for ease in the visual preprocessing stages of model building, including automatic adaptive mesh generation, and in the postprocessing stages of reviewing the solution results. A great abundance of literature has been presented on the subject, including many textbooks. A partial list of some textbooks, introductory and more comprehensive, is given at the end of this chapter.

Since the finite-element method is a numerical technique that discretizes the domain of a continuous structure, errors are inevitable. These errors are:

1 ***Computational errors.*** These are due to round-off errors from the computer floating-point calculations and the formulations of the numerical integration schemes that are employed. Most commercial finite-element codes concentrate on reducing these errors, and consequently the analyst generally is concerned with discretization factors.

2 ***Discretization errors.*** The geometry and the displacement distribution of a true structure continuously vary. Using a finite number of elements to model the structure introduces errors in matching geometry and the displacement distribution due to the inherent mathematical limitations of the elements.

For an example of discretization errors, consider the constant thickness, thin plate structure shown in Figure 19–2*a*. Figure 19–2*b* shows a finite-element model

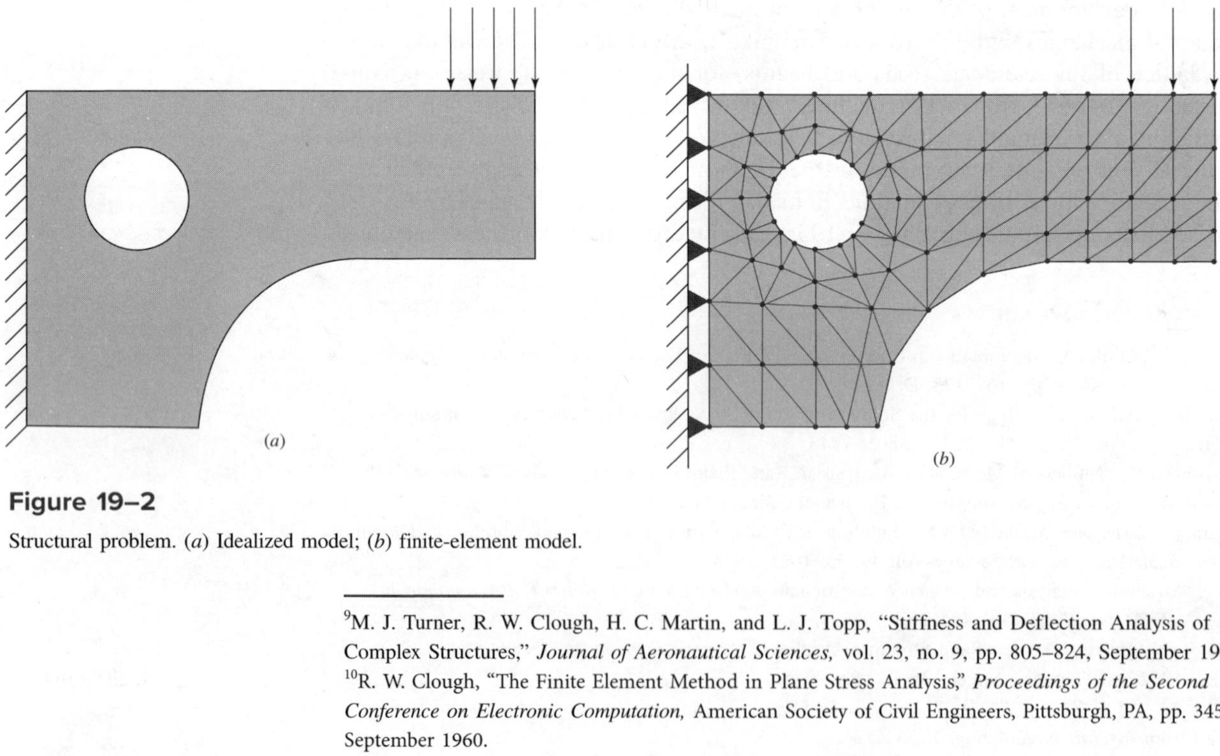

Figure 19–2

Structural problem. (*a*) Idealized model; (*b*) finite-element model.

[9]M. J. Turner, R. W. Clough, H. C. Martin, and L. J. Topp, "Stiffness and Deflection Analysis of Complex Structures," *Journal of Aeronautical Sciences,* vol. 23, no. 9, pp. 805–824, September 1956.

[10]R. W. Clough, "The Finite Element Method in Plane Stress Analysis," *Proceedings of the Second Conference on Electronic Computation,* American Society of Civil Engineers, Pittsburgh, PA, pp. 345–378, September 1960.

of the structure where three-node, plane stress, simplex triangular elements are employed. This element type has a flaw that creates two basic problems. The element has straight sides that remain straight after deformation. The strains throughout the plane stress triangular element are constant. The first problem, a geometric one, is the modeling of curved edges. Note that the surface of the model with a large curvature appears poorly modeled, whereas the surface of the hole seems to be reasonably modeled. The second problem, which is much more severe, is that the strains in various regions of the actual structure are changing rapidly, and the constant strain element will provide only an approximation of the average strain at the center of the element. So, in a nutshell, the results predicted by this model will be extremely poor. The results can be improved by significantly increasing the number of elements (increased mesh density). Alternatively, using a better element, such as an eight-node quadrilateral, which is more suited to the application, will provide the improved results. Because of higher-order interpolation functions, the eight-node quadrilateral element can model curved edges and provides a higher-order function for the strain distribution.

In Figure 19–2*b*, the triangular elements are shaded and the nodes of the elements are represented by the black dots. Forces and constraints can be placed only at the nodes. The nodes of the simplex triangular plane stress elements have only two degrees of freedom, corresponding to translation in the plane. Thus, the solid black, simple support triangles on the left edge represent the fixed support of the model. Also, the distributed load can be applied only to three nodes as shown. The modeled load has to be statically consistent with the actual load.

19–2 Element Geometries

Many geometric shapes of elements are used in finite-element analysis for specific applications. The various elements used in a general-purpose commercial FEA software code constitute what is referred to as the *element library* of the code. Elements can be placed in the following categories: *line elements, surface elements, solid elements,* and *special-purpose elements*. Table 19–1 provides some, but not all, of the types of elements available for finite-element analysis for structural problems. Not all elements support all degrees of freedom. For example, the 3-D truss element supports

Table 19–1 Sample Finite-Element Library

Element Type	None	Shape	Number of Nodes	Applications
Line	Truss	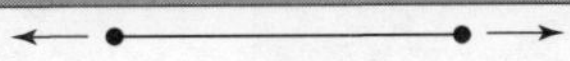	2	Pin-ended bar in tension or compression
	Beam	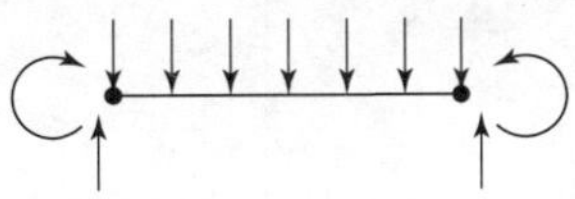	2	Bending
	Frame	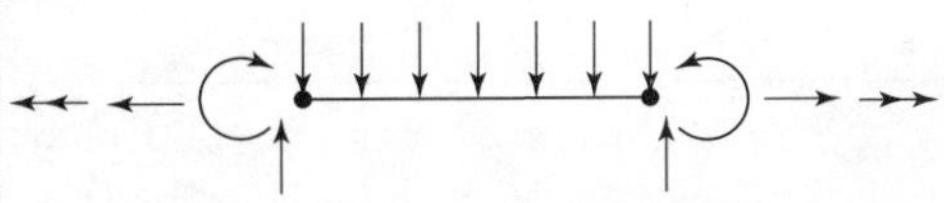	2	Axial, torsional, and bending. With or without load stiffening.

(*Continued*)

Table 19–1 Sample Finite-Element Library *(Continued)*

Element Type	None	Shape	Number of Nodes	Applications
Surface	4-node quadri-lateral	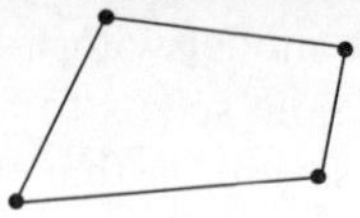	4	Plane stress or strain, axisymmetry, shear panel, thin flat plate in bending
	8-node quadri-lateral	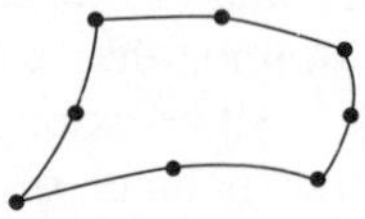	8	Plane stress or strain, thin plate or shell in bending
	3-node triangular	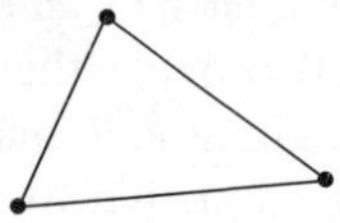	3	Plane stress or strain, axisymmetry, shear panel, thin flat plate in bending. Prefer quad where possible. Used for transitions of quads.
	6-node triangular	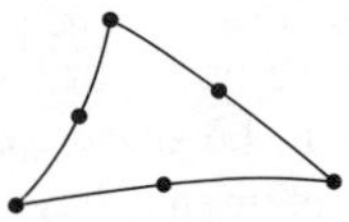	6	Plane stress or strain, axisymmetry, thin plate or shell in bending. Prefer quad where possible. Used for transitions of quads.
Solid[†]	8-node hexagonal (brick)	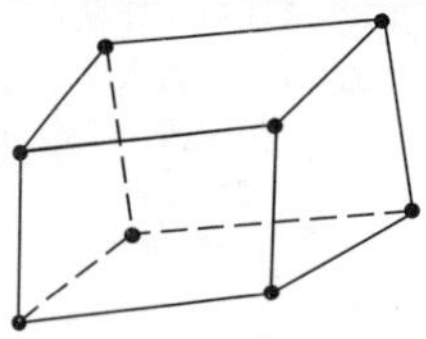	8	Solid, thick plate
	6-node pentagonal (wedge)	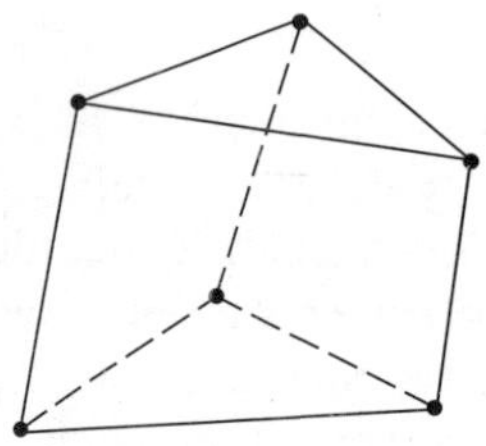	6	Solid, thick plate. Used for transitions.
	4-node tetrahedron (tet)	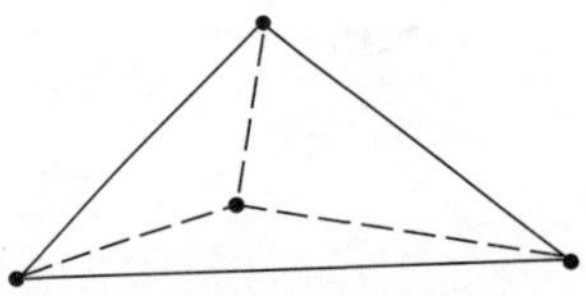	4	Solid, thick plate. Used for transitions.
Special purpose	Gap	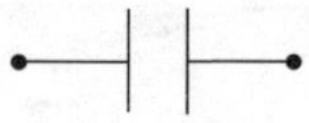	2	Free displacement for prescribed compressive gap
	Hook	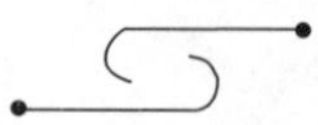	2	Free displacement for prescribed extension gap
	Rigid	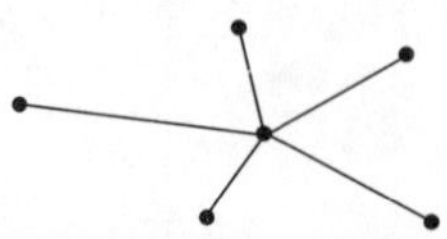	Variable	Rigid constraints between nodes

[†]These elements are also available with midside nodes.

only three translational degrees of freedom at each node. Connecting elements with differing dof's generally requires some manual modification. For example, consider connecting a truss element to a frame element. The frame element supports all six dof's at each node. A truss member, when connected to it, can rotate freely at the connection.

19–3 The Finite-Element Solution Process

We will describe the finite-element solution process on a very simple one-dimensional problem, using the linear truss element. A truss element is a bar loaded in tension or compression and is of constant cross-sectional area A, length l, and elastic modulus E. The basic truss element has two nodes, and for a one-dimensional problem, each node will have only one degree of freedom. A truss element can be modeled as a simple linear spring with a spring rate, given by Equation (4–4), as

$$k = \frac{AE}{l} \tag{19–1}$$

Consider a spring element (e) of spring rate k_e, with nodes i and j, as shown in Figure 19–3. Nodes and elements will be numbered. So, to avoid confusion as to what a number corresponds to, elements will be numbered within parentheses. Assuming all forces f and displacements u directed toward the right as positive, the forces at each node can be written as

$$\begin{aligned} f_{i,e} &= k_e(u_i - u_j) = k_e u_i - k_e u_j \\ f_{j,e} &= k_e(u_j - u_i) = -k_e u_i + k_e u_j \end{aligned} \tag{19–2}$$

The two equations can be written in matrix form as

$$\begin{Bmatrix} f_{i,e} \\ f_{j,e} \end{Bmatrix} = \begin{bmatrix} k_e & -k_e \\ -k_e & k_e \end{bmatrix} \begin{Bmatrix} u_i \\ u_j \end{Bmatrix} \tag{19–3}$$

Next, consider a two-spring system as shown in Figure 19–4*a*. Here we have numbered the nodes and elements. We have also labeled the forces at each node. However, these forces are the total external forces at each node, F_1, F_2, and F_3. If we draw separate free-body diagrams we will expose the internal forces as shown in Figure 19–4*b*.

Using Equation (19–3) for each spring gives

Element 1
$$\begin{Bmatrix} f_{1,1} \\ f_{2,1} \end{Bmatrix} = \begin{bmatrix} k_1 & -k_1 \\ -k_1 & k_1 \end{bmatrix} \begin{Bmatrix} u_1 \\ u_2 \end{Bmatrix} \tag{19–4a}$$

Element 2
$$\begin{Bmatrix} f_{2,2} \\ f_{3,2} \end{Bmatrix} = \begin{bmatrix} k_2 & -k_2 \\ -k_2 & k_2 \end{bmatrix} \begin{Bmatrix} u_2 \\ u_3 \end{Bmatrix} \tag{19–4b}$$

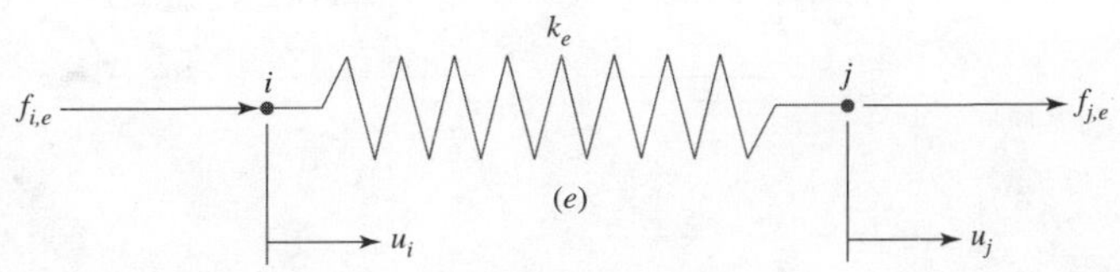

Figure 19–3
A simple spring element.

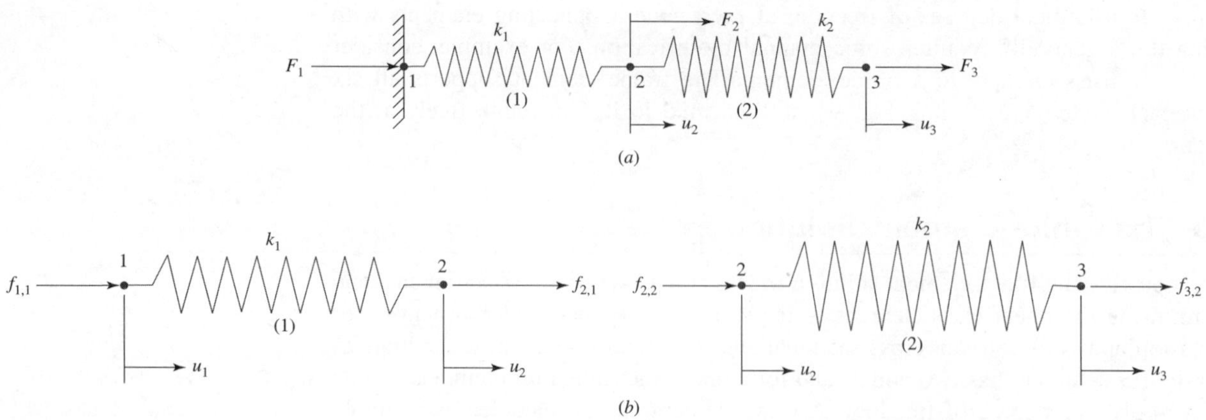

Figure 19–4

A two-element spring system. (*a*) System model; (*b*) separate free-body diagrams.

The total force at each node is the external force, $F_1 = f_{1,1}$, $F_2 = f_{2,1} + f_{2,2}$, and $F_3 = f_{3,2}$. Combining the two matrices in terms of the external forces gives

$$\begin{Bmatrix} f_{1,1} \\ f_{2,1} + f_{2,2} \\ f_3 \end{Bmatrix} = \begin{Bmatrix} F_1 \\ F_2 \\ F_3 \end{Bmatrix} = \begin{bmatrix} k_1 & -k_1 & 0 \\ -k_1 & (k_1 + k_2) & -k_2 \\ 0 & -k_2 & k_2 \end{bmatrix} \begin{Bmatrix} u_1 \\ u_2 \\ u_3 \end{Bmatrix} \tag{19–5}$$

If we know the displacement of a node, then the force at the node will be unknown. For example, in Figure 19–4*a*, the displacement of node 1 at the wall is zero, so F_1 is the unknown reaction force (note, up to this point, we have not applied a static solution of the system). If we do not know the displacement of a node, then we know the force. For example, in Figure 19–4*a*, the displacements at nodes 2 and 3 are unknown, and the forces F_2 and F_3 are to be specified. To see how the remainder of the solution process can be implemented, let us consider the following example.

EXAMPLE 19–1

Consider the aluminum step-shaft shown in Figure 19–5*a*. The areas of sections *AB* and *BC* are 64.5 mm^2 and 96.8 mm^2, respectively. The lengths of sections *AB* and *BC* are 250 mm and 300 mm, respectively. A force F = 4.5 kN is applied to *B*. Initially, a gap of $\varepsilon = 0.05$ mm exists between end *C* and the right rigid wall. Determine the wall reactions, the internal forces in the members, and the deflection of point *B*. Let $E = 70$ GPa and assume that end *C* hits the wall. Check the validity of the assumption.

Figure 19–5

(*a*) Step shaft; (*b*) spring model.

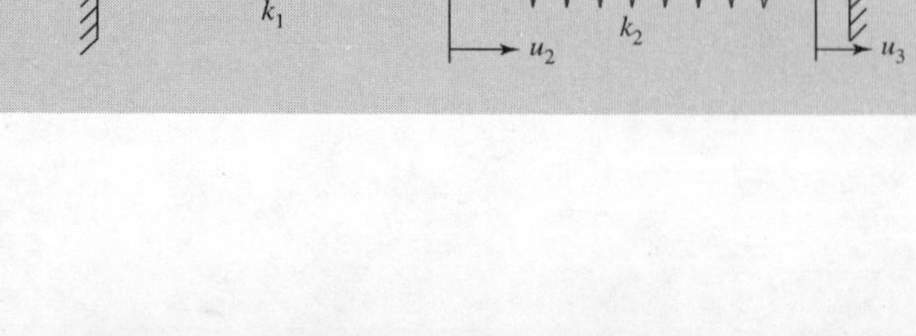

Solution

The step-shaft is modeled by the two-spring system of Figure 19–5*b* where

$$k_1 = \left(\frac{AE}{l}\right)_{AB} = \frac{64.5(70)10^3}{250} = 18\ 060\ \text{N/mm}$$

$$k_2 = \left(\frac{AE}{l}\right)_{BC} = \frac{96.8(70)10^3}{300} = 22\ 587\ \text{N/mm}$$

With $u_1 = 0$, $F_2 = 4.5$ kN and the assumption that $u_3 = \varepsilon = 0.05$ mm, Equation (19–5) becomes

$$\begin{Bmatrix} F_1 \\ 4500 \\ F_3 \end{Bmatrix} = \begin{bmatrix} 18\ 060 & -18\ 060 & 0 \\ -18\ 060 & 40\ 647 & -22\ 587 \\ 0 & -22\ 587 & 22\ 587 \end{bmatrix} \begin{Bmatrix} 0 \\ u_2 \\ 0.05 \end{Bmatrix} \tag{1}$$

For large problems, there is a systematic method of solving equations like Equation (1), called *partitioning* or *the elimination approach*.[11] However, for this simple problem, the solution is quite simple. From the second equation of the matrix equation

$$4500 = [-1(0) + 40\ 647u_2 - 22\ 587(0.05)]$$

or,

Answer
$$u_B = u_2 = \frac{4500 + 22\ 587(0.05)}{40\ 647} = 0.138\ \text{mm}$$

Since $u_B > \varepsilon$, it is verified that point C hits the wall.

The reactions at the walls are F_1 and F_3. From the first and third equations of matrix Equation (1),

Answer
$$F_1 = [-1(u_2)] = [-1(0.138)] = -2501\ \text{N}$$

and

Answer
$$F_3 = [-22\ 587u_2 + 22\ 587(0.05)]$$
$$= [-22\ 587(0.138) + 22\ 587(0.05)] = -1999\ \text{N}$$

Since F_3 is negative, this also verifies that C hits the wall. Note that $F_1 + F_3 = -2501 - 1999 = -4500$ N, balancing the applied force (with no statics equations necessary).

For internal forces, it is necessary to return to the individual (local) equations. From Equation (19–4*a*),

$$\begin{Bmatrix} f_{1,1} \\ f_{2,1} \end{Bmatrix} = \begin{bmatrix} k_1 & -k_1 \\ -k_1 & k_1 \end{bmatrix} \begin{Bmatrix} u_1 \\ u_2 \end{Bmatrix} = \begin{bmatrix} 18\ 060 & -18\ 060 \\ -18\ 060 & 18\ 060 \end{bmatrix} \begin{Bmatrix} 0 \\ 0.138 \end{Bmatrix} = \begin{Bmatrix} -2501 \\ 2501 \end{Bmatrix} \text{N}$$

Answer

Since $f_{1,1}$ is directed to the left and $f_{2,1}$ is directed to the right, the element is in tension, with a force of 2501 N. If the stress is desired, it is simply $\sigma_{AB} = f_{2,1}/A_{AB} = 2501/64.5 = 39$ MPa.

For element BC, from Equation (19.4*b*),

$$\begin{Bmatrix} f_{2,2} \\ f_{3,2} \end{Bmatrix} = \begin{bmatrix} k_2 & -k_2 \\ -k_2 & k_2 \end{bmatrix} \begin{Bmatrix} u_2 \\ u_3 \end{Bmatrix} = 10^5 \begin{bmatrix} 22\ 587 & -22\ 587 \\ -22\ 587 & 22\ 587 \end{bmatrix} \begin{Bmatrix} 0.138 \\ 0.05 \end{Bmatrix} = \begin{Bmatrix} 1999 \\ -1999 \end{Bmatrix} \text{N}$$

Answer

Since $f_{2,2}$ is directed to the right and $f_{3,2}$ is directed to the left, the element is in compression, with a force of 1999 N. If the stress is desired, it is simply $\sigma_{BC} = -f_{2,2}/A_{BC} = -1999/96.8 = -21$ MPa.

---

[11] See T. R. Chandrupatla and A. D. Belegundu, *Introduction to Finite Elements in Engineering,* 4th ed., Prentice Hall, Upper Saddle River, NJ, pp. 71–75, 2012.

19–4 Mesh Generation

The network of elements and nodes that discretize a region is referred to as a *mesh*. The *mesh density* increases as more elements are placed within a given region. *Mesh refinement* is when the mesh is modified from one analysis of a model to the next analysis to yield improved results. Results generally improve when the mesh density is increased in areas of high stress gradients and/or when geometric transition zones are meshed smoothly. Generally, but not always, the FEA results converge toward the exact results as the mesh is continuously refined. To assess improvement, in regions where high stress gradients appear, the structure can be remeshed with a higher mesh density at this location. If there is a minimal change in the maximum stress value, it is reasonable to presume that the solution has converged. There are three basic ways to generate an element mesh—manually, semiautomatically, or fully automatically.

1 ***Manual mesh generation.*** This is how the element mesh was created in the early days of the finite-element method. This is a very labor intensive method of creating the mesh, and except for some quick modifications of a model is it rarely done. *Note:* Care must be exercised in editing an input text file. With some FEA software, other files such as the preprocessor binary graphics file may not change. Consequently, the files may no longer be compatible with each other.

2 ***Semiautomatic mesh generation.*** Over the years, computer algorithms have been developed that enable the modeler to automatically mesh regions of the structure using well-defined boundaries. Since the modeler has to define these regions, the technique is deemed *semiautomatic*. The development of the many computer algorithms for mesh generation emanates from the field of computer graphics. If the reader desires more information on this subject, a review of the literature available from this field is recommended.

3 ***Fully automatic mesh generation.*** Many software vendors have concentrated their efforts on developing fully automatic mesh generation, and in some instances, with automatic *self-adaptive* mesh refinement. The obvious goal is to significantly reduce the modeler's preprocessing time and effort to arrive at a final well-constructed FEA mesh. Once the complete boundary of the structure is defined, without subdivisions as in semiautomatic mesh generation and with a minimum of user intervention, various schemes are available to discretize the region with *one element type*. For plane elastic problems the boundary is defined by a series of internal and external geometric lines and the element type to be automeshed would be the plane elastic element. For thin-walled structures, the geometry would be defined by three-dimensional surface representations and the automeshed element type would be the three-dimensional plate element. For solid structures, the boundary could be constructed by using *constructive solid geometry (CSG)* or *boundary representation (B-rep)* techniques. The finite-element types for automeshing would be the brick and/or tetrahedron element(s).

Automatic self-adaptive mesh refinement programs estimate the error of the FEA solution. On the basis of the error, the mesh is automatically revised and reanalyzed. The process is repeated until some convergence or termination criterion is satisfied.

Returning to the thin-plate model of Figure 19–2, the boundaries of the structure are constructed as shown in Figure 19–6*a*. The boundaries were then automeshed as

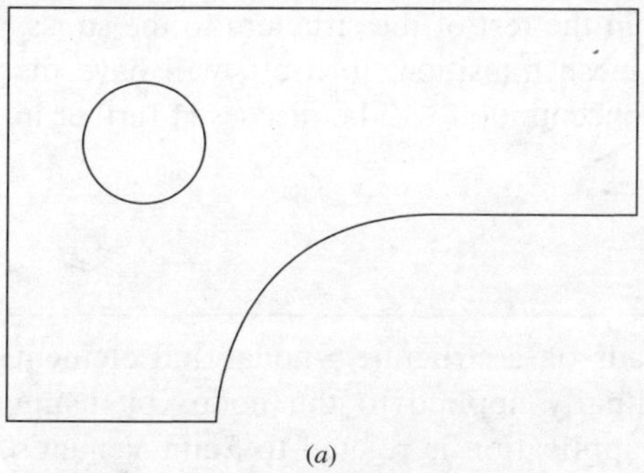

Figure 19–6

Automatic meshing the thin-plate model of Figure 19–2. (*a*) Model boundaries; (*b*) automesh with 294 elements and 344 nodes; (*c*) deflected (exaggerated scale) with stress contours; (*d*) automesh with 1008 elements and 1096 nodes, (*e*) deflected (exaggerated scale) with stress contours.

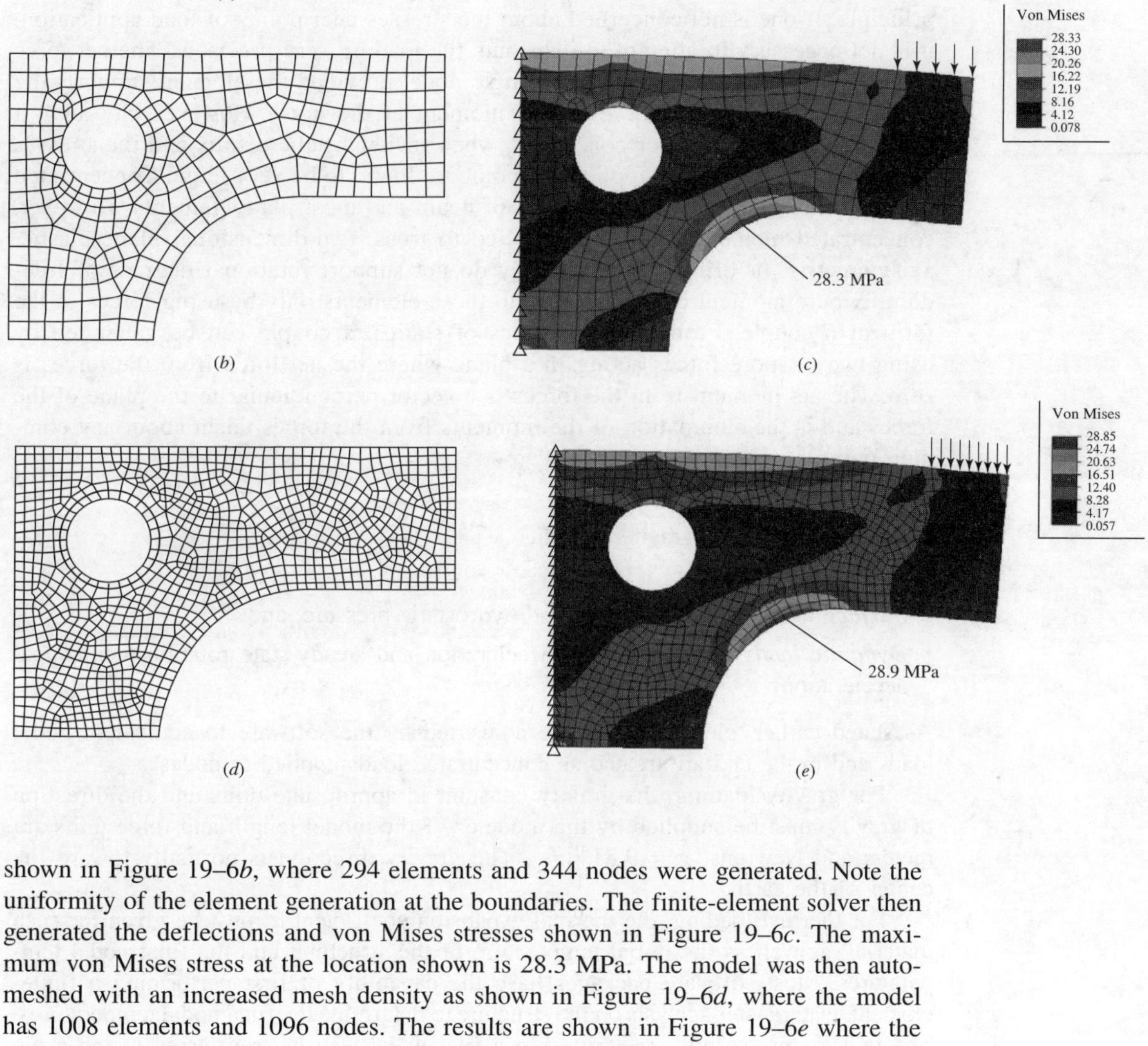

shown in Figure 19–6*b*, where 294 elements and 344 nodes were generated. Note the uniformity of the element generation at the boundaries. The finite-element solver then generated the deflections and von Mises stresses shown in Figure 19–6*c*. The maximum von Mises stress at the location shown is 28.3 MPa. The model was then automeshed with an increased mesh density as shown in Figure 19–6*d*, where the model has 1008 elements and 1096 nodes. The results are shown in Figure 19–6*e* where the maximum von Mises stress is found to be 28.9 MPa, which is only 1.8 percent higher. In all likelihood, the solution has nearly converged. *Note:* The stress contours of Figures 19–6*c* and *e* are better visualized in color.

When stress concentrations are present, it is necessary to have a very fine mesh at the stress-concentration region in order to get realistic results. What is important is that the mesh density needs to be increased only in the region around the stress concentration and that the transition mesh from the rest of the structure to the stress-concentration region be gradual. An abrupt mesh transition, in itself, will have the same effect as a stress concentration. Stress concentration will be discussed further in Section 19–7, Modeling Techniques.

19–5 Load Application

There are two basic forms of specifying loads on a structure—nodal and element loading. However, element loads are eventually applied to the nodes by using equivalent nodal loads. One aspect of load application is related to Saint-Venant's principle. If one is not concerned about the stresses near points of load application, it is not necessary to attempt to distribute the loading very precisely. The net force and/or moment can be applied to a single node, provided the element supports the dof associated with the force and/or moment at the node. However, the analyst should not be surprised, or concerned, when reviewing the results and the stresses in the vicinity of the load application point are found to be very large. Concentrated moments can be applied to the nodes of beam and most plate elements. However, concentrated moments cannot be applied to truss, two-dimensional plane elastic, axisymmetric, or brick elements. They do not support rotational degrees of freedom. A pure moment can be applied to these elements only by using forces in the form of a couple. From the mechanics of statics, a couple can be generated by using two or more forces acting in a plane where the net force from the forces is zero. The net moment from the forces is a vector perpendicular to the plane of the forces and is the summation of the moments from the forces taken about any common point.

Element loads include

- ***Static loads*** due to gravity (weight),
- ***Thermal effects,***
- ***Surface loads*** such as uniform and hydrostatic pressure, and
- ***Dynamic loads*** due to constant acceleration and steady-state rotation (centrifugal acceleration).

As stated earlier, element loads are converted by the software to equivalent nodal loads and in the end are treated as concentrated loads applied to nodes.

For gravity loading, the gravity constant in appropriate units and the direction of gravity must be supplied by the modeler. If the model length and force units are meters and Newtons, $g = 9.81\ \text{m/s}^2$. The gravity direction is normally toward the center of the earth.

For thermal loading, the thermal expansion coefficient α must be given for each material, as well as the initial temperature of the structure, and the final nodal temperatures. Most software packages have the capability of first performing a finite-element heat transfer analysis on the structure to determine the final nodal temperatures. The temperature results are written to a file, which can be transferred to the static stress analysis. Here the heat transfer model should have the same nodes and element type the static stress analysis model has.

Surface loading can generally be applied to most elements. For example, uniform or linear transverse line loads (force/length) can be specified on beams. Uniform and linear pressure can normally be applied on the edges of two-dimensional plane and axisymmetric elements. Lateral pressure can be applied on plate elements, and pressure can be applied on the surface of solid brick elements. Each software package has its unique manner in which to specify these surface loads, usually in a combination of text and graphic modes.

19–6 Boundary Conditions

The simulation of boundary conditions and other forms of constraint is probably the single most difficult part of the accurate modeling of a structure for a finite-element analysis. In specifying constraints, it is relatively easy to make mistakes of omission or misrepresentation. It may be necessary for the analyst to test different approaches to model esoteric constraints such as bolted joints, welds, etc., which are not as simple as the idealized pinned or fixed joints. Testing should be confined to simple problems and not to a large, complex structure. Sometimes, when the exact nature of a boundary condition is uncertain, only limits of behavior may be possible. For example, we have modeled shafts with bearings as being simply supported. It is more likely that the support is something between simply supported and fixed, and we could analyze both constraints to establish the limits. However, by assuming simply supported, the results of the solution are conservative for stress and deflections. That is, the solution would predict stresses and deflections larger than the actual.

For another example, consider beam 16 in Table A–9. The horizontal beam is uniformly loaded and is fixed at both ends. Although not explicitly stated, tables such as these assume that the beams are not restrained in the horizontal direction. That is, it is assumed that the beam can slide horizontally in the supports. If the ends were completely or partially restrained, a beam-column solution would be necessary.[12] With a finite-element analysis, a special element, a beam with stiffening, could be used.

Multipoint constraint equations are quite often used to model boundary conditions or rigid connections between elastic members. When used in the latter form, the equations are acting as elements and are thus referred to as *rigid elements.* Rigid elements can rotate or translate only rigidly.

Boundary elements are used to force specific nonzero displacements on a structure. Boundary elements can also be useful in modeling boundary conditions that are askew from the global coordinate system.

19–7 Modeling Techniques

With today's CAD packages and automatic mesh generators, it is an easy task to create a solid model and mesh the volume with finite elements. With today's computing speeds and with gobs of computer memory, it is very easy to create a model with extremely large numbers of elements and nodes. The finite-element modeling techniques of the past now seem passé and unnecessary. However, much unnecessary time can be spent on a very complex model when a much simpler model will do. The complex model may not even provide an accurate solution, whereas a simpler one

[12]See R. G. Budynas, *Advanced Strength and Applied Stress Analysis,* 2nd ed., McGraw Hill, New York, 1999, pp. 471–482.

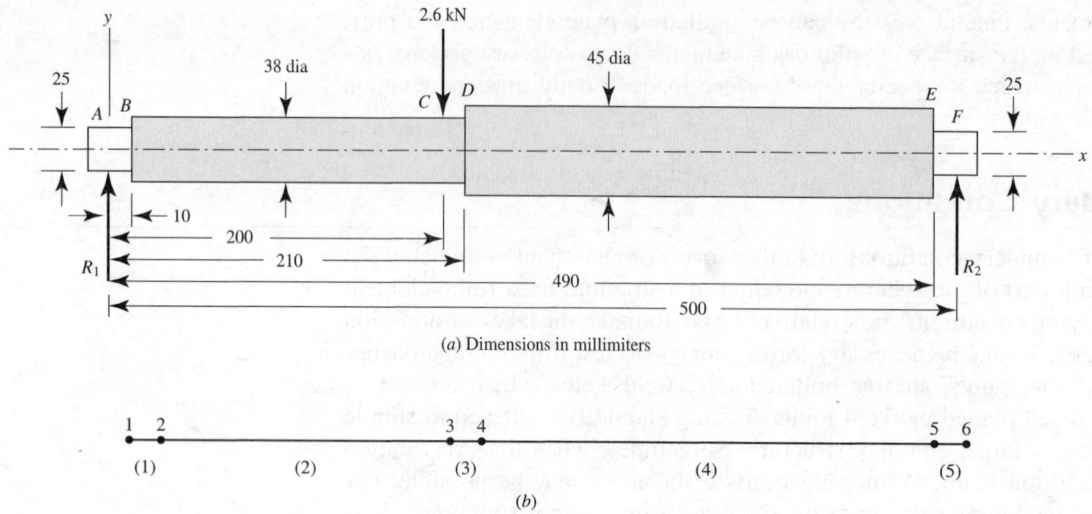

Displacements/rotations (degrees) of nodes

Node No.	x Translation	y Translation	z Translation	θ_x Rotation (deg)	θ_y Rotation (deg)	θ_z Rotation (deg)
1	0.000 E + 00	0.000 E + 00	0.000 E + 00	0.000 E + 00	0.000 E + 00	−9.293 E − 02
2	0.000 E + 00	−2.081 E − 02	0.000 E + 00	0.000 E + 00	0.000 E + 00	−9.127 E − 02
3	0.000 E + 00	−2.294 E − 01	0.000 E + 00	0.000 E + 00	0.000 E + 00	−7.580 E − 03
4	0.000 E + 00	−2.299 E − 01	0.000 E + 00	0.000 E + 00	0.000 E + 00	2.704 E − 02
5	0.000 E + 00	−1.482 E − 02	0.000 E + 00	0.000 E + 00	0.000 E + 00	6.506 E − 02
6	0.000 E + 00	0.000 E + 00	0.000 E + 00	0.000 E + 00	0.000 E + 00	6.520 E − 02

(c)

Figure 19–7

(*a*) Steel step shaft of Example 4–7; (*b*) finite-element model using five beam elements; (*c*) displacement results for FEA model.

will. What is important is what solution the analyst is looking for: deflections, stresses, or both?

For example, consider the steel step shaft of Example 4–7, repeated here as Figure 19–7*a*. Let the fillets at the steps have a radius of 0.5 mm. If only deflections and slopes were sought at the steps, a highly meshed solid model would not yield much more than the simple five-element beam model, shown in Figure 19–7*b*, would. The fillets at the steps, which could not be modeled easily with beam elements, would not contribute much to a difference in results between the two models. Nodes are necessary wherever boundary conditions, applied forces, and changes in cross section and/or material occur. The displacement results for the FEA model are shown in Figure 19–7*c*.

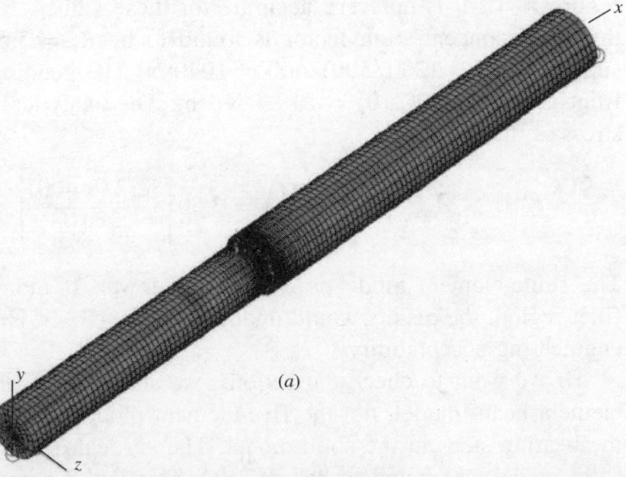

(a)

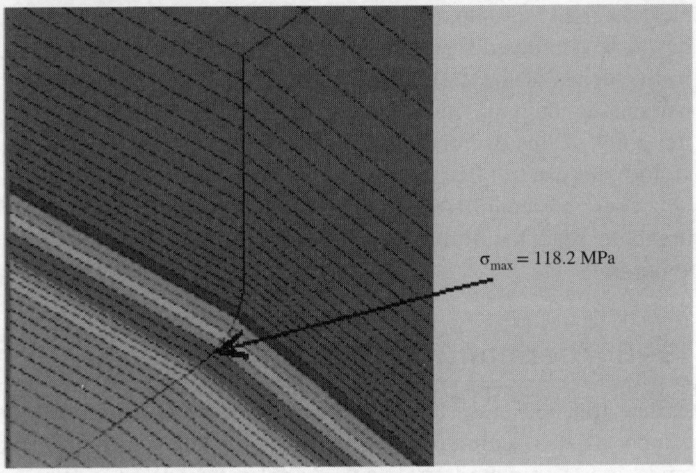

(b)

Figure 19–8

(*a*) Solid model of the step-shaft of Example 4–7 using 56 384 brick and tetrahedron elements; (*b*) view of stress contours at step, rotated 180° about x axis, showing maximum tension.

The FE model of Figure 19–7*b* is not capable of providing the stress at the fillet of the step at *D*. Here, a full-blown solid model would have to be developed and meshed, using solid elements with a high mesh density at the fillet as shown in Figure 19–8*a*. Here, the steps at the bearing supports are not modeled, as we are concerned only with the stress concentration at $x = 210$ mm. The brick and tetrahedron elements do not support rotational degrees of freedom. To model the simply supported boundary condition at the left end, nodes along the z axis were constrained from translating in the x and y directions. Nodes along the y axis were constrained from translating in the z direction. Nodes on the right end on an axis parallel with the z axis through the center of the shaft were constrained from translating in the y direction, and nodes on an axis parallel with the y axis through the center of the shaft were constrained from translating in the z direction. This ensures no rigid-body translation or rotation and no overconstraint at the ends. The maximum tensile stress at the fillet at the beam bottom is found to be $\sigma_{max} = 118.2$ MPa. Performing an analytical check at the step yields $D/d = 45/38 = 1.184$, and $r/d = 0.5/38 = 0.0132$.

Figure A–13–9 is not very accurate for these values. Resorting to another source,[13] the stress-concentration factor is found to be $K_t = 3.00$. The reaction at the right support is $R_F = (200/500)2600 = 1040$ N. The bending moment at the start of the fillet is $M = 1040(210) = 218.4$ N · m. The analytical prediction of the maximum stress is thus

$$\sigma_{\max} = K_t\left(\frac{32M}{\pi d^3}\right) = 3.00\left[\frac{32(218\ 400)}{\pi(38)^3}\right] = 121.6 \text{ MPa}$$

The finite-element model is 2.8 percent lower. If more elements were used in the fillet region, the results would undoubtedly be closer. However, the results are within engineering acceptability.

If we want to check deflections, we should compare the results with the three-element beam model, not the five-element model. This is because we did not model the bearing steps in the solid model. The vertical deflection, at $x = 210$ mm, for the solid model was found to be −0.2402. This is 4.3 percent higher in magnitude than the −0.2299 mm deflection for the three-element beam model. For slopes, the brick element does not support rotational degrees of freedom, so the rotation at the ends has to be computed from the displacements of adjacent nodes at the ends. This results in the slopes at the ends of $\theta_A = -0.0996°$ and $\theta_F = 0.0698°$; these are 6.7 and 6.6 percent higher in magnitude than the three-element beam model, respectively. However, the point of this exercise is, if deflections were the only result desired, which model would you use?

There are countless modeling situations which could be examined. The reader is urged to read the literature, and peruse the tutorials available from the software vendors.[14]

19–8 Thermal Stresses

A heat transfer analysis can be performed on a structural component including the effects of heat conduction, convection, and/or radiation. After the heat transfer analysis is completed, the same model can be used to determine the resulting thermal stresses. For a simple illustration, we will model a 254 × 102 mm, 6 mm-thick steel plate with a centered 25 mm-diameter hole. The plate is supported as shown in Figure 19–9*a*, and the temperatures of the ends are maintained at temperatures of 38°C and −18°C. Other than at the walls, all surfaces are thermally insulated. Before placing the plate between the walls, the initial temperature of the plate was −18°C. The thermal coefficient of expansion for steel is $\alpha_s = 11.7 \times 10^{-6}\ °\text{C}^{-1}$. The plate was meshed with 1312 two-dimensional elements, with the mesh refined along the border of the hole. Figure 19–9*b* shows the temperature contours of the steady-state temperature distribution obtained by the FEA. Using the same elements for a linear stress analysis, where the temperatures were transferred from the heat transfer

[13]See W. D. Pilkey and D. F. Pilkey, *Peterson's Stress-Concentration Factors,* 3rd ed. John Wiley & Sons, New York, 2008, Chart 3.11.

[14]See, for example, R. D. Cook, *Finite Element Modeling for Stress Analysis,* Wiley & Sons, New York, 1995; and R. G. Budynas, *Advanced Strength and Applied Stress Analysis,* 2nd ed., McGraw Hill, New York, 1999, Chapter 10.

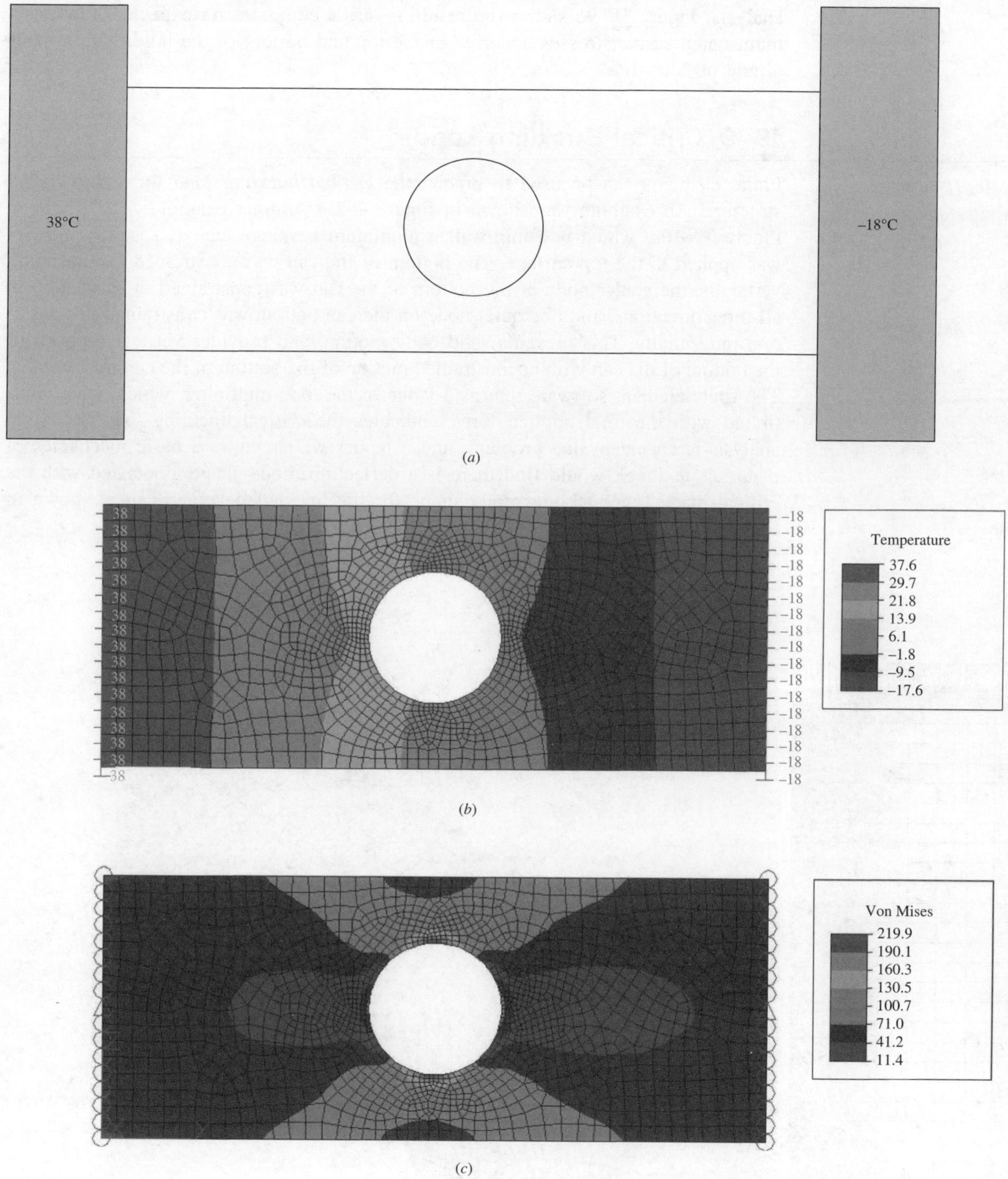

Figure 19–9

(*a*) Plate supported at ends and maintained at the temperatures shown; (*b*) steady-state temperature contours; (*c*) thermal von Mises stress contours where the initial temperature of the plate was −18°C.

analysis, Figure 19–9*c* shows the resulting stress contours. As expected, the maximum compressive stresses occurred at the top and bottom of the hole, with a magnitude of 220 MPa.

19–9 Critical Buckling Load

Finite elements can be used to predict the *critical buckling load* for a thin-walled structure. An example was shown in Figure 4–29. Another example can be seen in Figure 19–10*a*, which is a thin-walled aluminum beverage can. A specific pressure was applied to the top surface. The bottom of the can was constrained in translation vertically, the center node of the bottom of the can was constrained in translation in all three directions, and one outer node on the can bottom was constrained in translation tangentially. This prevents rigid-body motion, and provides vertical support for the bottom of the can with unconstrained motion of the bottom of the can horizontally. The finite element software returns a value of the load multiplier, which, when multiplied with the total applied force, indicates the critical buckling load. Buckling analysis is an eigenvalue problem, and a reader who reviews a basic mechanics of materials textbook would find there is a deflection mode shape associated with the critical load. The buckling mode shape for the buckled beverage can is shown in Figure 19–10*b*.

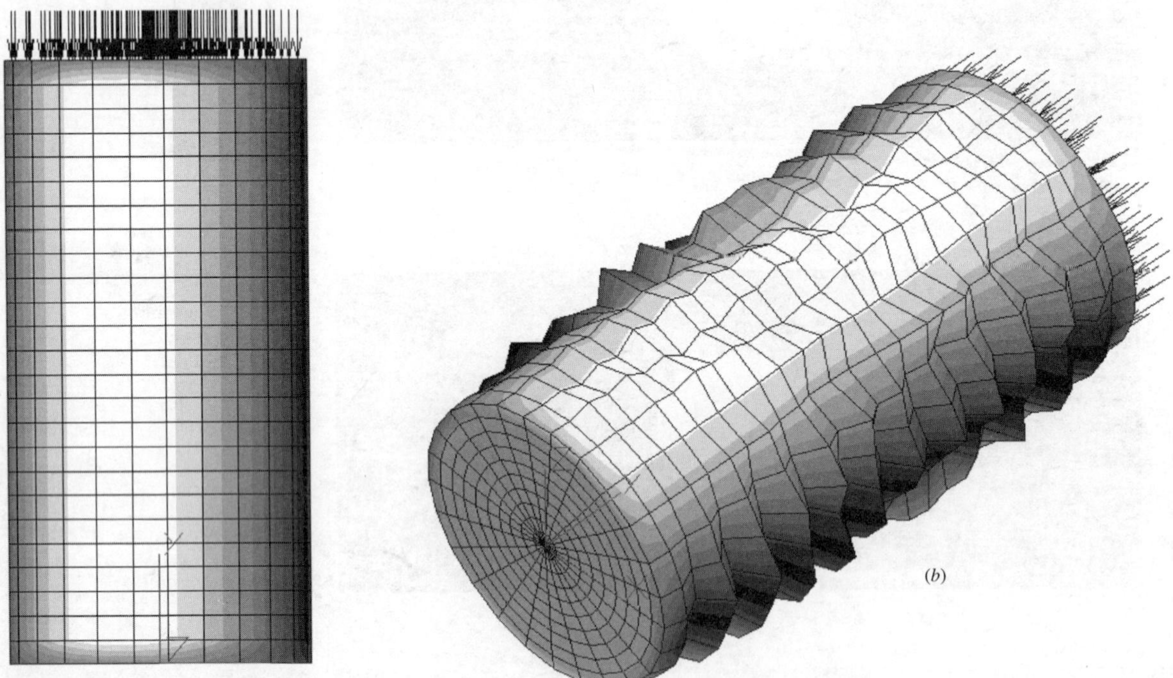

(*a*) (*b*)

Figure 19–10

(*a*) Thin-walled aluminum beverage container loaded vertically downward on the top surface; (*b*) isometric view of the buckled can (deflections greatly exaggerated).

19–10 Vibration Analysis

The design engineer may be concerned as to how a component behaves relative to dynamic input, which results in vibration. For vibration, most finite element packages start with a *modal analysis* of the component. This provides the natural frequencies and mode shapes that the component naturally vibrates at. These are called the eigenvalues and eigenvectors of the component. Next, this solution can be transferred (much the same as for thermal stresses) to solvers for forced vibration analyses, such as frequency response, transient impact, or random vibration, to see how the component's modes behave to dynamic input. The mode shape analysis is primarily based on stiffness and the resulting deflections. Thus, similar to static stress analysis, simpler models will suffice. However, if, when solving forced response problems, stresses are desired, a more detailed model is necessary (similar to the shaft illustration given in Section 19–7).

A modal analysis of the beam model without the bearing steps was performed for a 20-element beam model,[15] and the 56 384-element brick and tetrahedron model. Needless to say, the beam model took less than 9 seconds to solve, whereas the solid model took *considerably* longer. The first (fundamental) vibration mode was bending and is shown in Figure 19–11 for both models, together with the respective frequencies. The difference between the frequencies is about 1.9 percent. Further note that the mode shape is just that, a shape. The actual magnitudes of the deflections are unknown, only their relative values are known. Thus, any scale factor can be used to exaggerate the view of the deflection shape.

The convergence of the 20-element beam model was checked by doubling the number of elements. This resulted in no change.

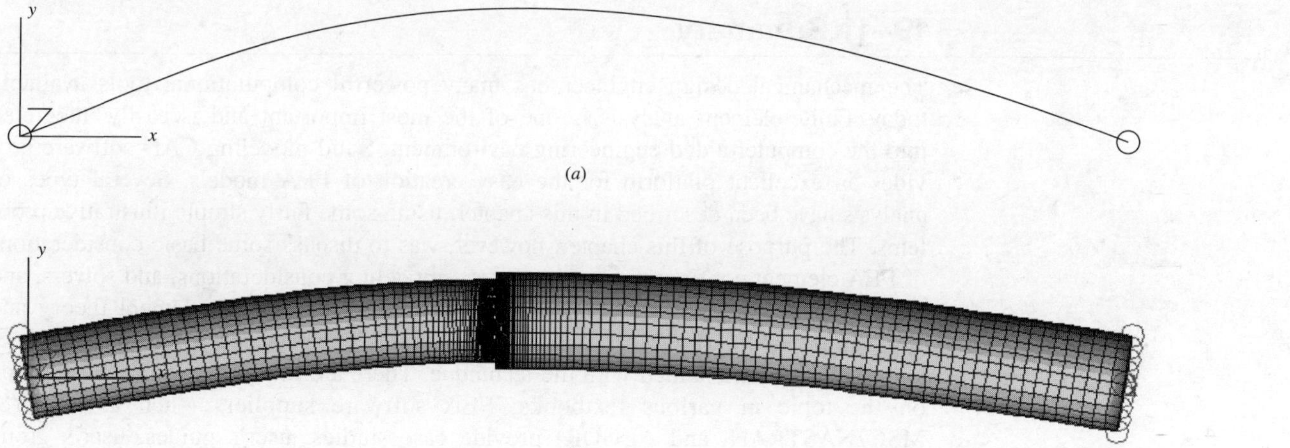

Figure 19–11

First free vibration mode of step beam. (*a*) Twenty-element beam model, $f_1 = 322$ Hz; (*b*) 56 384-element brick and tetrahedron model, $f_1 = 316$ Hz.

[15] For static deflection analysis, only three beam elements were necessary. However, because of mass distribution for the dynamics problem, more beam elements are necessary.

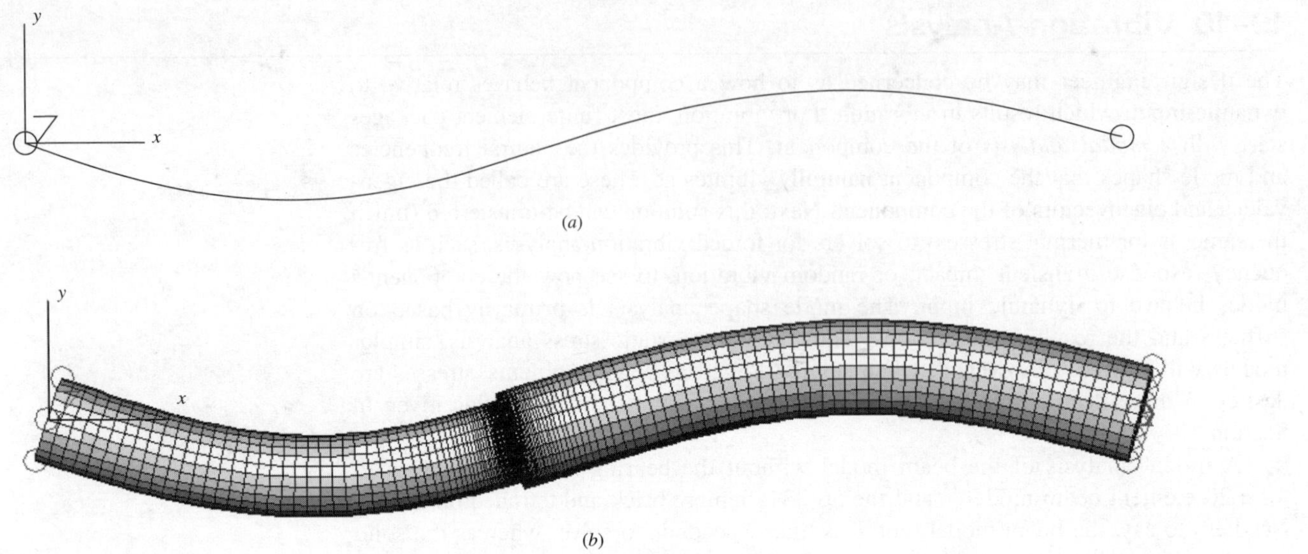

Figure 19–12

Second free-vibration mode of step beam. (*a*) Twenty-element beam model, f_2 = 1296 Hz; (*b*) 56 384-element brick and tetrahedron model, f_2 = 1249 Hz.

Figure 19–12 provides the frequencies and shapes for the second mode.[16] Here, the difference between the models is 3.6 percent.

As stated earlier, once the mode shapes are obtained, the response of the structure to various dynamic loadings, such as harmonic, transient, or random input, can be obtained. This is accomplished by using the mode shapes together with modal superposition. The method is called *modal analysis*.[17]

19–11 Summary

The mechanical design engineer has many powerful computational tools available today. Finite-element analysis is one of the most important and is easily integrated into the computer-aided engineering environment. Solid-modeling CAD software provides an excellent platform for the easy creation of FEA models. Several types of analysis have been described in this chapter, using some fairly simple illustrative problems. The purpose of this chapter, however, was to discuss some basic considerations of FEA element configurations, parameters, modeling considerations, and solvers, and not to necessarily describe complex geometric situations. Finite-element theory and applications is a vast subject, and will take years of experience before one becomes knowledgeable and skilled with the technique. There are many sources of information on the topic in various textbooks; FEA software suppliers (such as ANSYS, MSC/NASTRAN, and ALGOR) provide case studies, user's guides, user's group newsletters, tutorials, etc.; and the Internet provides many sources. Footnotes 11, 12, and 14 referenced some textbooks on FEA. Additional references are cited below.

[16]*Note:* Both models exhibited repeated frequencies and mode shapes for each bending mode. Since the beam and the bearing supports (boundary conditions) are axisymmetric, the bending modes are the same in all transverse planes. So, the second mode shown in Figure 19–12 is the next unrepeated mode.

[17]See S. S. Rao, *Mechanical Vibrations,* 5th ed., Pearson Prentice Hall, Upper Saddle River, NJ, 2010, Section 6.14.

Additional FEA References

K. J. Bathe, *Finite Element Procedures,* Prentice Hall, Englewood Cliffs, NJ, 1996.

R. D. Cook, D. S. Malkus, M. E. Plesha, and R. J. Witt, *Concepts and Applications of Finite Element Analysis*, 4th ed., Wiley, New York, 2001.

D. L. Logan, *A First Course in the Finite Element Method*, 4th ed., Nelson, a division of Thomson Canada Limited, Toronto, 2007.

J. N. Reddy, *An Introduction to the Finite Element Method,* 3rd ed., McGraw Hill, New York, 2002.

O. C. Zienkiewicz and R. L. Taylor, *The Finite Element Method,* 4th ed., Vols. 1 and 2, McGraw Hill, New York, 1989 and 1991.

PROBLEMS

The following problems are to be solved by FEA. It is recommended that you also solve the problems analytically, compare the two results, and explain any differences.

19–1 Solve Example 3–8.

19–2 For Example 3–13, apply a torque of 2700 N · m, and determine the maximum shear stress and angle of twist. Use 3 mm-thick plate elements.

19–3 The steel tube with the cross section shown is transmitting a torsional moment of 100 N · m. The tube wall thickness is 2.5 mm, all radii are $r = 6.25$ mm, and the tube is 500 mm long. For steel, let $E = 207$ GPa and $\nu = 0.29$. Determine the average shear stress in the wall and the angle of twist over the given length. Use 2.5-mm-thick plate elements.

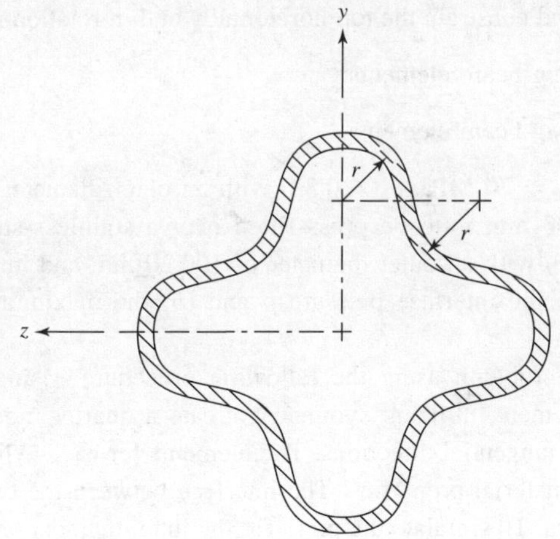

Problem 19–3

19–4 For Figure A–13–1, let $w = 50$ mm, $d = 7.5$ mm, and estimate K_t. Use 1/4 symmetry and 3 mm-thick 2-D elements.

19–5 For Figure A–13–3, let $w = 38$ mm, $d = 25$ mm, $r = 2.5$ mm, and estimate K_t. Use 1/4 symmetry and 3 mm-thick 2-D elements.

19–6 For Figure A–13–5, let $D = 75$ mm, $d = 50$ mm, $r = 6$ mm, and estimate K_t. Use 1/2 symmetry and 3 mm-thick 2-D elements.

19–7 Solve Problem 3–132, using solid elements. *Note:* You may omit the top part of the eyebolt above the applied force.

19–8 Solve Problem 3–142, using solid elements. *Note:* Since there is a plane of symmetry, a one-half model can be constructed. However, be very careful to constrain the plane of symmetry properly to assure symmetry without overconstraint.

19–9 Solve Example 4–11, with $F = 45$ N, $d = 3$ mm, $a = 12$ mm, $b = 25$ mm, $c = 50$ mm, $E = 207$ GPa, and $\nu = 0.29$, using beam elements.

19–10 Solve Example 4–13, modeling Figure 4–18*b* with 2-D elements of 50-mm thickness. Since this example uses symmetry, be careful to constrain the boundary conditions of the bottom horizontal surface appropriately.

19–11 Solve Problem 4–12, using beam elements.

19–12 Solve Problem 4–47, using beam elements. Pick a diameter, and solve for the slopes. Then, use Equation 7–18 to readjust the diameter. Use the new diameter to verify.

19–13 Solve Problem 4–69, using beam elements.

19–14 Solve Problem 4–94, using solid elements. Use a one-half model with symmetry. Be very careful to constrain the plane of symmetry properly to assure symmetry without overconstraint.

19–15 Solve Problem 4–95, using beam elements. Use a one-half model with symmetry. At the plane of symmetry, constrain translation and rotation.

19–16 Solve Problem 4–96, using beam elements. Model the problem two ways: (*a*) Model the entire wire form, using 200 elements. (*b*) Model half the entire wire form, using 100 elements and symmetry. That is, model the form from point *A* to point *C*. Apply half the force at the top, and constrain the top horizontally and in rotation in the plane.

19–17 Solve Problem 4–108, using beam elements.

19–18 Solve Problem 10–46, using beam elements.

19–19 An aluminum cylinder ($E_a = 70$ MPa, $\nu_a = 0.33$) with an outer diameter of 150 mm and inner diameter of 100 mm is to be press-fitted over a stainless-steel cylinder ($E_s = 190$ MPa, $\nu_s = 0.30$) with an outer diameter of 100.20 mm and inner diameter of 50 mm. Determine (*a*) the interface pressure *p* and (*b*) the maximum tangential stresses in the cylinders.

Note: Solve the press-fit problem, using the following procedure. Using the plane-stress two-dimensional element, utilizing symmetry, create a quarter model meshing elements in the radial and tangential directions. The elements for each cylinder should be assigned their unique material properties. The interface between the two cylinders should have common nodes. To simulate the press fit, the inner cylinder will be forced to expand thermally. Assign a coefficient of expansion and temperature increase, α and ΔT, respectively, for the inner cylinder. Do this according to the relation $\delta = \alpha \Delta T b$, where δ and b are the radial interference and the outer radius of the inner member, respectively. Nodes along the straight edges of the quarter model should be fixed in the tangential directions, and free to deflect in the radial direction.

20 Geometric Dimensioning and Tolerancing

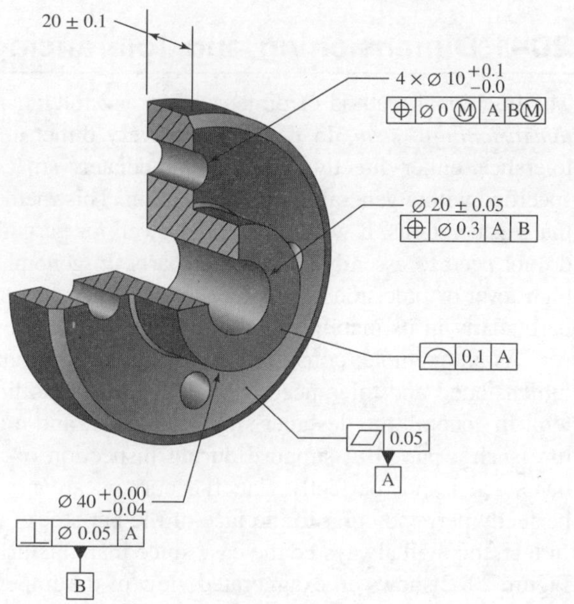

Chapter Outline

Early in the design process the designer works largely from a macro point of view, in which components are nominally sized to satisfy the design requirements, typically including control of stresses and deflections. But when it comes to issues of manufacturability, fits between components, and assembly of components, the designer must take a much closer look at the precise specification of the parts. Section 1–14 addressed some of the basic issues of dimensioning and tolerancing that a designer must consider. In this chapter, the focus is on a standardized method of defining part geometry that takes into consideration that no part is perfectly formed—nor should it need to be. The method known as *Geometric Dimensioning and Tolerancing* allows clarity for functionality, flexibility for manufacturing, and level of precision for inspection.

20–1 Dimensioning and Tolerancing Systems

The traditional method of dimensioning and tolerancing is referred to as the *coordinate dimensioning system.* In this system, every dimension is associated with a plus/minus tolerance, either directly specified immediately adjacent to the dimension, or implicitly specified with a general tolerance notation. This method of tolerancing has been used for many generations. It works reasonably well for parts that are not mass produced and that do not need to assemble with other parts. In general, this method is acceptable when a high level of precision is not needed. However, this system is lacking in many respects, particularly in its inability to address geometric issues of form and orientation.

As an example, consider the part shown in Figure 20–1. This simple part is fully dimensioned and toleranced according to the traditional coordinate dimensioning system. In general, the designer's intent is clear, and most machine shops could manufacture such a part. But suppose during inspection of an actual part, it is found that the bar stock is not perfectly flat, the corners are not perfectly square, the hole is not perfectly perpendicular to the face of the bar stock, and the hole is not perfectly round. In fact, this will always be the case since manufacturing can never achieve perfect form. Figure 20–2 shows an exaggerated view of the imperfections of the manufactured part. The problem is that every dimension may be within its tolerance, but the part may be

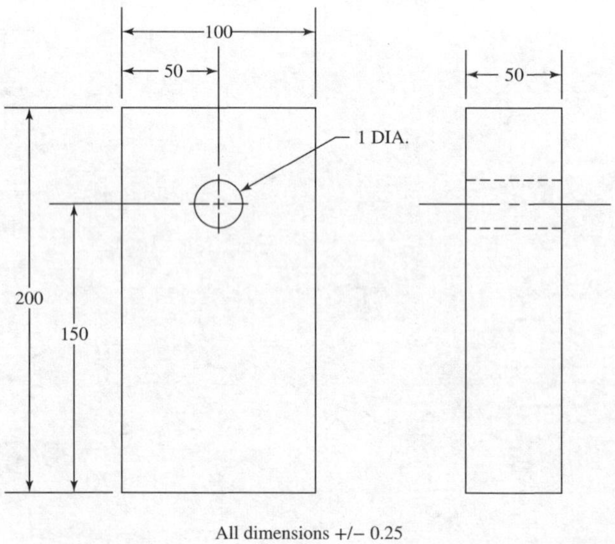

Figure 20–1

Part dimensioned with traditional coordinate dimensioning system.

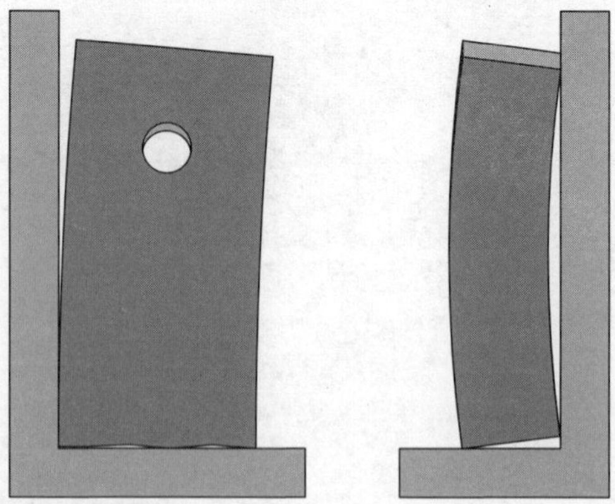

Figure 20–2

Exaggerated view of imperfections of a manufactured part.

unusable for its application owing to one of the geometric imperfections. A more challenging problem is that it is not even clear how to measure some of the dimensions. For example, the center of the hole is to be 50 mm from the edge of the part. If the hole is not perfectly round, how is its center defined? If the corner of the part is not square, from where should the 50 mm be measured? Bottom corner? Top corner? Closest edge? There is no defined correct answer with this dimensioning system.

For many applications these issues can be overlooked because typical manufacturing methods are deemed to be good enough. However, mass production calls for the most efficient and inexpensive operation allowable. The manufacturer legitimately needs to be able to cut corners as much as possible. This requires the part specification to define precisely how good is good enough. In fact, the designer should always view the task of geometry specification as simultaneously restricting and freeing the manufacturer—restricting within the necessary limits for functional requirements, and freeing from unnecessary levels of perfectness. Proper balance of these two aspects provides for cost-effective *and* functional parts.

Clearly, to address issues of functionality, manufacturability, interchangeability, and quality control requires parts that can be uniquely defined and consistently measured. This requires a dimensioning and tolerancing method that takes into account not only size, but geometric location, orientation, and form as well. This system is known as Geometric Dimensioning and Tolerancing (GD&T).

20–2 Definition of Geometric Dimensioning and Tolerancing

Geometric Dimensioning and Tolerancing (GD&T) is a comprehensive system of symbols, rules, and definitions for defining the nominal (theoretically perfect) geometry of parts and assemblies, along with the allowable variation in size, location, orientation, and form of the features of a part. It serves as a means of accurately representing the part for the purposes of design, manufacture, and quality control. GD&T is not new. It has been developing as a standard within industry since the 1940s. Today, most of the major manufacturing companies utilize GD&T. The part previously considered in Figure 20–1 is shown again in Figure 20–3 using GD&T terminology. Unfortunately,

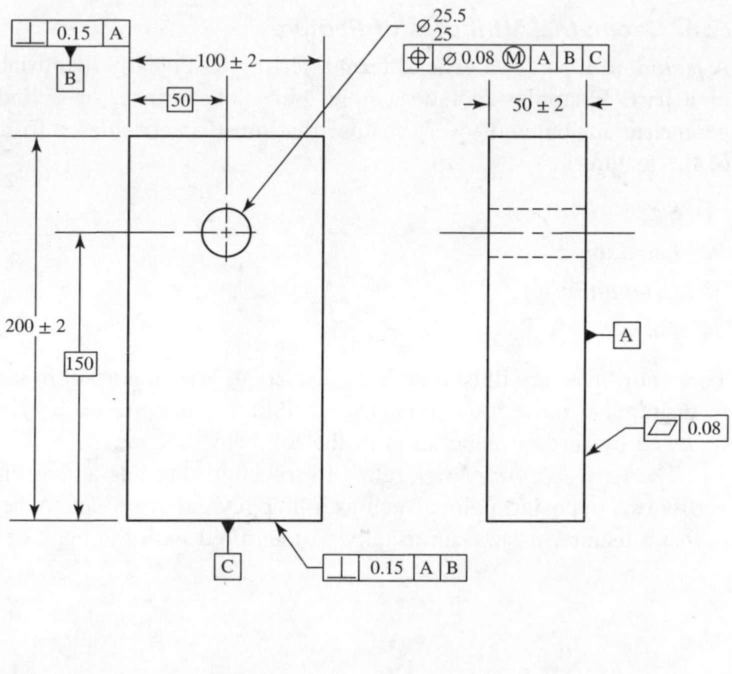

Figure 20–3

Part dimensioned and toleranced with GD&T terminology.

many mechanical engineers are not able to interpret the drawing. During the time when the standard was becoming most prevalent in manufacturing, most engineering schools were phasing out comprehensive drafting courses in favor of computerized CAD instruction. Consequently, GD&T is often missing from the engineering curriculum. A full understanding of GD&T is usually obtained through an intensive course or training program, readily available to practicing engineers. Some mechanical engineers will benefit from such a rigorous training. *All* mechanical engineers, however, should at least be familiar with the basic concepts and notation. The purpose of the coverage of GD&T in this chapter is to provide this foundational exposure that is essential for all machine designers. The coverage is not comprehensive. The focus is on the foundational concepts and the most commonly used notation. A first exposure to GD&T can feel overwhelming, as there are numerous concepts and terminology that all depend upon each other. This chapter is organized to help a beginner gradually build the most important concepts first, adding detail as needed, and culminating with a section on practical applications. Section 20–9 includes a glossary of some of the most important vocabulary terms used in GD&T, and should be used as a reference for clarification of terms while reading the rest of the chapter.

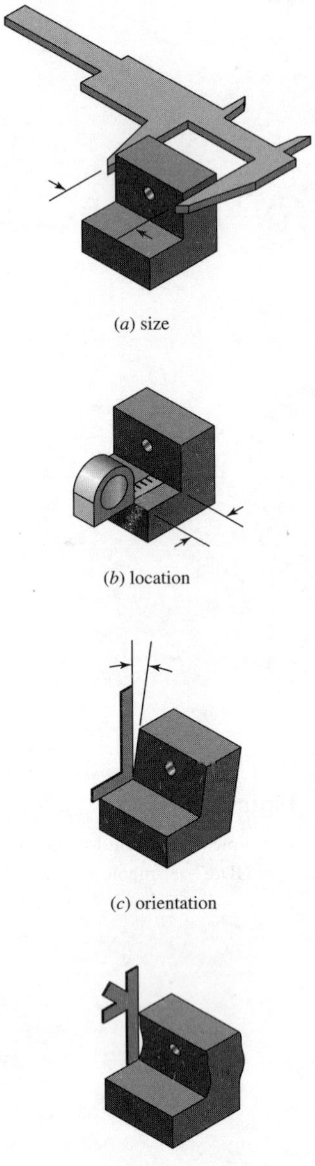

(*a*) size

(*b*) location

(*c*) orientation

(*d*) form

Figure 20–4

The four geometric attributes of a feature. (*a*) Size; (*b*) location; (*c*) orientation; (*d*) form.

GD&T Standards

GD&T is defined and controlled by standards to provide uniformity and clarity on a global scale. One widely utilized standard is published by the American Society of Mechanical Engineers as *ASME Y14.5–2009 Dimensioning and Tolerancing*. It is part of the broader set of ASME Y14 standards that cover all aspects of engineering drawings and terminology. The International Organization of Standards (ISO) also publishes a series of standards that have been more commonly used in European countries. The two standards have developed in parallel and are mostly similar in concept and terminology. The ASME standard tends to put more emphasis on *design intent* while the ISO standards have a greater emphasis on *metrology,* or the measurement of the resulting part. According to the ASME approach, the parts are defined primarily in a way to ensure that they will perform the desired function, without specifying what equipment or processes should be used to manufacture or inspect the parts. The ASME Y14.5–2009 standard is utilized in this textbook.

Four Geometric Attributes of Features

A *feature* of a part is a general term referring to a clearly identifiable physical portion of a part. Examples include a hole, pin, slot, surface, or cylinder. There are four geometric attributes of every feature that must be considered to define the geometry of the feature:

1 ***Size***

2 ***Location***

3 ***Orientation***

4 ***Form***

These attributes are illustrated in Figure 20–4. It is important in understanding GD&T to distinguish these four geometric attributes. They are each briefly described here, followed by further elaboration in the following sections.

The term *feature of size* refers to a feature that has a size that can be measured across two opposing points, such as a hole, cylinder, or slot. A helpful rule of thumb is that a feature of size can usually be measured with the head of a caliper tool, such

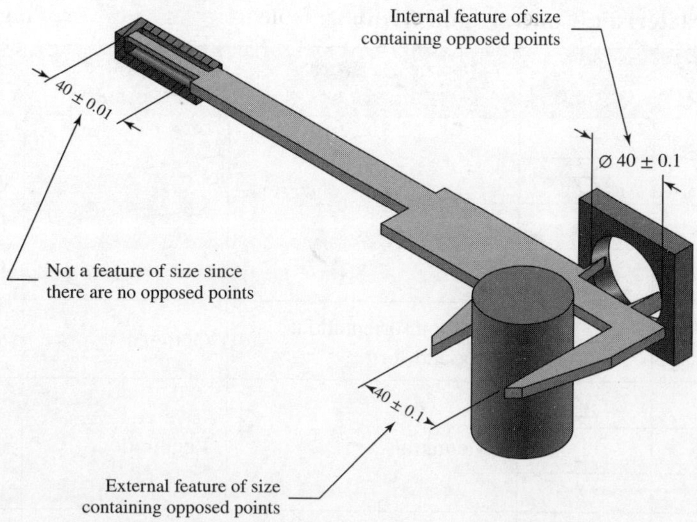

Figure 20–5

Measuring opposing points of features of size with a caliper head.

as illustrated in Figure 20–5. A dimension that is measured with the probe end of the caliper does not have opposed points, and is therefore not a feature of size. Such a dimension would be a *locating* dimension rather than a *size* dimension.

Location refers to the location of a feature with respect to some origin of measurement. *Orientation* refers to the angle of a feature or centerline of a feature with respect to some origin of measurement. It includes parallelism, perpendicularity, and angularity. *Form* refers to imperfections in the shape of a feature, and includes straightness, flatness, circularity, and cylindricity.

According to the GD&T standard, plus/minus tolerancing should only be directly applied to the size dimension of a feature of size. The other attributes (location, orientation, and form) are controlled through *geometric tolerancing,* described in Section 20–4.

Symbolic Language

The ASME Y14.5–2009 standard utilizes an international language of symbols that minimizes the need for, and potential confusion from, written notes on machine drawings. The symbols are classified into two categories—geometric characteristic and modifying. For reference, a brief introductory summary of the symbols will be given here. The usage of the symbols will be clarified in later sections.

Table 20–1 shows the 14 geometric characteristics and their symbols. These geometric characteristics are refinements of the geometric attributes (size, location, orientation, and form), and are each used to control a geometric tolerance of a feature. Each geometric characteristic has a symbol that is used on the drawing to specify a tolerance zone associated with a geometric characteristic. The geometric characteristic symbols are directly associated with the geometry of a feature, not with a size dimension. This is why they are referred to as geometric tolerancing. As shown in the table, the geometric characteristics are further subdivided into the types of tolerance defined by GD&T (form, profile, orientation, location, and runout), as well as the broader geometric attribute being controlled (size, location, orientation, and form). The table also indicates whether the symbol can be associated with a datum reference and any material condition modifiers, which will be clarified in later sections.

Table 20–1 Geometric Characteristic Controls and Symbols

Type of Tolerance	Geometric Characteristic	Symbol	Geometric Attribute Controlled	Datum Referencing?	Material Condition Modifier Allowed
Form	Straightness	⏤	Form	No	Ⓜ Ⓛ or RFS
	Flatness	▱			Ⓜ Ⓛ or RFS
	Circularity	○			RFS
	Cylindricity	⌭			RFS
Profile	Profile of a line	⌒	Location, orientation, size, and form	Optional	Ⓜ Ⓛ or RFS
	Profile of a surface	⌓			
Orientation	Angularity	∠	Orientation	Required	Ⓜ Ⓛ or RFS
	Perpendicularity	⊥			
	Parallelism	//			
Location	Position	⌖	Location and orientation of feature of size	Required	Ⓜ Ⓛ or RFS
	Concentricity	◎	Location of derived median points or planes		RFS
	Symmetry	⌯			RFS
Runout	Circular runout	↗	Location of cylinder	Required	RFS
	Total runout	⌰			RFS

Tables 20–2 and 20–3 show the modifying symbols. Table 20–2 includes the dimensioning modifiers and their symbols. These are used to modify or clarify the meaning of a dimension on the drawing. Table 20–3 includes the tolerance modifiers and their symbols. These are used in a feature control frame (to be defined later) to modify or clarify the tolerance specification.

Table 20–2 Dimensioning Modifiers and Symbols

Description	Symbol	Description	Symbol
Basic Dimension	⟵[98]⟶	Dimension Origin	⌱
Diameter	Ø	All Around	◄─○─
Spherical Diameter	SØ	All Over	◄─◎─
Radius	R	Independency	Ⓘ
Spherical Radius	SR	Continuous Feature	⟨CF⟩
Controlled Radius	CR	Counterbore	⌴
Square	□	Countersink	⌵
Reference	( )	Spotface	⌴SF
Arc Length	⌒	Depth	↧

Table 20–3 Tolerance Modifiers and Symbols

Description	Symbol
Maximum Material Condition (applied to tolerance) Maximum Material Boundary (applied to datum)	Ⓜ
Least Material Condition (applied to tolerance) Least Material Boundary (applied to datum)	Ⓛ
Translation	▷
Projected Tolerance Zone	Ⓟ
Free State	Ⓕ
Tangent Plane	Ⓣ
Statistical Tolerance	⟨ST⟩
Between	⟷
Unequally Disposed Profile	Ⓤ

20–3 Datums

Several concepts are foundational to the implementation of GD&T. In this section, the concept of datums will be explained, followed by specific methods of implementation in GD&T.

The geometric characteristics of features are defined and measured by relating them to clearly defined datums. A *datum* is an origin from which *location* or *orientation* of part features is established. Note that *size* dimensions and *form* control do not require an origin for measurement, and therefore do not need to be referenced to a datum. For a feature of a part to be manufactured or inspected, the entire part is located with respect to a datum reference frame. A *datum reference frame* is a set of up to three mutually perpendicular planes that are defined as the origin of measurement for locating the features of a part. The datum reference frame is idealized and geometrically perfect. It is necessary to consider its relationship to the nonideal physical part and the processing equipment. To do so, it is necessary to distinguish between several related terms, namely datum, datum reference frame, datum feature, and datum feature simulator.

A *datum feature* is a nonideal physical surface of the part that is specified in order to establish a theoretically exact datum. A datum feature is always a surface of the part that can be physically touched, not a centerline or other theoretical entity. Since the datum feature is not perfect, it is not directly used for measurements. Suppose a flat surface of a part is selected as a datum feature. The surface is an imperfect plane with localized hills and valleys, and is not perfectly flat. If the part is placed on a polished granite surface plate, a minimum of three high points on the surface of the datum feature will contact the nearly perfect plane of the polished surface plate. The surface plate serves as a *datum feature simulator* of the actual datum feature. A datum feature simulator is a precision embodiment, such as a surface plate, gauge pin, or machine tool bed, of the datum described by an imperfect datum feature. The datum feature simulator is often a physical gauging surface, but may also be simulated by "soft gauging" optical or probing methods. The *datum* itself is a theoretically exact point, axis, or plane derived from the datum feature simulator.

A recap of the relationship between the various datum terms may be helpful, using the example shown in Figure 20–6. An imperfect actual surface of a part, such

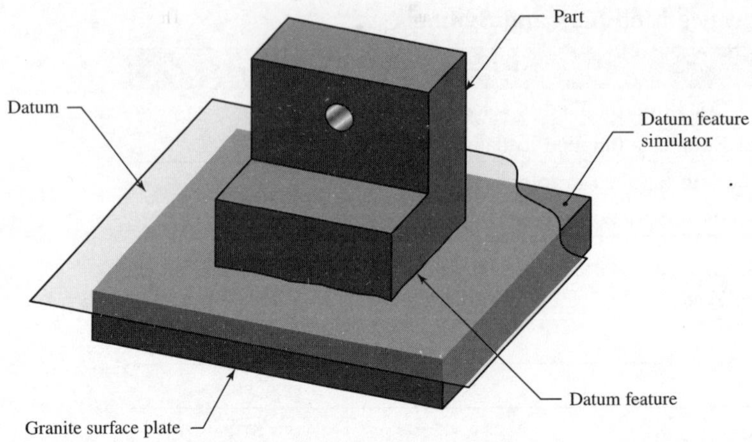

Figure 20–6

Example demonstrating datum terminology.

as the bottom surface, is designated as a datum feature. The datum feature (bottom surface) is placed in contact with a nearly perfect datum feature simulator (granite surface plate). A theoretical datum (true plane) is defined in association with the datum feature simulator. The process is repeated as necessary to define enough datums to obtain a three-plane datum reference frame. For example, if the back surface and one side surface are also selected as datum features, then the three-plane datum reference frame in Figure 20–7 may be obtained.

How, then, are locations and orientations of features handled through the process of design, manufacture, and inspection? The designer specifies datum features that are best suited for the functionality, manufacture, and inspection of the part. Locations and orientations are defined by the designer on the drawing with respect to the datum reference frame. They are actually manufactured with respect to a datum feature simulator inherent in the manufacturing equipment, such as the table surface of a milling machine. They are measured for quality control with respect to a datum feature simulator, such as a granite surface plate. Note that measurements of location and orientation are not made with respect to the actual surface of the datum feature, but from the datum feature simulator.

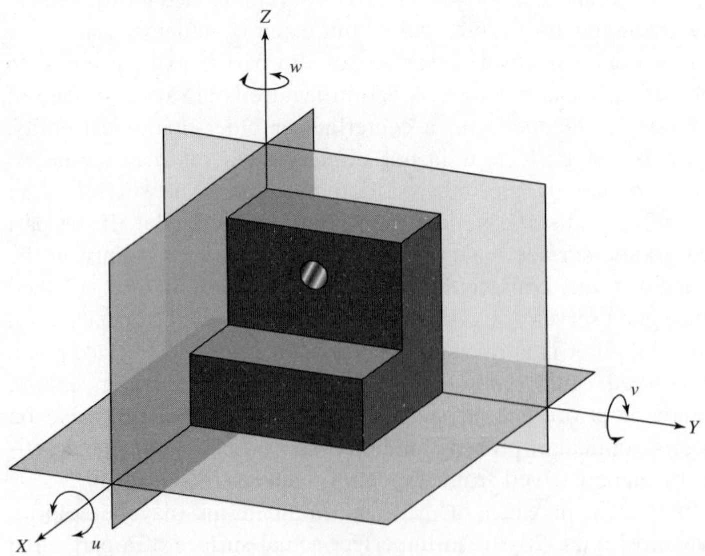

Figure 20–7

Example demonstrating a three-plane datum reference frame.

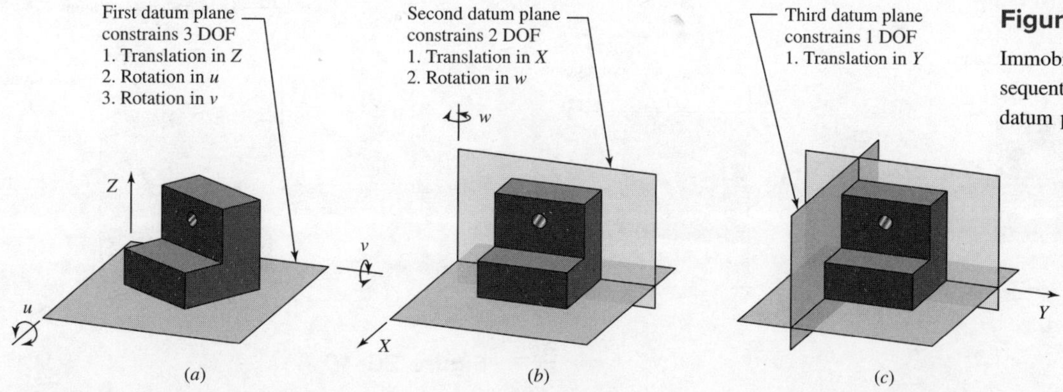

Figure 20–8

Immobilization of a part by sequential application of datum planes.

Immobilization of the Part

The selection of datum features can be thought of as selecting which surfaces of the part will be put into contact with datum feature simulators in order to immobilize the part for manufacture and inspection. The part floating in space has six degrees of freedom (three translations and three rotations). Each datum constrains some of the degrees of freedom in order to immobilize the part in a precise, repeatable location. Consider the process of immobilizing a part with three datum planes, demonstrated in Figure 20–8. First, let the bottom surface of the part be selected as a datum feature to be constrained by the first datum plane, as shown in Figure 20–8*a*. Remember that the datum feature is imperfect, so it may touch the datum plane in only a few places. Specifically, a minimum of three points of contact are required to prevent the part from rocking on the datum plane. This contact with the datum plane will constrain three degrees of freedom of motion of the part: translation in Z, rotation in u, and rotation in v. Next, let the back surface of the part be designated as the second datum feature, from which the second datum plane is derived, as shown in Figure 20–8*b*. Imagine holding the part in contact with the first datum plane and sliding it into contact with the second datum plane. It must touch in a minimum of two points to stabilize it with respect to the second datum plane. This will constrain an additional two degrees of freedom of motion: translation in X and rotation in w. Finally, let a side surface be designated as the third datum feature to define the third datum plane, as shown in Figure 20–8*c*. Maintaining contact of the part with the first two datum planes, and sliding it into contact with the third datum plane will result in a minimum of one point of contact with the third datum plane. This constrains the final degree of freedom: translation in Y. The part is now fully constrained in a precise, repeatable location.

Order of Datums

Notice that the order of application of the datum planes is important. Suppose the part from Figure 20–8 is constrained by the first datum plane as before, but then the order of application of the second and third datum planes is reversed. Figure 20–9*a* shows a top view of the part that has been constrained by the YZ plane first, then by the XZ plane. Figure 20–9*b* shows the same part with the order of application of the two datum planes reversed. The final position of the two parts is not the same. Because measurements are made from the datum planes, not from the edges of the part itself, the measured locations of features on the part are clearly dependent on the choice of datum features and the order of application of the resulting datum planes. It is necessary for the part drawings to specify clearly, for each feature to be located on the part,

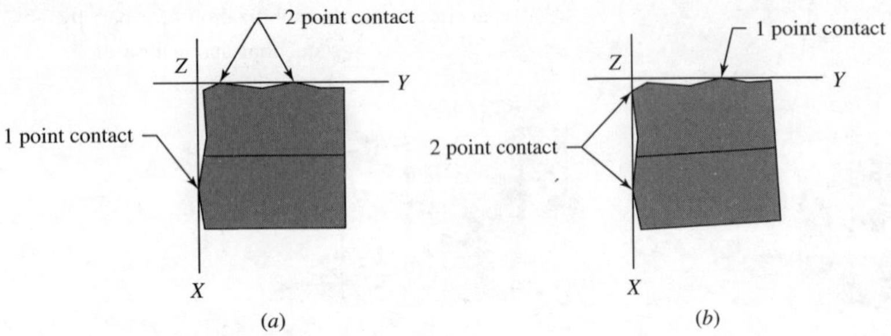

Figure 20–9

Comparison of order of application of datum planes. (*a*) *YZ* plane constrained first; (*b*) *XZ* plane constrained first.

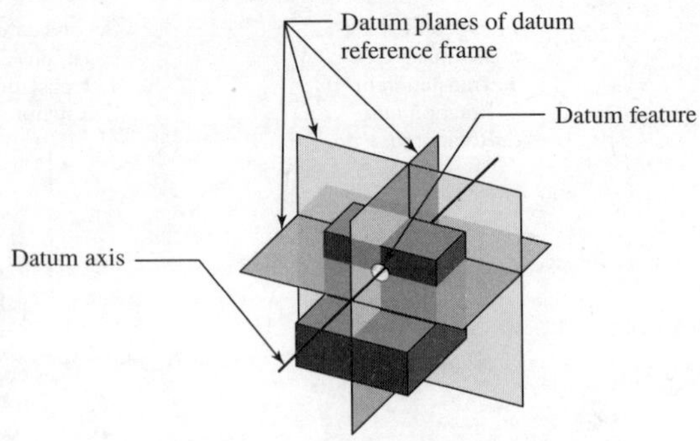

Figure 20–10

An example of a hole as a datum feature.

the datum features as well as the order of application of the resulting datum planes. The features do not all have to use the same datums and order of application.

Nonplanar Datum Features

So far, only planar datum features have been presented since they are the easiest to visualize the progression from the datum feature to the datum reference frame. Several other datum features are also provided for in the Y14.5 standard. In particular, cylindrical features such as shafts, bosses, and holes are often useful as datum features. Suppose in the part shown in Figure 20–10, the hole is selected as a datum feature. The actual surface of the hole is the datum feature; the center axis of the hole is the datum. The center axis defines the intersection of two perpendicular datum planes. In conjunction with another datum feature, say the back surface, the part is constrained and a datum reference frame is defined.

Actual Mating Envelopes

In the previous paragraph, it was stated that the center axis of the hole is the datum. This is a simplified statement of a more detailed concept that warrants a better explanation. Since the hole feature is imperfect in form (that is, it does not have a perfectly circular cross section, or a perfectly straight centerline, or a perfectly smooth surface), how is its theoretically perfect datum axis determined? To answer this question, a few GD&T terms will be introduced.

An *actual mating envelope* is a perfectly shaped counterpart of an imperfect feature of size, which can be contracted about an external feature, or expanded within an internal feature, so that it contacts the high points of the feature's surface. For example, Figure 20–11*a* shows an imperfect dowel pin (the feature of size) that is circumscribed by the smallest possible perfect cylinder (the actual mating envelope). The imperfect pin does not technically have a center axis. Instead, it has a collection of *derived median points* that represent the centroids of each cross section. When referring to the center axis of an imperfect feature such as the pin, what is actually meant is the theoretically perfect center axis of the theoretically perfect actual mating envelope of the pin. The same concept can be applied to a feature of size with an internal surface, such as the hole feature shown in Figure 20–11*b*.

Actual mating envelopes are categorized as *related* or *unrelated* to a datum. An *unrelated actual mating envelope* is sized to fit the feature without any constraint to any datum. In other words, it is free to float to find the best fit. A *related actual*

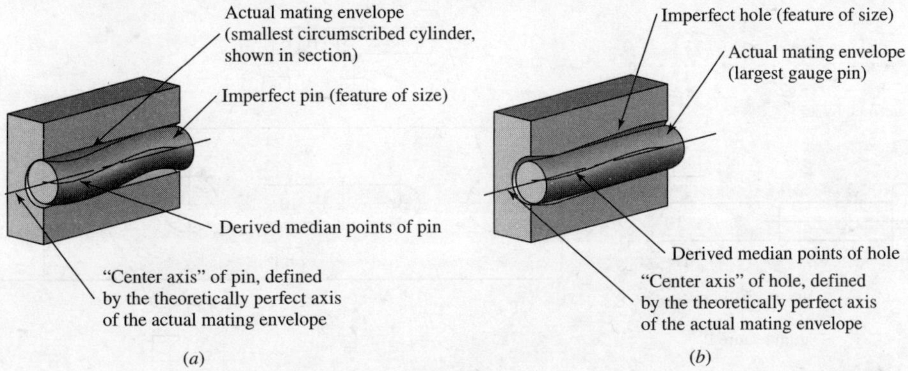

Figure 20–11

Definition of terms for an actual mating envelope. (*a*) External feature; (*b*) internal feature.

mating envelope is sized to fit the feature while maintaining some constraint in orientation or location with respect to a datum. For example, for the hole feature in Figure 20–10, the related actual mating envelope with respect to the back plane datum surface is the largest pin that can fit in the hole while being held perpendicular to the back datum plane.

Now, returning to the datum example in Figure 20–10, the datum axis corresponding to the datum feature (the hole) is defined by the actual mating envelope of the hole, that is, the largest cylinder that can fit within the hole. In practical implementation, this largest cylinder can be determined by physically inserting very precisely manufactured gauge pins of increasing size until the largest one is found. Alternatively, an expanding mandrel can be used. The largest gauge pin serves as the datum feature simulator (previously defined). In the case of an external datum feature, such as the surface of a shaft, the datum feature simulator is typically the jaws of a chuck or collet that is closed onto the surface. The center axis of the chuck is then the datum axis.

Datum Feature Symbol

On a drawing, a datum feature is defined symbolically by a capital letter enclosed in a square frame, attached to a leader line that terminates at the datum feature with a triangle. The triangle can be filled or unfilled. Any letter can be used, except I, O, and Q, which may be confused with numbers. The letters need not be assigned alphabetically as the precedence of the datums will be specified later as needed for each feature to be controlled.

The triangle may attach directly to the datum feature surface outline, point to it with a leader line, or attach to an extension line to the surface. All three of these methods are illustrated in Figure 20–12, where the datum features correlate to those previously demonstrated for the part in Figure 20–8.

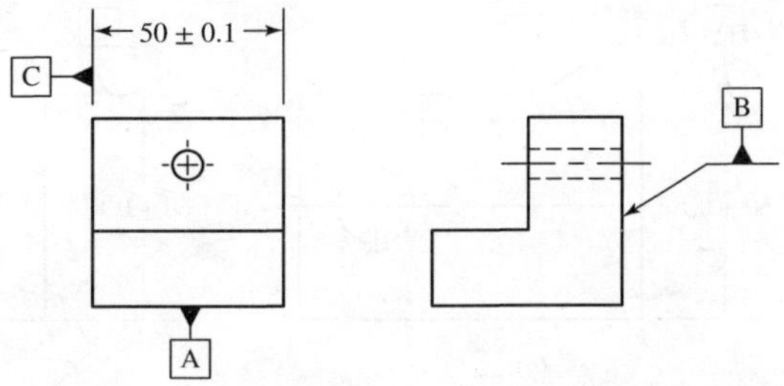

Figure 20–12

Three methods of designating a datum feature.

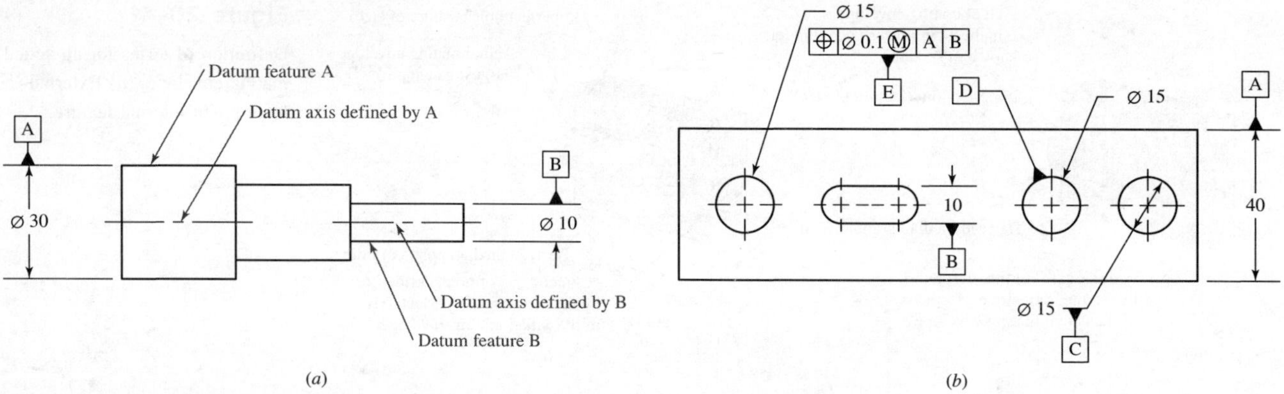

Figure 20–13

Methods of designating an axis or center plane of a feature of size as a datum. (*a*) Datums A and B are defined as the center axes of two different cylindrical surface features. (*b*) Datum symbols A and B are attached to width dimensions, defining the center planes as the datums. Datum symbols C, D, and E are attached to features with axes, defining the center axes as the datums.

If the datum is to be an axis or center plane of a feature of size, then the datum triangle is placed in-line with the dimension line of the feature of size. In the case of a cylinder it may be attached directly to the surface of the cylinder. The triangle may optionally replace one of the dimension arrowheads if both will not fit. Several examples are shown in Figure 20–13. A datum triangle always indicates the datum feature (a physical surface) from which the datum (a theoretical axis or center plane) is derived. Accordingly, the triangle is never placed directly on an axis, centerline, or center plane.

The datum triangle may also be attached to a feature control frame (to be defined in a later section) that controls the geometric tolerance of the datum feature. An example is included in Figure 20–13*b*.

Note that in the case of a feature of size, the subtle difference in placement of the triangle in Figures 20–14*a* and 20–14*b* has a significant difference in meaning. When the triangle is placed in-line with the dimension line, as in Figure 20–14*a*, the datum is the center plane of the feature of size. When the triangle is placed away from the dimension line, as in Figure 20–14*b*, the datum is the plane defined by the edge of the part. The center plane datum would likely be used if it was important that the hole be centered in the part, regardless of fluctuations in the overall width of the part. The edge datum has the advantage of simpler setup for manufacturing and inspecting, but because the hole location is controlled from one edge, the part may not be symmetric.

Figure 20–14

Different datum plane designation due to placement of datum symbol. (*a*) Symbol in-line with dimension; (*b*) symbol out-of-line with dimension.

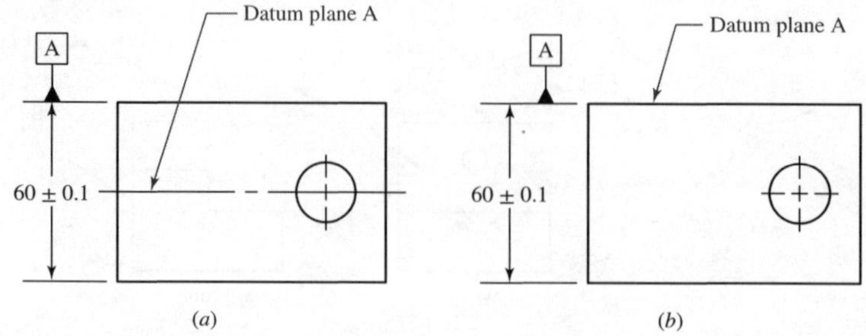

20–4 Controlling Geometric Tolerances

Tolerance Zones

GD&T generally only uses direct application of plus/minus tolerances when dimensioning a *feature of size*. A fundamental concept of GD&T is that in addition to any dimensional tolerance on a feature of size, the geometric shape and location of the surfaces must be controlled to stay within *tolerance zones*. Tolerance zones are defined in a variety of ways, such as by two parallel planes or concentric cylinders, to define limiting boundaries for the physical surfaces of the parts. The tolerance zones are defined relative to a theoretically exact location or shape. The actual shape and location of the part surfaces may vary from the theoretically exact location and shape, so long as they stay within the limiting boundaries of the tolerance zones.

There are special terms defined to represent the maximum and minimum boundaries of a feature of size. The *maximum material condition* (*MMC*) is the condition in which a feature of size contains the maximum amount of material within the stated limits of size. For an external feature, such as the outer surface of a shaft, the MMC is when the shaft diameter is at its maximum value allowed by the tolerance. For an internal feature, such as a hole, the MMC is when the hole is at its smallest diameter allowed by the tolerance. Similarly, the *least material condition* (*LMC*) is the condition in which a feature of size contains the least amount of material within the stated limits of size. This would correlate with the smallest allowed shaft diameter, or the largest allowed hole diameter. These terms are used extensively in GD&T.

The Y14.5 standard specifies a default tolerance zone for features of size through what is referred to as *Rule #1,* also known as the *envelope principle*. This rule states that when only a tolerance of size (i.e., a plus/minus tolerance) is specified for a feature of size, the limits of size prescribe the extent of which variation in its geometric form, as well as size, are allowed. Specifically, the envelope principle states that the surface of a feature of size may not extend beyond an envelope of perfect form at MMC. Consider a simple example of a dowel pin in Figure 20–15*a*. The limiting envelope is a perfectly formed pin (e.g., perfectly straight, perfectly circular cross section, etc.) that is at its largest allowed size, as shown in Figure 20–15*b*. An imperfect dowel pin would still meet the specifications so long as its diameter at any

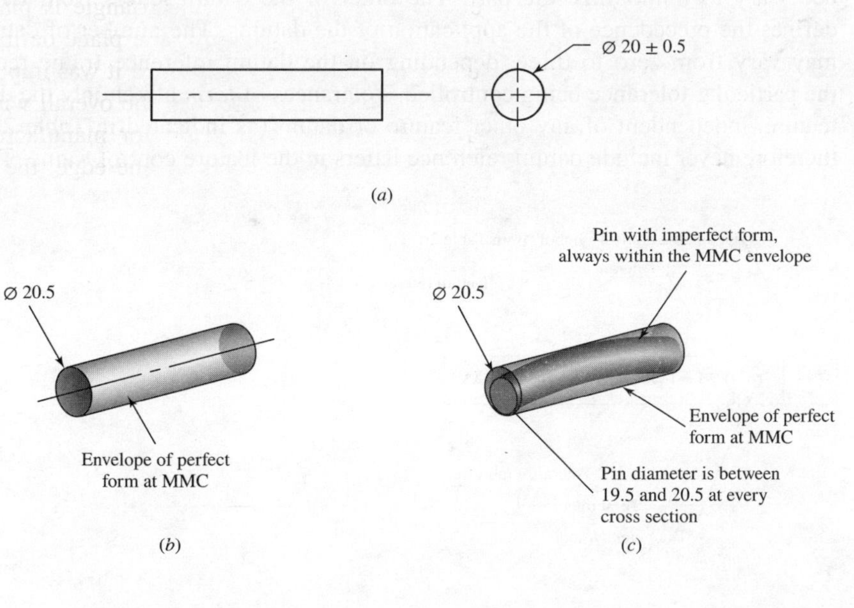

Figure 20–15

The envelope principle (Rule #1). (*a*) Size and tolerance specification on the drawing; (*b*) the envelope of perfect form at MMC; (*c*) an acceptable imperfect pin within the envelope.

location is within the allowed tolerance, and its surface does not exceed the envelope. An implication of this is that if the pin is manufactured at its MMC, then it must have perfect form. As the diameter is reduced from the MMC, the pin may deviate from perfect form, as shown in Figure 20–15*c*.

Other geometric controls (described in Section 20–5) may be specified when the default tolerance zone provided by Rule #1 is not sufficient to meet the requirements of the application.

Basic Dimensions

A theoretically exact location is specified with a *basic dimension*. A basic dimension is a theoretically exact dimension which does not have a tolerance directly associated with it, but is instead associated with a geometric control of a tolerance zone. When a basic dimension is used to locate a part feature, the feature itself must include a geometric control that defines a tolerance zone specifying the permissible variation from perfect form and location.

Basic dimensions are indicated on a drawing by enclosing the dimension within a rectangular box, or by a general note indicating that all untoleranced dimensions are basic.

Feature Control Frames

A geometric control is specified on a drawing in a *feature control frame*. A feature control frame is a rectangular box attached to a feature on a drawing, containing the necessary information to define the tolerance zone of the specified feature. The frame is subdivided into compartments in a specific order, as shown in the example in Figure 20–16.

The first compartment always contains one of the geometric control symbols from Table 20–1 to indicate what aspect of the feature is being controlled by the tolerancing information to follow.

The second compartment always contains a numerical value designating the total allowed tolerance. The tolerance value specified is always a total tolerance, that is, the entire range of the tolerance, not a plus/minus tolerance value from a midpoint. If the tolerance is circular or cylindrical, the diameter symbol will precede the specified tolerance.

The third and following compartments are used as needed to define the datum(s) necessary to immobilize the part. The order of the datum letters from left to right defines the precedence of the application of the datums. The number of datum letters may vary from zero to three, depending on the datum reference frame required for the particular tolerance being controlled. Tolerances of *form* affect only the designated feature, independent of any other feature or datum (as indicated in Table 20–1), and therefore never include datum reference letters in the feature control frame. Tolerances

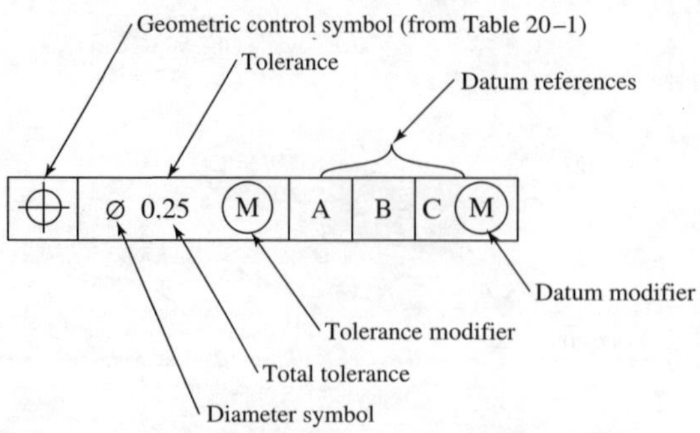

Figure 20–16

Example of a feature control frame.

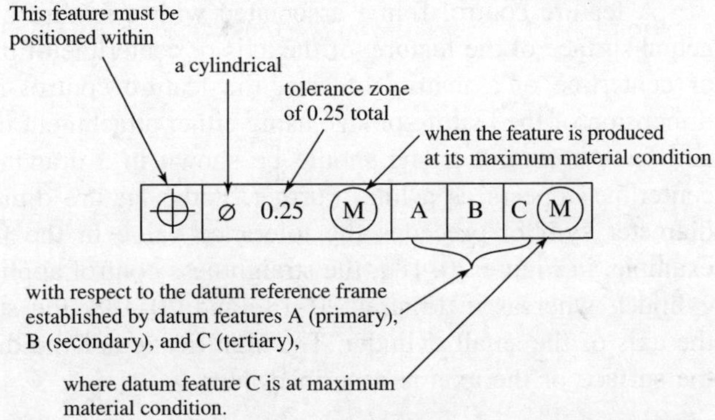

Figure 20–17

Reading a feature control frame.

of location, orientation, and runout always relate the designated feature to some other feature or datum, and therefore always require the specification of datum reference letters.

Modifiers from Table 20–3 can be included in a compartment immediately following a tolerance value, or after a datum specification. The effect of these modifiers is discussed in Section 20–6.

The feature control frame is read from left to right, as illustrated in Figure 20–17. If more than one geometric control is to be applied to the same feature, the feature control frames can be stacked, and are applied from top to bottom.

A feature control frame controls the feature to which it is attached. The four attachment methods are as follows:

1. A leader from the feature control frame points directly to the feature. See Figure 20–18*a*.
2. The feature control frame is attached to an extension line from a planar feature. See Figure 20–18*b*.
3. The feature control frame is placed below a leader-directed dimension or note pertaining to the feature. See Figure 20–18*c*.
4. The feature control frame is attached to an extension of the dimension line pertaining to a feature of size. See Figure 20–18*d*.

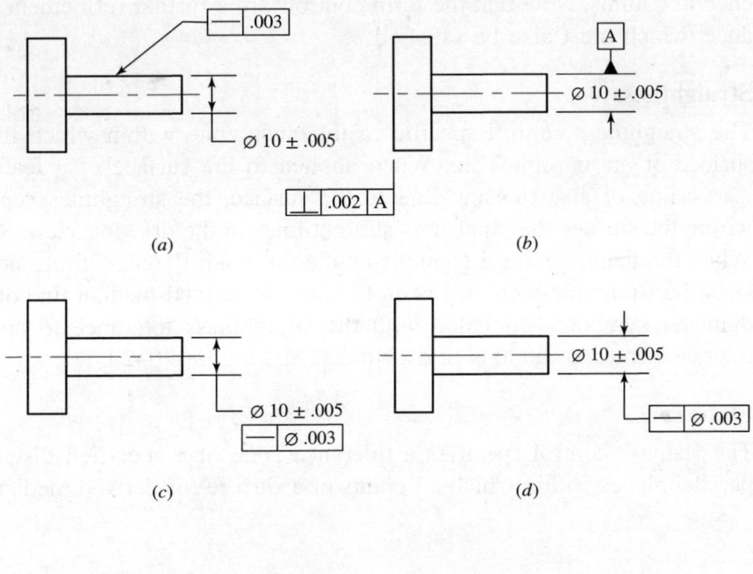

Figure 20–18

Four methods of attaching a feature control frame to a feature.

A feature control frame associated with a feature of size may control either the actual surface of the feature, or the axis or centerline of the feature. To control the axis or centerline of a feature of size, the feature control frame is associated with the dimension of the feature of size using either attachment method 3 or 4. The dimension and feature control frame should be shown in a drawing view in which the axis or centerline appears as a line. Further, if the feature dimension is for a diameter, the diameter symbol precedes the tolerance value in the feature control frame. As an example, in Figure 20–18*a*, the straightness control applies to the surface of the small cylinder, whereas in Figures 20–18*c* and 20–18*d*, the straightness control applies to the axis of the small cylinder. The significance of the difference between controlling the surface or the axis is described later.

20–5 Geometric Characteristic Definitions

Each of the geometric characteristic symbols in Table 20–1 is used to define a tolerance zone particular to a certain geometric characteristic. The geometric controls are categorized as controls of *form, orientation, profile, location,* and *runout*. Some of the controls are quite general and encompass most of the common needs, while some are very specific to a particular geometric need. For reference, a basic description will be given for each geometric control, followed by a broader discussion on practical implementation. The reader may find it helpful to only skim this section the first time through, then refer back to it as a reference when the details are needed for practical applications.

Form Controls

The four geometric characteristics that provide *form* control are

1 ***Straightness***

2 ***Flatness***

3 ***Circularity***

4 ***Cylindricity***

These control the form of an individual feature, independent of that feature's location or relationship to any other feature. Consequently, form controls never include reference to datums. Note that the form controls are a further refinement of any size tolerance, which must also be satisfied.

Straightness —

The straightness control specifies a tolerance zone within which line elements of a surface or an axis must lie. When applied to the surface of a feature with either a leader line or an extension line to the surface, the straightness applies to all lines within the surface that appear as straight lines in the drawing view. See Figure 20–19. When the feature control frame is applied beneath the size dimension of a feature of size, the straightness control is on the axis or derived median line of the feature. The diameter symbol is included with the straightness tolerance to apply a cylindrical tolerance zone to the axis of a cylinder. See Figure 20–20.

Flatness ⏥

The flatness control specifies a tolerance zone of a specified distance between two parallel planes within which all points of a surface (or derived median plane) must lie.

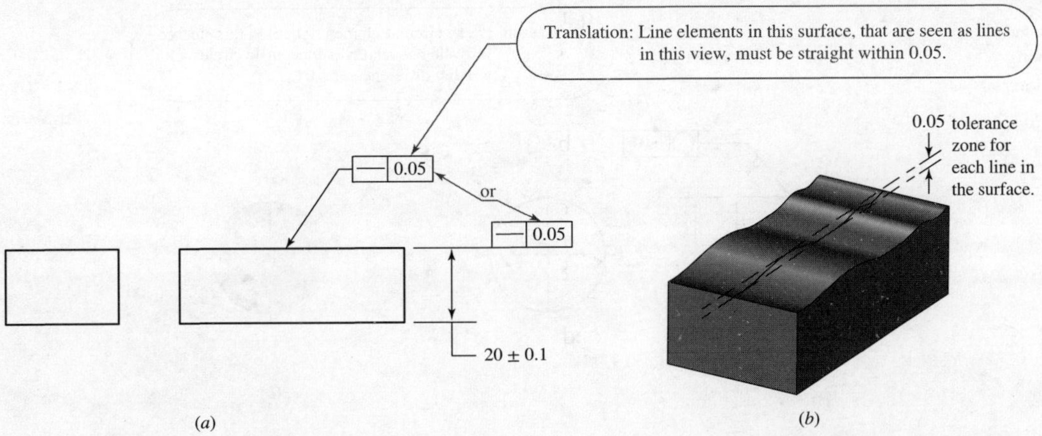

Figure 20–19

Straightness control applied to a surface. (*a*) Drawing specification; (*b*) Interpretation.

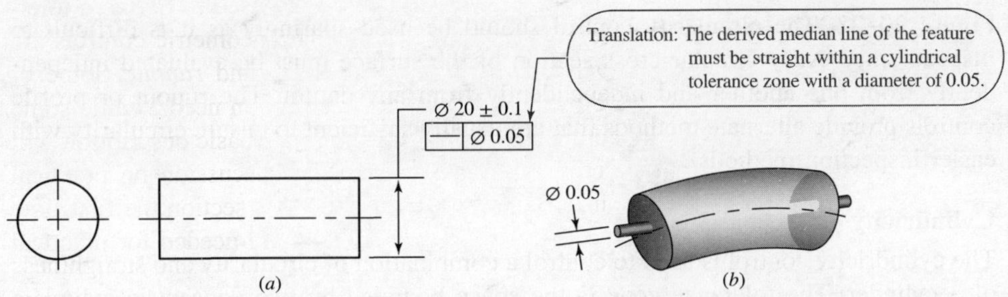

Figure 20–20

Straightness control applied to an axis of a feature. (*a*) Drawing specification; (*b*) Interpretation.

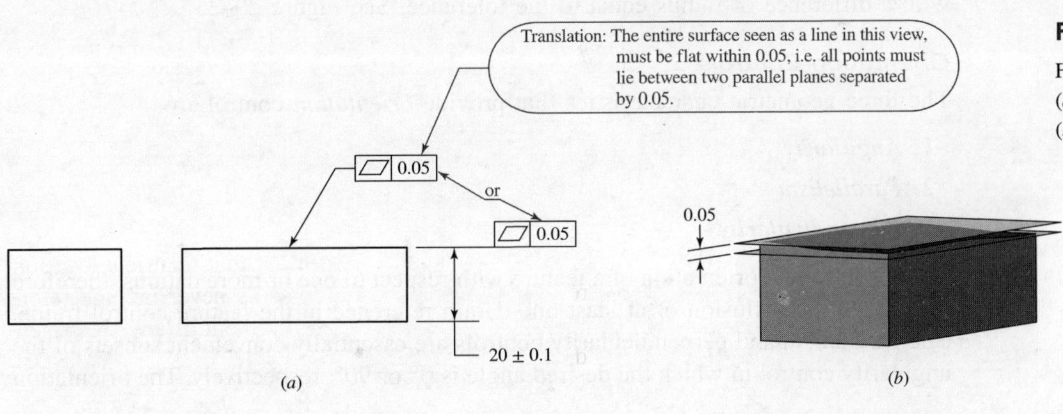

Figure 20–21

Flatness control. (*a*) Drawing specification; (*b*) Interpretation.

The feature control frame is applied with a leader line or an extension line to the surface in a drawing view in which the surface appears as a line. See Figure 20–21. If the feature control frame is applied beneath the size dimension of a feature of size, the flatness control is on the derived median plane rather than on the surface. The flatness control is often used to provide additional control of the primary datum feature to improve reproducibility of measurement.

Circularity ○

The circularity control is used to control the periphery of the circular cross sections of a cylinder, cone, or sphere. The tolerance zone is the annulus between two concentric circles within which each circular element of the surface must lie. See

Figure 20–22

Circularity control. (*a*) Drawing specification; (*b*) Interpretation.

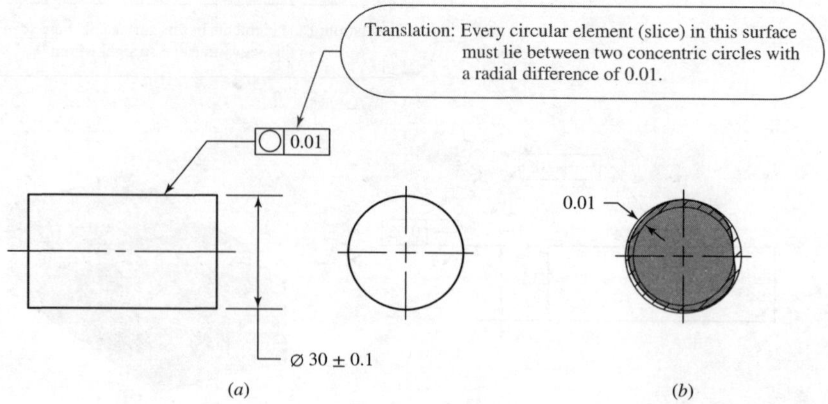

Figure 20–22. The circularity control should be used sparingly as it is difficult to inspect, since every circular cross section of the surface must be evaluated independently from one another and independently from any datum. The runout or profile controls provide alternate methods that are usually sufficient to ensure circularity with easier inspection methods.

Cylindricity ⌭

The cylindricity control is used to control a combination of circularity and straightness of a cylinder. The tolerance zone is the space bounded by two concentric cylinders with a difference of radius equal to the tolerance. See Figure 20–23.

Orientation Controls ∠ ⊥ //

The three geometric characteristics that provide *orientation* control are

1 ***Angularity***

2 ***Parallelism***

3 ***Perpendicularity***

These control the orientation of a feature with respect to one or more datums, therefore mandating the inclusion of at least one datum reference in the feature control frame. The parallelism and perpendicularity controls are essentially convenient subsets of the angularity control in which the desired angle is 0° or 90°, respectively. The orientation

Figure 20–23

Cylindricity control. (*a*) Drawing specification; (*b*) Interpretation.

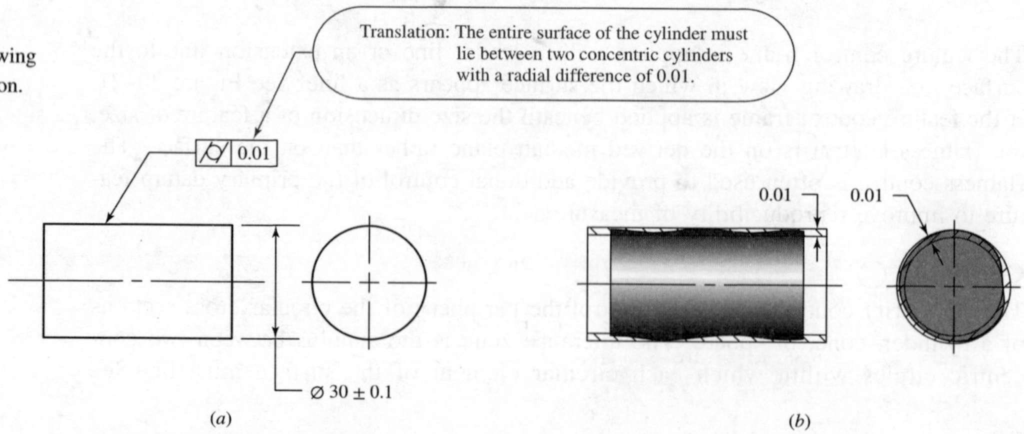

controls may be applied to surfaces, axes, or center planes. In the case of surfaces and planes, the tolerance zone is defined by two parallel planes oriented at the specified basic angle from the datum reference. See Figures 20–24 and 20–25. When controlling axes or center planes, the feature control frame is placed beneath the size dimension. When controlling axes, the diameter symbol is included before the tolerance, and the tolerance zone is cylindrical, as demonstrated in Figure 20–26.

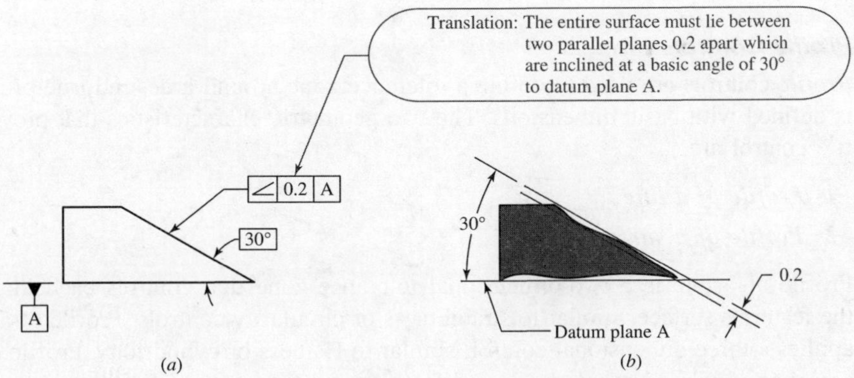

Figure 20–24

Angularity control.
(*a*) Drawing specification;
(*b*) Interpretation.

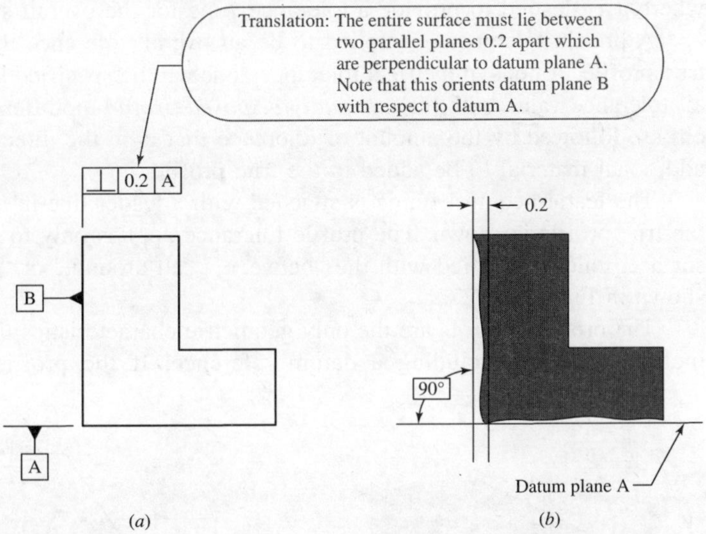

Figure 20–25

Perpendicularity control.
(*a*) Drawing specification;
(*b*) Interpretation.

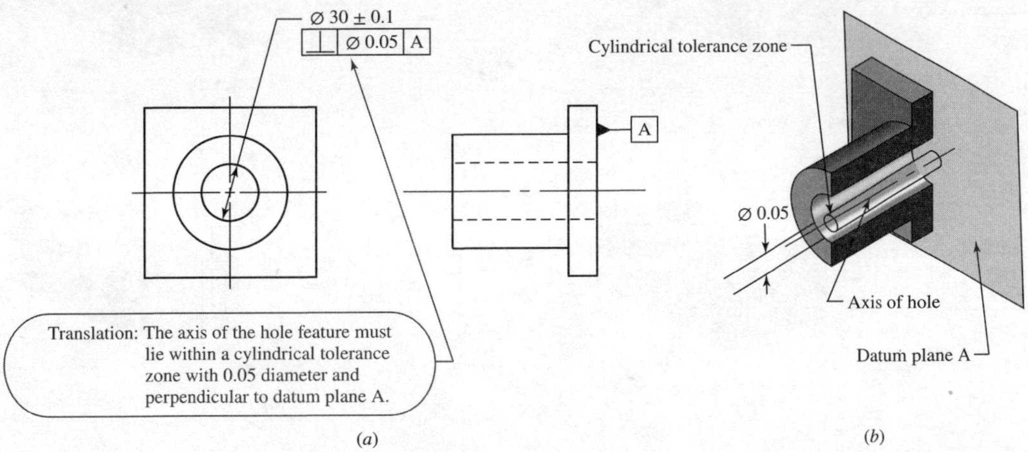

Figure 20–26

An orientation control applied to the axis of a feature.
(*a*) Drawing specification;
(*b*) Interpretation.

Orientation tolerances are constrained only in rotational degrees of freedom with respect to the referenced datum. Since the translational degrees of freedom are not constrained by orientation tolerances, an orientation tolerance zone cannot be used to locate a feature. It should only be used as a refinement of a tolerance that is doing the locating, such as position or profile of a surface. The most common use of the orientation controls is to orient a secondary or tertiary datum with respect to the primary datum plane.

Profile Controls ⌒ ⌓

Profile controls are used to define a tolerance zone around a desired true profile that is defined with basic dimensions. The two geometric characteristics that provide profile control are

1 ***Profile of a line***

2 ***Profile of a surface***

Profile of a line is a two-dimensional tolerance zone that controls each line within the feature's surface, similar to straightness or circularity controls. Profile of a surface applies a three-dimensional control, similar to flatness or cylindricity. Profile controls are often used for irregularly shaped features and for castings, forgings, or stampings where it is desired to provide a tolerance zone for the overall surface of the part.

A profile tolerance is implied to be an overall tolerance that is centered on the true profile. A nonsymmetrical tolerance zone can be specified by following the overall tolerance value with the *unequally disposed profile* modifier (an uppercase U in a circle), followed by the amount of tolerance that is in the direction that would allow additional material to be added to the true profile.

The feature control frame is attached with a leader line in a drawing view where the true profile is shown. The profile tolerance applies only to the individual feature surface, unless modified with the "between," "all around," or "all over" symbols, as shown in Figure 20–27.

The profile controls are the only geometric characteristics that have the option of including or not including a datum reference. If the profile tolerance does not

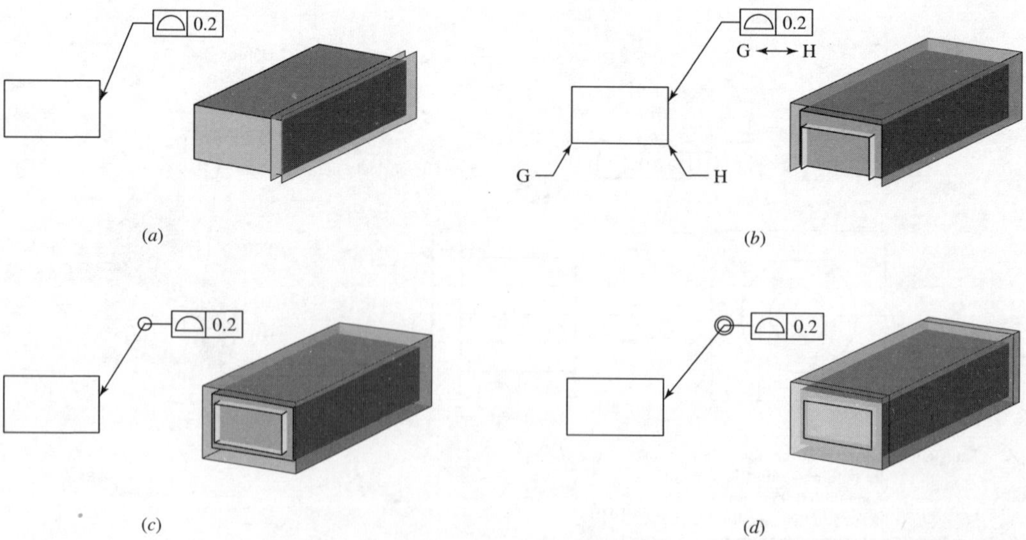

Figure 20–27

Application of the profile control. (*a*) Applied to a single surface; (*b*) applied between two designated points; (*c*) applied all around; (*d*) applied all over.

reference a datum, then the tolerance zone "floats" around the true profile, providing a form control of the surface, but not a location control. This option should be used sparingly, as it usually makes the inspection of the part more difficult.

If a datum is referenced, then the profile tolerance can simultaneously control size, form, orientation, and location of a feature. This general capability makes this control extremely useful as an overall default tolerance control. When the profile tolerance feature control frame is placed in a general note on the drawing, the tolerance applies to all features in the drawing unless otherwise specified. With this general note, other controls (e.g., flatness, perpendicularity, etc.) are needed only if a tighter control is desired than provided by the general profile tolerance.

Consider the example in Figure 20–28. The bottom surface is the datum feature that defines datum A. The flatness control establishes that when this surface is in contact with datum plane A, all points on the surface must be within the 0.05 tolerance zone. Datum plane B is defined to be exactly perpendicular to datum plane A, but the datum feature B (the actual surface), may vary in form and orientation so long as it stays within the 0.2 tolerance zone. The profile control on the curved surface indicates that the part is first placed into contact with datum plane A, then into contact with datum plane B. Then the ideal curved surface is defined with the basic dimensions. Then a tolerance zone is defined as the space between two curved surfaces centered around the true surface with a space of 0.1 between them. In addition, owing

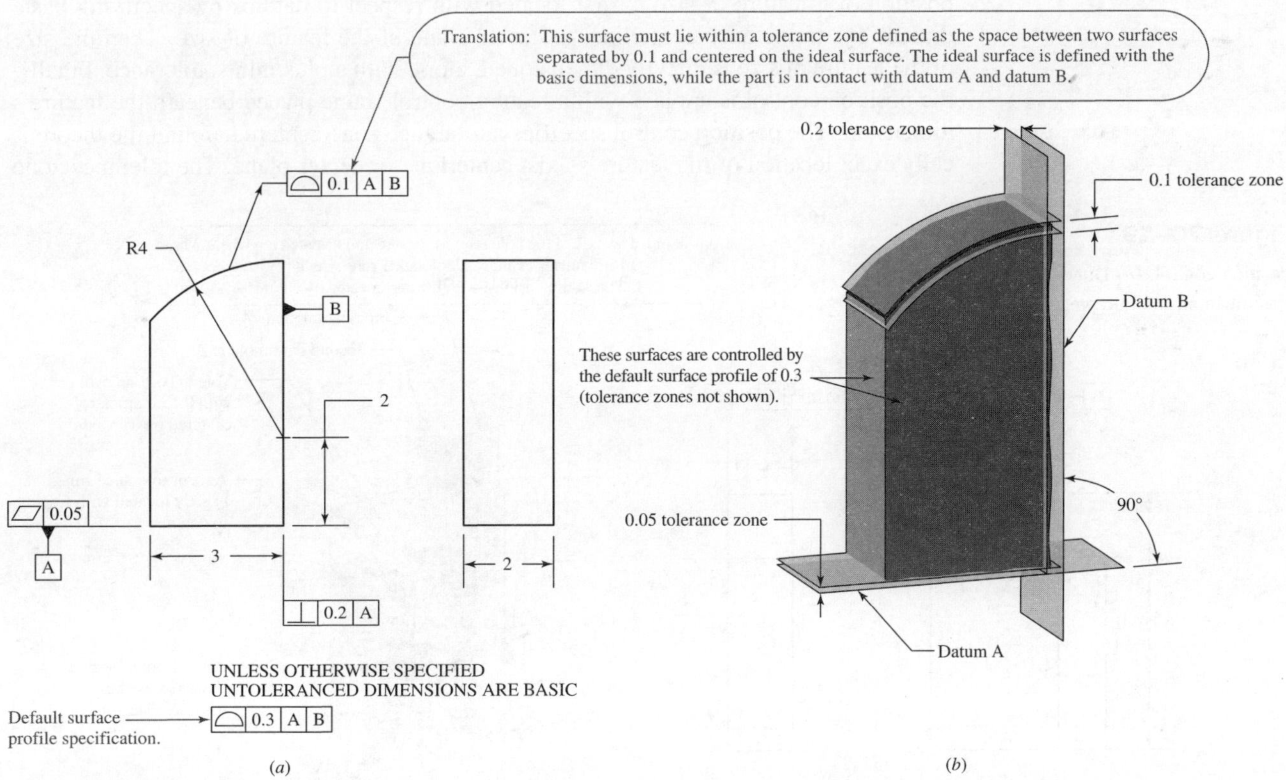

Figure 20–28

An example using a profile control note as an overall default tolerance control, with tighter controls added as needed. (*a*) Drawing specification; (*b*) Interpretation.

to the profile tolerance of 0.3 in the general note at the bottom of the drawing, all other surfaces that do not have a tighter tolerance specified will be within a profile tolerance zone of 0.3 centered on the basic (ideal) shape. Any deviation of size, form, orientation, and location is allowable so long as the surfaces remain within these tolerance zones.

Location Controls

There are three *location* controls:

1 ***Position***

2 ***Concentricity***

3 ***Symmetry***

The location controls are specified to control the location of a *feature of size,* such as a hole, slot, boss, or tab, with respect to a datum or another feature.

Position ⌖

The *position* control is one of the most effective and oft-used controls, as it incorporates most of the advantages of GD&T. The position control defines the allowed location (and orientation) of the axis, centerline, or center plane of a feature of size. It does not control the size or form of the feature.

The application of the position control is interpreted in Figure 20–29. The true position of a feature of size is first located with respect to datums by specifying basic dimensions to the axis, centerline, or center plane of the feature of size. Then the size of the feature of size is directly dimensioned, along with a plus/minus tolerance. Finally, the position control is applied with a feature control frame placed beneath the feature's dimension. The position control specifies a tolerance zone centered around the theoretically exact location of the feature's axis, centerline, or center plane. The tolerance zone

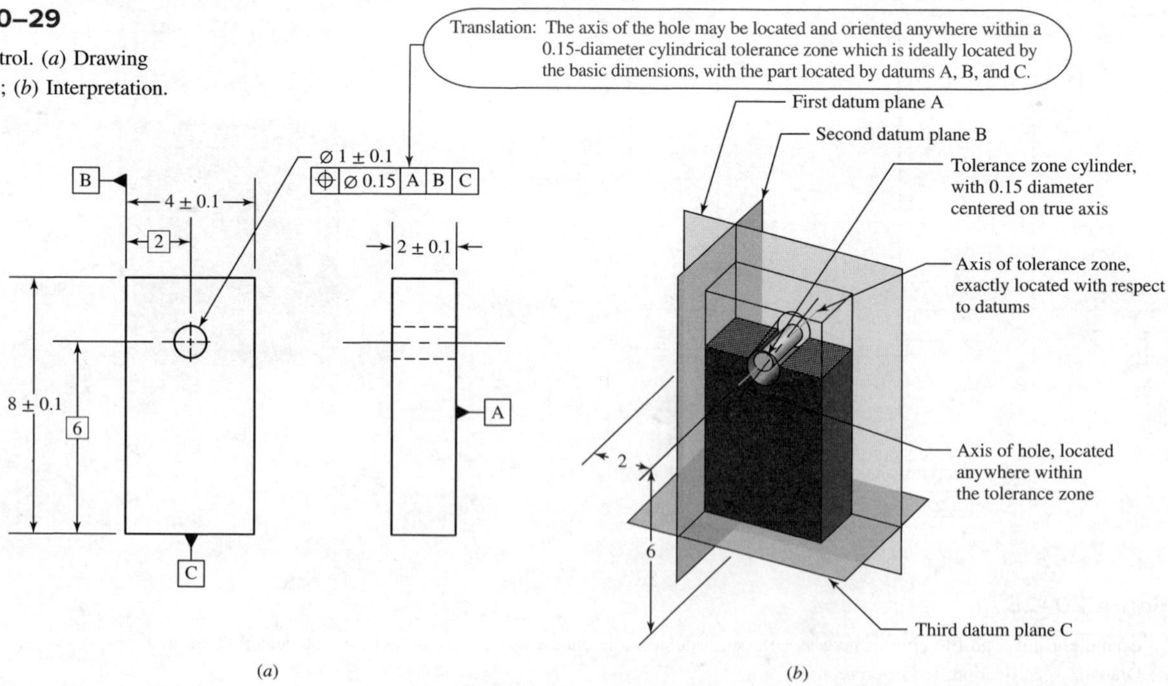

Figure 20–29

Position control. (*a*) Drawing specification; (*b*) Interpretation.

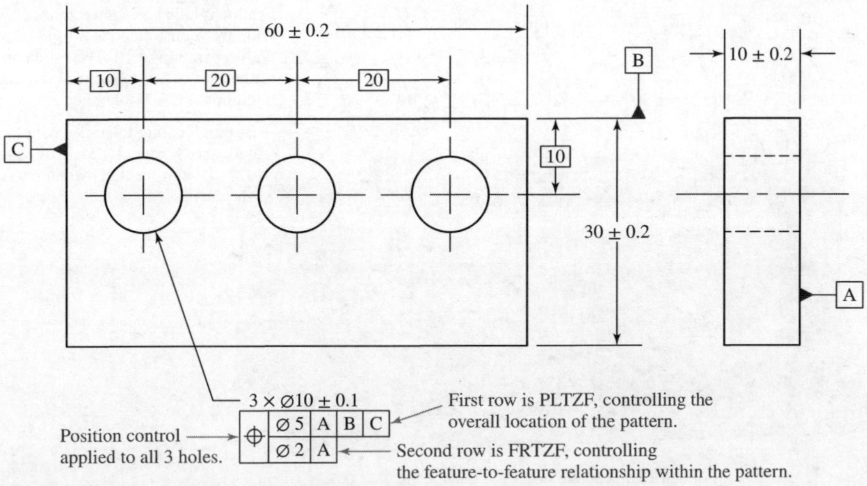

Figure 20–30

Drawing specification for position control applied to a group of features. The interpretation is shown in Figure 20–31.

is cylindrical if the diameter symbol precedes the tolerance; otherwise, it is the space between two parallel planes. The specification for the part shown in the machine drawing in Figure 20–29*a* leads to the interpretation shown in Figure 20–29*b*. Note that the hole's diameter may vary within its specified dimensional tolerance, while the hole's axis may be at any position and orientation within the cylindrical tolerance zone.

The position control also provides an excellent means of controlling the location of a group of features of size. A group of features is indicated to share the same dimension, dimensional tolerance, and position control by including the number of intended features preceding the specification of the size dimension, such as shown by the 3× in Figure 20–30.

Figure 20–30 also demonstrates the use of a composite control frame where there are two rows associated with the position control specification. This allows different tolerance specifications for the overall location of the pattern and the interrelation of features within the pattern. This is commonly needed when the feature pattern must mate with similarly spaced features on another part.

The first row in the composite control frame is referred to as a *Pattern-Locating Tolerance Zone Framework (PLTZF),* pronounced "plahtz." The PLTZF applies to the overall location of the pattern as a group with respect to the datums. The PLTZF defines tolerance zones for the centerline of each feature just like previously described for the position control.

The second row of the composite control frame adds additional constraint to the feature-to-feature relationship within the pattern, and is referred to as the *Feature-Relating Tolerance Zone Framework (FRTZF),* pronounced "fritz." The FRTZF is applicable to the basic dimensions between the features, but not to the basic dimensions locating the features with respect to the datums. The FRTZF defines another smaller tolerance zone for each feature, centered around the exact locations as defined by the basic dimensions between the features. The FRTZF tolerance zones may float anywhere within the PLTZF tolerance zones, as long as they maintain their relative positions with respect to each other. The actual feature centerlines must lie within the FRTZF tolerance zones, whose centerlines must in turn lie within the PLTZF tolerance zones.

Consider the example specified in Figure 20–30, where the tolerance specifications are larger than typical in order to allow the tolerance zones to be more easily visualized. Figure 20–31 shows the tolerance zones, without showing the actual holes, to minimize the clutter. The first row of the composite control frame specifies a

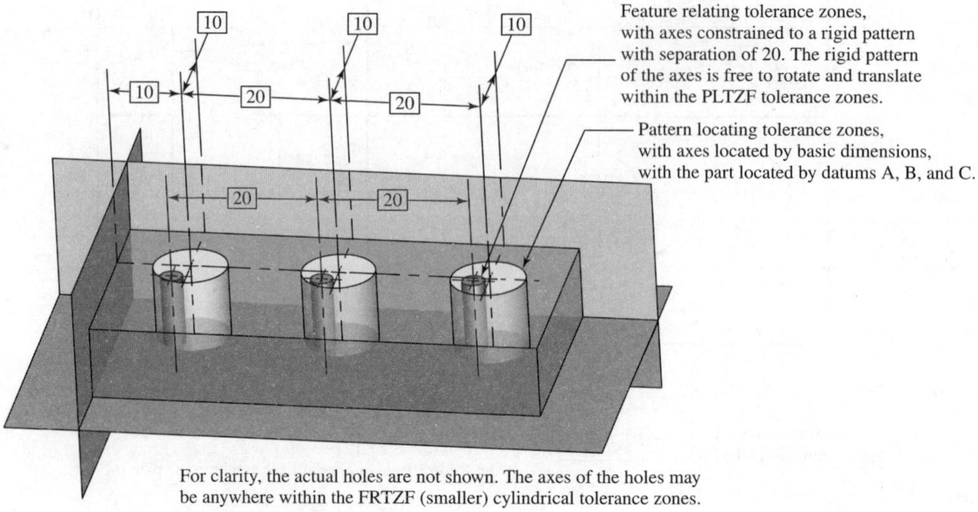

Figure 20–31

Interpretation of the PLTZF and FRTZF for a pattern of features located with a composite control frame, as specified in Figure 20–30.

PLTZF consisting of cylindrical tolerance zones each having a diameter of 5, centered around axes located by the basic dimensions from the datums. These are shown in Figure 20–31 as the larger cylinders. The second row specifies a FRTZF consisting of cylindrical tolerance zones each having a diameter of 2. The FRTZF tolerance zones form a rigid pattern that must maintain the relative distances of 20 with respect to each other, but which is free to rotate and translate within the cylindrical zones defined by the PLTZF. Think of the FRTZF tolerance zones as being rigidly connected to each other by a frame that maintains the basic dimensions between them. The entire frame may be translated and rotated to any position that keeps all of the FRTZF centerlines within the PLTZF tolerance zones.

When datum references are included in the FRTZF specification, they govern only the rotation of the FRTZF relative to the specified datums. In the example in Figure 20–30, since the second row references datum A, the FRTZF must be aligned with datum plane A, that is, the tolerance zone axes will be perpendicular to datum plane A. Likewise, if datum B had been specified, the tolerance zone axes would be required to lie in a plane that was parallel to datum plane B.

Concentricity ◎

Nominally, *concentricity* is the condition of the center axis of a surface of revolution, such as a cylinder, being congruent with a datum axis. The Y14.5 standard defines it more precisely as the condition where the median points of all diametrically opposed elements of a surface of revolution are within a cylindrical tolerance zone centered around a datum axis. This means the feature's center is not determined as a single straight line, but rather as a collection of all points obtained by finding the medians of all diameter measurements across the surface. This is extremely difficult and expensive to measure, so it is recommended that the concentricity control be rarely used. The preferred options for controlling concentric features are position, profile, and runout. The concentricity control might be warranted in cases where it is critical to control the axis rather than the surface of a feature, such as in dynamic balancing of a high-speed rotating part.

Symmetry ⌯

Symmetry is the condition where a feature has the same profile on either side of the center plane of a datum feature. In the Y14.5 standard, it is defined similarly to concentricity, except it applies to the center plane of a feature of size instead of the center axis of a surface of revolution. Because it is based on controlling a collection of all median points measured across the feature, rather than a single center plane of the feature, symmetry suffers from the same difficulties of measurement as concentricity. Consequently, the symmetry control should rarely be used. In most cases, symmetrical features are best controlled with position or profile controls.

Runout Controls ↗ ⌰

Runout controls the surface variation on a feature as it rotates around a datum axis. A simple example is the surface of a rotating shaft. There are two runout controls:

1 ***Circular runout***

2 ***Total runout***

Circular runout measures the radial variation of each circular section of a feature independently from each other. *Total runout* measures the runout of the entire surface of a cylindrical feature simultaneously.

Both types of runout are demonstrated in Figure 20–32. The cylinder on the left is defined as datum feature A. Chuck jaws clamped onto the surface of this datum feature serve as the datum feature simulator to define the centerline as the actual datum. Thus, when the chuck rotates, the part necessarily rotates around the datum centerline. The circular runout specification on the tapered feature requires that an indicator at any location along the feature must not move more than 0.01 units during a complete rotation of the part. Each point where the indicator is located should independently satisfy the runout tolerance control. The total runout specification on the right cylindrical feature requires that an indicator must not move more than 0.02 for all locations along the cylinder, measured in one setup. Another way of stating this is that the entire surface of the controlled feature must lie within the zone between two concentric cylinders that are separated radially by the stated tolerance of 0.02.

Circular runout can be applied to any surface of revolution since the measurements are made independently at each cross section. It is inherently controlling both concentricity and circularity. Total runout applies only to cylindrical features, since the diameter at each cross section must fit within the same tolerance zone. It is inherently controlling cylindricity, circularity, straightness, and surface profile. Total runout is a particularly

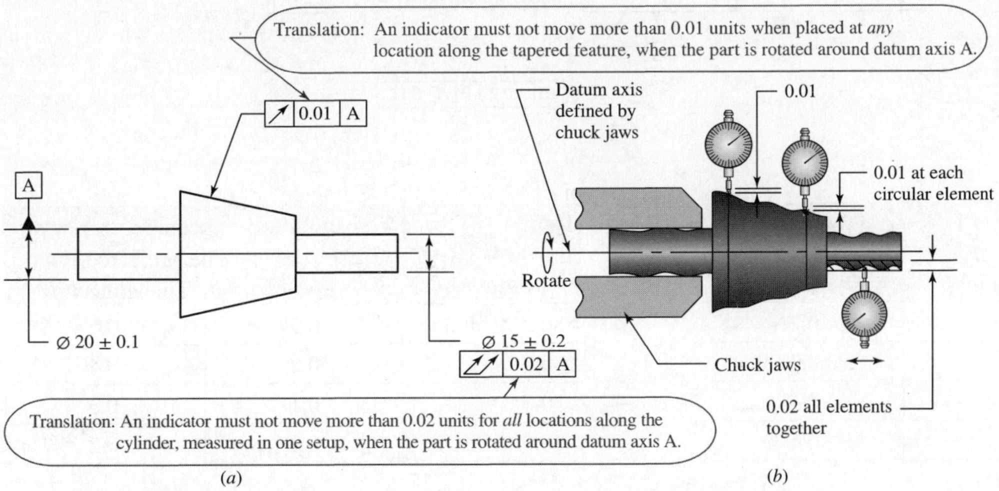

Figure 20–32

Circular runout and total runout.

(*a*) Drawing specification;

(*b*) Interpretation.

useful control for rotating shafts that carry bearings or gears that are sensitive to misalignment. Total runout can also be used effectively to control the coaxiality of multiple cylindrical surfaces by relating each cylinder to the same datum centerline.

The runout controls can also be applied to features constructed at right angles to a datum axis, such as the end-face of a cylinder. In this case, runout controls variations of perpendicularity (such as wobble) and flatness, measured while rotating the feature about the datum axis.

20–6 Material Condition Modifiers

Maximum material condition (MMC) and least material condition (LMC) can be applied as modifiers to most of the geometric controls that deal with a feature of size. The symbols for the modifiers, Ⓜ and Ⓛ, can be included in a feature control frame directly following the geometric tolerance value, and/or immediately following a datum reference.

When a geometric tolerance is modified with the maximum material condition modifier, it indicates that the stated tolerance value applies when the feature is produced at its MMC. If the feature is produced at a size with less material than its MMC, the deviation from the MMC is added to the allowed geometric tolerance value. The implication is that the value of the geometric tolerance is not constant, as it depends on the size of the actual produced part.

As an example, consider the part in Figure 20–33*a*, which contains both an external feature of size (the cylindrical boss) and an internal feature of size (the hole), each of which is located with a position control that is toleranced at MMC. The position

Figure 20–33

Application of MMC to a position tolerance. (*a*) Drawing specification; (*b*) summary of bonus tolerance for external feature; (*c*) summary of bonus tolerance for internal feature.

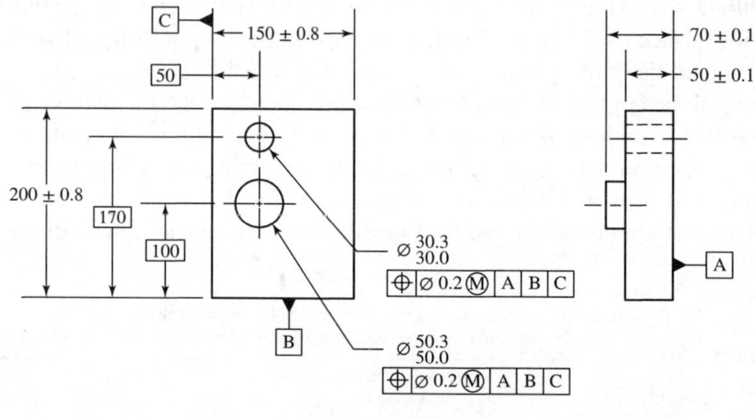

(*a*)

External Feature ϕ	Additional Tolerance (Bonus)	Tolerance Zone Cylinder ϕ
50.3 (MMC)	0.0	0.2
50.2	0.1	0.3
50.1	0.2	0.4
50.0 (LMC)	0.3	0.5

(*b*)

Internal Feature ϕ	Additional Tolerance (Bonus)	Tolerance Zone Cylinder ϕ
30.3 (LMC)	0.3	0.5
30.2	0.2	0.4
30.1	0.1	0.3
30.0 (MMC)	0.0	0.2

(*c*)

control on the external feature is interpreted as follows: "The center axis of this external cylindrical feature must be in position within a cylindrical tolerance zone of 0.2 diameter if the feature is produced at its maximum material condition of 50.3, where the position is specified by the basic dimensions with respect to the datum reference frame established by datum features A, B, and C." If the cylinder is produced with a diameter less than its MMC of 50.3, say at 50.2, then the amount of the deviation from the MMC, that is 0.1, is added to the specified geometric tolerance, providing a realized tolerance of 0.3. This increase in tolerance is traditionally referred to as a "bonus" tolerance. The table in Figure 20–33*b* shows how this bonus tolerance accumulates as the produced diameter of the cylinder is reduced from its MMC of 50.3 to its LMC of 50.0. Note that the bonus tolerance is applied to the tolerance associated with the geometric control (position, in this example), and not to the direct tolerance on the size of the feature.

This bonus tolerance is one of the significant advantages afforded by GD&T over the fixed tolerances provided by the traditional coordinate dimensioning. It is particularly useful for applications where mating parts need to fit with each other with a clearance for assembly. A minimum tolerance is guaranteed, but a bonus tolerance is made available to the manufacturer to reduce costs. This modifier would not be appropriate for an application where the fit between the mating parts is important, such as a press fit between a bearing and a shaft.

The same bonus tolerance concept is applicable for the hole in Figure 20–33*a*, except that the MMC condition for the internal feature is the *smallest* hole size. The table in Figure 20–33*c* shows how the bonus tolerance adds to the stated MMC tolerance as the hole size gets larger, moving from the MMC to the LMC.

The *least material condition* modifier works similarly to the MMC modifier, but in the opposite direction. When a geometric tolerance is modified with the LMC modifier, it indicates that the stated tolerance value applies when the feature is produced at its LMC. If the feature is produced at a size with more material than its LMC, the deviation from the LMC is added to the allowed geometric tolerance value. The LMC modifier is typically used for applications in which it is critical to maintain a minimum amount of material. Examples include the material between a hole and the edge of a part, or a wall thickness. LMC is also useful for specifying features on a casting that will later be machined, to ensure that enough material is left in the casting for the finish machining operation.

The MMC and LMC material condition modifiers can be similarly applied to most of the geometric tolerances, in particular to those that control features of size with a central axis or plane. If no material condition modifier is specified, the default material condition is known as *regardless of feature size (RFS)*. RFS means that the stated tolerance is applicable regardless of the feature size. In other words, no matter what size the feature is produced (within its tolerance), the geometric tolerance is fixed at the stated value. This is much more restrictive for manufacturing. It is warranted in applications where variable play between mating parts is not desirable, such as the press fit between a shaft and components like bearings and gears.

A fine detail to be noted is that in all the preceding discussion, the produced diameter of the feature is determined by the unrelated actual mating envelope, as defined in Section 20–3, as it is about the only practical way to determine a single value for the diameter of an imperfect feature.

The Ⓜ and Ⓛ symbols can also be applied immediately following a datum reference in the feature control frame, in particular when the datum is based on a feature of size. When applied to a datum, the symbols are referred to as *maximum material boundary (MMB)* and *least material boundary (LMB)*. A full explanation of the effect of applying MMB or LMB to a datum is beyond the scope of this chapter. In essence, it allows the part to float or shift relative to the datum reference frame as the produced datum feature of size deviates from its maximum or minimum material condition. Consequently, it does not change the tolerance on the considered feature, but just the relative position of the feature with respect to the datum reference frame.

20–7 Practical Implementation

The basic concept of GD&T is to define the ideal part, then specify the amount of variation that is acceptable. The allowable variation includes all four of the geometric attributes: size, location, orientation, and form. The size of a feature of size is directly dimensioned and toleranced with a plus/minus tolerance. The other three geometric attributes, broken down into more specific geometric characteristics, are controlled with geometric controls (Table 20–1). Some of the geometric controls are broad and inherently provide control of multiple characteristics. Anytime a boundary envelope is defined by *any* geometric control, it constrains *all* geometric variations from the ideal size and shape to fit within the envelope. For example, by applying profile of a surface to a feature, the bounding envelope that defines the allowable variation from the ideal profile also automatically controls orientation characteristics (e.g., parallelism and perpendicularity) and form characteristics (e.g., flatness and cylindricity). Consequently, most of the geometric characteristics can be controlled by a few controls, and refinements are only added as necessary.

A suggested general framework for implementation of GD&T comprises the following five steps:

1 Select the datum features.

2 Control the datum features.

3 Locate the features.

4 Size and locate the features of size.

5 Refine the orientation and the form of features, if needed.

Each step is elaborated in the following sections.

1 *Select the datum features.*

The datum features should be selected based on the functional use of the part first, rather than on the anticipated manufacturing method. The primary datum is usually the most critical to the part function, and of sufficient surface size to assure a stable setup for establishing the remaining datums. For mating parts, the corresponding interfacing features are usually selected as datum features. Remember that datum features identified with the datum symbol are only utilized when they are called out in the feature control frame of a feature being controlled. Though it is not unusual for every feature of a part to reference the same set of datums, it is not a requirement that they do so.

2 *Control the datum features.*

Though a datum is considered theoretically perfect, the physical datum feature is not. Consequently, the imperfect datum features need to be geometrically controlled just like any other feature. Sometimes the default controls are sufficient. However, since the datum features are used to stabilize the part for manufacturing and inspection, they may warrant additional consideration for that purpose, beyond what is needed just for the functionality of the part. If the primary datum feature is a flat surface, it may be helpful to consider a flatness or surface profile control on it. The secondary and tertiary datum features establish datum planes perpendicular to the primary datum plane. Consequently, it might be useful to use an orientation control (such as perpendicularity) on the secondary and tertiary datum features with respect to the primary datum. A default surface profile applied by a general note to the entire part might be sufficient for this.

When a feature of size is used as a datum feature, the size tolerance automatically provides form control through Rule #1 (see Section 20–4). Also, when the position control is used to locate a datum feature of size, it will automatically provide orientation control of the datum.

3 *Locate the features.*

All features have surfaces that need to be located with respect to appropriate datums. The best strategy for most situations is to use basic dimensions to locate the true position of each feature, accompanied by one or more appropriate geometric controls. A default surface profile control can be established with a general note.

4 *Size and locate the features of size.*

Features of size need to be sized as well as located. The following three steps are typical: (1) Locate the true position of the center axis or center plane of each feature of size with basic dimensions. (2) Directly dimension the feature's size, including a plus/minus tolerance. (3) Attach a position control to the feature's size callout to establish limits for location and orientation. For cylindrical features of size that are coaxial to a datum axis, the runout or surface profile controls may be used instead of the position control. The size tolerance on the feature of size automatically provides form control through Rule #1.

5 *Refine the orientation and the form of features, if needed.*

If any feature needs a tighter control in orientation or form than is provided by the previous steps, additional controls can be added.

The following example demonstrates this process.

EXAMPLE 20–1

Interpret and explain the GD&T notation for the part shown in Figure 20–3.

Solution

Since this part has already been drawn, the five steps will be used to organize the explanation rather than to make the decisions.

1 *Select the datum features.*

The datum features that have been identified with the datum symbols are the back face and the left and bottom edges. The functional needs of the part are not specified, but this choice of datums is pretty common for this type of simple rectangular plate. Note that the callout for datum B is not placed in-line with the dimension, thus indicating the datum is at the edge of the part rather than the center plane of the part. Since the features are located with respect to the datum, the choice of datum indicates that it is more important to maintain the distance of the hole from the edge rather than to ensure that it is centered. The back face makes a good primary datum (as called out in the position control of the hole), as it is of sufficient size to stabilize the part with respect to three degrees of freedom while establishing the other datums. It is also likely that the back surface will be in contact with a mating part.

2 *Control the datum features.*

A flatness specification ensures that the surface of datum feature A does not vary more than 0.003. This is a common control for a primary datum that is a planar surface, especially if it must fit with a mating surface. Datum feature B, the left surface of the part, is required to be perpendicular to the back face, within a tolerance zone between two parallel planes separated by a distance of 0.005. Note that the envelope generated by this orientation control also constrains the form of this surface (e.g., flatness and straightness). Datum feature C must be perpendicular to datum features A and B, within a tolerance zone of 0.005.

3 *Locate the features.*

All features of this part are features of size, which are handled in the next step.

4 *Size and locate the features of size.*

There are four features of size: the hole, and the height, width, and thickness of the plate. The plate features do not need to be located, since each dimension starts at one of the datum features. Consequently, the three plate dimensions simply need a directly toleranced dimension.

The hole feature is located by basic dimensions from datums B and C. The hole diameter is specified to be within the range from 1.000 to 1.002. The position control on the hole stipulates that the center axis of the hole must be within a cylindrical tolerance zone that has a diameter of 0.003 if the hole is produced at its MMC of 1.000. The tolerance zone may increase to as much as 0.005 as the produced hole diameter increases from 1.000 to 1.002. The tolerance zone is determined with the part immobilized with respect to the three datums, applied in order of the back surface datum A, then the side edge datum B, then the lower edge datum C. Changing this order would change the location and orientation of the hole's tolerance zone. Note that the hole's tolerance zone, in addition to providing a tolerance on the location of the axis, also limits the orientation of the hole's axis. A separate orientation control could have been stipulated if the same tolerance zone was not suitable for both position and orientation.

5 *Refine the orientation and the form of features, if needed.*

No additional controls are specified, so apparently no further refinement is necessary. It is always a good idea to consider how far from the true geometry the part could be produced and still be within specifications. For example, consider how out-of-flat the front face (opposite of datum A) could be. Since the thickness of the plate is a feature of size, the front face is controlled by Rule #1, that is, the size tolerance also establishes an envelope for form control. The size tolerance allows the thickness to vary between 1.9 and 2.1. Since the back datum face may vary only by 0.003, the majority of the size tolerance can be realized by a front face that is curved, bent, or wavy. The plate could be 1.9 thick on one edge and 2.1 thick on the other. If this is not acceptable, the Rule #1 control should be overruled by another more specific refinement.

There are many concepts of GD&T that have not been addressed in this brief introductory chapter. Significant training is necessary to gain competence in defining appropriate geometric controls to achieve the desired function. However, a fundamental ability to "read" a drawing is not out of reach. The next example will provide an opportunity to practice.

EXAMPLE 20–2

Interpret and explain the GD&T notation for the part shown in Figure 20–34.

Solution

Various aspects of the drawing are circled and annotated with note numbers that correlate with the explanations given here.

Note 1. The front face is defined as a datum feature and is used as the primary datum for several of the geometric controls on the drawing. Since the front face, rather than the back face, is selected as a datum, it is likely that this surface will fit against a mating surface. Functionally, this fit is apparently more important than the back face.

The flatness control is common for a surface that is a primary datum. The entire surface must fit within a tolerance zone defined by two parallel planes separated by a distance of 0.05.

Note 2. Consider the outer surface of the protruding cylinder, which is identified as a datum feature. Since it is a feature of size, the datum is the center axis. More precisely, the datum is the theoretically perfect center axis of the imperfect outer surface, as determined by the unrelated actual mating envelope. Note that if the bore through the center of the part had been chosen as the datum feature, the datum would also have been a center axis, but not precisely the same one. The fact that the outer surface is selected indicates it may have greater functional precedent over the bore. Perhaps the outer cylindrical surface will fit within the bore of a mating part. Even though the function of the part is not specified, the geometric requirements of the functional needs are clear, and the part should be manufactured and inspected accordingly.

Every feature of size must be controlled in all four geometric attributes: size, location, orientation, and form. Consider each in turn for the outer surface of the protruding cylinder.

The cylinder's size is directly specified. The diameter at any cross section must be within the stated range of 39.96 to 40.00.

The center axis of a feature of size must be located, usually by basic dimensions. In this case, since this is a datum feature, its own center axis becomes part of the definition of the origin of the datum reference frame. The feature's location is at the origin, so no further basic dimensions are needed to locate it.

Figure 20–34

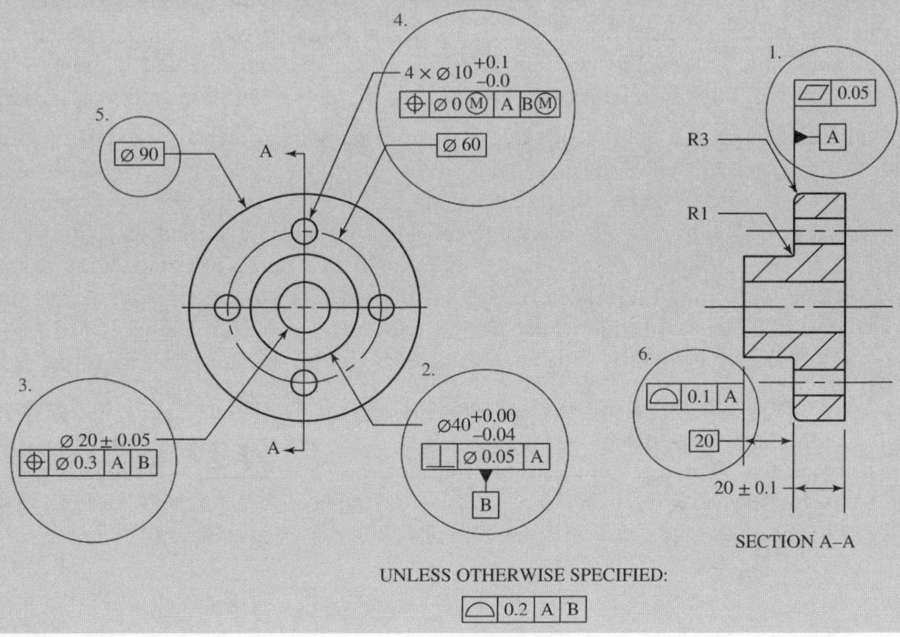

The orientation of this cylinder feature is controlled by the perpendicularity control. This control requires the center axis of the feature to lie within a cylindrical tolerance zone that has a diameter of 0.05. The tolerance zone is exactly perpendicular to datum plane A; the center axis of the feature is allowed to tip within the tolerance zone.

No specific form controls are specified for this feature. The form of its surface is controlled by default through Rule #1 for features of size. Rule #1 requires that the surface of the cylinder must be within the cylindrical envelope of perfect form at the MMC diameter of 40.00.

This feature of size is controlled in all aspects: size, location, orientation, and form.

Note 3. Now consider the center bore. This is a feature of size, so its size is directly specified and toleranced. Its form is controlled by Rule #1. It is located by implication (since no basic dimensions are given) at the origin of the datum reference frame. Its orientation and location is controlled by the position control. The position control defines a cylindrical tolerance zone that is first held perpendicular to datum plane A, then centered at the datum axis defined from datum feature B. Since there is no material modifier specified on the tolerance, the default condition of *regardless of feature size (RFS)* applies. Therefore, the tolerance zone diameter is a constant 0.3, regardless of what size the feature is actually produced.

Note 4. The four holes are defined collectively as a pattern. They are features of size, so their diameters are directly dimensioned and toleranced, and their surfaces must not exceed the envelope of perfect form at the MMC diameter of 10 (Rule #1). Their location is specified by the basic dimension of 60 for the bolt circle, as well as by the implication that they are spaced at 90 degrees around the bolt circle.

Control of orientation and location is provided by the position control. Specifically, the position control requires the center axis of each hole to be within a cylindrical tolerance zone. The datum callouts require this tolerance zone to be first held perpendicular to datum A, then centered at the true location as measured with respect to datum axis B. The Ⓜ modifier that accompanies datum B allows a little more leeway on the overall shift of the hole pattern if datum feature B is produced at a diameter less than its MMC.

The zero tolerance value in the position control does not mean that no tolerance is allowed, as it is accompanied by the Ⓜ modifier. This is known as *zero tolerancing* at MMC. The meaning is that the tolerance zone diameter is zero if the hole is produced at its maximum material diameter of 10, but grows to a diameter of 0.1 as the hole's diameter increases to its least material diameter of 10.1. Consequently, the axis of the hole would need to be perfectly located and oriented if the hole is produced at its MMC of 10, but can deviate from perfection if the hole is produced at a larger size.

Note 5. This cylindrical feature is a feature of size, and could have been directly dimensioned and toleranced. Instead, it is specified with a basic dimension of 90 with no tolerance or geometric control in sight. This does not mean it has to be perfect. It means that it is controlled by the default profile tolerance that is stated at the bottom of the drawing. The ideal cylindrical surface is first sized at the basic diameter of 90, and located with respect to datums A and B. Then a tolerance zone is centered around this ideal surface with a total tolerance of 0.2. In this case, the tolerance zone is the space between two concentric cylinders with diameters of 89.8 and 90.2 (a radial difference of 0.2). This tolerance zone controls all four geometric attributes of size, location, orientation, and form.

Note 6. The basic dimension of 20 locates the ideal surface of the front face of the protruding cylinder. The surface profile control requires that the actual front face lie within a tolerance zone consisting of the space of 0.1 between two planes centered around the ideal surface, where each plane is parallel to datum A. This effectively establishes the location of the face (a distance of 19.95 to 20.05 from datum A), as well as the flatness of the face, and the parallelism of the face with respect to datum A.

20–8 GD&T in CAD Models

Many industries are utilizing 3D computer-aided design (CAD) data for some or all of the engineering, manufacturing, and inspection phases of the product life cycle. The prominent standard regulating this is ASME Y14.41-2003, Digital Production Definition Data Practices. The standard addresses many aspects of the practices, requirements, and interpretation of CAD data. The standard defines the use of GD&T in a digital environment where specifications are embedded directly into the data set—not just visually, but functionally as well.

Most of the concepts of GD&T apply directly to digital models. The significant difference in CAD models versus 2D drawings is that the CAD model represents the ideal geometry of the part. Any dimension can be queried from the model and the exact (ideal) dimension can be obtained. In fact, the ideal data can be directly utilized in computer-aided manufacturing operations. This leads to a misconception that a part manufactured directly from the model on a computer numerical controlled (CNC) machine will be perfect. In fact, CNC manufacturing has the same requirements as manual manufacturing methods with regard to the need to specify and inspect geometric tolerances.

The Y14.41-2003 standard allows for *all* geometry data to be considered *basic,* unless superseded by a toleranced dimension or defined as a reference dimension. Geometric controls are applied to the 3D data models to control the features just the same as on the 2D drawings. Direct tolerancing is recommended only for features of size. Essentially, GD&T works the same on 3D models as in 2D drawings, except that in 3D models the basic dimensions do not have to be shown, since the default is that all dimensions are basic. Figure 20–35 shows the 3D solid model of the part used in Example 20–2, with appropriate GD&T elements embedded into the CAD data.

The ongoing transition to 3D digital representation brings with it the possibilities of tighter integration of the various processes of design, analysis, and manufacturing. For example, the embedded geometric dimensioning and tolerancing information can be directly accessed for tolerance analysis and process planning.

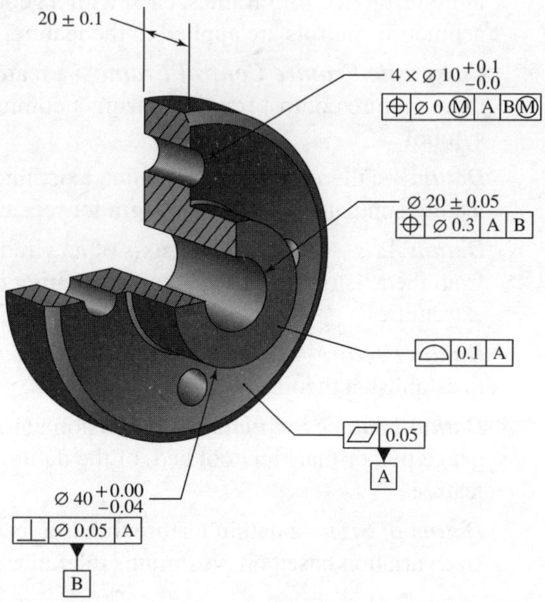

Figure 20–35

An example of GD&T applied to a 3D CAD model.

20–9 Glossary of GD&T Terms

Most of the concepts in GD&T are simple, but the vocabulary to describe them can seem overwhelming at first. That is because the vocabulary needs to be precise enough to be consistent and unambiguous. For convenient reference, some of the most commonly used terms in GD&T are summarized in this section.

Actual Mating Envelope—a perfectly shaped counterpart of an imperfect feature of size, which can be contracted about an external feature, or expanded within an internal feature, so that it contacts the high points of the feature's surface.

Actual Mating Envelope, Related—an actual mating envelope that is sized to fit the feature while maintaining some constraint in orientation or location with respect to a datum.

Actual Mating Envelope, Unrelated—an actual mating envelope that is sized to fit the feature without any constraint to any datum.

Axis—a line defining the center of a cylindrical feature, established from the theoretical axis of the unrelated actual mating envelope of the cylindrical feature's extremities.

Basic Dimension—a theoretically exact dimension that ideally locates and/or orients the tolerance zone of a feature. It does not have a tolerance directly associated with it, but is instead associated with a geometric control of a tolerance zone. Basic dimensions are denoted by a box around the dimension, or by a general note.

Bonus Tolerance—additional tolerance that applies to a feature as its size shifts from a stated material condition of MMC or LMC.

Center Plane—the theoretical plane located at the center of a noncylindrical feature of size, established from the center plane of the unrelated actual mating envelope of the feature's extremities.

Combined Feature Control Frame—a feature control frame made of two or more feature control frames, each with a geometric characteristic symbol. The geometric controls are applied to the feature, in order from top to bottom.

Composite Feature Control Frame—a feature control frame made of two or more feature control frames sharing a common geometric characteristic symbol.

Datum—a theoretically exact point, axis, line, or plane derived from a datum feature simulator, used as an origin for repeatable measurements.

Datum Axis—the theoretical axis of a cylindrical datum feature, established from the axis of the unrelated actual mating envelope of the cylindrical feature's extremities.

Datum Feature—an actual physical surface of the part that is specified in order to establish a theoretically exact datum.

Datum Feature Simulator—a precision embodiment, such as a surface plate, gauge pin, or machine tool bed, of the datum described by an imperfect datum feature.

Datum of Size—a datum feature that is a feature of size, and therefore subject to size variation based on plus/minus tolerances.

Datum Reference Frame—a set of up to three mutually perpendicular planes that are defined as the origin of measurement for locating the features of a part.

Derived Median Line—an imperfect "line" formed by the center points of all cross sections of a feature, where the cross sections are normal to the axis of the Unrelated Actual Mating Envelope of the feature.

Derived Median Plane—an imperfect "plane" formed by the center points of all line segments bounded by a feature, where the line segments are normal to the center plane of the unrelated actual mating envelope of the feature.

Envelope Principle—See definition for Rule #1, given later in this section.

Feature—a general term referring to a clearly identifiable physical portion of a part, such as a hole, pin, slot, surface, or cylinder.

Feature Control Frame—a rectangular box attached to a feature on a drawing, containing the necessary information to define the tolerance zone of the specified feature.

Feature of Size, Irregular—a directly toleranced feature or collection of features that may contain or be contained by an actual mating envelope.

Feature of Size, Regular—a cylindrical surface, a spherical surface, a circular element, or a set of two opposed parallel elements or surfaces that are associated with a directly toleranced dimension. A regular feature of size has a size that can be measured across two opposing points, and has a reproducible center point, axis, or center plane.

Feature-Relating Tolerance Zone Framework (FRTZF)—a tolerance zone framework that governs the positional relationship from feature to feature within a pattern of features. It is specified in the bottom row of a composite feature control frame.

Geometric Attributes—the four broad attributes (size, location, orientation, and form) that must be considered to geometrically define a feature. This term is not strictly defined by GD&T.

Geometric Characteristics—the 14 characteristics defined in Table 20–1 that are available to control some aspect of a geometric tolerance of a feature. A geometric characteristic symbol is the first item in any feature control frame.

Least Material Boundary (LMB)—the limit defined by a tolerance or combination of tolerances that exists on or inside the material of a feature(s). When applied as a modifier to a datum reference in a feature control frame with the Ⓛ symbol, it establishes the datum feature simulator at a boundary determined by the combined effects of size (least material), and all applicable geometric tolerances.

Least Material Condition (LMC)—the condition in which a feature of size contains the least amount of material within the stated limits of size (e.g., maximum hole diameter or minimum shaft diameter). This condition may be specified as a tolerance modifier in a feature control frame with the Ⓛ symbol.

Material Condition Modifier—a modifier symbol, Ⓜ or Ⓛ, applied to a geometric tolerance to indicate that the tolerance applies at maximum material condition or least material condition, respectively. The absence of a material condition modifier indicates that the tolerance applies at all material conditions, that is, regardless of feature size (RFS).

Maximum Material Boundary (MMB)—the limit defined by a tolerance or combination of tolerances that exists on or outside the material of a feature(s). When applied as a modifier to a datum reference in a feature control frame with the Ⓜ symbol, it establishes the datum feature simulator at a boundary determined by the combined effects of size (maximum material), and all applicable geometric tolerances.

Maximum Material Condition (MMC)—the condition in which a feature of size contains the maximum amount of material within the stated limits of size (e.g., minimum hole diameter or maximum shaft diameter). This condition may be specified as a tolerance modifier in a feature control frame with the Ⓜ symbol.

Pattern-Locating Tolerance Zone Framework (PLTZF)—a tolerance zone framework that governs the positional relationship between a pattern of features and the datum features. It is specified in the top row of a composite feature control frame.

Regardless of Feature Size (RFS)—indicates that the stated tolerance applies, regardless of the actual size to which the feature is produced. This is the default condition for a tolerance with no modifier symbol (i.e., Ⓜ or Ⓛ).

Regardless of Material Boundary (RMB)—indicates that a datum feature simulator progresses from MMB toward LMB until it makes maximum contact with the extremities of a feature(s). This is the default condition for a datum reference with no modifier symbol (i.e., Ⓜ or Ⓛ).

Rule #1—when only a tolerance of size (i.e., a plus/minus tolerance) is specified for a feature of size, the limits of size prescribe the extent of which variation in its geometric form, as well as size, are allowed. The surface of a feature of size may not extend beyond an envelope of perfect form at MMC. This rule is also referred to as the envelope principle.

Tolerance—the total amount a specific dimension is permitted to vary, between the maximum and minimum limits.

Tolerance Zone—a limiting boundary in which the actual feature must be contained.

Virtual Condition—a constant "worst case" boundary defined by the collective effects of a feature's size, geometric tolerance, and material condition.

PROBLEMS

20–1 In the traditional coordinate dimensioning system, which of the following is true? (Select one.)

i. Only "features of size" need to include a tolerance.
ii. Only dimensions that are important need to include a tolerance.
iii. Only dimensions that need to be tightly controlled need to include a tolerance.
iv. All dimensions should include a tolerance.

20–2 In GD&T, which type of dimension should generally be directly toleranced with a plus/minus tolerance?

20–3 What underlying purpose is emphasized by the *ASME Y14.5–2009* standard in the dimensioning and tolerancing of a part? (Select one.)

i. The method of manufacturing.
ii. The design intent.
iii. The inspection process.
iv. Equal attention to all of the above.

20–4 What is the term that refers to a feature which has a size that can be measured across two opposing points?

20–5 What are the four geometric attributes that must be considered to define the geometry of a feature of a part?

20–6 What are the four geometric characteristics that provide form control?

20–7 What are the three geometric characteristics that provide orientation control?

20–8 What are the three geometric characteristics that provide location control? Which of the three is most prominently used?

20–9 How is a basic dimension toleranced? (Select one.)

i. Basic dimensions receive the default tolerance specified in the title block.
ii. Basic dimensions are not toleranced.
iii. Basic dimensions receive the tolerance from an associated feature control frame.

20–10 For the part shown, identify all of the features of size.

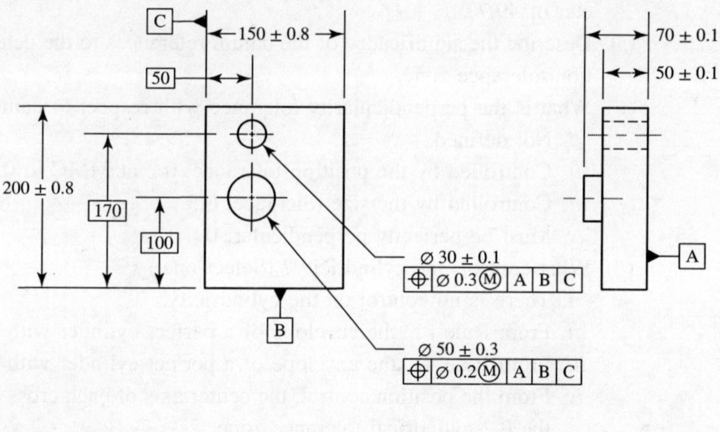

Problems 20–10 to 20–14

20–11 For the part shown, clearly identify each of the following, with labels and sketches on the drawing.

(*a*) Datum features A, B, and C.
(*b*) Datums A, B, and C.
(*c*) Datum reference frame based on datum features A, B, and C.

20–12 For the part shown, the ideal position of the cylindrical boss is located with the basic dimensions of 100 and 50. These basic dimensions are measured from which of the following? (Select one.)

i. The physical edges of the part.
ii. The high points on the physical edges of the part.
iii. The low points on the physical edges of the part.
iv. The theoretical datum planes B and C.

The basic dimensions are locating which of the following? (Select one.)

i. The physical location of the center axis of the boss.
ii. The center axis of the boss as determined by the actual mating envelope of the boss.
iii. The ideal location of the center axis of the tolerance zone specified by the position control.

If the part is produced and is being inspected, the location of the boss will be measured from which of the following? (Select one.)

i. The physical edges of the part.
ii. The high points on the physical edges of the part.
iii. The low points on the physical edges of the part.
iv. The theoretical datum planes B and C.
v. The datum feature simulators for datum features B and C.

20–13 For the part shown, answer the following questions with regard to the cylindrical boss.

(*a*) What are the maximum and minimum diameters allowed for the boss?
(*b*) What is the effect of the position tolerance of 0.2 on the diameters specified in part (*a*)?
(*c*) The position control defines a tolerance zone. Specifically what must stay within that tolerance zone?
(*d*) What is the diameter of the tolerance zone if the boss is produced with a diameter of 50.3?
(*e*) What is the diameter of the tolerance zone if the boss is produced with a diameter of 49.7?
(*f*) Describe the significance of the datum references to the determination of the position tolerance zone.
(*g*) What is the perpendicularity tolerance with respect to datum A? (Select one.)
i. Not defined.
ii. Controlled by the position tolerance; 0.2 at MMC to 0.8 at LMC.
iii. Controlled by the size tolerance; 0.3.
iv. Must be perfectly perpendicular; 0.
(*h*) What controls the cylindricity? (Select one.)
i. There is no control on the cylindricity.
ii. From Rule #1, the envelope of a perfect cylinder with diameter of 50.
iii. From Rule #1, the envelope of a perfect cylinder with diameter of 50.3.
iv. From the position control, the center axis of each cross section must be within the 0.2 cylindrical tolerance zone.

20–14 For the part shown, answer the following questions with regard to the hole.

(*a*) What are the maximum and minimum diameters allowed for the hole?
(*b*) What is the effect of the position tolerance of 0.3 on the diameters specified in part (*a*)?
(*c*) The position control defines a tolerance zone. Specifically what must stay within that tolerance zone?
(*d*) What is the diameter of the tolerance zone if the hole is produced with a diameter of 30.1?
(*e*) What is the diameter of the tolerance zone if the hole is produced with a diameter of 29.9?
(*f*) Describe the significance of the datum references to the determination of the position tolerance zone.

(*g*) What is the perpendicularity tolerance with respect to datum A? (Select one.)
  i. Not defined.
  ii. Controlled by the position tolerance; 0.3 at MMC to 0.5 at LMC.
  iii. Controlled by the size tolerance; 0.1.
  iv. Must be perfectly perpendicular; 0.

(*h*) What controls the cylindricity? (Select one.)
  i. There is no control on the cylindricity.
  ii. From Rule #1, the envelope of a perfect cylinder with diameter of 30.
  iii. From Rule #1, the envelope of a perfect cylinder with diameter of 29.9.
  iv. From the position control, the center axis of each cross section must be within the 0.3 cylindrical tolerance zone.

20–15 Describe how a center axis is determined for a physical hole that is not perfectly formed.

20–16 According to the envelope principle (Rule #1), the size tolerance applied to a feature of size controls the size and _____ of the feature. (Select one.)
  i. location
  ii. orientation
  iii. form
  iv. runout
  v. All of the above.

20–17 If a shaft diameter is dimensioned 20 ± 0.2, determine the diameter of the shaft at MMC and at LMC.

20–18 If a hole diameter is dimensioned 20 ± 0.2, determine the diameter of the hole at MMC and at LMC.

20–19 A hole diameter is dimensioned 20 ± 0.2. According to the form control provided by the envelope principle (Rule #1), is the limiting envelope a perfect cylinder with diameter of 19.8, 20.2, or both? Explain your answer.

20–20 A shaft diameter is dimensioned 20 ± 0.2. According to the form control provided by the envelope principle (Rule #1), is the limiting envelope a perfect cylinder with diameter of 19.8, 20.2, or both? Explain your answer.

20–21 The diameter of a cylindrical boss is dimensioned 25 ± 0.2. A position control is used to control the basic location of the boss. Specify the diameters allowed for the position tolerance zone if the boss is produced with diameters of 24.8, 25.0, and 25.2, for each of the following position tolerance specifications:
(*a*) ⌀0.1 (*b*) ⌀0.1Ⓜ (*c*) ⌀0.1Ⓛ

20–22 A hole diameter is dimensioned $32^{+0.4}_{-0.0}$. A position control is used to control the basic location of the hole. Specify the diameters allowed for the position tolerance zone if the hole is produced with diameters of 32.0, 32.2, and 32.4, for each of the following position tolerance specifications:
(*a*) ⌀0.3 (*b*) ⌀0.3Ⓜ (*c*) ⌀0.3Ⓛ

20–23 What is the name of the geometric characteristic that effectively controls a combination of circularity and straightness of a cylinder?

20–24 What is the name of the geometric characteristic that can be specified in a note to provide a default tolerance zone to control size, form, orientation, and location of all features that are not otherwise controlled?

20–25 Which geometric characteristics never reference datums? Why?

20–26 Answer the following questions regarding material condition modifiers.

(*a*) What are the three material condition modifiers?

(*b*) Which one is the default if nothing is specified?

(*c*) Which one(s) can provide "bonus" tolerance?

(*d*) Which of the following is increased by a bonus tolerance? (Select one.)

i. A size dimension.

ii. A ± tolerance of a size dimension.

iii. A basic dimension locating a feature.

iv. A size of a tolerance zone controlling a feature.

(*e*) To which of the following can a material condition modifier symbol be applied? (Select one.)

i. A size dimension.

ii. A ± tolerance of a size dimension.

iii. A tolerance of a geometric characteristic controlling a feature of size.

iv. A tolerance of a geometric characteristic controlling any feature.

(*f*) Which material condition modifier should be considered if the goal is to ensure a minimum clearance fit for a bolt in a hole, but to give greater manufacturing flexibility if the hole is produced with a greater clearance?

(*g*) Which material condition modifier should be considered if the goal is to provide a consistent press fit between interchangeable parts?

(*h*) Which material condition modifier should be considered if the goal is to ensure a minimum wall thickness for a casting, but to give greater manufacturing flexibility if the wall is produced with a greater thickness?

20–27 The drawing shown is of a mounting fixture to locate and orient a rod (not shown) through the large bore. The fixture will be bolted to a frame through the four bolt holes that are countersunk to recess the bolt heads. The bolt holes have too much clearance to properly align the rod, so the fixture will be aligned with two locating pins in the frame that will fit in the Ø6 hole and slot.

(*a*) Determine the minimum diameter allowed for the countersink.

(*b*) Determine the maximum depth allowed for the countersink.

(*c*) Determine the diameter of the bolt holes at MMC.

(*d*) Identify every feature that qualifies as a feature of size.

(*e*) The width of the base is specified with a basic dimension of 60, with no tolerance. (Note that as a feature of size, it could have had a tolerance directly specified.) What are the minimum and maximum allowed dimensions for the base width? Explain how they are determined.

(*f*) Describe the datum features A, B, and C. Describe their corresponding datums. Describe the datum reference frame that is defined by applying A, B, and C in that order. Describe how the part is stabilized by these datums. Explain why this is more appropriate for this application than using the edges of the base for datums B and C. (Notice that the basic dimensions are either measured from, or implied to be centered on, the datums of the datum reference frame.)

(*g*) If datum feature B is produced with a diameter of Ø6.00, what is the diameter of the tolerance zone in which its axis must lie? What if it is produced at Ø6.05?

(*h*) If the bolt holes are produced at Ø6.0, what is the diameter of the tolerance zones locating the bolt hole pattern with respect to the true position specified by the basic dimensions? What if the bolt holes are produced at Ø6.1?

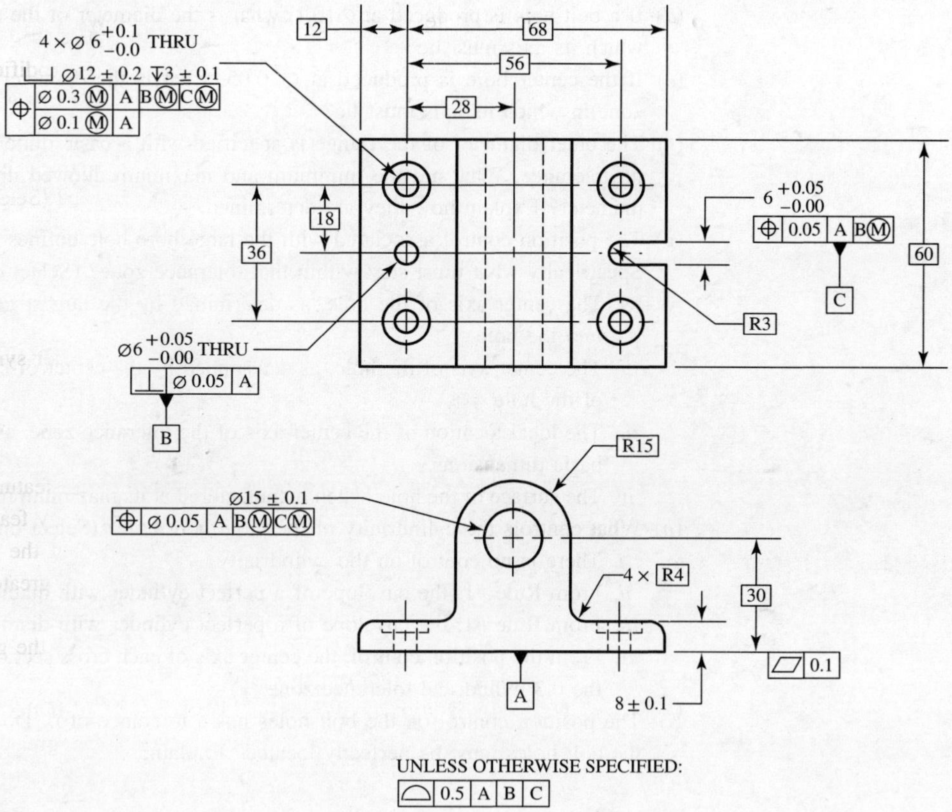

Problem 20–27

(*i*) If the bolt holes are produced at Ø6.0, what is the diameter of the tolerance zones locating the position of the bolt holes with respect to one another? What if the bolt holes are produced at Ø6.1?

(*j*) Explain why the Ⓜ modifier is appropriate for the bolt hole position tolerance.

(*k*) For the large bore, explain what provides control of each of the following: orientation, straightness of its center axis, and cylindricity of its surface.

(*l*) Assume the part is cast, and the casting operation can provide a surface profile tolerance of less than 0.5. Which surfaces can likely be left in the as-cast condition without compromising any of the requirements of the drawing? How would this change if the drawing were modified to use the edges of the base as datum features B and C, while still maintaining the functional goals for the alignment of the rod?

20–28 Answer the following questions for the drawing shown in Figure 20–34.

(*a*) Which of the following is *datum feature* B? (Select one.)

i. The surface of the bore hole.

ii. The center axis of the bore hole.

iii. The center axis of the largest gauge pin that will fit into the bore hole.

iv. The outer surface of the protruding cylinder.

v. The center axis of the outer surface of the protruding cylinder.

(*b*) What is the maximum diameter allowed for a bolt hole?

(*c*) What is the diameter of a bolt hole at its maximum material condition (MMC)?

(*d*) If a bolt hole is produced at ⌀10.1, what is the diameter of the tolerance zone in which its axis must lie?

(*e*) If the center bore is produced at ⌀20.05, what is the diameter of the tolerance zone in which its axis must lie?

(*f*) The outer diameter of the flange is specified with a basic dimension of 90, with no tolerance. What are the minimum and maximum allowed dimensions for the diameter? Explain how they are determined.

(*g*) The position control associated with the large bore hole defines a tolerance zone. Specifically what must stay within that tolerance zone? (Select one.)

i. The center axis of the hole, as determined by the largest gauge pin that fits into the hole.

ii. The center axis of the hole, as determined by the center of each cross section of the hole.

iii. The ideal location of the center axis of the tolerance zone, as specified by the basic dimensions.

iv. The surface of the hole, when manufactured at its maximum material condition.

(*h*) What controls the cylindricity of the large bore hole? (Select one.)

i. There is no control on the cylindricity.

ii. From Rule #1, the envelope of a perfect cylinder with diameter of 20.05.

iii. From Rule #1, the envelope of a perfect cylinder with diameter of 19.95.

iv. From the position control, the center axis of each cross section must be within the 0.3 cylindrical tolerance zone.

(*i*) The position control on the bolt holes has a tolerance of 0. Does this mean that the bolt holes must be perfectly located? Explain.

Useful Tables Appendix A

Appendix Outline

Table A–1 Standard SI Prefixes*†

Name	Symbol	Factor
exa	E	1 000 000 000 000 000 000 = 10^{18}
peta	P	1 000 000 000 000 000 = 10^{15}
tera	T	1 000 000 000 000 = 10^{12}
giga	G	1 000 000 000 = 10^{9}
mega	M	1 000 000 = 10^{6}
kilo	k	1 000 = 10^{3}
hecto‡	h	100 = 10^{2}
deka‡	da	10 = 10^{1}
deci‡	d	0.1 = 10^{-1}
centi‡	c	0.01 = 10^{-2}
milli	m	0.001 = 10^{-3}
micro	μ	0.000 001 = 10^{-6}
nano	n	0.000 000 001 = 10^{-9}
pico	p	0.000 000 000 001 = 10^{-12}
femto	f	0.000 000 000 000 001 = 10^{-15}
atto	a	0.000 000 000 000 000 001 = 10^{-18}

*If possible use multiple and submultiple prefixes in steps of 1000.

†Spaces are used in SI instead of commas to group numbers to avoid confusion with the practice in some European countries of using commas for decimal points.

‡Not recommended but sometimes encountered.

Table A–2 Conversion Factors A to Convert Input X to Output Y Using the Formula $Y = AX$*

Multiply Input X	By Factor A	To Get Output Y	Multiply Input X	By Factor A	To Get Output Y
British thermal unit, Btu	1055	joule, J	mass, lbf · s^2/in	175	kilogram, kg
Btu/second, Btu/s	1.05	kilowatt, kW	mile, mi	1.610	kilometer, km
calorie	4.19	joule, J	mile/hour, mi/h	1.61	kilometer/hour, km/h
centimeter of mercury (0°C)	1.333	kilopascal, kPa	mile/hour, mi/h	0.447	meter/second, m/s
centipoise, cP	0.001	pascal-second, Pa · s	moment of inertia, lbm · ft^2	0.0421	kilogram-meter^2, kg · m^2
degree (angle)	0.0174	radian, rad	moment of inertia, lbm · in^2	293	kilogram-millimeter^2, kg · mm^2
foot, ft	0.305	meter, m	moment of section (second moment of area), in^4	41.6	centimeter^4, cm^4
foot^2, ft^2	0.0929	meter^2, m^2	ounce-force, oz	0.278	newton, N
foot/minute, ft/min	0.0051	meter/second, m/s	ounce-mass	0.0311	kilogram, kg
foot-pound, ft · lbf	1.35	joule, J	pound, lbf†	4.45	newton, N
foot-pound/second, ft · lbf/s	1.35	watt, W	pound-foot, lbf · ft	1.36	newton-meter, N · m
foot/second, ft/s	0.305	meter/second, m/s	pound/foot^2, lbf/ft^2	47.9	pascal, Pa
gallon (U.S.), gal	3.785	liter, L	pound-inch, lbf · in	0.113	newton-meter, N · m
horsepower, hp	0.746	kilowatt, kW	pound/inch, lbf/in	175	newton/meter, N/m
inch, in	0.0254	meter, m	pound/inch^2, psi (lbf/in^2)	6.89	kilopascal, kPa
inch, in	25.4	millimeter, mm	pound-mass, lbm	0.454	kilogram, kg
inch^2, in^2	645	millimeter^2, mm^2	pound-mass/second, lbm/s	0.454	kilogram/second, kg/s
inch of mercury (32°F)	3.386	kilopascal, kPa	quart (U.S. liquid), qt	946	milliliter, mL
inch-pound, in · lbf	0.113	joule, J	section modulus, in^3	16.4	centimeter^3, cm^3
kilopound, kip	4.45	kilonewton, kN	slug	14.6	kilogram, kg
kilopound/inch^2, kpsi (ksi)	6.89	megapascal, MPa (N/mm^2)	ton (short 2000 lbm)	907	kilogram, kg
			yard, yd	0.914	meter, m

*Approximate.

†The U.S. Customary system unit of the pound-force is often abbreviated as lbf to distinguish it from the pound-mass, which is abbreviated as lbm.

Table A–3 Optional SI Units for Bending Stress $\sigma = Mc/I$, Torsion Stress $\tau = Tr/J$, Axial Stress $\sigma = F/A$, and Direct Shear Stress $\tau = F/A$

Bending and Torsion				Axial and Direct Shear		
M, T	I, J	c, r	σ, τ	F	A	σ, τ
N · m*	m^4	m	Pa	N*	m^2	Pa
N · m	cm^4	cm	MPa (N/mm^2)	N†	mm^2	MPa (N/mm^2)
N · m†	mm^4	mm	GPa	kN	m^2	kPa
kN · m	cm^4	cm	GPa	kN†	mm^2	GPa
N · mm†	mm^4	mm	MPa (N/mm^2)			

*Basic relation.

†Often preferred.

Table A–4 Optional SI Units for Bending Deflection $y = f(Fl^3/EI)$ or $y = f(wl^4/EI)$ and Torsional Deflection $\theta = Tl/GJ$

Bending Deflection					Torsional Deflection				
F, wl	l	I	E	y	T	l	J	G	θ
N*	m	m^4	Pa	m	N · m*	m	m^4	Pa	rad
kN†	mm	mm^4	GPa	mm	N · m†	mm	mm^4	GPa	rad
kN	m	m^4	Gpa	μm	N · mm	mm	mm^4	MPa (N/mm^2)	rad
N	mm	mm^4	kPa	m	N · m	cm	cm^4	MPa (N/mm^2)	rad

*Basic relation.

†Often preferred.

Table A–5 Physical Constants of Materials

Material	Modulus of Elasticity E GPa	Modulus of Rigidity G GPa	Poisson's Ratio ν	Unit Weight w kN/m^3
Aluminum (all alloys)	71.7	26.9	0.333	26.6
Beryllium copper	124.0	48.3	0.285	80.6
Brass	106.0	40.1	0.324	83.8
Carbon steel	207.0	79.3	0.292	76.5
Cast iron (gray)	100.0	41.4	0.211	70.6
Copper	119.0	44.7	0.326	87.3
Douglas fir	11.0	4.1	0.33	4.3
Glass	46.2	18.6	0.245	25.4
Inconel	214.0	75.8	0.290	83.3
Lead	36.5	13.1	0.425	111.5
Magnesium	44.8	16.5	0.350	17.6
Molybdenum	331.0	117.0	0.307	100.0
Monel metal	179.0	65.5	0.320	86.6
Nickel silver	127.0	48.3	0.322	85.8
Nickel steel	207.0	79.3	0.291	76.0
Phosphor bronze	111.0	41.4	0.349	80.1
Stainless steel (18-8)	190.0	73.1	0.305	76.0
Titanium alloys	114.0	42.4	0.340	43.4

Table A–6 Properties of Structural-Steel Equal Legs Angles*†

m = mass per meter, kg/m
A = area, cm^2
I = second moment of area, cm^4
k = radius of gyration, cm
y = centroidal distance, cm
Z = section modulus, cm^3

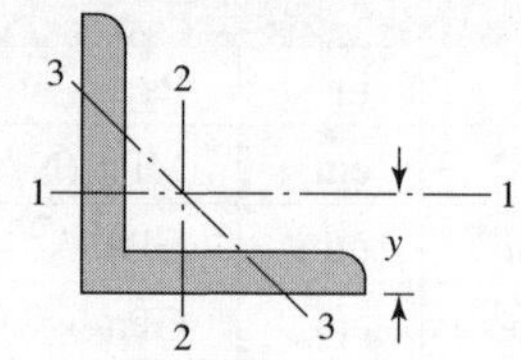

Table A–6 Properties of Structural-Steel Equal Legs Angles*†

Size, mm	m	A	I_{1-1}	k_{1-1}	Z_{1-1}	y	k_{3-3}
25 × 25 × 3	1.11	1.42	0.80	0.75	0.45	0.72	0.48
× 4	1.45	1.85	1.01	0.74	0.58	0.76	0.48
× 5	1.77	2.26	1.20	0.73	0.71	0.80	0.48
40 × 40 × 4	2.42	3.08	4.47	1.21	1.55	1.12	0.78
× 5	2.97	3.79	5.43	1.20	1.91	1.16	0.77
× 6	3.52	4.48	6.31	1.19	2.26	1.20	0.77
50 × 50 × 5	3.77	4.80	11.0	1.51	3.05	1.40	0.97
× 6	4.47	5.59	12.8	1.50	3.61	1.45	0.97
× 8	5.82	7.41	16.3	1.48	4.68	1.52	0.96
60 × 60 × 5	4.57	5.82	19.4	1.82	4.45	1.64	1.17
× 6	5.42	6.91	22.8	1.82	5.29	1.69	1.17
× 8	7.09	9.03	29.2	1.80	6.89	1.77	1.16
× 10	8.69	11.1	34.9	1.78	8.41	1.85	1.16
80 × 80 × 6	7.34	9.35	55.8	2.44	9.57	2.17	1.57
× 8	9.63	12.3	72.2	2.43	12.6	2.26	1.56
× 10	11.9	15.1	87.5	2.41	15.4	2.34	1.55
100 × 100 × 8	12.2	15.5	145	3.06	19.9	2.74	1.96
× 12	17.8	22.7	207	3.02	29.1	2.90	1.94
× 15	21.9	27.9	249	2.98	35.6	3.02	1.93
150 × 150 × 10	23.0	29.3	624	4.62	56.9	4.03	2.97
× 12	27.3	34.8	737	4.60	67.7	4.12	2.95
× 15	33.8	43.0	898	4.57	83.5	4.25	2.93
× 18	40.1	51.0	1050	4.54	98.7	4.37	2.92

*Metric sizes also available in sizes of 45, 70, 90, 120, and 200 mm.

†These sizes are also available in aluminum alloy.

Table A–7 Properties of Structural-Steel Channels*

a, b = size, mm
m = mass per meter, kg/m
t = web thickness, mm
A = area, cm^2
I = second moment of area, cm^4
k = radius of gyration, cm
x = centroidal distance, cm
Z = section modulus, cm^3

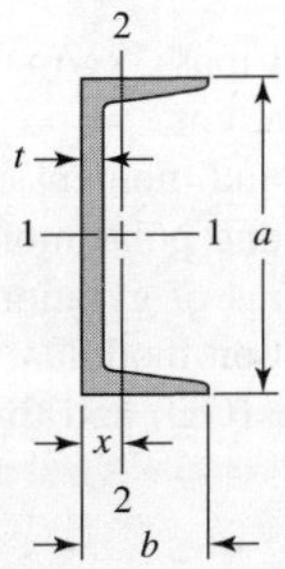

Table A–7 Properties of Structural-Steel Channels*

$a \times b$, mm	m	t	A	I_{1-1}	k_{1-1}	Z_{1-1}	I_{2-2}	k_{2-2}	Z_{2-2}	x
76 × 38	6.70	5.1	8.53	74.14	2.95	19.46	10.66	1.12	4.07	1.19
102 × 51	10.42	6.1	13.28	207.7	3.95	40.89	29.10	1.48	8.16	1.51
127 × 64	14.90	6.4	18.98	482.5	5.04	75.99	67.23	1.88	15.25	1.94
152 × 76	17.88	6.4	22.77	851.5	6.12	111.8	113.8	2.24	21.05	2.21
152 × 89	23.84	7.1	30.36	1166	6.20	153.0	215.1	2.66	35.70	2.86
178 × 76	20.84	6.6	26.54	1337	7.10	150.4	134.0	2.25	24.72	2.20
178 × 89	26.81	7.6	34.15	1753	7.16	197.2	241.0	2.66	39.29	2.76
203 × 76	23.82	7.1	30.34	1950	8.02	192.0	151.3	2.23	27.59	2.13
203 × 89	29.78	8.1	37.94	2491	8.10	245.2	264.4	2.64	42.34	2.65
229 × 76	26.06	7.6	33.20	2610	8.87	228.3	158.7	2.19	28.22	2.00
229 × 89	32.76	8.6	41.73	3387	9.01	296.4	285.0	2.61	44.82	2.53
254 × 76	28.29	8.1	36.03	3367	9.67	265.1	162.6	2.12	28.21	1.86
254 × 89	35.74	9.1	45.42	4448	9.88	350.2	302.4	2.58	46.70	2.42
305 × 89	41.69	10.2	53.11	7061	11.5	463.3	325.4	2.48	48.49	2.18
305 × 102	46.18	10.2	58.83	8214	11.8	539.0	499.5	2.91	66.59	2.66

*These sizes are also available in aluminum alloy.

Table A–8 Properties of Round Tubing

m = unit mass, kg/m
A = area, cm^2
I = second moment of area, cm^4
J = second polar moment of area, cm^4
k = radius of gyration, cm
Z = section modulus, cm^3
$d \times t$ = size (OD) and thickness, mm

Size, mm	m	A	I	k	Z	J
12 × 2	0.490	0.628	0.082	0.361	0.136	0.163
16 × 2	0.687	0.879	0.220	0.500	0.275	0.440
16 × 3	0.956	1.225	0.273	0.472	0.341	0.545
20 × 4	1.569	2.010	0.684	0.583	0.684	1.367
25 × 4	2.060	2.638	1.508	0.756	1.206	3.015
25 × 5	2.452	3.140	1.669	0.729	1.336	3.338
30 × 4	2.550	3.266	2.827	0.930	1.885	5.652
30 × 5	3.065	3.925	3.192	0.901	2.128	6.381
42 × 4	3.727	4.773	8.717	1.351	4.151	17.430
42 × 5	4.536	5.809	10.130	1.320	4.825	20.255
50 × 4	4.512	5.778	15.409	1.632	6.164	30.810
50 × 5	5.517	7.065	18.118	1.601	7.247	36.226

Table A–9 Shear, Moment, and Deflection of Beams

(*Note:* Force and moment reactions are positive in the directions shown; equations for shear force V and bending moment M follow the sign conventions given in Sec. 3–2.)

1 Cantilever—end load

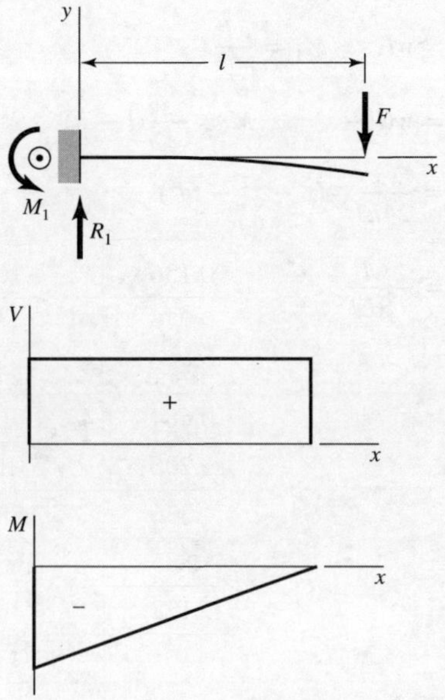

$$R_1 = V = F \qquad M_1 = Fl$$

$$M = F(x - l)$$

$$y = \frac{Fx^2}{6EI}(x - 3l)$$

$$y_{\max} = -\frac{Fl^3}{3EI}$$

2 Cantilever—intermediate load

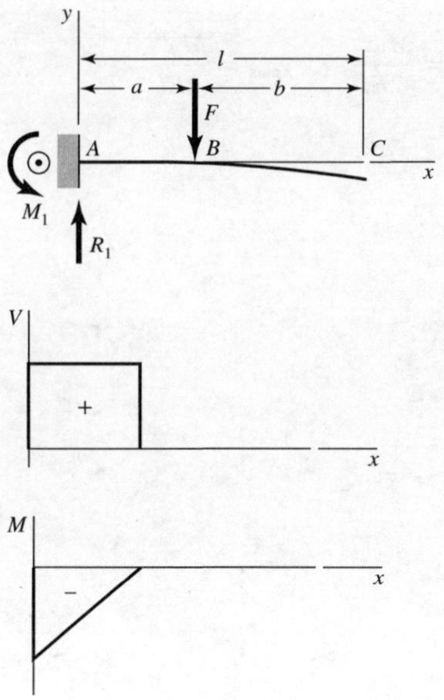

$$R_1 = V = F \qquad M_1 = Fa$$

$$M_{AB} = F(x - a) \qquad M_{BC} = 0$$

$$y_{AB} = \frac{Fx^2}{6EI}(x - 3a)$$

$$y_{BC} = \frac{Fa^2}{6EI}(a - 3x)$$

$$y_{\max} = \frac{Fa^2}{6EI}(a - 3l)$$

(*Continued*)

Table A–9 Shear, Moment, and Deflection of Beams ***(Continued)***

(*Note:* Force and moment reactions are positive in the directions shown; equations for shear force V and bending moment M follow the sign conventions given in Sec. 3–2.)

3 Cantilever—uniform load

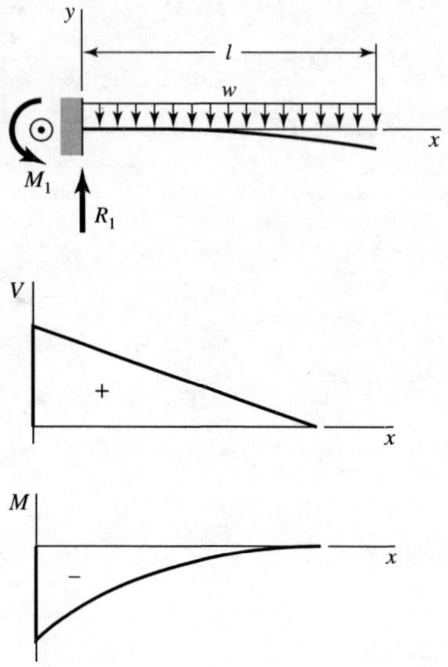

$$R_1 = wl \qquad M_1 = \frac{wl^2}{2}$$

$$V = w(l - x) \qquad M = -\frac{w}{2}(l - x)^2$$

$$y = \frac{wx^2}{24EI}(4lx - x^2 - 6l^2)$$

$$y_{max} = -\frac{wl^4}{8EI}$$

4 Cantilever—moment load

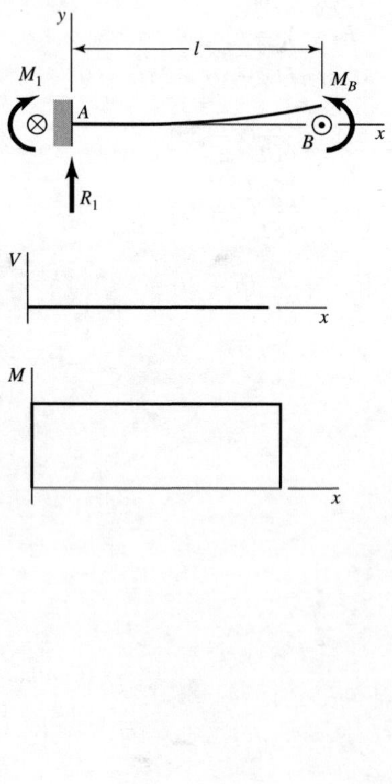

$$R_1 = V = 0 \qquad M_1 = M = M_B$$

$$y = \frac{M_B x^2}{2EI} \qquad y_{max} = \frac{M_B l^2}{2EI}$$

Table A–9 Shear, Moment, and Deflection of Beams *(Continued)*

(*Note:* Force and moment reactions are positive in the directions shown; equations for shear force V and bending moment M follow the sign conventions given in Sec. 3–2.)

5 Simple supports—center load

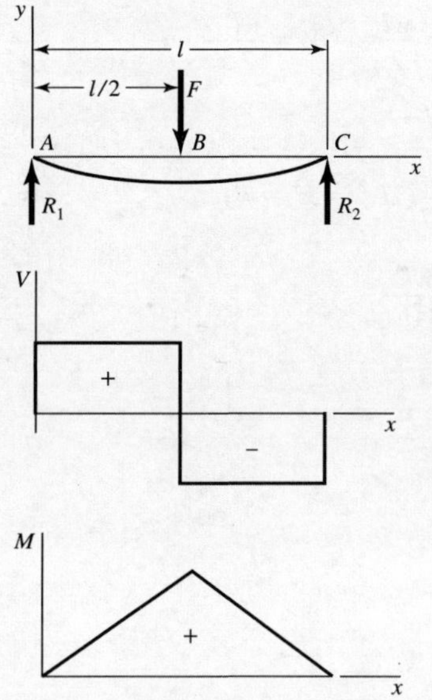

$$R_1 = R_2 = \frac{F}{2}$$

$$V_{AB} = R_1 \qquad V_{BC} = -R_2$$

$$M_{AB} = \frac{Fx}{2} \qquad M_{BC} = \frac{F}{2}(l - x)$$

$$y_{AB} = \frac{Fx}{48EI}(4x^2 - 3l^2)$$

$$y_{max} = -\frac{Fl^3}{48EI}$$

6 Simple supports—intermediate load

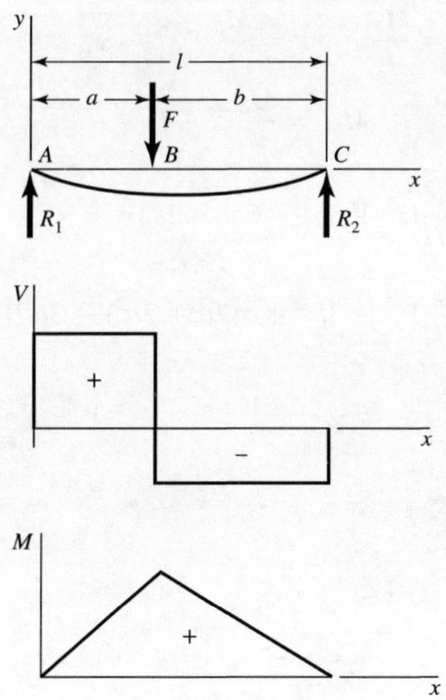

$$R_1 = \frac{Fb}{l} \qquad R_2 = \frac{Fa}{l}$$

$$V_{AB} = R_1 \qquad V_{BC} = -R_2$$

$$M_{AB} = \frac{Fbx}{l} \qquad M_{BC} = \frac{Fa}{l}(l - x)$$

$$y_{AB} = \frac{Fbx}{6EIl}(x^2 + b^2 - l^2)$$

$$y_{BC} = \frac{Fa(l - x)}{6EIl}(x^2 + a^2 - 2lx)$$

(*Continued*)

Table A–9 Shear, Moment, and Deflection of Beams (*Continued*)

(*Note:* Force and moment reactions are positive in the directions shown; equations for shear force V and bending moment M follow the sign conventions given in Sec. 3–2.)

7 Simple supports—uniform load

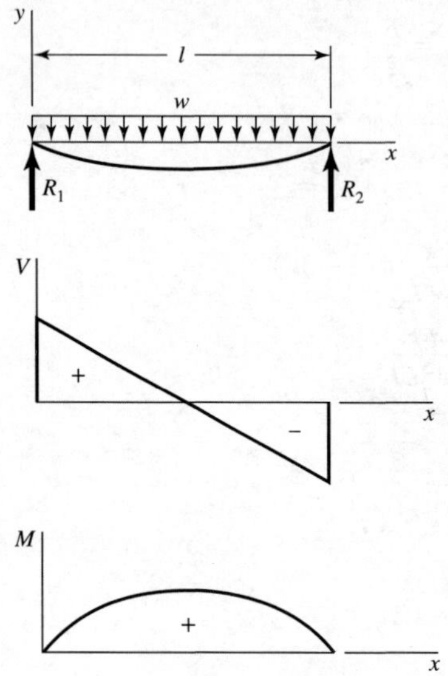

$$R_1 = R_2 = \frac{wl}{2} \qquad V = \frac{wl}{2} - wx$$

$$M = \frac{wx}{2}(l - x)$$

$$y = \frac{wx}{24EI}(2lx^2 - x^3 - l^3)$$

$$y_{max} = -\frac{5wl^4}{384EI}$$

8 Simple supports—moment load

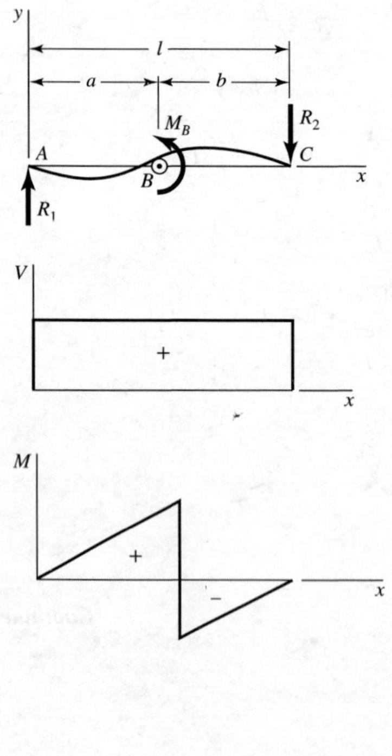

$$R_1 = R_2 = \frac{M_B}{l} \qquad V = \frac{M_B}{l}$$

$$M_{AB} = \frac{M_B x}{l} \qquad M_{BC} = \frac{M_B}{l}(x - l)$$

$$y_{AB} = \frac{M_B x}{6EIl}(x^2 + 3a^2 - 6al + 2l^2)$$

$$y_{BC} = \frac{M_B}{6EIl}[x^3 - 3lx^2 + x(2l^2 + 3a^2) - 3a^2 l]$$

Table A–9 Shear, Moment, and Deflection of Beams (*Continued*)

(*Note:* Force and moment reactions are positive in the directions shown; equations for shear force V and bending moment M follow the sign conventions given in Sec. 3–2.)

9 Simple supports—twin loads

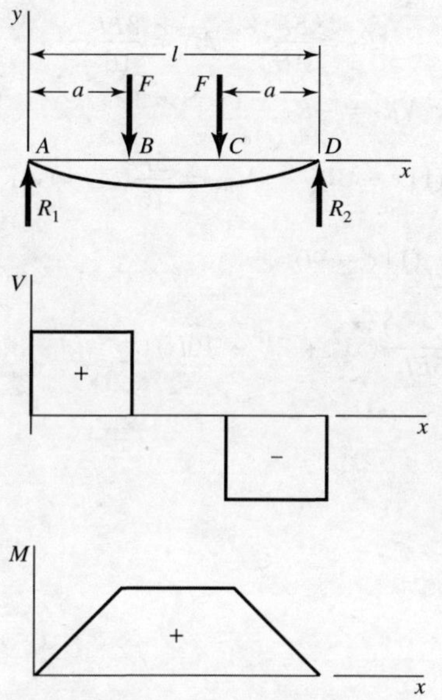

$$R_1 = R_2 = F \qquad V_{AB} = F \qquad V_{BC} = 0$$

$$V_{CD} = -F$$

$$M_{AB} = Fx \qquad M_{BC} = Fa \qquad M_{CD} = F(l - x)$$

$$y_{AB} = \frac{Fx}{6EI}(x^2 + 3a^2 - 3la)$$

$$y_{BC} = \frac{Fa}{6EI}(3x^2 + a^2 - 3lx)$$

$$y_{\max} = \frac{Fa}{24EI}(4a^2 - 3l^2)$$

10 Simple supports—overhanging load

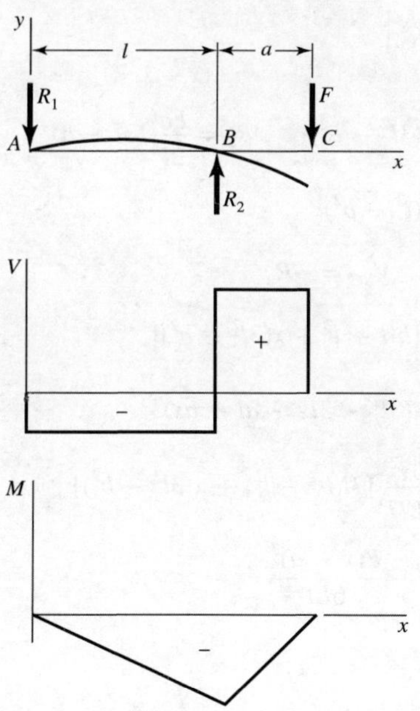

$$R_1 = \frac{Fa}{l} \qquad R_2 = \frac{F}{l}(l + a)$$

$$V_{AB} = -\frac{Fa}{l} \qquad V_{BC} = F$$

$$M_{AB} = -\frac{Fax}{l} \qquad M_{BC} = F(x - l - a)$$

$$y_{AB} = \frac{Fax}{6EIl}(l^2 - x^2)$$

$$y_{BC} = \frac{F(x - l)}{6EI}[(x - l)^2 - a(3x - l)]$$

$$y_C = -\frac{Fa^2}{3EI}(l + a)$$

(*Continued*)

Table A–9 Shear, Moment, and Deflection of Beams (*Continued*)

(*Note:* Force and moment reactions are positive in the directions shown; equations for shear force V and bending moment M follow the sign conventions given in Sec. 3–2.)

11 One fixed and one simple support—center load

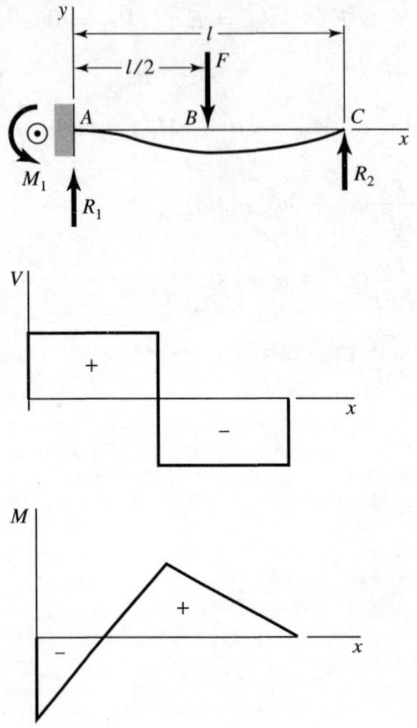

$$R_1 = \frac{11F}{16} \qquad R_2 = \frac{5F}{16} \qquad M_1 = \frac{3Fl}{16}$$

$$V_{AB} = R_1 \qquad V_{BC} = -R_2$$

$$M_{AB} = \frac{F}{16}(11x - 3l) \qquad M_{BC} = \frac{5F}{16}(l - x)$$

$$y_{AB} = \frac{Fx^2}{96EI}(11x - 9l)$$

$$y_{BC} = \frac{F(l - x)}{96EI}(5x^2 + 2l^2 - 10lx)$$

12 One fixed and one simple support—intermediate load

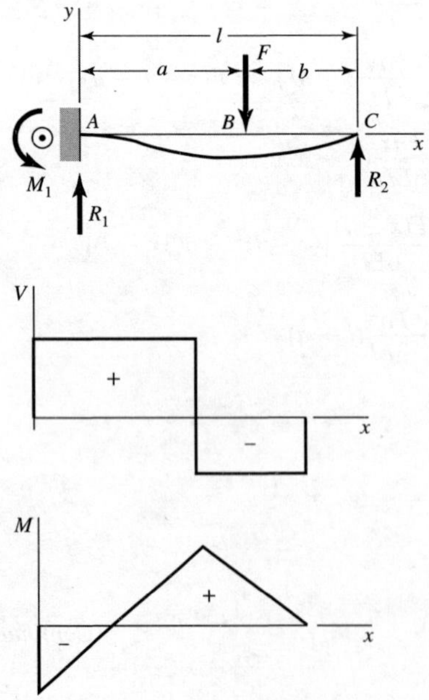

$$R_1 = \frac{Fb}{2l^3}(3l^2 - b^2) \qquad R_2 = \frac{Fa^2}{2l^3}(3l - a)$$

$$M_1 = \frac{Fb}{2l^2}(l^2 - b^2)$$

$$V_{AB} = R_1 \qquad V_{BC} = -R_2$$

$$M_{AB} = \frac{Fb}{2l^3}[b^2 l - l^3 + x(3l^2 - b^2)]$$

$$M_{BC} = \frac{Fa^2}{2l^3}(3l^2 - 3lx - al + ax)$$

$$y_{AB} = \frac{Fbx^2}{12EIl^3}[3l(b^2 - l^2) + x(3l^2 - b^2)]$$

$$y_{BC} = y_{AB} - \frac{F(x - a)^3}{6EI}$$

Table A–9 Shear, Moment, and Deflection of Beams *(Continued)*

(*Note:* Force and moment reactions are positive in the directions shown; equations for shear force V and bending moment M follow the sign conventions given in Sec. 3–2.)

13 One fixed and one simple support—uniform load

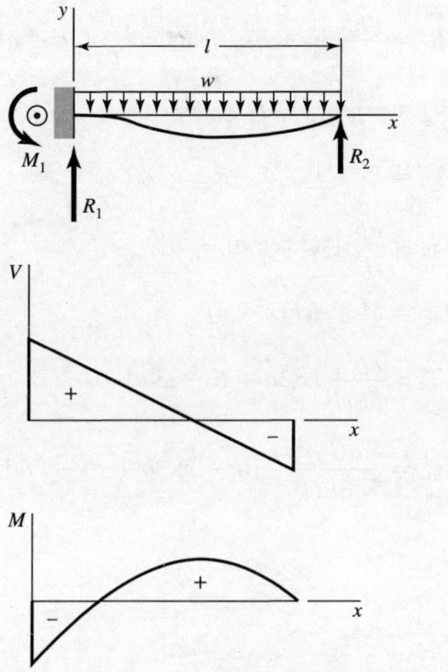

$$R_1 = \frac{5wl}{8} \qquad R_2 = \frac{3wl}{8} \qquad M_1 = \frac{wl^2}{8}$$

$$V = \frac{5wl}{8} - wx$$

$$M = -\frac{w}{8}(4x^2 - 5lx + l^2)$$

$$y = \frac{wx^2}{48EI}(l - x)(2x - 3l)$$

14 Fixed supports—center load

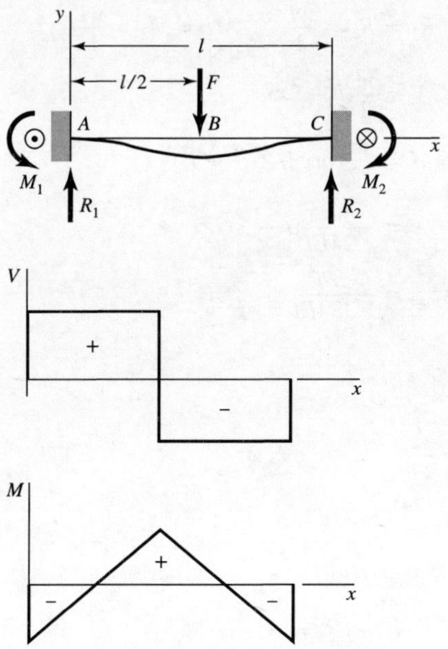

$$R_1 = R_2 = \frac{F}{2} \qquad M_1 = M_2 = \frac{Fl}{8}$$

$$V_{AB} = -V_{BC} = \frac{F}{2}$$

$$M_{AB} = \frac{F}{8}(4x - l) \qquad M_{BC} = \frac{F}{8}(3l - 4x)$$

$$y_{AB} = \frac{Fx^2}{48EI}(4x - 3l)$$

$$y_{max} = -\frac{Fl^3}{192EI}$$

(Continued)

Table A–9 Shear, Moment, and Deflection of Beams (*Continued*)

(*Note:* Force and moment reactions are positive in the directions shown; equations for shear force V and bending moment M follow the sign conventions given in Sec. 3–2.)

15 Fixed supports—intermediate load

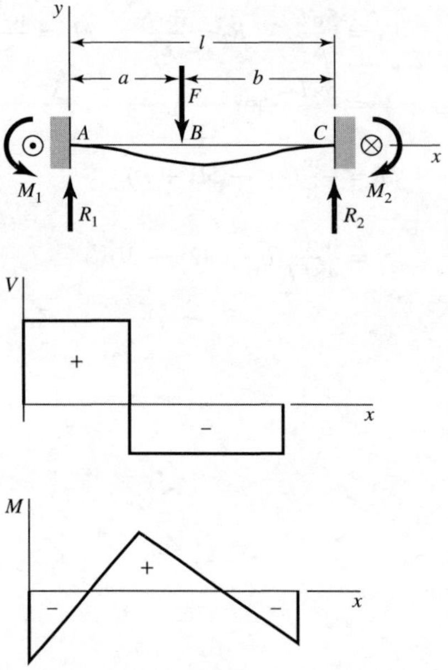

$$R_1 = \frac{Fb^2}{l^3}(3a + b) \qquad R_2 = \frac{Fa^2}{l^3}(3b + a)$$

$$M_1 = \frac{Fab^2}{l^2} \qquad M_2 = \frac{Fa^2b}{l^2}$$

$$V_{AB} = R_1 \qquad V_{BC} = -R_2$$

$$M_{AB} = \frac{Fb^2}{l^3}[x(3a + b) - al]$$

$$M_{BC} = M_{AB} - F(x - a)$$

$$y_{AB} = \frac{Fb^2x^2}{6EIl^3}[x(3a + b) - 3al]$$

$$y_{BC} = \frac{Fa^2(l - x)^2}{6EIl^3}[(l - x)(3b + a) - 3bl]$$

16 Fixed supports—uniform load

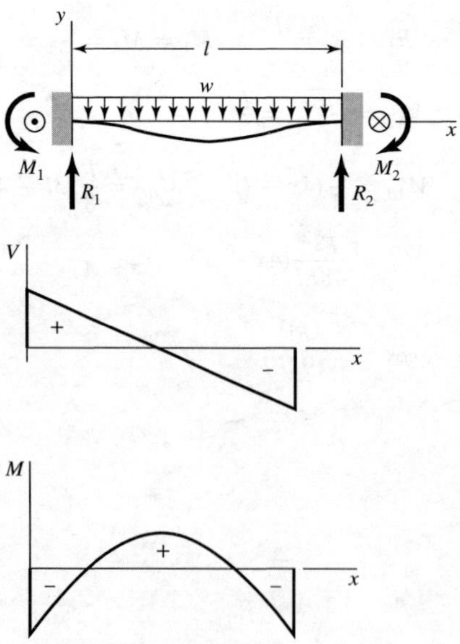

$$R_1 = R_2 = \frac{wl}{2} \qquad M_1 = M_2 = \frac{wl^2}{12}$$

$$V = \frac{w}{2}(l - 2x)$$

$$M - \frac{w}{12}(6lx - 6x^2 - l^2)$$

$$y = -\frac{wx^2}{24EI}(l - x)^2$$

$$y_{max} = -\frac{wl^4}{384EI}$$

Table A–10 Cumulative Distribution Function of Normal (Gaussian) Distribution

$$\Phi(z_\alpha) = \int_{-\infty}^{z_\alpha} \frac{1}{\sqrt{2\pi}} \exp\left(-\frac{u^2}{2}\right) du$$

$$= \begin{cases} \alpha & z_\alpha \le 0 \\ 1 - \alpha & z_\alpha > 0 \end{cases}$$

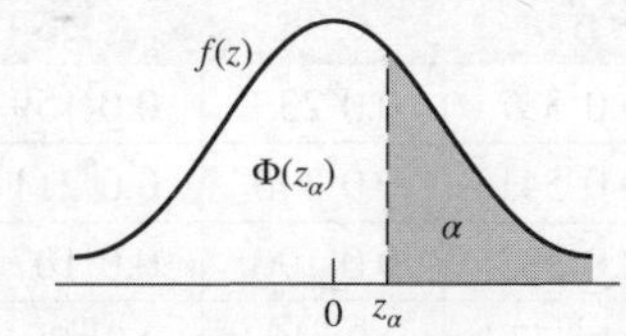

z_α	0.00	0.01	0.02	0.03	0.04	0.05	0.06	0.07	0.08	0.09
0.0	0.5000	0.4960	0.4920	0.4880	0.4840	0.4801	0.4761	0.4721	0.4681	0.4641
0.1	0.4602	0.4562	0.4522	0.4483	0.4443	0.4404	0.4364	0.4325	0.4286	0.4247
0.2	0.4207	0.4168	0.4129	0.4090	0.4052	0.4013	0.3974	0.3936	0.3897	0.3859
0.3	0.3821	0.3783	0.3745	0.3707	0.3669	0.3632	0.3594	0.3557	0.3520	0.3483
0.4	0.3446	0.3409	0.3372	0.3336	0.3300	0.3264	0.3238	0.3192	0.3156	0.3121
0.5	0.3085	0.3050	0.3015	0.2981	0.2946	0.2912	0.2877	0.2843	0.2810	0.2776
0.6	0.2743	0.2709	0.2676	0.2643	0.2611	0.2578	0.2546	0.2514	0.2483	0.2451
0.7	0.2420	0.2389	0.2358	0.2327	0.2296	0.2266	0.2236	0.2206	0.2177	0.2148
0.8	0.2119	0.2090	0.2061	0.2033	0.2005	0.1977	0.1949	0.1922	0.1894	0.1867
0.9	0.1841	0.1814	0.1788	0.1762	0.1736	0.1711	0.1685	0.1660	0.1635	0.1611
1.0	0.1587	0.1562	0.1539	0.1515	0.1492	0.1469	0.1446	0.1423	0.1401	0.1379
1.1	0.1357	0.1335	0.1314	0.1292	0.1271	0.1251	0.1230	0.1210	0.1190	0.1170
1.2	0.1151	0.1131	0.1112	0.1093	0.1075	0.1056	0.1038	0.1020	0.1003	0.0985
1.3	0.0968	0.0951	0.0934	0.0918	0.0901	0.0885	0.0869	0.0853	0.0838	0.0823
1.4	0.0808	0.0793	0.0778	0.0764	0.0749	0.0735	0.0721	0.0708	0.0694	0.0681
1.5	0.0668	0.0655	0.0643	0.0630	0.0618	0.0606	0.0594	0.0582	0.0571	0.0559
1.6	0.0548	0.0537	0.0526	0.0516	0.0505	0.0495	0.0485	0.0475	0.0465	0.0455
1.7	0.0446	0.0436	0.0427	0.0418	0.0409	0.0401	0.0392	0.0384	0.0375	0.0367
1.8	0.0359	0.0351	0.0344	0.0336	0.0329	0.0322	0.0314	0.0307	0.0301	0.0294
1.9	0.0287	0.0281	0.0274	0.0268	0.0262	0.0256	0.0250	0.0244	0.0239	0.0233
2.0	0.0228	0.0222	0.0217	0.0212	0.0207	0.0202	0.0197	0.0192	0.0188	0.0183
2.1	0.0179	0.0174	0.0170	0.0166	0.0162	0.0158	0.0154	0.0150	0.0146	0.0143
2.2	0.0139	0.0136	0.0132	0.0129	0.0125	0.0122	0.0119	0.0116	0.0113	0.0110
2.3	0.0107	0.0104	0.0102	0.00990	0.00964	0.00939	0.00914	0.00889	0.00866	0.00842
2.4	0.00820	0.00798	0.00776	0.00755	0.00734	0.00714	0.00695	0.00676	0.00657	0.00639
2.5	0.00621	0.00604	0.00587	0.00570	0.00554	0.00539	0.00523	0.00508	0.00494	0.00480
2.6	0.00466	0.00453	0.00440	0.00427	0.00415	0.00402	0.00391	0.00379	0.00368	0.00357
2.7	0.00347	0.00336	0.00326	0.00317	0.00307	0.00298	0.00289	0.00280	0.00272	0.00264
2.8	0.00256	0.00248	0.00240	0.00233	0.00226	0.00219	0.00212	0.00205	0.00199	0.00193
2.9	0.00187	0.00181	0.00175	0.00169	0.00164	0.00159	0.00154	0.00149	0.00144	0.00139

(Continued)

Table A–10 Cumulative Distribution Function of Normal (Gaussian) Distribution* ***(Continued)***

Z_α	0.00	0.01	0.02	0.03	0.04	0.05	0.06	0.07	0.08	0.09
3	0.00135	0.0^3968	0.0^3687	0.0^3483	0.0^3337	0.0^3233	0.0^3159	0.0^3108	0.0^4723	0.0^4481
4	0.0^4317	0.0^4207	0.0^4133	0.0^5854	0.0^5541	0.0^5340	0.0^5211	0.0^5130	0.0^6793	0.0^6479
5	0.0^6287	0.0^6170	0.0^7996	0.0^7579	0.0^7333	0.0^7190	0.0^7107	0.0^8599	0.0^8332	0.0^8182
6	0.0^9987	0.0^9530	0.0^9282	0.0^9149	$0.0^{10}777$	$0.0^{10}402$	$0.0^{10}206$	$0.0^{10}104$	$0.0^{11}523$	$0.0^{11}260$

z_α	−1.282	−1.643	−1.960	−2.326	−2.576	−3.090	−3.291	−3.891	−4.417
$F(z_\alpha)$	0.10	0.05	0.025	0.010	0.005	0.001	0.0005	0.0001	0.000005
$R(z_\alpha)$	0.90	0.95	0.975	0.990	0.995	0.999	0.9995	0.9999	0.999995

*The superscript on a zero after the decimal point indicates how many zeros there are after the decimal point. For example, $0.0^4481 = 0.000\ 048\ 1$.

Table A–11 A Selection of International Tolerance Grades—Metric Series

(Size Ranges Are for *Over* the Lower Limit and *Including* the Upper Limit. All Values Are in Millimeters)

Basic Sizes	Tolerance Grades					
	IT6	IT7	IT8	IT9	IT10	IT11
0–3	0.006	0.010	0.014	0.025	0.040	0.060
3–6	0.008	0.012	0.018	0.030	0.048	0.075
6–10	0.009	0.015	0.022	0.036	0.058	0.090
10–18	0.011	0.018	0.027	0.043	0.070	0.110
18–30	0.013	0.021	0.033	0.052	0.084	0.130
30–50	0.016	0.025	0.039	0.062	0.100	0.160
50–80	0.019	0.030	0.046	0.074	0.120	0.190
80–120	0.022	0.035	0.054	0.087	0.140	0.220
120–180	0.025	0.040	0.063	0.100	0.160	0.250
180–250	0.029	0.046	0.072	0.115	0.185	0.290
250–315	0.032	0.052	0.081	0.130	0.210	0.320
315–400	0.036	0.057	0.089	0.140	0.230	0.360

Source: Preferred Metric Limits and Fits, ANSI B4.2-1978. See also BSI 4500.

Table A–12 Fundamental Deviations for Shafts—Metric Series

(Size Ranges Are for *Over* the Lower Limit and *Including* the Upper Limit. All Values Are in Millimeters)

Basic Sizes	Upper-Deviation Letter					Lower-Deviation Letter				
	c	d	f	g	h	k	n	p	s	u
0–3	−0.060	−0.020	−0.006	−0.002	0	0	+0.004	+0.006	+0.014	+0.018
3–6	−0.070	−0.030	−0.010	−0.004	0	+0.001	+0.008	+0.012	+0.019	+0.023
6–10	−0.080	−0.040	−0.013	−0.005	0	+0.001	+0.010	+0.015	+0.023	+0.028
10–14	−0.095	−0.050	−0.016	−0.006	0	+0.001	+0.012	+0.018	+0.028	+0.033
14–18	−0.095	−0.050	−0.016	−0.006	0	+0.001	+0.012	+0.018	+0.028	+0.033
18–24	−0.110	−0.065	−0.020	−0.007	0	+0.002	+0.015	+0.022	+0.035	+0.041
24–30	−0.110	−0.065	−0.020	−0.007	0	+0.002	+0.015	+0.022	+0.035	+0.048
30–40	−0.120	−0.080	−0.025	−0.009	0	+0.002	+0.017	+0.026	+0.043	+0.060
40–50	−0.130	−0.080	−0.025	−0.009	0	+0.002	+0.017	+0.026	+0.043	+0.070
50–65	−0.140	−0.100	−0.030	−0.010	0	+0.002	+0.020	+0.032	+0.053	+0.087
65–80	−0.150	−0.100	−0.030	−0.010	0	+0.002	+0.020	+0.032	+0.059	+0.102
80–100	−0.170	−0.120	−0.036	−0.012	0	+0.003	+0.023	+0.037	+0.071	+0.124
100–120	−0.180	−0.120	−0.036	−0.012	0	+0.003	+0.023	+0.037	+0.079	+0.144
120–140	−0.200	−0.145	−0.043	−0.014	0	+0.003	+0.027	+0.043	+0.092	+0.170
140–160	−0.210	−0.145	−0.043	−0.014	0	+0.003	+0.027	+0.043	+0.100	+0.190
160–180	−0.230	−0.145	−0.043	−0.014	0	+0.003	+0.027	+0.043	+0.108	+0.210
180–200	−0.240	−0.170	−0.050	−0.015	0	+0.004	+0.031	+0.050	+0.122	+0.236
200–225	−0.260	−0.170	−0.050	−0.015	0	+0.004	+0.031	+0.050	+0.130	+0.258
225–250	−0.280	−0.170	−0.050	−0.015	0	+0.004	+0.031	+0.050	+0.140	+0.284
250–280	−0.300	−0.190	−0.056	−0.017	0	+0.004	+0.034	+0.056	+0.158	+0.315
280–315	−0.330	−0.190	−0.056	−0.017	0	+0.004	+0.034	+0.056	+0.170	+0.350
315–355	−0.360	−0.210	−0.062	−0.018	0	+0.004	+0.037	+0.062	+0.190	+0.390
355–400	−0.400	−0.210	−0.062	−0.018	0	+0.004	+0.037	+0.062	+0.208	+0.435

Source: Preferred Metric Limits and Fits, ANSI B4.2-1978. See also BSI 4500.

Table A–13 Charts of Theoretical Stress-Concentration Factors K_t^*

Figure A–13–1

Bar in tension or simple compression with a transverse hole. $\sigma_0 = F/A$, where $A = (w - d)t$ and t is the thickness.

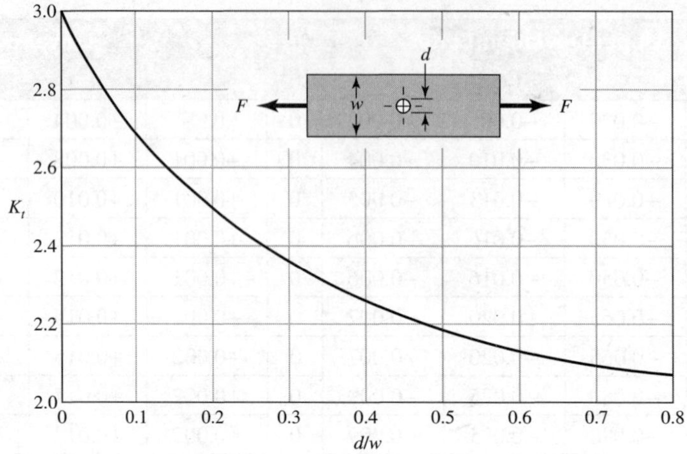

Figure A–13–2

Rectangular bar with a transverse hole in bending. $\sigma_0 = Mc/I$, where $I = (w - d)h^3/12$.

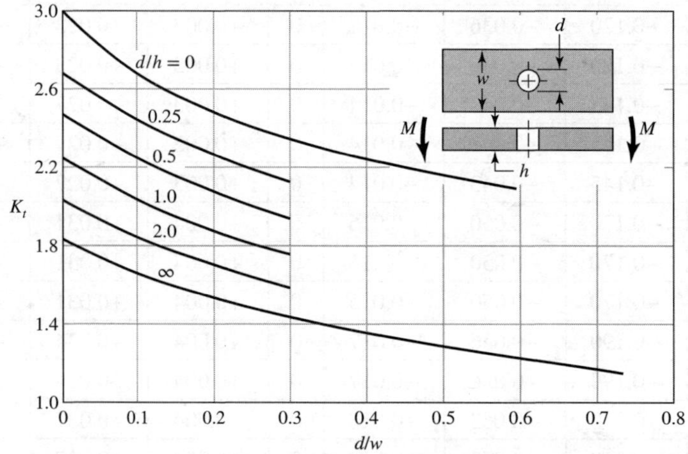

Figure A–13–3

Notched rectangular bar in tension or simple compression. $\sigma_0 = F/A$, where $A = dt$ and t is the thickness.

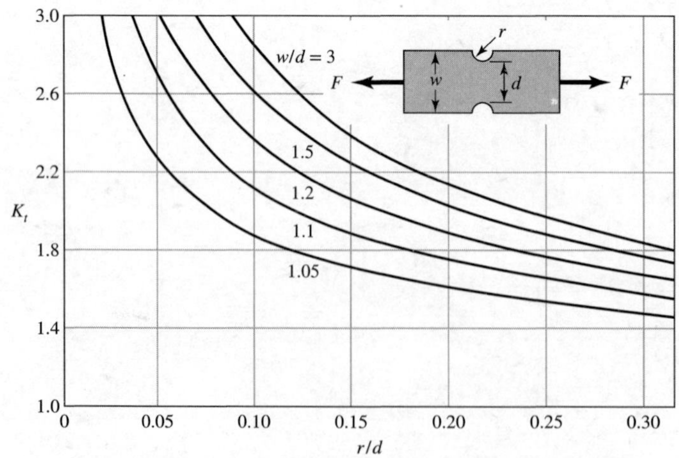

Table A–13 Charts of Theoretical Stress-Concentration Factors K_t^* (*Continued*)

Figure A–13–4

Notched rectangular bar in bending. $\sigma_0 = Mc/I$, where $c = d/2$, $I = td^3/12$, and t is the thickness.

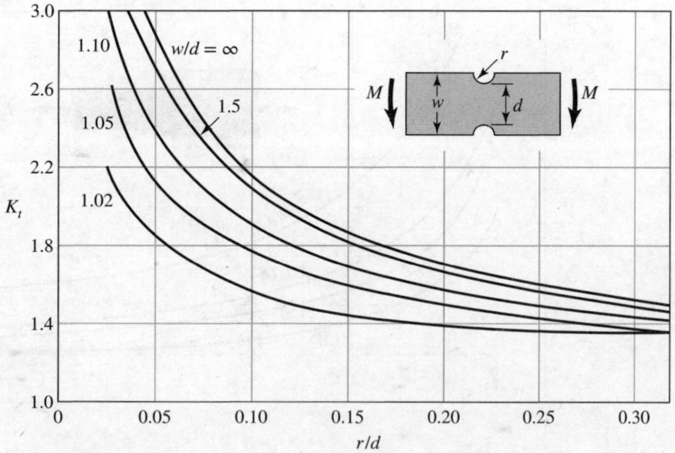

Figure A–13–5

Rectangular filleted bar in tension or simple compression. $\sigma_0 = F/A$, where $A = dt$ and t is the thickness.

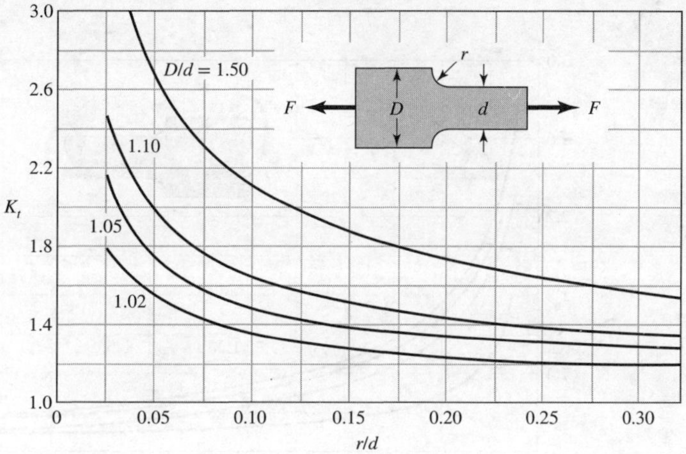

Figure A–13–6

Rectangular filleted bar in bending. $\sigma_0 = Mc/I$, where $c = d/2$, $I = td^3/12$, and t is the thickness.

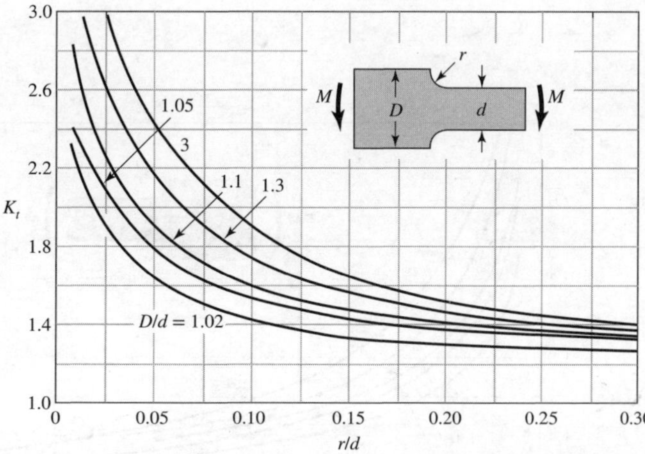

(*Continued*)

*Factors from Peterson, R. E. "Design Factors for Stress Concentration." *Machine Design 23*, no. 2 (February 1951): 169; no. 3, March 1951, p. 161; no. 5, May 1951, p. 159; no. 6, June 1951, p. 173; no. 7, July 1951, p. 155. Reprinted with permission from Machine Design, a Penton Media Inc. publication.

Table A–13 Charts of Theoretical Stress-Concentration Factors K_t^* (*Continued*)

Figure A–13–7

Round shaft with shoulder fillet in tension. $\sigma_0 = F/A$, where $A = \pi d^2/4$.

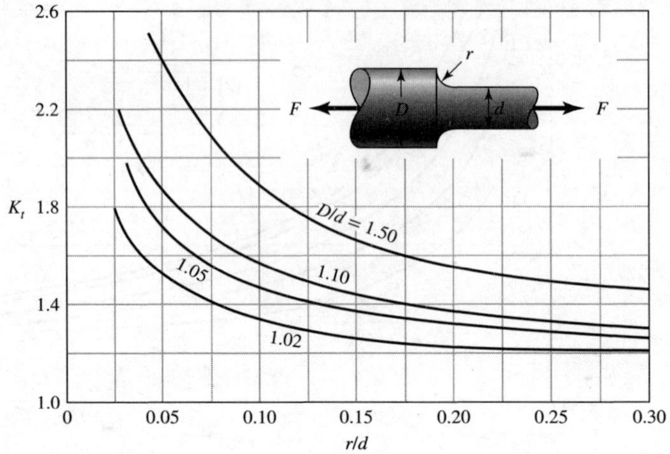

Figure A–13–8

Round shaft with shoulder fillet in torsion. $\tau_0 = Tc/J$, where $c = d/2$ and $J = \pi d^4/32$.

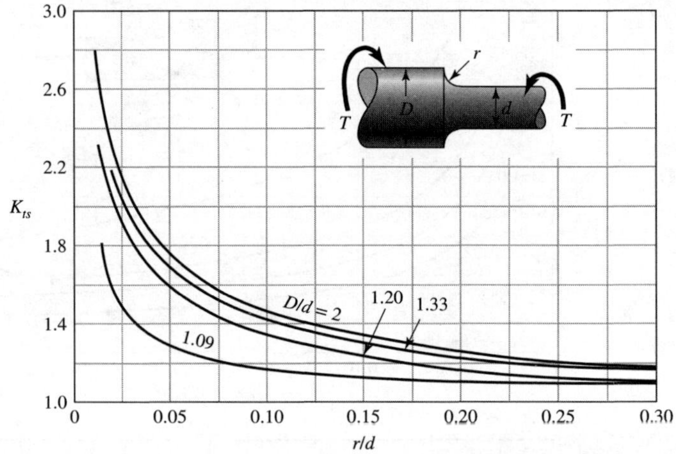

Figure A–13–9

Round shaft with shoulder fillet in bending. $\sigma_0 = Mc/I$, where $c = d/2$ and $I = \pi d^4/64$.

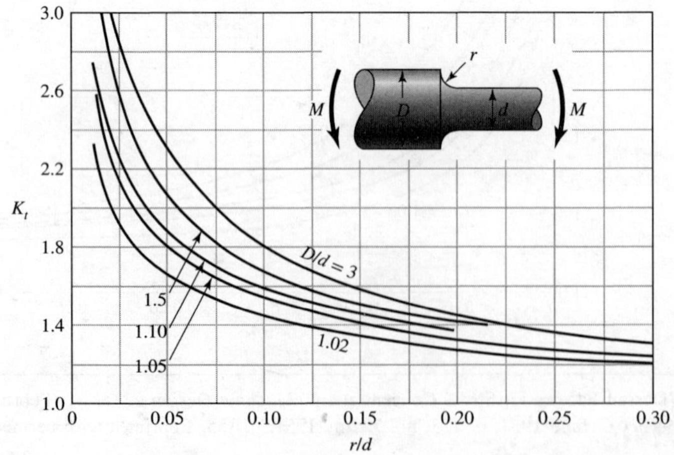

Table A–13 Charts of Theoretical Stress-Concentration Factors K_t^* (*Continued*)

Figure A–13–10

Round shaft in torsion with transverse hole.

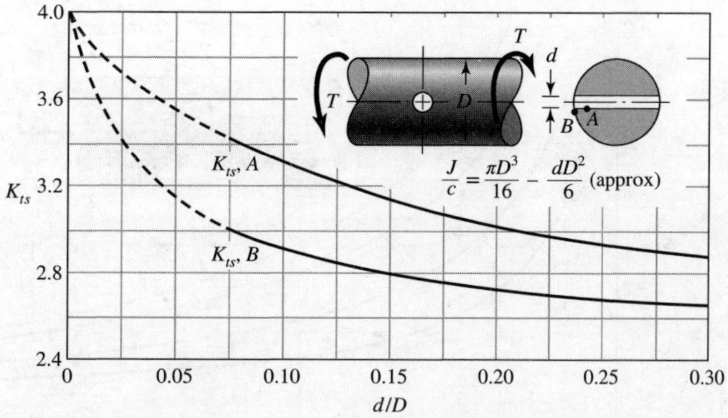

Figure A–13–11

Round shaft in bending with a transverse hole. $\sigma_0 = M/[(\pi D^3/32) - (dD^2/6)]$, approximately.

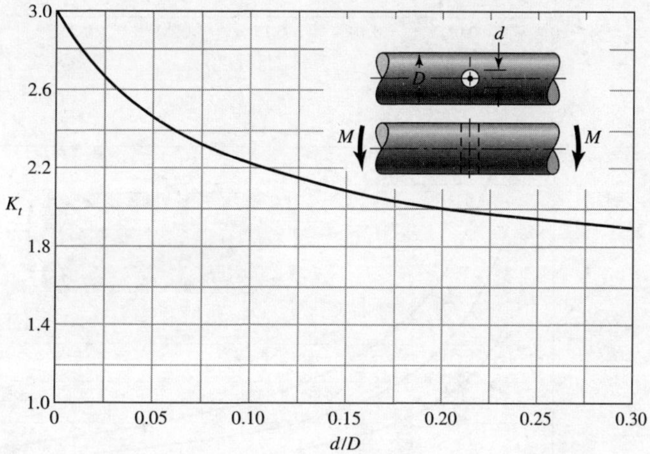

Figure A–13–12

Plate loaded in tension by a pin through a hole. $\sigma_0 = F/A$, where $A = (w - d)t$. When clearance exists, increase K_t 35 to 50 percent. (*Source:* Frocht, M. M. and H. N. Hill. "Stress-Concentration Factors around a Central Circular Hole in a Plate Loaded through a Pin in Hole." *Journal of Applied Mechanics* 7, no. 1 (March 1940): A–5.)

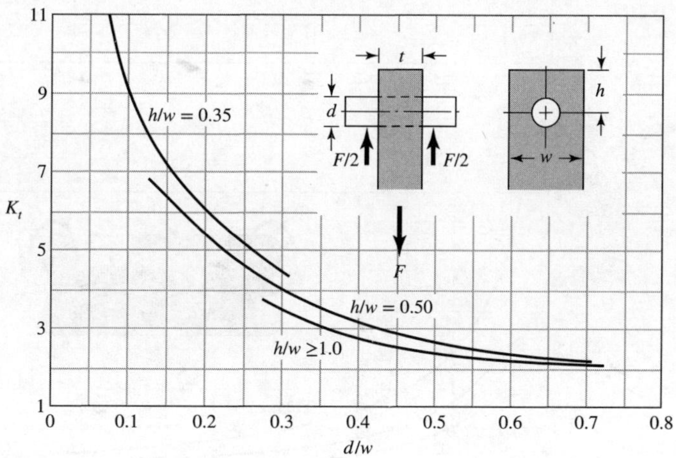

(*Continued*)

*Factors from Peterson, R. E. "Design Factors for Stress Concentration." *Machine Design* 23, no. 2 (February 1951): 169; no. 3, March 1951, p. 161; no. 5, May 1951, p. 159; no. 6, June 1951, p. 173; no. 7, July 1951, p. 155. Reprinted with permission from Machine Design, a Penton Media Inc. publication.

Table A–13 Charts of Theoretical Stress-Concentration Factors K_t^* (*Continued*)

Figure A–13–13

Grooved round bar in tension. $\sigma_0 = F/A$, where $A = \pi d^2/4$.

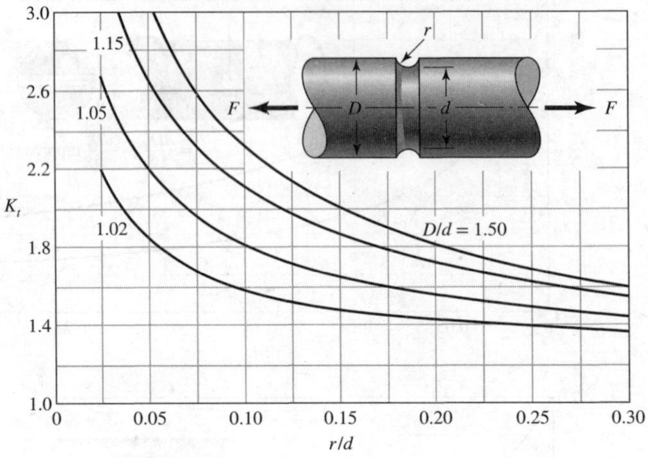

Figure A–13–14

Grooved round bar in bending. $\sigma_0 = Mc/I$, where $c = d/2$ and $I = \pi d^4/64$.

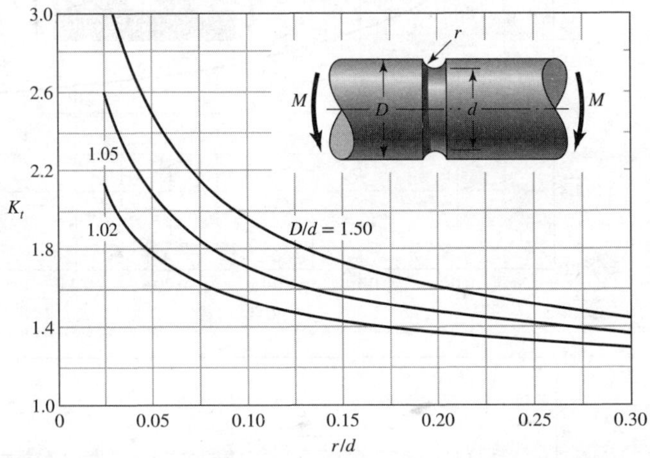

Figure A–13–15

Grooved round bar in torsion. $\tau_0 = Tc/J$, where $c = d/2$ and $J = \pi d^4/32$.

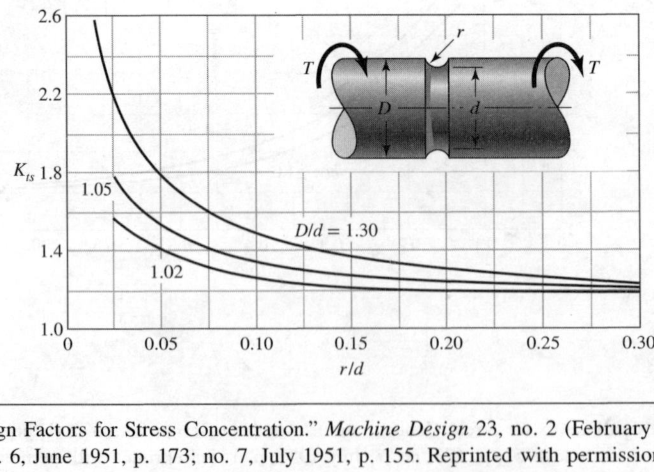

*Factors from Peterson, R. E. "Design Factors for Stress Concentration." *Machine Design* 23, no. 2 (February 1951): 169; no. 3, March 1951, p. 161; no. 5, May 1951, p. 159; no. 6, June 1951, p. 173; no. 7, July 1951, p. 155. Reprinted with permission from Machine Design, a Penton Media Inc. publication.

Table A–13 Charts of Theoretical Stress-Concentration Factors K_t^* (*Continued*)

Figure A–13–16

Round shaft with flat-bottom groove in bending and/or tension.

$$\sigma_0 = \frac{4F}{\pi d^2} + \frac{32M}{\pi d^3}$$

(*Source:* Adapted from Pilkey, W. D. and D. F. Pilkey. *Peterson's Stress Concentration Factors*. 3rd ed. Hoboken, NJ: John Wiley & Sons, 2008, 115.)

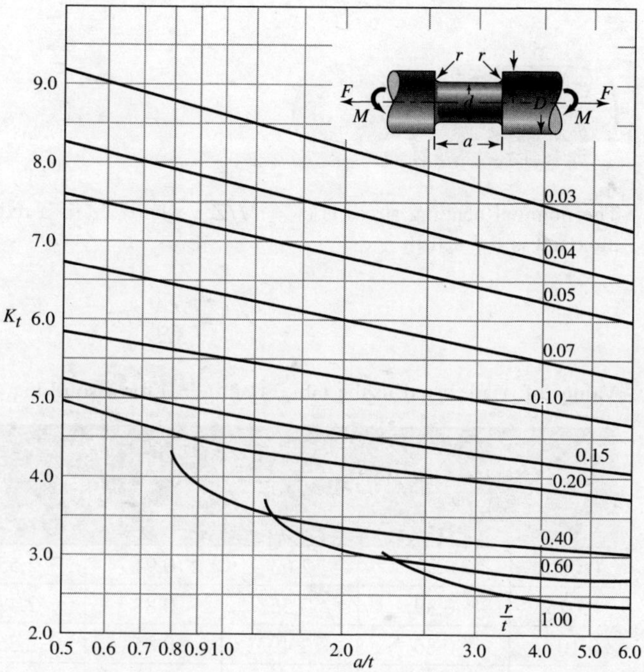

Figure A–13–17

Round shaft with flat-bottom groove in torsion.

$$\tau_0 = \frac{16T}{\pi d^3}$$

(*Source:* Adapted from Pilkey, W. D. and D. F. Pilkey. *Peterson's Stress Concentration Factors*. 3rd ed. Hoboken, NJ: John Wiley & Sons, 2008, 133.)

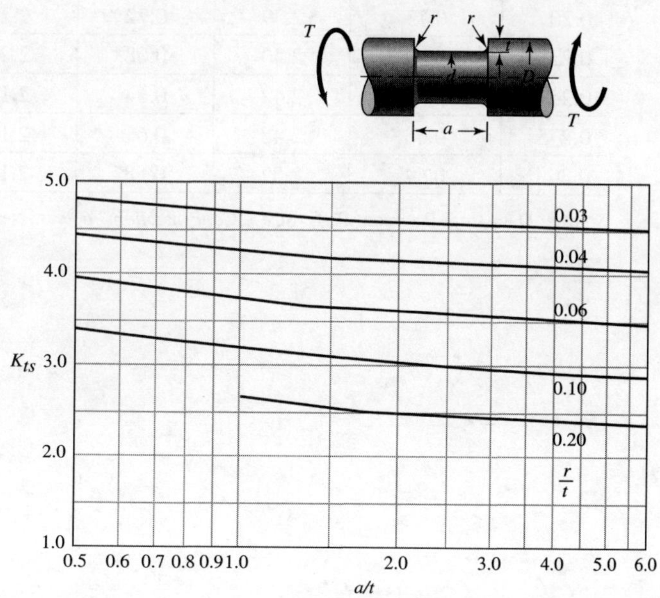

Table A–14 Approximate Stress-Concentration Factor K_t of a Round Bar or Tube with a Transverse Round Hole and Loaded in Bending

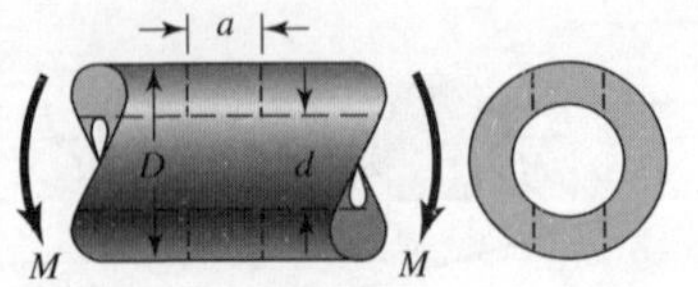

The nominal bending stress is $\sigma_0 = M/Z_{net}$ where Z_{net} is a reduced value of the section modulus and is defined by

$$Z_{net} = \frac{\pi A}{32D}(D^4 - d^4)$$

Values of A are listed in the table. Use $d = 0$ for a solid bar.

	d/D					
	0.9		0.6		0	
a/D	A	K_t	A	K_t	A	K_t
0.050	0.92	2.63	0.91	2.55	0.88	2.42
0.075	0.89	2.55	0.88	2.43	0.86	2.35
0.10	0.86	2.49	0.85	2.36	0.83	2.27
0.125	0.82	2.41	0.82	2.32	0.80	2.20
0.15	0.79	2.39	0.79	2.29	0.76	2.15
0.175	0.76	2.38	0.75	2.26	0.72	2.10
0.20	0.73	2.39	0.72	2.23	0.68	2.07
0.225	0.69	2.40	0.68	2.21	0.65	2.04
0.25	0.67	2.42	0.64	2.18	0.61	2.00
0.275	0.66	2.48	0.61	2.16	0.58	1.97
0.30	0.64	2.52	0.58	2.14	0.54	1.94

Source: Data from Peterson, R. E. *Stress Concentration Factors*. New York: Wiley, 1974, 146, 235.

(Continued)

Table A–14 Approximate Stress-Concentration Factors K_{ts} for a Round Bar or Tube Having a Transverse Round Hole and Loaded in Torsion *(Continued)*

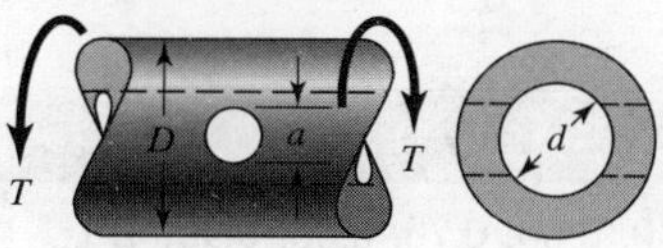

The maximum stress occurs on the inside of the hole, slightly below the shaft surface. The nominal shear stress is $\tau_0 = TD/2J_{net}$, where J_{net} is a reduced value of the second polar moment of area and is defined by

$$J_{net} = \frac{\pi A(D^4 - d^4)}{32}$$

Values of A are listed in the table. Use $d = 0$ for a solid bar.

	d/D									
	0.9		0.8		0.6		0.4		0	
a/D	A	K_{ts}	A	K_{ts}	A	K_{ts}	A	K_{ts}	A	K_{ts}
0.05	0.96	1.78							0.95	1.77
0.075	0.95	1.82							0.93	1.71
0.10	0.94	1.76	0.93	1.74	0.92	1.72	0.92	1.70	0.92	1.68
0.125	0.91	1.76	0.91	1.74	0.90	1.70	0.90	1.67	0.89	1.64
0.15	0.90	1.77	0.89	1.75	0.87	1.69	0.87	1.65	0.87	1.62
0.175	0.89	1.81	0.88	1.76	0.87	1.69	0.86	1.64	0.85	1.60
0.20	0.88	1.96	0.86	1.79	0.85	1.70	0.84	1.63	0.83	1.58
0.25	0.87	2.00	0.82	1.86	0.81	1.72	0.80	1.63	0.79	1.54
0.30	0.80	2.18	0.78	1.97	0.77	1.76	0.75	1.63	0.74	1.51
0.35	0.77	2.41	0.75	2.09	0.72	1.81	0.69	1.63	0.68	1.47
0.40	0.72	2.67	0.71	2.25	0.68	1.89	0.64	1.63	0.63	1.44

Source: Data from Peterson, R. E. *Stress Concentration Factors*. New York: Wiley, 1974, 148, 244.

Table A–15 Preferred Sizes and Renard (R-Series) Numbers

(When a choice can be made, use one of these sizes; however, not all parts or items are available in all the sizes shown in the table.)

Millimeters
0.05, 0.06, 0.08, 0.10, 0.12, 0.16, 0.20, 0.25, 0.30, 0.40, 0.50, 0.60, 0.70, 0.80, 0.90, 1.0, 1.1, 1.2, 1.4, 1.5, 1.6, 1.8, 2.0, 2.2, 2.5, 2.8, 3.0, 3.5, 4.0, 4.5, 5.0, 5.5, 6.0, 6.5, 7.0, 8.0, 9.0, 10, 11, 12, 14, 16, 18, 20, 22, 25, 28, 30, 32, 35, 40, 45, 50, 60, 80, 100, 120, 140, 160, 180, 200, 250, 300

Renard Numbers*
1st choice, R5: 1, 1.6, 2.5, 4, 6.3, 10
2d choice, R10: 1.25, 2, 3.15, 5, 8
3d choice, R20: 1.12, 1.4, 1.8, 2.24, 2.8, 3.55, 4.5, 5.6, 7.1, 9
4th choice, R40: 1.06, 1.18, 1.32, 1.5, 1.7, 1.9, 2.12, 2.36, 2.65, 3, 3.35, 3.75, 4.25, 4.75, 5.3, 6, 6.7, 7.5, 8.5, 9.5

*May be multiplied or divided by powers of 10.

Table A–16 Geometric Properties

Part 1 Properties of Sections

$A =$ area

$G =$ location of centroid

$I_x = \int y^2\, dA =$ second moment of area about x axis

$I_y = \int x^2\, dA =$ second moment of area about y axis

$I_{xy} = \int xy\, dA =$ mixed moment of area about x and y axes

$J_G = \int r^2\, dA = \int (x^2 + y^2)\, dA = I_x + I_y$

$=$ second polar moment of area about axis through G

$k_x^2 = I_x/A =$ squared radius of gyration about x axis

Rectangle

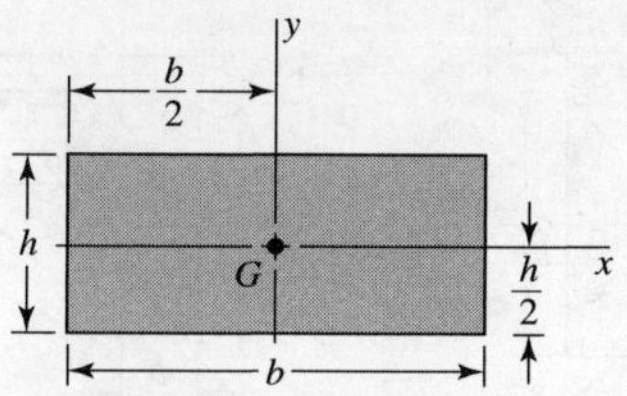

$A = bh \quad I_x = \dfrac{bh^3}{12} \quad I_y = \dfrac{b^3h}{12} \quad I_{xy} = 0$

Circle

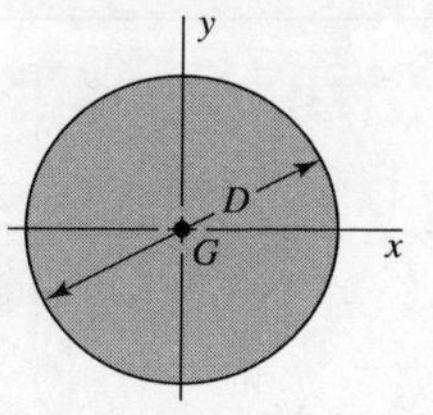

$A = \dfrac{\pi D^2}{4} \quad I_x = I_y = \dfrac{\pi D^4}{64} \quad I_{xy} = 0 \quad J_G = \dfrac{\pi D^4}{32}$

Hollow circle

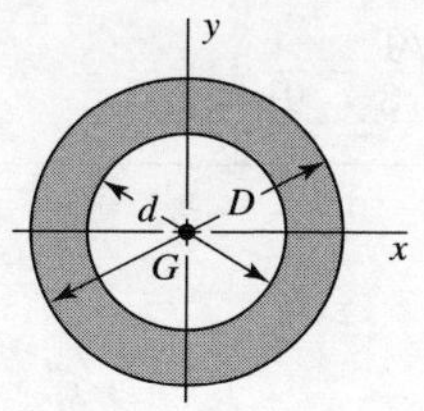

$A = \dfrac{\pi}{4}(D^2 - d^2) \quad I_x = I_y = \dfrac{\pi}{64}(D^4 - d^4) \quad I_{xy} = 0 \quad J_G = \dfrac{\pi}{32}(D^4 - d^4)$

Table A–16 Geometric Properties *(Continued)*

Right triangles

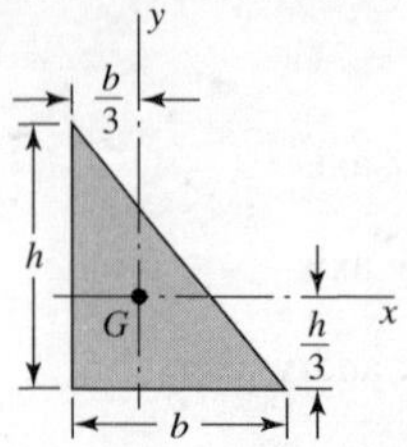

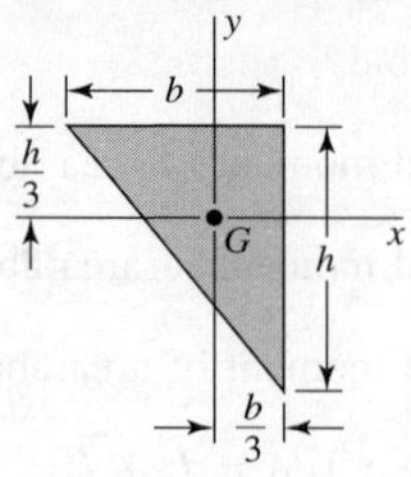

$$A = \frac{bh}{2} \quad I_x = \frac{bh^3}{36} \quad I_y = \frac{b^3h}{36} \quad I_{xy} = \frac{-b^2h^2}{72}$$

Right triangles

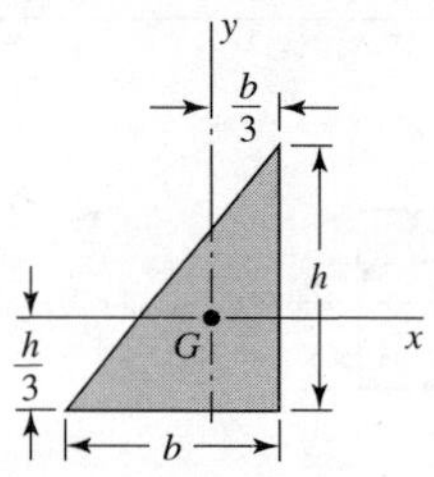

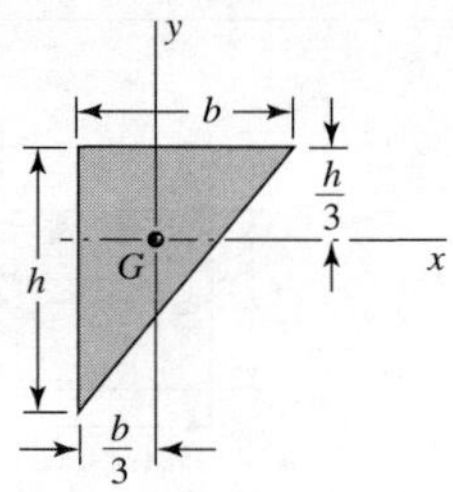

$$A = \frac{bh}{2} \quad I_x = \frac{bh^3}{36} \quad I_y = \frac{b^3h}{36} \quad I_{xy} = \frac{b^2h^2}{72}$$

Quarter-circles

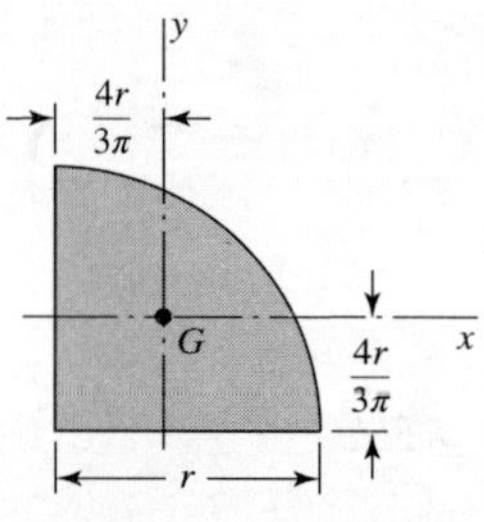

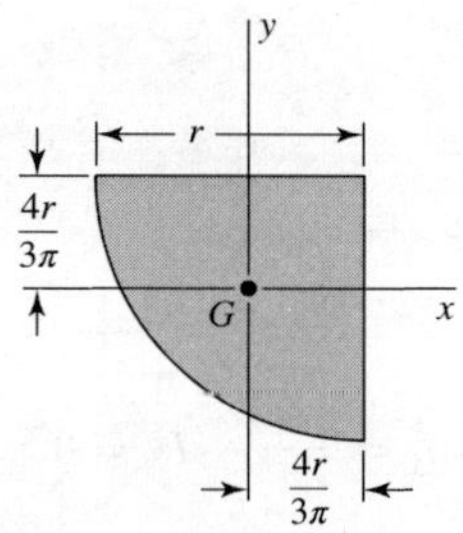

$$A = \frac{\pi r^2}{4} \quad I_x = I_y = r^4\left(\frac{\pi}{16} - \frac{4}{9\pi}\right) \quad I_{xy} = r^4\left(\frac{1}{8} - \frac{4}{9\pi}\right)$$

Quarter-circles

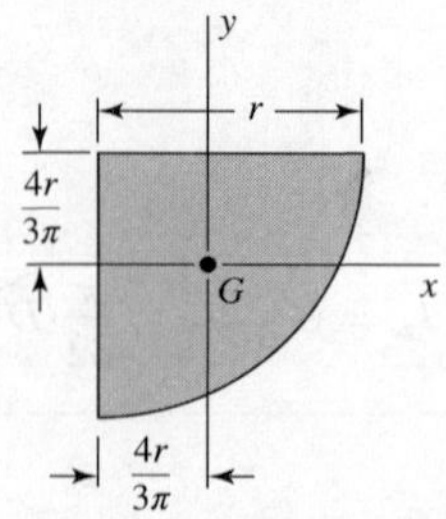

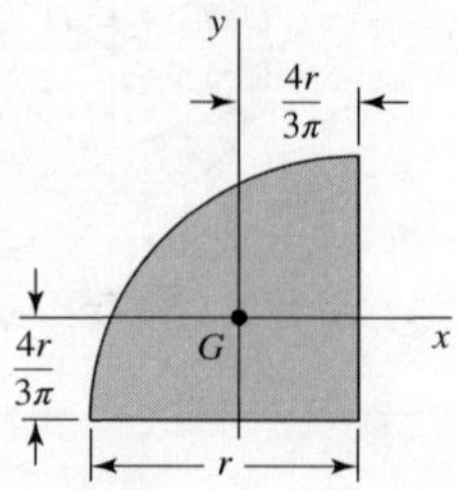

$$A = \frac{\pi r^2}{4} \quad I_x = I_y = r^4\left(\frac{\pi}{16} - \frac{4}{9\pi}\right) \quad I_{xy} = r^4\left(\frac{4}{9\pi} - \frac{1}{8}\right)$$

(Continued)

Table A–16 Geometric Properties *(Continued)*

Part 2 Properties of Solids (ρ = Mass Density, Mass per Unit Volume)

Rods

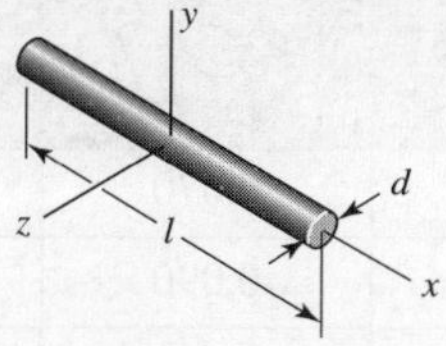

$$m = \frac{\pi d^2 l \rho}{4} \qquad I_y = I_z = \frac{ml^2}{12}$$

Round disks

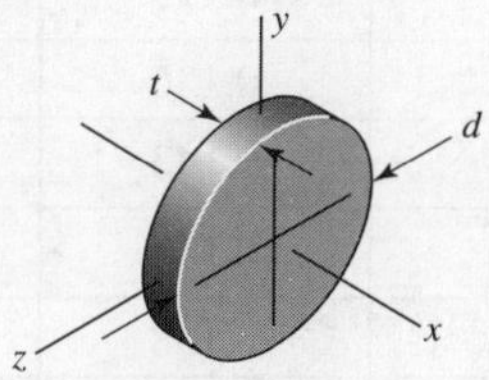

$$m = \frac{\pi d^2 t \rho}{4} \qquad I_x = \frac{md^2}{8} \qquad I_y = I_z = \frac{md^2}{16}$$

Rectangular prisms

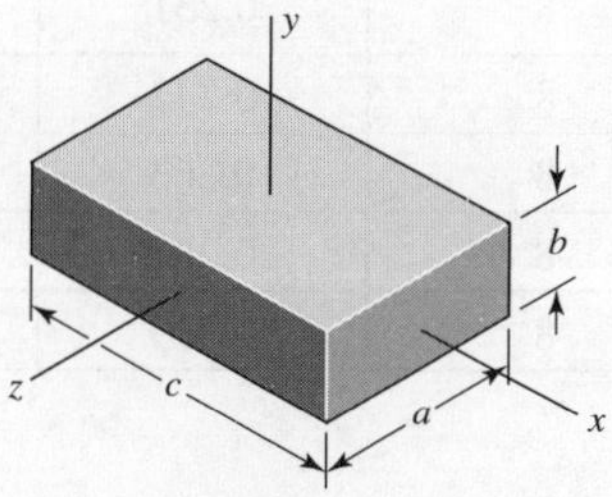

$$m = abc\rho \qquad I_x = \frac{m}{12}(a^2 + b^2) \qquad I_y = \frac{m}{12}(a^2 + c^2) \qquad I_z = \frac{m}{12}(b^2 + c^2)$$

Cylinders

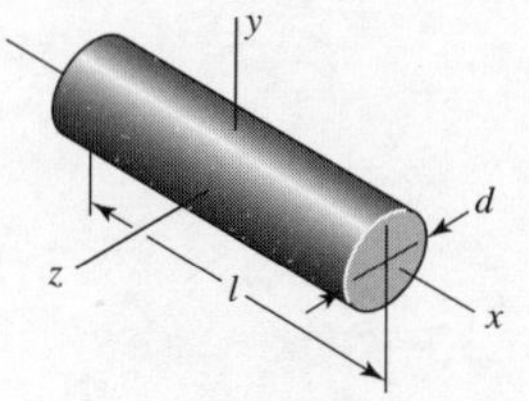

$$m = \frac{\pi d^2 l \rho}{4} \qquad I_x = \frac{md^2}{8} \qquad I_y = I_z = \frac{m}{48}(3d^2 + 4l^2)$$

Hollow cylinders

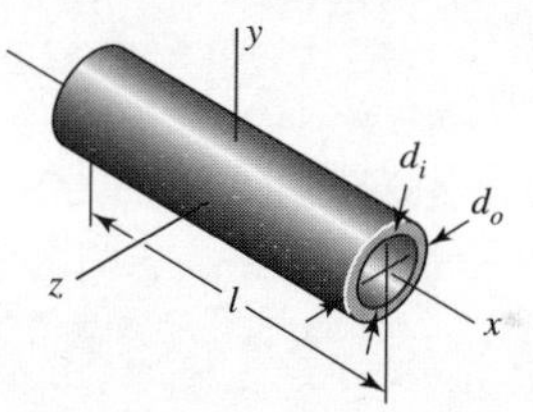

$$m = \frac{\pi (d_o^2 - d_i^2) l \rho}{4} \qquad I_x = \frac{m}{8}(d_o^2 + d_i^2) \qquad I_y = I_z = \frac{m}{48}(3d_o^2 + 3d_i^2 + 4l^2)$$

Table A–17 American Standard Pipe

Nominal Size, in	Outside Diameter, in	Threads per inch	Wall Thickness, in		
			Standard No. 40	Extra Strong No. 80	Double Extra Strong
$\frac{1}{8}$	0.405	27	0.070	0.098	
$\frac{1}{4}$	0.540	18	0.090	0.122	
$\frac{3}{8}$	0.675	18	0.093	0.129	
$\frac{1}{2}$	0.840	14	0.111	0.151	0.307
$\frac{3}{4}$	1.050	14	0.115	0.157	0.318
1	1.315	$11\frac{1}{2}$	0.136	0.183	0.369
$1\frac{1}{4}$	1.660	$11\frac{1}{2}$	0.143	0.195	0.393
$1\frac{1}{2}$	1.900	$11\frac{1}{2}$	0.148	0.204	0.411
2	2.375	$11\frac{1}{2}$	0.158	0.223	0.447
$2\frac{1}{2}$	2.875	8	0.208	0.282	0.565
3	3.500	8	0.221	0.306	0.615
$3\frac{1}{2}$	4.000	8	0.231	0.325	
4	4.500	8	0.242	0.344	0.690
5	5.563	8	0.263	0.383	0.768
6	6.625	8	0.286	0.441	0.884
8	8.625	8	0.329	0.510	0.895

Table A–18 Dimensions and Mass (kg/m) of European Standard Pipes by Outside Diameter and Wall Thickness

Outside Diameter, mm	Wall Thickness, mm							
	1	1.2	1.5	2	2.5	3	3.2	3.6
6	0.125	–	–	–	–	–	–	–
8	0.175	–	0.244	–	–	–	–	–
10	0.225	0.264	0.319	–	–	–	–	–
12	0.275	0.325	0.394	–	–	–	–	–
13	0.3	–	–	–	–	–	–	–
14	0.326	0.385	0.47	0.601	–	–	–	–
15	0.351	–	0.507	0.537	–	–	–	–
16	0.376	0.445	0.545	0.701	–	–	–	–
17.2	0.406	0.481	0.59	0.761	–	–	–	–
18	0.406	0.505	0.62	0.801	–	–	–	–
19.05	0.452	0.536	0.659	0.854	–	–	–	–
20	0.476	0.565	0.695	0.901	–	–	–	–
21.3	0.508	0.604	0.744	0.967	1.094	1.375	–	–
22	0.526	0.625	0.77	1.002	–	–	–	–
23	–	–	0.88	–	–	–	–	–
24	0.576	0.685	0.845	–	–	–	–	–
25	0.601	0.715	0.883	1.152	–	–	–	–
25.4	0.611	0.727	0.898	1.172	1.434	1.683	–	–
26.9	–	0.772	0.954	1.247	1.527	1.795	–	–
28	0.676	0.805	0.995	1.302	1.596	–	–	–
29	–	–	1.033	–	–	–	–	–
30	0.726	0.865	1.07	1.402	1.722	–	–	–
32	0.776	0.925	1.146	1.502	1.847	2.178	–	–
33.7	0.819	0.977	1.209	1.588	1.953	2.306	–	–
34	0.826	0.986	1.221	1.603	–	–	–	–
35	0.851	1.016	1.258	1.653	2.035	–	–	–
38.1	0.929	1.109	1.375	1.808	2.229	2.637	–	–
40	0.977	1.166	1.446	1.903	2.348	2.779	–	–
42.4	–	1.238	1.536	2.023	2.498	2.96	3.141	3.498
43	–	–	1.559	–	–	–	–	–
44.5	–	1.301	1.615	2.128	2.629	3.117	–	–
45	–	1.316	1.634	2.153	2.661	–	–	–
48.3	–	1.145	1.758	2.319	2.867	3.403	3.614	4.029
50	1.227	1.466	1.822	2.404	2.974	3.531	–	–
50.8	1.247	1.49	1.852	2.444	3.024	3.591	–	–
51	1.252	1.496	1.859	2.454	–	–	–	–
52	1.277	1.526	1.897	2.504	–	–	–	–
53	–	–	1.934	–	–	–	–	–

Table A–19 Deterministic ASTM Minimum Tensile and Yield Strengths for Some Hot-Rolled (HR) and Cold-Drawn (CD) Steels

[The strengths listed are estimated ASTM minimum values in the size range 18 to 32 mm. These strengths are suitable for use with the design factor defined in Sec. 1–10, provided the materials conform to ASTM A6 or A568 requirements or are required in the purchase specifications. Remember that a numbering system is not a specification.]

1 UNS No.	2 SAE and/or AISI No.	3 Processing	4 Tensile Strength, MPa	5 Yield Strength, MPa	6 Elongation in 50 mm, %	7 Reduction in Area, %	8 Brinell Hardness
G10060	1006	HR	300	170	30	55	86
		CD	330	280	20	45	95
G10100	1010	HR	320	180	28	50	95
		CD	370	300	20	40	105
G10150	1015	HR	340	190	28	50	101
		CD	390	320	18	40	111
G10180	1018	HR	400	220	25	50	116
		CD	440	370	15	40	126
G10200	1020	HR	380	210	25	50	111
		CD	470	390	15	40	131
G10300	1030	HR	470	260	20	42	137
		CD	520	440	12	35	149
G10350	1035	HR	500	270	18	40	143
		CD	550	460	12	35	163
G10400	1040	HR	520	290	18	40	149
		CD	590	490	12	35	170
G10450	1045	HR	570	310	16	40	163
		CD	630	530	12	35	179
G10500	1050	HR	620	340	15	35	179
		CD	690	580	10	30	197
G10600	1060	HR	680	370	12	30	201
G10800	1080	HR	770	420	10	25	229
G10950	1095	HR	830	460	10	25	248

Source: Data from 1986 SAE Handbook, p. 2.15.

Table A–20 Mean Mechanical Properties of Some Heat-Treated Steels

[These are typical properties for materials normalized and annealed. The properties for quenched and tempered (Q&T) steels are from a single heat. Because of the many variables, the properties listed are global averages. In all cases, data were obtained from specimens of diameter 13 mm, machined from 25 mm rounds, and of gauge length 50 mm. Unless noted, all specimens were oil-quenched.]

1 AISI No.	2 Treatment	3 Temperature °C	4 Tensile Strength MPa	5 Yield Strength, MPa	6 Elongation, %	7 Reduction in Area, %	8 Brinell Hardness
1030	Q&T*	205	848	648	17	47	495
	Q&T*	315	800	621	19	53	401
	Q&T*	425	731	579	23	60	302
	Q&T*	540	669	517	28	65	255
	Q&T*	650	586	441	32	70	207
	Normalized	925	521	345	32	61	149
	Annealed	870	430	317	35	64	137
1040	Q&T	205	779	593	19	48	262
	Q&T	425	758	552	21	54	241
	Q&T	650	634	434	29	65	192
	Normalized	900	590	374	28	55	170
	Annealed	790	519	353	30	57	149
1050	Q&T*	205	1120	807	9	27	514
	Q&T*	425	1090	793	13	36	444
	Q&T*	650	717	538	28	65	235
	Normalized	900	748	427	20	39	217
	Annealed	790	636	365	24	40	187
1060	Q&T	425	1080	765	14	41	311
	Q&T	540	965	669	17	45	277
	Q&T	650	800	524	23	54	229
	Normalized	900	776	421	18	37	229
	Annealed	790	626	372	22	38	179
1095	Q&T	315	1260	813	10	30	375
	Q&T	425	1210	772	12	32	363
	Q&T	540	1090	676	15	37	321
	Q&T	650	896	552	21	47	269
	Normalized	900	1010	500	9	13	293
	Annealed	790	658	380	13	21	192
1141	Q&T	315	1460	1280	9	32	415
	Q&T	540	896	765	18	57	262

(Continued)

Table A–20 Mean Mechanical Properties of Some Heat-Treated Steels *(Continued)*

[These are typical properties for materials normalized and annealed. The properties for quenched and tempered (Q&T) steels are from a single heat. Because of the many variables, the properties listed are global averages. In all cases, data were obtained from specimens of diameter 13 mm, machined from 25 mm rounds, and of gauge length 50 mm. Unless noted, all specimens were oil-quenched.]

1 AISI No.	2 Treatment	3 Temperature °C	4 Tensile Strength MPa	5 Yield Strength, MPa	6 Elongation, %	7 Reduction in Area, %	8 Brinell Hardness
4130	Q&T*	205	1630	1460	10	41	467
	Q&T*	315	1500	1380	11	43	435
	Q&T*	425	1280	1190	13	49	380
	Q&T*	540	1030	910	17	57	315
	Q&T*	650	814	703	22	64	245
	Normalized	870	670	436	25	59	197
	Annealed	865	560	361	28	56	156
4140	Q&T	205	1770	1640	8	38	510
	Q&T	315	1550	1430	9	43	445
	Q&T	425	1250	1140	13	49	370
	Q&T	540	951	834	18	58	285
	Q&T	650	758	655	22	63	230
	Normalized	870	1020	655	18	47	302
	Annealed	815	655	417	26	57	197
4340	Q&T	315	1720	1590	10	40	486
	Q&T	425	1470	1360	10	44	430
	Q&T	540	1170	1080	13	51	360
	Q&T	650	965	855	19	60	280

*Water-quenched

Source: Data from *ASM Metals Reference Book*. 2nd ed. Metals Park, Ohio: American Society for Metals, 1983, 217.

Table A–21 Results of Tensile Tests of Some Metals*

Number	Material	Condition	Yield Strength S_y MPa	Ultimate Strength S_u MPa	True Fracture Strength $\tilde{\sigma}_f$ MPa	Strain-Strengthening Coefficient σ_0 MPa	Strain-Strengthening Exponent m	True Fracture Strain $\tilde{\varepsilon}_f$
1018	Steel	Annealed	220	341	628†	620	0.25	1.05
1020	Steel	HR	290	456	772†	793	0.22	0.90
1144	Steel	Annealed	358	646	898†	992	0.14	0.49
1212	Steel	HR	193	424	729†	758	0.24	0.85
1045	Steel	HR	414	638	896†	965	0.14	0.58
1045	Steel	Q&T 600°F	1520	1580	2380	1880†	0.041	0.81
4142	Steel	Q&T 600°F	1720	1930	2340	1760†	0.048	0.43
4340	Steel	HR	910	1041	1344†	1448	0.09	0.45
303	Stainless steel	Annealed	241	601	1520†	1410	0.51	1.16
304	Stainless steel	Annealed	276	568	1600†	1270	0.45	1.67
2011	Aluminum alloy	T6	169	324	325†	620	0.28	0.10
2024	Aluminum alloy	T4	296	446	533†	689	0.15	0.18
7075	Aluminum alloy	T6	542	593	706†	882	0.13	0.18

*Values from one or two heats and believed to be attainable using proper purchase specifications. The fracture strain may vary as much as 100 percent.

†Derived value.

Source: Data from Datalro, J. "Solid Materials." In *Standard Handbook of Machine Design*, edited by Joseph E. Shigley, Charles R. Mischlre, and Thomas H. Brown, Jr. (eds.-in-chief). 3rd ed. New York: McGraw Hill, 2004, 32.49–32.52.

Table A–22 Mean Monotonic and Cyclic Stress-Strain Properties of Selected Steels

Grade (a)	Orientation (e)	Description (f)	Hardness HB	Tensile Strength S_{ut} MPa	Reduction in Area %	True Fracture Strain $\tilde{\varepsilon}_f$	Modulus of Elasticity E GPa	Fatigue Strength Coefficient σ'_f MPa	Fatigue Strength Exponent b	Fatigue Ductility Coefficient ε'_f	Fatigue Ductility Exponent c
A538A (b)	L	STA	405	1515	67	1.10	185	1655	−0.065	0.30	−0.62
A538B (b)	L	STA	460	1860	56	0.82	185	2135	−0.071	0.80	−0.71
A538C (b)	L	STA	480	2000	55	0.81	180	2240	−0.07	0.60	−0.75
AM-350 (c)	L	HR, A		1315	52	0.74	195	2800	−0.14	0.33	−0.84
AM-350 (c)	L	CD	496	1905	20	0.23	180	2690	−0.102	0.10	−0.42
Gainex (c)	LT	HR sheet		530	58	0.86	200	805	−0.07	0.86	−0.65
Gainex (c)	L	HR sheet		510	64	1.02	200	805	−0.071	0.86	−0.68
H-11	L	Ausformed	660	2585	33	0.40	205	3170	−0.077	0.08	−0.74
RQC-100 (c)	LT	HR plate	290	940	43	0.56	205	1240	−0.07	0.66	−0.69
RQC-100 (c)	L	HR plate	290	930	67	1.02	205	1240	−0.07	0.66	−0.69
10B62	L	Q&T	430	1640	38	0.89	195	1780	−0.067	0.32	−0.56
1005-1009	LT	HR sheet	90	360	73	1.3	205	580	−0.09	0.15	−0.43
1005-1009	LT	CD sheet	125	470	66	1.09	205	515	−0.059	0.30	−0.51
1005-1009	L	CD sheet	125	415	64	1.02	200	540	−0.073	0.11	−0.41
1005-1009	L	HR sheet	90	345	80	1.6	200	640	−0.109	0.10	−0.39
1015	L	Normalized	80	415	68	1.14	205	825	−0.11	0.95	−0.64
1020	L	HR plate	108	440	62	0.96	205	895	−0.12	0.41	−0.51
1040	L	As forged	225	620	60	0.93	200	1540	−0.14	0.61	−0.57
1045	L	Q&T	225	725	65	1.04	200	1225	−0.095	1.00	−0.66
1045	L	Q&T	410	1450	51	0.72	200	1860	−0.073	0.60	−0.70
1045	L	Q&T	390	1345	59	0.89	205	1585	−0.074	0.45	−0.68
1045	L	Q&T	450	1585	55	0.81	205	1795	−0.07	0.35	−0.69
1045	L	Q&T	500	1825	51	0.71	205	2275	−0.08	0.25	−0.68
1045	L	Q&T	595	2240	41	0.52	205	2725	−0.081	0.07	−0.60
1144	L	CDSR	265	930	33	0.51	195	1000	−0.08	0.32	−0.58
1144	L	DAT	305	1035	25	0.29	200	1585	−0.09	0.27	−0.53

1541F	L	Q&T forging	290	950	49	0.68	205	1275	−0.076	0.68	−0.65
1541F	L	Q&T forging	260	890	60	0.93	205	1275	−0.071	0.93	−0.65
4130	L	Q&T	258	895	67	1.12	220	1275	−0.083	0.92	−0.63
4130	L	Q&T	365	1425	55	0.79	200	1695	−0.081	0.89	−0.69
4140	L	Q&T, DAT	310	1075	60	0.69	200	1825	−0.08	1.2	−0.59
4142	L	DAT	310	1060	29	0.35	200	1450	−0.10	0.22	−0.51
4142	L	DAT	335	1250	28	0.34	200	1250	−0.08	0.06	−0.62
4142	L	Q&T	380	1415	48	0.66	205	1825	−0.08	0.45	−0.75
4142	L	Q&T and deformed	400	1550	47	0.63	200	1895	−0.09	0.50	−0.75
4142	L	Q&T	450	1760	42	0.54	205	2000	−0.08	0.40	−0.73
4142	L	Q&T and deformed	475	2035	20	0.22	200	2070	−0.082	0.20	−0.77
4142	L	Q&T and deformed	450	1930	37	0.46	200	2105	−0.09	0.60	−0.76
4142	L	Q&T	475	1930	35	0.43	205	2170	−0.081	0.09	−0.61
4142	L	Q&T	560	2240	27	0.31	205	2655	−0.089	0.07	−0.76
4340	L	HR, A	243	825	43	0.57	195	1200	−0.095	0.45	−0.54
4340	L	Q&T	409	1470	38	0.48	200	2000	−0.091	0.48	−0.60
4340	L	Q&T	350	1240	57	0.84	195	1655	−0.076	0.73	−0.62
5160	L	Q&T	430	1670	42	0.87	195	1930	−0.071	0.40	−0.57
52100	L	SH, Q&T	518	2015	11	0.12	205	2585	−0.09	0.18	−0.56
9262	L	A	260	925	14	0.16	205	1040	−0.071	0.16	−0.47
9262	L	Q&T	280	1000	33	0.41	195	1220	−0.073	0.41	−0.60
9262	L	Q&T	410	565	32	0.38	200	1855	−0.057	0.38	−0.65
950C (d)	LT	HR plate	159	565	64	1.03	205	1170	−0.12	0.95	−0.61
950C (d)	L	HR bar	150	565	69	1.19	205	970	−0.11	0.85	−0.59
950X (d)	L	Plate channel	150	440	65	1.06	205	625	−0.075	0.35	−0.54
950X (d)	L	HR plate	156	530	72	1.24	205	1005	−0.10	0.85	−0.61
950X (d)	L	Plate channel	225	695	68	1.15	195	1055	−0.08	0.21	−0.53

Notes: (a) AISI/SAE grade, unless otherwise indicated. (b) ASTM designation. (c) Proprietary designation. (d) SAE HSLA grade. (e) Orientation of axis of specimen, relative to rolling direction; L is longitudinal (parallel to rolling direction); LT is long transverse (perpendicular to rolling direction). (f) STA, solution treated and aged; HR, hot rolled; CD, cold drawn; Q&T, quenched and tempered; CDSR, cold drawn strain relieved; DAT, drawn at temperature; A, annealed.

Source: Data from *ASM Metals Reference Book*. 2nd ed. Metals Park, Ohio: American Society for Metals, 1983, 217.

Table A–23 Mechanical Properties of Three Non-Steel Metals

(a) Typical Properties of Gray Cast Iron

ASTM Number	Tensile Strength S_{ut}, MPa	Compressive Strength S_{uc}, MPa	Shear Modulus of Rupture S_{su}, MPa	Modulus of Elasticity, MPa		Endurance Limit* S_e, MPa	Brinell Hardness H_B	Fatigue Stress-Concentration Factor K_f
				Tension†	Torsion			
20	151	572	179	66–96	26–38	69	156	1.00
25	179	668	220	79–102	31–41	79	174	1.05
30	213	751	275	89–113	35–45	96	201	1.10
35	251	854	334	100–118	40–47	110	212	1.15
40	293	965	344	110–137	44–53	127	235	1.25
50	361	1130	503	129–157	49–55	148	262	1.35
60	430	1292	610	140–162	53–58	169	302	1.50

*Polished or machined specimens.

†The modulus of elasticity of cast iron in compression corresponds closely to the upper value in the range given for tension and is a more constant value than that for tension.

Note: The American Society for Testing and Materials (ASTM) numbering system for gray cast iron is such that the numbers correspond to the *minimum tensile strength* in MPa. Thus an ASTM No. 20 cast iron has a minimum tensile strength of 20 MPa. Note particularly that the tabulations are *typical* of several heats.

Table A–23 Mechanical Properties of Three Non-Steel Metals *(Continued)*

(b) Mechanical Properties of Some Aluminum Alloys

[These are *typical* properties for sizes of about 13 mm; similar properties can be obtained by using proper purchase specifications. The values given for fatigue strength correspond to $50(10^7)$ cycles of completely reversed stress. Aluminum alloys do not have an endurance limit. Yield strengths were obtained by the 0.2 percent offset method.]

Aluminum Association Number	Temper	Yield Strength S_y MPa	Tensile Strength S_u MPa	Fatigue Strength S_f MPa	Elongation in 50 mm %	Brinell Hardness H_B
Wrought:						
2017	O	70	179	90	22	45
2024	O	76	186	90	22	47
	T3	345	482	138	16	120
3003	H12	117	131	55	20	35
	H16	165	179	65	14	47
3004	H34	186	234	103	12	63
	H38	234	276	110	6	77
5052	H32	186	234	117	18	62
	H36	234	269	124	10	74
Cast:						
319.0*	T6	165	248	69	2.0	80
333.0†	T5	172	234	83	1.0	100
	T6	207	289	103	1.5	105
335.0*	T6	172	241	62	3.0	80
	T7	248	262	62	0.5	85

*Sand casting.

†Permanent-mold casting.

(c) Mechanical Properties of Some Titanium Alloys

Titanium Alloy	Condition	Yield Strength S_y MPa	Tensile Strength S_{ut} MPa	Elongation in 50 mm %	Hardness (Brinell or Rockwell)
Ti-35A†	Annealed	210	275	30	135 HB
Ti-50A†	Annealed	310	380	25	215 HB
Ti-0.2 Pd	Annealed	280	340	28	200 HB
Ti-5 Al-2.5 Sn	Annealed	760	790	16	36 HRC
Ti-8 Al-1 Mo-1 V	Annealed	900	965	15	39 HRC
Ti-6 Al-6 V-2 Sn	Annealed	970	1030	14	38 HRC
Ti-6 Al-4V	Annealed	830	900	14	36 HRC
Ti-13 V-11 Cr-3 Al	Sol. + aging	1207	1276	8	40 HRC

†Commercially pure alpha titanium.

Table A–24 Decimal Equivalents of Wire and Sheet-Metal Gauges* (All Sizes Are Given in Millimeters)

Name of Gauge:	American or Brown & Sharpe	Birmingham or Stubs Iron Wire	United States Standard†	Manufacturers Standard	Steel Wire or Washburn & Moen	Music Wire	Stubs Steel Wire	Twist Drill
Principal Use:	Nonferrous Sheet, Wire, and Rod	Tubing, Ferrous Strip, Flat Wire, and Spring Steel	Ferrous Sheet and Plate, 75.4 kN/m^3	Ferrous Sheet	Ferrous Wire Except Music Wire	Music Wire	Steel Drill Rod	Twist Drills and Drill Steel
7/0			12.7		12.446			
6/0	14.732		11.906		11.722	0.012		
5/0	13.119		11.112		10.935	0.127		
4/0	11.684	11.532	10.319		10.003	0.152		
3/0	10.388	10.795	9.525		9.207	0.178		
2/0	9.266	9.652	8.731		8.407	0.203		
0	8.252	8.636	7.937		7.785	0.229		
1	7.348	7.62	7.144		7.188	0.254	5.766	5.791
2	6.543	7.214	6.747		6.667	0.279	5.563	5.613
3	5.816	6.579	6.35	6.073	6.19	0.305	5.385	5.41
4	5.189	6.045	5.953	5.695	5.723	0.33	5.258	5.309
5	4.620	5.588	5.556	5.314	5.258	0.356	5.182	5.22
6	4.115	5.156	5.159	4.935	4.87	0.406	5.105	5.182
7	2.908	4.572	4.762	4.554	4.496	0.457	5.055	5.105
8	3.264	4.191	4.366	4.175	4.115	0.508	5.004	5.055
9	2.906	3.759	3.969	3.797	3.759	0.559	4.928	4.987
10	2.588	3.404	3.572	3.416	3.429	0.61	4.851	4.915
11	2.305	3.048	3.175	3.038	3.061	0.66	4.775	4.851
12	2.052	2.768	2.778	2.657	2.68	0.737	4.699	4.8
13	1.828	2.413	2.381	2.278	2.324	0.787	4.623	4.699
14	1.628	2.108	1.984	1.897	2.032	0.838	4.57	4.623
15	1.449	1.829	1.786	1.709	1.829	0.889	4.52	4.572
16	1.29	1.651	1.587	1.499	1.587	0.94	4.445	4.696
17	1.15	1.473	1.429	1.367	1.372	0.991	4.37	4.394

(Continued)

Table A–24 Decimal Equivalents of Wire and Sheet-Metal Gauges* (All Sizes Are Given in Millimeters) *(Continued)*

Name of Gauge:	American or Brown & Sharpe	Birmingham or Stubs Iron Wire	United States Standard[†]	Manufacturers Standard	Steel Wire or Washburn & Moen	Music Wire	Stubs Steel Wire	Twist Drill
Principal Use:	Nonferrous Sheet, Wire, and Rod	Tubing, Ferrous Strip, Flat Wire, and Spring Steel	Ferrous Sheet and Plate, 75.4 kN/m^3	Ferrous Sheet	Ferrous Wire Except Music Wire	Music Wire	Steel Drill Rod	Twist Drills and Drill Steel
18	1.024	1.245	1.27	1.265	1.206	1.041	4.267	4.305
19	0.912	1.067	1.111	1.062	1.041	1.092	4.166	4.216
20	0.812	0.889	0.952	0.912	0.884	1.143	4.089	4.089
21	0.723	0.813	0.873	0.836	0.805	1.194	3.988	4.039
22	0.644	0.711	0.794	0.759	0.726	1.245	3.937	3.988
23	0.573	0.635	0.714	0.683	0.655	1.245	3.886	3.912
24	0.511	0.559	0.635	0.607	0.584	1.397	3.835	3.861
25	0.455	0.508	0.556	0.531	0.518	1.499	3.759	3.797
26	0.405	0.457	0.476	0.455	0.46	1.6	3.708	3.734
27	0.361	0.406	0.437	0.417	0.439	1.702	3.632	3.658
28	0.321	0.356	0.397	0.378	0.411	1.803	3.531	3.556
29	0.286	0.33	0.357	0.343	0.381	1.905	3.404	3.454
30	0.255	0.305	0.318	0.305	0.356	2.032	3.226	3.264
31	0.227	0.254	0.278	0.267	0.335	2.159	3.048	3.048
32	0.202	0.229	0.258	0.246	0.325	2.286	2.921	2.946
33	0.18	0.203	0.238	0.229	0.3	2.413	2.845	2.87
34	0.16	0.178	0.128	0.208	0.264		2.794	2.819
35	0.143	0.127	0.198	0.19	0.241		2.743	2.794
36	0.127	0.102	0.179	0.17	0.241		2.692	2.705
37	0.113		0.169	0.163	0.216		2.616	2.642
38	0.101		0.159	0.152	0.203		2.565	2.578
39	0.09				0.19		2.515	2.527
40	0.08				0.178		2.464	2.489

*Specify sheet, wire, and plate by stating the gauge number, the gauge name, and the decimal equivalent in parentheses.

[†]Reflects present average and weights of sheet steel.

Table A–25 Dimensions of Square and Hexagonal Bolts

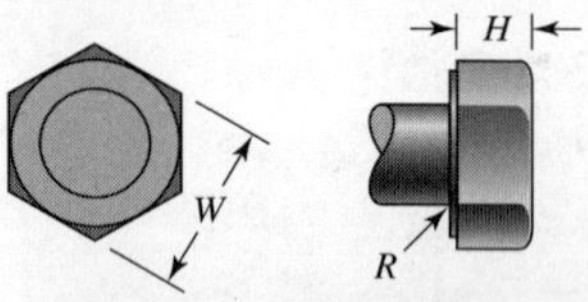

Nominal Size, mm	Head Type										
	Square		Regular Hexagonal			Heavy Hexagonal			Structural Hexagonal		
	W	H	W	H	R_{min}	W	H	R_{min}	W	H	R_{min}
M5	8	3.58	8	3.58	0.2						
M6			10	4.38	0.3						
M8			13	5.68	0.4						
M10			16	6.85	0.4						
M12			18	7.95	0.6	21	7.95	0.6			
M14			21	9.25	0.6	24	9.25	0.6			
M16			24	10.75	0.6	27	10.75	0.6	27	10.75	0.6
M20			30	13.40	0.8	34	13.40	0.8	34	13.40	0.8
M24			36	15.90	0.8	41	15.90	0.8	41	15.90	1.0
M30			46	19.75	1.0	50	19.75	1.0	50	19.75	1.2
M36			55	23.55	1.0	60	23.55	1.0	60	23.55	1.5

Table A–26 Dimensions of Hexagonal Cap Screws and Heavy Hexagonal Screws (W = Width across Flats; H = Height of Head; See Figure in Table A–25)

		Type of Screw		
Nominal Size, mm	Minimum Fillet Radius	Cap W	Heavy W	Height H
M5	0.2	8		3.65
M6	0.3	10		4.15
M8	0.4	13		5.50
M10	0.4	16		6.63
M12	0.6	18	21	7.76
M14	0.6	21	24	9.09
M16	0.6	24	27	10.32
M20	0.8	30	34	12.88
M24	0.8	36	41	15.44
M30	1.0	46	50	19.48
M36	1.0	55	60	23.38

Table A–27 Dimensions of Hexagonal Nuts

		Height H		
Nominal Size, mm	Width W	Regular Hexagonal	Thick or Slotted	JAM
M5	8	4.7	5.1	2.7
M6	10	5.2	5.7	3.2
M8	13	6.8	7.5	4.0
M10	16	8.4	9.3	5.0
M12	18	10.8	12.0	6.0
M14	21	12.8	14.1	7.0
M16	24	14.8	16.4	8.0
M20	30	18.0	20.3	10.0
M24	36	21.5	23.9	12.0
M30	46	25.6	28.6	15.0
M36	55	31.0	34.7	18.0

Table A–28 Dimensions of Metric Plain Washers (All Dimensions in Millimeters)

Washer Size*	Minimum ID	Maximum OD	Maximum Thickness	Washer Size*	Minimum ID	Maximum OD	Maximum Thickness
1.6 N	1.95	4.00	0.70	10 N	10.85	20.00	2.30
1.6 R	1.95	5.00	0.70	10 R	10.85	28.00	2.80
1.6 W	1.95	6.00	0.90	10 W	10.85	39.00	3.50
2 N	2.50	5.00	0.90	12 N	13.30	25.40	2.80
2 R	2.50	6.00	0.90	12 R	13.30	34.00	3.50
2 W	2.50	8.00	0.90	12 W	13.30	44.00	3.50
2.5 N	3.00	6.00	0.90	14 N	15.25	28.00	2.80
2.5 R	3.00	8.00	0.90	14 R	15.25	39.00	3.50
2.5 W	3.00	10.00	1.20	14 W	15.25	50.00	4.00
3 N	3.50	7.00	0.90	16 N	17.25	32.00	3.50
3 R	3.50	10.00	1.20	16 R	17.25	44.00	4.00
3 W	3.50	12.00	1.40	16 W	17.25	56.00	4.60
3.5 N	4.00	9.00	1.20	20 N	21.80	39.00	4.00
3.5 R	4.00	10.00	1.40	20 R	21.80	50.00	4.60
3.5 W	4.00	15.00	1.75	20 W	21.80	66.00	5.10
4 N	4.70	10.00	1.20	24 N	25.60	44.00	4.60
4 R	4.70	12.00	1.40	24 R	25.60	56.00	5.10
4 W	4.70	6.00	2.30	24 W	25.60	72.00	5.60
5 N	5.50	11.00	1.40	30 N	32.40	56.00	5.10
5 R	5.50	15.00	1.75	30 R	32.40	72.00	5.60
5 W	5.50	20.00	2.30	30 W	32.40	90.00	6.40
6 N	6.65	13.00	1.75	36 N	38.30	66.00	5.60
6 R	6.65	18.80	1.75	36 R	38.30	90.00	6.40
6 W	6.65	25.40	2.30	36 W	38.30	110.00	8.50
8 N	8.90	18.80	2.30				
8 R	8.90	25.40	2.30				
8 W	8.90	32.00	2.80				

N = narrow; R = regular; W = wide.

*Same as screw or bolt size.

Table A–29 Gamma Function*

Values of $\Gamma(n) = \int_0^{\infty} e^{-x} x^{n-1}\,dx$

n	$\Gamma(n)$	n	$\Gamma(n)$	n	$\Gamma(n)$	n	$\Gamma(n)$
1.00	1.000 00	1.25	.906 40	1.50	.886 23	1.75	.919 06
1.01	.994 33	1.26	.904 40	1.51	.886 59	1.76	.921 37
1.02	.988 84	1.27	.902 50	1.52	.887 04	1.77	.923 76
1.03	.983 55	1.28	.900 72	1.53	.887 57	1.78	.926 23
1.04	.978 44	1.29	.899 04	1.54	.888 18	1.79	.928 77
1.05	.973 50	1.30	.897 47	1.55	.888 87	1.80	.931 38
1.06	.968 74	1.31	.896 00	1.56	.889 64	1.81	.934 08
1.07	.964 15	1.32	.894 64	1.57	.890 49	1.82	.936 85
1.08	.959 73	1.33	.893 38	1.58	.891 42	1.83	.939 69
1.09	.955 46	1.34	.892 22	1.59	.892 43	1.84	.942 61
1.10	.951 35	1.35	.891 15	1.60	.893 52	1.85	.945 61
1.11	.947 39	1.36	.890 18	1.61	.894 68	1.86	.948 69
1.12	.943 59	1.37	.889 31	1.62	.895 92	1.87	.951 84
1.13	.939 93	1.38	.888 54	1.63	.897 24	1.88	.955 07
1.14	.936 42	1.39	.887 85	1.64	.898 64	1.89	.958 38
1.15	.933 04	1.40	.887 26	1.65	.900 12	1.90	.961 77
1.16	.929 80	1.41	.886 76	1.66	.901 67	1.91	.965 23
1.17	.936 70	1.42	.886 36	1.67	.903 30	1.92	.968 78
1.18	.923 73	1.43	.886 04	1.68	.905 00	1.93	.972 40
1.19	.920 88	1.44	.885 80	1.69	.906 78	1.94	.976 10
1.20	.918 17	1.45	.885 65	1.70	.908 64	1.95	.979 88
1.21	.915 58	1.46	.885 60	1.71	.910 57	1.96	.983 74
1.22	.913 11	1.47	.885 63	1.72	.912 58	1.97	.987 68
1.23	.910 75	1.48	.885 75	1.73	.914 66	1.98	.991 71
1.24	.908 52	1.49	.885 95	1.74	.916 83	1.99	.995 81
						2.00	1.000 00

*For $n > 2$, use the recursive formula

$$\Gamma(n) = (n-1)\Gamma(n-1)$$

For example, $\Gamma(5.42) = 4.42(3.42)\,(2.42)\,\Gamma(1.42) = 4.42(3.42)\,(2.42)\,(0.886\,36) = 32.4245$

†For large positive values of x, $\Gamma(x)$ can be expressed by the asymptotic series based on Stirling's approximation

$$\Gamma(x) \approx x^x e^{-x}\sqrt{\frac{2\pi}{x}}\left(1 + \frac{1}{12x} + \frac{1}{288x^2} - \frac{139}{51\,840x^3} - \frac{571}{2\,488\,320x^4}\right)$$

Source: Data from Beyer, H. William (ed.). *Handbook of Tables for Probability and Statistics*. 2nd ed. Boca Raton, Florida: CRC Press, 1966.

Answers to Selected Problems — Appendix B

B–1 Chapter 1

1–8 $P = 100$ units

1–11 (a) $e_1 = 0.005\ 751\ 311\ 1$, $e_2 = 0.008\ 427\ 124\ 7$, $e = 0.014\ 178\ 435\ 8$, (b) $e_1 = -0.004\ 248\ 688\ 9$, $e_2 = -0.001\ 572\ 875\ 3$, $e = -0.005\ 821\ 564\ 2$

1–13 (a) $\bar{x} = 122.9$ kcycles, $\hat{s}_x = 30.3$ kcycles, (b) 27

1–15 $L_{10} = 84.1$ kcycles

1–18 $\bar{n} = 1.32$, $d = 31.9$ mm

1–20 $\bar{n} = 1.17$, $R = 94.9$ percent

1–25 $D_o = 14.43 \pm 0.29$ mm

1–30 (a) $\sigma = 8$ MPa, (b) $I = 13.0(10^3)$ mm^4, (c) $I = 20.4(10^3)$ mm^4, (d) $\tau = 62.2$ MPa

B–2 Chapter 2

2–2 (a) Maximize yield strength: Q&T at 425°C, (b) Maximize elongation: Q&T at 650°C

2–12 $S_u = 935$ MPa

2–25 Aluminum alloys have greatest potential followed closely by high carbon heat-treated steel. Warrants further discussion.

2–33 Steel, titanium, aluminum alloys, and composites

B–3 Chapter 3

3–3 $R_O = 1.39$ kN, $R_A = 1.6$ kN

3–7 $R_1 = -0.91$ kN, $R_2 = 6.91$ kN, $M_1 = 1.09$ kN·m, $M_2 = -4$ kN·m

3–15 (a) $\sigma_1 = 22$ MPa, $\sigma_2 = -12$ MPa, $\sigma_3 = 0$ MPa, $\phi_p = 14.0°$ cw, $\tau_1 = 17$ MPa, $\sigma_{ave} = 5$ MPa, $\phi_s = 31.0°$ ccw,
(b) $\sigma_1 = 18.6$ MPa, $\sigma_2 = 6.4$ MPa, $\sigma_3 = 0$ MPa, $\phi_p = 27.5°$ ccw, $\tau_1 = 6.10$ MPa, $\sigma_{ave} = 12.5$ MPa, $\phi_s = 17.5°$ cw,
(c) $\sigma_1 = 26.2$ MPa, $\sigma_2 = 7.78$ MPa, $\sigma_3 = 0$ MPa, $\phi_p = 69.7°$ ccw, $\tau_1 = 9.22$ MPa, $\sigma_{ave} = 17$ MPa, $\phi_s = 24.7°$ ccw,
(d) $\sigma_1 = 25.8$ MPa, $\sigma_2 = -15.8$ MPa, $\sigma_3 = 0$ MPa, $\phi_p = 72.4°$ cw, $\tau_1 = 20.8$ MPa, $\sigma_{ave} = 5$ MPa, $\phi_s = 27.4°$ ccw

3–22 $\sigma_1 = 90$ MPa, $\sigma_2 = 0$ MPa, $\sigma_3 = 0$ MPa, $\tau_{max} = 45$ MPa

3–26 $\delta = 5.86$ mm

3–27 $\sigma_x = 382$ MPa, $\sigma_y = -37.4$ MPa

3–34 $\sigma_{max} = 84.3$ MPa, $\tau_{max} = 5.63$ MPa

3–39 model c: $\sigma = 88.4$ MPa, $\tau = 23.6$ MPa, model d: $\sigma = 141.5$ MPa, $\tau = 23.6$ MPa, model e: $\sigma = 88.4$ MPa, $\tau = 23.6$ MPa

3–44 (b) $(\tau_{max})_A = 75.1$ MPa, $(\tau_{max})_B = 53.1$ MPa, $(\tau_{max})_C = 116.8$ MPa

3–50 (a) $r = 6$ mm, 126.25° CCW from the y axis, (b) $\sigma_x = 112.3$ MPa, $\tau_{xz} = 35.4$ MPa, (d) $\sigma_1 = 122.6$ MPa, $\sigma_2 = 0$ MPa, $\sigma_3 = -10.2$ MPa, $\tau_{max} = 66.4$ MPa

3–55 (a) $\tau_{max} = 180.4$ MPa, (b) $\tau_{max} = 179.2$ MPa, (c) $\theta = 1.74°$, (d) $\theta = 1.74°$

3–57 (a) $T = 204$ N·m, (b) $T = 52.5$ N·m

3–61 (a) $T_1 = 1.47$ N·m, $T_2 = 7.45$ N·m, $T_3 = 0$ N·m, $T = 8.92$ N·m, (b) $\theta_1 = 0.348$ rad/m

3–67 $H = 55.5$ kW

3–77 (a) $T_1 = 2880$ N, $T_2 = 432$ N, (b) $R_C = 1794$ N, $R_O = 3036$ N, (d) $\sigma = 263$ MPa, $\tau = 57.7$ MPa, (e) $\sigma_1 = 275$ MPa, $\sigma_2 = -12.1$ MPa, $\tau_{max} = 144$ MPa

3–81 (a) $F_B = 22\ 810$ N, (b) $R_{Cy} = 8319$ N, $R_{Cz} = -10\ 830$ N, $R_{Oy} = 5083$ N, $R_{Oz} = 494$ N, (d) $\sigma = 333.9$ MPa, $\tau = 126.3$ MPa, (e) $\sigma_1 = 376$ MPa, $\sigma_2 = -42.4$ MPa, $\tau_{max} = 209$ MPa

3–105 $x_{min} = 8.3$ mm

3–110 $p_o = 82.8$ MPa

3–118 $\sigma_t = 1.85$ MPa

3–121 $\delta_{max} = 0.021$ mm, $\delta_{min} = 0.0005$ mm, $p_{max} = 65.2$ MPa, $p_{min} = 1.55$ MPa

3–131 $\sigma_i = 300$ MPa, $\sigma_o = -195$ MPa

3–139 $\sigma_i = 64.6$ MPa, $\sigma_o = -21.7$ MPa

3–144 $\sigma_{max} = 352F^{1/3}$ MPa, $\tau_{max} = 106F^{1/3}$ MPa

3–147 $z = 0.0604$ mm, $\tau_{max} = 181$ MPa

3–152 $\sigma_x = -35.0$ MPa, $\sigma_y = -22.9$ MPa, $\sigma_z = -96.9$ MPa, $\tau_{max} = 37.0$ MPa

B–4 Chapter 4

4–10 $y_{max} = -25.4$ mm, $\sigma_{max} = -163$ MPa

4–13 $y_O = y_C = -3.72$ mm, $y|_{x=550mm} = 1.11$ mm

4–16 $d_{min} = 32.3$ mm

4–24 $y_A = -7.99$ mm, $\theta_A = -0.0304$ rad

4–28 $y_A = -3.735$ mm, $z_A = 1{,}791$ mm, $(\theta_A)_y = -0.00489$ rad, $(\theta_A)_z = -0.00507$ rad

4–30 $(\theta_O)_z = 0.0131$ rad, $(\theta_C)_z = -0.0191$ rad

4–32 $(\theta_O)_y = 0.00316$ rad, $(\theta_O)_z = 0.00680$ rad, $(\theta_C)_y = 6.04(10^{-5})$ rad, $(\theta_C)_z = -0.00558$ rad

4–36 $d = 62.0$ mm

4–40 $d = 100.9$ mm

4–46 $d = 38.1$ mm, $y_{max} = -0.0678$ mm

4–52 $y_B = -4.18$ mm

4–56 $y_A = 0.784$ mm

4–58 $k = 8.10$ N/mm

4–83 $y_b = -4.18$ mm

4–89 (a) $(y_D)_{AB} = -0.150$ mm, (b) $(y_D)_{BC} = 0$ mm, (c) $(y_D)_{BD} = 0$ mm, (d) $y_D = -0.150$ mm

4–92 $\delta = 0.0338$ mm

4–95 $F = 11.65$ N

4–96 $H = 9.55$ N, $\delta_P = 0.224$ mm

4–105 $\delta = 6.067$ mm

4–112 $R_B = 1.6$ kN, $R_O = 2.4$ kN, $\delta_A = 0.0223$ mm

4–122 $R_C = -2378$ N, $F_{BE} = 4189$ N, $F_{DF} = -189.2$ N, $C_1 = 1.036(10^7)$ N · mm², $C_2 = -7.243(10^8)$ N · mm², $y_A = -0.167$ mm, $y_B = -0.0201$ mm, $y_D = -1.18(10^{-3})$ mm

4–127 (a) $t = 11$mm, (b) No

4–136 $F_{max} = 930$ N, $\delta_{max} = 29.06$ mm

B–5 Chapter 5

5–1 (a) MSS: $n = 3.5$, DE: $n = 3.5$, (b) MSS: $n = 3.5$, DE: $n = 4.04$, (c) MSS: $n = 1.94$, DE: $n = 2.13$, (d) MSS: $n = 3.07$, DE: $n = 3.21$, (e) MSS: $n = 3.34$, DE: $n = 3.57$

5–2 (a) MSS: $n = 3.3$, DE: $n = 3.5$, (b) MSS: $n = 1.75$, DE: $n = 2.02$, (c) MSS: $n = 3.5$, DE: $n = 4.04$, (d) MSS: $n = 2.33$, DE: $n = 2.65$, (e) MSS: $n = 3.5$, DE: $n = 4.04$

5–7 (a) $n = 3.03$

5–17 (a) $n = 1.81$

5–39 (a) $n = 1.22$, (b) $n = 1.41$

5–51 MSS: $n = 1.29$, DE: $n = 1.32$

5–55 MSS: $n = 0.88$, DE: $n = 0.93$

5–58 MSS: $n = 3.63$, DE: $n = 3.91$

5–70 (a) $p_i = 36.1$ MPa, (b) $p_i = 41.6$ MPa

5–80 $F_x = 2\pi fT/(0.2d)$

5–81 (a) $F_i = 16.7$ kN, (b) $p_i = 111.3$ MPa, (c) $\sigma_t = 185.5$ MPa, $\sigma_r = -111.3$ MPa, (d) $\tau_{max} = 148.4$ MPa, $\sigma' = 259.7$ MPa, (e) MSS: $n = 1.52$, DE: $n = 1.73$

5–87 $n_o = 1.84$, $n_i = 1.80$

5–89 $n = 1.91$

5–97 (a) $F = 958$ kN, (b) $F = 329.4$ kN

B–6 Chapter 6

6–1 $S_e = 429$ MPa

6–4 $N = 46\ 400$ cycles

6–6 $S_f = 624$ MPa

6–25 $n_y = 3.32$, using Goodman: $n_f = 0.59$, $N = 25\ 000$ cycles

6–29 (a) $F_{min} = 48.3$ N, $F_{max} = 144.9$ N, (b) $n_f = 3.33$

6–35 $n_f = 0.81$, $N = 716\ 000$ cycles, yielding is not predicted

6–38 $n_f = 0.51$

6–42 $n_f = 0.36$

6–45 $n_f = 1.68$

6–59 (a) $n_2 = 7\ 000$ cycles, (b) $n_2 = 10\ 000$ cycles

6–61 (a) $n_3 = 184\ 100$ cycles, (b) $N_3 = 33\ 610$ cycles

B–7 Chapter 7

7–1 (a) DE-Goodman: $d = 27.27$ mm, (b) DE-Morrow: $d = 26.68$ mm, (c) DE-Gerber: $d = 25.85$ mm, (d) DE-SWT: $d = 27.99$ mm

7–4 (a) $d = 43.0$ mm, (b) $d = 57$ mm

7–33 (a) $\omega = 883$ rad/s, (b) $d = 50$ mm, (c) $\omega = 1766$ rad/s (doubles)

7–35 (b) $\omega = 466$ rad/s $= 4450$ rev/min

7–40 14 mm square key, 50 mm long, AISI 1020 CD

7–41 $d_{\min} = 14.989$ mm, $d_{\max} = 15.000$ mm, $D_{\min} = 15.000$ mm, $D_{\max} = 15.018$ mm

7–47 (a) $d_{\min} = 35.043$ mm, $d_{\max} = 35.059$ mm, $D_{\min} = 35.000$ mm, $D_{\max} = 35.025$ mm, (b) $p_{\min} = 35.1$ MPa, $p_{\max} = 115$ MPa, (c) Shaft: $n_y = 3.4$, hub: $n_y = 1.9$, (d) Assuming $f = 0.8$, $T = 2700$ N·m

B–8 Chapter 8

8–1 (a) Thread depth 2.5 mm, thread width 2.5 mm, $d_m = 22.5$ mm, $d_r = 20$ mm, $l = p = 5$ mm

8–4 $T_R = 15.85$ N·m, $T_L = 7.83$ N·m, $e = 0.251$

8–11 (a) $L = 45$ mm, (b) $k_b = 874.6$ MN/m, (c) $k_m = 3116.5$ MN/m

8–19 (a) $L = 60$ mm, (b) $k_b = 292.1$ MN/m, (c) $k_m = 692.5$ MN/m

8–25 From Eqs. (8–20) and (8–22), $k_m = 2762$ MN/m. From Eq. (8–23), $k_m = 2843$ MN/m

8–33 $L = 55$ mm, $n_p = 1.29$, $n_L = 11.1$, $n_0 = 11.8$

8–37 $n_p = 1.29$, $n_L = 10.7$, $n_0 = 12.0$

8–41 Bolt sizes of diameters 8, 10, 12, and 14 mm were evaluated and all were found acceptable. For $d = 8$ mm, $k_m = 854$ MN/m, $L = 50$ mm, $k_b = 233.9$ MN/m, $C = 0.215$, $N = 20$ bolts, $F_i = 6.18$ kN, $P = 2.71$ kN/bolt, $n_p = 1.22$, $n_L = 3.53$, $n_0 = 2.90$

8–46 (a) $T = 823$ N·m, (b) $n_p = 1.10$, $n_L = 17.7$, $n_0 = 57.7$

8–52 (a) Goodman: $n_f = 7.55$, (b) Gerber: $n_f = 11.4$, (c) ASME-elliptic: $n_f = 9.73$

8–56 Goodman: $n_f = 11.9$

8–64 $n_p = 1.24$, $n_L = 4.62$, $n_0 = 5.39$, $n_f = 4.75$

8–69 $F = 60.9$ kN

8–72 $n = 2.72$

8–76 $F = 2.32$ kN based on channel bearing

8–77 Bolt shear, $n = 0.728$; bolt bearing, $n = 1.07$; member bearing, $n = 0.534$; member bending, $n = 1.91$

B–9 Chapter 9

9–1 $F = 49.5$ kN

9–5 $F = 51.2$ kN

9–9 $F = 31.1$ kN

9–13 $\tau = 141$ MPa

9–19 (a) $F = 6.06$ kN, (b) $F = 3.29$ kN

9–25 $F = 12.1$ kN

9–33 $F = 12.8$ kN

9–40 $F = 5.04$ kN

9–43 All-around square, four beads each $h = 6$ mm, 75 mm long, Electrode E6010

9–56 $\tau_{\max} = 45.3$ MPa

9–60 $F = 61.2$ kN

B–10 Chapter 10

10–2 (a) $L_0 = 162.8$ mm, (b) $F_s = 167.9$ N, (c) $k = 1.314$ N/mm, (d) $(L_0)_{cr} = 149.9$ mm, spring needs to be supported

10–5 (a) $k = 3.333$ N/mm, (b) OD $= 44$ mm, (c) $N_t = 11.6$ coils, (d) $L_s = 50.4$ mm, (e) $n_s = 4.02$

10–15 Spring is solid safe, $n_s = 1.51$

10–16 Spring is not solid safe, ($n_s < 1.2$), $L_0 = 68.2$ mm

10–21 (a) $C = 9$, (b) $p = 7.8$ mm/coil, (c) $y_s = 144$ mm, (d) $F_s = 189.45$ kN, (e) $\tau_s = 555.8$ MPa, (f) $n_s = 1.22$, (g) Exact, $k = 1.316$ N/mm; approximate, $k = 1.324$ N/mm, 0.61% higher than the exact.

10–27 With $d = 2$ mm, $L_0 = 48$ mm, $k = 4.286$ N/mm, $D = 13.25$ mm, $N_a = 15.9$ coils, $n_s = 2.63 > 1.2$, ok. No other d works.

10–43 $\Sigma = 31.3°$ (see Fig. 10–9), $F_{\max} = 87.3$ N

B–11 Chapter 11

11–2 $x_D = 525$, $F_D = 3.0$ kN, $C_{10} = 24.3$ kN, 02–35 mm deep-groove ball bearing, $R = 0.920$

11–8 $x_D = 456$, $C_{10} = 145$ kN

11–10 $C_{10} = 20$ kN

11–17 $C_{10} = 26.1$ kN

11–23 (a) $F_e = 5.34$ kN, (b) $\mathcal{L}_D = 444$ h

11–26 60 mm deep-groove

11–31 (a) $C_{10} = 204$ kN

11–35 $C_{10} = 5.7$ kN, 02–12 mm deep-groove ball bearing

11–44 $l_2 = 0.267(10^6)$ rev

11–49 $F_{RA} = 35.4$ kN, $F_{RB} = 17.0$ kN

B–12 Chapter 12

12–1 $c_{\min} = 0.015$ mm, $r = 12.5$ mm, $r/c = 833$, $N_j = 18.3$ rev/s, $S = 0.182$, $h_0/c = 0.3$, $rf/c = 5.4$, $Q/(rcNl) = 5.1$, $Q_s/Q = 0.81$, $h_0 = 0.0045$ mm, $H_{\text{loss}} = 11.2$ W, $Q = 219$ mm^3/s, $Q_s = 177$ mm^3/s

12–2 $h_0 = 0.023$ mm, $p_{\max} = 1.31$ MPa, $Q = 1.23(10^3)$ mm^3/s

12–9 $h_0 = 0.011$ mm, $H = 48.1$ W, $Q = 1426$ mm^3/s, $Q_s = 1012$ mm^3/s

12–19 Approx: $\mu = 45.7$ mPa·s, Fig. 12–3: $\mu = 39$ mPa·s

B–13 Chapter 13

13–2 400 rev/min, $p = 3\pi$ mm, $C = 112.5$ mm

13–3 $N_G = 64$ teeth, $d_G = 384$ mm, $d_P = 96$ mm, $C = 240$ mm

13–7 (a) $p_n = 7.854$ mm, $p_t = 9.069$ mm, $p_x = 15.71$ mm, (b) $m_t = 2.887$ mm, $\Phi_n = 22.80°$, (c) $a = 2.5$ mm, $b = 3.125$ mm, $d_P = 54.85$ mm, $d_G = 164.6$ mm

13–10 (a) 13, (b) 15, 45, (c) 18

13–15 (a) $p_n = 3\pi$ mm, $p_t = 10.40$ mm, $p_x = 22.30$ mm, (b) $m_t = 3.310$ mm, $\phi_t = 21.88°$, (c) $d_p = 59.58$ mm, $d_G = 105.92$ mm

13–17 $e = 4/51$, $n_d = 47.06$ rev/min cw

13–25 (a) $n_A = 300$ rev/min, (b) $n_6 = 600$ rev/min

13–30 $n_A = 68.57$ rev/min cw

13–45 (a) $N_{Pmin} = 11$, (b) $m = 27.3$ mm/tooth, (c) $F_A = 11.40$ kN, $F_B = 25.16$ kN

13–46 (a) $N_F = 15$ teeth, $N_C = 30$ teeth , (b) $m = 8.33$ mm/tooth, (c) $T = 100$ N · m, (d) $W_t = 1600$ N, W_r =583 N, $W = 1700$ N

13–55 $\mathbf{F}_{2a} = -1119\mathbf{i} - 407.3\mathbf{j} - 299.8\mathbf{k}$ N, $\mathbf{F}_{3b} = 711.7\mathbf{i} - 711.7\mathbf{j}$ N, $\mathbf{F}_{4c} = 407.3\mathbf{i} + 1119\mathbf{j} + 299.8\mathbf{k}$ N

13–57 A is the thrust bearing, $T = 31.85$ N · m, $\mathbf{F}_A = 318.5\mathbf{i} - 1946\mathbf{j} - 5119\mathbf{k}$ N, $\mathbf{F}_B = 318.5\mathbf{i} + 613\mathbf{j}$ N

B–14 Chapter 14

14–3 $\sigma = 86.6$ MPa

14–4 $\sigma = 32.6$ MPa

14–6 $F = 28$ mm

14–10 $m = 2$ mm, $F = 25$ mm

14–15 $\sigma_c = -617$ MPa

14–18 $W^t = 16\,390$ N, $H = 94.3$ kW (pinion bending); $W^t = 3469$ N, $H = 20.0$ kW (pinion and gear wear)

B–15 Chapter 15

15–5 $H = 3.92$ kW

15–6 $H = 3.92$ kW

B–16 Chapter 16

16–1 (a) Right shoe: $p_a = 734.5$ kPa cw rotation, (b) Right shoe: $T = 277.6$ N·m; left shoe: 144.4 N·m; total $T = 422$ N·m, (c) RH shoe: $R^x = -1.007$ kN, $R^y = 4.13$ kN, $R = 4.25$ kN, LH shoe: $R^x = 570$ N, $R^y = 751$ N, $R = 959$ N

16–2 RH shoe: $p_a = 852.4$ kPa, $T = 263$ N·m; LH shoe: $p_a = 479.1$ kPa, T = 148 N·m

16–4 (a) $F = 5.70$ kN, (b) $T = 1750$ N·m, (c) $R = 5.90$ kN

16–8 $a' = 1.209r$, $a = 1.170r$

16–11 $P_1 = 10.85$ kN, $P_2 = 2.64$ kN, $T = 1.437$ kN·m

16–21 Key: $\tau = 13.2$ MPa, $\sigma_b = -26.5$ MPa; Jaws: $\tau = 0.869$ MPa, $\sigma_b = -22.6$ MPa

16–23 $C_s = 0.08$, $t = 143$ mm

16–26 (b) $I_e = I_M + I_p + n^2 I_p + I_L/n^2$, (c) $I_e = 10 + 1 + 10^2(1) + 100/10^2 = 112$

16–27 (c) $n^* = 2.430$, $m^* = 4.115$, which are independent of I_L

B–20 Chapter 20

20–13 Partial answers: (a) 50.3, 49.7, (b) No effect, (c) the center axis of the boss, as determined by the related actual mating envelope, (d) 0.2, (e) 0.8

20–15 Hint: Read about actual mating envelopes.

20–17 20.2, 19.8

20–21 (a) 0.1, 0.1, 0.1, (b) 0.5, 0.3, 0.1,(c) 0.1, 0.3, 0.5

A

B

C

D

E

F

G

H

M

N

O

P

Q

R

S

T

U

V